Engineering Design and Graphics with SolidWorks® 2014

James D. Bethune

PEARSON

Boston Columbus Indianapolis New York San Francisco Upper Saddle River
Amsterdam Cape Town Dubai London Madrid Milan Munich Paris Montreal Toronto Delhi
Mexico City São Paulo Sydney Hong Kong Seoul Singapore Taipei Tokyo

Executive Editor: Lisa McClain
Production Editor: Katerina Malone
Cover Designer: Mimi Heft
Full-Service Project Management:
Mohinder Singh/Aptara®, Inc.
Composition: Aptara®, Inc.
Printer/Binder: Edwards Brothers
Cover Printer: Edwards Brothers
Text Font: Bookman

Credits and acknowledgments borrowed from other sources and reproduced, with permission, in this textbook appear on the appropriate page within the text. Unless otherwise stated, all artwork has been provided by the author.

Disclaimer:
The publication is designed to provide tutorial information about SolidWorks® and/or other Dassault Systèmes SolidWorks Corp computer programs. Every effort has been made to make this publication complete and as accurate as possible. The reader is expressly cautioned to use any and all precautions necessary, and to take appropriate steps to avoid hazards, when engaging in the activities described herein.

Neither the author nor the publisher makes any representations or warranties of any kind, with respect to the materials set forth in this publication, express or implied, including without limitation any warranties of fitness for a particular purpose or merchantability. Nor shall the author or the publisher be liable for any special, consequential, or exemplary damages resulting, in whole or in part, directly or indirectly, from the reader's use of, or reliance upon, this material or subsequent revisions of this material.

Library of Congress Control Number: 2014942492

10 9 8 7 6 5 4 3 2 1

ISBN 10: 0-321-99399-3
ISBN 13: 978-0-321-99399-1

Preface

This book shows and explains how to use SolidWorks® 2014 to create engineering drawings and designs. Emphasis is placed on creating engineering drawings including dimensions and tolerances and using standard parts and tools. Each chapter contains step-by-step sample problems that show how to apply the concepts presented in the chapter.

The book contains hundreds of projects of various degrees of difficulty specifically designed to reinforce the chapter's content. The idea is that students learn best by doing. In response to reviewers' requests, some more difficult projects have been included.

Chapter 1 and **2** show how to set up a part document and how to use the SolidWorks **Sketch** tools. **Sketch** tools are used to create 2D part documents that can then be extruded into 3D solid models. The chapters contain an explanation of how SolidWork's colors are used and of how shapes can be fully defined. The usage of mouse gestures, S key, and origins is also included. The two chapters include 43 projects using both inches and millimeters for students to use for practice in applying the various **Sketch** tools.

Chapter 3 shows how to use the **Features** tools. **Features** tools are used to create and modify 3D solid models. In addition, reference planes are covered, and examples of how to edit existing models are given.

Chapter 4 explains how to create and interpret orthographic views. Views are created using third-angle projection in compliance with ANSI standards and conventions. The differences between first-angle and third-angle projections are demonstrated. Five exercise problems are included to help students learn to work with the two different standards. Also included are section views, auxiliary views, and broken views. Several of the projects require that a 3D solid model be drawn from a given set of orthographic views to help students develop visualization skills.

Chapter 5 explains how to create assembly drawings using the **Assembly** tools (**Mate**, exploded **View**) and how to document assemblies using the **Drawing Documents** tools. Topics include assembled 3D solid models, exploded isometric drawings, and bills of materials (BOMs). Assembly numbers and part numbers are discussed. Both the **Animate Collapse/Explode** and **Motion Study** tools are demonstrated. In addition, the title, release, and revision blocks are discussed. An explanation of how to use **Interference Detection** is given.

Chapter 6 shows how to create and design with threads and fasteners. Both ANSI inch and ANSI metric threads are covered. The **Design Library** is presented, and examples are used to show how to select and size screws and other fasteners for assembled parts.

Chapter 7 covers dimensioning and is in compliance with ANSI standards and conventions. There are extensive visual examples of dimensioned shapes and features that serve as references for various dimensioning applications.

Chapter 8 covers tolerances. Both linear and geometric tolerances are included. This is often a difficult area to understand, so there are many examples of how to apply and how to interpret the various types of tolerances. Standard tolerances as presented in the title block are demonstrated. Many of the figures have been updated.

Chapter 9 explains bearings and fit tolerances. The **Design Library** is used to create bearing drawings, and examples show how to select the correct interference tolerance between bearings and housing, and clearance tolerances between bearings and shafts.

Chapter 10 presents gears. Gear terminology, gear formulas, gear ratios, and gear creation using the SolidWorks **Toolbox** are covered. The chapter relies heavily on the **Design Library**. Keys, keyways, and set screws are discussed. Both English and metric units are covered. There is an extensive sample problem that shows how to draw a support plate for mating gears and how to create an assembly drawing for gear trains. The exercise problems at the end of the chapter are supplemented with two large gear assembly exercises.

Chapter 11 covers belts and pulleys. Belts, pulleys, sprockets, and chains are drawn. All examples are based on information from the **Design Library**. There are several sample problems.

Chapter 12 covers cams. Displacement drawings are defined. The chapter shows how to add hubs and keyways to cams and then insert the cams into assembly drawings. The chapter also shows how to add springs to followers.

Chapter 13 is a new chapter. It includes two large project-type problems. They can be used as team projects to help students learn to work together to share and compile files, or they can be used as end-of-the-semester individual projects. This chapter can be found on the web as a supplement to the Instructor's Manual at http://pearsonhighered.com/irc. Instructors may distribute to students.

The **Appendix** includes fit tables for use with projects in the text. Clearance, locational, and interference fits are included for both inch and millimeter values.

Download Instructor Resources from the Instructor Resource Center

To access supplementary materials online, instructors need to request an instructor access code. Go to www.pearsonhighered.com/irc to register for an instructor access code. Within 48 hours of registering, you will receive a confirming e-mail including an instructor access code. Once you have received your code, locate your text in the online catalog and click on the Instructor Resources button on the left side of the catalog product page. Select a supplement, and a login page will appear. Once you have logged in, you can access instructor material for all Prentice Hall textbooks. If you have any difficulties accessing the site or downloading a supplement, please contact Customer Service at http://247pearsoned.custhelp.com/.

Acknowledgments

I would like to acknowledge the reviewers of this text: Peggy Condon-Vance, Penn State Berks; Lisa Richter, Macomb Community College; Julie Korfhage, Clackamas Community College; Max P. Gassman, Iowa State University; Paul E. Lienard, Northeastern University; and Hossein Hemati, Mira Costa College.

Thanks to editor Lisa McClain. Thanks to my family—David, Maria, Randy, Hannah, Will, Madison, Jack, Luke, Sam, and Ben.

A special thanks to Cheryl.

James D. Bethune

Contents

chapterone

Getting Started

CHAPTER OBJECTIVES

- Learn how to create a sketch
- Learn how to create a file/part
- Learn how to create a solid model
- Learn how to edit angular and circular shapes
- Learn how to draw holes
- Learn how to use **Sketch** tools
- Learn how to change units of a part

1-1 Introduction

SolidWorks is a *parametric modeler*. A solid modeler uses dimensions, parameters, and relationships to define and drive 3D shapes. Solid modelers make it easy to edit and modify parts as they are constructed. This capability is ideal for creating new designs.

Parametric modelers use dimensions to drive the shapes. For example, to create a line of a defined length, a line is first sketched, then the length dimension is added. The line will assume the length of the dimension. If the dimension is changed, the length of the line will change to match the new dimension.

When using *non-parametric modelers*, a line is drawn and a dimension added. The dimension will define the length of the existing line but not drive it. If the length of the line is changed, the dimensions will not change. A new dimension is required to define the length of the line.

This chapter will show you how to start a **New** drawing and introduce the **Line**, **Circle**, and **Edit** tools. The **Smart Dimension** tool will be used to define and edit lines and circles. Line colors and relationships will also be introduced.

1-2 Starting a New Drawing

Figure 1-1 shows the initial SolidWorks screen.

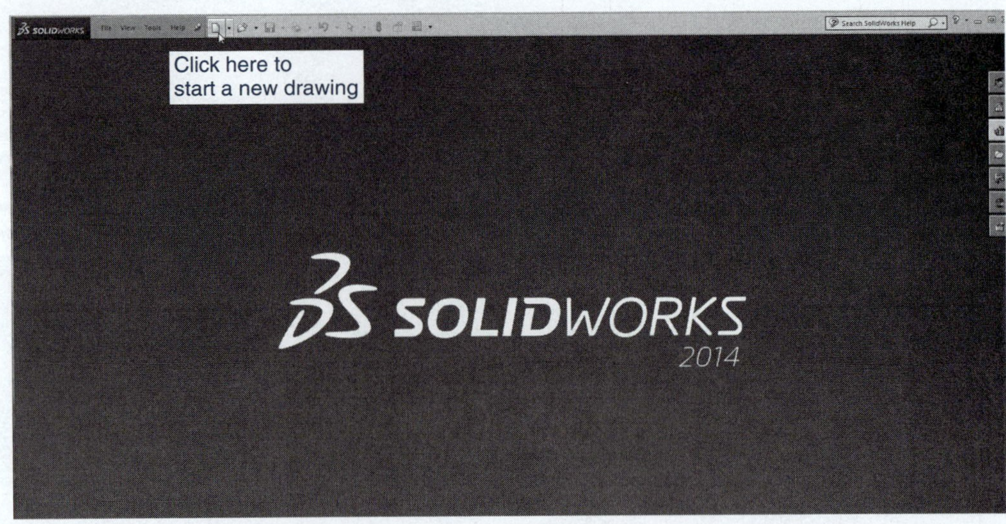

To Start a New Drawing

1 Click the **New** tool icon at the top of the drawing screen.

A new drawing screen will appear. See Figure 1-2. The **New SolidWorks Document** dialog box will appear. SolidWorks can be used to create three types of documents: **Part**, **Assembly**, and **Drawing**.

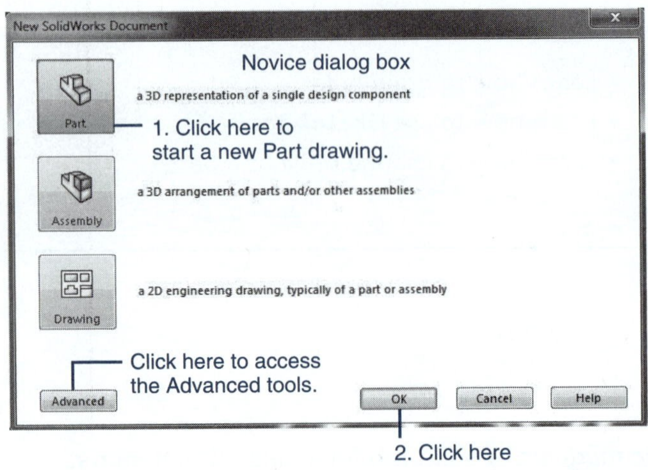

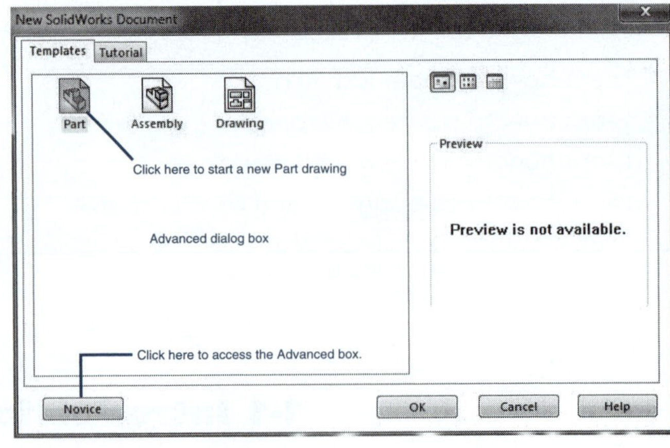

Figure 1-2

There are two versions of the **New SolidWorks Document** dialog box: Novice and Advanced. The Advanced version includes Tutorials. Either version can be used to access the **Part Document** area.

Part drawings are 3D solid models of individual parts.

Assembly drawings are used to create drawings of assemblies that contain several Part drawings.

Drawing drawings are used to create orthographic views of the Part and Assembly drawings. Dimensions and tolerances can be applied to Drawing drawings.

2 Click the **Part** tool and then click the **OK** box.

The **Part** drawing screen will appear. See Figure 1-3. Note the different areas of the screen. The **Features** tab is currently activated, so the

Features tools are displayed. Each tool icon on the Features toolbar is accompanied by its name. These names can be removed and the toolbar condensed to expand the size of the drawing screen. For clarity these named tools will be included in the first few chapters of the book so you gain enough knowledge of the tools to work without their names.

Figure 1-3

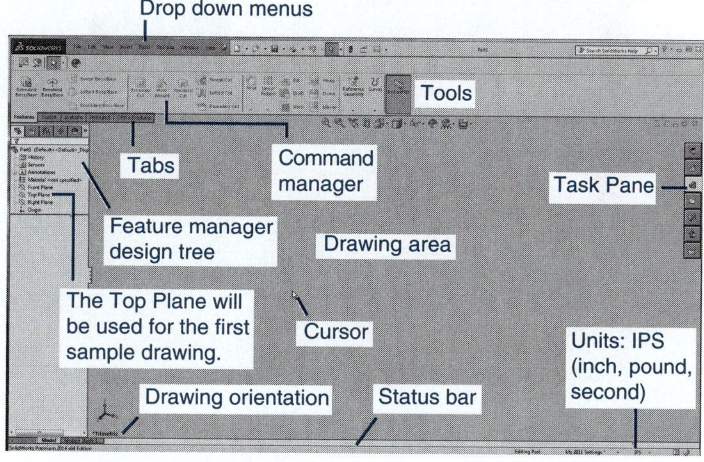

To Select a Drawing Plane

SolidWorks uses one of three basic planes to define a drawing: **Front**, **Top**, and **Right**. These planes correspond to the planes used to define orthographic views that will be explained in Chapter 4. The **Top** plane will be used to demonstrate the first few tools.

3 Define the plane on which the part will be created.

4 Click the **Top Plane** option in the **Feature manager** box on the left side of the drawing screen.

See Figure 1-4. An outline of the **Top** plane will appear using the **Trimetric** orientation, that is, a 3D orientation.

Figure 1-4

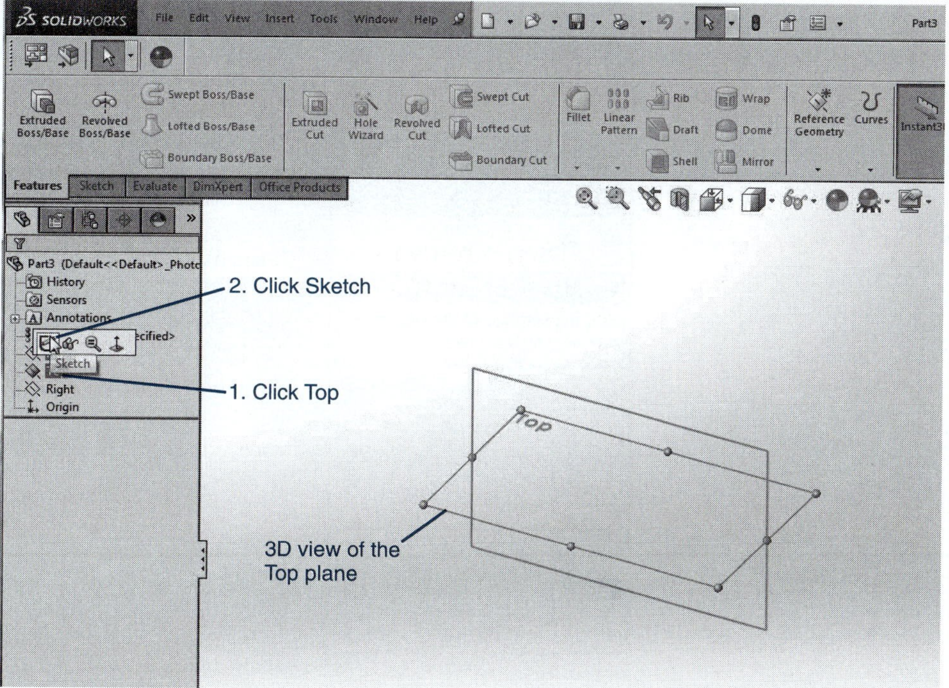

Figure 1-4
(*Continued*)

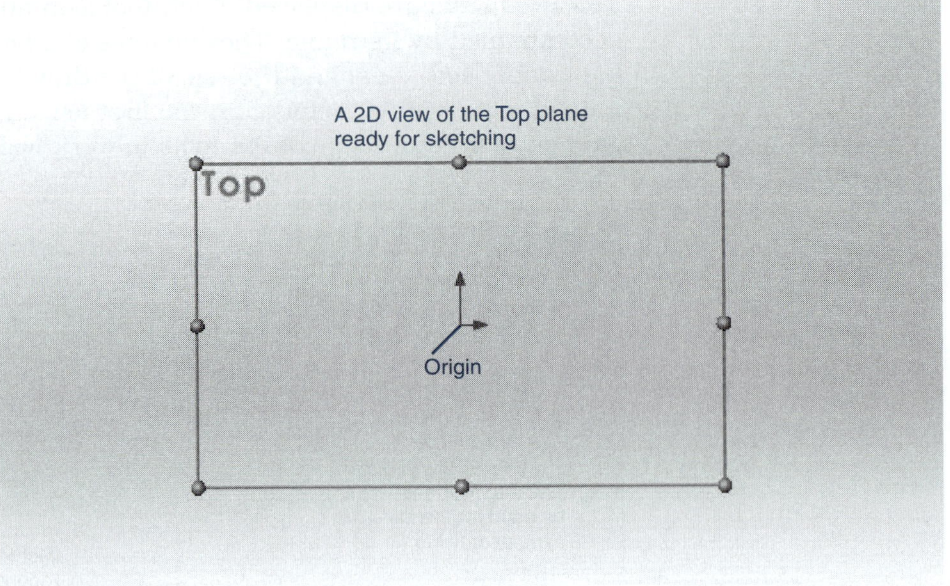

5 Click the **Sketch** tool.

The **Top** plane's orientation will change to a 2D view. The **Top** plane appears as a rectangle because the view is taken at 90° to the plane. This means that all 2D shapes drawn on the plane will appear as true shapes.

6 Click the **Line** tool.

With the **Line** tool activated, locate the cursor on the origin. The origin is indicated by the two red arrows spaced 90° apart. See Figure 1-5.

Figure 1-5

Two icons will appear on the screen: the **Line tool** icon indicating that the **Line** tool is active, and the **Coincident relationship** icon indicating that the origin and starting point for the line are on the same point.

7 Move the cursor away from the origin horizontally to the right.

As you move the cursor away from the origin a distance, an angle value will appear. See Figure 1-6. The distance is as measured from the origin or starting point for the line and the angle is based on the SolidWorks definition of 0° as a horizontal line to the left of the starting point. We are drawing to the right, so the angular value is 180°.

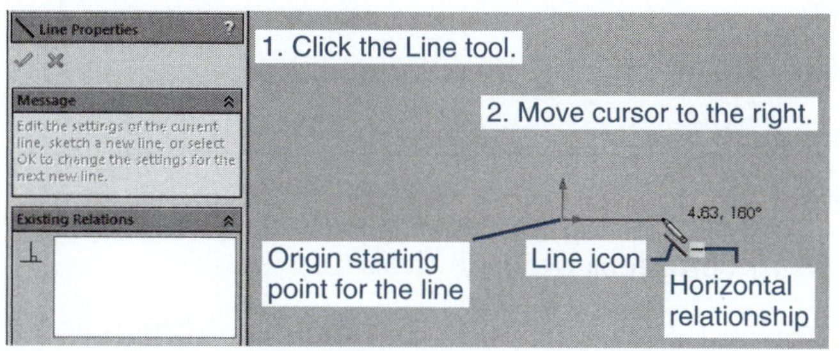

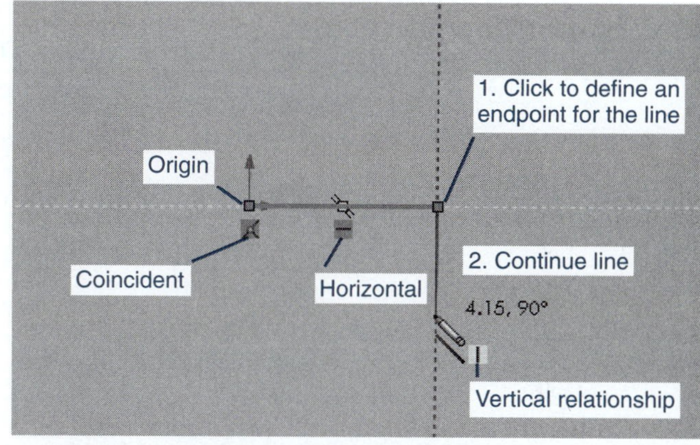

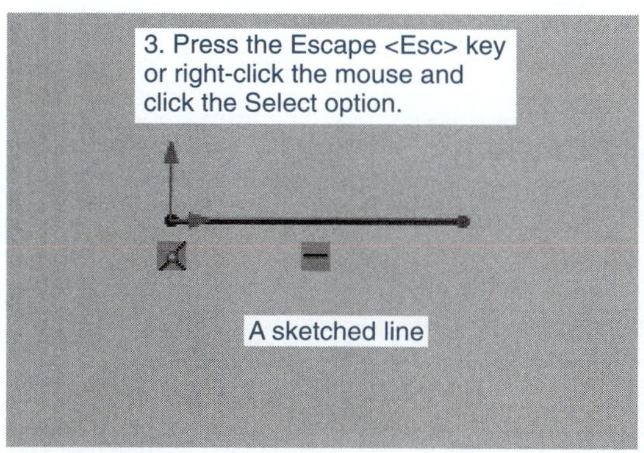

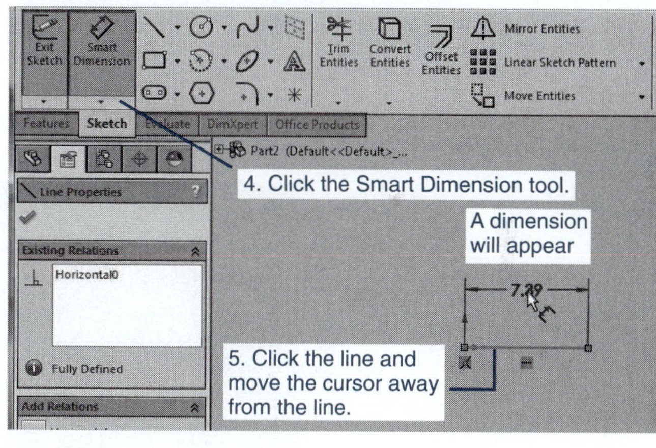

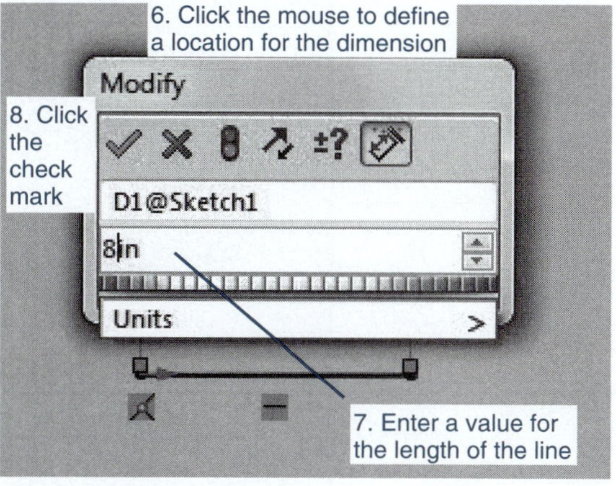

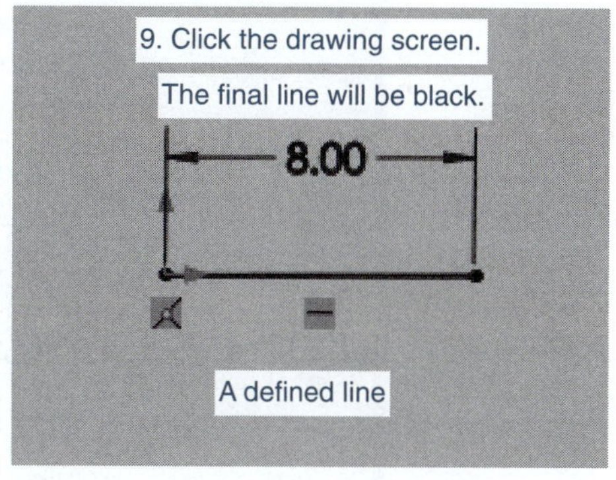

Figure 1-6

Two other icons will also appear: the **Line** tool icon and the horizontal relationship icon.

8 Click the mouse to define the endpoint of the line.

9 Move the cursor vertically downwards. Do not click the mouse.

A new line will be drawn using the endpoint of the horizontal line as the starting point for the vertical line. Distance and angle values will appear based on the new starting point, and the Line and vertical relationship icons will appear.

10 Press the Escape **<Esc>** key or right-click the mouse and click the **Select** option.

11 Click the **Smart Dimension** tool, click the line, and move the cursor away from the line.

A dimension will appear.

12 Click the mouse to define the location of the dimension.

The **Modify** dialog box will appear.

13 Enter a distance value for the line and click the green **OK** check mark.

14 Click the drawing screen to complete the line drawing.

The dimension can be moved by locating the cursor on the dimension, pressing and holding the mouse button, and dragging the cursor.

15 Click the **File** tab located at the top of the screen.

See Figures 1-7 and 1-8.

Figure 1-7

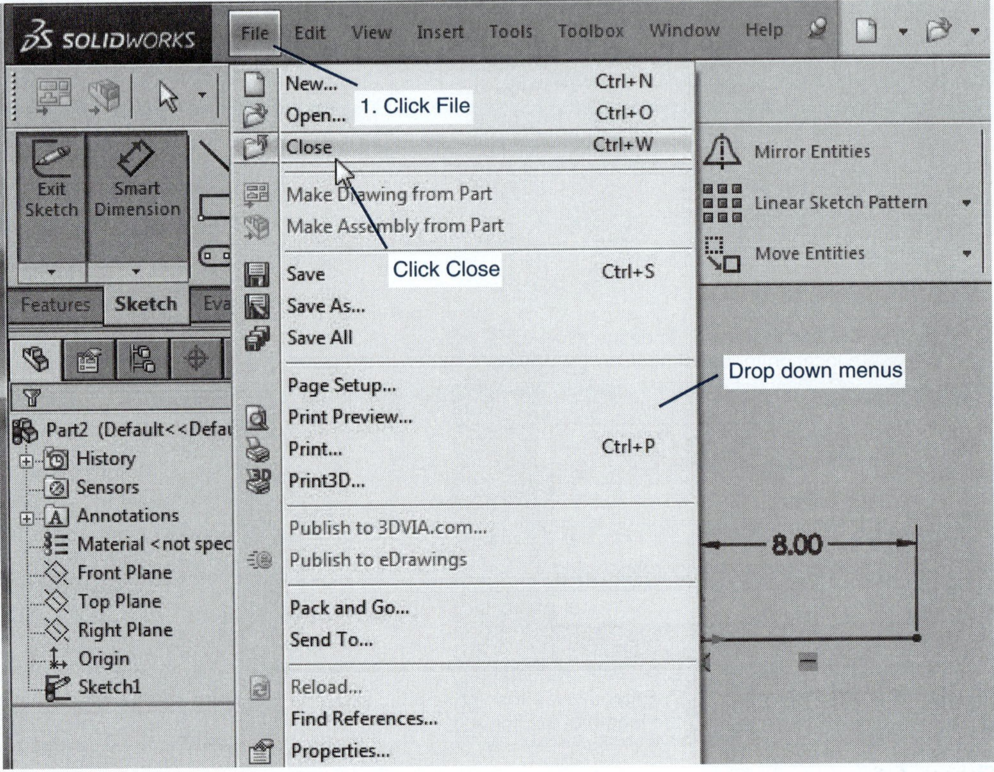

Figure 1-8

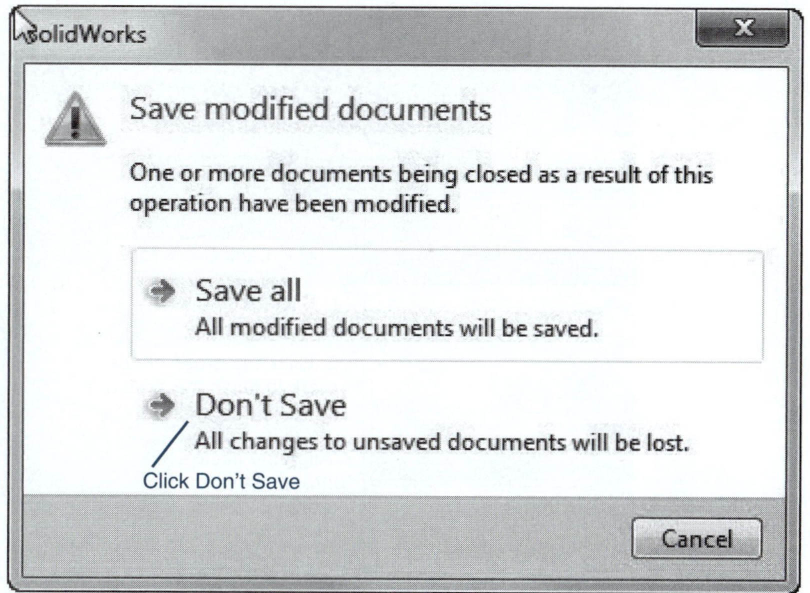

16 Click the **Don't Save** option.

The screen will return to the original SolidWorks screen.

1-3 SolidWorks Colors

As you work with SolidWorks you will notice that the lines change colors. These color changes serve to let you know the status of the sketch being drawn. There are four basic colors.

BLACK = Fully Defined

BLUE = Under Defined

RED = Over Defined

YELLOW = Redundant

1-4 Creating a Fully Defined Circle

In this section we will sketch a circle to help understand the difference between a fully defined and an under defined **Part**.

Start a **New Part** drawing and click the **Top Plane** tool as defined in Figure 1-4.

1 Click the **Sketch** tab. (It may already be activated.)

2 Click the **Circle** tool.

3 Locate the cursor on the origin, click the mouse, and drag the cursor away from the origin center point.

Note that the Coincident relationship symbol appears next to the origin, indicating that the center point of the circle is located on the origin.

4 Click the mouse to define a sketch radius for the circle.

This is a temporary radius, that is, a sketched radius, and is not the final radius. The circle will be blue, indicating that it is not fully defined. See Figure 1-9.

Figure 1-9

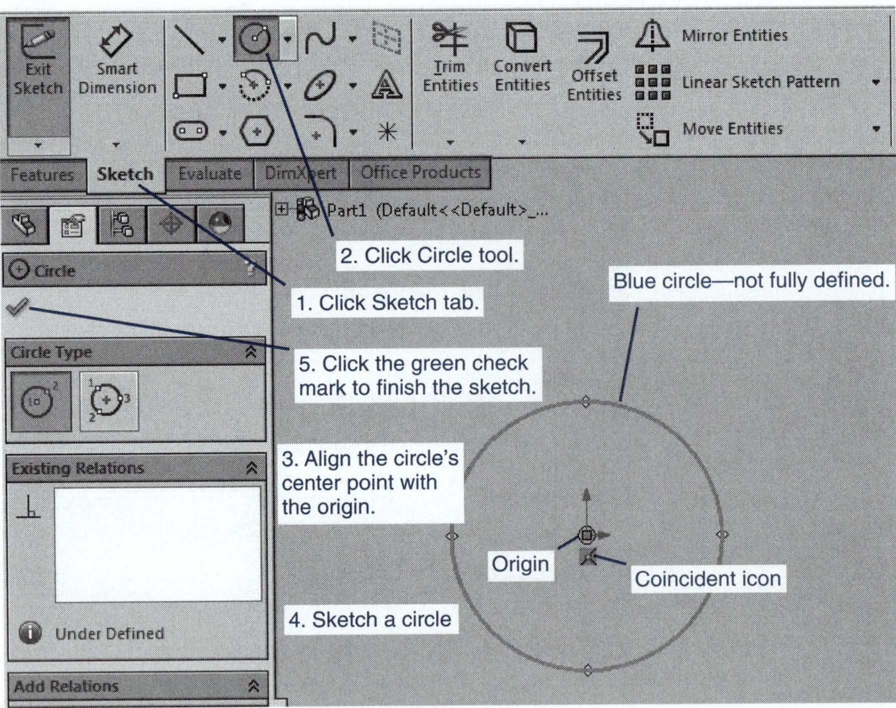

5 Click the **Smart Dimension** tool on the **Sketch** panel.

6 Click the circle and move the cursor away from the circle.

A dimension will appear. See Figure 1-10.

Figure 1-10

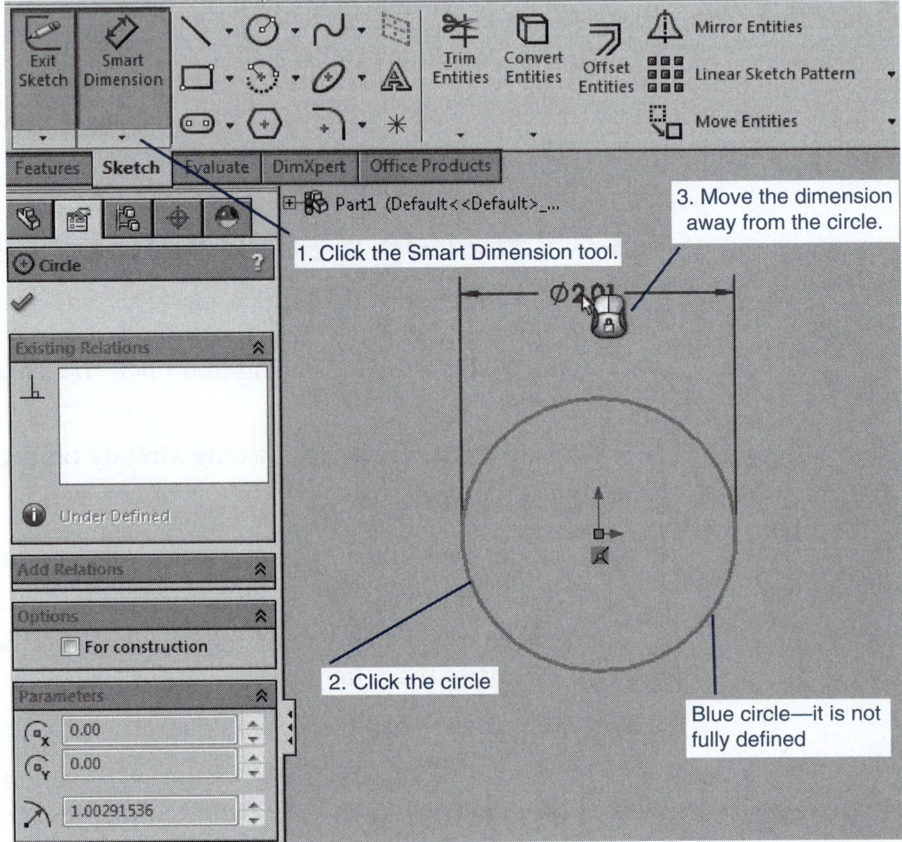

7 Select a location for the dimension and click the mouse.

The circle will initially be blue, not fully defined, until the mouse is clicked, locating the circle's dimension. When the mouse is clicked, the circle will turn black; it is now fully defined. We know the circle's diameter and location.

When the mouse is clicked, the **Modify** dialog box will appear. The sketched diameter value will be listed in the box. This sketched diameter value is now the circle's diameter until we enter a new value.

8 Enter a diameter value for the circle.

In this example a value of 2.00 was entered.

9 Click the green check mark in the **Modify** box to enter the diameter value.

10 Click the green **OK** check mark in the **Manager** area to finish defining the circle.

To Change an Existing Dimension

1 Double click the **2.00** dimension.

The **Modify** dialog box will reappear.

2 Enter a new value.

In this example a value of **3.00** was entered. See Figure 1-11. The circle's diameter will change to 3.00 and the circle's color will remain black. The circle still is fully defined.

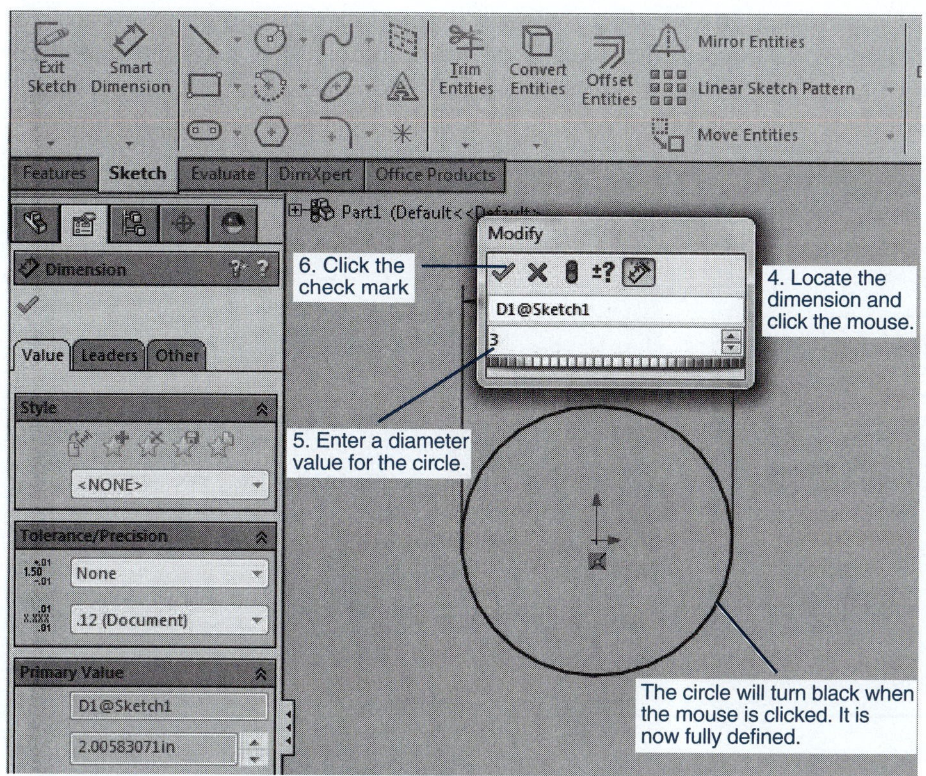

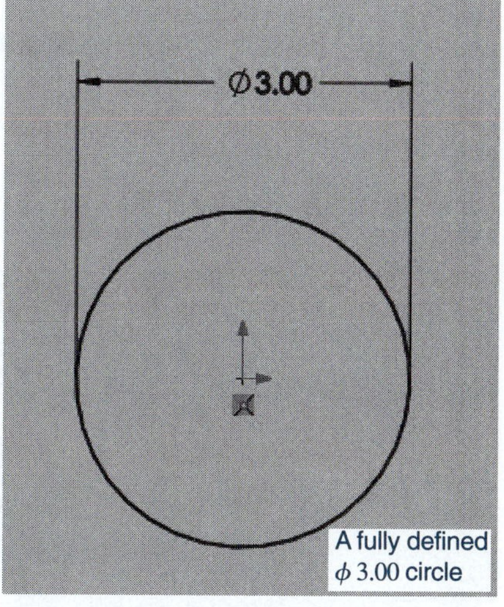

Figure 1-11

Note that the words **Fully Defined** appear at the bottom of the screen. The circle is fully defined because both its diameter and location are known. The location was fully defined when we located the circle's center point on the origin. Every circle needs a locational value and a diameter value to be fully defined. The locational value may be linear, an X and Y component value, or polar, an angular and radius value.

Fully Defined Entities

To help understand when an entity is fully defined, sketch two circles, one with its center point on the origin and one with its center point not on the origin. See Figure 1-12. Both circles are under defined because the diameter values have not been defined. Both circles are sketched circles.

Figure 1-12

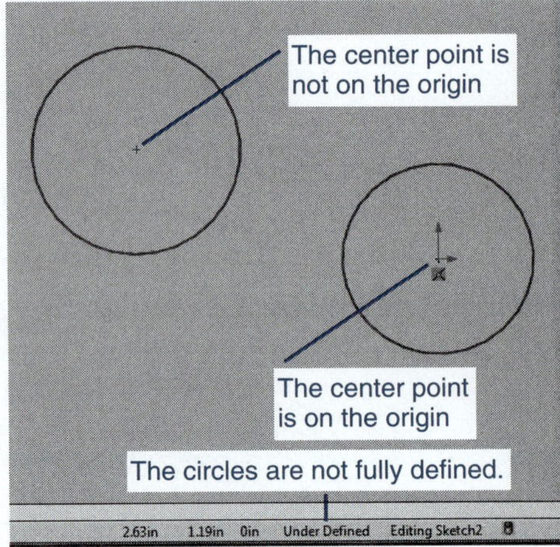

Use the **Smart Dimension** tool and define both their diameters as **Ø2.00**. The circle located on the origin will be black. It is fully defined. Both its diameter and location are known. The circle with its center point not located on the origin will remain blue. It is not fully defined. Its location is unknown. See Figure 1-13.

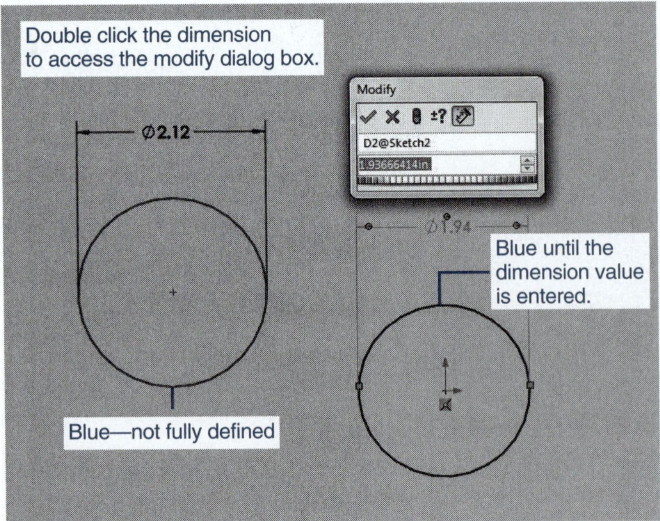

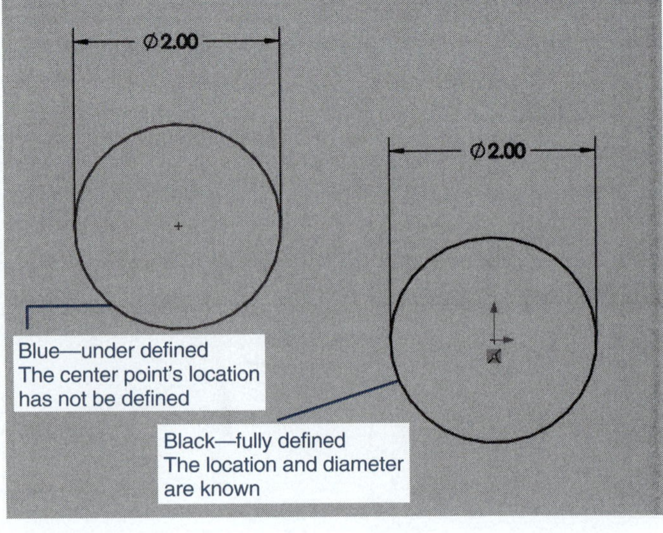

Figure 1-13

Figure 1-14 shows the two Ø2.00 circles again. This time, dimensions have been added to the circle not located on the origin. The dimensions define the circle's center point relative to the origin. It is now fully defined. Its color will change to black.

Figure 1-14

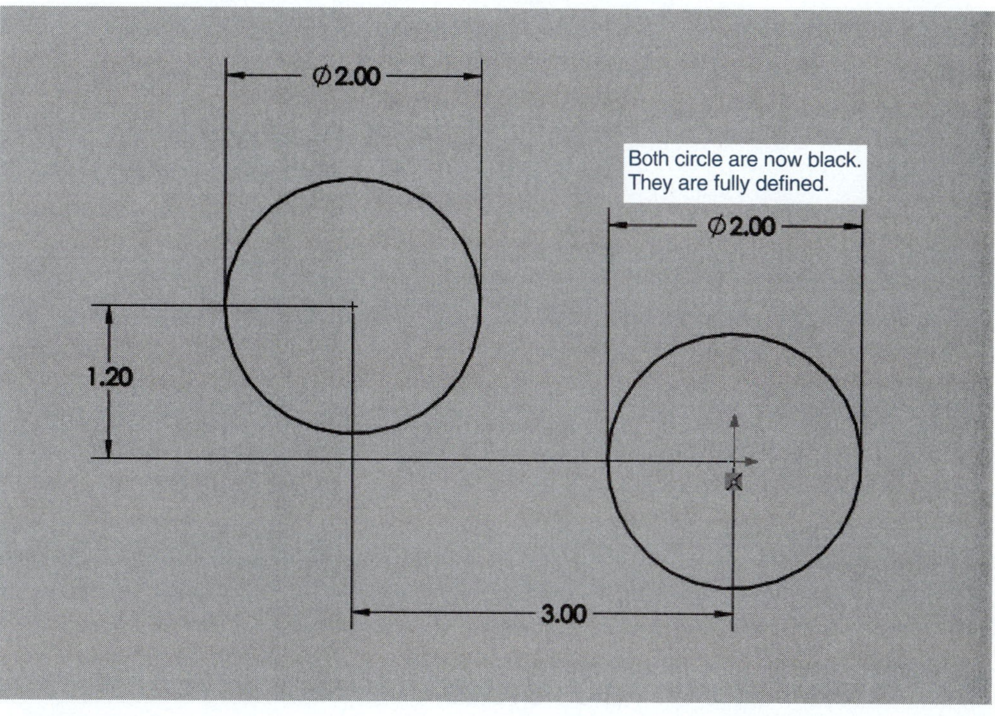

NOTE

Always include the origin as part of a 2D sketch.

Figure 1-15 shows the two Ø2.00 circles with an extra dimension. The 1.20 vertical dimension is not needed to define the location of the hole not centered on the origin. A 1.20 vertical dimension already exists. The 1.20 dimension is redundant, so the drawing lines change to yellow.

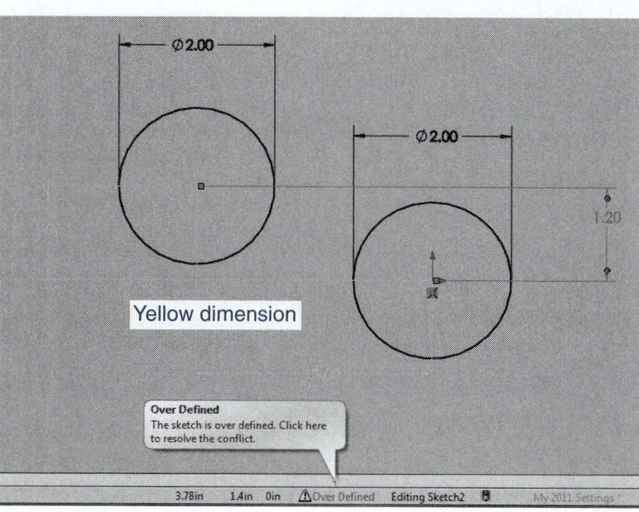

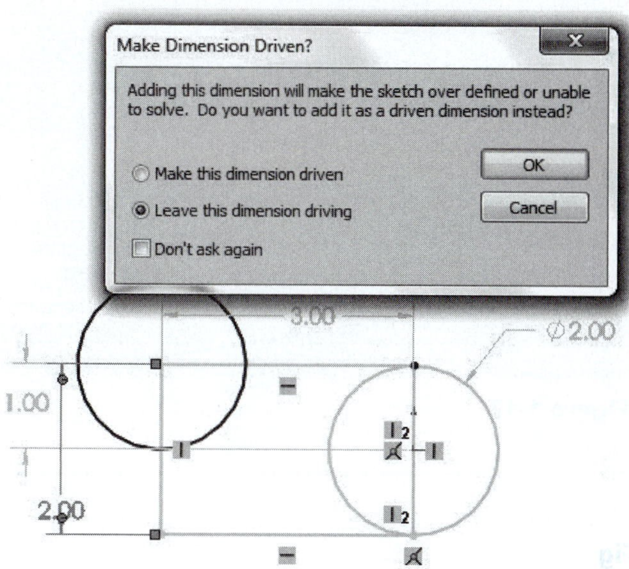

Figure 1-15

Figure 1-15 also shows the **Make Dimension Driven?** dialog box. A driving dimension drives the shape and/or location of the object. If the driving dimension is changed, the shape or location will change. Driven dimensions are reference dimensions. They are sometimes added to a drawing for clarity. For example, a reference dimension could be used to show the overall value of a string of smaller dimensions. See Chapter 7. In this example it would be better to delete the extra 1.20 dimension. If you save it on the drawing, click the **Make this dimension driven** option and click **OK**. It will appear as a gray color. See Figure 1-16.

Figure 1-16

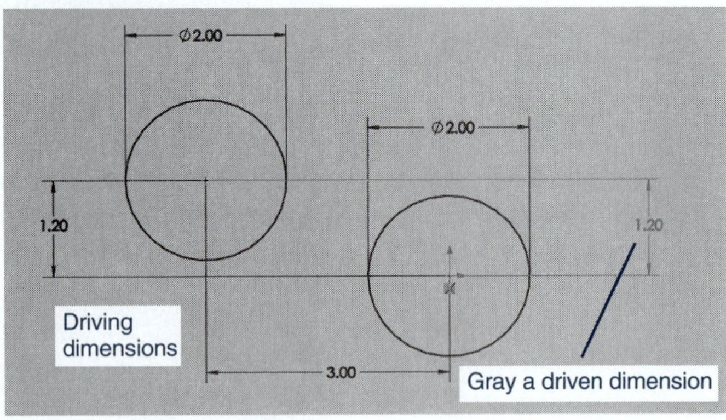

1-5 Units

This book will present examples and exercise problems using English units (inches) and Metric units (millimeters). Figure 1-17 shows the dimensioned circles created in the previous section. Note the letters **IPS** to the right of the **Fully Defined** callout. IPS stands for Inches, Pounds, and Seconds, the current units.

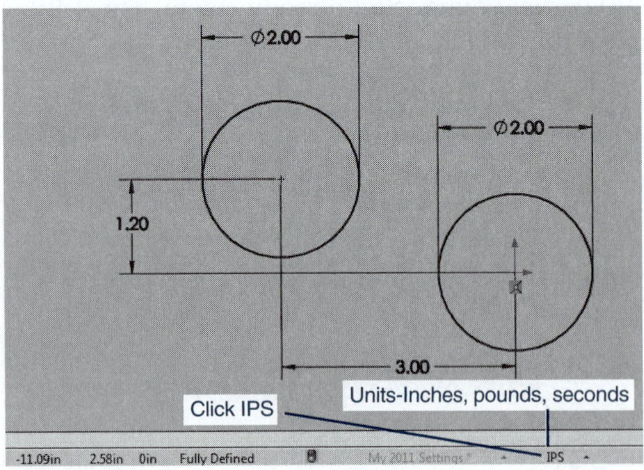

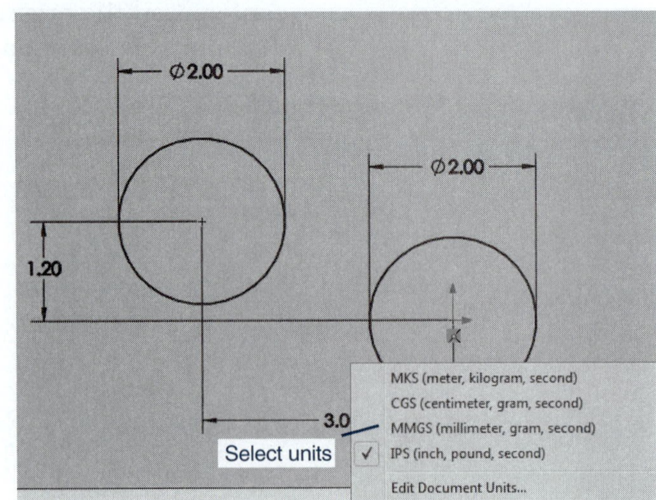

Figure 1-17

Figure 1-17
(*Continued*)

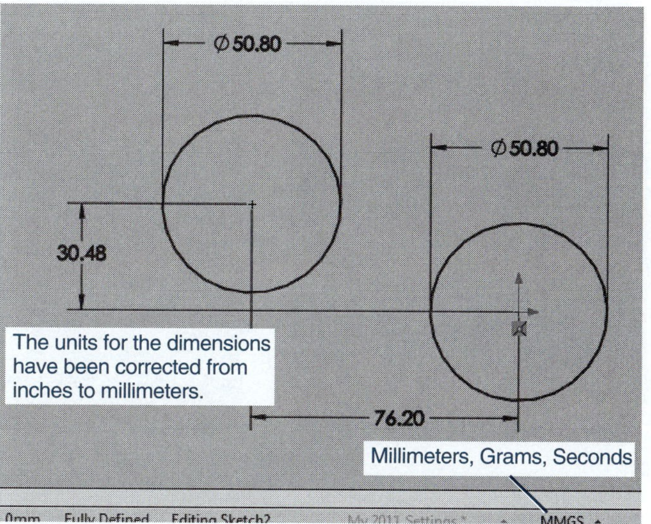

To Change Units

1 Click the **IPS** callout at the bottom of the screen.

2 Select the desired units.

In this example millimeters (**MMGS**) was selected. MMGS stands for Millimeter, Grams, and Seconds.

The letters **MMGS** will appear at the bottom of the screen, indicating that the drawing units are now millimeters.

> **NOTE**
>
> The converted millimeter dimensions are not whole numbers as were the inch units. It is better to do a drawing in either inches or millimeters from the beginning and not to convert units as a drawing is created. This helps prevent round off error.

1-6 Line

To Sketch a Line

See Figure 1-18. The example was created on the **Top** plane using the **Line** tool. The units are inches.

1 Start a **New Part** drawing, click the **Top** plane option, and click the **Sketch** tool.

See Figure 1-18. The outline rectangle for the **Top** plane will rotate to the *Normal* orientation, that is, you are looking at the plane from a 90° orientation. This means that any shape drawn on the plane will be a true shaped line. This concept will be covered in Chapter 4 on orthographic views.

2 Click the **Line** tool.

3 Click the origin and move the cursor to the right.

4 Click the mouse to define the line's endpoint.

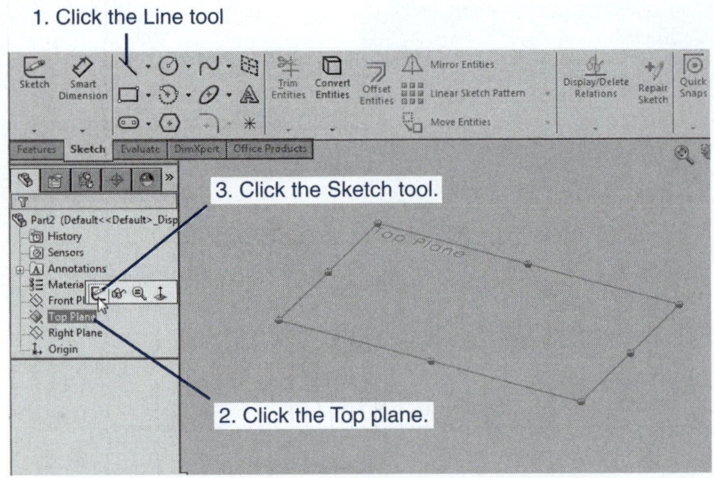
1. Click the Line tool
3. Click the Sketch tool.
2. Click the Top plane.
Top Plane
Part2 (Default<<Default>_Disp
History
Sensors
Annotations
Front Pl
Top Plane
Right Plane
Origin

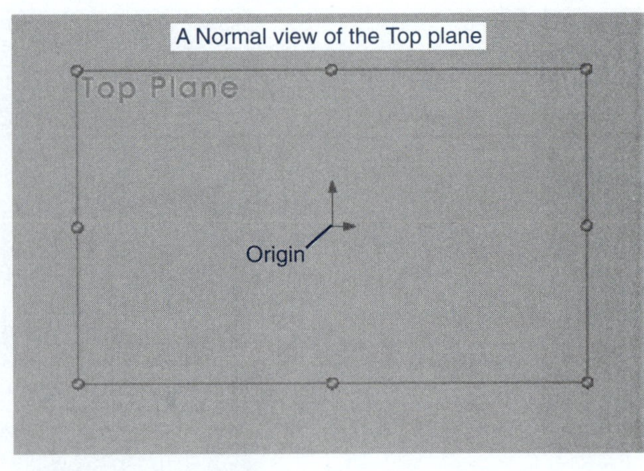
A Normal view of the Top plane
Top Plane
Origin

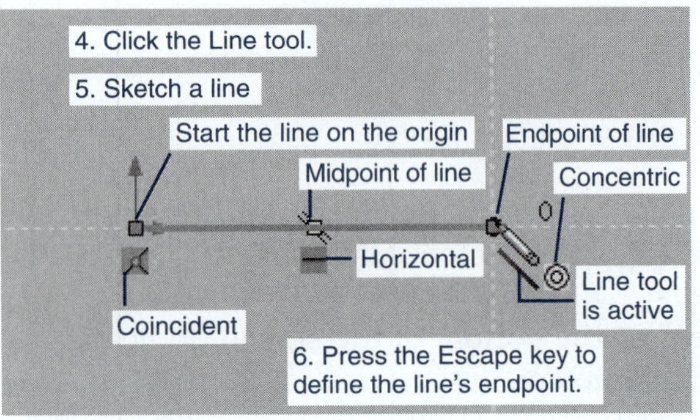
4. Click the Line tool.
5. Sketch a line
Start the line on the origin
Endpoint of line
Midpoint of line
Concentric
Horizontal
Line tool is active
Coincident
6. Press the Escape key to define the line's endpoint.

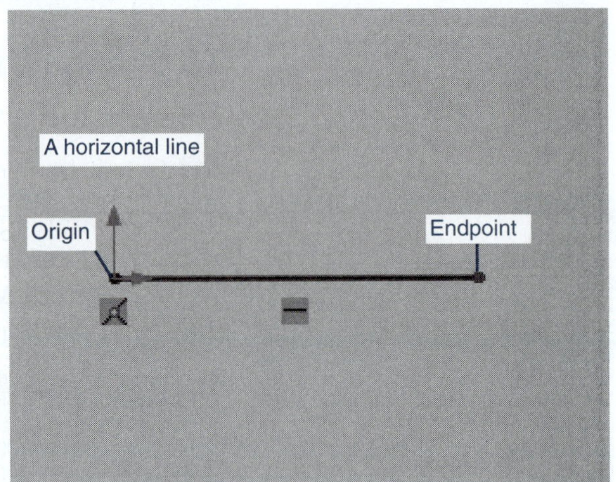
A horizontal line
Origin
Endpoint

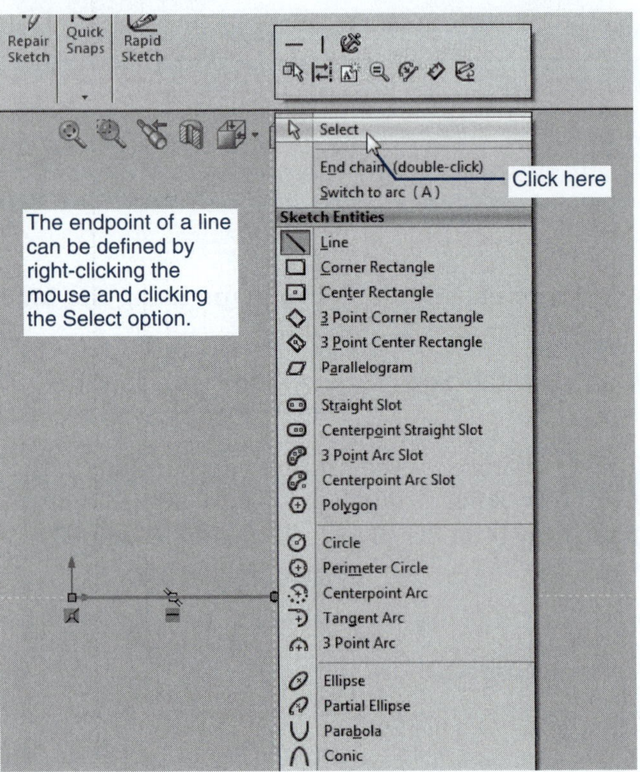
Repair Sketch
Quick Snaps
Rapid Sketch
Select
End chain (double-click)
Switch to arc (A)
Click here
The endpoint of a line can be defined by right-clicking the mouse and clicking the Select option.
Sketch Entities
Line
Corner Rectangle
Center Rectangle
3 Point Corner Rectangle
3 Point Center Rectangle
Parallelogram
Straight Slot
Centerpoint Straight Slot
3 Point Arc Slot
Centerpoint Arc Slot
Polygon
Circle
Perimeter Circle
Centerpoint Arc
Tangent Arc
3 Point Arc
Ellipse
Partial Ellipse
Parabola
Conic

Figure 1-18

5 Press the **<Esc>** key or right-click the mouse and click the **Select** option.

Note that two relationships are defined: coincident and horizontal. The starting point was located on the origin, so they are coincidental and the line is drawn in the horizontal direction. Note also that the line is not fully defined because its length has not been defined.

Releasing the mouse button will define the length of the sketched line, but you are still in the **Sketch** mode. If you click the mouse again, a new line will begin.

To Exit the Sketch Mode

1 Click the **Exit Sketch** icon on the **Sketch** panel or click the **Exit Sketch** icon that appears in the upper right corner of the drawing screen.

See Figure 1-19.

Figure 1-19

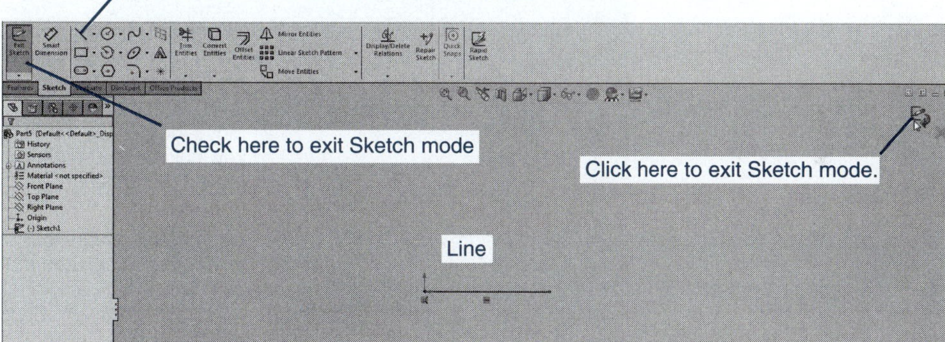
Click the Line tool icon to exit the Line mode.
Check here to exit Sketch mode
Click here to exit Sketch mode.
Line

To Reenter the Sketch Mode

Once you have created a sketch and left the **Sketch** mode, you can return to work on the sketch by using the **Edit Sketch** mode. See Figure 1-20.

Figure 1-20

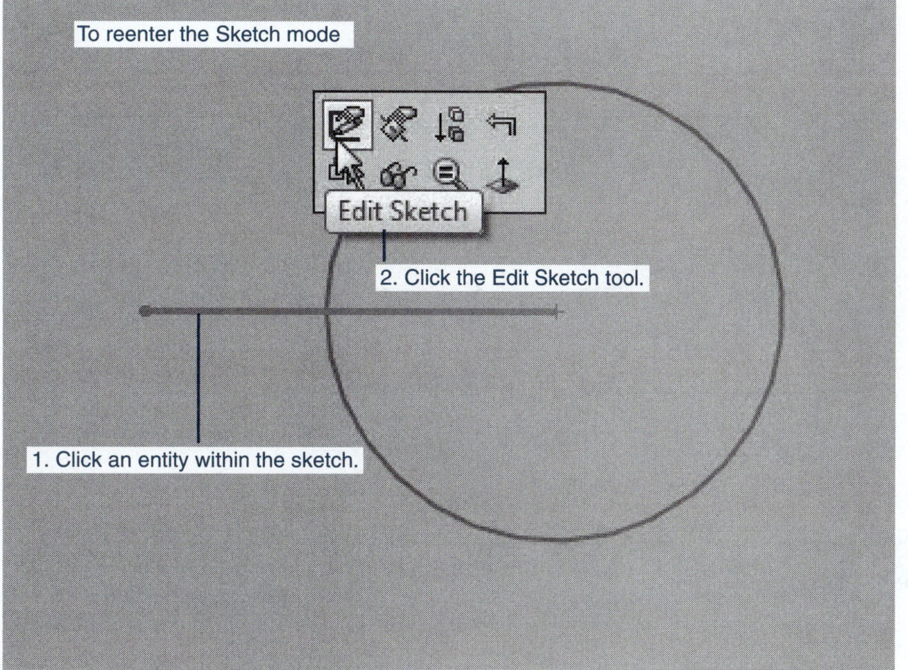
To reenter the Sketch mode
Edit Sketch
2. Click the Edit Sketch tool.
1. Click an entity within the sketch.

1. Click an entity in the existing sketch.
2. Click the **Edit Sketch** tool.

1-7 Moving Around the Drawing Screen

SolidWorks includes several methods that allow you to move entities about the screen. Entities can be moved, zoomed, or reorientated. Figure 1-21 shows the line created in the previous section.

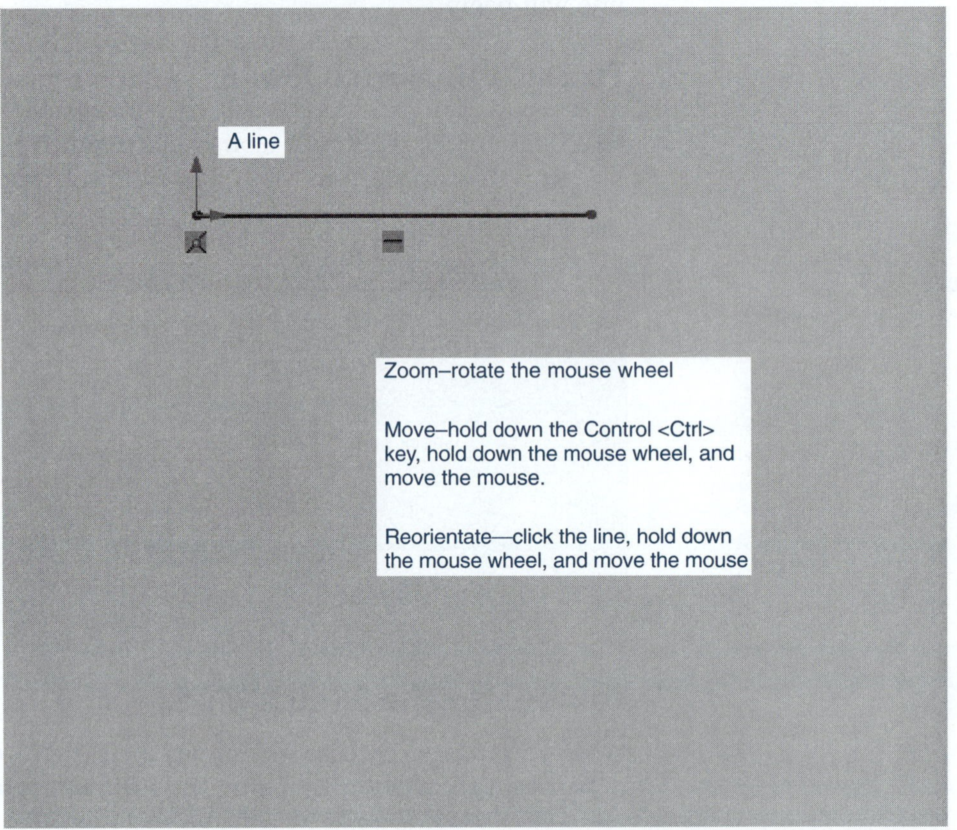

Figure 1-21

To Zoom the Line

1. Rotate the mouse wheel.

   The line will increase and decrease in length.

To Move the Line

1. Hold down the Control **<Ctrl>** key; press and hold down the mouse wheel.
2. Move the mouse around.

   The line will follow the mouse movement.

To Reorientate the Line

1. Click the line.
2. Hold down the mouse wheel and move the mouse.

   The mouse's orientation will follow the mouse movement.

1-8 Orientation

The line created in the previous sections was created in the **Top view** orientation. As you work on a sketch the orientation may change. There are three ways to return the sketch's orientation to its original orientation.

To Return to the Top View Orientation – View Selector

1 Click the **View Orientation** tool at the top of the drawing screen.

The **View Selector** cube will appear. See Figure 1-22. If the cube does not appear, click the **View Selector** icon on the **View Orientation** tool panel.

Figure 1-22

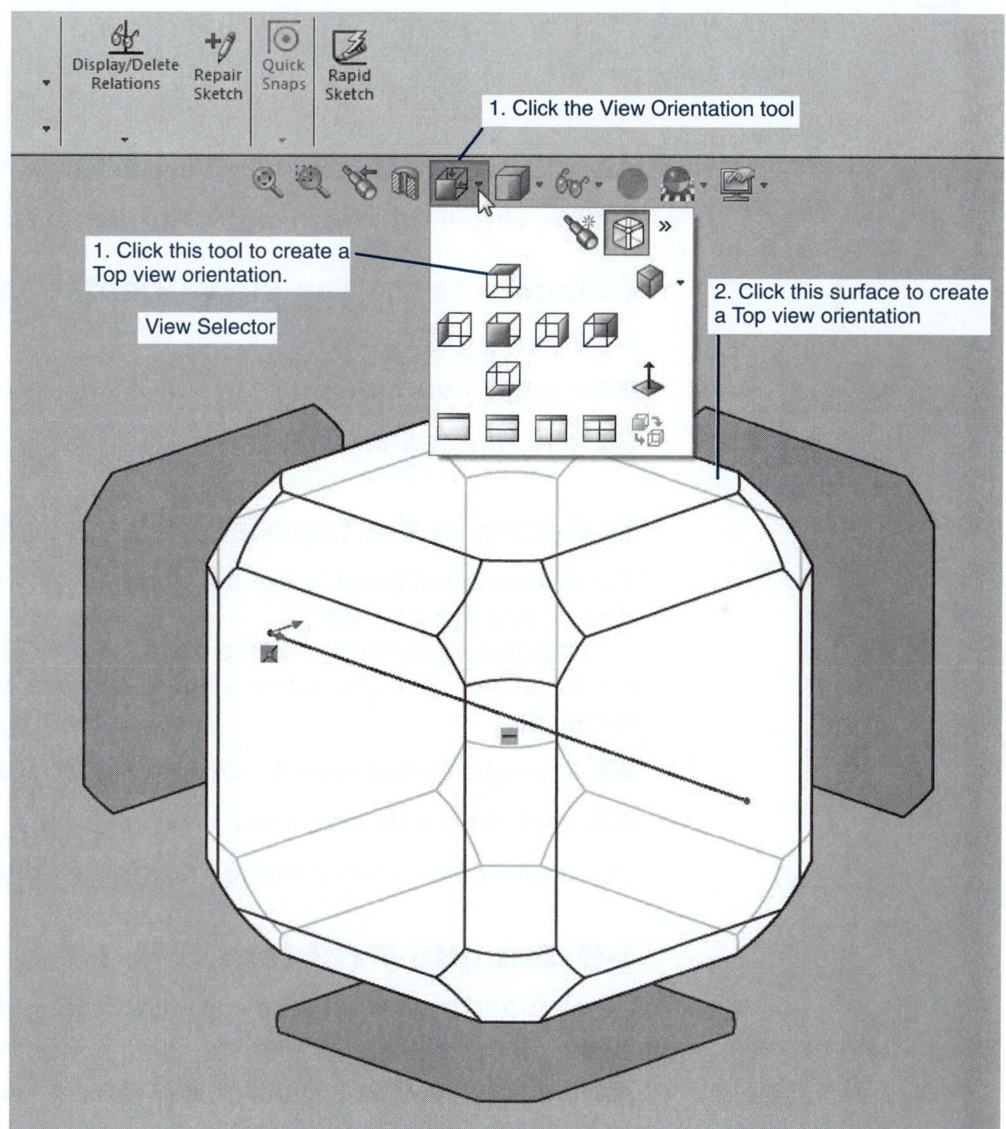

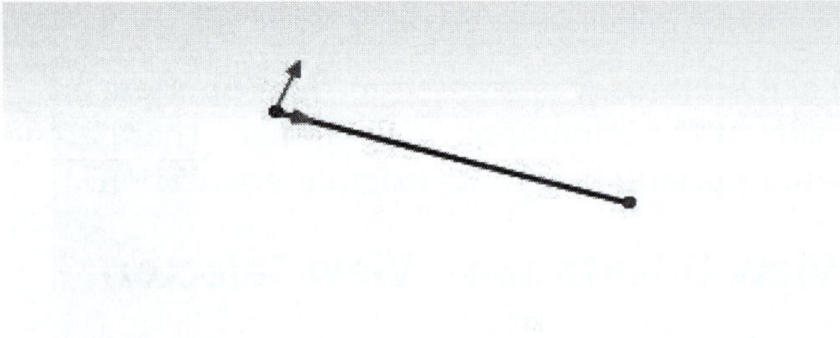

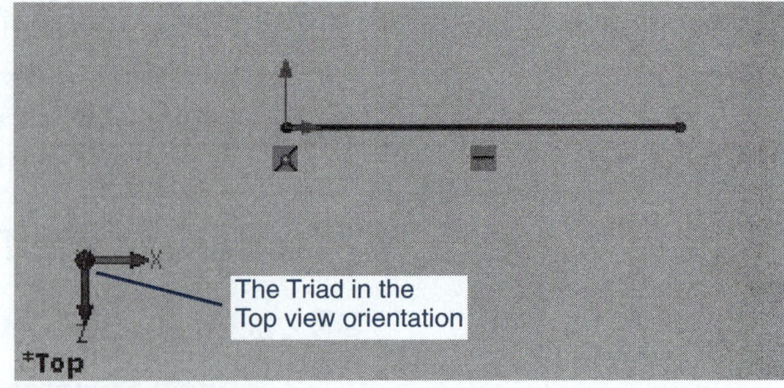

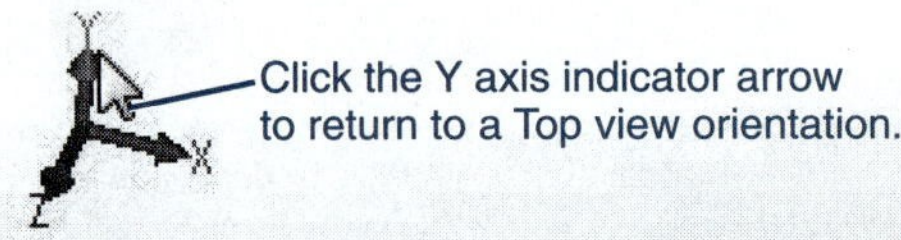

Figure 1-22
(Continued)

2 Click the top surface of the **View Selector** cube.

The sketch will return to the **Top view** orientation.

To Return to the Top View Orientation – Top View

See Figure 1-22.

1 Click the **View Orientation** tool at the top of the drawing screen.

2 Click the **Top View** tool.

To Return to the Top View Orientation – Orientation Triad

The **Orientation Triad** is located in the lower left corner of the drawing screen. See Figure 1-22.

SolidWorks defines the **Top** plane as the XZ plane. The Y axis is 90° to the XZ plane, so a view taken along the Y axis will generate a top view of the plane.

1 Move the cursor onto the **Orientation Triad**.

2 Click the Y axis indicator arrow.

The triad will reorientate to the **Top view** orientation.

1-9 Sample Problem SP1-1

Figure 1-23 shows a 2D shape sketched using the **Line** tool. The dimensions are in millimeters. This section will explain how to draw the shape.

1 Start a **New Part** document, select the **Front** plane, and create a **Sketch** plane.

See Figure 1-24.

2 Define the dimensional units as millimeters, **MMGS**.

See Figure 1-25.

3 Click the **Line** tool.

4 Select the origin as the starting point for the first line.

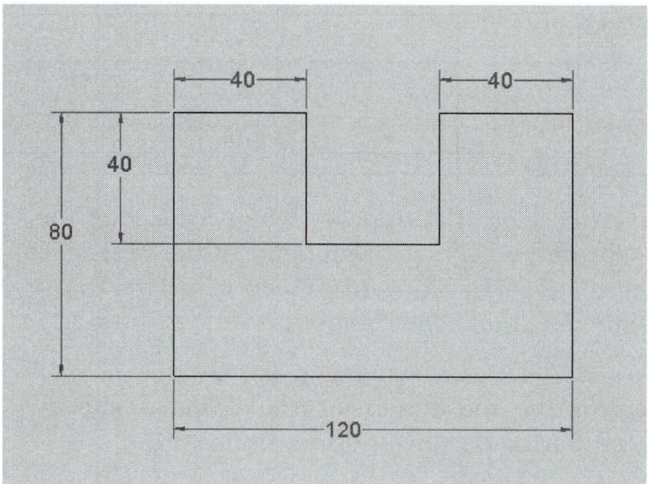

Figure 1-23

Figure 1-24

Figure 1-25

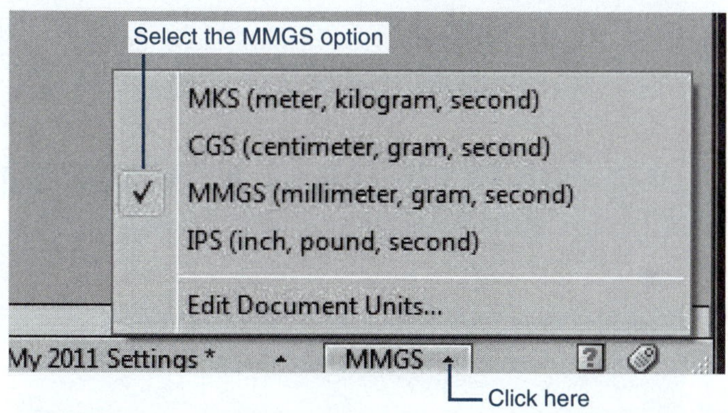

NOTE

The line command will generate a series of chain lines, where the endpoint of a sketched line becomes the starting point for the next line, until the line's endpoint is defined by pressing the <Esc> key or right-clicking the mouse and clicking the **Select** option.

See Figure 1-26.

Figure 1-26

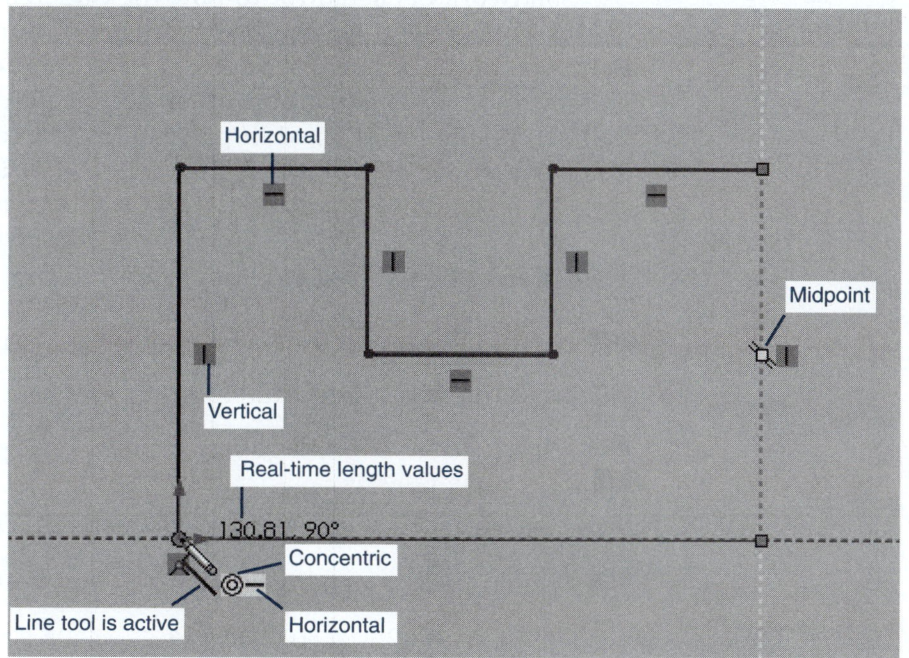

5 Sketch the general shape as shown.

> **HINT**
> Make each line slightly larger than the stated dimension. Exact values are not required. Use the real-time length values to estimate the length of the longer lines.

Note the double circle relation icon that appears when the end of the last horizontal line drawn is located on the starting point of the first line. This is the **Concentric** relation icon. The Concentric icon indicates that the two points occupy the same location. The midpoint of the right-side vertical line is also defined.

6 Click the **Smart Dimension** tool and dimension the shape as shown by clicking each line and entering the given dimensional value.

See Figure 1-27.

Figure 1-27

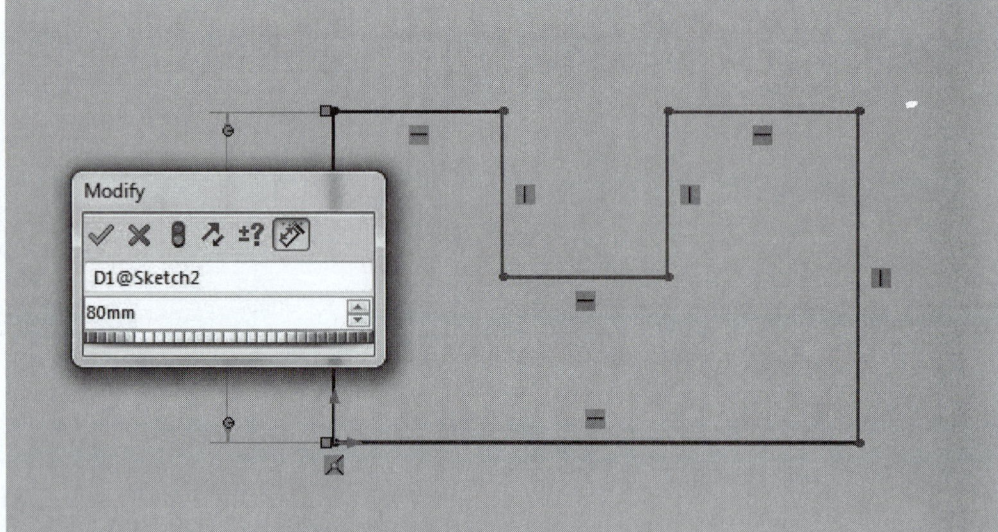

SolidWorks is sensitive to how the dimensions are entered. See Figure 1-28. Note that when the vertical 40 dimension was added to the right side of the shape the adjacent horizontal 40 line moved upwards. This means that the two horizontal 40 lines are no longer aligned. The right 40 line must be fixed in place so that it remains aligned with the other horizontal 40 line when the vertical 40 dimension is added. The vertical 40 dimension will then move the bottom of the slot downwards.

To Fix a Line in Place

1 Use the **Undo** tool to remove the vertical 40 dimension.

2 Click the right horizontal 40 line.

3 Click the **Make Fixed** tool.

The **Make Fixed** tool's icon is an anchor. When the **Make Fixed** tool is activated, an anchor icon will appear below the line.

4 Double click the vertical dimension and enter a value of **40**.

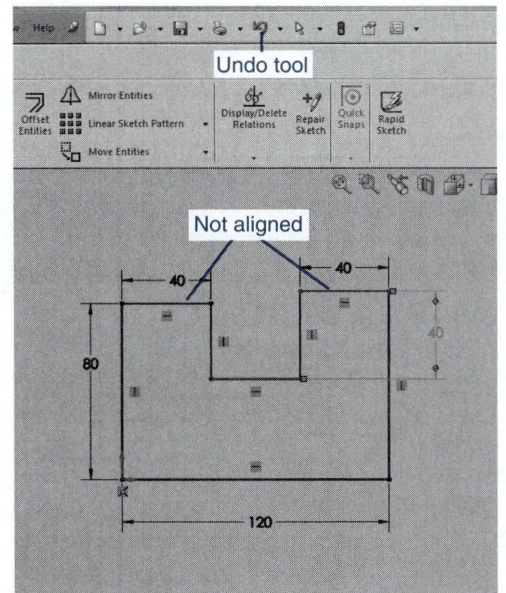

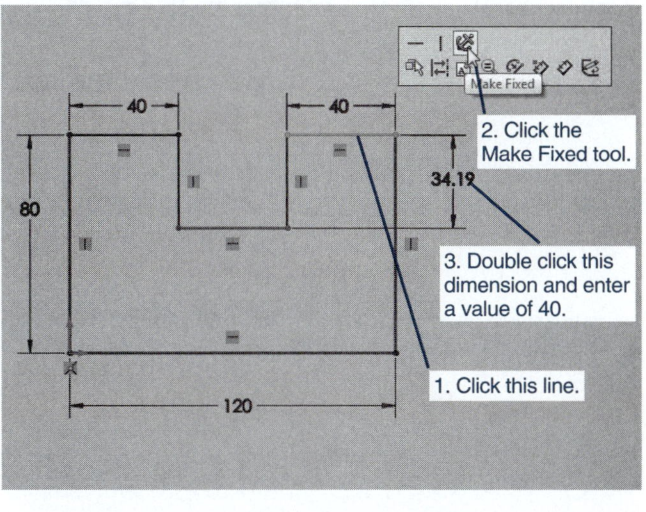

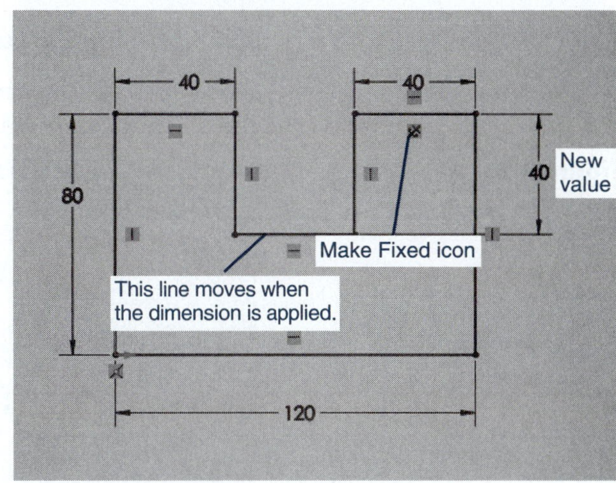

Figure 1-28

The horizontal line will move, accepting the 40 dimensional changes. The two horizontal lines remain aligned.

Figure 1-29 shows the shape dimensioned as it was presented in Figure 1-23. The status bar shows that the shape is **Under Defined**. The top horizontal 40 line is blue.

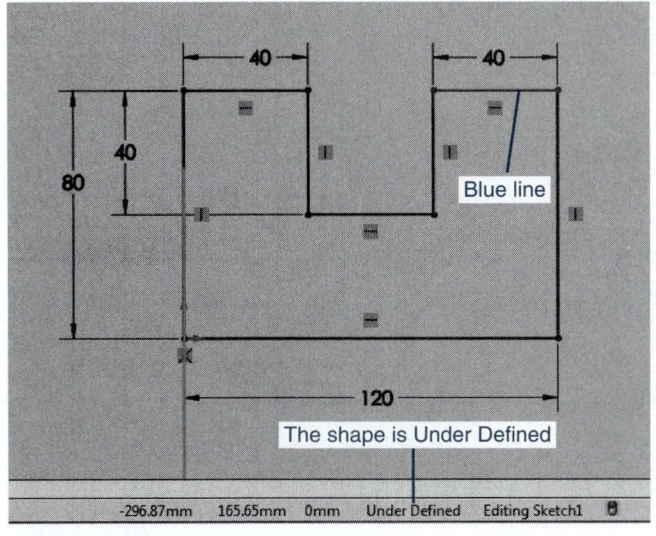

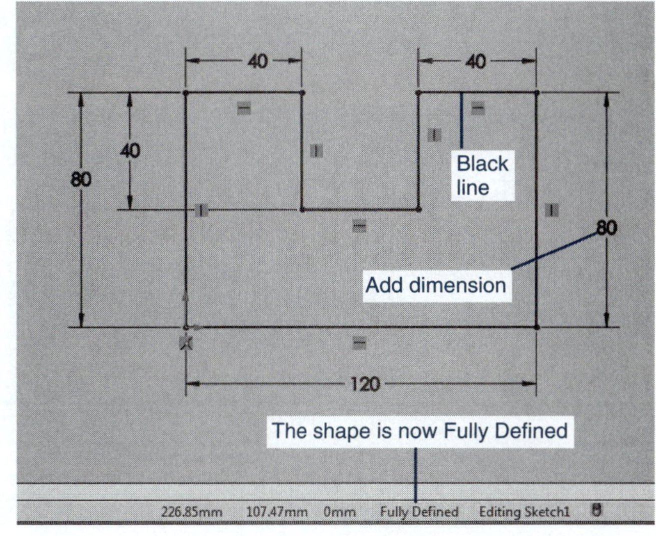

Figure 1-29

To Fully Define the Shape

1 Add another vertical 80 dimension as shown.

The shape is now **Fully Defined**.

> **NOTE**
> 2D shapes should always be fully defined before creating 3D models.

1-10 Creating 3D Models

The fully defined shape shown in Figure 1-29 can now be used to create a 3D model.

To Create a 3D Model

1 Click the **Features** tab.

2 Click the **Extruded Boss/Base** tool.

See Figure 1-30. The shape will change orientation to the **Trimetric** format. The sketch was created on the **Front** plane.

Figure 1-30

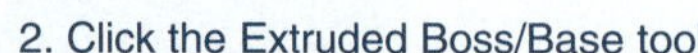

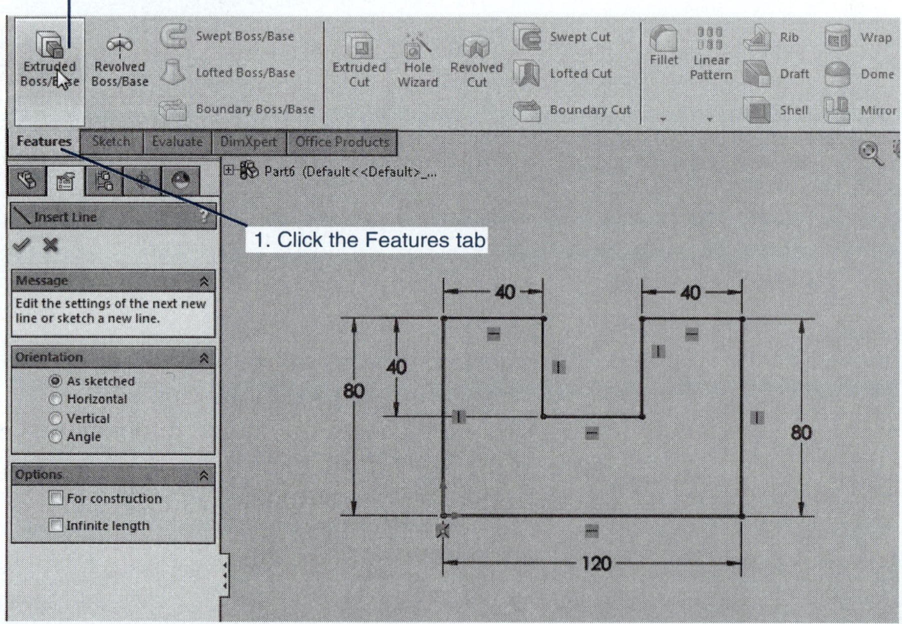

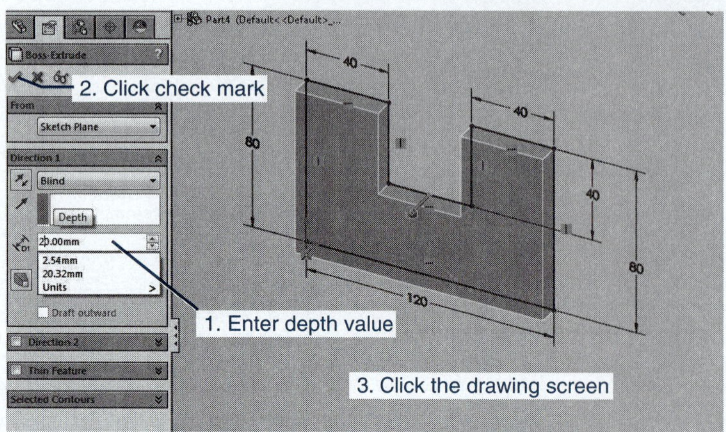

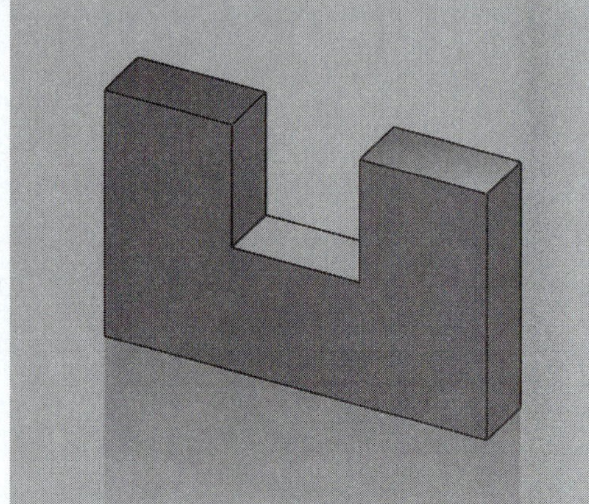

3 Define the depth as **20 mm**.

4 Click the green check mark.

5 Click the drawing screen.

1-11 Saving a Document

See Figure 1-31.

Figure 1-31

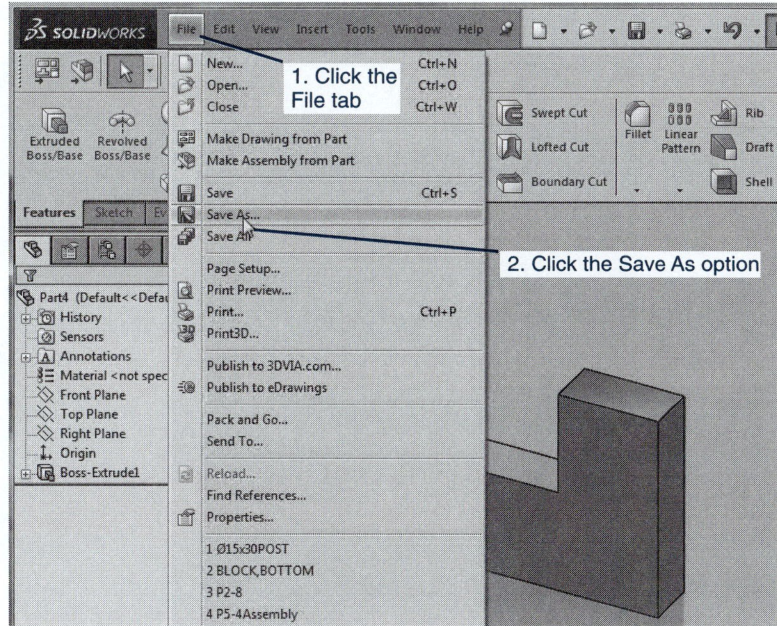

To Save a Document

1 Click the **File** tab at the top of the drawing screen.

A drop-down menu will appear.

2 Click the **Save As** tool.

The **Save As** dialog box will appear. See Figure 1-32.

Figure 1-32

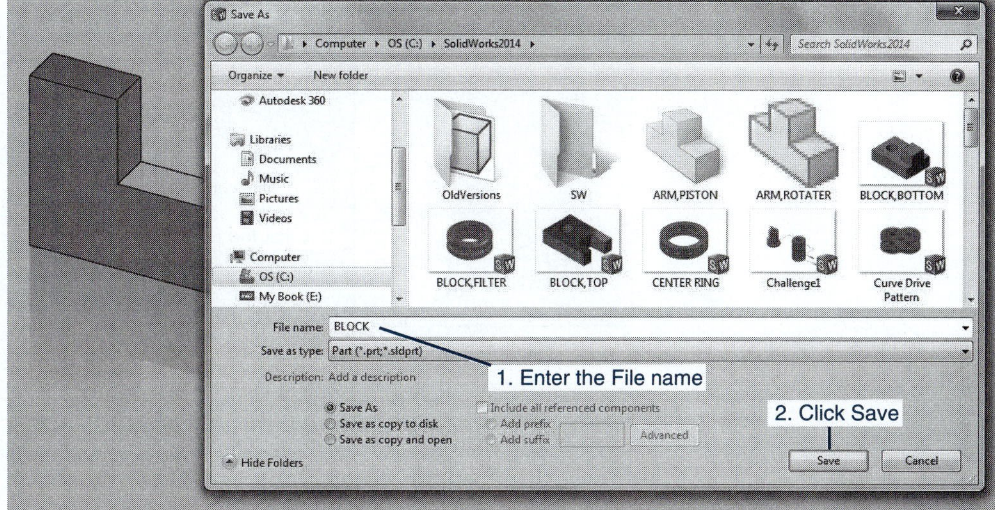

3 Enter the **File name**.

In this example the name **BLOCK** was used.

4 Click the **Save** box.

1-12 Lines and Angles – Sample Problem SP1-2

Figure 1-33 shows a 2D shape that includes two angles. The dimensions are in inches. This section will show how to create the shape.

Figure 1-33

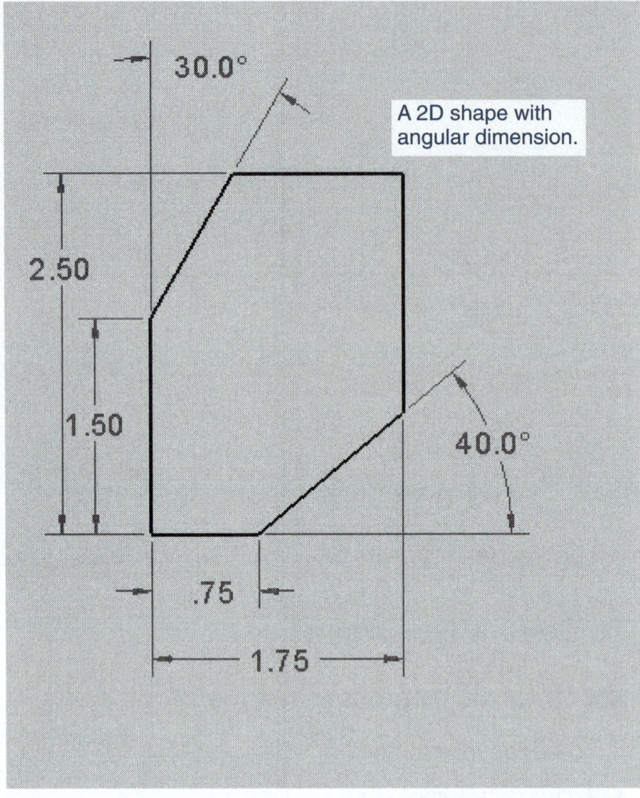

1 Click the **Sketch** tab, the **Front Plane**, and the **Sketch** tool. See Figure 1-34.

Figure 1-34

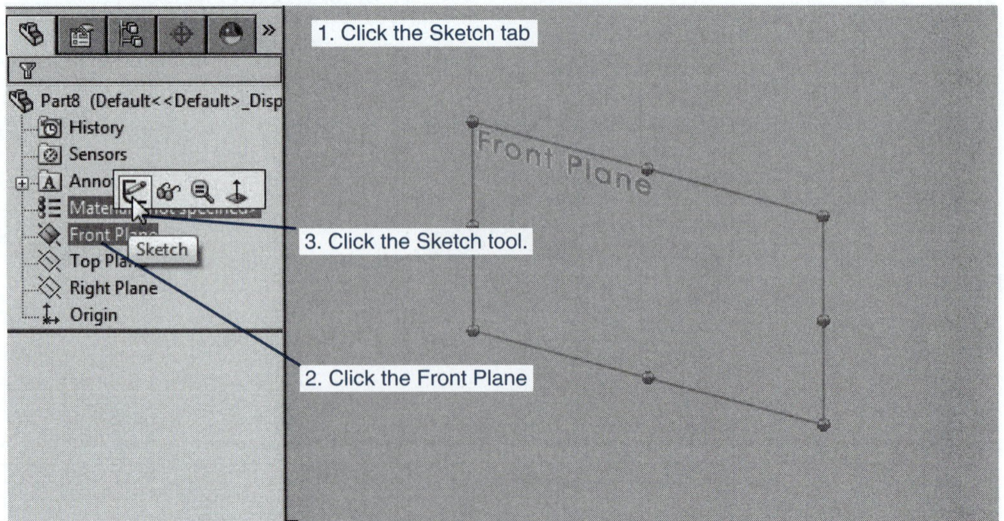

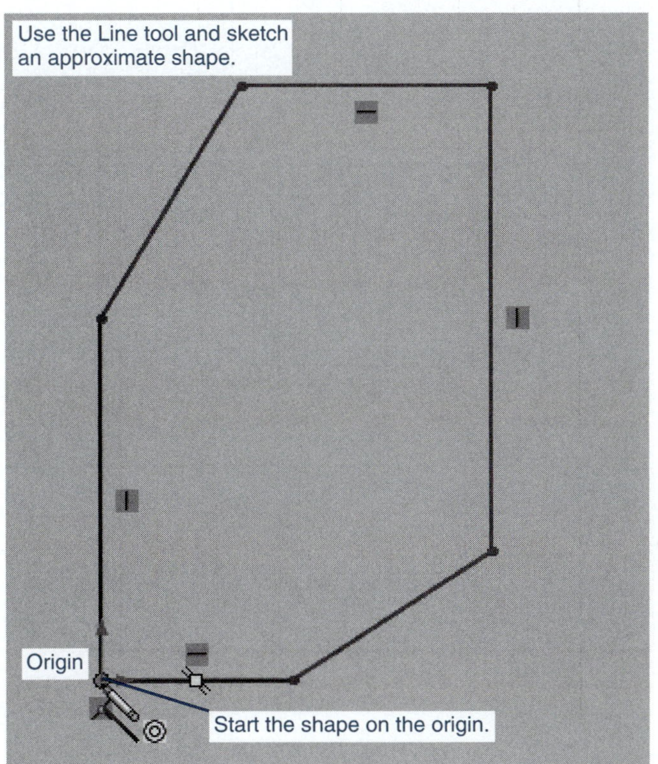
Use the Line tool and sketch an approximate shape.
Origin
Start the shape on the origin.

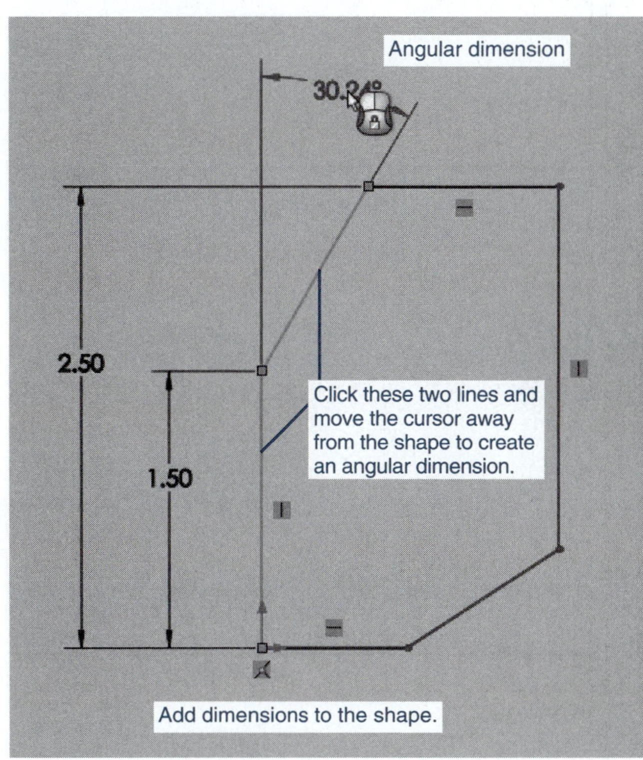
Angular dimension
2.50
1.50
Click these two lines and move the cursor away from the shape to create an angular dimension.
Add dimensions to the shape.

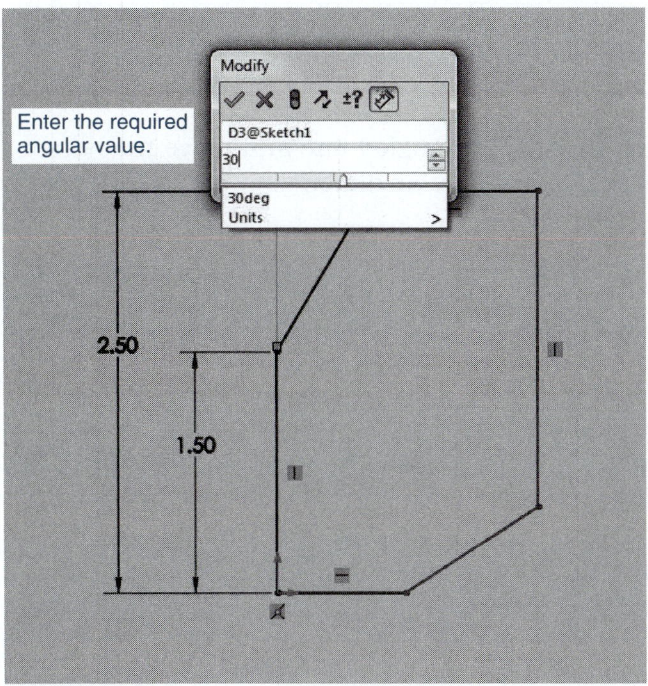
Enter the required angular value.
Modify
D3@Sketch1
30
30deg
Units
2.50
1.50

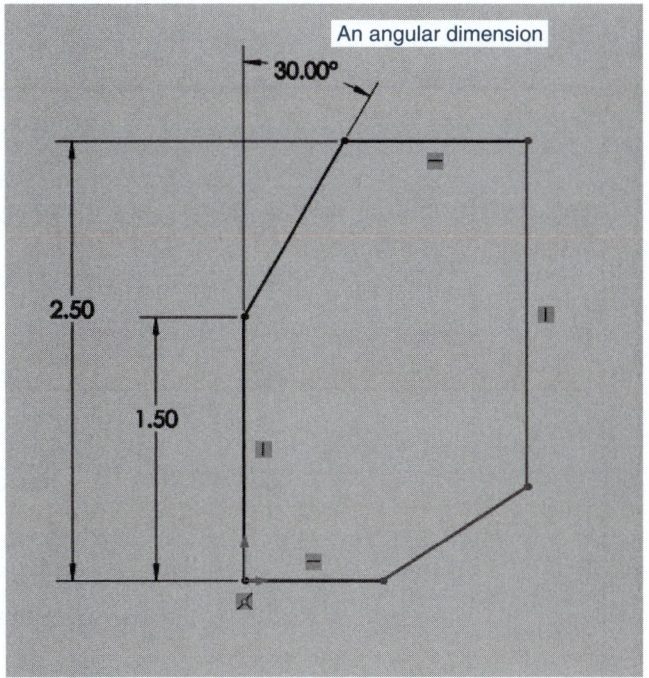
An angular dimension
30.00°
2.50
1.50

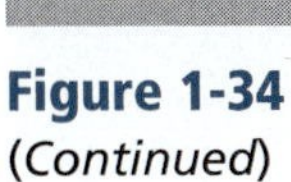
Figure 1-34
(*Continued*)

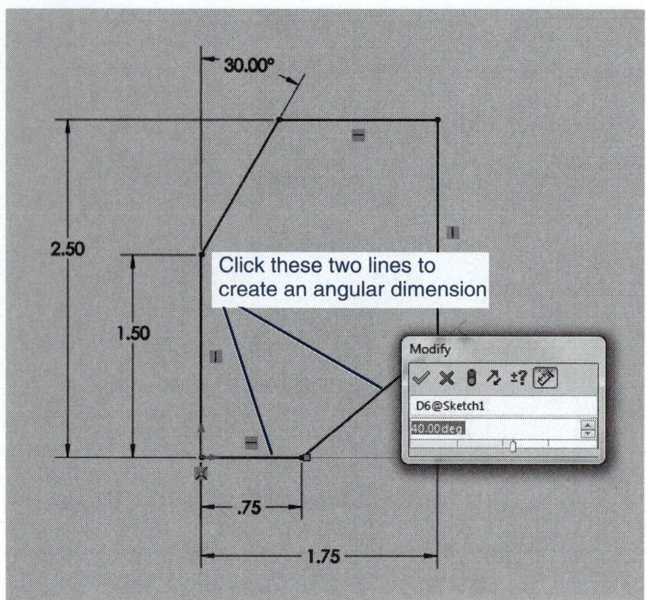

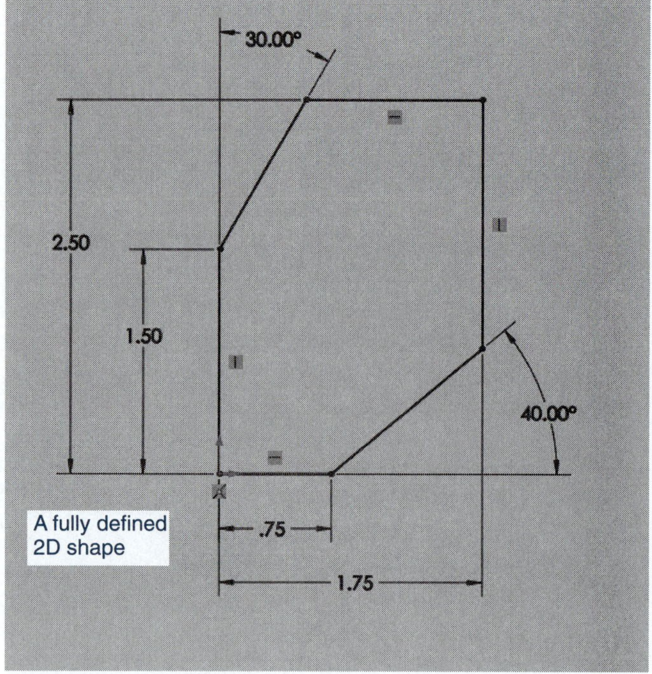

Figure 1-34
(Continued)

2 Use the **Line** tool and sketch the approximate shape.

Start the first line of the shape on the origin. Sketch the shape slightly larger than the final shape.

3 Add dimensions to the shape.

4 Click the left vertical line and the left angled line and move the cursor away from the shape to create an angular dimension.

5 Select a location for the dimension and click the mouse.

6 Enter the angle value.

In this example the value is **30°**.

7 Complete the remaining dimensions.

8 Ensure that the shape is fully defined.

9 Click the **Features** tab, the **Extruded Boss/Base tool**, and define the depth.

In this example, a depth of **0.50** was entered. See Figure 1-35.

10 Click the green **OK** check mark and then click the drawing screen.

See Figure 1-36.

Figure 1-35

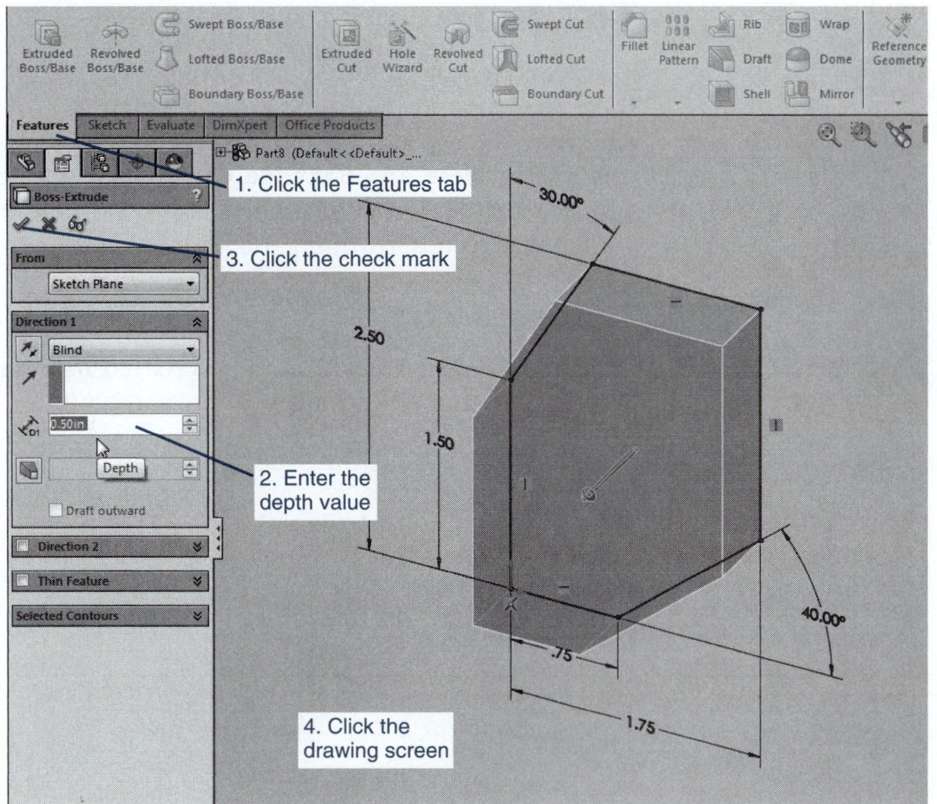

Figure 1-36

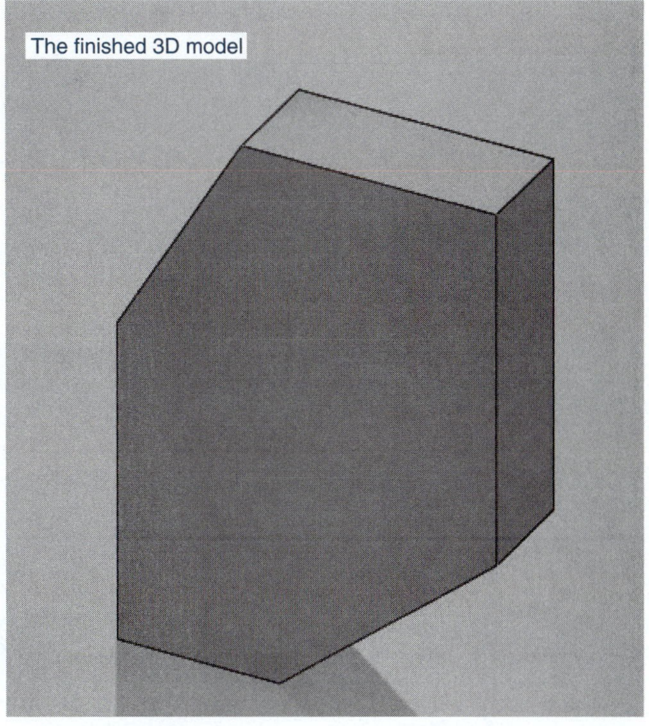

1-13 Holes

There are several different ways to create holes using SolidWorks. Most holes are created using the **Hole Wizard** tool. Hole Wizard is explained in Chapter 3. For purposes of this introductory chapter, holes will be created using the **Circle** and **Extrude Cut** tools. A circle will be created, then cut through the 3D shape. All holes will be simple through holes; that is, they will go completely through the shape.

To Create a Hole

Figure 1-37 shows the 3D shape created in Sample Problem SP1-1. Two Ø20.0 holes have been added.

Figure 1-37

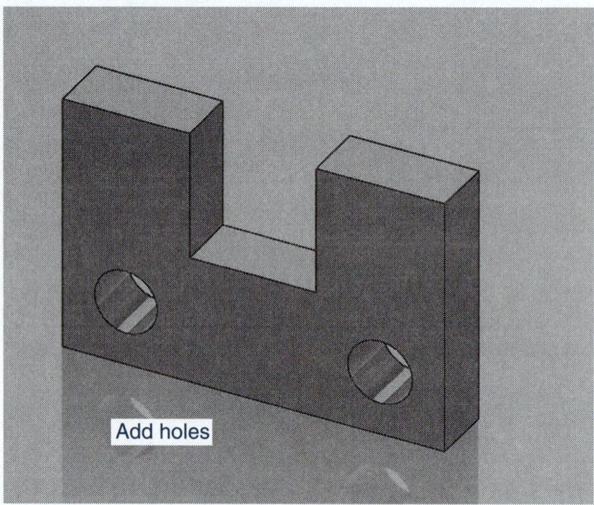

1. Click the **File** tool heading at the top of the screen and click the **Open** option, or click the **Open** tool.

2. Locate and click the **BLOCK** file created and saved in the last section.

See Figure 1-38. In this example the file was located on the C: drive under the file heading **SolidWorks 2014**.

Figure 1-38

Folder location

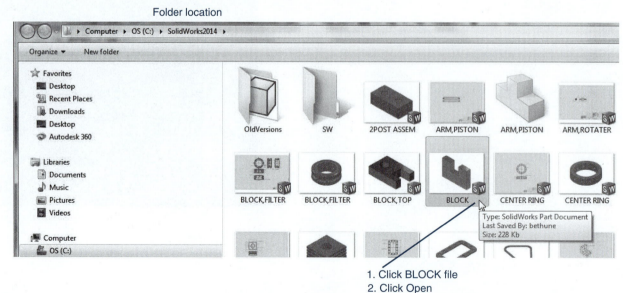

1. Click BLOCK file
2. Click Open

3 Click the **BLOCK** file, and click **Open**.

The BLOCK will appear on the screen. See Figure 1-39.

Figure 1-39

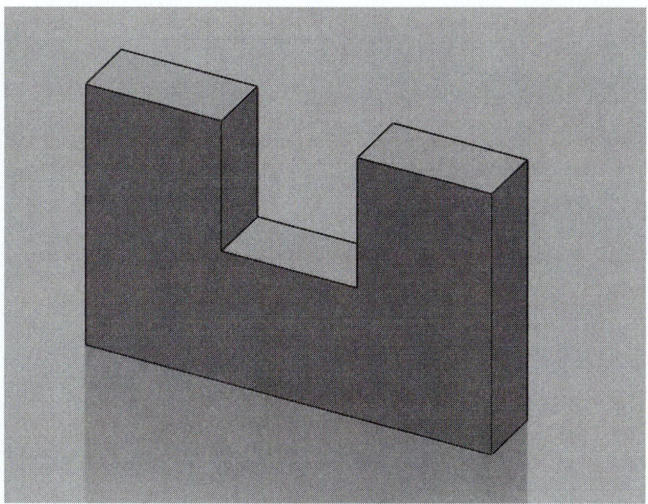

4 Click the **View Orientation** tool and select the **Normal** option.

This will create a view from an orientation point 90° to the surface. This is called a *normal* view. See Figure 1-40.

Figure 1-40

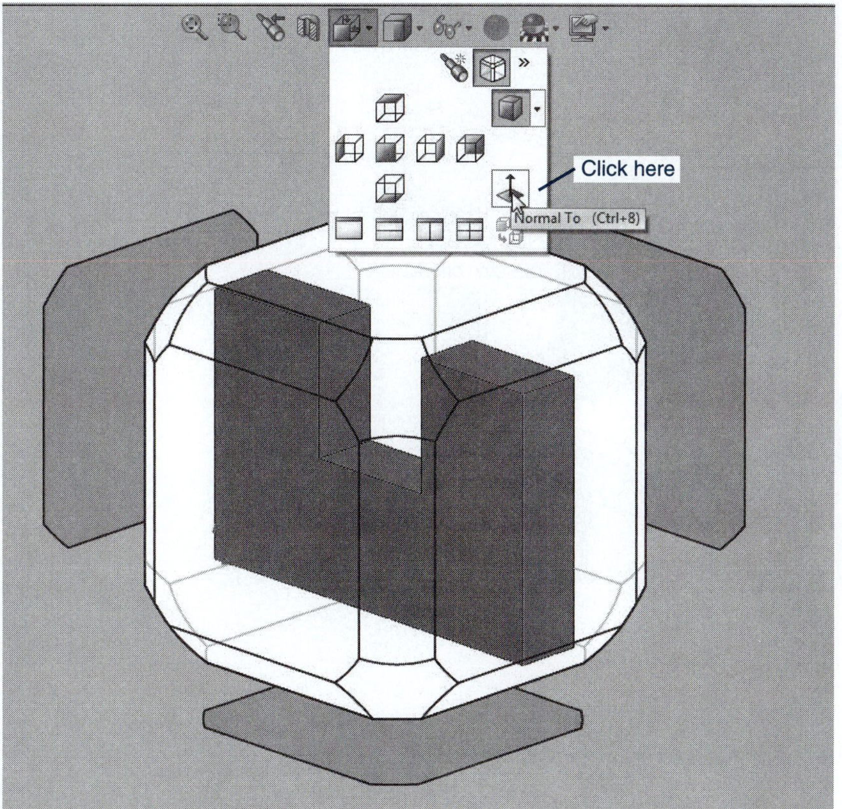

5 Right-click the front surface of the **BLOCK** and click the **Sketch** tool. See Figure 1-41.

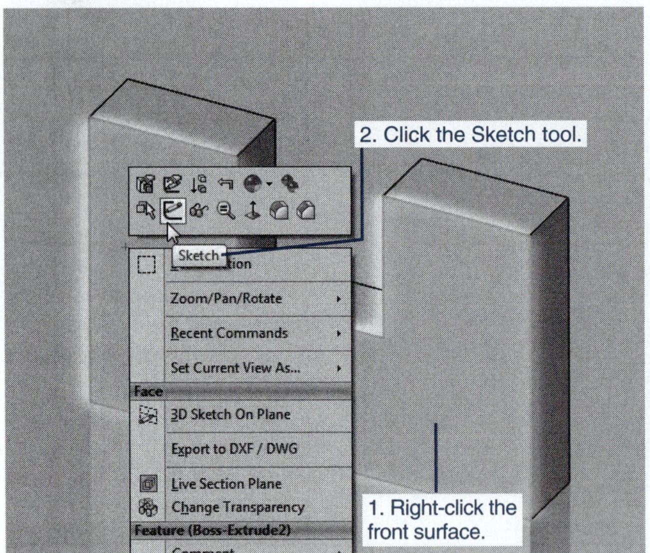

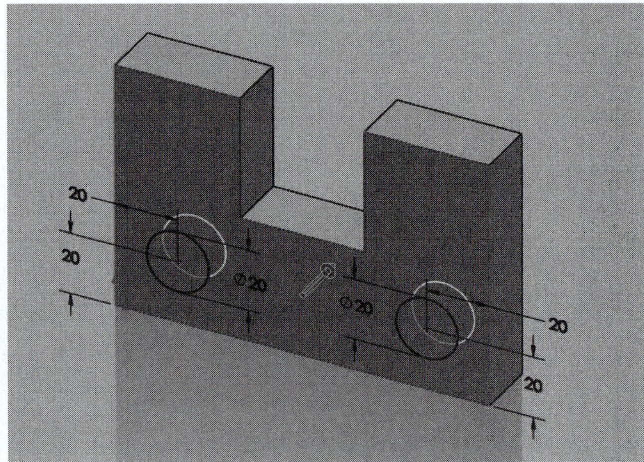

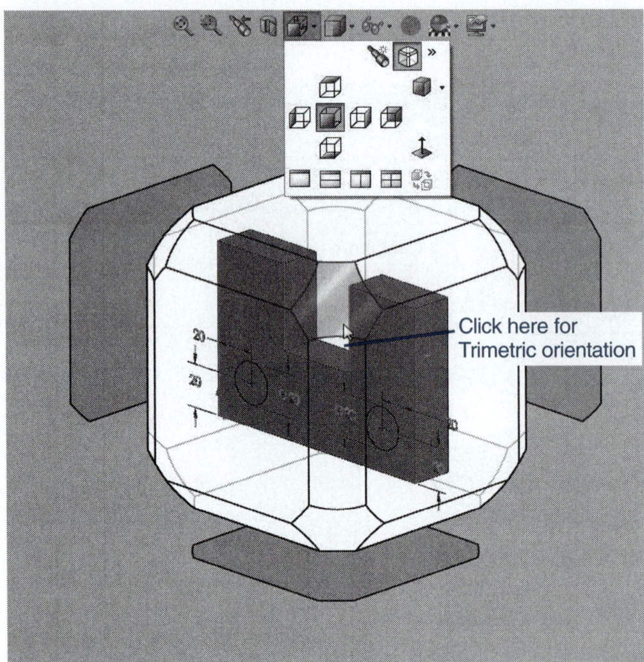

Figure 1-41

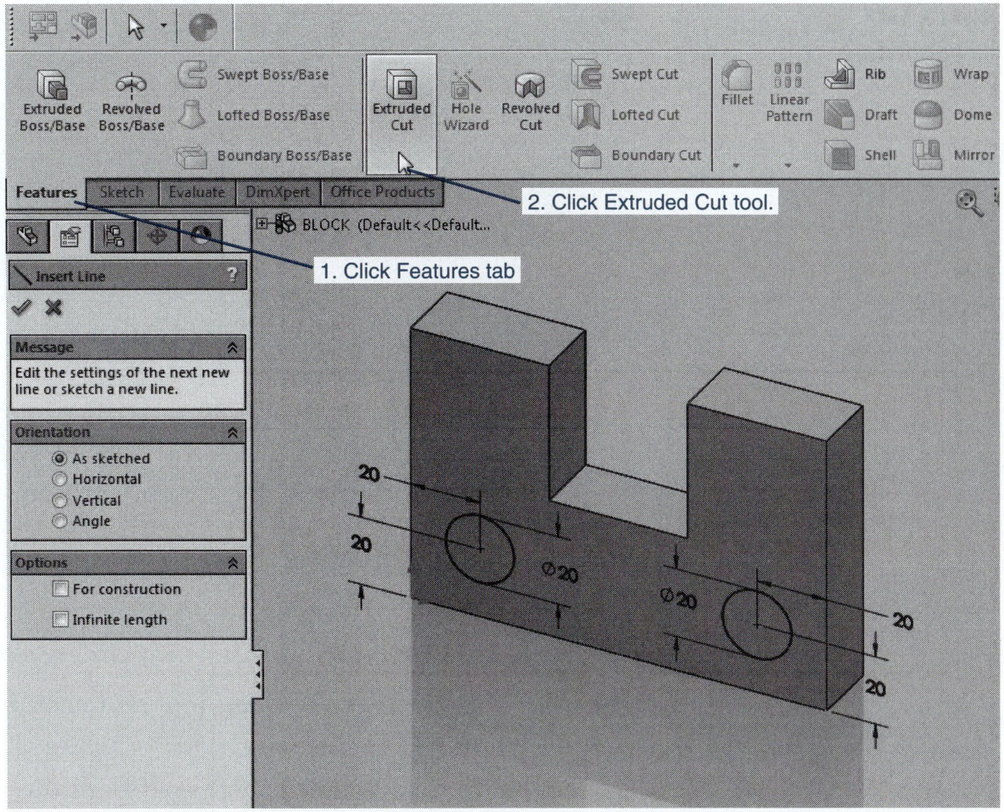

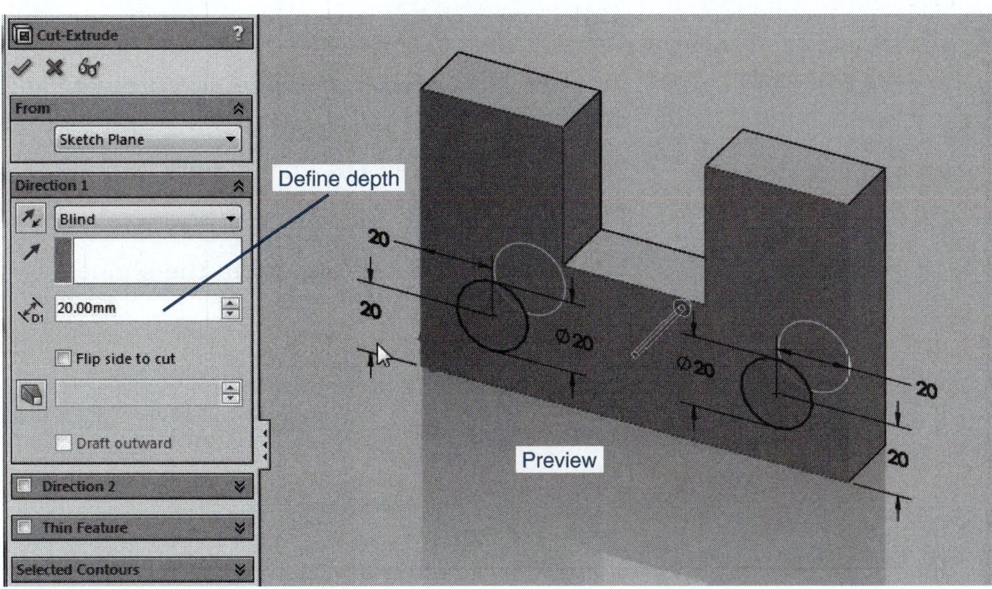

Figure 1-41
(Continued)

Figure 1-41
(Continued)

6 Click the **Circle** tool and add two circles using the given dimensions.

7 Click the **View Orientation** tool and click the **Trimetric** orientation option.

The **Trimetric** option is accessed from the **View Selector** cube as shown in Figure 1-41.

8 Click the **Features** tab, and the **Extruded Cut** tool.

9 Set the cut depth for **20**.

The **Cut-Extrude** tool should automatically select the two circles. If it does not, click the circles. A preview should appear.

10 Click the green **OK** check mark.

11 Click the drawing screen.

The holes should appear in the shape.

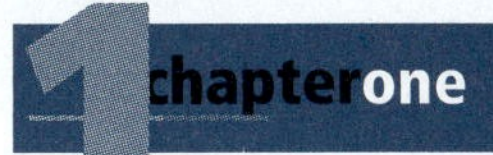

Chapter Project

Project 1-1:

Sketch the shapes shown in Figures P1-1 through P1-18. Create 3D models using the specified thickness values.

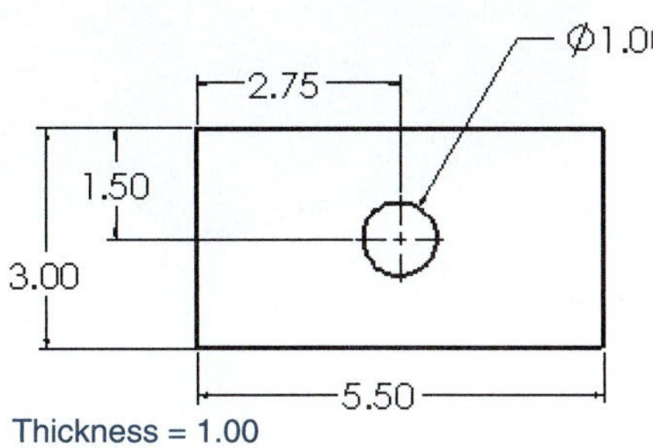

Figure P1-1
INCHES

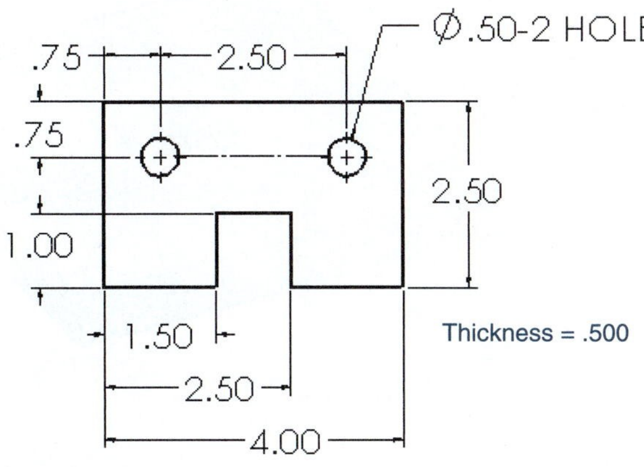

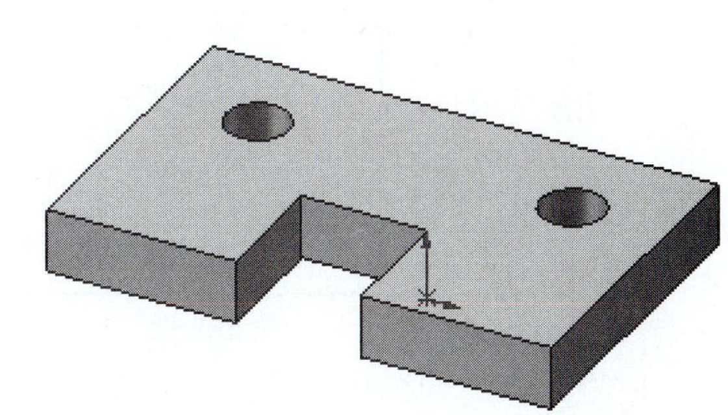

Figure P1-2
INCHES

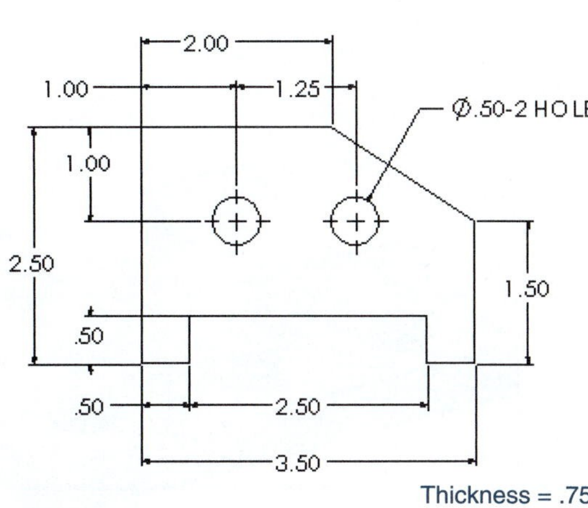

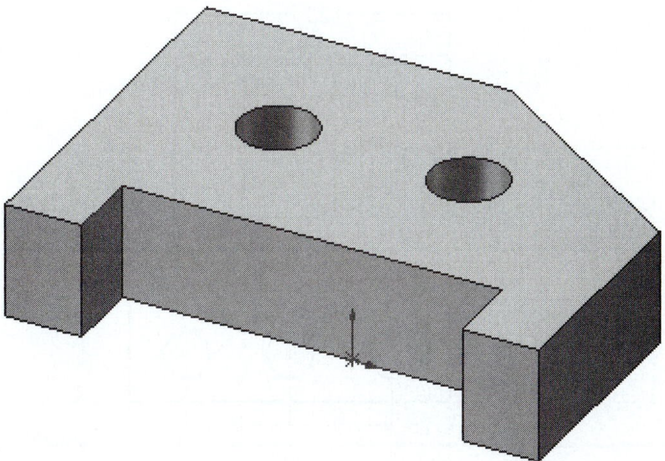

Figure P1-3
INCHES

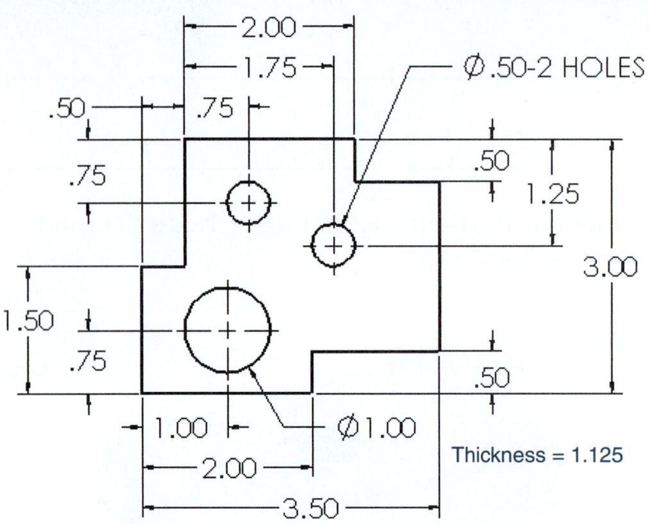

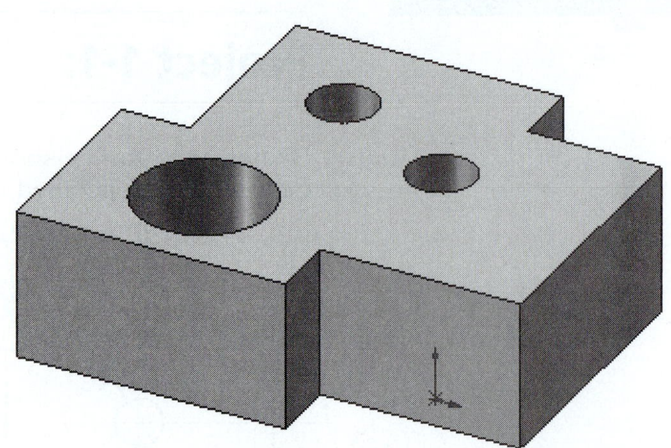

Figure P1-4
INCHES

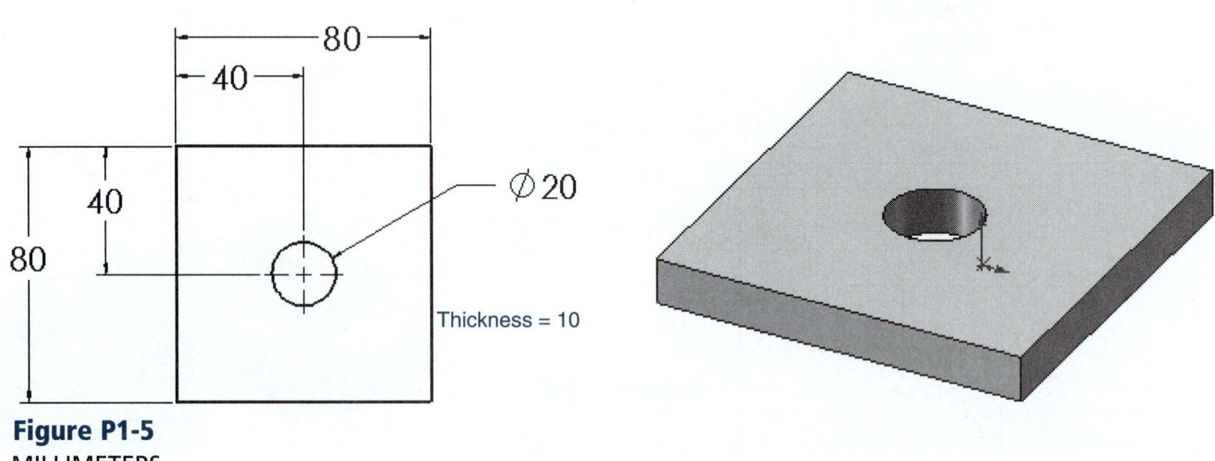

Figure P1-5
MILLIMETERS

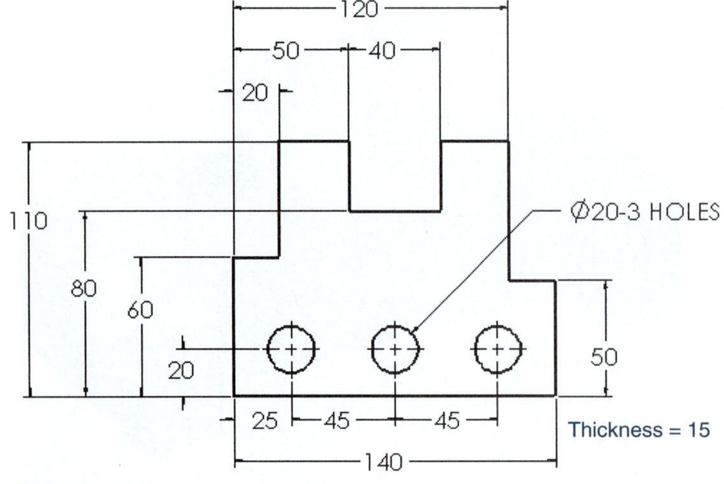

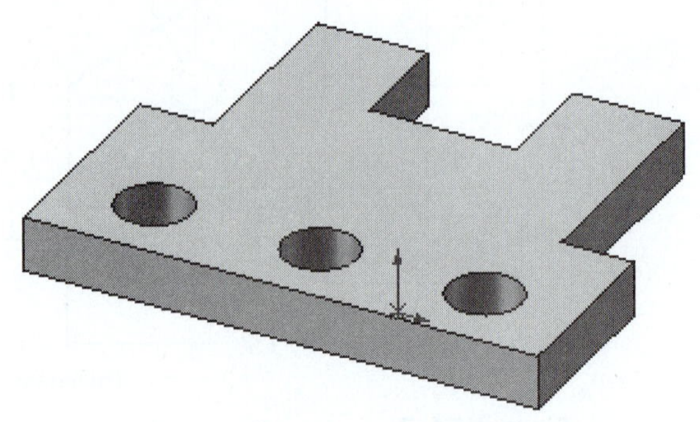

Figure P1-6
MILLIMETERS

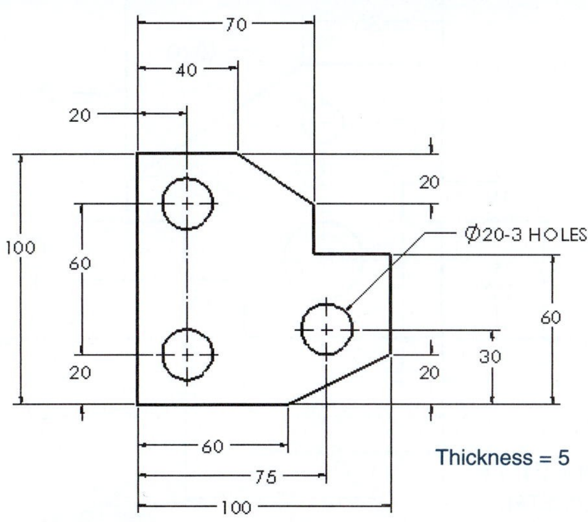

Figure P1-7
MILLIMETERS

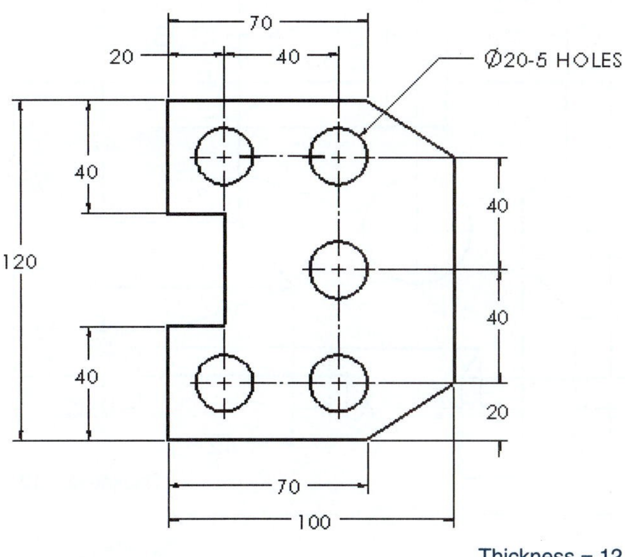

Figure P1-8
MILLIMETERS

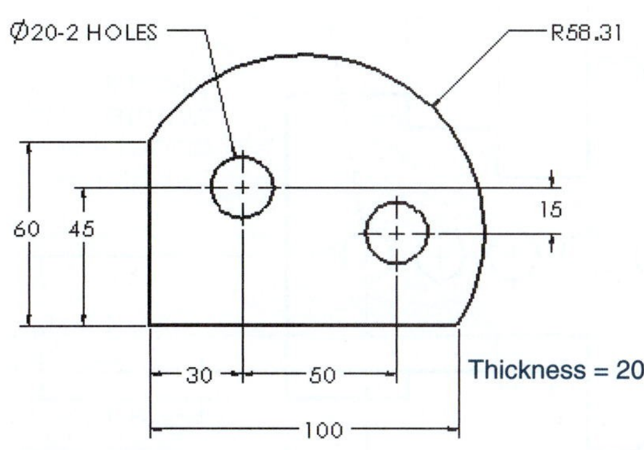

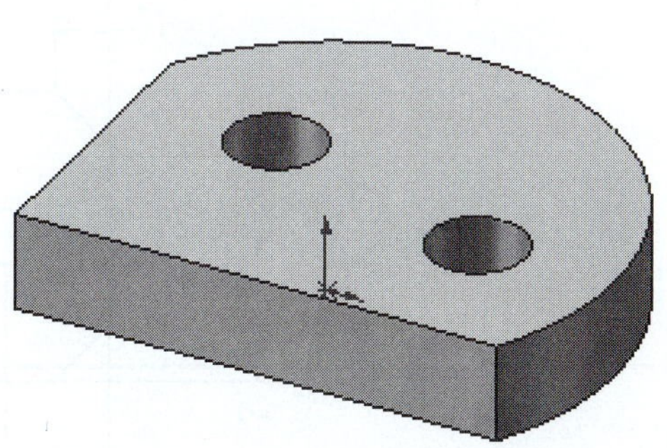

Figure P1-9
MILLIMETERS

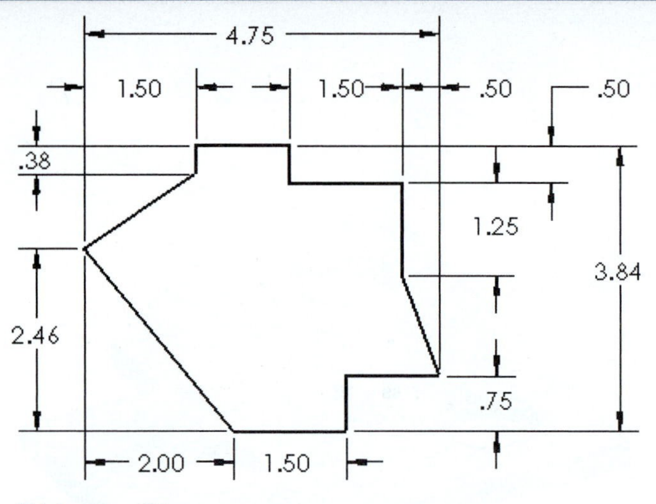

Thickness = .60

Figure P1-10
INCHES

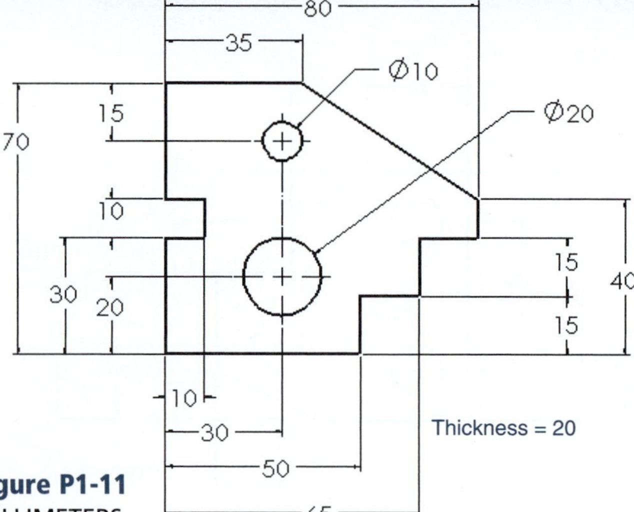

Thickness = 20

Figure P1-11
MILLIMETERS

Figure P1-12
MILLIMETERS

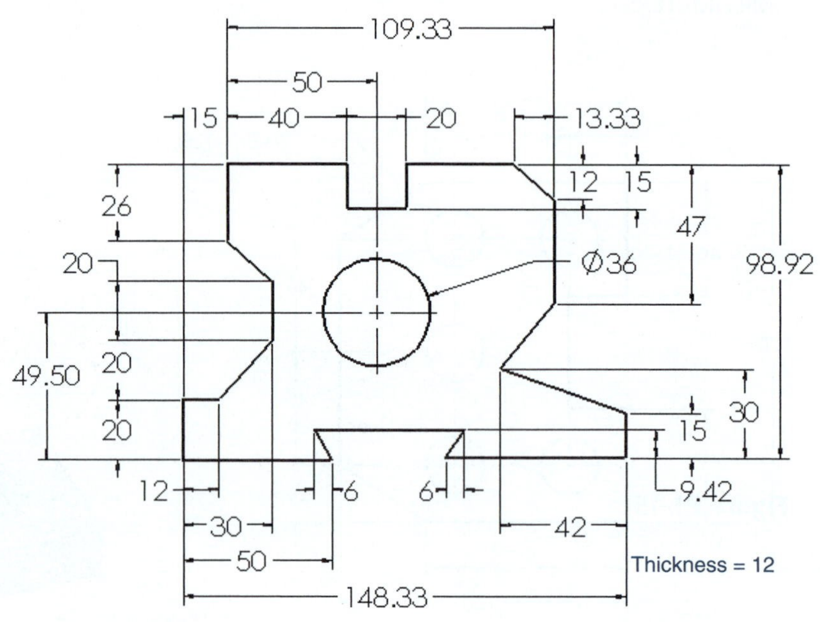

Thickness = 12

Figure P1-13
INCHES

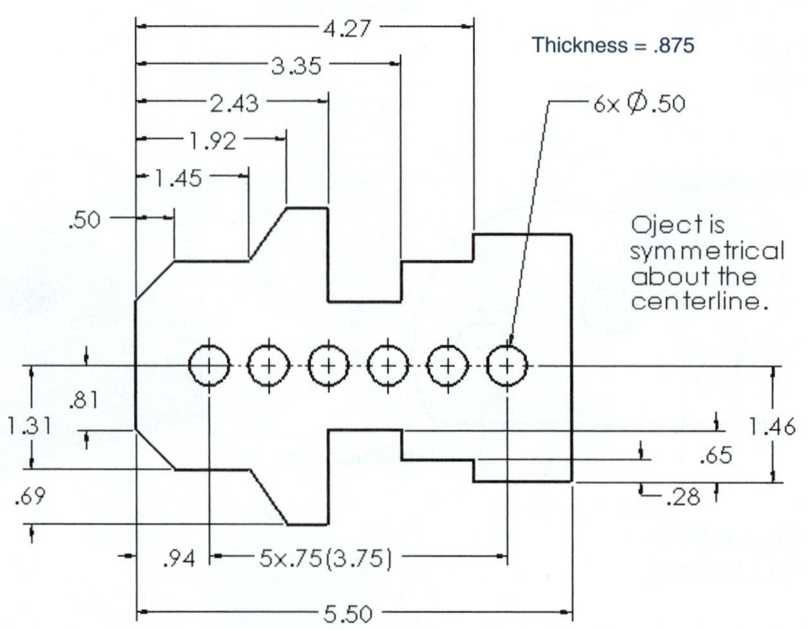

Thickness = .875

Figure P1-14
MILLIMETERS

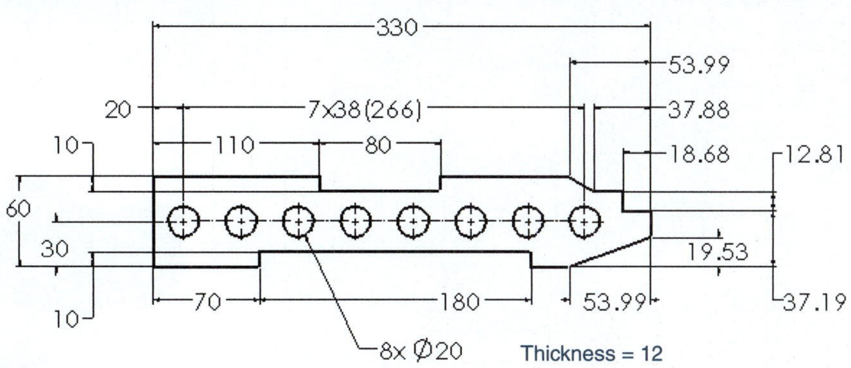

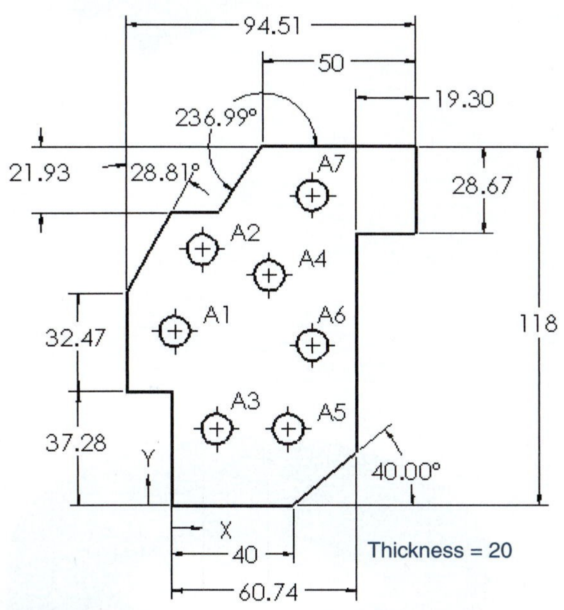

TAG	X LOC	Y LOC	SIZE
A1	1.22	57.14	Ø10
A2	10.27	84.04	Ø10
A3	15	25	Ø10
A4	32.38	75.51	Ø10
A5	38.51	25	Ø10
A6	46.50	52.61	Ø10
A7	46.50	101.88	Ø10

Figure P1-15
MILLIMETERS

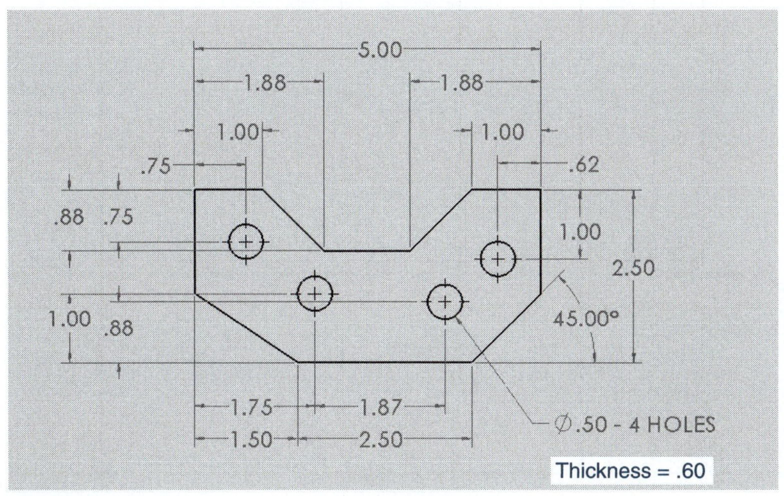

Figure P1-16
INCHES

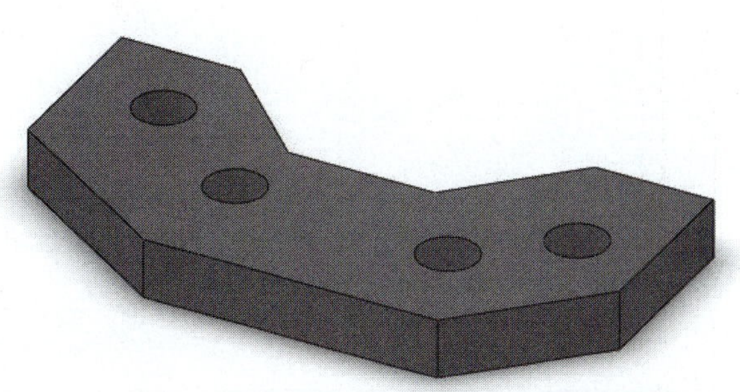

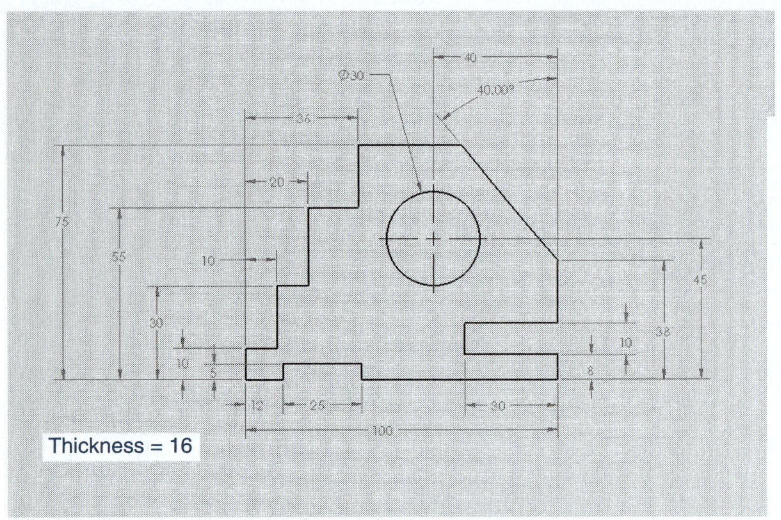

Figure P1-17
MILLIMETERS

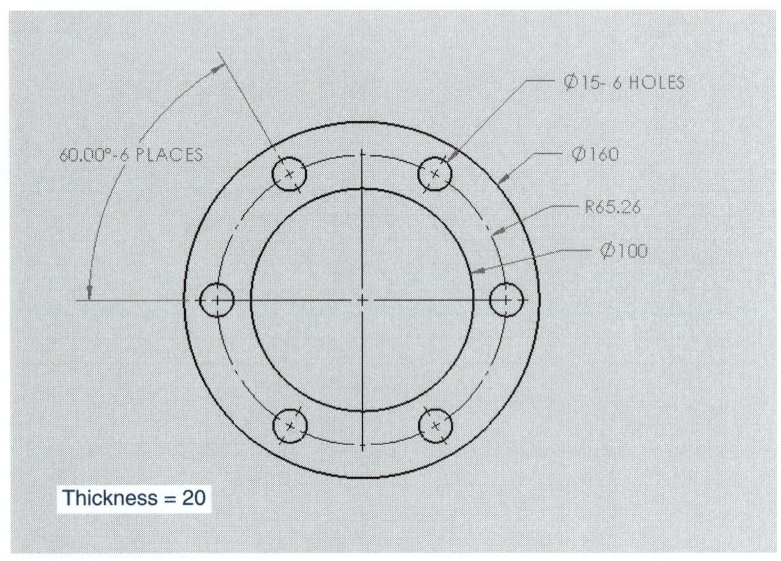

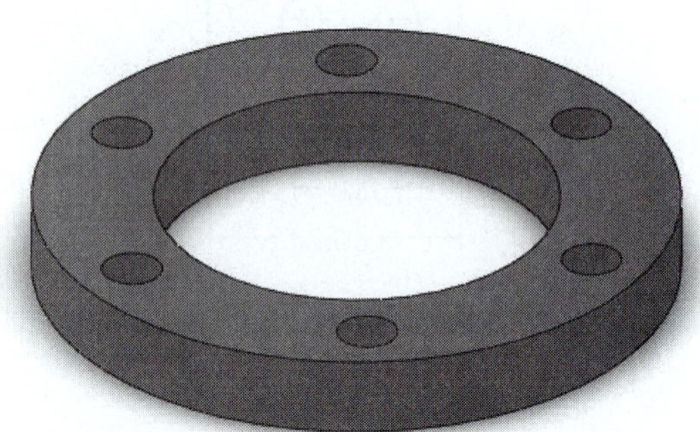

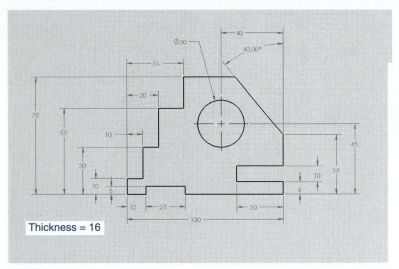

Figure P1-18
MILLIMETERS

chapter two

Sketch Entities and Tools

CHAPTER OBJECTIVES

- Learn how to create 2D sketches
- Learn how to use most of the sketch tools
- Learn how to create more complex shapes by combining individual sketch tools

2-1 Introduction

Chapter 2 explains and demonstrates the tools on the **Sketch** panel. Figure 2-1 shows the **Sketch** panel. Most of the sketch tools are initially explained and demonstrated individually. They are then combined in Sample Problems to show how they can be used together to create more complicated sketches.

Figure 2-1

The Sketch Panel

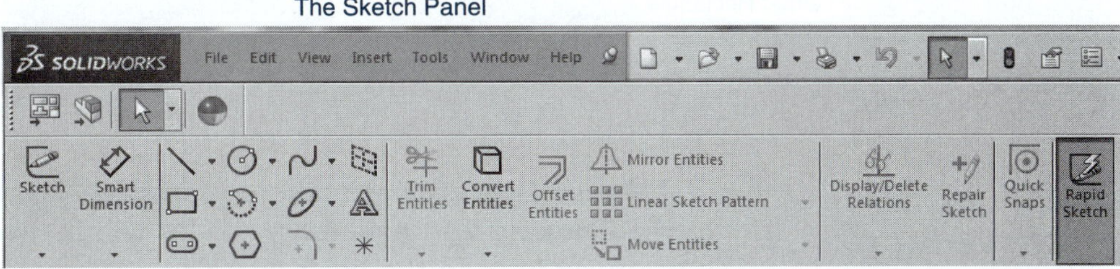

Most of the sections include only one tool so they can be referenced at a later time if a refresher is needed. The **Line** tool was covered in Chapter 1.

The **Circle** and **Rectangle** tools are presented in reference to an origin to create fully defined parts. However, many times sketches are added into an existing part that is already fully defined and referenced to the origin.

The added sketches are referenced to the existing part to create fully defined parts. Many of the sketching tools will be presented without reference to an origin.

2-2 Mouse Gestures and the S Key

As you work with SolidWorks you may find yourself constantly using the same sketching tools repeatedly. The **Mouse Gestures** and **S Key** tools can be customized to help speed up the application of repetitive tools.

Mouse Gestures

Mouse Gestures is a SolidWorks feature that enables you to work faster. Selective commands, those you use most often, are linked to the movement of the mouse so that you can activate these commands by simply moving the mouse.

Figure 2-2 shows the default mouse gesture settings for the **Sketch** commands. Four tools are shown on the wheel: **Smart Dimension**, **Circle**, **Line**, and **Corner Rectangle**. These are the default tools settings. Up to eight tools can be added to the mouse gesture wheel.

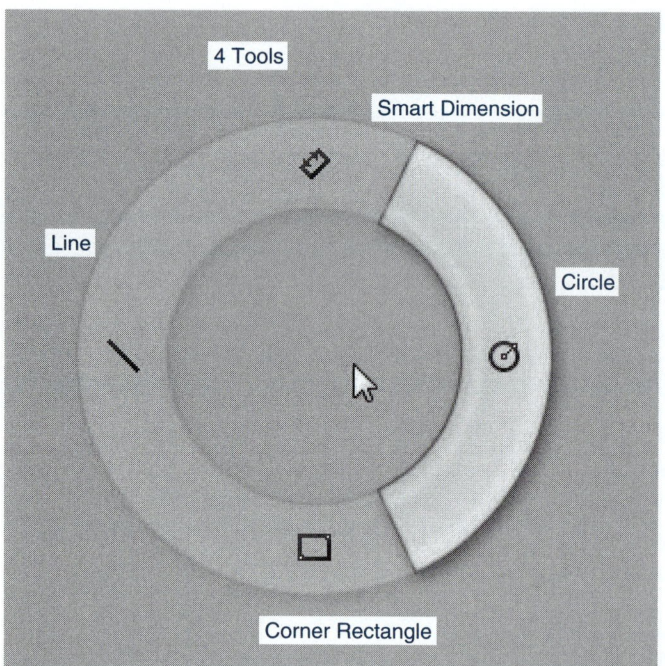

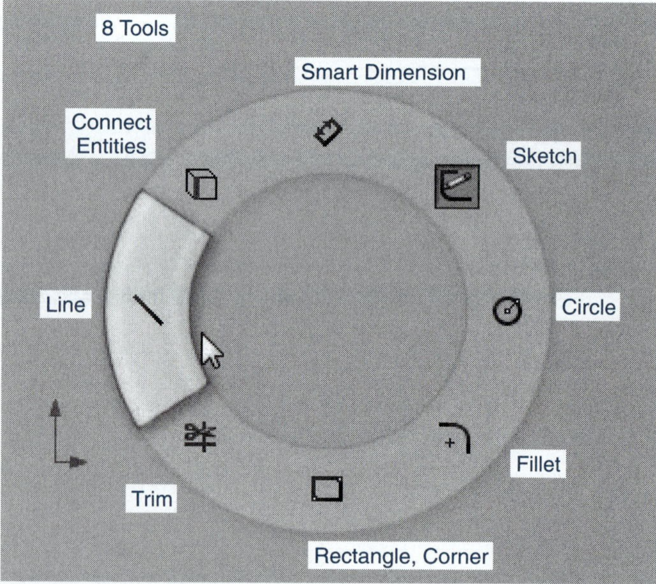

Figure 2-2

Mouse Gestures are different for each drawing mode. The tools listed in the **Drawing** and **Assembly** modes will be used when these subjects are introduced later in the book.

To Use Mouse Gestures. Say you want to activate the line command and the default settings are in place.

1. Press and hold the right mouse button.
2. Move the mouse slightly to the left.

The **Line** tool will be activated. Lines can now be drawn as if you had clicked the **Line** tool on the **Sketch** panel.

The Default Mouse Gestures Settings

1. Click the **Tools** heading at the top of the screen.
2. Click the **Customize** option.

   See Figure 2-3. The **Customize** dialog box will appear.

Figure 2-3

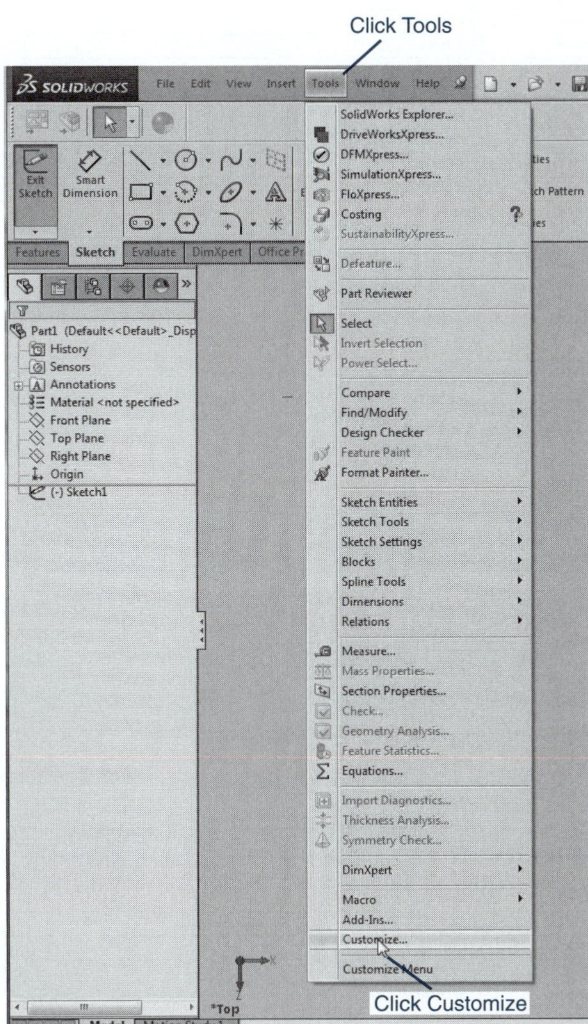

3. Click the **Mouse Gestures** tab.

   See Figure 2-4.

4. Click the box next to **Show only commands with mouse gestures assigned**.

The far right column with the heading **Sketch** shows the four default mouse gestures and the mouse motion needed to activate the tools. For example, a mouse motion to the left activates the **Line** tool.

If you click the **8 gestures** button, the eight default mouse gestures settings will appear. See Figure 2-5.

Figure 2-4

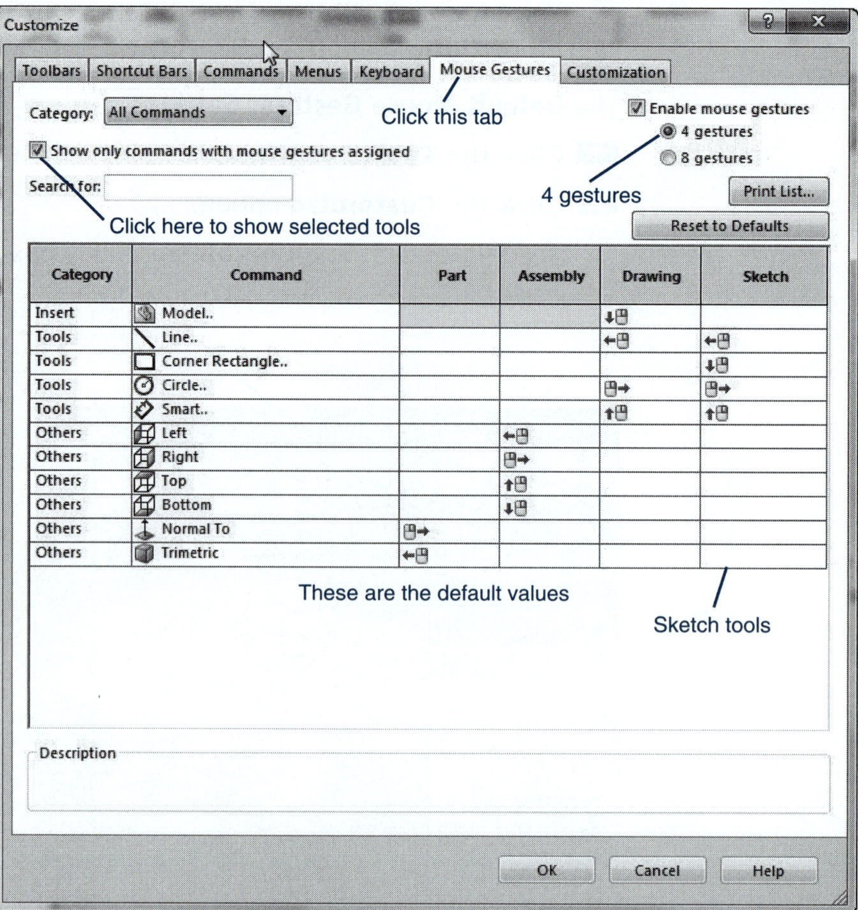

Figure 2-5

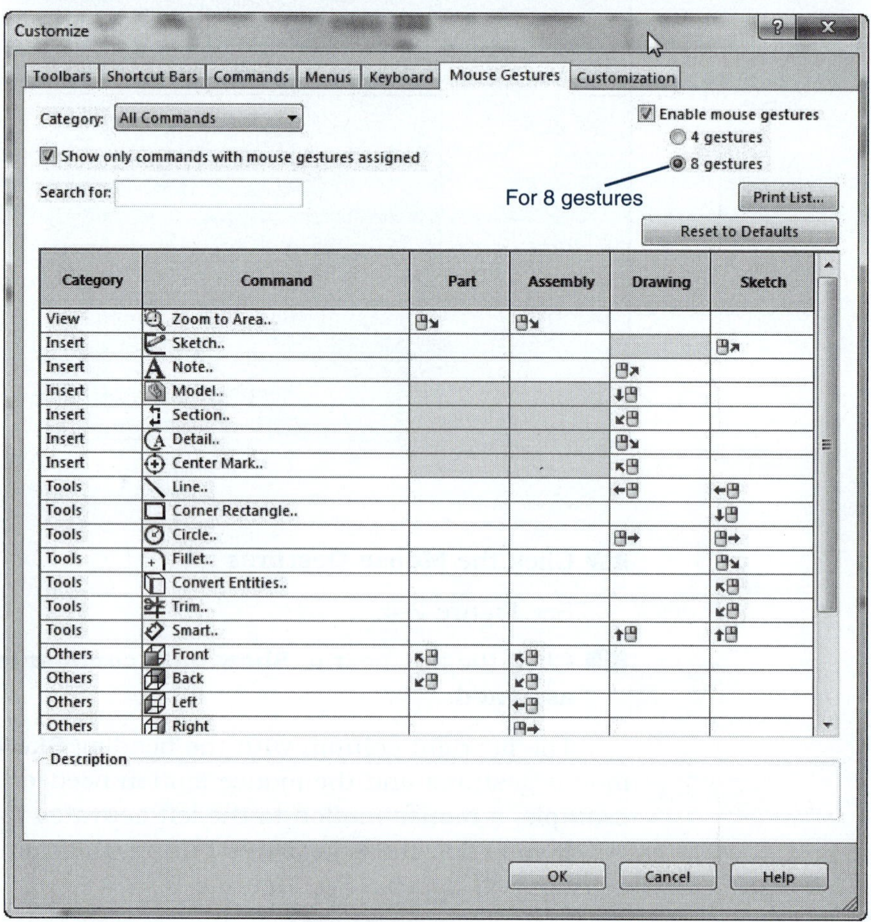

To Change Mouse Gestures. The **Mouse Gestures** settings can be changed.

See Figure 2-6.

Figure 2-6

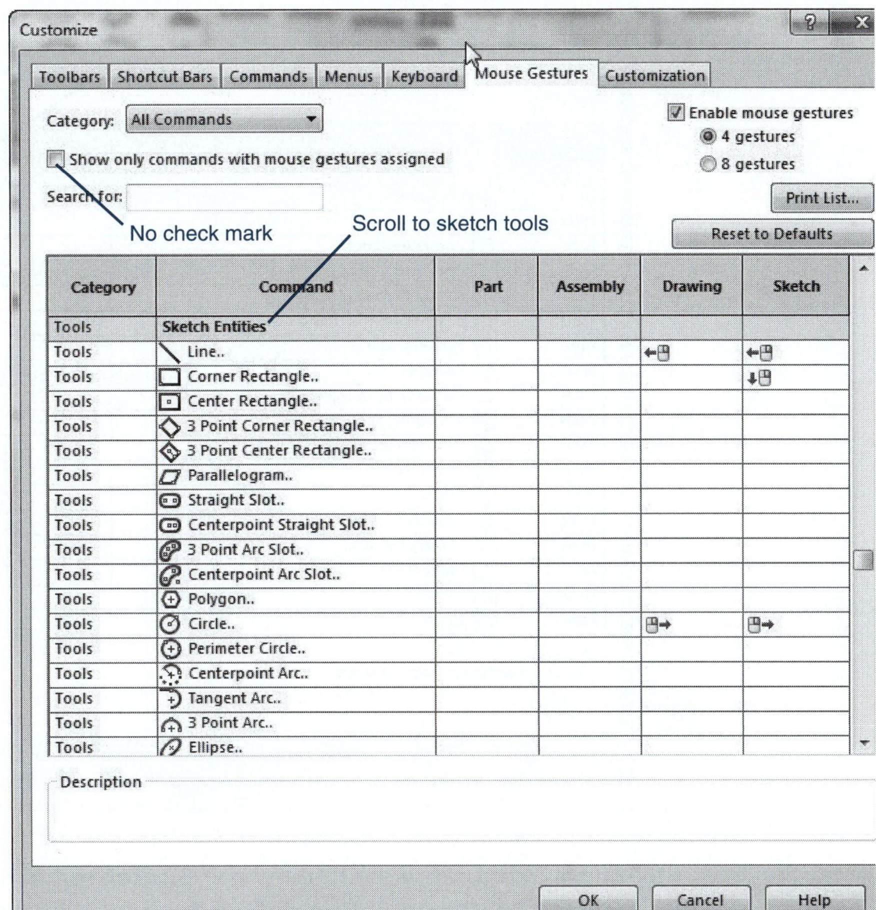

1. Remove the check mark from the **Show only commands with mouse gestures assigned** box.
2. Scroll down to the **Sketch Entities** tools.

Say we wish to change the mouse's left motion to activate the **Center Rectangle** tool instead of the **Line** tool. See Figure 2-7.

3. Click the arrowhead on the right of the **Center Rectangle** command line.
4. Scroll and click the left mouse motion icon.

   The left mouse motion button will appear on the Center Rectangle line.
5. Click **OK**.

> **NOTE**
>
> When you assign the left mouse motion to the **Center Rectangle** tool, the left mouse motion will be *removed* from the **Line** tool. If you want to have the **Line** tool as one of the mouse gestures you must assign the **Line** tool to another mouse gesture.

S Key

Another SolidWorks tool that helps to increase your working speed is the *S Key* option. The tools listed on the **S Key** dialog box may be customized.

Figure 2-7

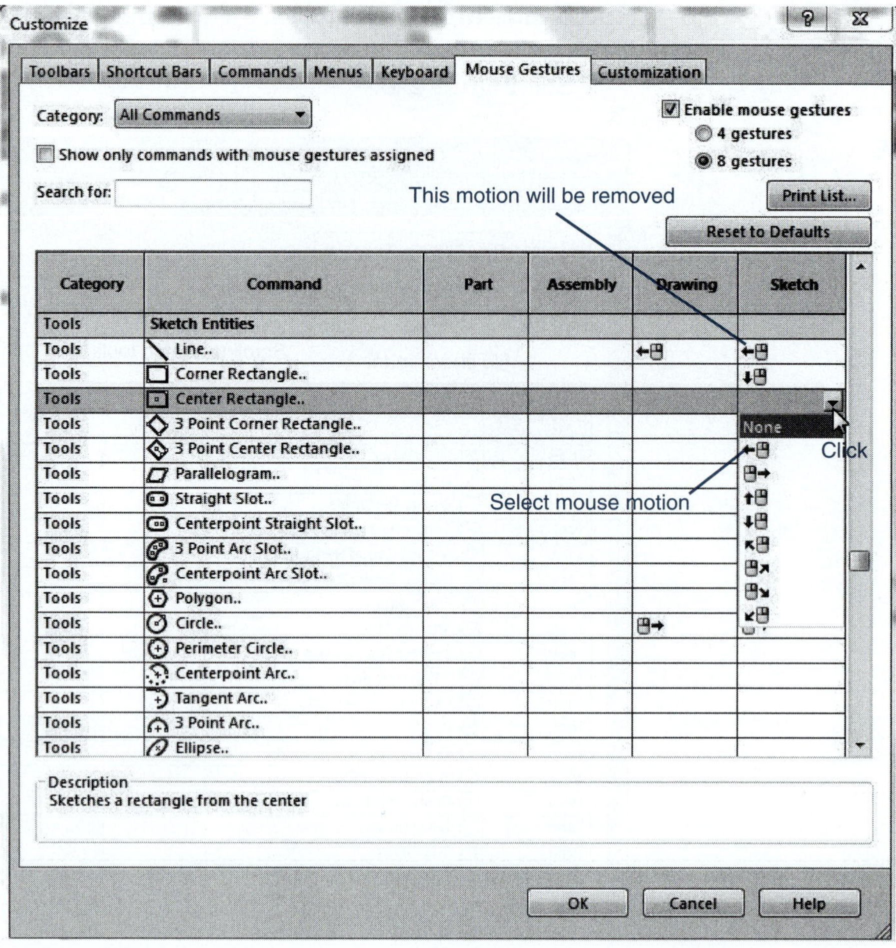

For this chapter we will show the **Sketching** S Key. Other S Key tool listings will be shown as the different modes are explained. The tools listed on any S Key can be customized to your personal preference.

To Activate the S Key. See Figure 2-8.

Figure 2-8

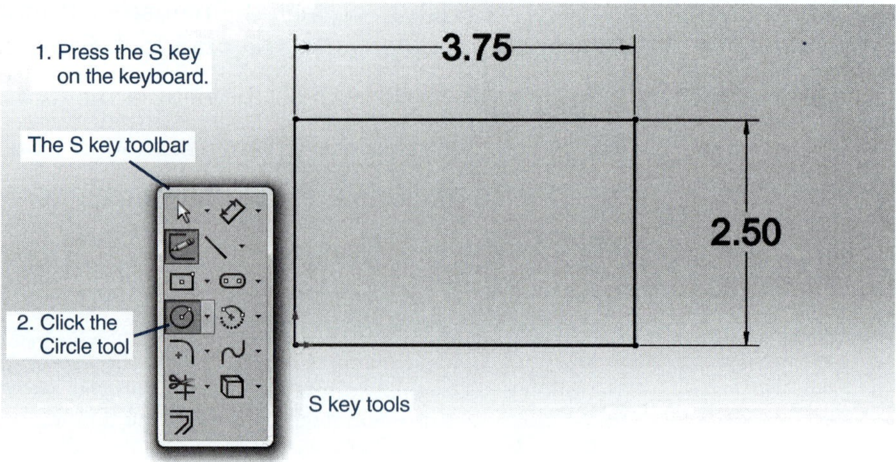

1 Press the **S** key on the keyboard.

The **S Key** toolbar will appear.

2. Click the desired tool.

   The tool will activate. See Figure 2-9.

Figure 2-9

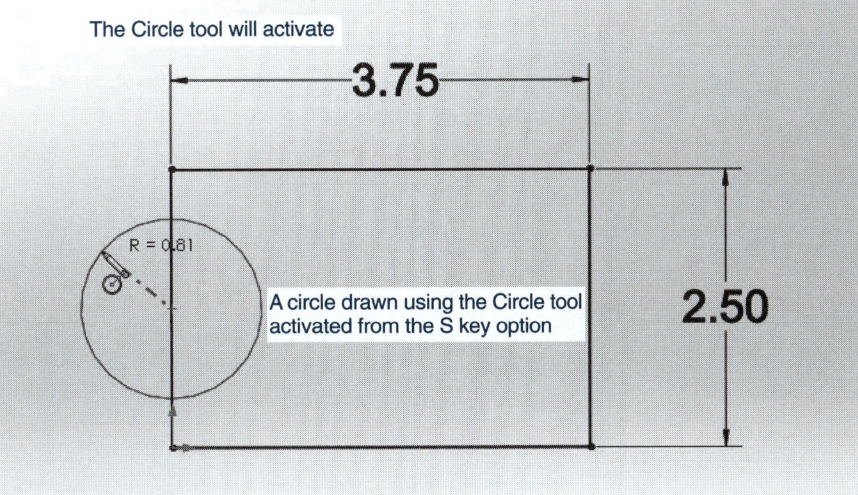
The Circle tool will activate
3.75
R = 0.81
A circle drawn using the Circle tool activated from the S key option
2.50

To Customize the S Key Shortcut Toolbar

1. Press the **S** key on the keyboard.
2. Right-click the **S Key** toolbar.

   See Figure 2-10.

Figure 2-10

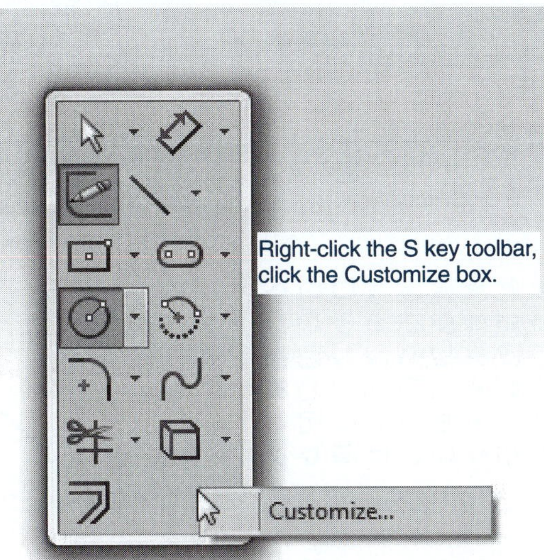
Right-click the S key toolbar, click the Customize box.
Customize...

3. Click the **Customize . . .** callout.

   The **Customize** dialog box will appear. See Figure 2-11.
4. Select the **Sketch Shortcut** tool.
5. Activate the **Sketch Buttons**.
6. Select the tool to be added and click and drag the tools icon from the **Buttons** box to the **Sketch Shortcut . . .** box, the **S Key** box, and release the mouse button.
7. Click **OK**.

   The tool icon will appear in the box. See Figure 2-12.

Figure 2-11

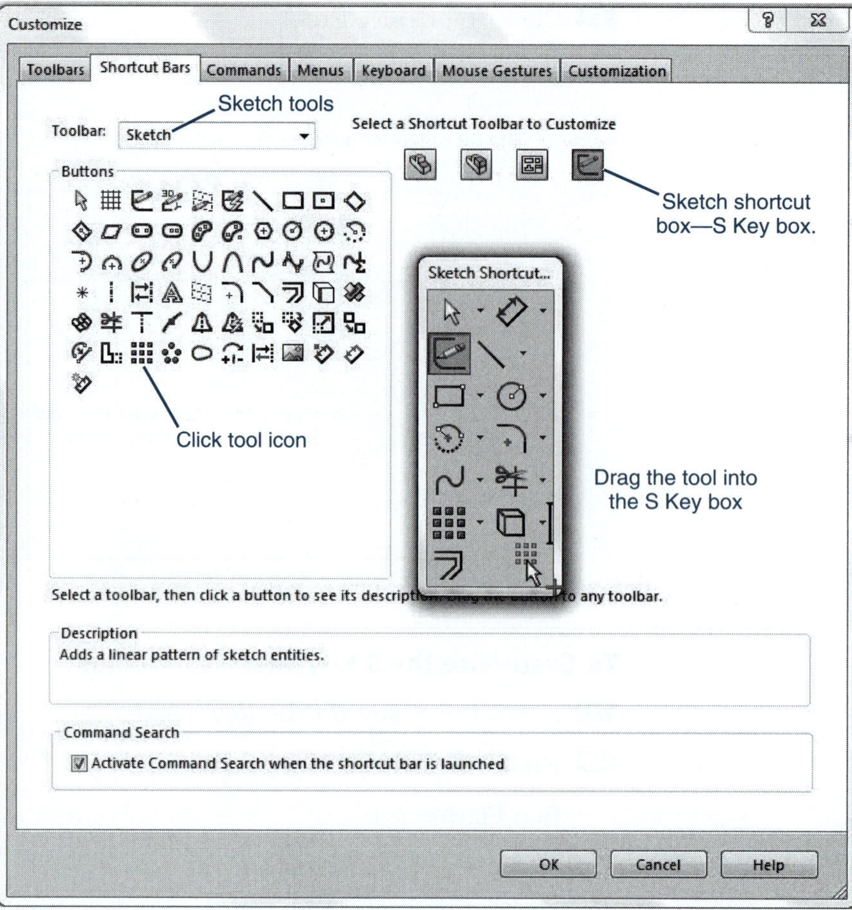

Figure 2-12

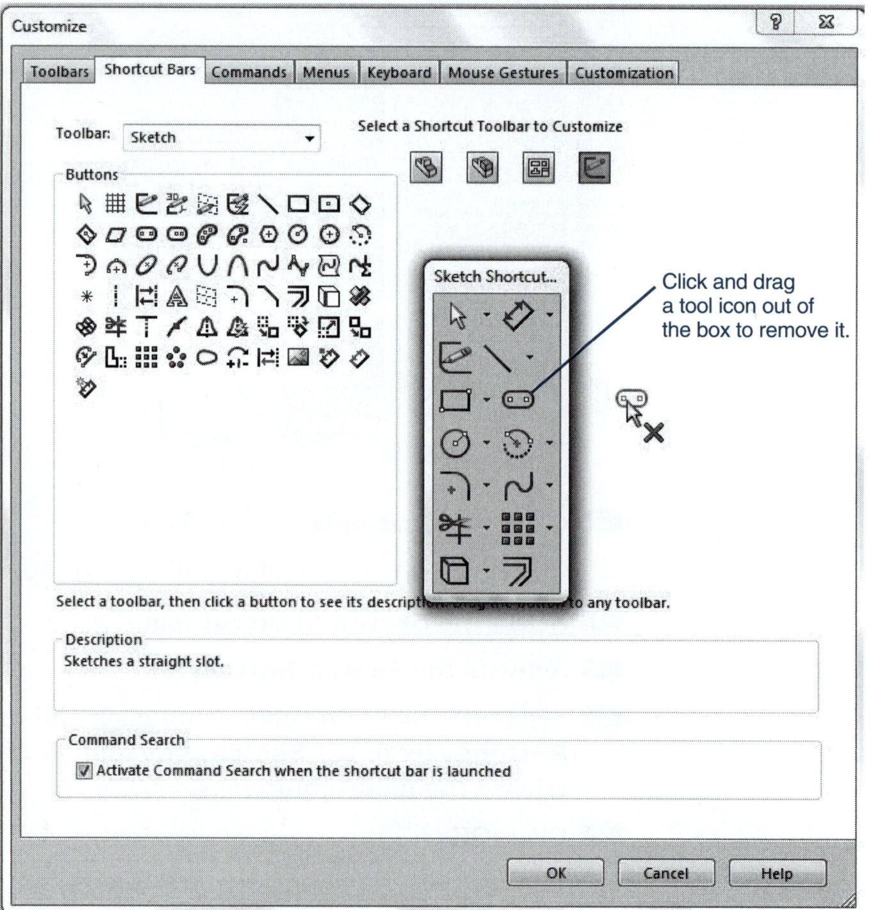

To Remove a Tool from the S Key Box

1 Click and drag the selected tool out and away from the box.

2 Release the mouse button.

3 Click **OK**.

2-3 Origins

All sketches must be referenced to the drawing's origin in order to be fully defined. See Section 1-4. The easiest way to create a fully defined sketch is to start it on the origin. The drawing's origin may not always be visible.

To Show the Origin

1 Click the **View** heading at the top of the drawing screen.

2 Click the **Origins** tool.

See Figure 2-13. The icon next to the **Origins** tool heading will become shaded when the tool is on.

Figure 2-13

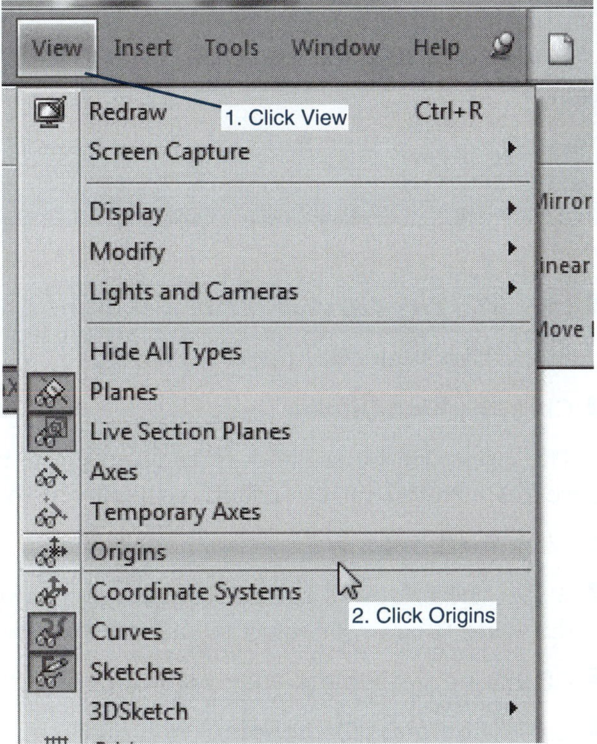

The drawing's origin icon will appear on the screen. It will remain visible on the screen until removed by clicking the shaded tool heading created when the **Origins** tool was turned on.

2-4 Circle

A circle is sketched using the **Circle** tool and then sized, that is, given a diameter value, using the **Smart Dimension** tool. There are two circle tools: **Circle** and **Perimeter Circle**.

To Sketch a Circle

1 Click the **New** tool at the top of the screen, click the **Part** tool on the **New SolidWorks Document** dialog box, and click **OK**.

See Figure 2-14. The dimensions for this example are in inches.

Figure 2-14

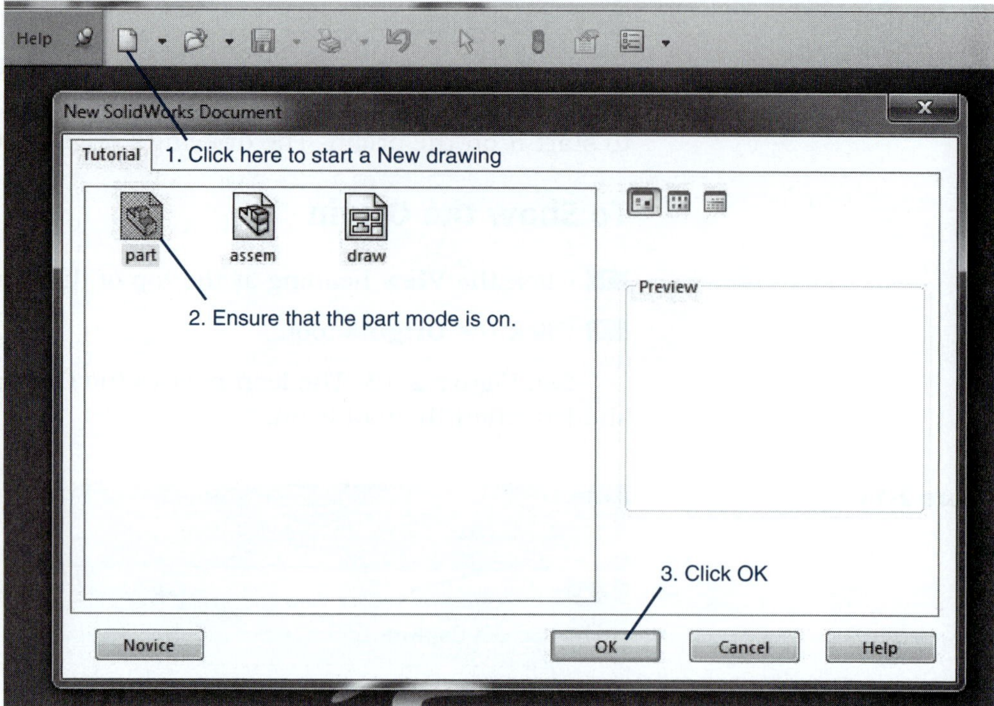

2 Click the **Top drawing plane** tool in the **Document Properties** box.

A small toolbar will appear.

3 Click the **Sketch** tool.

The circle will be created on the Top plane in the Sketch mode. See Figure 2-15. Note that the origin is displayed in the center of the Top plane.

4 Click the **Circle** tool.

5 Select the origin as the circle's centerpoint by clicking the origin and dragging the cursor away from the centerpoint.

6 Click an approximate edge point for the circle.

7 Click the **Smart Dimension** tool, click the circle, and define the circle's diameter.

8 Right-click the mouse and click the **Select** option.

For this example a diameter of Ø2.50 was selected. The circle is black, meaning it is fully defined. The orientation triad in the lower left corner of the screen shows the XZ or top plane orientation.

To Sketch a Perimeter Circle Using Three Points

A perimeter circle is a circle drawn using three points. These points may be three individual points or points tangent to an existing object.

Figure 2-15

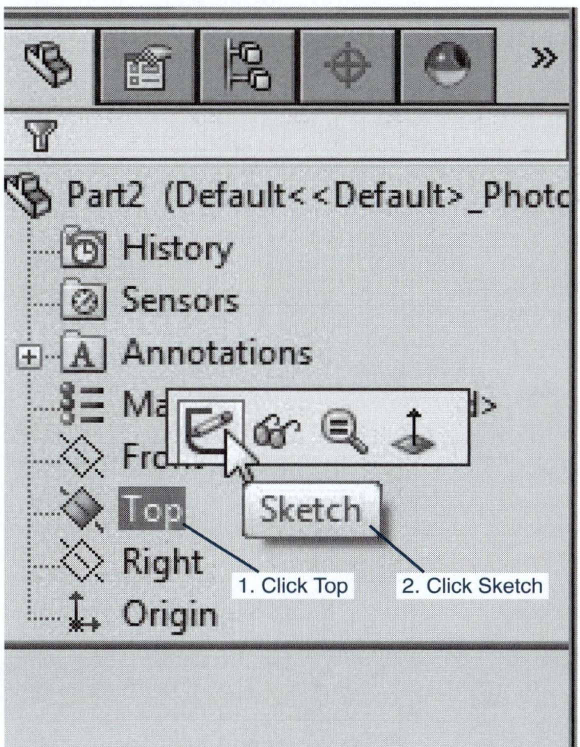

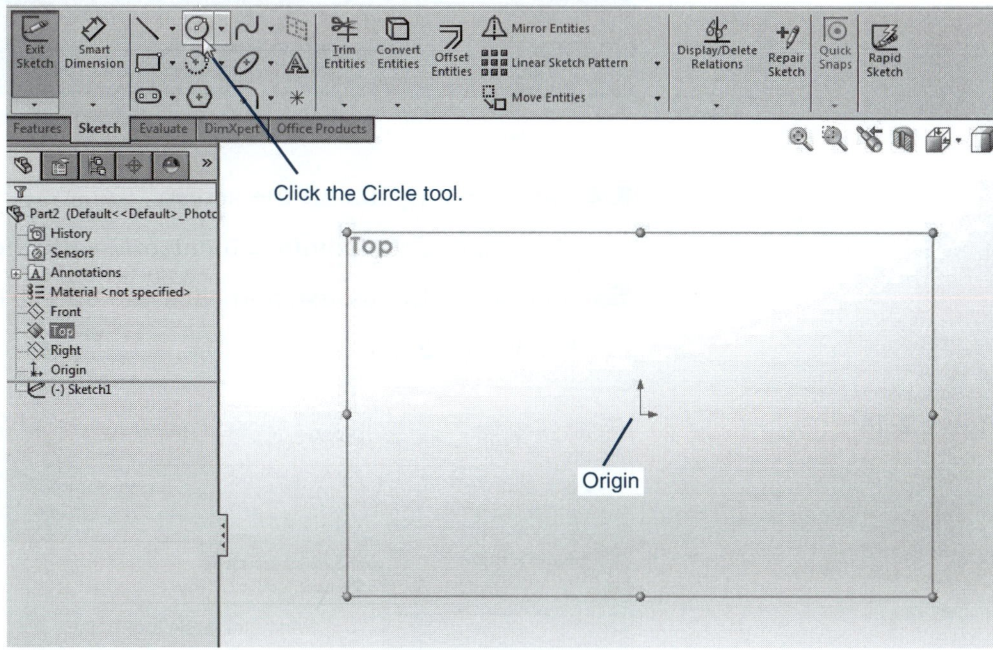

1 Create a **New Part** document.

2 Click the **Top drawing plane** tool in the **Document Properties** box.

3 Click the **Sketch** tool.

4 Set the **Units** for inches.

5 Click the **Perimeter Circle** tool.

The **Perimeter Circle** tool is a flyout from the **Circle** tool. See Figure 2-17.

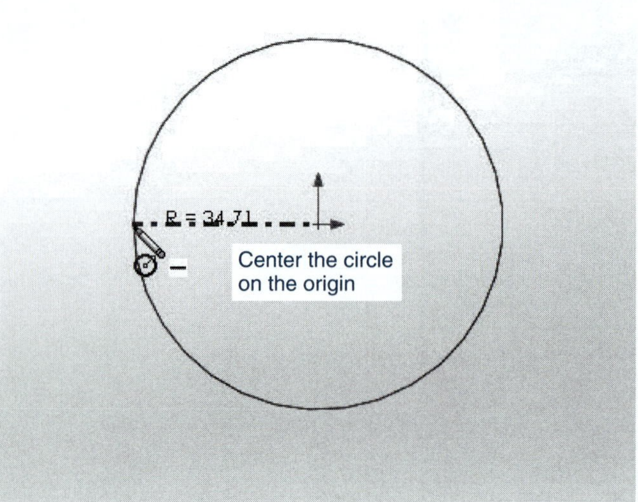

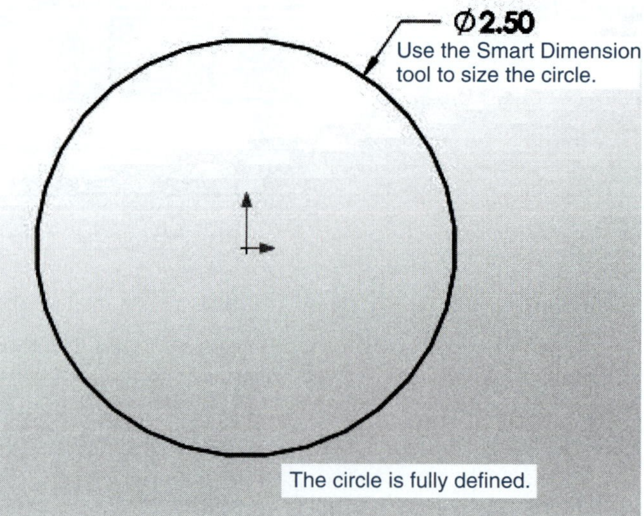

Figure 2-16

Figure 2-17

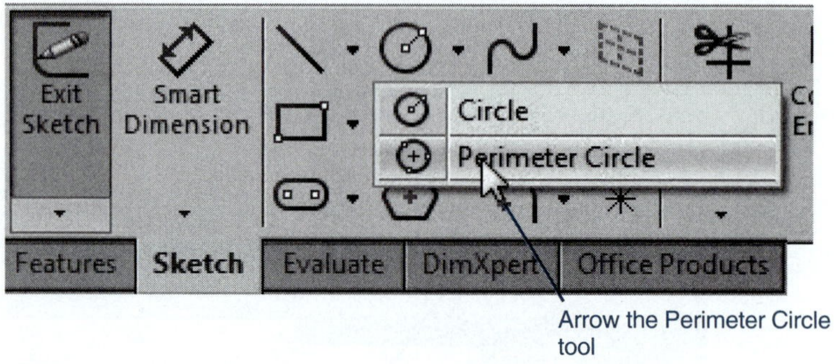

6 Click three points on the screen.

Make one of the points coincidental with the origin.

7 Right-click the mouse and click the **Select** option.

See Figure 2-18.

Figure 2-18

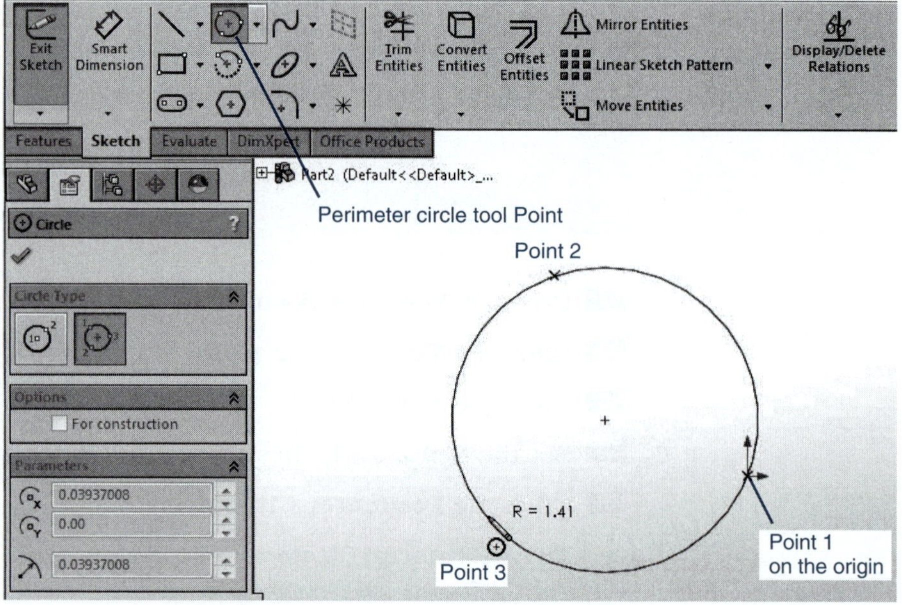

Figure 2-18
(Continued)

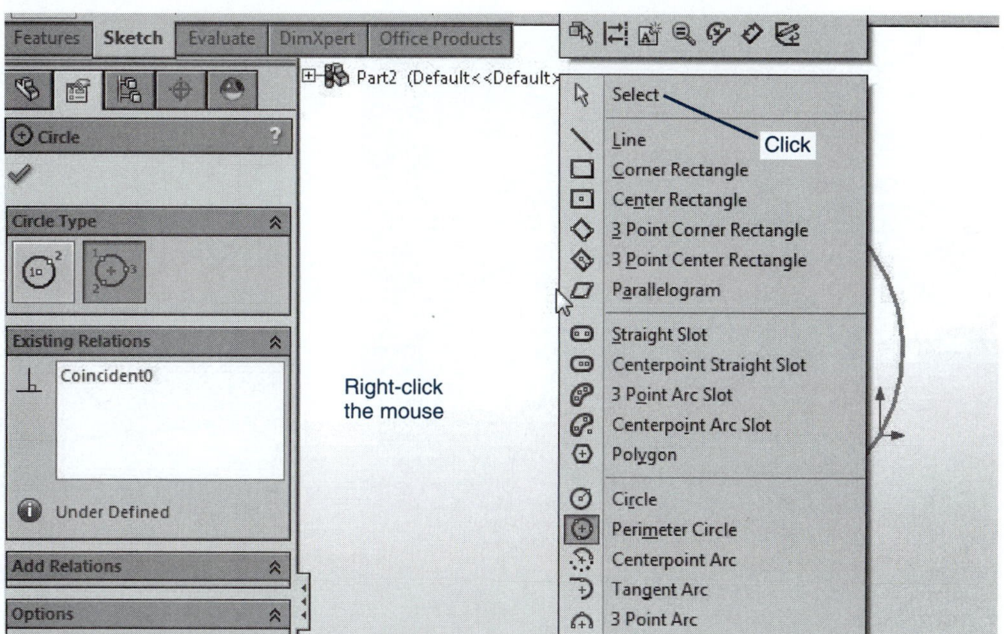

To Sketch a Perimeter Circle Tangent to Three Lines

Figure 2-19 shows three randomly drawn straight lines. Draw a circle tangent to each line.

Figure 2-19

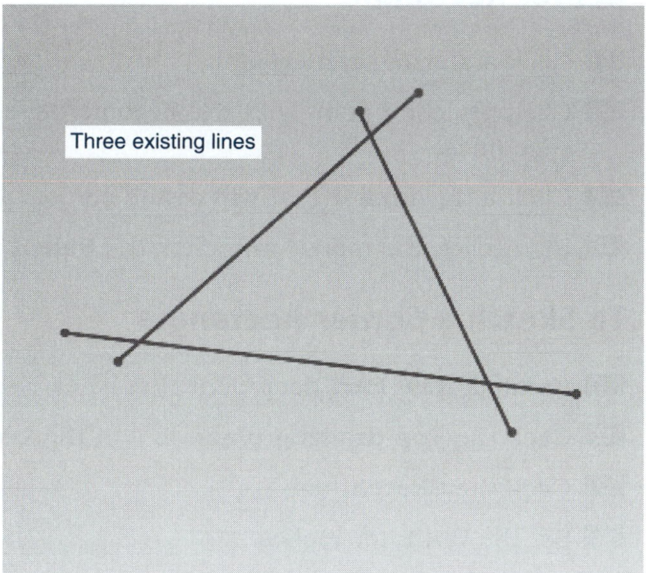

1. Click the **Perimeter Circle** tool.
2. Click the approximate centerpoint of each of the three lines.

   A circle will appear tangent to all three lines.
3. Right-click the mouse and click the **Select** option.

   See Figure 2-20.

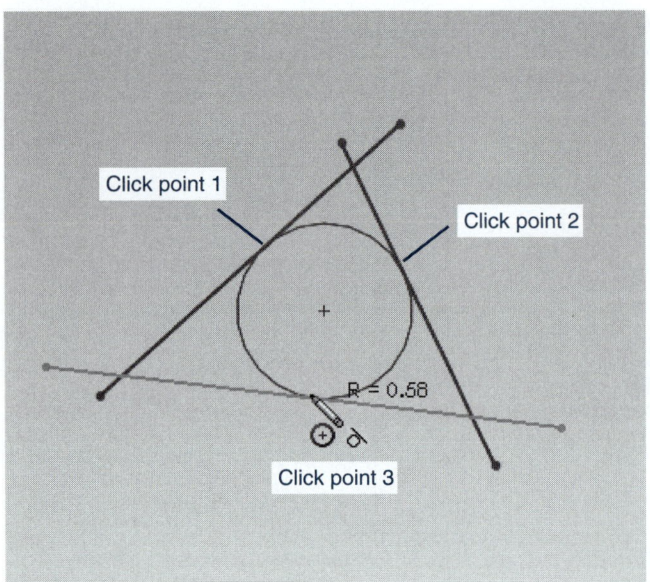

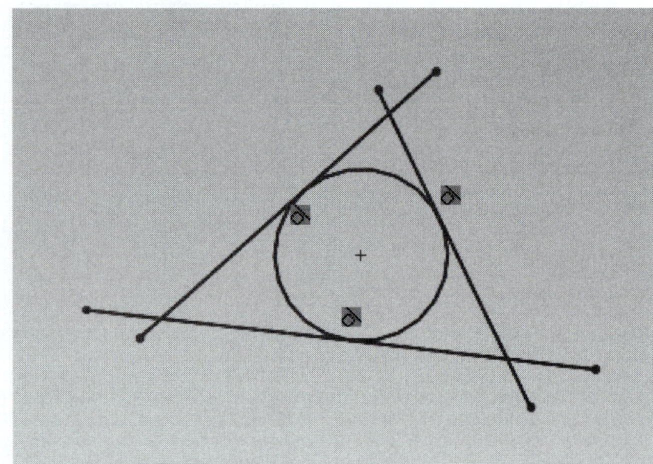

Figure 2-20

2-5 Rectangle

There are five different methods that can be used to create a rectangle, including the **Corner Rectangle** tool explained in the previous section. Figure 2-22 shows the four other methods.

Center Rectangle

1. Click a starting centerpoint.
2. Click a second point that will become the centerpoint of one of the edge lines.
3. Click a third point that will define one of the corner points.
4. Right-click the mouse and click the **Select** option.

To Sketch a Corner Rectangle

1. Create a **New Part** document.
2. Click the **Top drawing plane** tool in the **Document Properties** box.
3. Click the **Sketch** tool.
4. Set the **Units** for inches.
5. Click the **Corner Rectangle** tool.

   See Figure 2-21.
6. Select the origin for the first corner point.
7. Move the cursor and select the second corner point.
8. Click the **Smart Dimension** tool and add dimensions to the rectangle.
9. Right-click the mouse and click the **Select** option.

   The rectangle is fully defined.

Figure 2-21

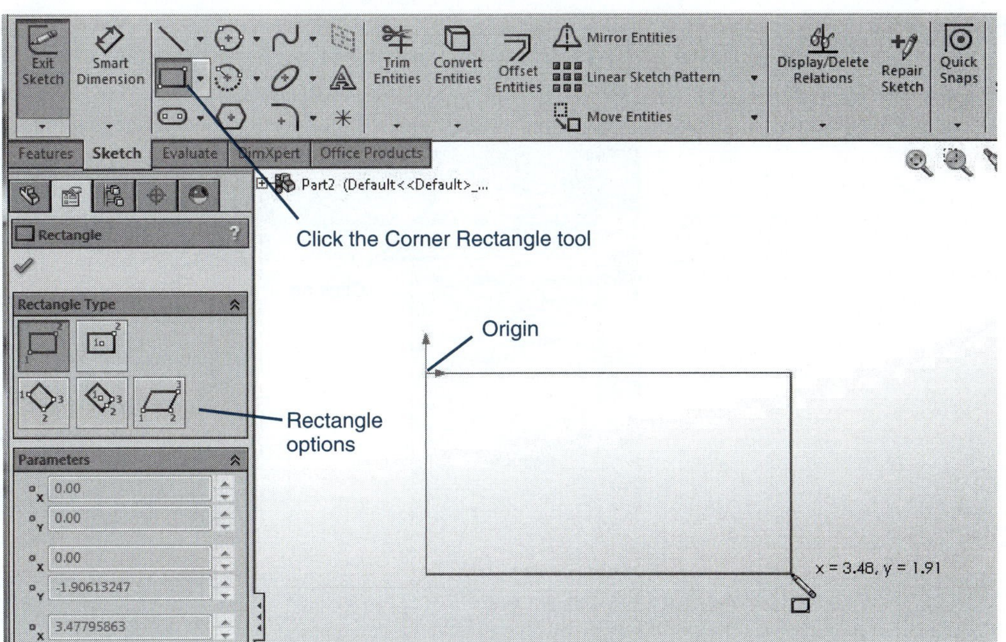

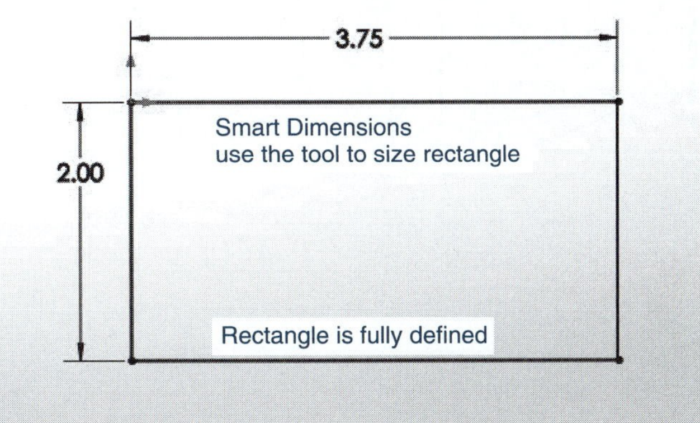

To Sketch a Center Rectangle

1. Access the **Sketching** tools, define a plane, and the inch units.
2. Click the **Center Rectangle** tool.

   The **Center Rectangle** tool is a flyout from the **Corner Rectangle** tool icon. See Figure 2-22.
3. Select the origin as the centerpoint for the rectangle.
4. Move the cursor away from the centerpoint and select a corner point.

   See Figure 2-23.
5. Use the **Smart Dimension** tool and define the rectangle's size.

To Sketch a 3 Point Corner Rectangle

1. Access the **Sketching** tools, define a plane, and the inch units.
2. Click the **3 Point Rectangle** tool.

Figure 2-22

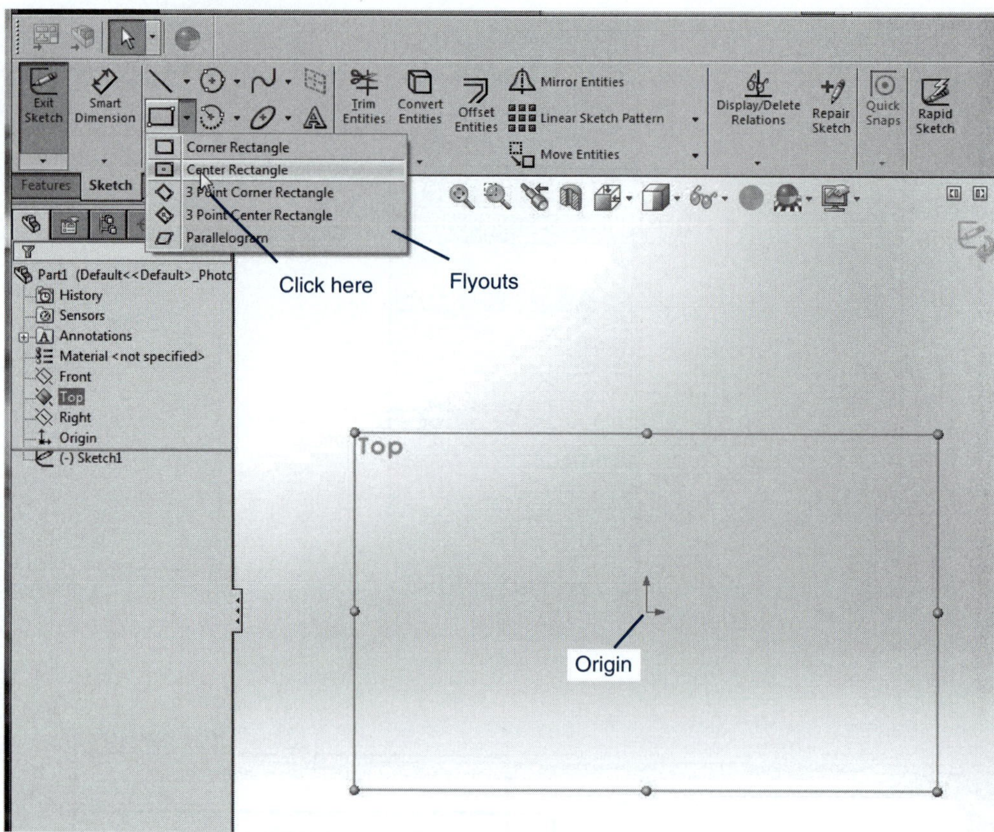

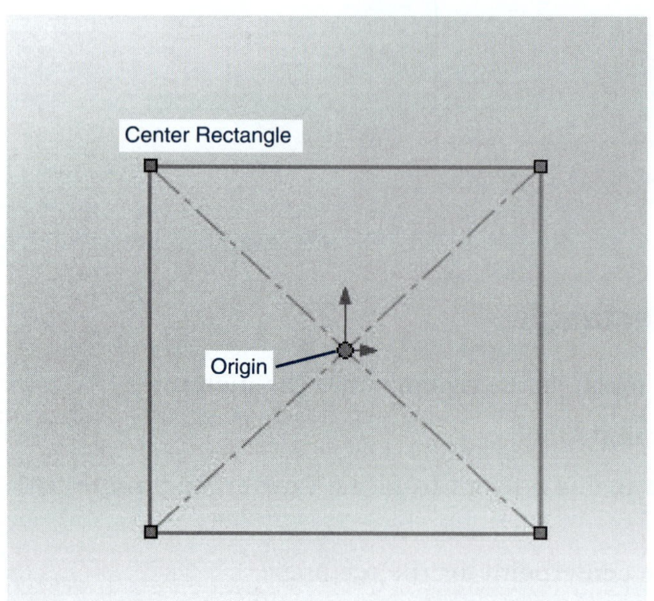

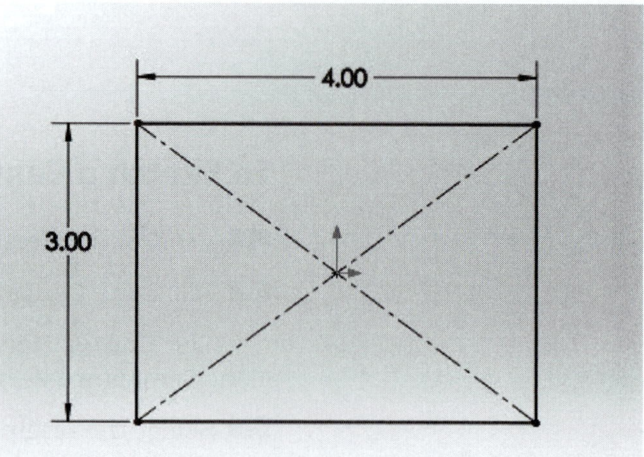

Figure 2-23

The **3 Point Corner Rectangle** tool is a flyout from the **Corner Rectangle** tool icon. See Figure 2-24.

3 Locate the first point on the origin.

4 Move the cursor horizontally and define a second point.

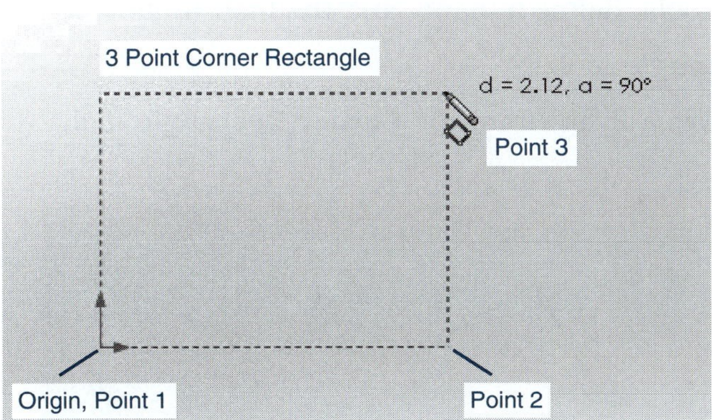

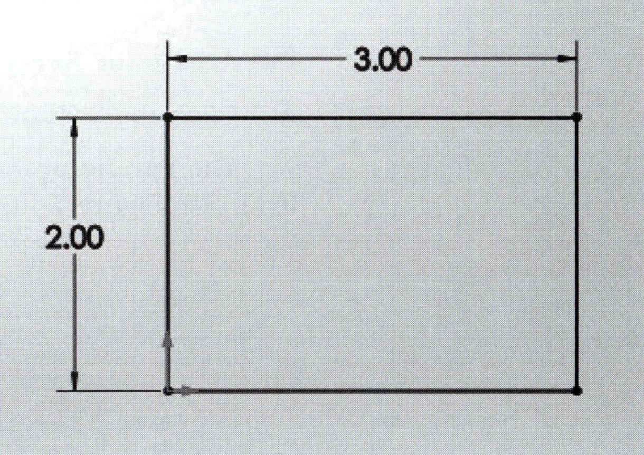

Figure 2-24

5 Move the cursor vertically and define a third point.

6 Click the green **OK** check mark.

7 Use the **Smart Dimension** tool and define the rectangle's size.

To Sketch a 3 Point Center Rectangle

1 Access the **Sketching** tools, define a plane, and the inch units.

2 Click the **3 Point Rectangle** tool.

The **3 Point Center Rectangle** tool is a flyout from the **Corner Rectangle** tool icon. See Figure 2-25.

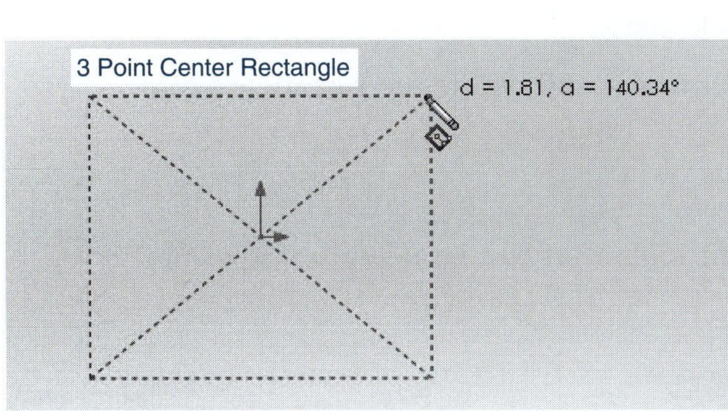

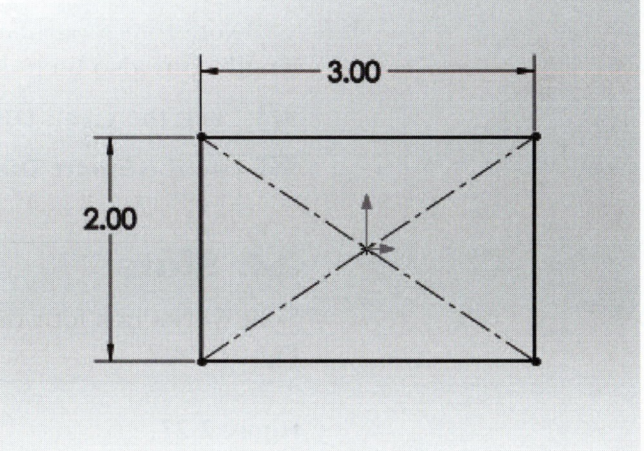

Figure 2-25

3 Locate the first point on the origin.

4 Move the cursor horizontally and define a second point.

5 Move the cursor vertically and define a third point.

6 Click the green **OK** check mark.

7 Use the **Smart Dimension** tool and define the rectangle's size.

To Sketch a Parallelogram

1. Access the **Sketching** tools, define a plane, and the inch units.
2. Click the **3 Point Rectangle** tool.

The **Parallelogram** tool is a flyout from the **Corner Rectangle** tool icon. See Figure 2-26.

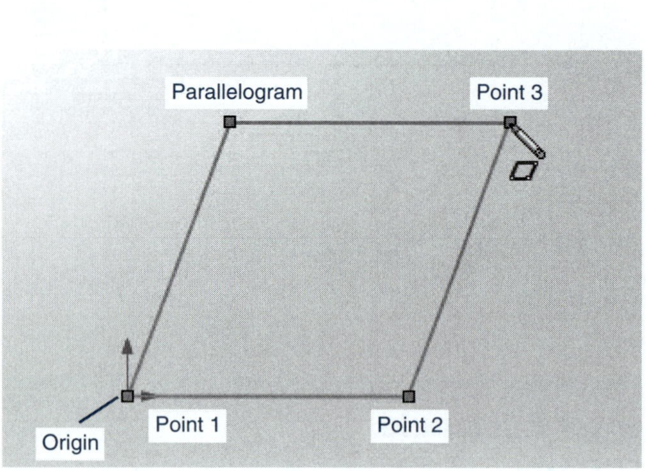

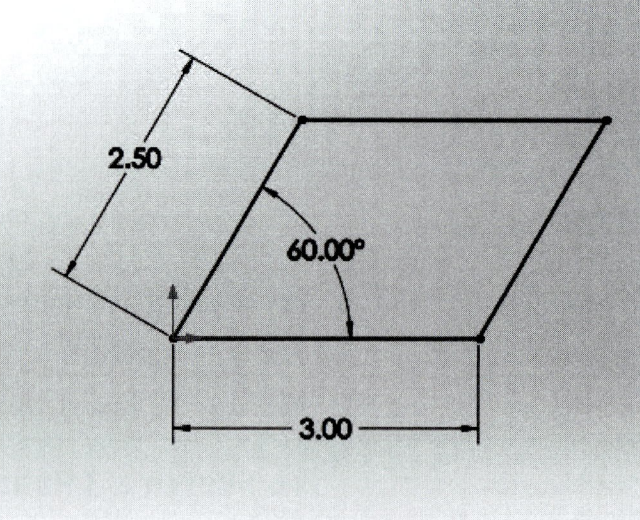

Figure 2-26

3. Locate the first point on the origin.
4. Move the cursor horizontally and define a second point.
5. Move the cursor vertically and define a third point.

Note that a parallelogram has four sides that are parallel and equal in length, but also includes an angle other than 90° between sides.

6. Click the green **OK** check mark.
7. Use the **Smart Dimension** tool and define the rectangle's size.

2-6 Slots

SolidWorks has four different tools that can be used to draw slots. See Figure 2-27.

Figure 2-27

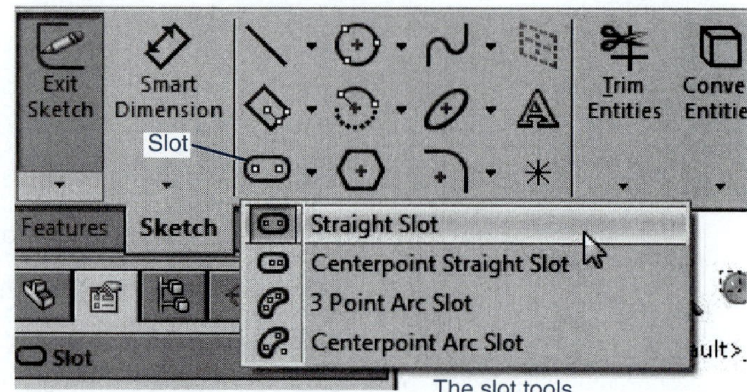

The **Slot** tool can be used to draw both internal and external slot shapes. Figure 2-28 shows examples of both internal and external slot shapes.

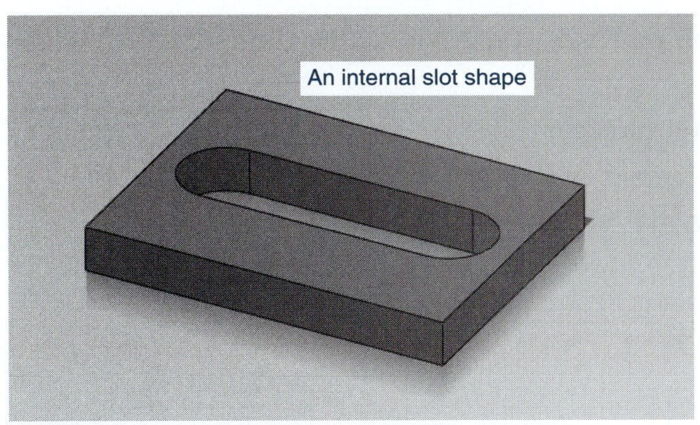

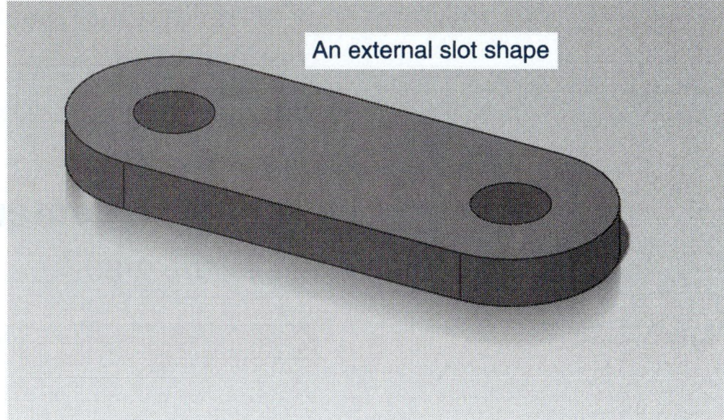

Figure 2-28

To Draw a Straight Slot

See Figure 2-29.

Figure 2-29

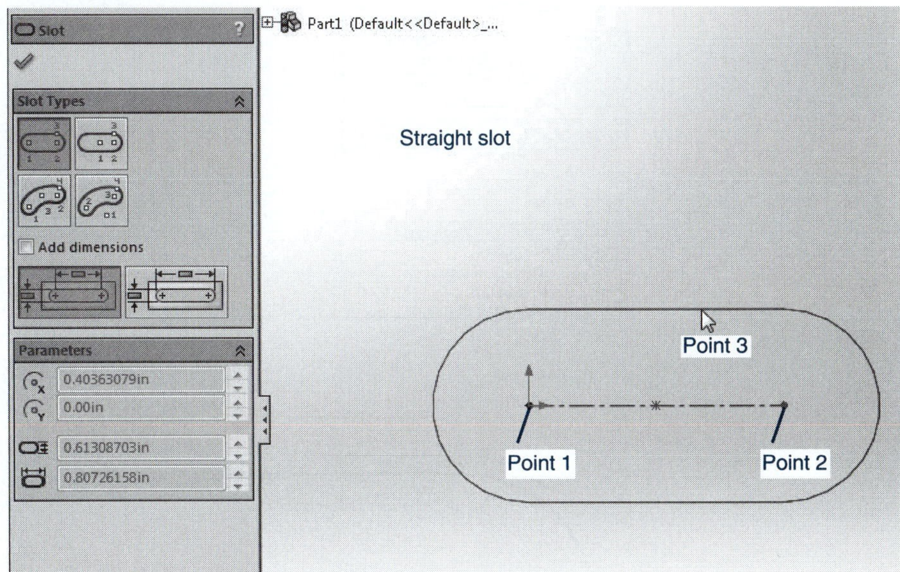

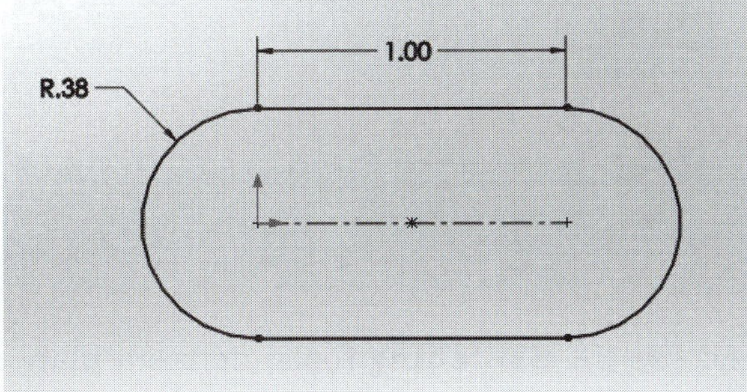

1. Access the **Sketching** tools, define a plane, and the inch units.
2. Click the **Straight Slot** tool.
3. Click a starting point.
4. Move the cursor horizontally and define a second point.

The distance between points 1 and 2 will define the slot's centerline.

5 Move the cursor and select a third point.

The distance between points 2 and 3 will define the radius of the slot's rounded ends.

6 Add dimensions as required.

To Draw a Centerpoint Straight Slot

See Figure 2-30.

Figure 2-30

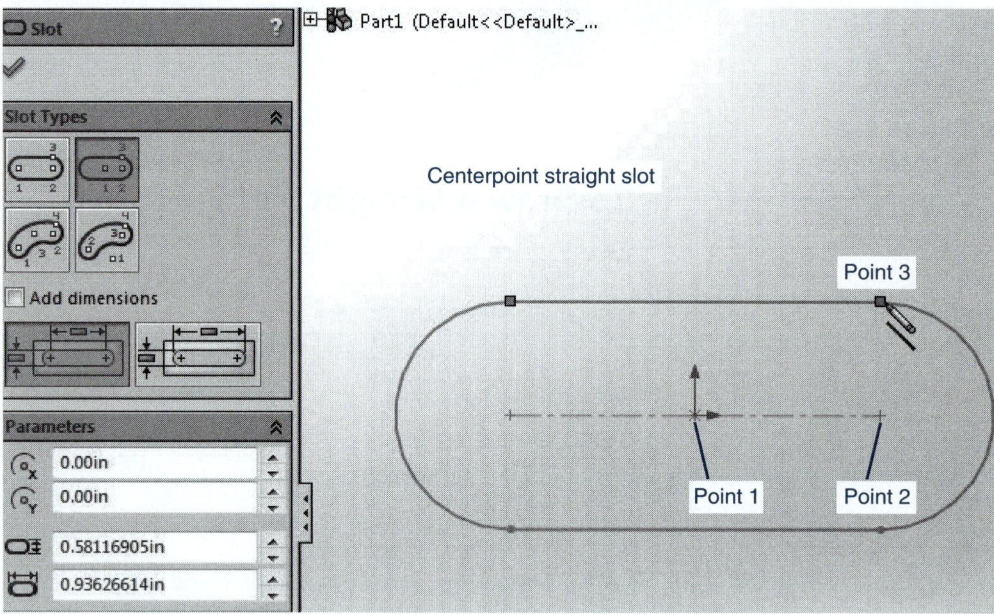

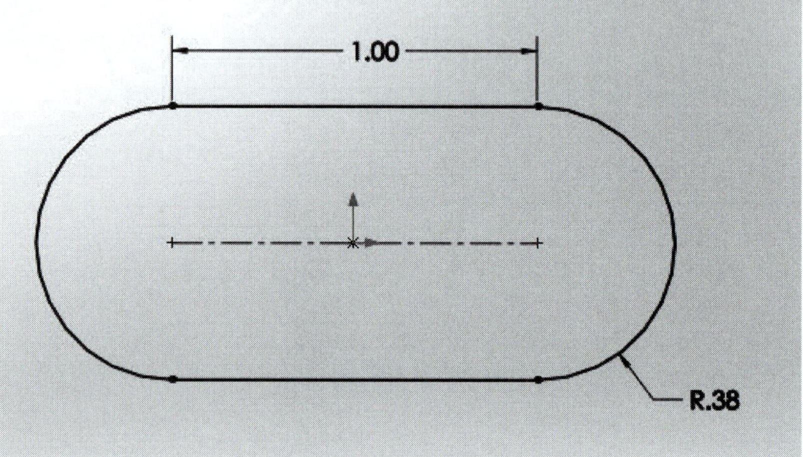

1 Access the **Sketching** tools, define a plane, and the inch units.

2 Click the **Centerpoint Straight Slot** tool.

3 Click a starting point.

4 Move the cursor horizontally and define a second point.

5 Move the cursor and select a third point.

6 Add dimensions as required.

To Draw a 3 Point Arc Slot

See Figure 2-31.

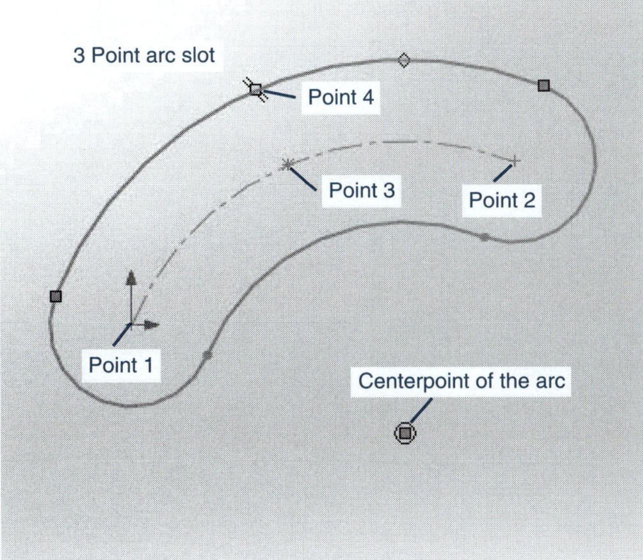

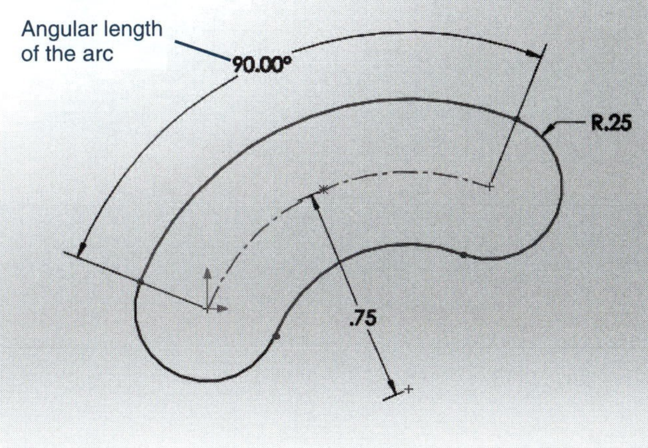

Figure 2-31

1. Access the **Sketching** tools, define a plane, and the inch units.
2. Click the **3 Point Arc Slot** tool.
3. Click a starting point.
4. Move the cursor and define point 2.
5. Move the cursor and define point 3.

   The angular distance between points 1 and 2 will define the angular length of the arc. The distance between point 3 and the arc's centerpoint will define the arc's radius.

6. Move the cursor and define point 4.

   The distance between points 3 and 4 will define the radius for the slot's rounded ends.

7. Add dimensions as required.

To Draw a Centerpoint Arc Slot

See Figure 2-32.

1. Access the **Sketching** tools, define a plane, and the inch units.
2. Click the **Centerpoint Arc Slot** tool.
3. Click a starting point.
4. Move the cursor and define point 2.

   The distance between the starting point, point 1, and point 2 will define the radius of a circle. The circle will be used to define the arc. The radius of the arc will be equal to the radius of the circle.

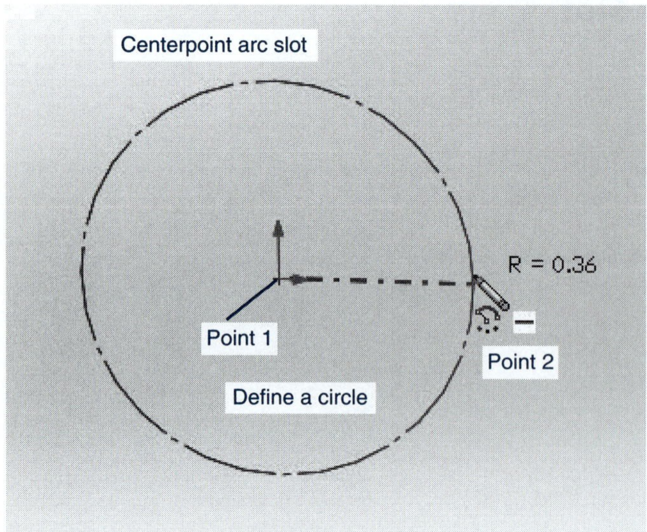

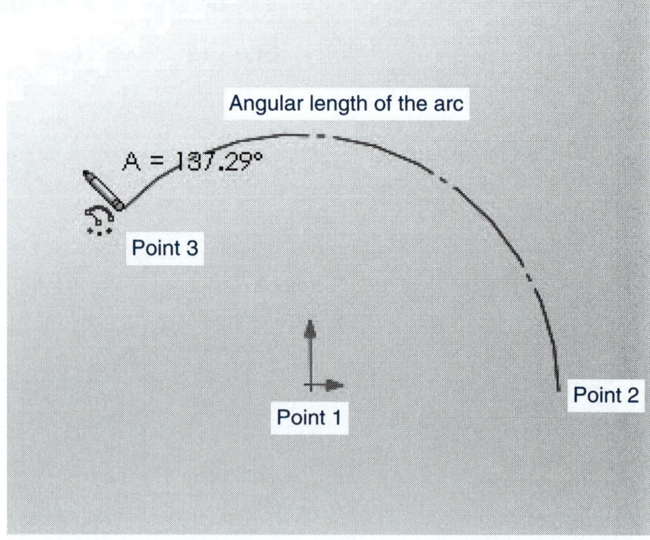

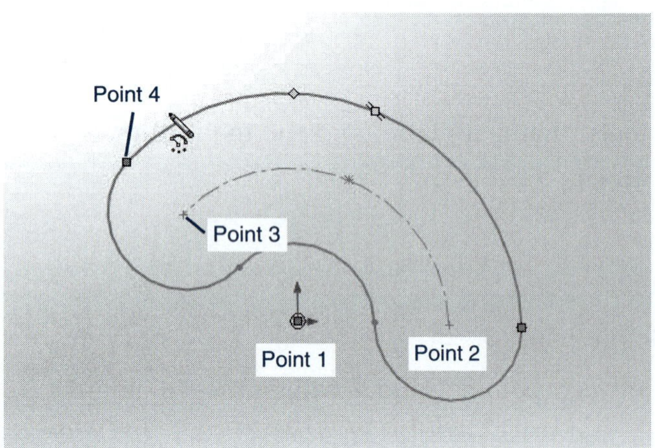

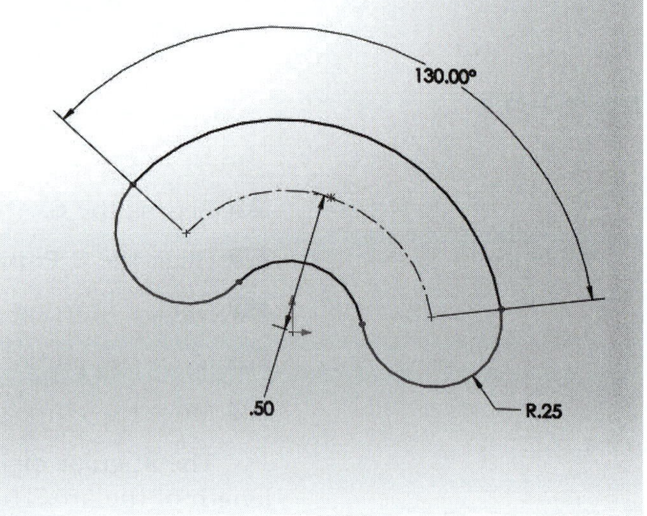

Figure 2-32

5 Move the cursor and define point 3.

The angular distance between points 2 and 3 will define the angular length of the arc.

6 Move the cursor and define point 4.

The distance between points 3 and 4 will define the radius for the slot's rounded ends.

7 Add dimensions as required.

2-7 Perimeter Circle

The **Circle** tool was presented in Section 1-4.

Perimeter circles are often drawn tangent to existing lines or arcs. Figure 2-33 shows a set of perpendicular lines and a circle. In this example a perimeter circle will be drawn tangent to the lines and circle.

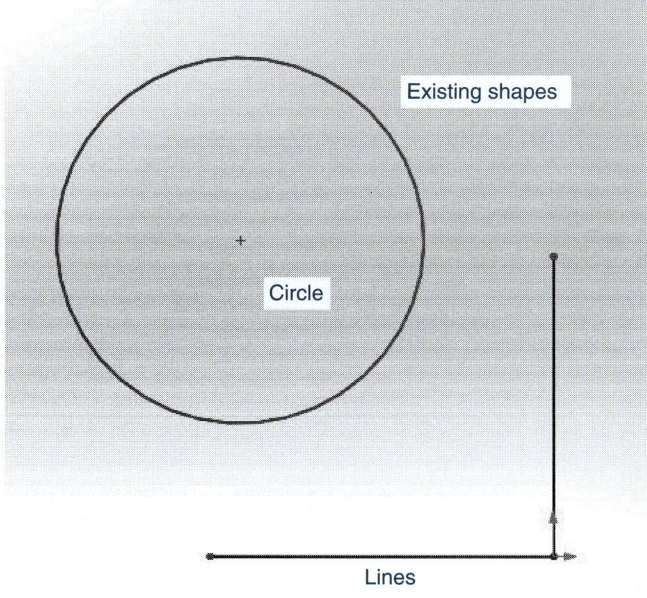

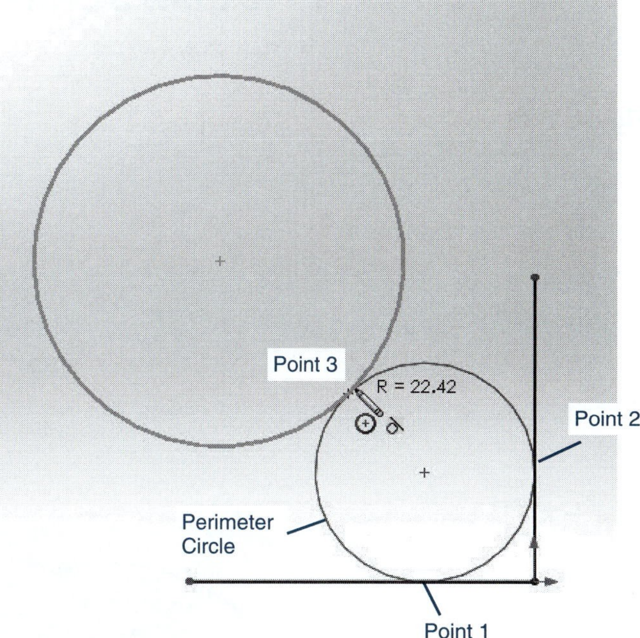

Figure 2-33

To Draw a Perimeter Circle

1. Access the **Sketching** tools, define a plane, and the inch units.
2. Click the **Perimeter Circle** tool.
3. Click a starting point, point 1.

   In this example the horizontal line was selected. Any point on the line is acceptable but it is good practice to select a point on the line near the approximate location of the tangent point.
4. Move the cursor and define point 2.

   In this example the vertical line was selected.
5. Move the cursor and select point 3.

   In this example the circle was selected.
6. Right-click the mouse and click the **Select** option.

2-8 Arcs

SolidWorks has three **Arc** tools: **Centerpoint Arc**, **Tangent Arc**, and **3 Point Arc**. See Figure 2-34.

Figure 2-34

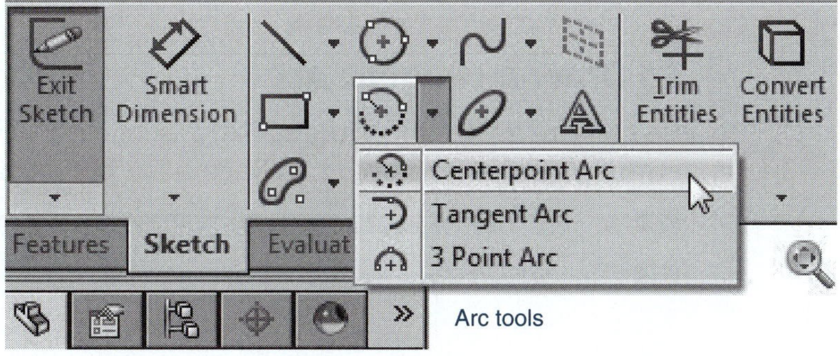

To Draw a Centerpoint Arc

See Figure 2-35.

Figure 2-35

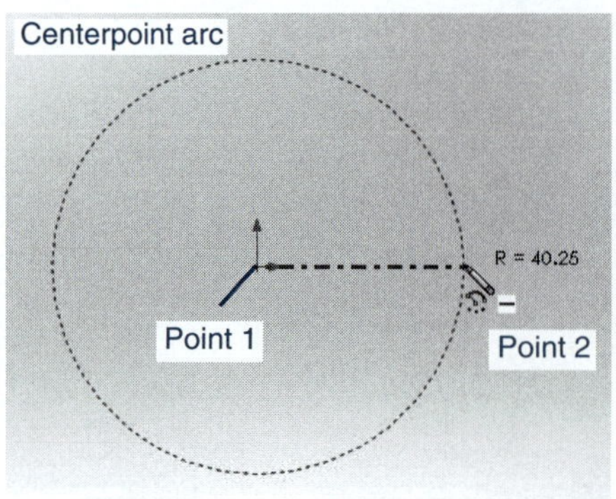

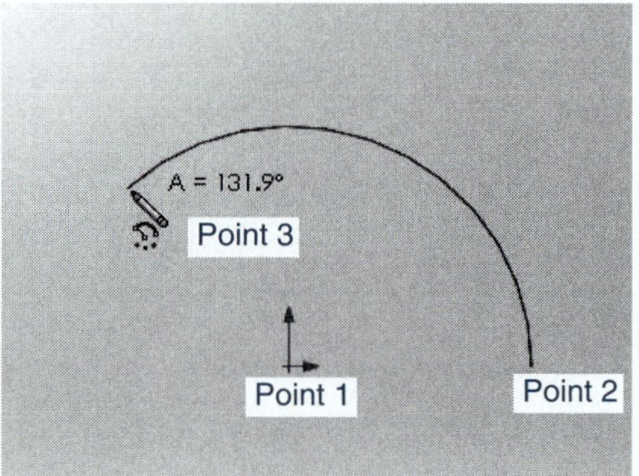

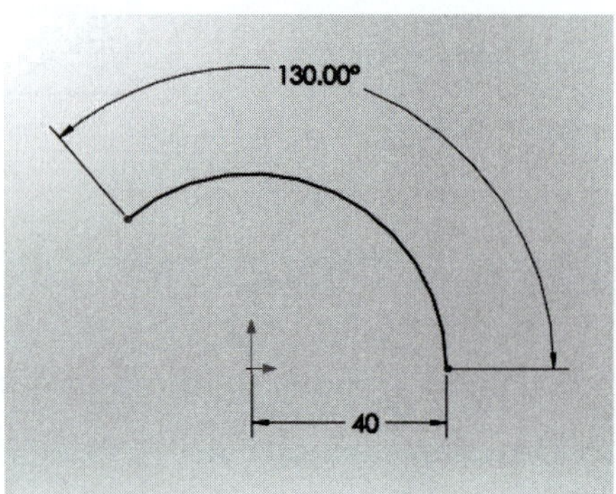

1. Access the **Sketching** tools, define a plane, and the millimeter units.
2. Click the **Centerpoint Arc** tool.
3. Click a starting point, point 1.
4. Move the cursor and define point 2.

   The distance between the starting point, point 1, and point 2 will define the radius of a circle. The circle will be used to define the arc. The radius of the arc will be equal to the radius of the circle.

5. Move the cursor and define point 3.

   The angular distance between points 2 and 3 will define the angular length of the arc.

6. Add dimensions as required.

To Draw a Tangent Arc

See Figure 2-36. The **Tangent Arc** is used to draw an arc between existing entities. In this example two lines have been drawn.

1. Access the **Sketching** tools, define a plane, and the millimeter units.
2. Click the **Tangent Arc** tool.

Figure 2-36

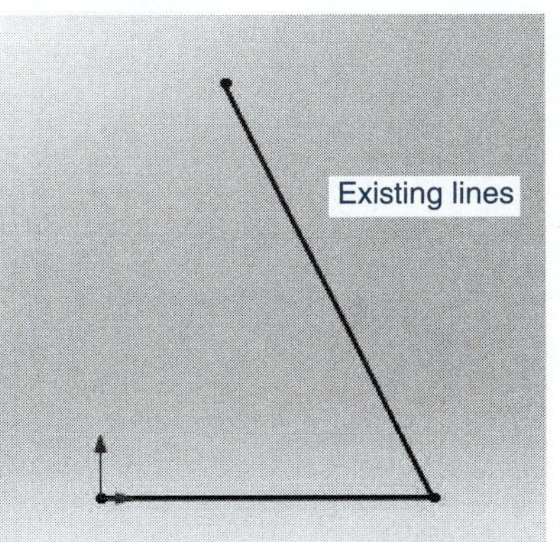

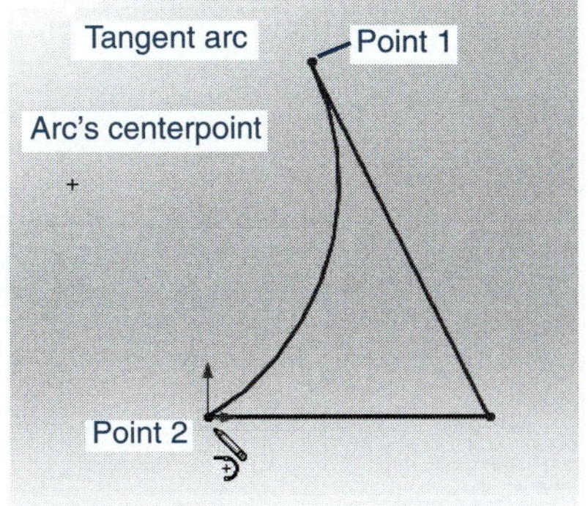

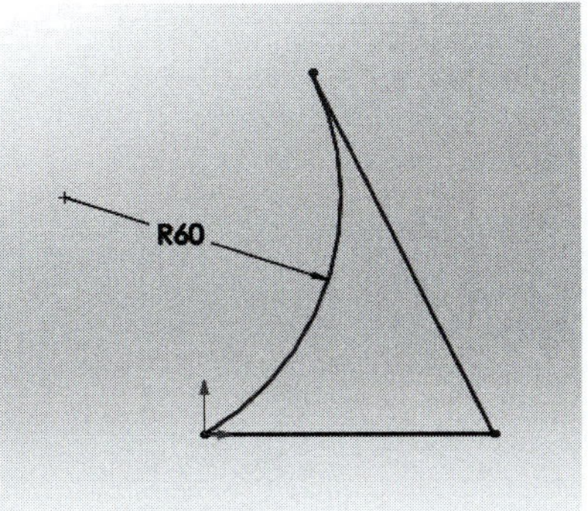

3 Click a starting point, point 1.

In this example the endpoint of the slanted line was selected.

4 Move the cursor and define point 2.

In this example the endpoint of the horizontal line was selected.

5 Add dimensions as required.

The radius of the arc is controlled by the radius dimensional value.

Figure 2-37 shows two parallel lines. The **Tangent Arc** tool was used to add a rounded end segment to the right end. Note that the radius value is **10** or half the distance between the two lines.

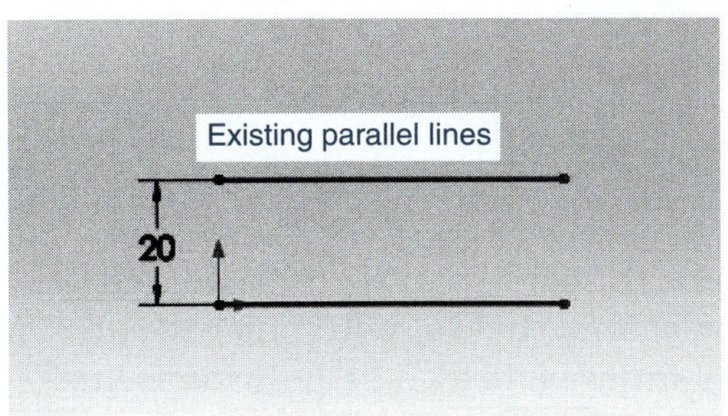

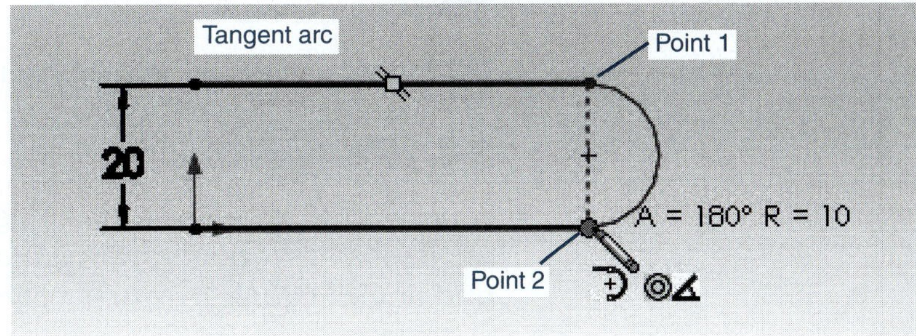

Figure 2-37

Figure 2-37
(*Continued*)

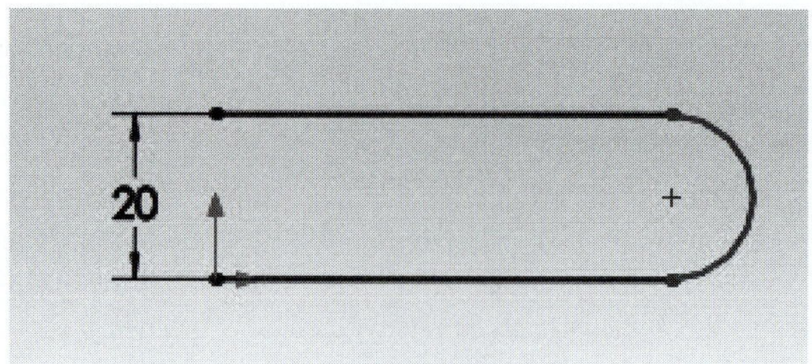

To Draw a 3 Point Arc

See Figure 2-38.

Figure 2-38

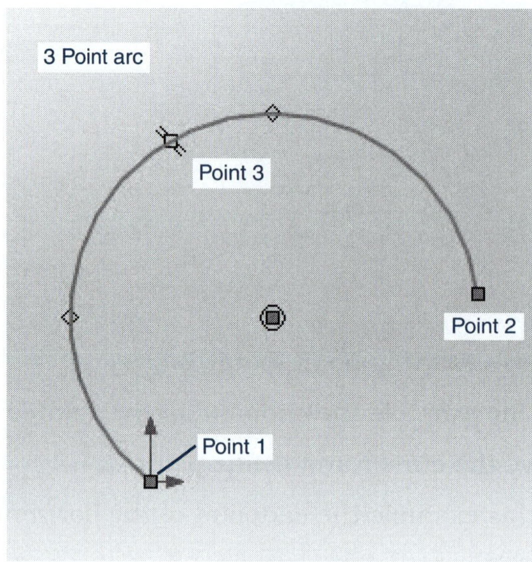

1. Access the **Sketching** tools, define a plane, and the millimeter units.
2. Click the **3 Point Arc** tool.
3. Click a starting point, point 1.
4. Move the cursor and define point 2.
5. Move the cursor and define point 3.
6. Add dimensions as required.

2-9 Polygons

A polygon is any closed plane figure with more than three sides and angles. SolidWorks draws regular polygons, that is, polygons with all sides and angles equal. Irregular polygons must be drawn as individual line segments.

To Draw a Hexagon

A hexagon is a polygon with six sides. For this example a regular hexagon with six equal sides and angles will be drawn. See Figure 2-39.

Figure 2-39

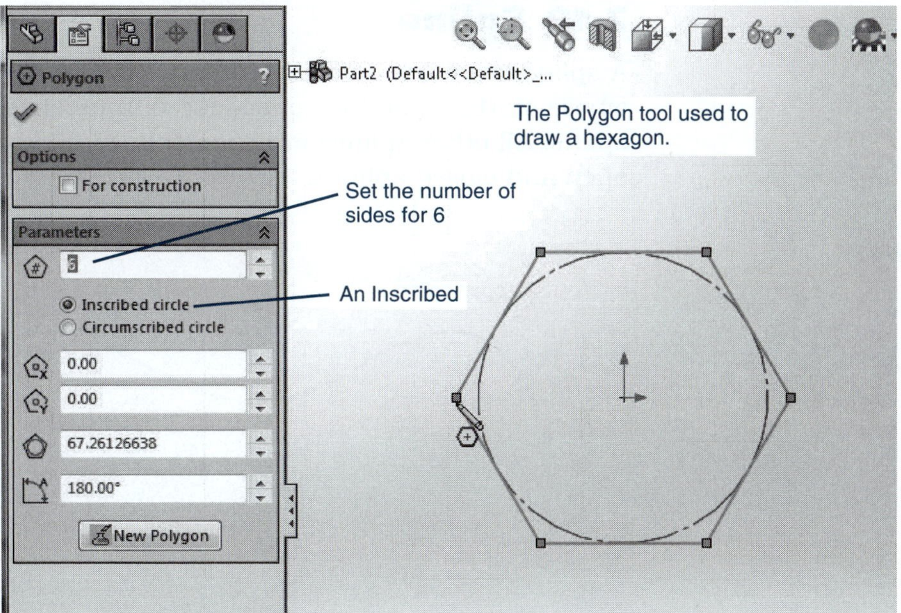

1. Access the **Sketching** tools, define a plane, and the millimeter units.
2. Click the **Polygon** tool.
3. Set the **Parameters** value for **6** and ensure that the **Inscribed circle** button is active. This will create a circumscribed hexagon.
4. Click a starting point, point 1.
5. Move the cursor and define point 2.
6. Add dimensions as required.

Figure 2-40 shows a circumscribed and an inscribed hexagon. The circumscribed hexagon is drawn around and tangent to a circle. An inscribed hexagon is drawn within a circle.

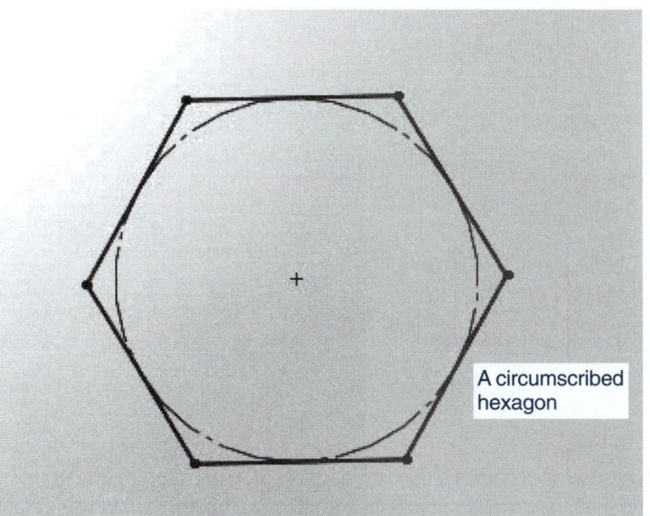

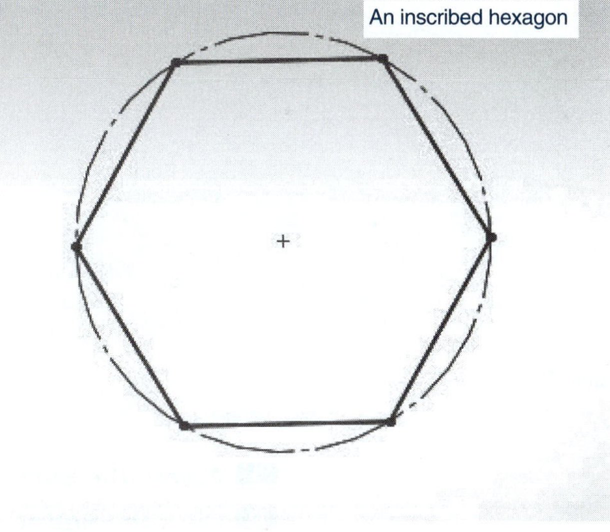

Figure 2-40

2-10 Spline

A spline is a curved line that intersects several defined points. A spline whose starting point is coincident with its endpoint is called a closed spline. All other splines are open splines. Figure 2-41 shows examples of open and closed splines.

Figure 2-41

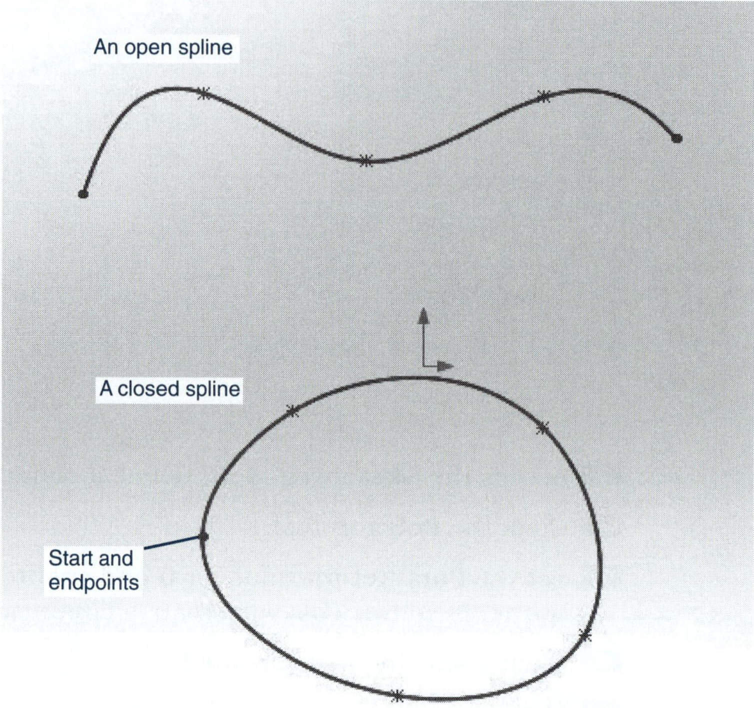

To Draw a Spline

See Figure 2-42.

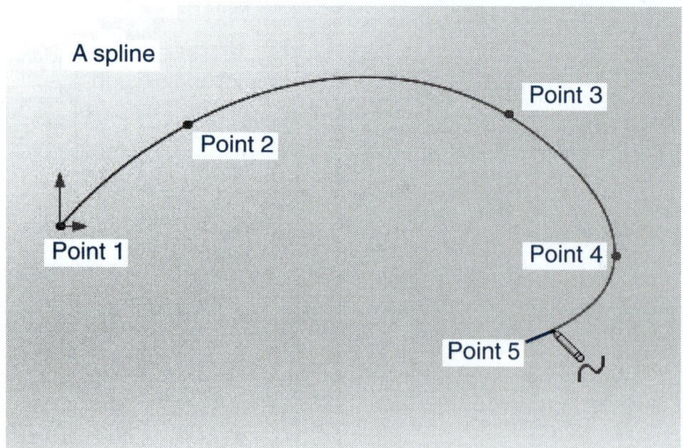

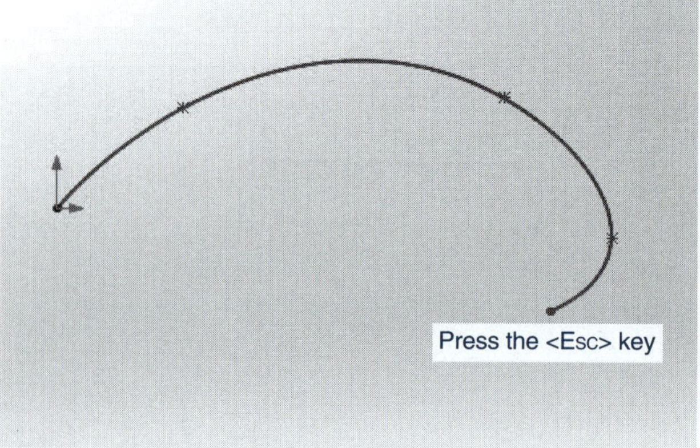

Figure 2-42

1. Access the **Sketching** tools, define a plane, and the millimeter units.
2. Click the **Spline** tool.
3. Click a starting point, point 1.
4. Move the cursor and define point 2.

5 Move the cursor and define point 3.

6 Move the cursor and define point 4.

7 Move the cursor and define point 5.

8 Press the <**Esc**> key.

The points used to create a spline may be defined using dimensions.

To Edit a Spline

See Figure 2-43.

Figure 2-43

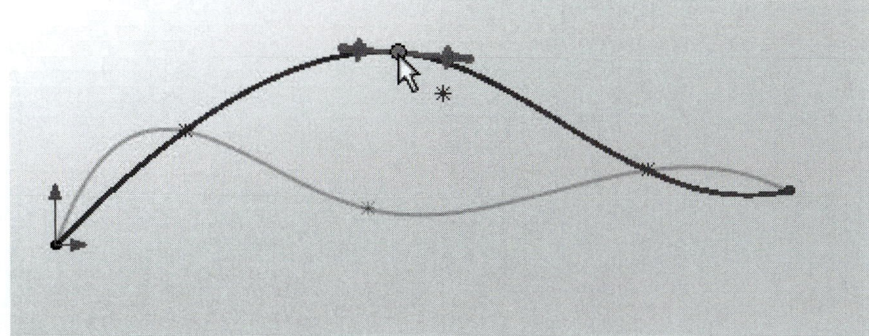

To edit a spline, click and drag the defining points.

1 Select a point on the spline.

2 Click and drag the point to a new location.

3 Click the green **OK** check mark.

A spline may also be edited by editing the dimensions used to define the spline's points.

2-11 Ellipse

There are four tools associated with the **Ellipse** tool: **Ellipse**, **Partial Ellipse**, **Parabola**, and **Conic**. See Figure 2-44. For technical drawings an ellipse is defined by its centerpoint location, its major axis, and its minor axis.

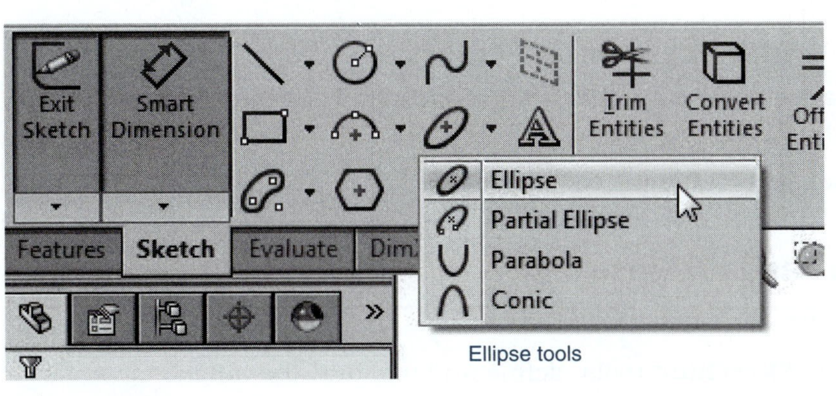

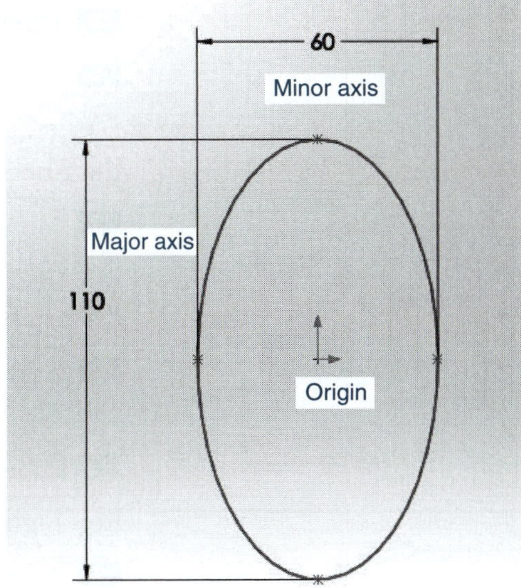

Figure 2-44

To Draw an Ellipse

See Figure 2-45.

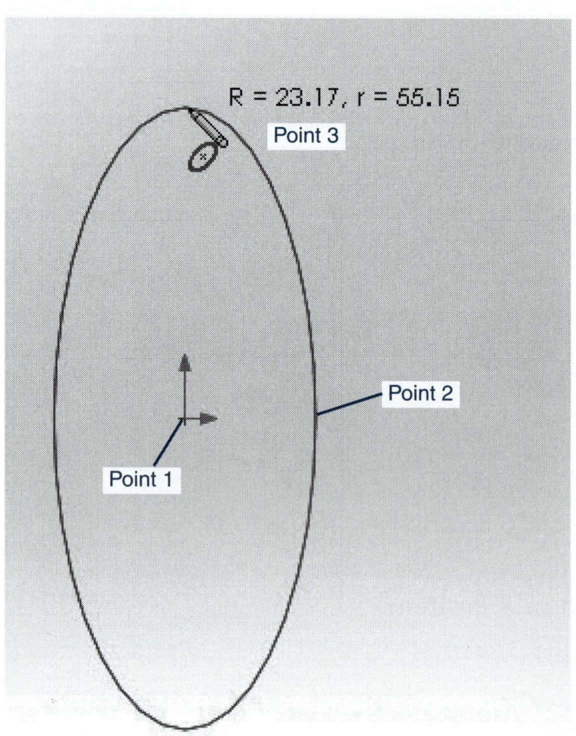

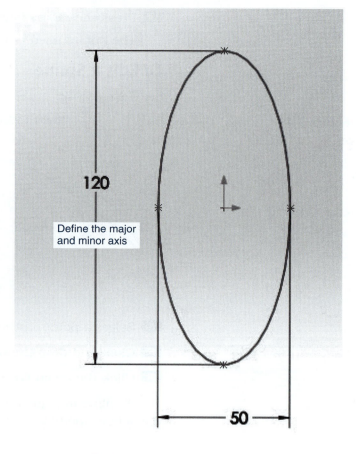

Figure 2-45

1. Access the **Sketching** tools, define a plane, and the millimeter units.
2. Click the **Ellipse** tool.
3. Click a starting point, point 1.
4. Move the cursor and define point 2.

   The distance between points 1 and 2 defines half the minor axis distance.
5. Move the cursor and define point 3.

   The distance between points 1 and 3 defines half the major axis distance.
6. Add dimensions as required.

To Draw a Partial Ellipse

See Figure 2-46.

1. Access the **Sketching** tools, define a plane, and the millimeter units.
2. Click the **Partial Ellipse** tool.

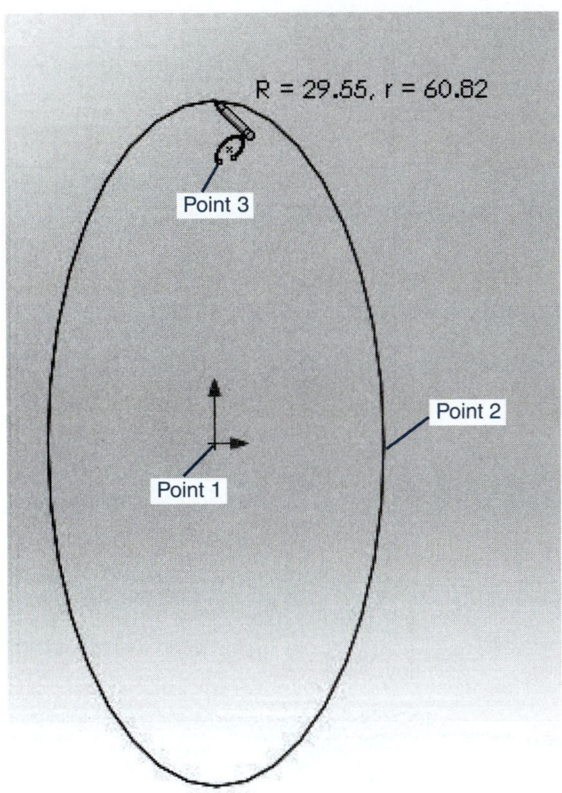

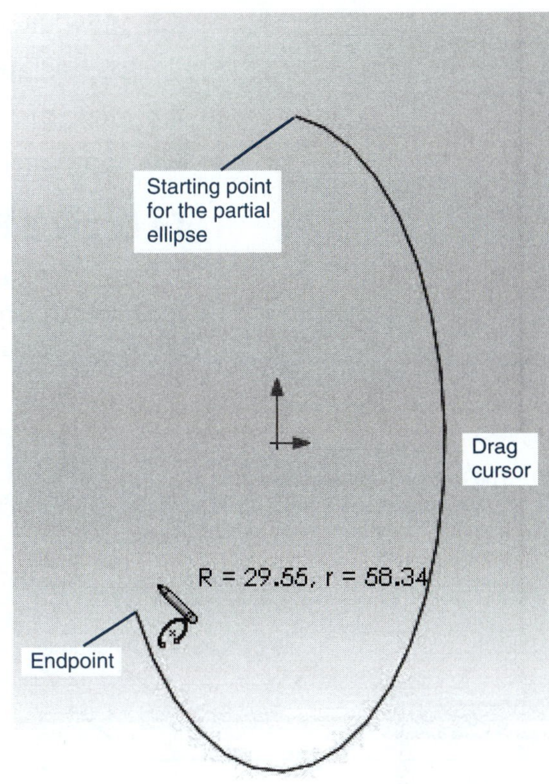

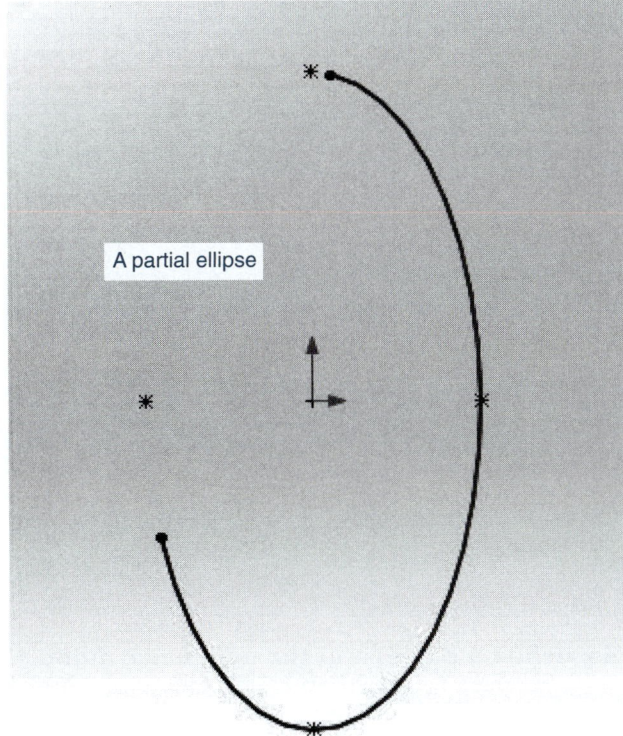

Figure 2-46

3. Click a starting point, point 1.
4. Move the cursor and define point 2.

   The distance between points 1 and 2 defines half the minor axis distance.

5 Move the cursor and define point 3.

The distance between points 1 and 3 defines half the major axis distance.

6 Move the cursor to define the partial ellipse.

7 Add dimensions as necessary.

To Draw a Parabola

A parabola is defined as the path of a moving point that is always equidistant from a fixed point, the focus, and a fixed straight line, the directrix. See Figure 2-47.

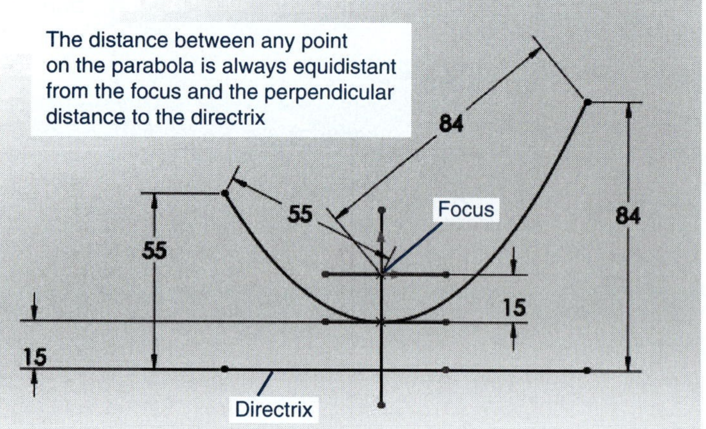

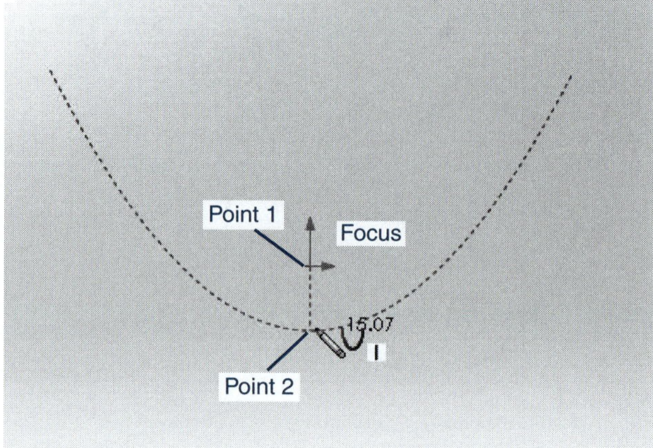

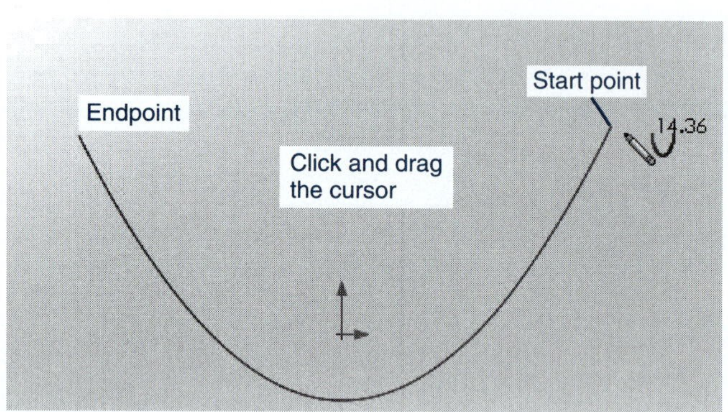

Figure 2-47

1 Access the **Sketching** tools, define a plane, and the millimeter units.

2 Click the **Parabola** tool.

3 Click a starting point, point 1.

In this example the selected point 1 becomes the focus.

4 Move the cursor and define point 2.

The distance between points 1 and 2 defines half the distance to the directrix.

5 Move the cursor and define a start point for the parabola.

6 Click and drag the cursor to define the size of the parabola.

7 Add dimensions as required.

The distance between the focus and the directrix is two times the distance between the focus and the closest point of the parabola.

Conic Section

A conic section is the intersection of a plane with a cone. Ellipses, parabolas, and hyperbolas are all conic sections. See Figure 2-48.

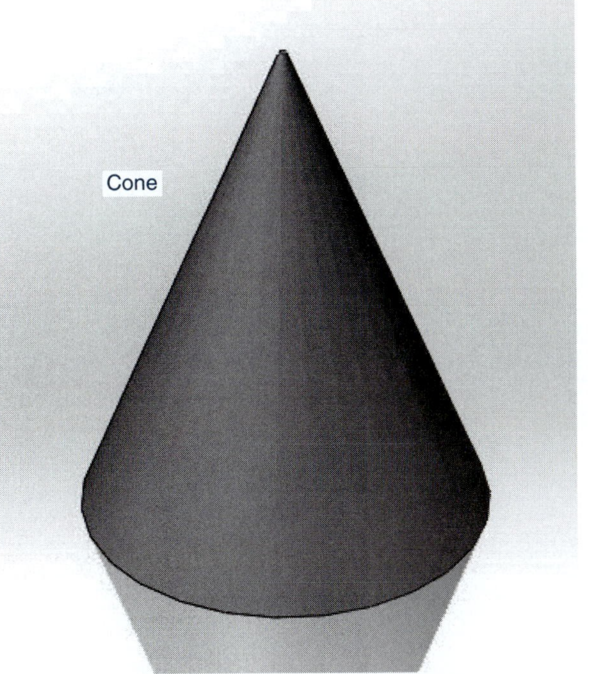
Cone

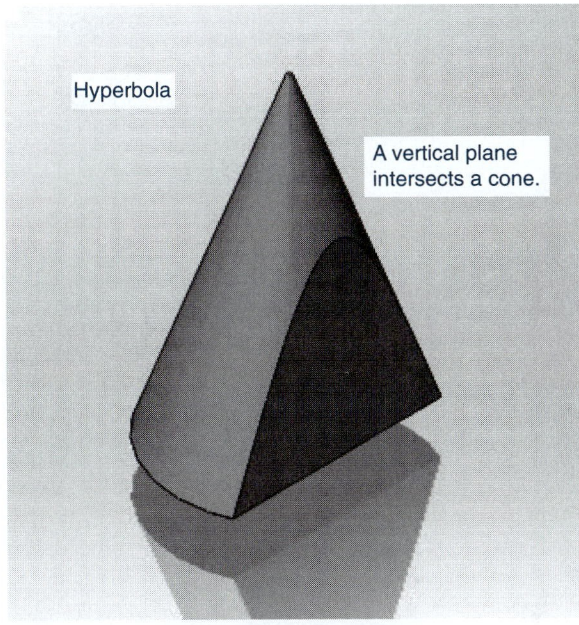
Hyperbola
A vertical plane intersects a cone.

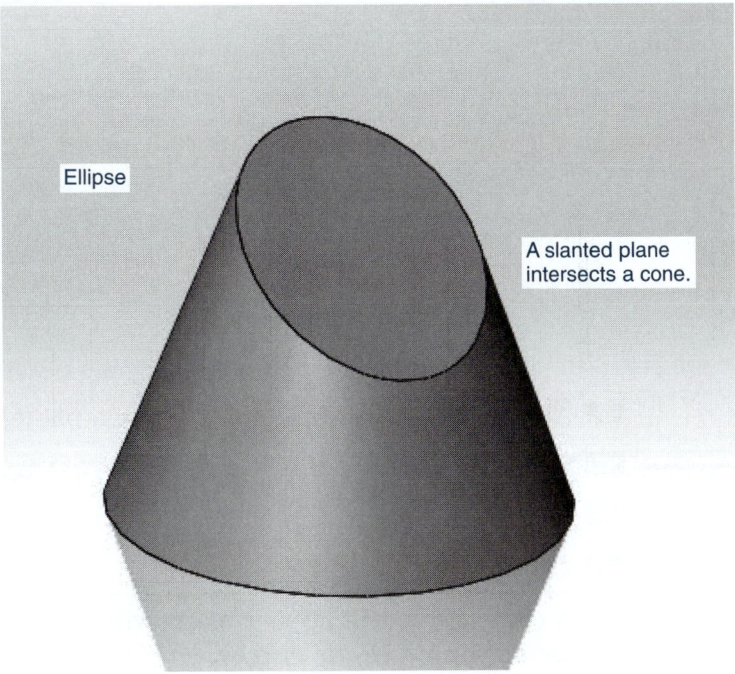
Ellipse
A slanted plane intersects a cone.

Figure 2-48

To Draw a Conic

Conic curves can be used to create smooth curves between existing endpoints. See Figure 2-49.

Figure 2-49

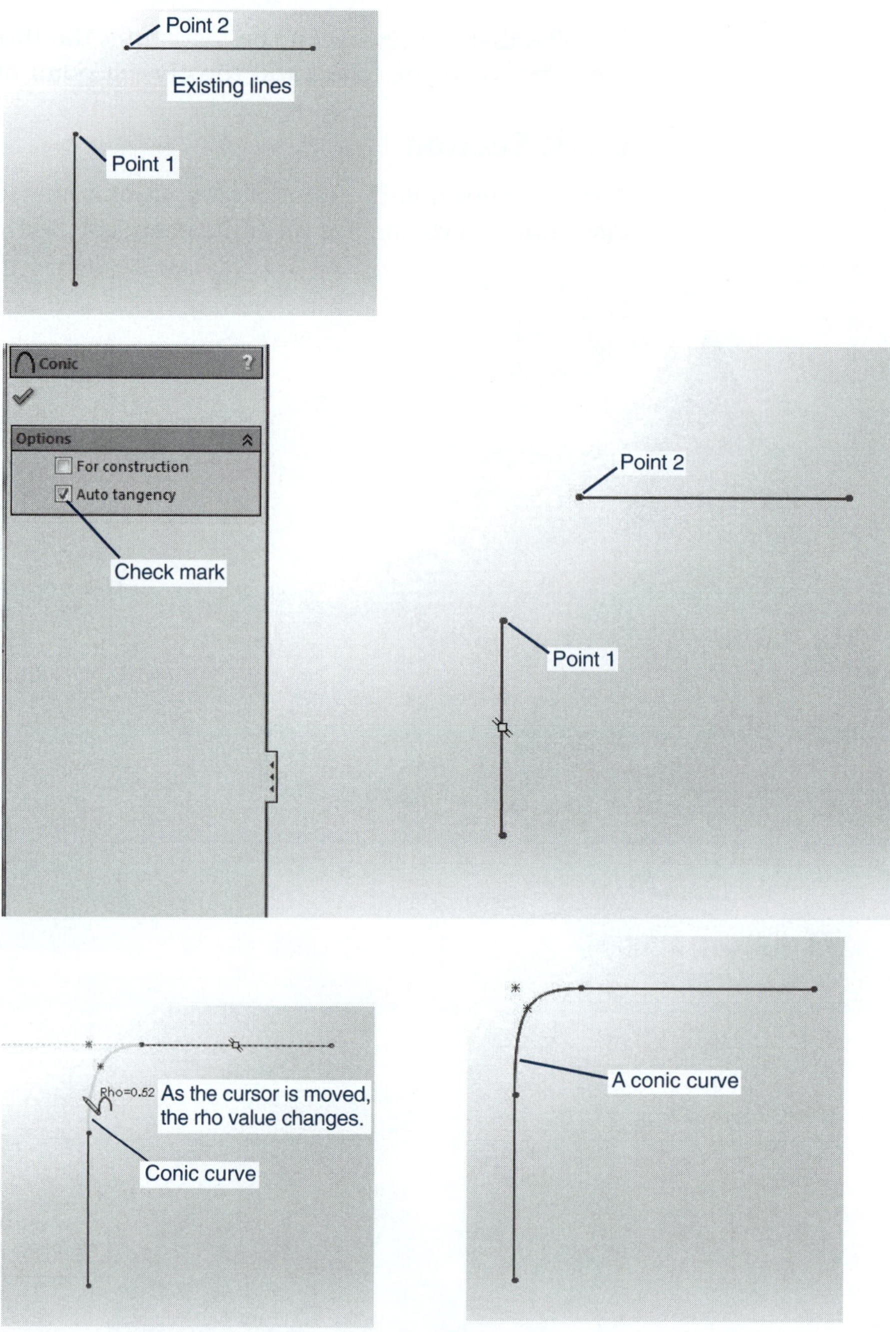

1. Access the **Sketching** tools, define a plane, and the millimeter units.
2. Click the **Conic** tool.
3. Ensure that there is a check mark in the **Auto tangency** box.
4. Click points 1 and 2.

   A conic curve will appear along with a rho value. The rho value will change as you move the cursor.

5. Select an appropriate conic shape and click the green **OK** check mark.

2-12 Fillets and Chamfers

A fillet is a rounded corner or edge and a chamfer is a beveled corner or edge. Both the **Fillet** and **Chamfer** commands can be applied to 2D sketches of 3D models. Figure 2-50 shows a sketch that includes a fillet and a chamfer.

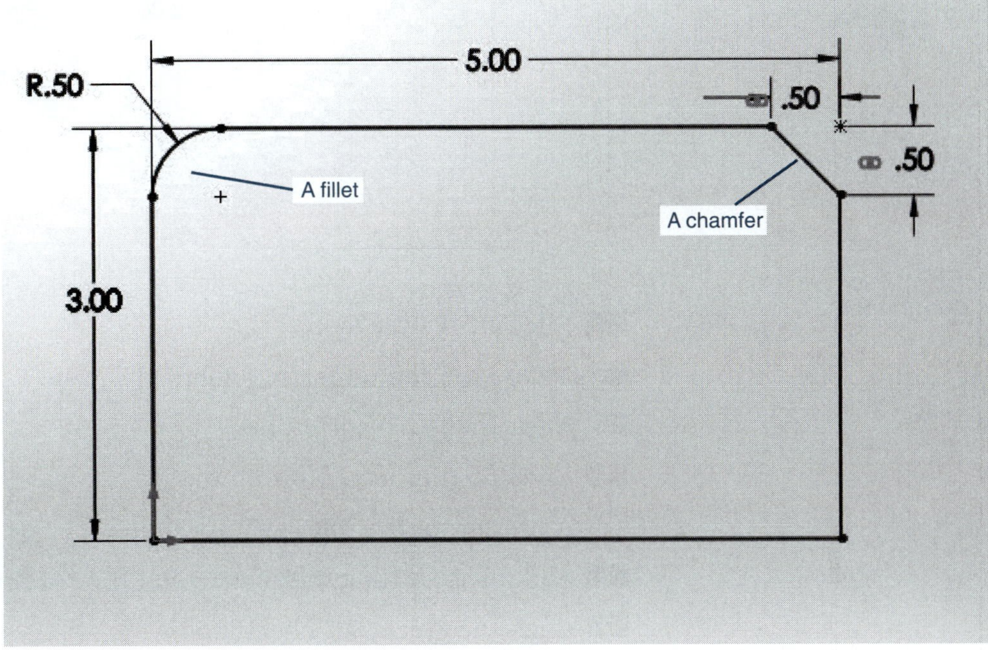

Figure 2-50

To Draw a Fillet

See Figure 2-51.

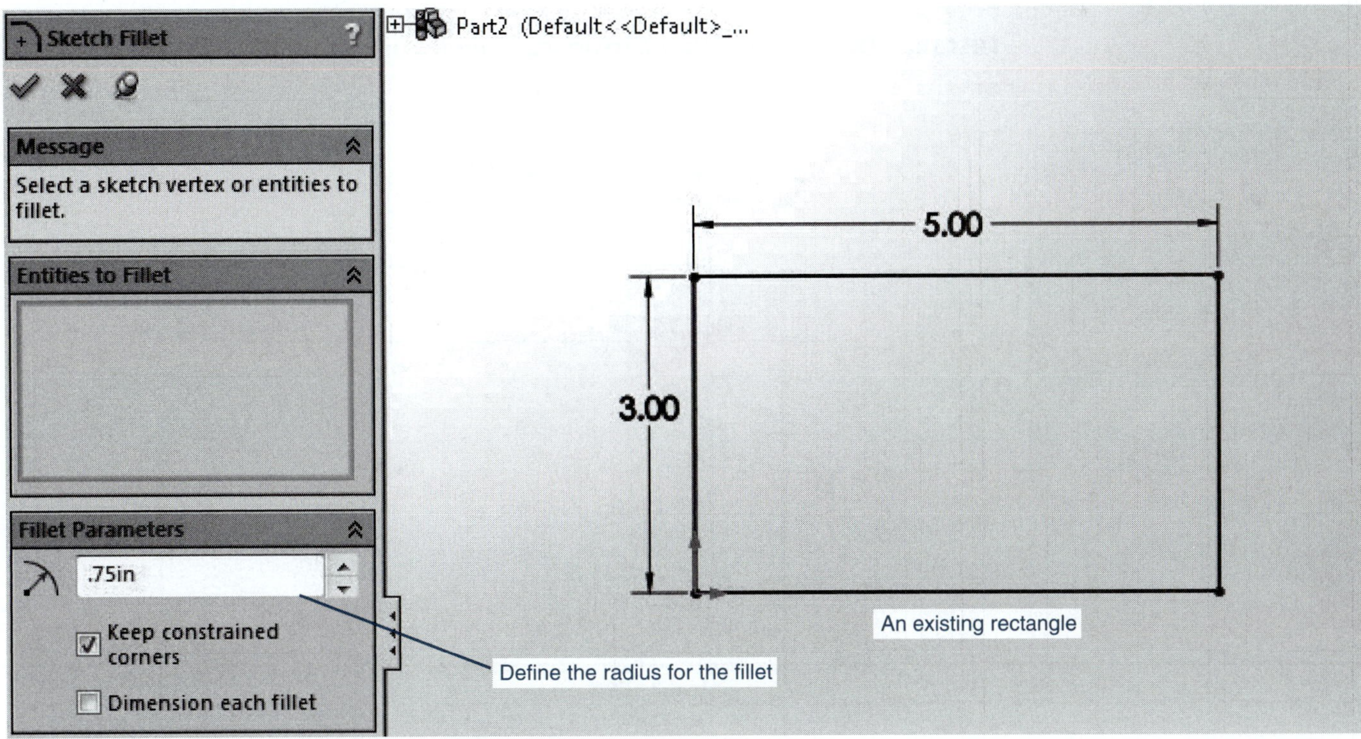

Figure 2-51

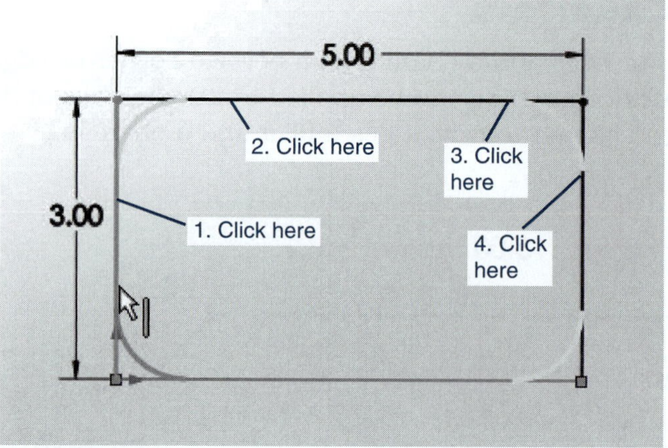

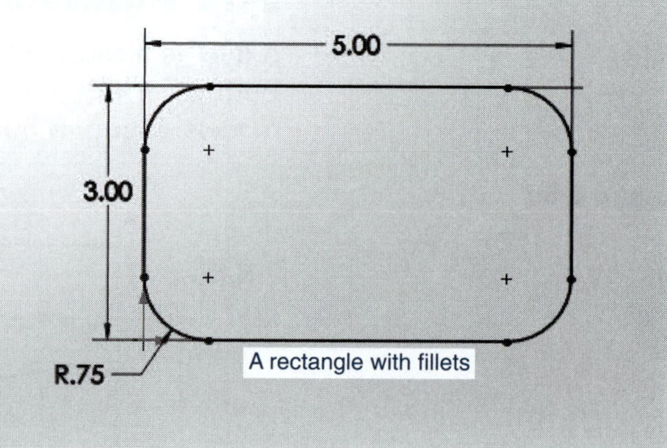

Figure 2-51
(*Continued*)

1 Click the **Fillet** tool.

2 Define the radius for the fillet.

3 Click the left vertical line.

4 Click the top horizontal line.

A preview of the fillet will appear.

5 Add fillets to the other corners of the rectangle.

6 Click the green **OK** check mark.

To Draw a Chamfer

There are three different ways to define a chamfer: Angle-distance, Distance-distance – equal distance, Distance-distance – not equal distance. Figure 2-52 shows an existing 3.00 × 5.00 rectangle.

Distance-Distance – Equal Distance. See Figure 2-52.

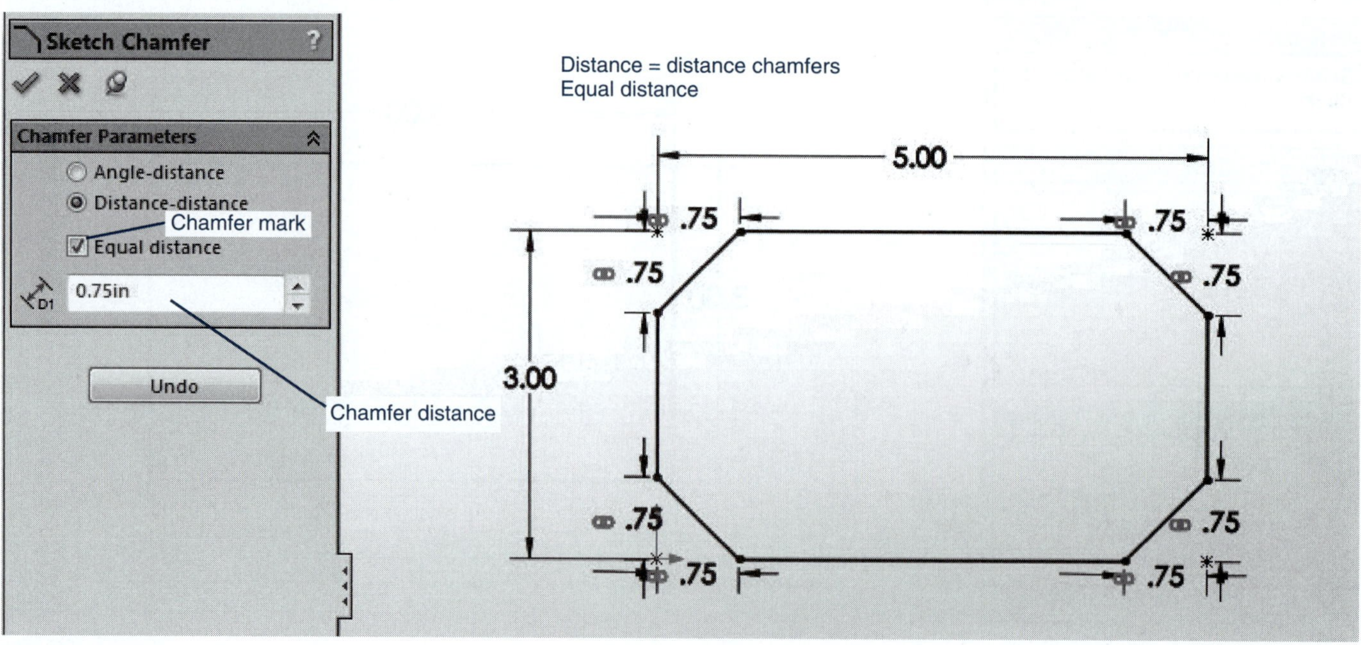

Figure 2-52

1. Click the **Chamfer** tool.

   The **Chamfer** tool is a flyout from the **Fillet** tool.
2. Define the size of the chamfer.
3. Click the left vertical line.
4. Click the top horizontal line.

   A preview of the chamfer will appear.
5. Add a .75 × .75 chamfer to the other corners.
6. Click the green **OK** check mark.

Angle-Distance. See Figure 2-53.

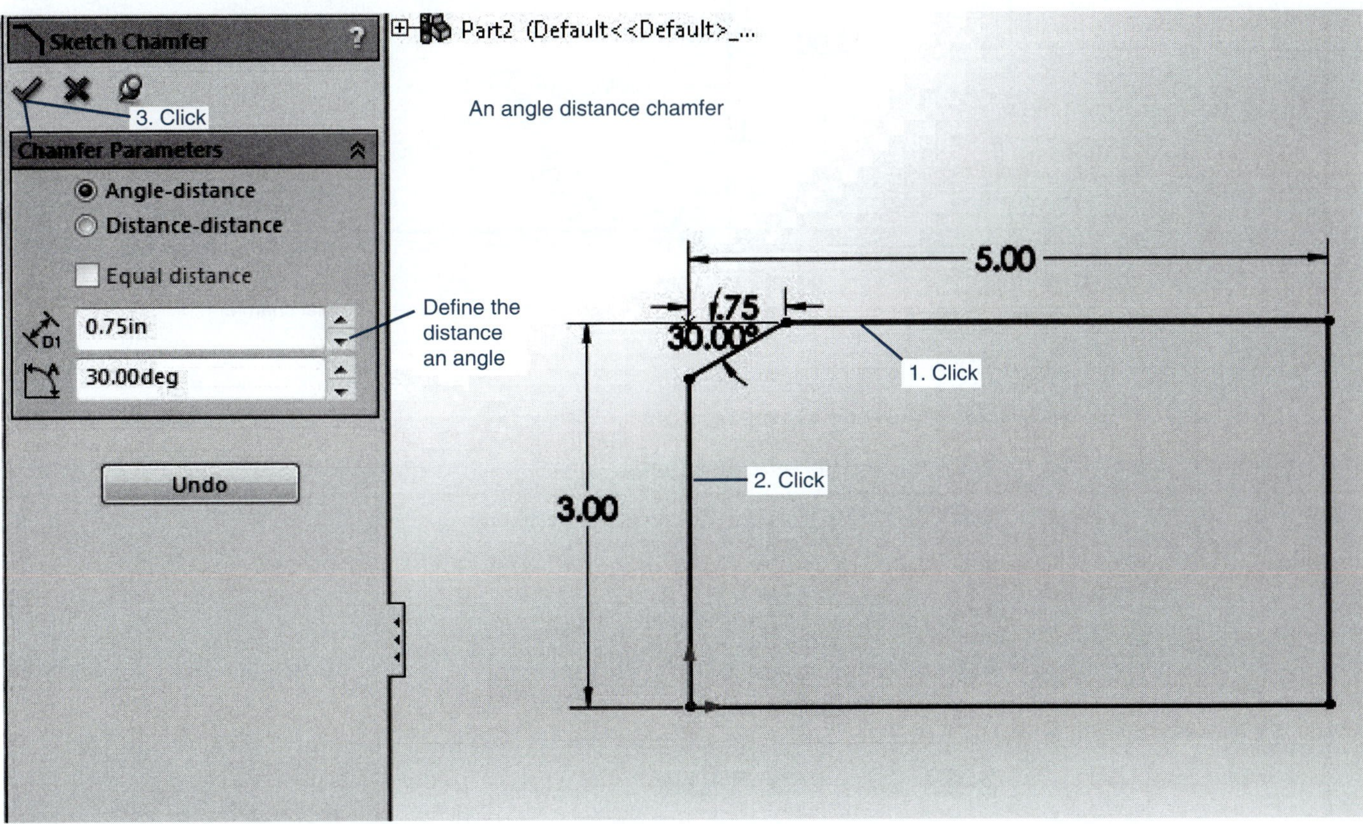

Figure 2-53

1. Click the **Chamfer** tool.
2. Define the chamfer's angle and distance.
3. Click the top horizontal line.
4. Click the left vertical line.

   A preview of the chamfer will appear.
5. Click the green **OK** check mark.

Note that an angle-distance chamfer created with a 45° angle will generate two equal distances.

Distance-Distance – Not Equal. See Figure 2-54.

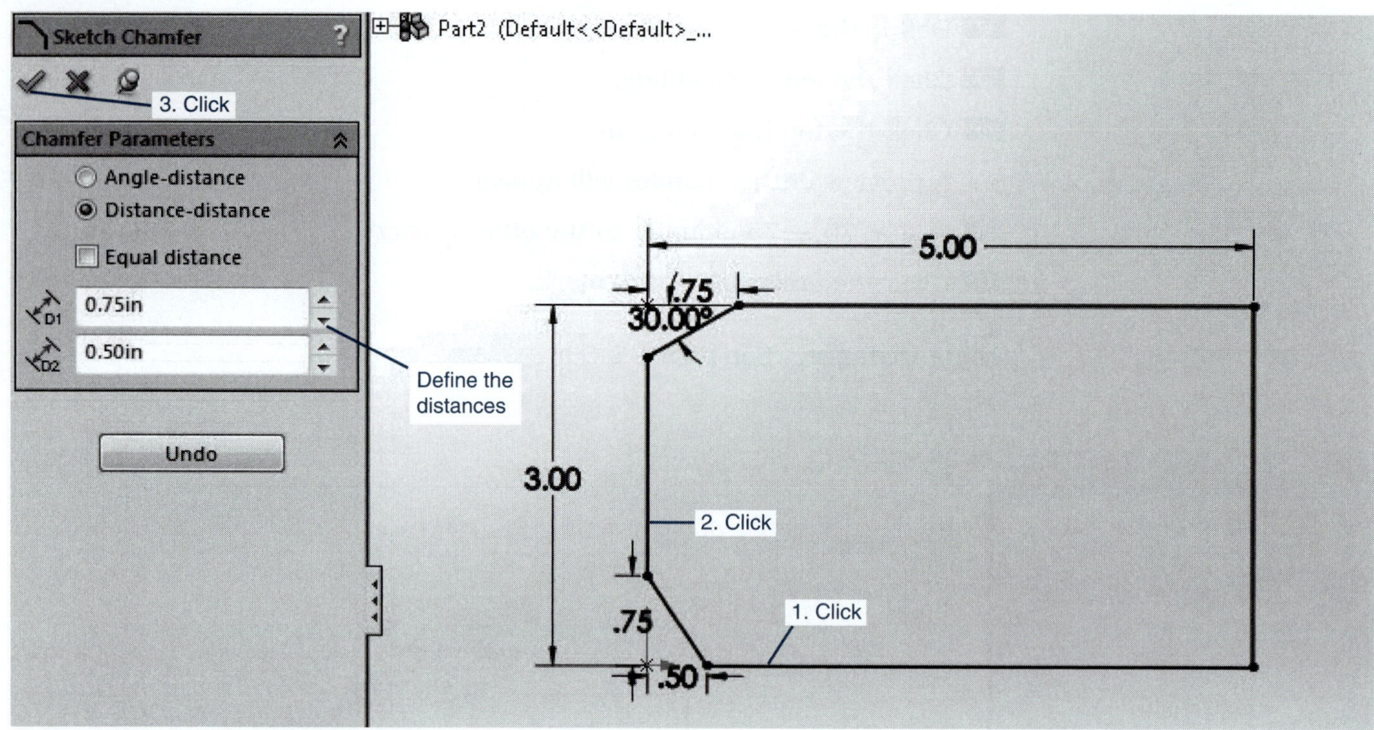

A Distance-distance–not equal chamfer

Figure 2-54

1. Click the **Chamfer** tool.
2. Define the two distances.
3. Click the bottom horizontal line.
4. Click the left vertical line.

   A preview of the chamfer will appear.
5. Click the green **OK** check mark.

2-13 Sketch Text

The **Sketch Text** tool is used to add text to a sketch. The default font for SolidWorks text is **Century Gothic**.

To Add Text

See Figure 2-55.

1. Click the **Text** tool.
2. Type the text in the text box in the **Sketch Text** manager box.

The **Text** will appear on the screen. The text can be moved by clicking the cursor in a different location.

Figure 2-55

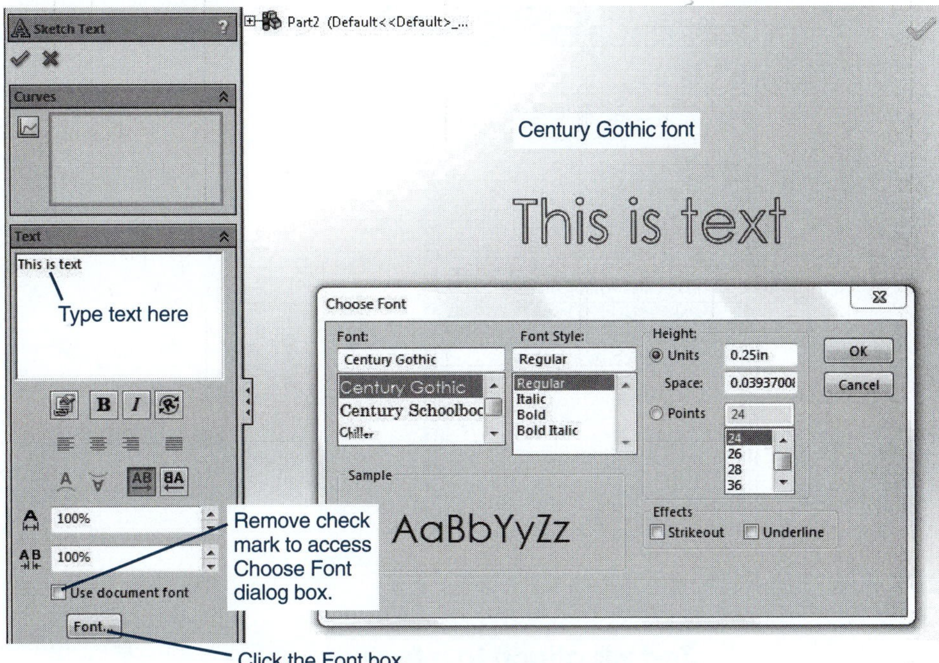

To Change the Font and Size of Text

1 Remove the check mark from the **Use document font** box and click the **Font** box.

The **Choose Font** dialog box will appear.

2 Define the new font and text height.

3 Add the text.

In this example **Arial Bold** font with a height of **0.375** was selected. See Figure 2-56.

Figure 2-56

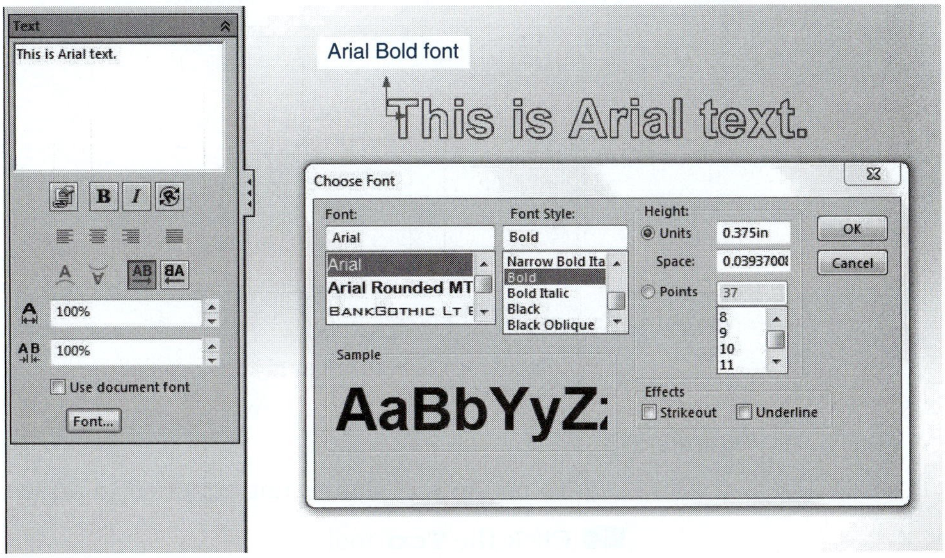

4 Click the green **OK** check mark.

The Text tool will automatically attach text to an existing feature. See Figure 2-57.

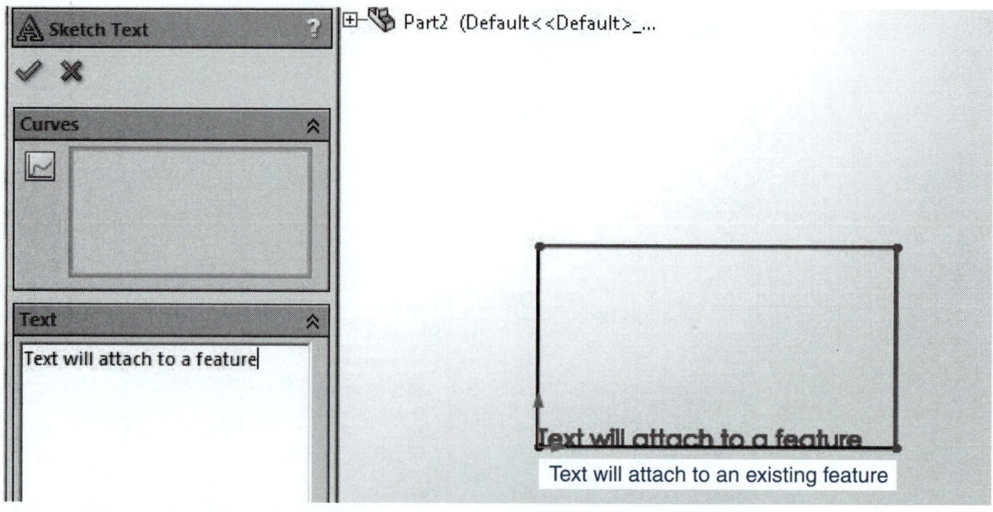

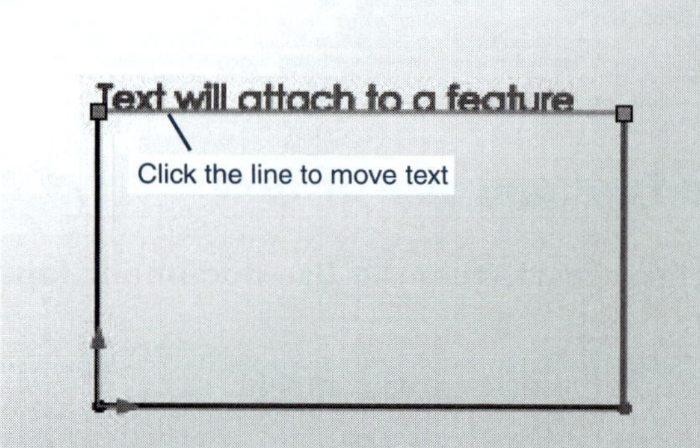

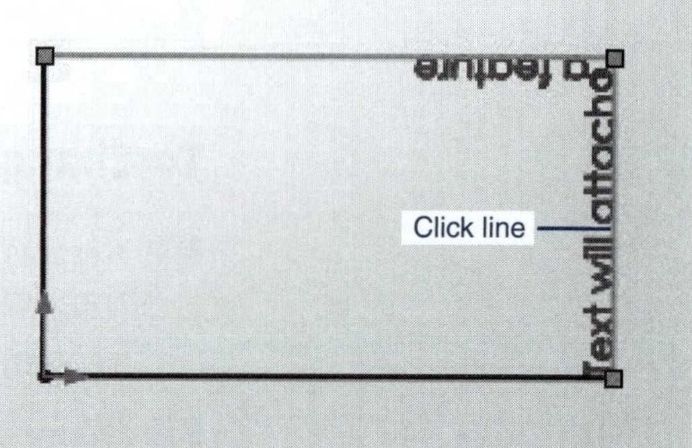

Figure 2-57

To create text that is not attached to an existing sketch.

1. Click the **Text** tool.
2. Use the cursor and click the point within the drawing screen.
3. Type the text.
4. Click the green **OK** check mark.

2-14 Point

The **Point** tool is used to create points anywhere on the drawing screen. Points are often helpful during the construction process of a sketch. Figure 2-58 shows a Point. It was created by clicking the **Point** tool and selecting a location on the drawing screen. The point can be precisely located using dimensions or by using the changing **Parameters** values listed in the **Point** manager box.

Figure 2-58

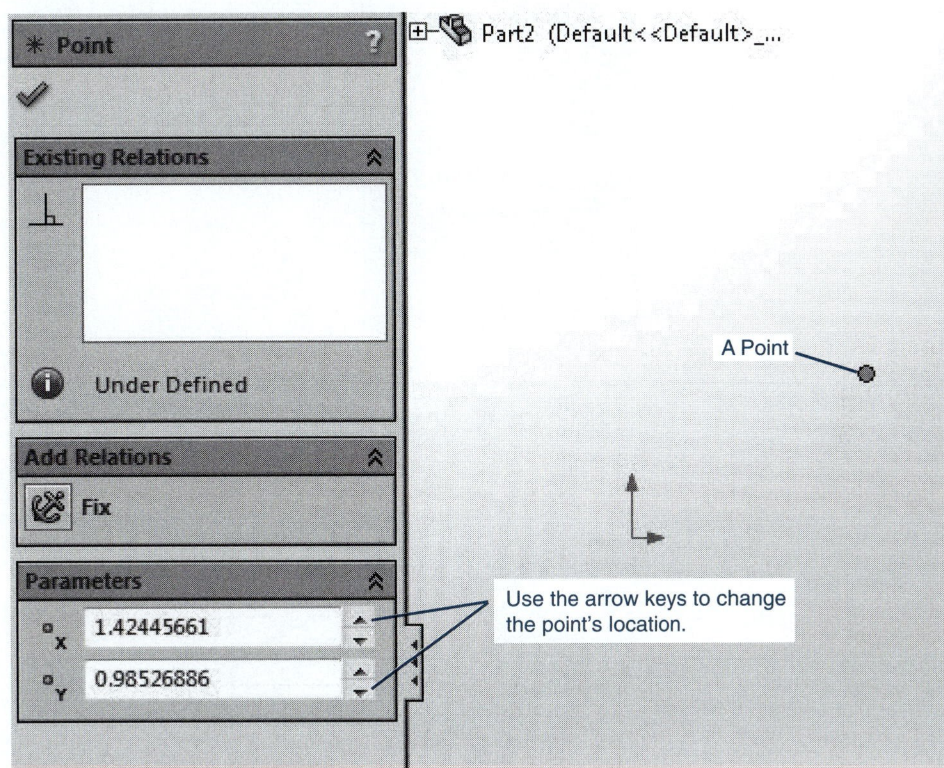

2-15 Trim Entities

The **Trim Entities** tool is used to remove unwanted entities from an existing sketch. Figure 2-59 shows a line passing through a circle and a rectangle. Say we wish to remove the portions of the line that pass through the circle and the rectangle.

To Use Trim Entities

1. Click the **Trim Entities** tool.
2. Locate the cursor on the portion of the line within the circle.
3. Click the mouse.

   The line segment will be removed.
4. Locate the cursor on the portion of the line that passes through the rectangle and click the mouse.
5. Click the green **OK** check mark.

Figure 2-59

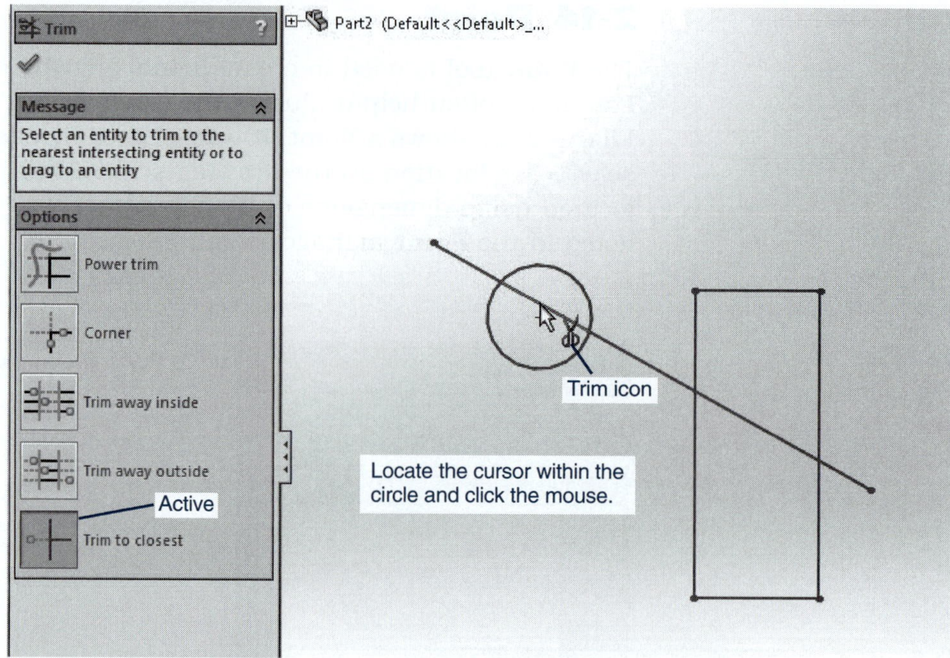

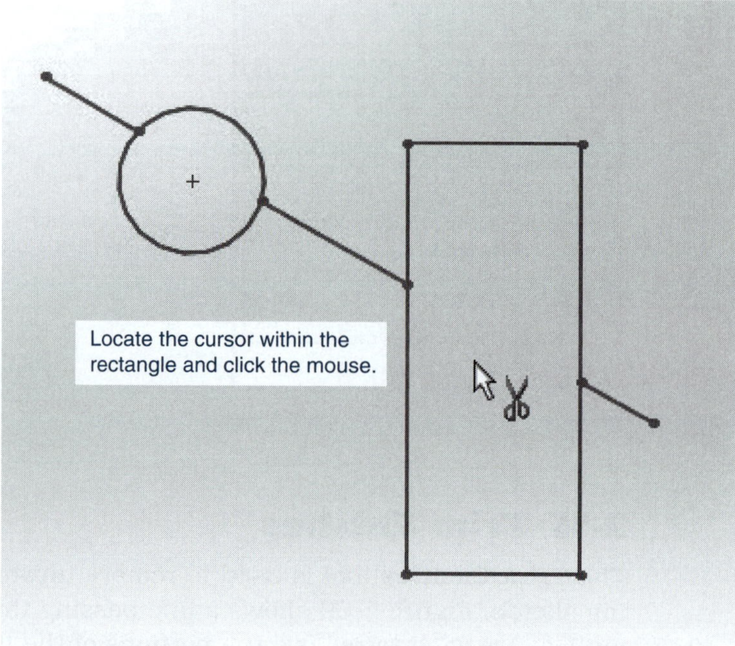

2-16 Extend Entities

The **Extend Entities** tool is used to extend existing lines to new lengths.

Figure 2-60 shows an existing 1.75 × 4.00 rectangle. We wish to extend the horizontal length to 6.00.

To Extend Entities in a Sketch

1. Sketch a vertical line about 6.00 from the left vertical line of the rectangle.
2. Use the **Smart Dimension** tool and define the location of the line as **6.00**.
3. Click the **Extend Entities** tool.

Figure 2-60

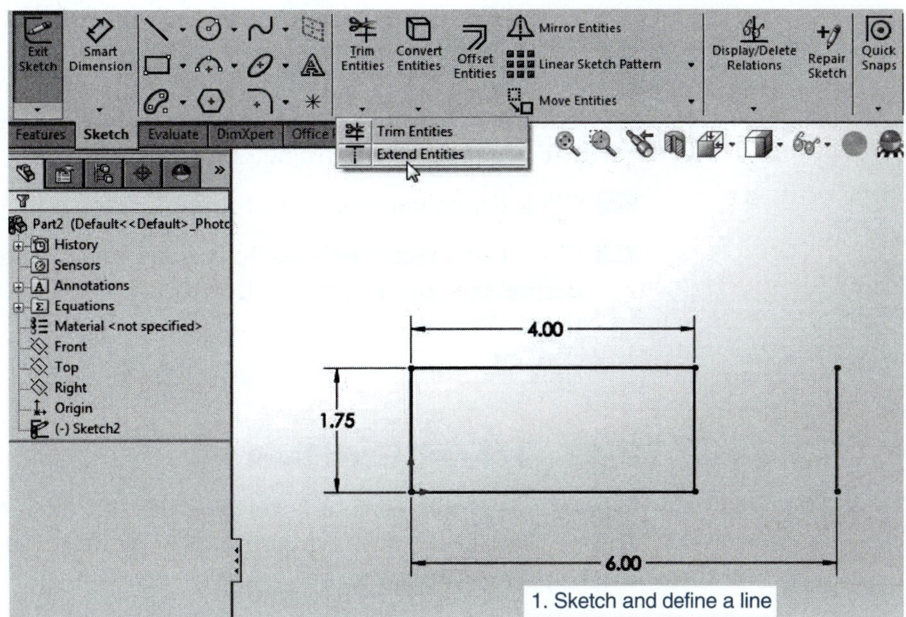
Exit Sketch
Smart Dimension
Trim Entities
Convert Entities
Offset Entities
Mirror Entities
Linear Sketch Pattern
Move Entities
Display/Delete Relations
Repair Sketch
Quick Snaps
Features
Sketch
Evaluate
DimXpert
Trim Entities
Extend Entities
Part2 (Default<<Default>_Photo
History
Sensors
Annotations
Equations
Material <not specified>
Front
Top
Right
Origin
(-) Sketch2
4.00
1.75
6.00
1. Sketch and define a line

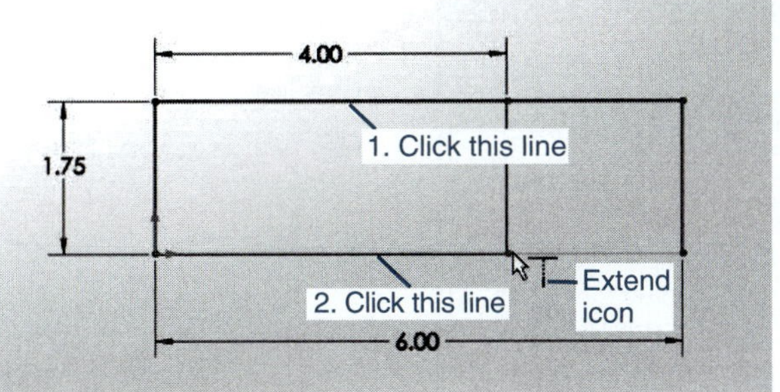
4.00
1.75
1. Click this line
2. Click this line
Extend icon
6.00

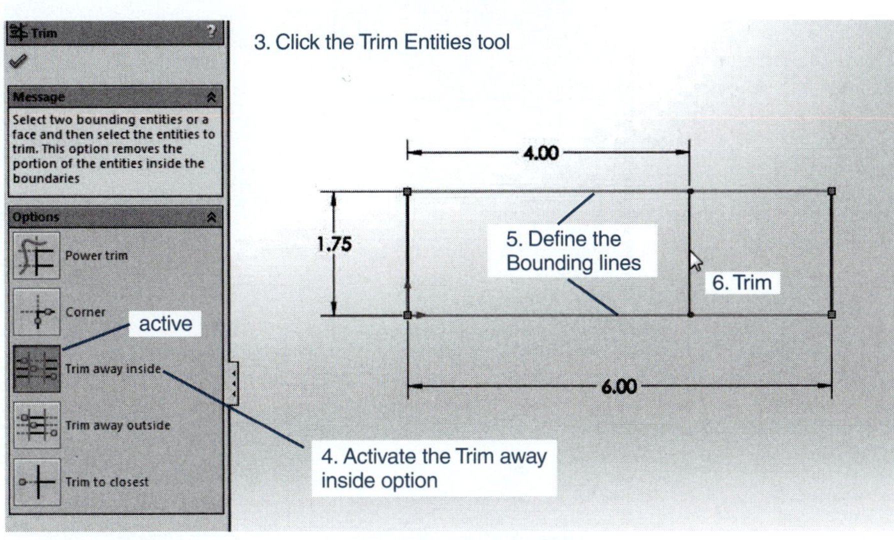
Trim
Message
Select two bounding entities or a face and then select the entities to trim. This option removes the portion of the entities inside the boundaries
Options
Power trim
Corner
active
Trim away inside
Trim away outside
Trim to closest
3. Click the Trim Entities tool
4.00
1.75
5. Define the Bounding lines
6. Trim
6.00
4. Activate the Trim away inside option

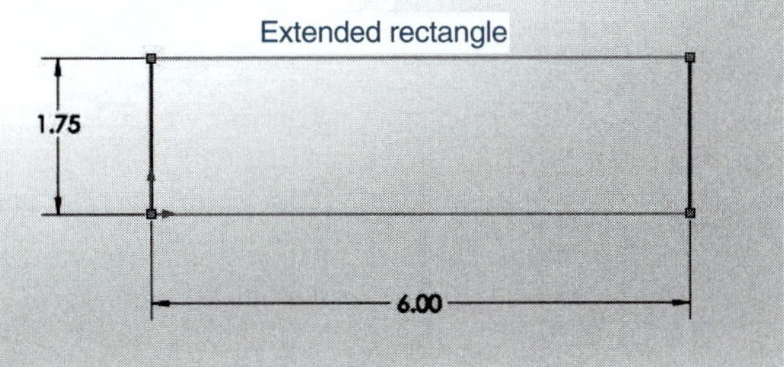
Extended rectangle
1.75
6.00

The cursor will appear with the **Extend Entities** icon attached.

4 Click the top horizontal line.

The line will extend to the vertical line located at 6.00 from the left vertical edge of the rectangle.

5 Click the lower horizontal line.

6 Click the **Trim Entities** tool, select the **Trim away inside** option, define the top and lower horizontal lines as bounding lines, and trim the **4.00** line.

Note that the dimension also is removed.

2-17 Offset Entities

The **Offset Entities** tool is used to sketch entities parallel to an existing entity. Figure 2-61 shows a line. Say we wish to draw a line parallel to the existing line .75 away.

Figure 2-61

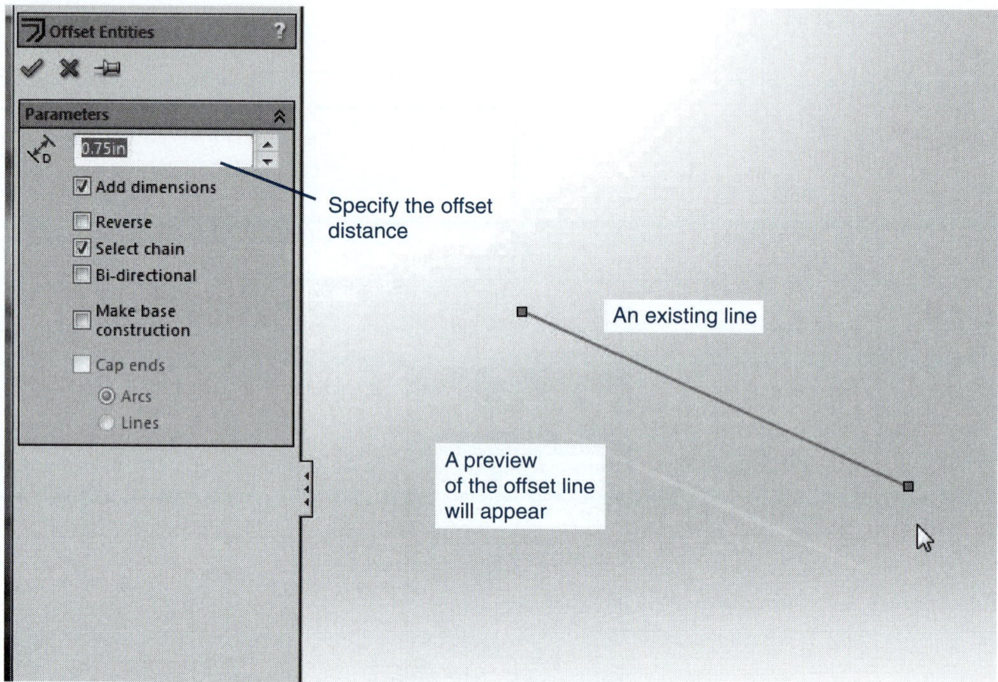

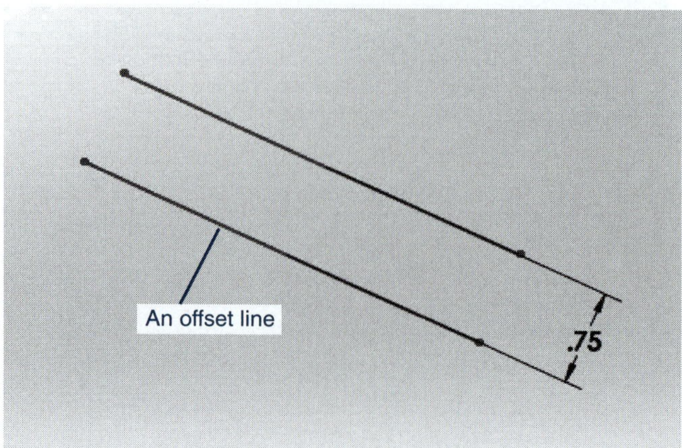

To Draw an Offset Line

1. Click the **Offset Entities** tool.
2. Specify the offset distance.
3. Click the existing line.
4. Ensure that the offset line is on the correct side of the existing line.
5. Click the green **OK** check mark.

The **Offset Entities** tool can be used to offset existing enclosed shapes. See Figure 2-62.

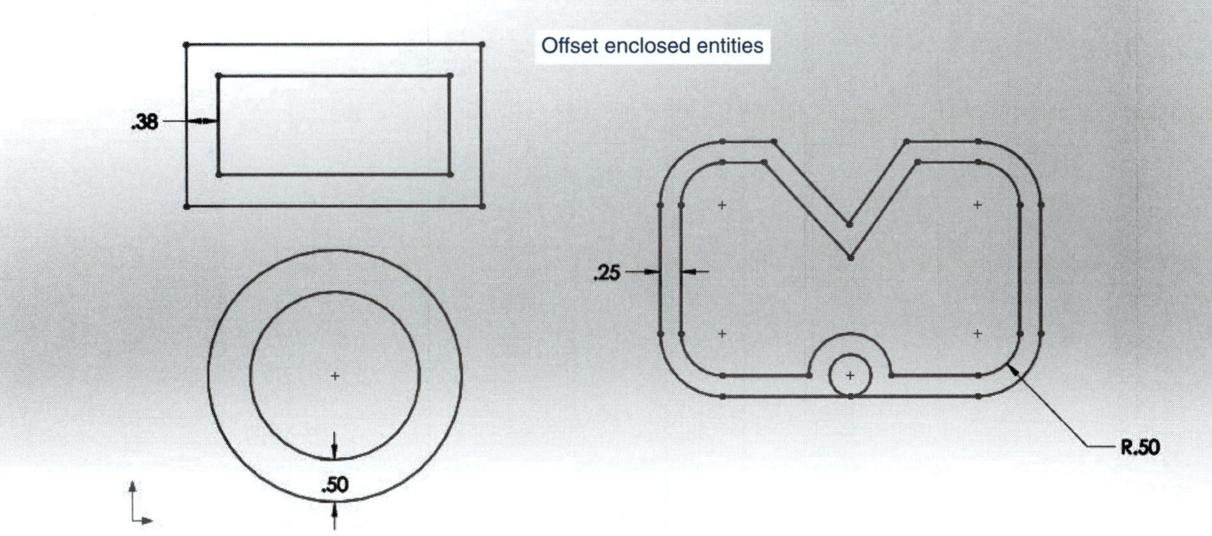

Figure 2-62

2-18 Mirror Entities

The **Mirror Entities** tool is used to create mirror images of entities. A mirror image is different from a copy. Figure 2-63 shows both a mirror image and a copy of the same shape. Note the differences.

Figure 2-63

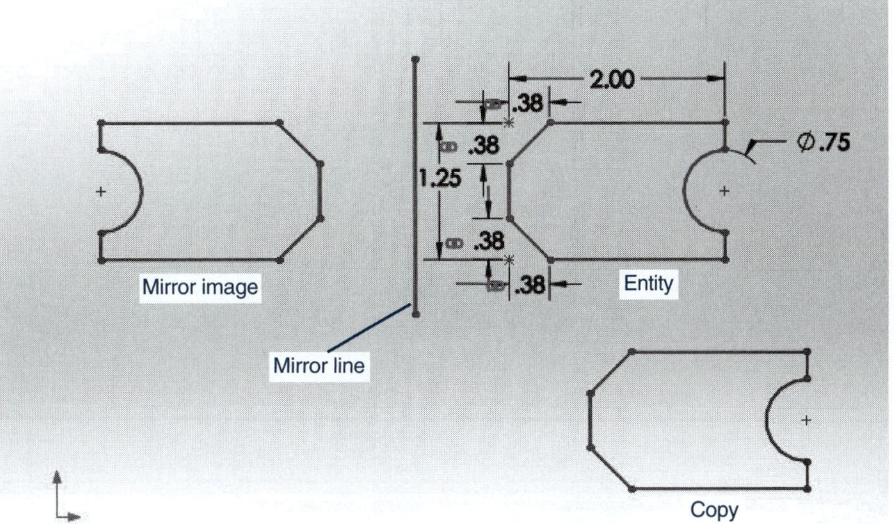

To Create a Mirror Entity

Figure 2-64 shows a circular shaped entity that includes a circular cutout.

1. Click the **Mirror Entity** tool.
2. Select all the lines in the entity.

   A listing will appear in the **Entities to mirror** box.
3. Click the **Mirror about** box.
4. Click the mirror line.

   A preview of the mirrored image will appear.
5. Click the green **OK** check mark.

Figure 2-64

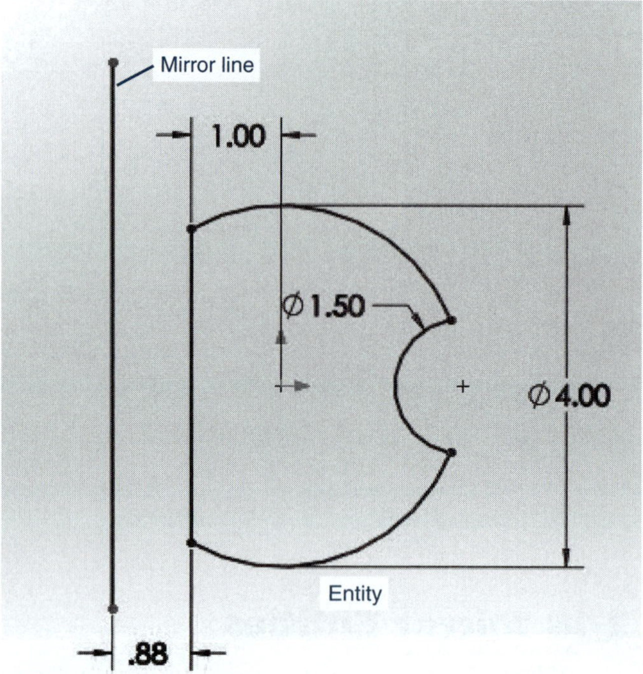

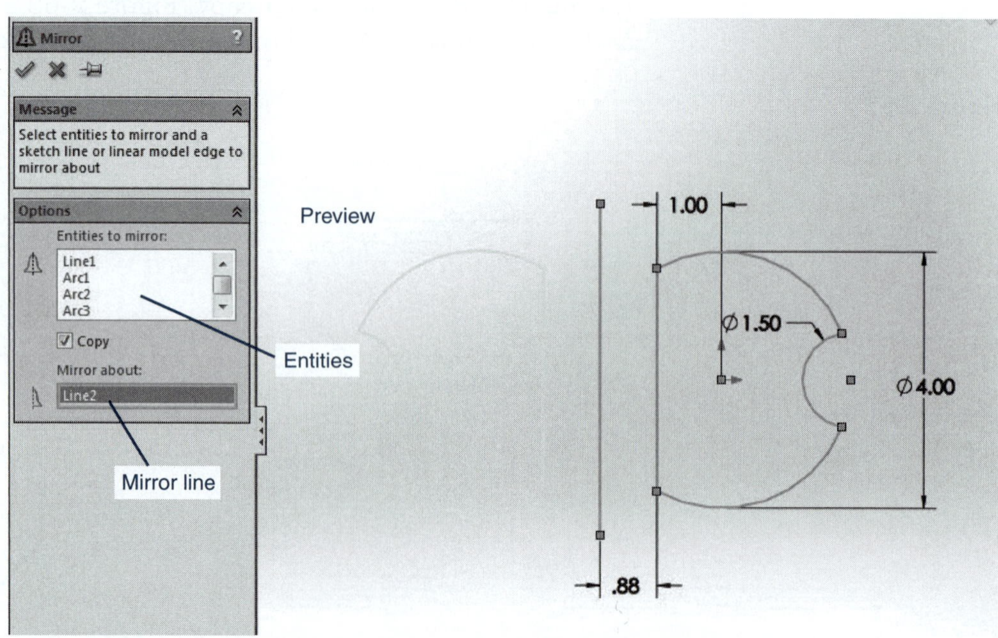

Figure 2-64
(*Continued*)

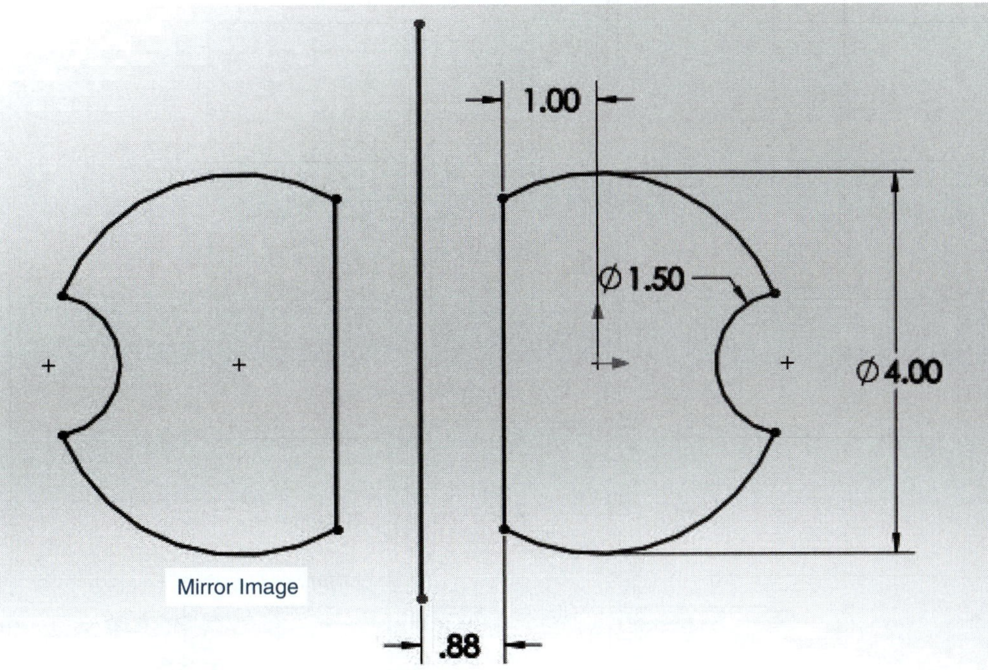

Figure 2-65

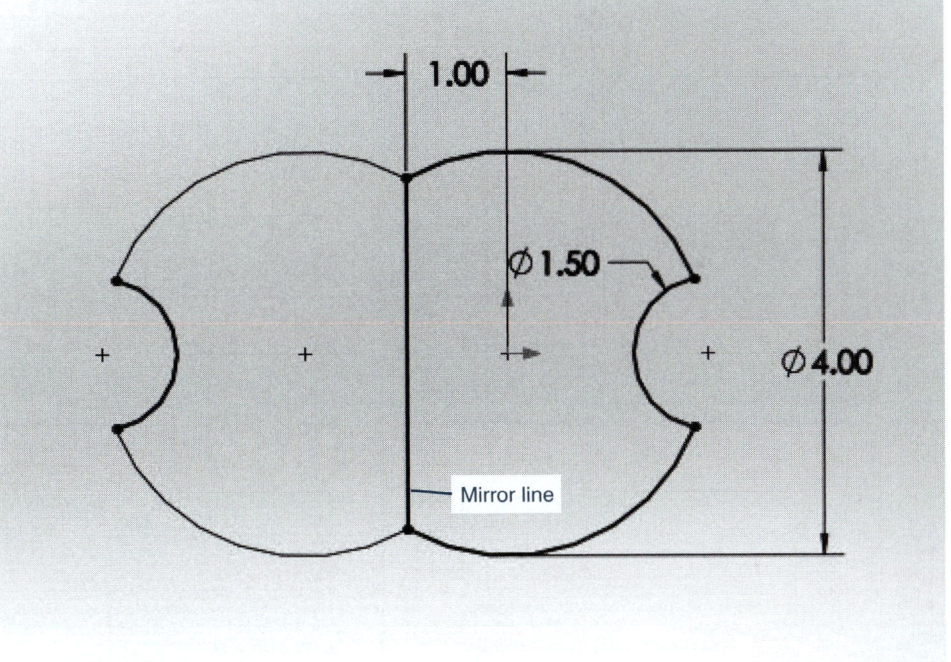

A line within the entity can be used as a mirror line. Figure 2-65 shows an entity mirrored about one of its edge lines.

2-19 Linear Sketch Pattern

The **Linear Sketch Pattern** tool is used to create patterns of sketched entities in the X and Y directions. Both **Linear** and **Circular** patterns can also be created from 3D models.

Figure 2-66 shows an entity. Say we wish to create a 3 × 5 linear pattern of the entity.

Figure 2-66

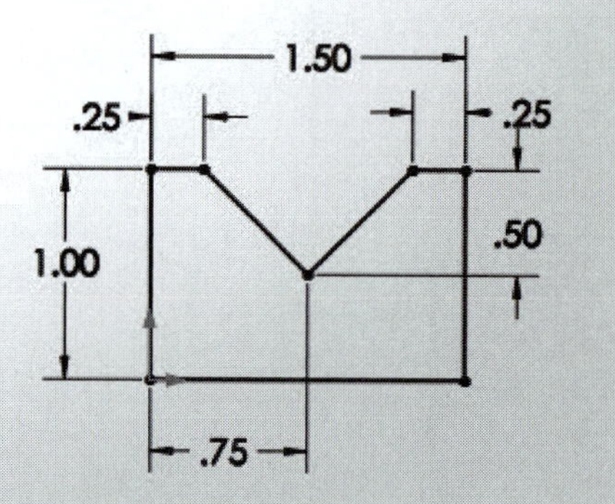

1.50
.25
.25
1.00
.50
.75

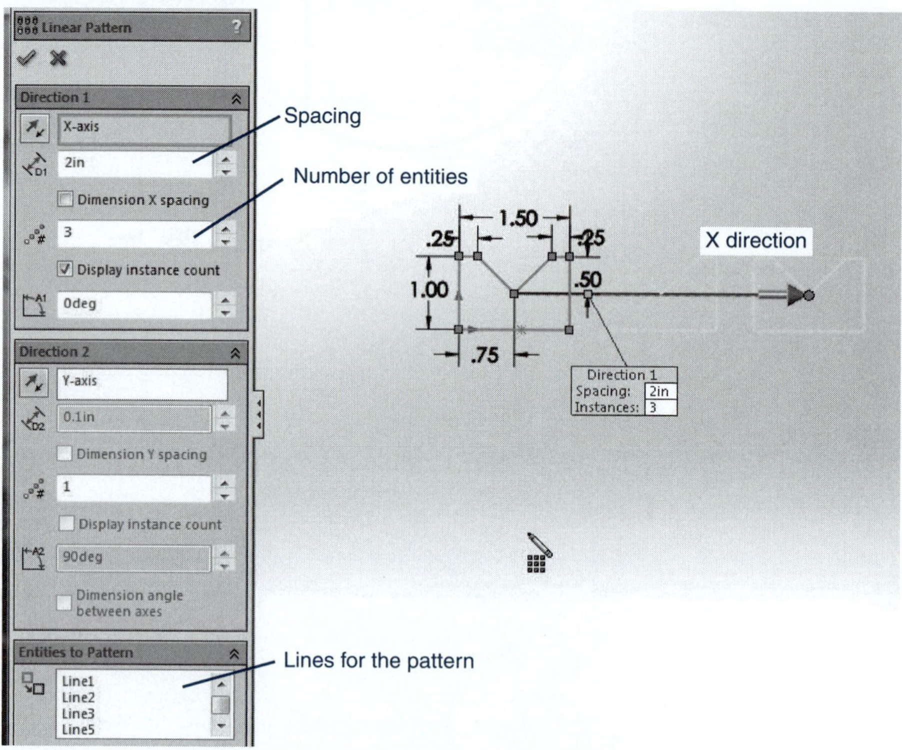

Linear Pattern
Direction 1
X-axis
2in
Dimension X spacing
3
Display instance count
0deg
Direction 2
Y-axis
0.1in
Dimension Y spacing
1
Display instance count
90deg
Dimension angle between axes
Entities to Pattern
Line1
Line2
Line3
Line5
Spacing
Number of entities
Lines for the pattern
X direction
1.50
.25
.25
1.00
.50
.75
Direction 1
Spacing: 2in
Instances: 3

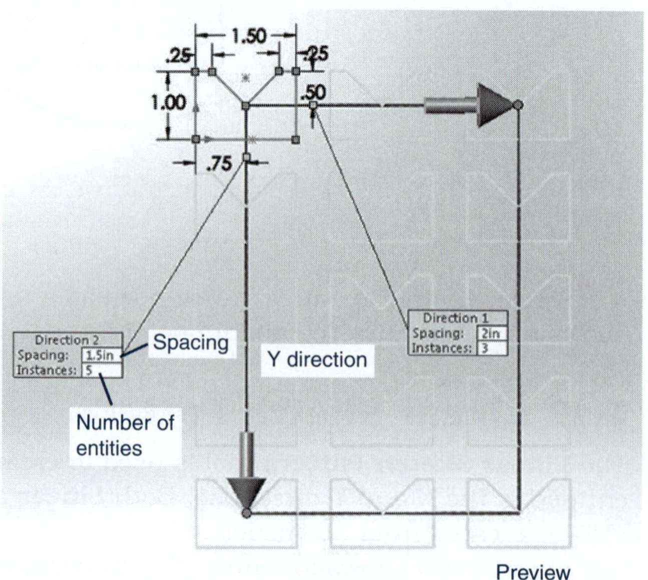

1.50
.25
.25
1.00
.50
.75
Direction 2
Spacing: 1.5in
Instances: 5
Spacing
Y direction
Direction 1
Spacing: 2in
Instances: 3
Number of entities
Preview

Figure 2-66
(Continued)

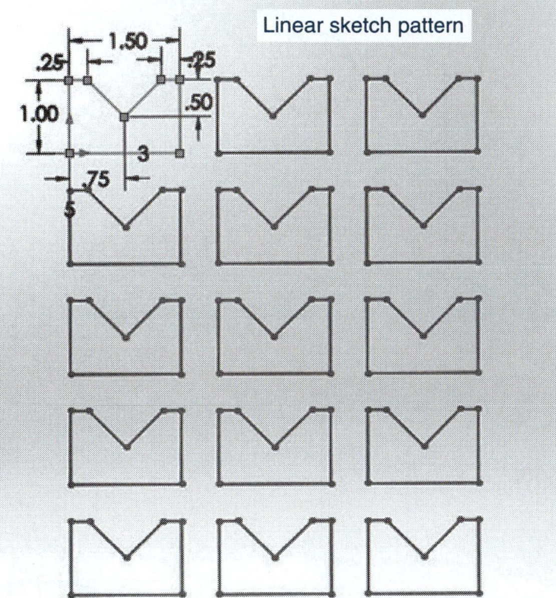

To Create a Linear Sketch Pattern

1. Click the **Linear Sketch Pattern** tool.
2. Click all the lines in the entity.
3. Ensure that the X-axis direction is active.

   The box will have a blue background.
4. Set the spacing distance for **2.00**.

   The 2.00 spacing distance is derived from the shapes' 1.50 width plus .50 clearance between each object in the X-direction.
5. Set the number of objects to **3**.
6. Click the Y-axis box.
7. Set the spacing distance for **1.50**.

   The 1.50 spacing distance is derived from the shapes' 1.00 height plus .50 clearance between each object in the Y-direction.
8. Set the number of objects to **5**.

   A preview will appear.
9. Click the green **OK** check mark.

2-20 Circular Sketch Pattern

The **Circular Sketch Pattern** tool is used to create circular patterns about a centerpoint. To demonstrate the **Circular Sketch Pattern** tool we will create a bolt circle. Figure 2-67 shows a Ø.75 circle located 2.00 from a fixed centerpoint.

Figure 2-67

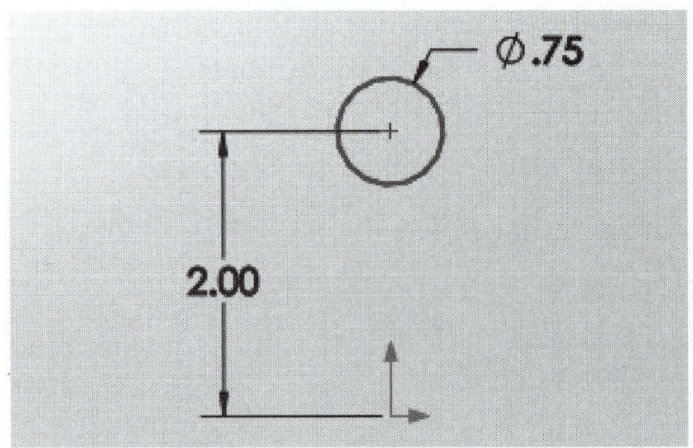

Ø.75
2.00

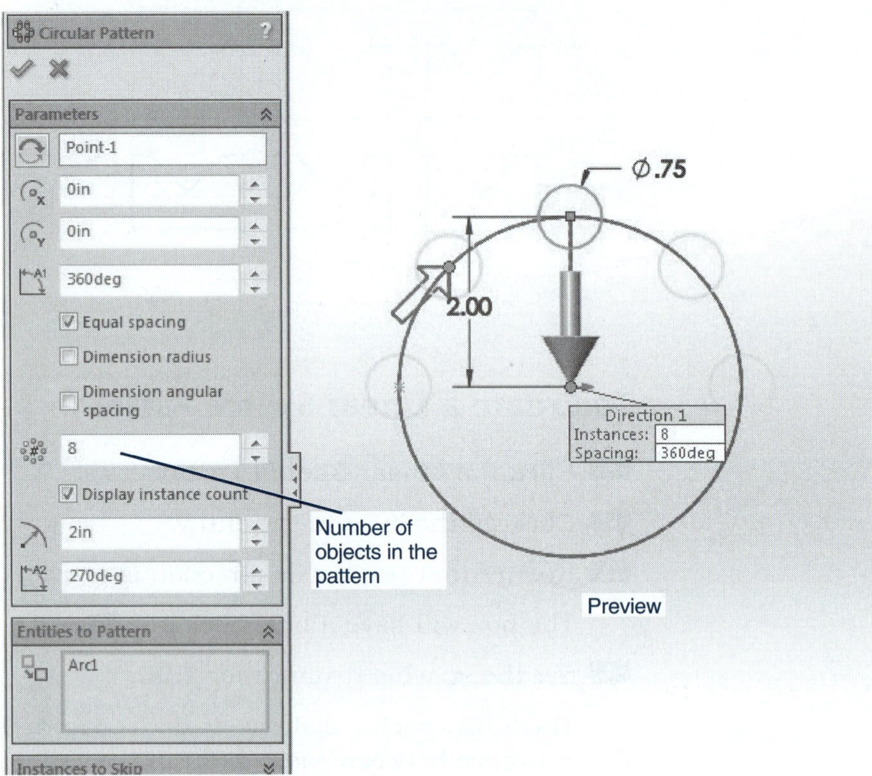

Circular Pattern
Parameters
Point-1
0in
0in
360deg
Equal spacing
Dimension radius
Dimension angular spacing
8
Display instance count
2in
270deg
Entities to Pattern
Arc1
Instances to Skip
Number of objects in the pattern
Ø.75
2.00
Direction 1
Instances: 8
Spacing: 360deg
Preview

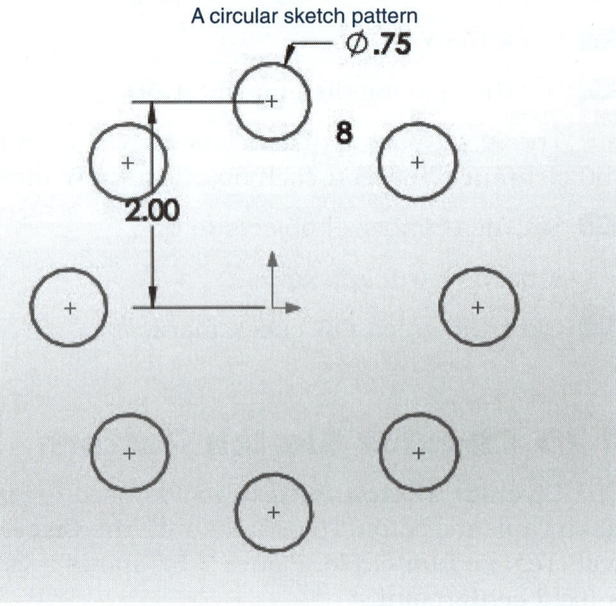

A circular sketch pattern
Ø.75
8
2.00

To Create a Circular Sketch Pattern

1 Click the **Circular Sketch Pattern** tool.

The **Circular Sketch Pattern** tool is a flyout from the **Linear Sketch Pattern** tool.

2 Click the edge of the Ø.75 circle.

3 Set the number of objects in the pattern to **8**.

4 Ensure that the **Spacing** is **360°**.

5 Click the green **OK** check mark.

2-21 Move Entities

The **Move Entities** tool is used to move an entity from one location to another. See Figure 2-68.

To Move an Entity

1 Click the **Move Entities** tool.

2 Click all the lines in the entity.

A listing of the selected lines and arcs will appear in the **Entities to Move** box.

3 Click the **Start point** box.

It should turn blue indicating that it is active.

4 Select a **Start** point.

In this example the lower left corner of the entity was selected.

5 Click the **Start** point and move the cursor away from the entity.

A preview of the entity will follow the cursor.

6 Select a new location for the entity.

7 Click the green **OK** check mark.

Figure 2-68

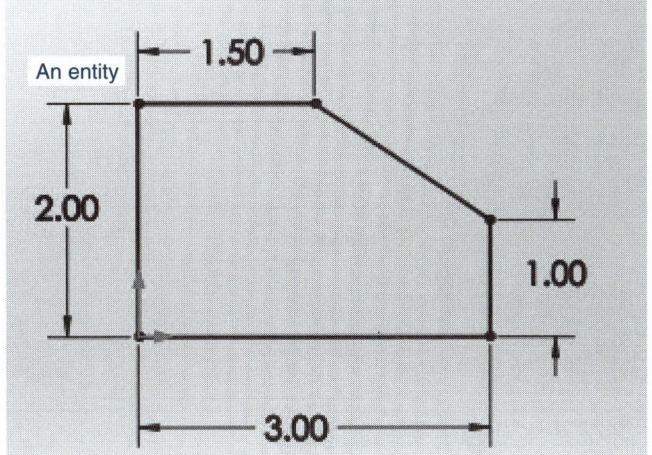

Figure 2-68
(*Continued*)

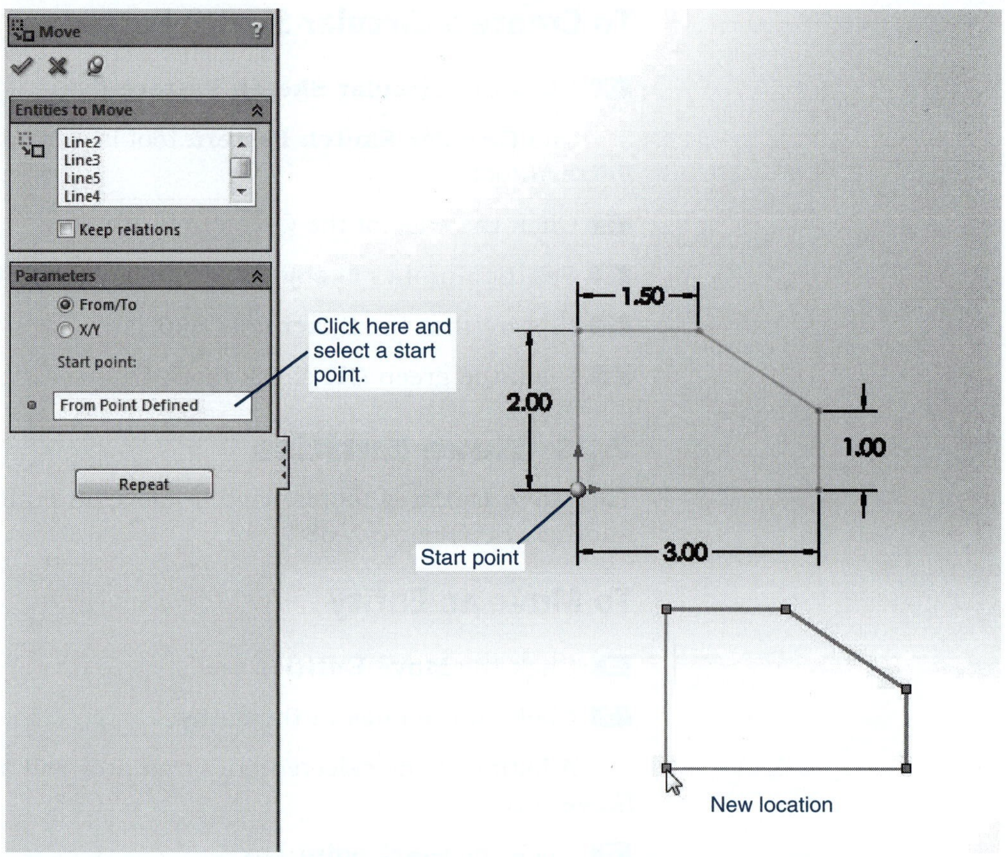

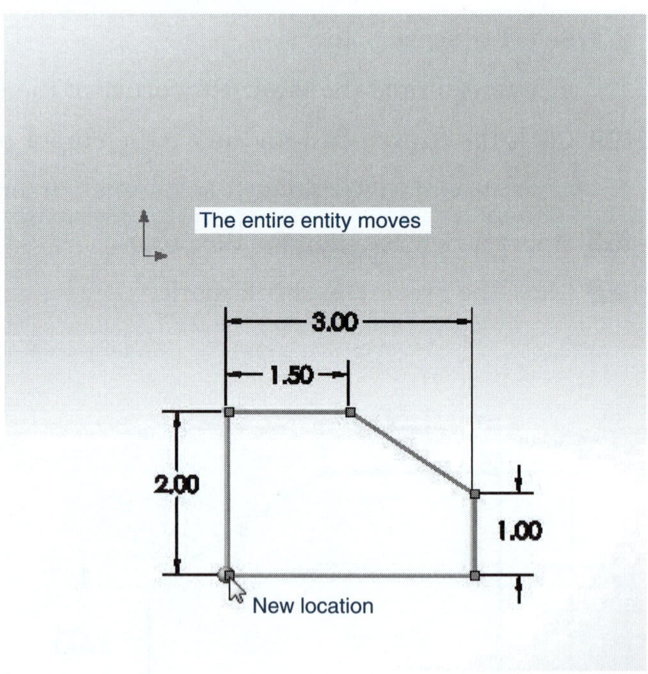

2-22 Copy Entities

The **Copy Entities** tool is used to create a duplication of an existing entity. The existing entity remains in place as the copy is created. See Figure 2-69.

Figure 2-69

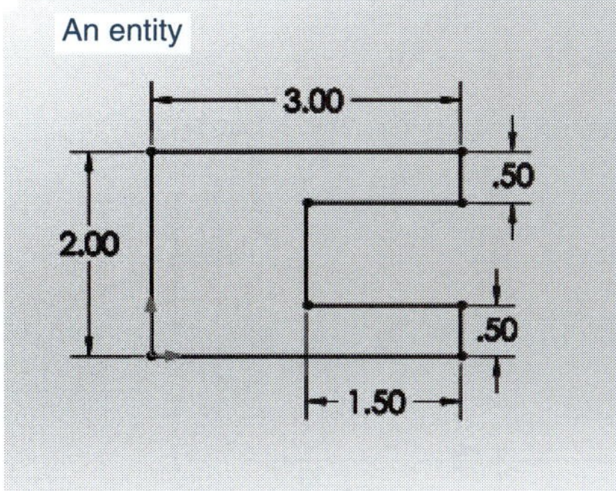

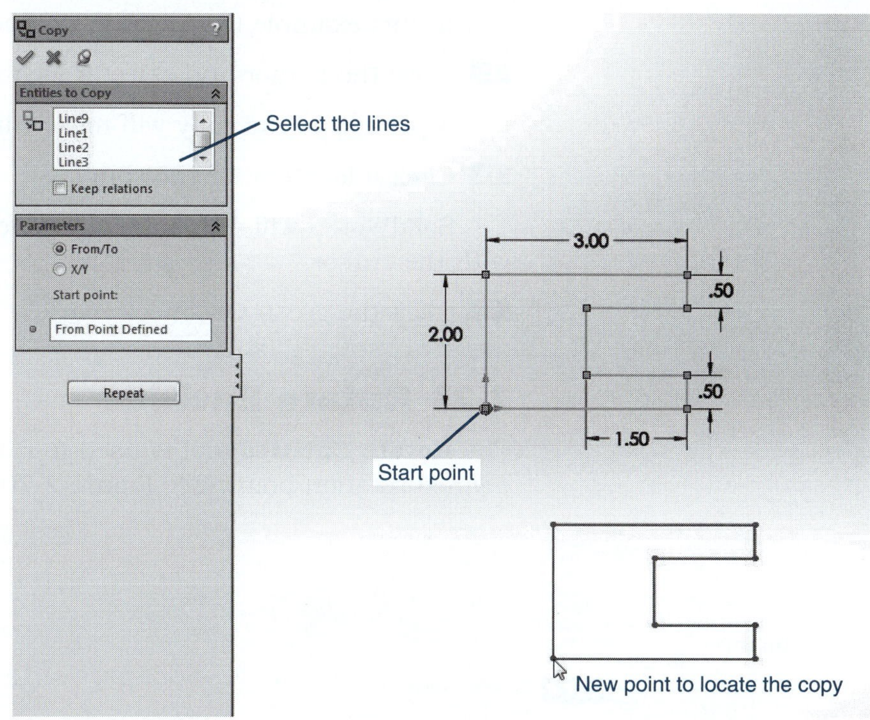

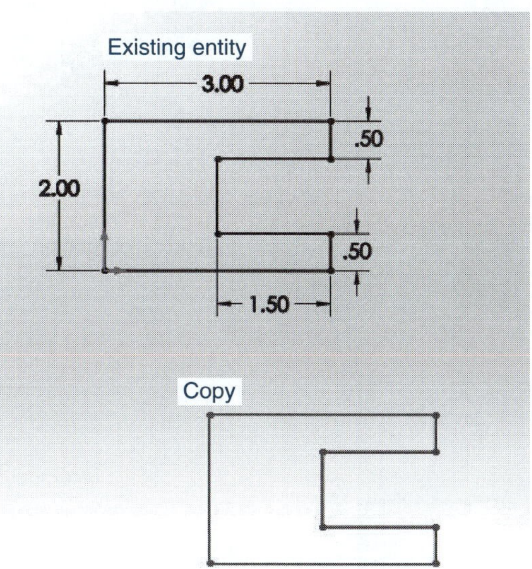

To Copy an Entity

1. Click the **Copy Entities** tool.

   The **Copy Entities** tool is a flyout from the **Move Entities** tool.

2. Select all the lines in the entity.
3. Click the **Start point** box.

   The **Start point** box will turn blue when it is active.

4. Click a selected **Start** point.

In this example the lower left corner was selected.

5 Move the cursor and select a location for the copy.

A preview of the copy will move with the cursor.

6 Click a location for the copy.

SolidWorks will automatically generate another copy that will move with the cursor.

7 Click the green **OK** check mark.

2-23 Rotate Entities

The **Rotate Entities** tool is used to rotate an existing entity about a defined rotation point. See Figure 2-70.

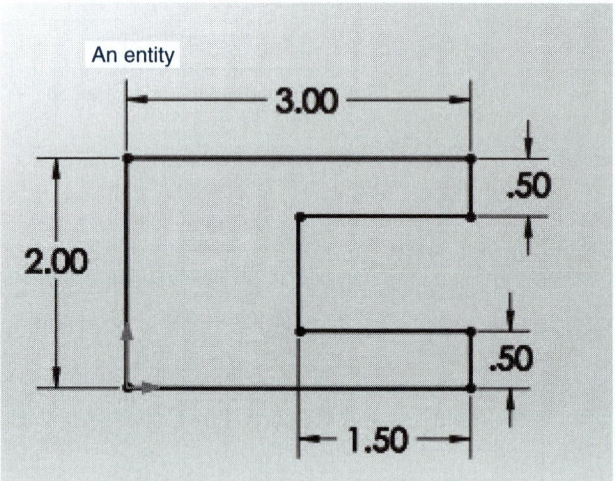

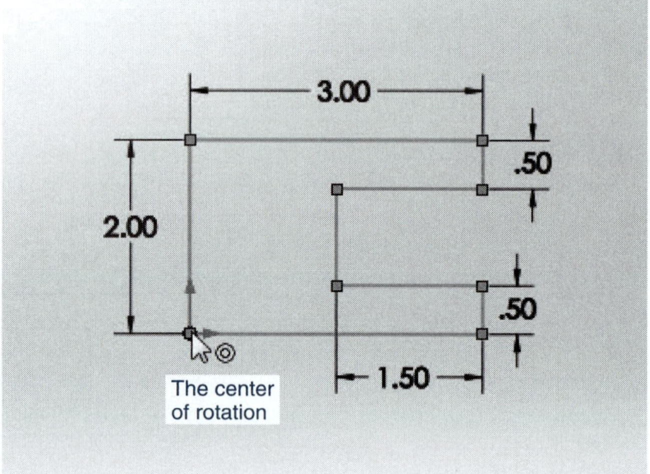

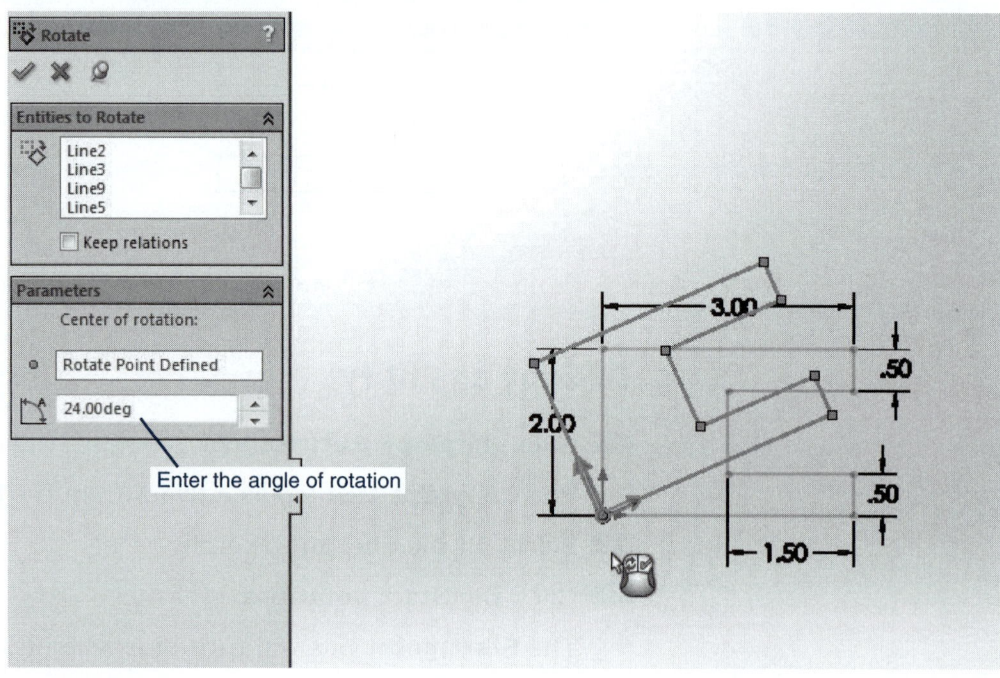

Figure 2-70

Figure 2-70
(*Continued*)

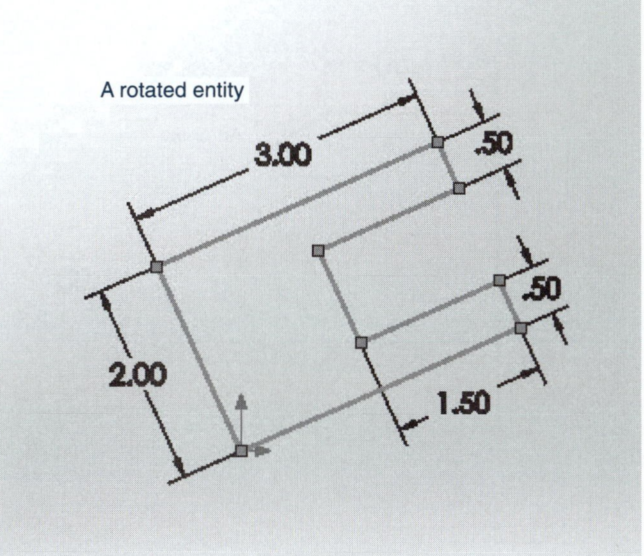

Chapter 2

To Rotate an Entity

1. Click the **Rotate Entities** tool.
2. Select all the lines in the entity.
3. Click the **Center of rotation** box.

   The box will have a blue background when it is active.
4. Enter an angular value for the rotation or move the cursor to select a rotation angle.

   SolidWorks defines the counterclockwise direction as the positive angular direction. A horizontal line to the right is 0.0°.
5. Click the green **OK** check mark.

2-24 Scale Entities

The **Scale Entities** tool is used to change the overall size of an entity while maintaining the proportions of the original entity. The **Scale Entities** tool includes a **Copy** option. If the **Copy** option is activated when the scaled drawing is created, the original drawing will be retained. If the **Copy** option is off (no check mark) when a scaled drawing is made, the original drawing will be deleted.

See Figure 2-71.

To Create a Scale Entity

1. Click the **Scale Entities** tool.

   The **Scale Entities** tool is a flyout from the **Move Entities** tool.
2. Select all the lines in the entity.
3. Define the scale factor.

Figure 2-71

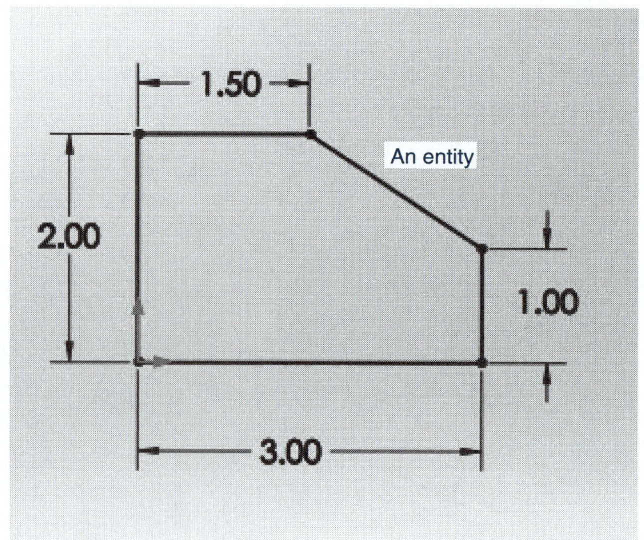

1.50
An entity
2.00
1.00
3.00

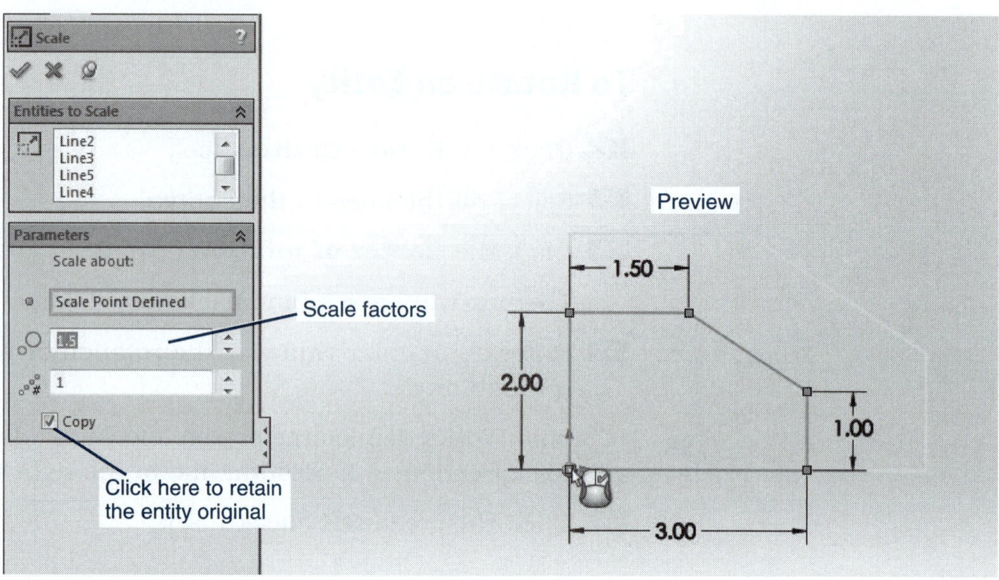

Scale
Entities to Scale
Line2
Line3
Line5
Line4
Parameters
Scale about:
Scale Point Defined
1.5
1
Copy
Scale factors
Click here to retain the entity original
Preview
1.50
2.00
1.00
3.00

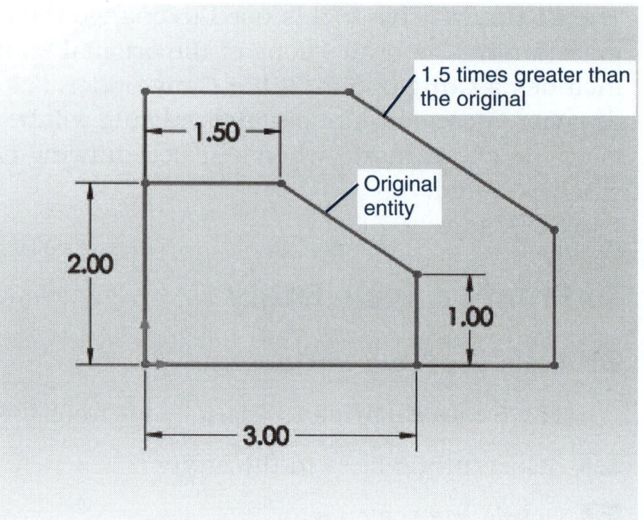

1.5 times greater than the original
1.50
Original entity
2.00
1.00
3.00

In this example a factor of **1.5** was selected.

4 Click the **Copy** box (check mark).

Checking the **Copy** box will cause the original drawing to remain when the scaled copy is created.

5 Define the **Scale** point.

In this example the lower left corner was selected. A preview will appear.

6 Click the green **OK** check mark.

2-25 Stretch Entities

The **Stretch Entities** tool is used to extend the length of some of the lines in an entity. See Figure 2-72.

Figure 2-72

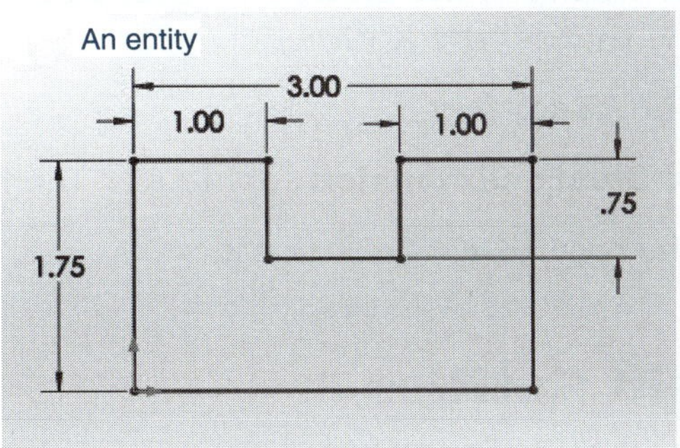

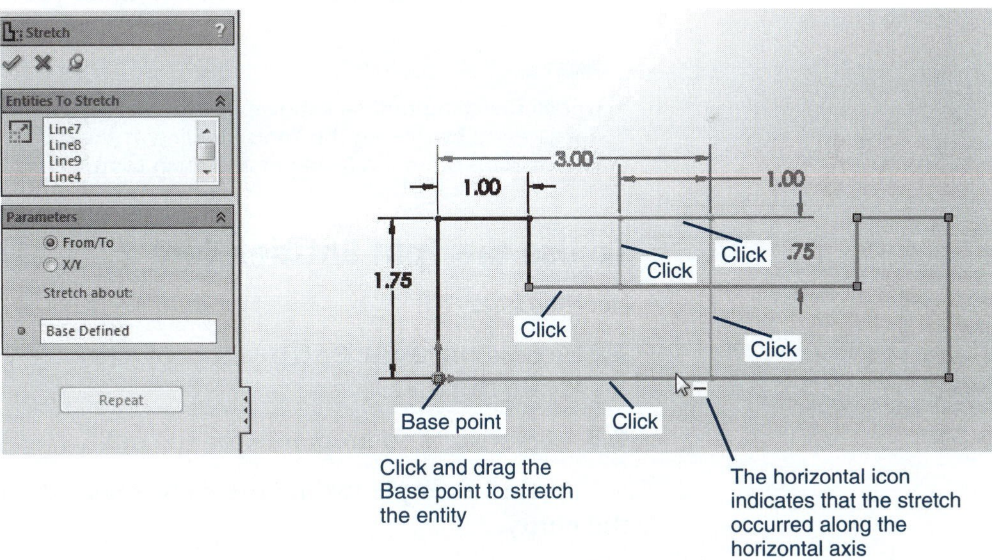

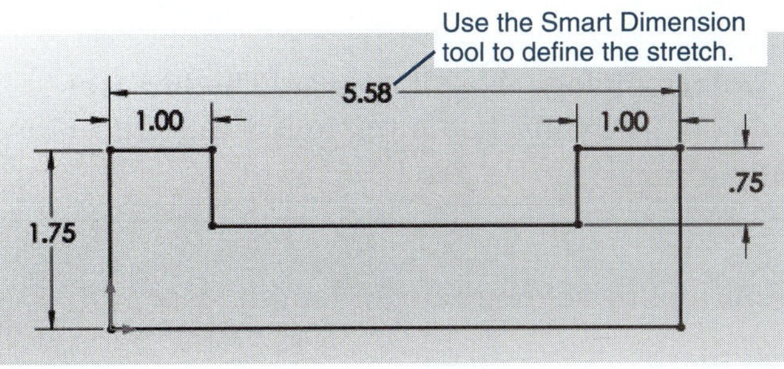

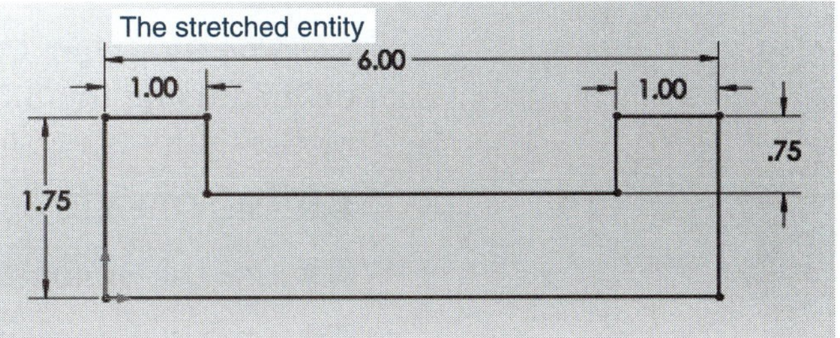

To Stretch an Entity

1 Click the **Stretch Entities** tool.

The **Stretch Entities** tool is a flyout from the **Move Entities** tool.

2 Click only the lines that will be involved in the stretch.

In this example five lines were selected.

3 Click the **Stretch about** box.

The box will turn blue when active.

4 Define a base point.

For this example the lower left corner was selected.

5 Click and drag the **Base** point to stretch the entity.

For this example, the stretch drag was along the horizontal lower line of the entity, that is, the X direction.

6 Press the **<Esc>** key.

7 Use the **Smart Dimension** tool to size the final stretch distance.

8 Click the green **OK** check mark.

2-26 Split Entities

The **Split Entities** tool is used to trim away internal segments of an existing entity or to split an entity into two or more parts by specifying split points.

> **NOTE**
>
> Not all sketching tools are shown on the Sketching panel. Additional sketch tools are accessed by clicking the **Tools** heading at the top of the screen, and clicking the **Sketch Tools** option. A listing of all sketch tools will appear. See Figure 2-73.

To Use the Split Entities Tool

See Figure 2-74.

1 Access the **Split Entities** tool by clicking **Tools**, **Sketch Tools**, and **Split Entities**.

2 Click two random points for the split.

In this example two points were selected on the top horizontal line of the entity.

3 Right-click the line segment between the two points and click the **Delete** option.

The **Sketcher Confirm Delete** dialog box will appear. The line was defined using the 3.50 dimension. The Split has interrupted this definition, so the entity is not fully defined.

4 Click the **Yes** option.

5 Use the **Smart Dimension** tool to define the split length.

6 Click the green **OK** check mark.

Figure 2-73

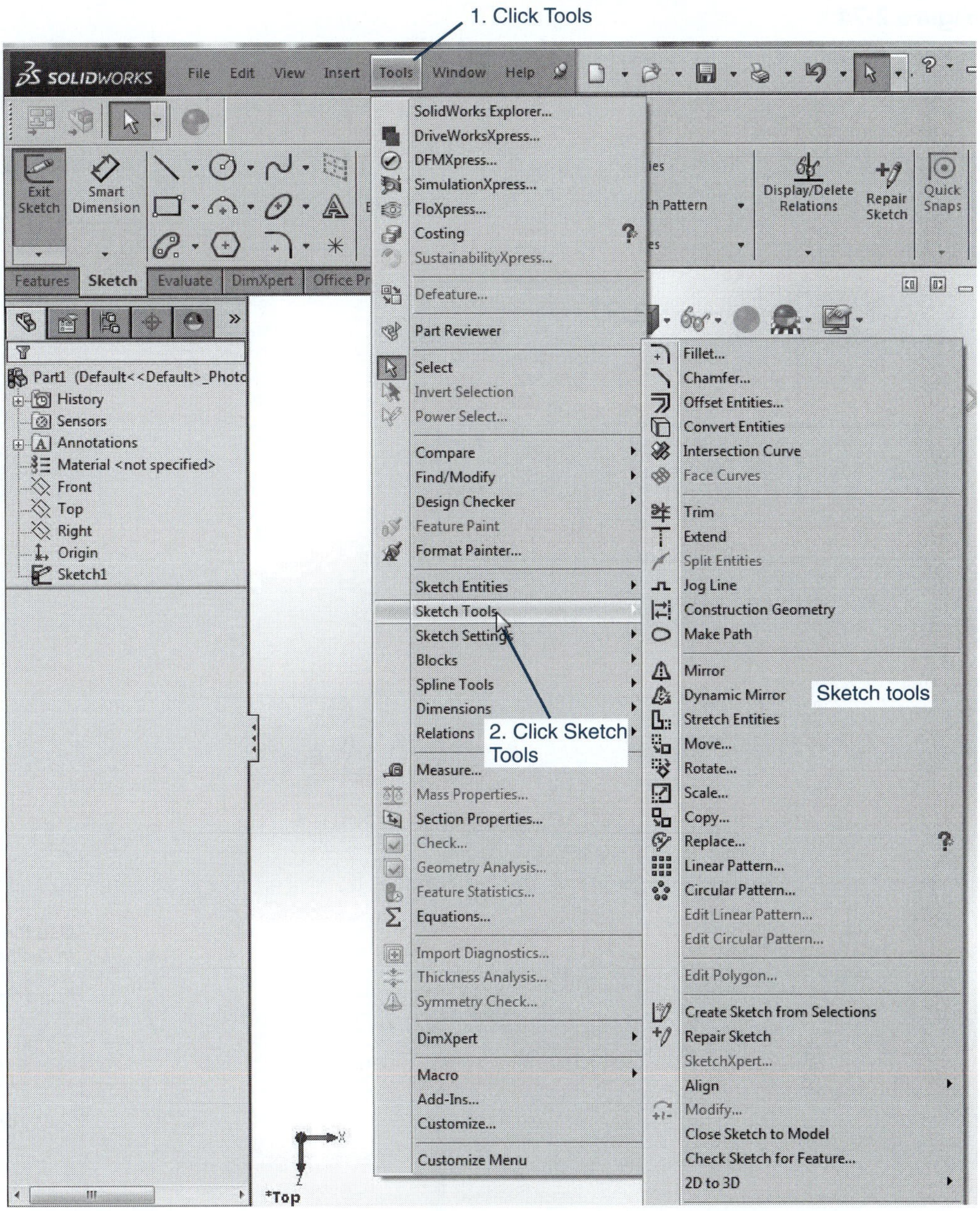

1. Click Tools

2. Click Sketch Tools

Sketch tools

Figure 2-74

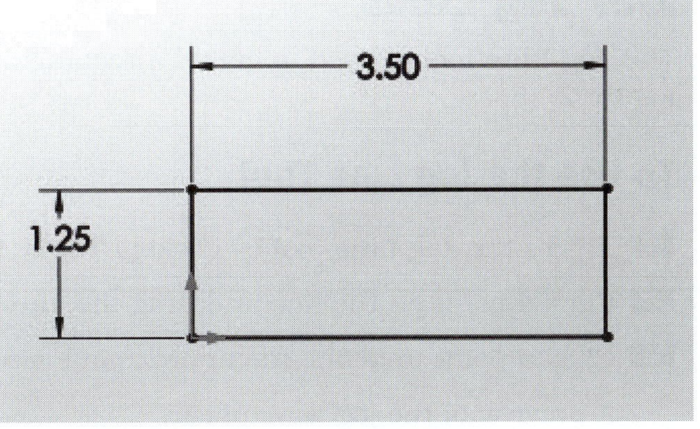

Figure 2-74
(*Continued*)

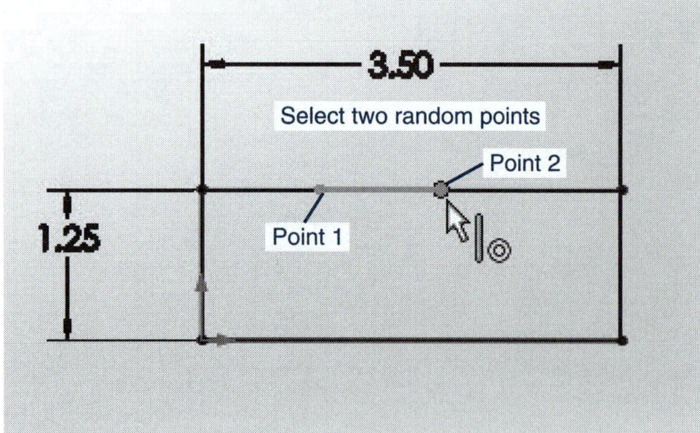

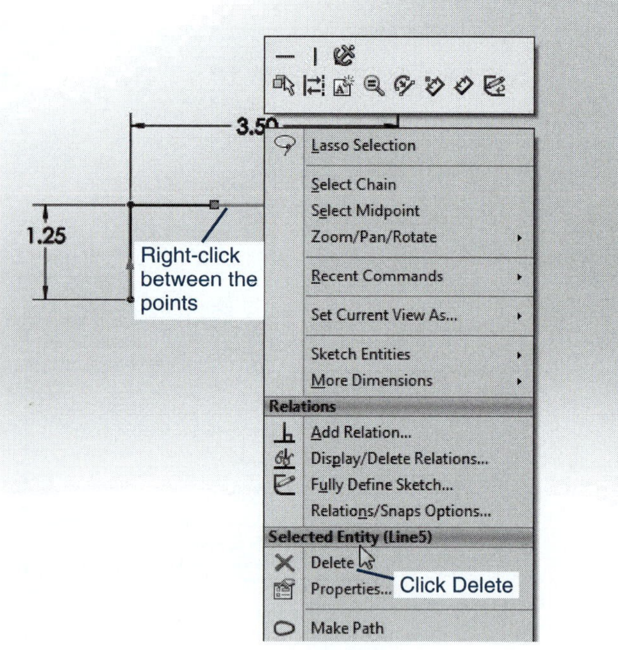

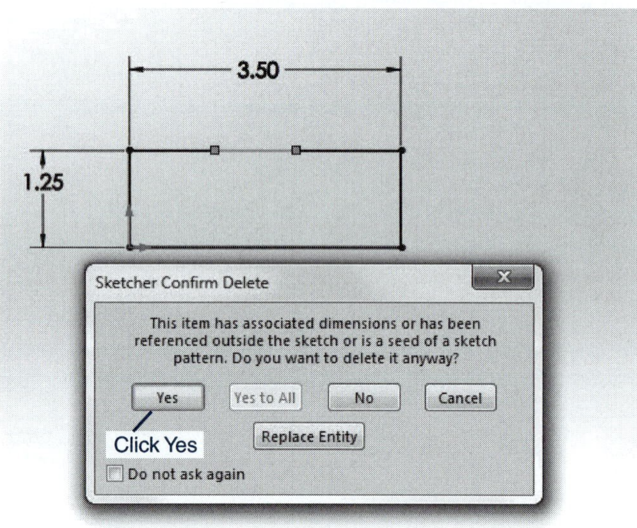

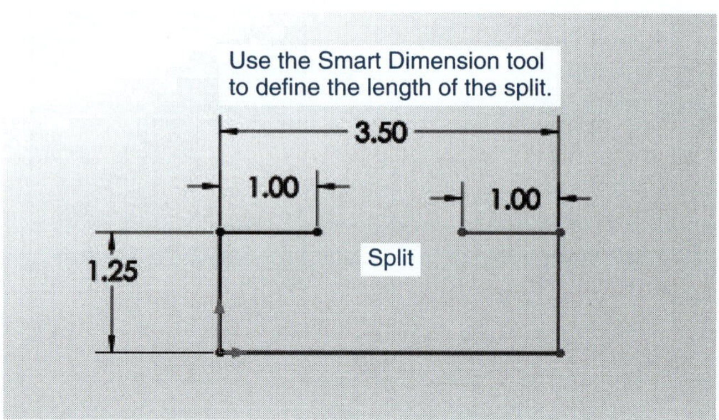

2-27 Jog Lines

The **Jog Line** tool is used to create a rectangular shape (jog) in a line. See Figure 2-75.

To Use the Jog Line Tool

1. Access the **Jog Line** tool by clicking **Tools**, **Sketch Tools**, and **Jog Line**.
2. Click a point on the line and drag the cursor away from the line.
3. Click a point to define the approximate size of the jog.

   A preview of the jog will appear.

Figure 2-75

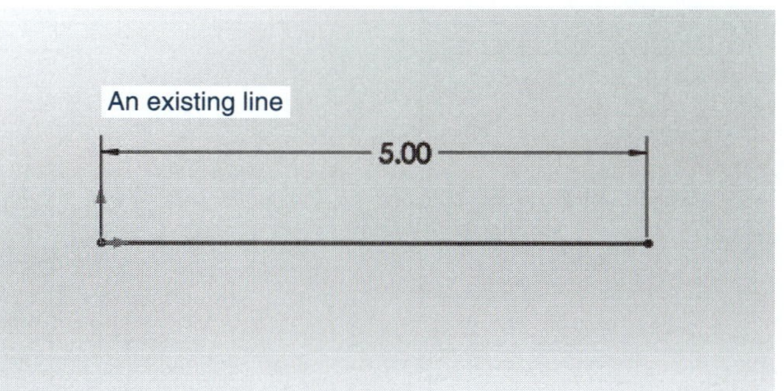

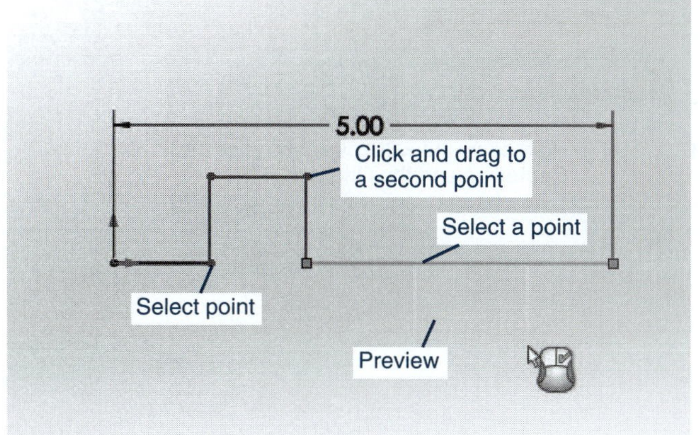

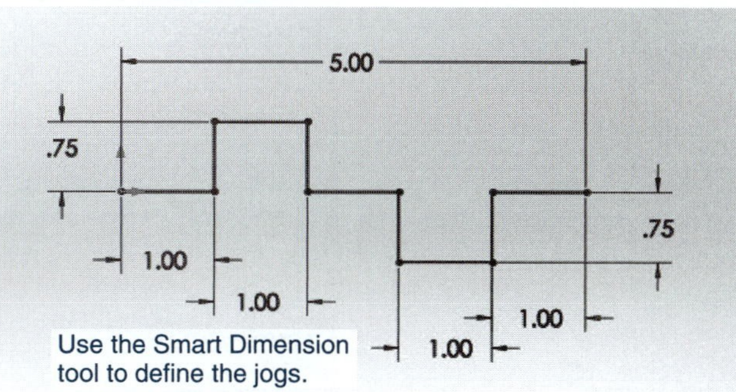

4 Repeat the process as necessary.

5 Use the **Smart Dimension** tool to define the location and depth of the jog.

2-28 Centerline

The **Centerline** tool is used to help define and locate the center of entities.

Figure 2-76 shows an existing 2.00 × 4.00 rectangle. It is not centered on the origin, and none of its lines touch the origin. The **Centerline** tool can be used to center the rectangle on the origin.

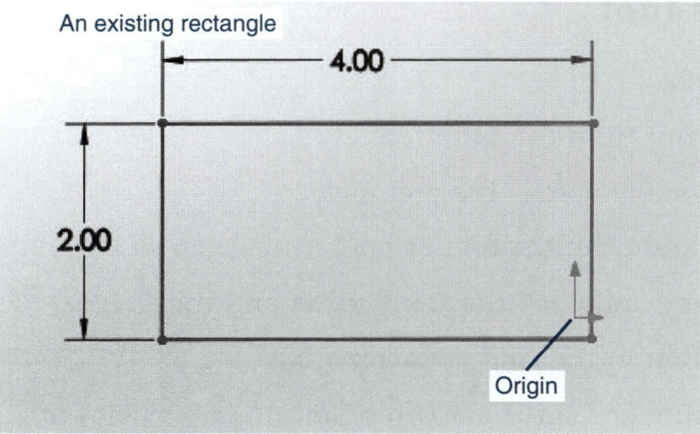

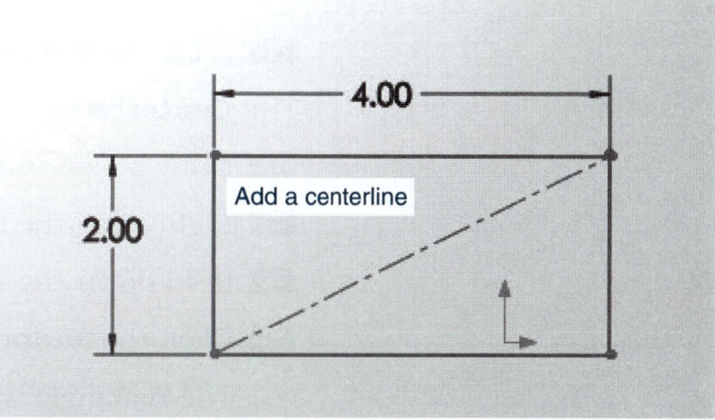

Figure 2-76

Figure 2-76
(Continued)

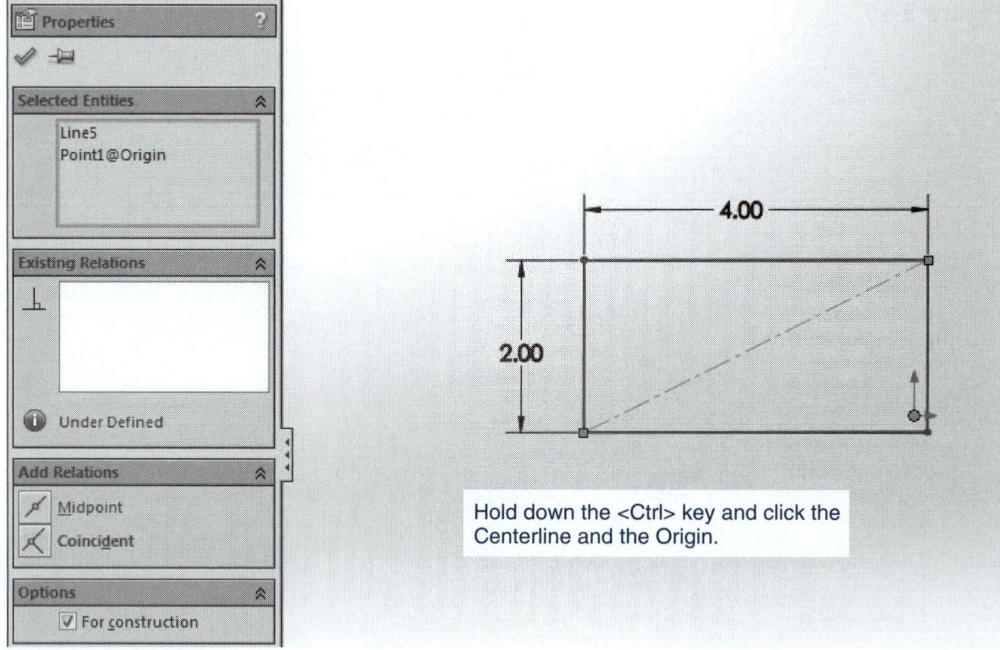

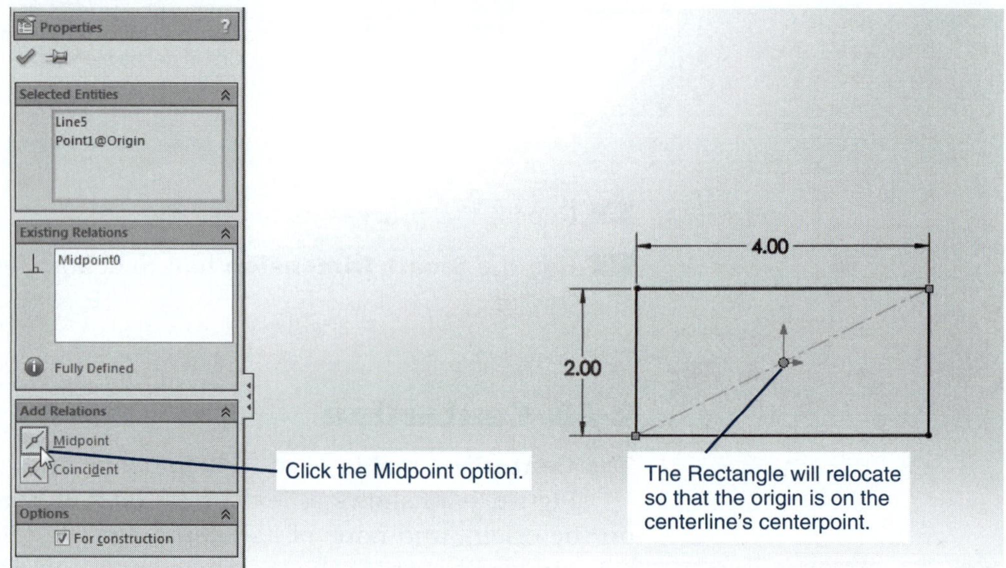

To Use the Centerline Tool

1 Click the **Centerline** tool.

The **Centerline** tool is a flyout from the **Line** tool.

2 Draw a centerline diagonally across the rectangle.

3 Right-click the mouse and click the **Select** option.

4 Hold down the **<Ctrl>**key and click the **Centerline** and the **Origin**.

5 Click the **Midpoint** option in the **Add Relations** box.

The rectangle will relocate so that the origin is on the centerpoint of the centerline.

2-29 Sample Problem SP2-1

Draw the shape shown in Figure 2-77.

1 Start a **New Part** document, use ANSI Standards, inch units, and create the sketch on the Top plane.

2 Use the **Corner Rectangle** and **Smart Dimension** tools and draw a 2.00 × 4.00 rectangle with its lower left corner coincidental with the origin.

Figure 2-77

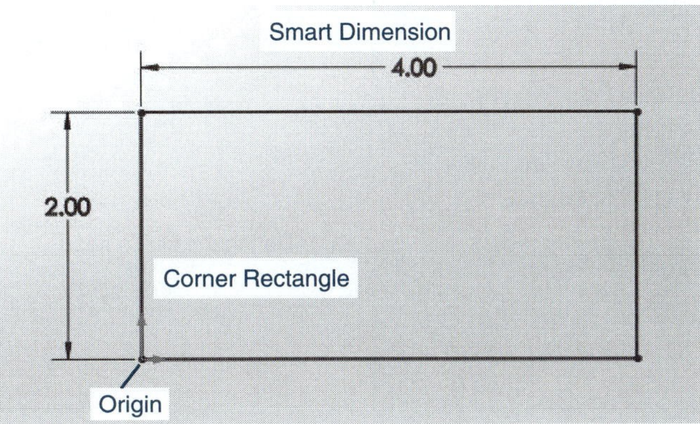

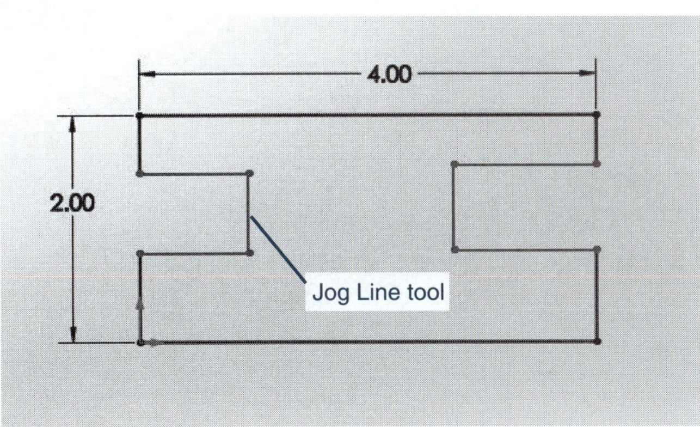

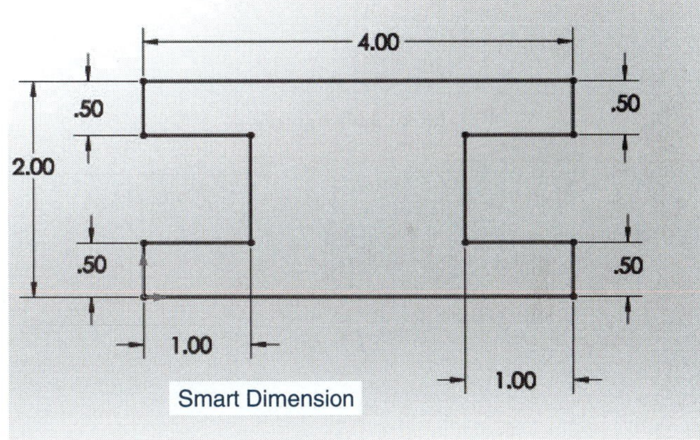

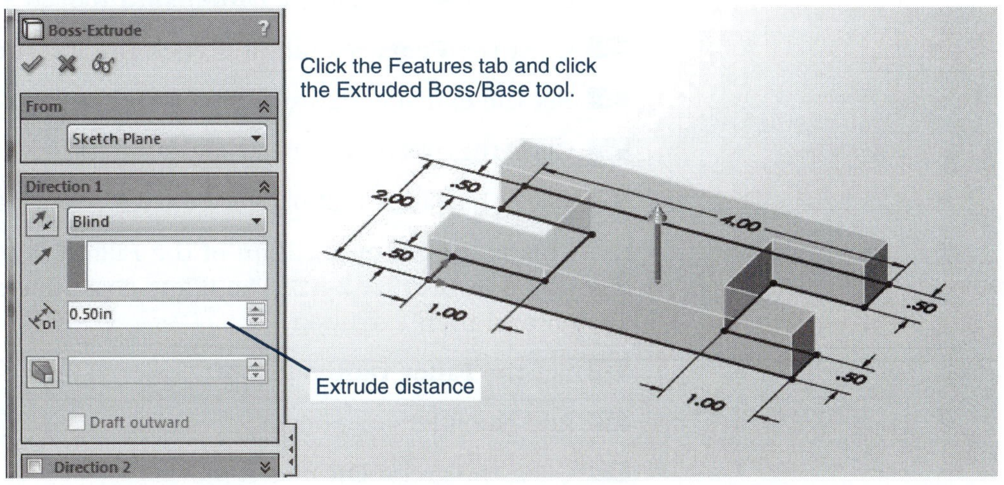

Figure 2-77
(Continued)

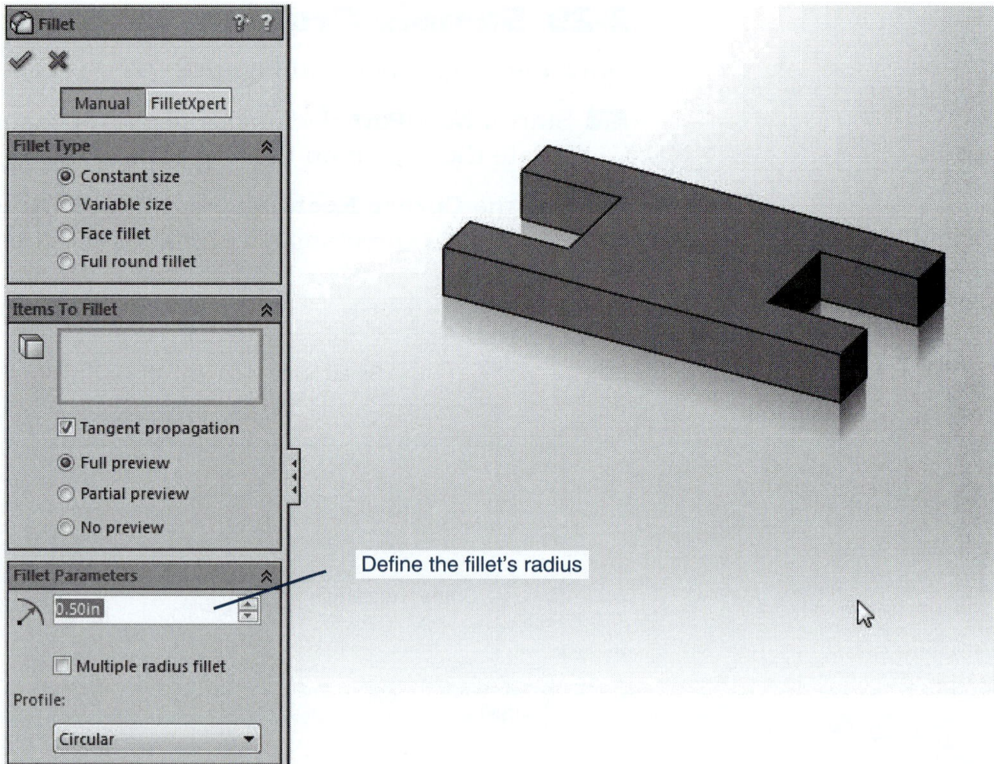

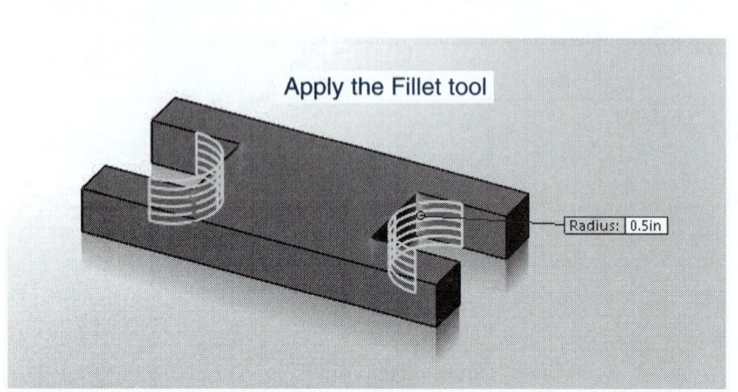

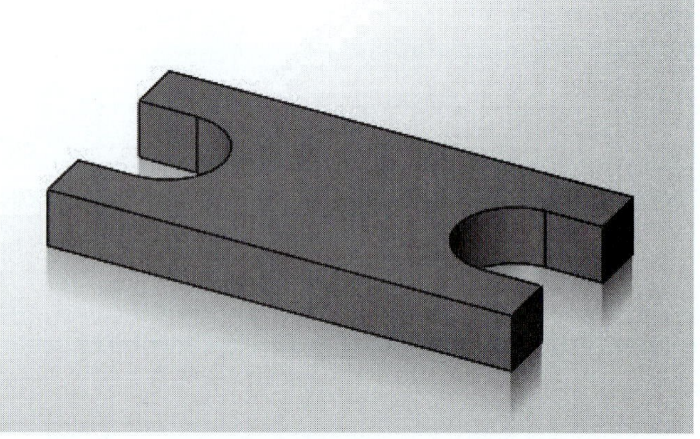

3 Use the **Jog Line** tool and add jogs at each end on the rectangle as shown. Use the **Smart Dimension** tool to size the jogs.

4 Click the **Features** tab and click the **Extruded Boss/Base** tool.

5 Set the extrude distance for **0.50**.

6 Click the green **OK** check mark.

7 Click the **Fillet** tool on the **Features** tool panel.

This is the 3D application of the Fillet tool. The 2D application was discussed in Section 2-12. The fillets could have been added to the 2D sketch and then extruded.

8 Set the **Fillet radius** for **.50**.

9 Add the fillets as shown.

10 Click the green **OK** check mark.

2-30 Sample Problem SP2-2

A circular pattern can be created using any shape. Figure 2-78 shows a slot shape located within a large circle. A circular pattern can be created from the slot shape.

1 Start a **New Part** document, use ANSI Standards, inch units, and create the sketch on the Top plane.

2 Sketch a Ø7.50 circle centered on the origin and add the slot shape as shown.

3 Click the **Circular Sketch Pattern** tool.

The **Circular Sketch Pattern** tool is a flyout from the **Linear Sketch Pattern** tool.

Figure 2-78

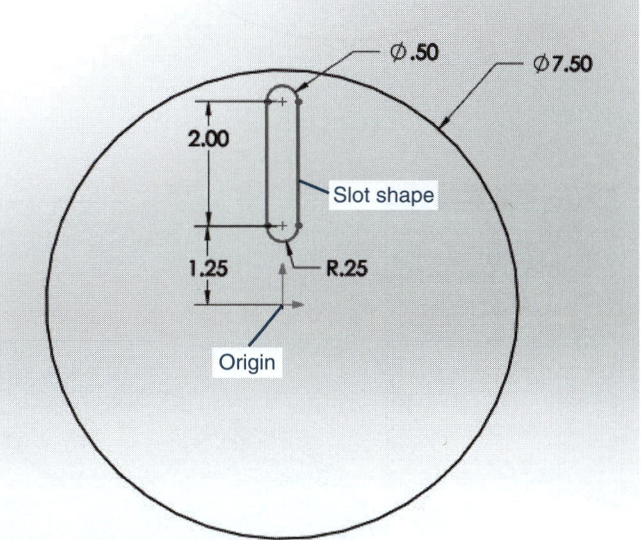

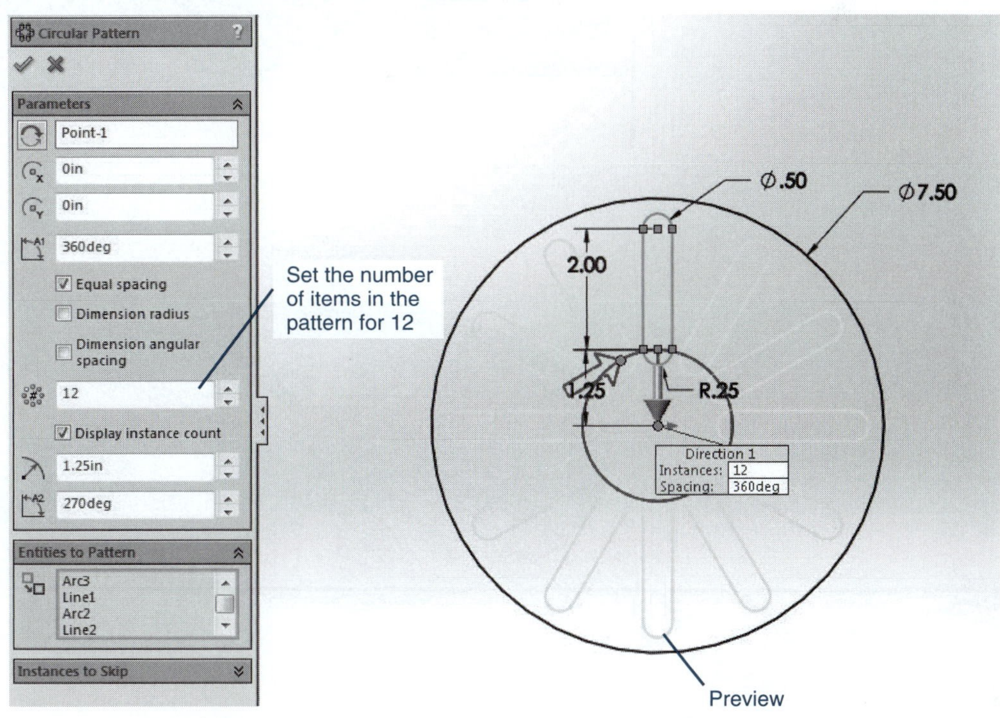

Figure 2-78
(*Continued*)

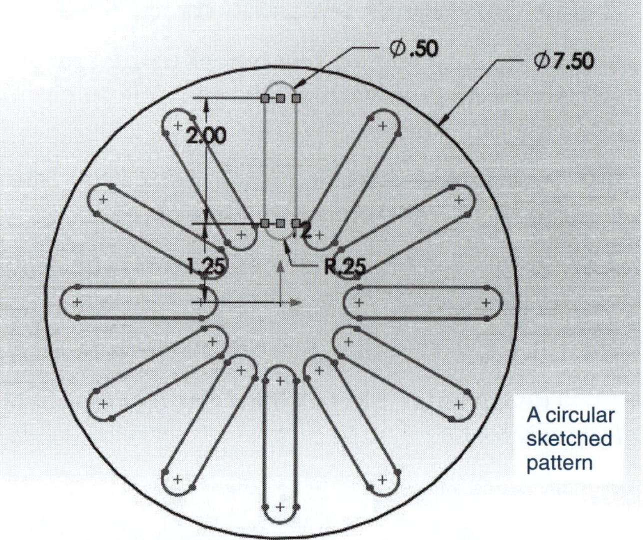

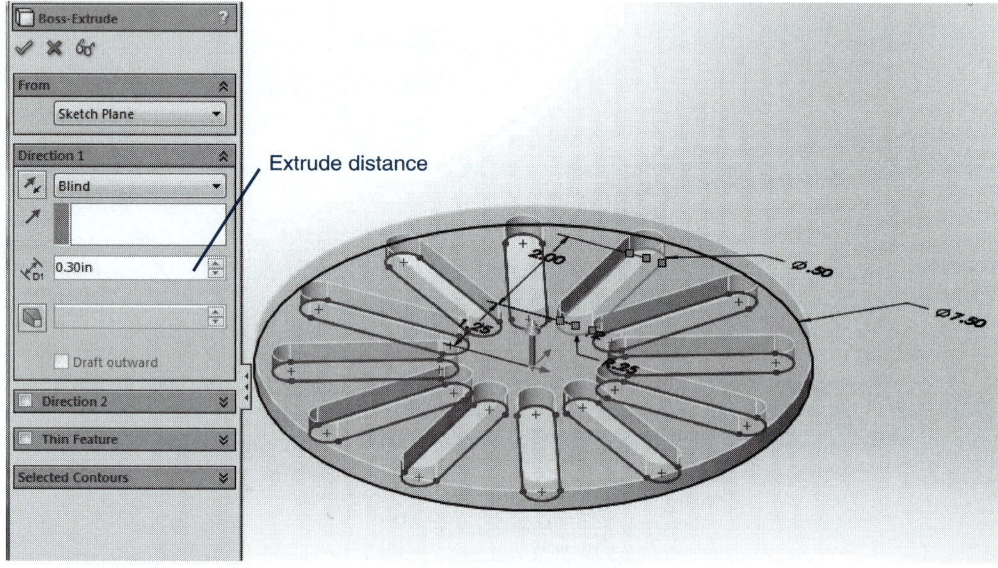

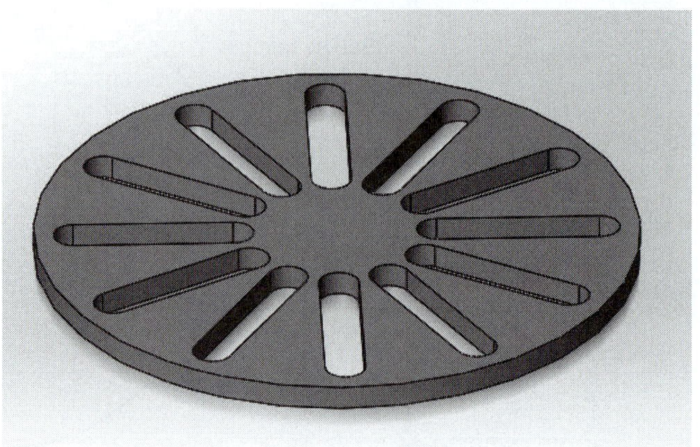

4 Click all the lines and arc in the slot shape.

5 Set the number of items in the pattern for **12**.

A preview will appear.

6 Click the green **OK** check mark.

7 Click the **Features** tab and click the **Extruded Boss/Base** tool.

8 Set the extrude distance for **0.3**.

9 Click the green **OK** check mark.

The model in this sample problem could have been created by first creating a 3D disk, sketching the slot shape on the disk, and cutting out the circular pattern.

2-31 Sample Problem SP2-3

Figure 2-79 shows a shape that includes fillets and rounded shapes. Draw the shape and extrude it to a thickness of .375 inch.

Figure 2-79

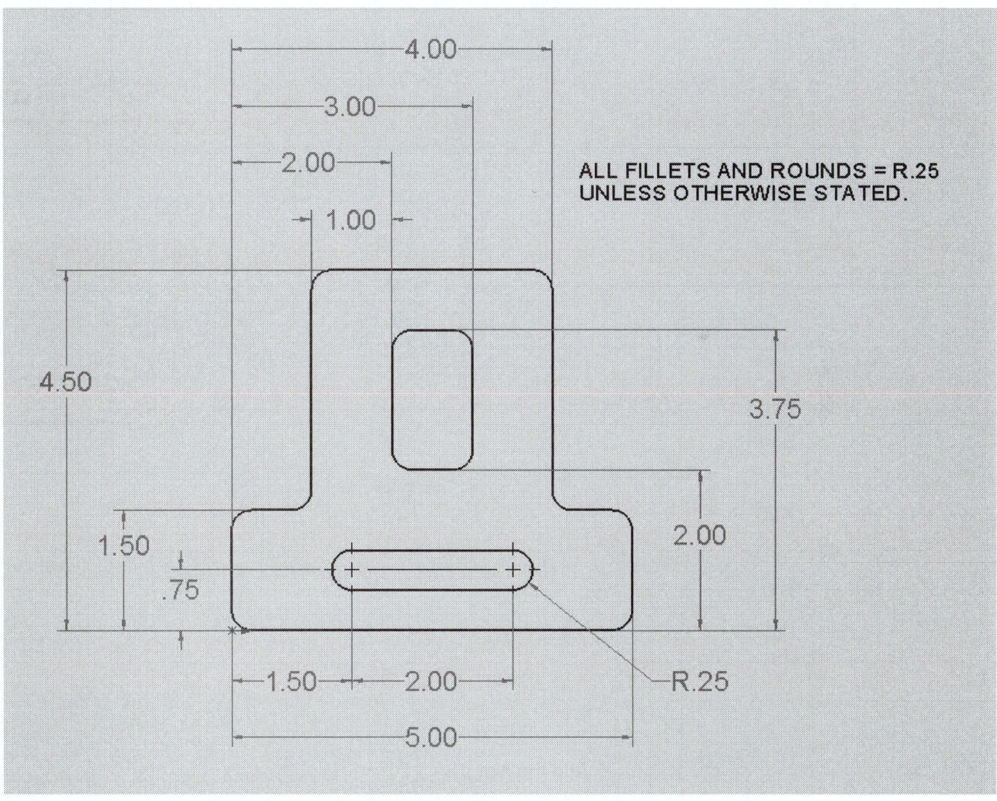

1 Start a **New Part** document, use ANSI Standards, inch units, and create the sketch on the Top plane.

See Figure 2-80.

2 Sketch the approximate shape of the part.

3 Use the **Smart Dimension** tool and size the part.

4 Use the **Fillet** tool and add fillets with a 0.25 radius as shown.

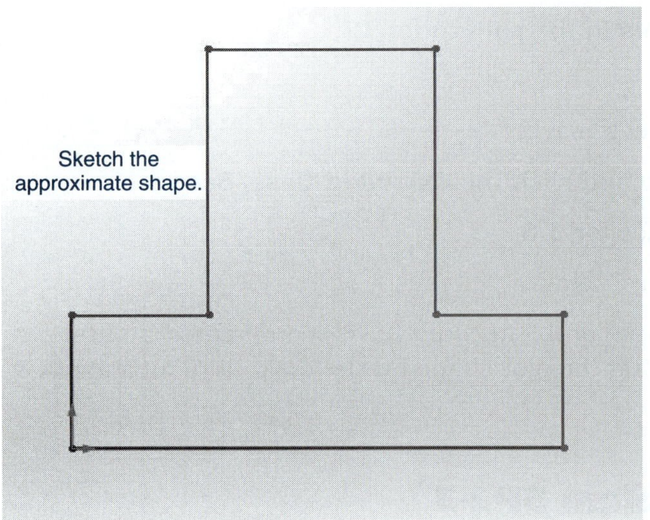
Sketch the
approximate shape.

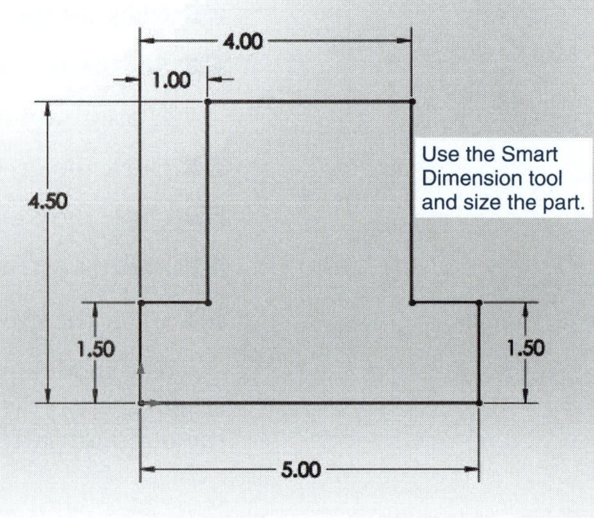
4.00
1.00
4.50
Use the Smart
Dimension tool
and size the part.
1.50
1.50
5.00

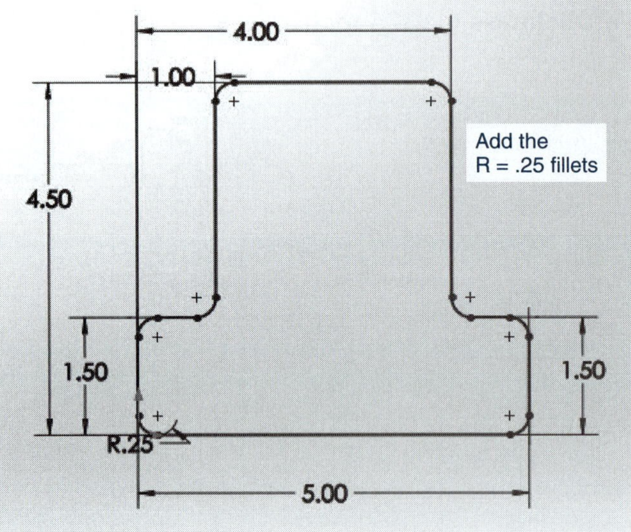
4.00
1.00
4.50
Add the
R = .25 fillets
1.50
1.50
R.25
5.00

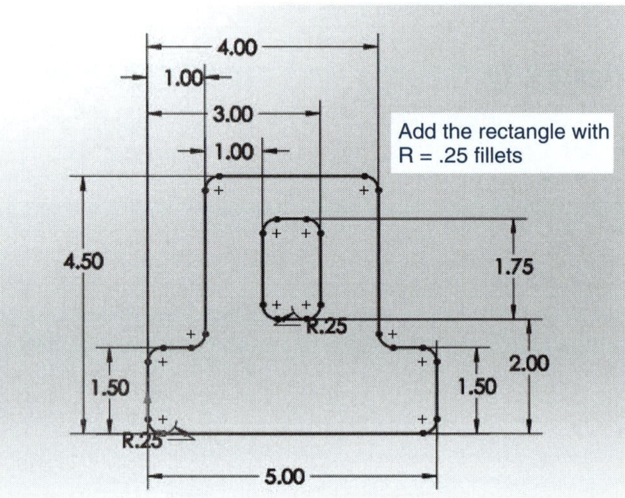
4.00
1.00
3.00
1.00
Add the rectangle with
R = .25 fillets
4.50
1.75
R.25
2.00
1.50
1.50
R.25
5.00

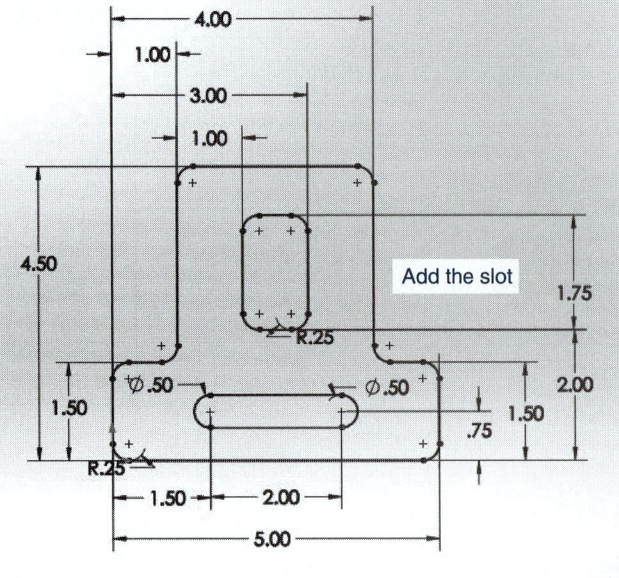
4.00
1.00
3.00
1.00
4.50
Add the slot
1.75
R.25
Ø.50
Ø.50
2.00
1.50
1.50
.75
R.25
1.50
2.00
5.00

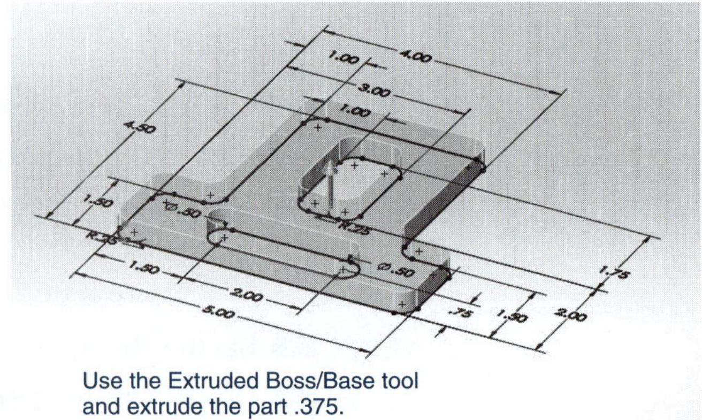
Use the Extruded Boss/Base tool
and extrude the part .375.

Figure 2-80

Figure 2-80
(Continued)

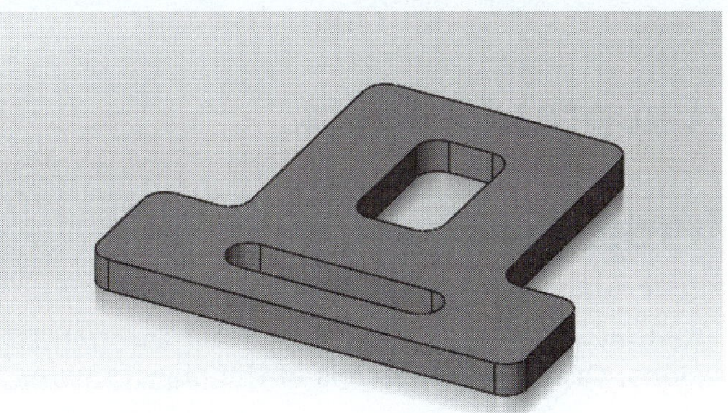

5. Use the **Corner Rectangle** and **Smart Dimension** tools and add the 1.00 × 1.75 rectangle as shown.
6. Use the **Fillet** tool to add R0.25 fillets inside the 1.00 × 1.75 rectangle.
7. Use the **Circle**, **Line**, **Trim**, and **Smart Dimension** tools to draw and locate the slot as shown.
8. Click the **Features** tab and the **Extruded Boss/Base** tool.
9. Set the extrude distance for **0.375**.
10. Click the green **OK** check mark.

Chapter Projects

Project 2-1:

Redraw the objects in Figures P2-1 through P2-24 using the given dimensions. Create solid models of the objects using the specified thicknesses.

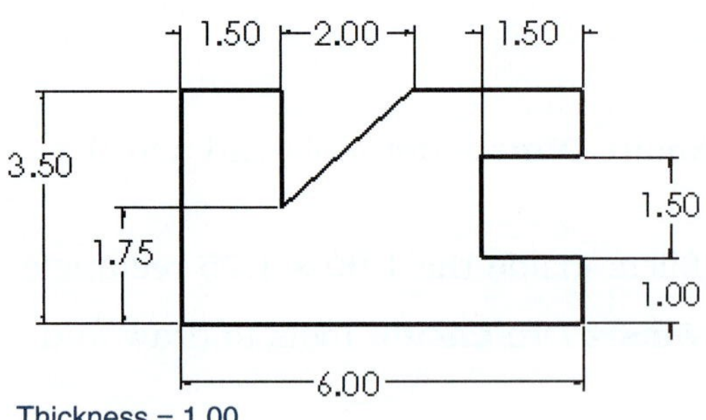

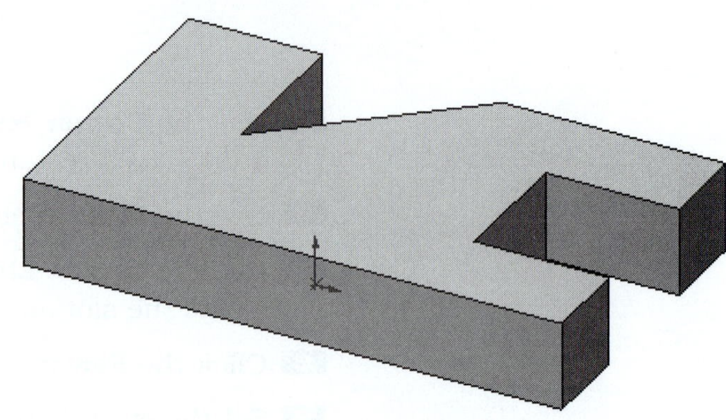

Thickness = 1.00

Figure P2-1
INCHES

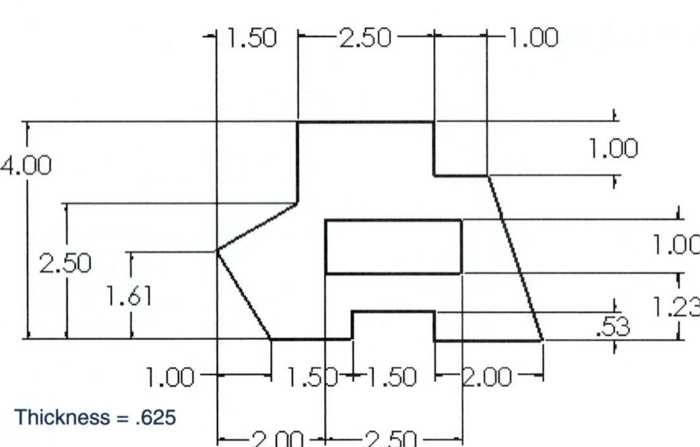

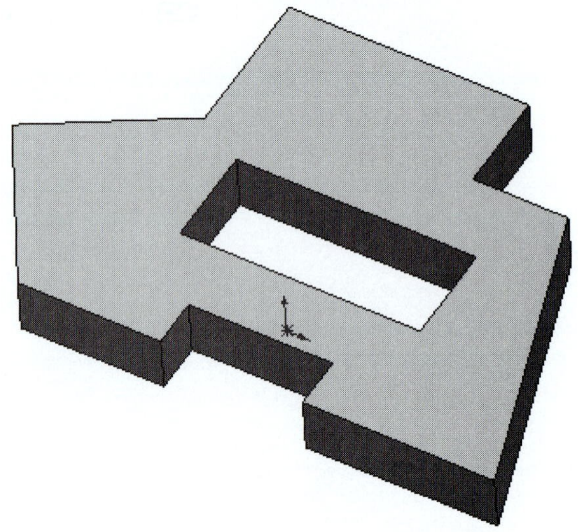

Thickness = .625

Figure P2-2
INCHES

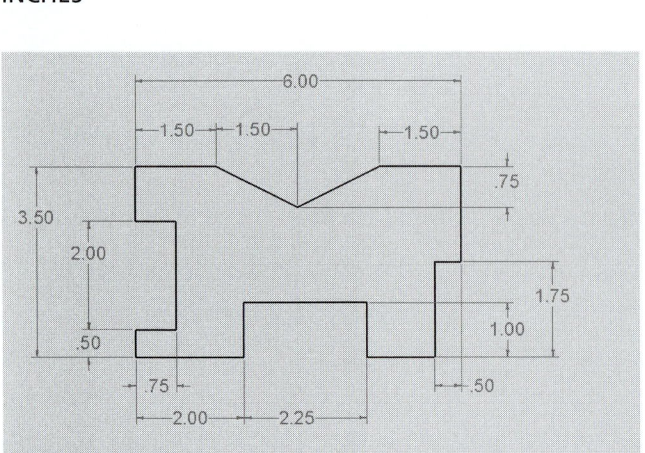

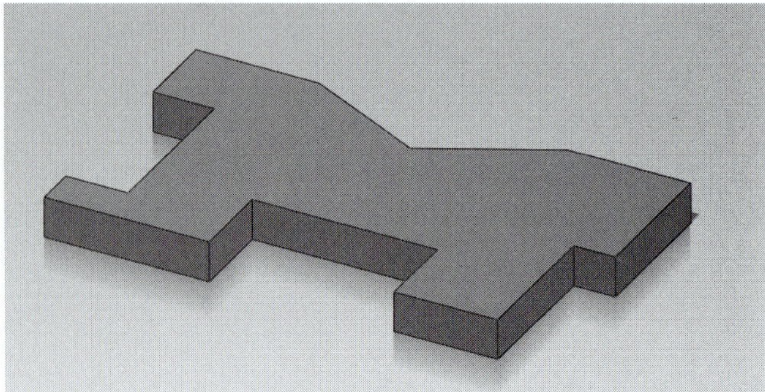

Figure P2-3
MILLIMETERS

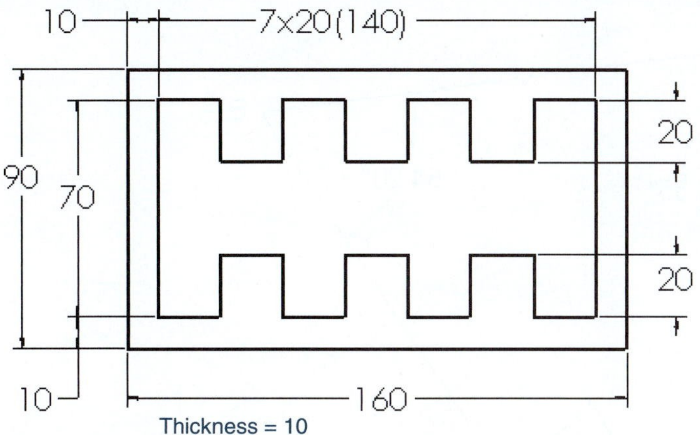

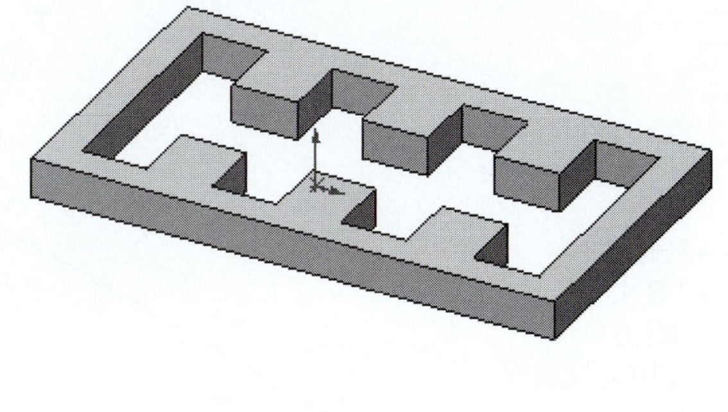

Figure P2-4
MILLIMETERS

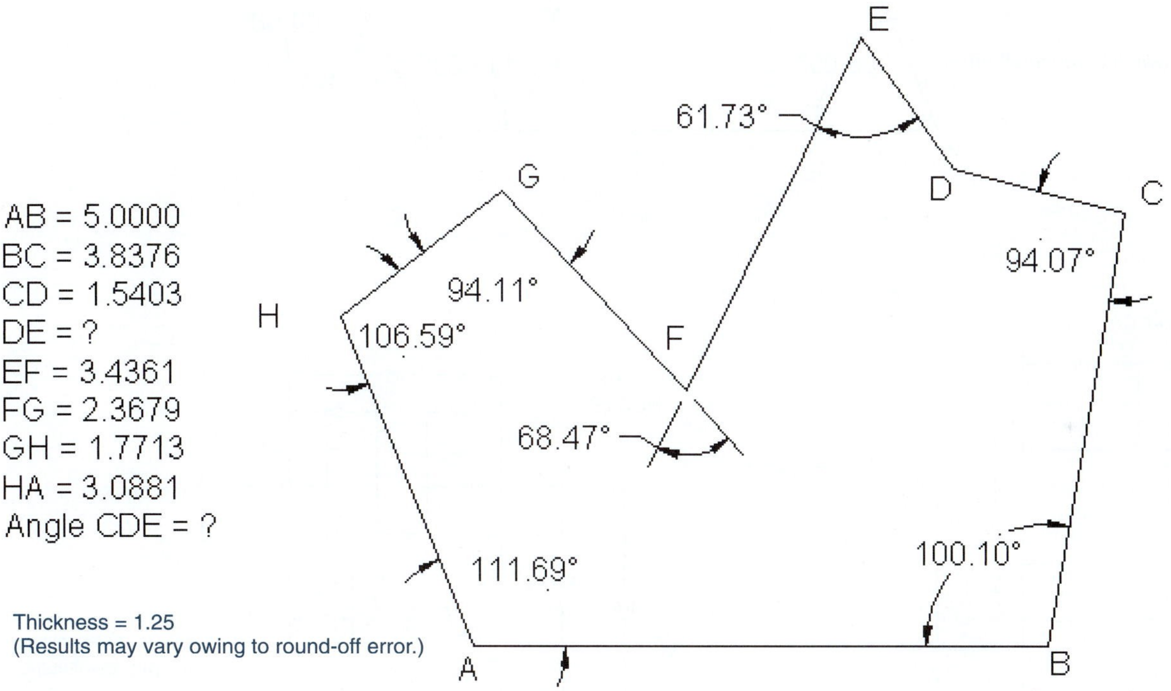

Figure P2-5
INCHES

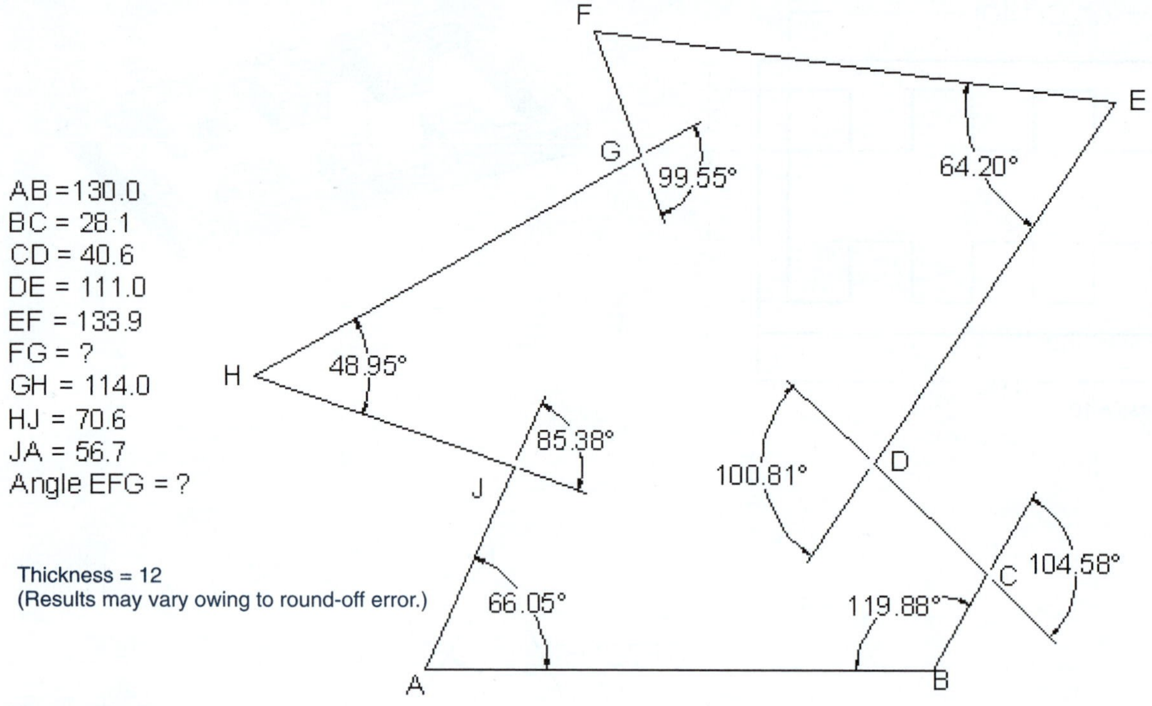

Figure P2-6
MILLIMETERS

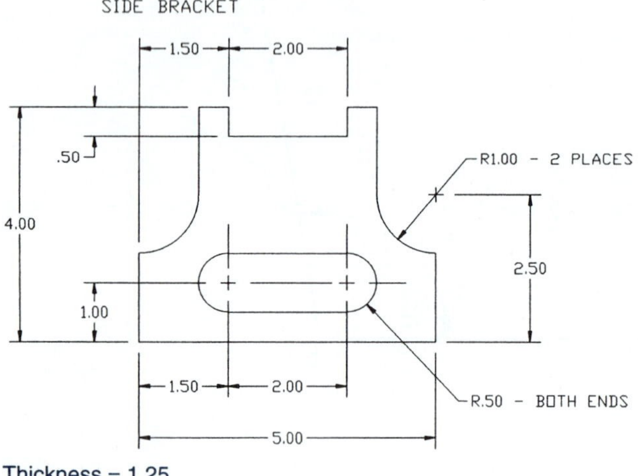

Thickness = 1.25

Figure P2-7
INCHES

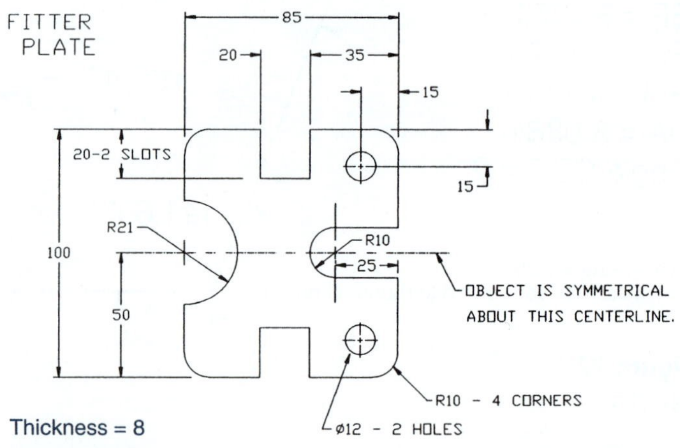

Thickness = 8

Figure P2-8
MILLIMETERS

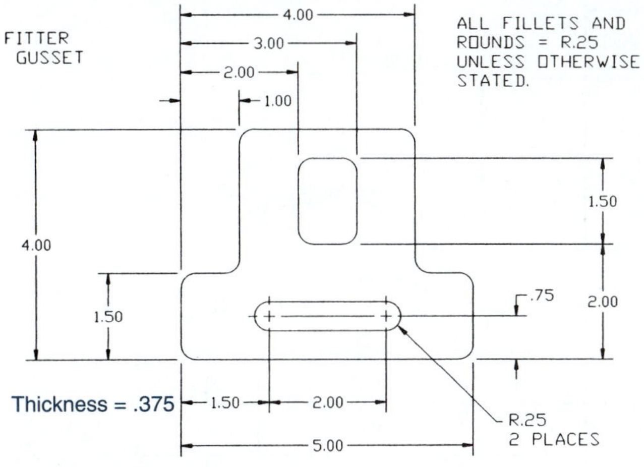

Figure P2-9
INCHES

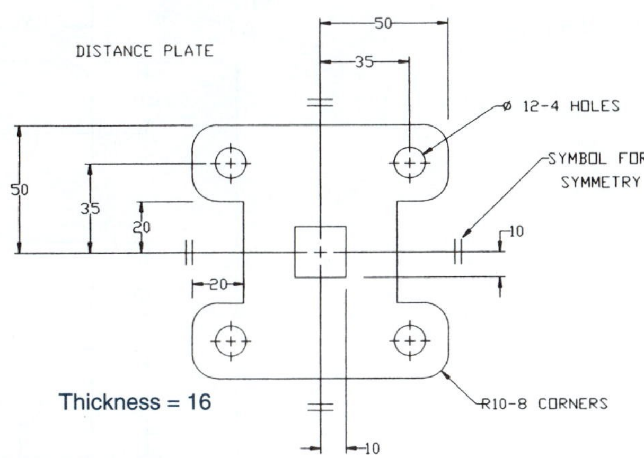

Figure P2-10
MILLIMETERS

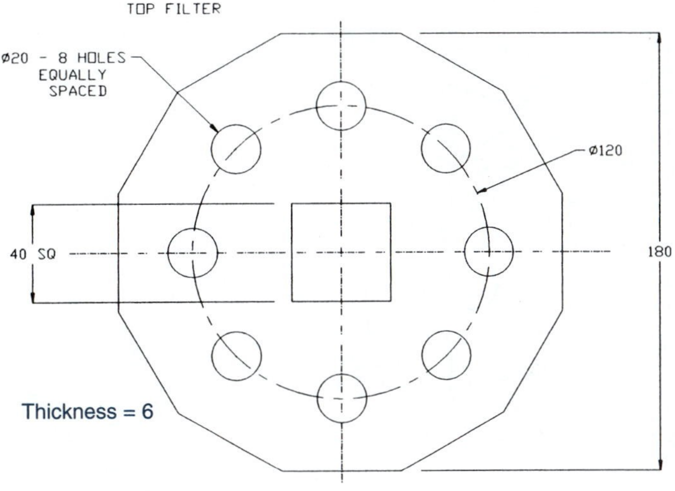

Figure P2-11
MILLIMETERS

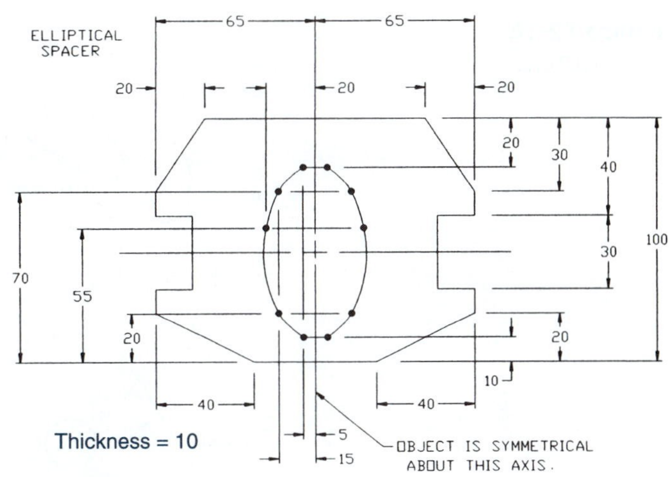

Figure P2-12
MILLIMETERS

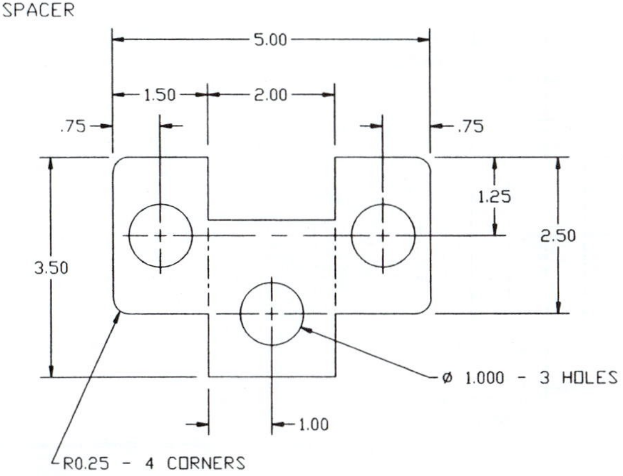

Figure P2-13
INCHES

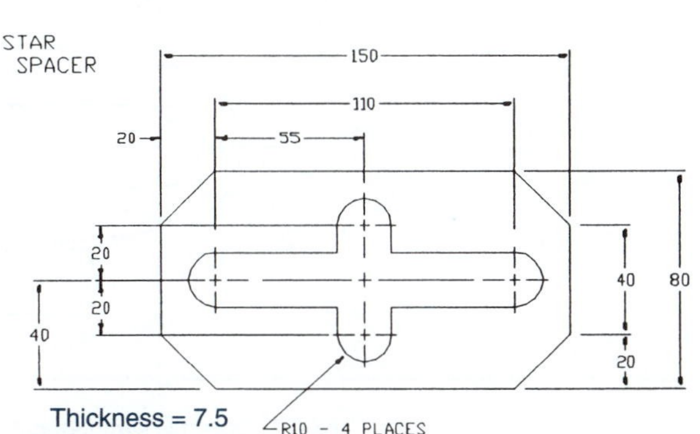

Figure P2-14
MILLIMETERS

Figure P2-15
MILLIMETERS

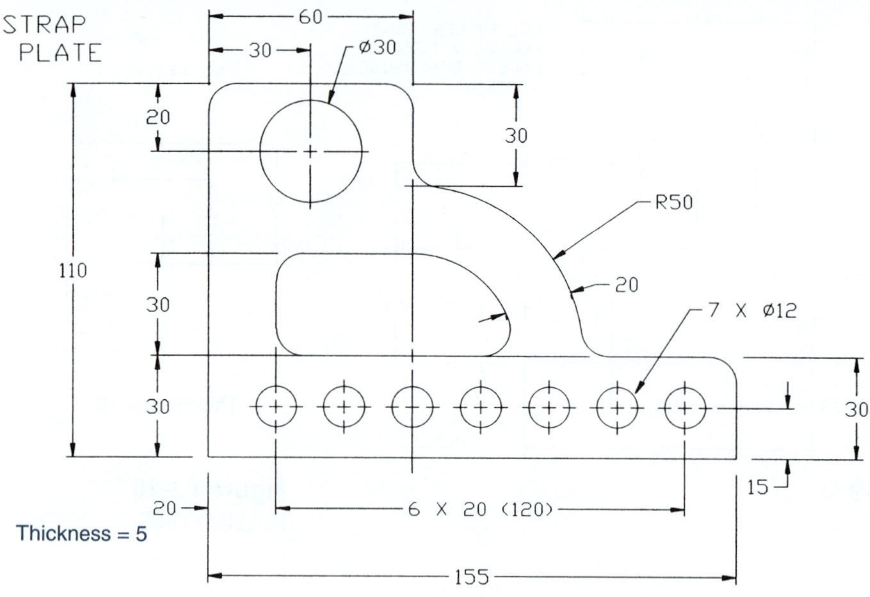

Thickness = 5

Figure P2-16
MILLIMETERS

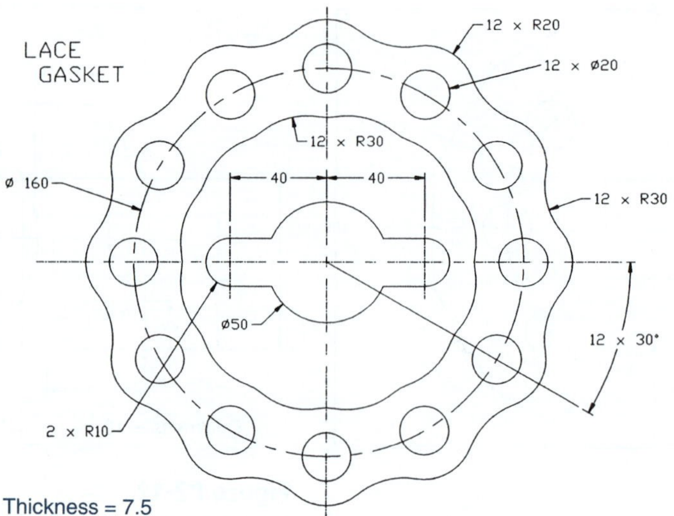

Thickness = 7.5

Figure P2-17
INCHES

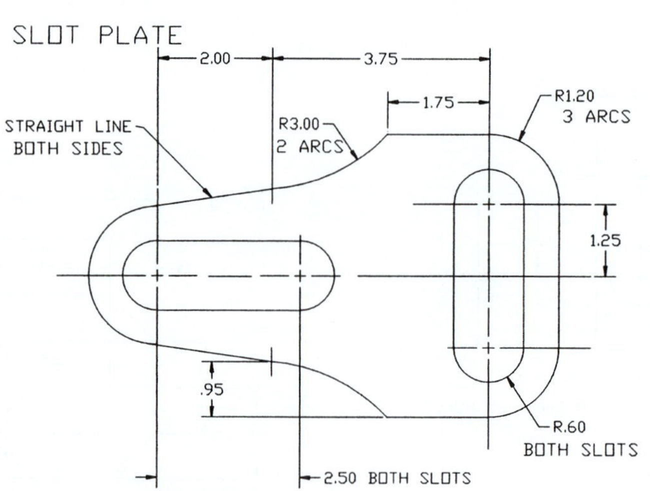

NOTE: Object is symmetrical about its horizontal centerline

Thickness = .875

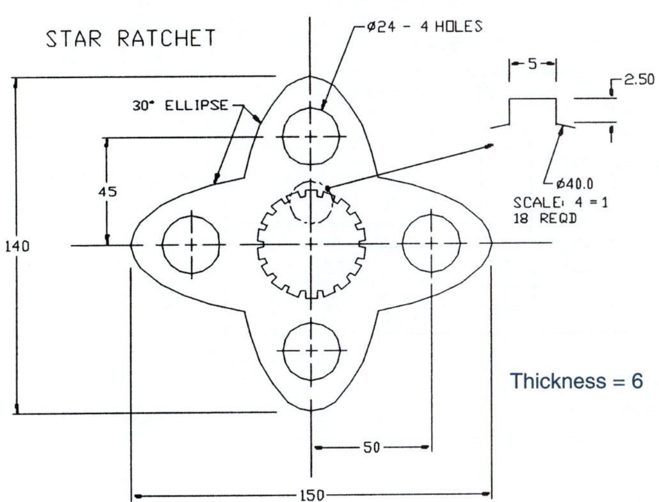

Figure P2-18
MILLIMETERS

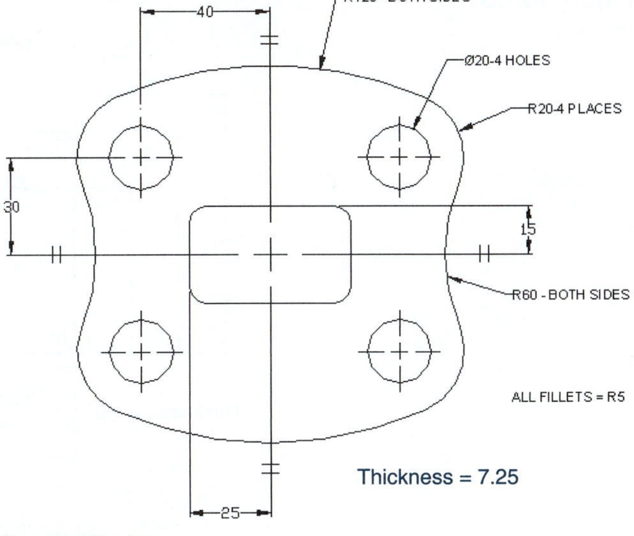

Figure P2-19
MILLIMETERS

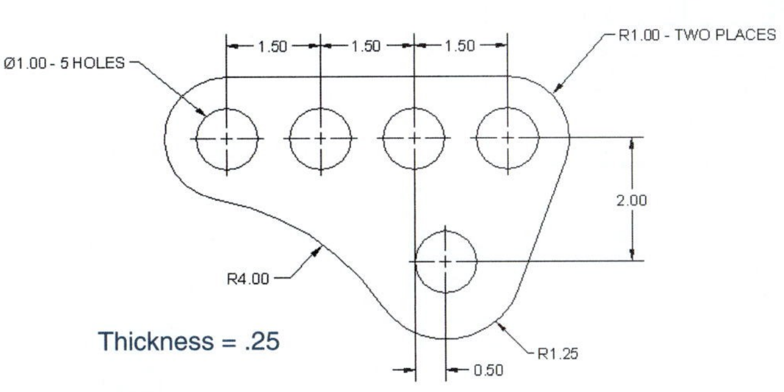

Figure P2-20
INCHES

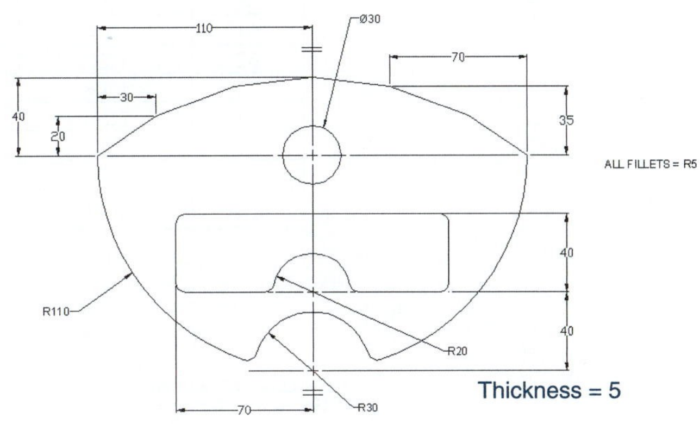

Figure P2-21
MILLIMETERS

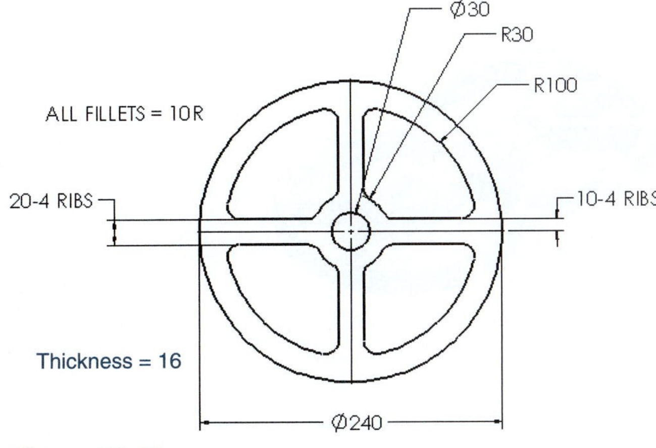

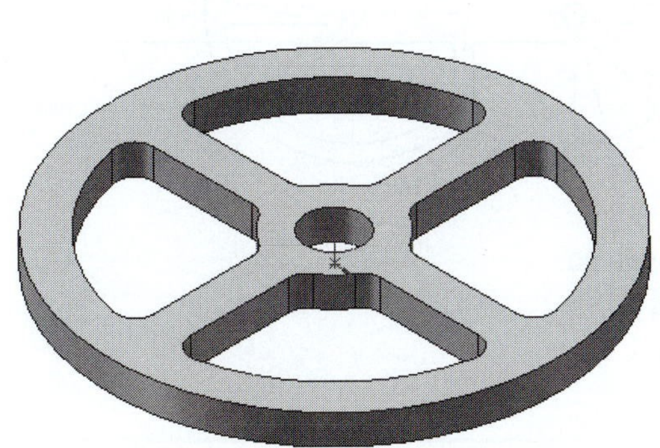

Figure P2-22
MILLIMETERS

Figure P2-23
INCHES

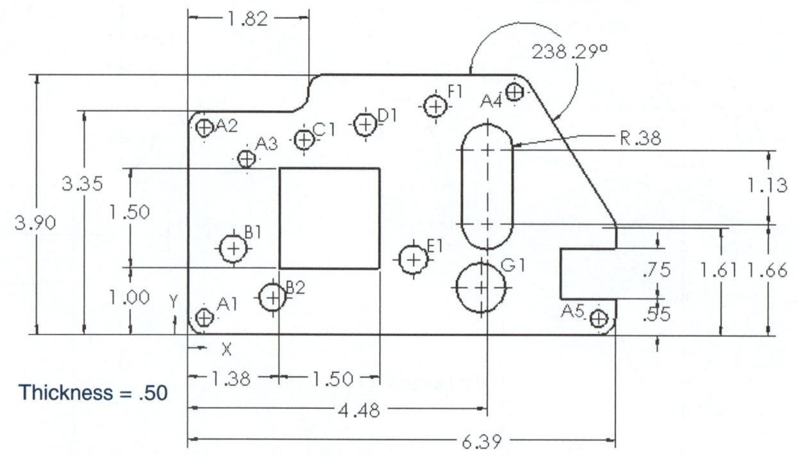

TAG	X LOC	Y LOC	SIZE
A1	.25	.25	Ø.25
A2	.25	3.10	Ø.25
A3	.88	2.64	Ø.25
A4	4.88	3.65	Ø.25
A5	6.14	.25	Ø.25
B1	.68	1.29	Ø.40
B2	1.28	.56	Ø.40
C1	1.76	2.92	Ø.29
D1	2.66	3.16	Ø.33
E1	3.38	1.13	Ø.42
F1	3.71	3.43	Ø.32
G1	4.39	.71	Ø.73

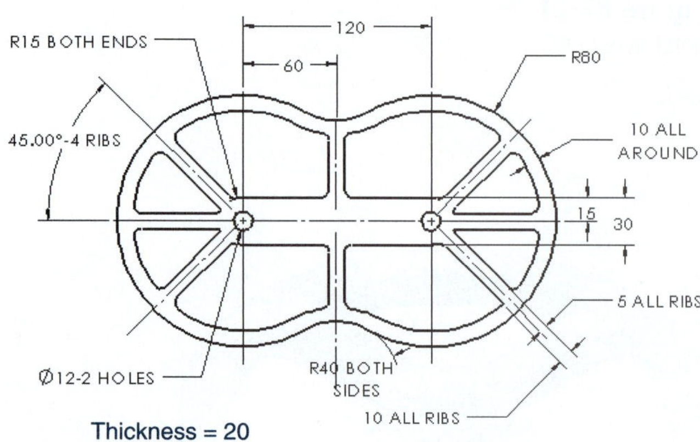

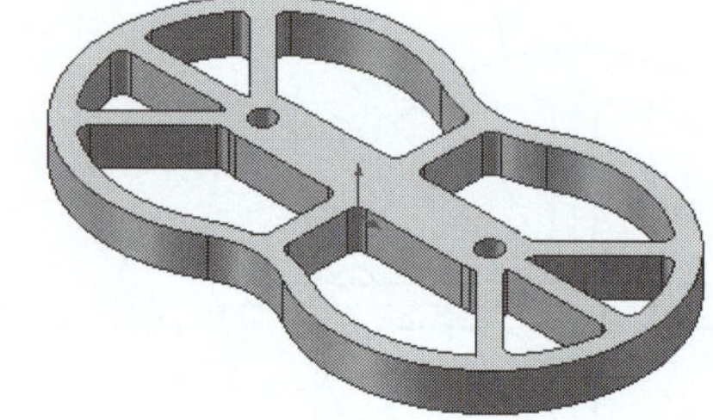

Figure P2-24
MILLIMETERS

Project 2-2:

Draw a retaining ring based on the following dimensions and as assigned by your instructor.

Figure P2-25

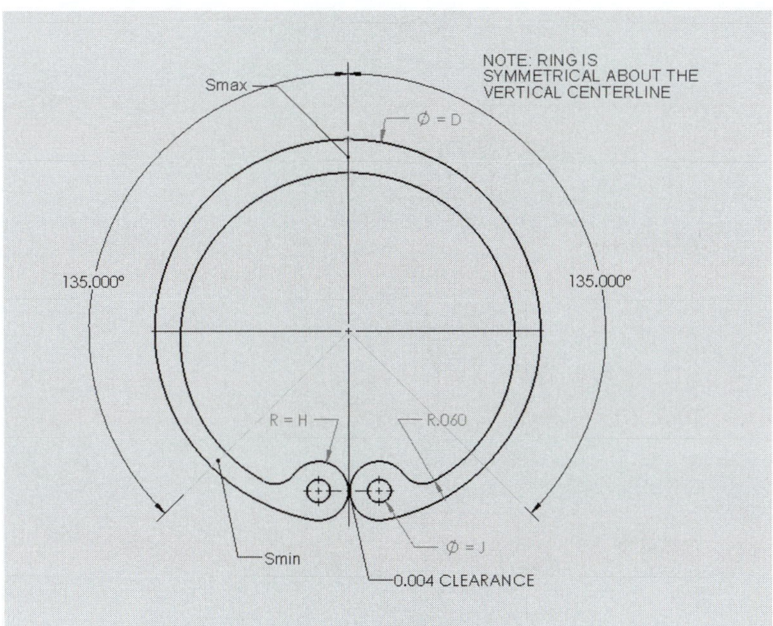

Retaing Ring - Internal - Inches

PART NO	ØD	Smax	Smin	H	A	ØJ	Thk
BU-25	.25	.025	.015	.065	.030	.031	.020
BU-50	.50	.053	.035	.114	.042	.047	.035
BU-75	.75	.070	.040	.142	.055	.060	.040
BU100	1.00	.091	.052	.155	.060	.060	.042
BU125	1.25	.120	.062	.180	.070	.075	.050
BU150	1.50	.127	.066	.180	.070	.075	.050

Retaining Ring - Internal - Millimeters

PART NO	ØD	Smax	Smin	H	A	ØJ	Thk
MBU-20	20	2.3	1.9	4.1	2.0	2.0	1.0
MBU-30	30	3.0	2.3	4.8	2.0	2.0	1.2
MBU-40	40	3.9	3.0	5.8	2.5	2.5	1.7
MBU-50	50	4.6	3.8	6.5	2.5	2.5	2.0
MBU-60	60	5.4	4.3	7.3	2.5	2.5	2.0
MBU-70	70	6.2	5.2	7.8	3.0	3.0	2.5

3 chapter three

Features

CHAPTER OBJECTIVES

- Learn about the **Features** tools
- Learn how to draw 3D objects
- Learn how to use **Features** tools to create objects

3-1 Introduction

This chapter introduces the **Features** tools. **Features** tools are used to create 3D models. A drawing usually starts with a 2D sketch as described in Chapter 2 and Features tools are applied to create a 3D model. For example, the **Extruded Boss/Base** features tool can be applied to a rectangle to create a rectangular prism or box. Figure 3-1 shows the **Features** tool panel.

Figure 3-1

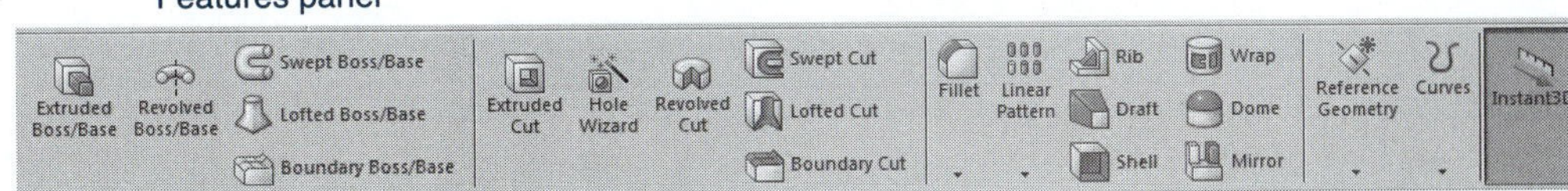

3-2 Extruded Boss/Base

The **Extruded Boss/Base** tool is used to add thickness or height to an existing 2D sketch. In this section we will first sketch a 2D rectangle, then extrude it into a box-like shape. All dimensions are in millimeters.

To Use the Extruded Boss/Base Tool

1 Click the **New** tool, create a new **Part** document, click the **Top Plane** option, and click the **Sketch** tab.

2 Define the drawing units for **MMGS** (Millimeters, grams, seconds).

The drawing units are referenced at the bottom of the drawing screen. In this example **Custom** units are in place. See Figure 3-2.

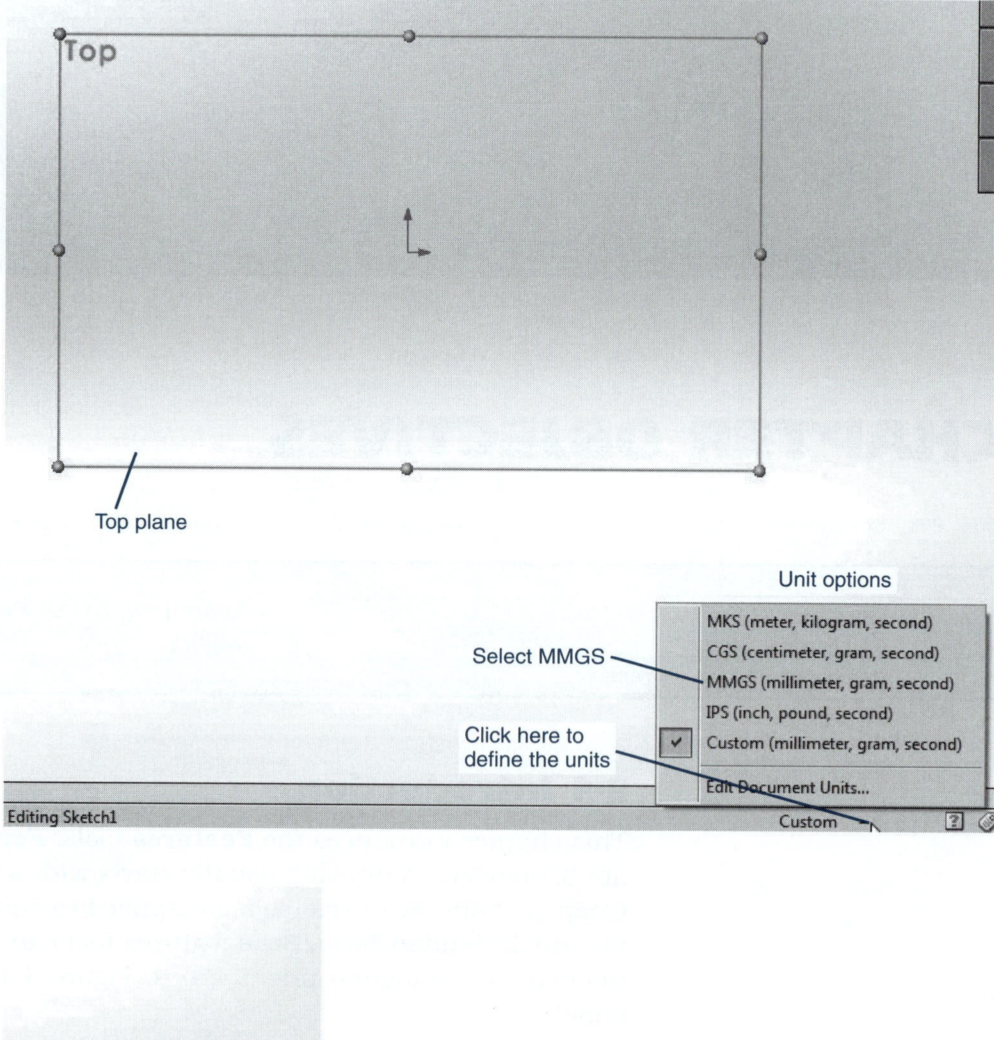

Figure 3-2

3 Click the arrowhead next to the **Custom** heading and select the **MMGS** option.

4 Use the **Corner Rectangle** tool to sketch a **60** × **100** rectangle in the top plane with one corner located on the origin. Use the **Smart Dimension** tool to size the rectangle.

See Figure 3-3.

Figure 3-3

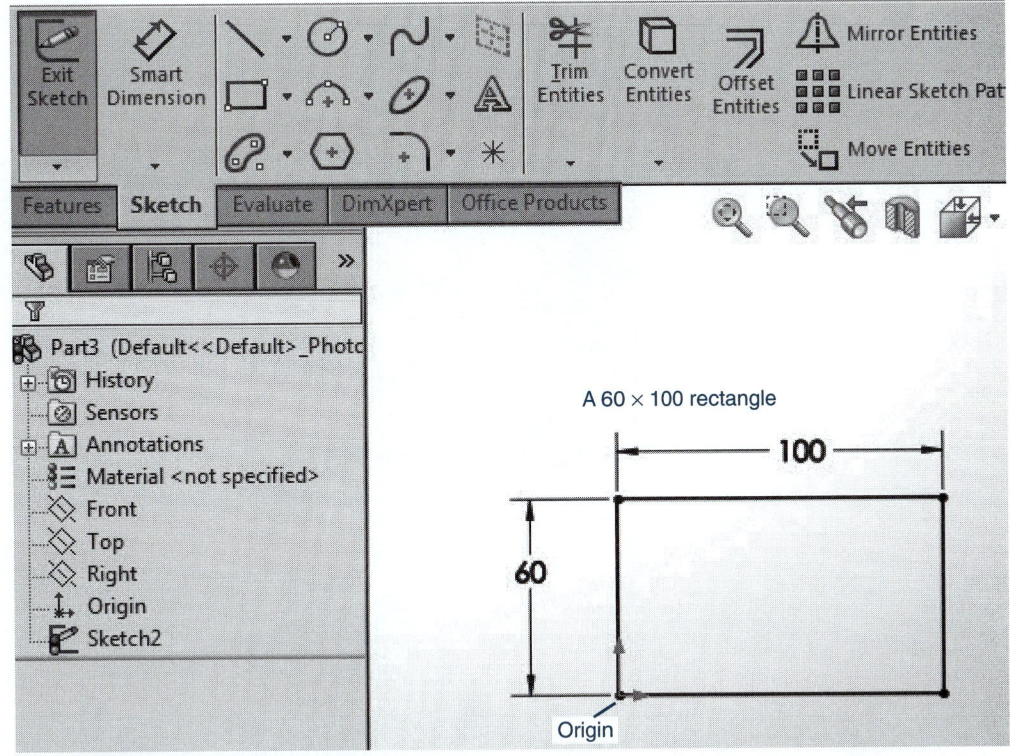

Exit Sketch
Smart Dimension
Trim Entities
Convert Entities
Offset Entities
Mirror Entities
Linear Sketch Pat
Move Entities
Features
Sketch
Evaluate
DimXpert
Office Products
Part3 (Default<<Default>_Photo
History
Sensors
Annotations
Material <not specified>
Front
Top
Right
Origin
Sketch2
A 60 × 100 rectangle
100
60
Origin

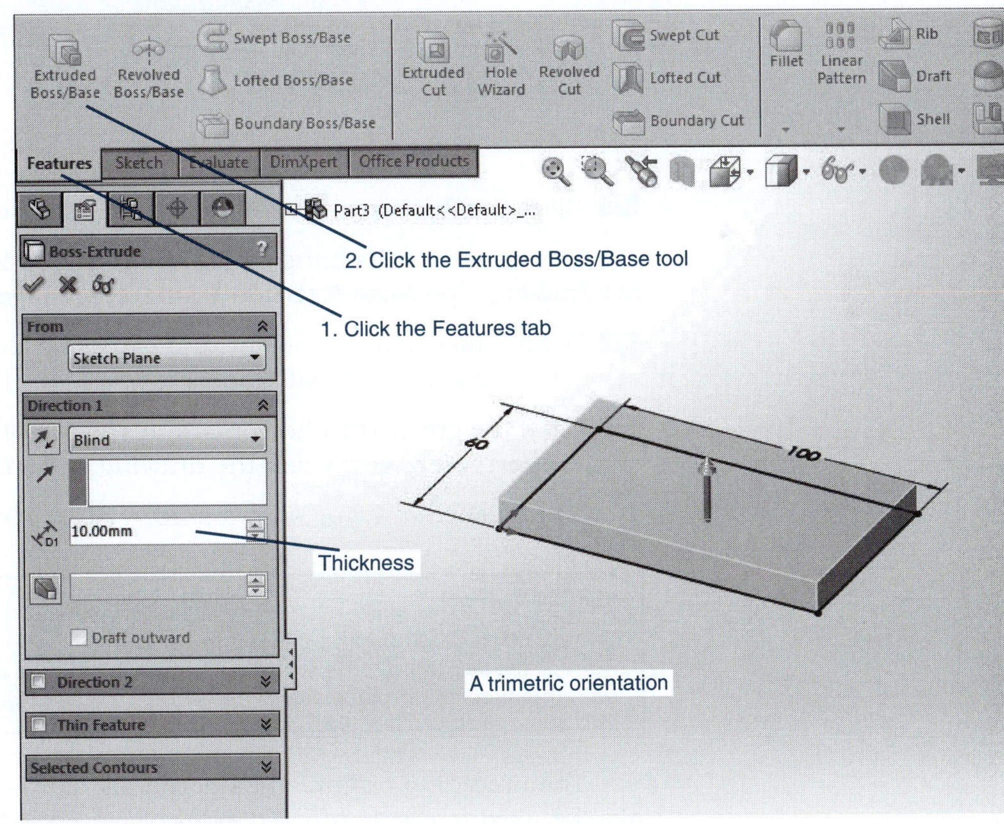

Extruded Boss/Base
Revolved Boss/Base
Swept Boss/Base
Lofted Boss/Base
Boundary Boss/Base
Extruded Cut
Hole Wizard
Revolved Cut
Swept Cut
Lofted Cut
Boundary Cut
Fillet
Linear Pattern
Rib
Draft
Shell
Features
Sketch
Evaluate
DimXpert
Office Products
Part3 (Default<<Default>_...
Boss-Extrude
From
Sketch Plane
Direction 1
Blind
10.00mm
Draft outward
Direction 2
Thin Feature
Selected Contours
2. Click the Extruded Boss/Base tool
1. Click the Features tab
Thickness
60
100
A trimetric orientation

Figure 3-3
(*Continued*)

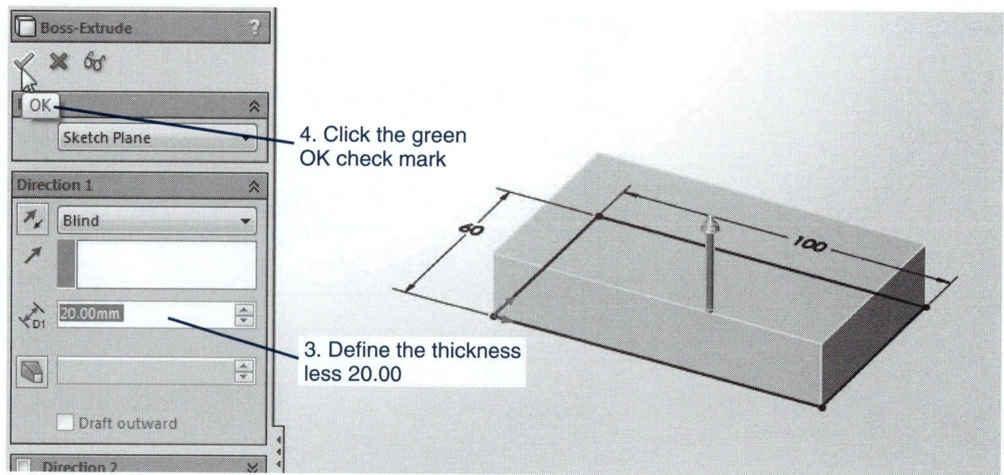

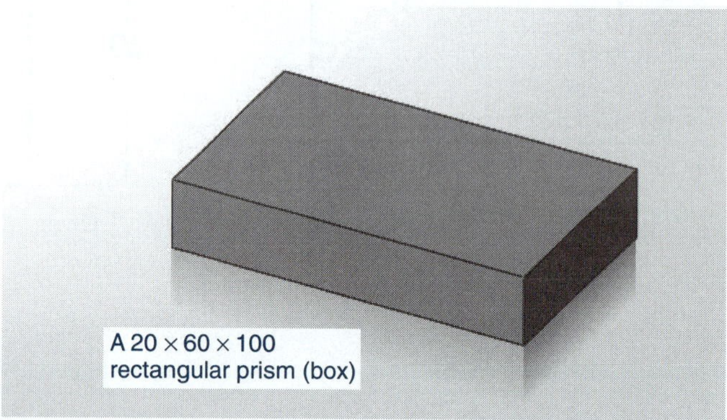

5 Click the **Features** tab, and click the **Extruded Boss/Base** tool.

The drawing will change orientation from a 2D top view to a Trimetric 3D drawing. The **Boss-Extrude PropertyManager** will appear.

6 Define the extrusion height, the extrusion distance, as **20.00mm**. A real-time preview will appear.

7 Click the green **OK** check mark at the top of the **Boss-Extrude PropertyManager**. Click the drawing screen.

The finished drawing shows a 20 × 60 × 100 box.

> **TIP**
> The extrusion depth may be defined by entering a value or by using the arrows at the right of the **Depth** box. The extrusion depth may also be changed by clicking and dragging the direction arrow in the center of the box. See Figure 3-4.

The preceding example has perpendicular sides. The **Extrude** tool may also be used to create tapered sides. Tapered sides are called ***draft sides***.

To Create Inward Draft Sides

1 Sketch a **60** × **100** rectangle as described in the previous section.

2 Click the **Features** tab and click the **Extruded Boss/Base** tool.

Figure 3-4

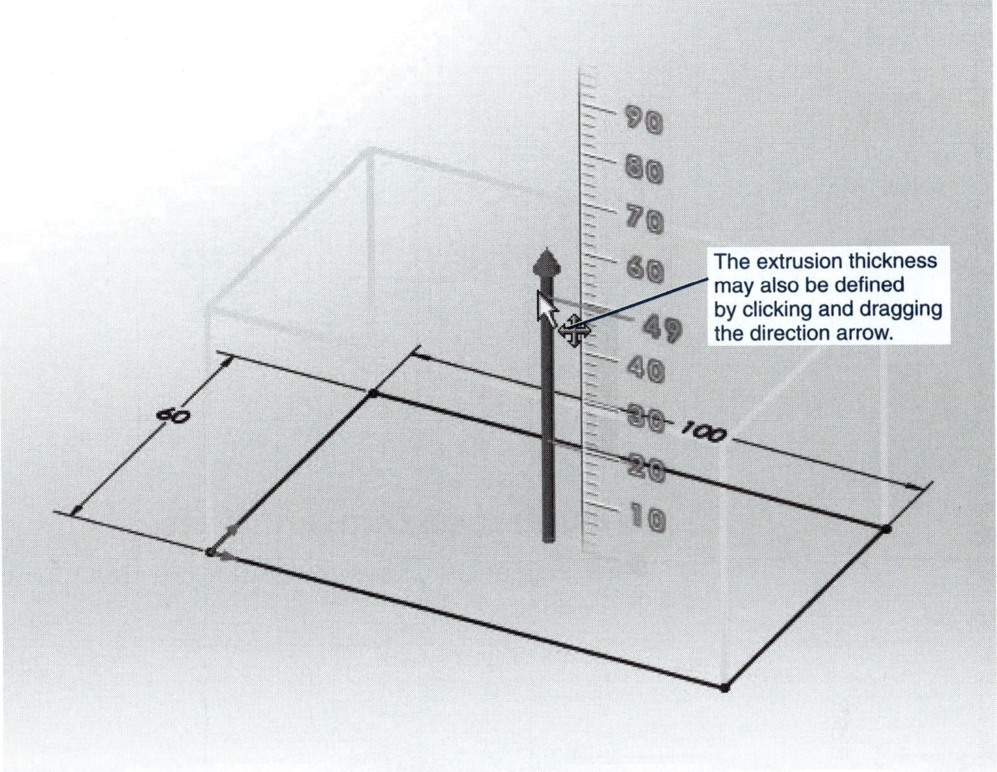

3 Click the **Draft On/Off** tool.

4 Enter the draft angle value.

In this example a **15°** value was entered. See Figure 3-5.

5 Click the green **OK** check mark at the top of the **Boss-Extrude PropertyManager** to complete the object.

The shape now has a 15° inward draft.

Figure 3-5

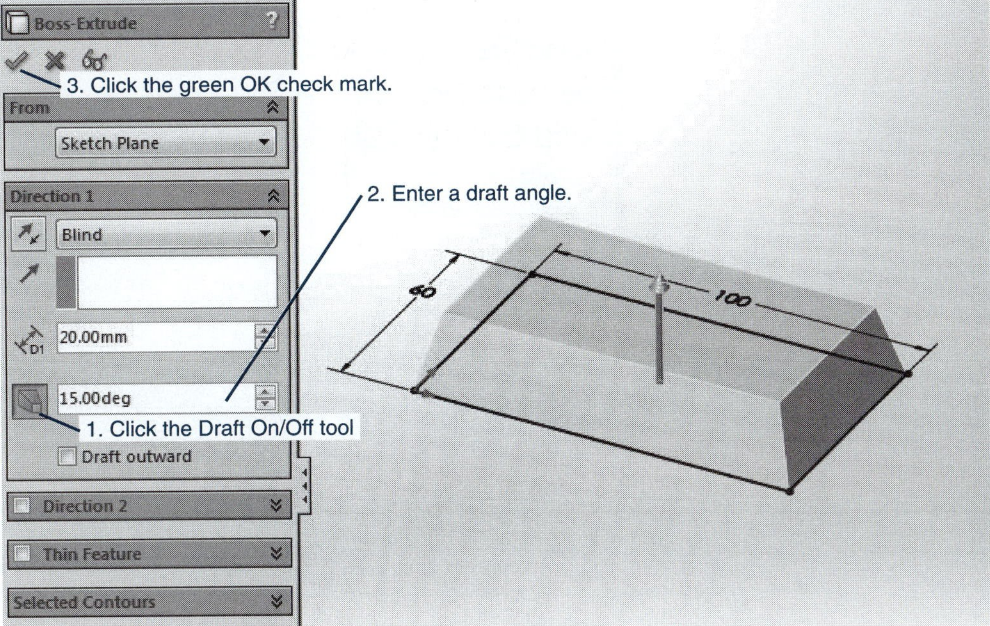

figure 3-5
(*continued*)

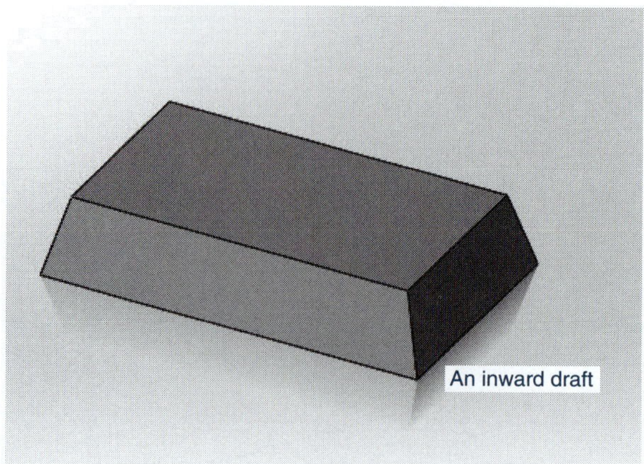

To Create an Outward Draft

1 Repeat the same procedure, but this time check the **Draft outward** box.

See Figure 3-6.

Figure 3-6

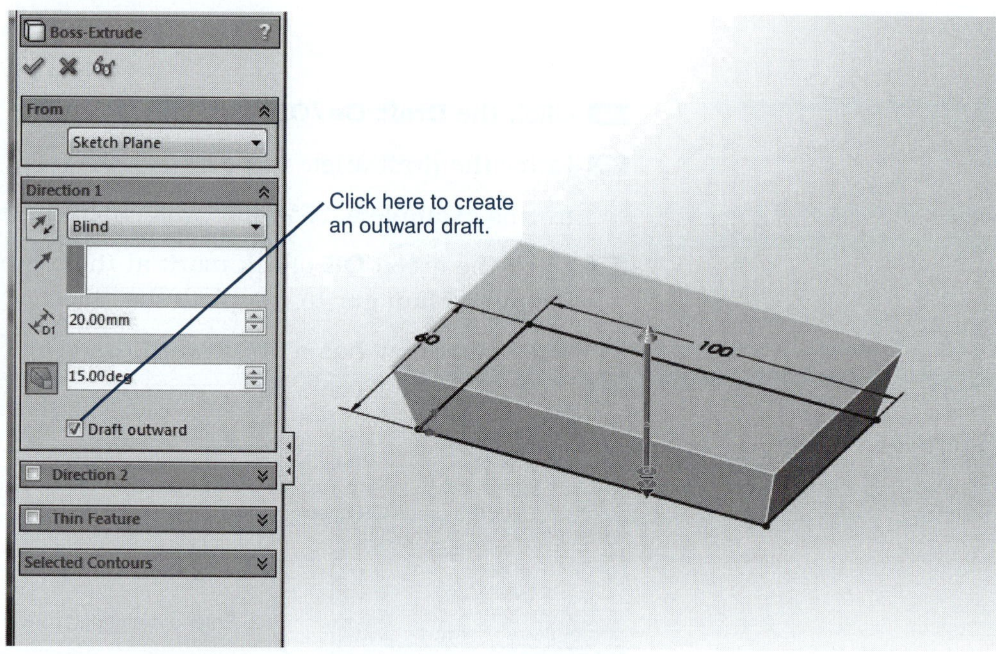

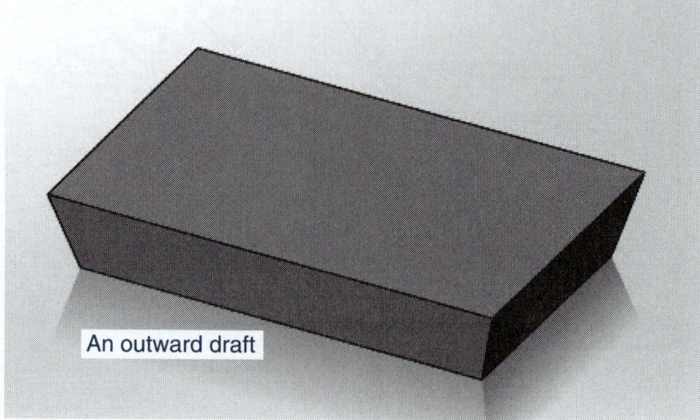

3-3 Sample Problem SP3-1

This section shows how to draw a solid 3D model of an L-bracket using the **Extruded Boss/Base** tool.

1 Draw a **60** × **100** rectangle and extrude it to a depth of **20mm** as explained in Section 3-2.

See Figure 3-7.

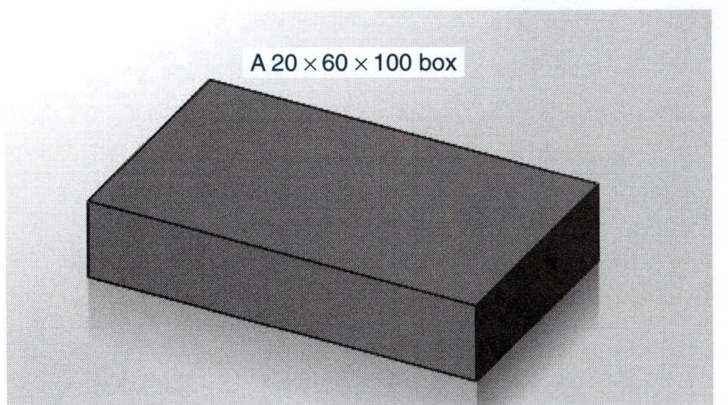

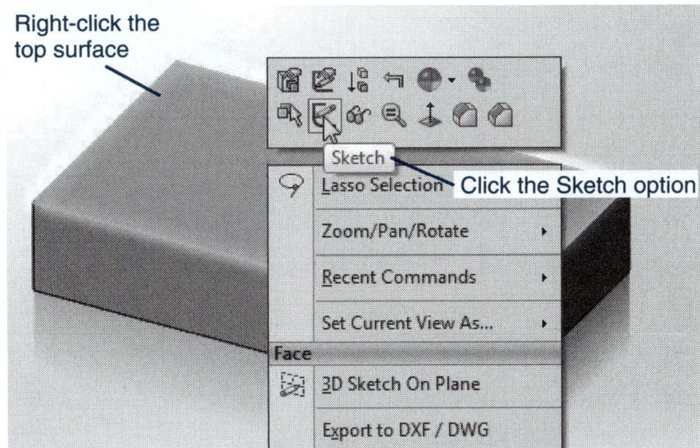

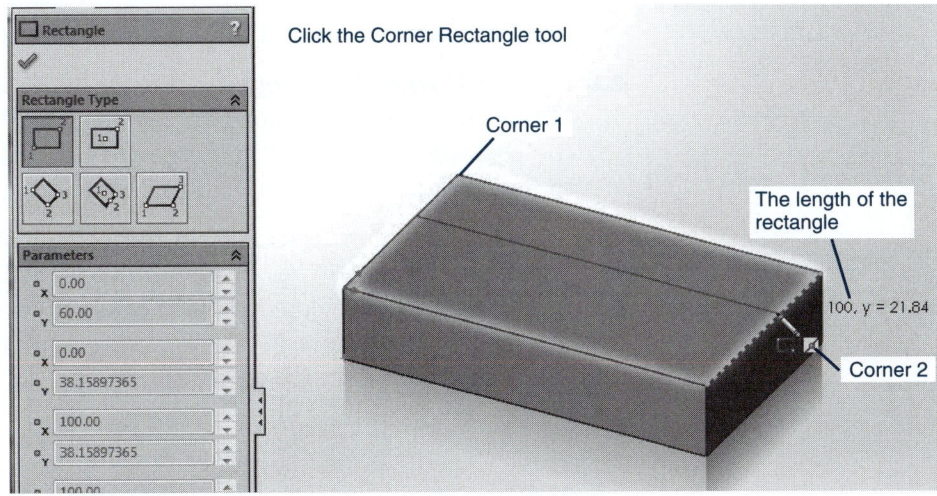

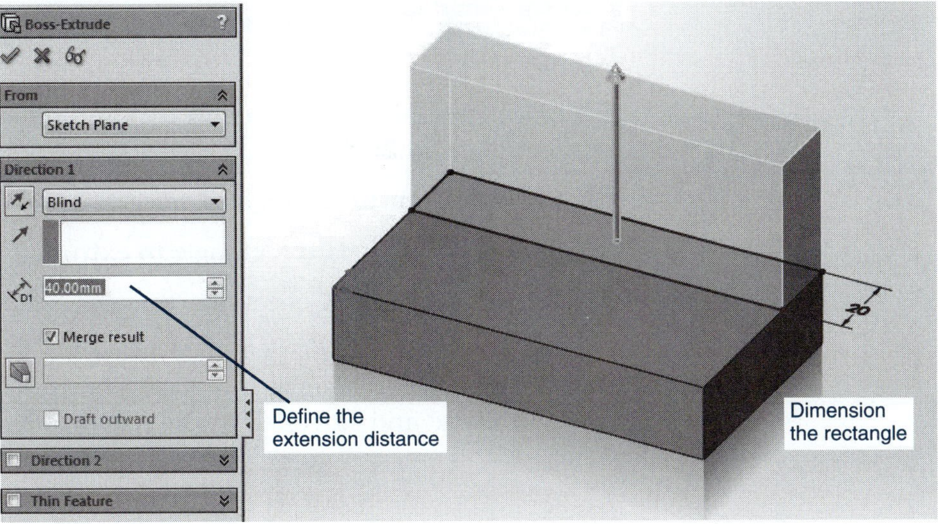

Figure 3-7

Figure 3-7
(*Continued*)

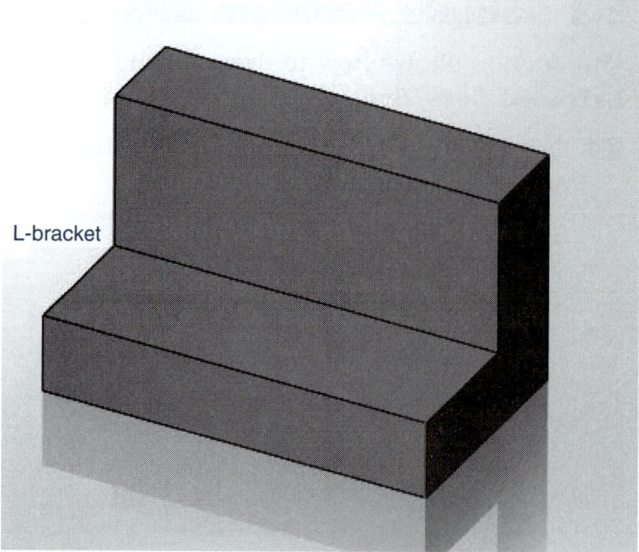

2 Locate the cursor on the top surface of the box and right-click the mouse. Select the **Sketch** tool.

The 2D sketch tools will be displayed across the top of the screen.

3 Use the **Corner Rectangle** tool to draw a rectangle on the top surface of the box. Use the upper left corner of the box as one corner point of the rectangle and drag the cursor to the right edge of the top surface to locate the second corner point for the rectangle.

SolidWorks will grab the upper left corner of the rectangle when the cursor is located near it. An orange fill circle will appear indicating that the corner has been selected. SolidWorks will also grab the edge line when the cursor is located near it. The edge line will appear as an orange broken line indicating that it has been selected.

NOTE

The corner points and edge lines will change color when they are activated. The part is currently under defined.

4 Select the **Smart Dimension** tool and dimension the width of the rectangle as **20.0**.

5 Click the **Features** tab, and select the **Extruded Boss/Base** tool.

6 Select the **20** × **100** rectangle to extrude to a depth of **40.00mm**.

7 Click the green **OK** check mark in the **Boss-Extrude PropertyManager**.

SolidWorks offers many different ways to create the same shape. Figure 3-8 shows the same L-bracket shown in Figure 3-7 created using the right plane and extruded in the negative X direction.

Figure 3-8

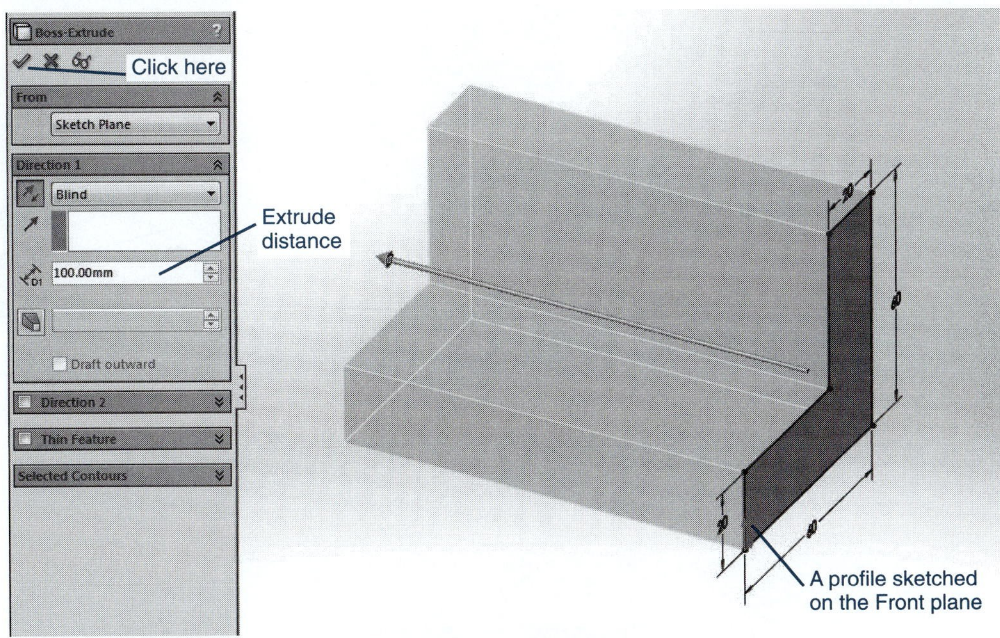

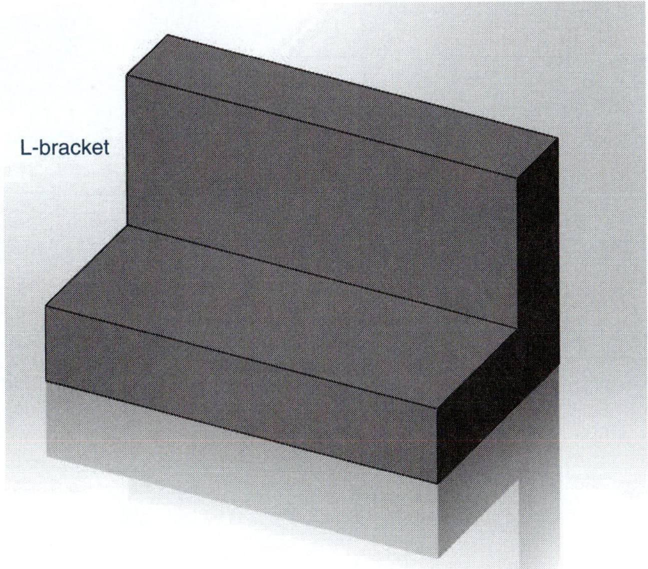

3-4 Extruded Cut

This section will add a cutout to the L-bracket drawn in Section 3-3 using the **Extruded Cut** tool. See Figure 3-9.

1. Locate the cursor on the lower front horizontal surface and right-click the mouse.
2. Click the **Sketch** option.
3. Use the **Corner Rectangle** and **Smart Dimension** tools to draw and size a rectangle as shown.

Locate the first point of the **Corner** rectangle on the front edge line. SolidWorks will grab the front edge line. The line will change to an orange broken line indicating that it has been selected.

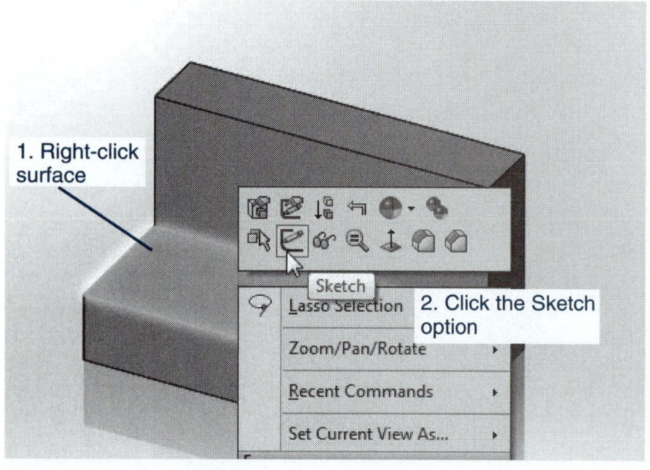

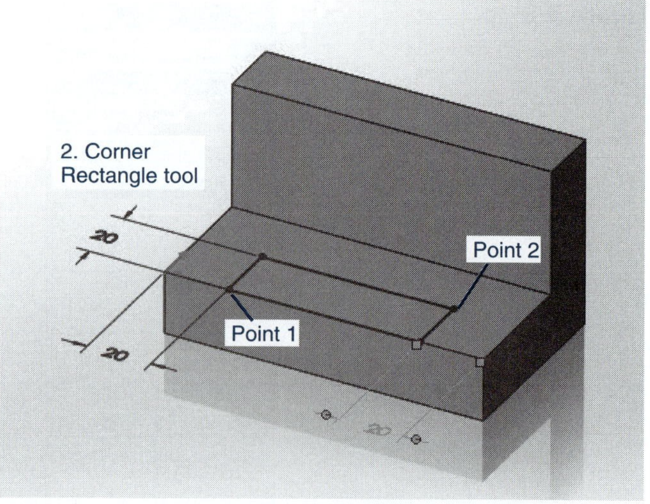

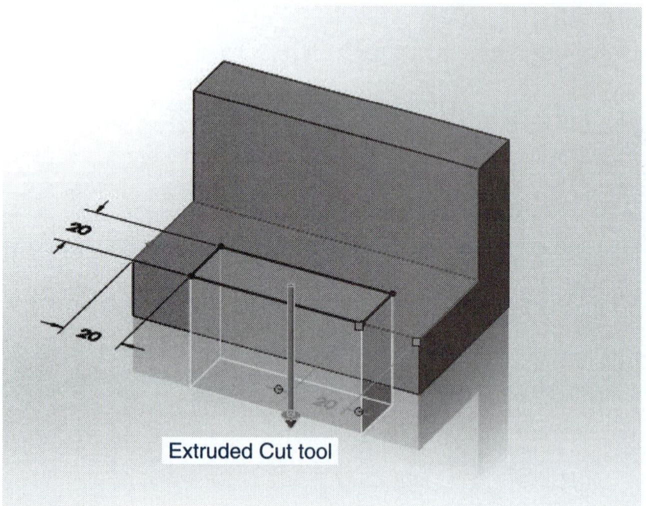

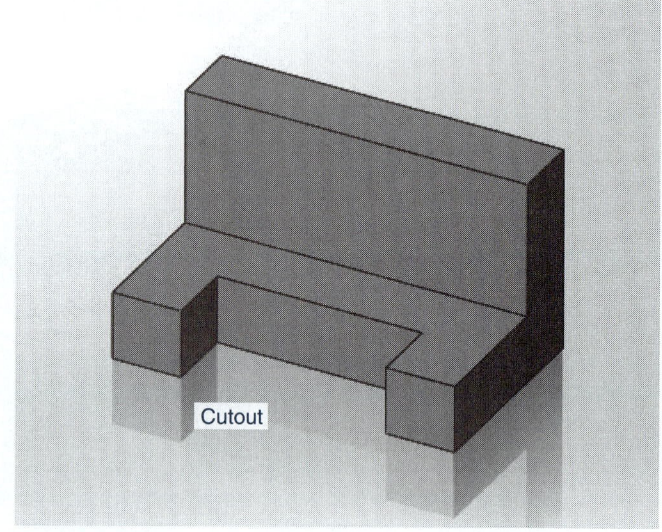

Figure 3-9

4 Click the **Features** tab, then click the **Extruded Cut** tool.

A preview will appear. Ensure that the **Extruded-Cut** extends beyond the lower surface of the bracket.

5 Click the green **OK** check mark in the **Cut-Extrude PropertyManager**.

3-5 Hole Wizard

This section will add a hole to the L-bracket created in Section 3-3 using the **Hole Wizard** tool. The **Hole Wizard** tool is used to add nine different type holes and slots to an existing 3D model. The hole types include simple holes and threaded holes among others. See Figure 3-10.

1 Click the **Features** tab, then click the **Hole Wizard** tool.

2 Select the **Hole** option.

Figure 3-10

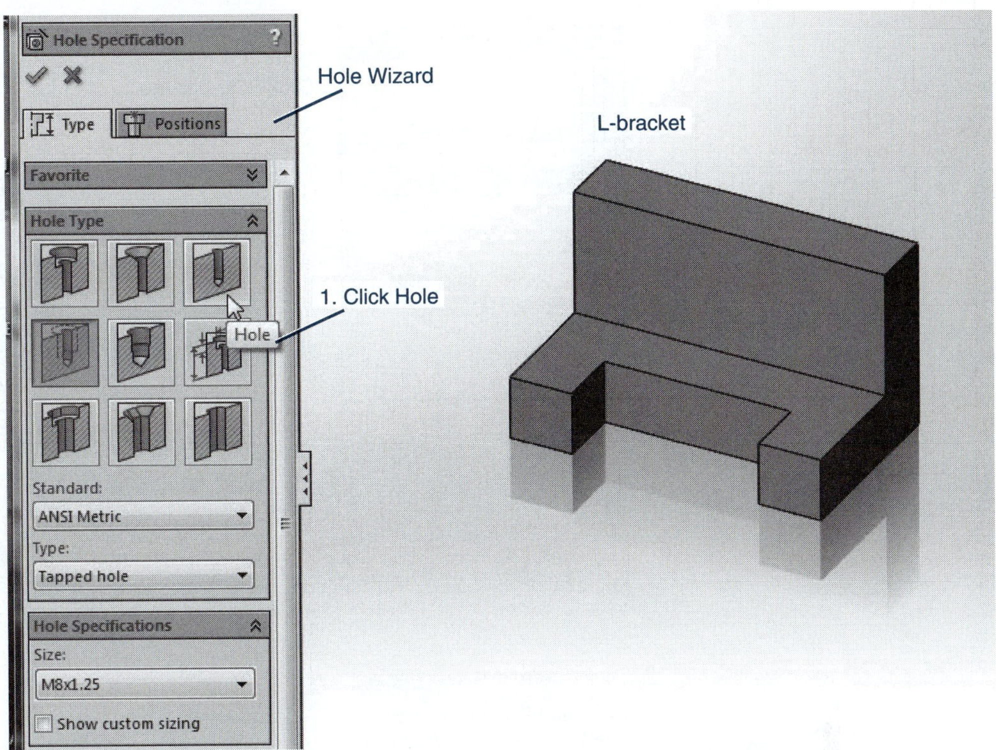
Hole Specification
Type
Positions
Hole Wizard
Favorite
Hole Type
Hole
1. Click Hole
Standard:
ANSI Metric
Type:
Tapped hole
Hole Specifications
Size:
M8x1.25
Show custom sizing
L-bracket

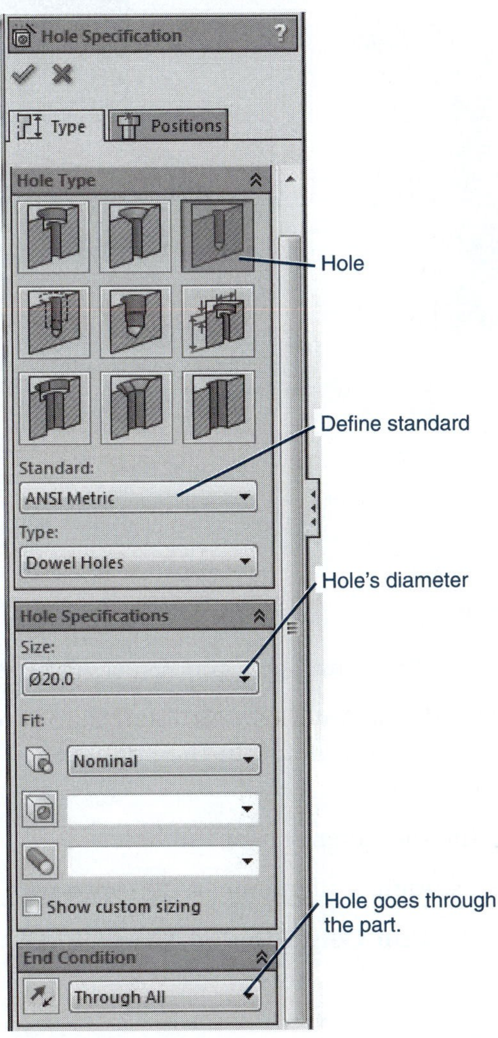
Hole Specification
Type
Positions
Hole Type
Hole
Define standard
Standard:
ANSI Metric
Type:
Dowel Holes
Hole's diameter
Hole Specifications
Size:
Ø20.0
Fit:
Nominal
Show custom sizing
Hole goes through the part.
End Condition
Through All

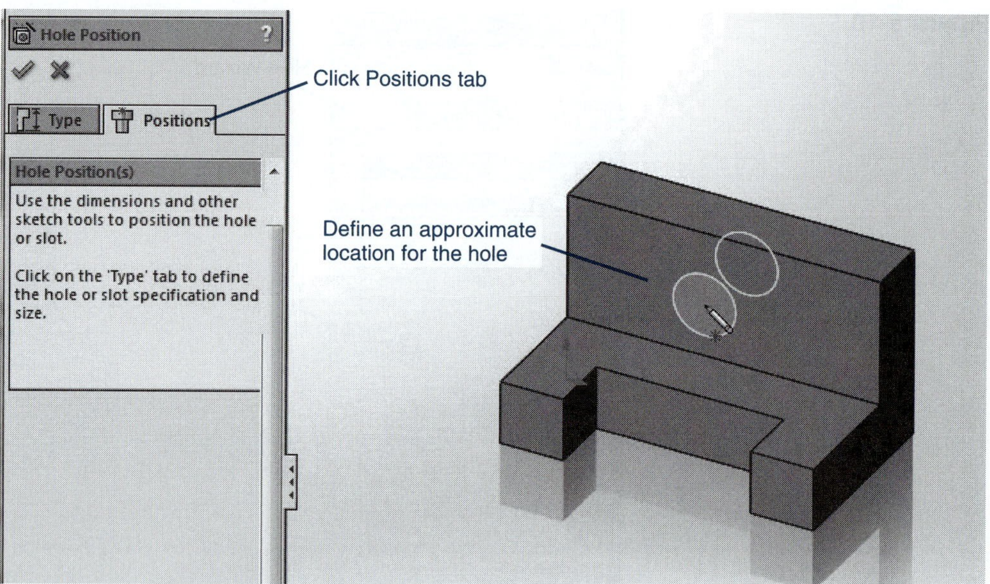

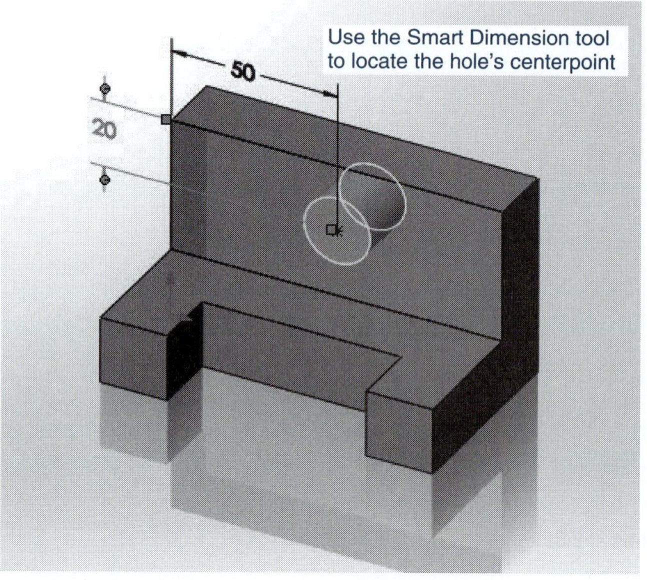

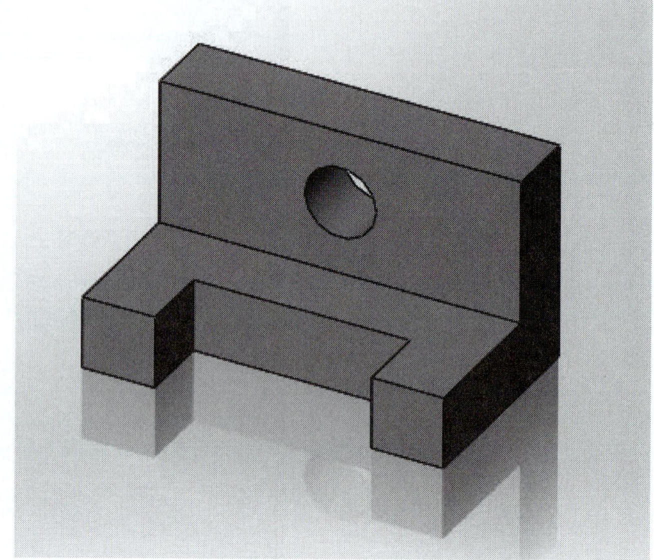

Figure 3-10
(*Continued*)

The **Hole** option will generate a clear hole.

3 Access the **Standard** box and select **ANSI Metric**.

ANSI stands for American National Standards Institute. ANSI standards will be covered in detail in the chapters on orthographic views, dimensions, and tolerances.

4 Define the diameter size of the hole.

In this example a diameter of **20.0mm** was selected.

5 Define the **End Condition** as **Through All**.

6 Click the **Positions** tab on the **Hole Wizard PropertyManager**.

7 Click an approximate location for the hole's centerpoint.

A preview of the hole will appear.

8 Use the **Smart Dimension** tool to locate the hole's centerpoint.

9 Click the green **OK** check mark and click the drawing screen.

The hole will be added to the L-bracket.

10 Save the L-bracket, as it will be used in later sections.

The hole created in Figure 3-10 is a ***through hole***, that is, it goes completely through the object. Holes that do not go completely through are called ***blind holes***. Note that the **Hole Wizard PropertyManager** shows a conical point at the bottom of the hole. Holes created using an extruded cut circle (see Section 1-13) will not have this conical endpoint. Blind holes created using a drill should include the conical point. For this reason, blind holes should, with a few exceptions, be created using the **Hole Wizard** tool.

3-6 A Second Method of Creating a Hole

Holes may also be created using the **Circle** tool and then applying the **Extruded Cut** tool. This method should be used only for through holes. Figure 3-11 shows the L-bracket created in Section 3-3.

1 Right-click the top front plane as shown and click the **Sketch** tool.

2 Use the **Circle** tool and sketch a circle.

3 Use the **Smart Dimension** tool and locate and size the circle.

4 Click the **Features** tab and select the **Extruded Cut** tool. Click the circle if necessary.

5 Click the green **OK** check mark and click the drawing screen.

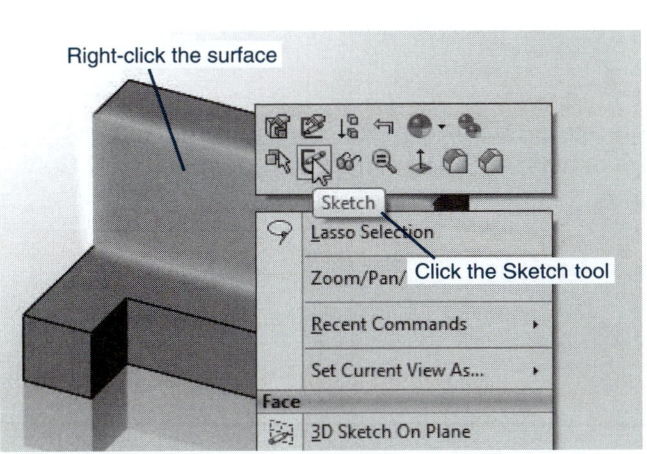

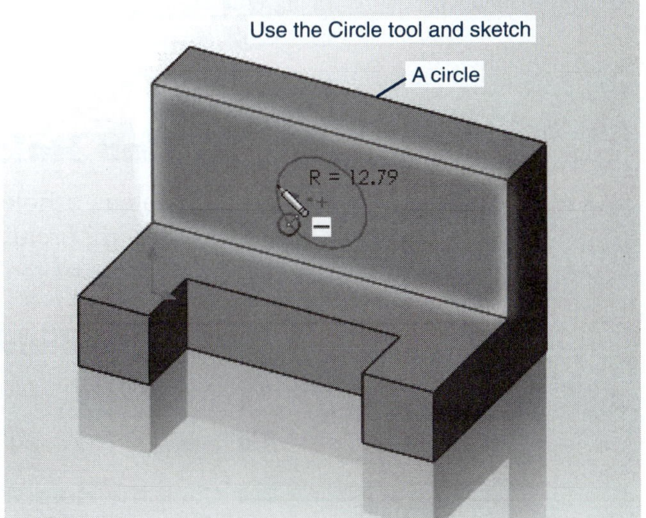

Figure 3-11

Figure 3-11
(*Continued*)

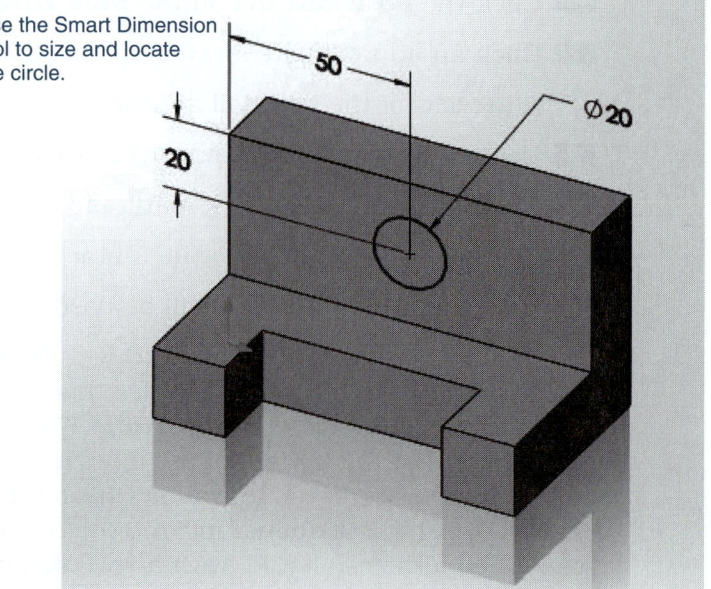
Use the Smart Dimension tool to size and locate the circle.
50
Ø20
20

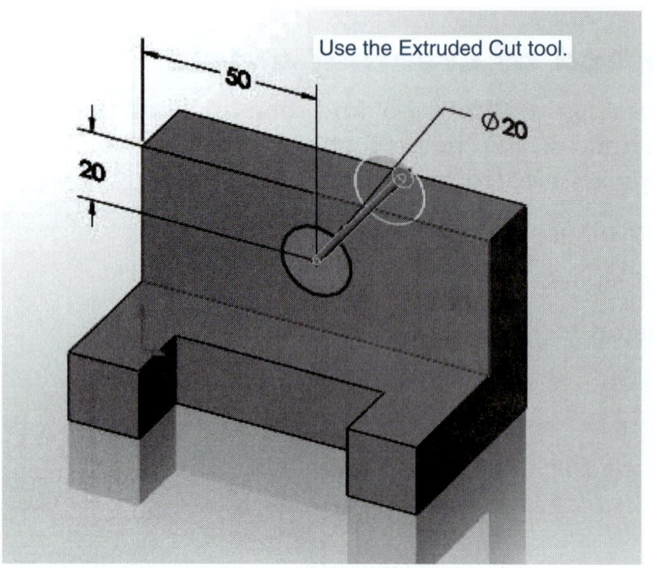
Use the Extruded Cut tool.
50
Ø20
20

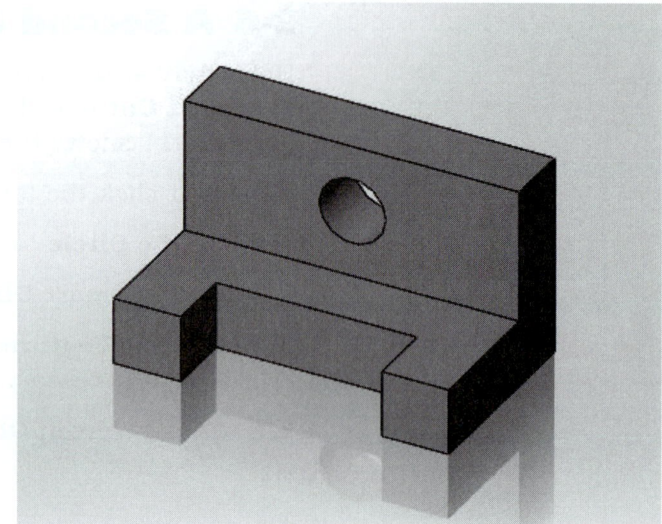

3-7 Blind Holes

A blind hole is a hole that does not go completely through a part. Most blind holes are created using a twist drill that includes a conical end tip. The **Hole Wizard** tool includes a conical point on blind holes.

To Create a Blind Hole – Inches

See Figure 3-12.

1. Draw a 1.00 × 1.50 × 1.25 box.
2. Click the **Hole Wizard** tool and click the **Hole** option.
3. Set the **Standard** for **ANSI Inch**, the **Hole size** for **1/2**, the **End Condition** for **Blind**, and the hole depth for **0.75**.

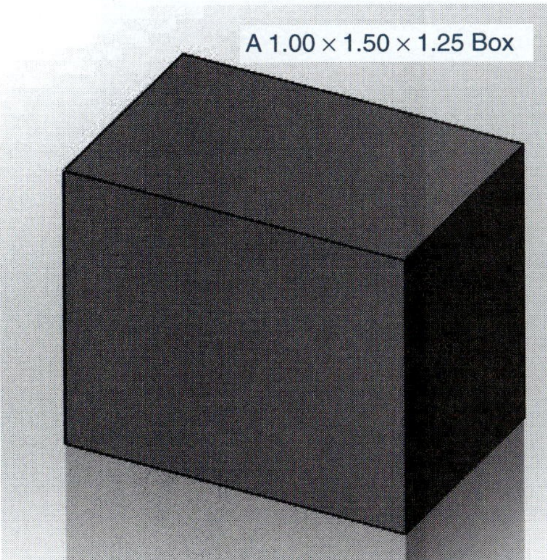
A 1.00 × 1.50 × 1.25 Box

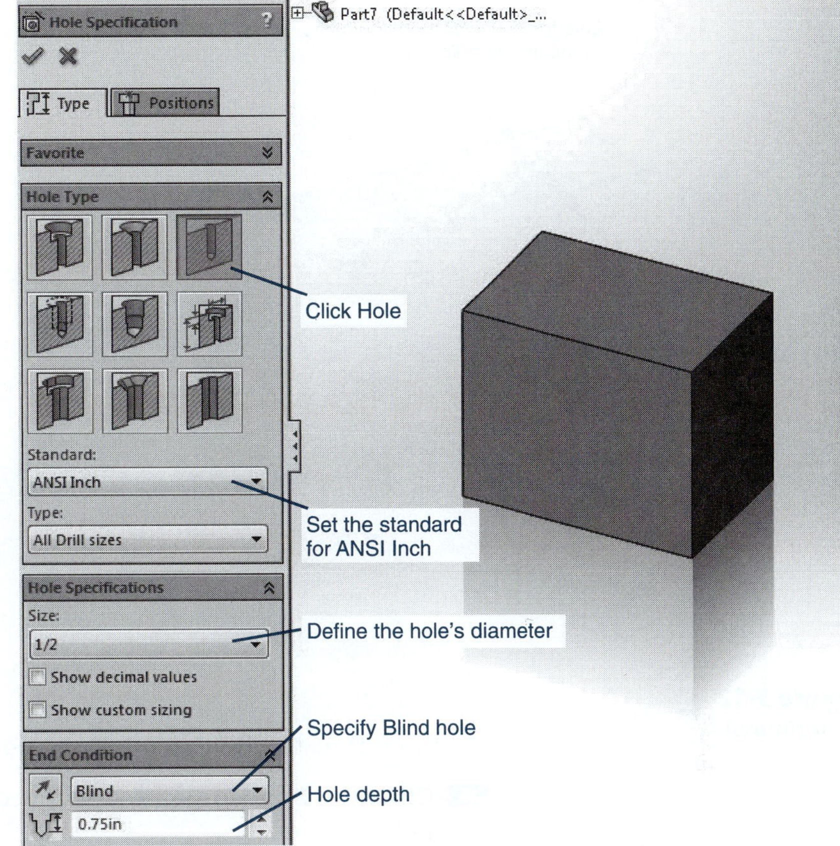
Hole Specification
Type
Positions
Favorite
Hole Type
Click Hole
Standard:
ANSI Inch
Set the standard for ANSI Inch
Type:
All Drill sizes
Hole Specifications
Size:
1/2
Define the hole's diameter
Show decimal values
Show custom sizing
Specify Blind hole
End Condition
Blind
Hole depth
0.75in
Part7 (Default<<Default>_...

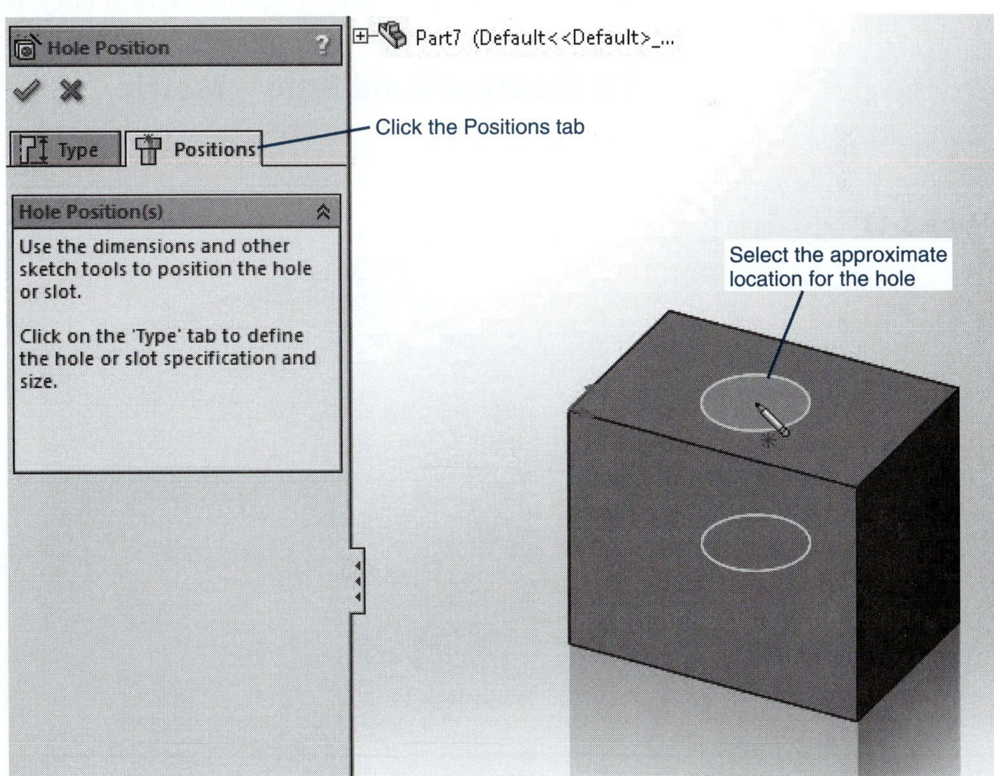
Hole Position
Type
Positions
Click the Positions tab
Hole Position(s)
Use the dimensions and other sketch tools to position the hole or slot.
Click on the 'Type' tab to define the hole or slot specification and size.
Part7 (Default<<Default>_...
Select the approximate location for the hole

Figure 3-12

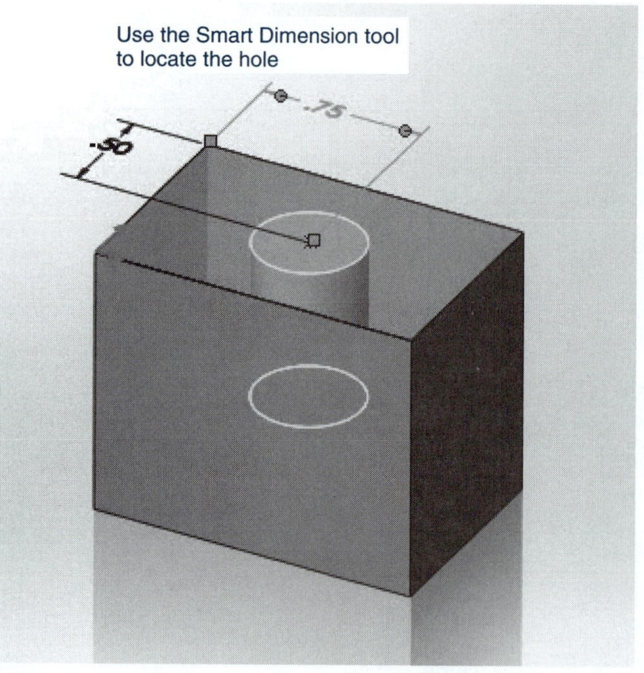

Figure 3-12
(*Continued*)

Note that the hole depth does not include the conical endpoint.

4. Click the **Positions** tab and select an approximate location for the hole.
5. Use the **Smart Dimension** tool to locate the hole.
6. Click the green **OK** check mark and click the drawing screen.

To Create a Blind Hole – Metric

See Figure 3-13.

Figure 3-13

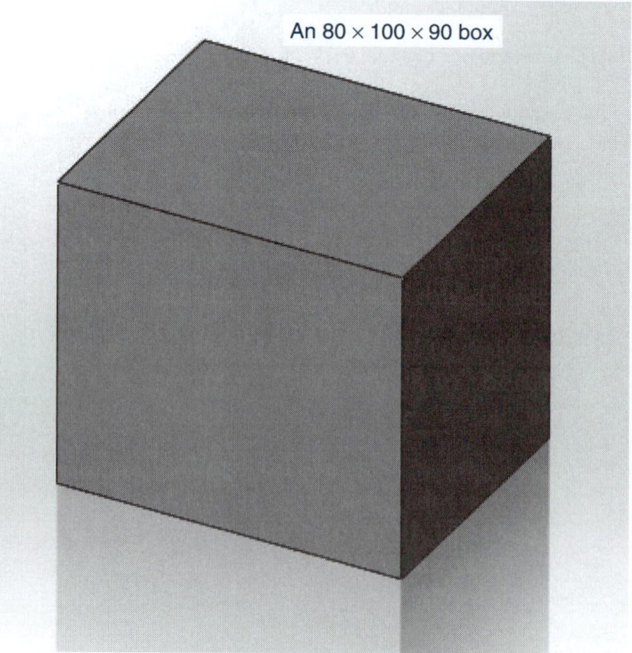

Figure 3-13
(*Continued*)

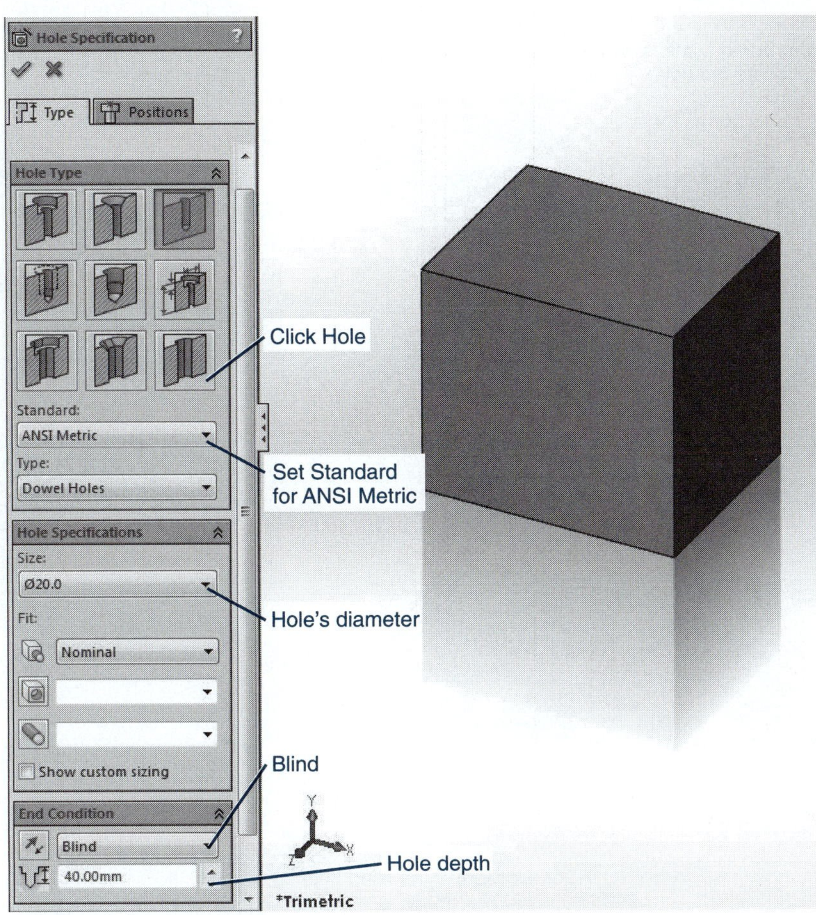

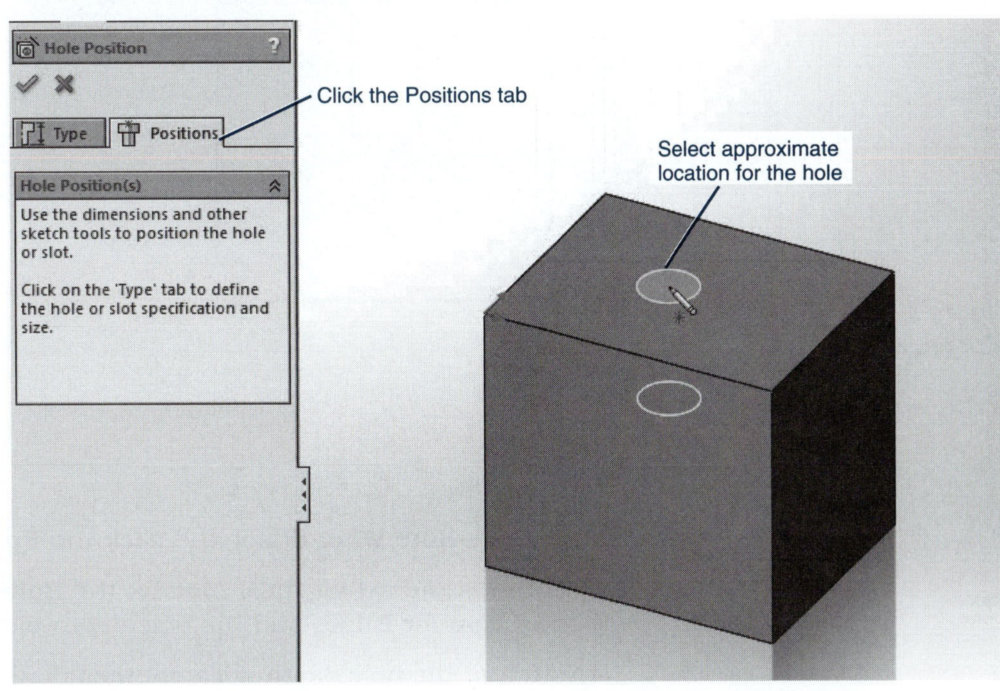

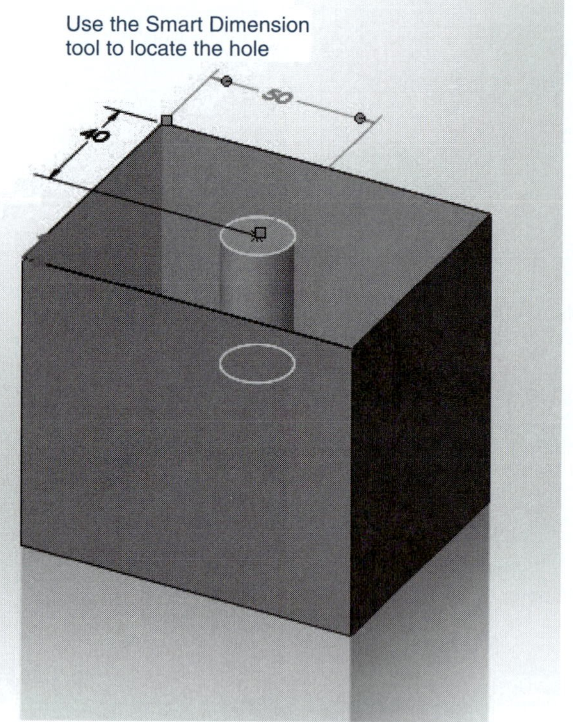

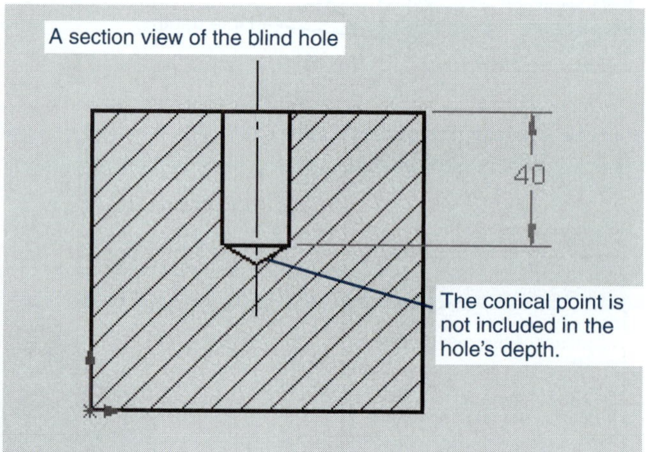

Figure 3-13
(*Continued*)

1. Draw a 80 × 100 × 90 box.
2. Click the **Hole Wizard** tool and click the **Hole** option.
3. Set the **Standard** for **ANSI Metric**, the **Hole size** for **Ø20.0**, the **End Condition** for **Blind**, and the hole depth for **40**.

   Note that the hole depth does not include the conical endpoint.
4. Click the **Positions** tab and select an approximate location for the hole.
5. Use the **Smart Dimension** tool to locate the hole.
6. Click the green **OK** check mark and click the drawing screen.

Figure 3-13 also shows a section view of the box. Note that the hole is blind and that it ends with a conical point. The specified hole depth does not include the conical point.

3-8 Fillet

A ***fillet*** is a rounded corner. Specifically, convex corners are called ***rounds***, and concave corners are called ***fillets***, but in general, all rounded corners are called ***fillets***.

SolidWorks can draw four types of fillets: constant radius, variable radius, face fillets, and full round fillets. Figure 3-14 shows the L-bracket used previously to demonstrate the **Features** tools. It will be used in this section to demonstrate **Fillet** tools.

Figure 3-14

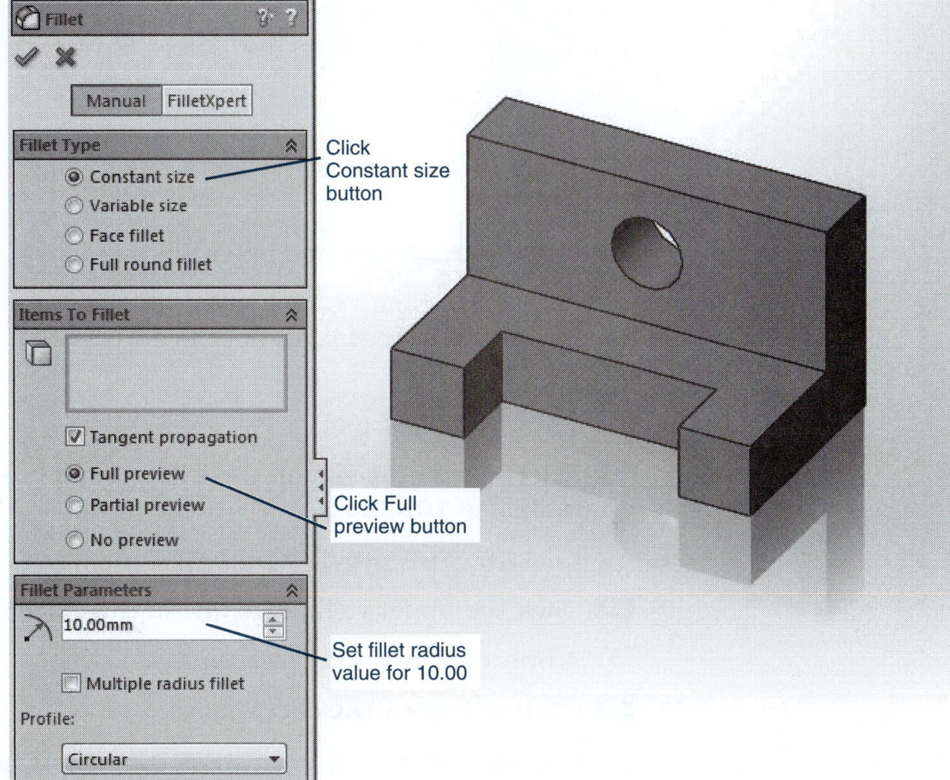

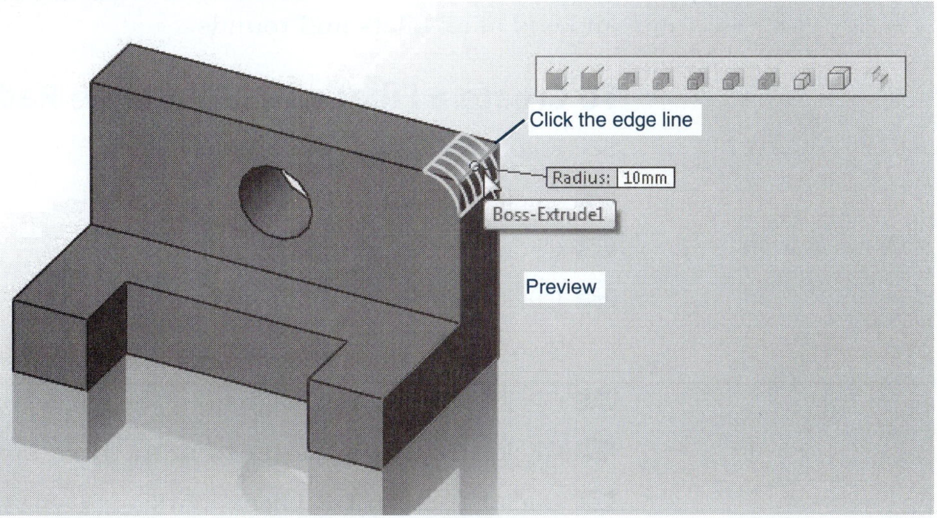

Figure 3-14
(*Continued*)

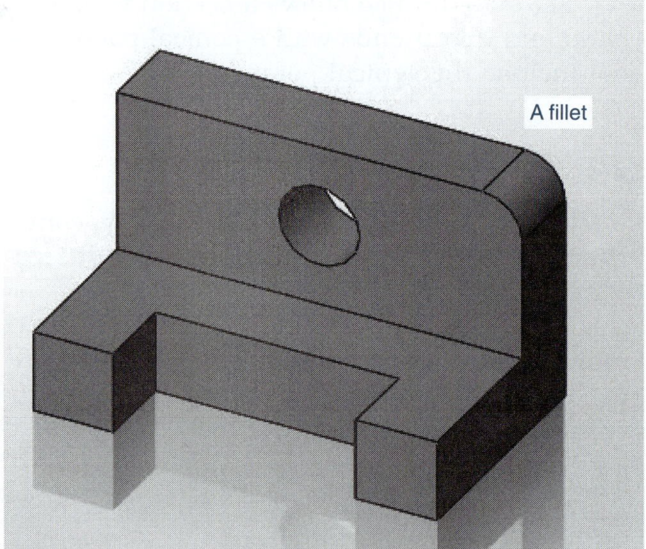

There is also a **2D Fillet** tool available on the **Sketch** tool panel. See Section 2-12.

1 Use the **Open** tool and open the L-bracket drawing.

If the L-bracket was not saved after the previous section, use the dimensions and procedures specified in Section 3-4 to re-create the bracket.

2 Click the **Fillet** tool.

3 Select **Constant size** in the **Fillet Type** box and define the fillet's radius as **10.00mm**.

Click the **Full preview** button.

4 Click the upper right edge line of the object.

A preview of the fillet will appear.

5 Click the green **OK** check mark.

Note that there is a line across the top surface of the part. This is a tangency line. SolidWorks includes tangency lines to show the beginning and end of all fillets and rounds.

To Create a Fillet with a Variable Radius

See Figure 3-15.

1 Click the **Fillet** tool.

2 Click the **Variable size** button.

3 Click the edge line shown in Figure 3-15.

Two boxes will appear on the screen, one at each end of the edge line.

4 Click the word **Unassigned** in the left box and enter a value of **15**.

5 Click the word **Unassigned** in the right box and enter a value of **5**.

6 Click the green **OK** check mark.

Figure 3-15

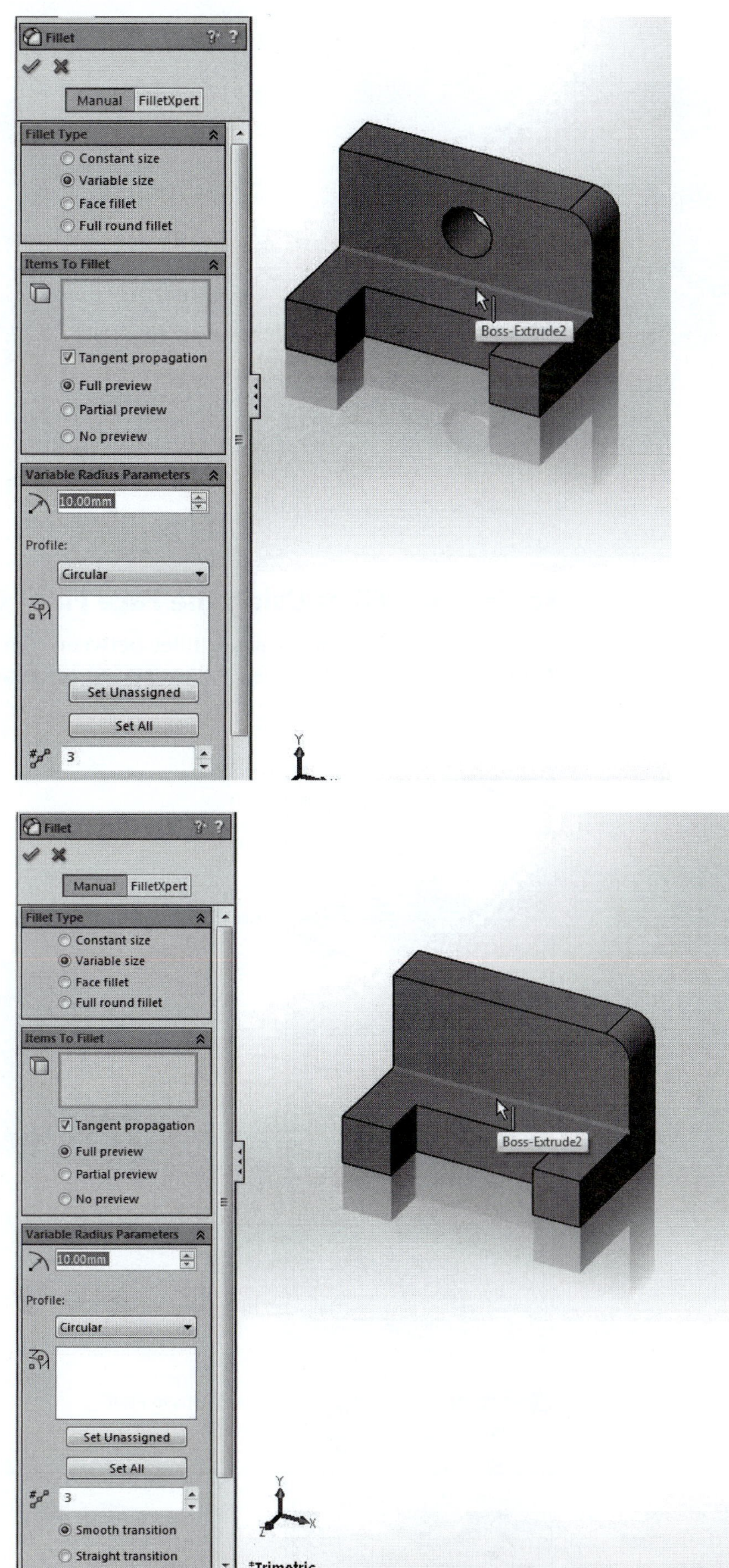

Fillet
Manual
FilletXpert
Fillet Type
Constant size
Variable size
Face fillet
Full round fillet
Items To Fillet
Tangent propagation
Full preview
Partial preview
No preview
Variable Radius Parameters
10.00mm
Profile:
Circular
Set Unassigned
Set All
3
Boss-Extrude2
Fillet
Manual
FilletXpert
Fillet Type
Constant size
Variable size
Face fillet
Full round fillet
Items To Fillet
Tangent propagation
Full preview
Partial preview
No preview
Variable Radius Parameters
10.00mm
Profile:
Circular
Set Unassigned
Set All
3
Smooth transition
Straight transition
Boss-Extrude2
*Trimetric

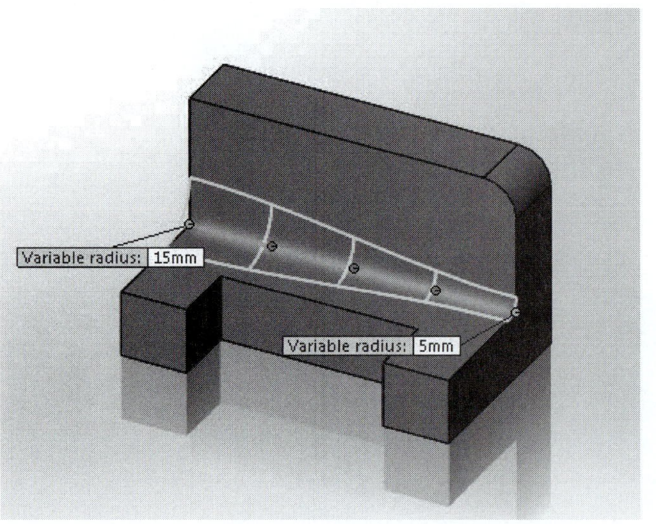

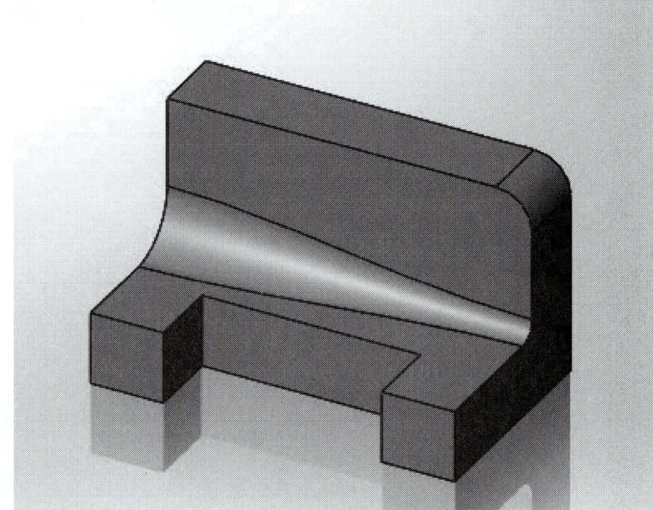

Figure 3-15
(*Continued*)

To Create a Fillet Using the Face Fillet Option

The **Face fillet** option draws a fillet between two faces (surfaces), whereas **Fillet** uses an edge between two surfaces to draw a fillet.

See Figure 3-16.

Figure 3-16

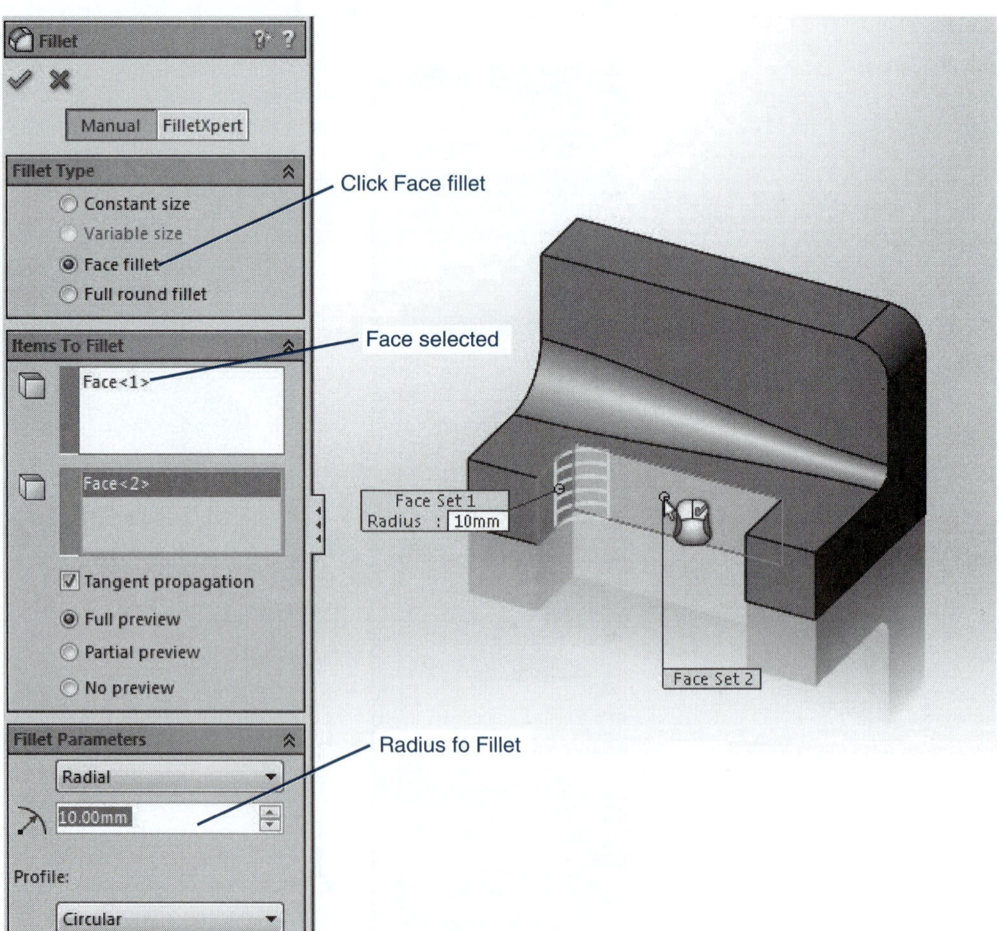

Figure 3-16
(Continued)

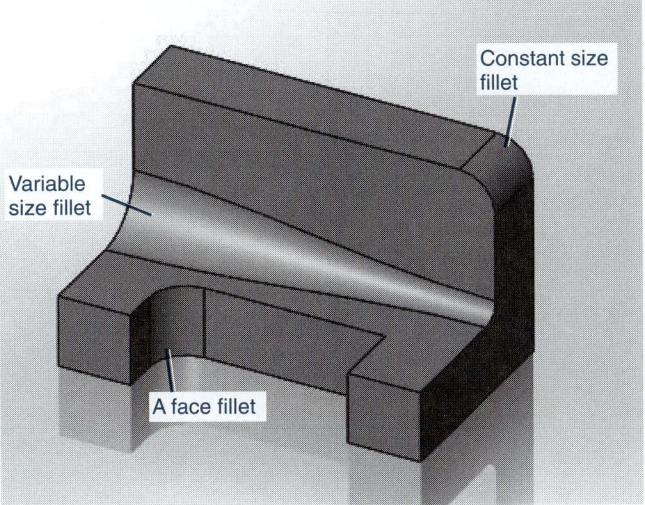

1. Click the **Fillet** tool.
2. Click the **Face fillet** option.

Two boxes will appear in the **Items To Fillet** box. They will be used to define the two faces of the fillet. The blue box is the active box and is awaiting an input.

3. Define the fillet radius as **10.00mm**.
4. Define **Face 1** as shown.

   The label **Face<1>** will appear in the blue box.

5. Click the lower box in the **Items To Fillet** box (it will turn blue) and define **Face 2** by clicking the surface as shown.
6. Click the green **OK** check mark and click the drawing screen.

To Create a Fillet Using the Full Round Fillet Option

See Figure 3-17.

1. Use the **Undo** tool and remove the fillets created previously.

   This will return the original L-bracket shape.

2. Click the **Fillet** tool and click the **Full round fillet** button.

Three boxes will appear. These boxes will be used to define Side Face 1, the Center Face, and Side Face 2. The top box is blue, indicating it is awaiting an input.

In this example, the default fillet radius value of **10** will be used.

> **TIP**
> Objects can be rotated by holding down the mouse wheel and moving the cursor.

3. Rotate the object so that the back surface can be selected.
4. Click the back surface.

Figure 3-17

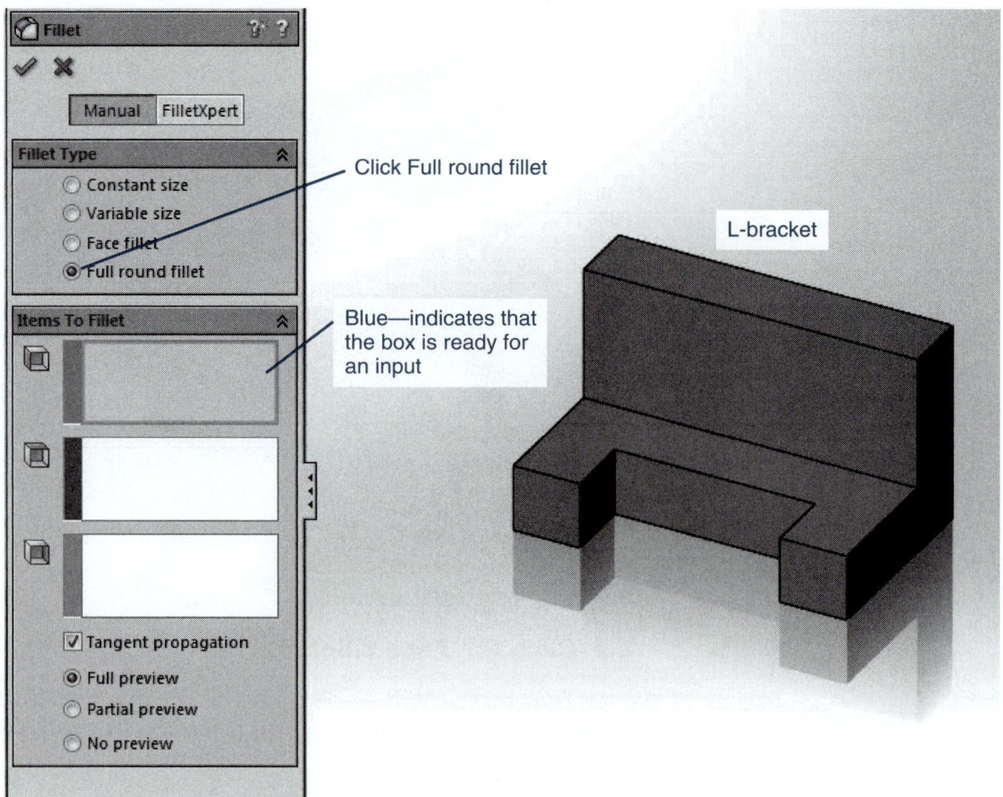
Fillet
Manual
FilletXpert
Fillet Type
Constant size
Variable size
Face fillet
Full round fillet
Items To Fillet
Tangent propagation
Full preview
Partial preview
No preview
Click Full round fillet
L-bracket
Blue—indicates that the box is ready for an input

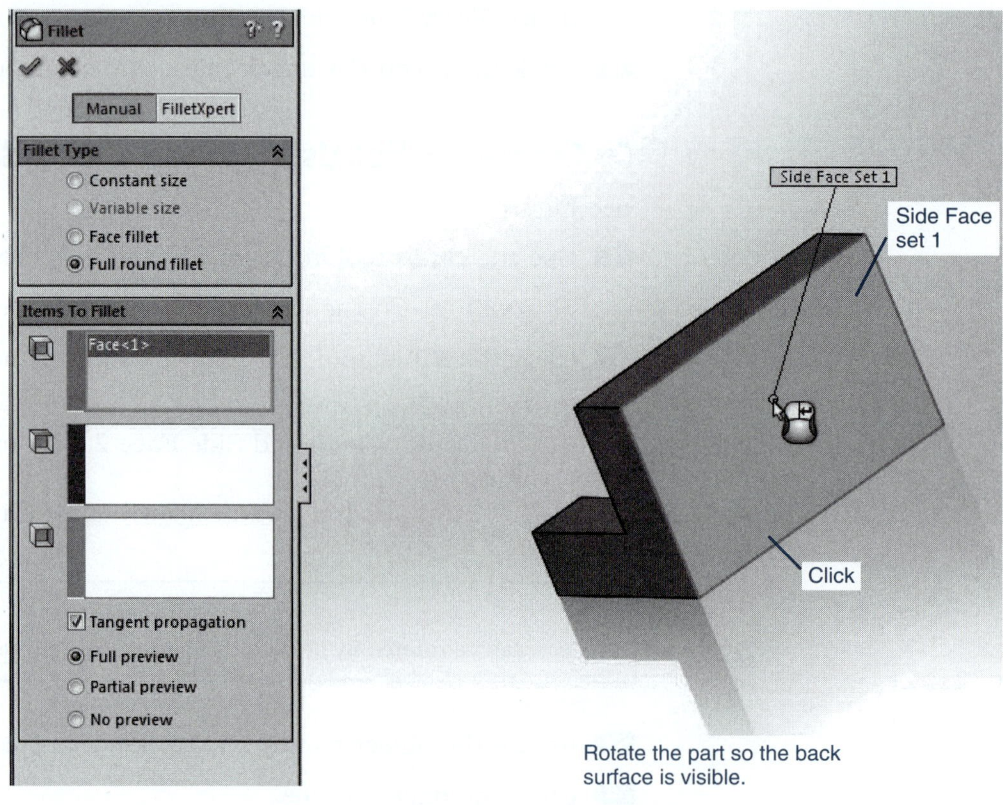
Fillet
Manual
FilletXpert
Fillet Type
Constant size
Variable size
Face fillet
Full round fillet
Items To Fillet
Face<1>
Tangent propagation
Full preview
Partial preview
No preview
Side Face Set 1
Side Face set 1
Click
Rotate the part so the back surface is visible.

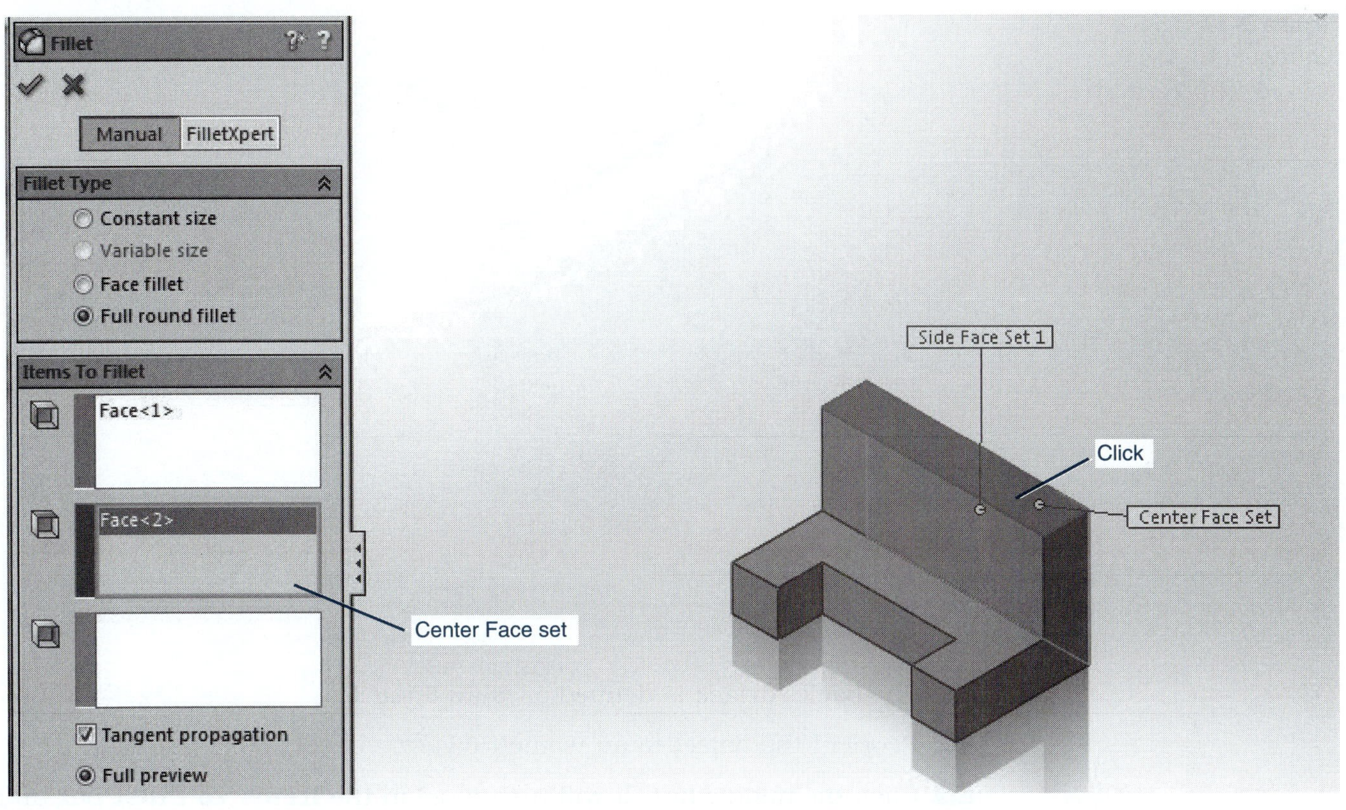

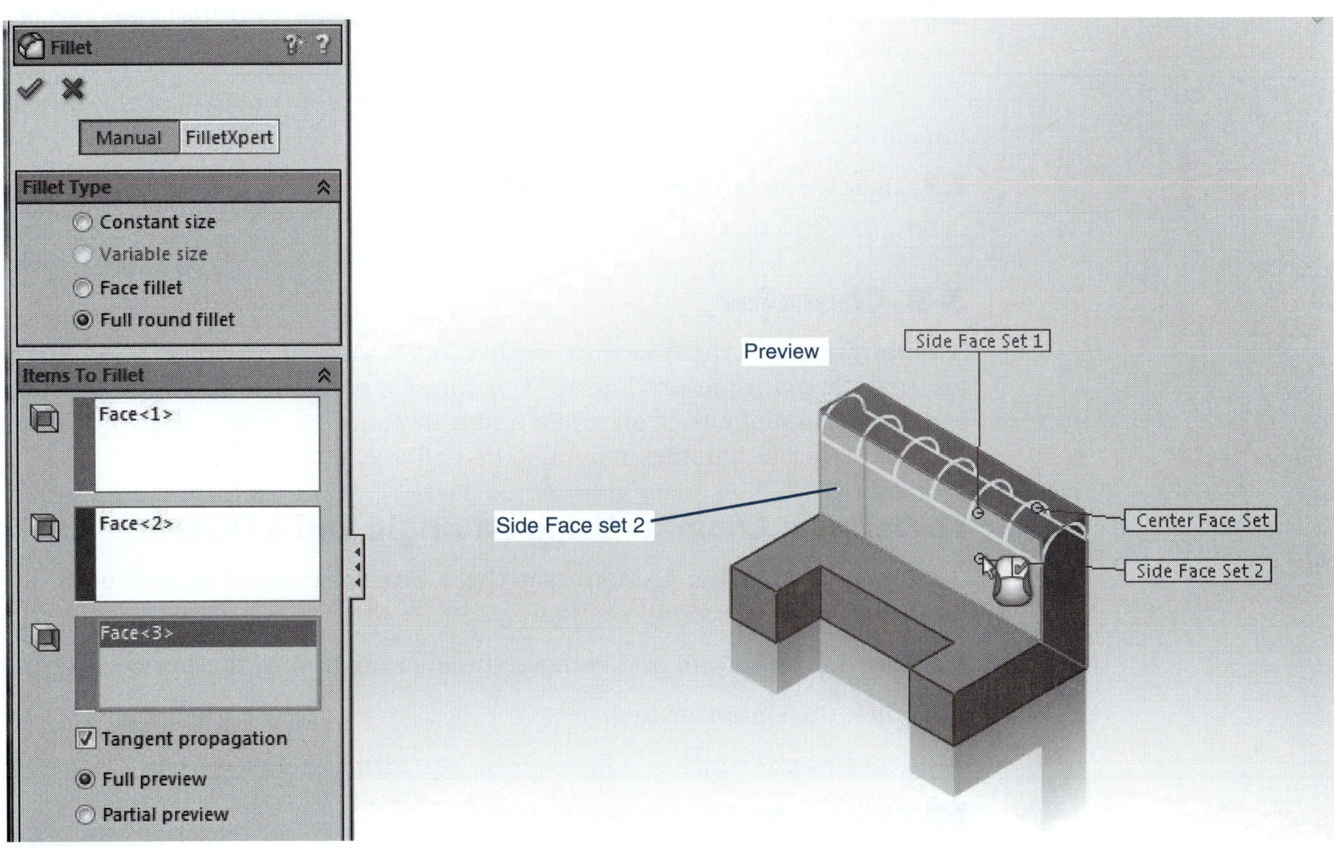

Figure 3-17
(*Continued*)

Figure 3-17
(*Continued*)

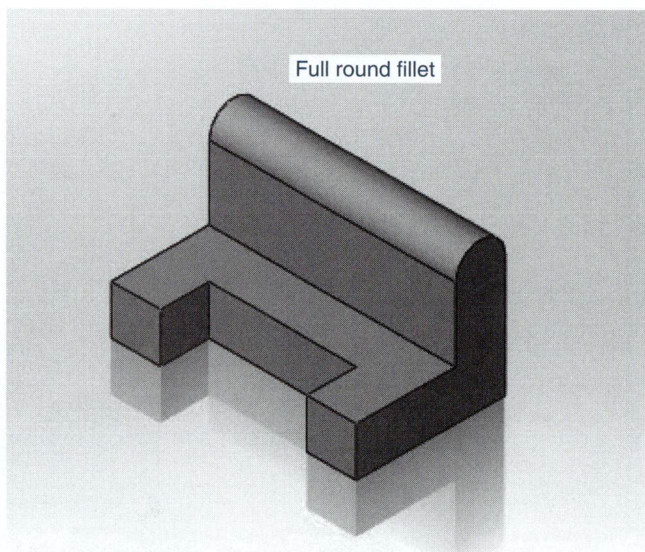

The back surface is defined as **Side Face 1**.

5 Reorient the object to an isometric view.

6 Click the middle box (it will turn blue) in the **Items To Fillet** box and click the top surface of the object.

The top surface is defined as the **Center Face**.

7 Click the lower of the three boxes in the **Items To Fillet** box and click the front surface of the object as shown.

The front surface is defined as **Side Face 2**. A preview of the fillet will appear.

8 Click the green **OK** check mark.

3-9 Chamfer

A ***chamfer*** is a slanted surface added to a corner of an object. Chamfers are usually manufactured at 45° but may be made at any angle. Chamfers are defined using either an angle and a distance (5 × 45°) or two distances (5 × 5). A vertex chamfer may also be defined.

To Define a Chamfer Using an Angle and a Distance

See Figure 3-18. This section uses the L-bracket created in Section 3-3 and used in the previous section.

1 Use the **Undo** tool and remove the fillet created in the previous section.

2 Click the **Chamfer** tool.

The **Chamfer** tool is a flyout from the **Fillet** tool.

3 Click the **Angle distance** button.

4 Define the chamfer distance as **5** and accept the **45°** default value.

5. Click the top side edge line as shown.

6. Click the green **OK** check mark.

Figure 3-18

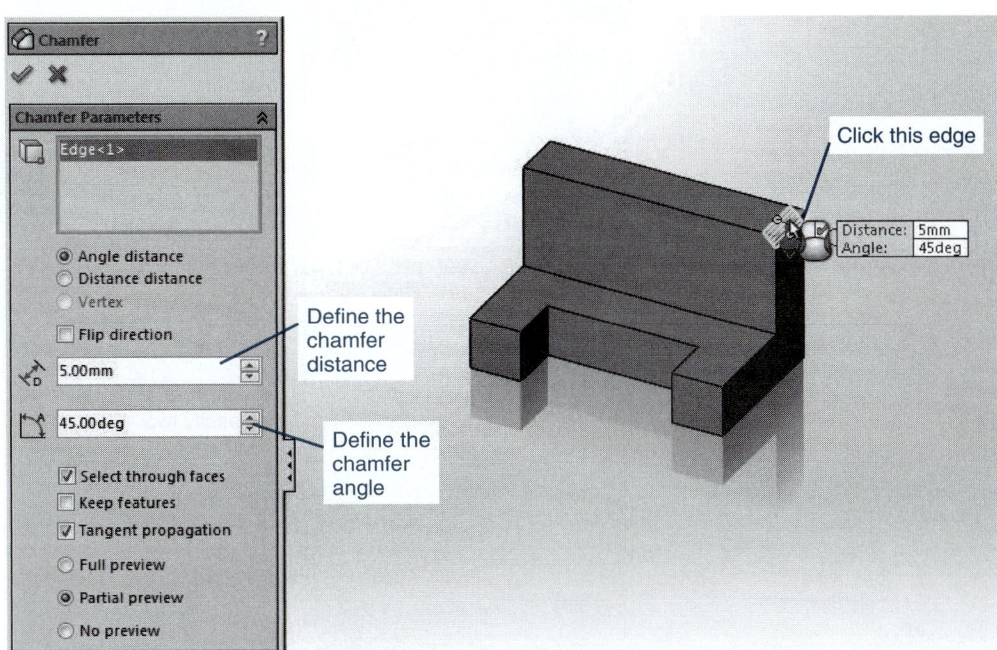

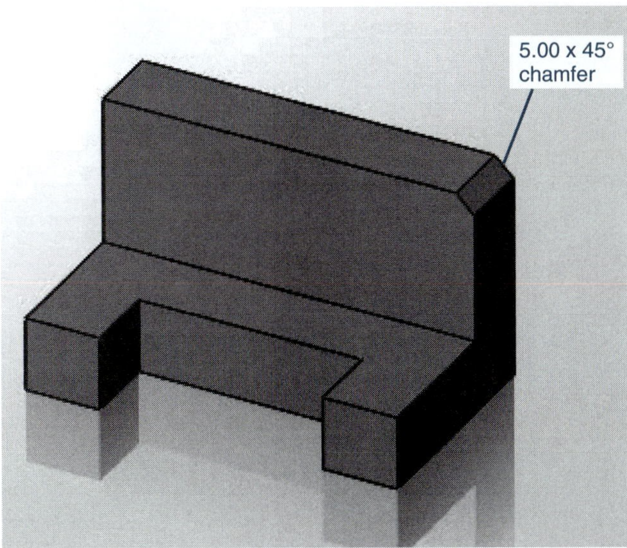

To Define a Chamfer Using Two Distances

See Figure 3-19.

1. Click the **Chamfer** tool.

2. Click the **Distance distance** button.

3. Define the two distances as **5.00mm** each.

In this example the two distances are equal. Distances of different lengths may be used.

Figure 3-19

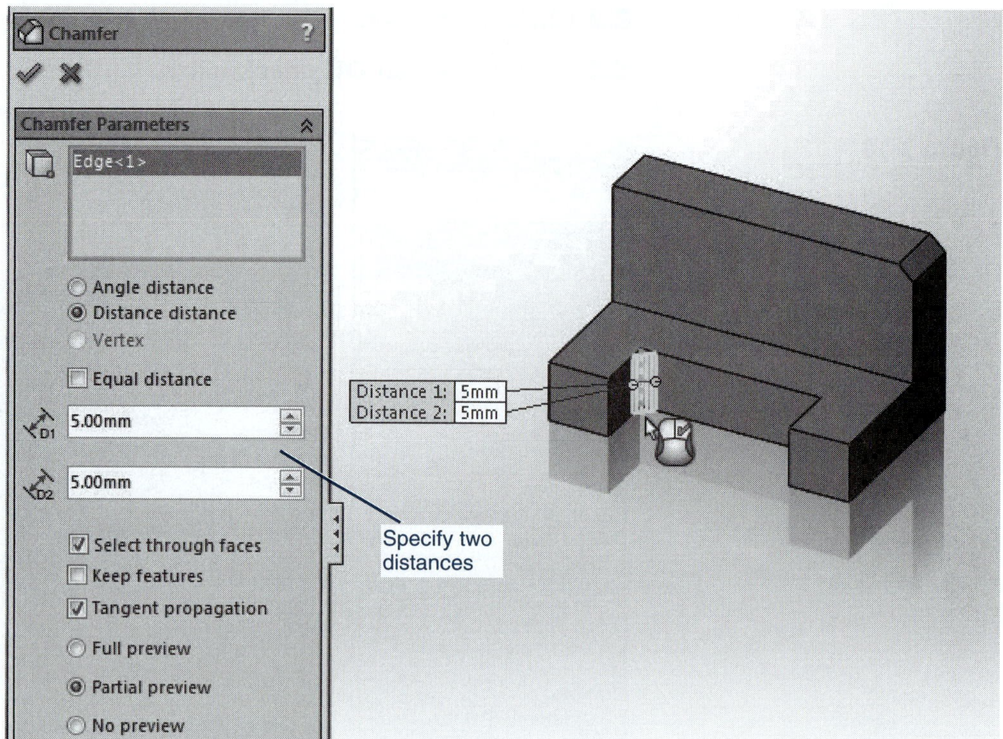

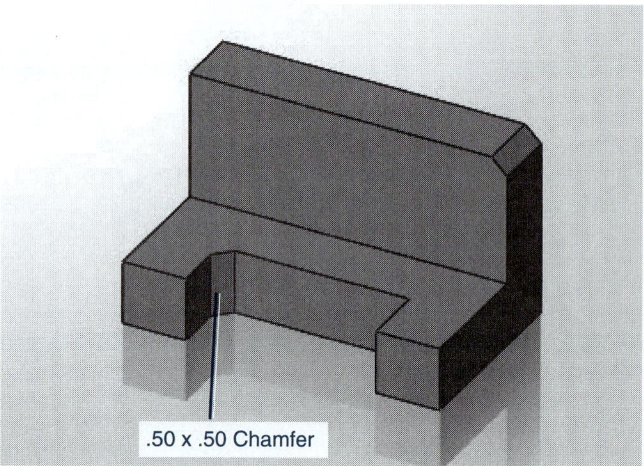

4 Click the inside vertical line as shown.

5 Click the green **OK** check mark.

The two distances need not be equal.

To Define a Vertex Chamfer

See Figure 3-20.

1 Click the **Chamfer** tool.

2 Click the **Vertex** button.

Three distance boxes will appear.

3 Define the three distances.

Figure 3-20

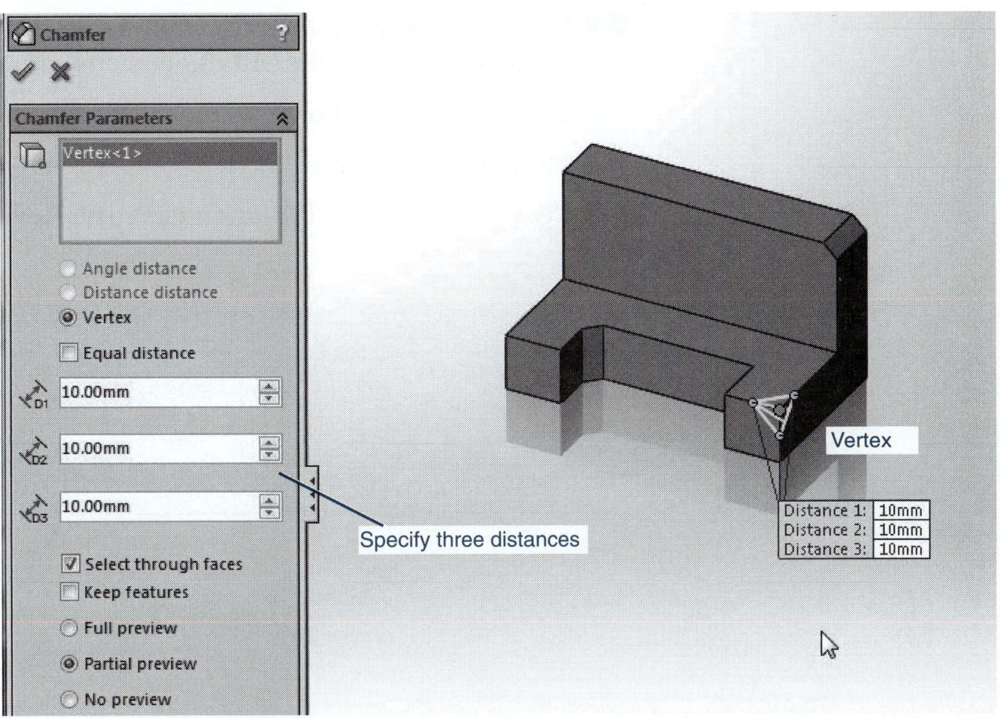

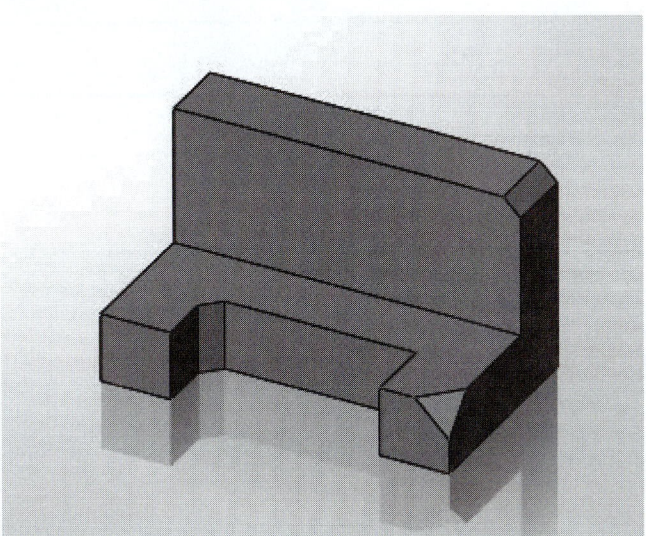

In this example three equal distances of **10.00mm** were used. The three distances need not be equal.

4 Click the lower top corner point as shown.

5 Click the green **OK** check mark.

3-10 Revolved Boss/Base

The **Revolved Boss/Base** tool rotates a contour about an axis line. See Figure 3-21.

1 Start a new drawing, click the **Sketch** tool, and click the **Top Plane** option.

2. Use the **Sketch** tools to draw a line on the screen and then draw a 2D shape next to the line. All dimensions are in inches.

   Draw the shape shown using the given dimensions.

3. All dimensions are in inches.

4. Click the **Features** tab, then click the **Revolved Boss/Base** tool.

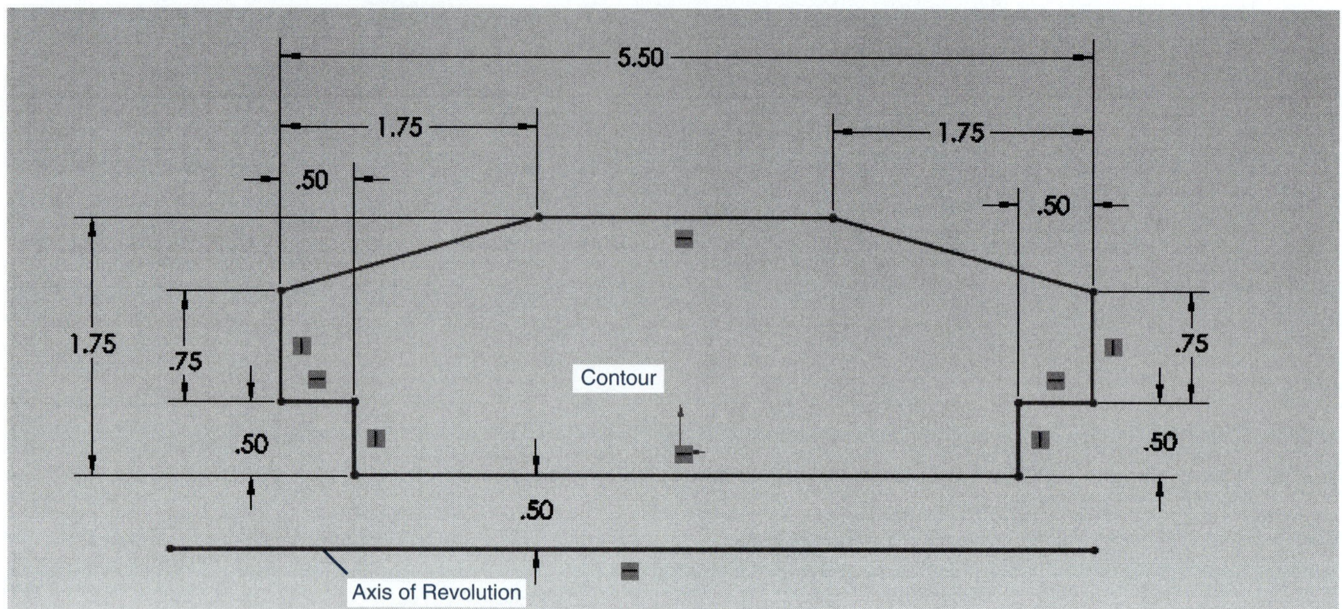

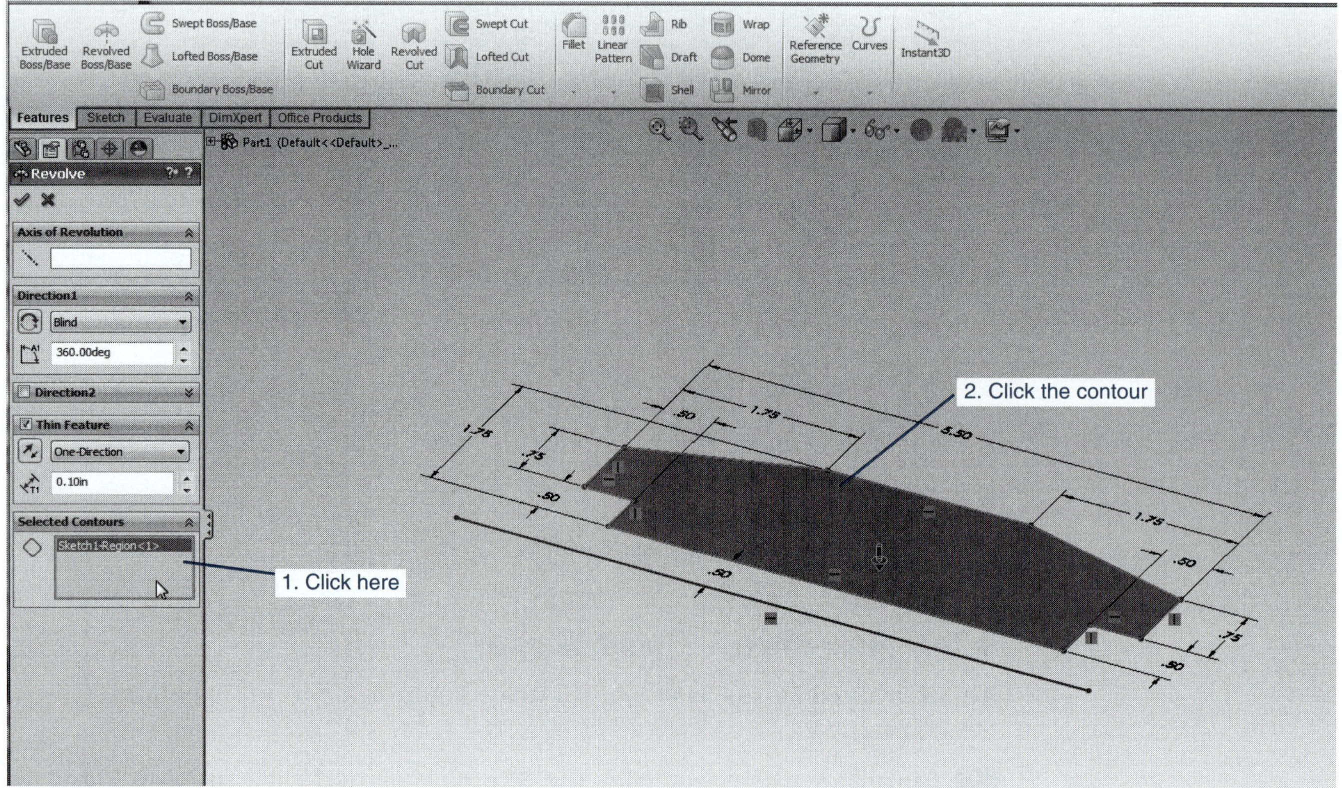

Figure 3-21

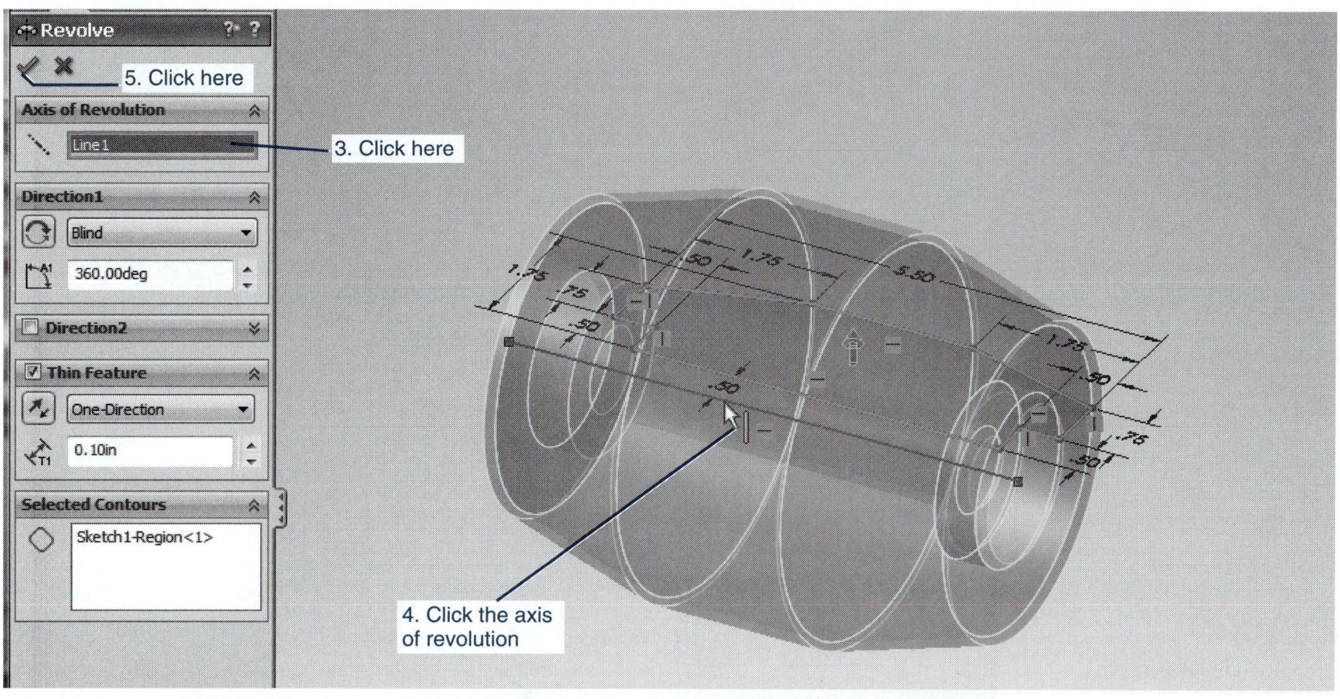

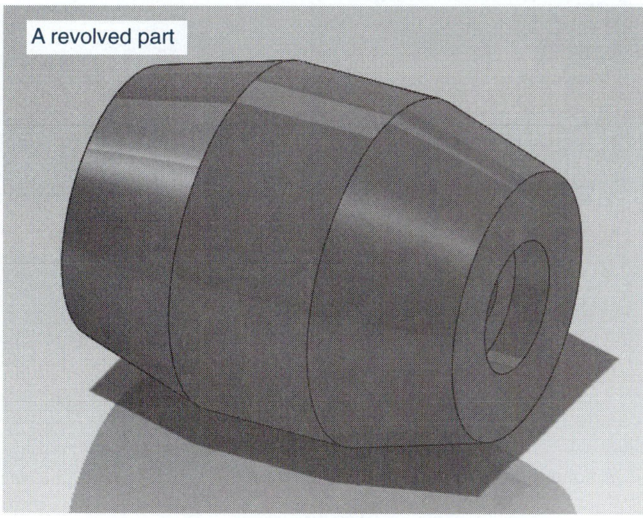

figure 3-21
(*continued*)

5 Click the **Selected Contours** box, then click the contour on the screen.

6 Click the axis box at the top of the **Revolve PropertyManager**.

> **NOTE**
> If the **Selected Contours** box is already shaded, it means that it has been activated automatically. Click the contour directly.

7 Click the axis line on the screen.

A preview of the revolved object will appear.

8 Click the green **OK** check mark.

Figure 3-22 shows an example of a sphere created using the **Revolve** tool. A **Centerline** was used as the axis of revolution.

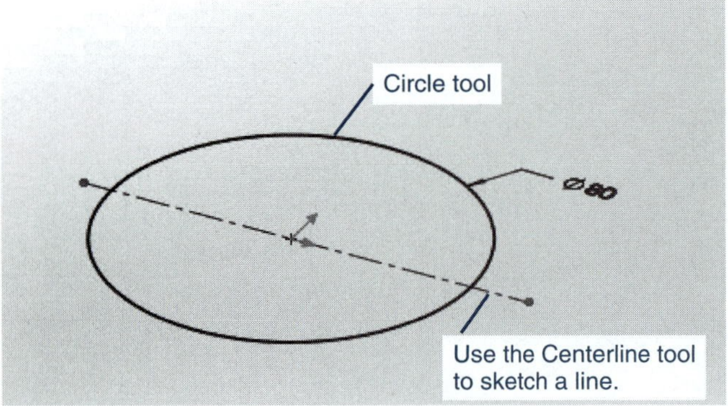

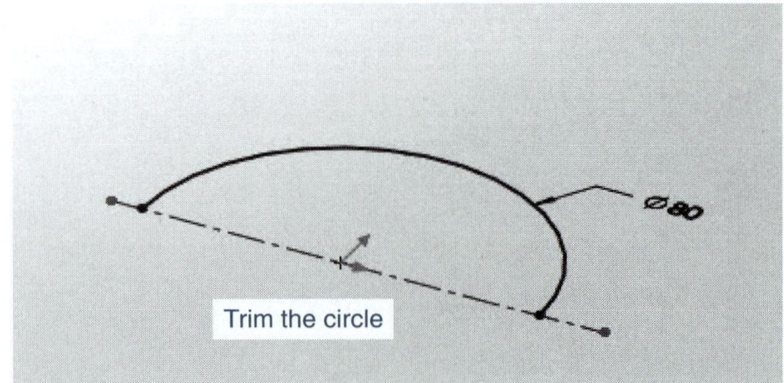

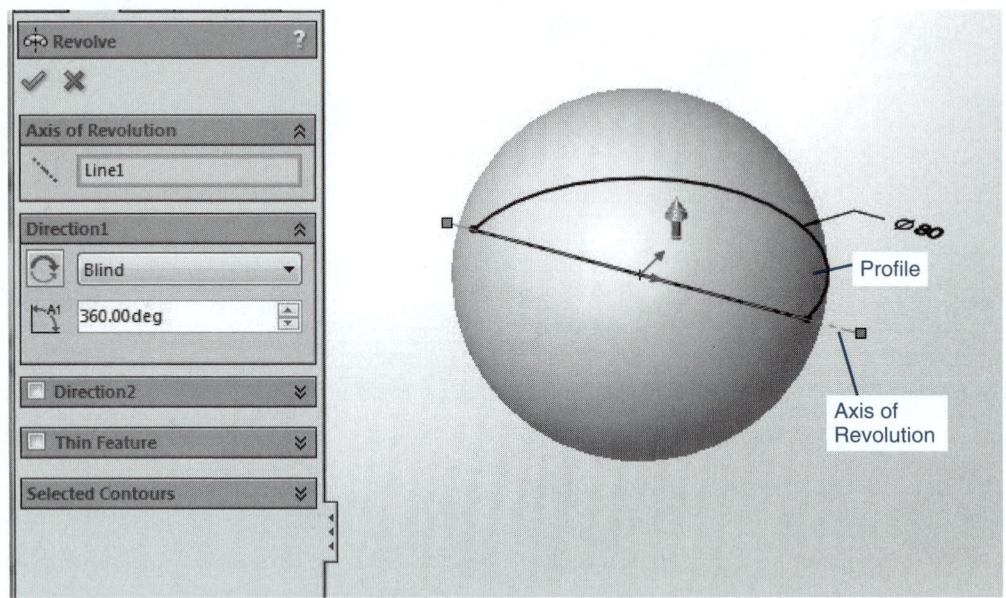

A sphere created using the Revolve tool.

Figure 3-22

3-11 Revolved Cut

The **Revolved Cut** tool is used to cut revolved sections out of objects. Figure 3-23 shows a 40 × 60 × 100 box. All dimensions are in millimeters.

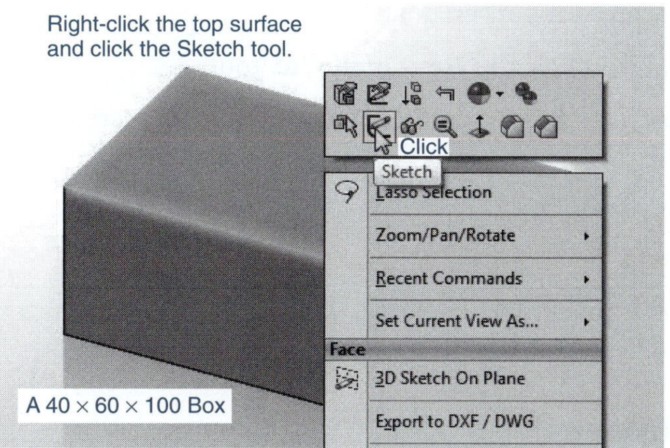

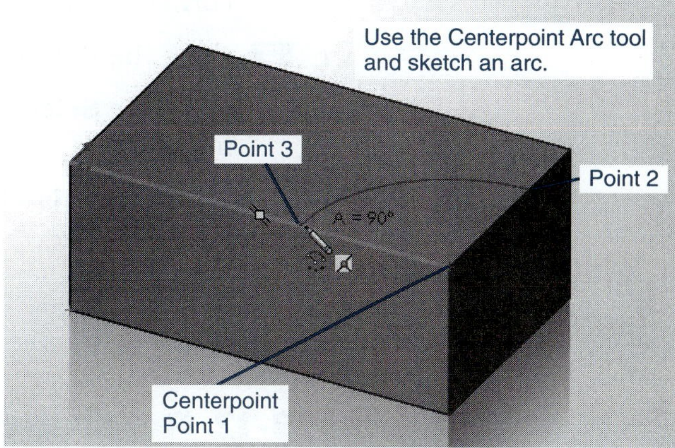

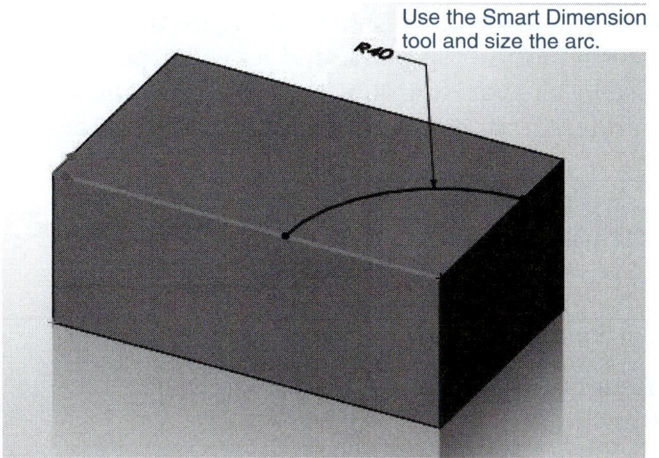

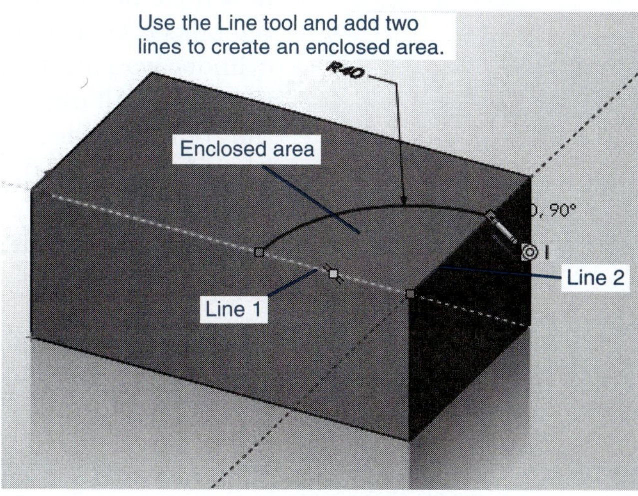

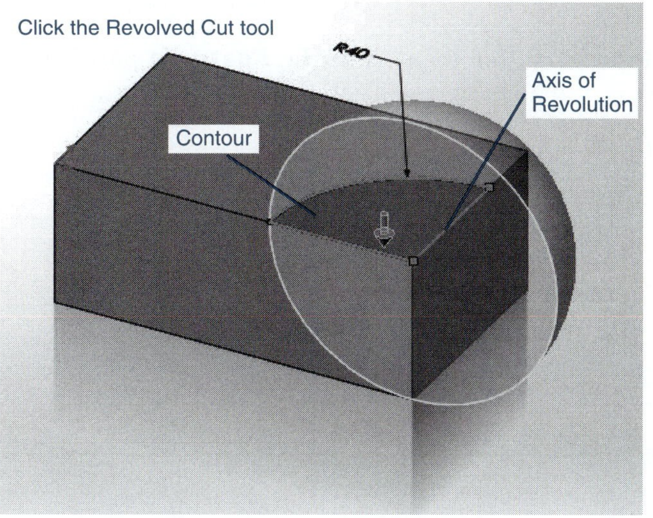

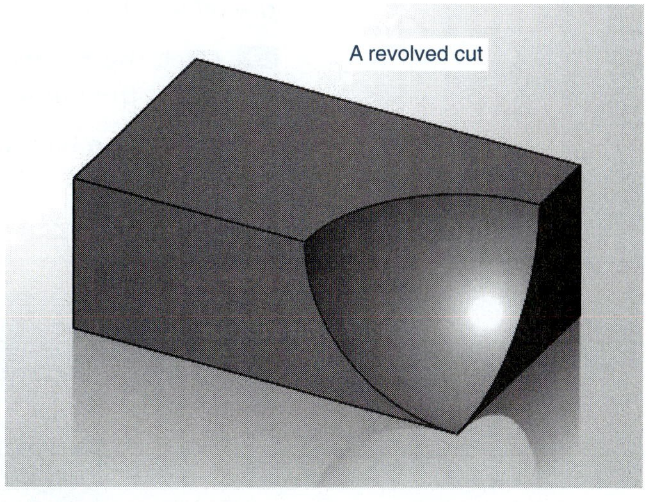

Figure 3-23

1. Create a new sketch plane (**Sketch**) on the top surface of the box and use the **Centerpoint Arc** tool to draw an arc of radius **40** centered about the lower corner of the top surface as shown.
2. Use the **Line** tool to draw two lines from this arc's endpoints to the corner point of the box, creating an enclosed area.
3. Click the **Features** tab and click the **Revolved Cut** tool.
4. Click the **Selected Contours** box and click the enclosed area created in Step 2.
5. Click the **Axis of Revolution** box.
6. Click the right edge line of the contour.
7. Click the green **OK** check mark and click the drawing screen.

3-12 Reference Planes

Reference planes are planes that are not part of an existing object. Until now if we needed a new sketch plane, we selected an existing surface on the object. Consider the Ø3.0 × 3.50 cylinder shown in Figure 3-24. The cylinder was drawn with its base on the top plane and its centerpoint at the origin. All dimensions are in inches. How do we create a hole through the rounded sides of the cylinder? If we right-click the rounded surface, no **Sketch** tool will appear.

To Create a Reference Plane

See Figure 3-24.

1. Click **Right Plane** in the **FeatureManager**.
2. Right-click the **Right Plane** feature and click **Show** to ensure that the right plane is visible. (It will probably already be on.)

The Right plane is located at the cylinder's origin. The **Reference Geometry** tool is used to create a second plane parallel to the Right plane. We can offset this plane from the Right plane and create a sketch plane on the offset plane.

3. Click the **Reference Geometry** tool located in the **Features** panel and select the **Plane** option.

The **Plane** box will appear.

Figure 3-24

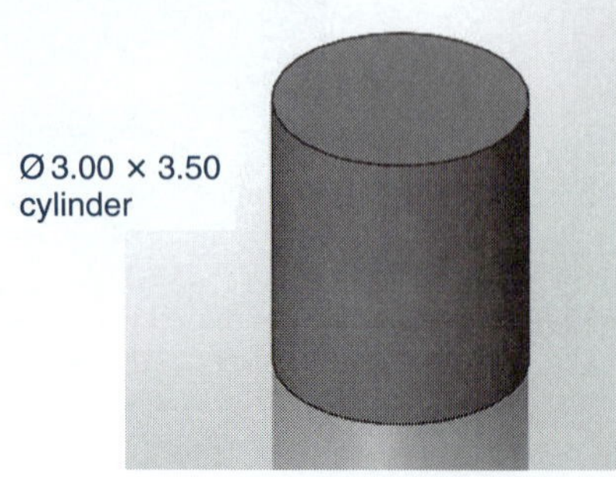

Figure 3-24
(Continued)

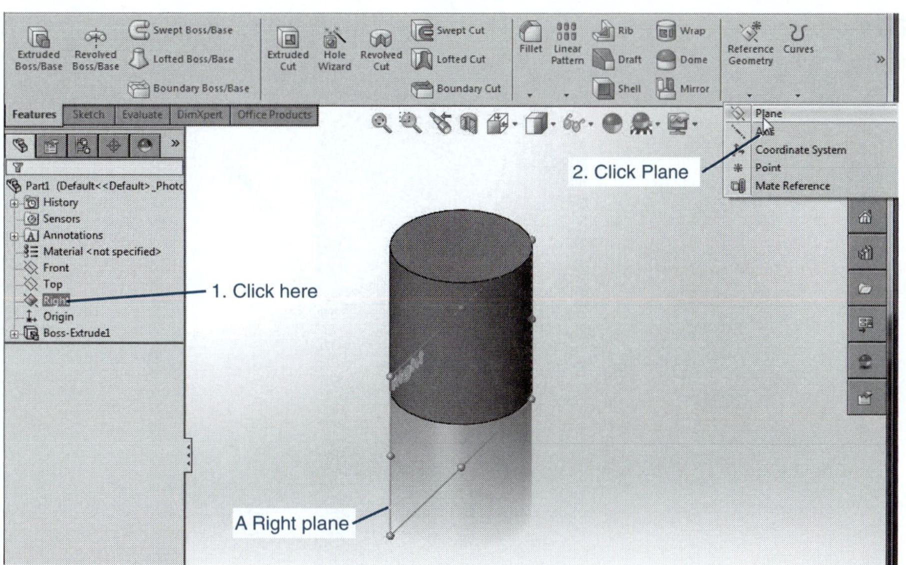
1. Click here
2. Click Plane
A Right plane

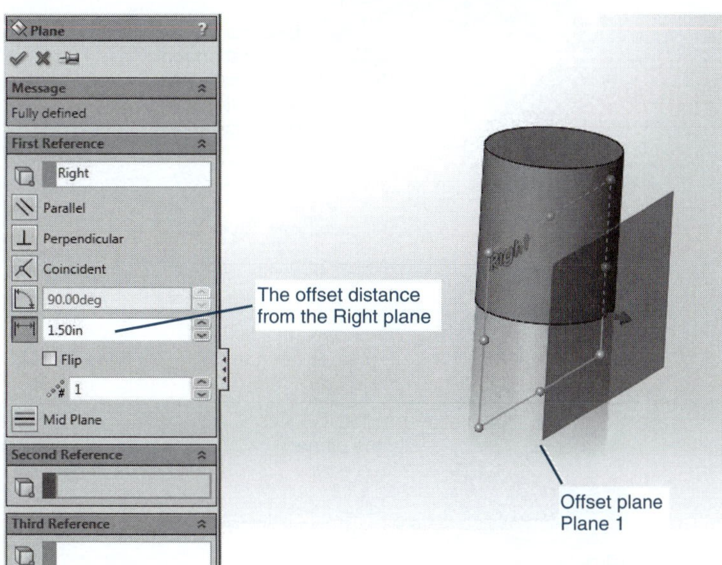
The offset distance from the Right plane
Offset plane Plane 1

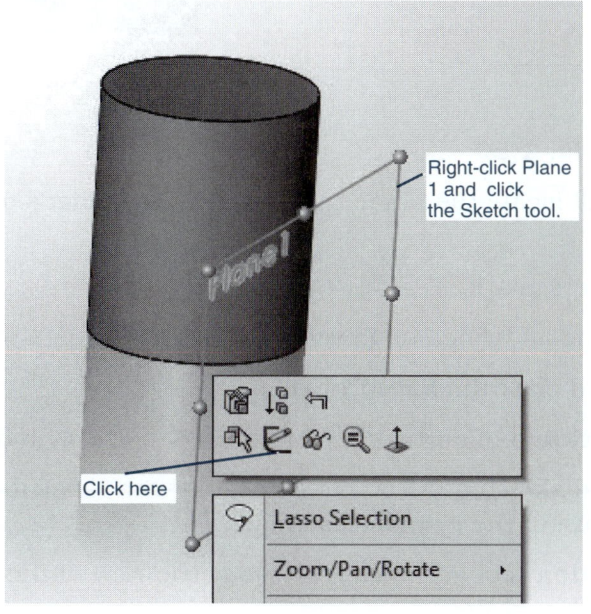
Right-click Plane 1 and click the Sketch tool.
Click here
Lasso Selection
Zoom/Pan/Rotate

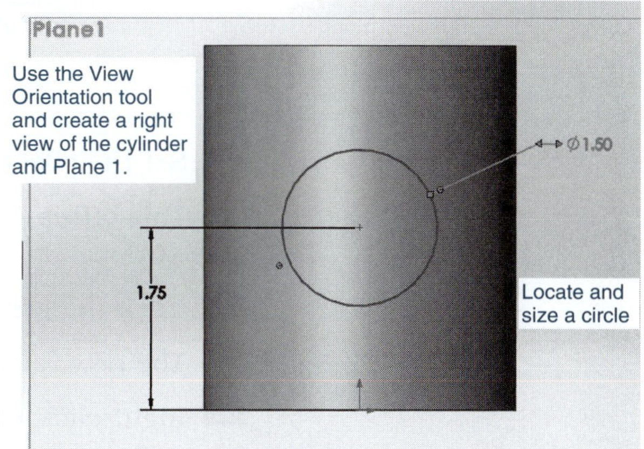
Plane1
Use the View Orientation tool and create a right view of the cylinder and Plane 1.
Ø1.50
1.75
Locate and size a circle

Figure 3-24
(*Continued*)

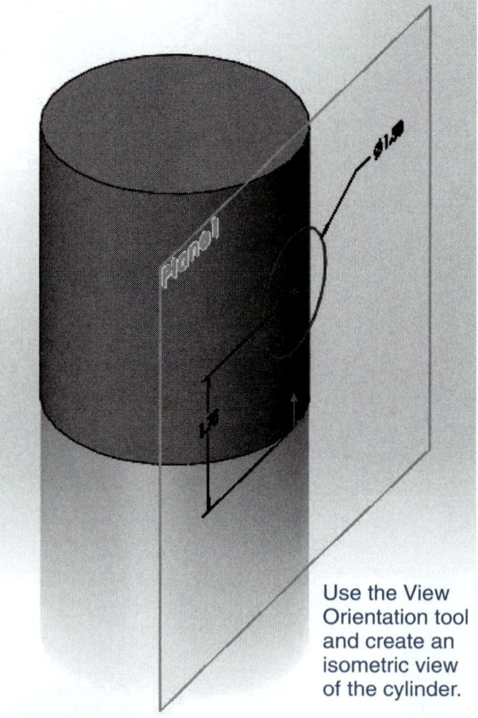
Ø1.50
Plane1
1.75
Use the View Orientation tool and create an isometric view of the cylinder.

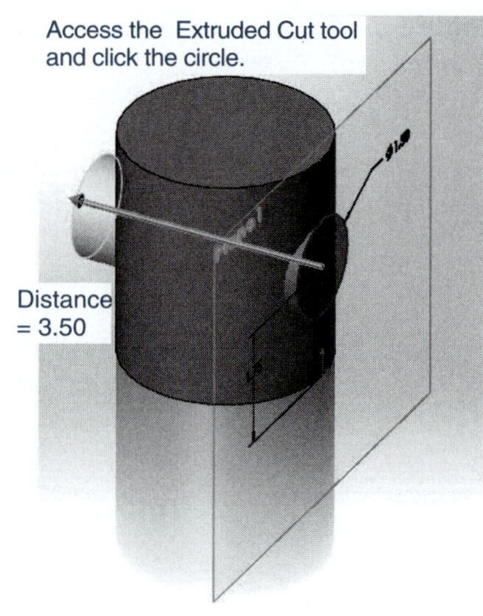
Access the Extruded Cut tool and click the circle.
Distance = 3.50

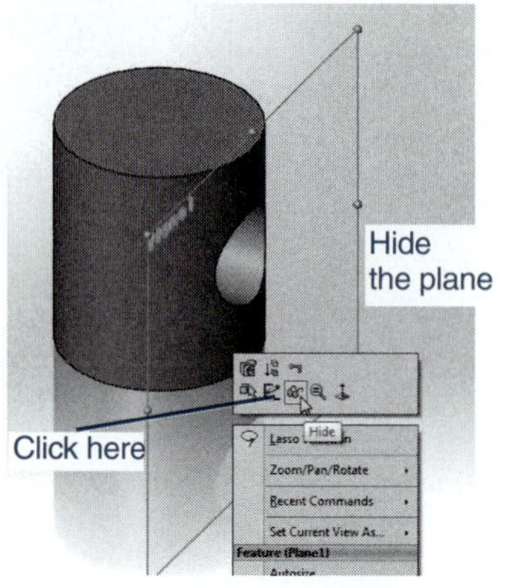
Hide the plane
Click here
Zoom/Pan/Rotate
Recent Commands
Set Current View As...

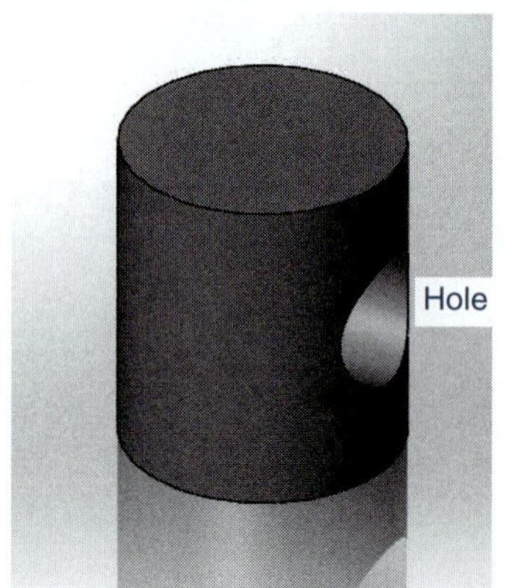
Hole

4 Set the offset distance for **1.50**.

This offset distance will locate the plane tangent to the cylinder's outside surface.

5 Click the green **OK** check mark.

The new reference plane is defined as **Plane 1**.

6 Right-click **Plane 1** and click the **Sketch** tool.

7 Use the **View Orientation** tool and create a right view of the cylinder.

8 Use the **Circle** tool and sketch a Ø1.50 circle located with its centerpoint 1.75 from the base of the cylinder as shown.

9 Use the **View Orientation** tool and create an isometric view of the cylnder.

10 Click the **Features** tab and click the **Extruded Cut** tool.

11 Define the length of the cutting cylinder as **3.50** to assure that it will pass completely through the Ø3.00 cylinder.

12 Click the green **OK** check mark.

13 Hide **Plane 1** by right-clicking on the planes and selecting the **Hide** option.

3-13 Lofted Boss/Base

The **Lofted Boss/Base** tool is used to create a shape between two planes, each of which contains a defined shape. Before drawing a lofted shape we must first draw two shapes on two different planes. In this example a square is lofted to a circle.

See Figure 3-25.

Figure 3-25

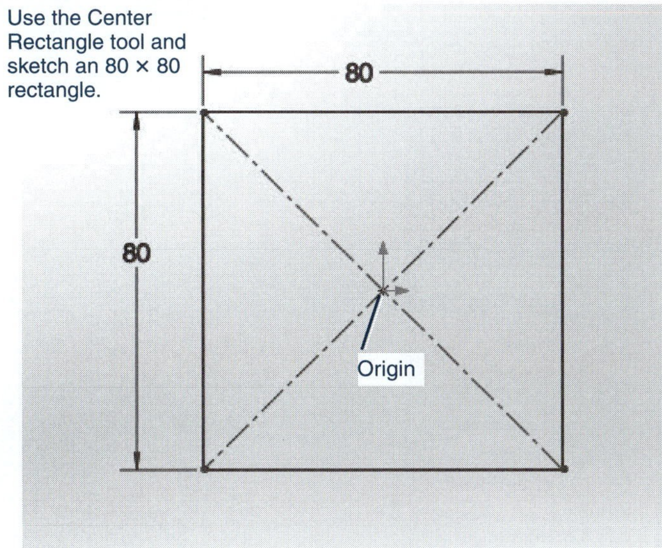
Use the Center Rectangle tool and sketch an 80 × 80 rectangle.
80
80
Origin

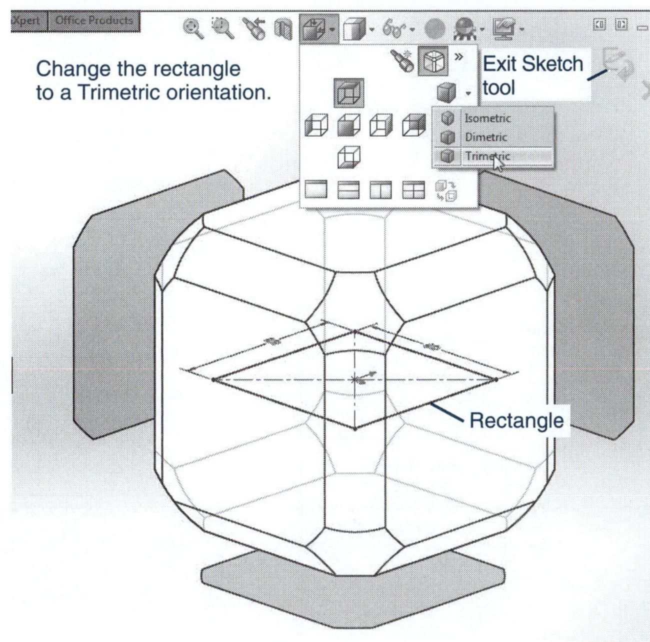
Change the rectangle to a Trimetric orientation.
Exit Sketch tool
Isometric
Dimetric
Trimetric
Rectangle

Figure 3-25
(Continued)

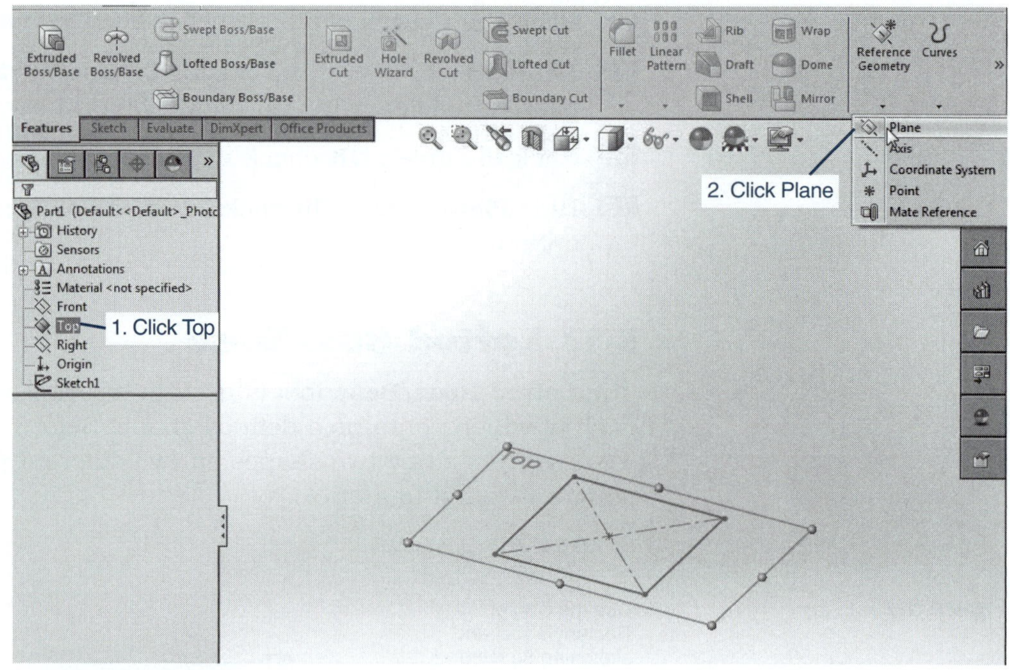

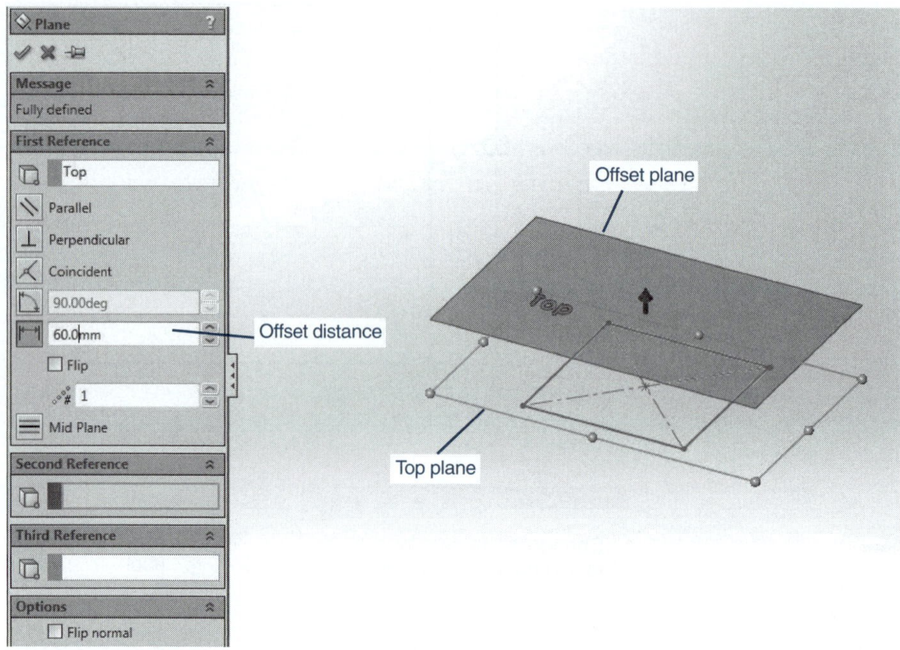

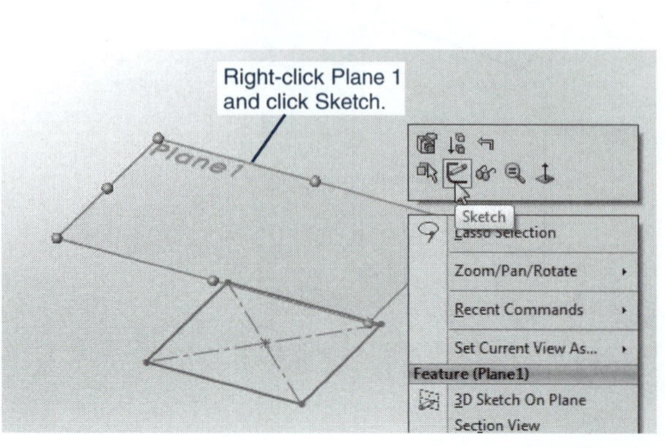

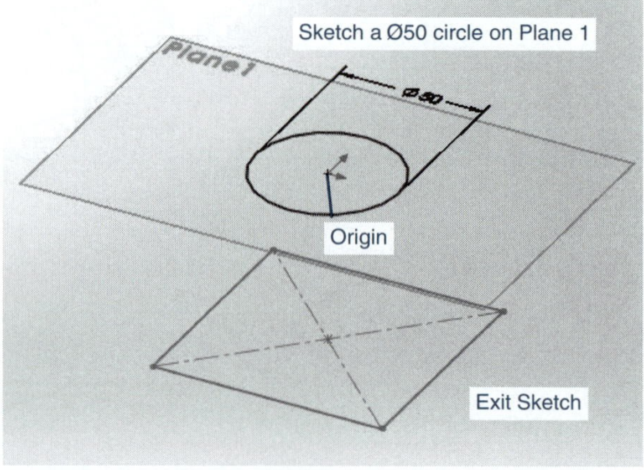

Figure 3-25
(Continued)

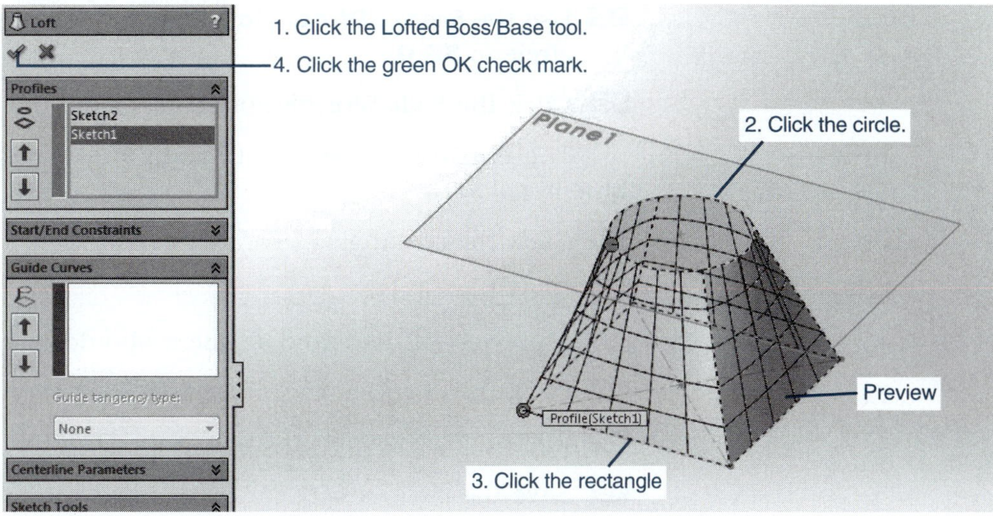

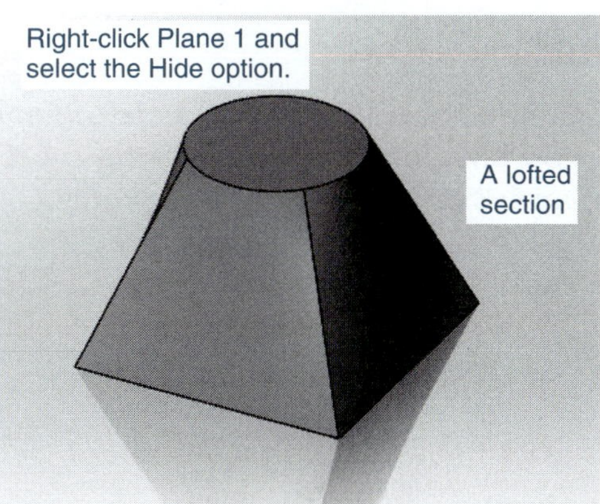

1. Create a new drawing, select the **Top Plane** option, click the **Sketch** tool, and set the units for millimeters **(MMGS)**.
2. Use the **Center Rectangle** tool and sketch a rectangle about the origin.
3. Use the **Smart Dimension** tool and create an **80** × **80** square.

   The square is centered about the origin.
4. Click the **Exit Sketch** option.
5. Click the **View Orientation** tool and select a **Trimetric** orientation.
6. Click the **Features** tab.
7. Click the **Top** plane option, click the **Reference Geometry** tool, and select the **Plane** option.

   A new plane, called **Plane 1**, will appear.
8. Offset **Plane 1 60** from the top plane.
9. Right-click the mouse and click **OK**.
10. Right-click **Plane 1** and click the **Sketch** option.
11. Use the **Circle** tool and sketch a circle centered about the origin in Plane 1.

12 Use the **Smart Dimension** tool and dimension the diameter of the circle to **50.0**.

13 Click the **Exit Sketch** option and click the **Lofted Boss/Base** tool.

The **Profiles** box should turn on automatically; that is, it should be blue in color.

14 Click the rectangle.

15 Click the circle.

A preview of the lofted segment should appear.

16 Click the check mark in the **Loft PropertyManager** box.

17 Right-click **Plane 1** and select the **Hide** option.

18 Save the part as **LOFT**.

3-14 Shell

The **Shell** tool is used to hollow out existing solid parts. Figure 3-26 shows a 35 × 40 × 60 box. It was created using the **Corner Rectangle, Smart Dimension,** and **Extruded Boss/Base** tools.

Figure 3-26

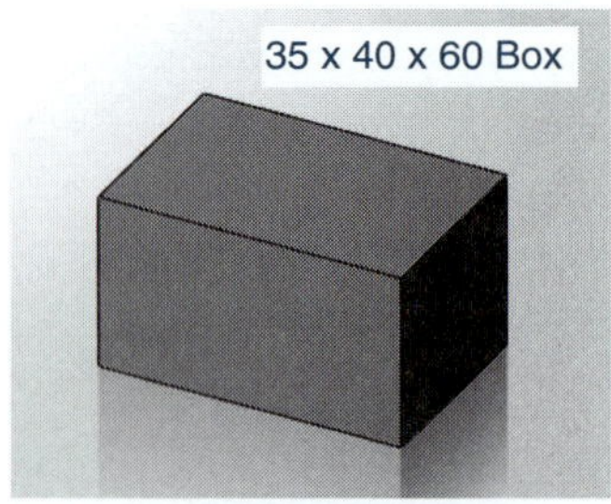

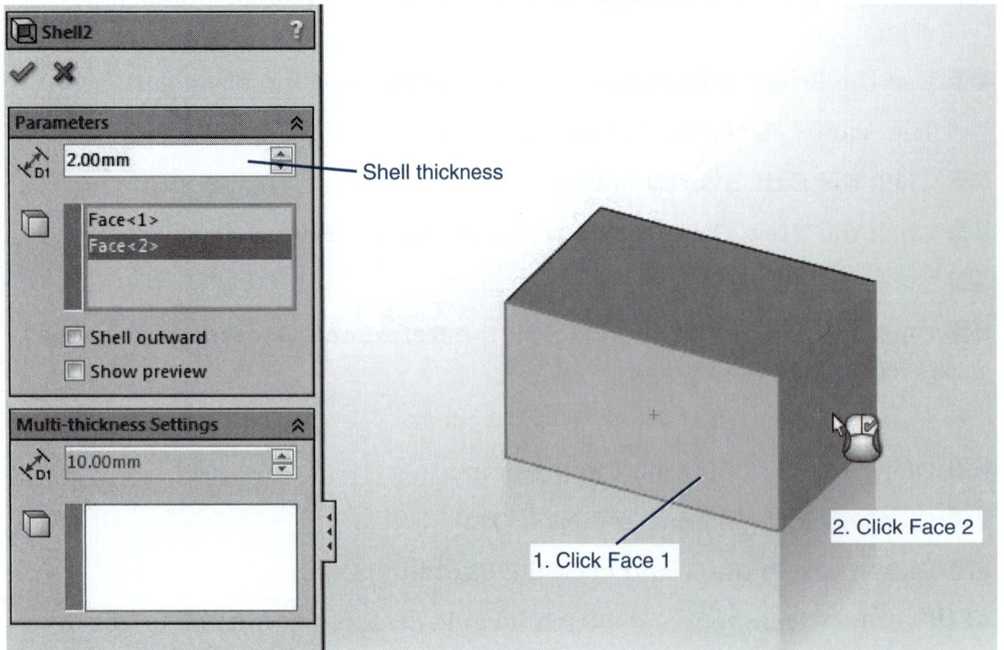

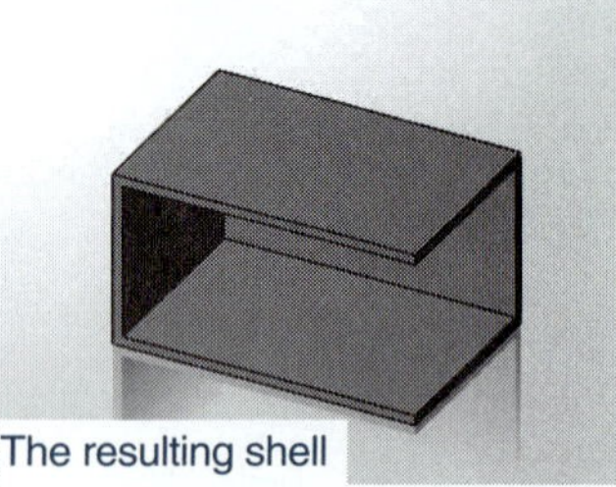

The **Shell** tool will be applied to the box.

1. Click the **Features** tab and click the **Shell** tool.
2. Define the shell thickness.

   In this example, a thickness of **2.00** was selected.
3. Click the two faces of the box as indicated.
4. Click the green **OK** check mark.

Figure 3-27 shows two more examples of how the **Shell** tool can be applied to parts.

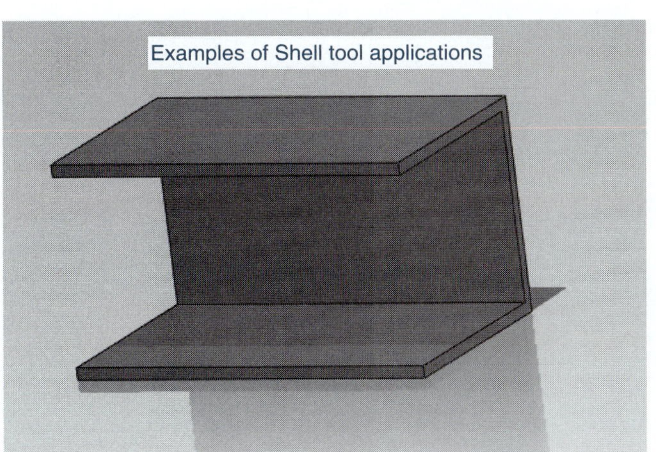

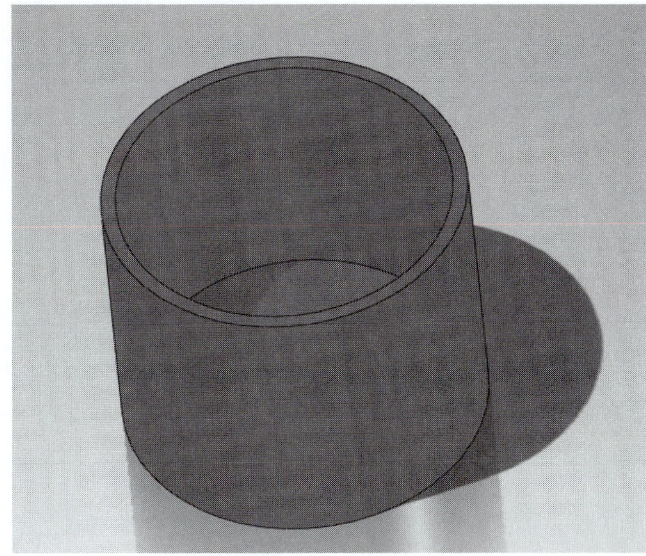

Figure 3-27

3-15 Swept Boss/Base

The **Swept Boss/Base** tool is used to sweep a profile along a path line. As with the **Lofted Boss/Base** tool, existing shapes must be present before the **Swept Boss/Base** tool can be applied. In this example, a Ø.50-inch circle will be swept along an arc with a 2.50-inch radius for 120°. See Figure 3-28.

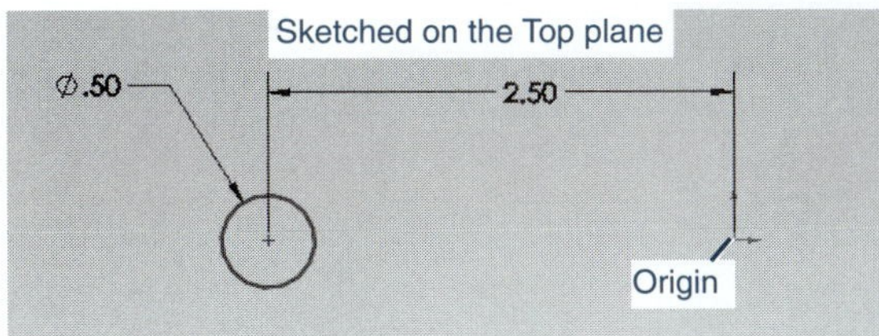

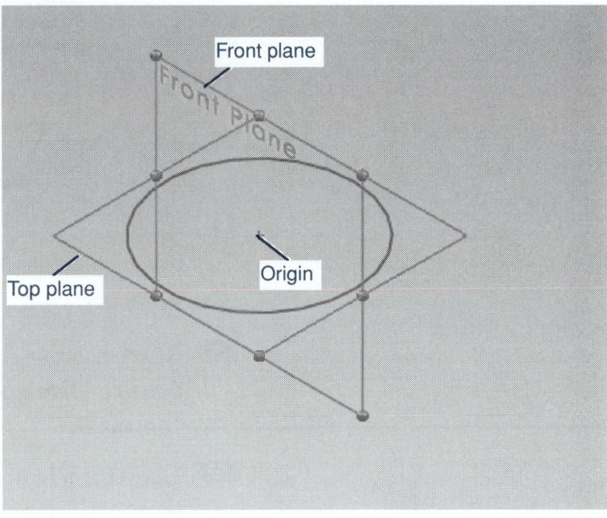

Figure 3-28

Figure 3-28
(Continued)

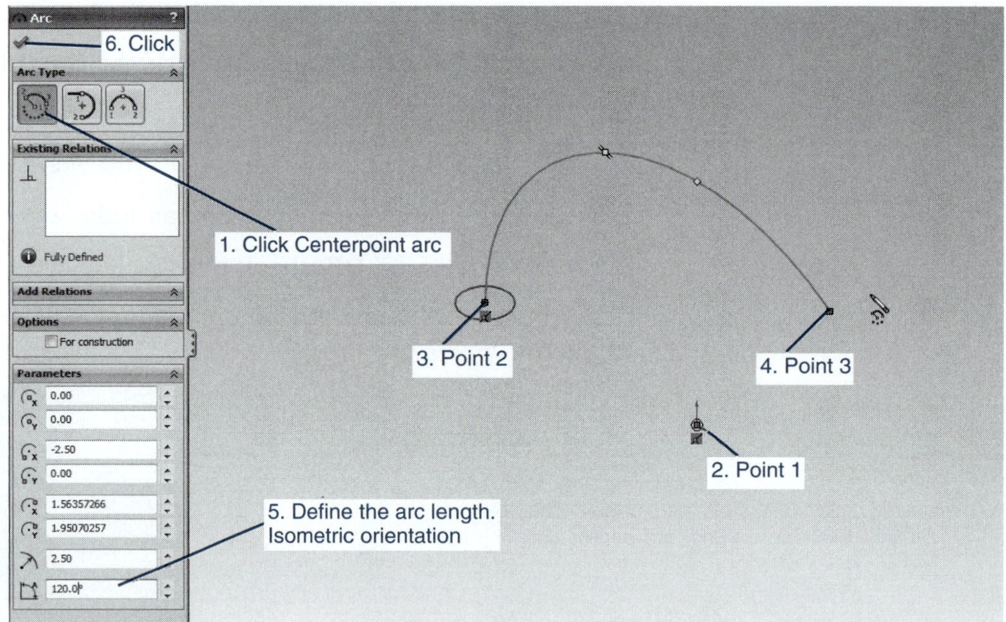

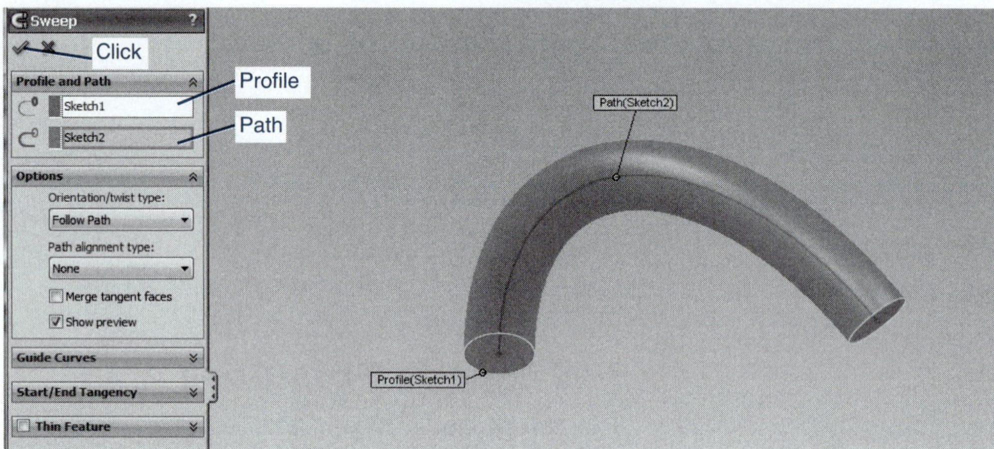

1 Start a **New** drawing and click the **Top Plane** tool. Use the **Circle** and **Smart Dimension** tools to draw a **Ø.50**-inch circle **2.50** inches from the origin.

2 Use the **View Orientation** tool and change the drawing screen to an isometric orientation.

Chapter 3

3. Right-click the mouse and click the **Select** option.
4. Click **Exit Sketch** tool.

We are now going to create a new sketch on a different sketch plane, so we must exit the top plane sketch plane.

5. Right-click the **Front** tool in the **Feature Manager Design** tree and click the **Sketch** option.
6. Use the **Centerpoint Arc** tool to draw an arc with a **2.50** radius with the origin as its centerpoint and an arc length of approximately 120°. Define the arc's length as **120°** in the **Parameters** box.
7. Click the green **OK** check mark.
8. Click the **Exit Sketch** option.

This completes the second sketch. The **Swept Boss/Base** tool can now be applied.

9. Click the **Features** tool, then click the **Swept Boss/Base** tool.
10. Select the circle as **Sketch 1** (the profile) and the arc as **Sketch 2** (the path).

The path area will automatically be selected.

11. Click the green **OK** check mark.

NOTE
Figure 3-29 shows an object hollowed out using the **Shell** tool.

Figure 3-29

3-16 Draft

The **Draft** tool is used to create slanted surfaces. Figure 3-30 shows a 30 × 50 × 60 box. In this example a 15° slanted surface will be added to the top surface.

Figure 3-30

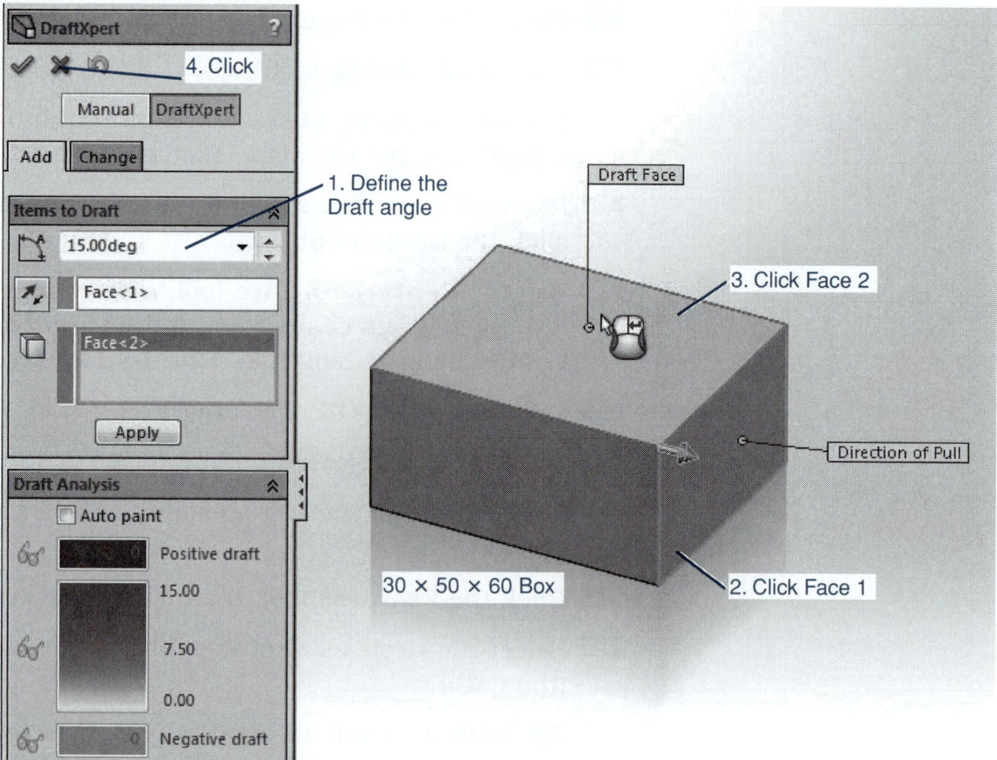

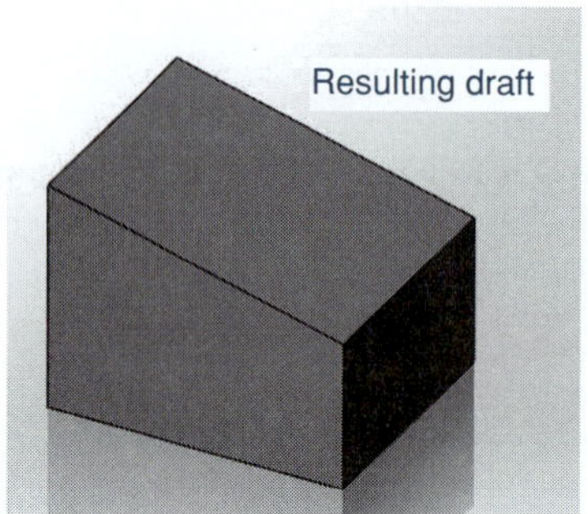

1 Draw a **30** × **50** × **60** box on the top plane.

2 Click the **Draft** tool.

3 Define the **Direction of Pull** by clicking the right vertical face of the box.

4 Define the **Draft angle** as **15°**.

5 Select the draft face by clicking the top surface of the box.

The draft angle will be applied to the draft face relative to the 90° angle between the two faces.

6 Click the green **OK** check mark.

Figure 3-31 shows a slanted surface created by making the top surface the direction of pull and the front surface the draft face. Note how the sequence of the draft face selection affects the resulting slanted surface.

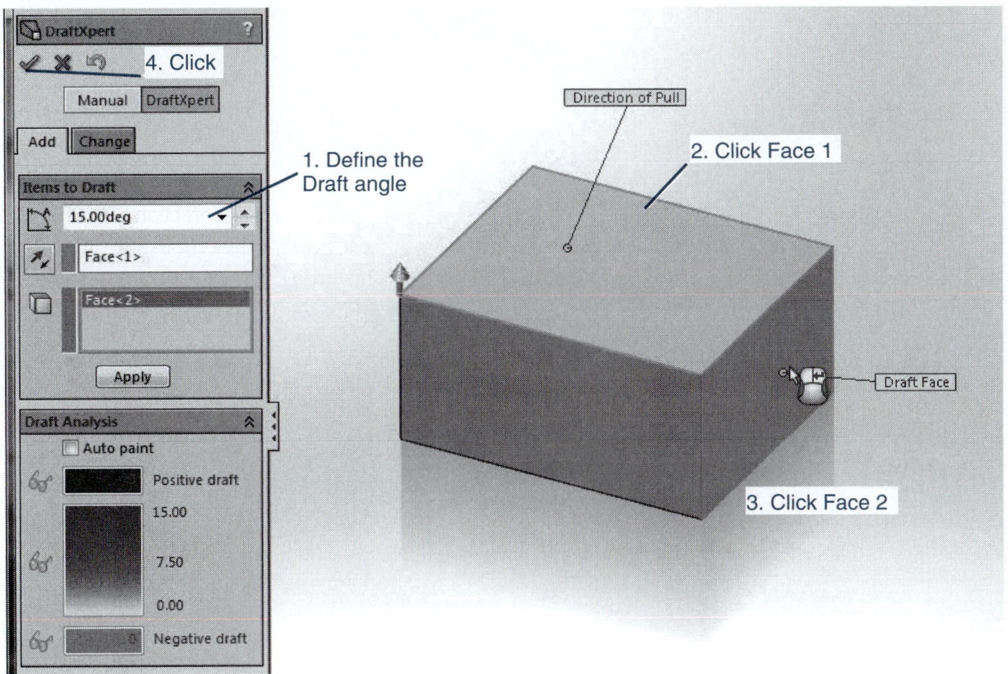

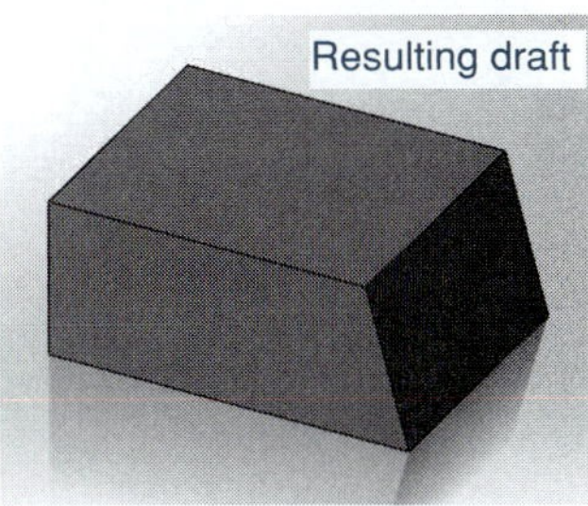

Figure 3-31

3-17 Linear Sketch Pattern

The **Linear Sketch Pattern** tool is used to create rectangular patterns based on a given object.

Figure 3-32 shows a 10 × 15 × 20 box located on a 5 × 80 × 170 base. The box is located 10 from each edge of the base as shown.

1 Draw the **Base** and **Box** as shown.

2 Click the **Linear Sketch Pattern** tool.

> **NOTE**
>
> The box should be selected automatically, but if is not, use the **Features to Pattern** tool to select the box.

Figure 3-32

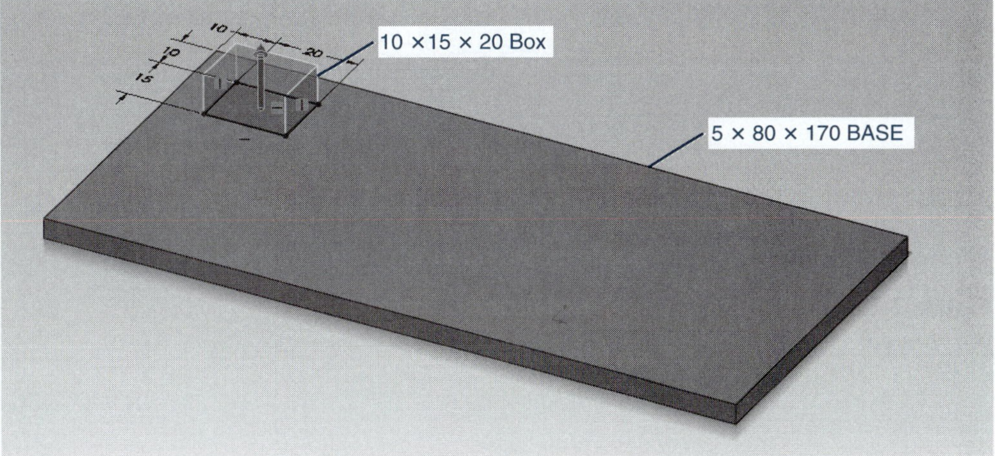

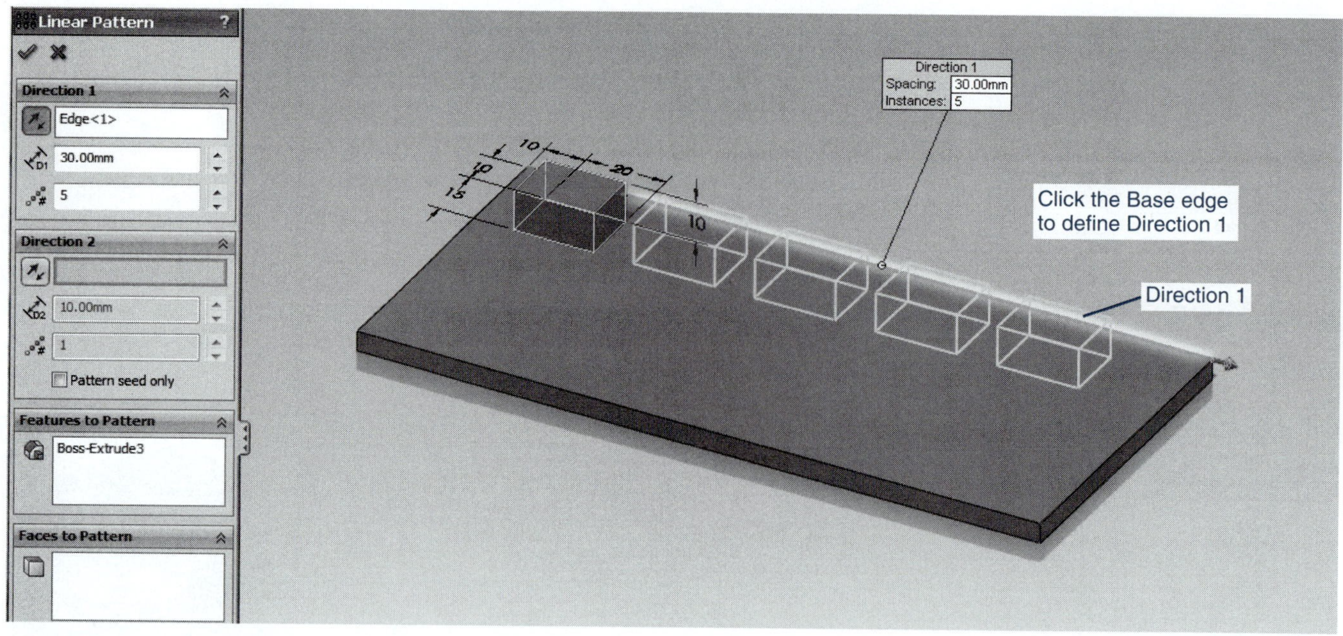

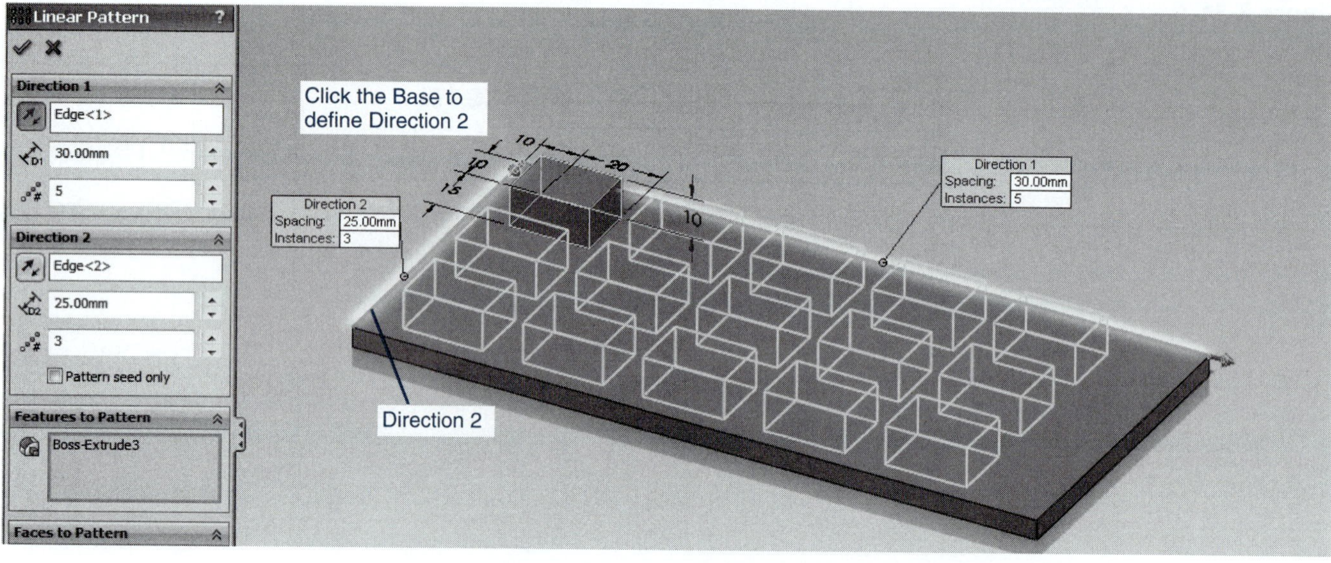

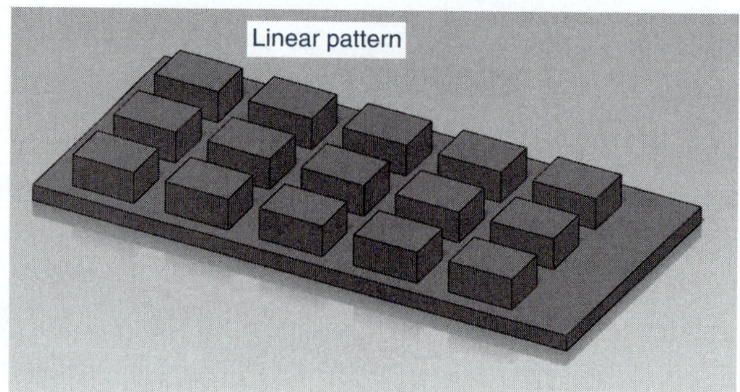

Figure 3-32
(Continued)

3 Define **Direction 1** by clicking the back top line as shown.

4 Define the spacing as **30.00mm**.

Spacing is the distance between two of the objects in the pattern as measured from the same point on each object; for example, the distance from the lower front corner on one object to the lower front corner of the next object.

5 Define the number of columns in the pattern **(Instances)** as **5**.

6 Define **Direction 2** by clicking the left edge of the base.

7 Define the spacing for **Direction 2** as **25** and the number of rows in the pattern **(Instances)** as **3**.

A preview of the pattern will appear.

8 Click the green **OK** check mark.

3-18 Circular Sketch Pattern

The **Circular Sketch Pattern** tool is used to create circular patterns about an origin. See Figure 3-33.

Figure 3-33

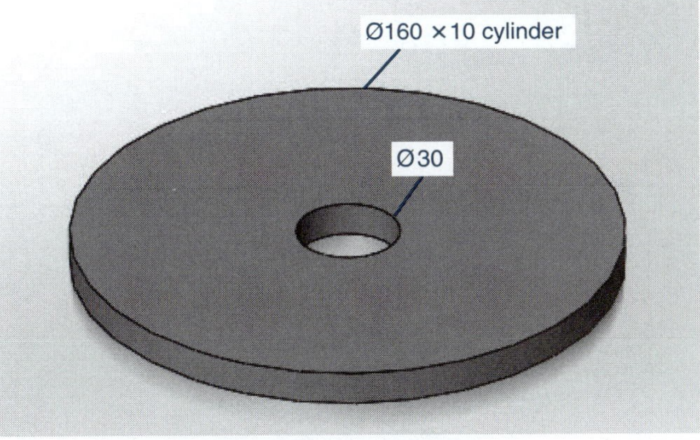

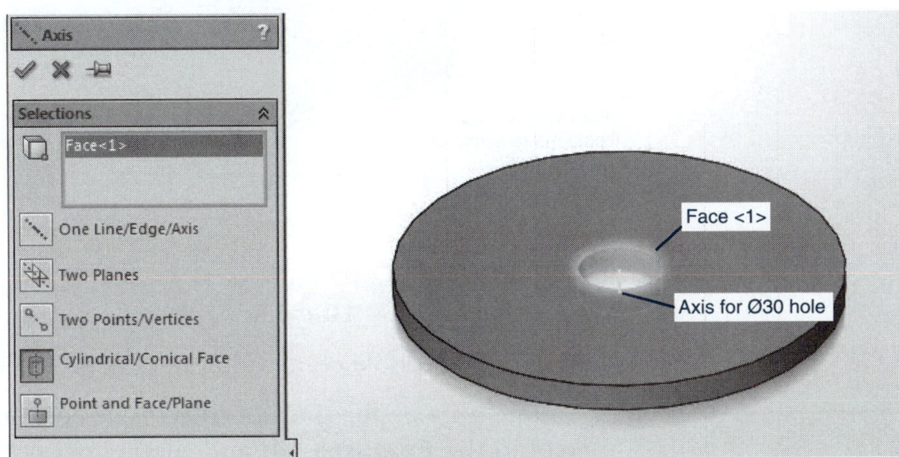

Figure 3-33
(Continued)

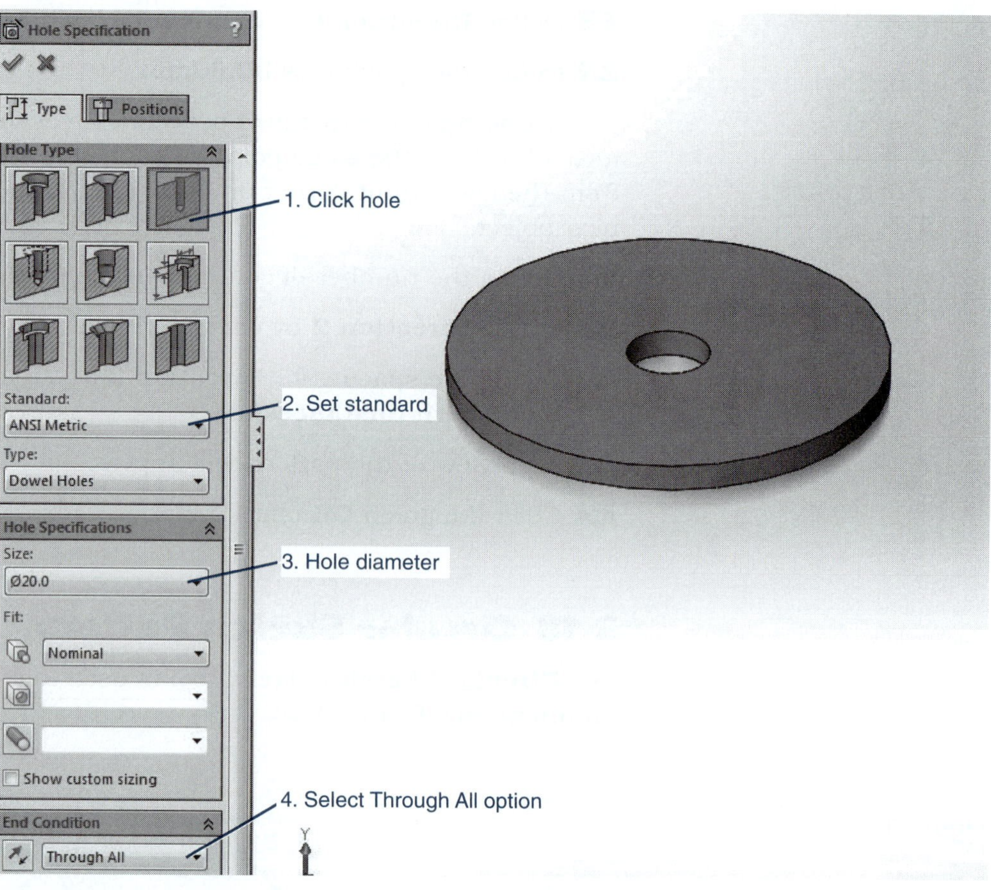

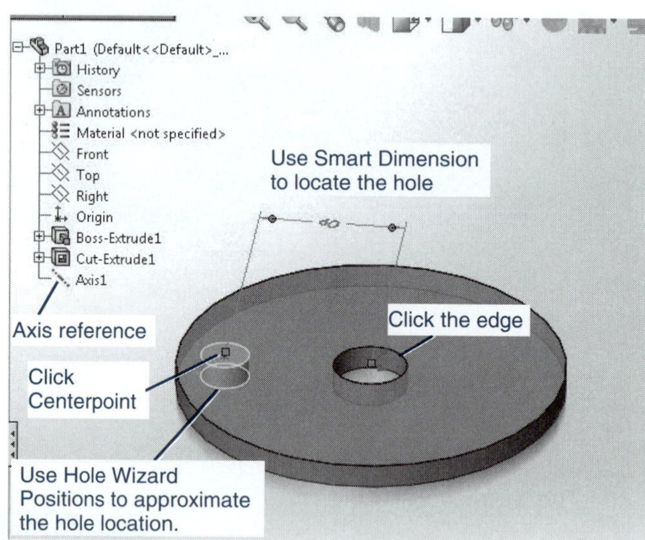

1 Draw a **Ø160** × **10** cylinder.

2 Draw a **Ø30** hole centered about the Ø160 cylinder's origin.

3 Create an axis for the Ø30 hole by accessing the **Reference Geometry** tool in the **Features** tab and then clicking the **Axis** option.

4 Click the **Cylindrical/Conical Face** option in the **Axis PropertyManager** and click the inside surface of the Ø30 hole.

Figure 3-33
(Continued)

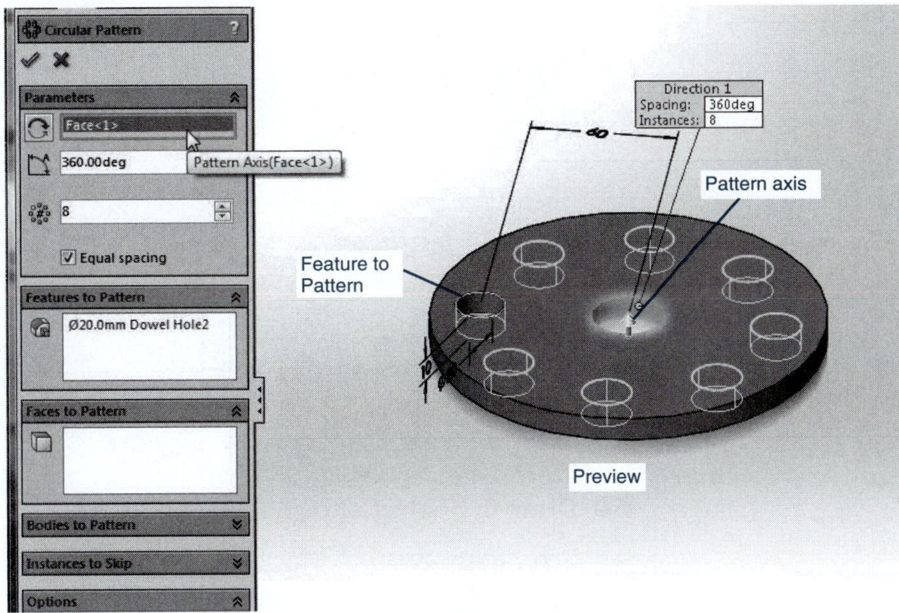

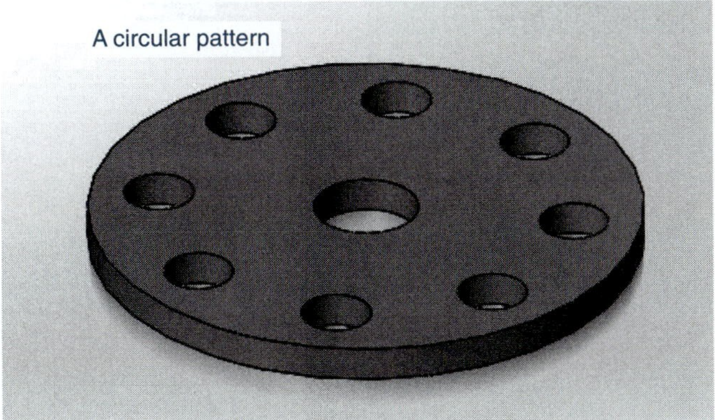

The word **Axis** should appear on the screen and in the **Design** tree.

5 Use the **Hole Wizard** tool and create a **Ø20.0** through hole **60** from the cylinder's centerpoint by clicking the centerpoint of the Ø20 hole and the edges of the Ø30 hole.

6 Click the **Circular Pattern** tool.

The **Circular Pattern** tool is a flyout from the **Linear Pattern** tool.

7 Access the **Features** tab and select the **Circular Pattern** tool.

8 Define the number of features **(Instances)** in the pattern for **8**, and click the **Equal spacing** option. Define the axis of the Ø30 hole as the axis and the Ø20 hole as the **Feature to Pattern**.

A preview will appear.

9 Click the green **OK** check mark.

3-19 Mirror

The **Mirror** tool is used to create mirror images of features. A mirror image is not the same as a copy. In this section we will mirror the object shown in Figure 3-34.

Figure 3-34

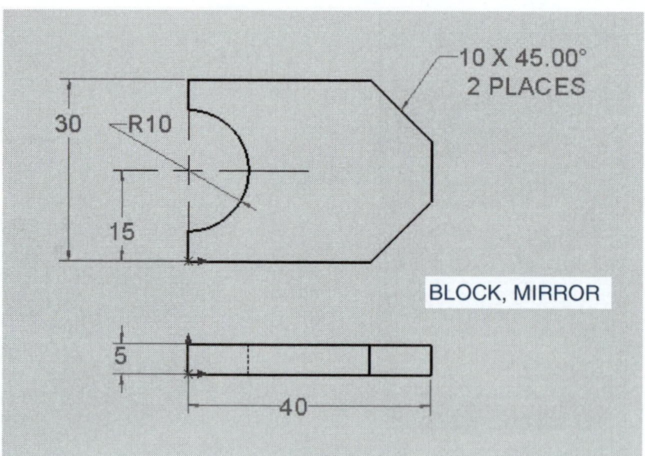

1. Draw the object shown in Figure 3-34.

   See Figure 3-35.

Figure 3-35

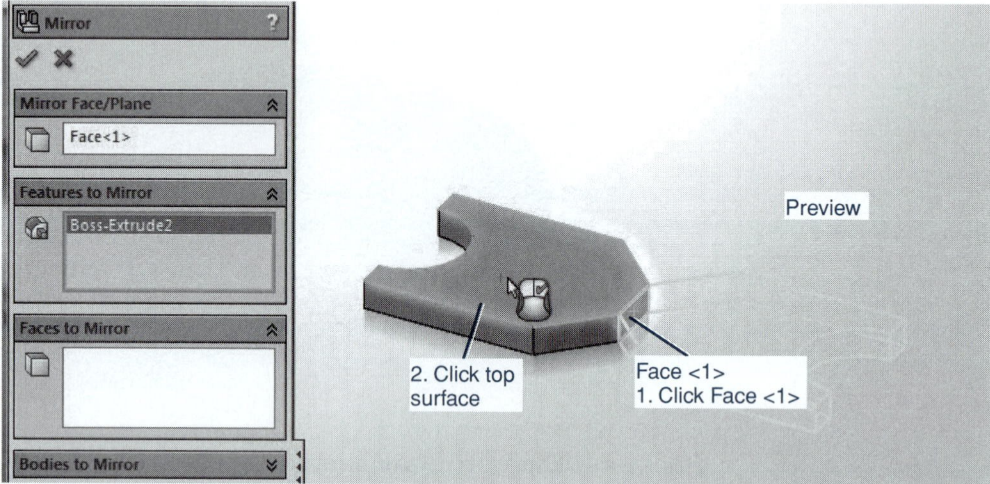

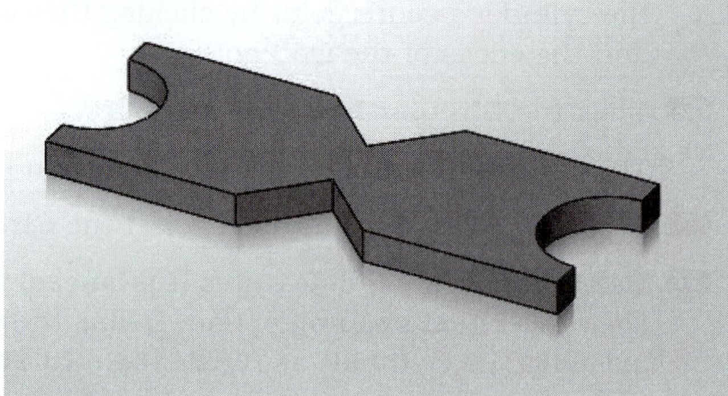

2. Access the **Features** tab, and click the **Mirror** tool.
3. Click the **Mirror Face/Plane** box, then click the right surface plane as shown.

   The box will be blue when it is active.
4. Click the **Features to Mirror** box, then click the object.

A preview will appear.

5 Click the green **OK** check mark.

3-20 Helix Curves and Springs

SolidWorks allows you to draw springs by drawing a helix and then sweeping a circle along the helical path.

To Draw a Helix

See Figure 3-36. All dimensions are in inches.

Figure 3-36

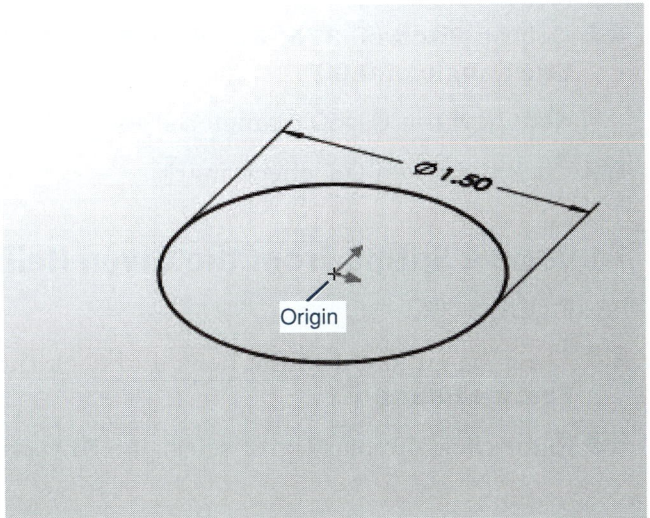

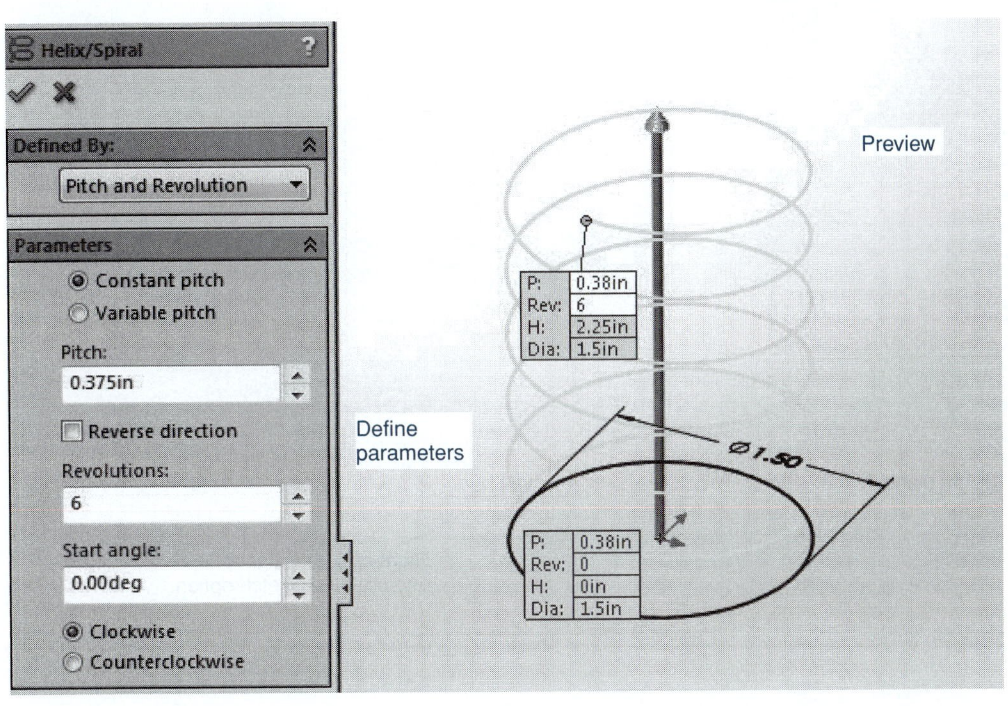

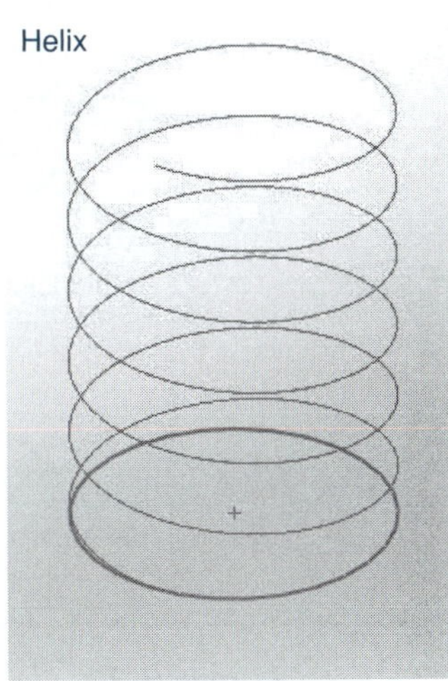

1. Start a new drawing, select the **Top Plane**, and click the **Sketch** tab.
2. Draw a **Ø1.50** circle.

   In this example the circle was centered about the origin. The diameter of this circle will determine the diameter of the helix.

3. Click the **View Orientation** tool and select the **Trimetric** view orientation.
4. Click the **Features** tab, and click the **Helix** tool.

   The **Helix and Spiral** tool is a flyout from the **Curves** tool on the **Features** panel.

   The **Helix/Spiral PropertyManager** box will appear along with a preview of the default helix.

5. Enter a **Pitch** of **.375, Revolutions** of **6**, **Clockwise** direction, and **Start angle** of **0.00°**.

   Note how the Ø1.50 diameter sizes the helix.

6. Click the green **OK** check mark.

To Draw a Spring from the Given Helix

See Figure 3-37.

1. Click the endpoint of the helix and click the **Right Plane** option in the **FeatureManager**.
2. Right-click the plane and select the **Sketch** option.

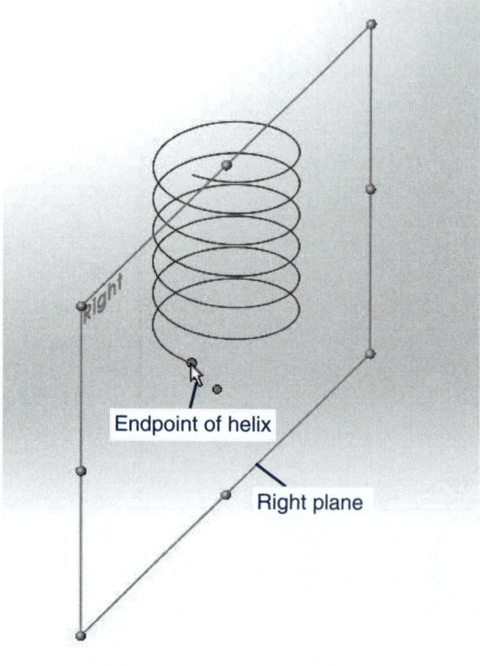

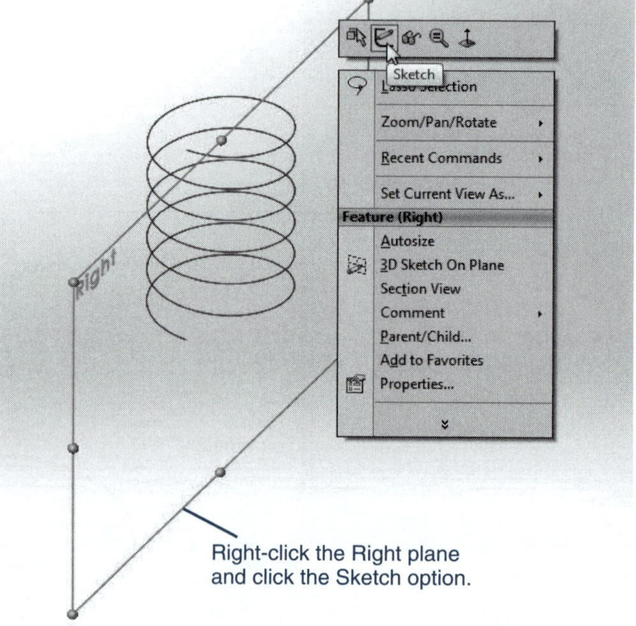

Figure 3-37

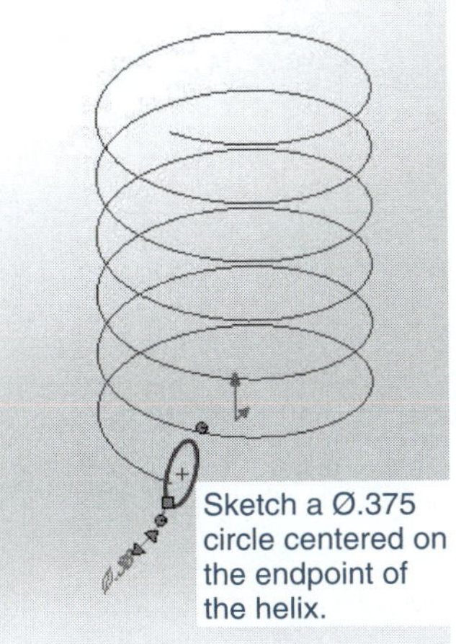

Sketch a Ø.375 circle centered on the endpoint of the helix.

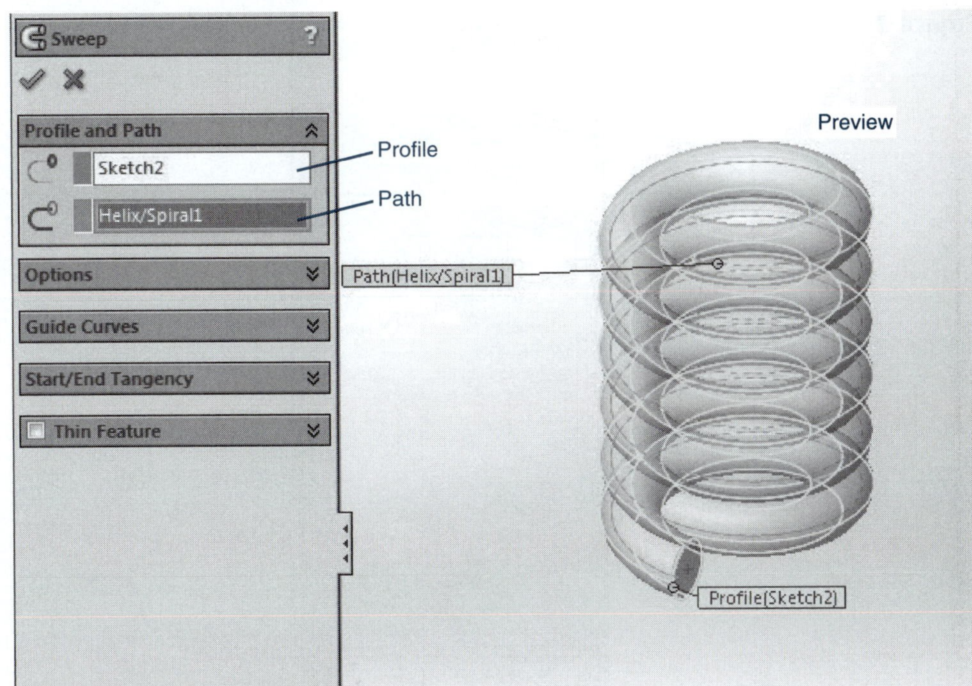

Figure 3-37
(*Continued*)

3. Draw a **Ø.375** circle on the front plane centered on the endpoint of the helix.

   The diameter of the circle defines the wire diameter for the spring.

4. Exit the **Sketch** mode, click the **Features** tab, and click the **Swept Boss/Base** tool.

5. Click the circle as the profile and the helix as the path.

   A preview will appear.

6. Click the green **OK** check mark.

7. Save the spring.

3-21 Compression Springs

Compression springs are designed to accept forces that squeeze them together. They often include ground ends that help them accept the loads while maintaining their position; that is, they don't pop out when the load is applied.

Figure 3-38 shows a spring. It has the following parameters. Dimensions are in inches.

Diameter = 1.00

Pitch = .25

Number of coils = 10

Start angle = 0.00

Wire diameter = .125

It was created using the procedure described in Section 3-20.

Figure 3-38

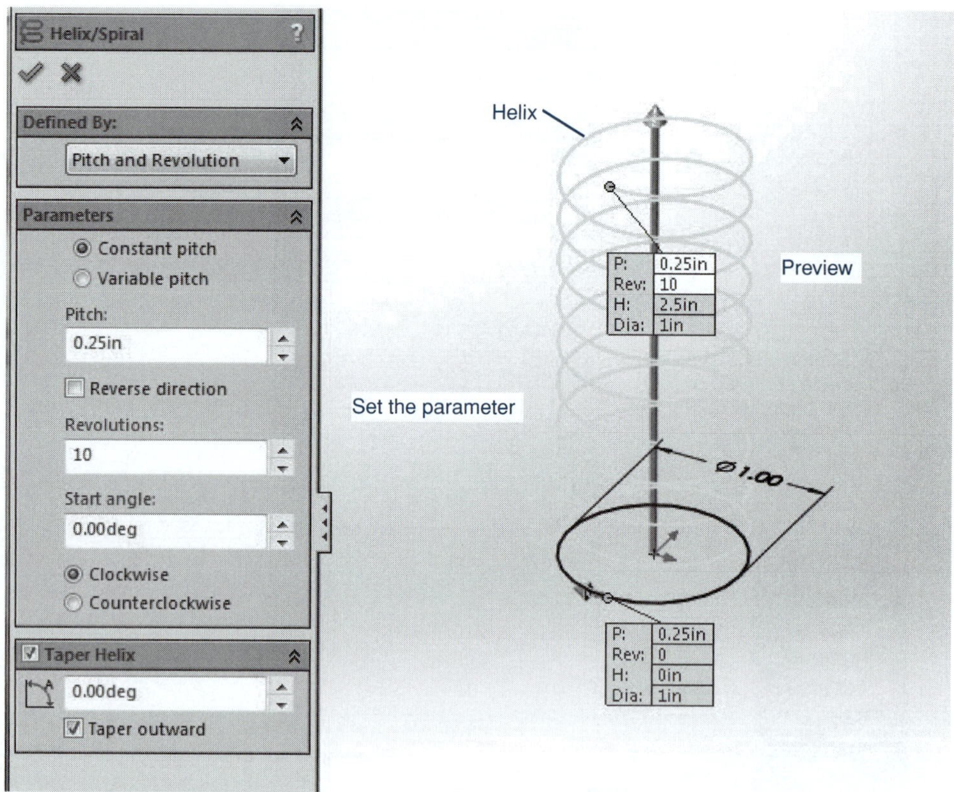

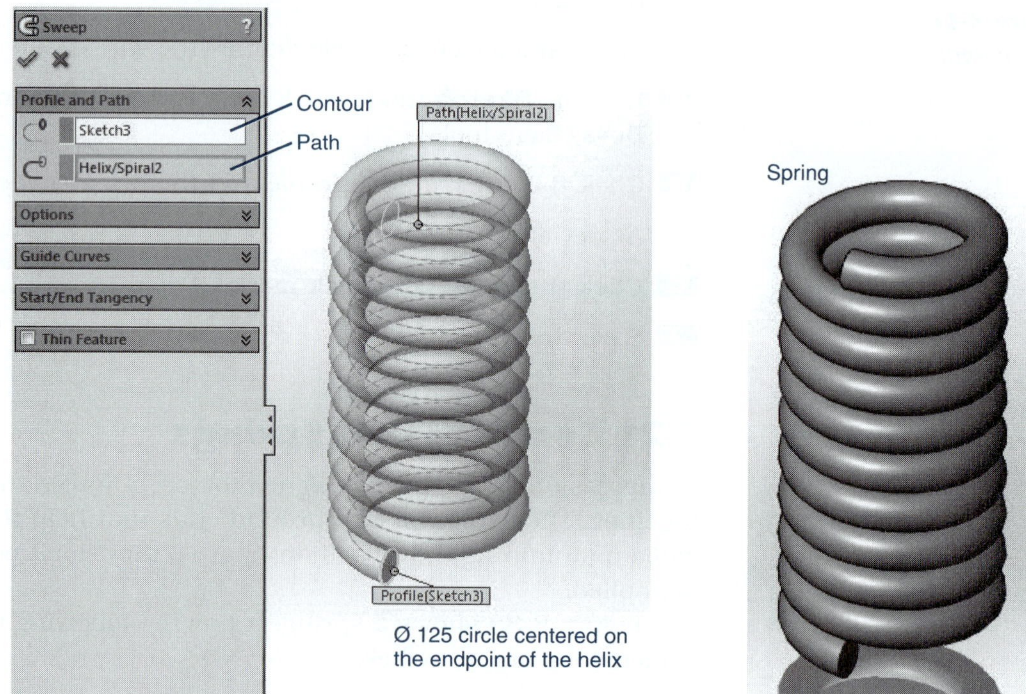

To Create Ground Ends

See Figure 3-39.

1 Orient the spring in the **Trimetric** orientation.

2 Click the **Top Plane** option in the **Feature Manager Design** tree and click the **Plane** option in the **Reference Geometry** tool on the **Features** tab.

Figure 3-39

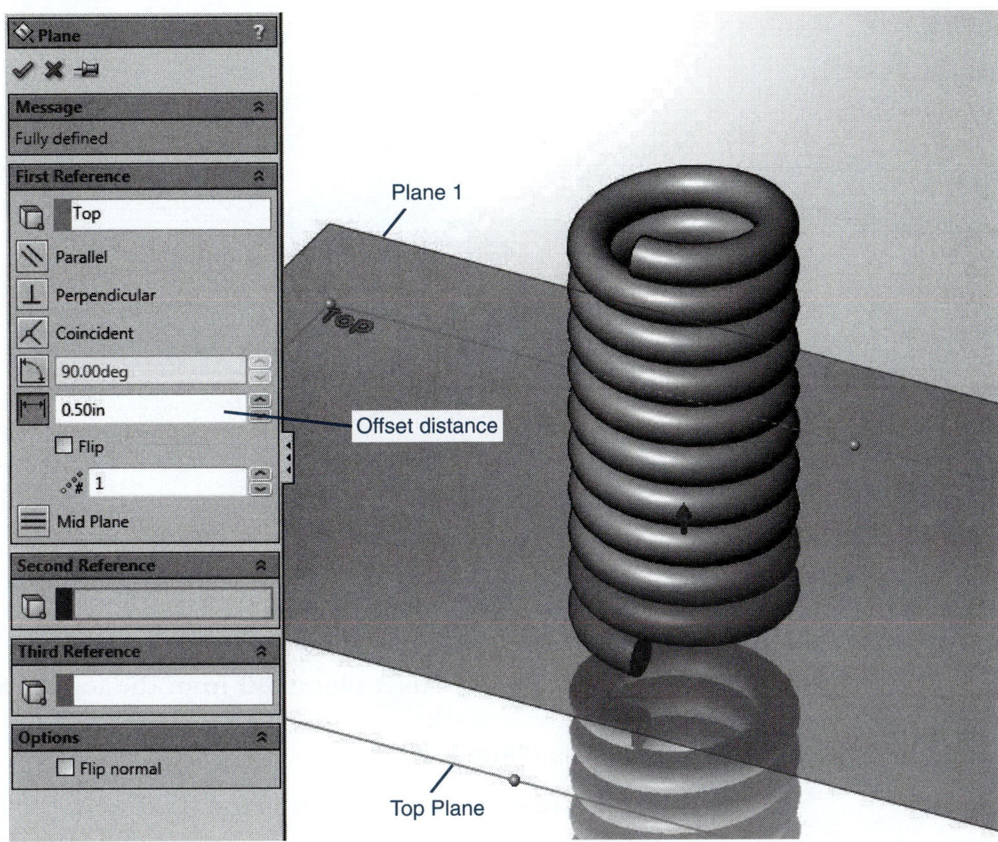
Plane
Message
Fully defined
First Reference
Top
Parallel
Perpendicular
Coincident
90.00deg
0.50in
Flip
1
Mid Plane
Second Reference
Third Reference
Options
Flip normal
Plane 1
Offset distance
Top Plane

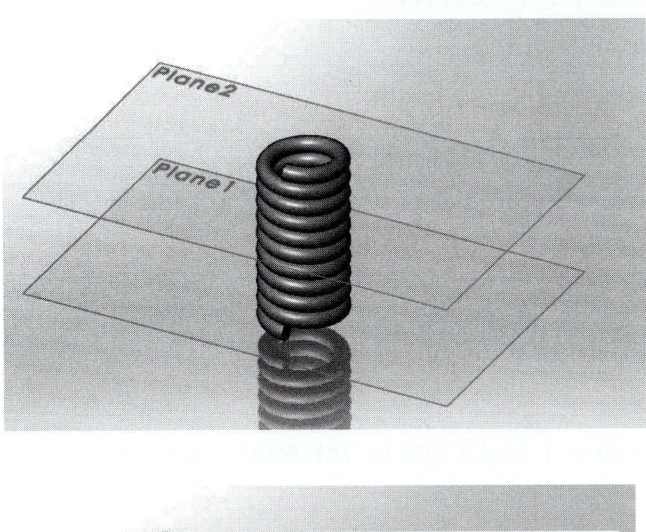
Plane2
Plane1

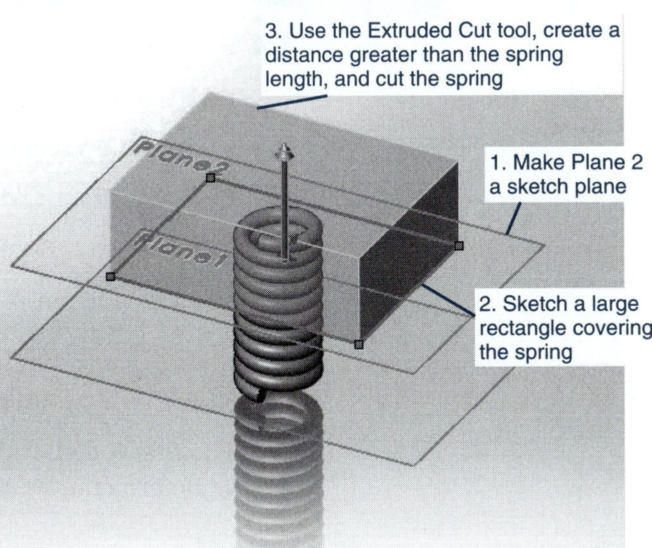
3. Use the Extruded Cut tool, create a distance greater than the spring length, and cut the spring
1. Make Plane 2 a sketch plane
2. Sketch a large rectangle covering the spring
Plane2
Plane1

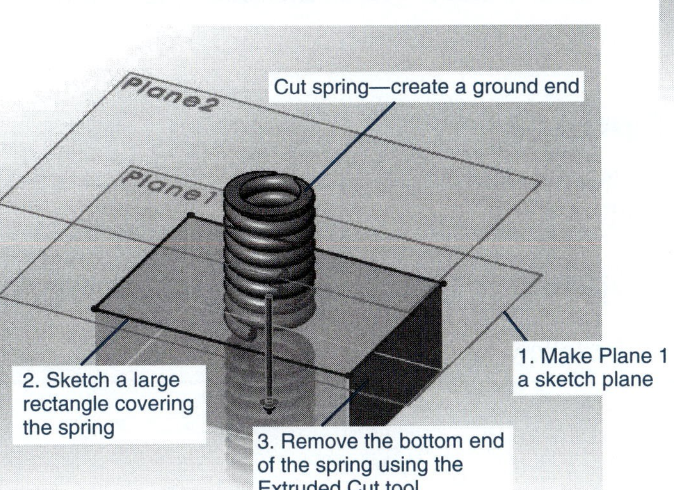
Plane2
Plane1
Cut spring—create a ground end
1. Make Plane 1 a sketch plane
2. Sketch a large rectangle covering the spring
3. Remove the bottom end of the spring using the Extruded Cut tool

Figure 3-39
(Continued)

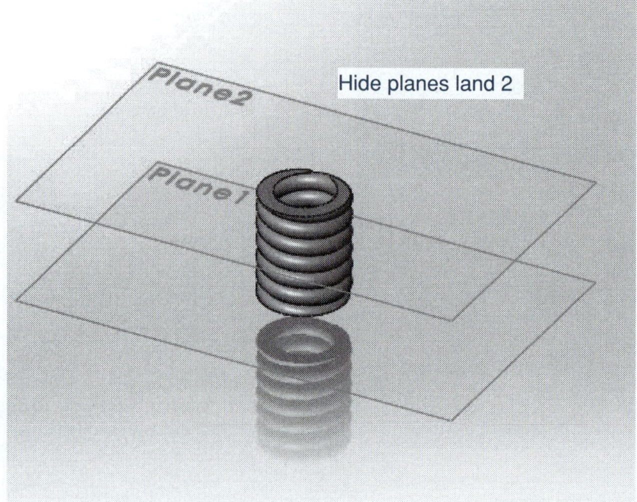

3 Create an offset plane **.50** from the top plane and click the green **OK** check mark.

This is Plane 1.

4 Create a second offset plane **2.00** from the top plane. This is Plane 2.

5 Right-click **Plane 2** and select the **Sketch** option.

6 Sketch a large rectangle on **Plane 2**.

Any size larger than the spring is acceptable.

7 Click the **Features** tab and select the **Extruded Cut** tool.

8 Create a distance that exceeds the end of the spring.

9 Click the green **OK** check mark.

The top portion of the spring will be cut off, simulating a ground end.

10 Repeat the procedure for the bottom end of the spring.

11 Hide Planes 1 and 2.

The distance between Planes 1 and 2 will be the final height of the spring before compression.

3-22 Torsional Springs

Torsional springs are designed to accept a twisting load. They usually include extensions. See Figure 3-40.

To Draw a Torsional Spring

1 Draw a spring using the following parameters. All dimensions are in inches. See Figure 3-41.

Figure 3-40

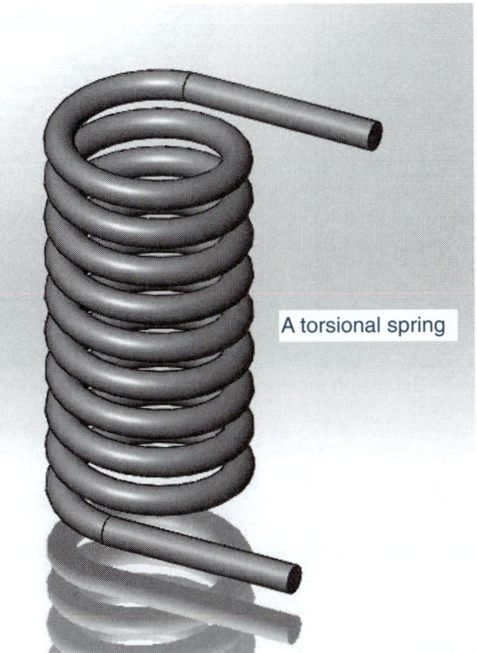
A torsional spring

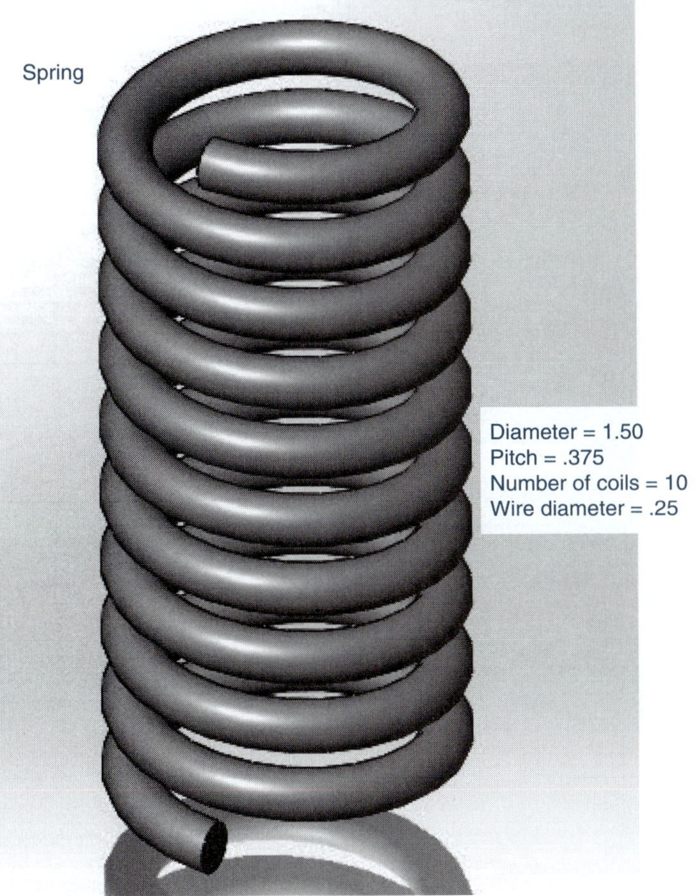
Spring
Diameter = 1.50
Pitch = .375
Number of coils = 10
Wire diameter = .25

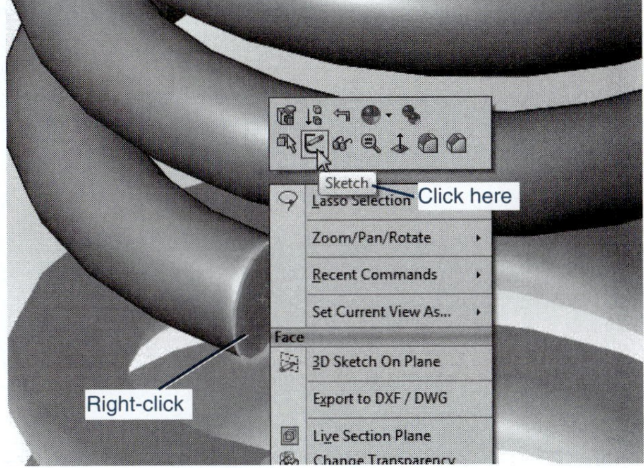
Sketch
Click here
Lasso Selection
Zoom/Pan/Rotate
Recent Commands
Set Current View As...
Face
3D Sketch On Plane
Export to DXF / DWG
Live Section Plane
Right-click

Figure 3-41

Figure 3-41
(*Continued*)

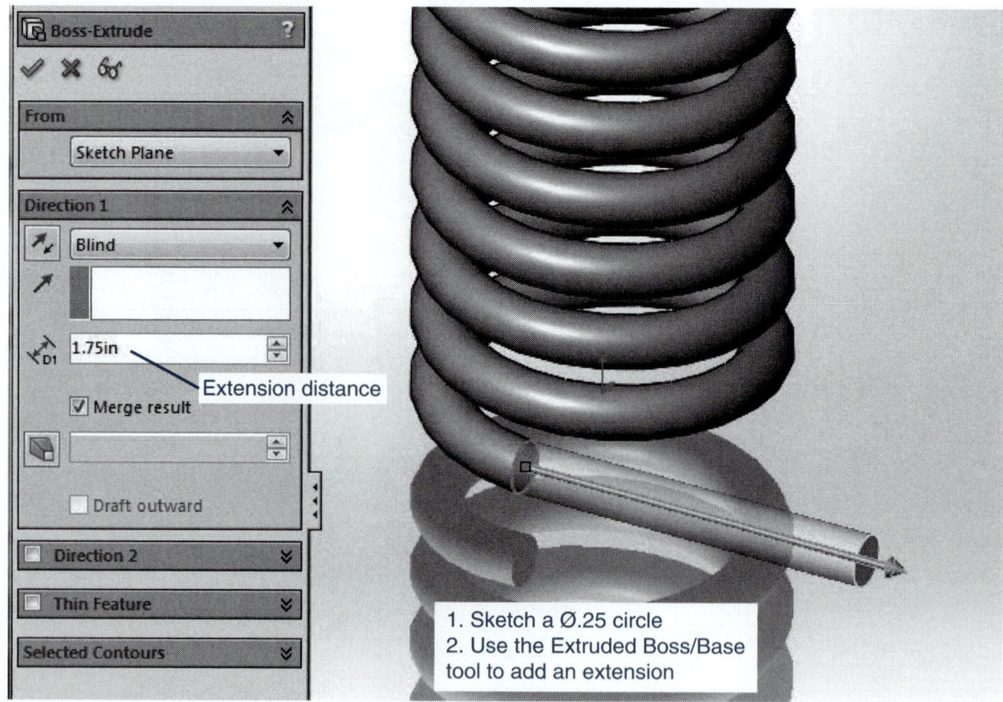

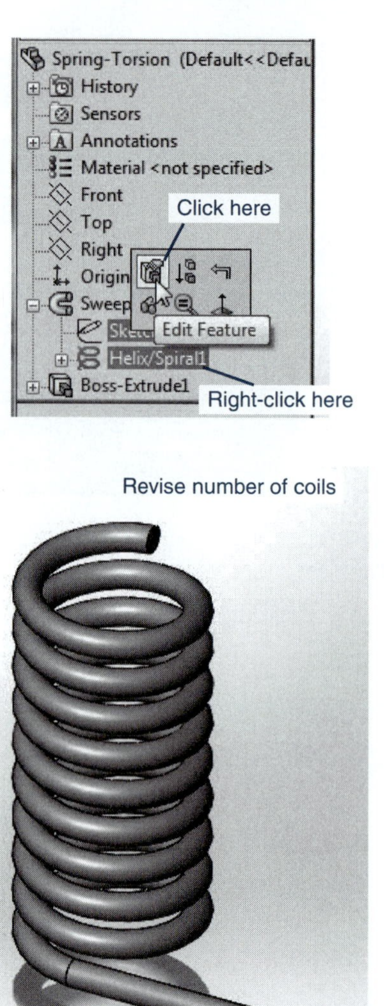

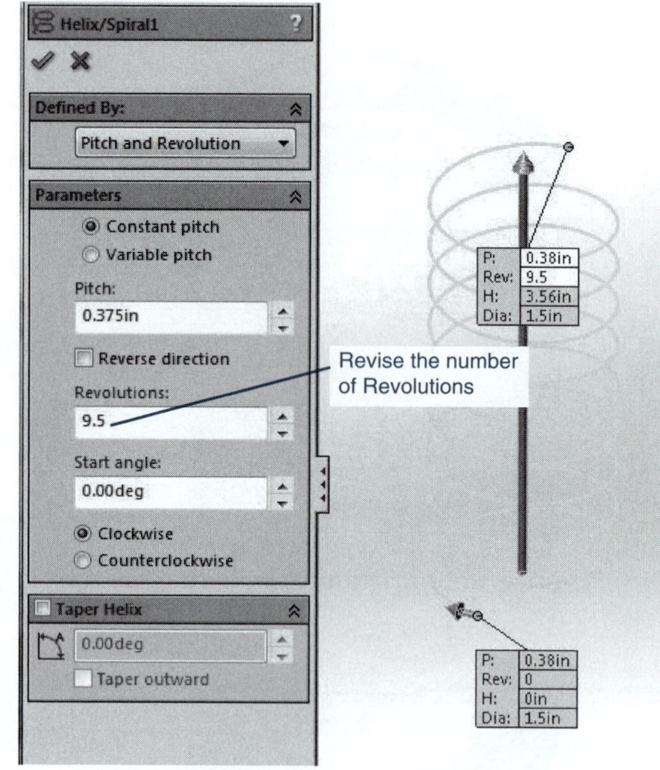

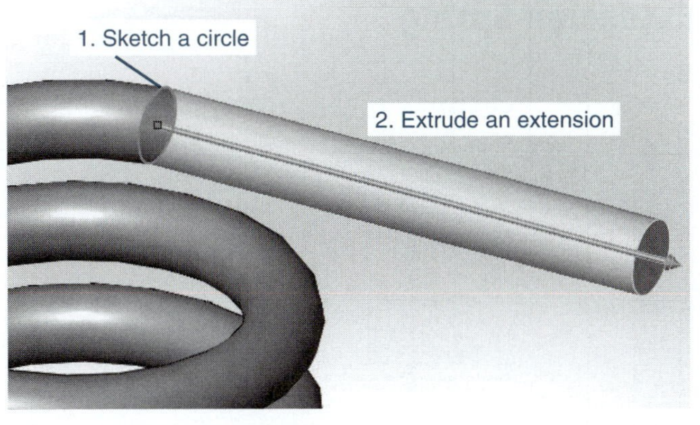

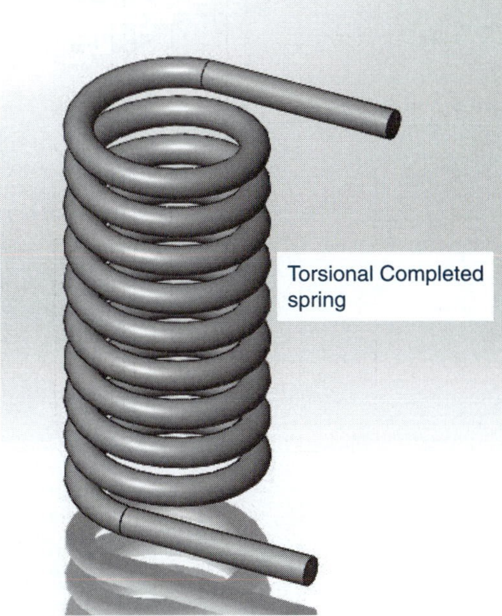

Figure 3-41
(*Continued*)

Diameter = 1.50

Pitch = .375

Number of coils = 10

Start angle = 0°

Wire diameter = .25

2 Zoom the bottom endpoint of the spring, right-click the end surface, and select the **Sketch** tool.

3 Click the **Sketch** tab and sketch a circle that exactly matches the existing end diameter.

In this example **Ø.250**, the wire diameter, was used.

4 Click the **Features** tab and select the **Extruded Boss/Base** tool.

5 Extrude the circle a distance of **1.75** and click the green **OK** check mark.

6 Right-click the **Helix/Spiral** heading in the **Manager** box and select the **Edit Feature** option.

7 Change the **Revolutions** value to **9.5** and click the green **OK** check mark.

8 Right-click the new top end surface of the spring, create a sketch plane, sketch a Ø.25 circle, and draw a **1.75** extension as shown.

9 Click the green **OK** check mark and save the spring.

3-23 Extension Springs

Extension springs are designed for loads that pull them apart, that is, tension loads. Extension springs usually have hooklike ends.

To Draw an Extension Spring

See Figure 3-42.

Figure 3-42

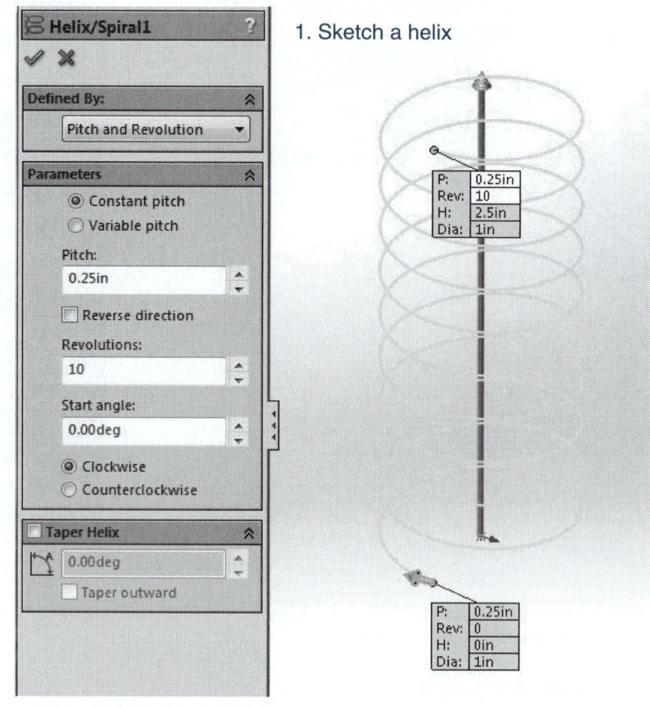

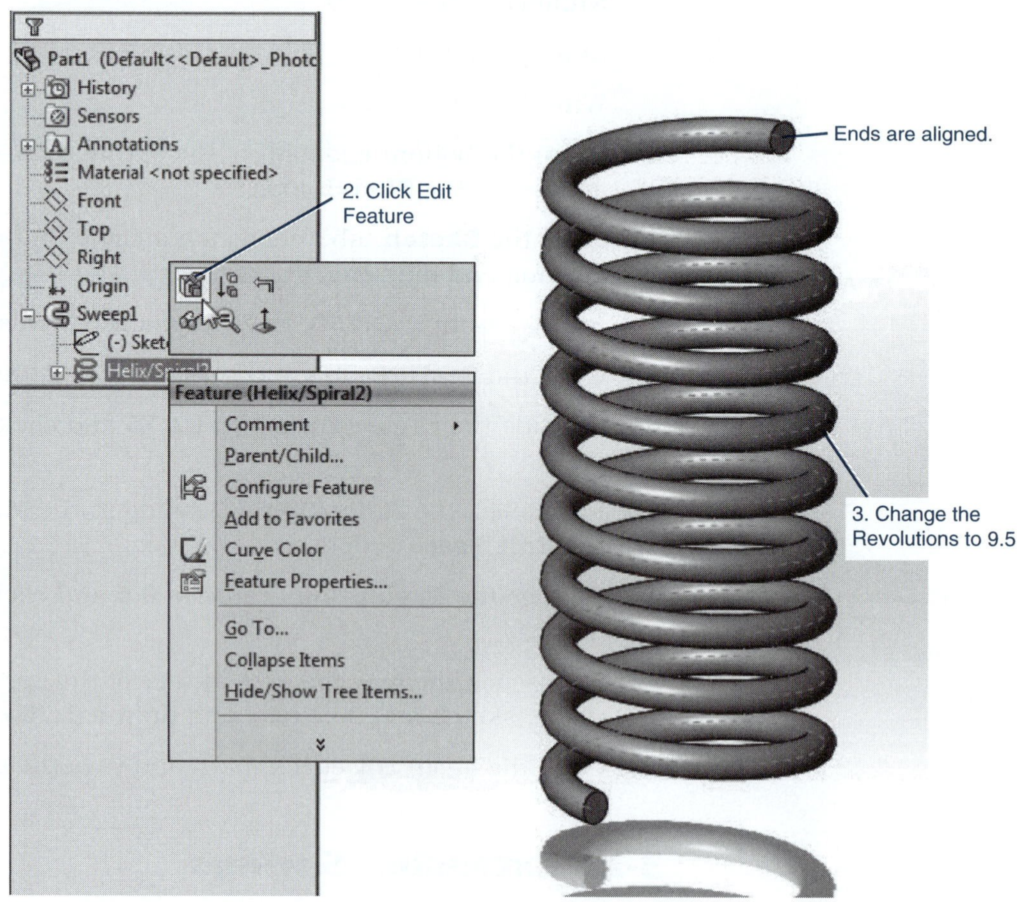

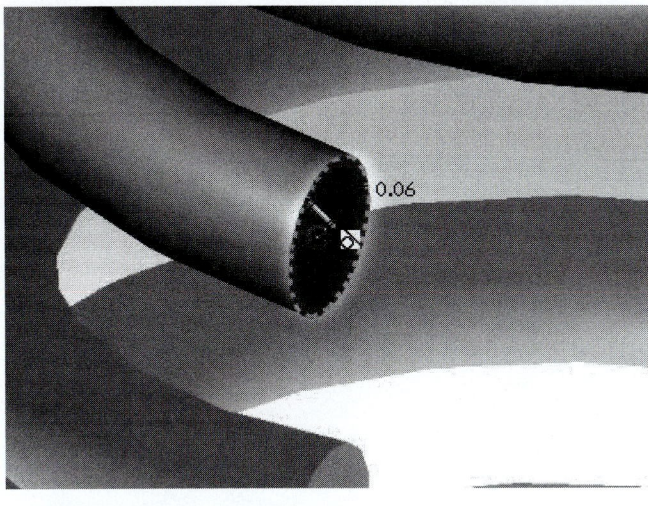
0.06

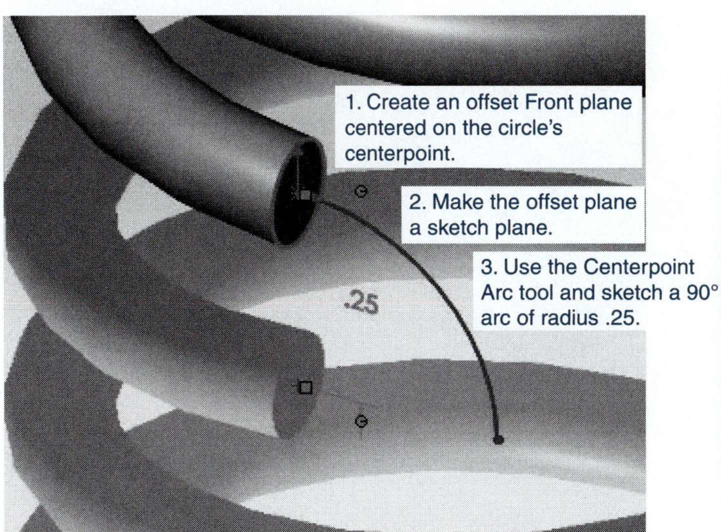
1. Create an offset Front plane centered on the circle's centerpoint.
2. Make the offset plane a sketch plane.
3. Use the Centerpoint Arc tool and sketch a 90° arc of radius .25.
.25

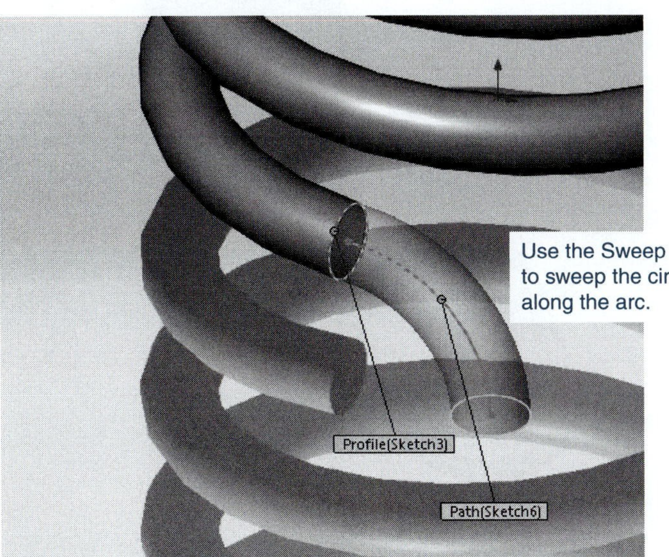
Use the Sweep tool to sweep the circle along the arc.
Profile(Sketch3)
Path(Sketch6)

2. Create an offset Right plane through the centerpoint of the circle.

3. Make the offset plane a sketch plane.

1. Sketch a circle on the end surface.

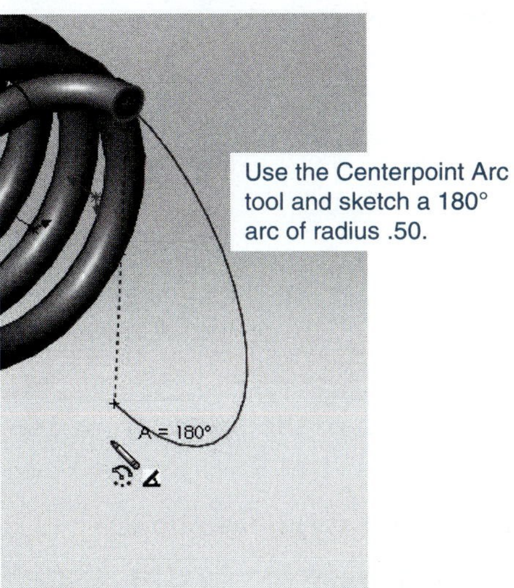
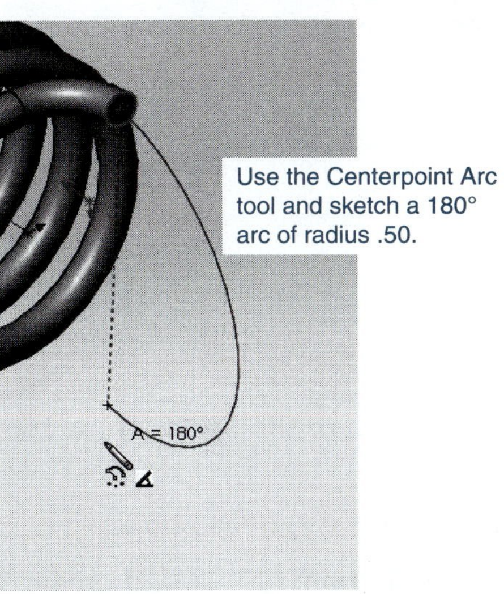
Use the Centerpoint Arc tool and sketch a 180° arc of radius .50.
A = 180°

Figure 3-42
(*Continued*)

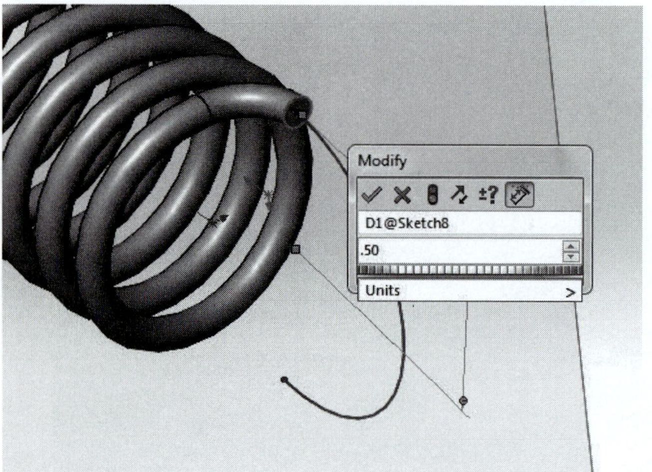

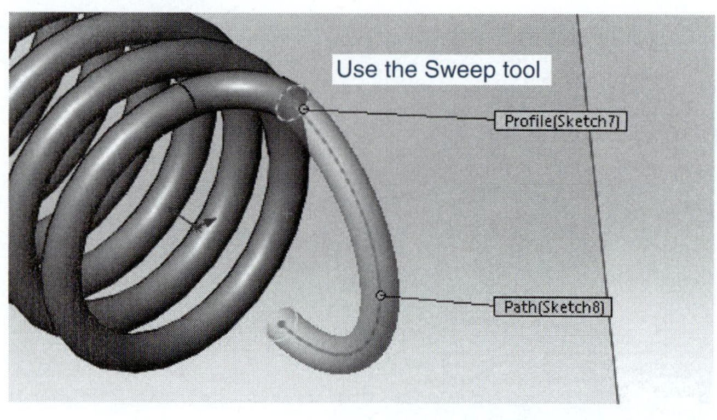

Figure 3-42
(Continued)

1. Draw a spring as defined below. All dimensions are in inches.

   Diameter = 1.00

   Pitch = .25

   Number of coils = 10

   Start angle = 0.00

   Wire diameter = .125

   Clockwise

2. Align the ends of the spring by editing the pitch of the spring from **10** to **9.5**.
3. Sketch a Ø.125 circle on the end surface of the spring as shown.
4. Use the **Plane** option under the **Reference Geometry** tool and create an offset **Front** plane offset **.50**, that is, through the centerpoint of the circle created in Step 1.

5. Right-click the mouse and select the **Sketch** tool, turning the offset plane into a sketch plane.
6. Use the **Centerpoint Arc** tool and sketch an arc from the center of the sketch circle a distance of **90°**.

   Use the **Smart Dimension** tool to define the radius of the arc as **.25**.
7. Use the **Swept Boss/Base** tool and sweep the circle along the arc path; click the green **OK** check mark. Hide the planes and the sketches.
8. Sketch a circle on the arc's end surface.
9. Use the **Plane** option under the **Reference Geometry** tool and create an offset **Right** plane through the centerpoint of the circle created in Step 3.
10. Use the **Centerpoint Arc** tool and sketch a **180°** arc with a radius of .50.
11. Use the **Swept Boss/Base** tool and sweep the circle through the arc.
12. Click the **OK** check mark.
13. Hide the planes.
14. Repeat the procedure for the other end surface of the spring.

    The spring shown represents one possible type of extension spring.

3-24 Wrap

The **Wrap** tool is used to wrap text or other shapes around a surface. There are three options: embossed, text that stands out from a face; debossed, text that is embedded on a face; and scribed, text that is written directly on the face.

To Create Debossed Text

See Figure 3-43.

1. Draw a **Ø3.00** × **2.50** cylinder based on the top plane centered on the origin.
2. Use the **Plane** option of the **Reference Geometry** tool and create an offset right plane tangent to the outside edge of the cylinder. The offset distance will be 1.50 from the origin. This is Plane 1.
3. Change the orientation of the cylinder to the right plane.
4. Right-click **Plane 1** and click the **Sketch** option.
5. Click the **Text** tool and add text to the plane.
6. Use the cursor or the **Smart Dimension** tool to locate the text.
7. Change the drawing orientation to **Trimetric**.
8. Exit the **Sketch** mode.
9. Click the **Wrap** tool.
10. Click the **Deboss** option.

Figure 3-43

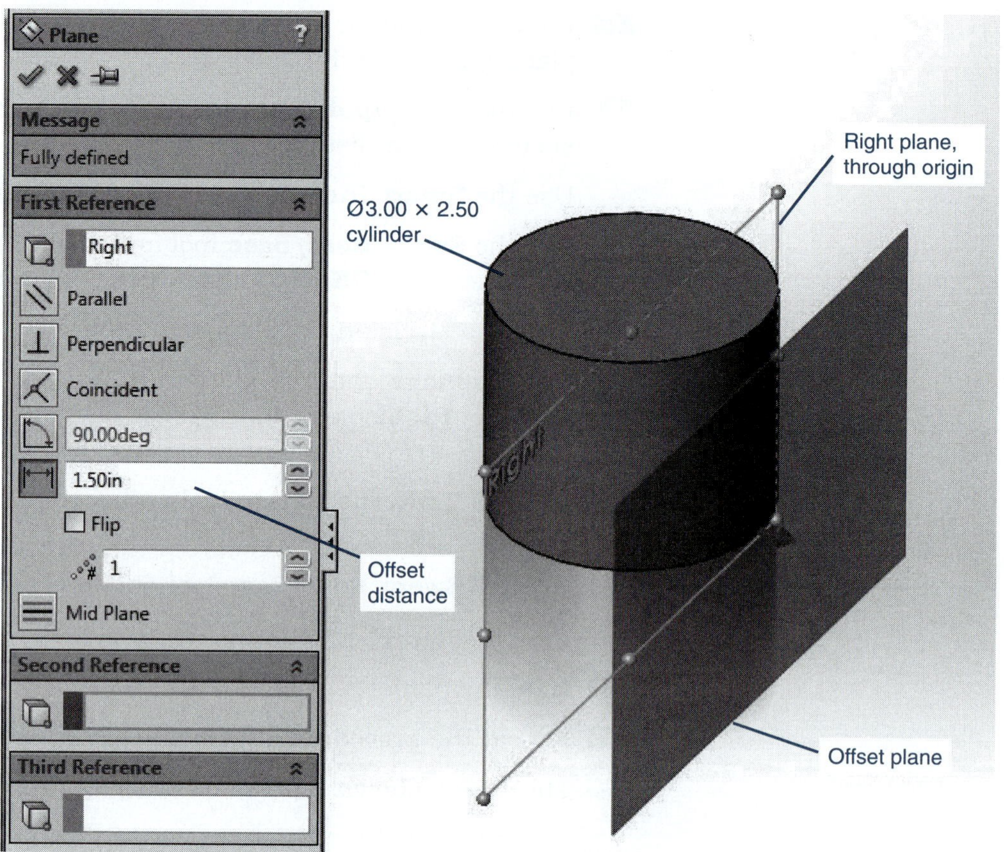
Plane
Message
Fully defined
First Reference
Right
Parallel
Perpendicular
Coincident
90.00deg
1.50in
Flip
1
Mid Plane
Second Reference
Third Reference
Ø3.00 × 2.50
cylinder
Right plane,
through origin
Offset
distance
Offset plane

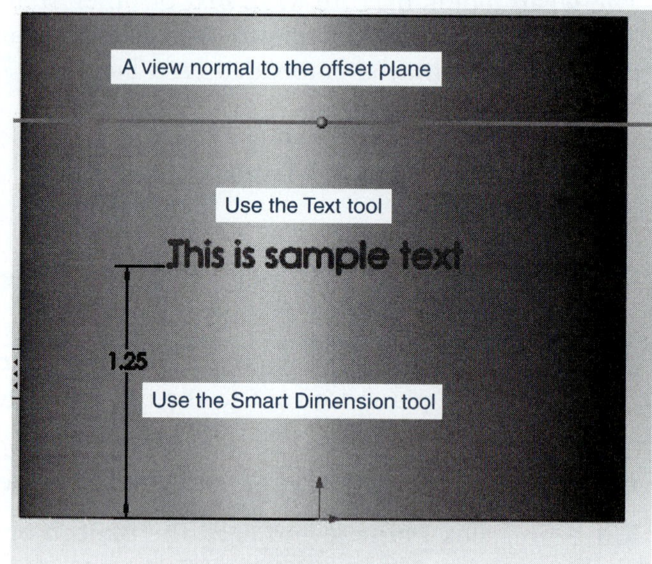
A view normal to the offset plane
Use the Text tool
This is sample text
1.25
Use the Smart Dimension tool

Figure 3-43
(Continued)

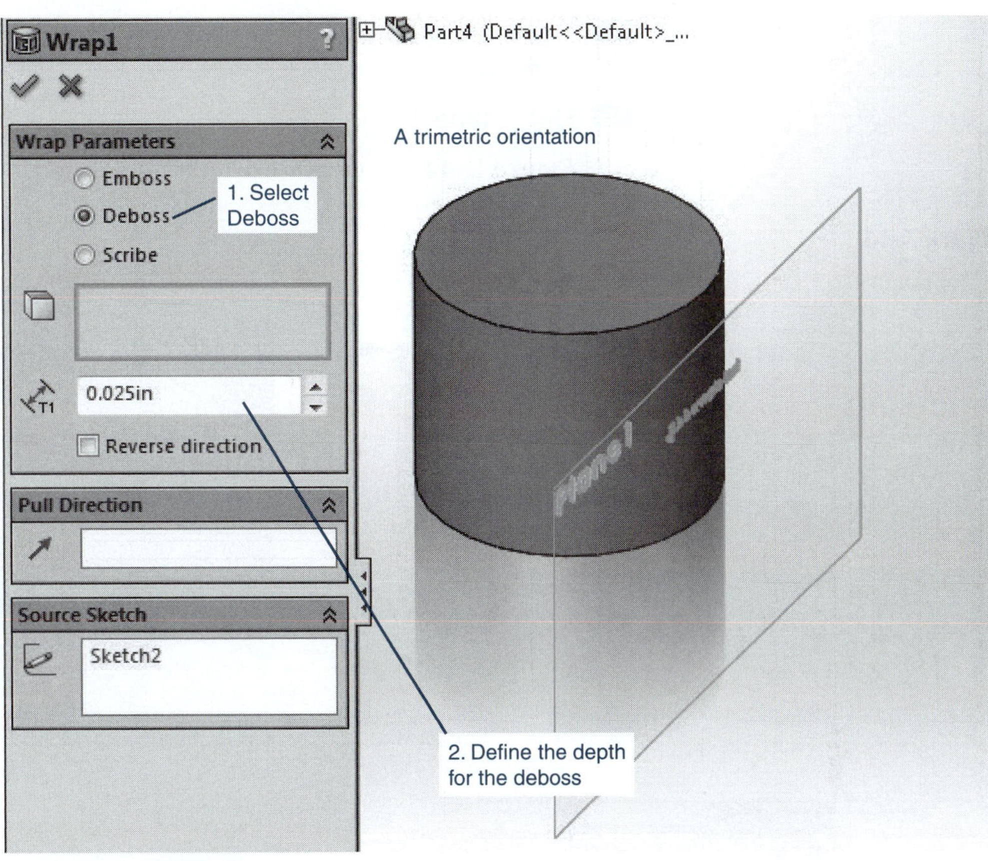

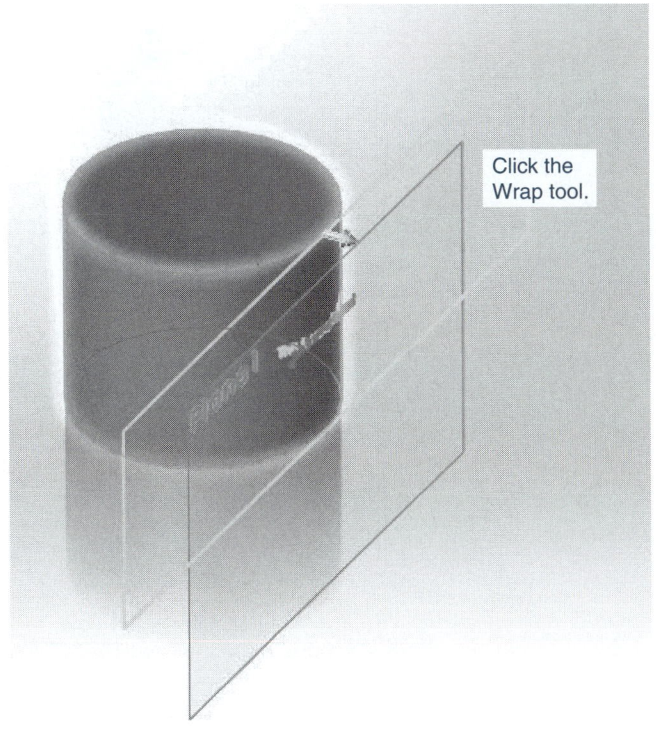

11 Click the edge surface of the cylinder to define the **Face for Wrap Sketch**.

12 Hide Plane 1.

Figure 3-44 shows an example of scribed text.

Figure 3-44

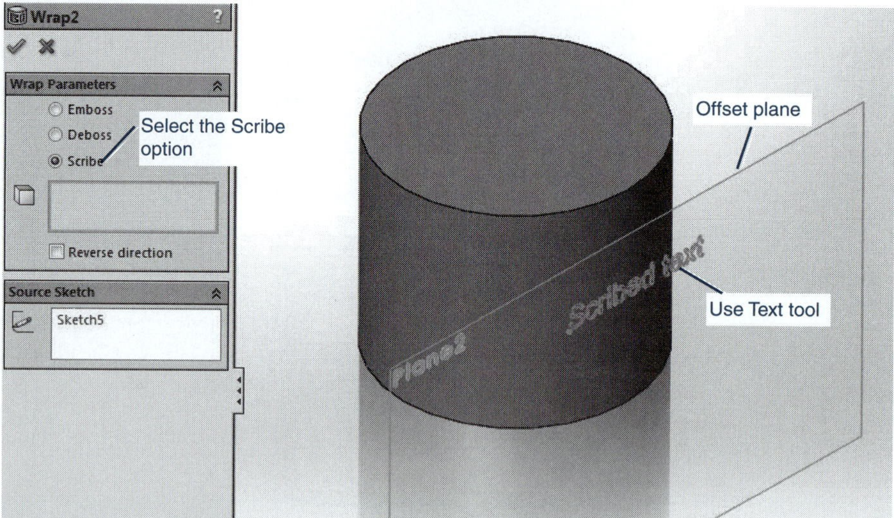

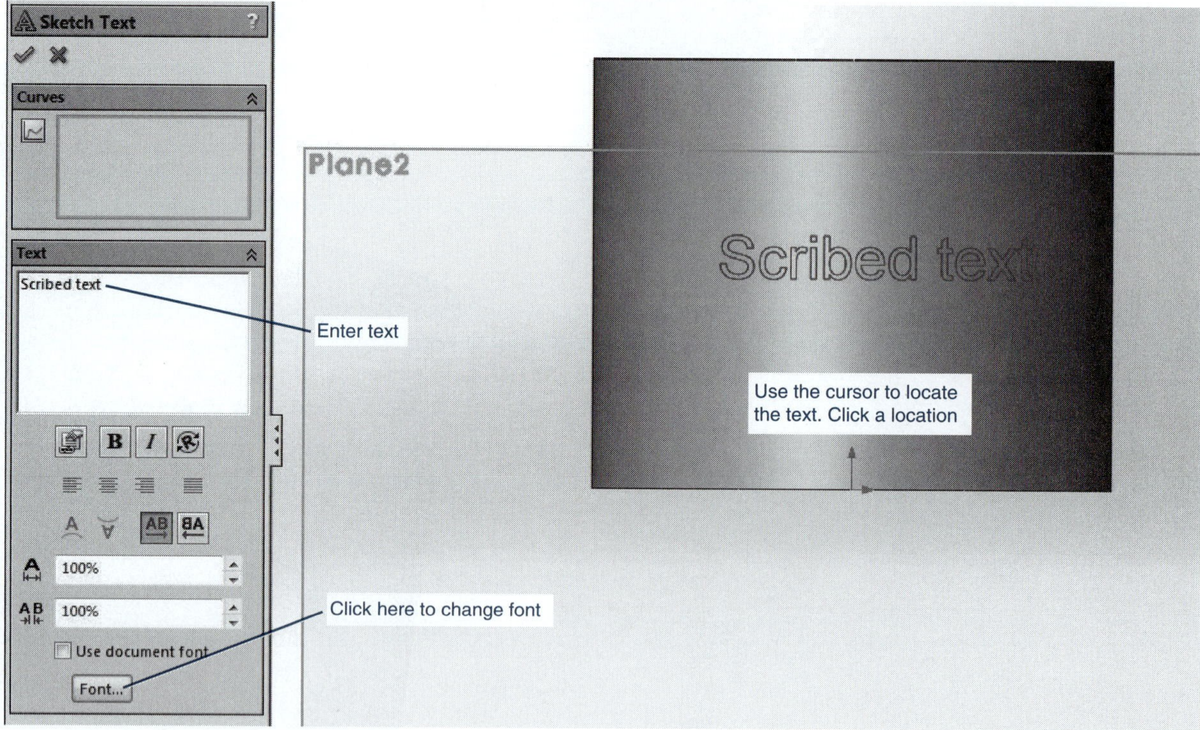

3-25 Editing Features

SolidWorks allows you to edit existing models. This is a very powerful feature in that you can easily make changes to a completed model without having to redraw the entire model.

> **TIP**
> The **Edit Sketch** tool is used to edit shapes created using the **Sketch** tools such as holes. The **Edit Features** tool is used to edit shapes created using the **Features** tools such as a cut or extrusion.

Figure 3-45 shows the L-bracket originally created in Sections 3-3 through 3-6. The finished object may be edited. In this example, the hole's diameter and the size of the cutout will be changed.

Figure 3-45

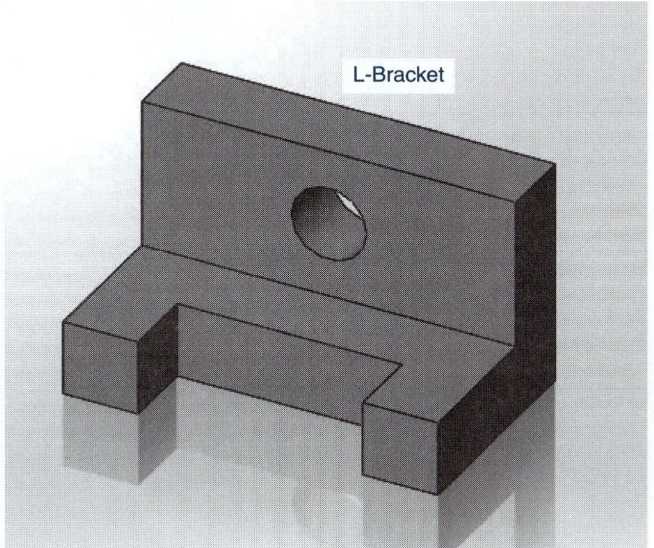

> **TIP**
> The hole will be highlighted when selected.

To Edit the Hole

See Figure 3-46.

1. Right-click the hole on the L-bracket.

   A dialog box will appear.

2. Select the **Edit Sketch** option.

   The sketch used to create the hole will appear.

3. Double-click the Ø20 dimension.

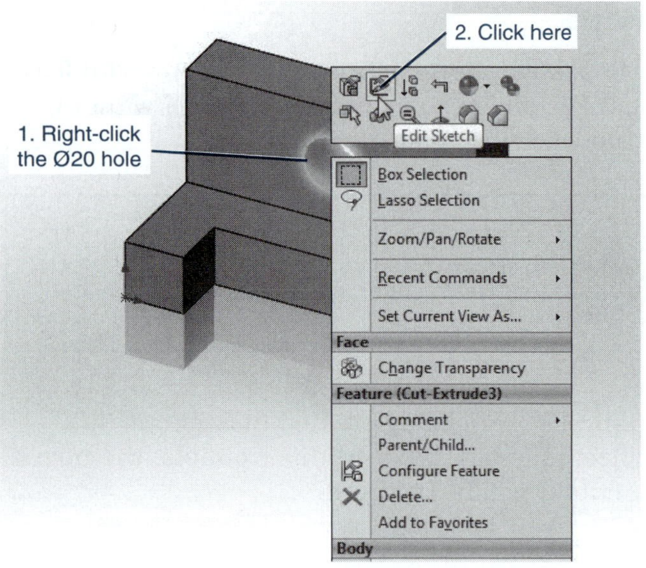

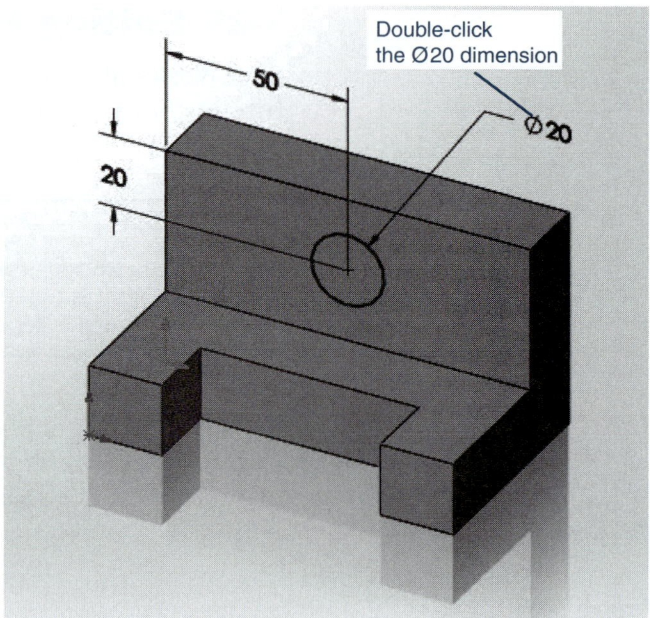

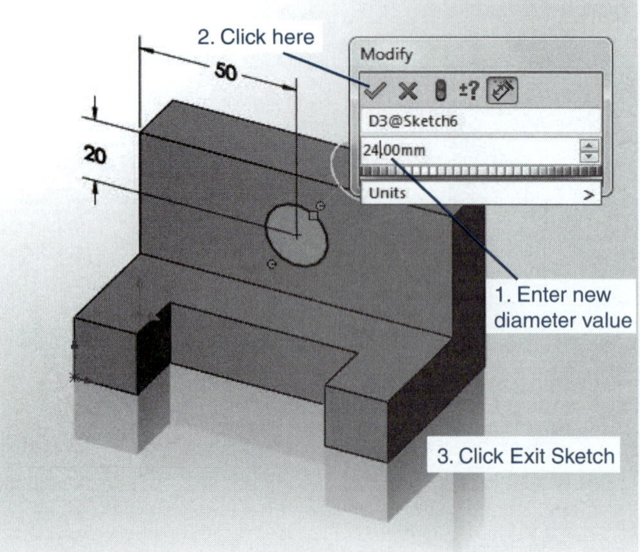

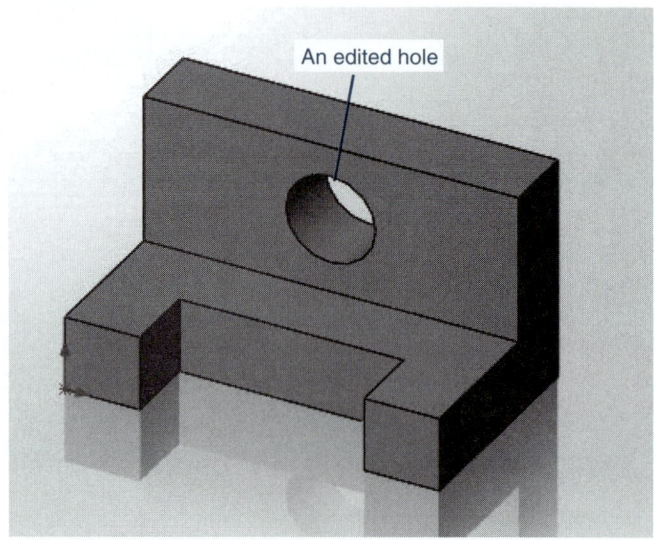

Figure 3-46

4 Change the dimension to **Ø24.00** and click the green **OK** check mark.

5 Click the **Exit Sketch** tool.

To Edit the Cutout

See Figure 3-47.

1 Right-click the back surface of the cutout.

The surface will be highlighted when selected. A dialog box will appear.

2 Click the **Edit Sketch** option.

3 Double-click the two **20** dimensions that define the length of the cutout and change their value to **30**.

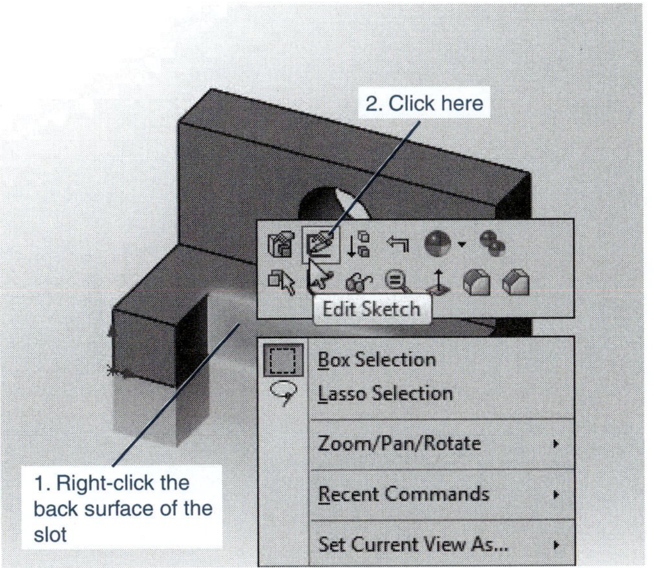

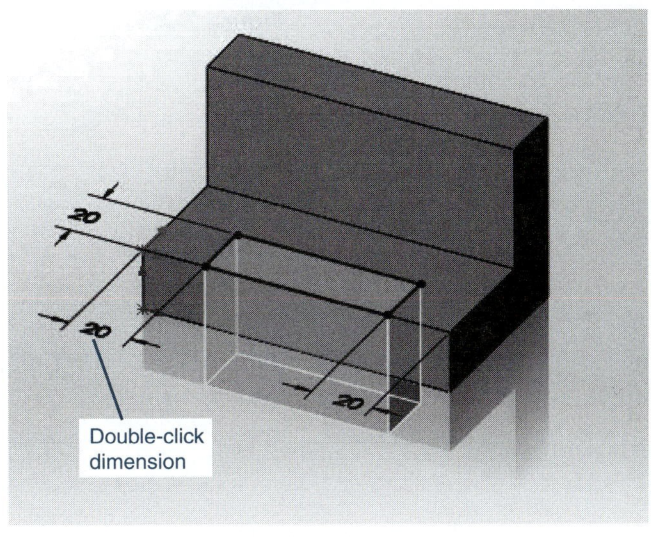

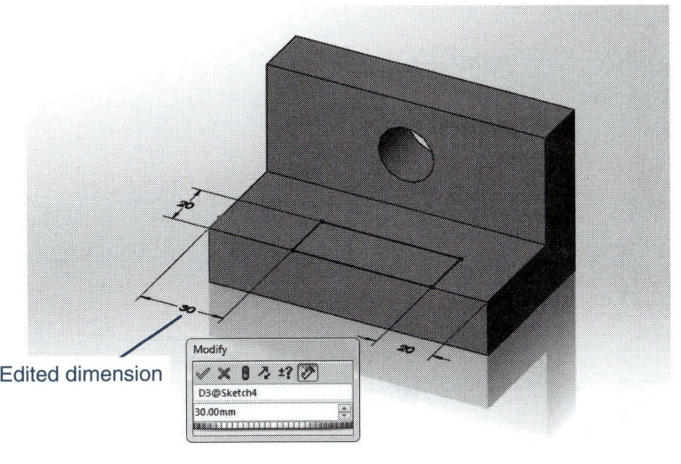

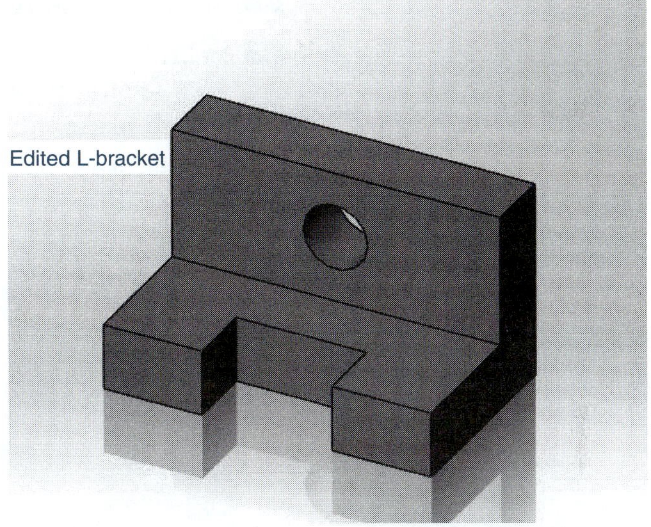

Figure 3-47

4 Click the green **OK** check mark.

5 Click the **Exit Sketch** option.

3-26 Sample Problem SP3-2

Figure 3-48 shows a cylindrical object with a slanted surface, a cutout, and a blind hole. Figure 3-49 shows how to draw the object. The procedures presented in Figure 3-43 represent one of several possible ways to create the object.

Figure 3-48

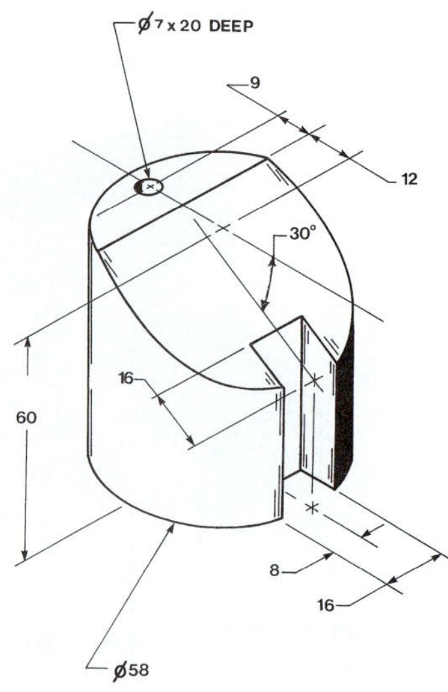

To Draw a Cylinder

1. Start a new **Part** document.
2. Define the units as millimeters **(MMGS)**, the Overall drafting standards should be **ANSI**, access the top plane, and make the top plane a sketch plane.
3. Draw a **Ø58** circle and extrude it to **60** centered on the origin.

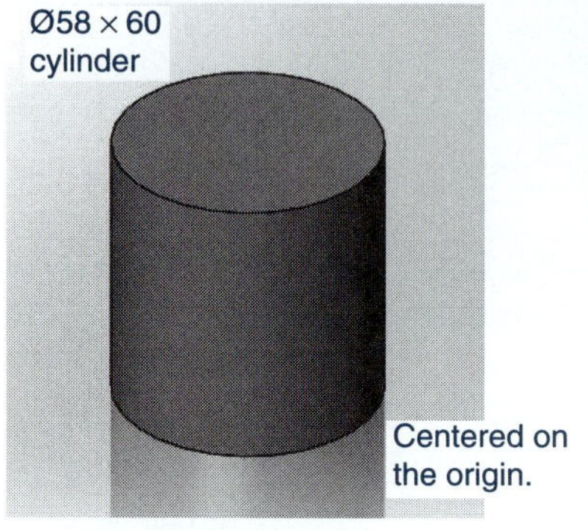

Figure 3-49

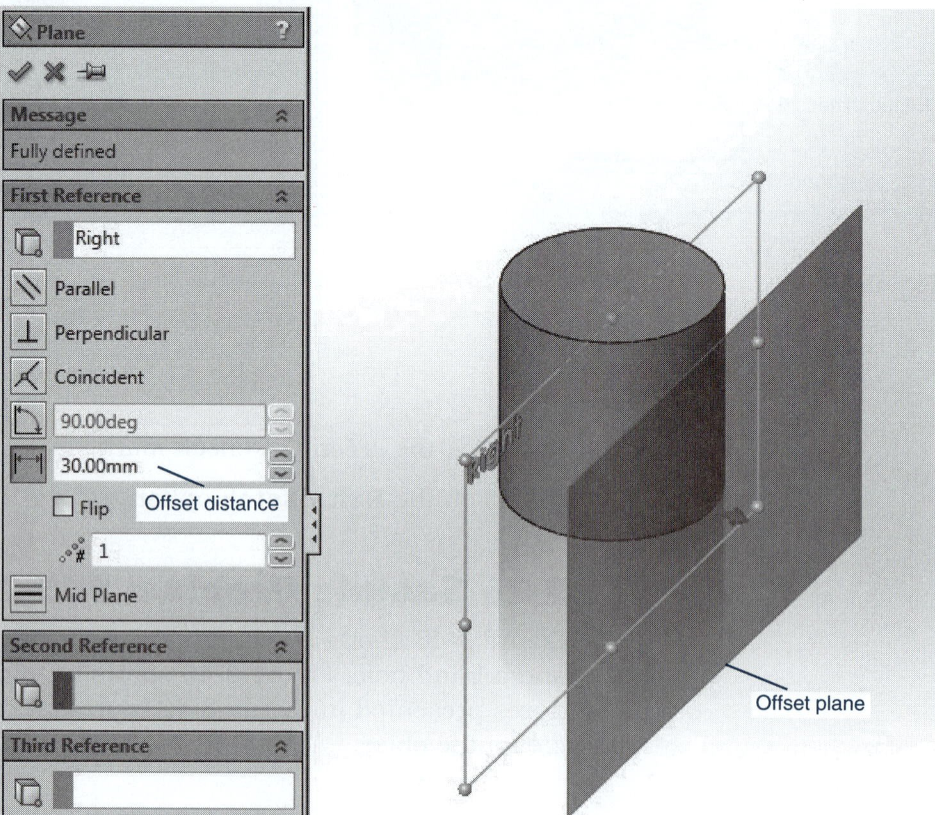

Figure 3-49
(*Continued*)

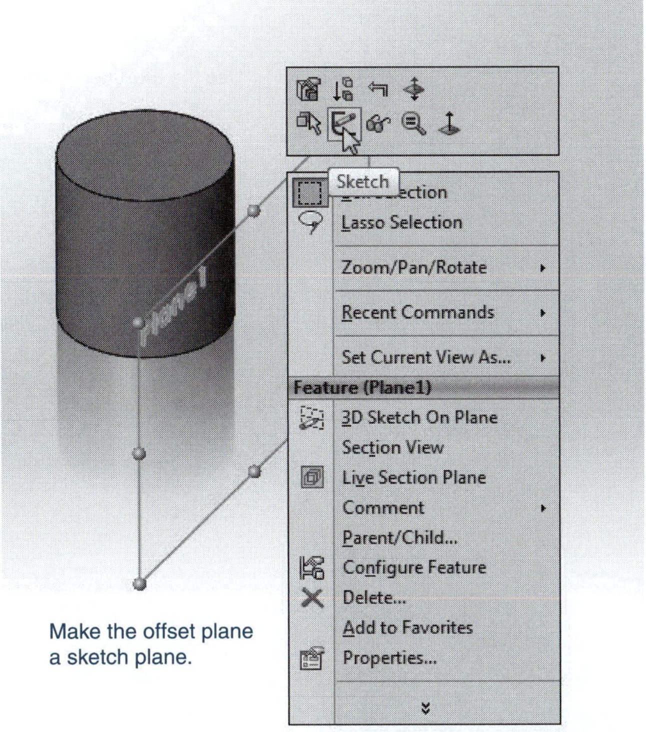
Sketch
Lasso Selection
Zoom/Pan/Rotate
Recent Commands
Set Current View As...
Feature (Plane1)
3D Sketch On Plane
Section View
Live Section Plane
Comment
Parent/Child...
Configure Feature
Delete...
Add to Favorites
Properties...
Plane1
Make the offset plane a sketch plane.

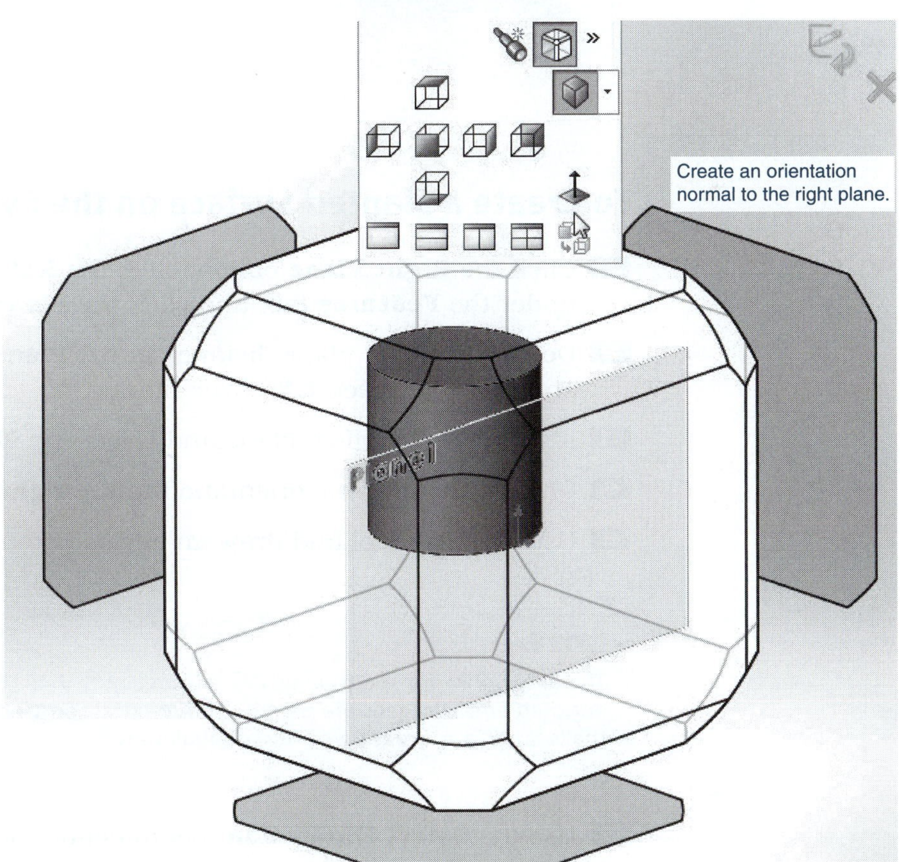
Create an orientation normal to the right plane.
Plane1

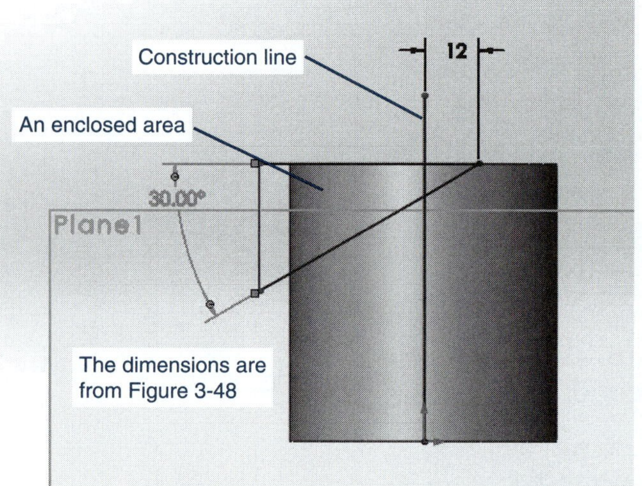

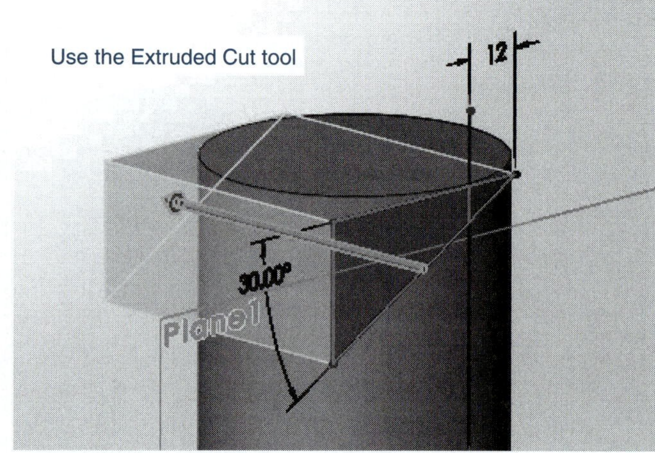

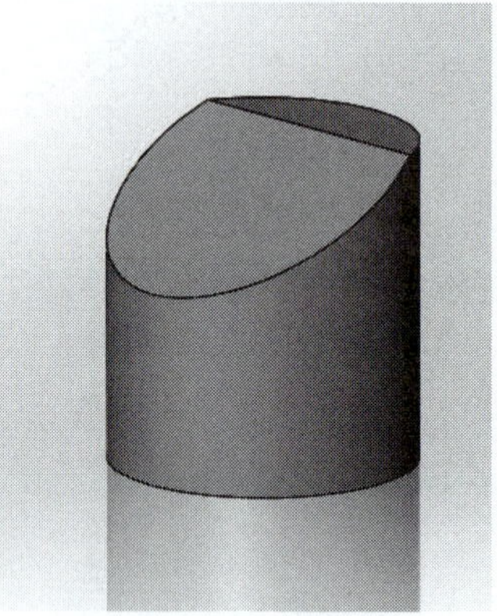

Figure 3-49
(*Continued*)

To Create a Slanted Surface on the Cylinder

1. Click the **Right Plane** option, click the **Reference Geometry** tool under the **Features** tab, and click the **Plane** option.
2. Define the offset plane distance in the **Plane PropertyManager** as **30**, and click the green **OK** check mark.
3. Right-click the offset plane and select the **Sketch** option.
4. Change the drawing orientation to the **right plane**.
5. Use the **Line** tool and draw an enclosed triangular shape.

> **NOTE**
>
> The dimensions for the triangle came from Figure 3-48. The triangle must be an enclosed area. No gaps are permitted. A vertical construction line was added to help in the location and creation of the triangular area.

6. Use the **Smart Dimension** tool to define the size and location of the triangle.
7. Change the drawing orientation to **Dimetric** and click the **Extruded Cut** tool in the **Features** tab.

8 Set the length of the cut to **60.00mm** and click the **OK** check mark.

9 Right-click the offset plane and click the **Hide** option.

To Add the Vertical Slot

See Figure 3-50.

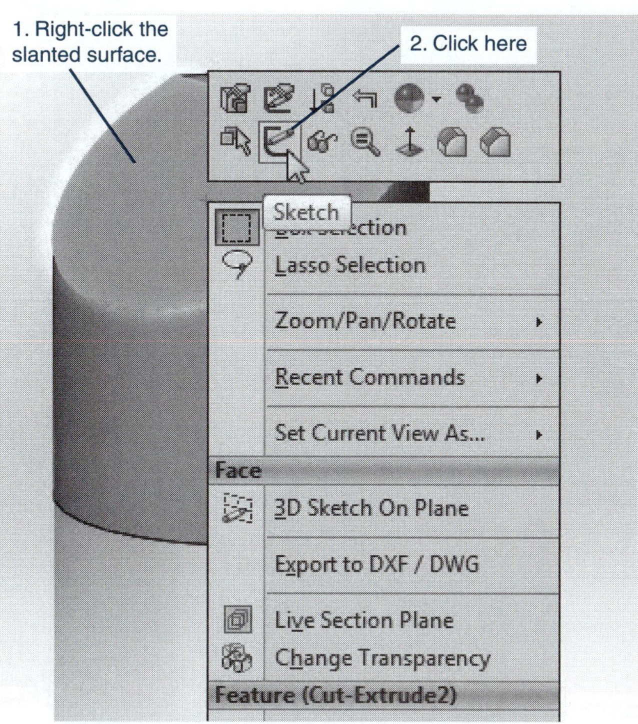

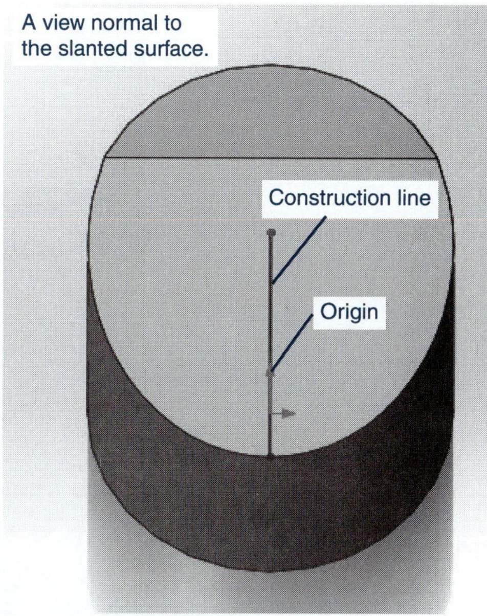

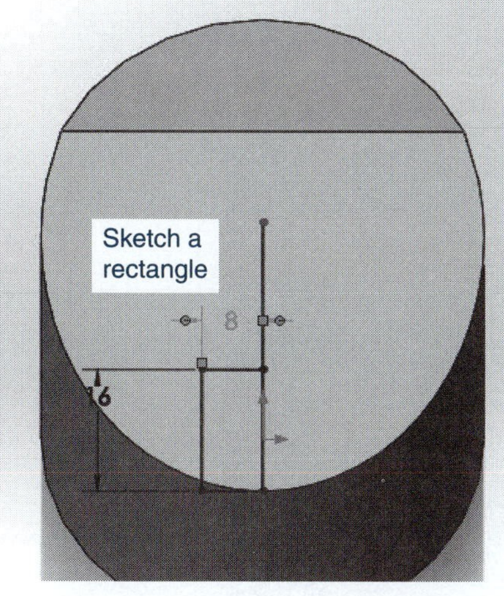

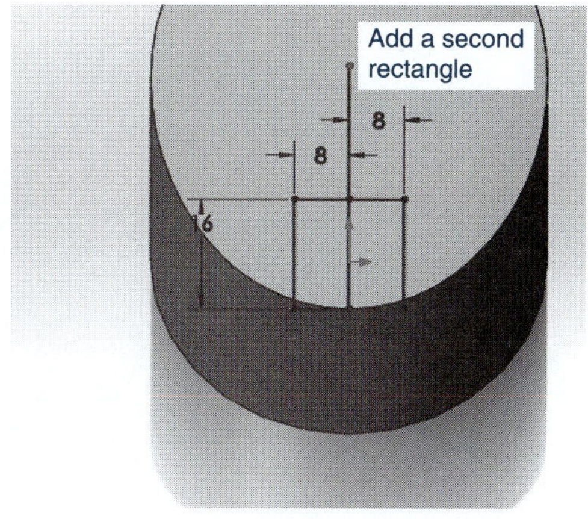

Figure 3-50

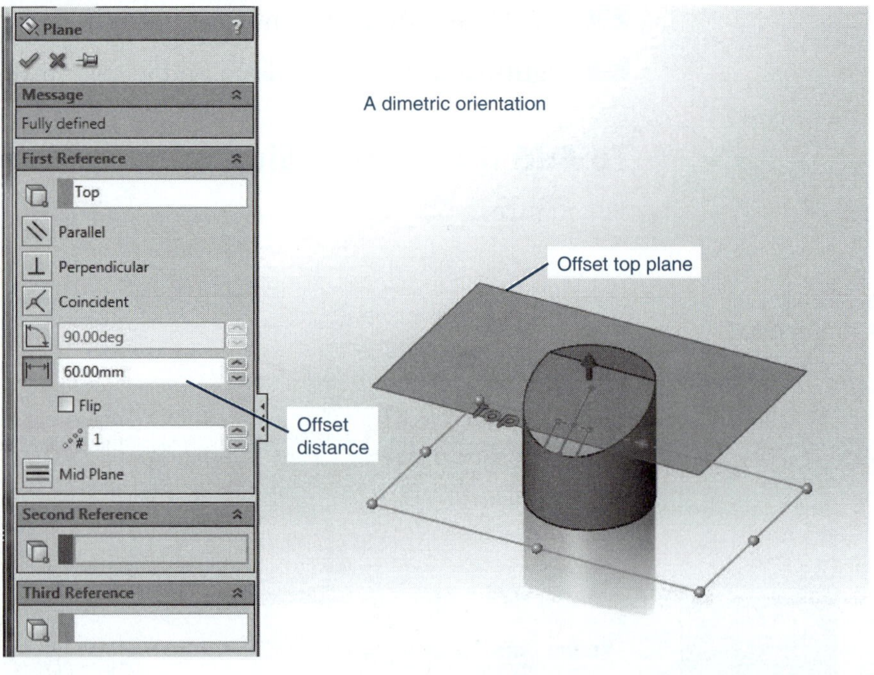
Plane
Message
Fully defined
First Reference
Top
Parallel
Perpendicular
Coincident
90.00deg
60.00mm
Flip
1
Mid Plane
Second Reference
Third Reference
A dimetric orientation
Offset top plane
Offset
distance

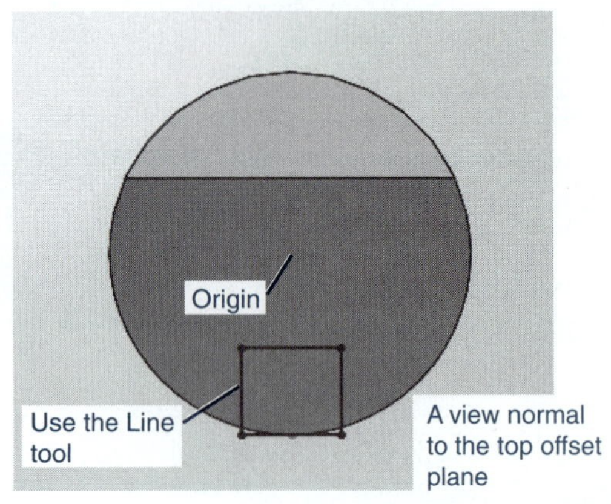
Origin
Use the Line
tool
A view normal
to the top offset
plane

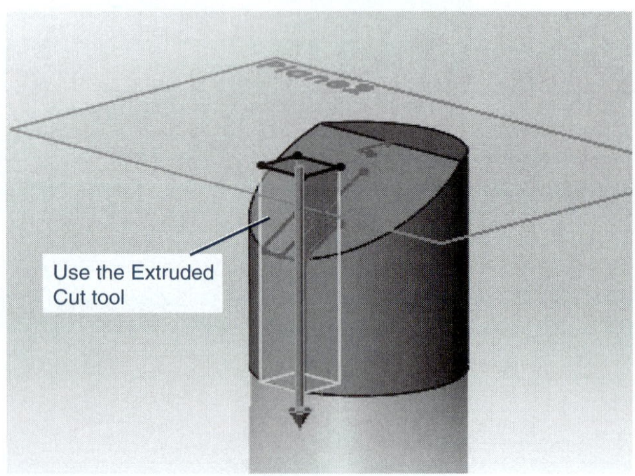
Use the Extruded
Cut tool

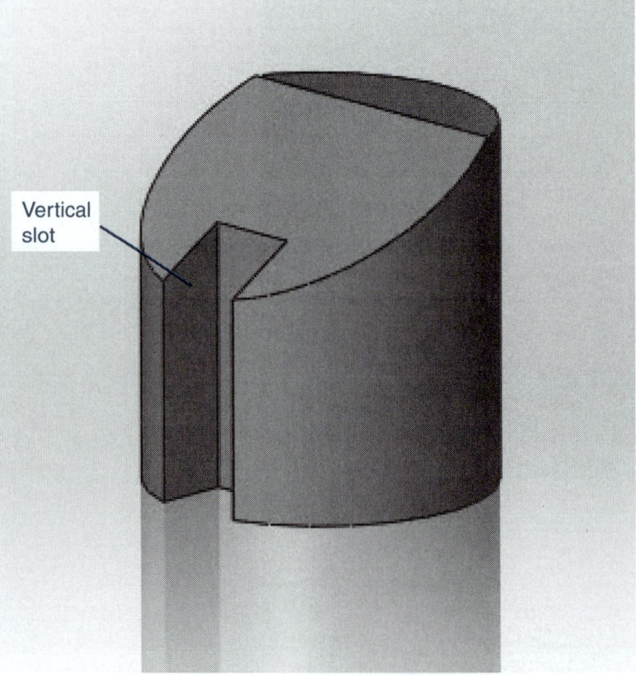
Vertical
slot

Figure 3-50
(*Continued*)

1. Right-click the slanted surface and click the **Sketch** tool.
2. Click the **View Orientation** tool and click the **Normal to View** tool, or click **<Ctrl-8>**.
3. Use the **Line** tool and sketch a vertical construction line through the origin. Start the line on the edge of the slanted surface.
4. Use the **Rectangle** tool on the **Sketch** toolbar and draw an **8 × 16** rectangle as shown. Use the **Smart Dimension** tool to size the rectangle.
5. Draw a second **8 × 16** rectangle as shown.
6. Change the drawing orientation to a dimetric view.
7. Exit the sketch.
8. Click the **Top Plane** option, then use the **Plane** option in the **Reference Geometry** tool on the **Features** tab and create an offset top plane **60** from the base of the cylinder.

> **NOTE**
>
> The **Extruded Cut** tool will extrude a shape perpendicular to the plane of the shape. In this example the plane is slanted, so the extrusion would not be vertical, as required. The rectangle is projected into the top offset plane and the extrusion tool applied there.

9. Right-click the 60 offset plane, select the **Sketch** option, and change the drawing orientation to the top view.
10. Use the **Line** tool and sketch a rectangle on the offset plane over the projected view of the 16 × 16 rectangle on the slanted plane, right-click the mouse, and click the **Select** option.
11. Change the drawing orientation to a dimetric view.
12. Use the **Extruded Cut** tool on the **Features** tab to cut out the slot.
13. Hide the 60 offset plane and hide the 16 × 16 rectangle on the slanted surface.
14. Click the green **OK** check mark.

To Add the Ø8 Hole

See Figure 3-51.

1. Right-click the flat top surface of the object and create a sketch plane.
2. Change the orientation to a view normal to the flat top surface.
3. Use the **Point** tool and sketch a point on the flat portion of the top surface. Use the origin to center the point.
4. Use the **Smart Dimension** tool and locate the point according to the given dimensions, in this example 9.00mm.
5. Change to a trimetric orientation and exit the sketch.

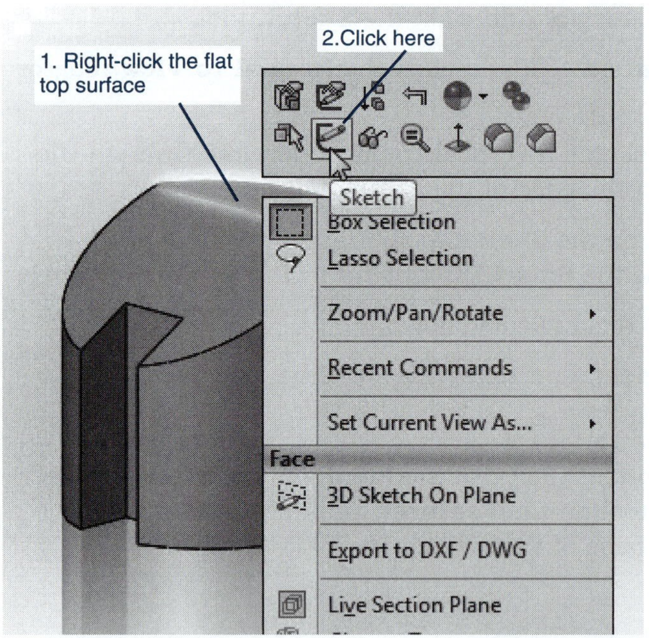
1. Right-click the flat top surface
2.Click here
Sketch
Box Selection
Lasso Selection
Zoom/Pan/Rotate
Recent Commands
Set Current View As...
Face
3D Sketch On Plane
Export to DXF / DWG
Live Section Plane

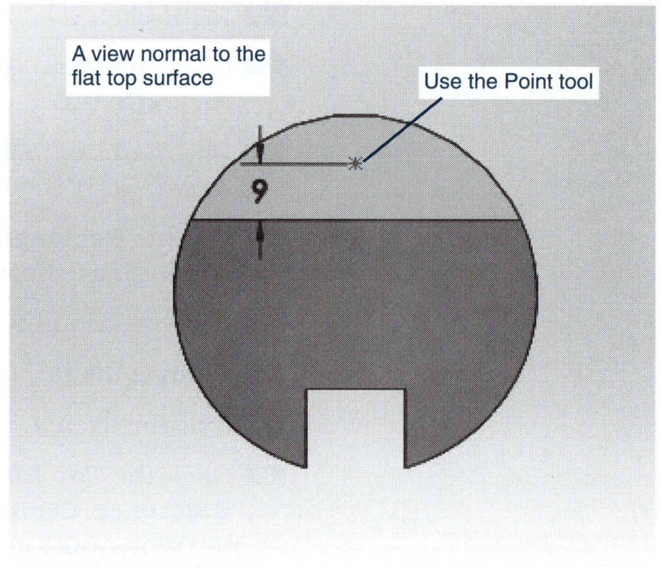
A view normal to the flat top surface
Use the Point tool
9

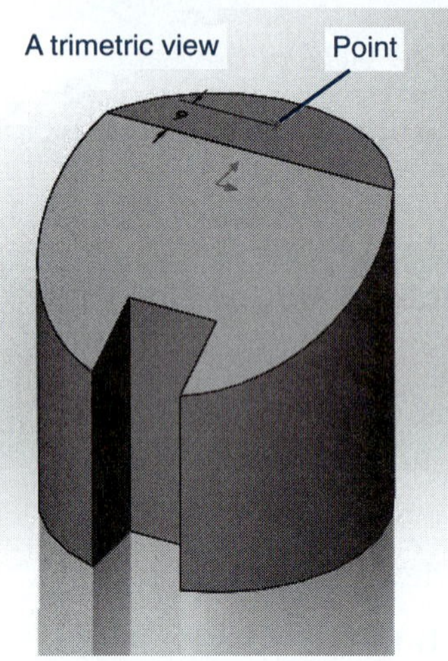
A trimetric view
Point

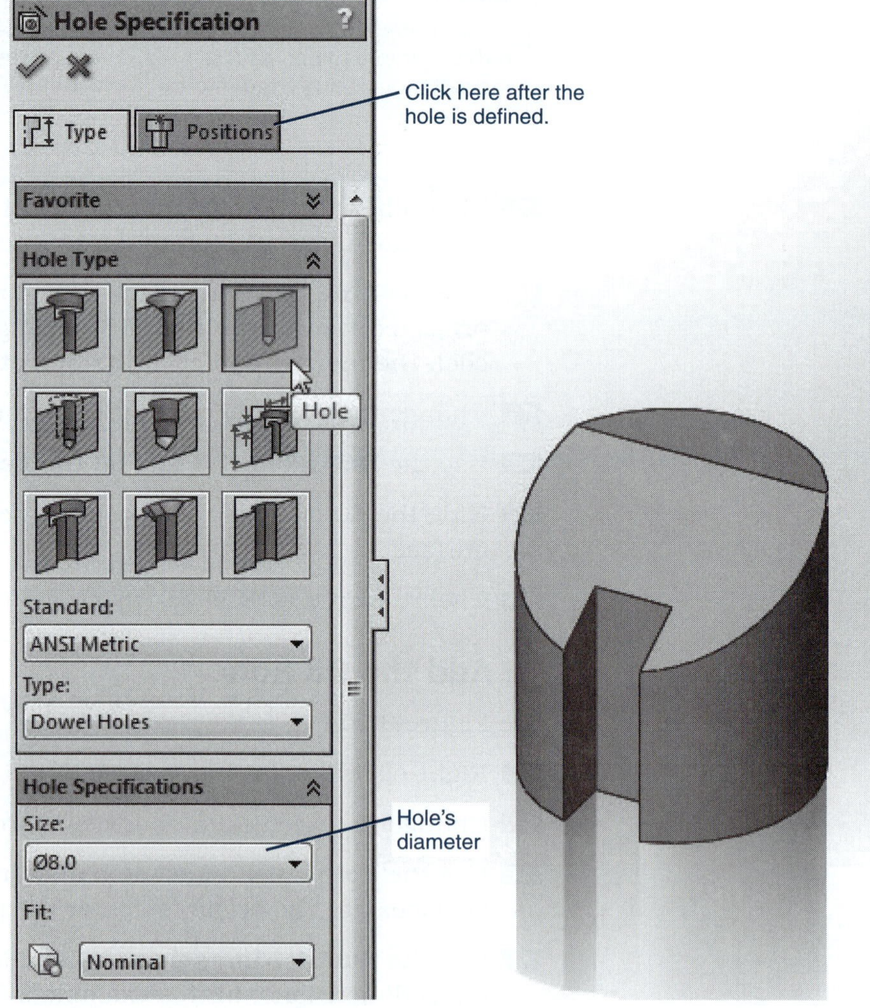
Hole Specification
Type
Positions
Click here after the hole is defined.
Favorite
Hole Type
Hole
Standard:
ANSI Metric
Type:
Dowel Holes
Hole Specifications
Size:
Ø8.0
Hole's diameter
Fit:
Nominal

Figure 3-51

Figure 3-51
(Continued)

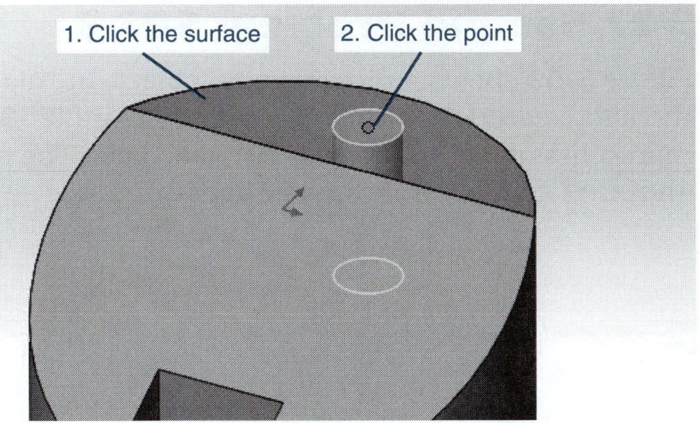

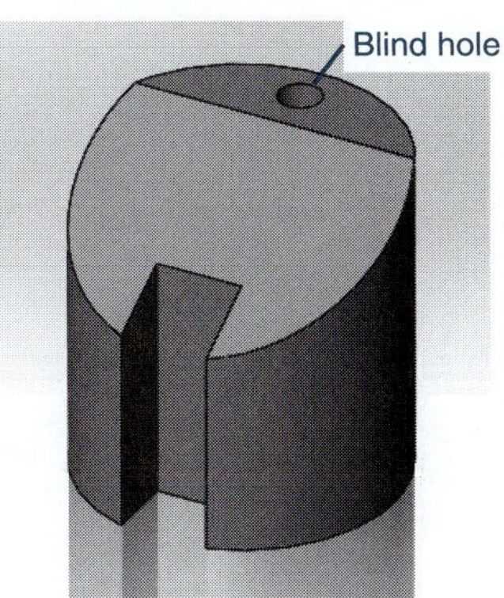

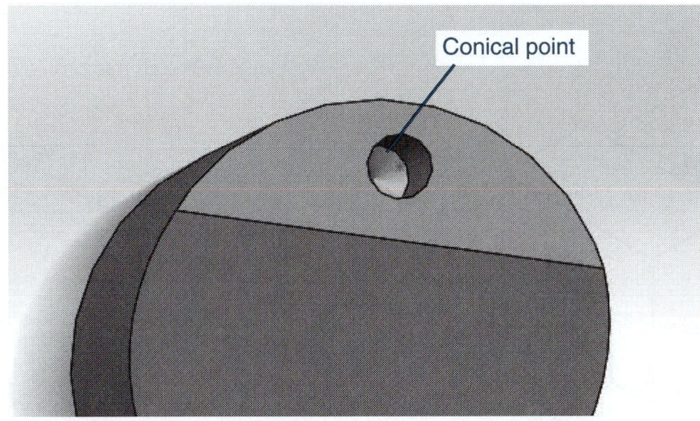

Chapter 3

NOTE

There are two ways to draw blind holes (holes that do not go all the way through): draw a circle and use the **Extruded Cut** tool to remove material, or use the **Hole Wizard**. In this example the **Hole Wizard** tool is used because it will generate a conical-shaped bottom to the hole. Conical-shaped hole bottoms result from using a twist drill, which has a conical-shaped cutting end.

6 Click the **Hole Wizard** tool on the **Features** tab.

7 Click the **Hole** option and define the hole's diameter and depth.

In this example the hole's diameter is 8.00 and the depth is 20. Note that the hole is defined as a blind hole and that the hole's depth does not include the conical point.

8 Click the **Positions** tab in the **Hole Wizard PropertyManager**.

9 Click the flat top surface, then click the point.

10 Click the **OK** check mark.

11 Rotate the drawing orientation and verify that the hole has a conical-shaped bottom.

3-27 Sample Problem SP3-3

Figure 3-52 shows a dimensioned object. In this example we will start with the middle section of the object. See Figure 3-53. The solution presented represents one of many possible solutions. The solution uses metric units and ANSI Overall drafting standards.

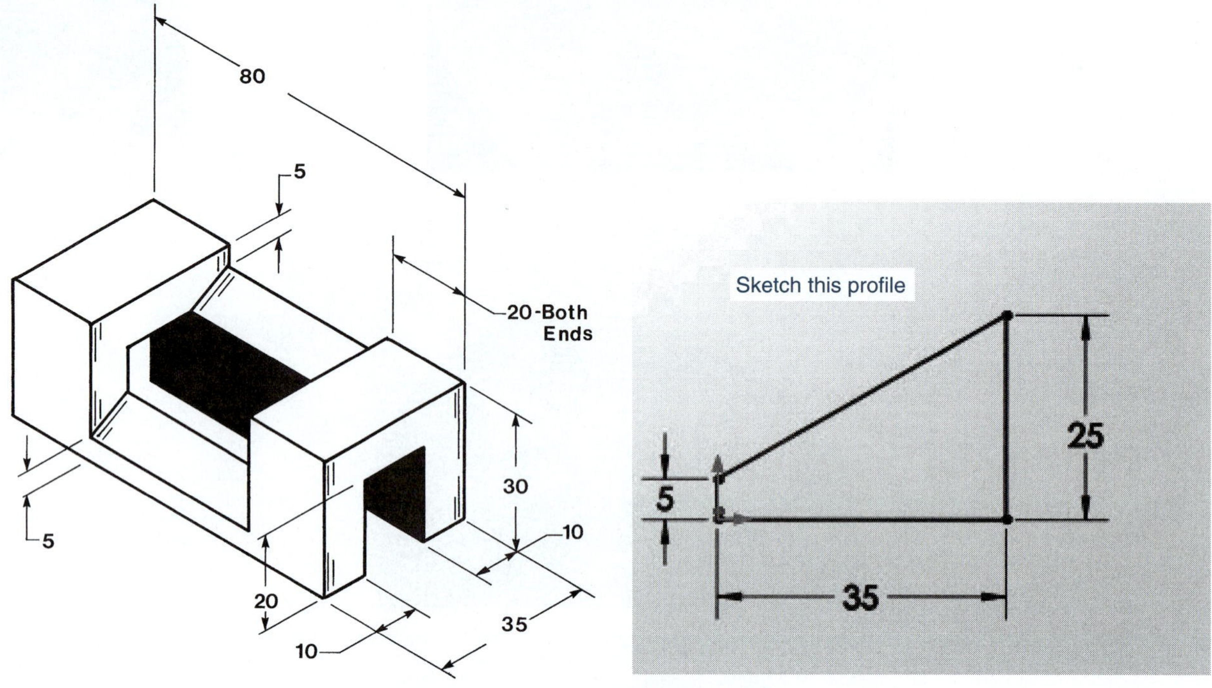

Figure 3-52

Figure 3-53

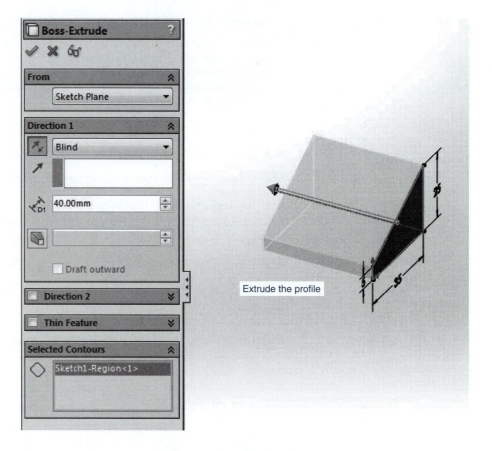

Boss-Extrude
From
Sketch Plane
Direction 1
Blind
40.00mm
Draft outward
Direction 2
Thin Feature
Selected Contours
Sketch1-Region<1>
Extrude the profile

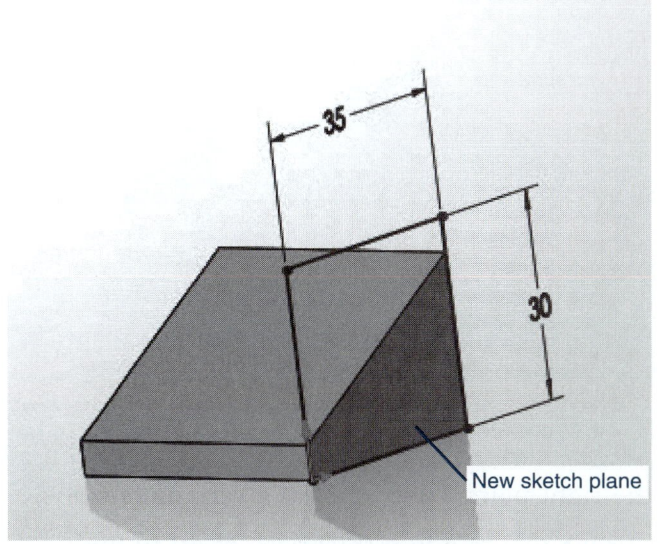

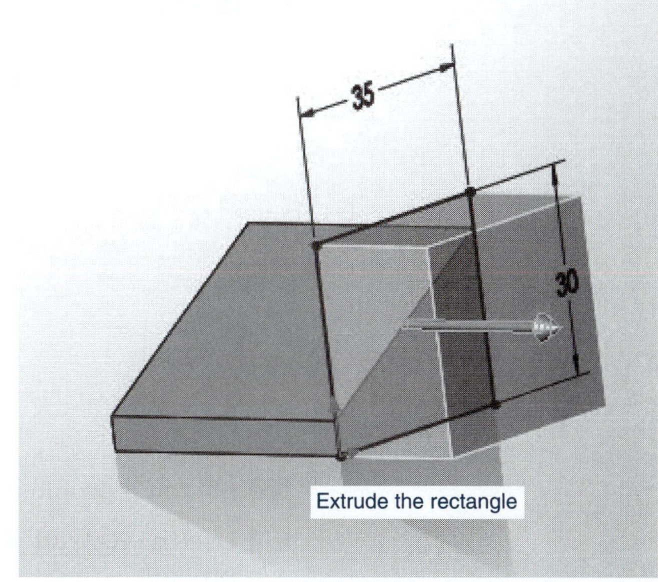

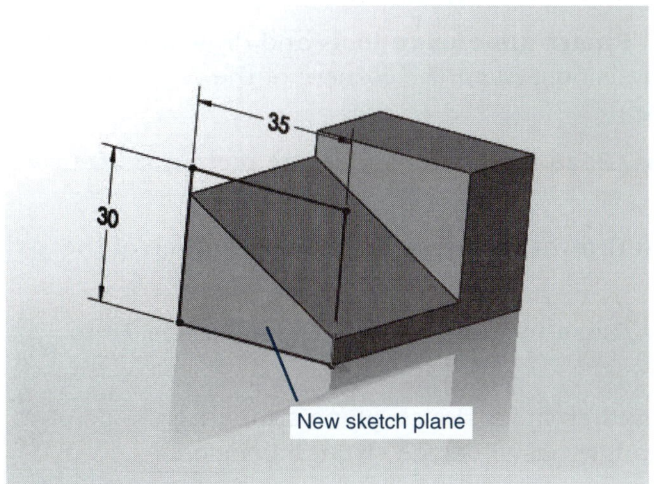

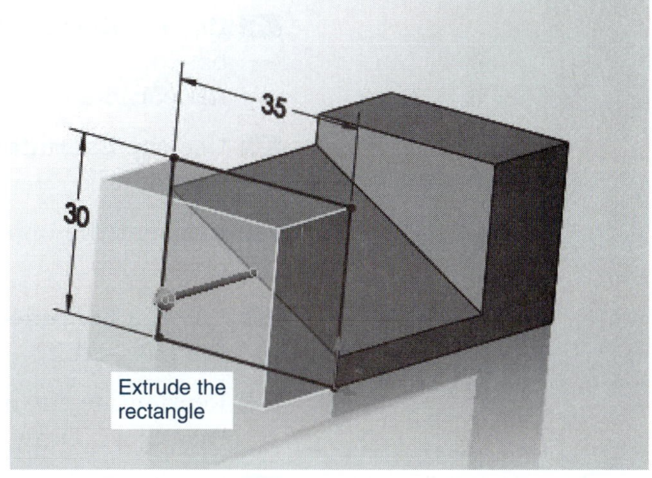

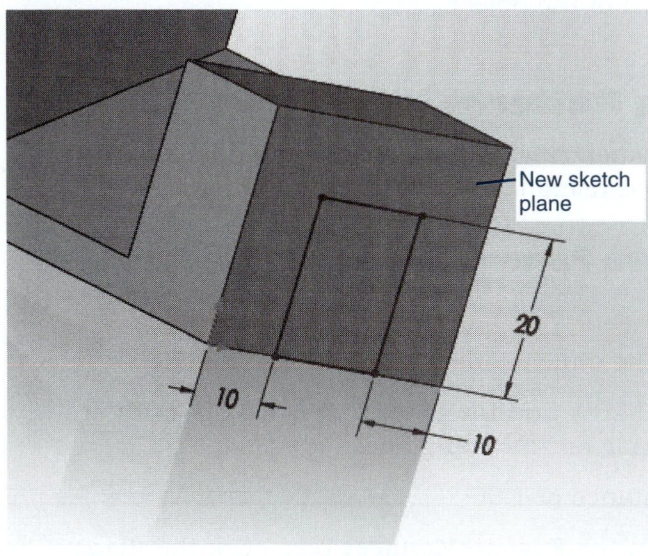

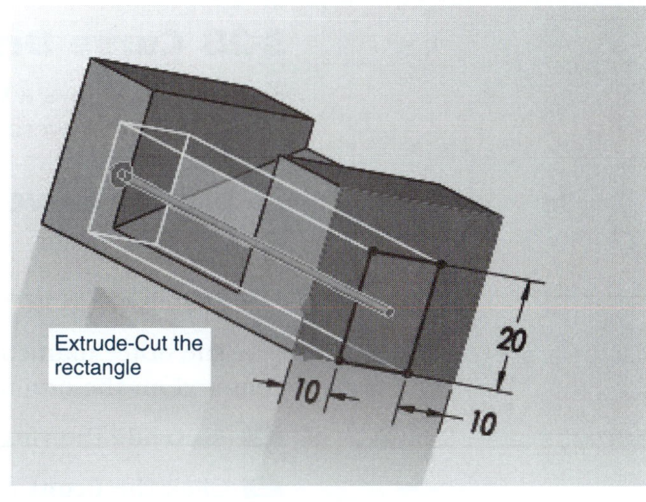

Figure 3-53
(*Continued*)

Figure 3-53
(*Continued*)

1. Sketch a profile using the right plane based on the given dimensions.
2. Use the **Extruded Boss/Base** tool to add a thickness of **40** to the profile.
3. Right-click the right surface of the object and select the **Sketch** option.
4. Use the **Rectangle** and **Smart dimension** tools and draw a rectangle based on the given dimensions. Align the corners of the rectangle with the corners of the object.
5. Use the **Extruded Boss/Base** tool and extrude the rectangle **20** to the right.
6. Reorient the object and draw a rectangle on the left surface of the object.
7. Use the **Extruded Boss/Base** tool and extrude the rectangle **20** to the left.
8. Reorientate the object and create a sketch plane on the right side of the object. Draw a rectangle based on the given dimensions.
9. Use the **Extruded Cut** tool on the **Features** tab and cut out the rectangle over the length of the object.

3-28 Curve Driven Patterns

Figure 3-54 shows a Ø4.00 inch ring with 12 holes though its side surfaces. The holes were created using the **Curve Driven Pattern** tool.

To Use the Curve Driven Pattern Tool – Example 1

See Figure 3-55.

1. Sketch the ring using the dimensions shown in Figure 3-54.

   The outer diameter is Ø4.00, and the inner ring is Ø3.00. Both are centered on the origin working on the **Top** plane.

2. **Extrude** the ring a distance of 1.00.
3. Click the **Front** plane in the **FeatureManager** box, click the **Plane** option under the **Reference Geometry** tool and create an offset plane tangent to the front edge of the ring.

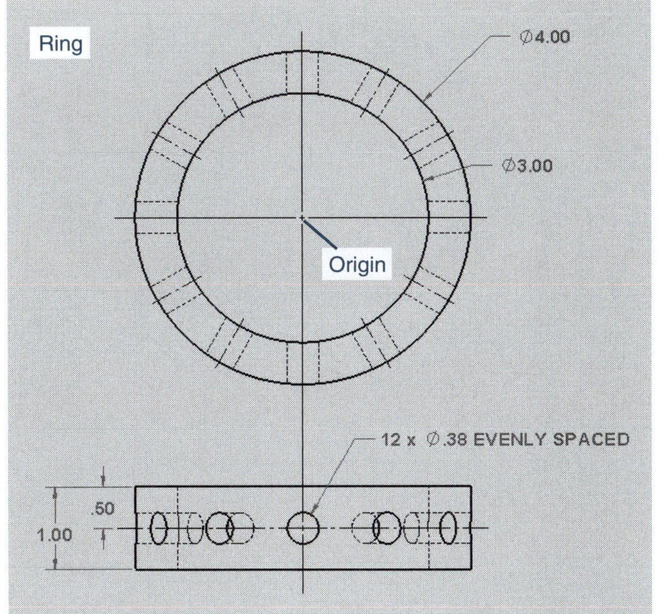

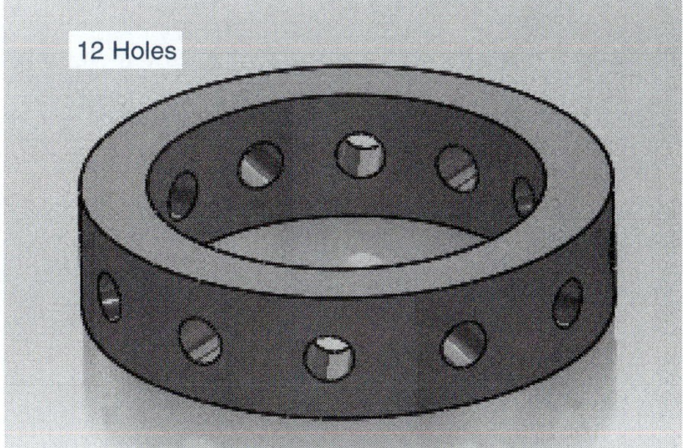

Figure 3-54

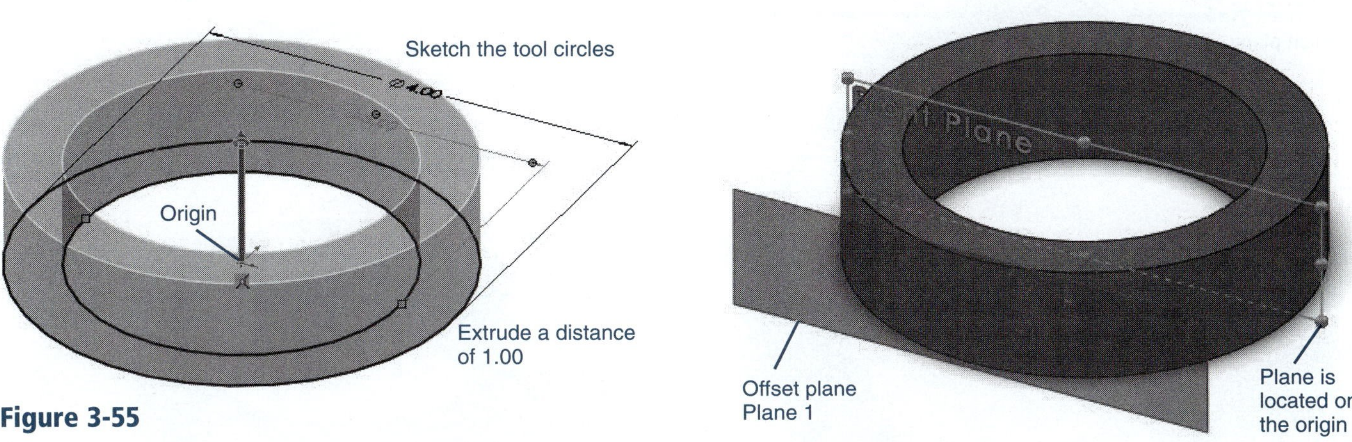

Figure 3-55

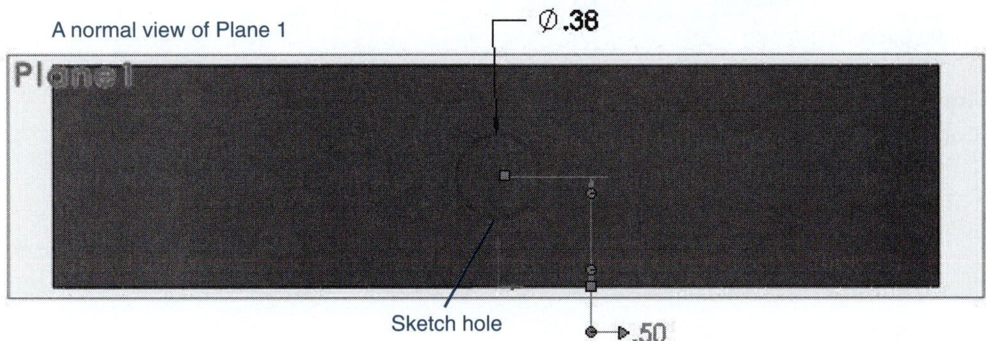

Chapter 3

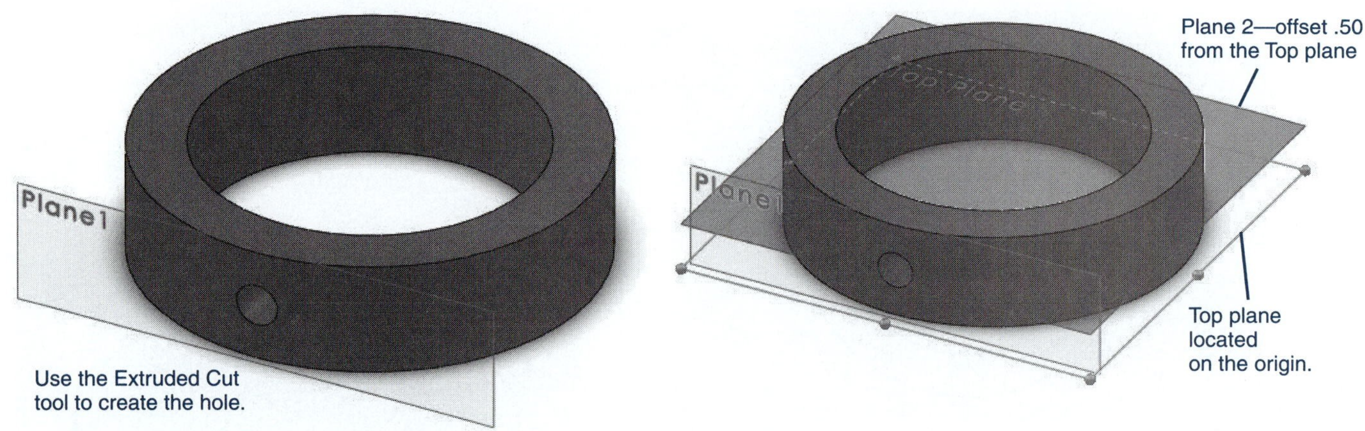
Plane1
Use the Extruded Cut
tool to create the hole.
Plane 2—offset .50
from the Top plane
Top plane
located
on the origin.

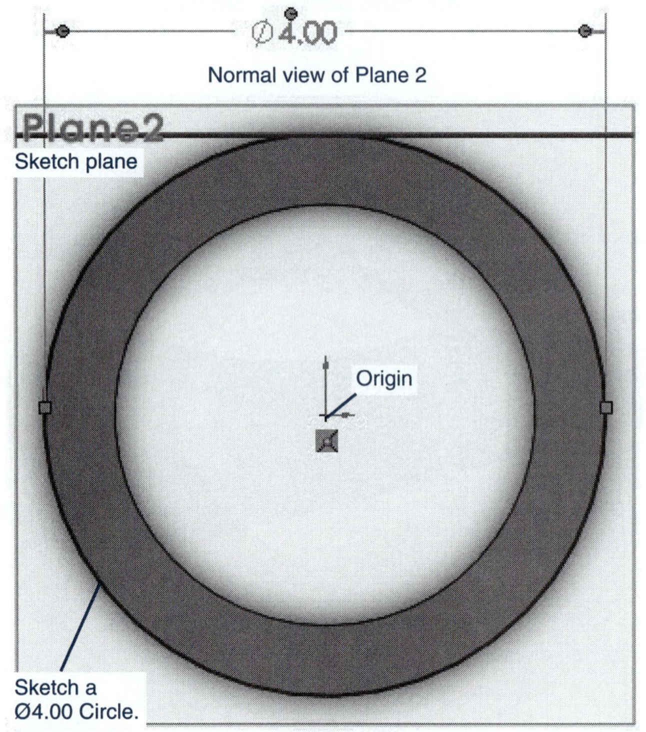
Ø4.00
Normal view of Plane 2
Plane2
Sketch plane
Origin
Sketch a
Ø4.00 Circle.

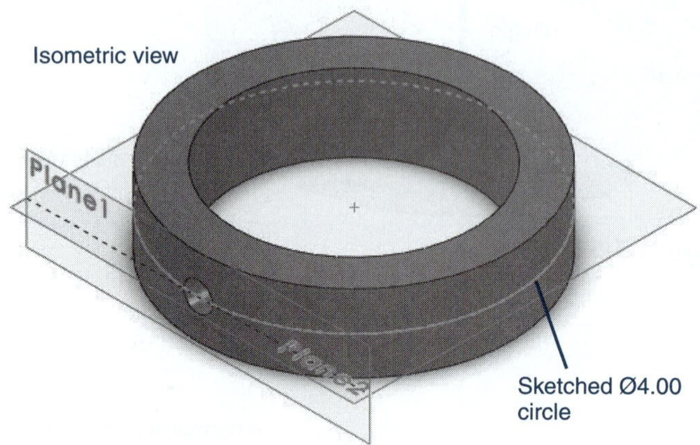
Isometric view
Plane1
Plane2
Sketched Ø4.00
circle

Figure 3-55
(*Continued*)

Figure 3-55
(*Continued*)

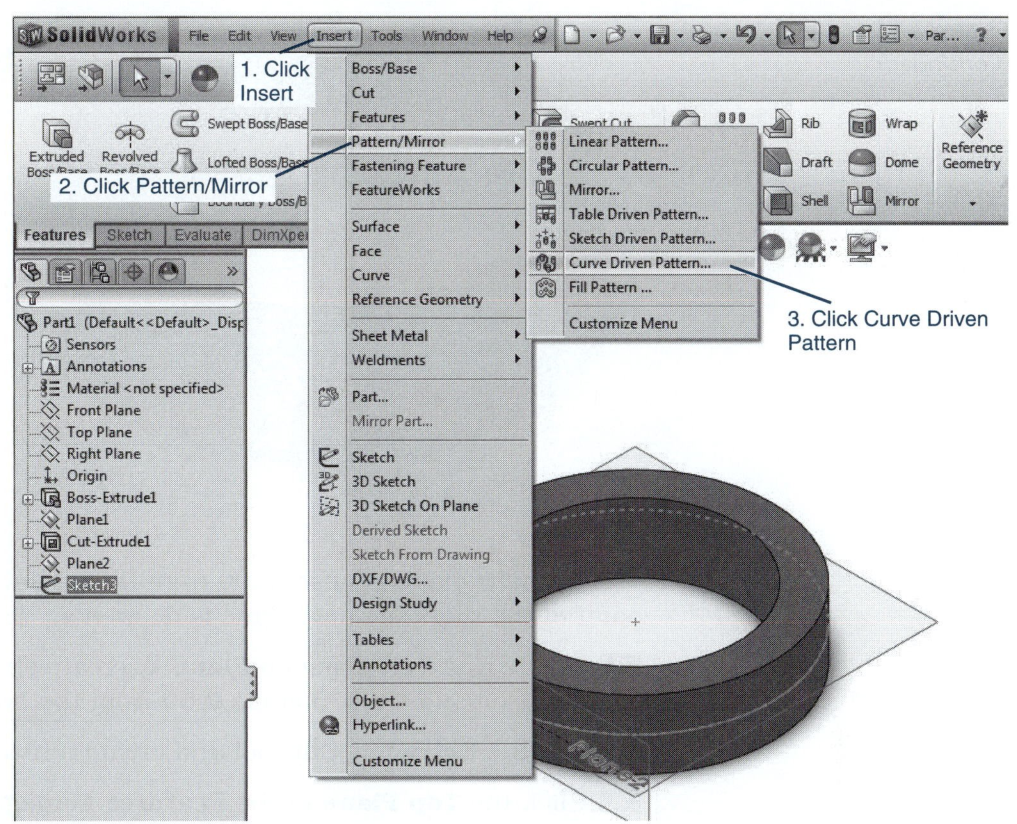

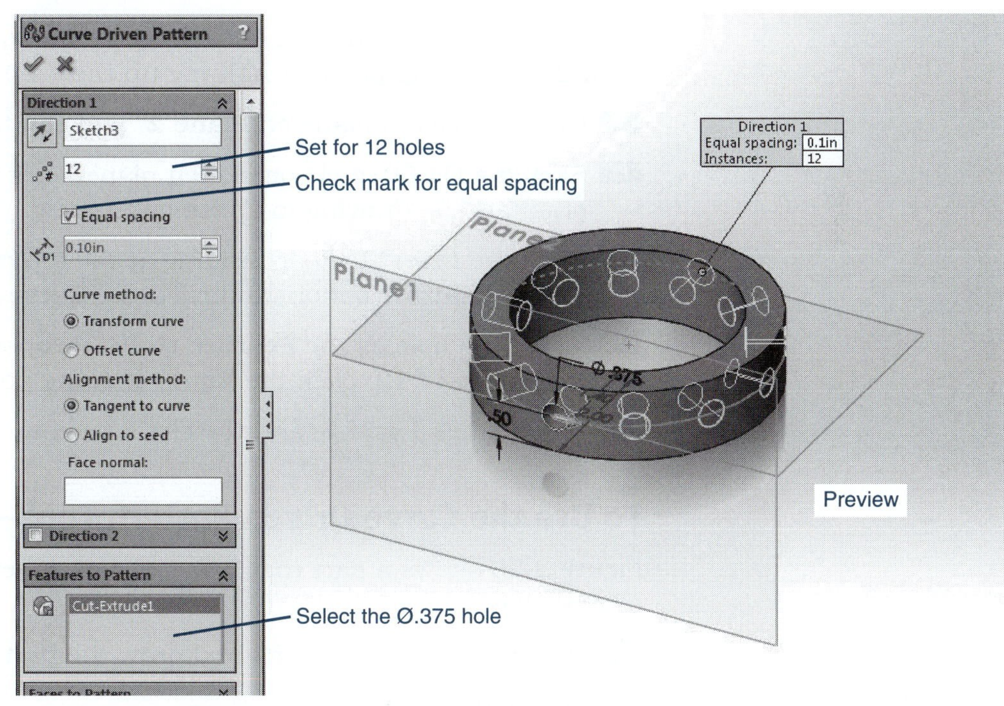

Figure 3-55
(*Continued*)

The new plane is offset 2.00 from the origin, that is, from the centerpoint of the ring. The new plane is defined as Plane 1 in this example.

4 Create a sketch plane on **Plane 1**, create a normal view to the plane, and sketch a Ø0.375 circle **0.50** from the bottom edge of the ring.

5 Use the **Extruded Cut** tool and create a hole from the circle.

6 Click the **Top Plane** in the **Features Manager** box, click the **Plane** option under the **Reference Geometry** tool and create an offset Plane **0.50** above the initial Top Plane used to create the ring.

In this example this plane is defined as Plane 2. **Plane 2** is offset **.50** from the initial top plane or halfway up the 1.00 thickness of the ring.

7 Create a sketch plane on **Plane 2**, and create a normal view.

8 Sketch a Ø4.00 circle on sketch plane, click the **Exit Sketch** icon, and orientate the drawing to an isometric view.

9 Click the **Insert** toolbar heading at the top of the screen, click the **Pattern/Mirror** option, and click the **Curve Driven Pattern** tool.

10 Select the hole as the **Feature to Pattern**, set the **Number of Instances** to **12**, click the **Equal spacing** box, and click **OK**.

11 Hide Planes 1 and 2, and the circle used to define the pattern.

To Use the Curve Driven Pattern Tool – Example 2

Figure 3-56 shows a part that has 12 holes offset 10 from the part's outer edge surface.

1 Use the given dimensions and draw the part as shown in Figure 3-56.

See Figure 3-57.

2 Define a sketch plane on the top surface of the part and use the **Offset** tool and create a curve offset **10** from the part's outer edge.

Figure 3-57 shows a normal view of the top surface. The offset curve is created using the **Offset** tool. Both the arcs and fillets can be offset to create a continuous curve.

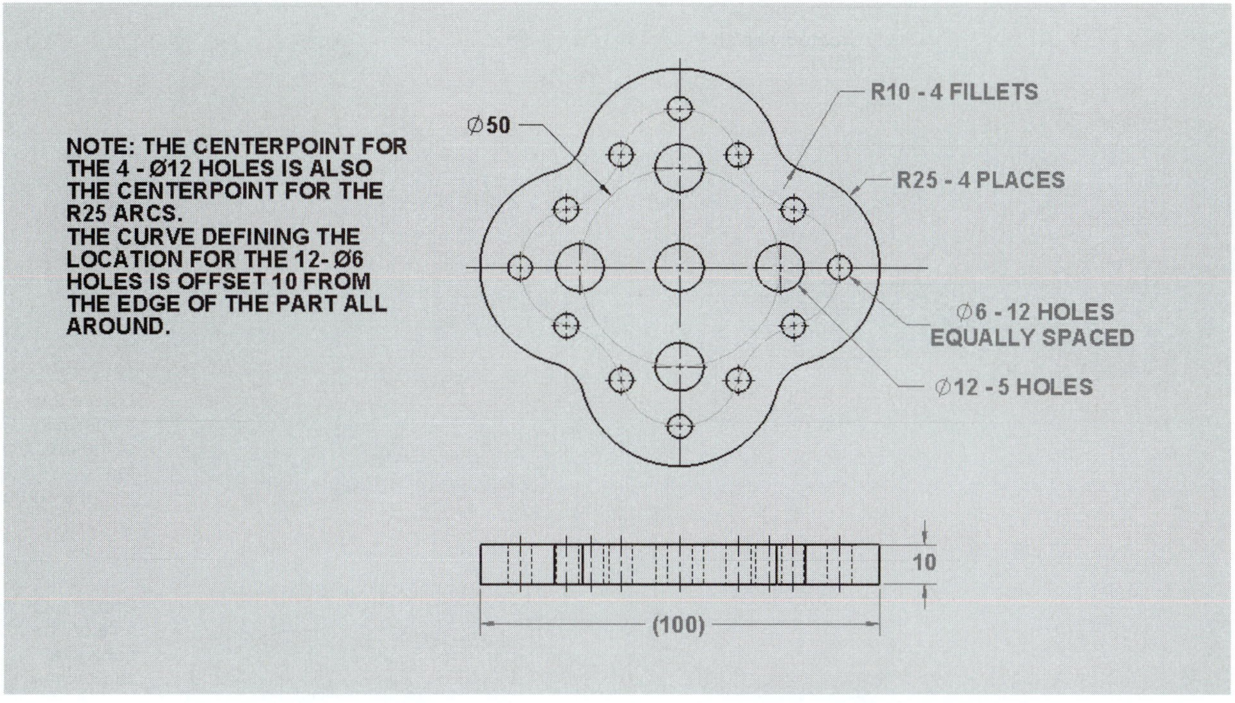
NOTE: THE CENTERPOINT FOR THE 4 - Ø12 HOLES IS ALSO THE CENTERPOINT FOR THE R25 ARCS.
THE CURVE DEFINING THE LOCATION FOR THE 12- Ø6 HOLES IS OFFSET 10 FROM THE EDGE OF THE PART ALL AROUND.
Ø50
R10 - 4 FILLETS
R25 - 4 PLACES
Ø6 - 12 HOLES EQUALLY SPACED
Ø12 - 5 HOLES
10
(100)

Figure 3-56

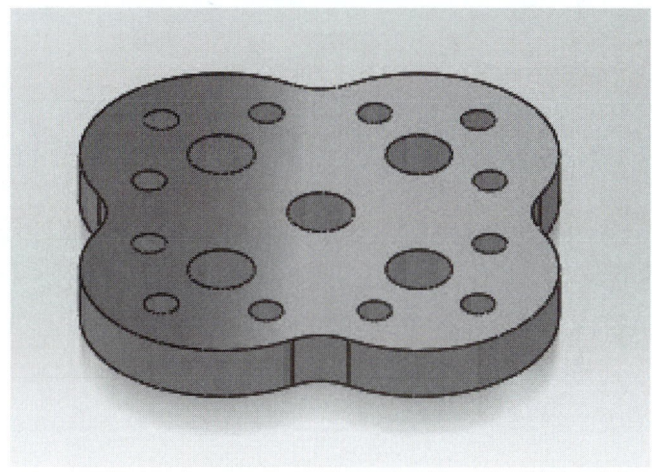

Figure 3-57

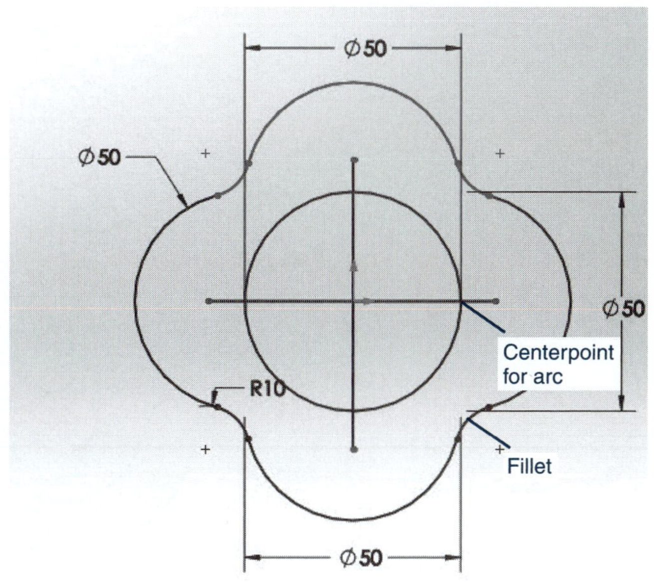
Ø50
Ø50
Ø50
Centerpoint for arc
R10
Fillet
Ø50

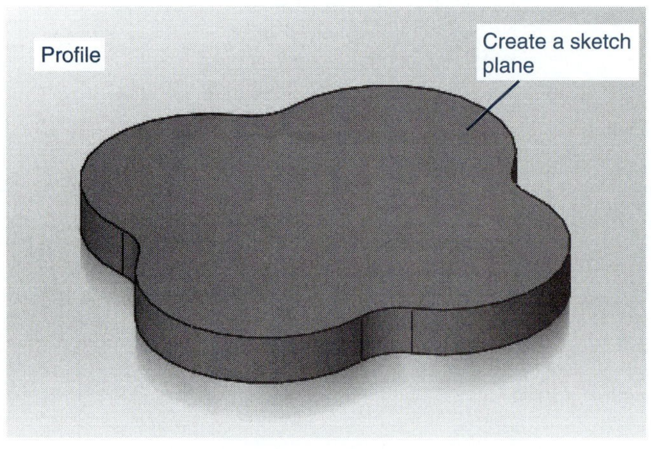
Profile
Create a sketch plane

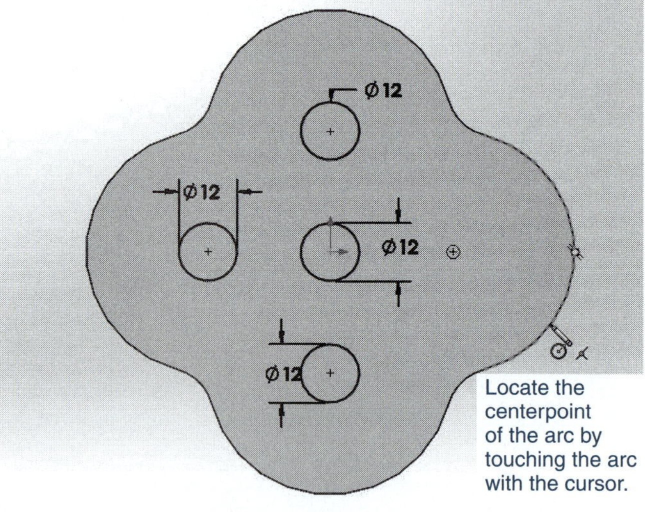
Ø12
Ø12
Ø12
Ø12
Locate the centerpoint of the arc by touching the arc with the cursor.

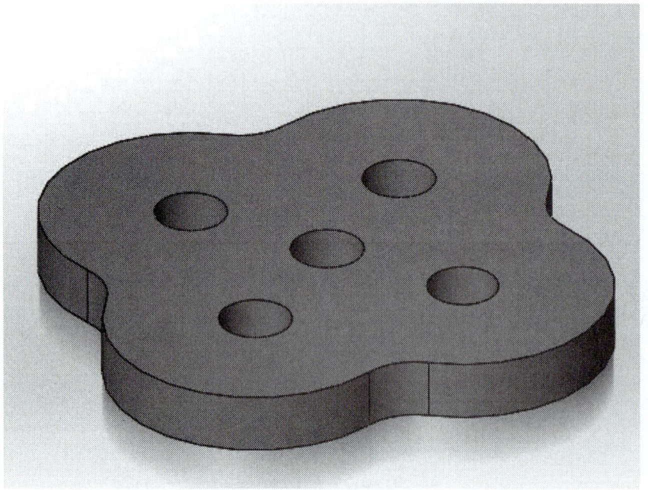

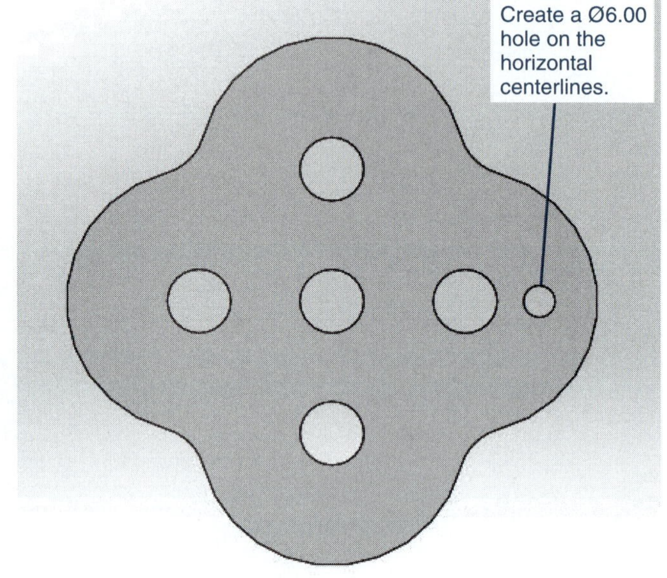
Create a Ø6.00 hole on the horizontal centerlines.

Figure 3-57
(*Continued*)

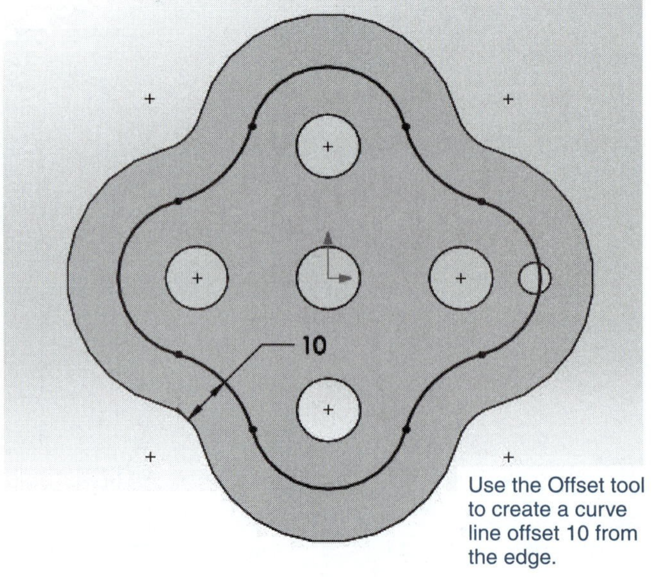
10
Use the Offset tool to create a curve line offset 10 from the edge.

Figure 3-57
(Continued)

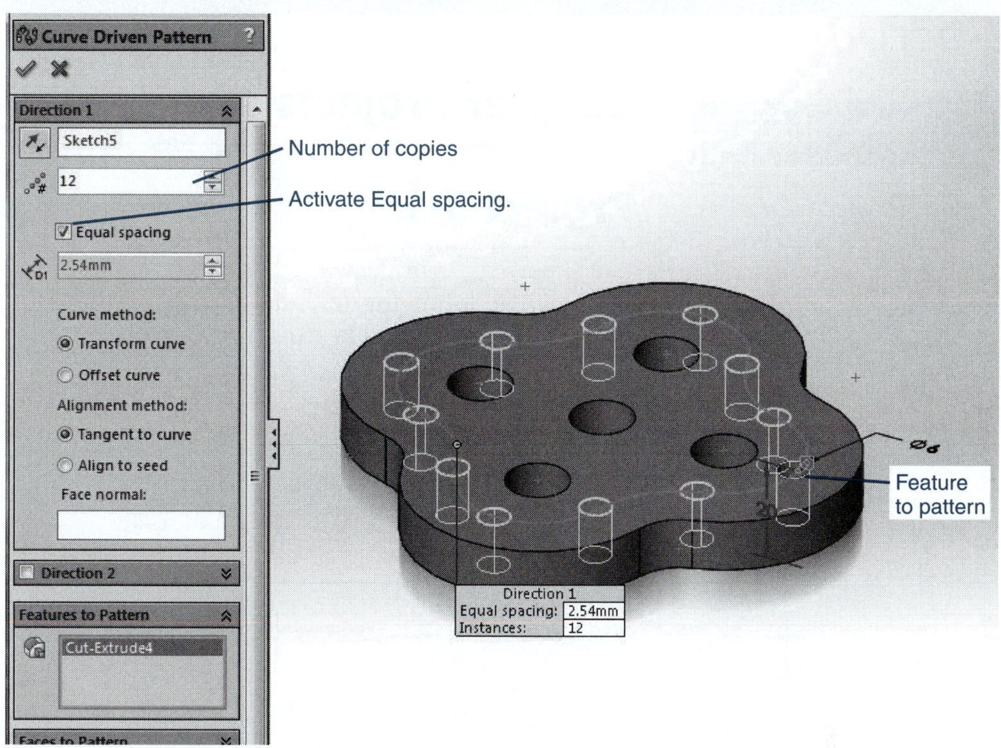

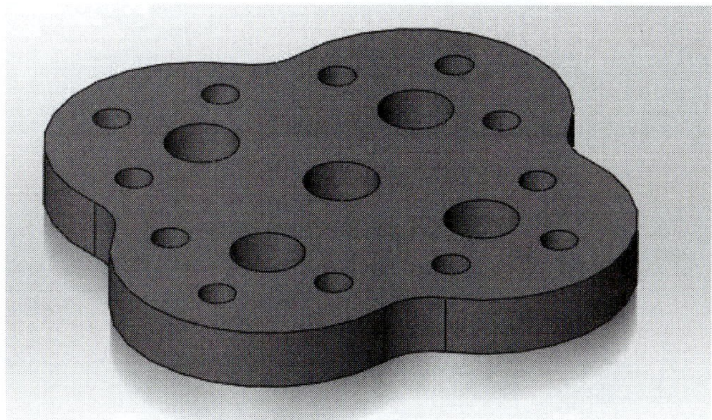

3. Draw a Ø6 hole centered on the intersection of the offset curve and the horizontal center line of the part. The centerpoint is 40 from the part's origin.

4. Click the **Insert** toolbar heading at the top of the screen, click the **Pattern/Mirror** option, and click the **Curve Driven** pattern tool.

5. Select the hole as the **Feature to Pattern**, the offset curve as the **Path**, and set the **Number of Instances** to **12**, click the **Equal spacing** box, and click **OK**.

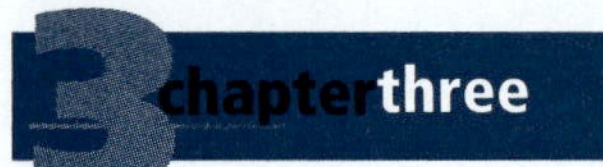

Chapter Projects

Project 3-1:

Redraw the following objects as solid models based on the given dimensions. Make all models from mild steel.

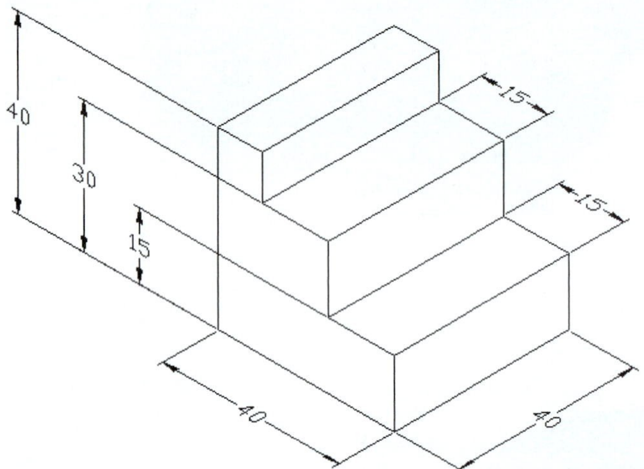

Figure P3-1
MILLIMETERS

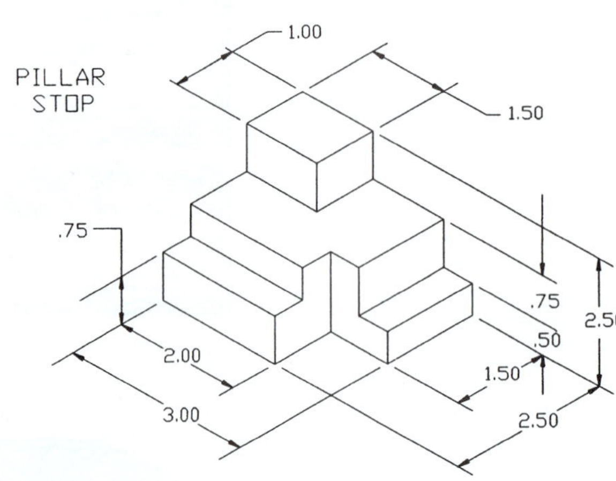

Figure P3-2
INCHES

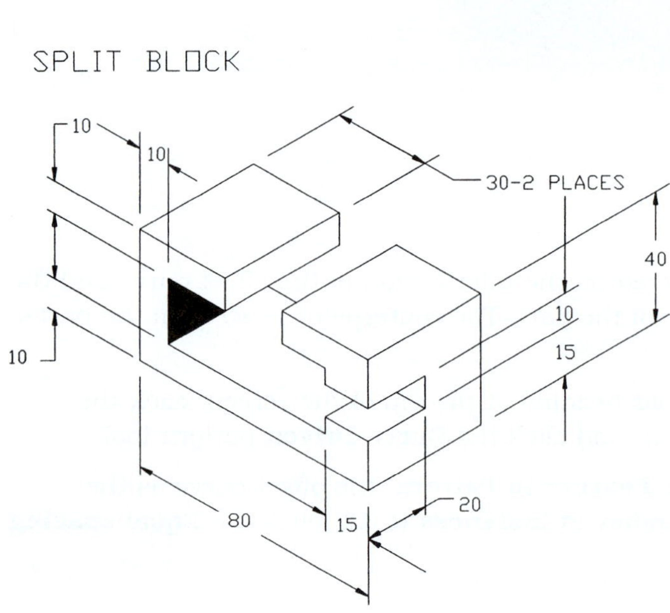

Figure P3-3
MILLIMETERS

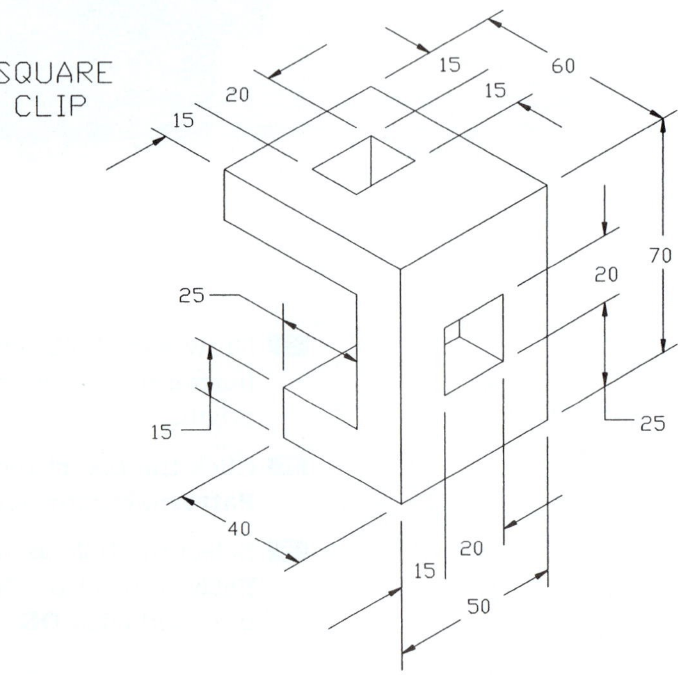

Figure P3-4
MILLIMETERS

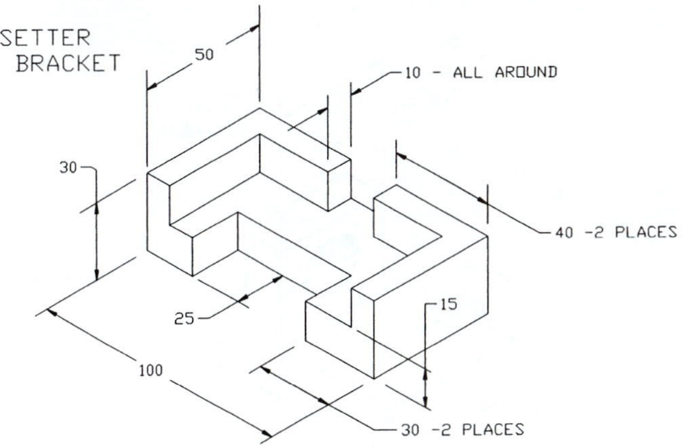

Figure P3-5
MILLIMETERS

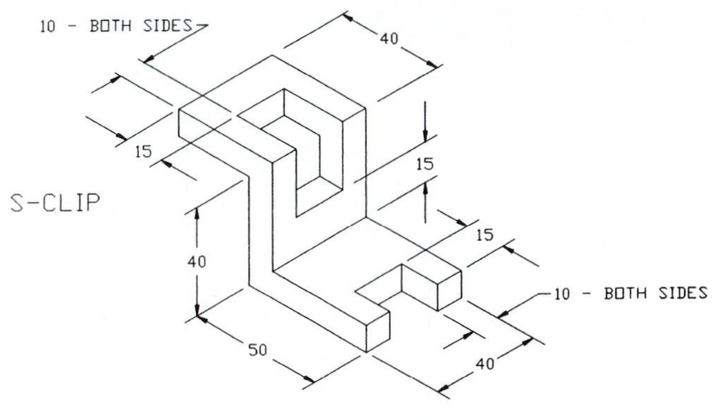

Figure P3-6
MILLIMETERS

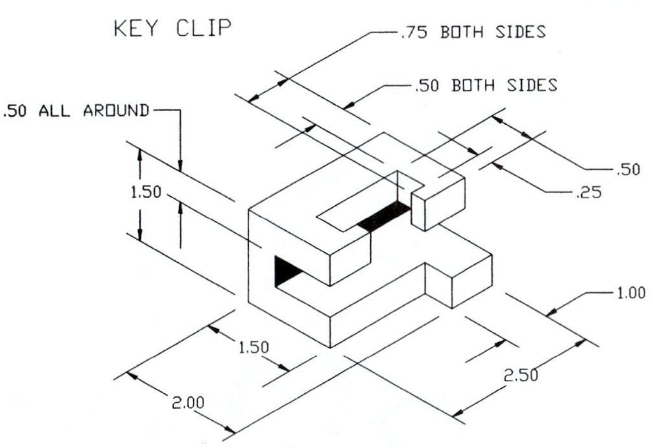

Figure P3-7
INCHES

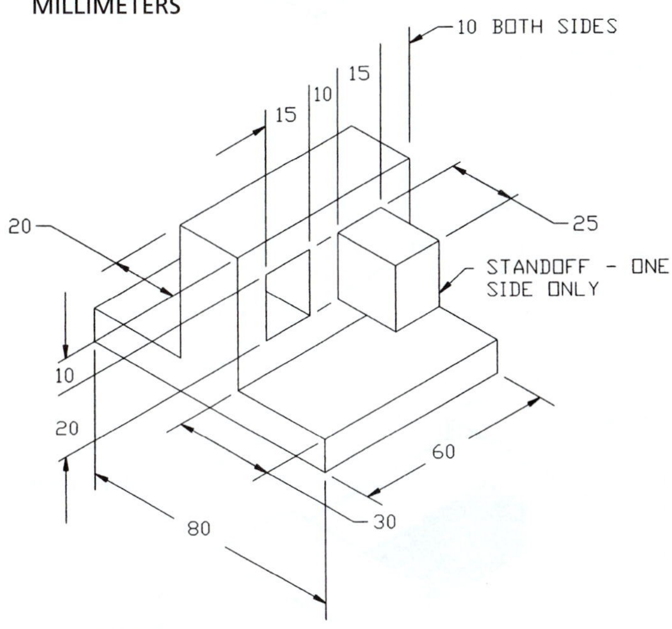

Figure P3-8
MILLIMETERS

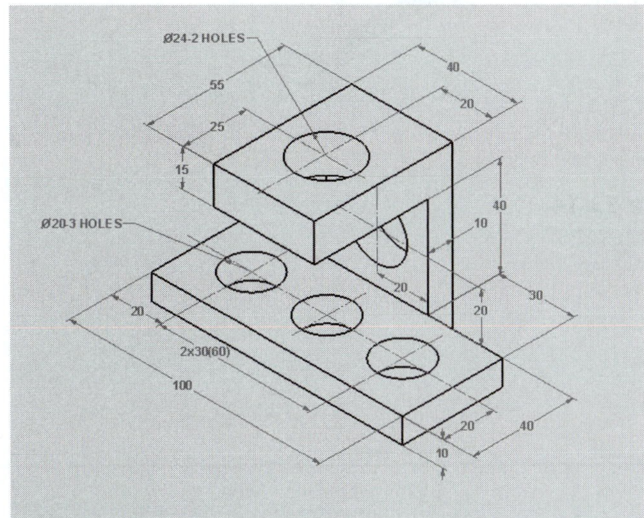

Figure P3-9
MILLIMETERS

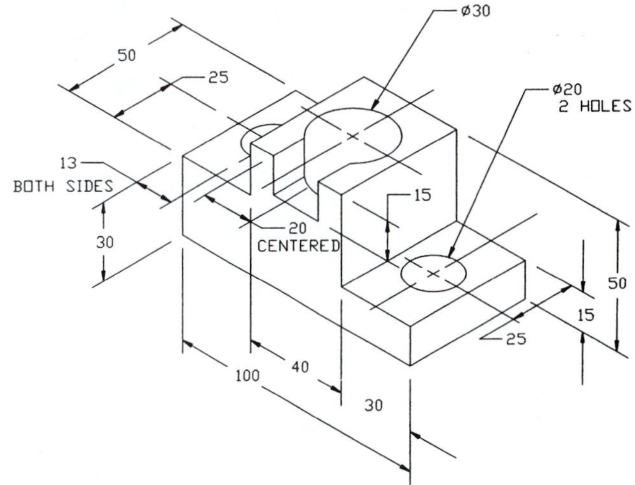

Figure P3-10
MILLIMETERS

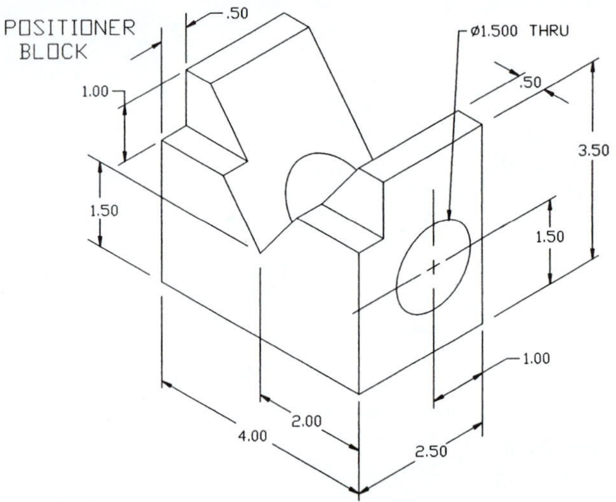

Figure P3-11
INCHES

Figure P3-12
MILLIMETERS

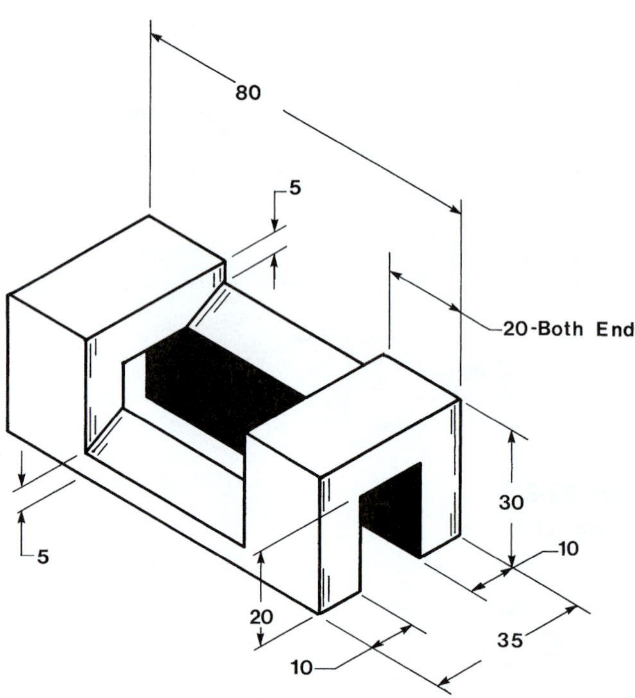

Figure P3-13
MILLIMETERS

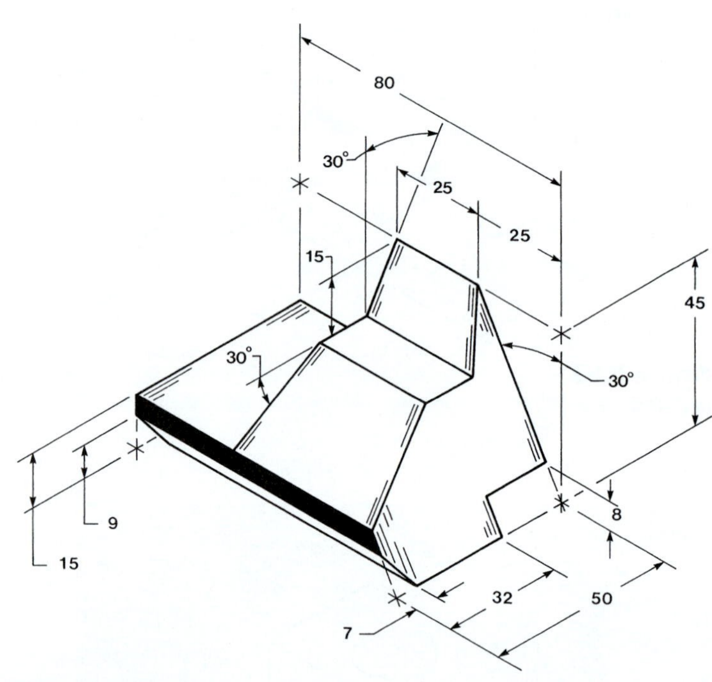

Figure P3-14
MILLIMETERS

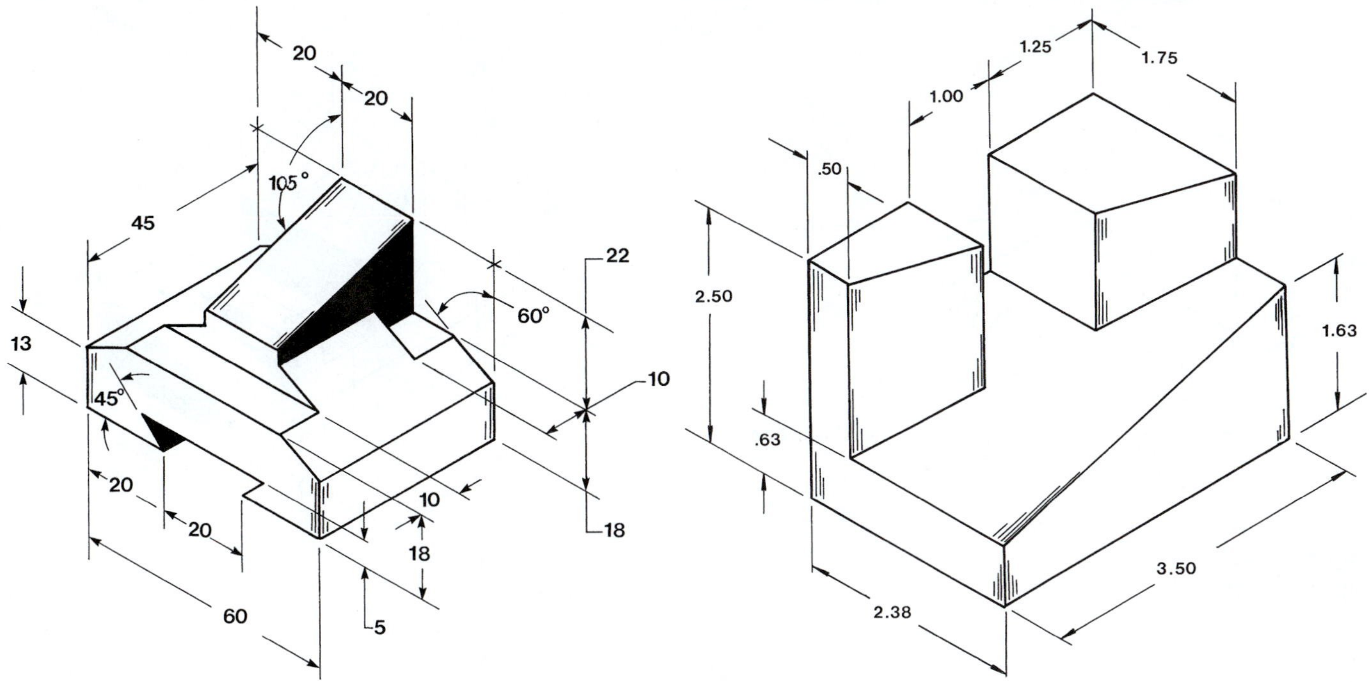

Figure P3-15
MILLIMETERS

Figure P3-16
INCHES

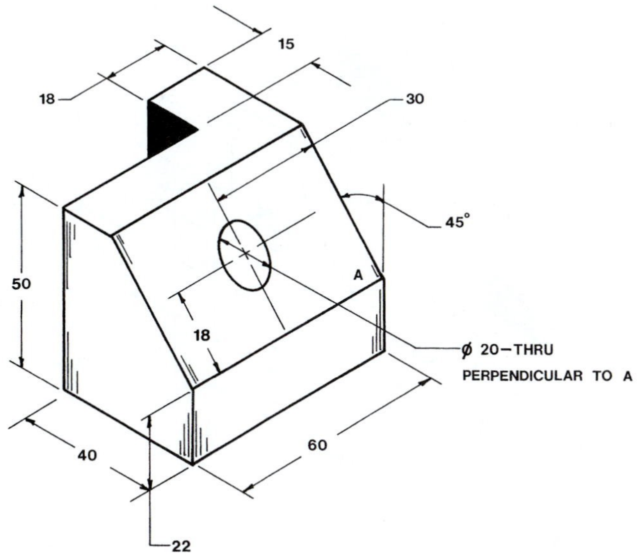

15
18
30
45°
50
A
18
ϕ 20—THRU
PERPENDICULAR TO A
40
60
22

Figure P3-17
MILLIMETERS

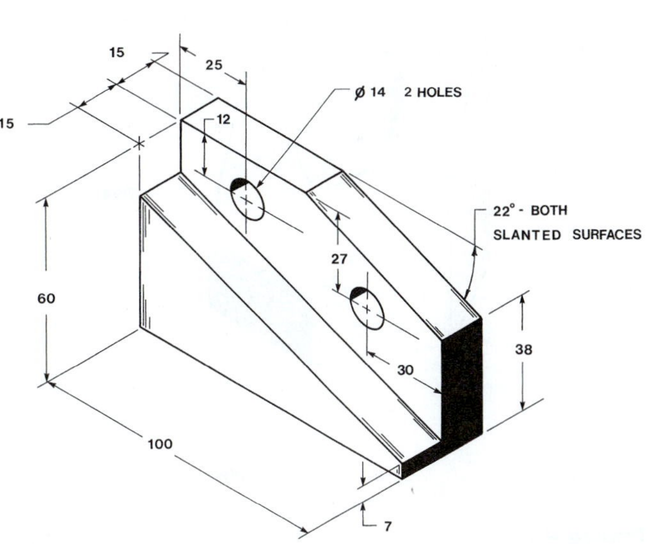

Figure P3-18
MILLIMETERS

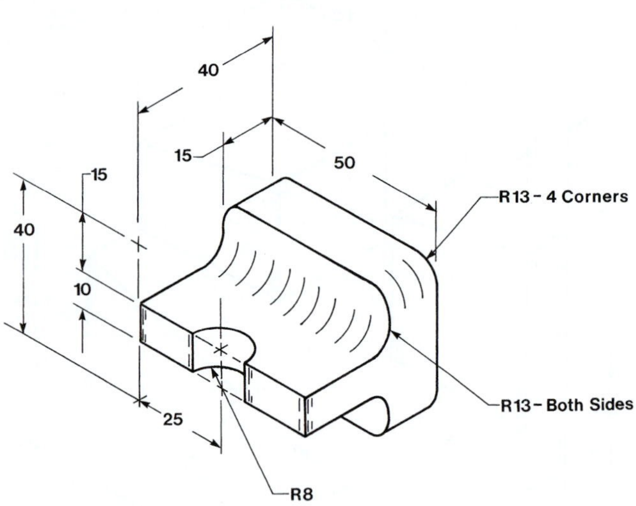

Figure P3-19
MILLIMETERS

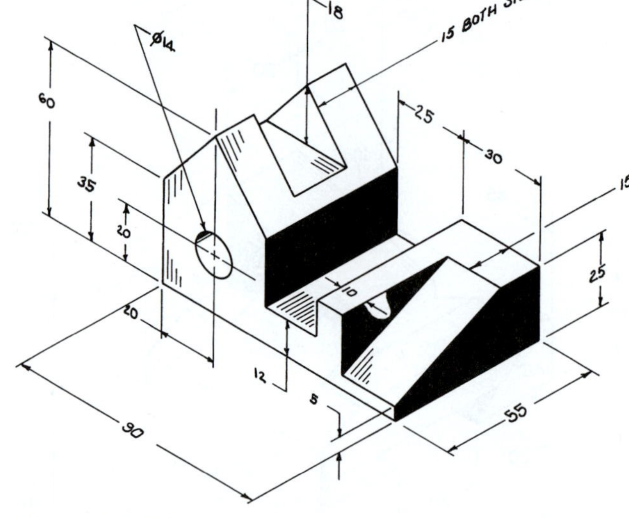

Figure P3-20
MILLIMETERS

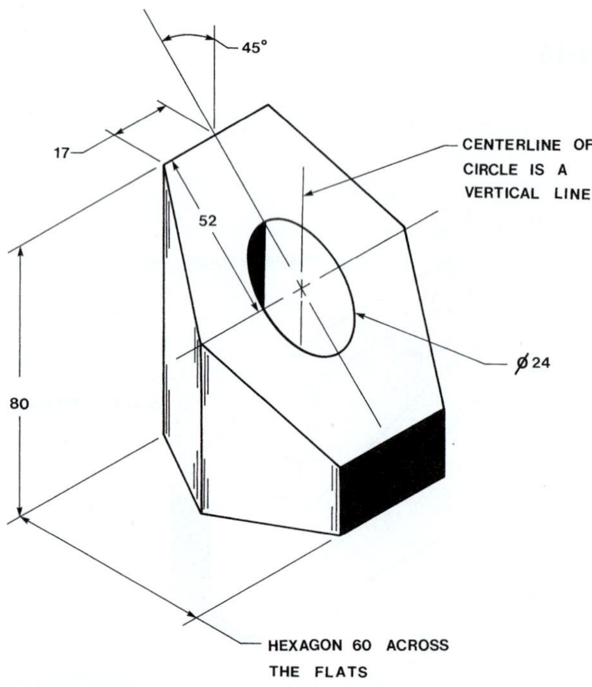

Figure P3-21
MILLIMETERS

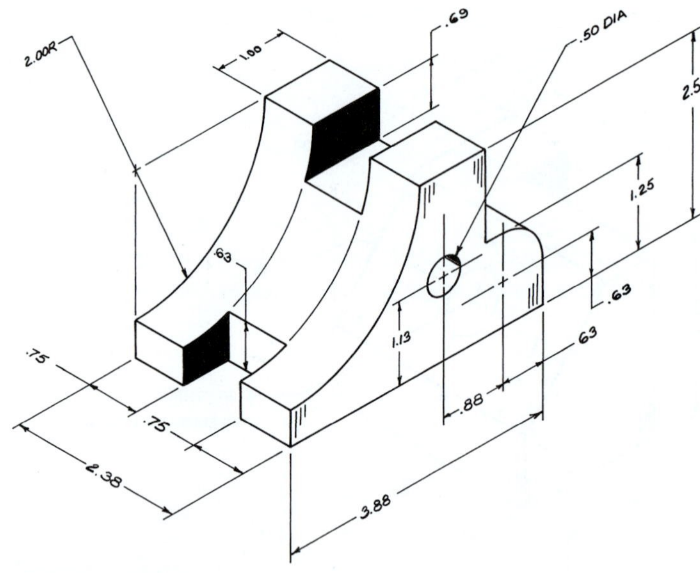

Figure P3-22
INCHES

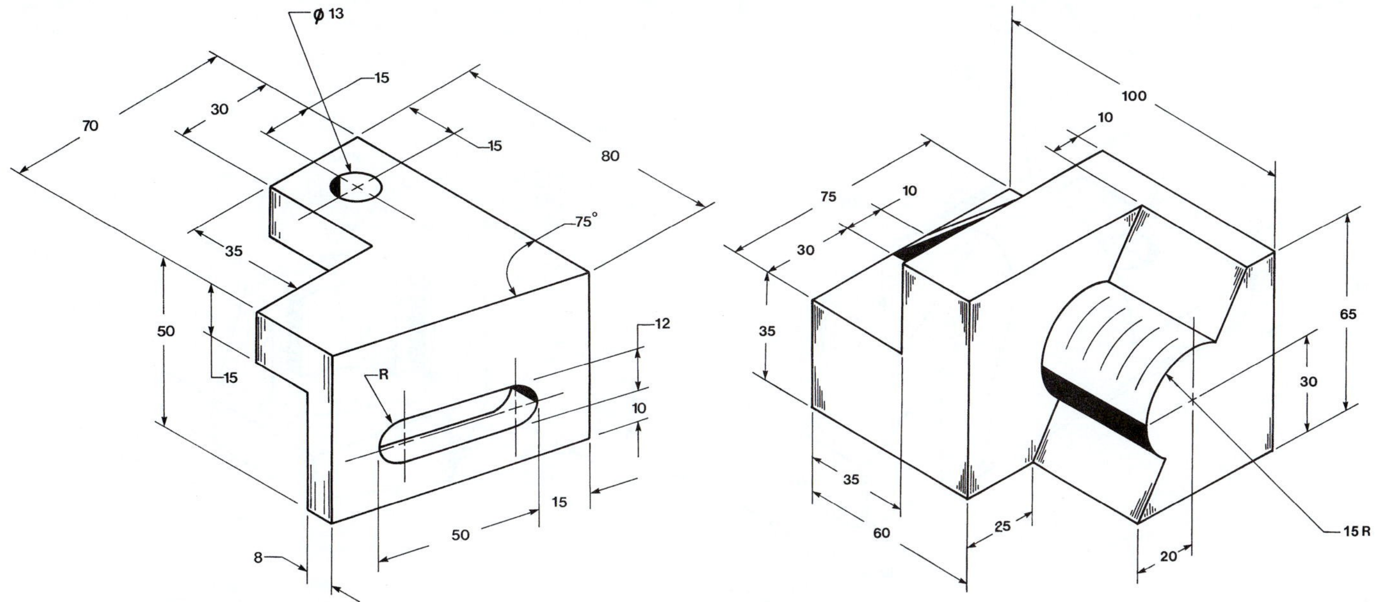

Figure P3-23
MILLIMETERS

Figure P3-24
MILLIMETERS

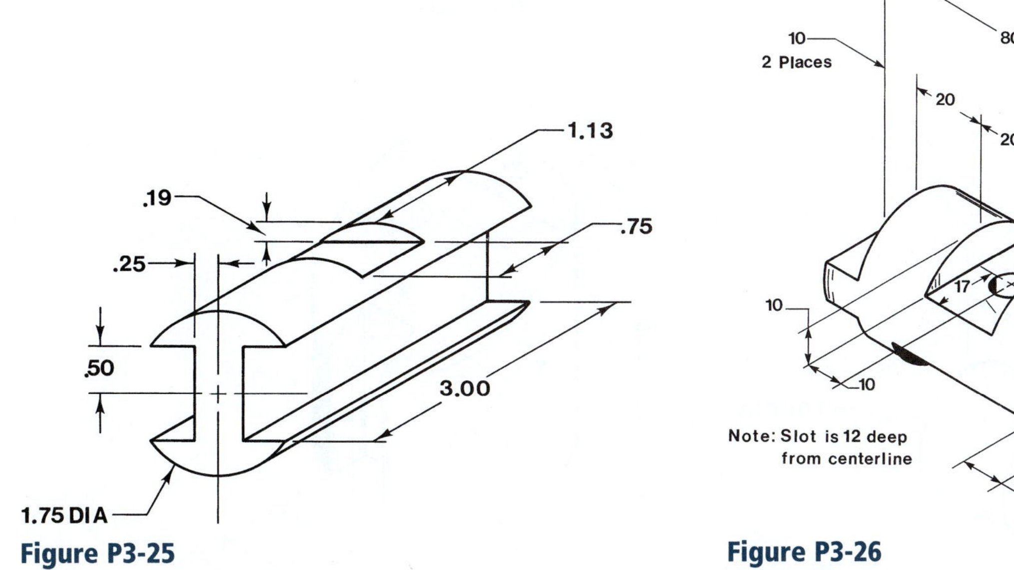

Figure P3-25
INCHES (SCALE: 4 = 1)

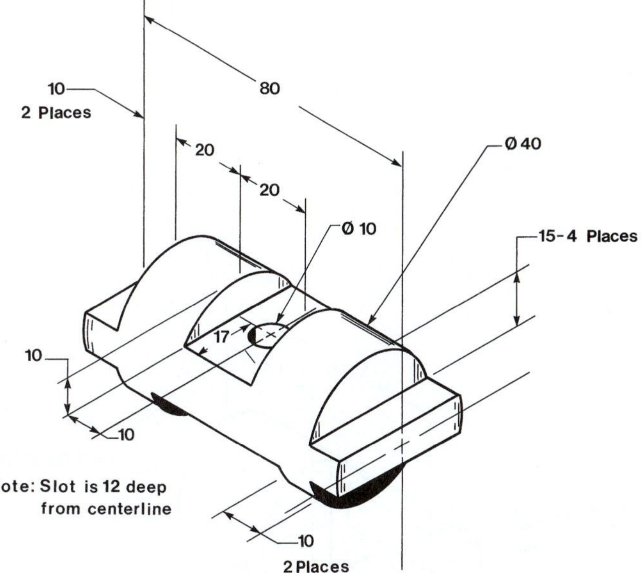

Figure P3-26
MILLIMETERS

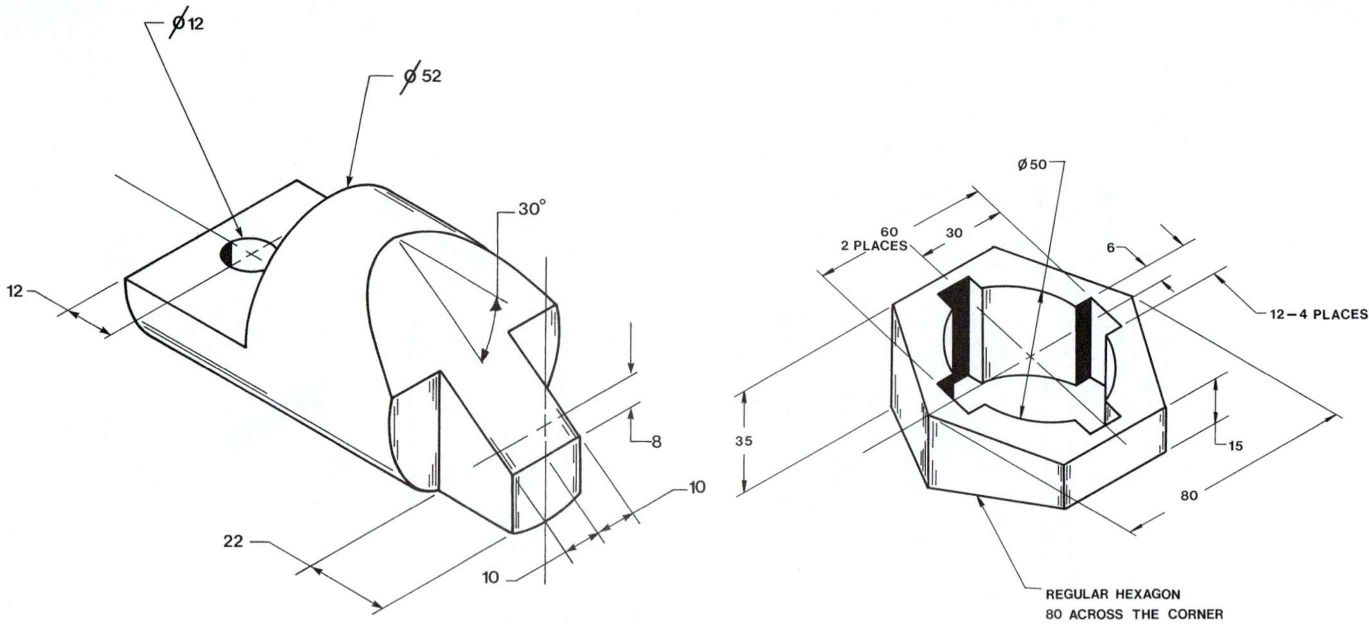

Figure P3-27
MILLIMETERS

Figure P3-28
MILLIMETERS

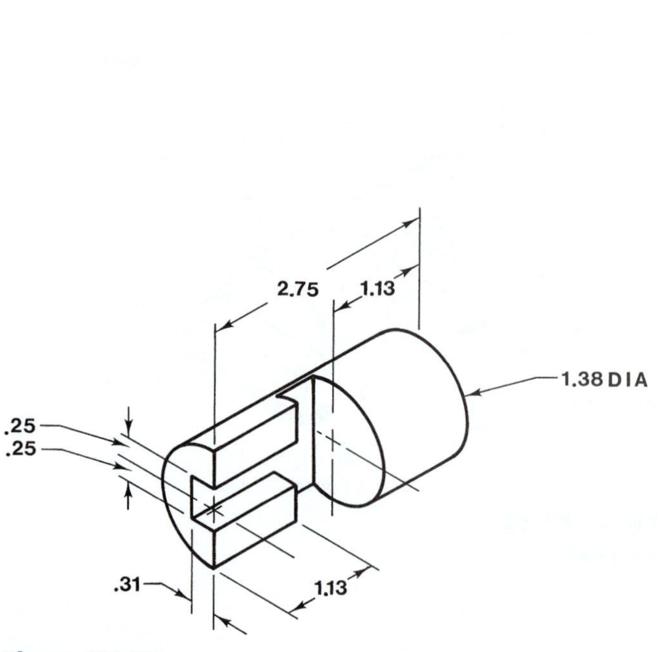

Figure P3-29
INCHES (SCALE: 4 = 1)

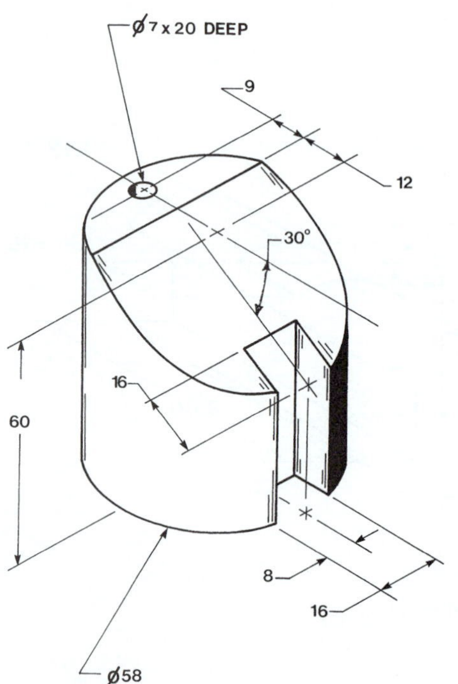

Figure P3-30
MILLIMETERS (SCALE: 2 = 1)

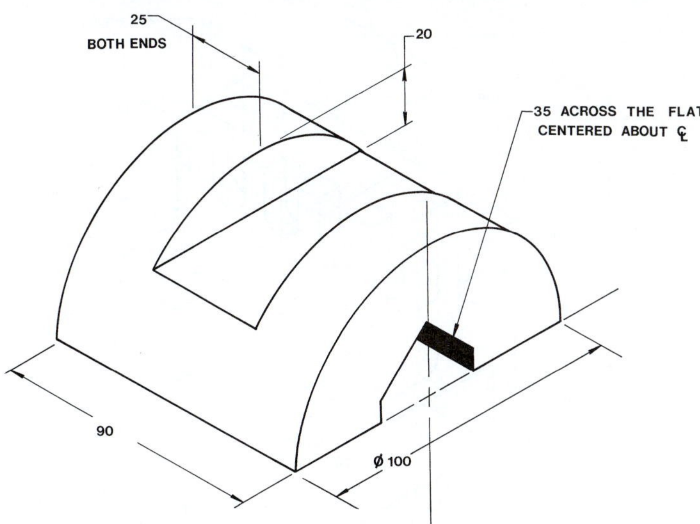

Figure P3-31
MILLIMETERS

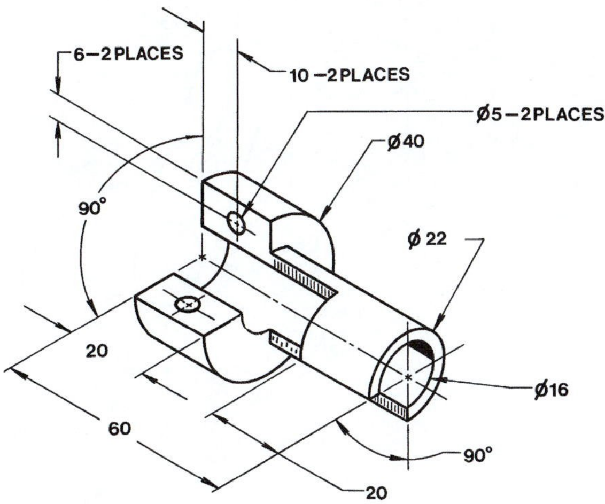

Figure P3-32
MILLIMETERS

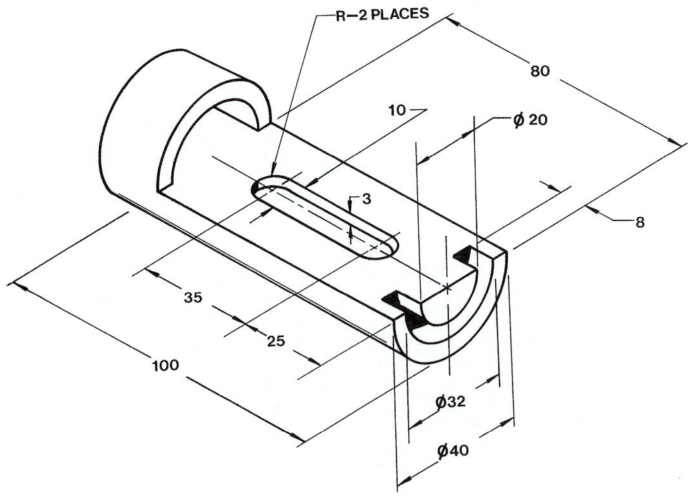

Figure P3-33
MILLIMETERS

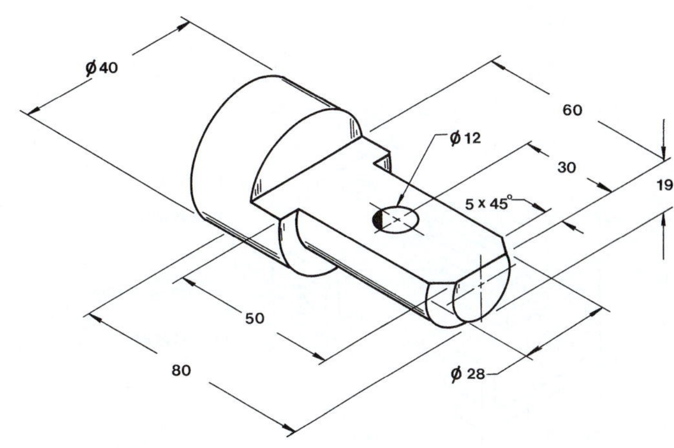

Figure P3-34
MILLIMETERS

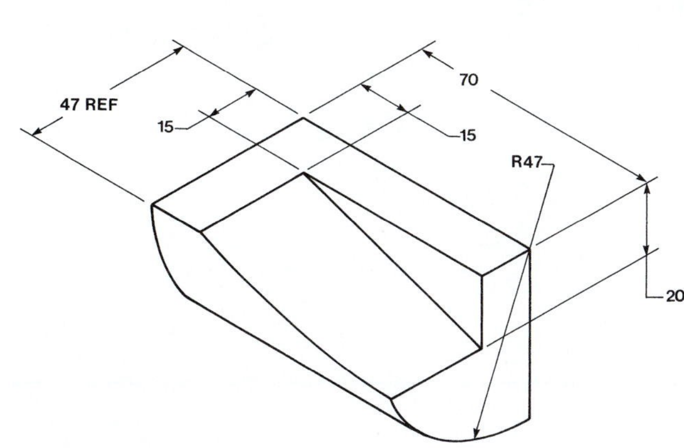

Figure P3-35
MILLIMETERS

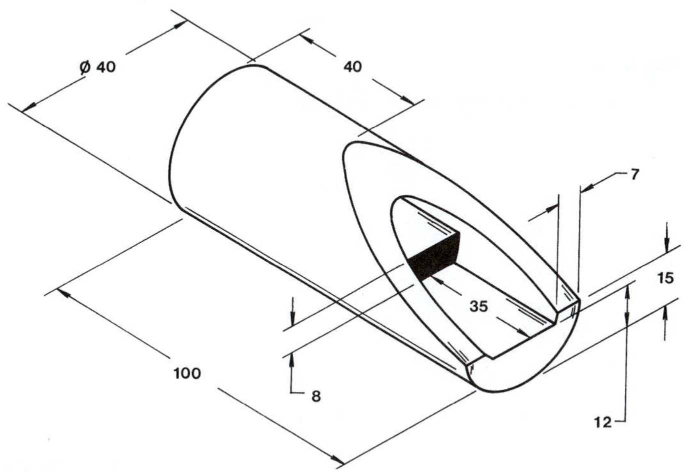

Figure P3-36
MILLIMETERS

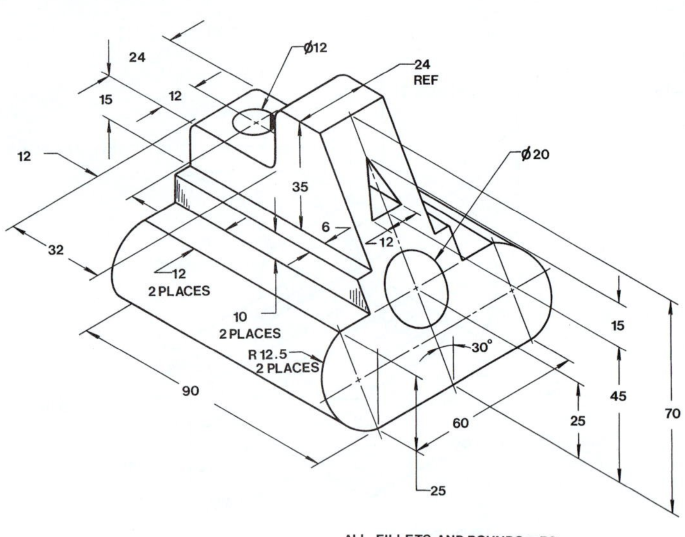

Figure P3-37
MILLIMETERS

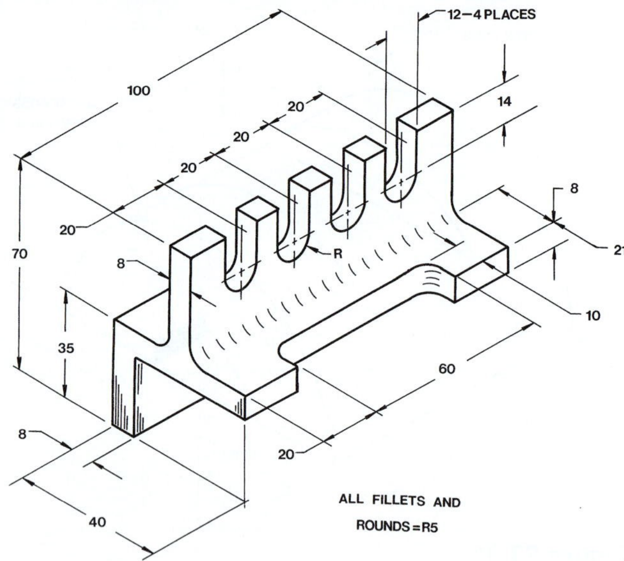

Figure P3-38
MILLIMETERS

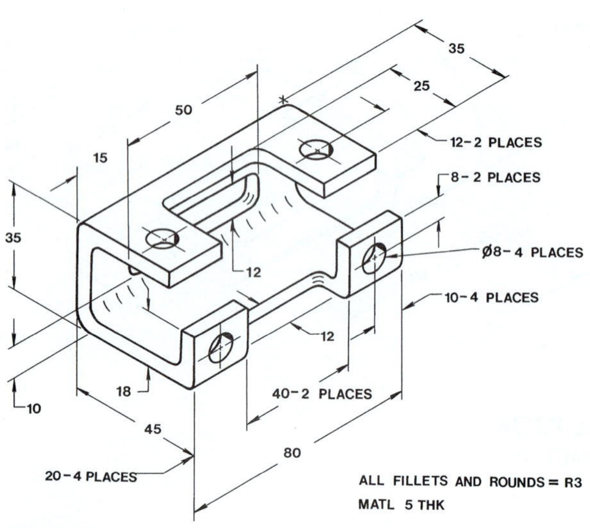

Figure P3-39
MILLIMETERS (CONSIDER A SHELL)

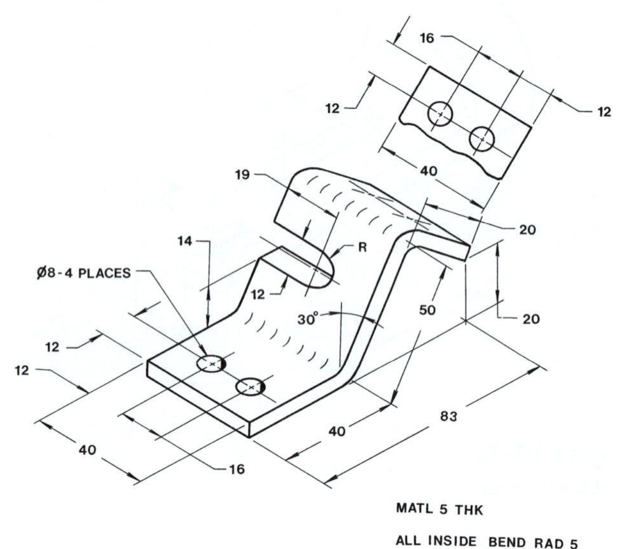

Figure P3-40
MILLIMETERS

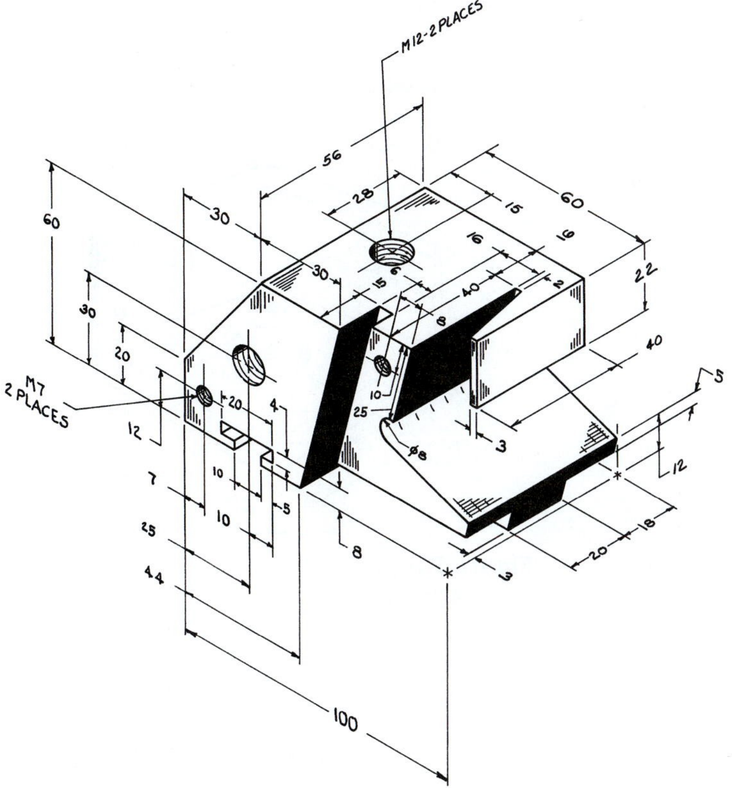

Figure P3-41
MILLIMETERS

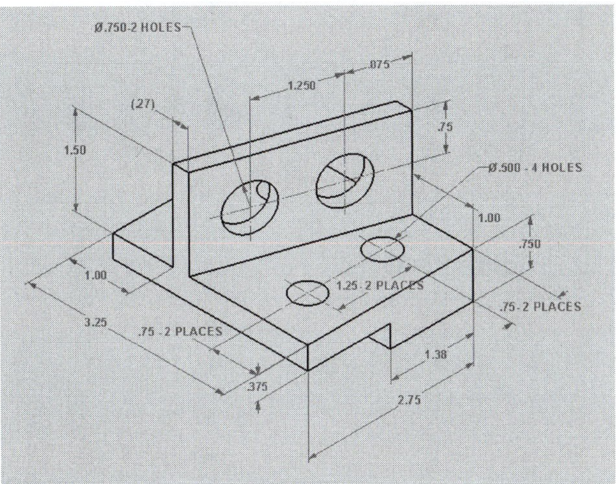

Figure P3-42
INCHES

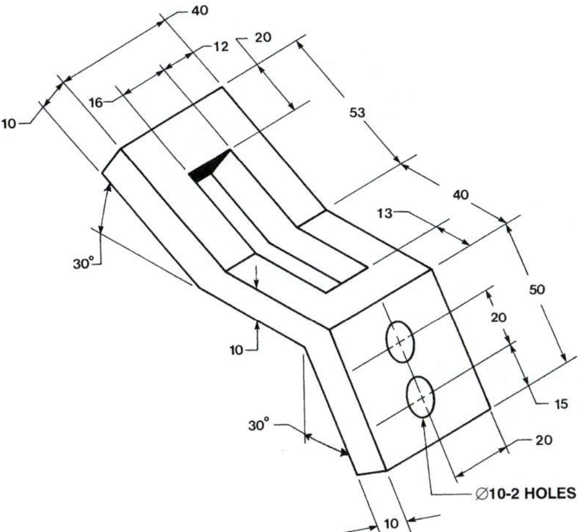

Figure P3-43
MILLIMETERS

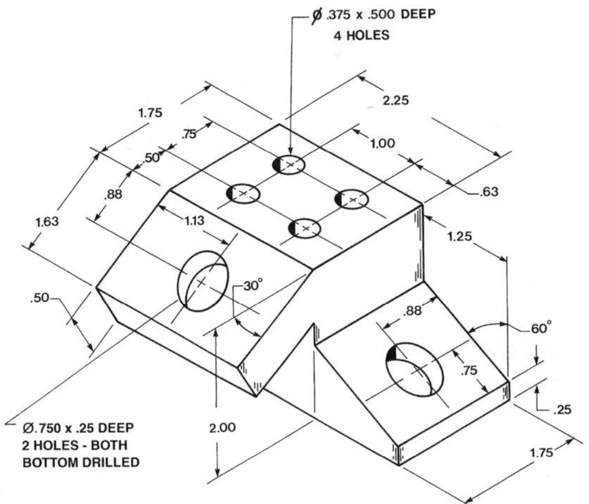

Figure P3-44
INCHES

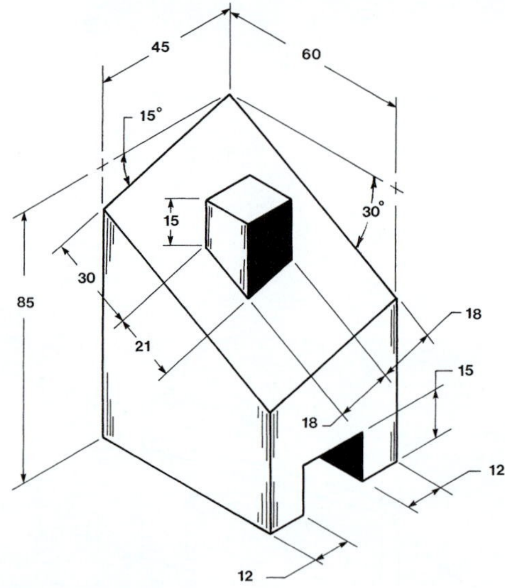

Figure P3-45
MILLIMETERS

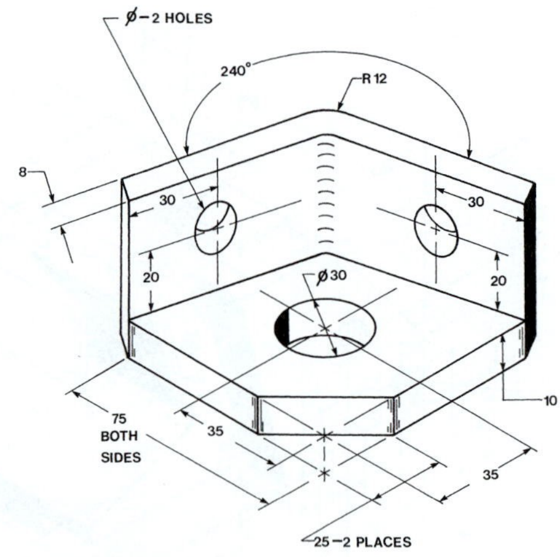

Figure P3-46
MILLIMETERS

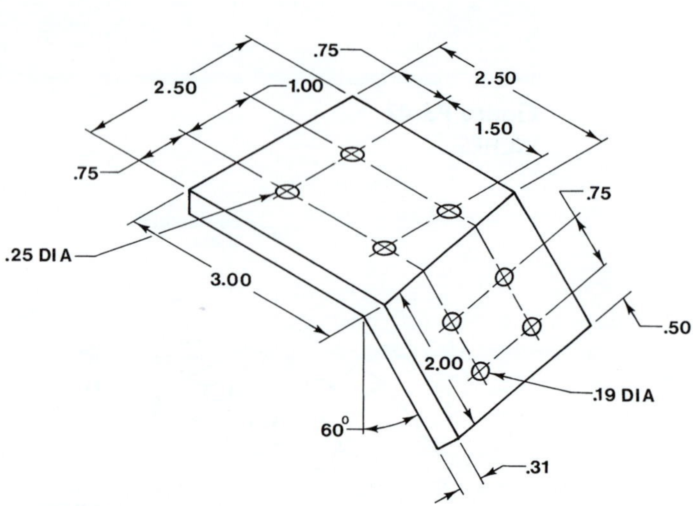

Figure P3-47
INCHES

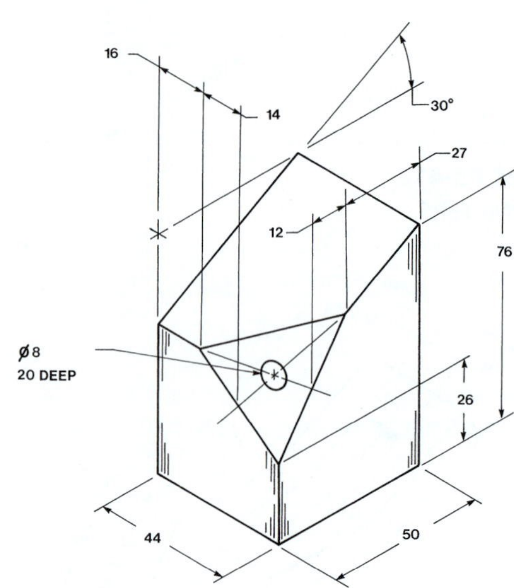

Figure P3-48
MILLIMETERS

Project 3-2:

A. Draw the following spring.

Diameter = 2.00

Wire diameter = .125

Pitch = .375

Revolutions = 16

B. Grind both ends to create a spring 2.00 long in its unloaded position.

Project 3-3:

A. Draw the following spring.

Diameter = 25

Wire diameter = 5 × 5 Square

Pitch = 6

Revolutions = 8

Figure P3-49
INCHES

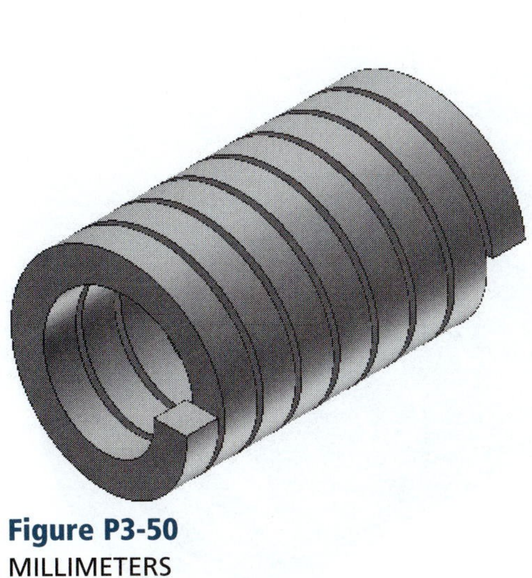

Figure P3-50
MILLIMETERS

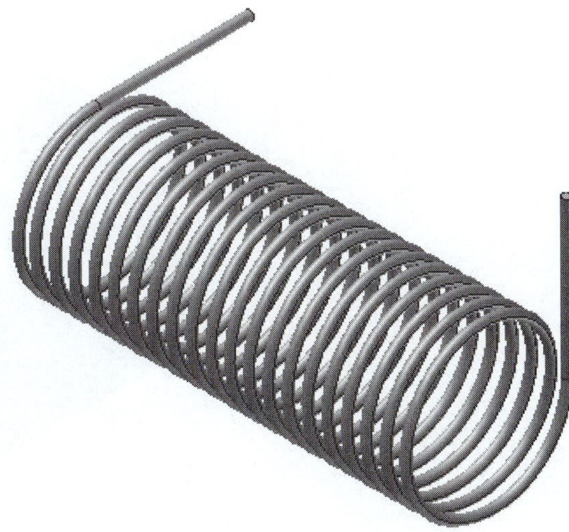

Figure P3-51
INCHES

Project 3-4:

A. Draw the following torsional spring.

Diameter = .500

Wire diameter = .06

Pitch = .125

Revolutions = 20

Extension lengths = 1.00, 90° apart

Project 3-5:

A. Draw the following torsional spring.

Diameter = 12.00

Wire diameter = 4.0

Pitch = 6.0

Revolutions = 18

Extension lengths = 15, 180° apart

Figure P3-52
MILLIMETERS

Project 3-6:

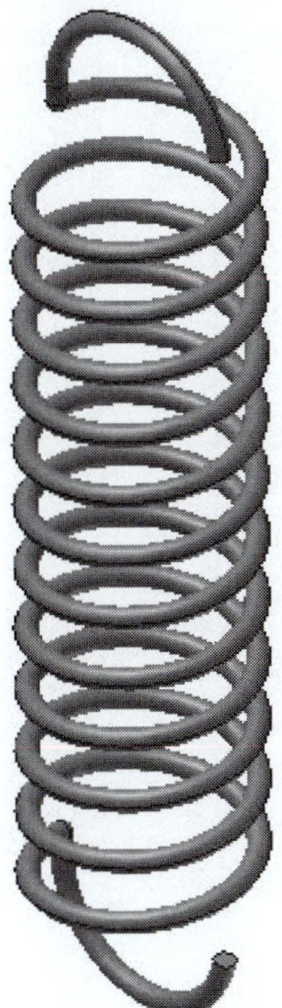

Figure P3-53
INCHES

A. Draw the following extension spring.

Diameter = 1.00

Wire diameter = .0938

Pitch = .180

Revolutions = 12

Extension radius = .125

Hook radius = .50

Project 3-7:

A. Draw the following extension spring.

Diameter = 30

Wire diameter = 6

Pitch = 30

Revolutions = 10

Extension radius = 6

Hook radius = 12

Project 3-8:

Draw a Ø4.00 × 3.00 cylinder and deboss on one of the following. Use a font style and size of your choice, or as assigned by your instructor.

A. Your name and address

B. Your school's name and address

Figure P3-54
MILLIMETERS

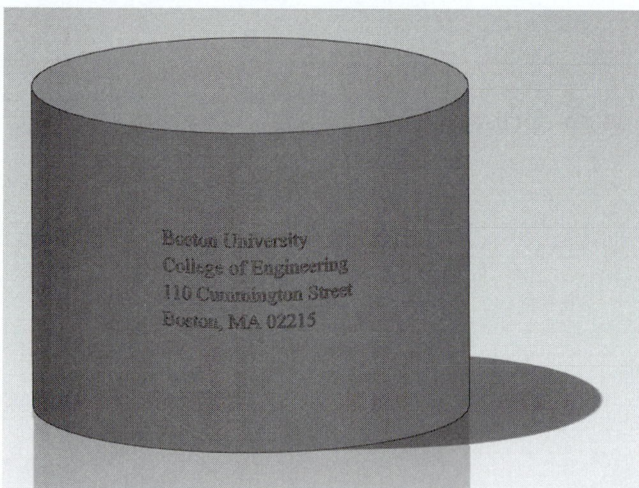

Figure P3-55
INCHES

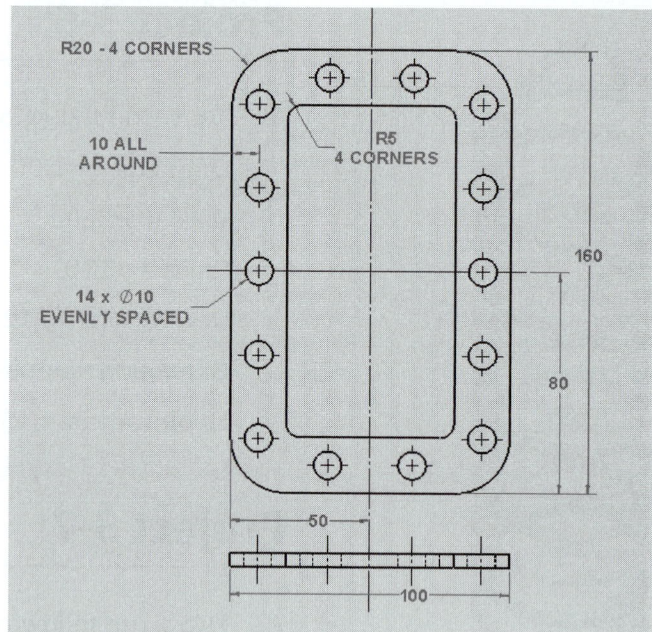

Figure P3-56
MILLIMETERS

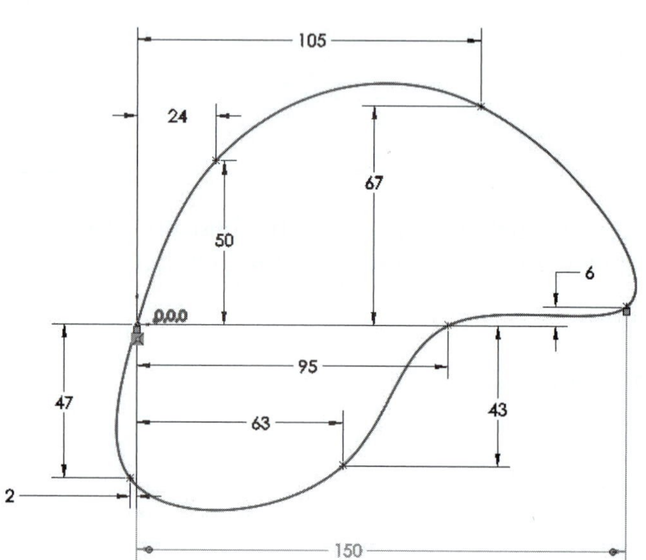

Figure P3-57A
MILLIMETERS

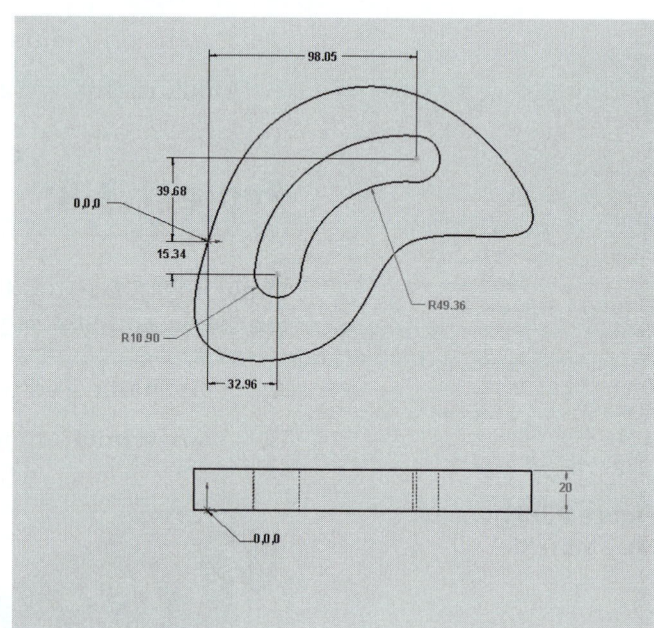

Figure P3-57B
MILLIMETERS

4 chapter four Orthographic Views

CHAPTER OBJECTIVES

- Learn about orthographic views
- Learn ANSI standards and conventions
- Learn how to draw section and auxiliary views

4-1 Introduction

Orthographic views are two-dimensional views used to define a three-dimensional model. More than one orthographic view is needed to define a model unless the model is of uniform thickness. Standard practice calls for three orthographic views, a front, top, and side view, although more or fewer views may be used as needed.

There are two sets of standards used to define the projection and placement of orthographic views: the American National Standards Institute (ANSI) and the International Organization for Standardization (ISO). The ANSI calls for orthographic views to be created using third-angle projection and is the accepted method for use in the United States. See the American Society of Mechanical Engineers publication ASME Y14.3-2003. Some countries, other than the United States, use first-angle projection. See ISO publication 128-30.

This chapter will present orthographic views using third-angle projections as defined by ANSI. However, there is so much international commerce happening today that you should be able to work in both conventions as you should be able to work in inches or millimeters.

Figure 4-1 shows a three-dimensional model and three orthographic views created using third-angle projection and three orthographic views created using first-angle projection. Note the differences and similarities. The front view in both projections is the same. The top views are the same but are in different locations. The third-angle projection presents a right-side view, while the first-angle projection presents a left-side view.

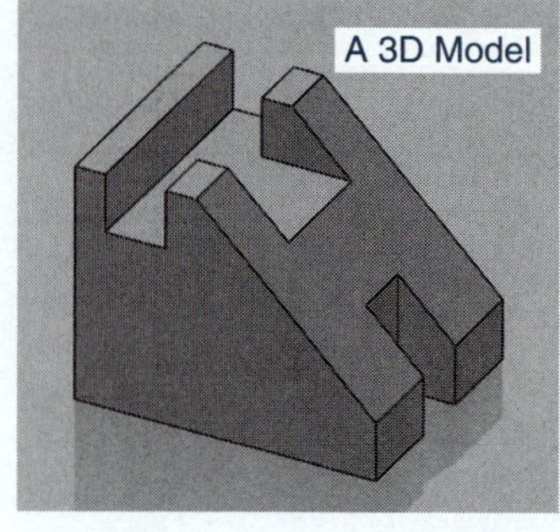

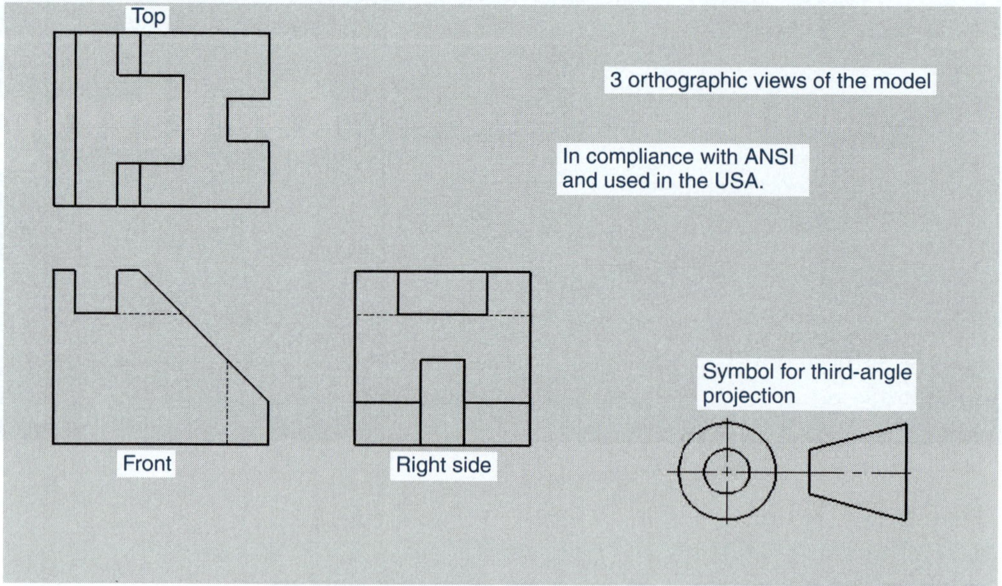

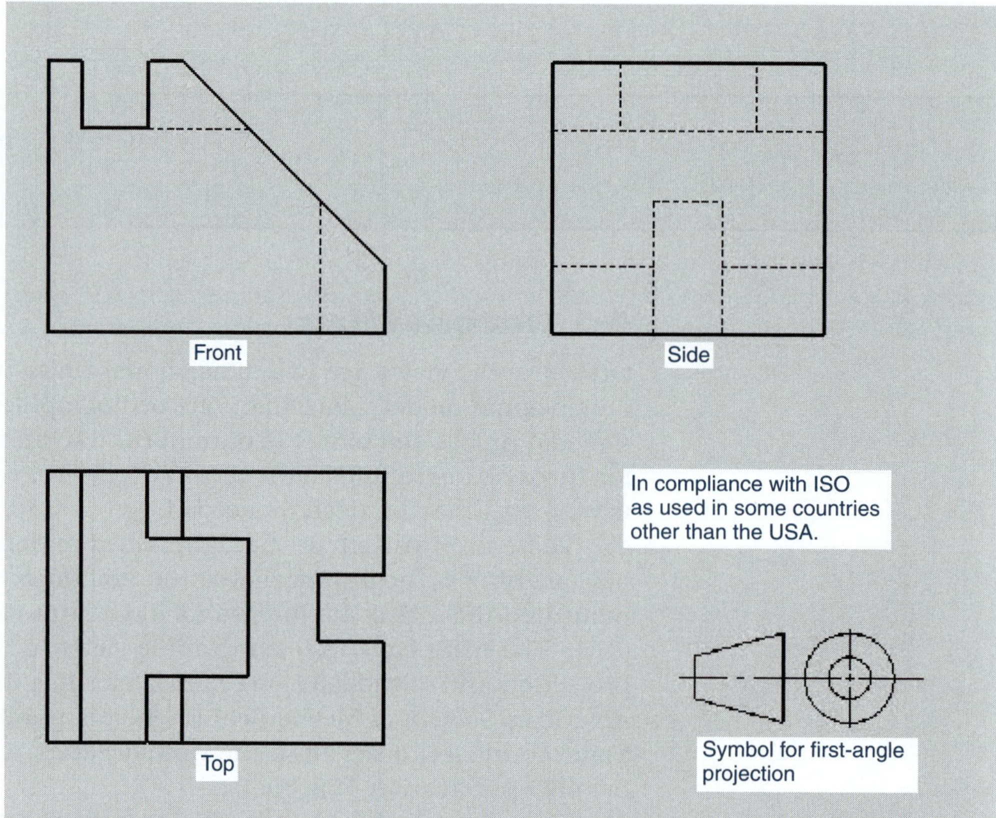

Figure 4-1

Figure 4-2 shows the drawing symbols for first- and third-angle projections. These symbols can be added to a drawing to help the reader understand which type of projection is being used. These symbols were included in the projections presented in Figure 4-1.

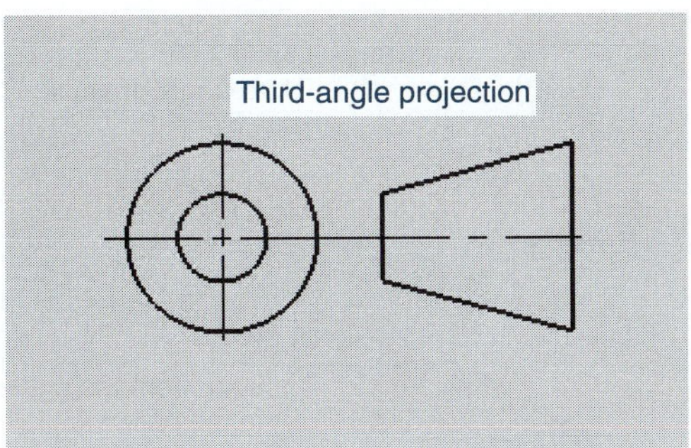

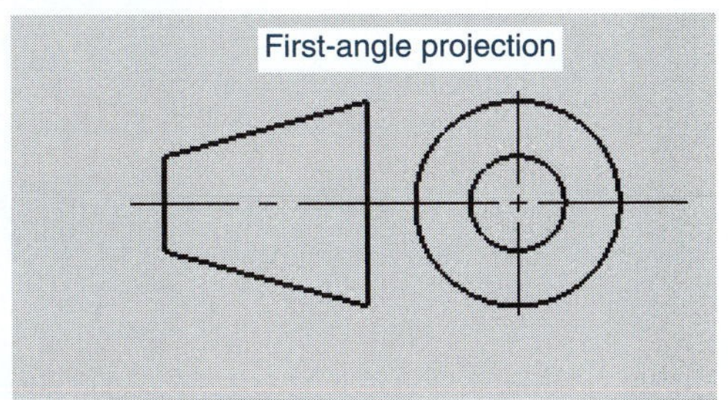

Figure 4-2

4-2 Third- and First-Angle Projections

Figure 4-3 shows an object with front and top orthographic views created using third-angle and first-angle projections. For third-angle projections the orthographic view is projected on a plane located between the viewer's position and the object. For first-angle projections the orthographic view is projected on a plane located beyond the object. The front and top views for third- and first-angle projections appear the same, but they are located in different positions relative to the front view.

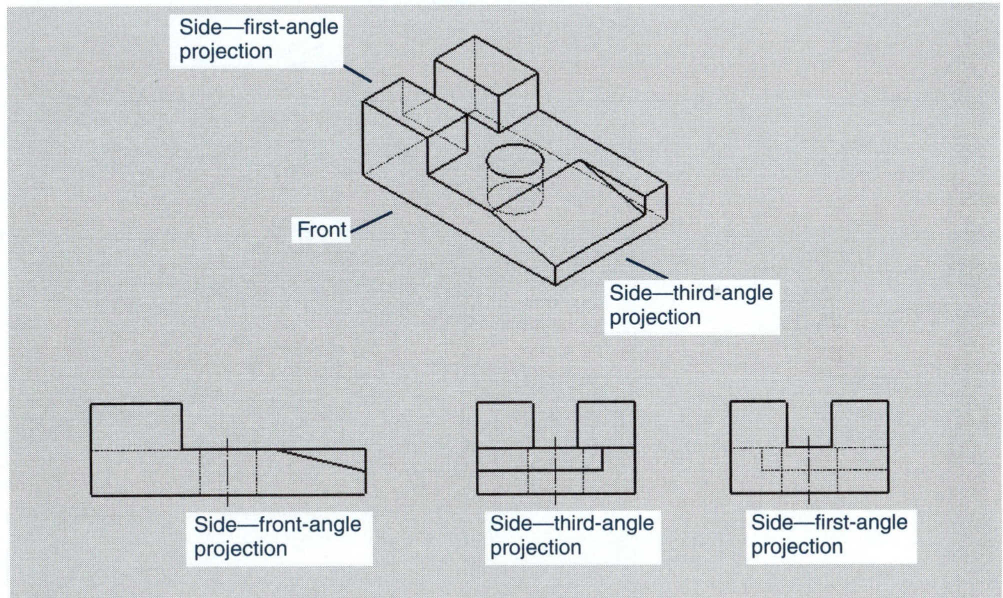

Figure 4-3

The side orthographic views are different for third- and first-angle projections. Third-angle projections use a right-side view. First-angle projections use a left-side view. Figures 4-4 and 4-5 show the side views for two different objects. For third-angle projections, the viewer is located on the right side of the object and creates the side orthographic view on a plane located between the view position and the object. For first-angle projections the viewer is located on the left side of the object and creates the side orthographic view on a plane located beyond the object.

To help understand the difference between side view orientations for third- and first-angle projections, locate your right hand with the heel facing down and the thumb facing up. Rotate your hand so that the palm is facing up—this is the third-angle projection orientation. Return to the

Figure 4-4

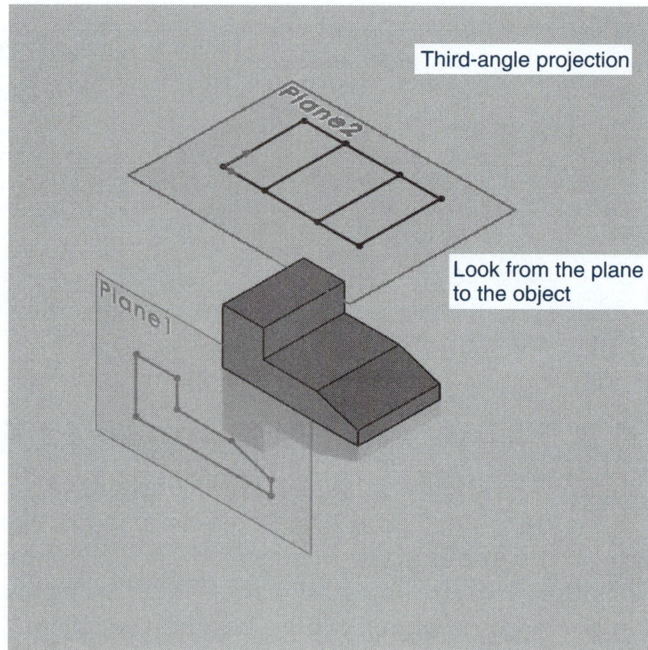

Figure 4-5

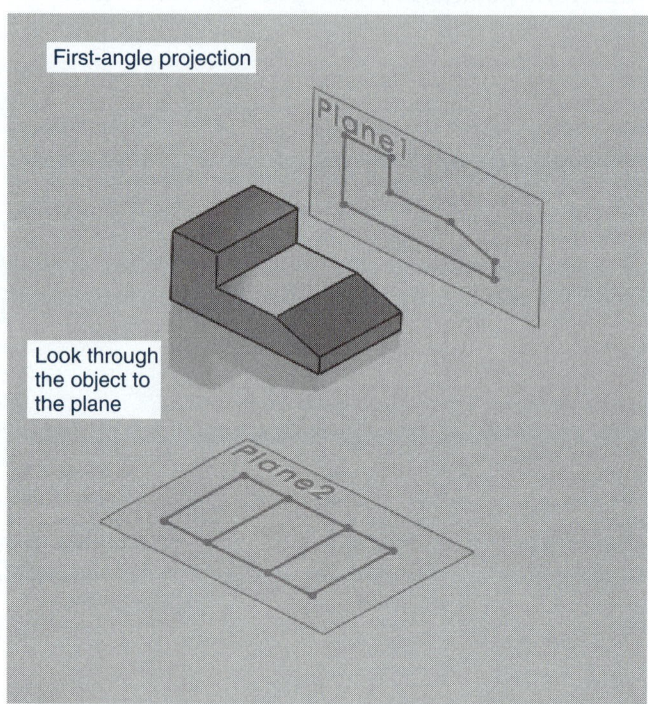

thumb-up position. Rotate your hand so that the palm is down—this is the first-angle view orientation.

4-3 Fundamentals of Orthographic Views

Figure 4-6 shows an object with its front, top, and right-side orthographic views projected from the object. The views are two-dimensional, so they show no depth. Note that in the projected right plane there are three rectangles. There is no way to determine which of the three is closest and which is farthest away if only the right-side view is considered. All views must be studied to analyze the shape of the object.

Figure 4-7 shows three orthographic views of a book. After the views are projected they are positioned as shown. The positioning of views relative to one another is critical. The views must be aligned and positioned as shown.

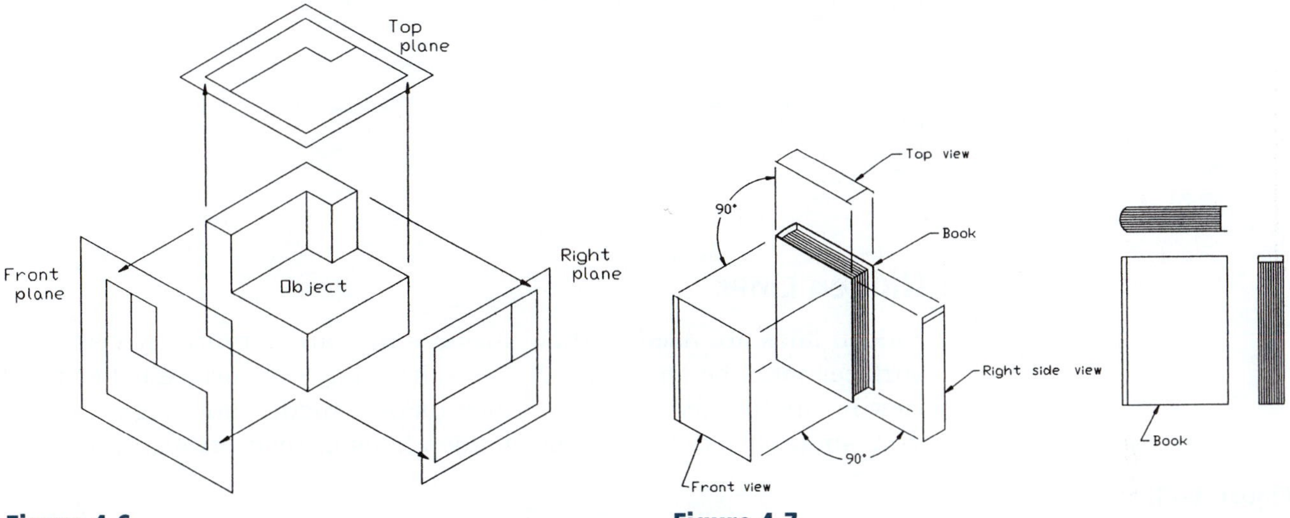

Figure 4-6

Figure 4-7

Normal Surfaces

Normal surfaces are surfaces that are at 90° to each other. Figures 4-8, 4-9, and 4-10 show objects that include only normal surfaces and their orthographic views.

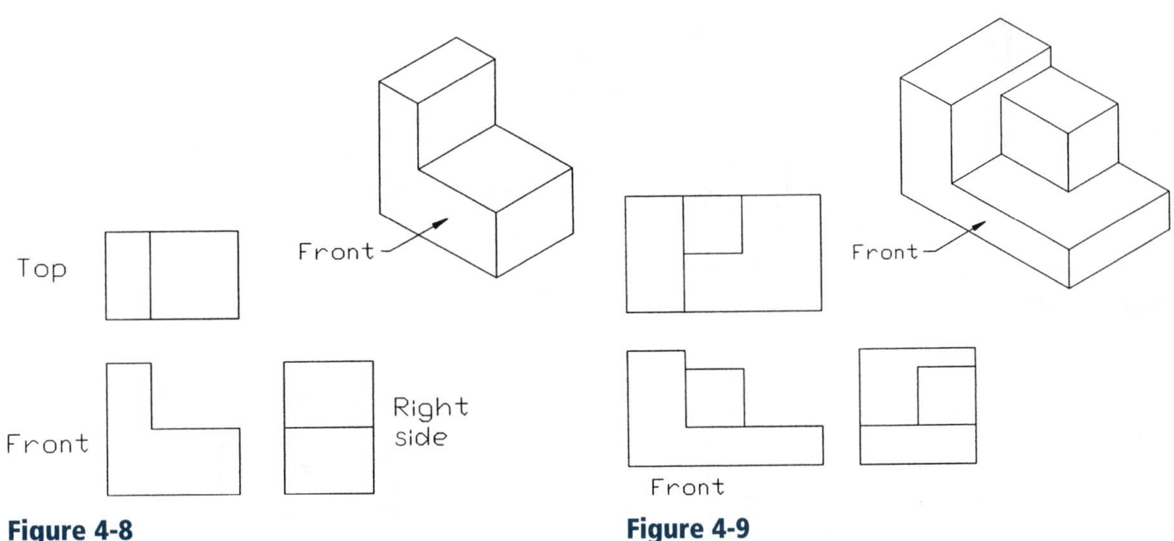

Figure 4-8

Figure 4-9

Figure 4-10

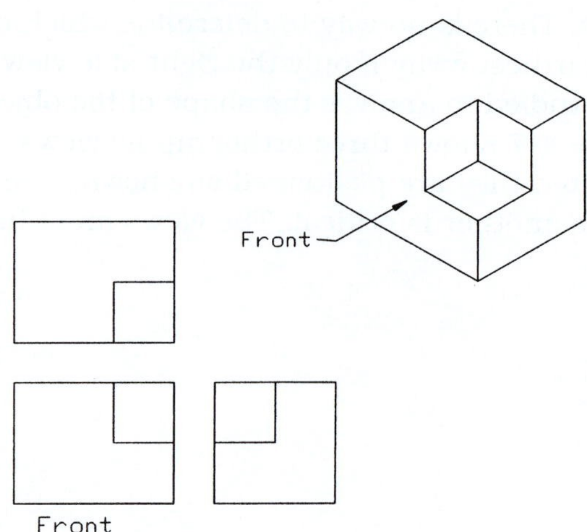

Hidden Lines

Hidden lines are used to show surfaces that are not directly visible. All surfaces must be shown in all views. If an edge or surface is blocked from view by another feature, it is drawn using a hidden line. Figures 4-11 and 4-12 show objects that require hidden lines in their orthographic views.

Figure 4-11

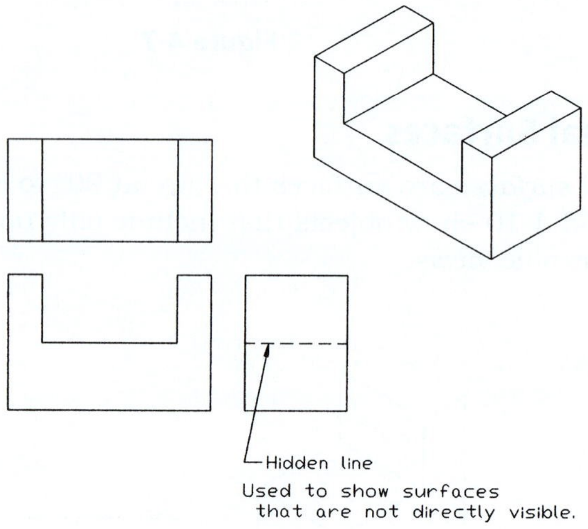

Figure 4-12

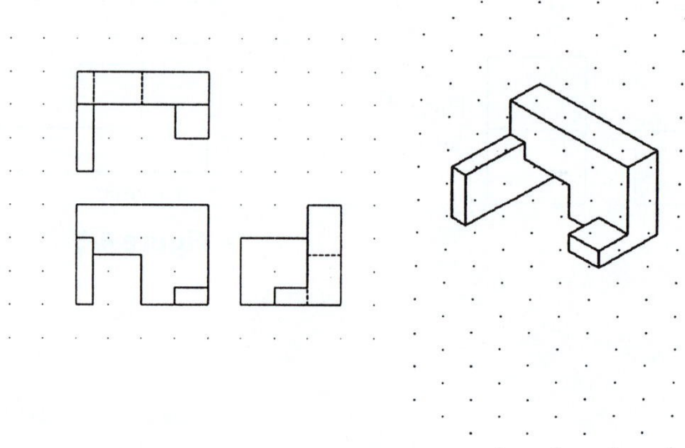

Figure 4-13 shows an object that contains an edge line, **A-B**. In the top view, line A-B is partially hidden and partially visible. The hidden portion of the line is drawn using a hidden-line pattern, and the visible portion of the line is drawn using a solid line.

Figures 4-14 and 4-15 show objects that require hidden lines in their orthographic views.

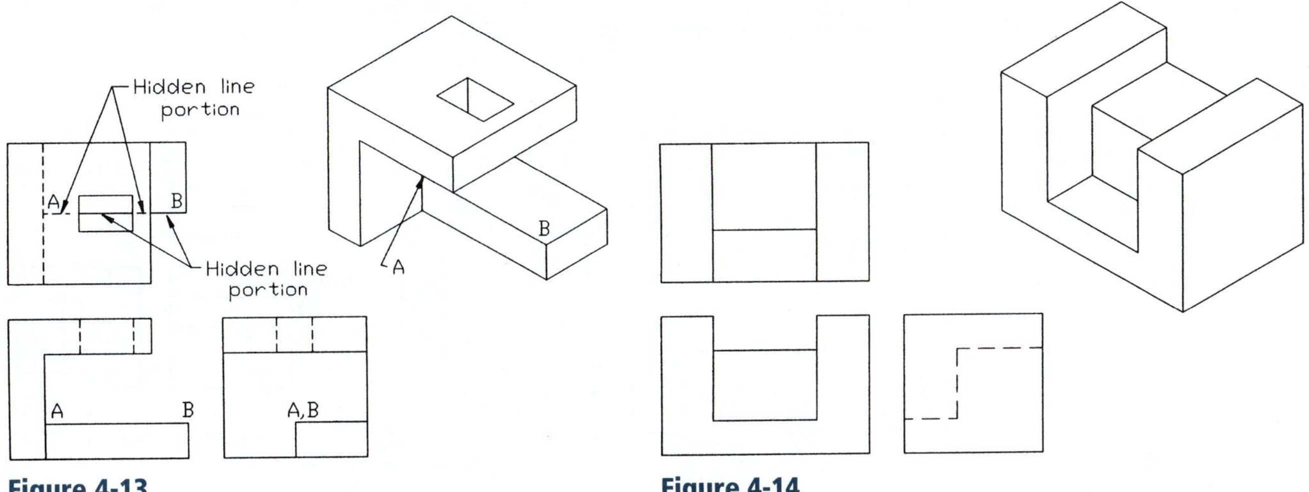

Figure 4-13

Figure 4-14

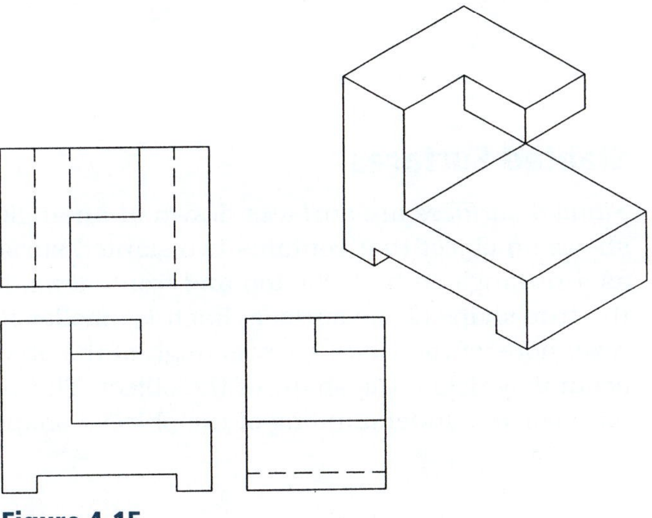

Figure 4-15

Precedence of Lines

It is not unusual for one type of line to be drawn over another type of line. Figure 4-16 shows two examples of overlap by different types of lines. Lines are shown on the views in a prescribed order of precedence. A solid line (object or continuous) takes precedence over a hidden line, and a hidden line takes precedence over a centerline.

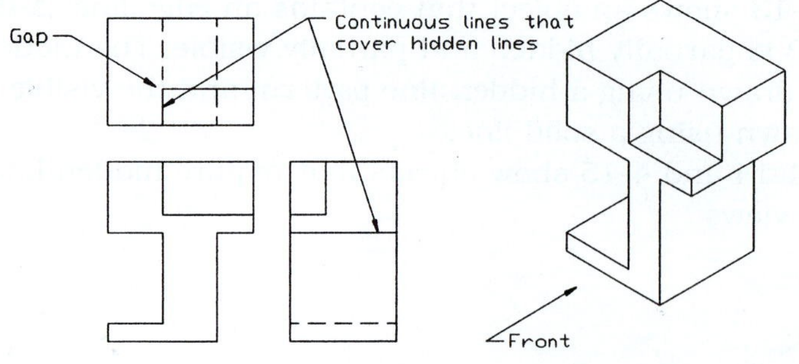

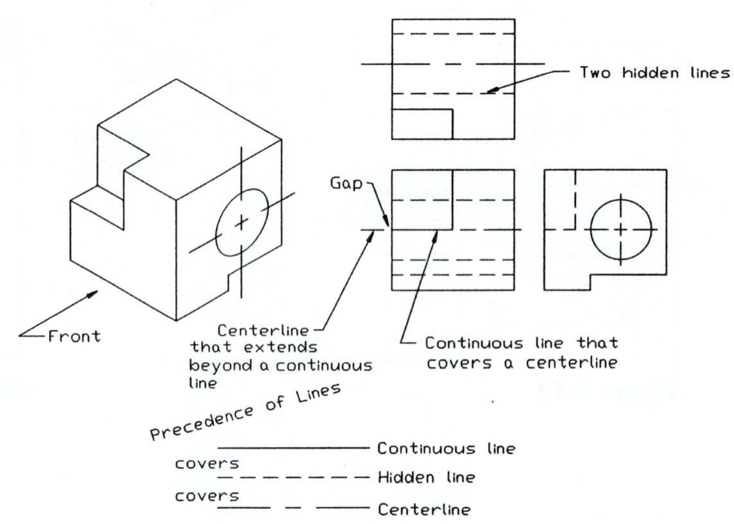

Figure 4-16

Slanted Surfaces

Slanted surfaces are surfaces drawn at an angle to each other. Figure 4-17 shows an object that contains two slanted surfaces. Surface **ABCD** appears as a rectangle in both the top and front views. Neither rectangle represents the true shape of the surface. Each is smaller than the actual surface. Also, none of the views show enough of the object to enable the viewer to accurately define the shape of the object. The views must be used together for a correct understanding of the object's shape.

Figure 4-17

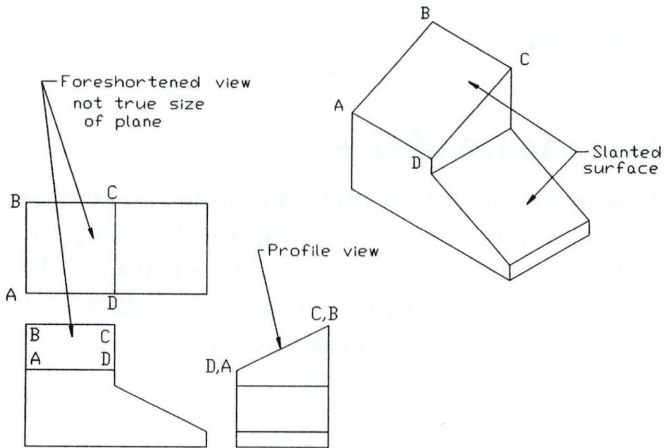

Figures 4-18 and 4-19 show objects that include slanted surfaces. Projection lines have been included to emphasize the importance of correct view location. Information is projected between the front and top views using vertical lines and between the front and side views using horizontal lines.

Figure 4-18

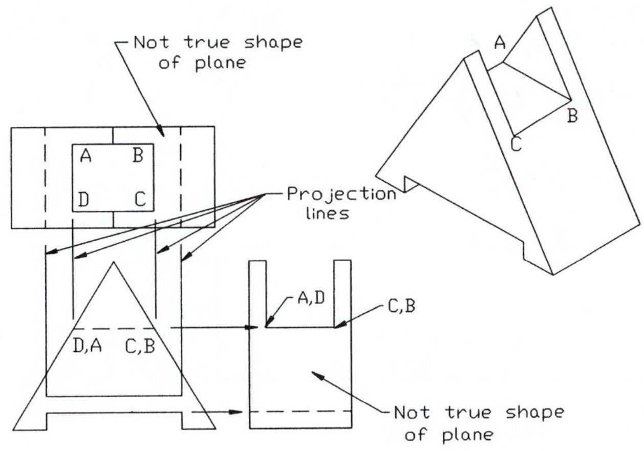

Figure 4-19

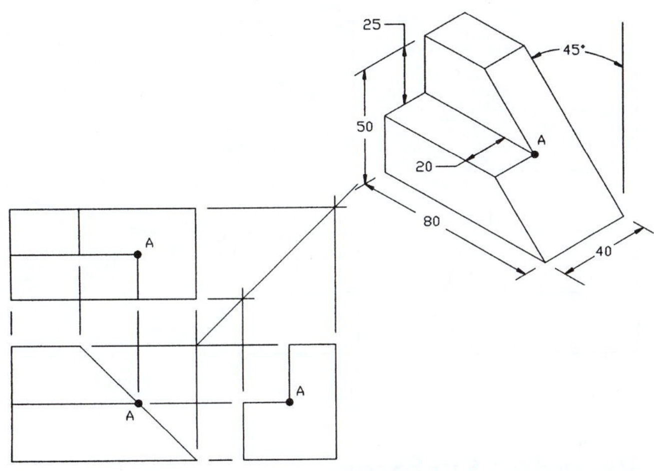

Compound Lines

A *compound line* is formed when two slanted surfaces intersect. Figure 4-20 shows an object that includes a compound line.

Figure 4-20

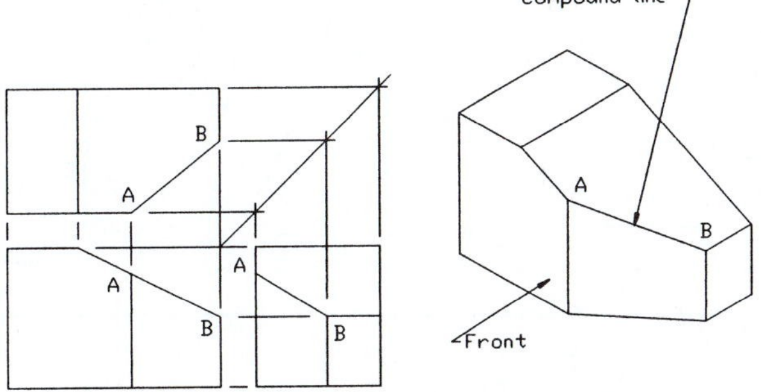

Oblique Surfaces

An *oblique surface* is a surface that is slanted in two different directions. Figures 4-21 and 4-22 show objects that include oblique surfaces.

Figure 4-21

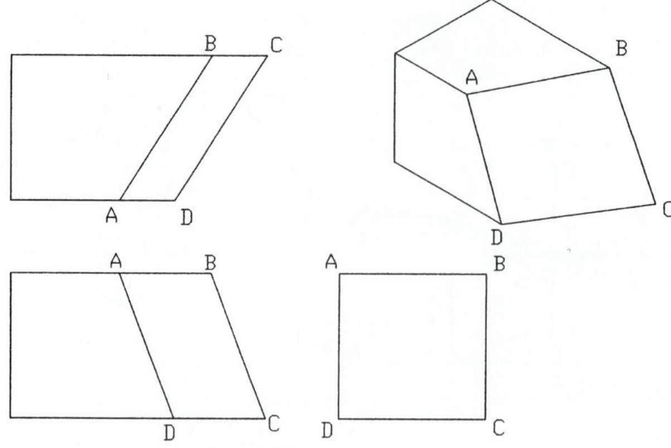

Figure 4-22

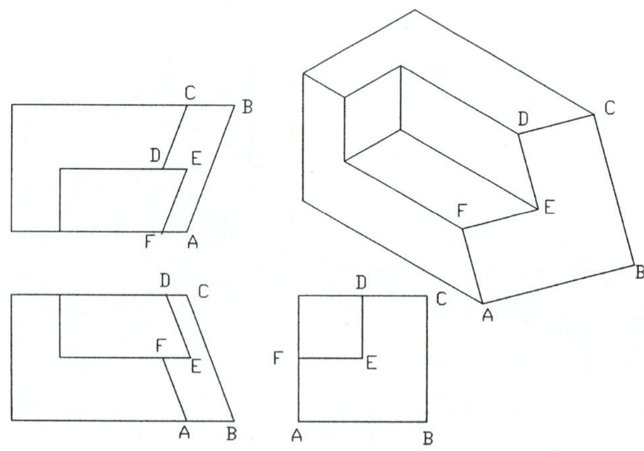

Rounded Surfaces

Figure 4-23 shows an object with two rounded surfaces. Note that as with slanted surfaces, an individual view is insufficient to define the shape of a surface. More than one view is needed to accurately define the surface's shape.

Figure 4-23

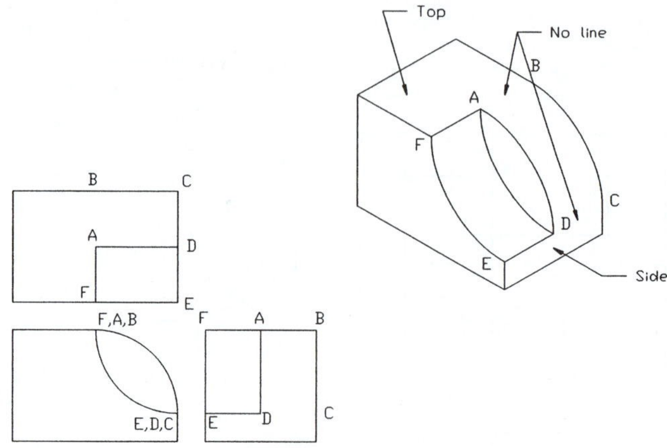

Chapter 4

Convention calls for a smooth transition between rounded and flat surfaces; that is, no lines are drawn to indicate the tangency. SolidWorks includes a line to indicate tangencies between surfaces in the isometric drawings created using the multiview options but does not include them in the orthographic views. Tangency lines are also not included when models are rendered.

Figure 4-24 shows the drawing conventions for including lines for rounded surfaces. If a surface includes no vertical portions or no tangency, no line is included.

Figure 4-25 shows an object that includes two tangencies. Each is represented by a line. Note in Figure 4-20 that SolidWorks will add tangent lines to the 3D model. These lines will not appear in the orthographic views.

Figure 4-24

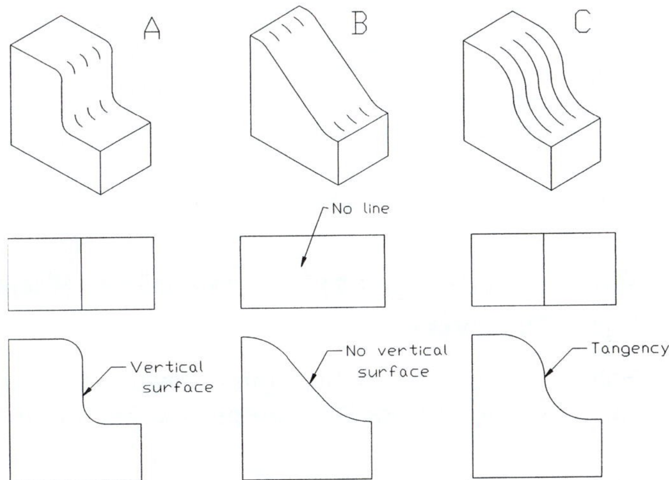

Figure 4-25

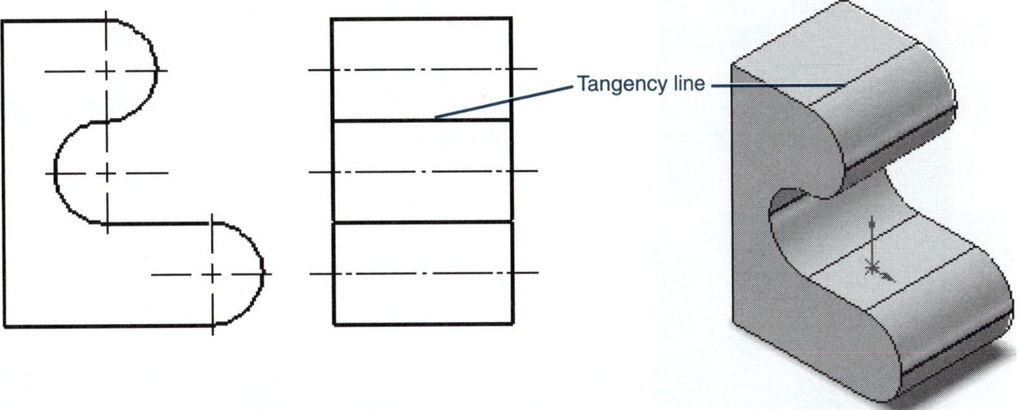

Figure 4-26 shows two objects with similar configurations; however, the boxlike portion of the lower object blends into the rounded portion exactly on its widest point, so no line is required.

Figure 4-26

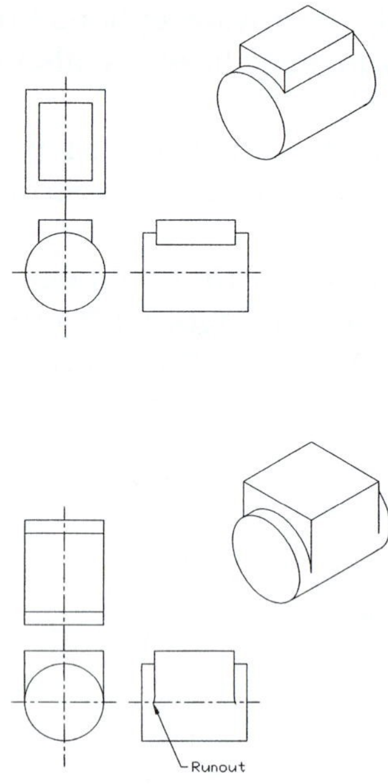

4-4 Drawing Orthographic Views Using SolidWorks

SolidWorks creates orthographic views using the drawing tools found on the **New SolidWorks Document** box. See Figure 4-27.

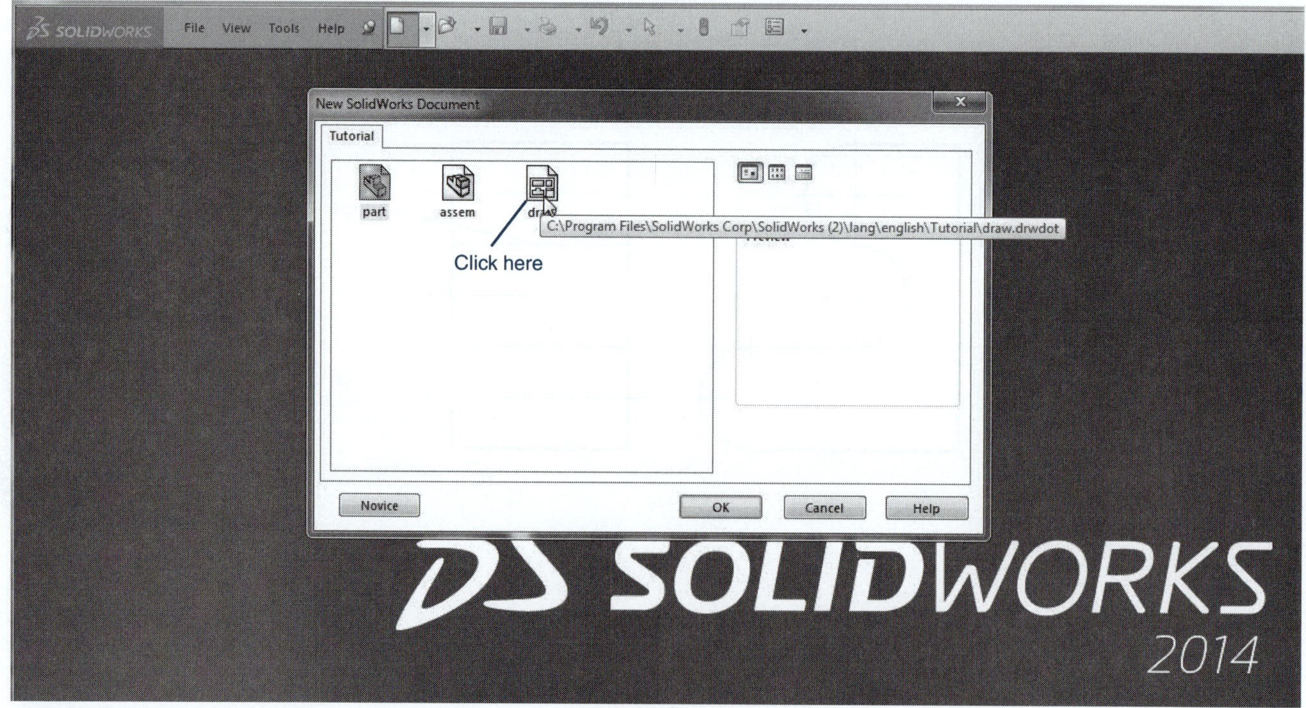

Figure 4-27

1 Start a new drawing by clicking the **New** tool.

2 Click the **Drawing** icon on the **New SolidWorks Document** box.

3 Click **OK**.

The **Sheet Format/Size** box will appear. See Figure 4-28. Accept the **A (ANSI) Landscape** format.

Figure 4-28

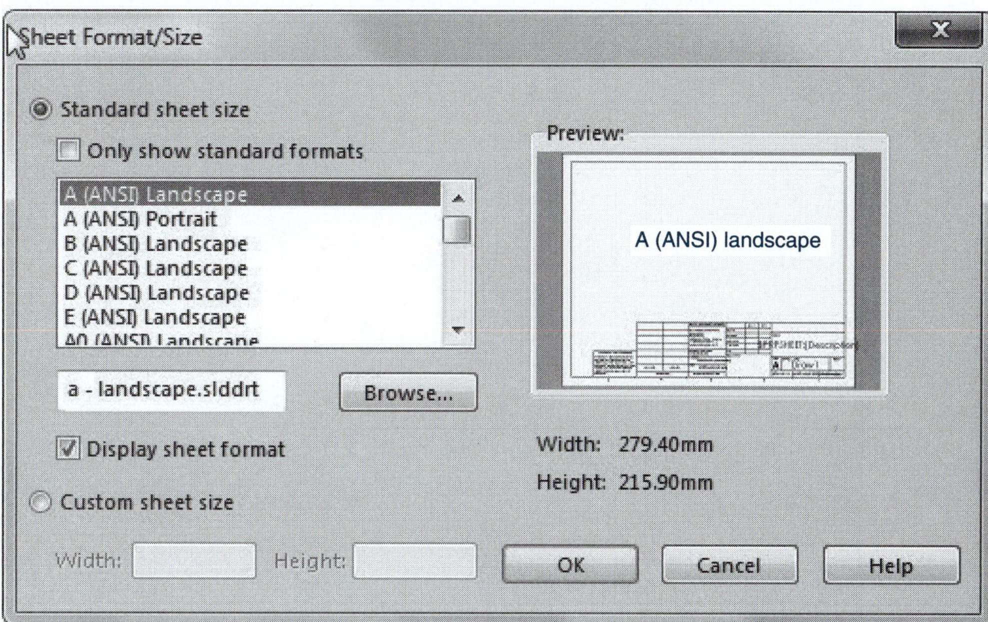

Standard Drawing Sheet Sizes Inches
A = 8.5 × 11
B = 11 × 17
C = 17 × 22
D = 22 × 34
E = 34 × 44

Standard Drawing Sheet Sizes Millimeters
A4 = 210 × 297
A3 = 297 × 420
A2 = 420 × 594
A1 = 594 × 841
A0 = 841 × 1189

TIP

Drawing sheets, that is, the paper drawings are printed on, are manufactured in standard sizes. For example, in the English unit system an A-size drawing sheet is 8.5 × 11 in. In the metric unit system an A4-size drawing sheet is 210 × 297 mm.

A listing of standard sheet sizes is shown in Figure 4-28.

4 Click **OK**.

A drawing template will appear. See Figure 4-29. The template includes a title block, a release block, a tolerance block, and two other blocks. The template format can be customized, but in this example the default template will be used. The title block will be explained in the next section.

Figure 4-29

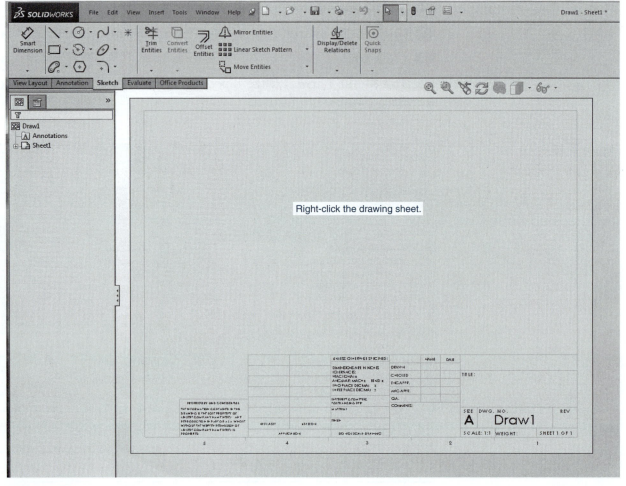

5 Move the cursor into the drawing area and right-click the mouse.

6 Select the **Properties** option.

See Figure 4-30.

Figure 4-30

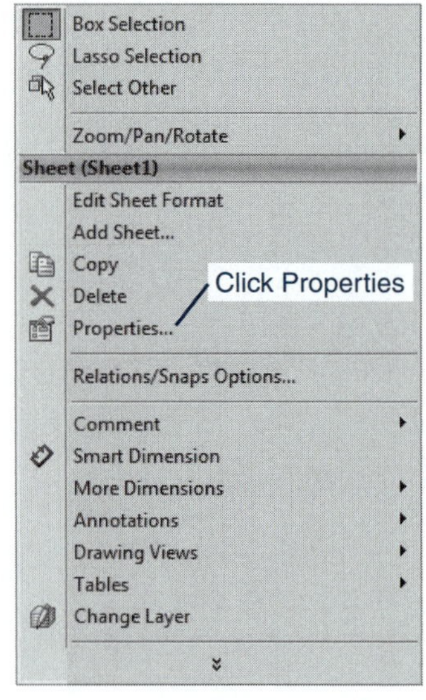

The **Sheet Properties** dialog box will appear. See Figure 4-31.

Figure 4-31

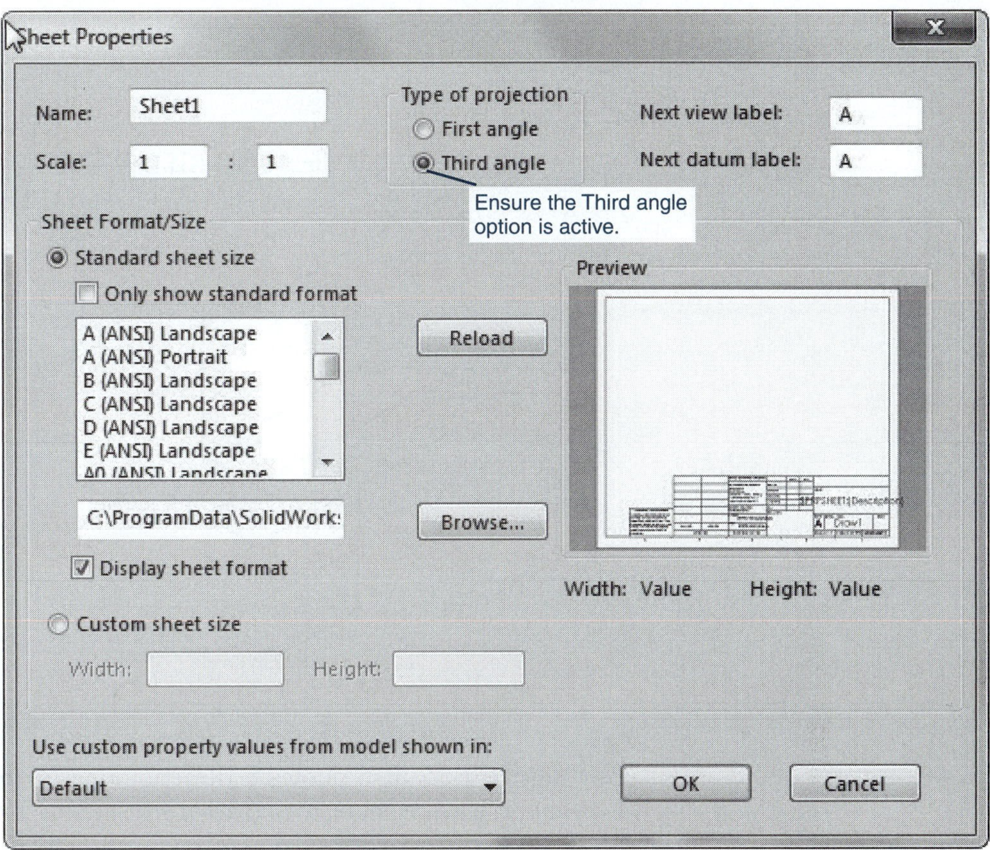

7 Click the **Third angle** button.

8 Click **OK**.

Third-angle projection is the format preferred by U.S. companies in compliance with ANSI (American National Standards Institute) standards. First-angle projection is used by countries that are in compliance with ISO (International Organization for Standardization). Figure 4-32 shows an L-bracket drawn in both first- and third-angle projection. Compare the differences in the projected views.

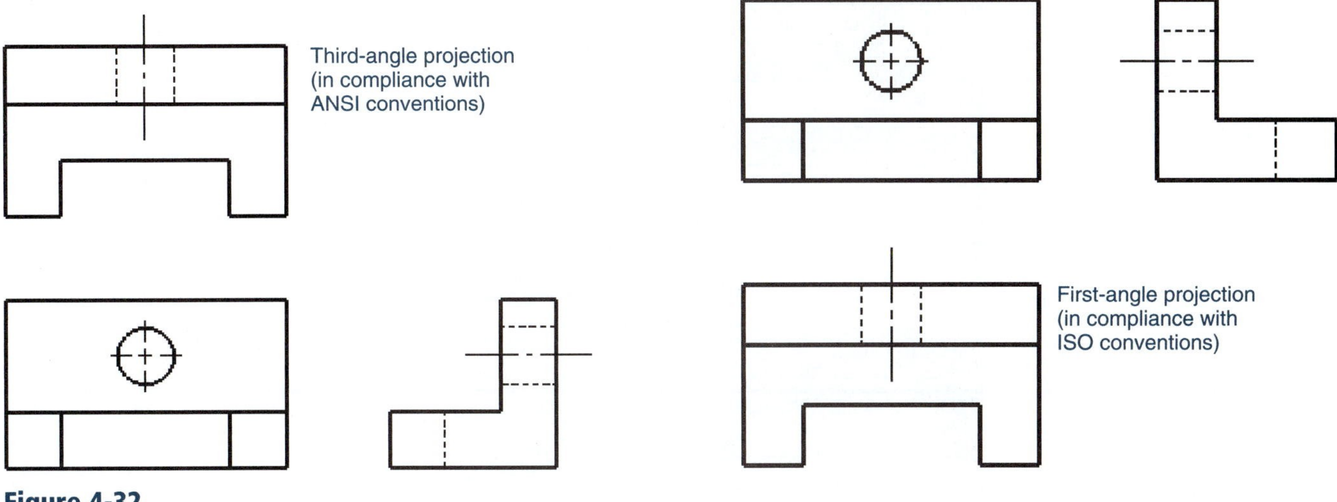

Figure 4-32

Figure 4-32
(*Continued*)

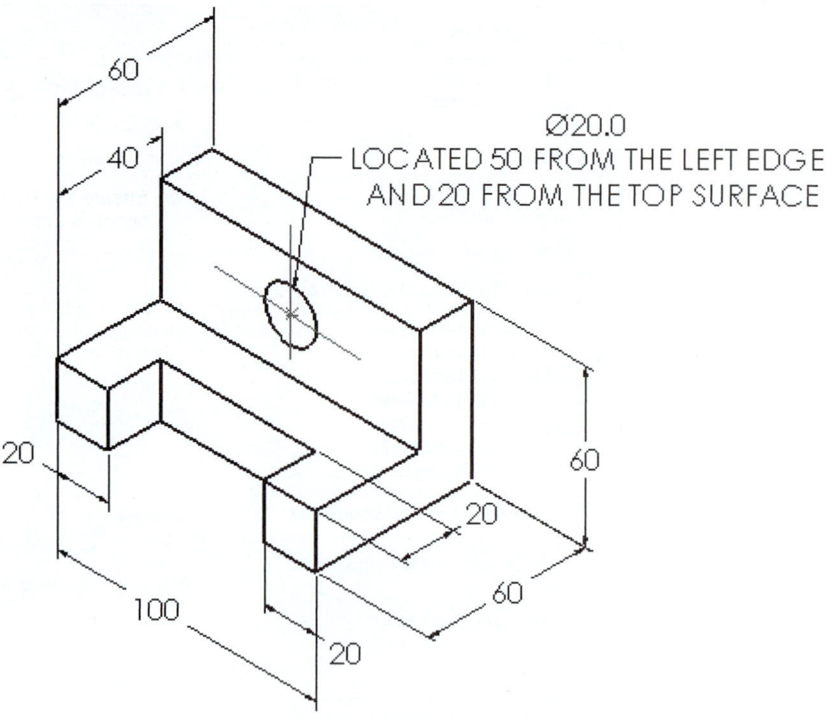

Figure 4-32 also shows a dimensioned isometric drawing of the L-bracket. The bracket was drawn in Section 3-3. If you have not previously drawn the bracket, do so now and save it as **L-bracket**.

9 Click the **View Layout** tab.

10 Click the **Standard 3 View** tool located on the **View Layout** panel.

See Figure 4-33. The **Standard 3 View PropertyManager** will appear on the left side of the screen.

Figure 4-33

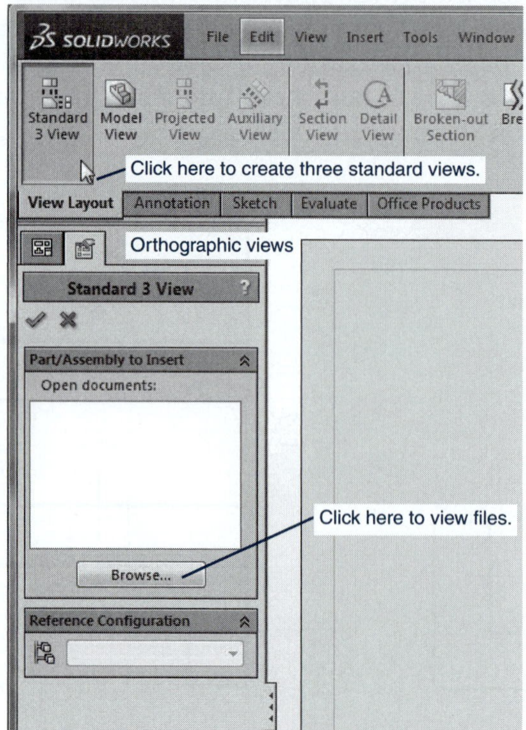

11 Click the **Browse . . .** box.

The **Open** box will appear. See Figure 4-34.

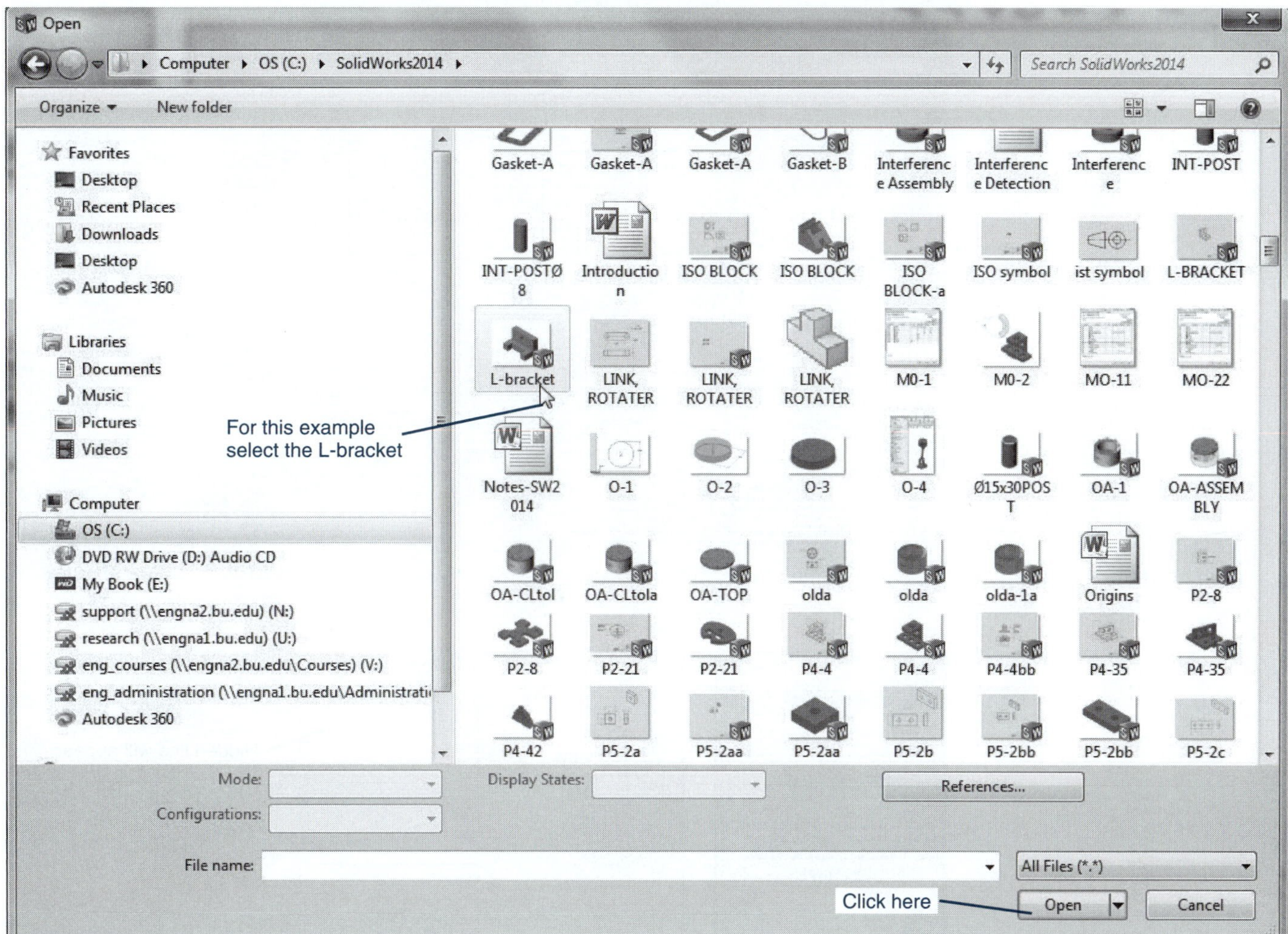

Figure 4-34

12 Select the **L-BRACKET** file. A rectangle will appear on the screen representing the views.

13 Select the L-bracket, and click **Open**.

Three orthographic views will appear on the screen. They include no hidden lines. The hidden lines must be added. See Figure 4-35.

14 Click the top orthographic view and select the **Hidden Lines Visible** tool in the **Display Style** box of the **Drawing View PropertyManager**.

The hidden lines will appear in the top view.

15 Click the right-side view, then click the **Hidden Lines Visible** tool to add hidden lines to the right-side view.

Figure 4-35

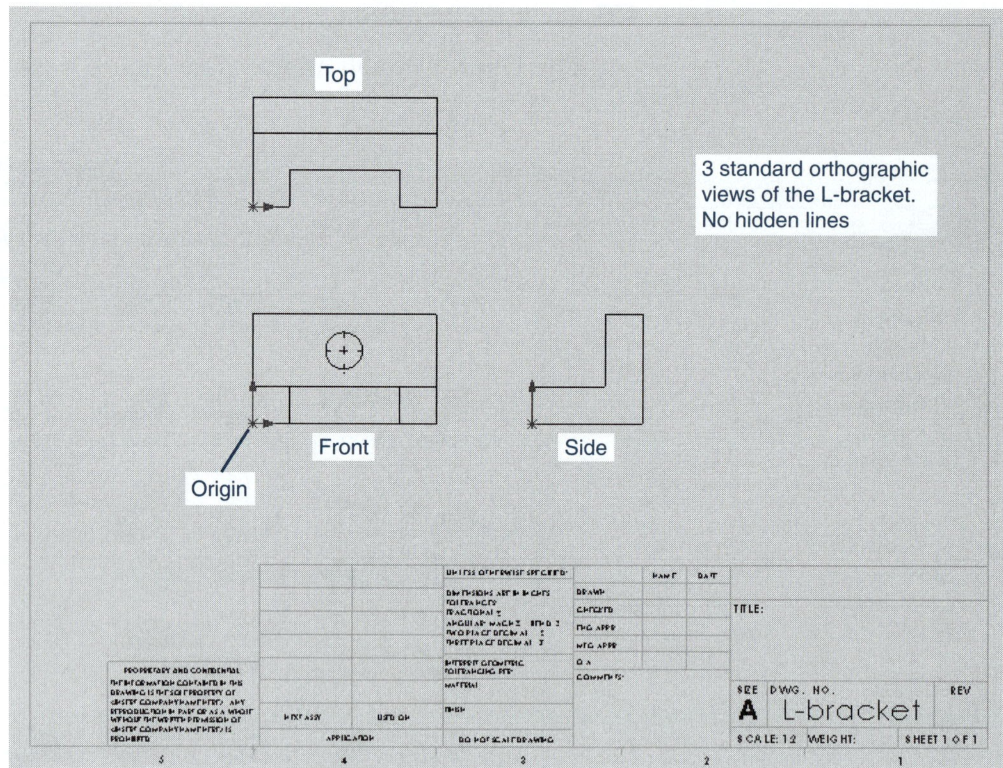
Top
3 standard orthographic views of the L-bracket. No hidden lines
Front
Side
Origin
TITLE:
SIZE DWG. NO.
A L-bracket
REV
SCALE: 1:2 WEIGHT:
SHEET 1 OF 1

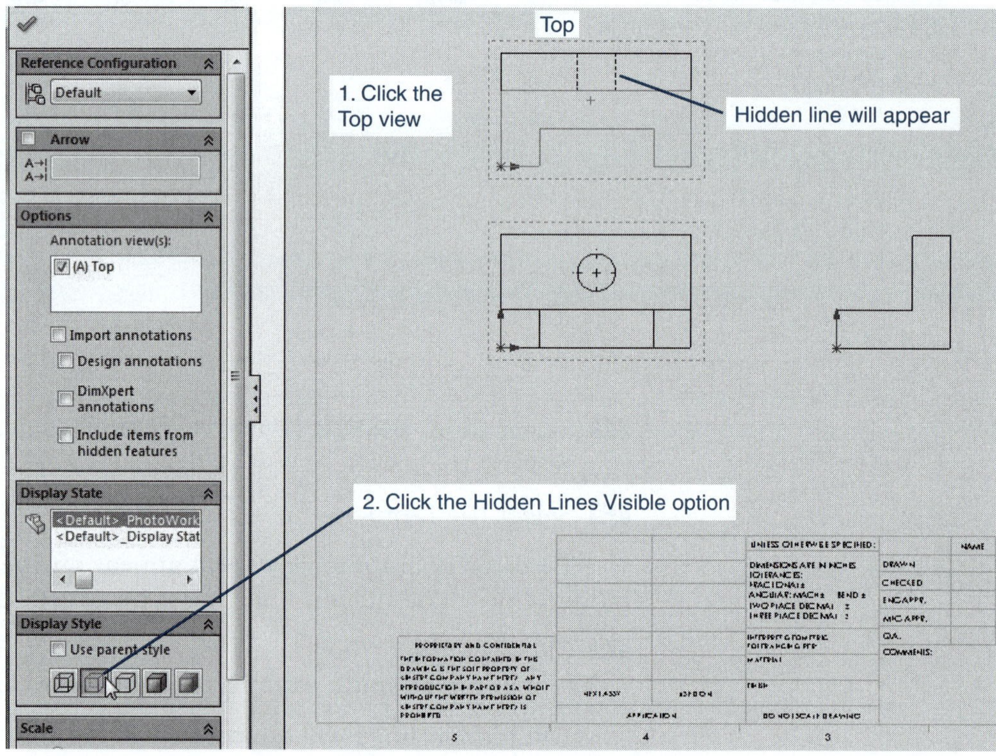
Reference Configuration
Default
Arrow
Options
Annotation view(s):
(A) Top
Import annotations
Design annotations
DimXpert annotations
Include items from hidden features
Display State
<Default>_PhotoWork
<Default>_Display Stat
Display Style
Use parent style
Scale
Top
1. Click the Top view
Hidden line will appear
2. Click the Hidden Lines Visible option

Figure 4-35
(Continued)

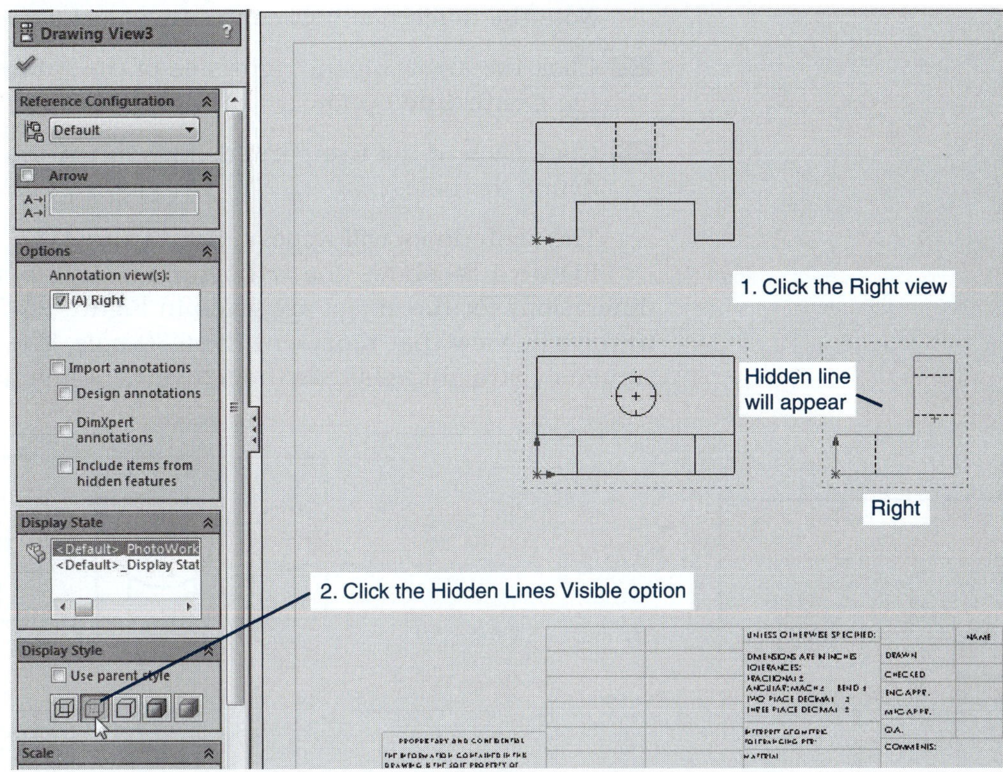

Notice in the top and right-side views that there are no centerlines for the hole. Centerlines are added using the **Centerline** option found on the **Annotation** tab. See Figure 4-36. The circular view of the hole will automatically generate a set of perpendicular centerlines.

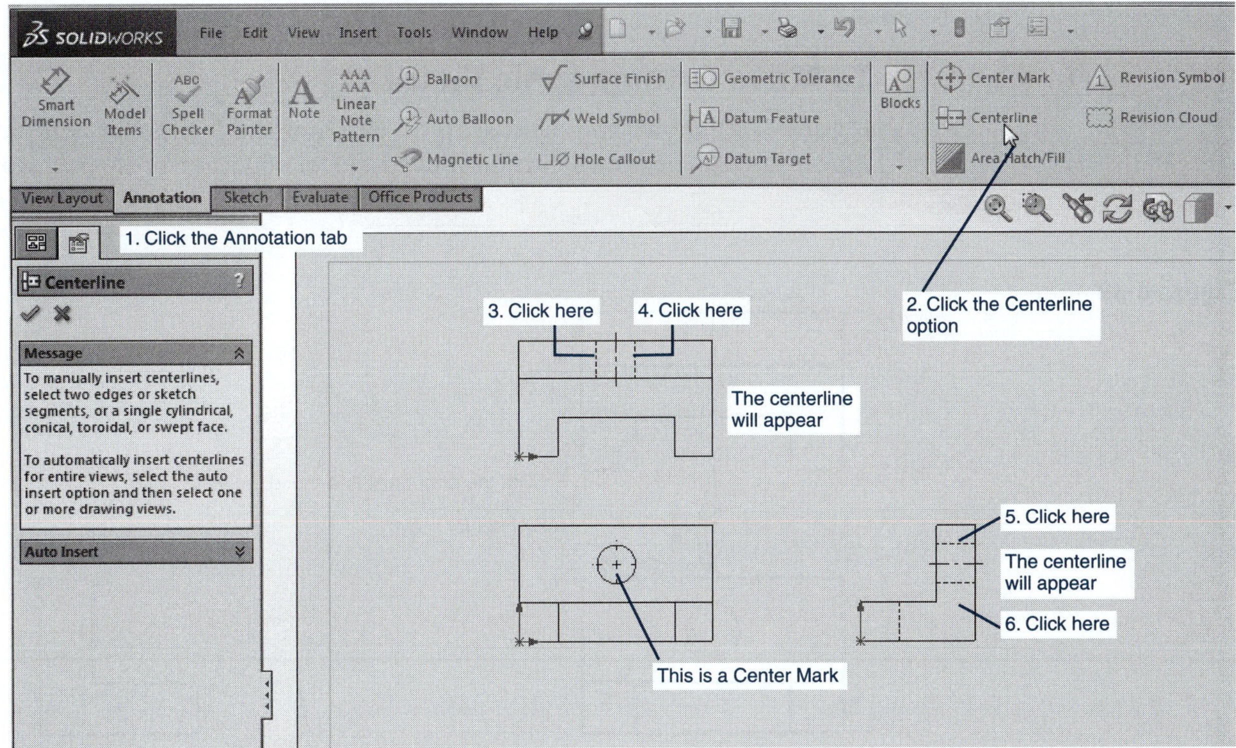

Figure 4-36

Note the difference between center marks and centerline tools.

16 Click the arrow on the right side of the **Annotation** panel and select the **Centerline** option.

17 Click each of the two parallel lines in the top and side views that define the hole.

The centerlines will appear. See Figure 4-37.

Figure 4-38 shows the orthographic views of another object. The dimensions for the object are given in Figure P4-23. Note the hidden lines in the side view that represent the Ø30 hole. The right vertical line is continuously straight, whereas the left vertical line has a step. Why?

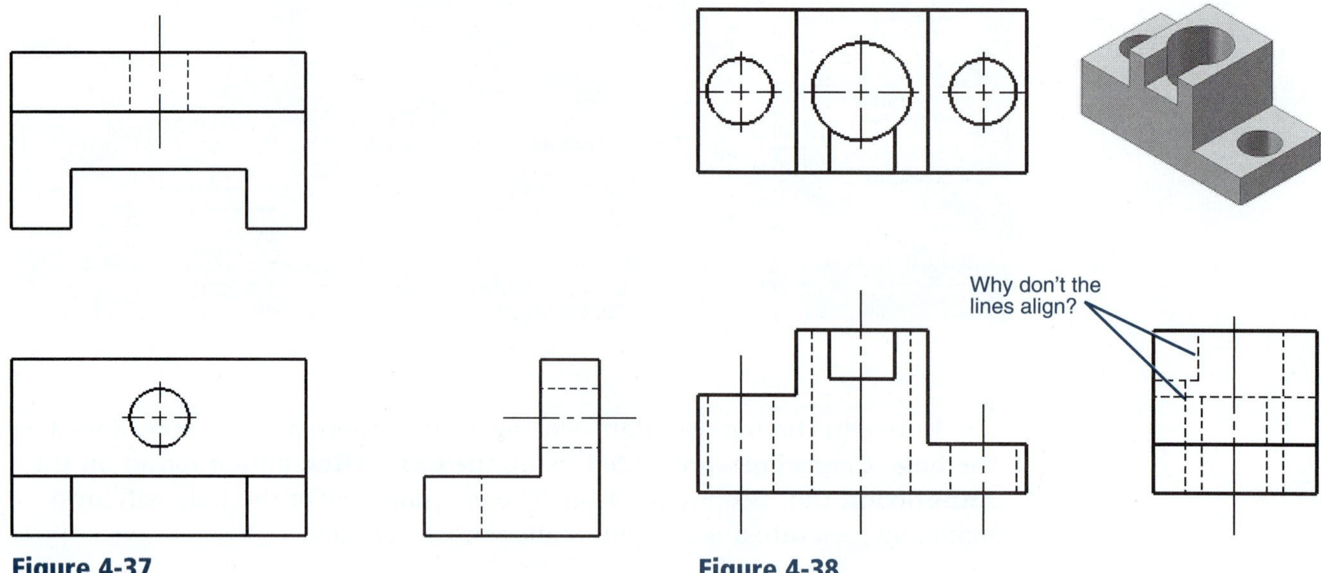
Why don't the lines align?

Figure 4-37

Figure 4-38

To Move Orthographic Views

Figure 4-39 shows the orthographic views of the L-bracket generated for Figure 4-37. The views can be moved closer together or farther apart.

Figure 4-39

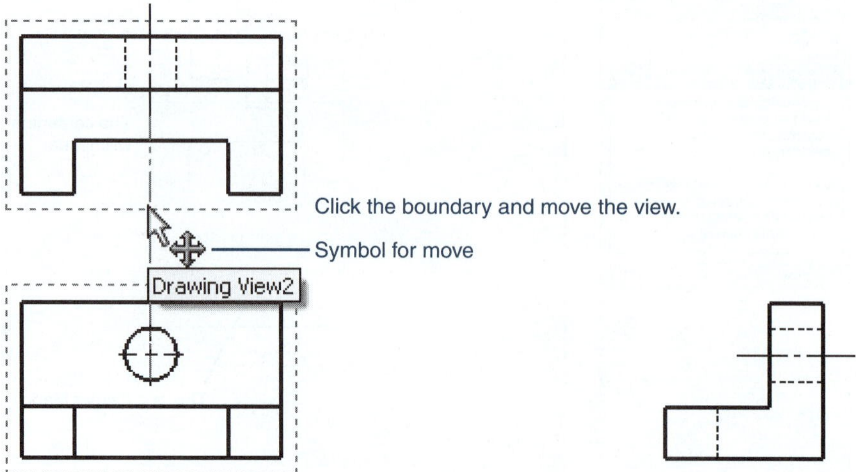
Click the boundary and move the view.
Symbol for move
Drawing View2

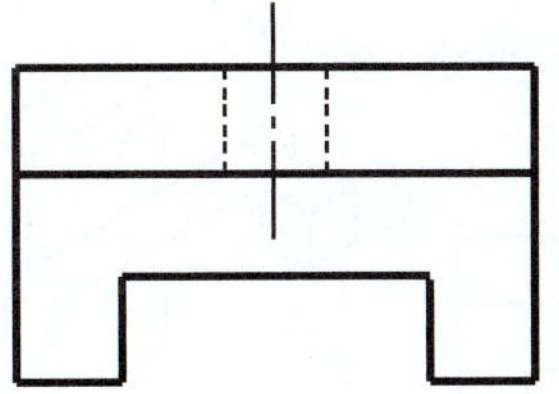

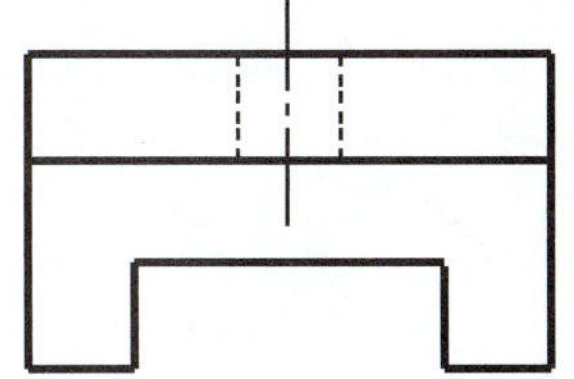

Views are now closer together.

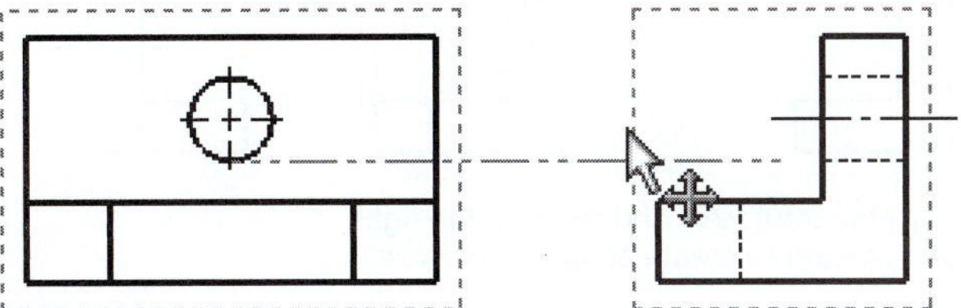
Click the boundary and move the view.

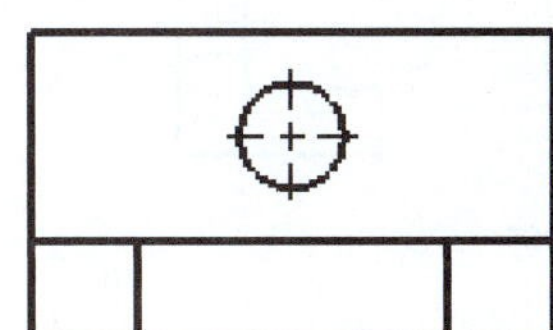

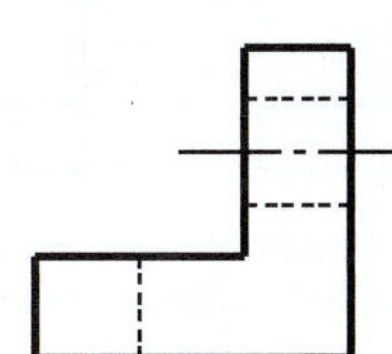

Figure 4-39
(*Continued*)

1. Move the cursor into the area of the top view.

   A red boundary line will appear.

2. Click and hold one of the boundary lines.
3. Drag the view to a new location.

To Create Other Views

The **Standard 3 View** tool will generate front, top, and right-side orthographic views of an object. These views are considered the standard three views. Other orthographic views and isometric views can be generated.

1. Click the **Projected View** tool.

   The **Projected View** tool is one of the **View Layout** tools.

2. Click the front view and move the cursor to the left of the front view, creating a new orthographic view.

   In this example a left-side view was created. Add hidden lines and centerlines as needed.

3. Click the left view in its new location.
4. Press the <Esc> key or click the green **OK** check mark.

   See Figure 4-40.

5. Use the **Centerline** tool to add a centerline to the hole in the left-side view.
6. Click the **Projected View** tool and click the front view again.
7. Move the cursor to the right and upward.
   An isometric view will appear.
8. Click the isometric view in its new location.
9. Press the **<Esc>** key or click the green **OK** check mark.

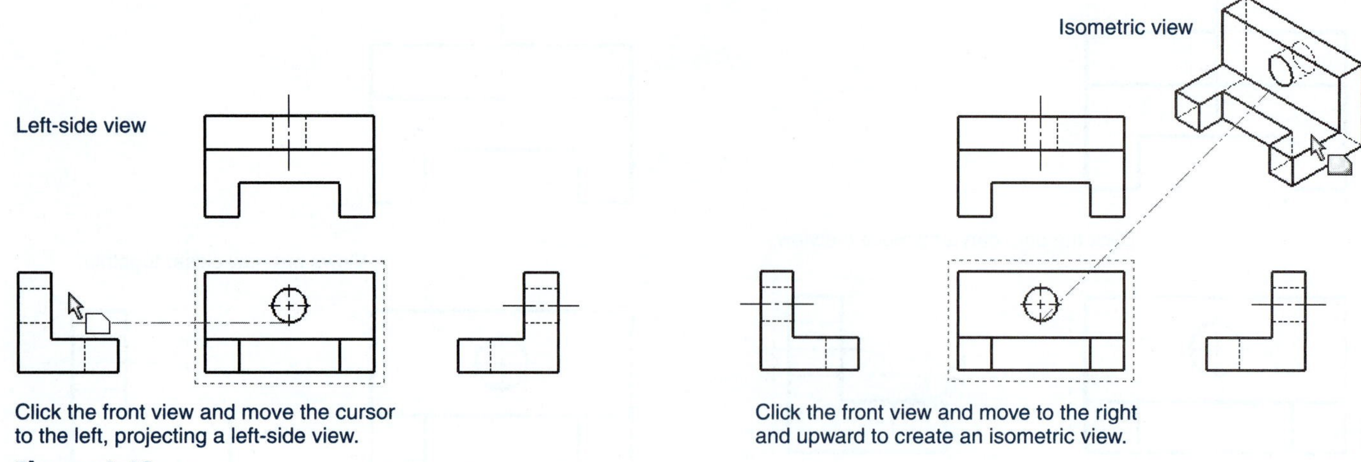
Left-side view
Isometric view
Click the front view and move the cursor to the left, projecting a left-side view.
Click the front view and move to the right and upward to create an isometric view.

Figure 4-40

4-5 Section Views

Some objects have internal surfaces that are not directly visible in normal orthographic views. ***Section views*** are used to expose these surfaces. Section views do not include hidden lines.

Any material cut when a section view is defined is hatched using section lines. There are many different styles of hatching, but the general style is evenly spaced 45° lines. This style is defined as ANSI 31 and will be applied automatically by SolidWorks.

Figure 4-41 shows a three-dimensional view of an object. The object is cut by a cutting plane. ***Cutting planes*** are used to define the location of the section view. Material to one side of the cutting plane is removed, exposing the section view.

Figure 4-42 shows the same object presented in Figure 4-41 using two orthographic views and a section view. The cutting plane is represented by a cutting plane line. The cutting plane line is defined as A-A, and the section view is defined as view A-A.

Figure 4-41

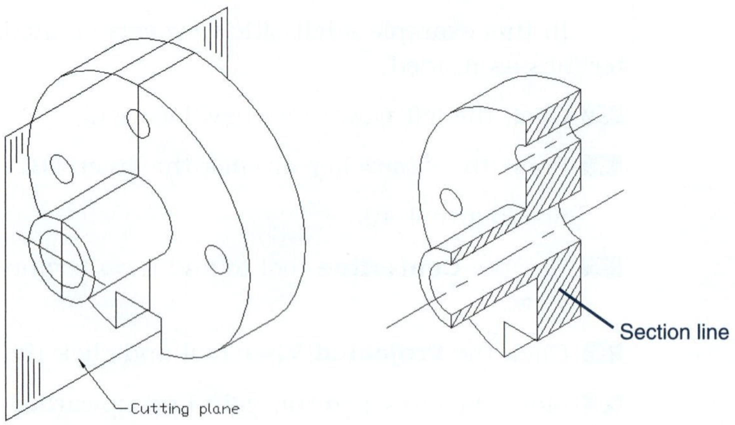
Cutting plane
Section line

Figure 4-42

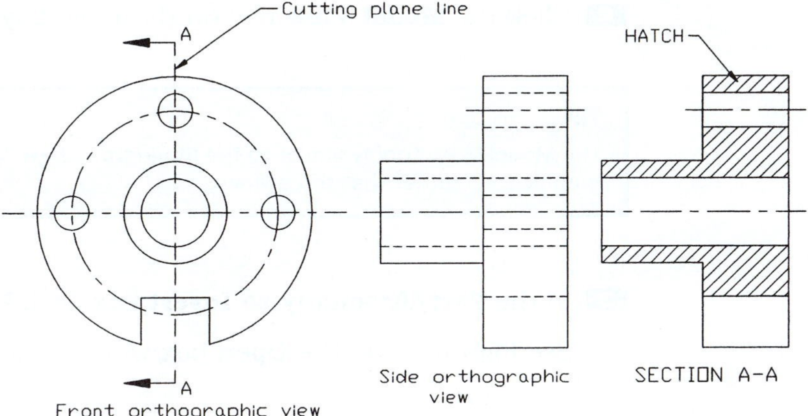

All surfaces directly visible must be shown in a section view. In Figure 4-43 the back portion of the object is not affected by the section view and is directly visible from the cutting plane. The section view must include these surfaces. Note how the rectangular section blocks out part of the large hole. No hidden lines are used on section views.

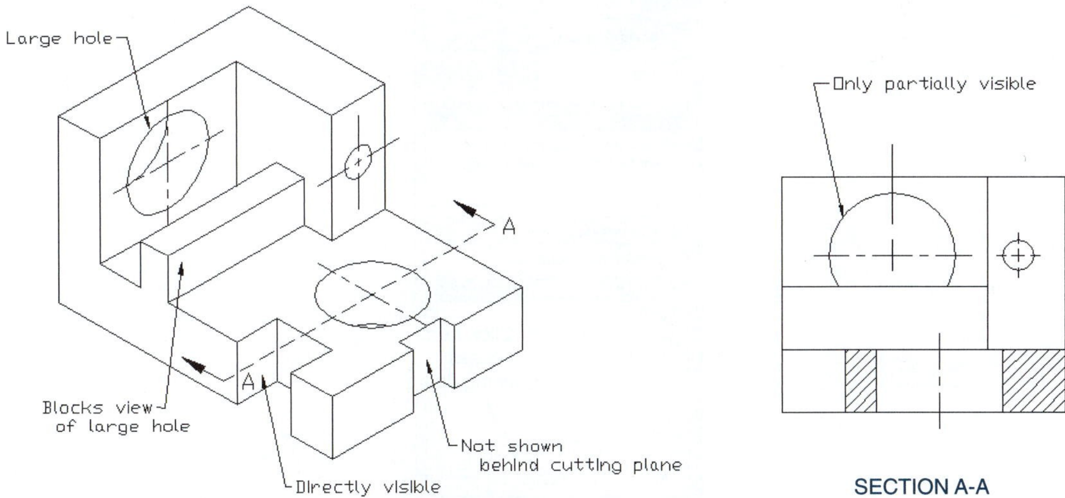

Figure 4-43

4-6 Drawing a Section View Using SolidWorks

This section will show how to draw a section view of an existing model. In this example, the model pictured in Figure 4-38 is used. This is the same as Figure P4-22 in the Chapter Projects.

1 Start a new drawing using the **Drawing** format.

See the previous section on how to create orthographic views using SolidWorks. Select the **A (ANSI) Landscape** format and select **Third angle** projection.

NOTE

See Figures 4-29 to 4-32 for an explanation of how to specify the third-angle format.

2 Click the **Model View** tool on the **View Layout** panel.

> **TIP**
>
> The **Model View** tool is similar to the **Standard 3 View** tool but creates only one orthographic view rather that three views.

3 In the **Part/Assembly to Insert** box click **Browse. . . .**

See Figure 4-44. The **Open** box will appear. See Figure 4-45.

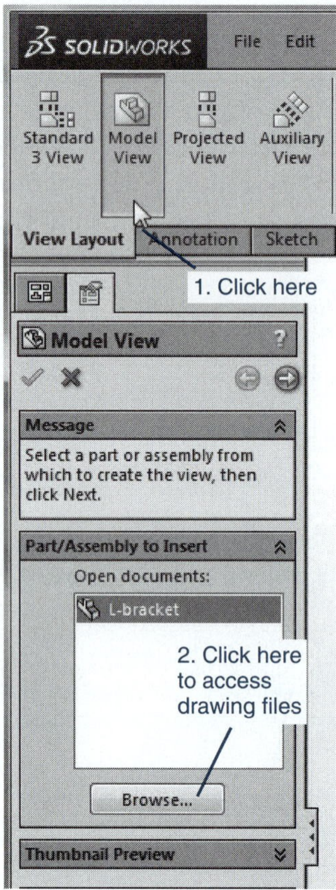

Figure 4-44

4 Click the model to be used to draw orthographic views, and click **Open**.

In this example the model is called **BLOCK, 3 HOLE**. The dimensions for the BLOCK, 3 HOLE can be found in Figure P4-22.

A rectangular outline will appear defining the boundaries of the orthographic view. By default, this will be a front view. In this example we want a top view.

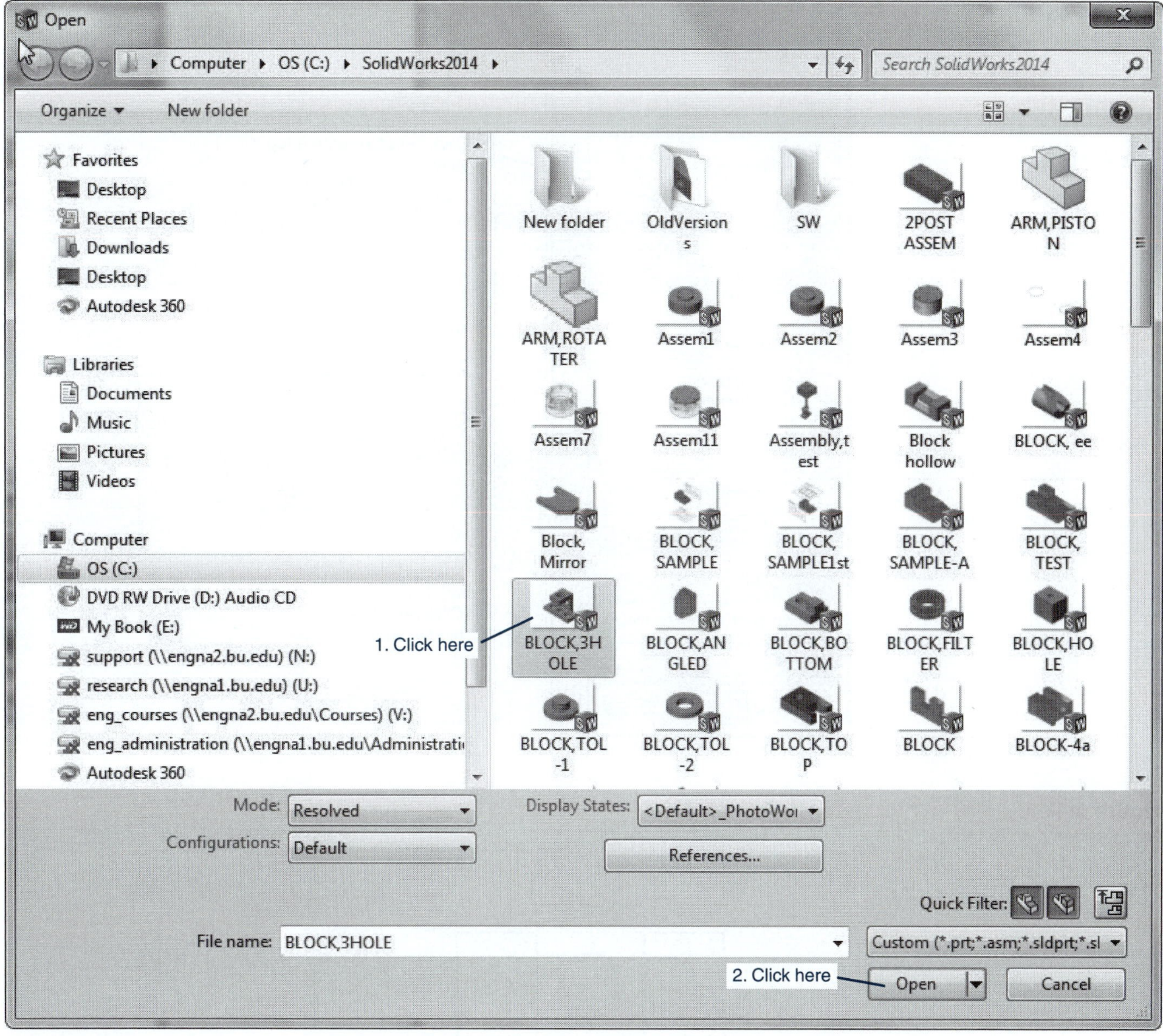

Figure 4-45

5 Click the **Top** view tool.

See Figure 4-46.

6 Locate the top orthographic view on the drawing screen and click the mouse.

7 Add a center mark to the Ø30 hole.

See Figure 4-47.

8 Click the **View Layout** tab and click the **Section View** tool. The orthographic view will be outlined by a dotted line.

Figure 4-46

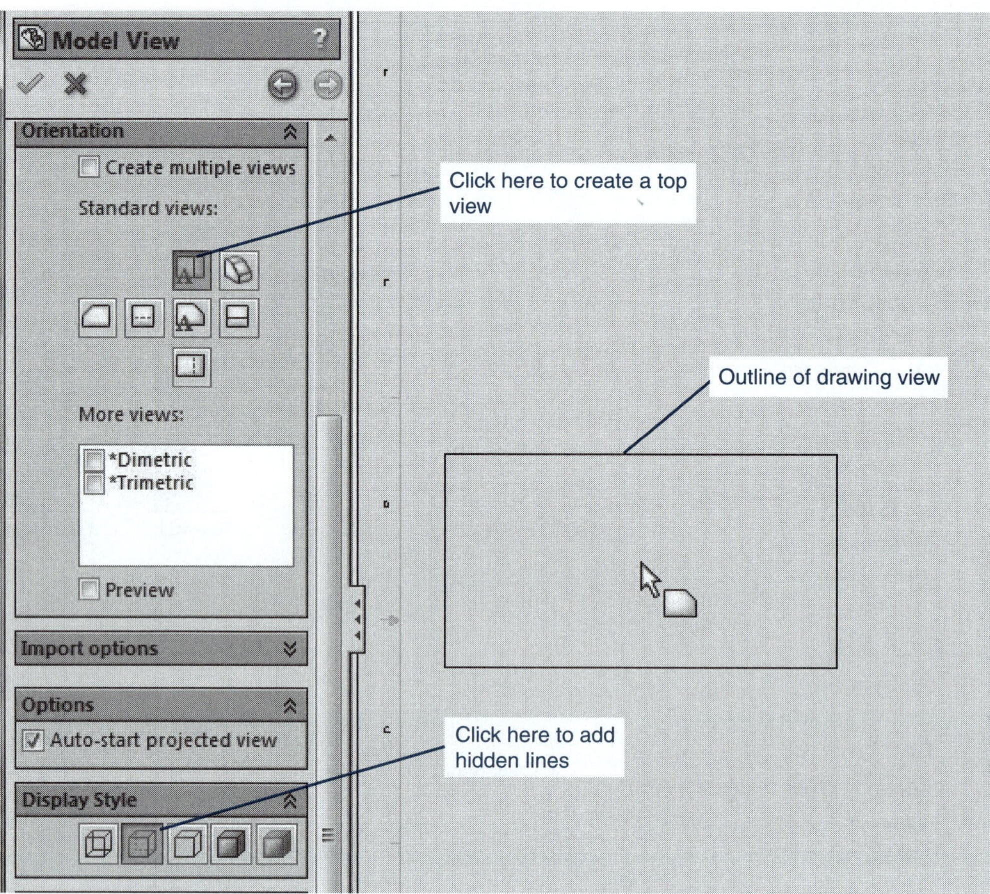

Figure 4-47

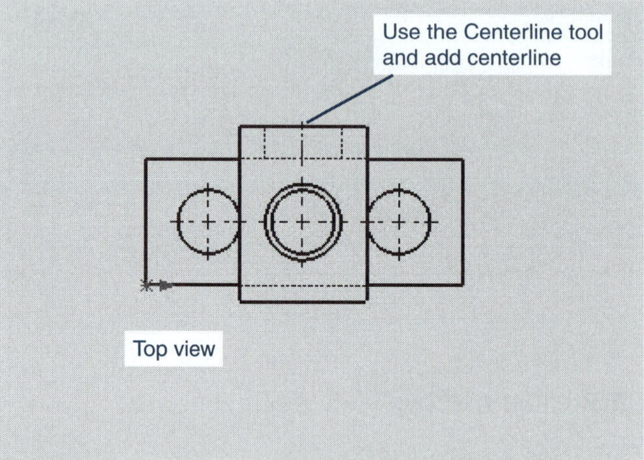

NOTE

If more than one view was present on the screen, you would first have to select which view you wanted to be used to create the section view.

9 Select a horizontal cutting plane line.

10 Define the location of the cutting plane line by moving the cursor to the approximate midpoint of the left vertical line of the orthographic view.

The system will automatically jump to the line's midpoint. A filled square icon will appear.

11 Click the green **OK** check mark.

12 Move the cursor downward.

The section view will appear and move with the cursor. See Figure 4-48.

Figure 4-48

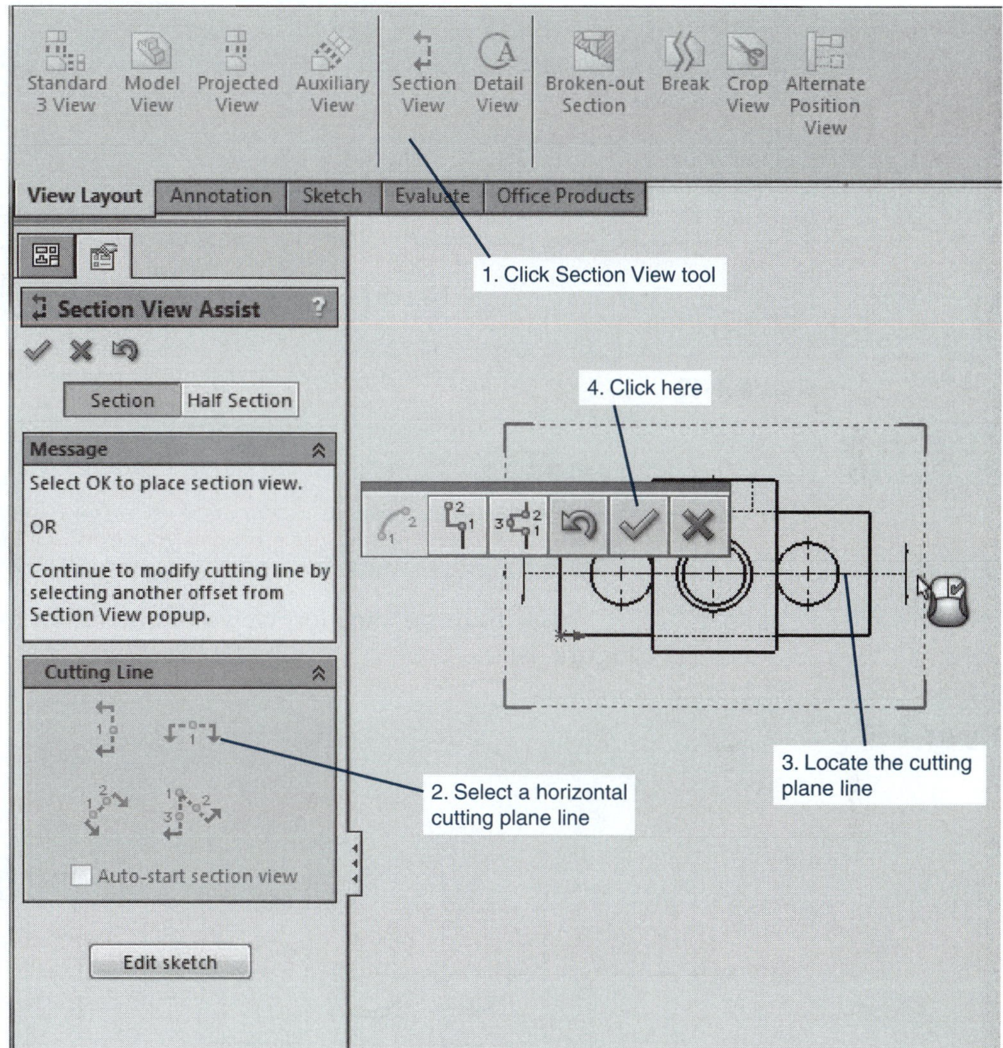

13 Select an appropriate location and click the mouse.

14 Click the **Flip direction** box if necessary.

See Figure 4-49.

15 Add centerlines.

16 Click the green **OK** check mark.

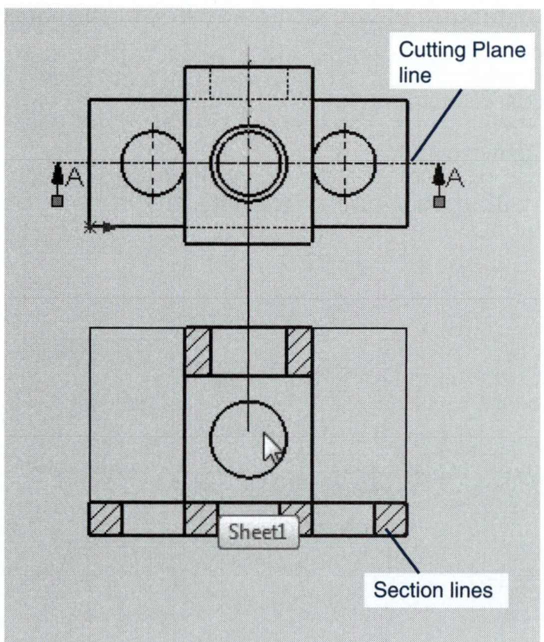

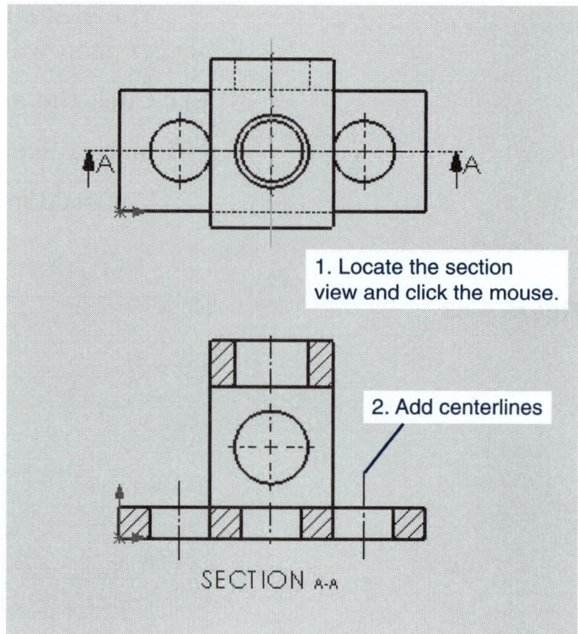

Figure 4-49

> **NOTE**
>
> Section views are always located behind the arrows; that is, the arrows point away from the section view. Think of the arrows as your eyes looking at the section view.

More than one section view may be taken from the same model. See Figure 4-50.

Figure 4-50

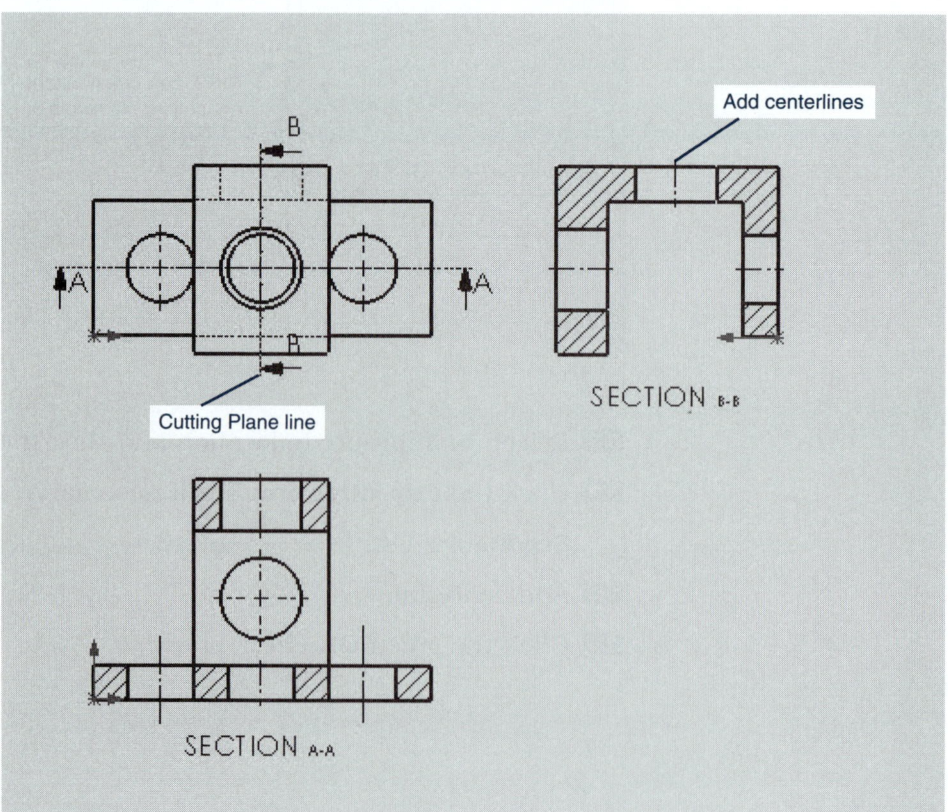

The section views shown in Figure 4-50 use a hatching pattern made from evenly spaced 45° lines. This is the most commonly used hatch pattern for section views and is designated as ANSI 31 in the ANSI hatch patterns. SolidWorks can also draw section views using one of five different styles. See Figure 4-51.

Figure 4-51

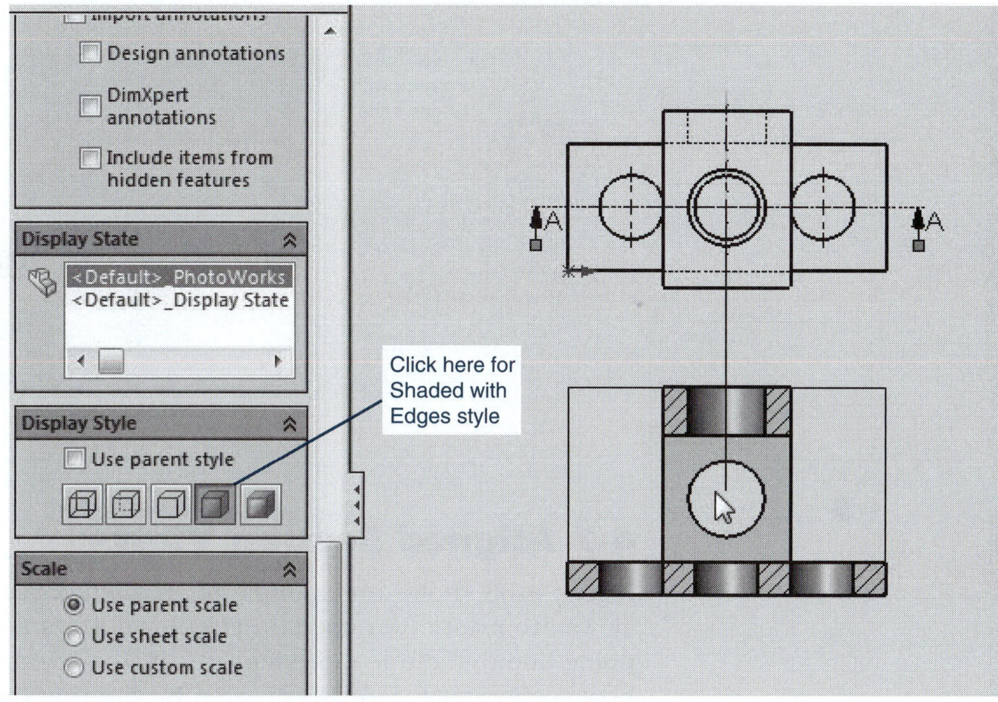

To Change the Style of a Section View

1 Move the cursor into the area of the section view and right-click the mouse.

A listing of tools will appear, and the **Display Style** box will appear.

2 Click the mouse again in the section view area to remove the list of tools.

3 Click one of the boxes in the **Display Style** box.

Figure 4-52 shows two of the styles available: shaded with edge lines, and shaded. The hidden lines removed style is used for all other illustrations in this chapter.

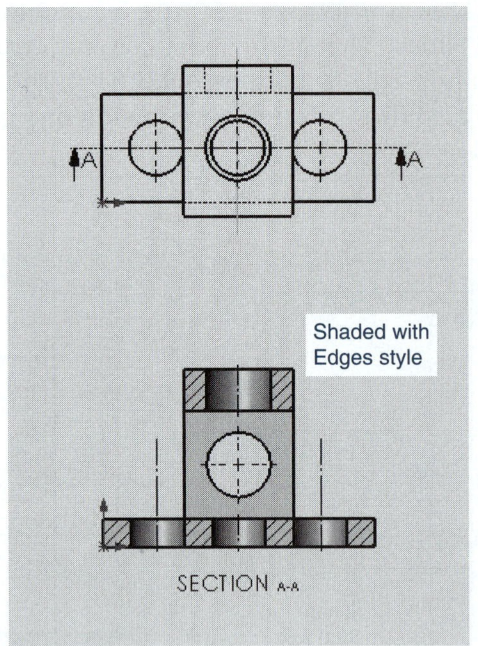

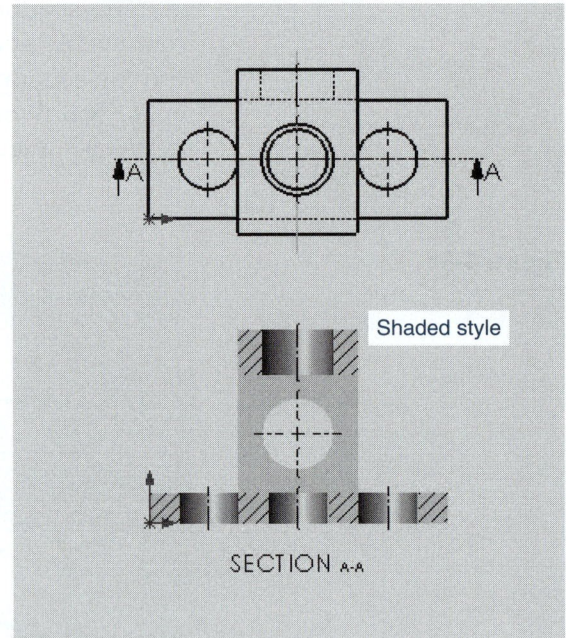

Figure 4-52

4-7 Aligned Section Views

Figure 4-53 shows an example of an aligned section view. Aligned section views are most often used on circular objects and use an angled cutting plane line to include more features in the section view. Note in Figure 4-48 how section view A-A was created by rotating the cutting plane into a vertical position before projecting the section view.

Figure 4-53 shows an aligned section view created using SolidWorks. The aligned section view was created as follows.

1. Start a new drawing using the **Drawing** format and click on the model for the aligned section view.

Figure 4-53

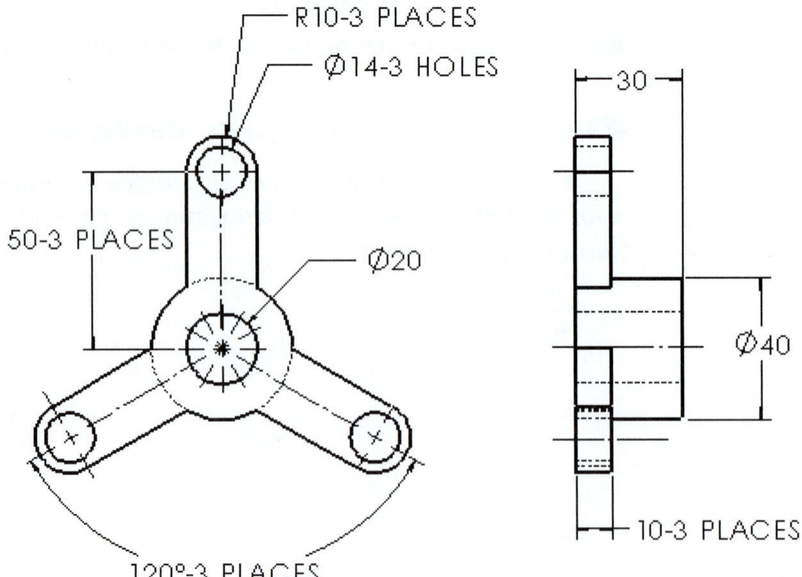

Figure 4-53
(Continued)

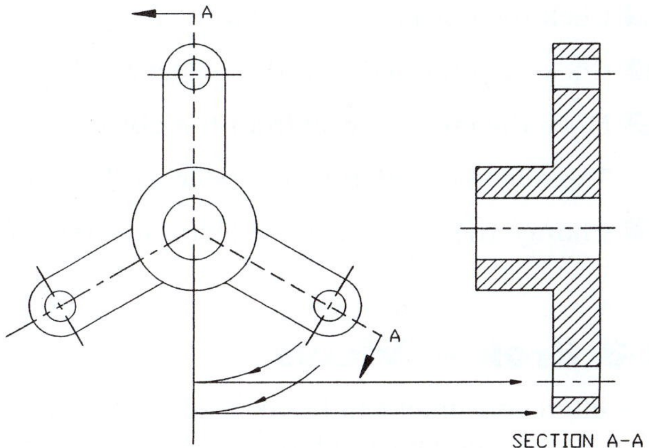

The model was drawn previously using the given dimensions.

2 Access the **Aligned Section View** tool.

The **Aligned Section View** tool is a flyout from the **Section View** tool.

3 Click the edge of the object as shown in Figure 4-54.

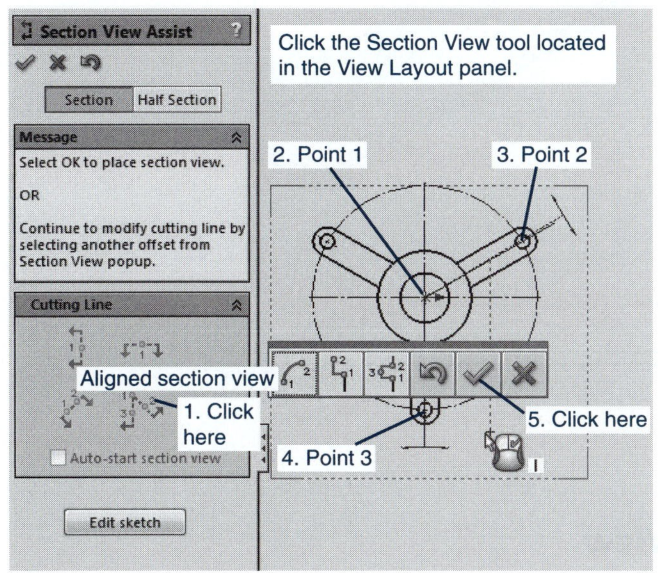

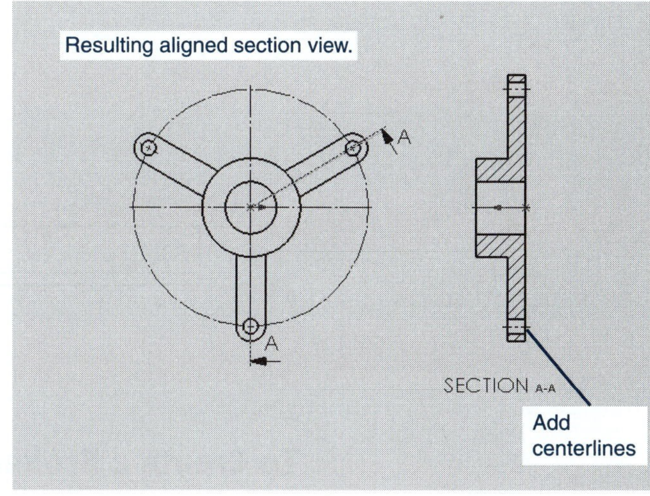

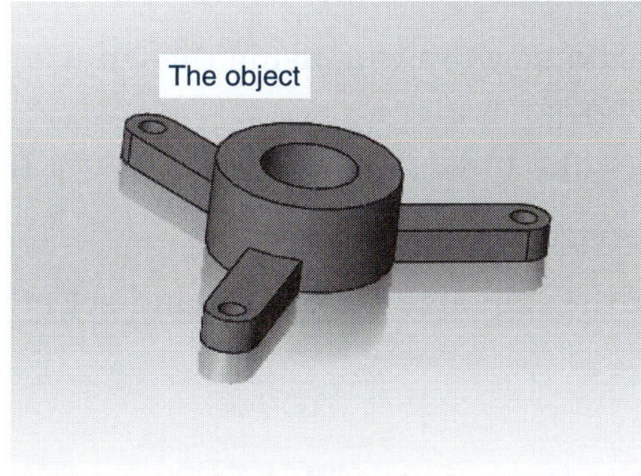

Figure 4-54

4 Click the object's centerpoint.

5 Click the other edge of the object as shown.

6 Move the cursor away from the object.

The aligned section will appear as the cursor is moved.

7 Change the scale of the object if desired and add centerlines.

4-8 Broken Views

It is often convenient to break long continuous shapes so that they take up less drawing space. Figure 4-55 shows a long L-bracket that has a continuous shape; that is, its shape is constant throughout its length.

Figure 4-55

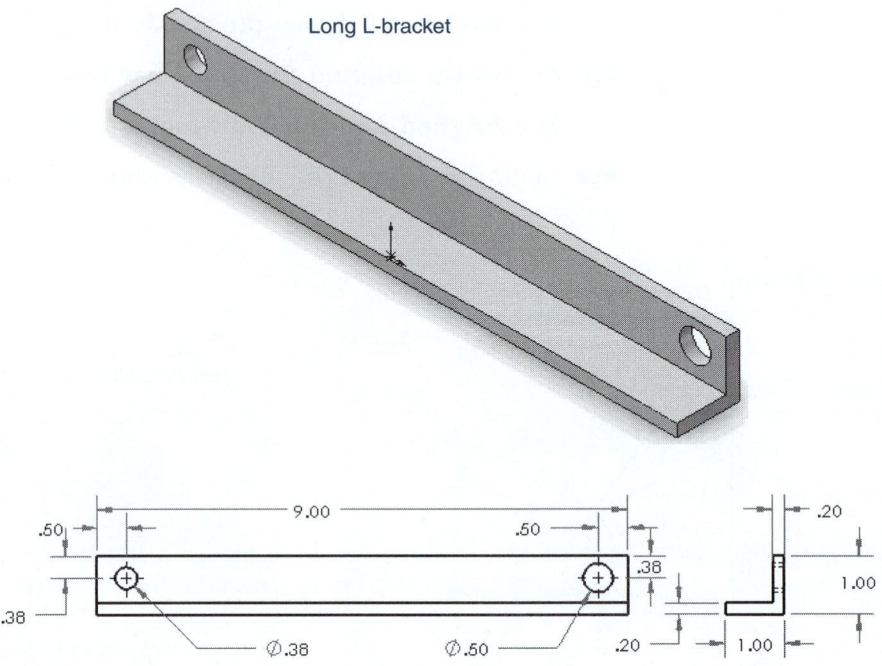

To Create a Broken View

1 Draw a model of the long L-bracket using the dimensions shown in Figure 4-55. Save the model.

2 Start a new drawing using the **Drawing** format and click on the long L-bracket.

3 Click the **View Layout** tab, and click the **Break** tool.

See Figure 4-56.

4 Set the **Gap size** for **0.25in** and select the **Small Zig Zag Cut** style.

Figure 4-56

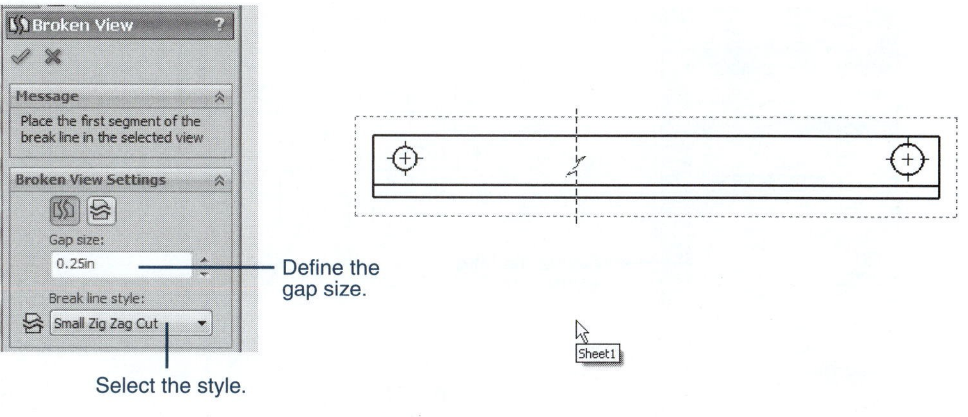

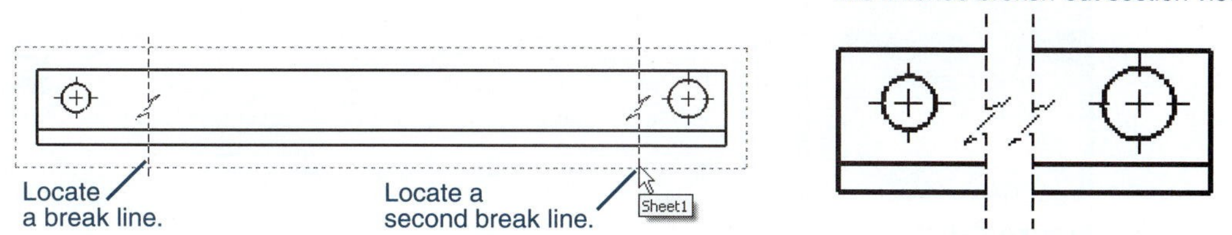

5. Move the cursor onto the long L-bracket and click a location for the first break line.
6. Click a location for the second break line.

   The area between the break lines will be removed.
7. Click the green **OK** check mark.

   If the break is not satisfactory, undo the break and insert a new one.

4-9 Detail Views

A *detail view* is used to clarify specific areas of a drawing. Usually, an area is enlarged so that small details are easier to see.

To Draw a Detail View

1. Create a **Part** drawing for the model shown in Figure P4-7. Save the model.
2. Start a new drawing using the **Drawing** format and create front and top orthographic views of a model. Use third-angle projection.
3. Click the **Detail View** tool.

   See Figure 4-57.

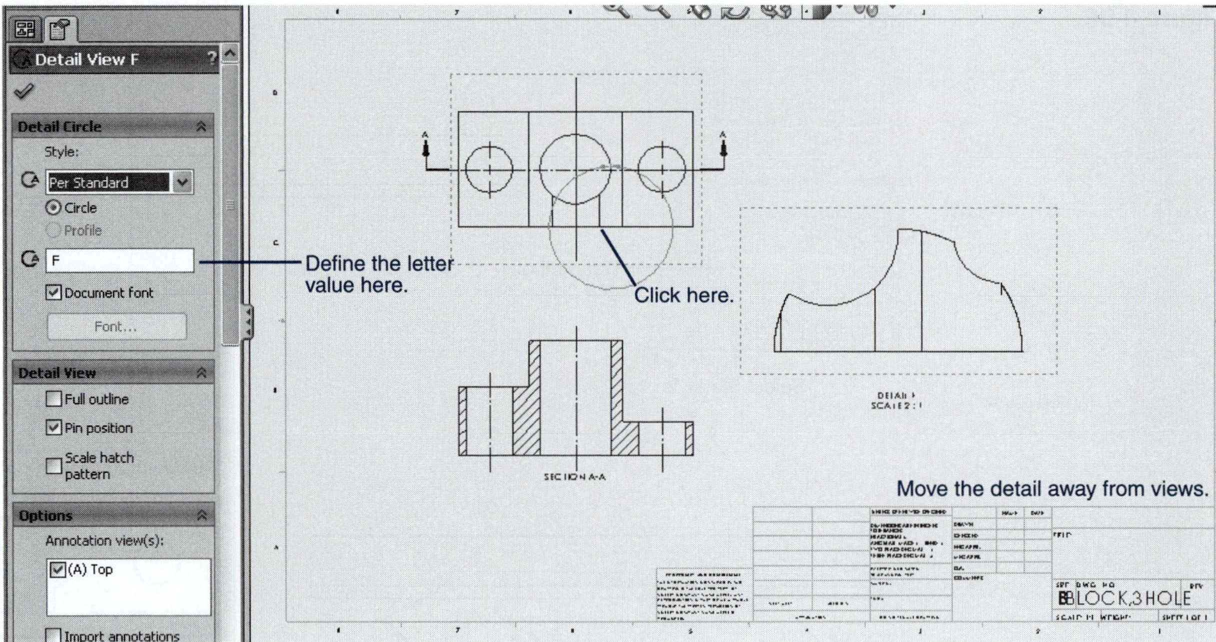

Figure 4-57

4 Locate the centerpoint for a circle that will be used to define the area for the detail view by clicking a point.

In this example the intersection of the top view's front edge line and the right edge line of the slot were selected.

5 Move the cursor away from the point.

A detail view will appear.

6 When the circle is big enough to enclose all the area you wish to display in the detail view, click the mouse.

7 Move the cursor away from the views.

8 Select a location for the detail view and click the mouse.

The scale of the detail view can be changed by changing the values in the **Scale** box. The callout letters are changed using the **Detail Circle** box.

4-10 Auxiliary Views

Auxiliary views are orthographic views used to present true-shaped views of a slanted surface. In Figure 4-58 neither the front nor the side view shows a true shape of the slanted surface. A top view would show a foreshortened view. Only a view taken 90° to the surface will show its true shape.

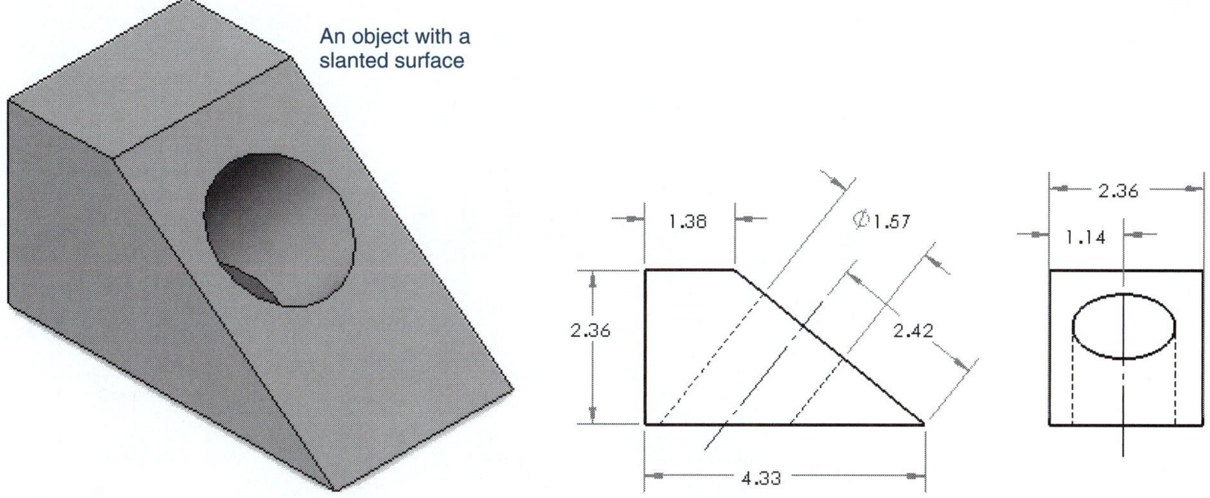

Figure 4-58

To Draw an Auxiliary View

1. Draw the model with a slanted surface shown in Figure 4-58.
2. Start a new drawing using the **Drawing** format and create orthographic views of the model.

   In this example a front and a right-side view were drawn. See Figure 4-59.

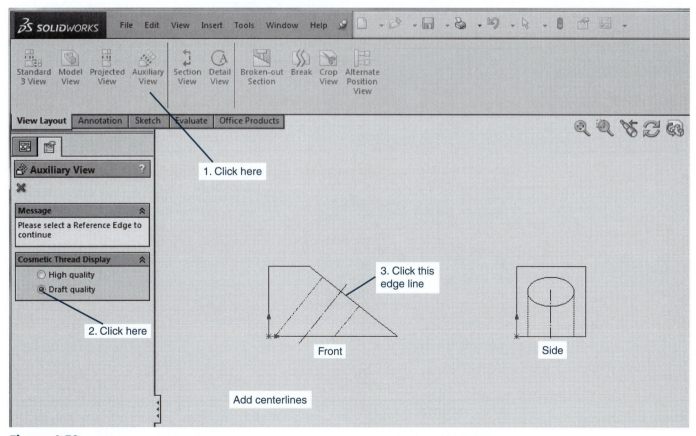

Figure 4-59

Figure 4-59
(*Continued*)

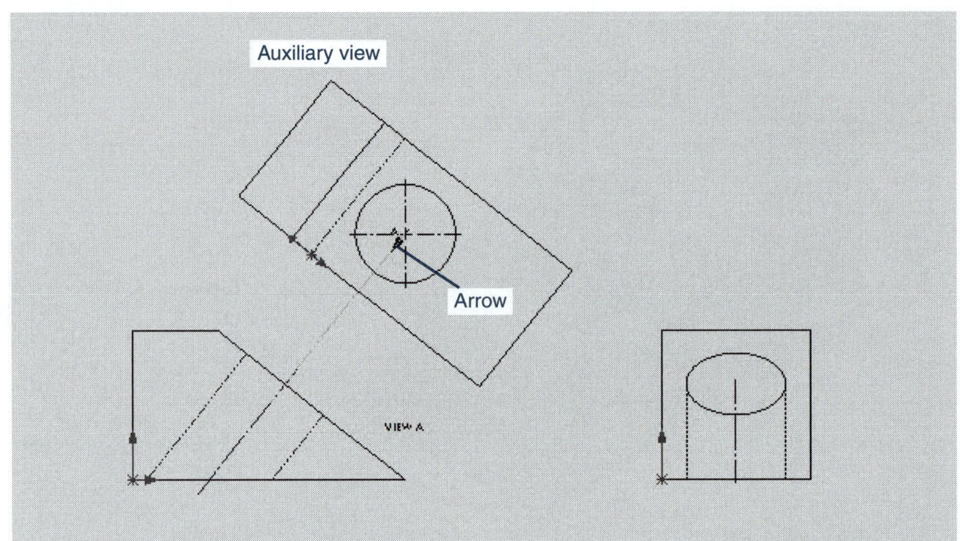

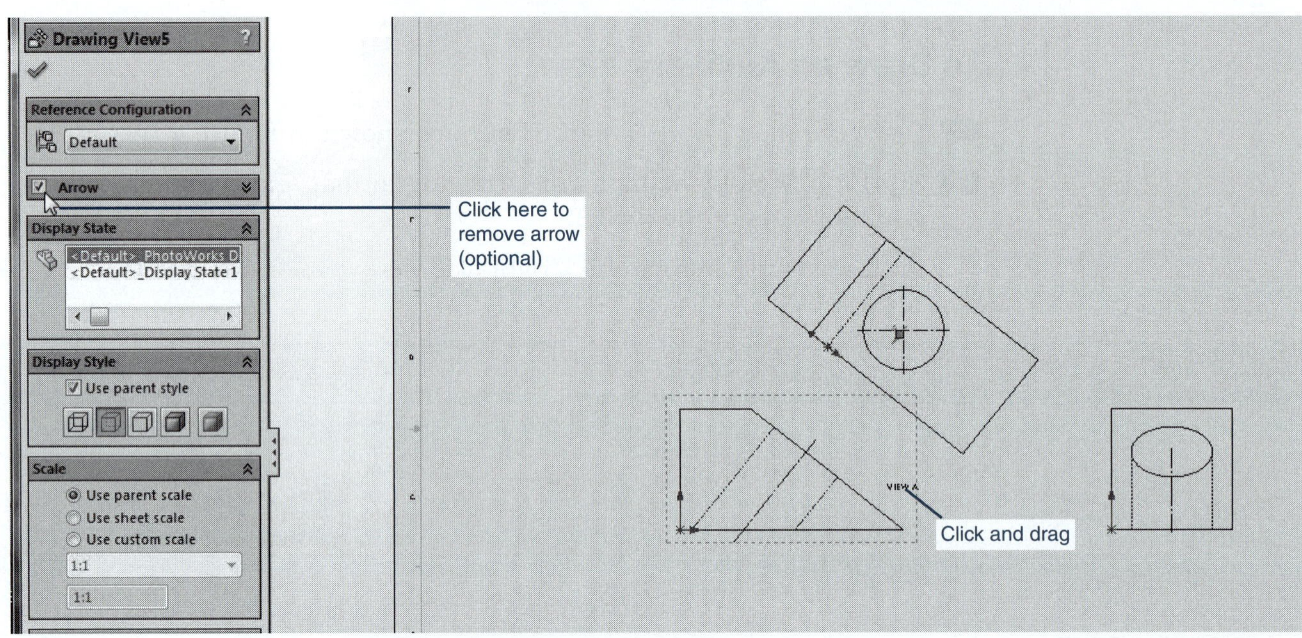

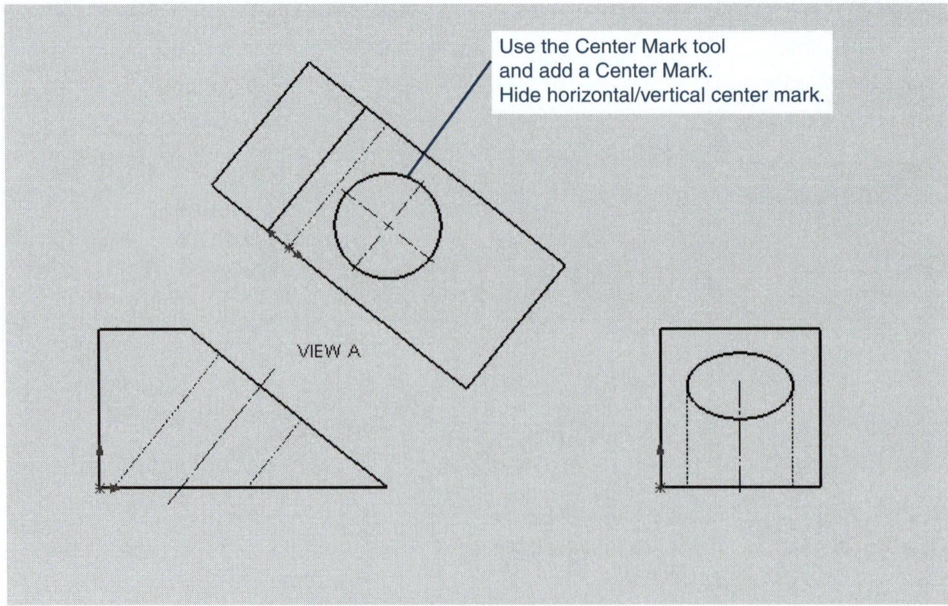

3. Click the **View Layout** tab, then click the **Auxiliary View** tool.
4. Click the slanted edge line in the front view.
5. Move the cursor away from the slanted edge line.
6. Select a location for the auxiliary view and click the mouse.
7. Adjust the cutting plane line as shown.

Figures 4-60 and 4-61 show additional examples of auxiliary views.

Figure 4-60

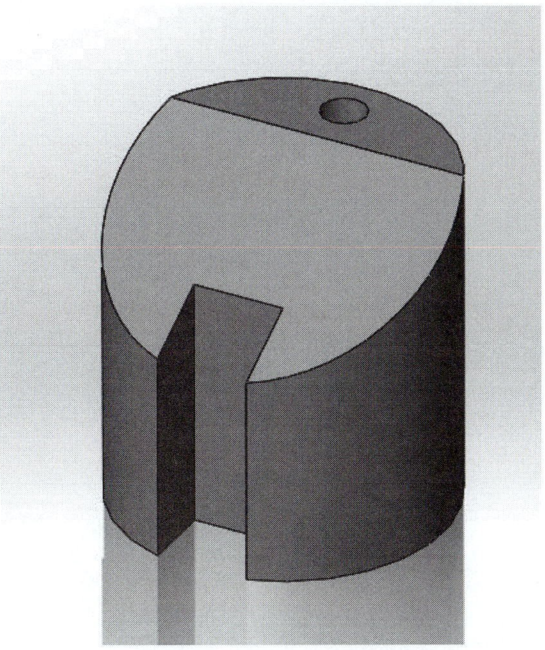

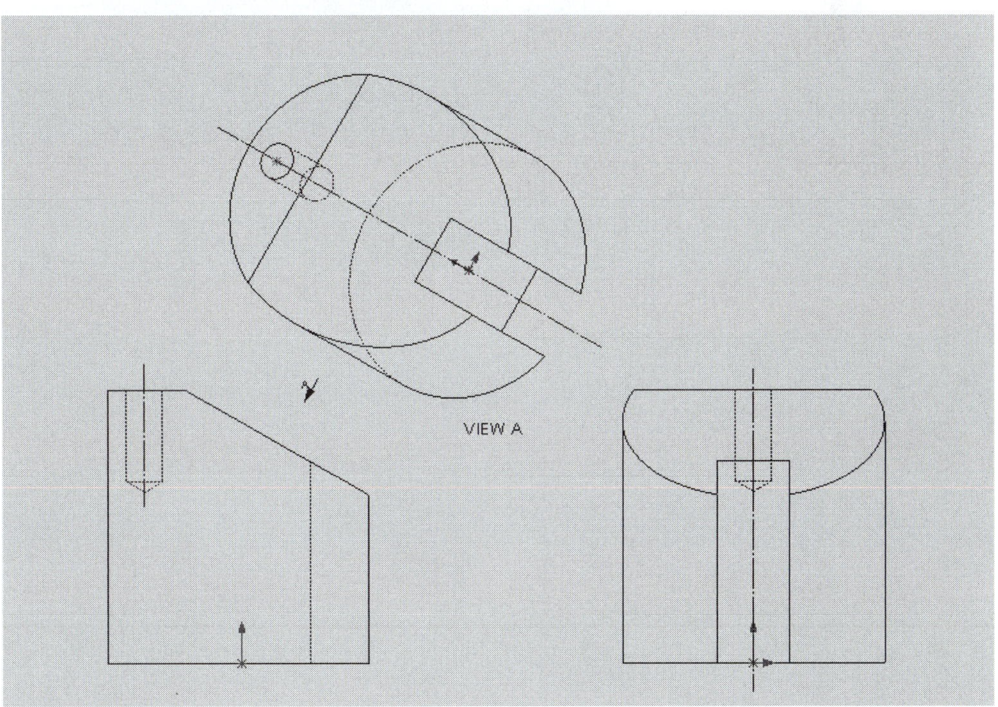

Figure 4-61

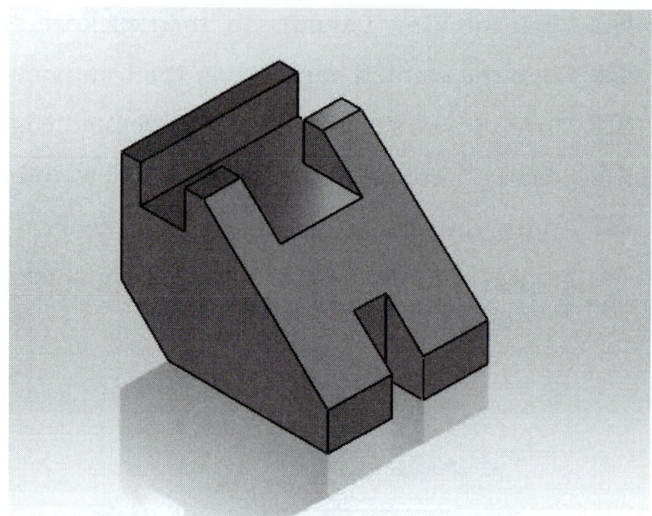

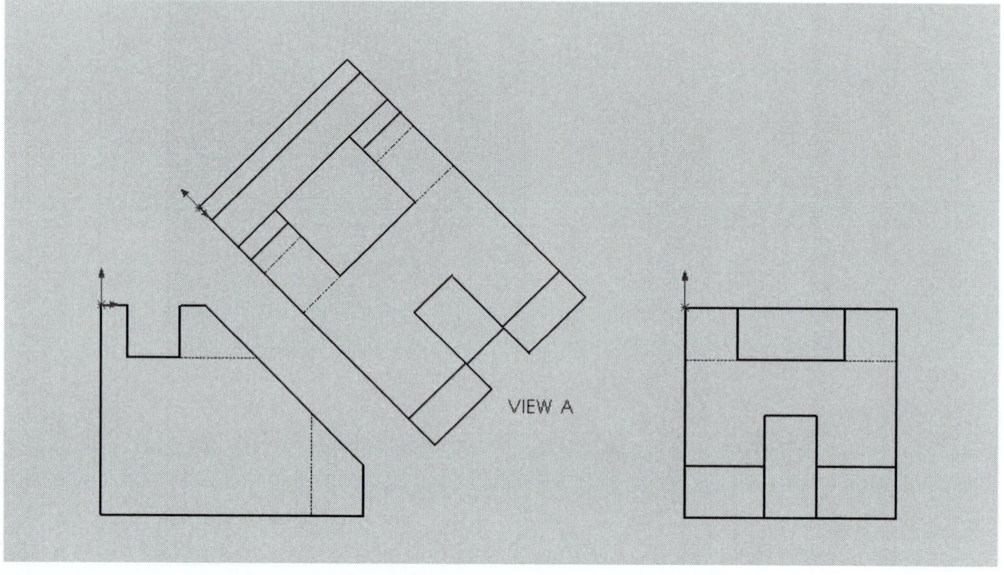

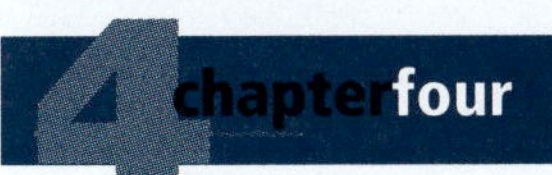

Chapter Projects

Draw a front, top, and right-side orthographic view of each of the objects in Figures P4-1 through P4-94. Do not include dimensions.

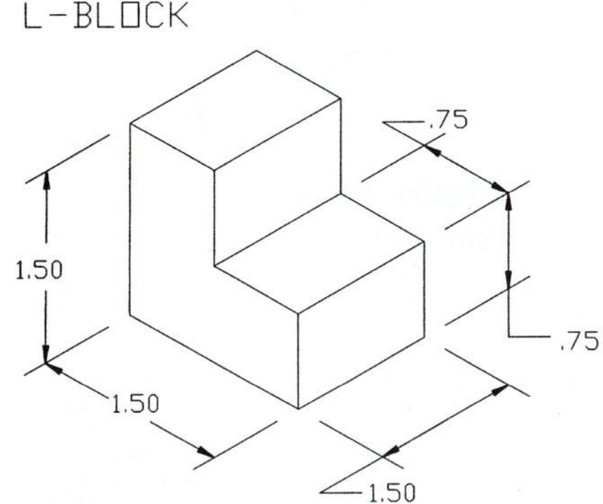

Figure P4-1
INCHES

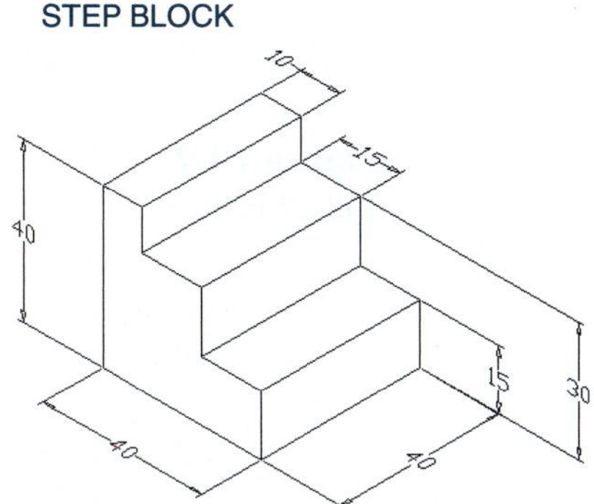

Figure P4-2
MILLIMETERS

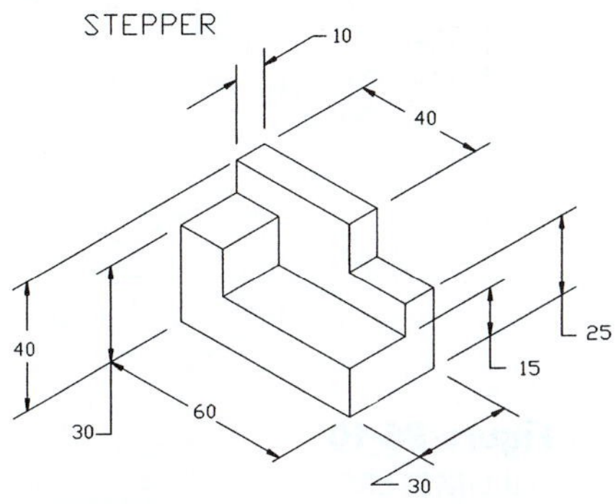

Figure P4-3
MILLIMETERS

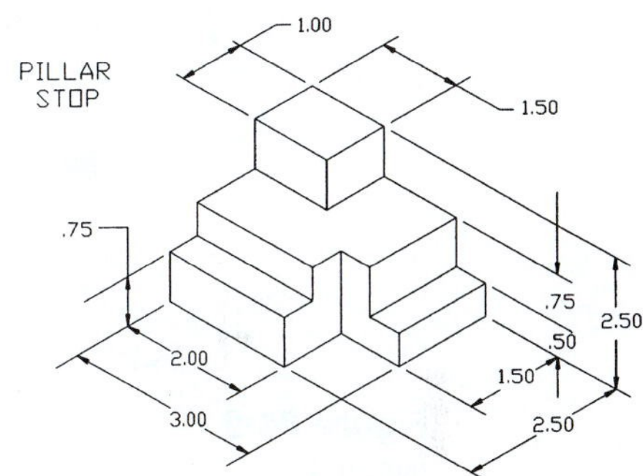

Figure P4-4
INCHES

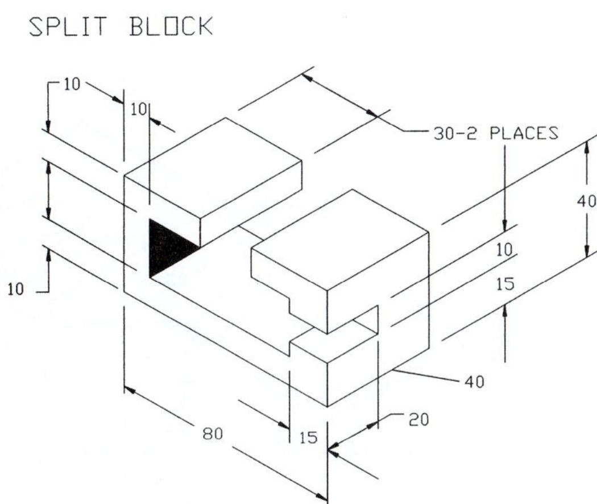

Figure P4-5
MILLIMETERS

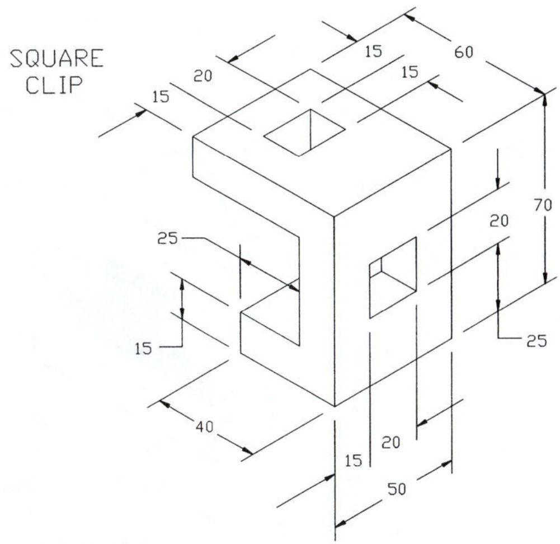

Figure P4-6
MILLIMETERS

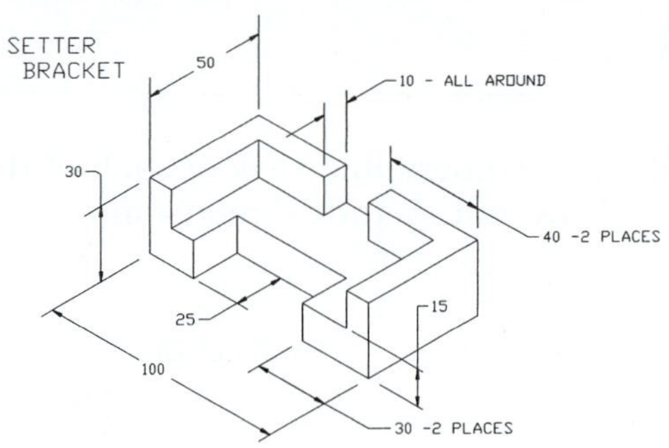

Figure P4-7
MILLIMETERS

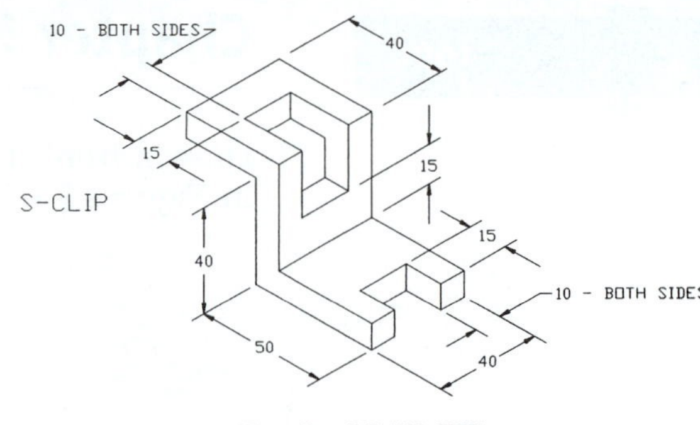

Figure P4-8
MILLIMETERS

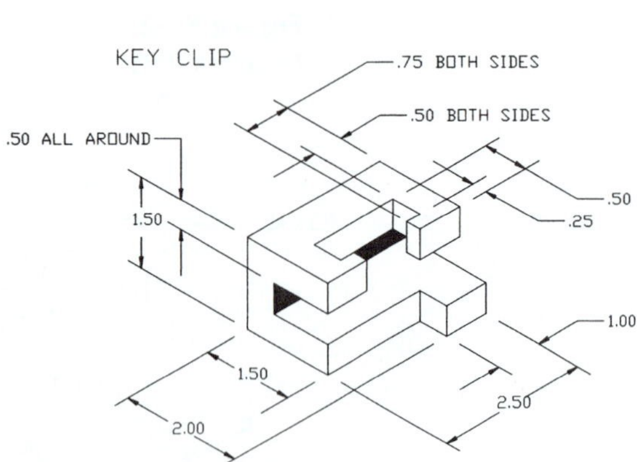

Figure P4-9
INCHES

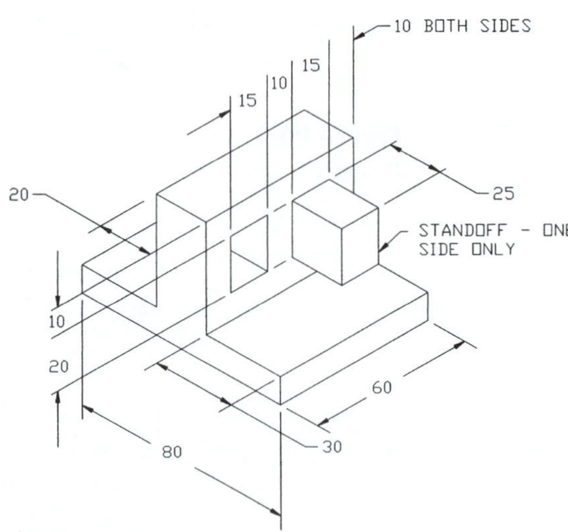

Figure P4-10
MILLIMETERS

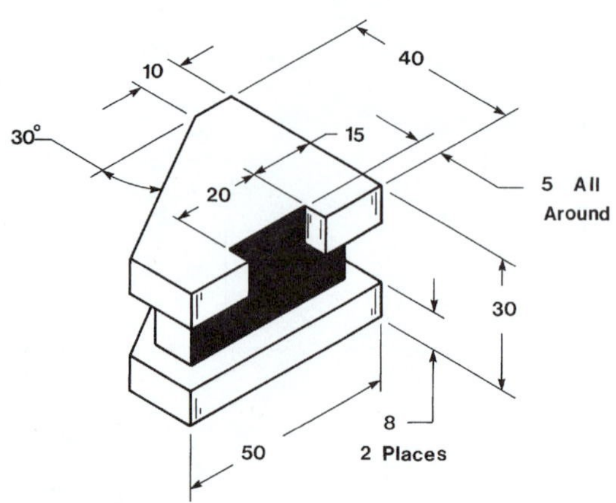

Figure P4-11
MILLIMETERS

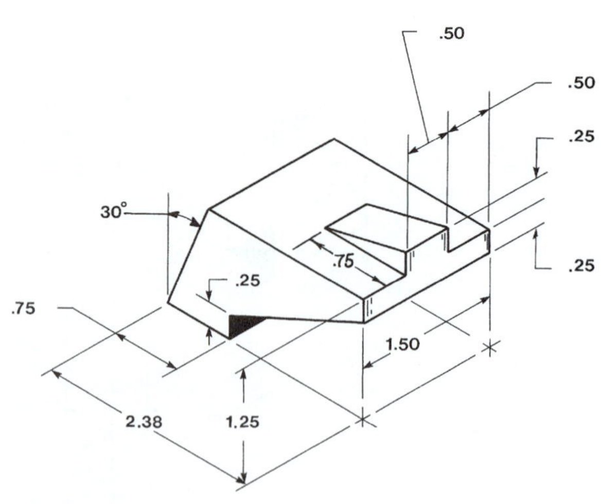

Figure P4-12
INCHES

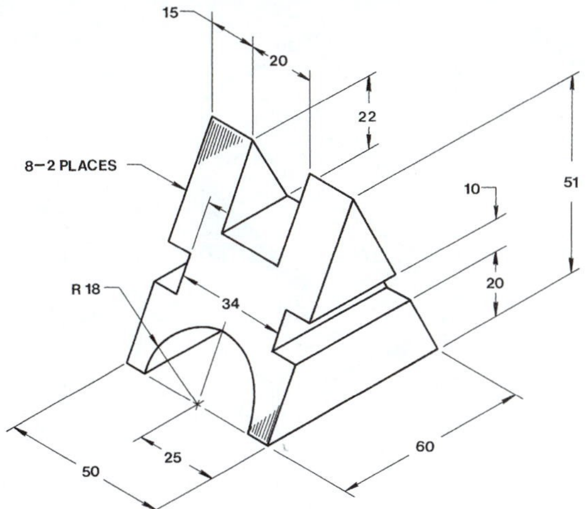

Figure P4-13
MILLIMETERS

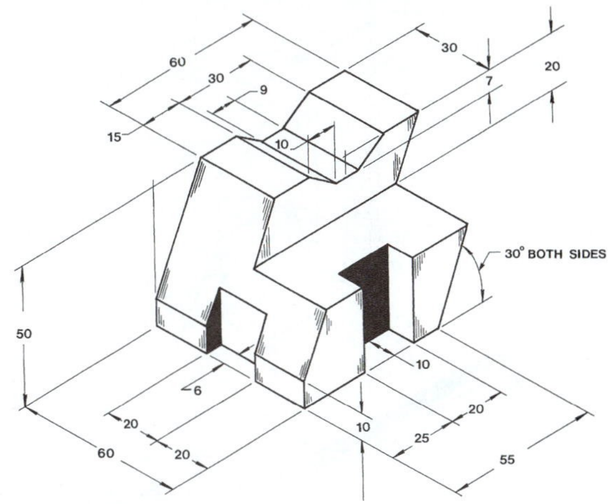

Figure P4-14
MILLIMETERS

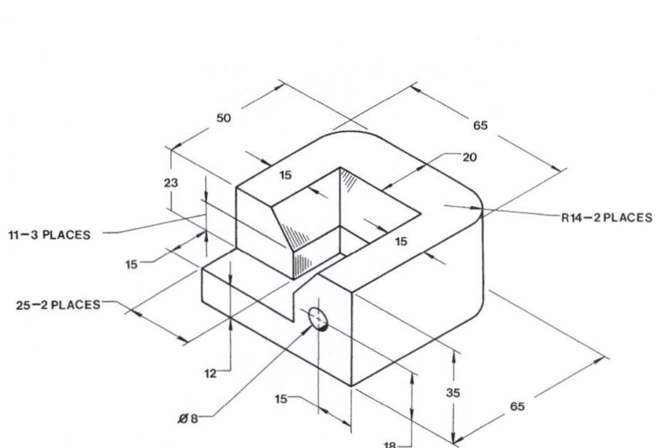

Figure P4-15
MILLIMETERS

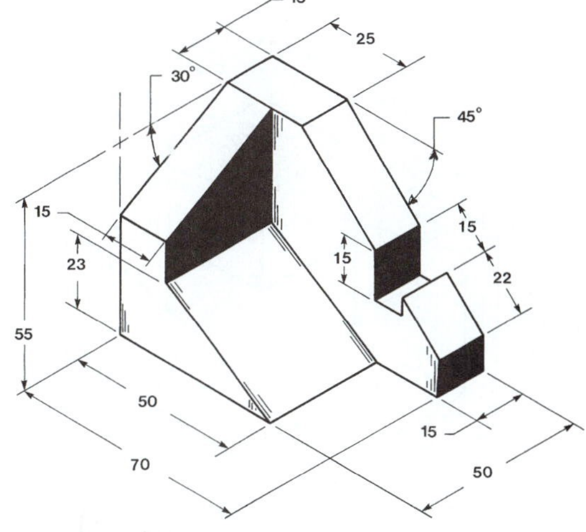

Figure P4-16
MILLIMETERS

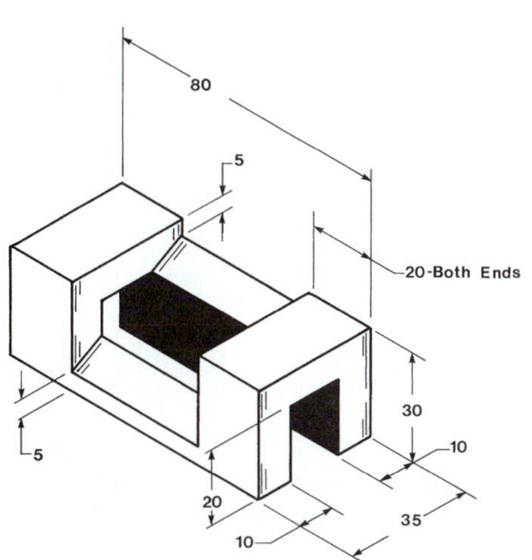

Figure P4-17
MILLIMETERS

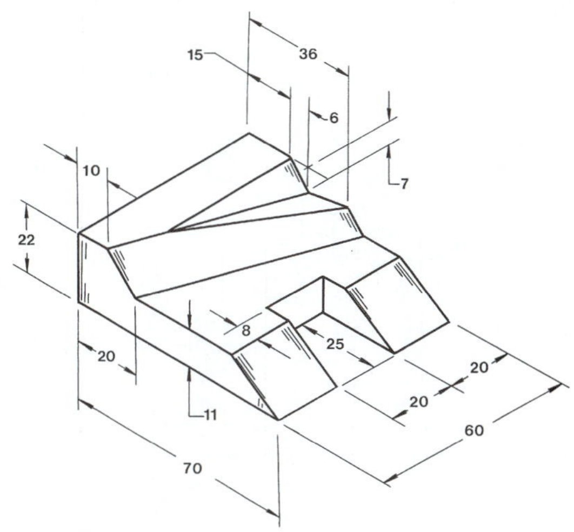

Figure P4-18
MILLIMETERS

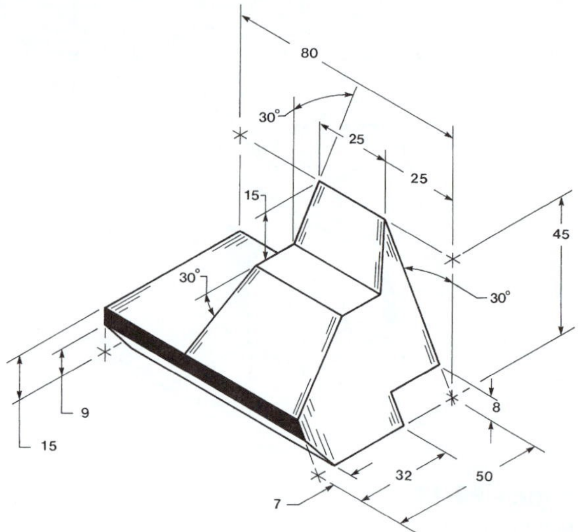

Figure P4-19
MILLIMETERS

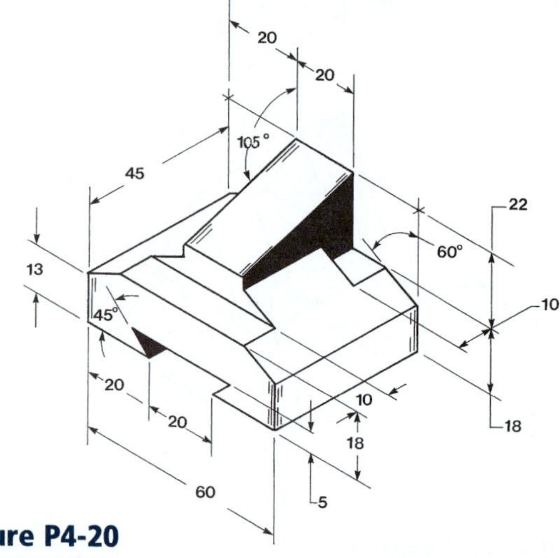

Figure P4-20
MILLIMETERS

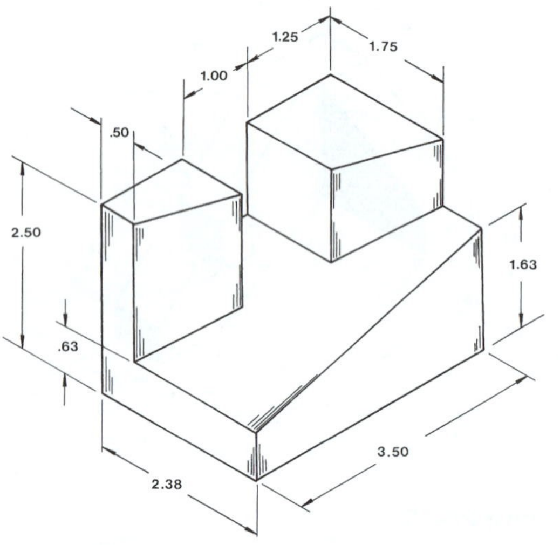

Figure P4-21
INCHES

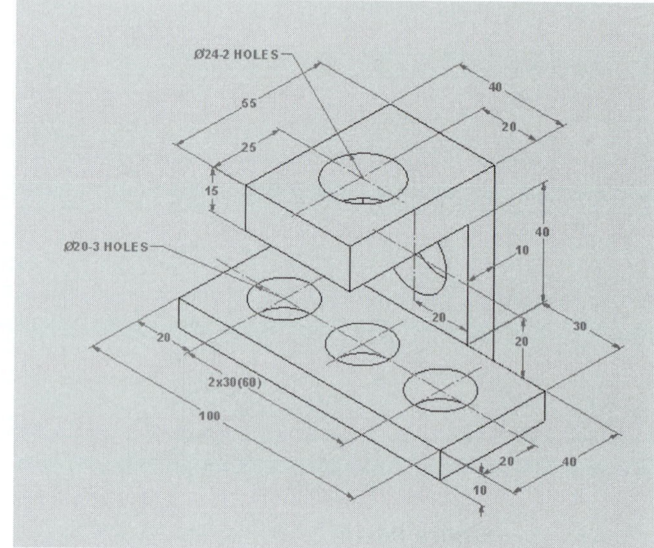

Figure P4-22
MILLIMETERS

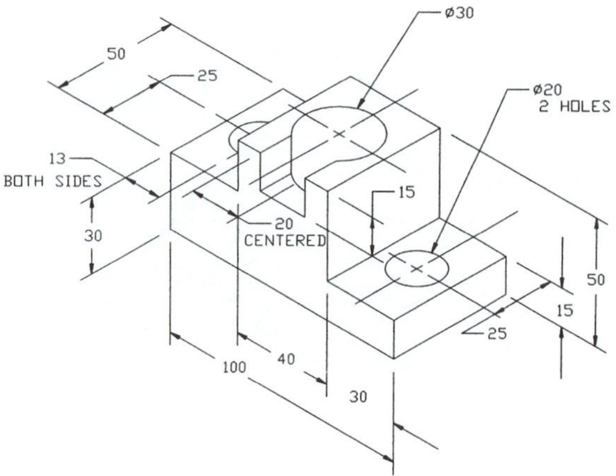

Figure P4-23
MILLIMETERS

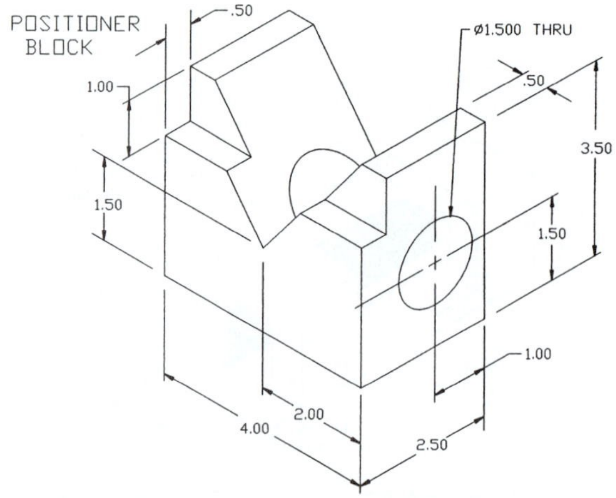

Figure P4-24
INCHES

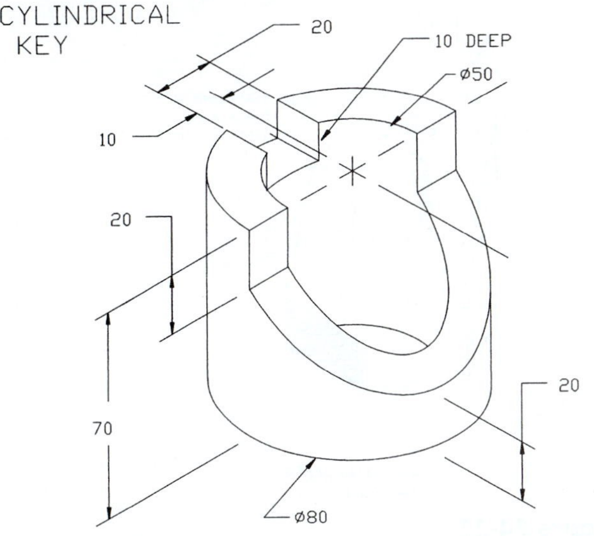

Figure P4-25
MILLIMETERS

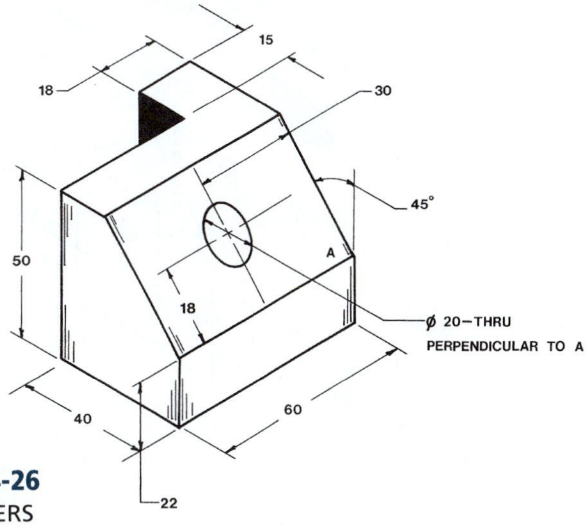

Figure P4-26
MILLIMETERS

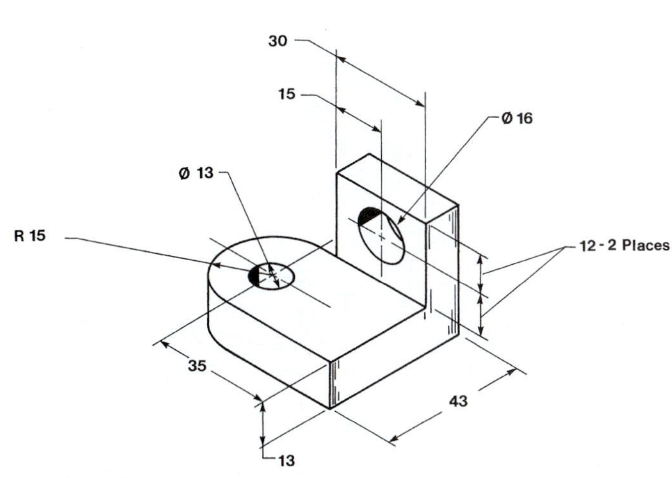

Figure P4-27
MILLIMETERS

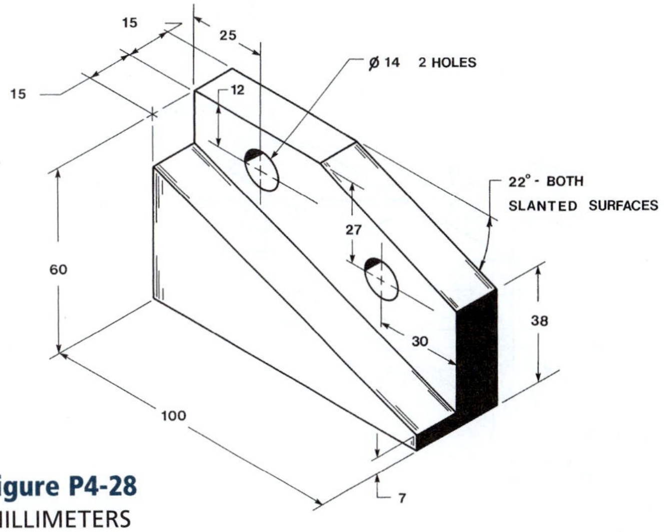

Figure P4-28
MILLIMETERS

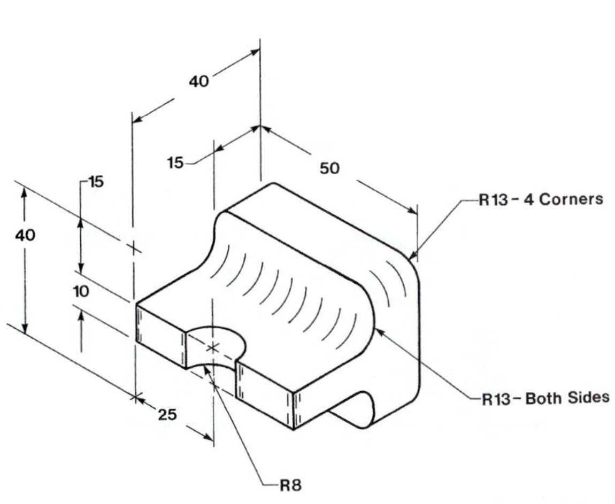

Figure P4-29
MILLIMETERS

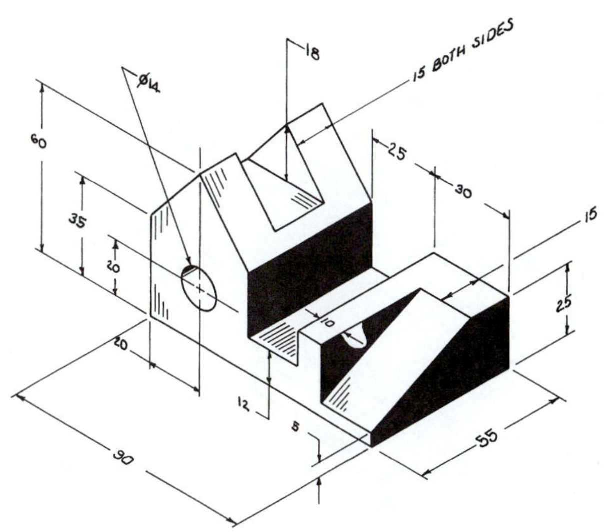

Figure P4-30
MILLIMETERS

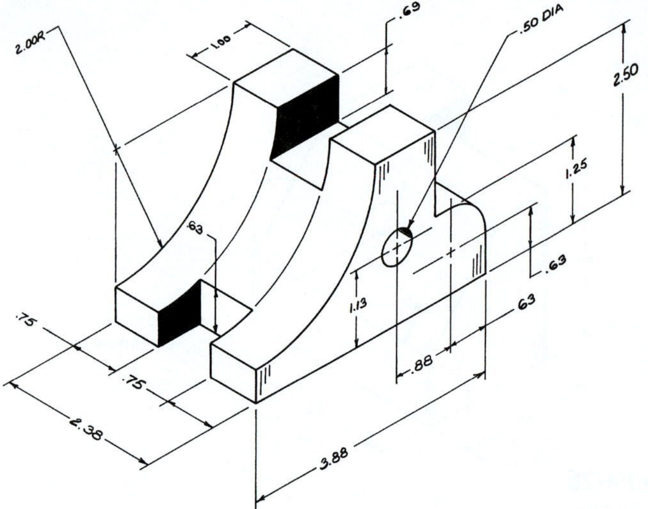

Figure P4-31
INCHES

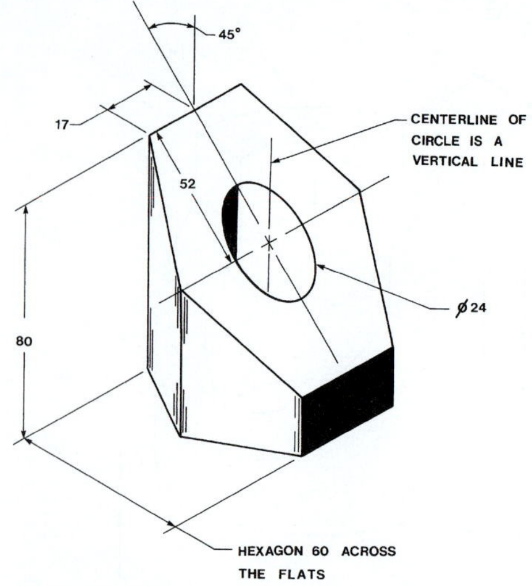

Figure P4-32
MILLIMETERS

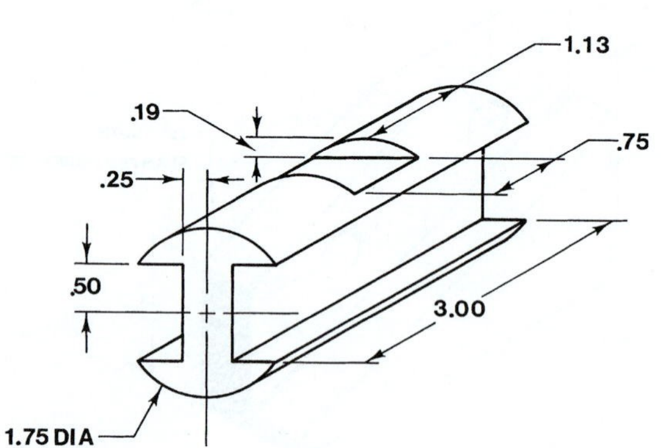

Figure P4-33
INCHES

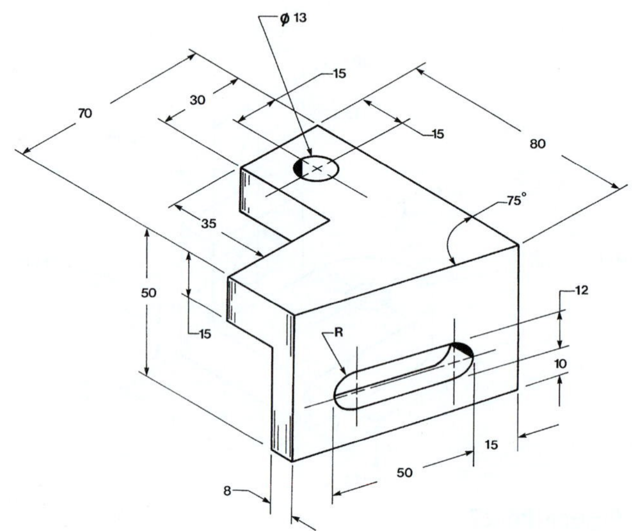

Figure P4-34
MILLIMETERS

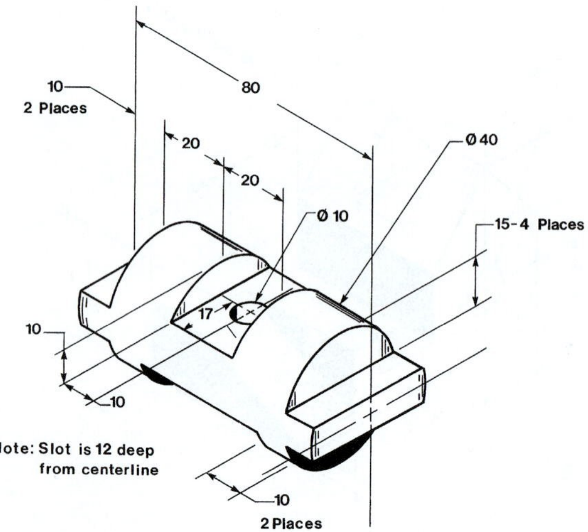

Figure P4-35
MILLIMETERS

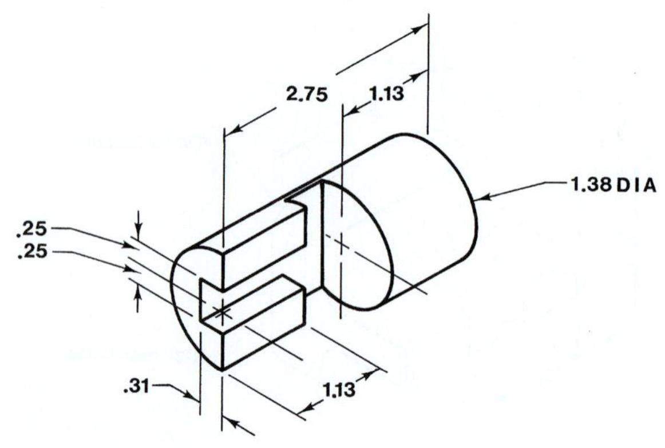

Figure P4-36
INCHES

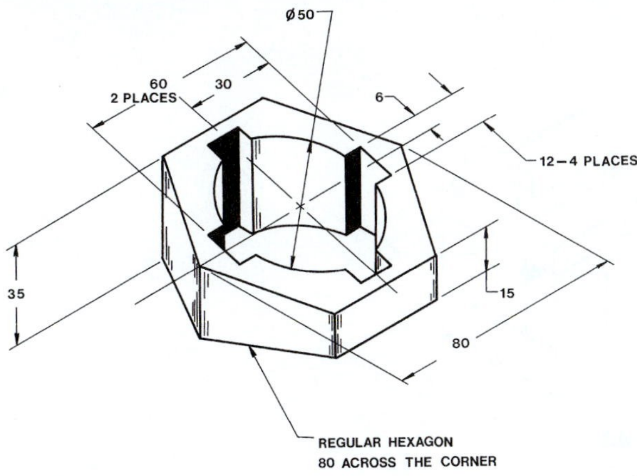

Figure P4-37
MILLIMETERS

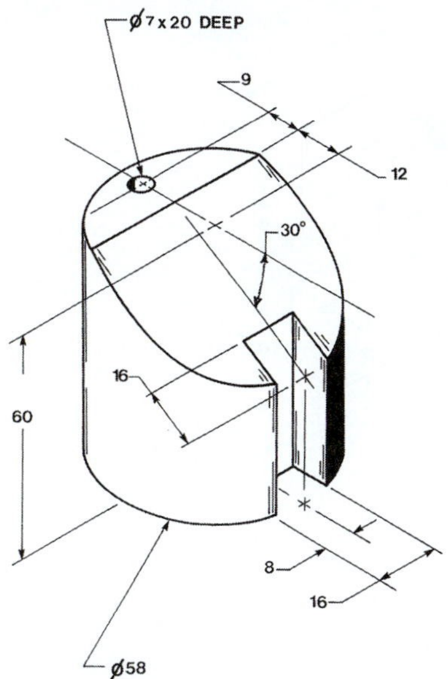

Figure P4-38
MILLIMETERS

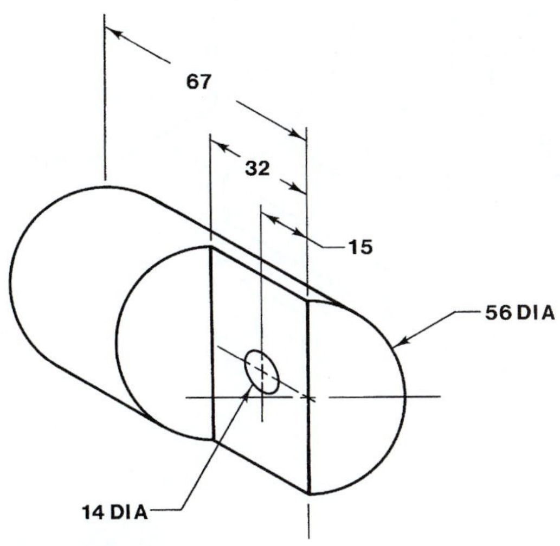

Figure P4-39
MILLIMETERS

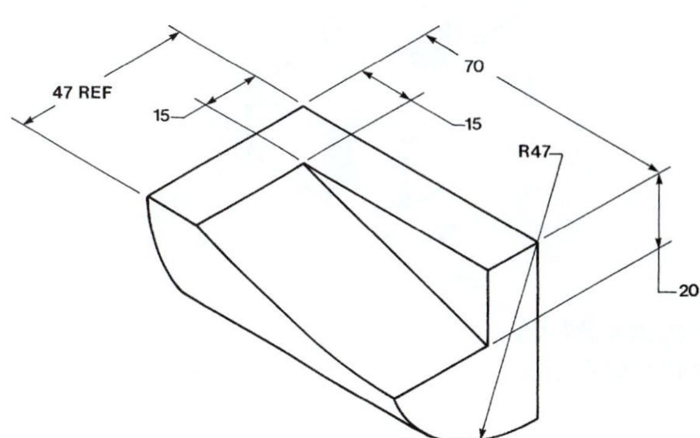

Figure P4-40
MILLIMETERS

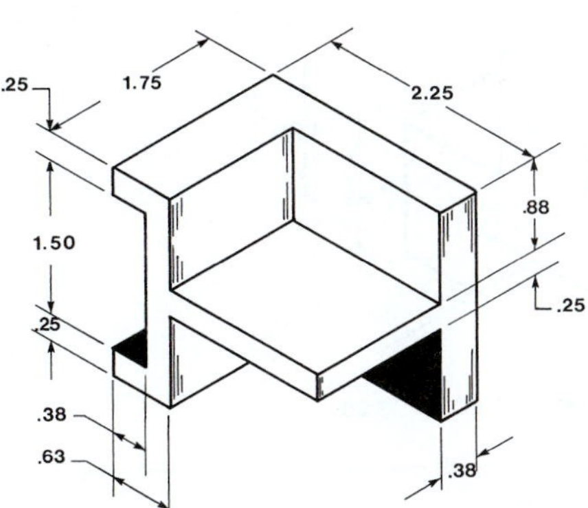

Figure P4-41
INCHES

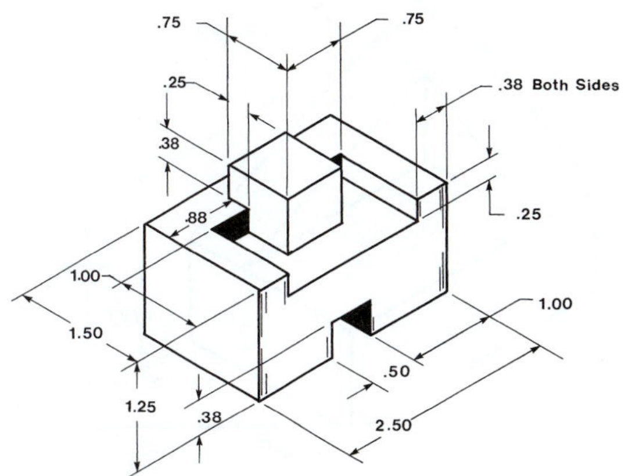

Figure P4-42
INCHES

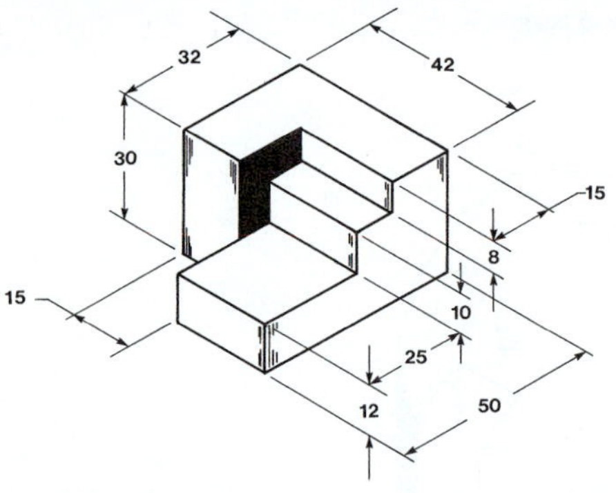

Figure P4-43
MILLIMETERS

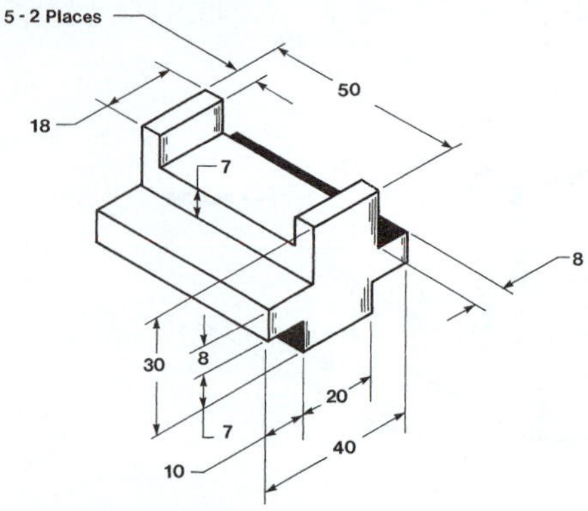

Figure P4-44
MILLIMETERS

NOTE: THE SLOT IS 15 LONG

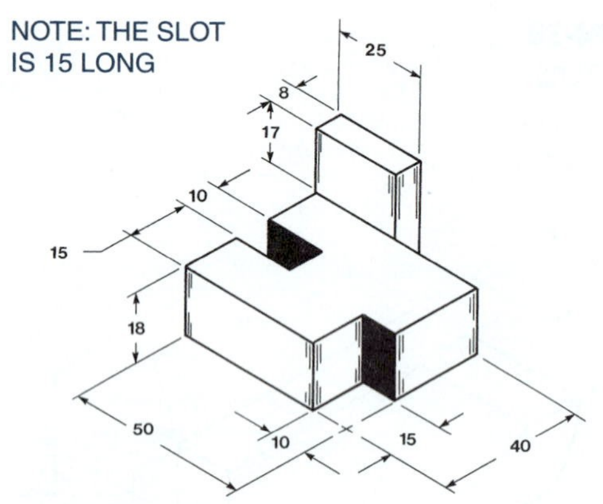

Figure P4-45
MILLIMETERS

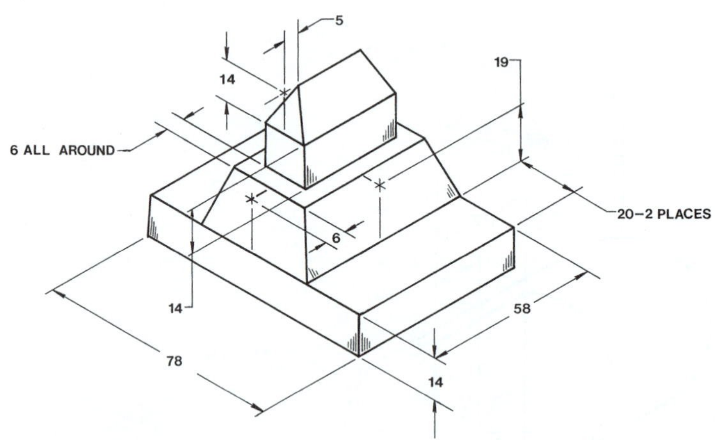

Figure P4-46
MILLIMETERS

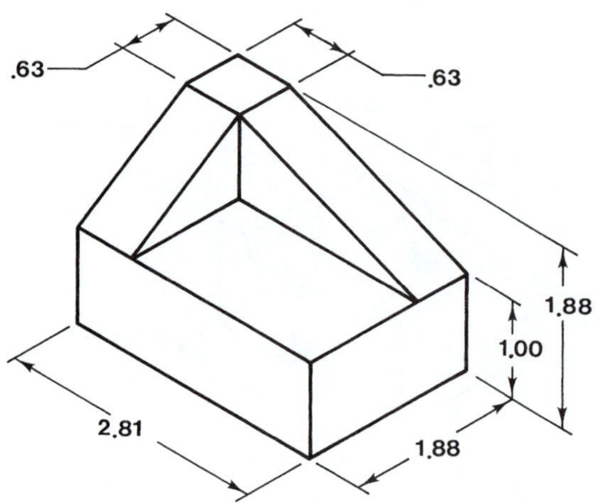

Figure P4-47
INCHES

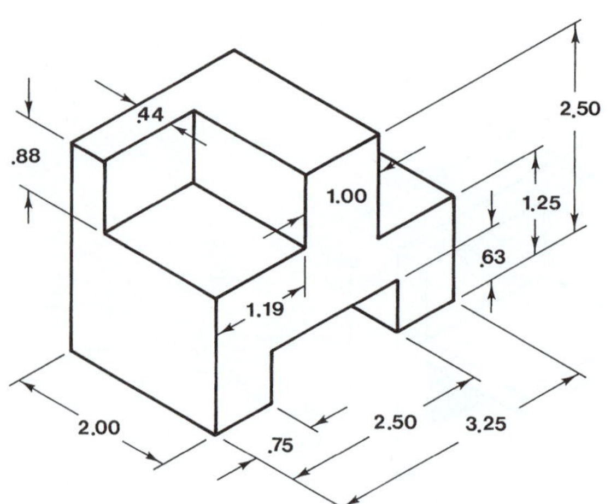

Figure P4-48
INCHES

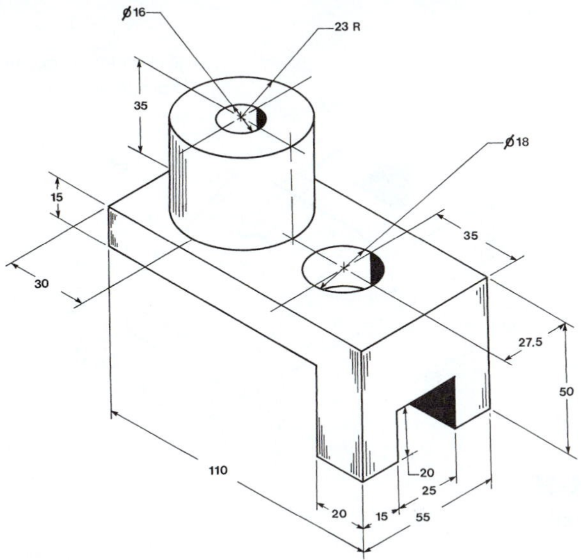

Figure P4-49
MILLIMETERS

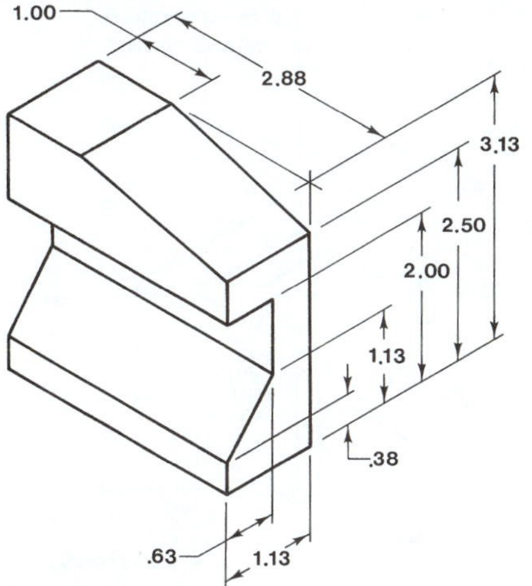

Figure P4-50
INCHES

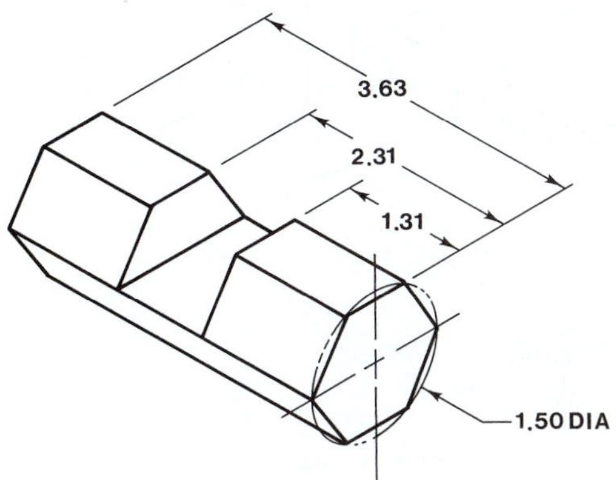

Figure P4-51
INCHES

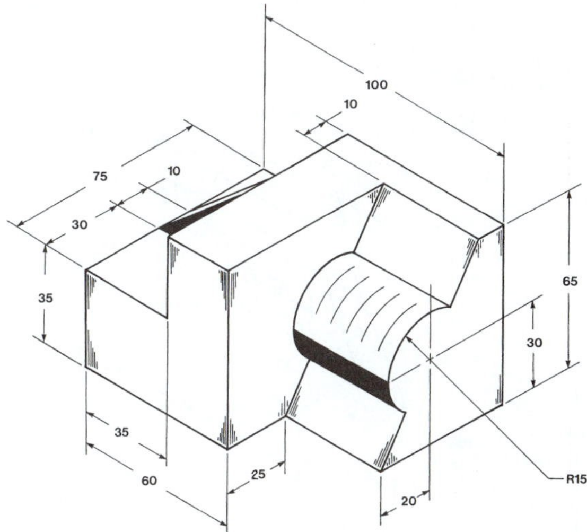

Figure P4-52
MILLIMETERS

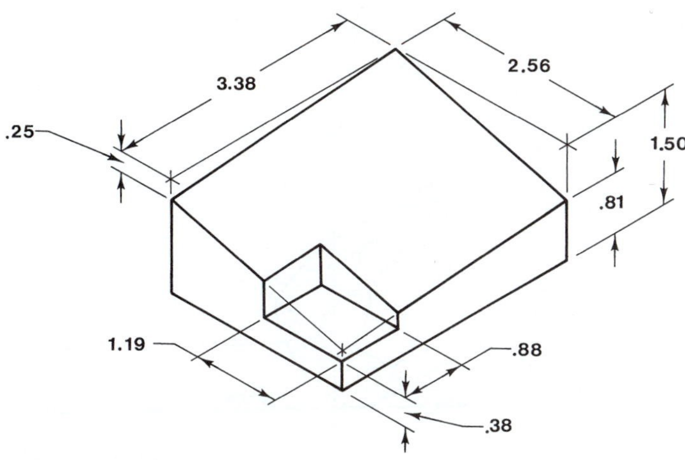

Figure P4-53
INCHES

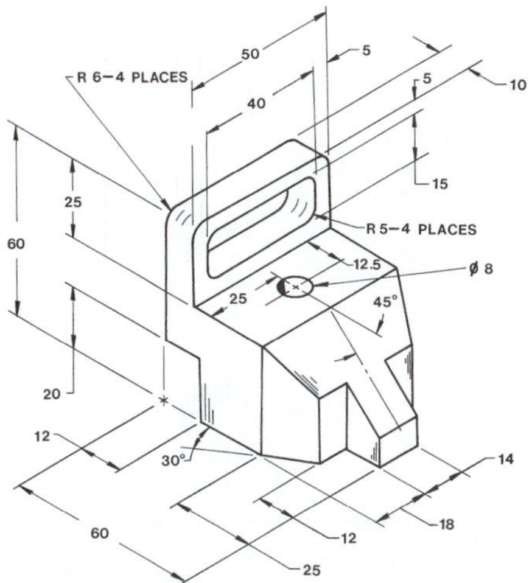

Figure P4-54
MILLIMETERS

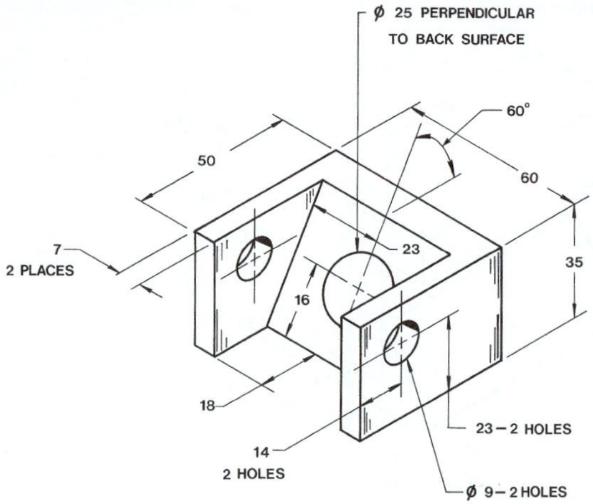

Figure P4-55
MILLIMETERS

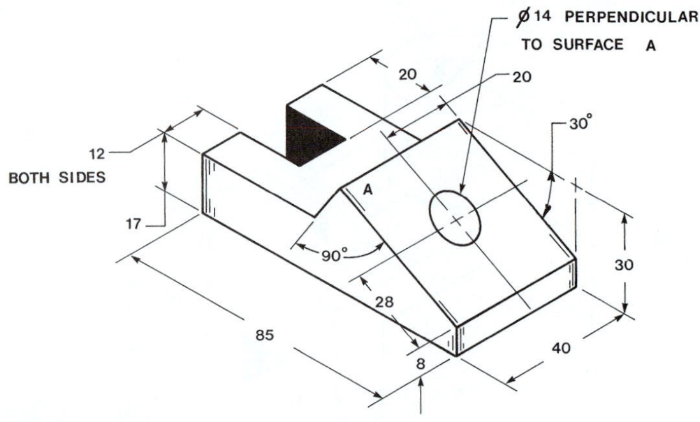

Figure P4-56
MILLIMETERS

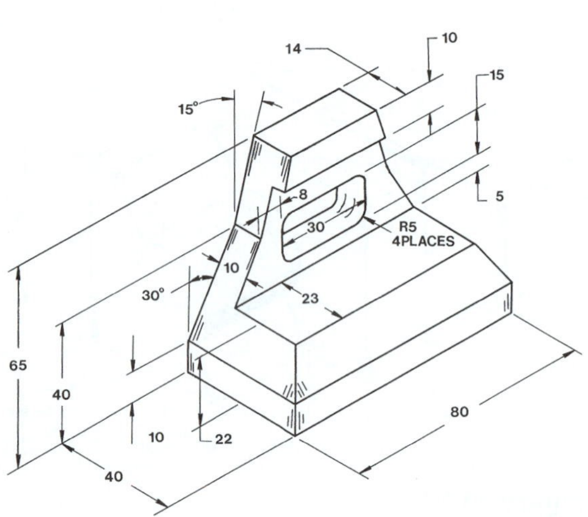

Figure P4-57
MILLIMETERS

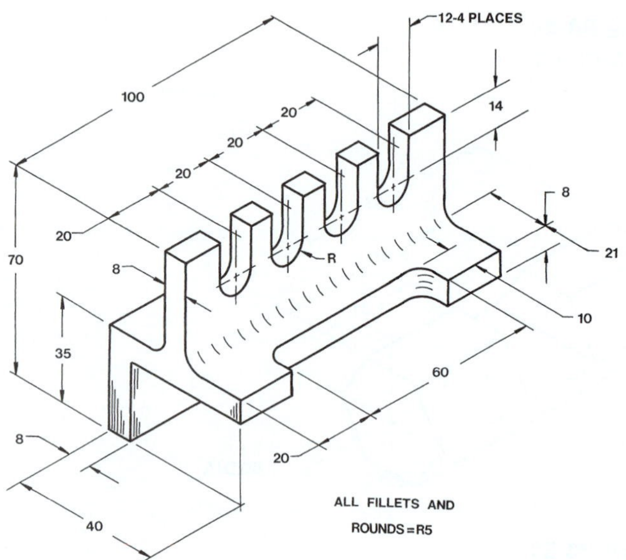

Figure P4-58
MILLIMETERS

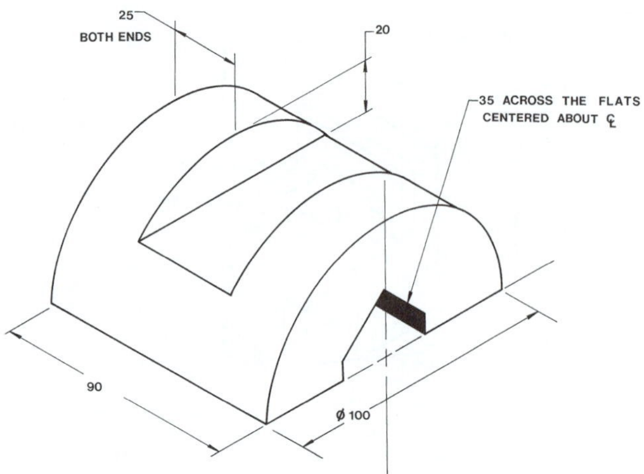

Figure P4-59
MILLIMETERS

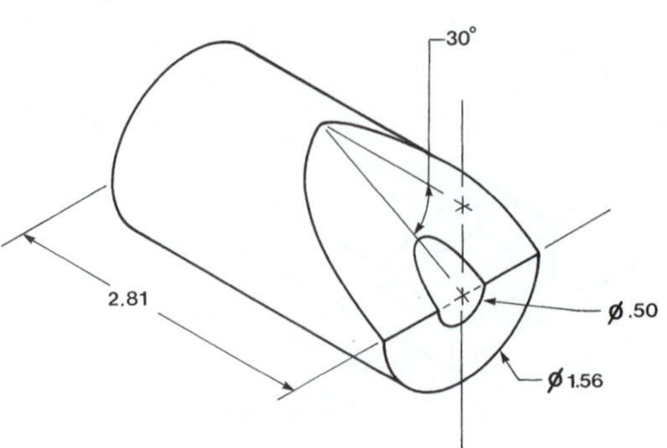

Figure P4-60
INCHES

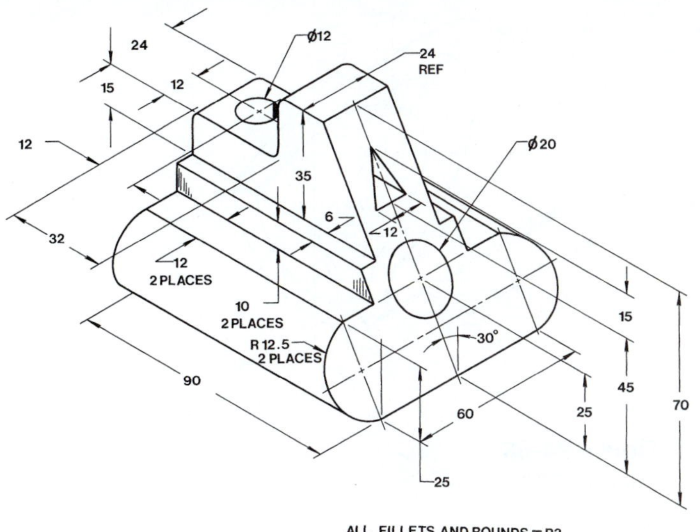

Figure P4-61
MILLIMETERS

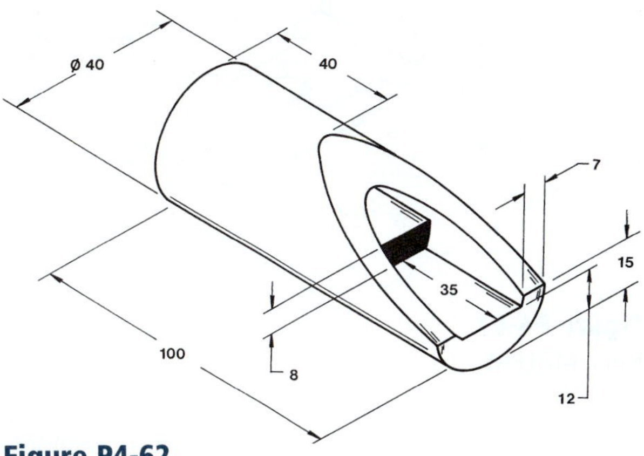

Figure P4-62
MILLIMETERS

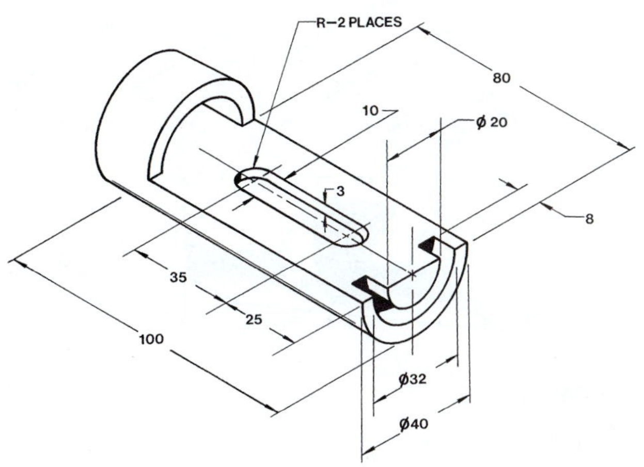

Figure P4-63
MILLIMETERS

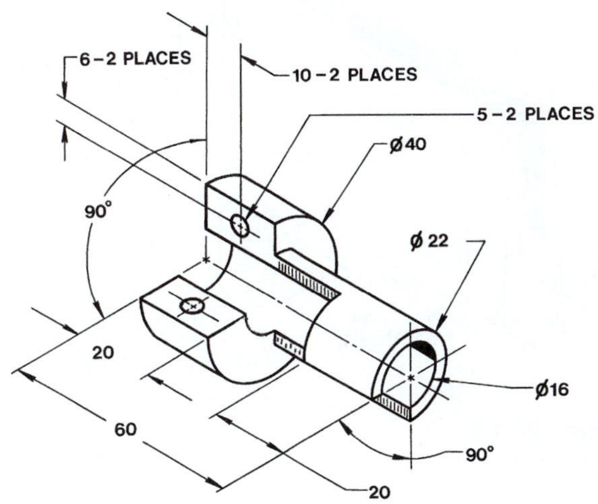

Figure P4-64
MILLIMETERS

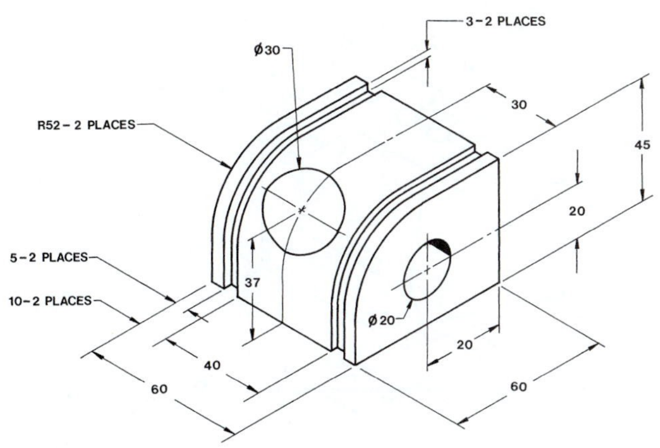

Figure P4-65
MILLIMETERS

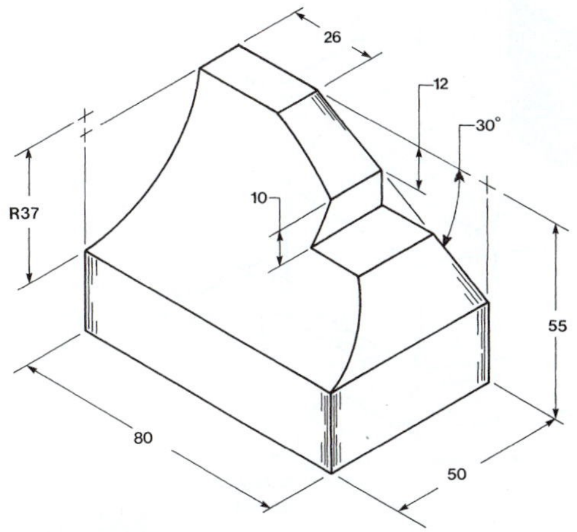

Figure P4-66
MILLIMETERS

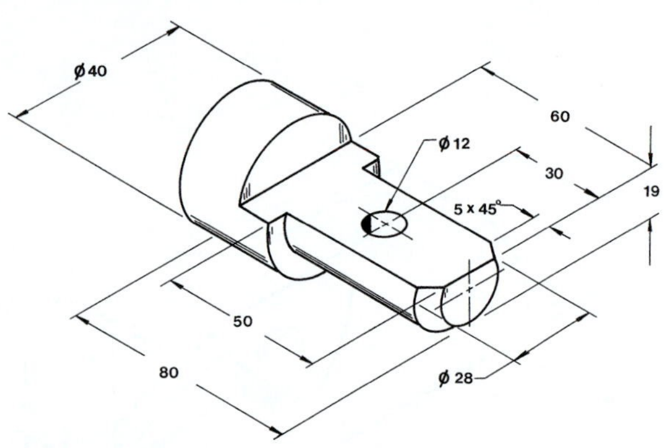

Figure P4-67
MILLIMETERS

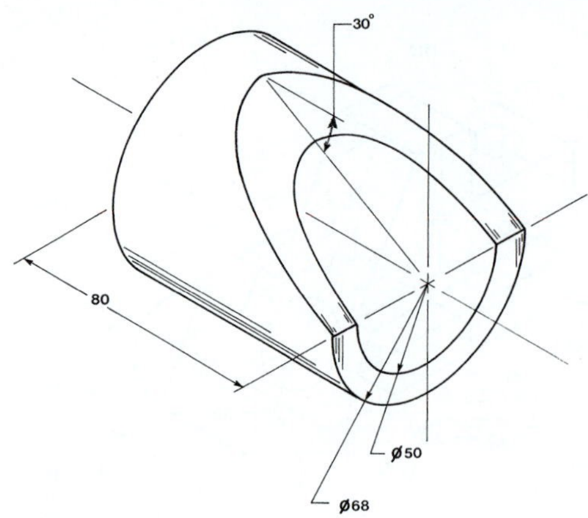

Figure P4-68
MILLIMETERS

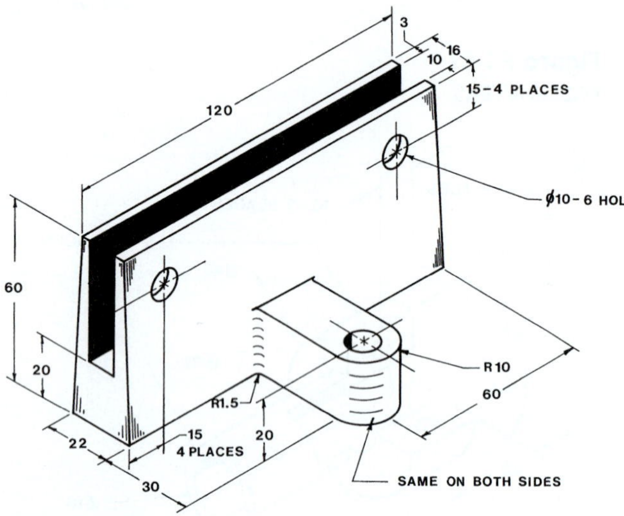

Figure P4-69
MILLIMETERS

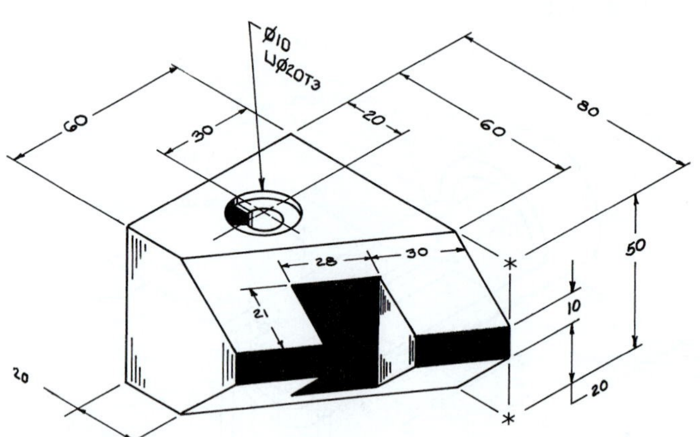

Figure P4-70
MILLIMETERS

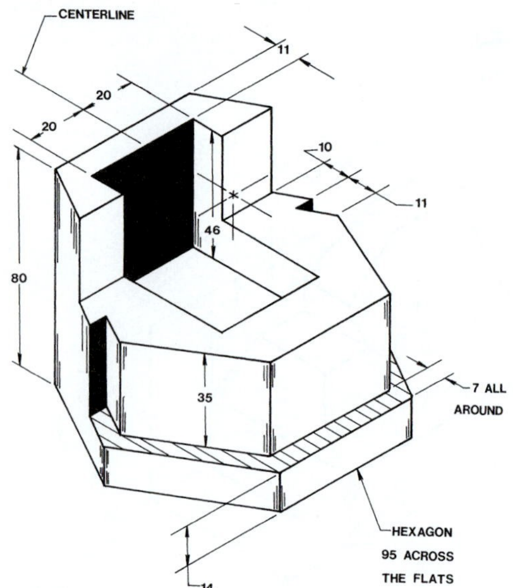

Figure P4-71
MILLIMETERS

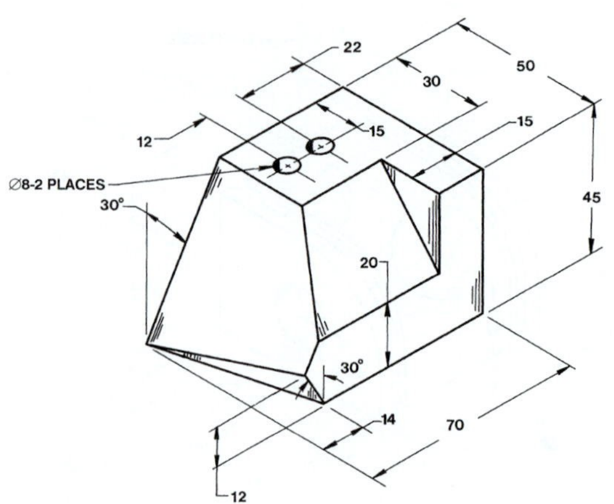

Figure P4-72
MILLIMETERS

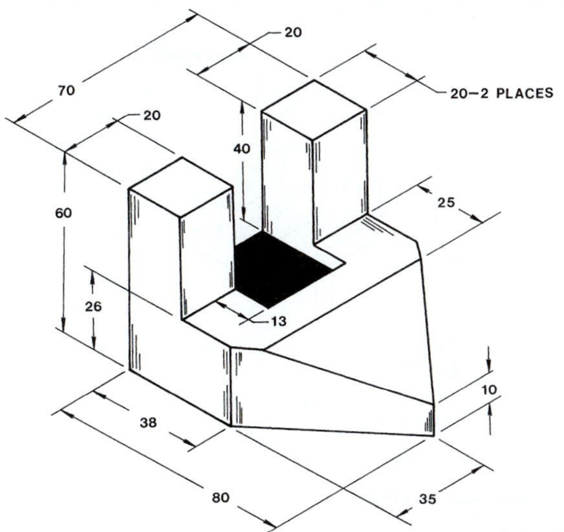

Figure P4-73
MILLIMETERS

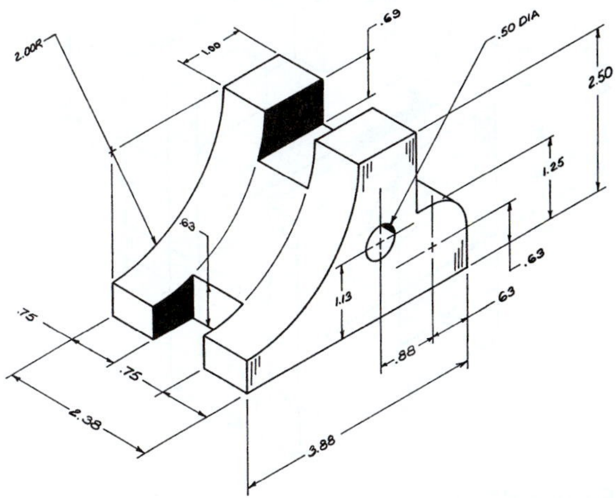

Figure P4-74
INCHES

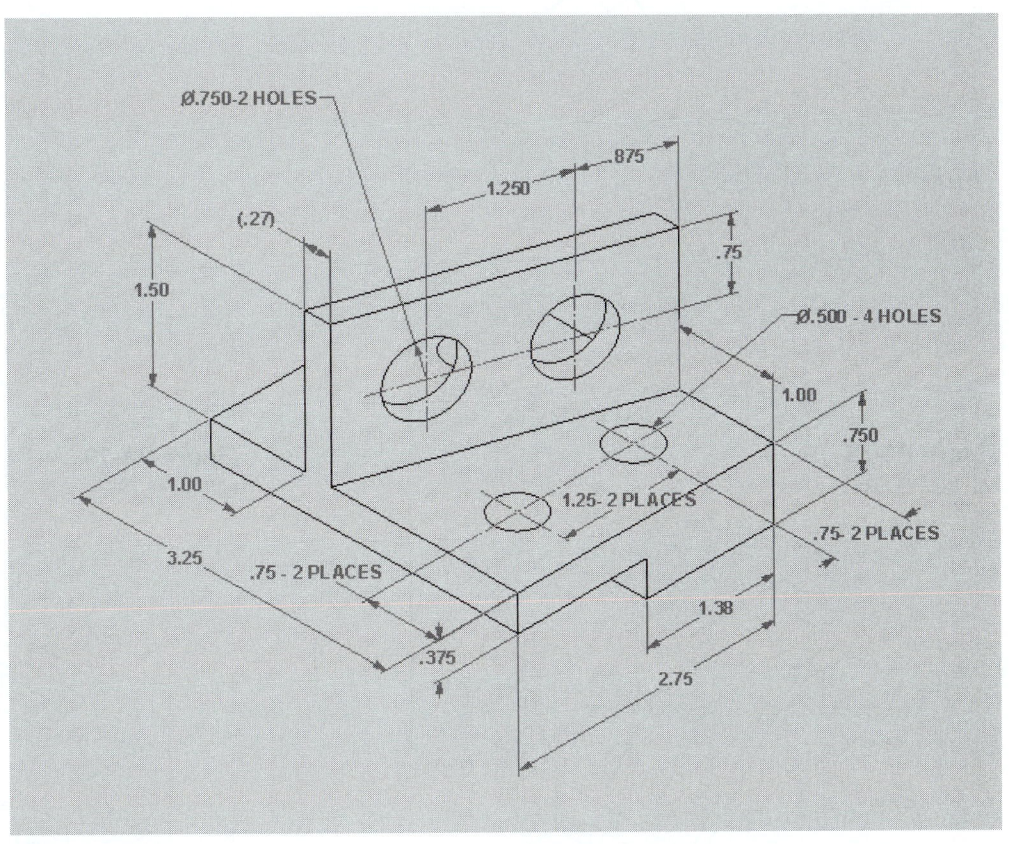

Figure P4-75
INCHES

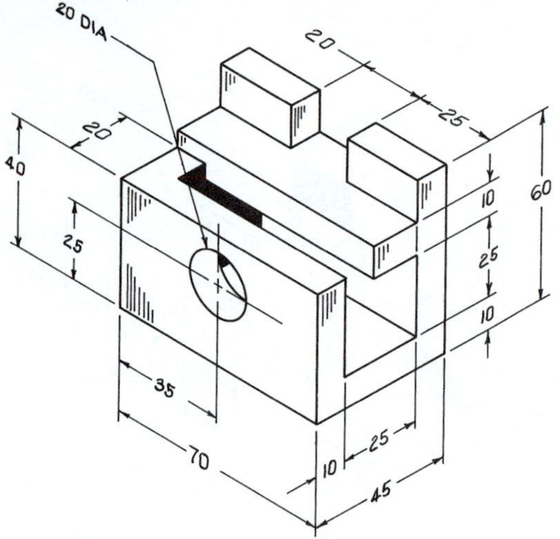

Figure P4-76
MILLIMETERS

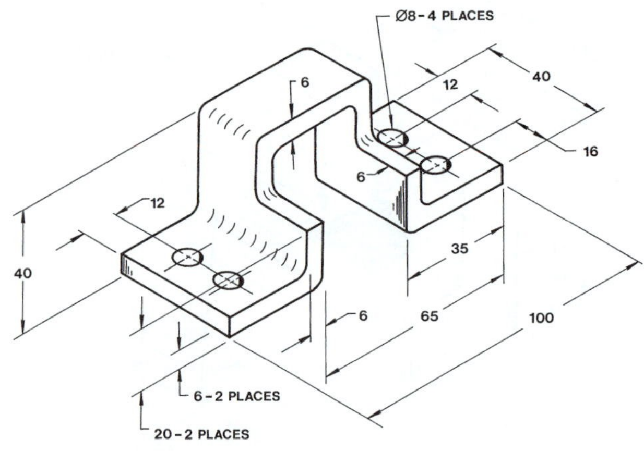

Figure P4-77
MILLIMETERS

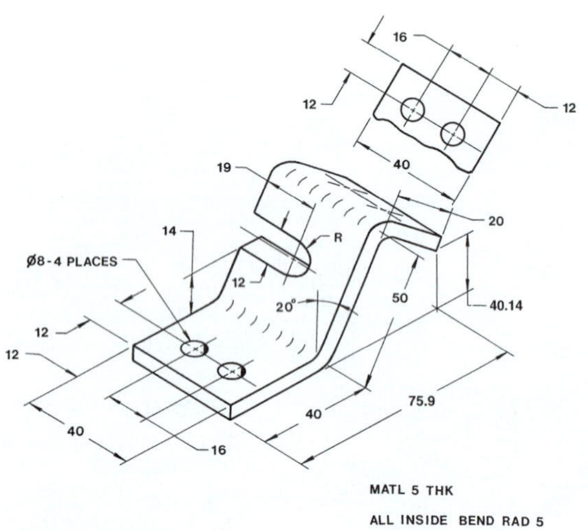

Figure P4-78
MILLIMETERS

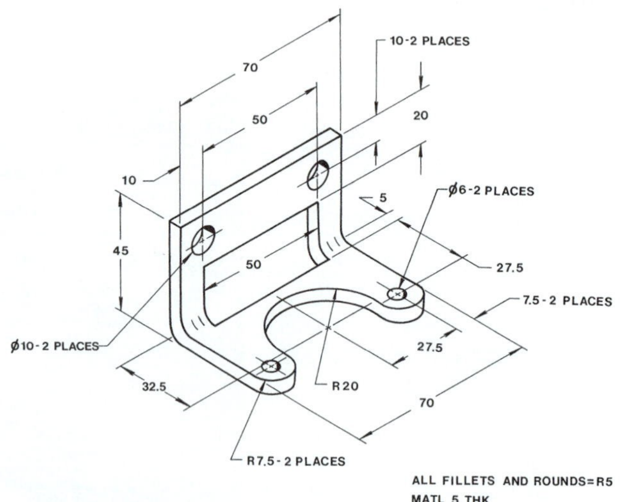

Figure P4-79
MILLIMETERS

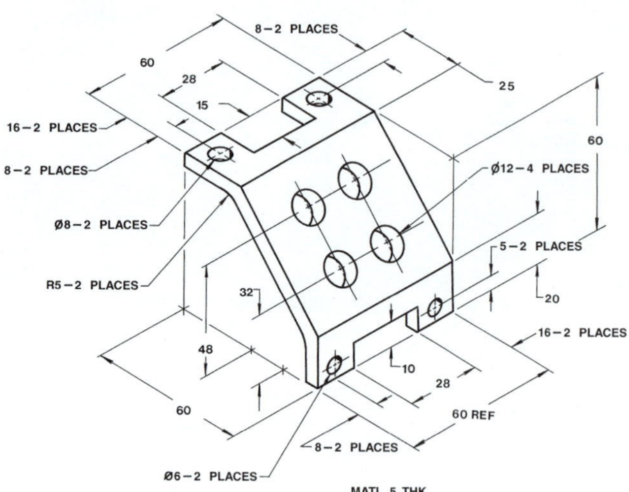

Figure P4-80
MILLIMETERS

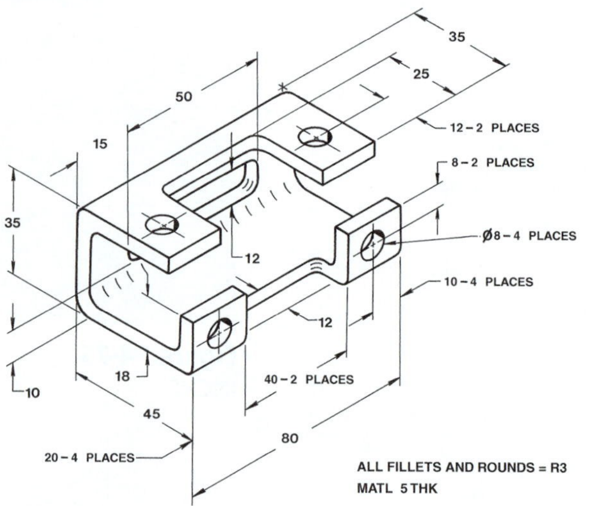

Figure P4-81
MILLIMETERS

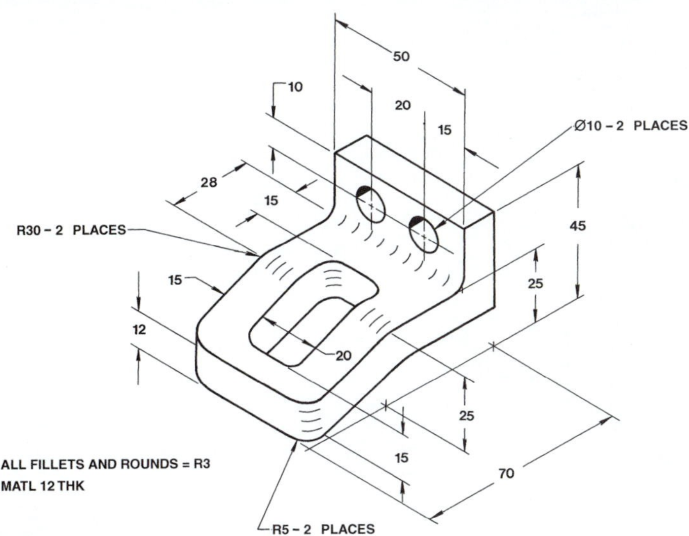

Figure P4-82
MILLIMETERS

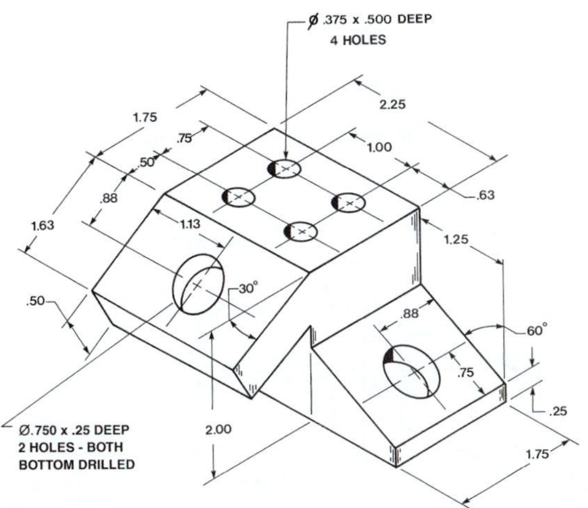

Figure P4-84
INCHES

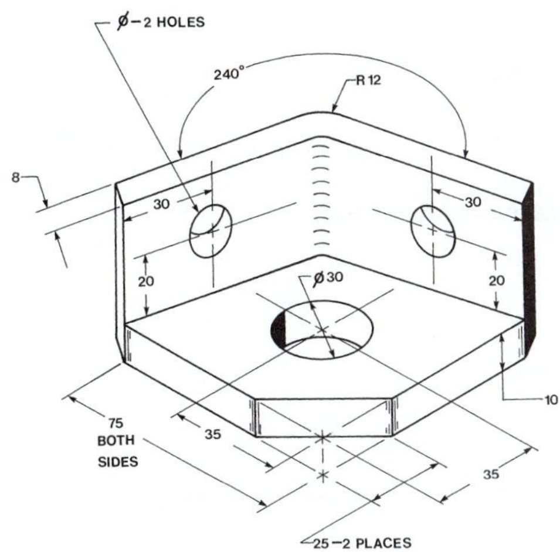

Figure P4-86
MILLIMETERS

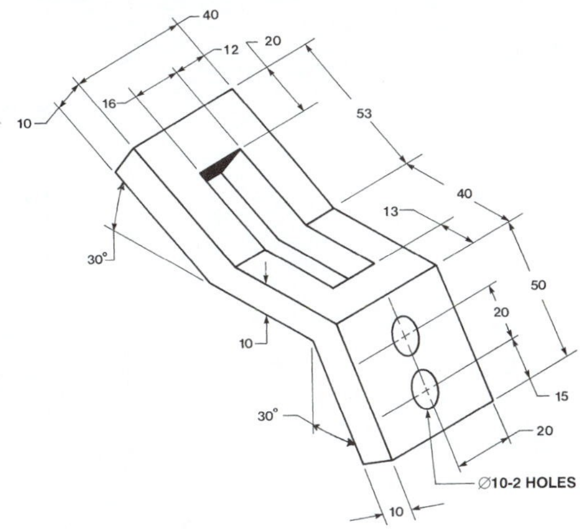

Figure P4-83
MILLIMETERS

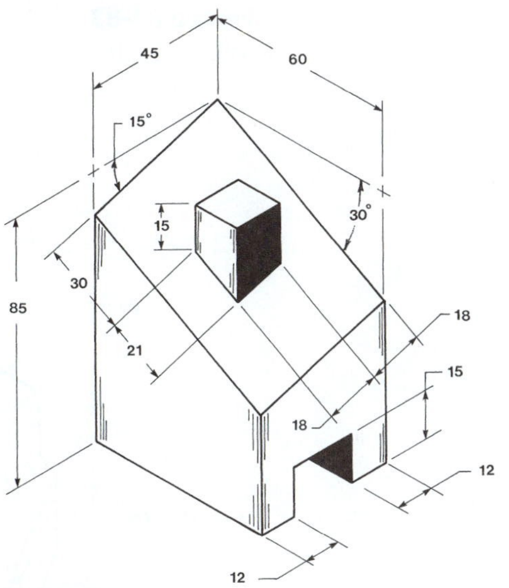

Figure P4-85
MILLIMETERS

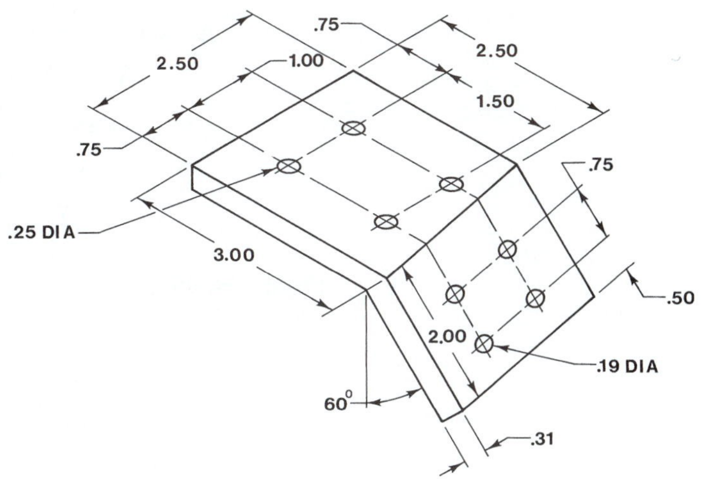

Figure P4-87
INCHES

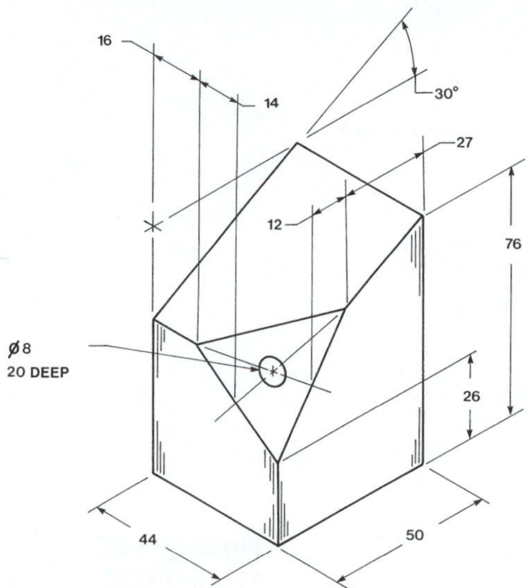

Figure P4-88
MILLIMETERS

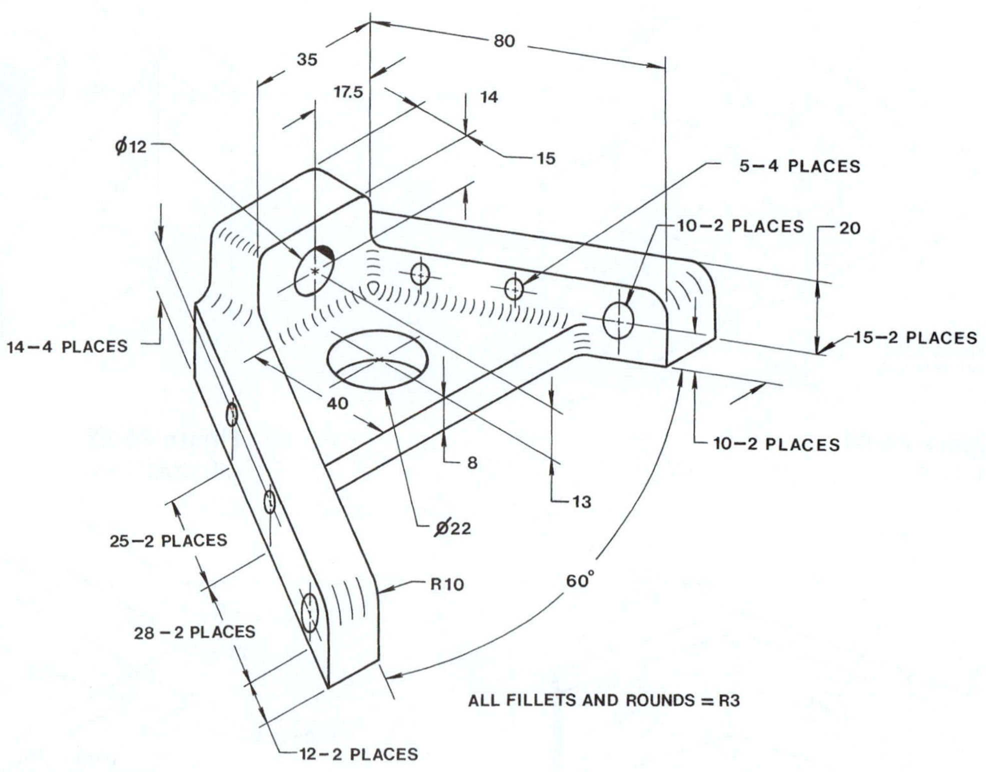

Figure P4-89
MILLIMETERS

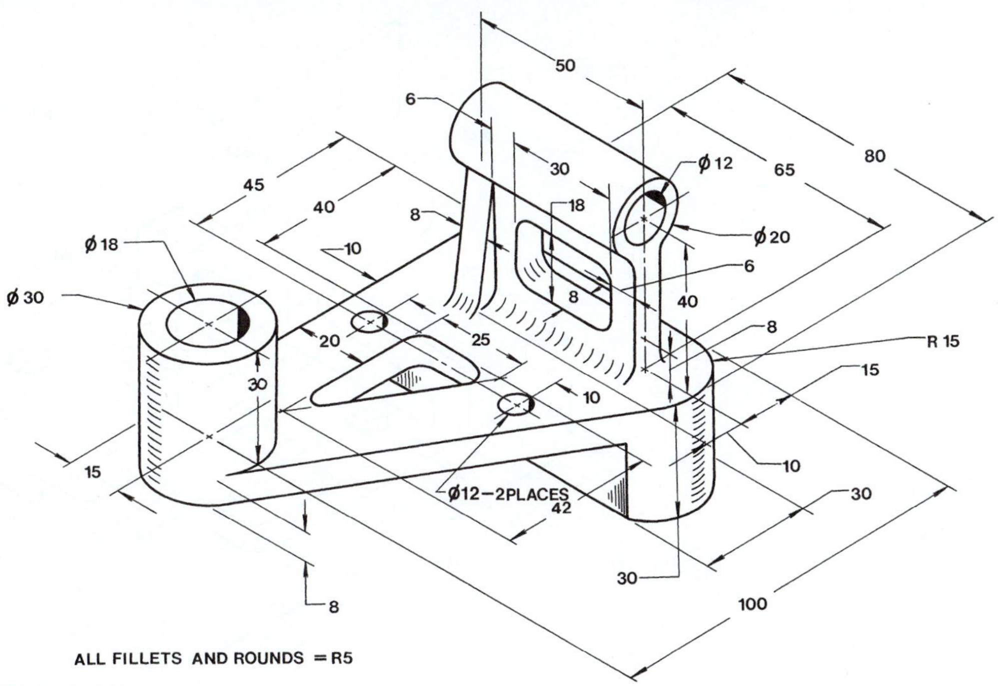

Figure P4-90
MILLIMETERS

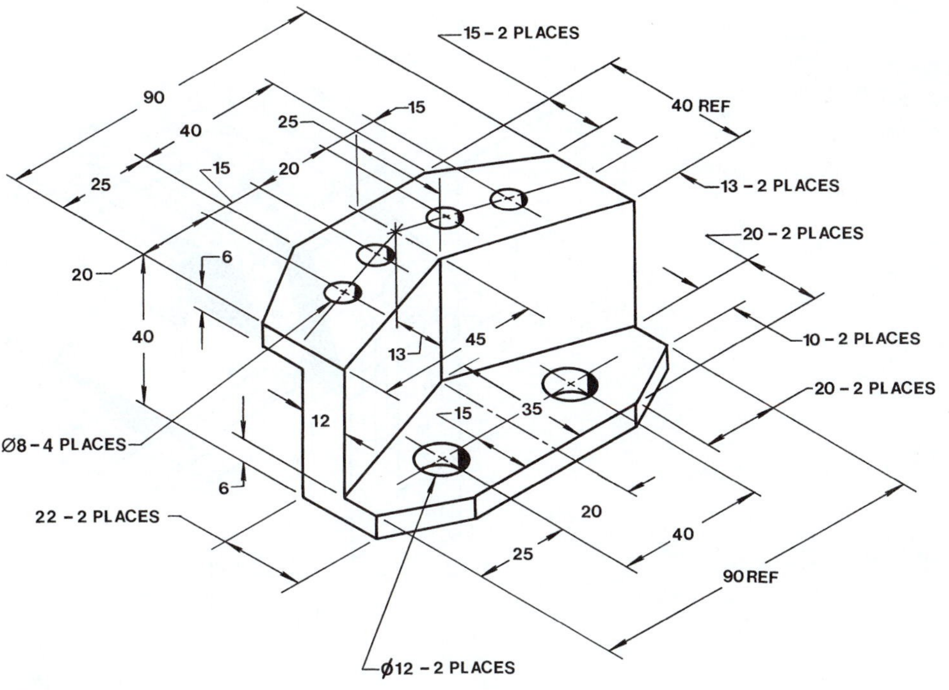

Figure P4-91
MILLIMETERS

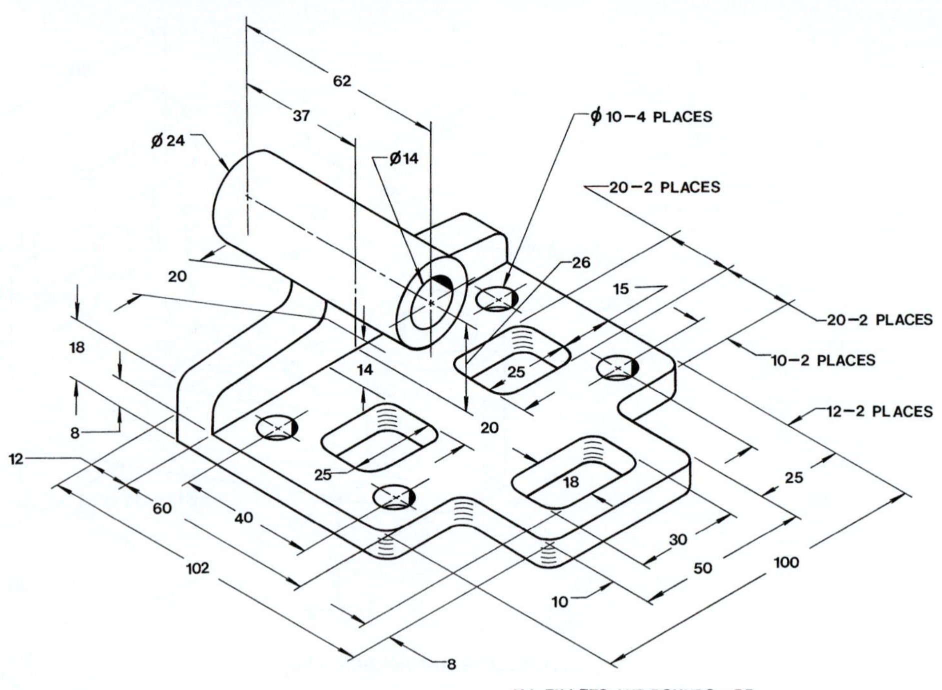

Figure P4-92
MILLIMETERS

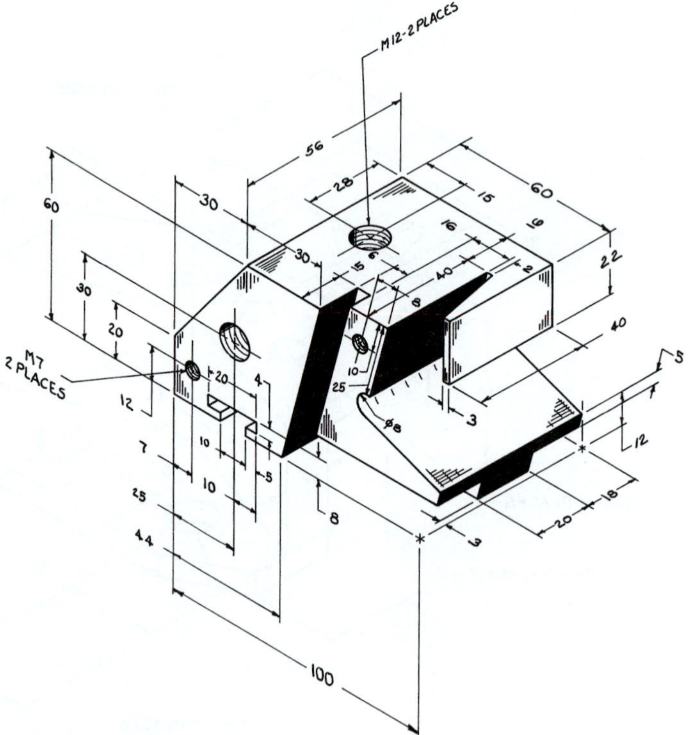

Figure P4-93
MILLIMETERS

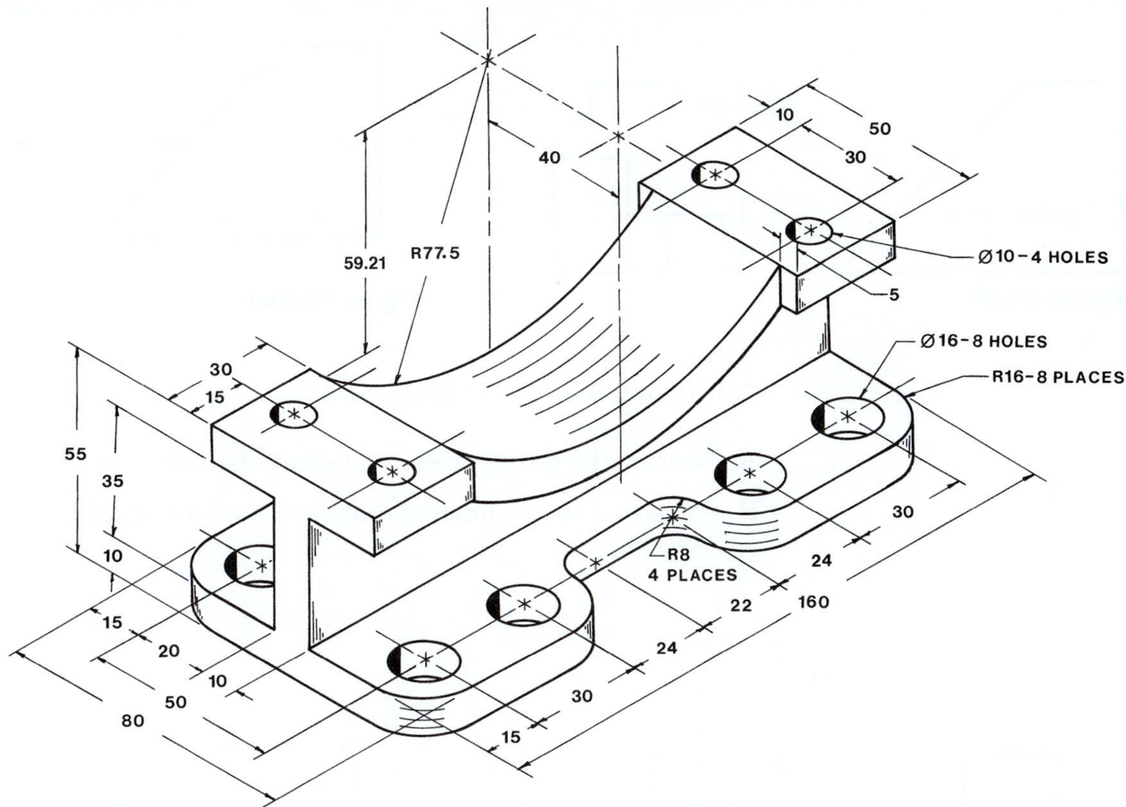

Figure P4-94
MILLIMETERS

For Figures P4-95 through P4-100:

A. Sketch the given orthographic views, and add the top view so that the final sketch includes a front, top, and right-side view.

B. Prepare a three-dimensional sketch of the object.

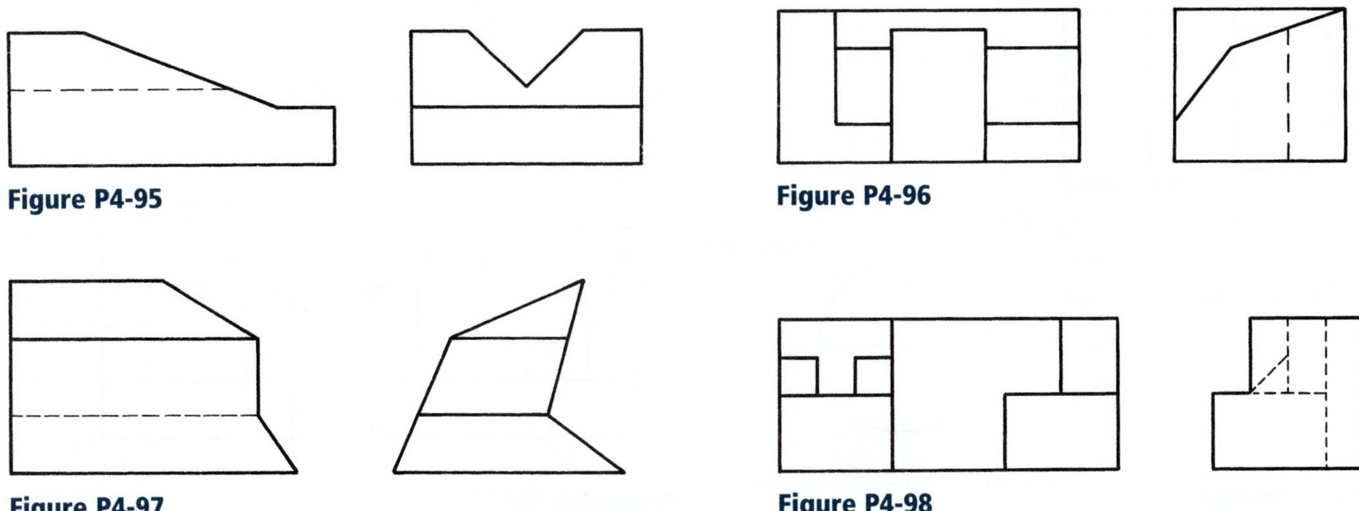

Figure P4-95

Figure P4-96

Figure P4-97

Figure P4-98

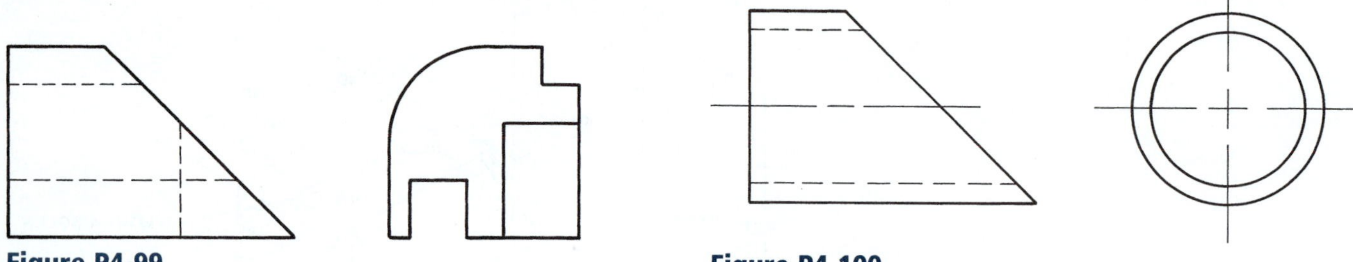

Figure P4-99

Figure P4-100

For Figures P4-101 through P4-128:

A. Redraw the given views, and draw the third view.

B. Prepare a three-dimensional sketch of the object.

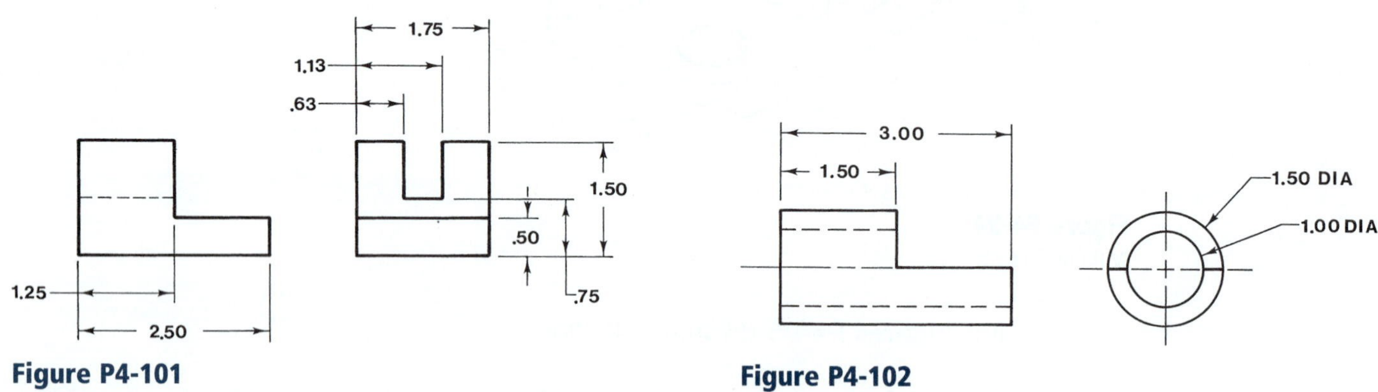

Figure P4-101
INCHES

Figure P4-102
INCHES

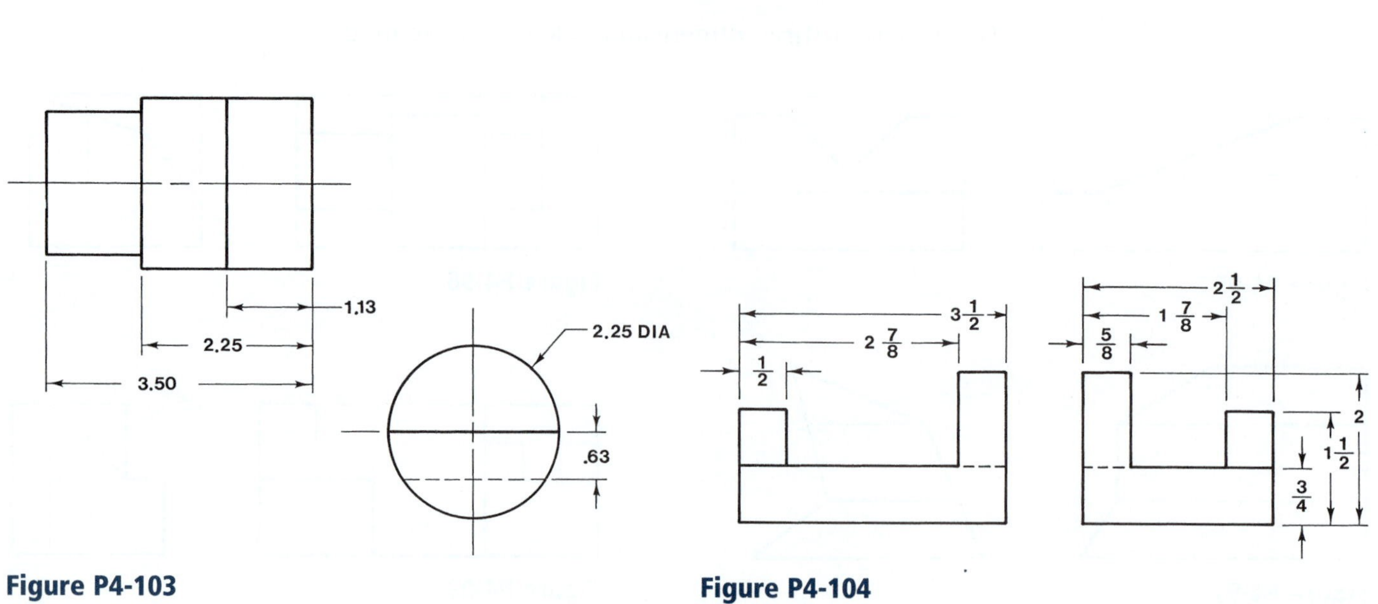

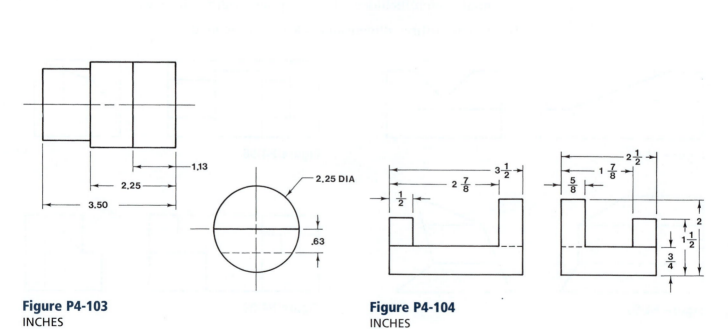

Figure P4-103
INCHES

Figure P4-104
INCHES

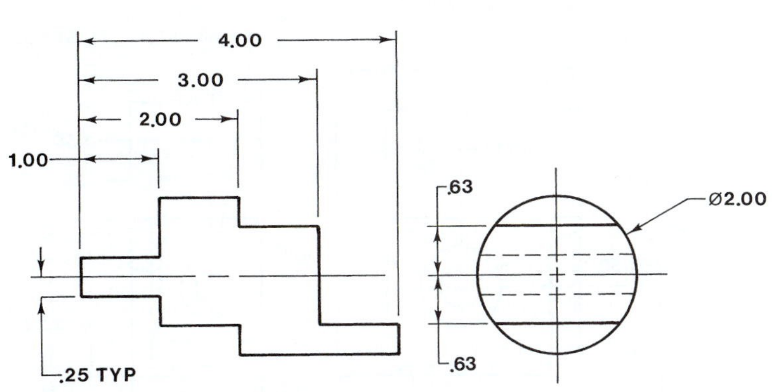

Figure P4-105
INCHES

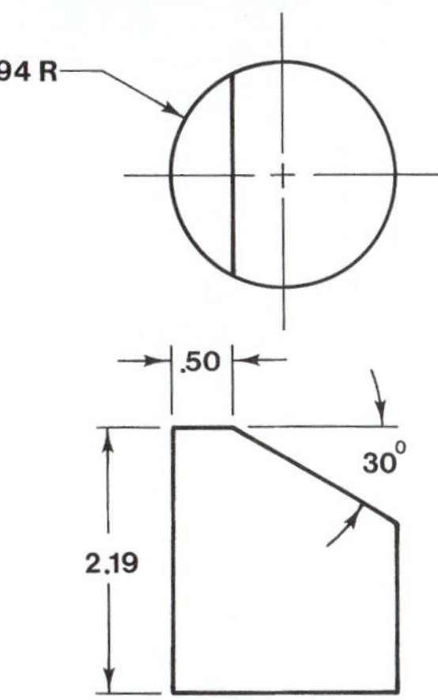

Figure P4-106
INCHES

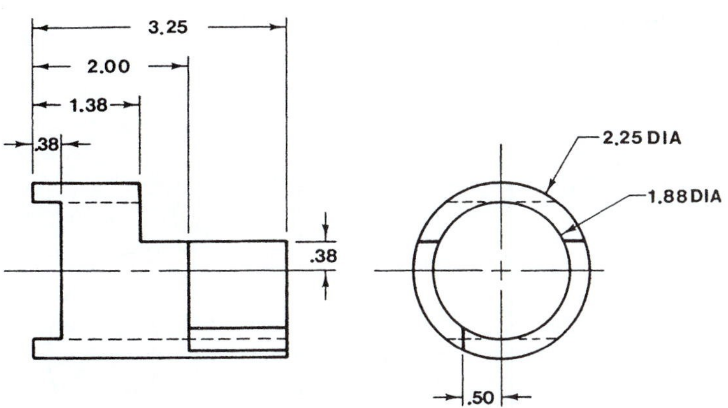

Figure P4-107
INCHES

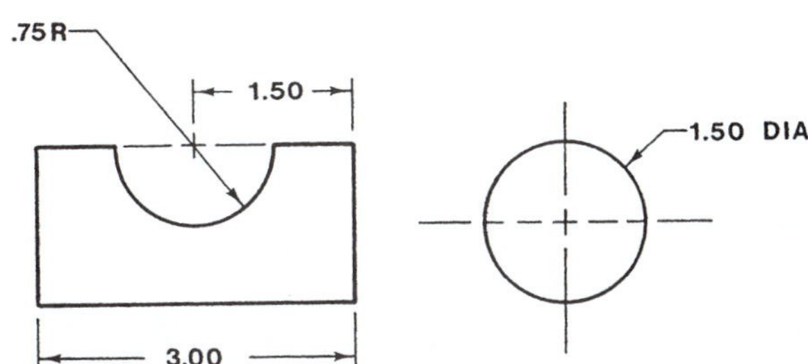

Figure P4-108
INCHES

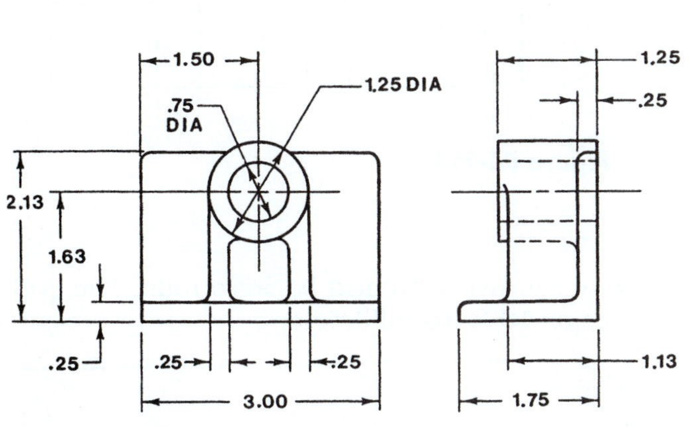

Figure P4-109
INCHES

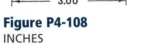

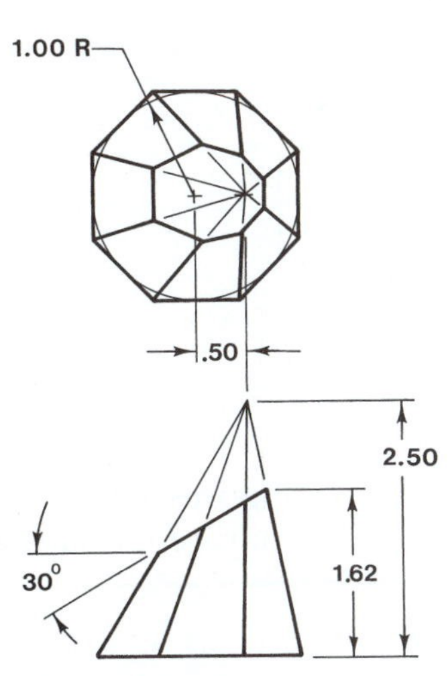

Figure P4-110
INCHES

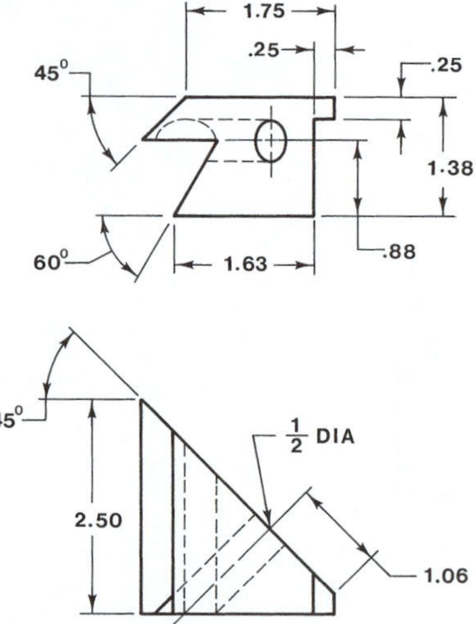

Figure P4-111
INCHES

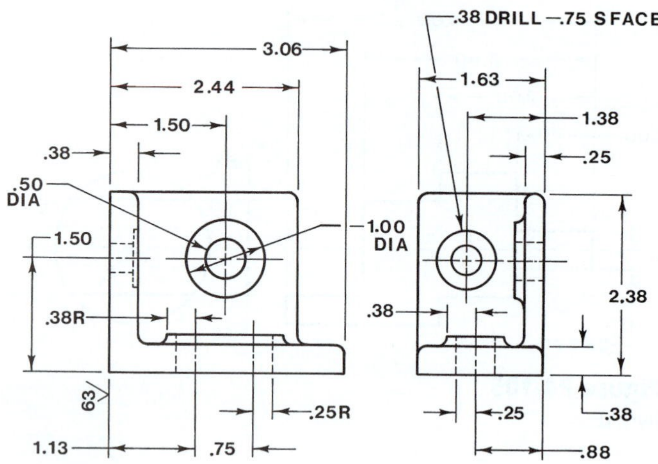

Figure P4-112
INCHES

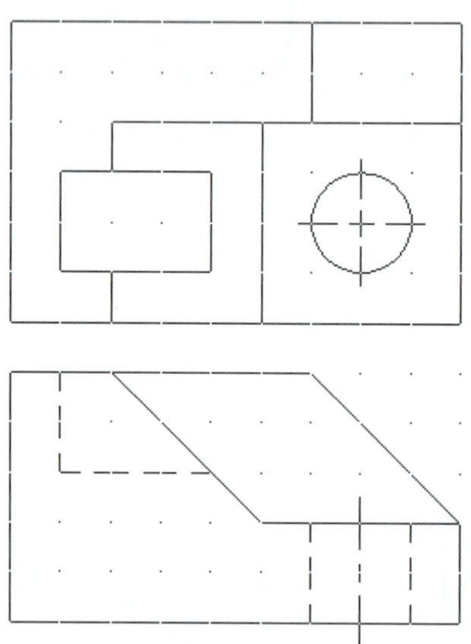

Figure P4-113

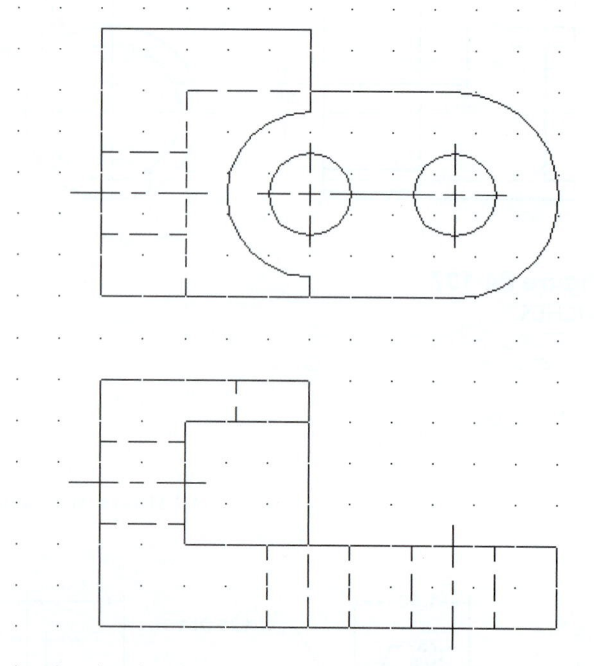

Figure P4-114

Figures P4-113 through P4-128 are drawn with grid backgrounds. The grid may be either 0.50 × 0.50 inches or 10 × 10 millimeters.

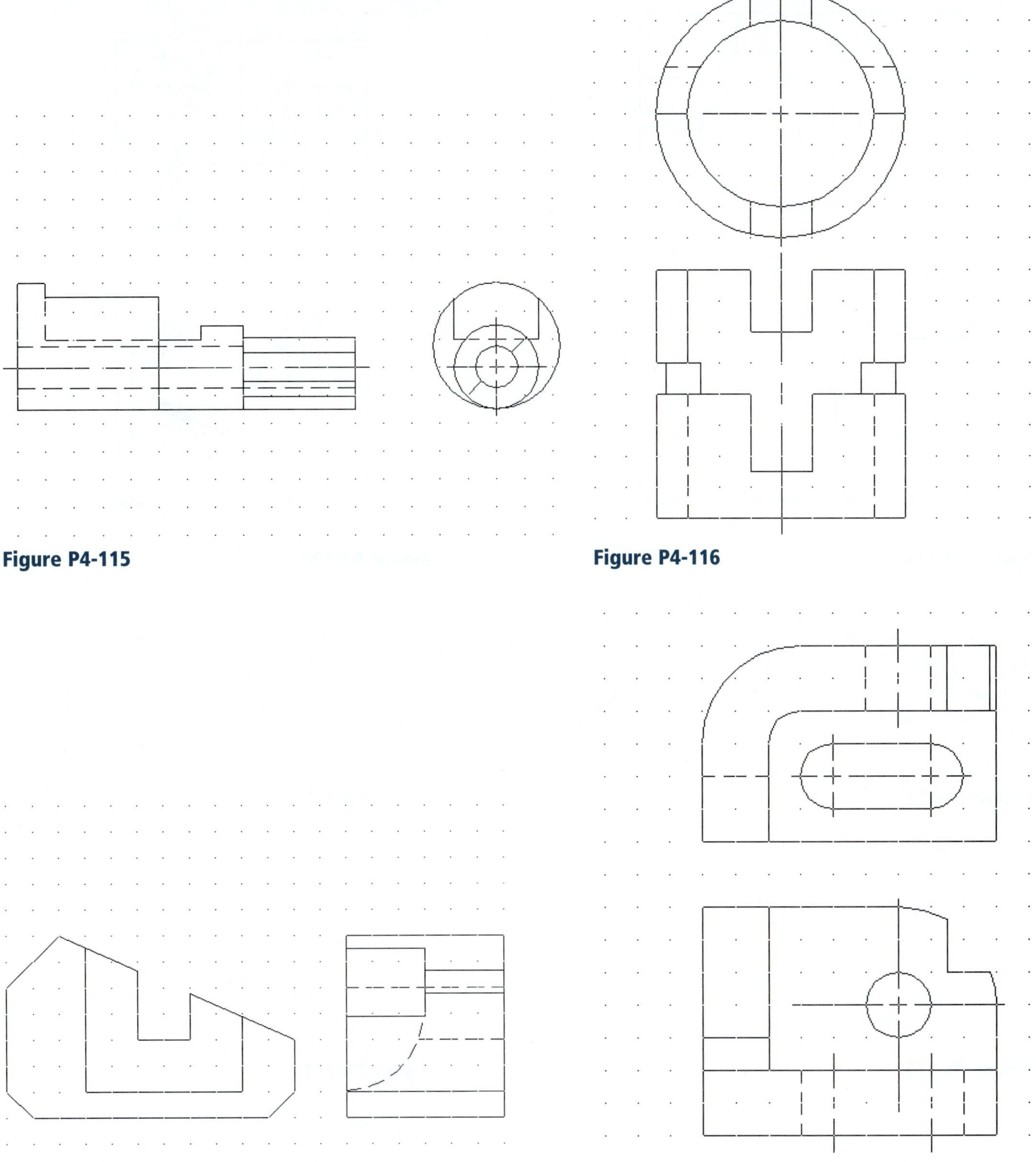

Figure P4-115

Figure P4-116

Figure P4-117

Figure P4-118

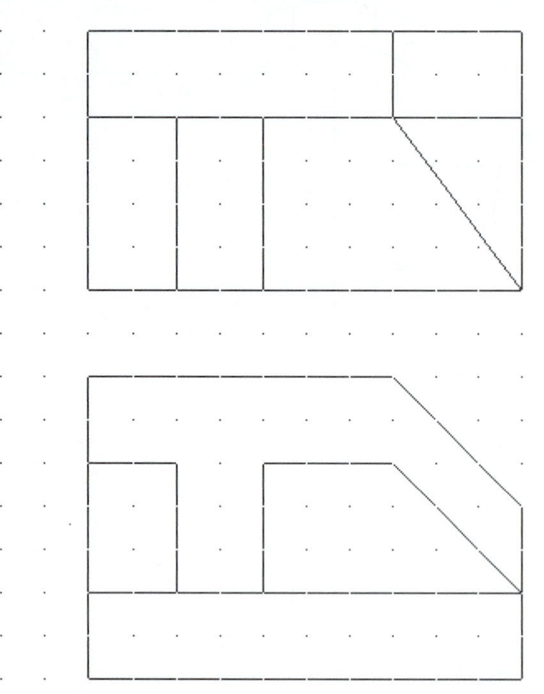

Figure P4-119

Figure P4-120

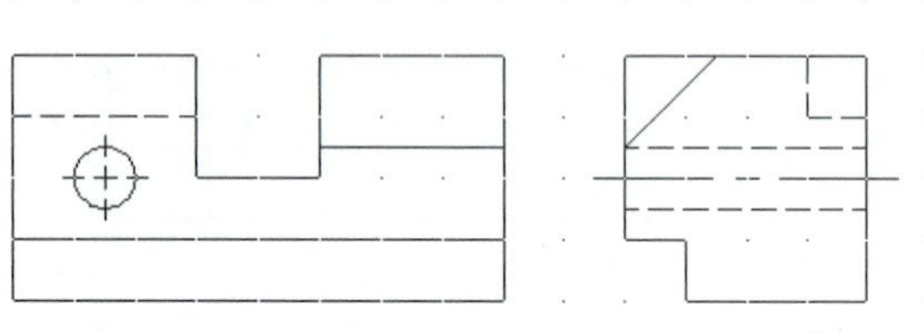

Figure P4-121

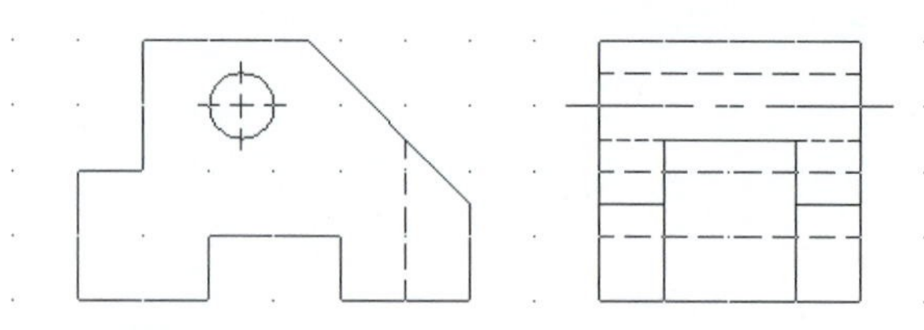

Figure P4-122

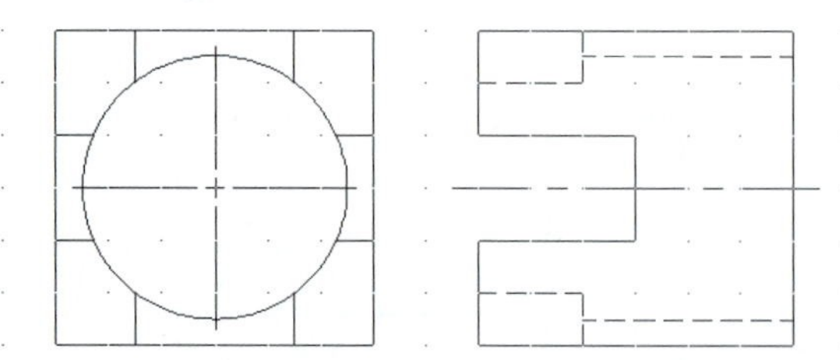

Figure P4-123

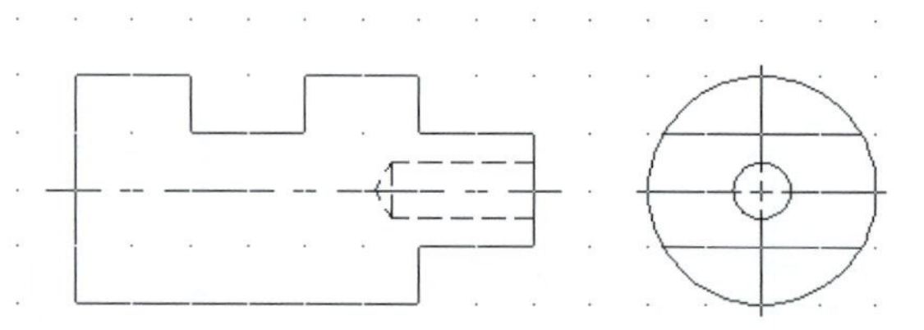

Figure P4-124

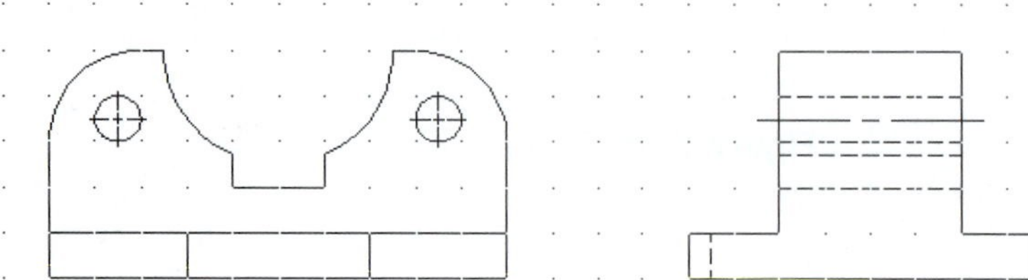

Figure P4-125

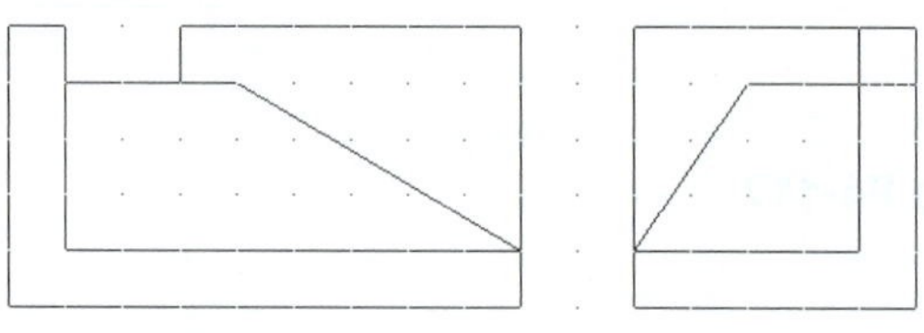

Figure P4-126

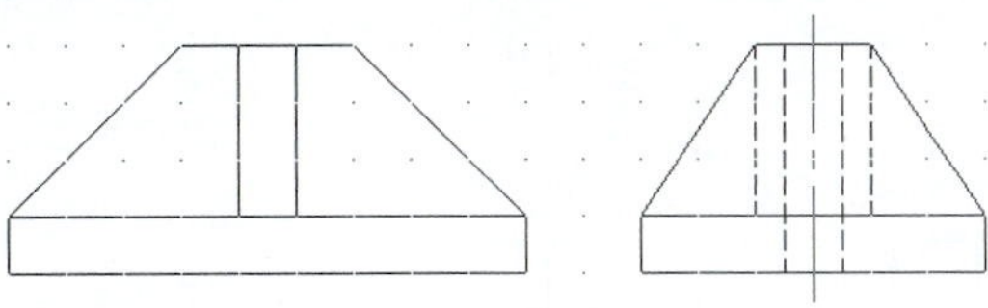

Figure P4-127

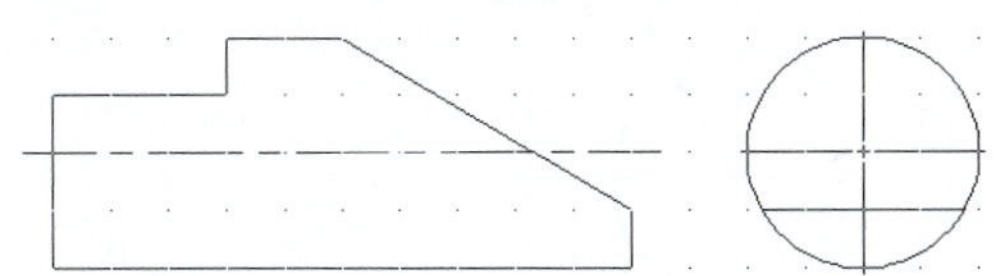

Figure P4-128

Draw the complete front, top, and side views of the two intersecting objects given in Figures P4-129 through P4-134 on the basis of the given complete and partially complete orthographic views.

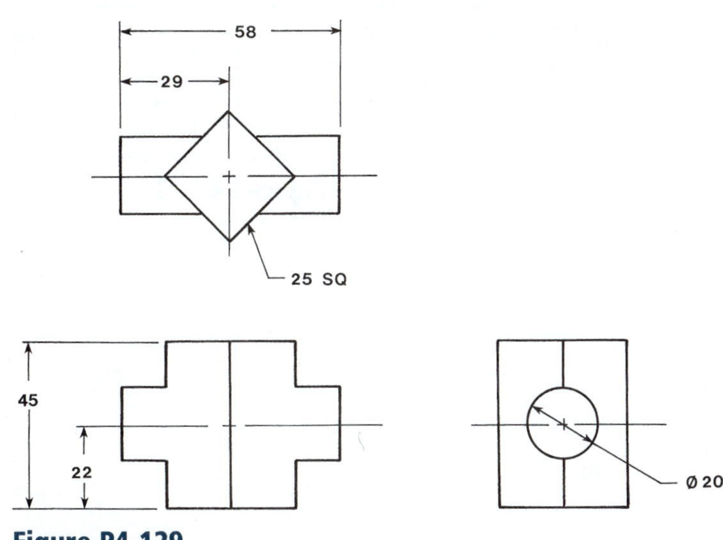

Figure P4-129
MILLIMETERS

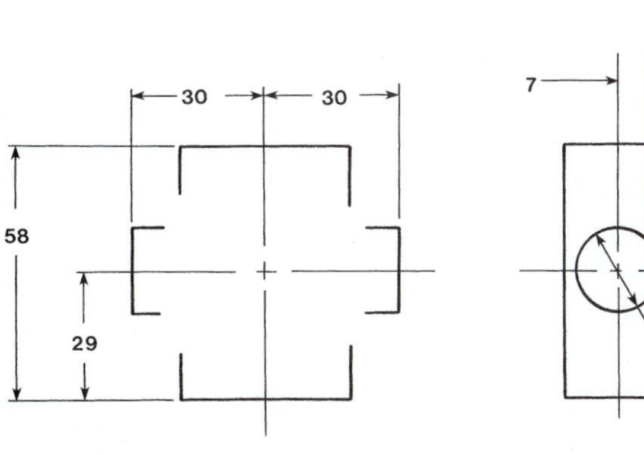

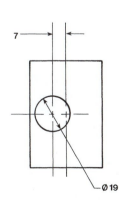

Figure P4-130
MILLIMETERS

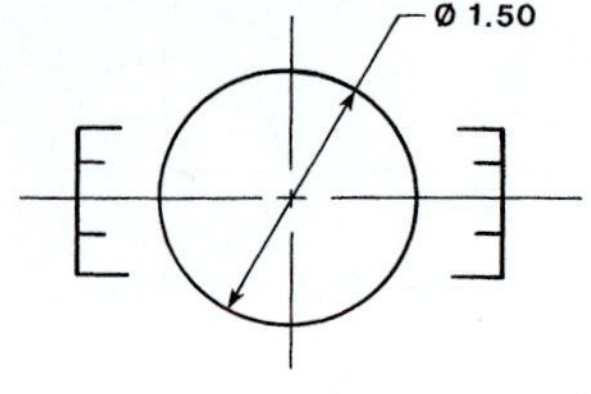

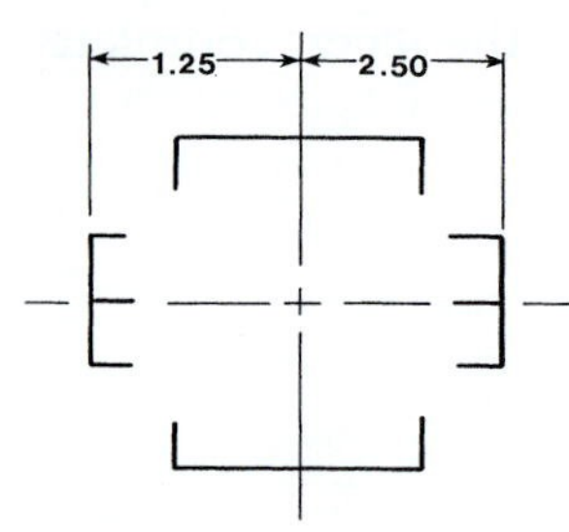

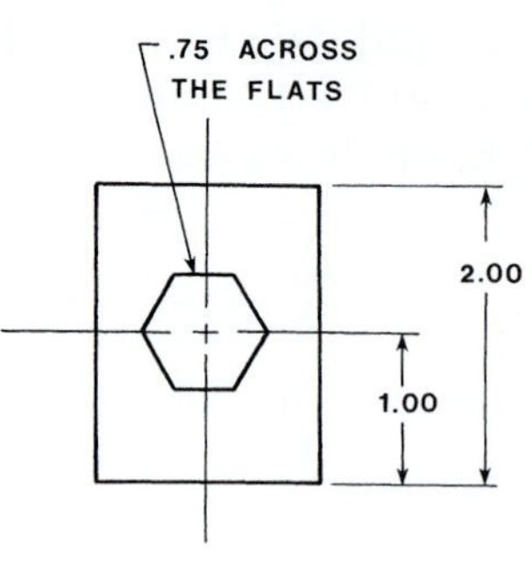

Figure P4-131
INCHES

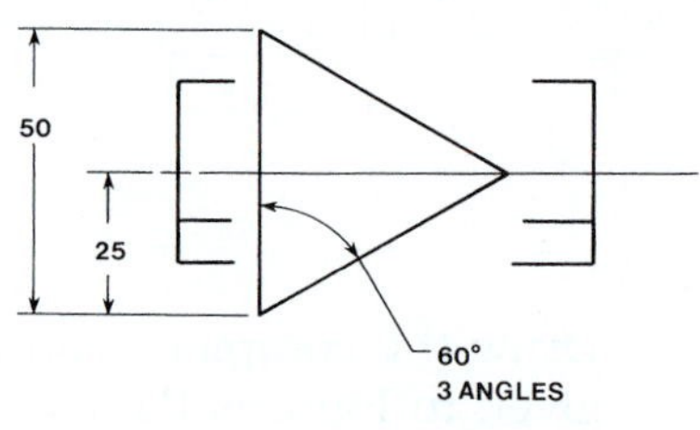

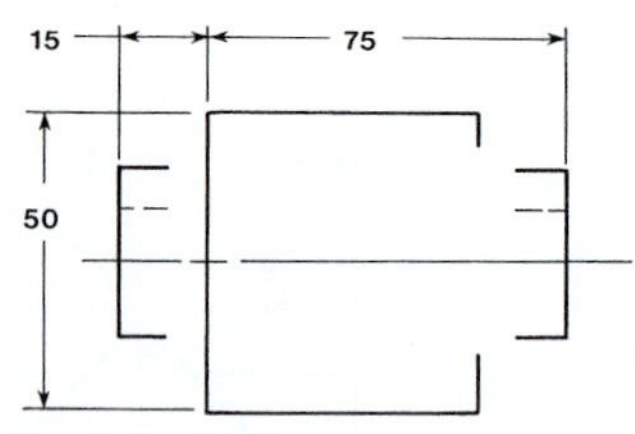

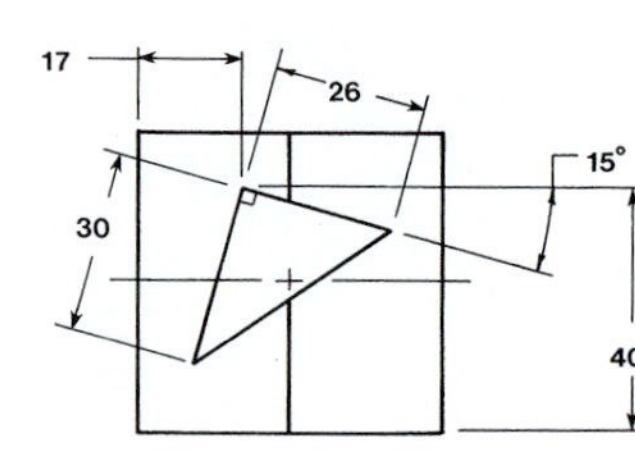

Figure P4-132
INCHES

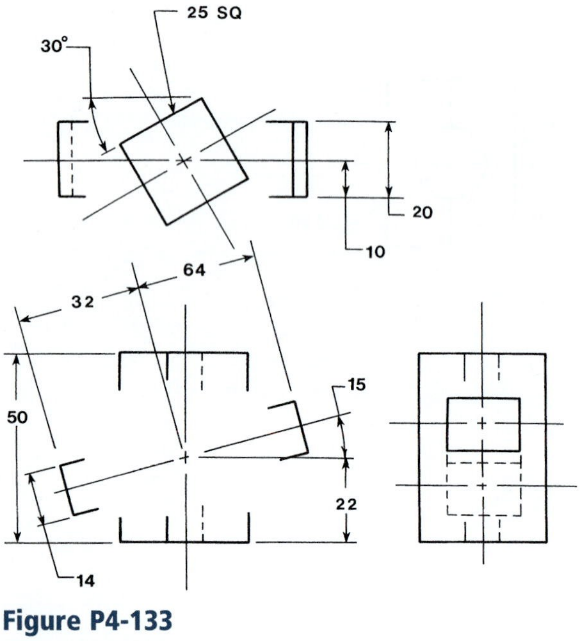

Figure P4-133
MILLIMETERS

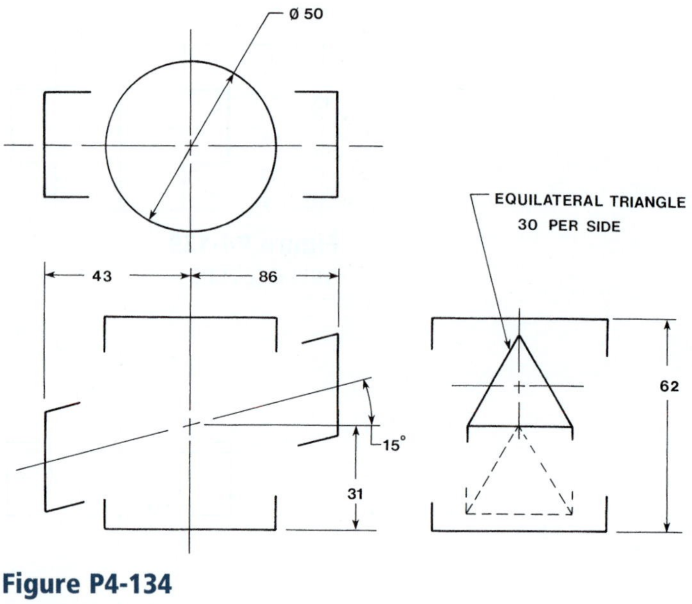

Figure P4-134
MILLIMETERS

Draw the front, top, and side orthographic views of the objects given in Figures P4-135 through P4-138 on the basis of the partially complete isometric drawings.

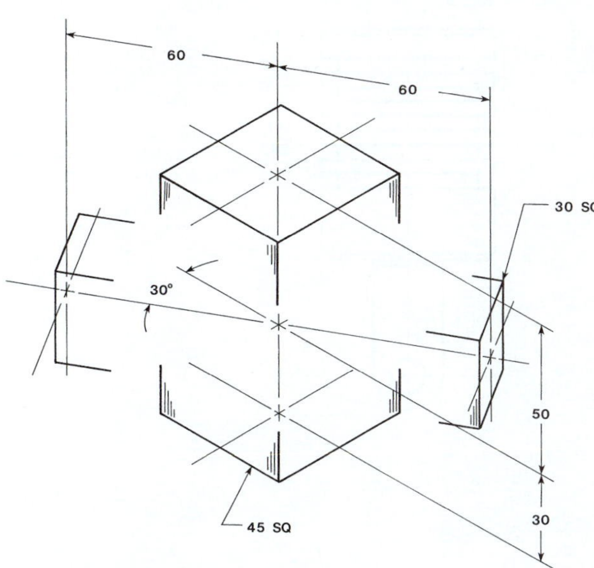

Figure P4-135
MILLIMETERS

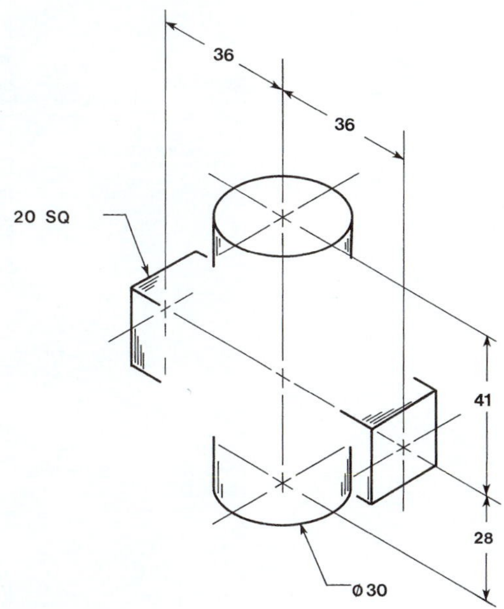

Figure P4-136
MILLIMETERS

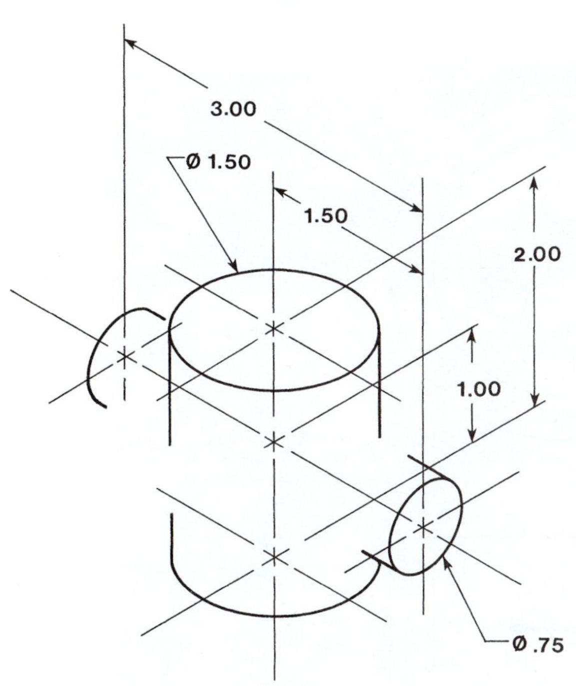

Figure P4-137
INCHES

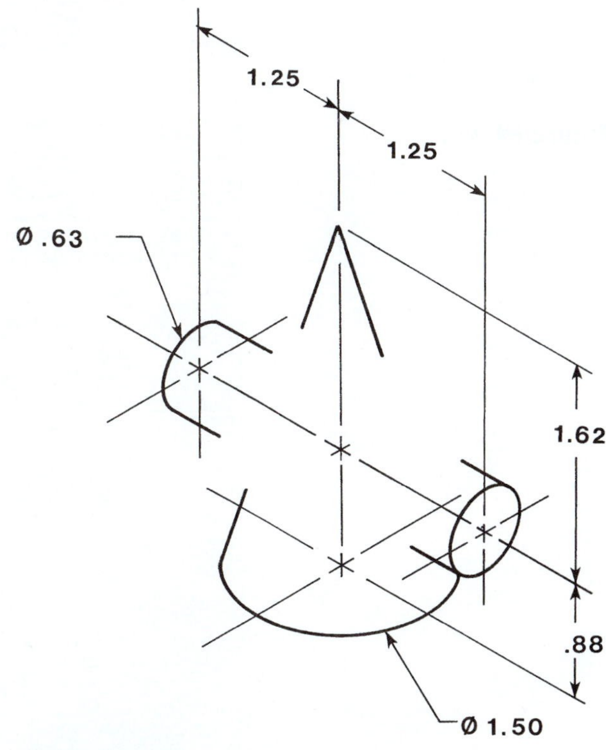

Figure P4-138
INCHES (SCALE: 5=1)

Draw the followings as 3D models using the given dimensions.

Figure P4-139

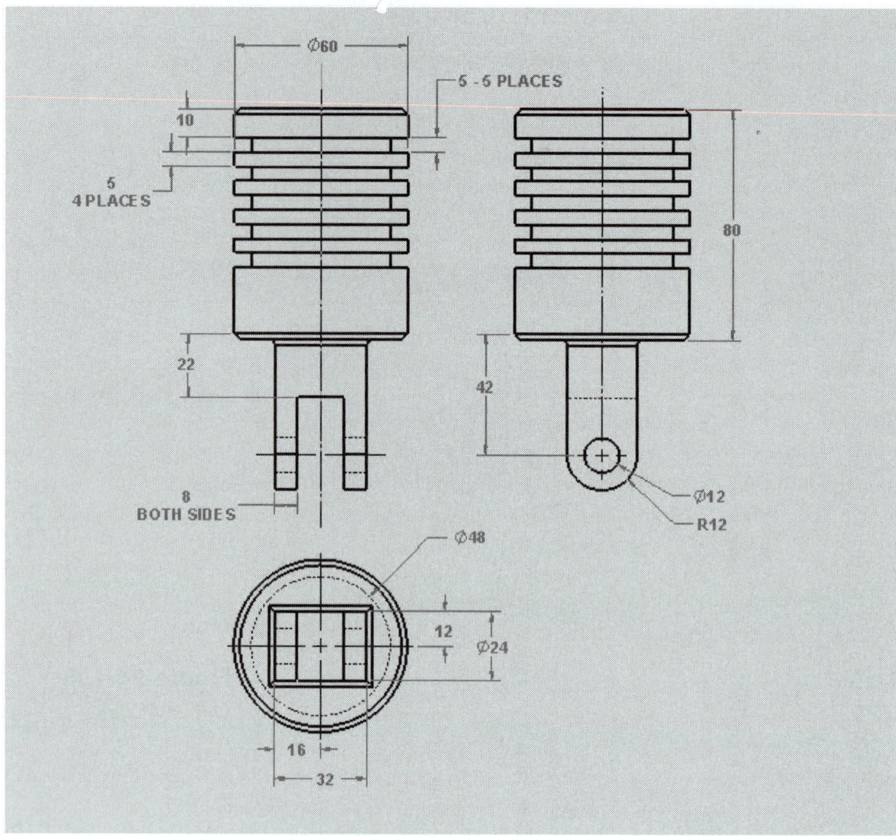

Figure P4-140

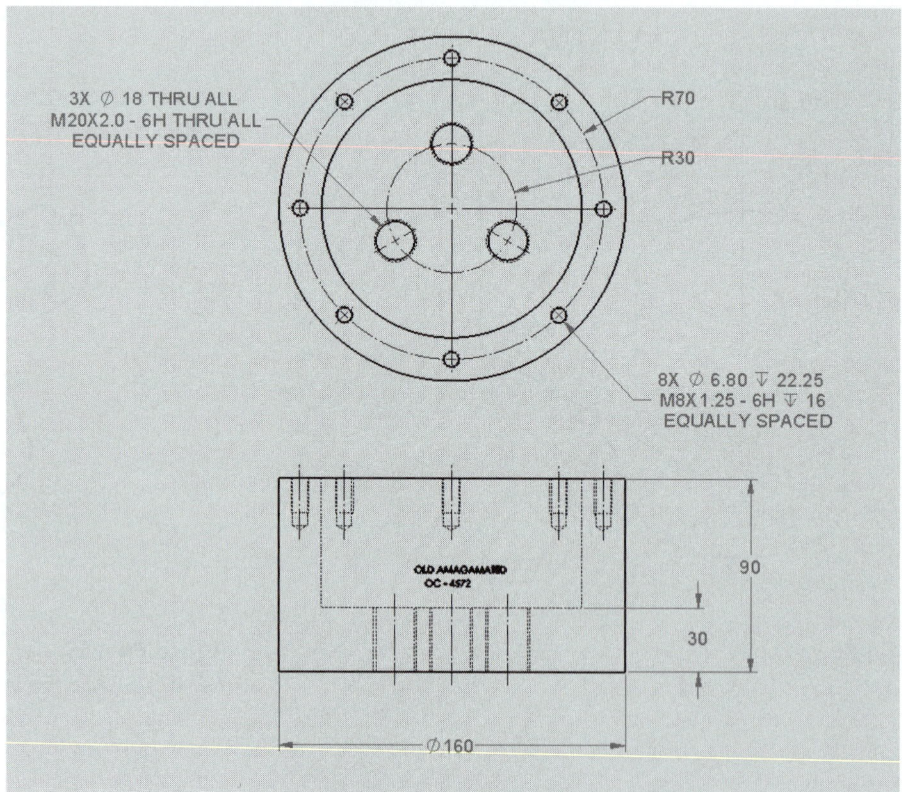

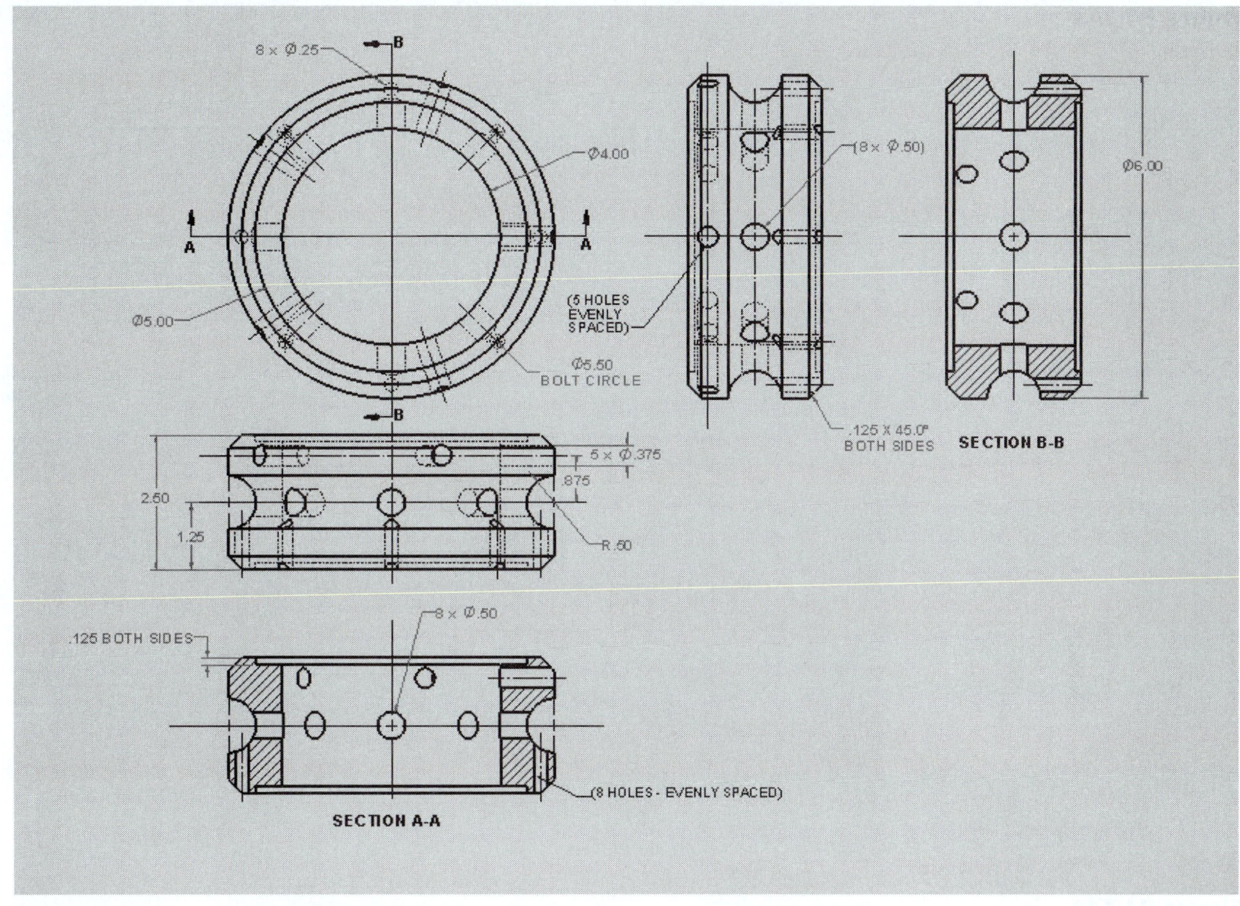

Figure P4-141

Figures P4-142 through P4-146 are presented using first-angle projection per ISO standards.

Draw the objects as

a. 3D models

b. Orthographic views using third-angle projections per ANSI standards.

Figure P4-142

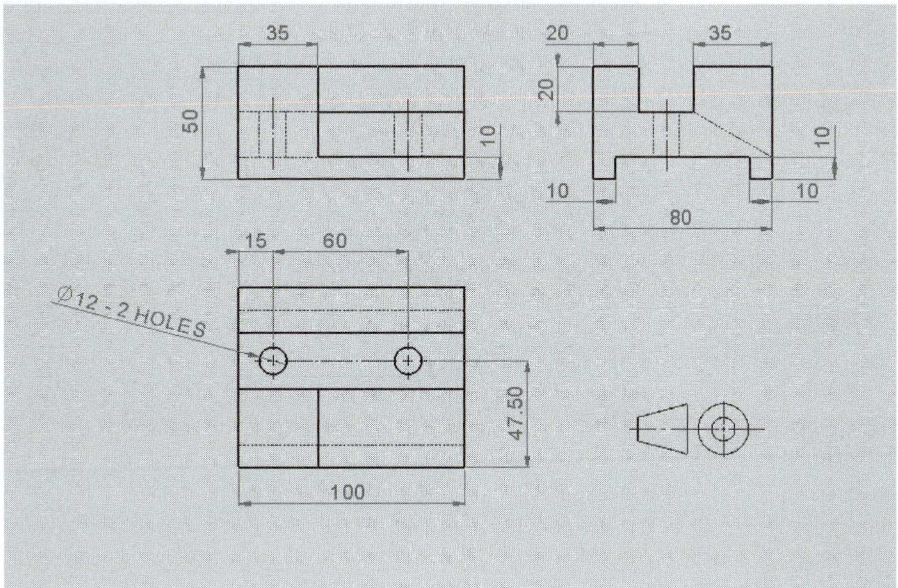

Figure P4-143

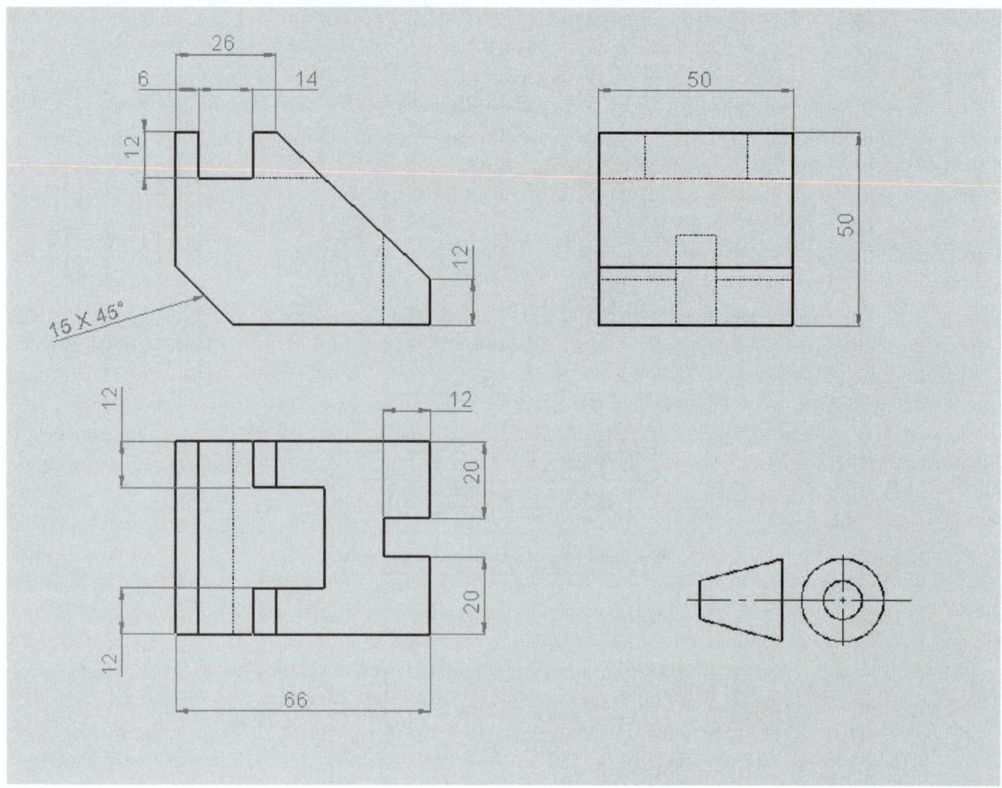

Figure P4-144

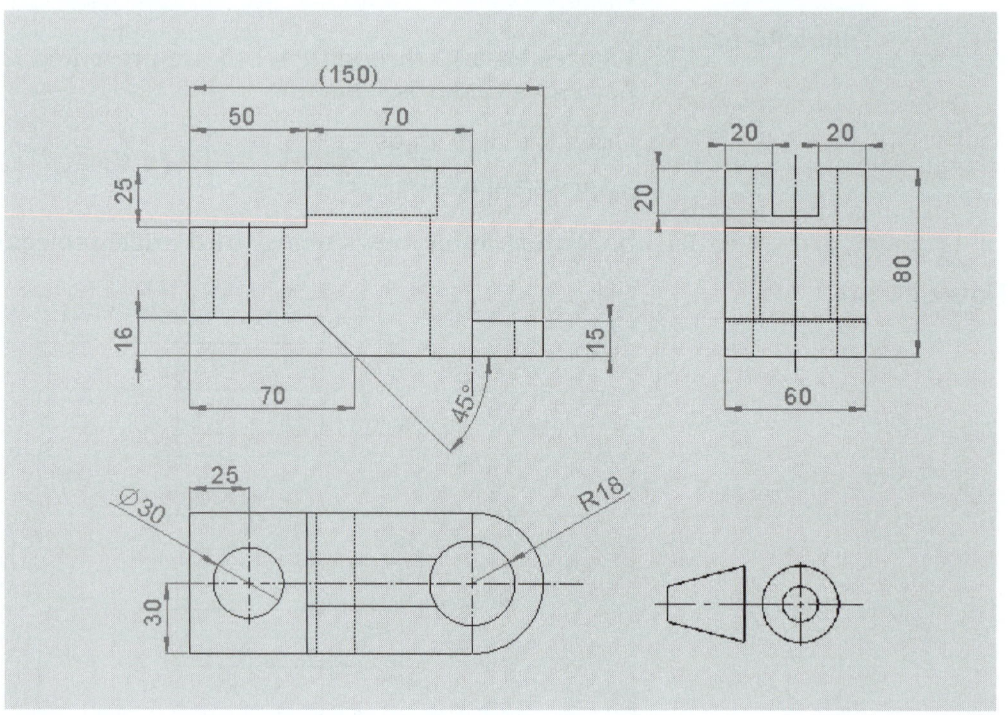

Figure P4-145

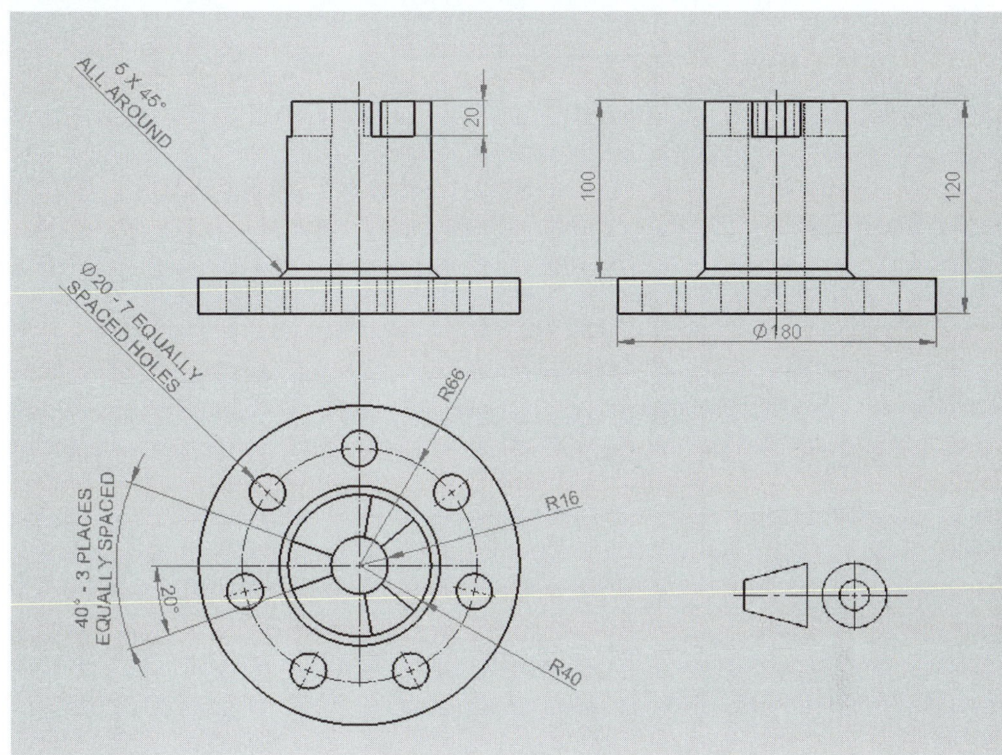

Figure P4-146

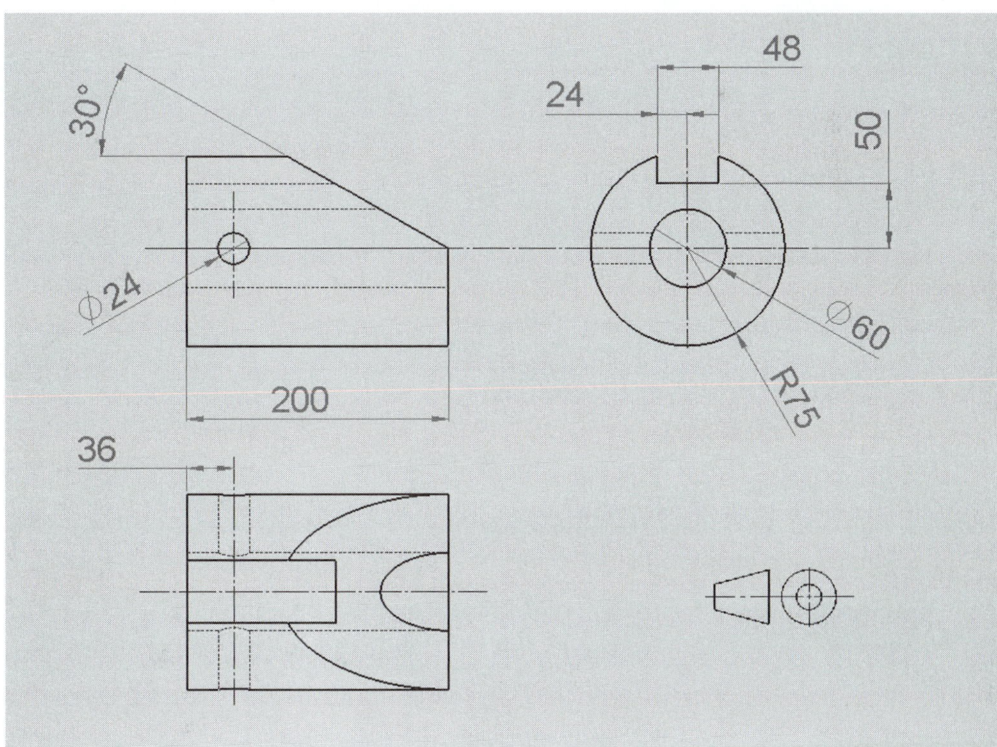

5 chapterfive
Assemblies

CHAPTER OBJECTIVES

- Learn how to create assembly drawings
- Learn how to create exploded assembly drawings
- Learn how to create a parts list
- Learn how to animate an assembly
- Learn how to edit a title block

5-1 Introduction

This chapter introduces the **Assembly** tools. These tools are used to create assembly drawings. Assembly drawings can be exploded to form isometric assembly drawings that when labeled and accompanied by a parts list become working drawings. Assembly drawings can be animated.

5-2 Starting an Assembly Drawing

Figure 5-1 shows a test block. The overall dimensions for the block are 80 × 80 × 80 mm. The cutout is 40 × 40 × 80 mm. The test block will be used to help introduce the **Assembly** tools. Draw the block and save it as **BLOCK,TEST**. Draw the Test Block so that one of its corners is on the origin. If the origin icon does not appear on your drawing screen, click the **View** tab and click the **Origin** option. See Figure 5-2.

1. Start a **New** drawing.
2. Select the **Assembly** format.
3. Click **OK**.

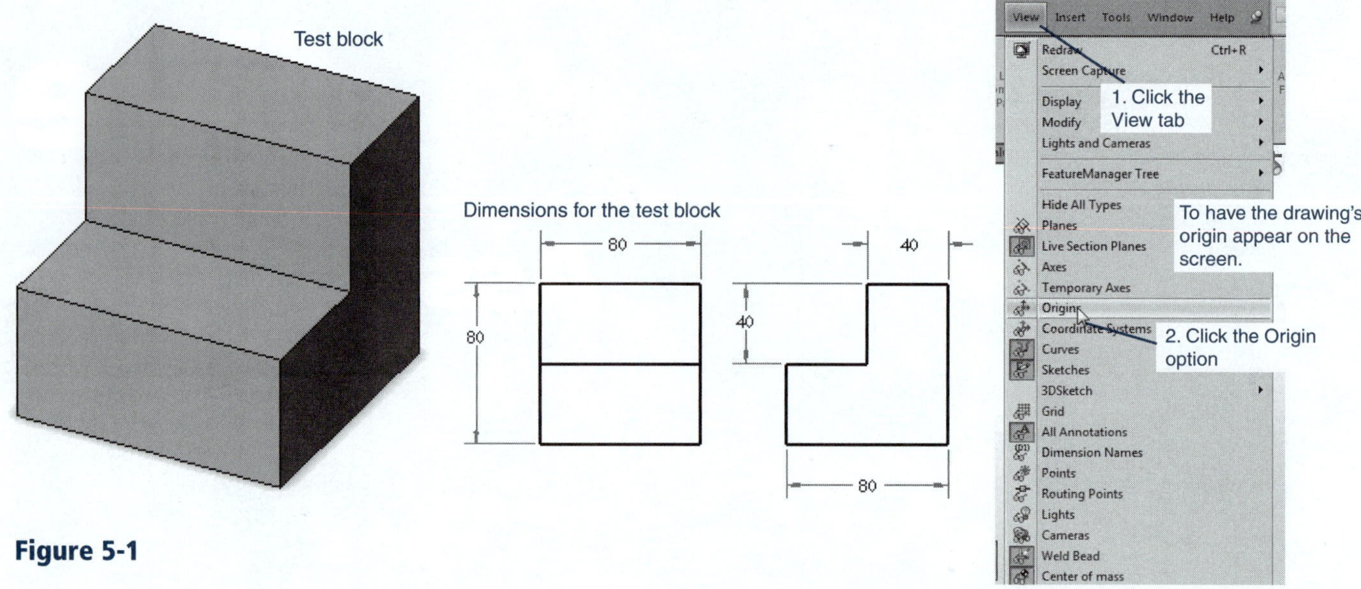

Figure 5-1

Figure 5-2

See Figure 5-3.

The **Begin Assembly PropertyManager** will appear. See Figure 5-4.

4 Click the **Browse. . .** box.

The **Open** box will appear. See Figure 5-5.

5 Click **BLOCK,TEST**, then click **Open**.

The test block will appear on the screen. See Figure 5-6.

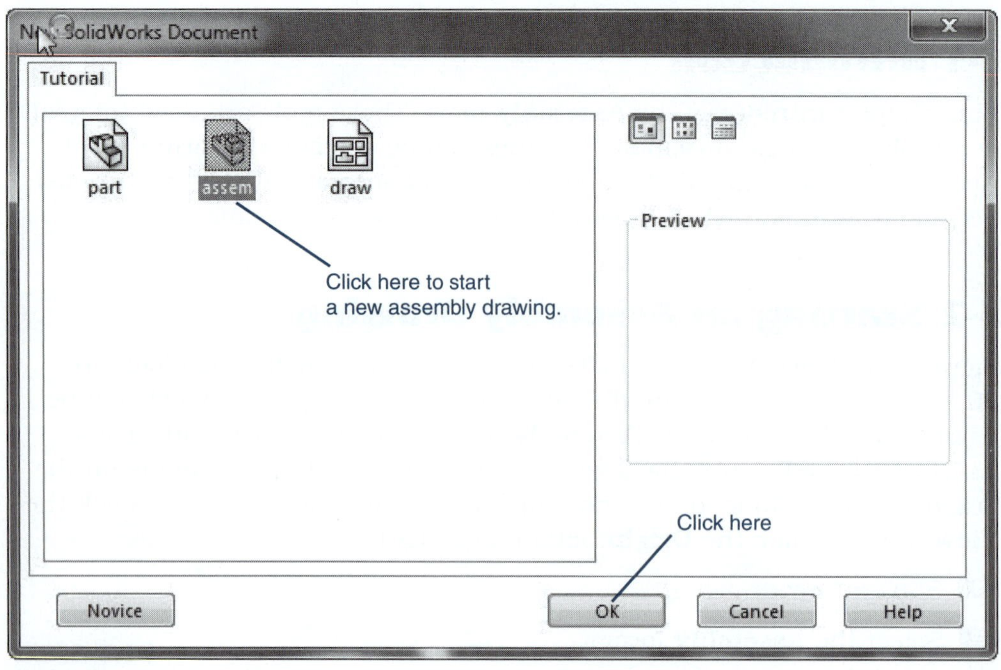

Figure 5-3

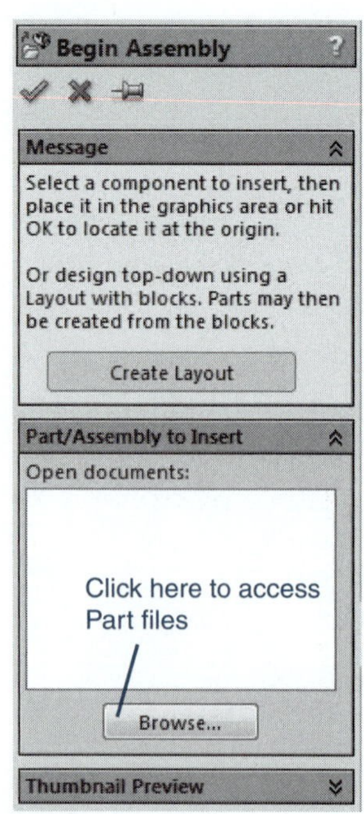

Figure 5-4

Figure 5-5

Click the Block, Test icon.
Click here

Figure 5-6

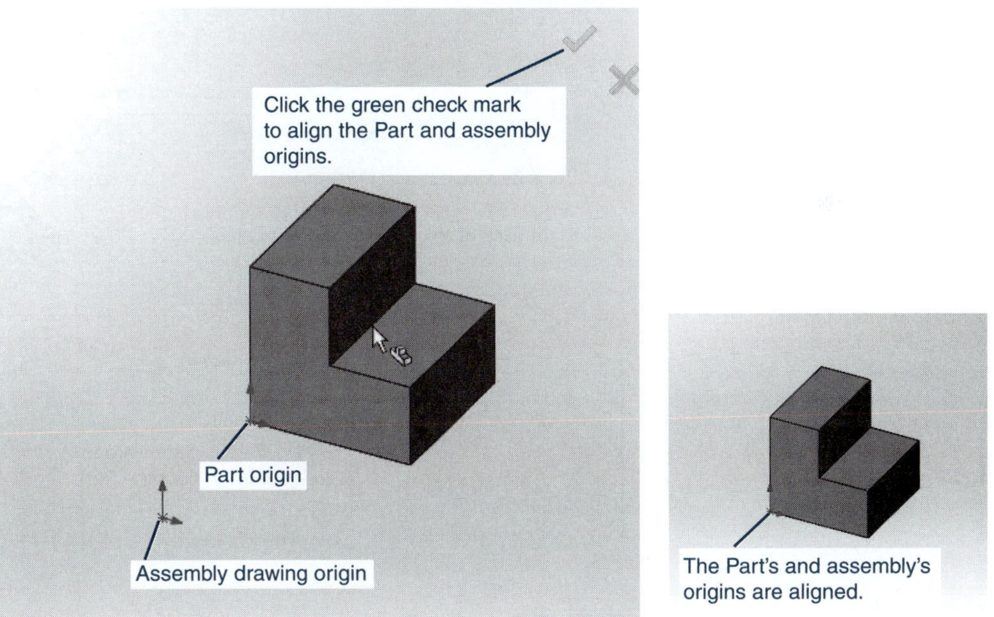
Click the green check mark to align the Part and assembly origins.
Part origin
Assembly drawing origin
The Part's and assembly's origins are aligned.

NOTE

It is good practice to anchor the assembly to the assembly drawing's origin.

Align the origin of the Test Block with the assembly drawing's origin.

6 When the Test Block appears on the screen, click the **green check mark** in the upper right corner of the drawing screen.

The Part's origin and the assembly drawing's origin will align.

7 Click the **Insert Components** tool, click the **Browser box**, select the **BLOCK,TEST**, and insert a second block.

NOTE

Parts is an assembly drawing. Do not use hidden lines.

See Figure 5-7. The first block inserted is fixed in place. In any assembly drawing the first component will automatically be fixed in place. Note the **(f)** notation to the left of **BLOCK,TEST<1>**. See Figure 5-7. As the assembly is created, components will move to the fixed first component.

TIP

To remove the fixed condition, locate the cursor on the **(f) BLOCK,TEST<1>** callout, right-click the mouse, and click the **Float** option. To return the block to the fixed condition or to fix another component, right-click the component name callout and select the **Fix** option.

Figure 5-7

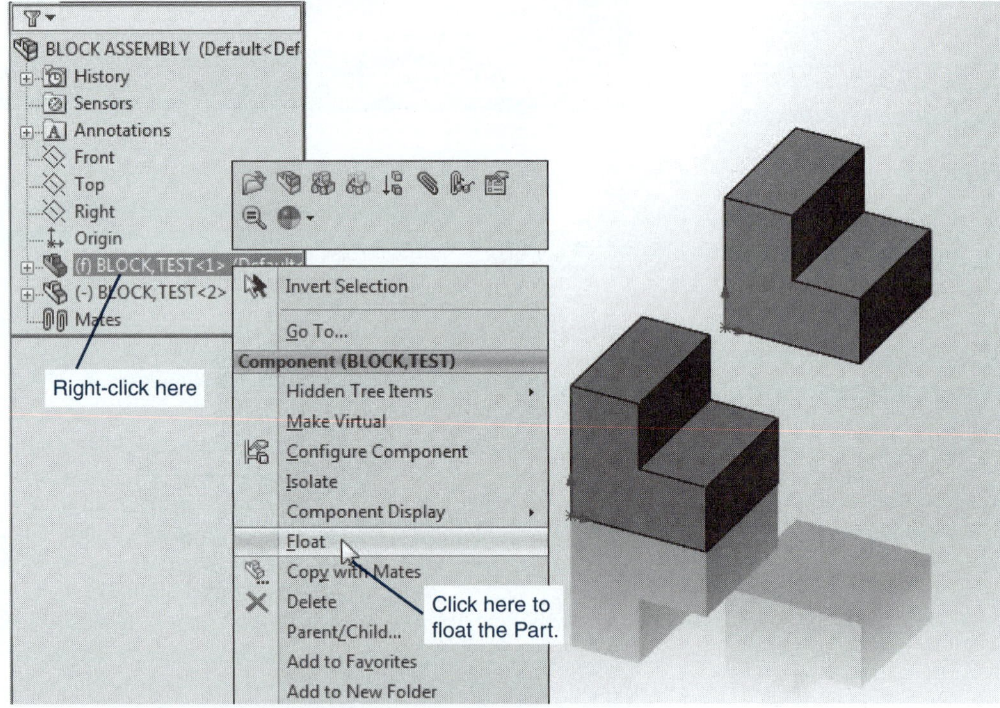

5-3 Move Component

See Figure 5-8.

1 Click the **Move Component** tool.

2 Click the second block inserted and hold down the left mouse button.

3 While holding down the left button, move the block around the screen by moving the mouse.

4 Release the button and click the green **OK** check mark.

Figure 5-8

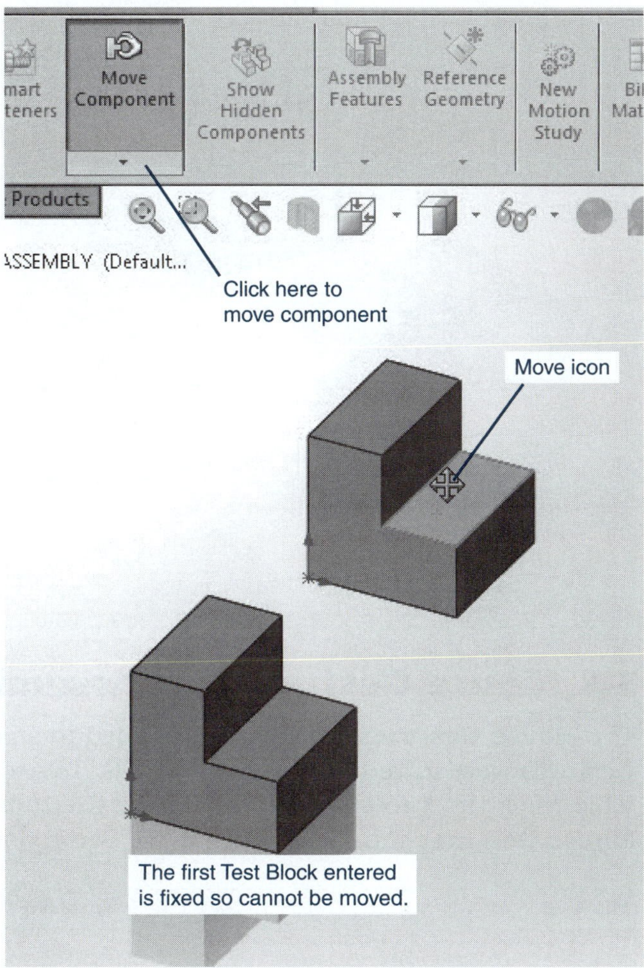

NOTE

If you try to apply the **Move** tool to the fixed block, the block will not move. The **Float** tool must be applied before the block can be moved.

The entire assembly can be moved by pressing and holding the **<Ctrl>** key down and holding down the mouse wheel while moving the cursor. The mouse wheel can be used to zoom the assembly.

5-4 Rotate Component

See Figure 5-9.

1. Click the **Rotate Component** tool.

   The **Rotate Component** tool is a flyout from the **Move Component** tool.

2. Click the second block inserted and hold down the left mouse button.
3. While holding down the left button, rotate the block around the screen by moving the mouse.
4. Release the button and click the green **OK** check mark.
5. Use the **Undo** tool to return the block to its original position.

Figure 5-9

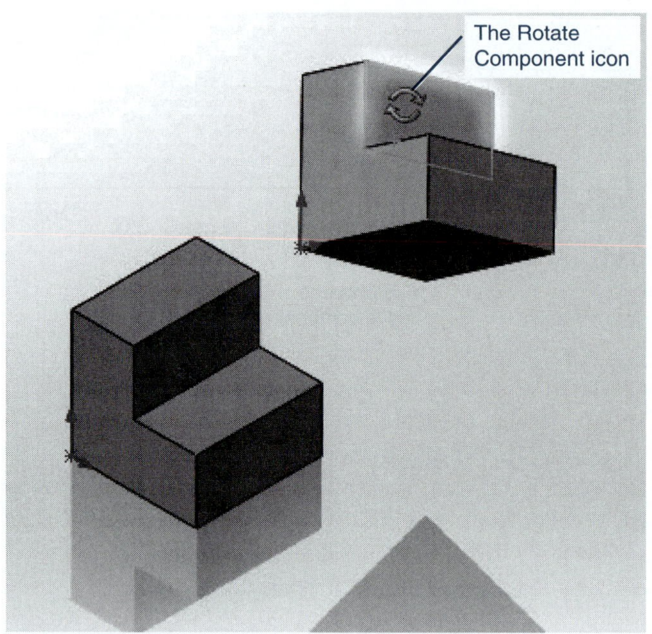

5-5 Mouse Gestures for Assembly Drawings

The **Mouse Gestures** tools can be applied to assembly drawings. Mouse Gestures were introduced in Section 2-2. To access the **Mouse Gestures** tools, click the **Tools** tab and select the **Customize** option. Click the **Mouse Gestures** tab. See Figure 5-10. Scroll down the listings, watching

Figure 5-10

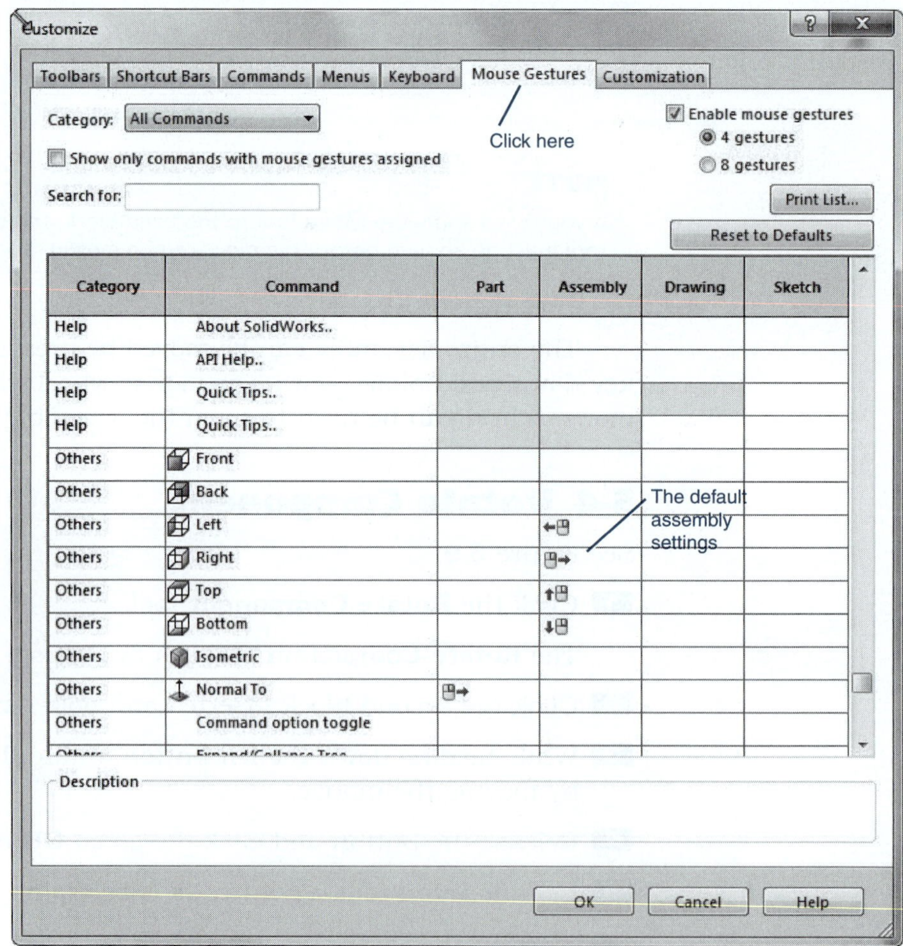

the Assembly column. The default settings for Assembly drawings are the Left, Right, Top, and Bottom orientation tools. These may be changed as preferred. Up to eight tools may be added to the **Mouse Gestures** tool.

To access the mouse gestures, press and hold the right mouse button. The **mouse gesture** wheel will show the selected settings. While still holding the right mouse button down, move the cursor onto the selected tool and release the button. The selected tool will be applied to the assembly. See Figure 5-11.

Figure 5-11

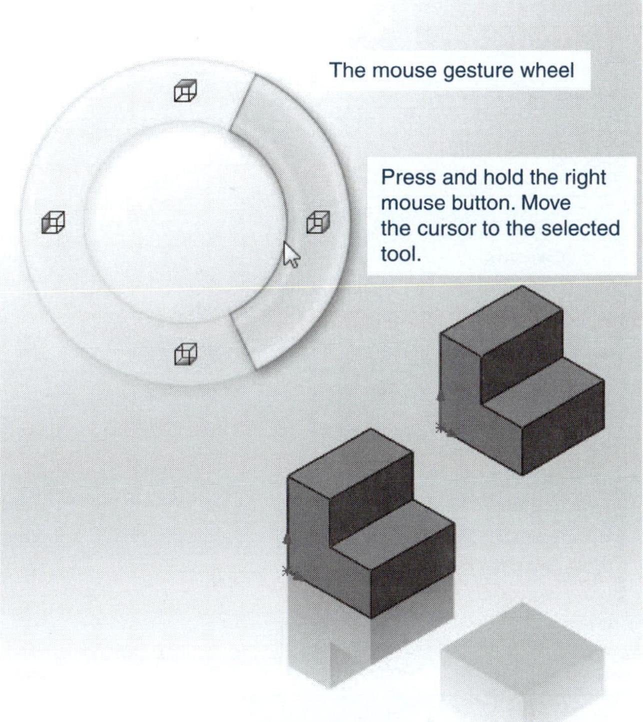

5-6 Mate

The **Mate** tool is used to align components to create assembly drawings. See Figure 5-12.

To Create the First Assembly

Mate the two test blocks side by side.

1. Click the **Mate** tool.
2. Click the upper right edge of the first block inserted.

   See Figure 5-13. Note that the first block is listed in the **Mate Selections** box after it is selected and that the **Coincident** option became active automatically.

3. Click the upper left edge on the second block.

Figure 5-12

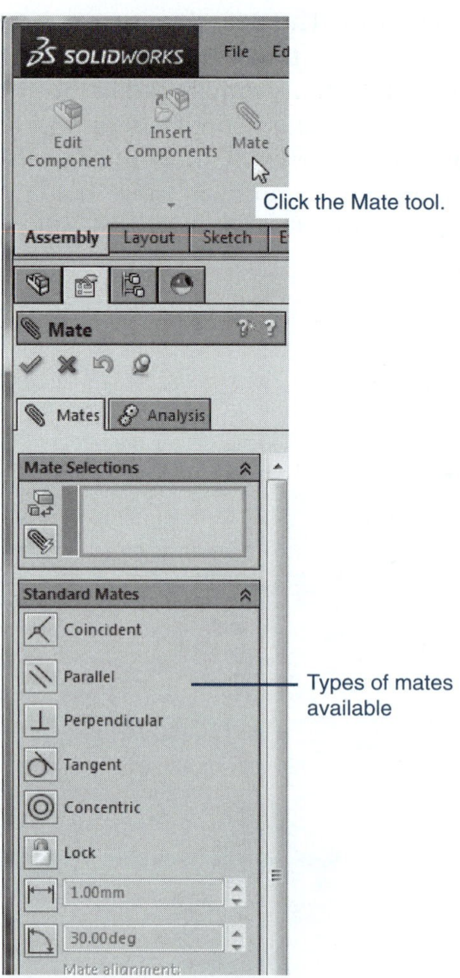

Figure 5-13

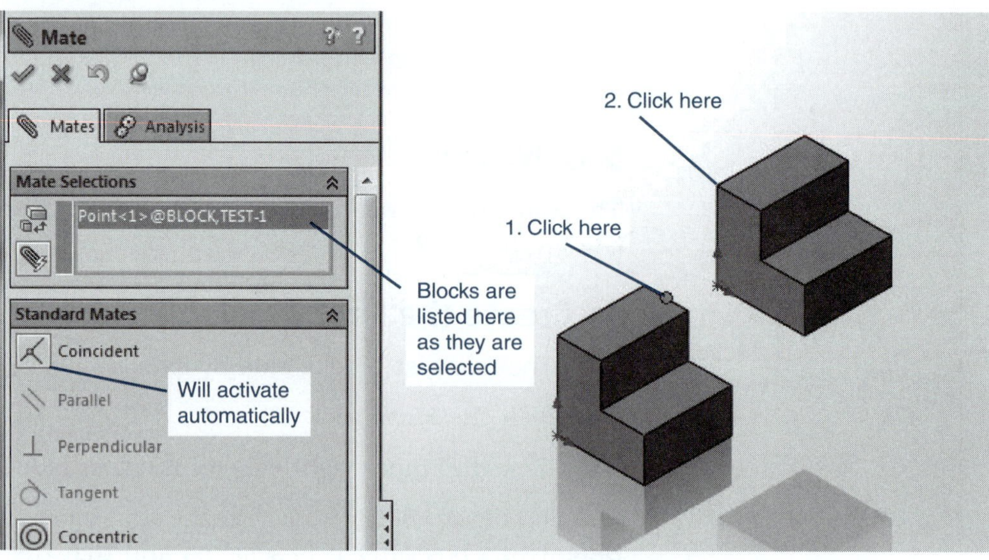

The edges will align. See Figure 5-14.

4 Click the green **OK** check mark to clear the tools.

The **Mate Selections** box should be clear.

Figure 5-14

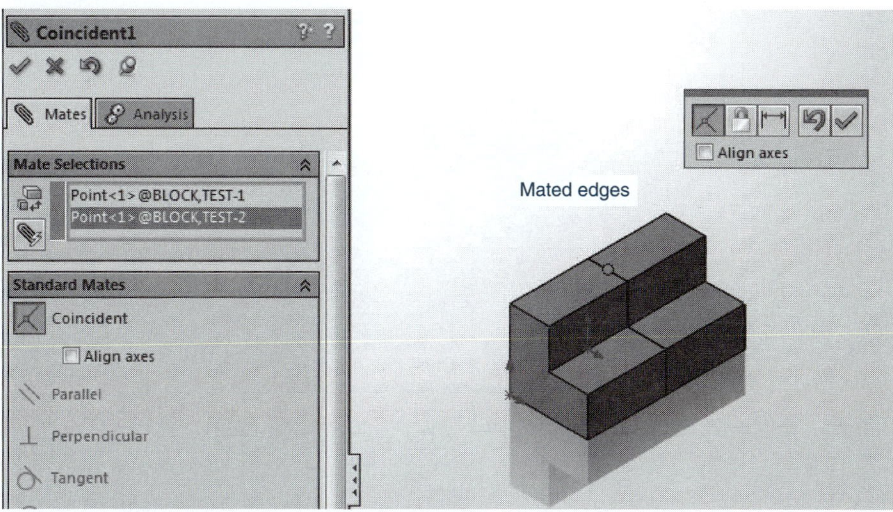

To Create a Second Assembly

Mate the two test blocks face-to-face.

1 Use the **Undo** tool and return the blocks to their original positions.

See Figure 5-15. The second test block must be rotated into a different position relative to the first test block.

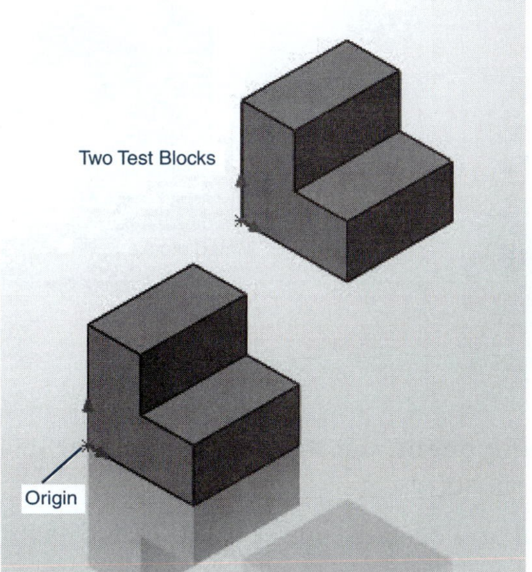

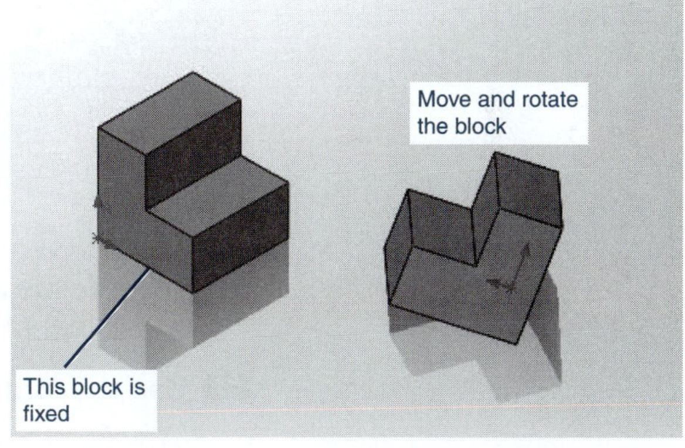

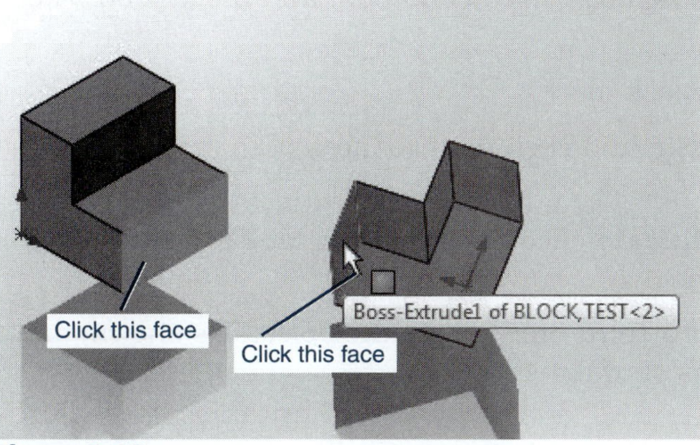

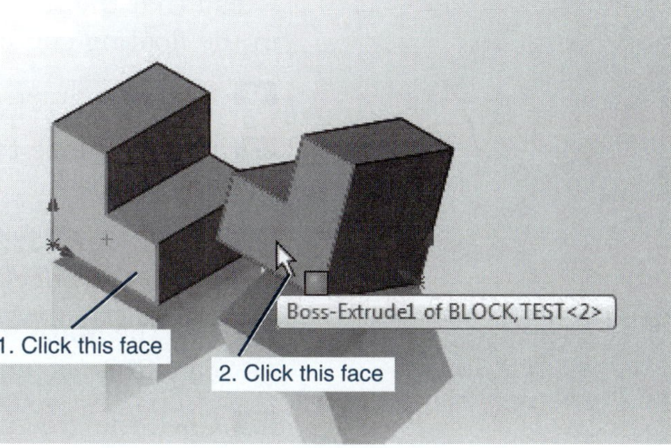

Figure 5-15

Figure 5-15
(continued)

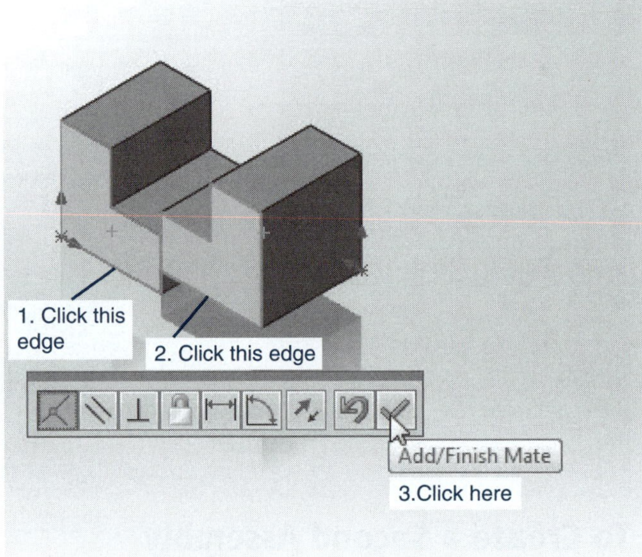

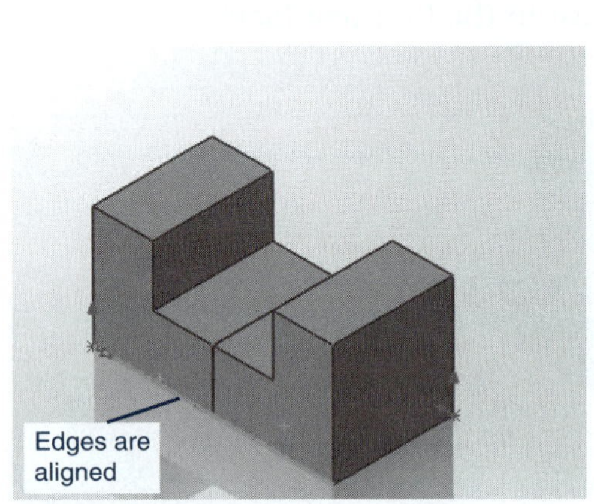

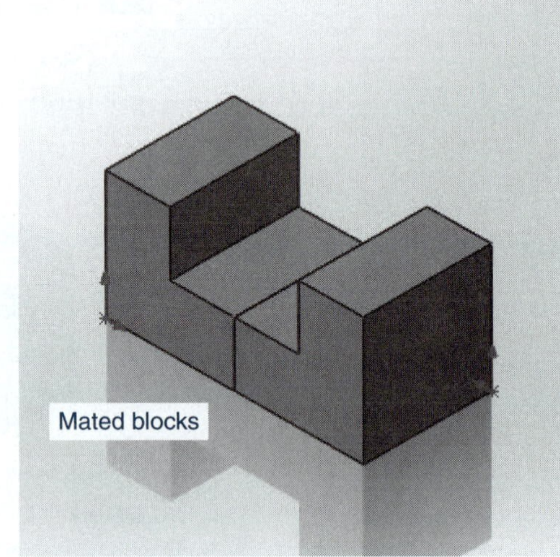

2 Use the **Move Component** and **Rotate Component** tools and position the second block as shown.

3 Use the **Mate** tool and click the lower front face of each block.

The second block will rotate relative to the first block. See Figure 5-16. Recall that the first block is in the fixed condition, and the second block is in the floating condition.

4 Click the green **OK** check mark.

5 Use the **Mate** tool again and click the two faces of the test block as shown.

6 Click the **Add/Finish Mate** check mark on the toolbar, then click the green **OK** check mark.

7 Use the **Mate** tool for a third time, rotate the objects as shown, and click the two edges as shown.

8 Click the green **OK** check mark.

Figure 5-16

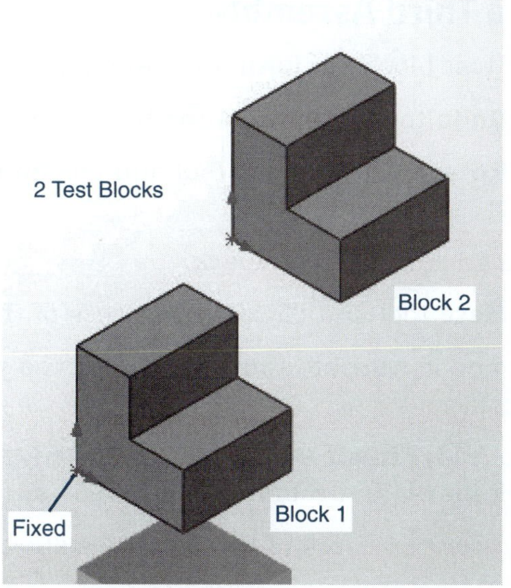
2 Test Blocks
Block 2
Fixed
Block 1

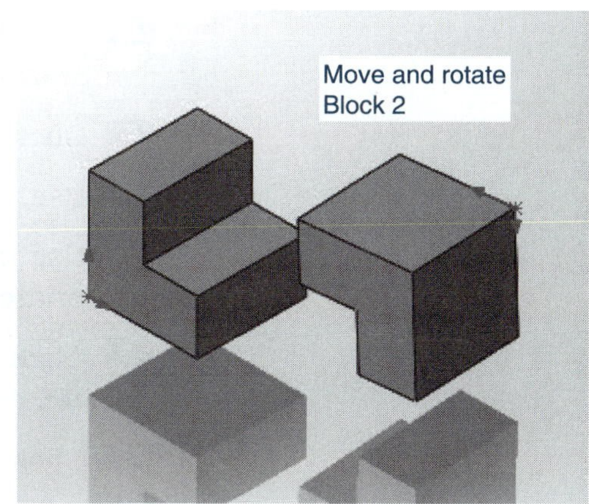
Move and rotate
Block 2

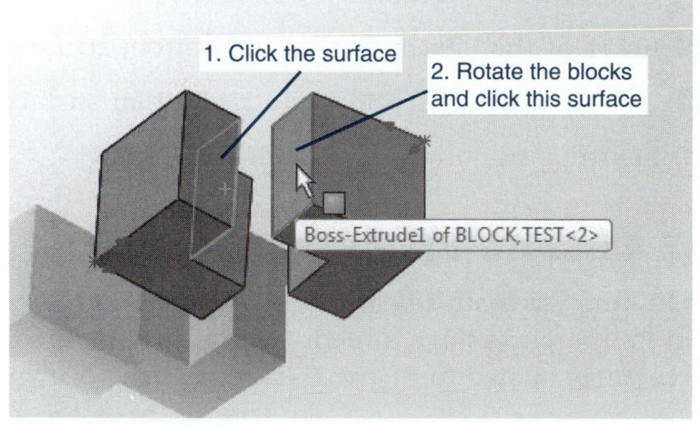
1. Click the surface
2. Rotate the blocks
and click this surface
Boss-Extrude1 of BLOCK,TEST<2>

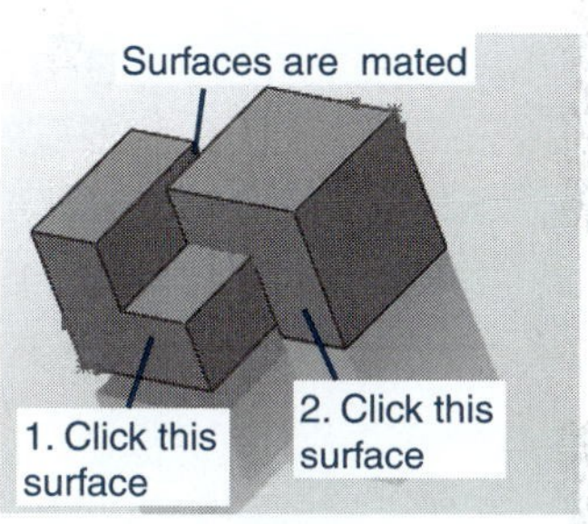
Surfaces are mated
1. Click this
surface
2. Click this
surface

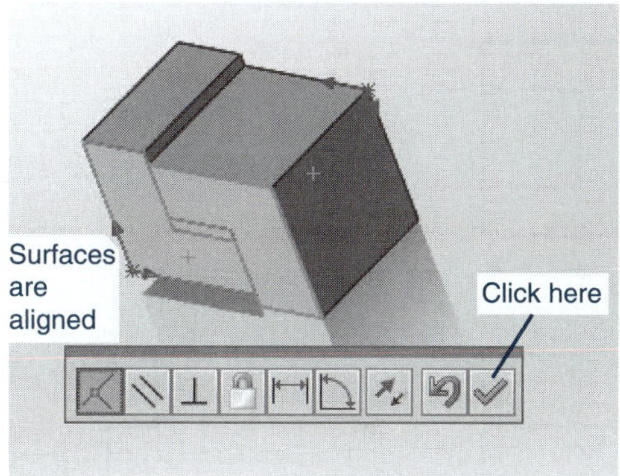
Surfaces
are
aligned
Click here

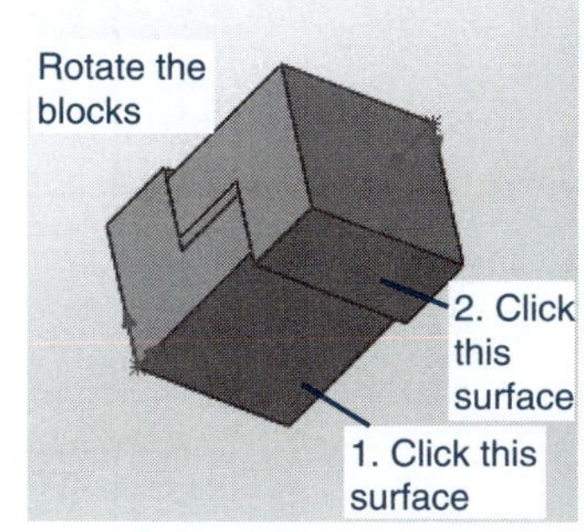
Rotate the
blocks
2. Click
this
surface
1. Click this
surface

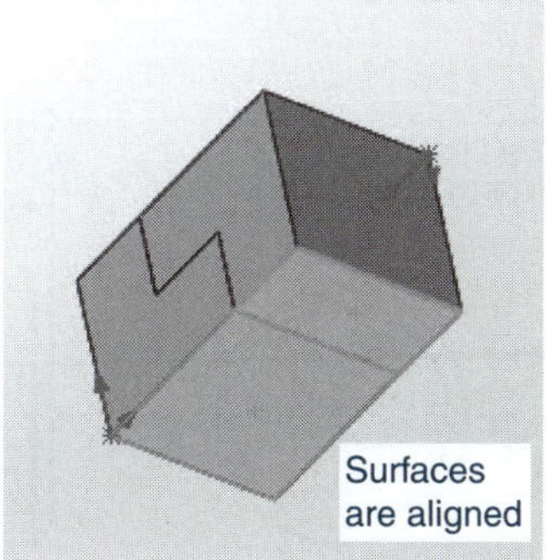
Surfaces
are aligned

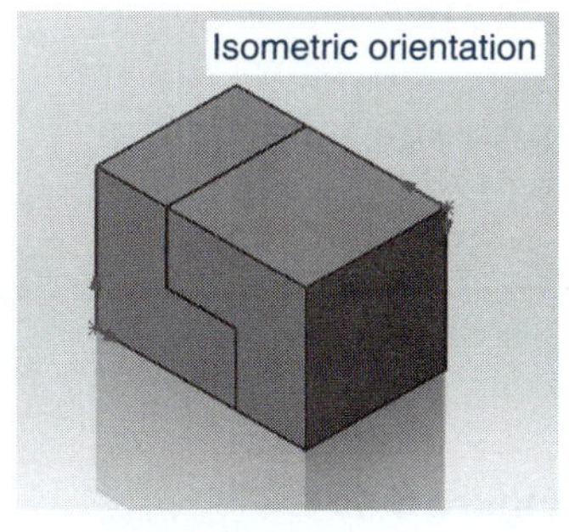
Isometric orientation

To Create a Third Assembly

Mate the two test blocks to form a rectangular prism.

1. Use the **Undo** tool and return the blocks to their original positions.
2. Use the **Rotate Component** tool and position the second test block as shown.
3. Click the green **OK** check mark.
4. Use the **Mate** tool and click the surfaces of the blocks as shown.

   Use the mouse wheel to rotate the blocks so you can click the appropriate surfaces.
5. Click the **Add/Finish Mate** check mark on the toolbar, click the green **OK** check mark, and rotate the blocks as shown.
6. Click the front surfaces of the blocks as shown.
7. Click the **Add/Finish Mate** check mark on the toolbar, then click the green **OK** check mark.
8. Reorient the blocks so that the bottom surfaces are visible.
9. Use the **Mate** tool and click the bottom surfaces of the blocks.

   Figure 5-16 shows the finished assembly.

5-7 Bottom-up Assemblies

Bottom-up assemblies are assemblies that are created for existing parts; that is, the parts have already been drawn as models. In this example the three parts shown in Figure 5-17 have been drawn and saved.

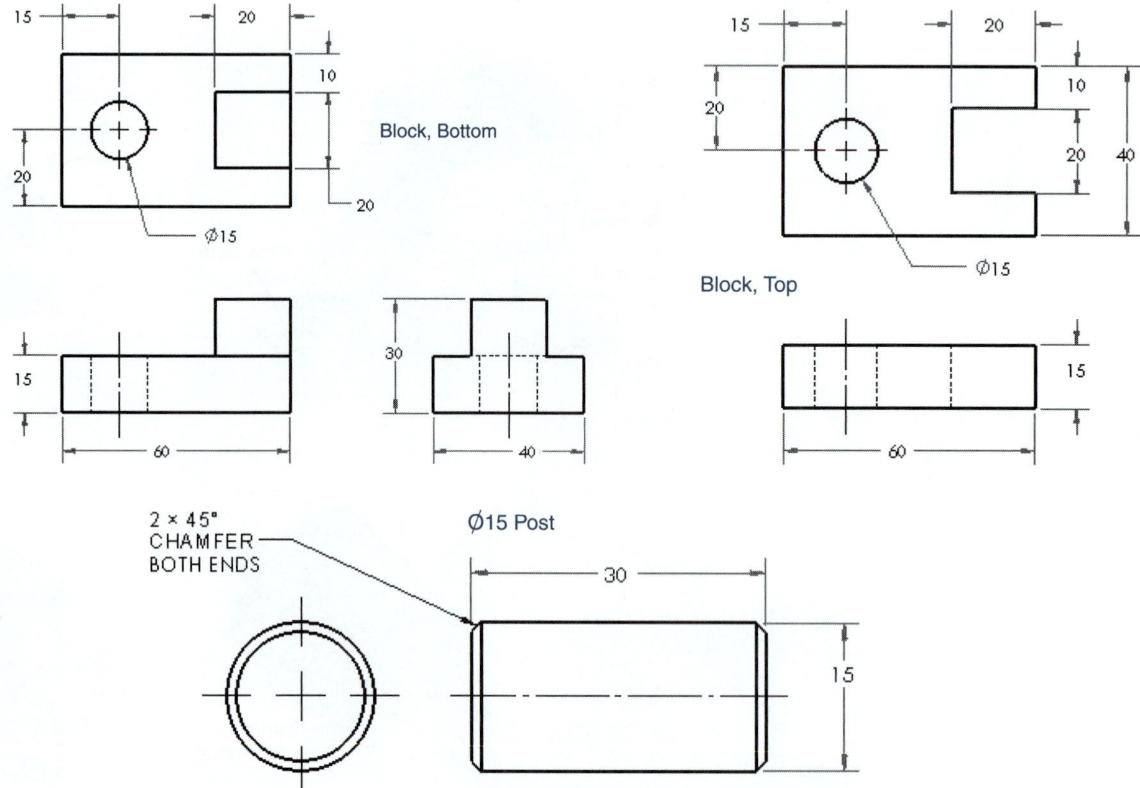

Figure 5-17

1 Start a new drawing and select the **Assembly** format.

2 Click the **Browse. . .** box, then click the **Block, Bottom** component.

See Figure 5-18. The Block, Bottom will appear on the screen.

Figure 5-18

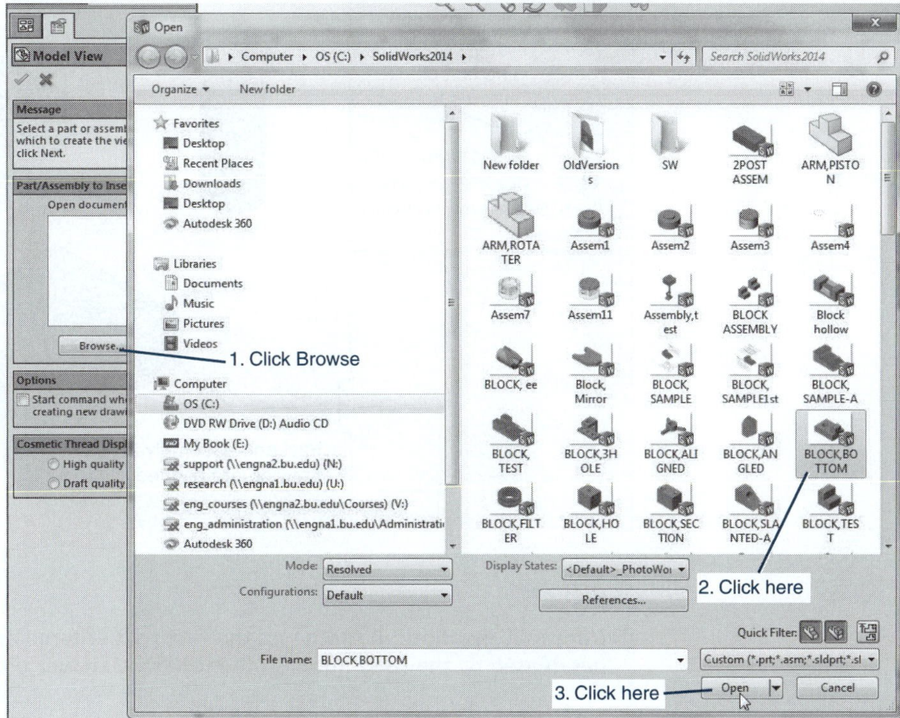

3 Click the **green check mark** in the upper right corner of the drawing screen to align the origin of the block with the origin of the assembly drawing.

See Figure 5-19.

Figure 5-19

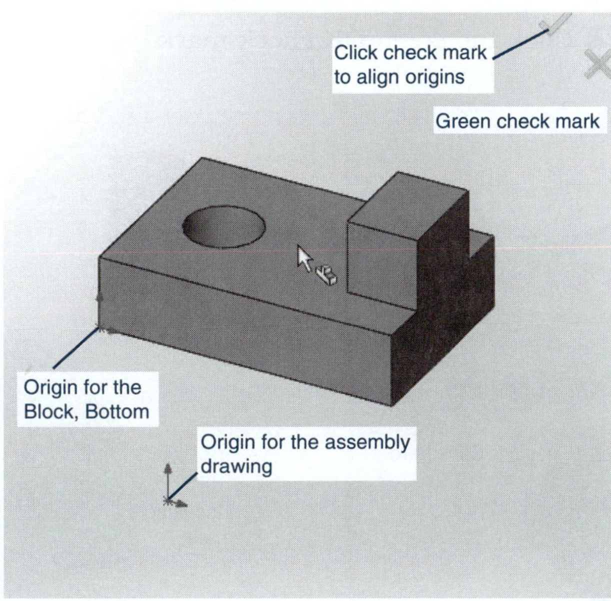

4 Click the **Insert Component** tool, click the **Browse. . .** box, and select **Block, Top**.

5 Repeat the sequence and select **Ø15 Post**.

See Figure 5-20.

Figure 5-20

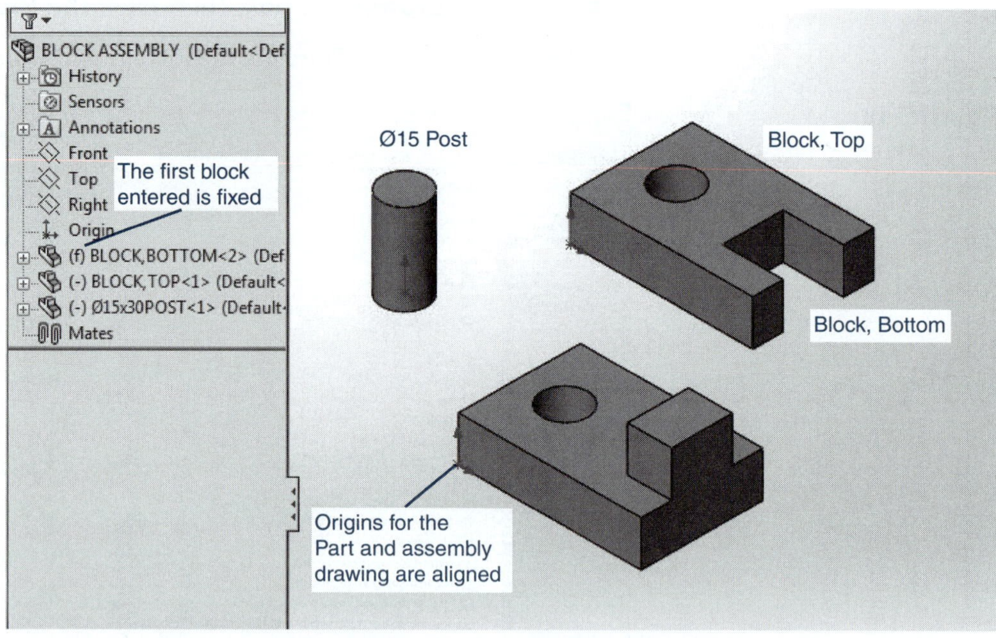

> **NOTE**
>
> Note that the **Block, Bottom** was the first part entered and is fixed in its location, as designated by the **(f)** symbol in the **PropertyManager** box.

6 Use the **Mate** tool and click the center point of the edge line of the **Block, Bottom** and the **Block, Top** as shown.

See Figure 5-21. A dot will appear when the cursor is on the centerpoint of the edge. If the blocks do not align, use the **Mate** tool again to align the blocks by clicking their end surfaces.

7 Click the green **OK** check mark.

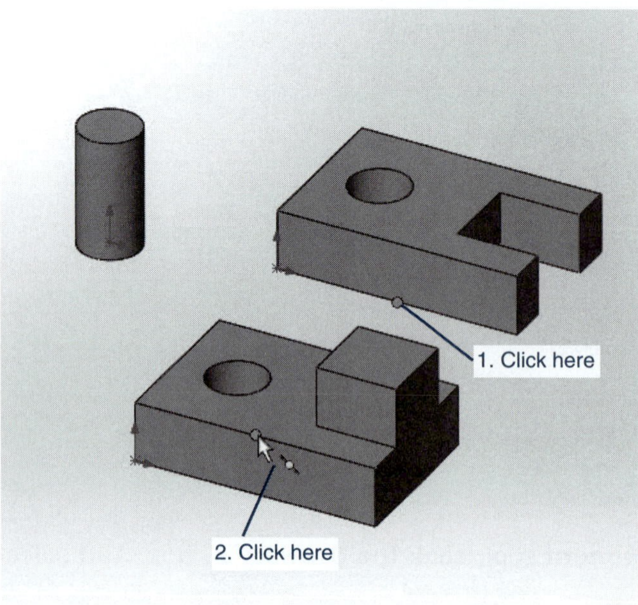

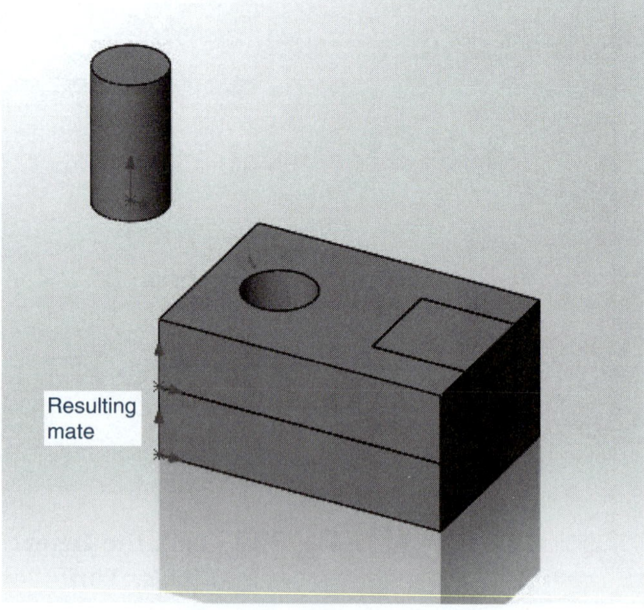

Figure 5-21

8 Click the **Mate** tool and click the **Concentric** option.

See Figure 5-22.

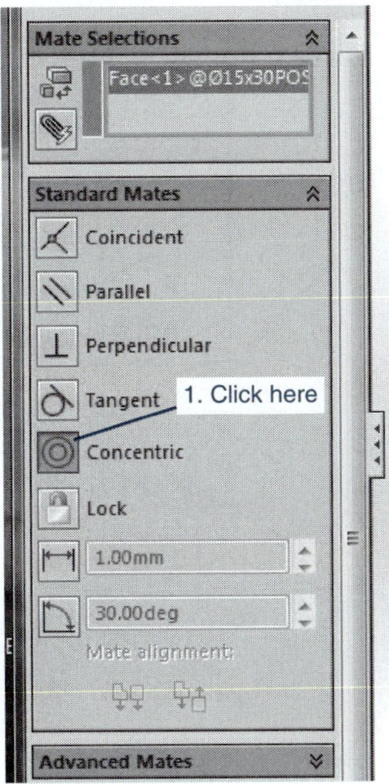

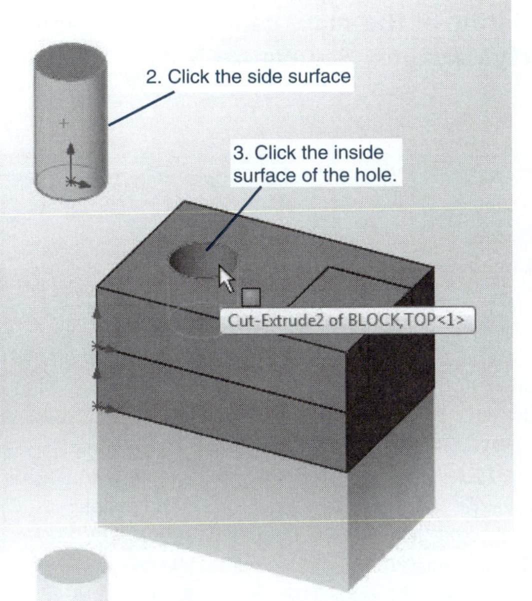

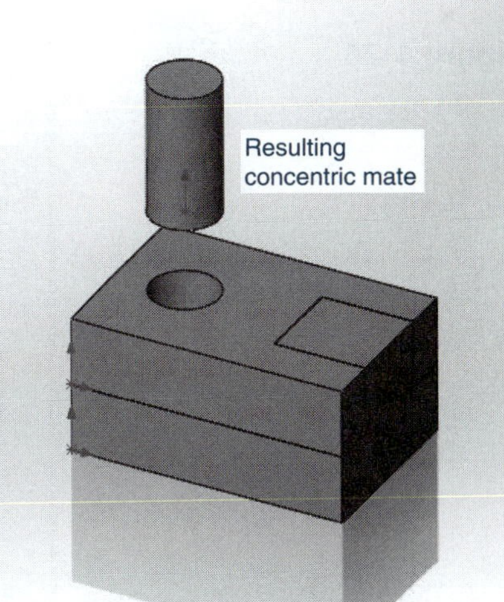

Figure 5-22

9 Click the side of the **Ø15 Post** and the inside of the hole in the **Block, Top**.

> **NOTE**
>
> Click the surface of the components, not their edge lines. Clicking the edge lines would produce different results.

10 Click the **Add/Finish Mate** check mark on the toolbar, then click the green **OK** check mark.

11 Use the **Mate** tool and click the top surface of the **Ø15 Post** and the top surface of the **Block, Top**.

See Figure 5-23.

Figure 5-23

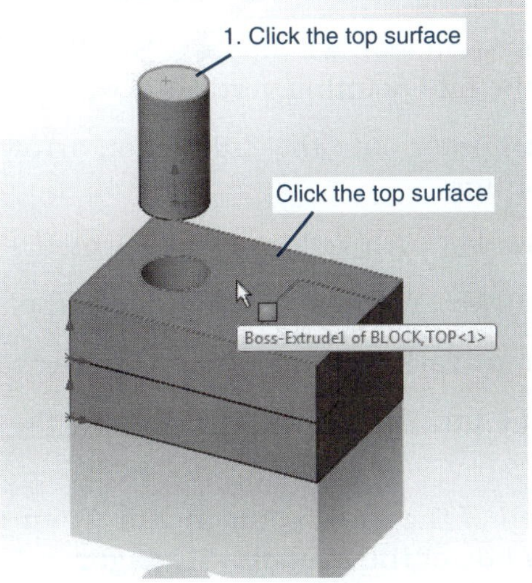

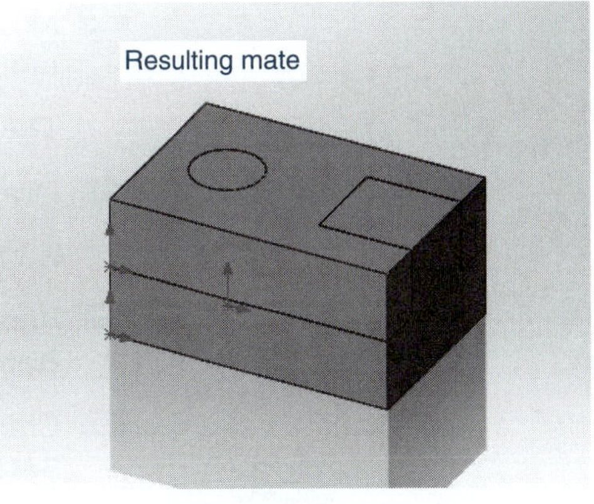

12 Click the **Add/Finish Mate** check mark on the toolbar, then click the green **OK** check mark.

13 Save the assembly as **Block Assembly**.

A listing of the mate can be seen by clicking the **Mates** heading in the **PropertyManager**. See Figure 5-24.

Figure 5-24

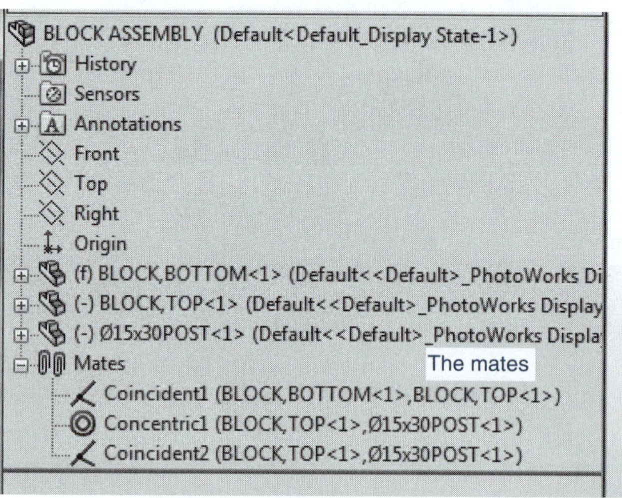

The origin icons can be removed from view using the **View, Origins** tools. For examples in this book the origins will be shown.

5-8 Creating an Exploded Isometric Assembly Drawing

An exploded isometric assembly drawing shows the components of an assembly pulled apart. This makes it easier to see how the parts fit together.

1 Click the **Exploded View** tool.

See Figure 5-25.

2 Click the top surface of the **Ø15 Post**.

An axis system icon will appear. See Figure 5-26. The arrow in the Z-direction (the one pointing vertically) will initially be green.

3 Move the cursor onto the Z-direction arrow and hold down the left mouse button.

The arrow will turn yellow when selected.

4 Drag the Ø15 Post to a location above the assembly as shown.

A real-time scale will appear as you drag the post.

5 Repeat the procedure and drag the **Block, Top** away from the **Block, Bottom**.

6 Click each of the three components. Their names should appear in the **Settings** box of the **PropertyManager**.

Figure 5-25

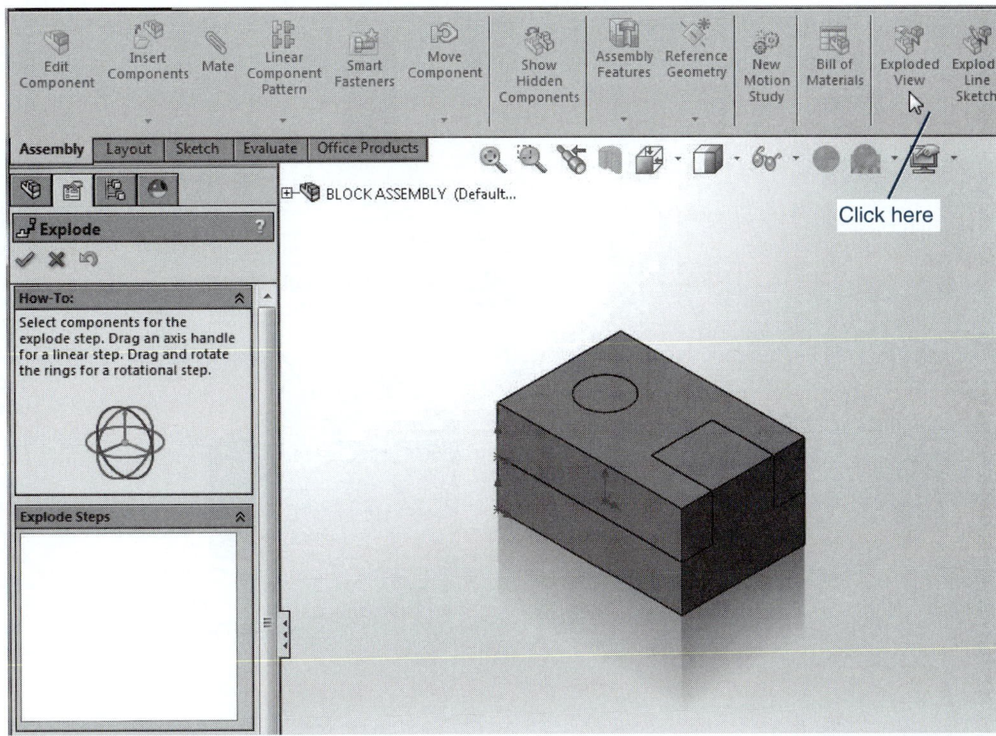
Edit Component
Insert Components
Mate
Linear Component Pattern
Smart Fasteners
Move Component
Show Hidden Components
Assembly Features
Reference Geometry
New Motion Study
Bill of Materials
Exploded View
Explode Line Sketch
Assembly
Layout
Sketch
Evaluate
Office Products
BLOCK ASSEMBLY (Default...
Explode
How-To:
Select components for the explode step. Drag an axis handle for a linear step. Drag and rotate the rings for a rotational step.
Explode Steps
Click here

Figure 5-26

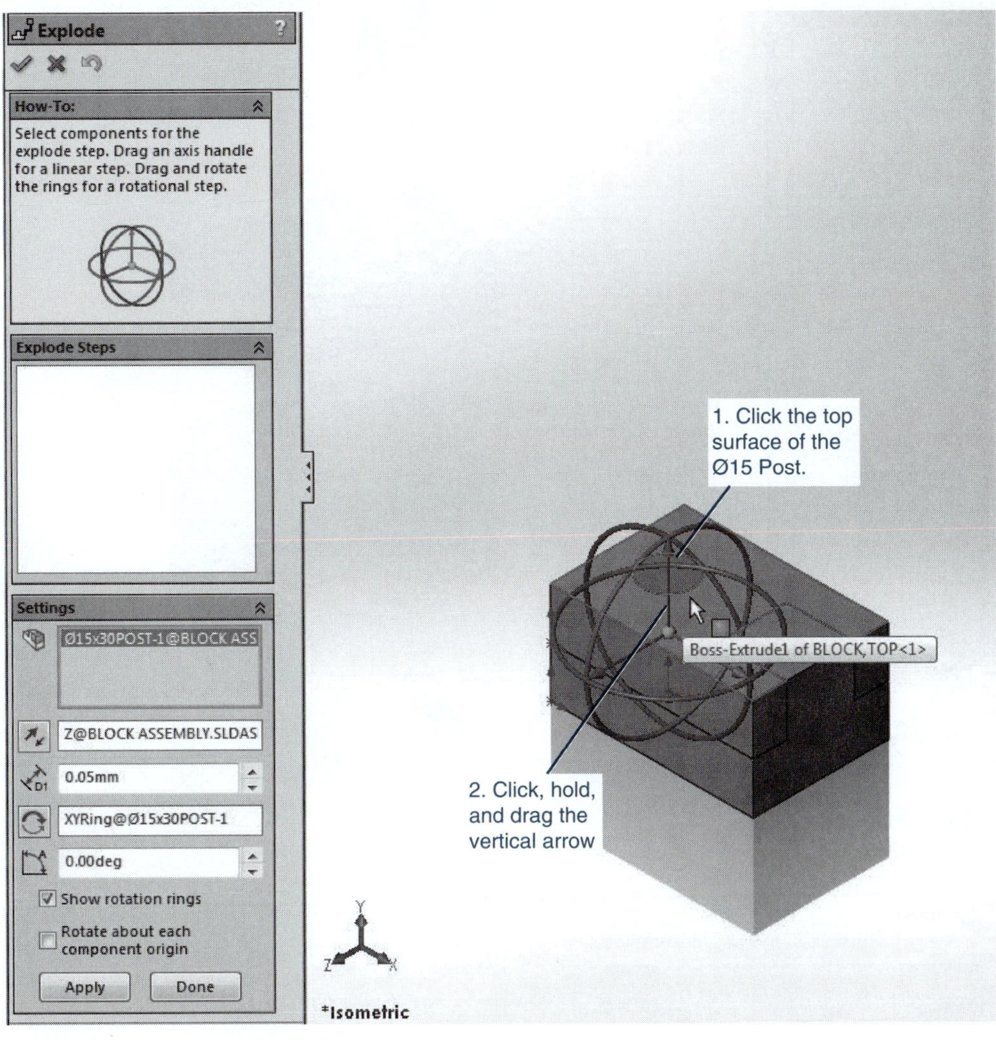
Explode
How-To:
Select components for the explode step. Drag an axis handle for a linear step. Drag and rotate the rings for a rotational step.
Explode Steps
Settings
Ø15x30POST-1@BLOCK ASS
Z@BLOCK ASSEMBLY.SLDAS
0.05mm
XYRing@Ø15x30POST-1
0.00deg
Show rotation rings
Rotate about each component origin
Apply
Done
1. Click the top surface of the Ø15 Post.
Boss-Extrude1 of BLOCK,TOP<1>
2. Click, hold, and drag the vertical arrow
*Isometric

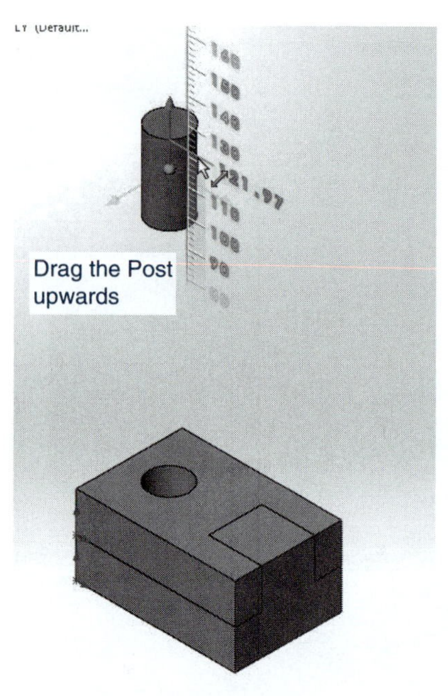
Drag the Post upwards

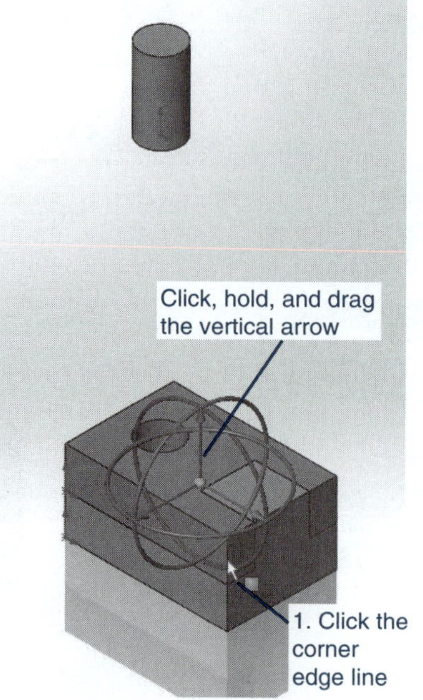
Click, hold, and drag the vertical arrow
1. Click the corner edge line

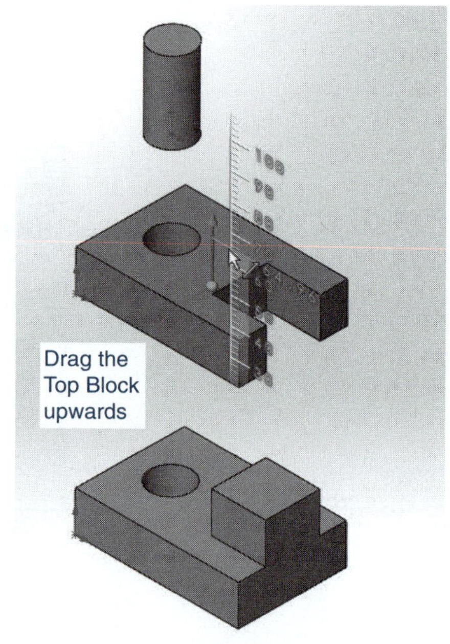
Drag the Top Block upwards

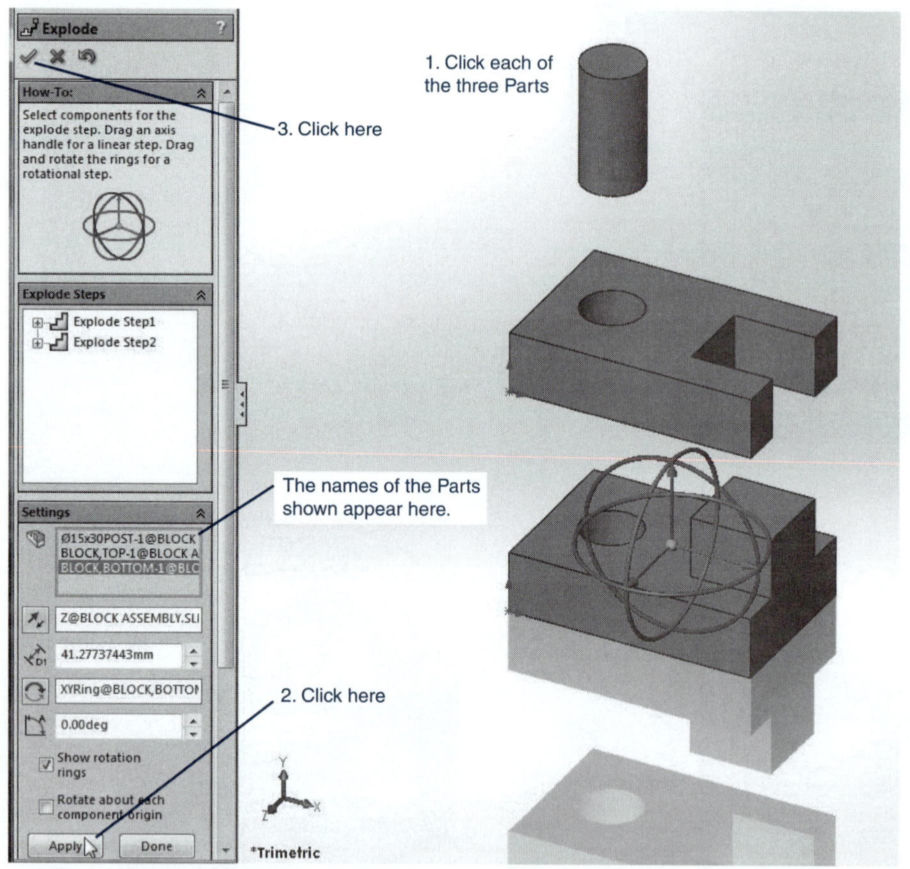
Explode
How-To:
Select components for the explode step. Drag an axis handle for a linear step. Drag and rotate the rings for a rotational step.
Explode Steps
Explode Step1
Explode Step2
Settings
Ø15x30POST-1@BLOCK
BLOCK,TOP-1@BLOCK A
BLOCK,BOTTOM-1@BLO
Z@BLOCK ASSEMBLY.SLI
41.27737443mm
XYRing@BLOCK,BOTTOI
0.00deg
Show rotation rings
Rotate about each component origin
Apply
Done
*Trimetric
1. Click each of the three Parts
3. Click here
The names of the Parts shown appear here.
2. Click here

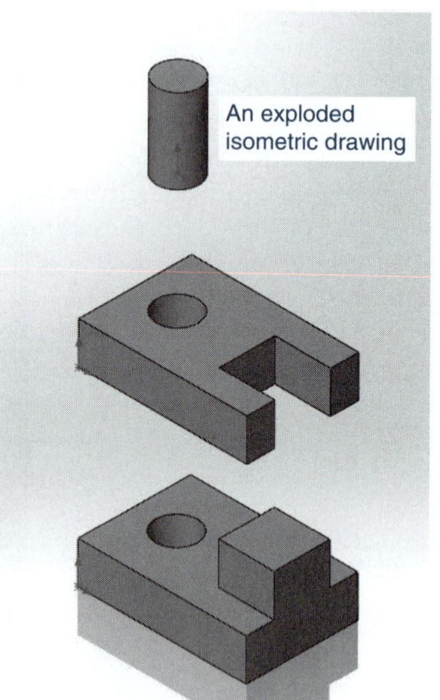
An exploded isometric drawing

Figure 5-26
(*Continued*)

7 Click the **Apply** box in the **Settings** box.

8 Click the **green check mark**.

9 Save the assembly as **Block Assembly**.

Figure 5-26 shows the final assembly.

5-9 Creating an Exploded Isometric Drawing Using the Drawing Format

1 Create a new drawing using the **Drawing** format.

2 Select the **A-(ANSI)Landscape** sheet format.

3 Click the **Browse. . .** box in the **Model View PropertyManager**.

See Figure 5-27.

4 Select **Block Assembly;** click **Open**.

5 Set the **Orientation** for **Isometric** and the **Display Style** for **Hidden Lines Removed**.

6 Move the cursor into the drawing area.

Figure 5-27

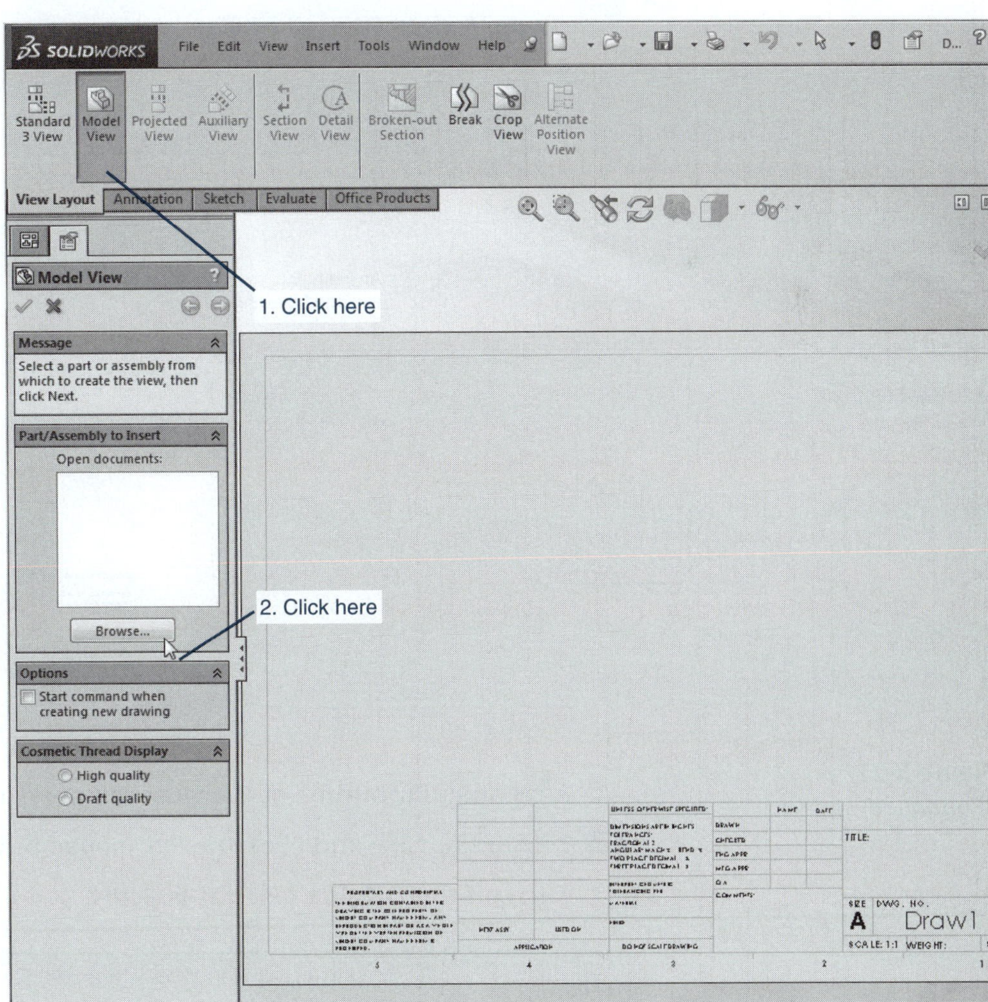

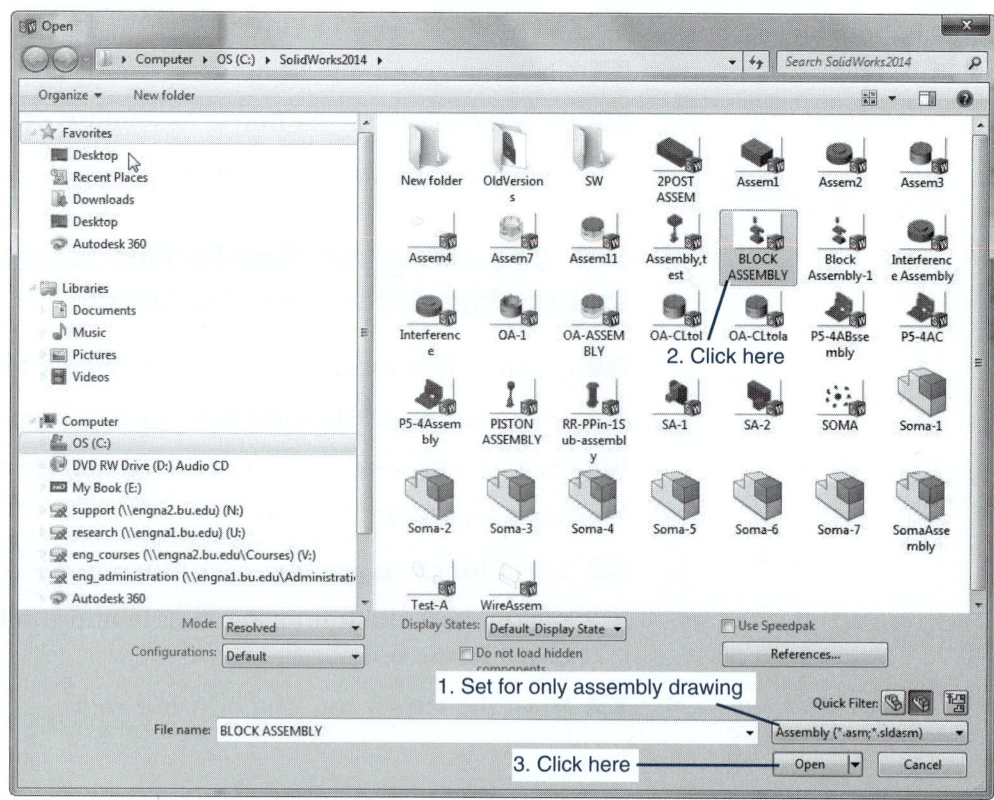

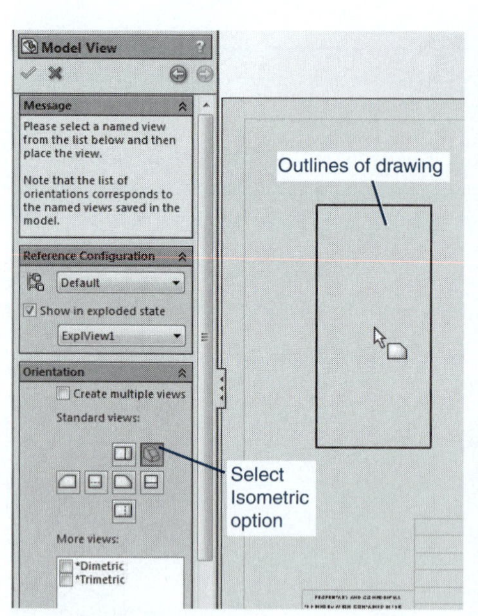

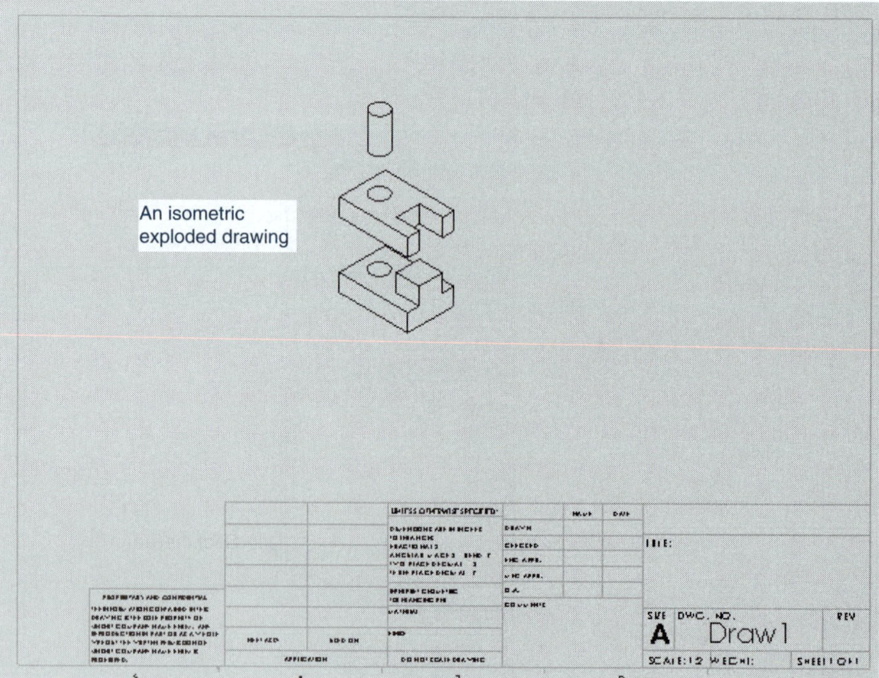

Figure 5-27
(Continued)

A rectangular outline of the view will appear.

7 Locate the view and click the left mouse button.

Turn off the **Origins** tool if it is active.

> **TIP**
> As a general rule, hidden lines are not included on isometric drawings.

5-10 Assembly Numbers

Assembly numbers are numbers that identify a part within an assembly. They are different from part numbers. A part number identifies a specific part, and the part number is unique to that part. A part has only one part number but may have different assembly numbers in different assemblies.

Assembly numbers are created using the **Balloon** and **Auto Balloon** tools located on the **Annotation** tool panel. See Figure 5-28.

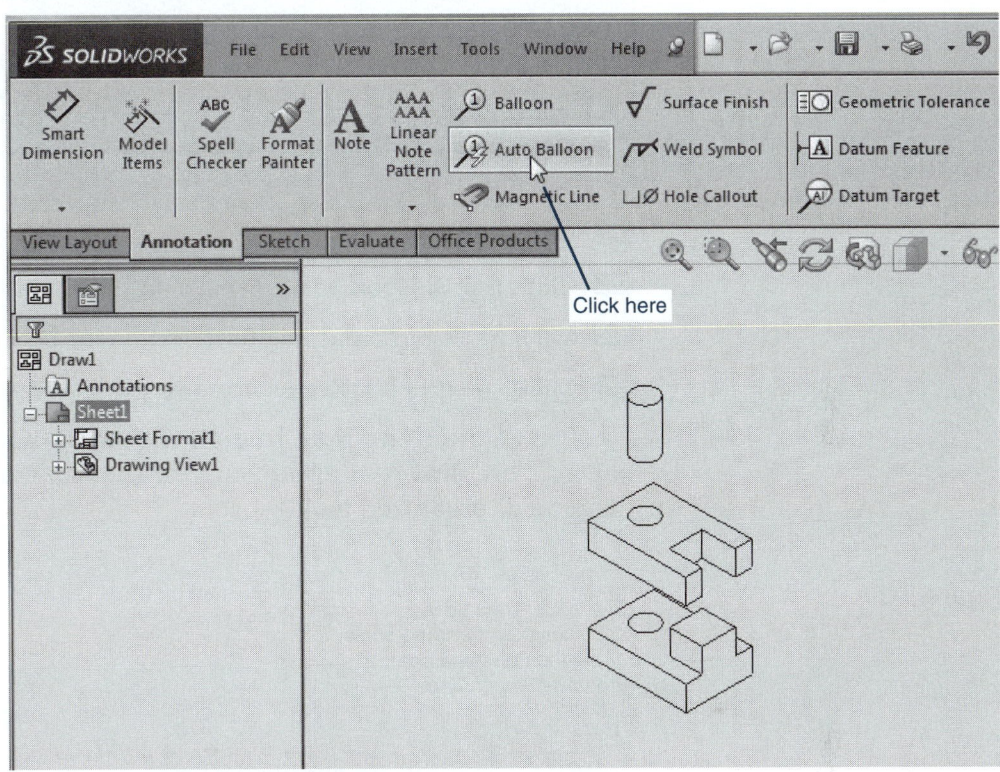

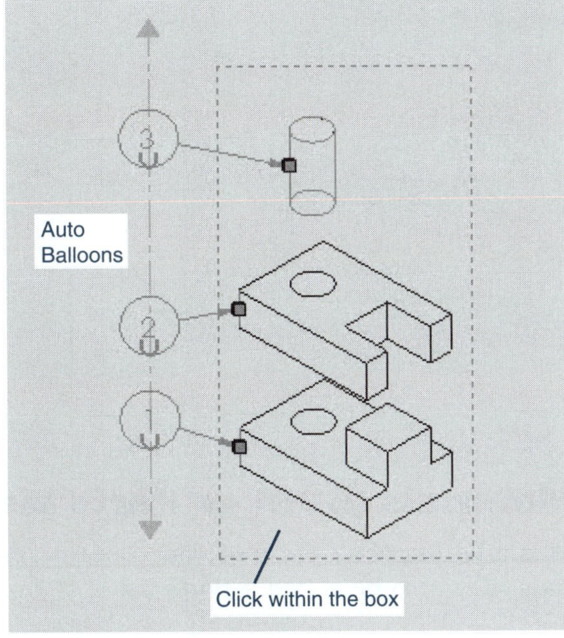

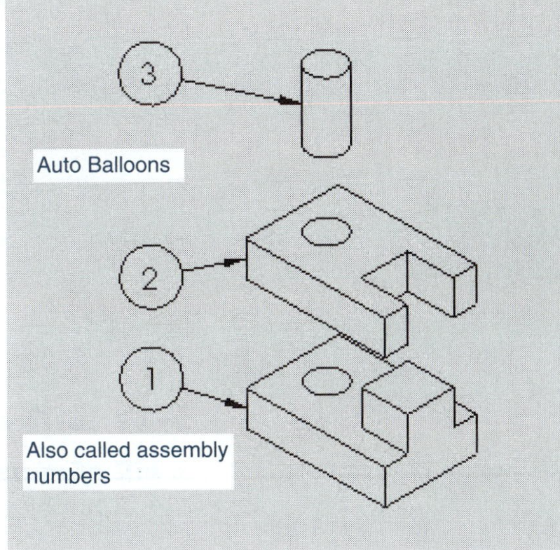

Figure 5-28

1. Click the **Annotation** tab.
2. Click the **Auto Balloon** tool.
3. Move the cursor into the **Block Assembly** area (a red outline will appear) and click the mouse.

The **Auto Balloon** arrangements may not always be the best presentation. The balloons may be rearranged or applied individually.

> **TIP**
>
> Balloons can be moved by first clicking them. They will change color. Click and hold either the balloon or the box that will appear on the arrow. Drag either the balloon or the arrow to a new location.

4. Undo the auto balloons.
5. Click the **Balloon** tool.
6. Click each part and locate the balloon.
7. Click the green **OK** check mark.

See Figure 5-29. Note that all the leader lines from the parts to the balloons are drawn at approximately the same angle. This gives the drawing a neat, organized look.

Figure 5-29

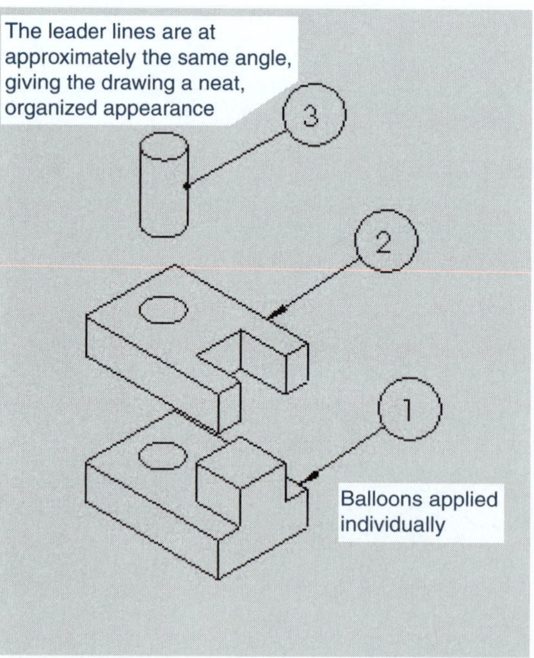

5-11 Bill of Materials (BOM or Parts List)

A ***bill of materials*** is a listing of all parts included in an assembly drawing. A bill of materials may also be called a *parts list*.

1 To access the **Bill of Materials** tool, click the **Annotation** tab, then the **Tables** and **Bill of Materials** tools.

See Figure 5-30.

2 Click the area of the **Block Assembly**.

A box will appear around the Block Assembly. Click within that box.

3 Click the green **OK** check mark.

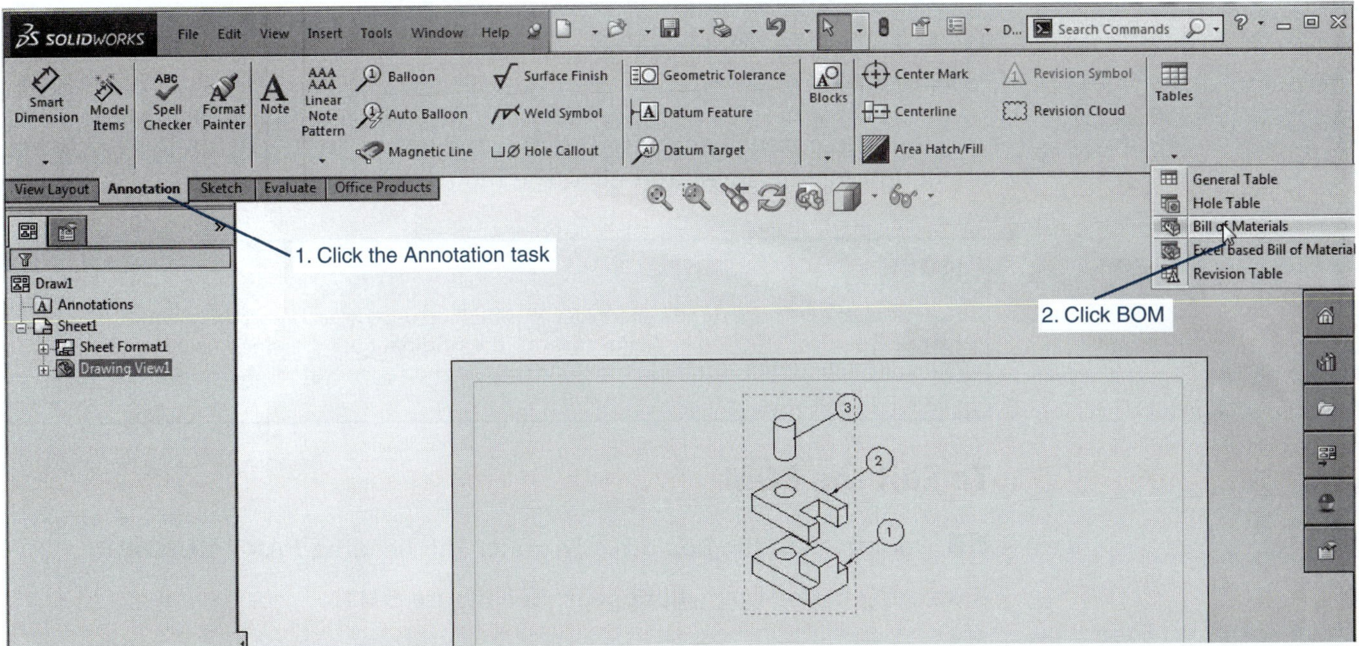

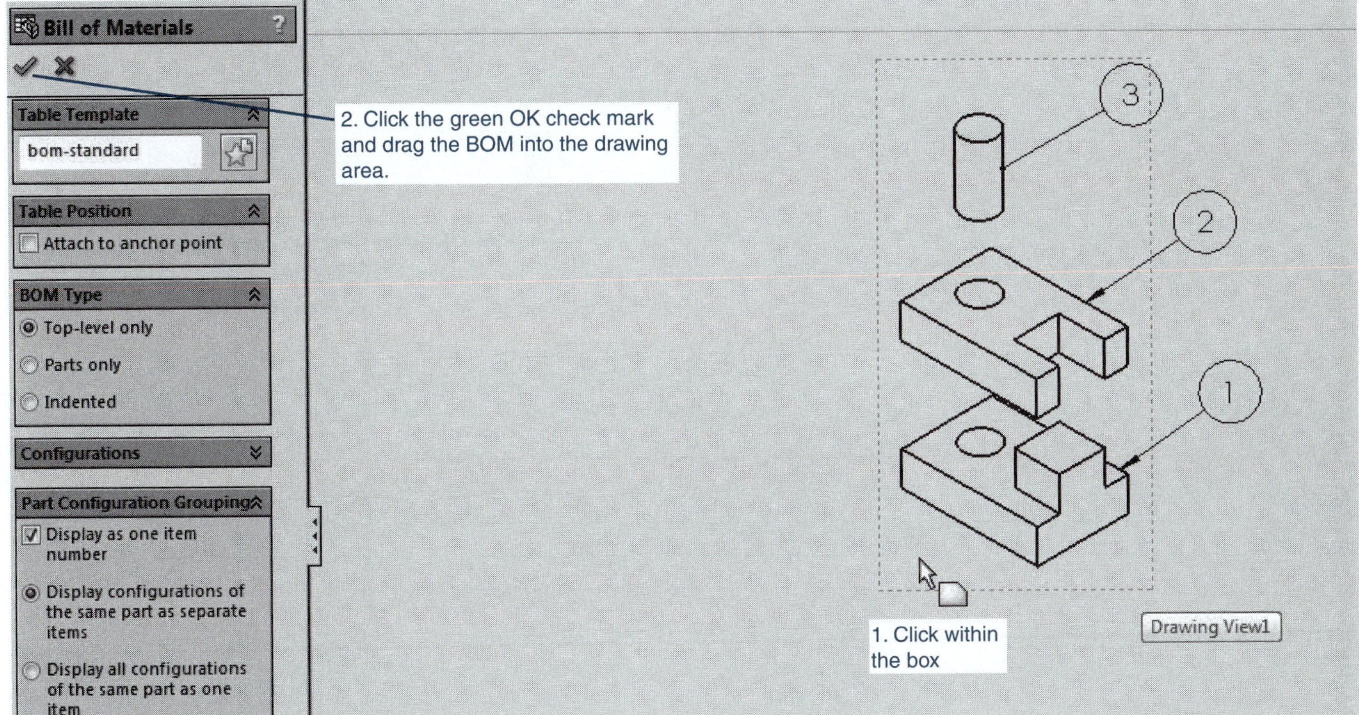

Figure 5-30

Figure 5-30
(*Continued*)

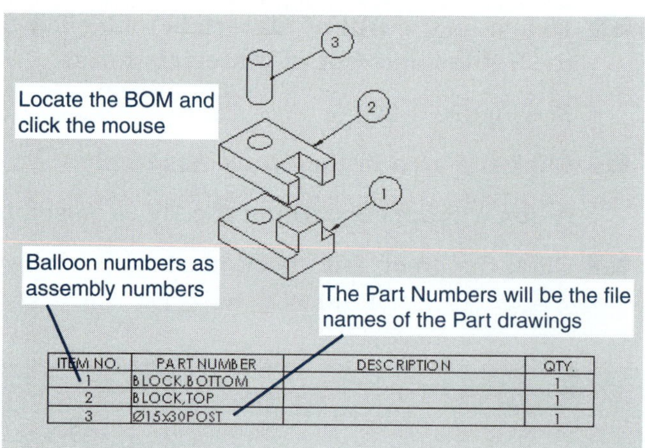
Locate the BOM and click the mouse
Balloon numbers as assembly numbers
The Part Numbers will be the file names of the Part drawings
ITEM NO.
PART NUMBER
DESCRIPTION
QTY.
BLOCK,BOTTOM
BLOCK,TOP
Ø15x30POST

Pull the cursor back into the drawing area. The BOM will follow. Select a location for the BOM and click the mouse.

> **NOTE**
>
> Note that the information in the **PART NUMBER** column is each part's file name. These names are directly linked by SolidWorks to the original part drawings. They can be manually edited, but if the assembly is changed and regenerated, the original part names will appear.

To Edit the BOM

1 Double-click the box directly under the heading **PART NUMBER**.

A warning dialog box will appear. See Figure 5-31.

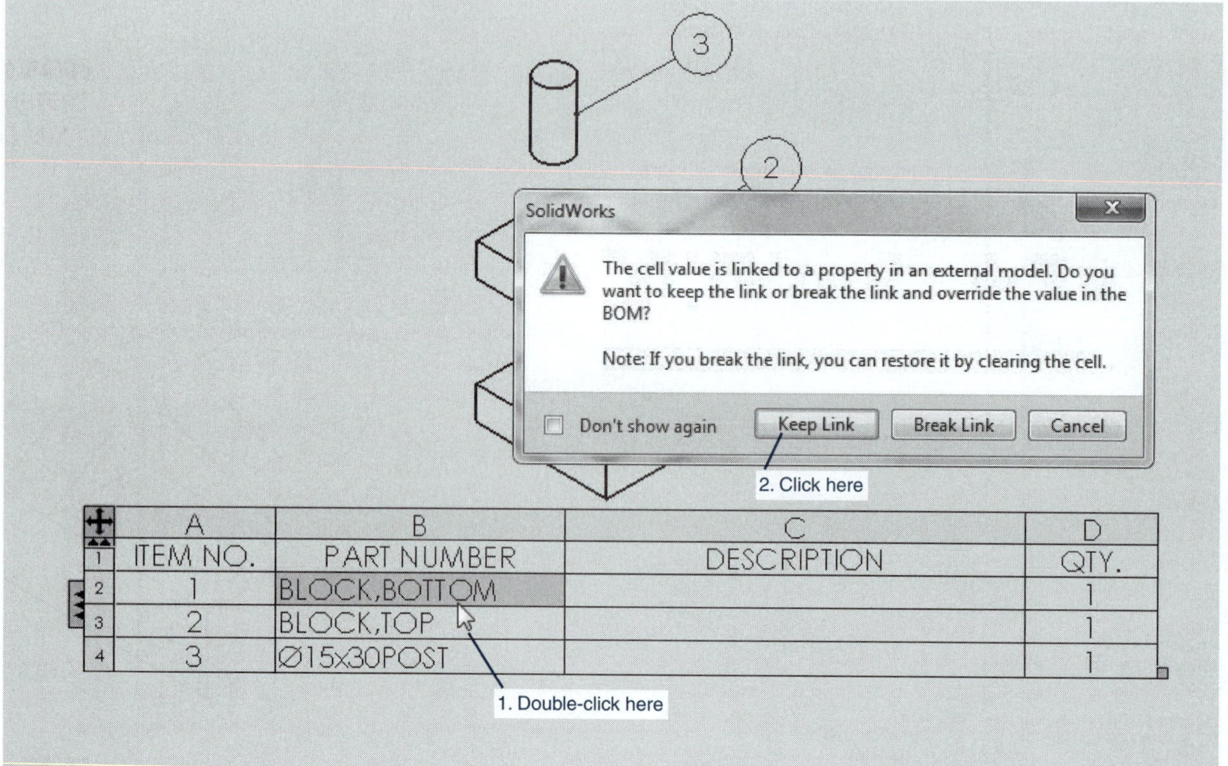
SolidWorks
The cell value is linked to a property in an external model. Do you want to keep the link or break the link and override the value in the BOM?
Note: If you break the link, you can restore it by clearing the cell.
Don't show again
Keep Link
Break Link
Cancel
2. Click here
ITEM NO.
PART NUMBER
DESCRIPTION
QTY.
BLOCK,BOTTOM
BLOCK,TOP
Ø15x30POST
1. Double-click here

Figure 5-31

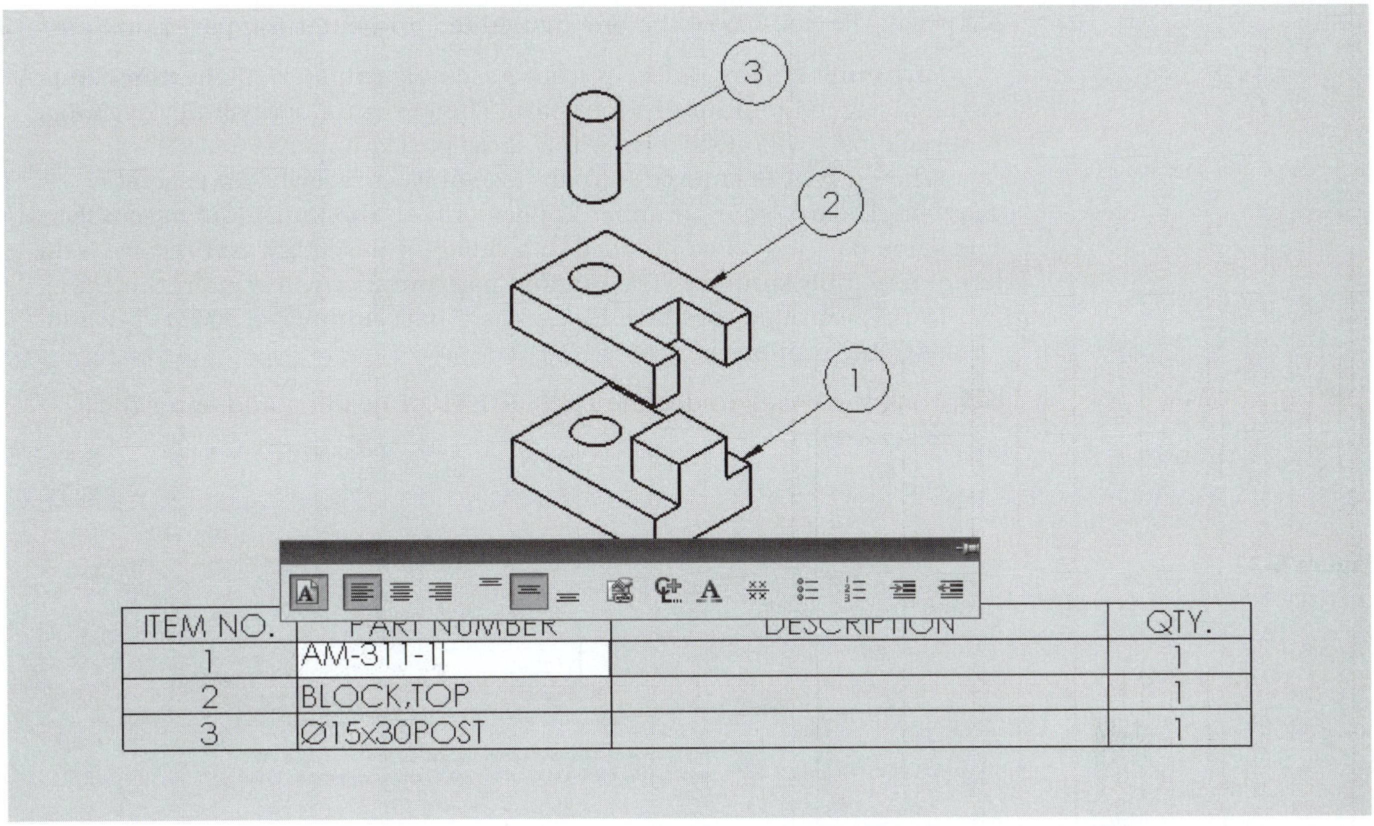

ITEM NO.	PART NUMBER	DESCRIPTION	QTY.
1	AM-311-1		1
2	BLOCK,TOP		1
3	Ø15x30POST		1

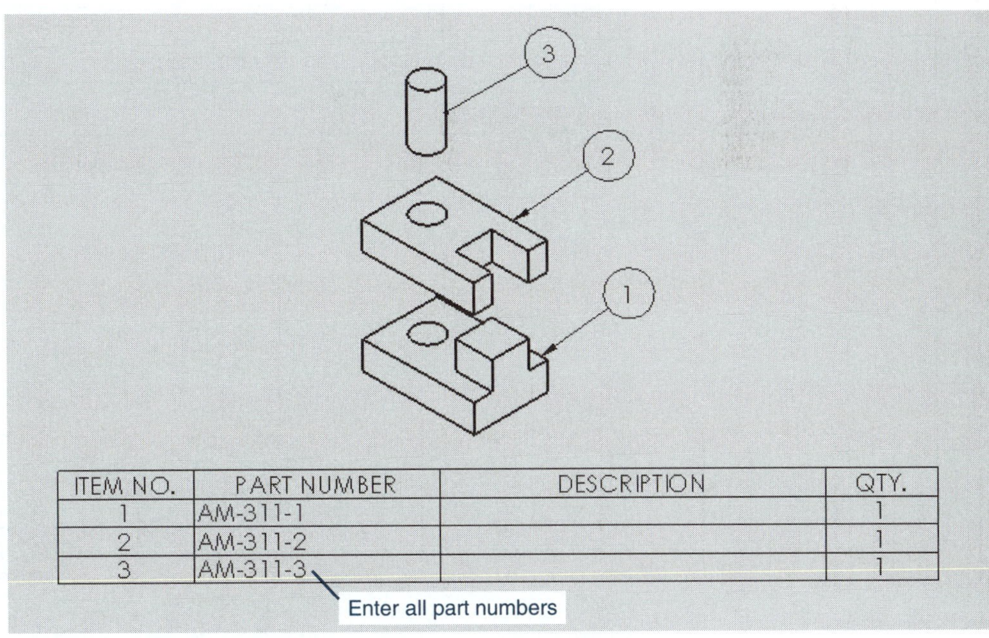

3
2
1
ITEM NO.
PART NUMBER
DESCRIPTION
QTY.
1
AM-311-1
1
2
AM-311-2
1
3
AM-311-3
1
Enter all part numbers

Figure 5-31
(*Continued*)

2 Click the **Keep Link** box.

The **Formatting** dialog box will appear.

3 Type in a part number.

4 Click the **Left Justify** option if necessary.

5 Click the box below the one just edited and enter the part numbers.

Part numbers are different from Assembly numbers (Item numbers). A part is assigned to an individual part. The part number will never change. A specific part will always have the same part number.

When a part is entered into an assembly, it receives an assembly number. The assembly number applies only to the individual assembly. The same part may also be used in a different assembly and receive a different assembly number. The original part number will remain the same.

In this example the Base Block has a part number of AM-311-1 and an assembly number of 1.

6 Click the boxes under the **DESCRIPTION** heading and enter the descriptions.

See Figure 5-32.

Figure 5-32

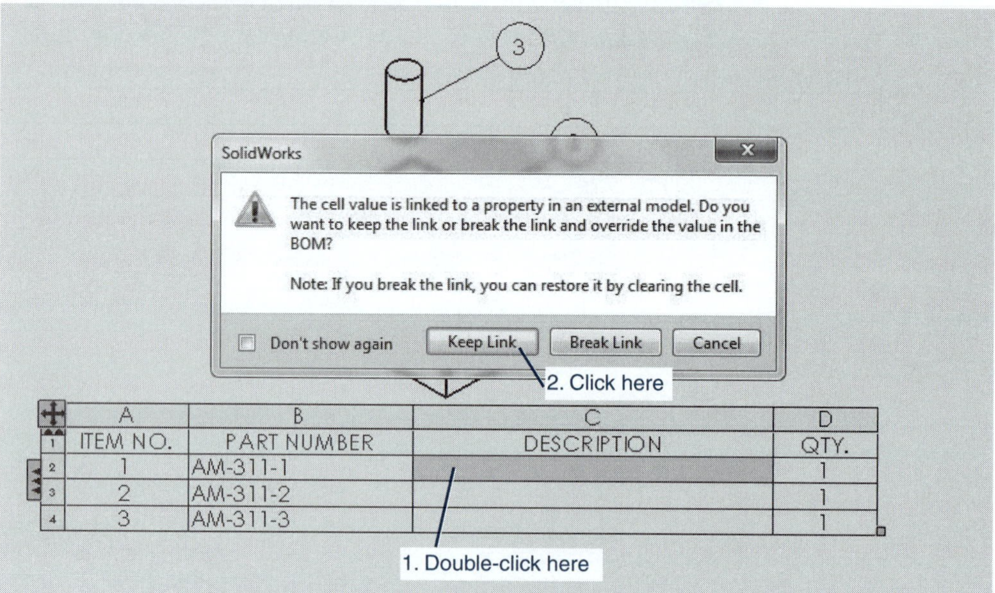

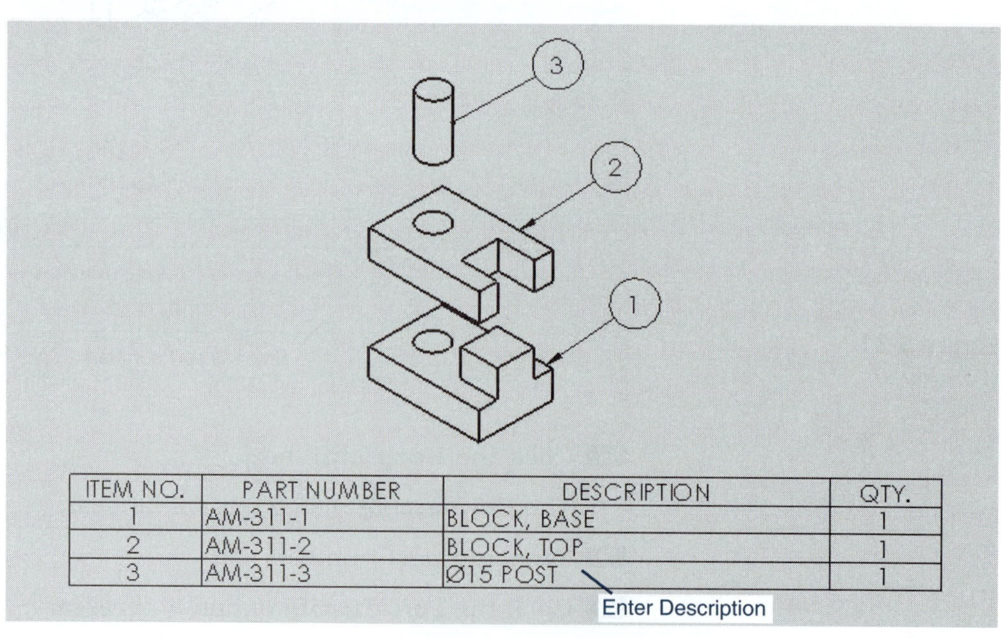

To Add Columns to the BOM

1 Right-click the **BLOCK, BASE** box in the **Description** column.

See Figure 5-33.

2 Click the **Insert** option, then click **Column Right**.

A new column will appear to the right of the BLOCK, BASE box and a dialog box will appear above the new column.

3 Click the **Property name** box and select the **SW-Title (Title)** option.

4 Swipe and remove the **Title** entry and enter a new column heading.

In this example the heading **MATERIAL** was entered.

5 Enter the required materials for each part.

In this example a material specification of SAE 1020 was added. SAE 1020 is a type of mild steel.

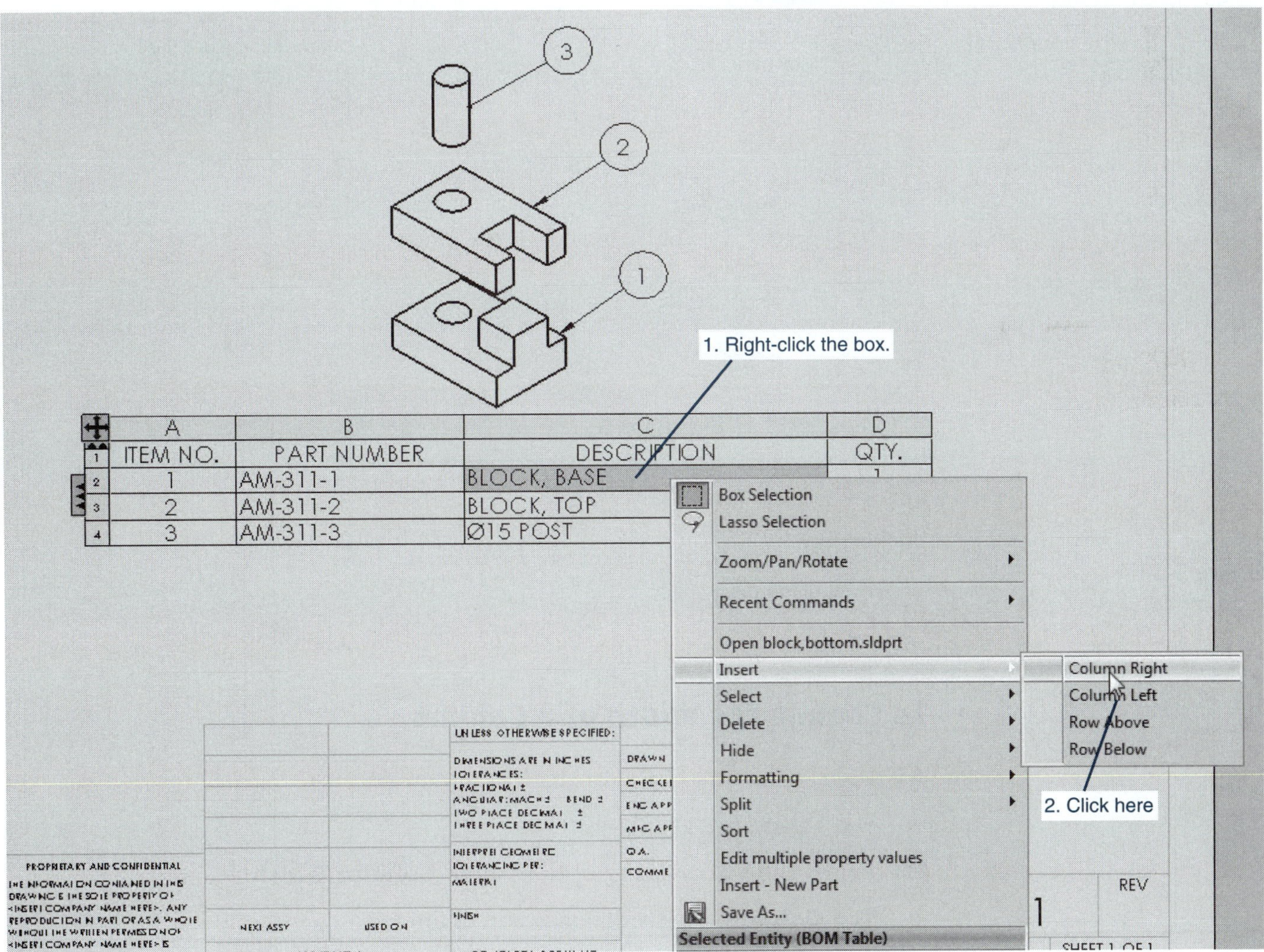

Figure 5-33

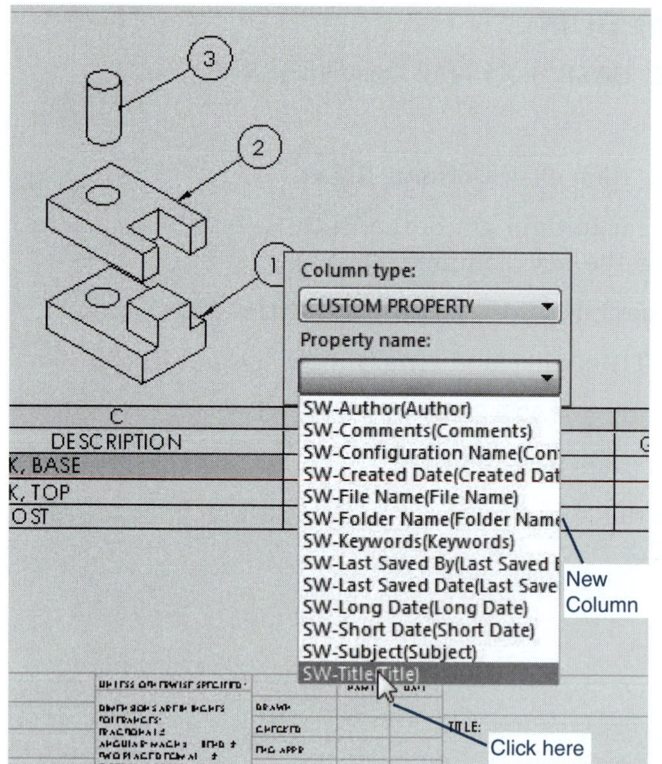

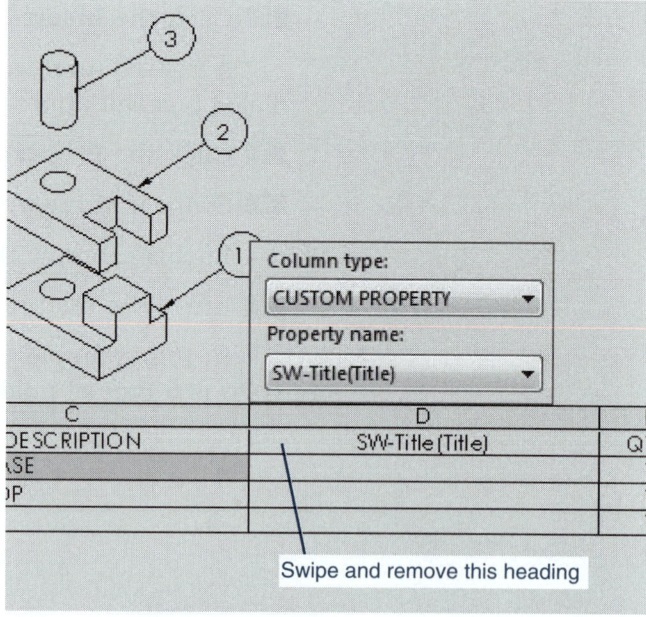

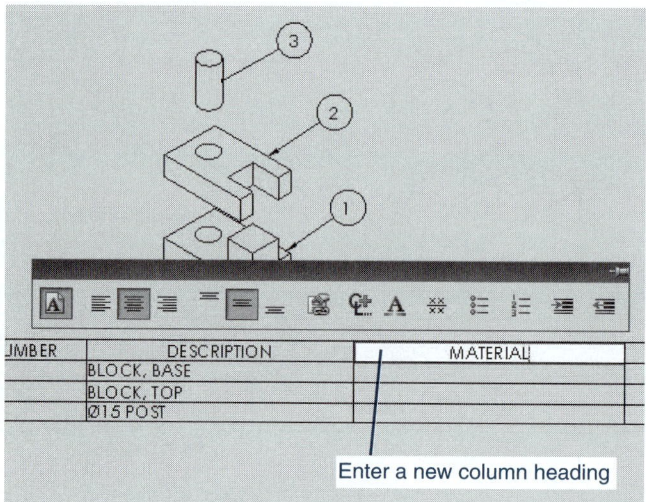

Figure 5-33
(Continued)

To Change the Width of a Column

1 Right-click one of the boxes in the new column.

See Figure 5-34.

2 Select the **Formatting** option; click **Column Width**.

The **Column Width** dialog box will appear.

3 Enter a new value.

In this example a value of **1.75in** was entered. Figure 5-34 shows the new column width. Edit the other columns if necessary.

Figure 5-34

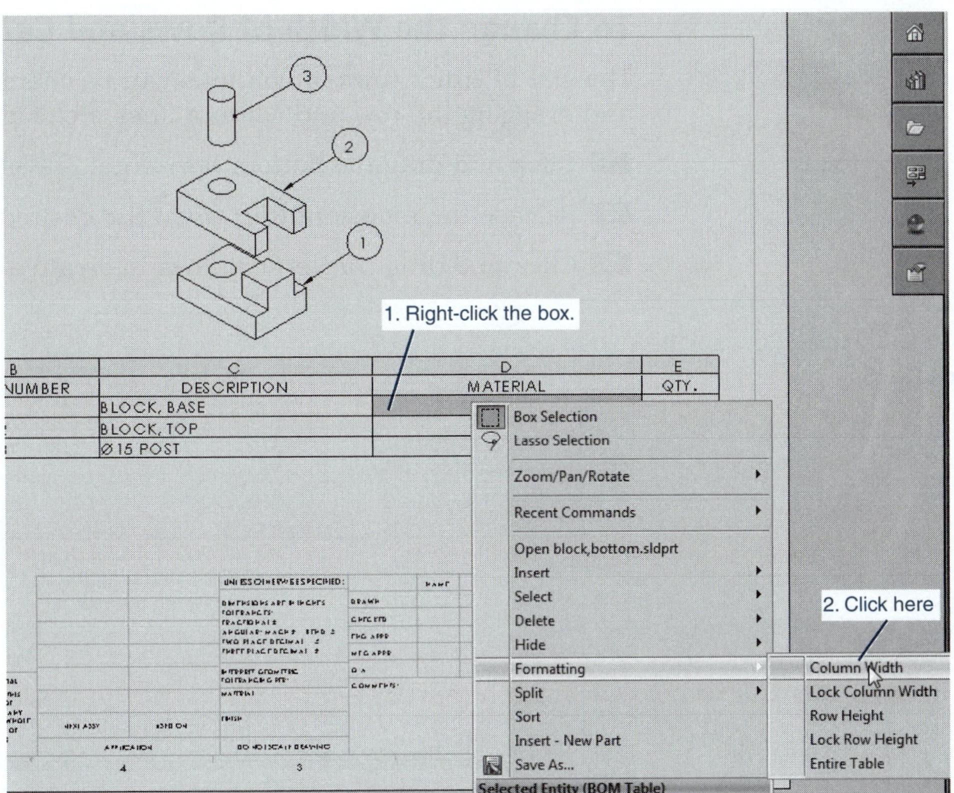

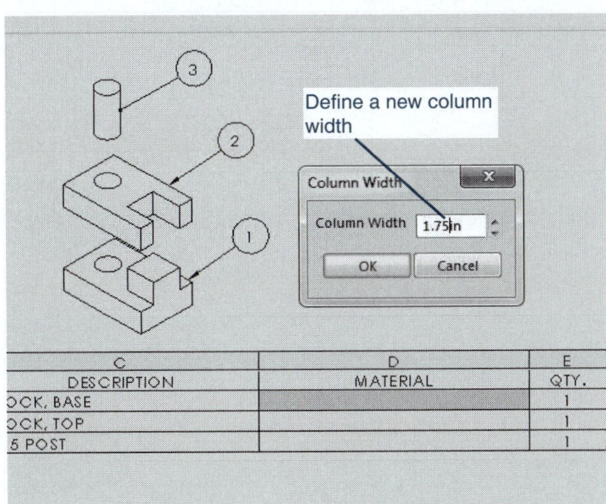

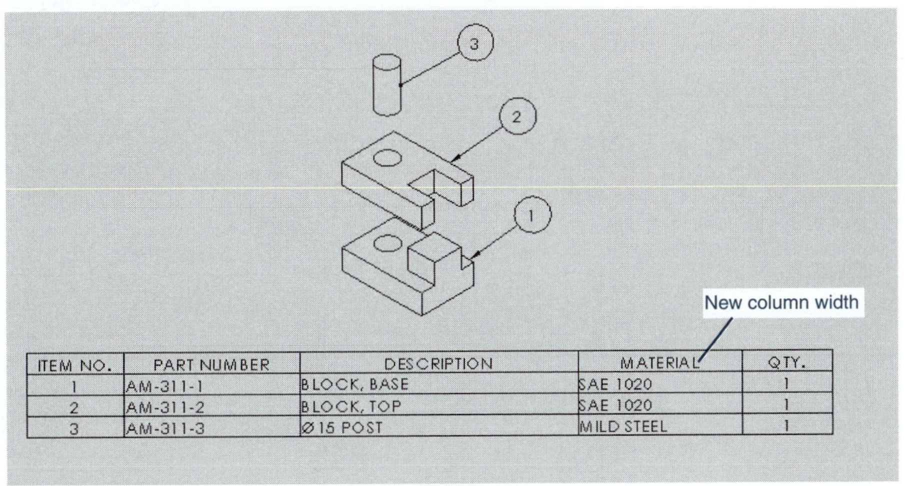

ITEM NO.	PART NUMBER	DESCRIPTION	MATERIAL	QTY.
1	AM-311-1	BLOCK, BASE	SAE 1020	1
2	AM-311-2	BLOCK, TOP	SAE 1020	1
3	AM-311-3	Ø15 POST	MILD STEEL	1

To Change the Width of Rows and Columns

The size of either rows or columns can be changed in real time by clicking and dragging the row and column lines. See Figure 5-35.

1. Click and drag the bottom horizontal line of the BOM.
2. Release the mouse button when the desired size is reached.
3. Click and drag the vertical lines to create a new format for the BOM.

Figure 5-35

A	B	C	D	E
ITEM NO.	PART NUMBER	DESCRIPTION	MATERIAL	QTY.
1	AM-311-1	BLOCK, BASE	SAE 1020	1
2	AM-311-2	BLOCK, TOP	SAE 1020	1
3	AM-311-3	Ø15 POST	MILD STEEL	1

Click and drag the line to widen the row

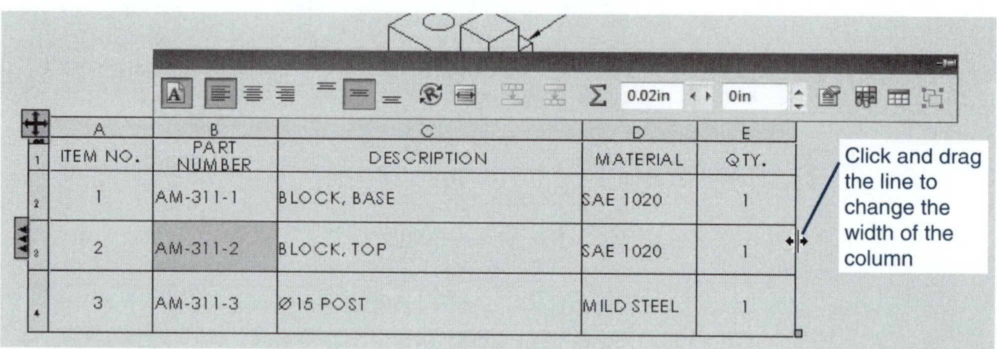

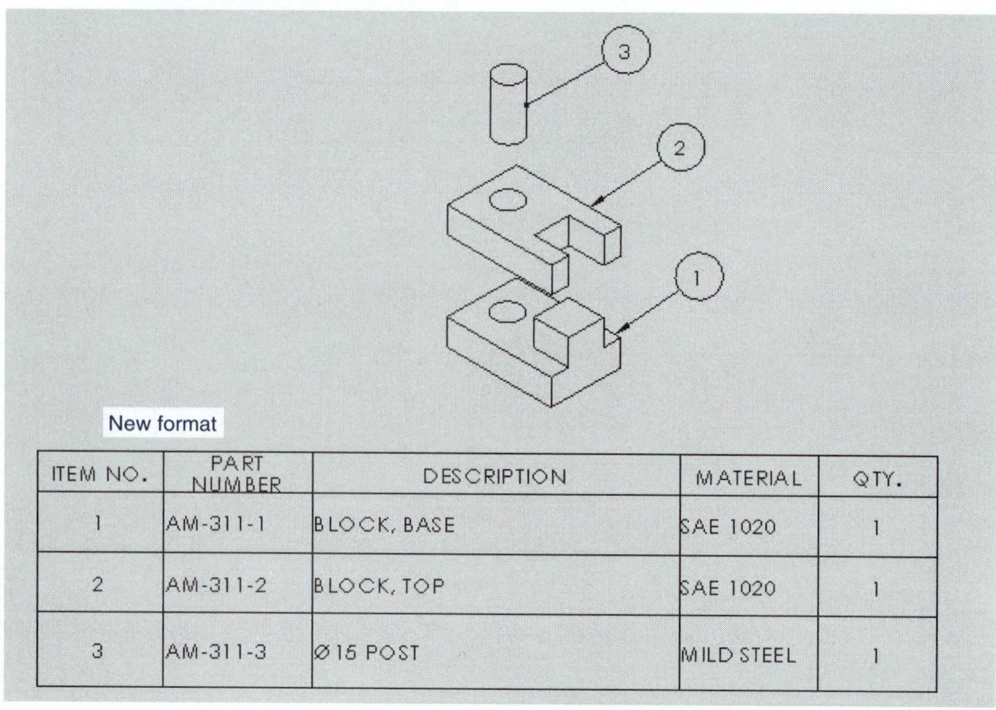

ITEM NO.	PART NUMBER	DESCRIPTION	MATERIAL	QTY.
1	AM-311-1	BLOCK, BASE	SAE 1020	1
2	AM-311-2	BLOCK, TOP	SAE 1020	1
3	AM-311-3	Ø15 POST	MILD STEEL	1

To Change the BOM's Font

The text font can be changed. In Figure 5-36 the font of the column heads was changed from the default SolidWorks font Century Gothic to Times New Roman and made bold.

Figure 5-36

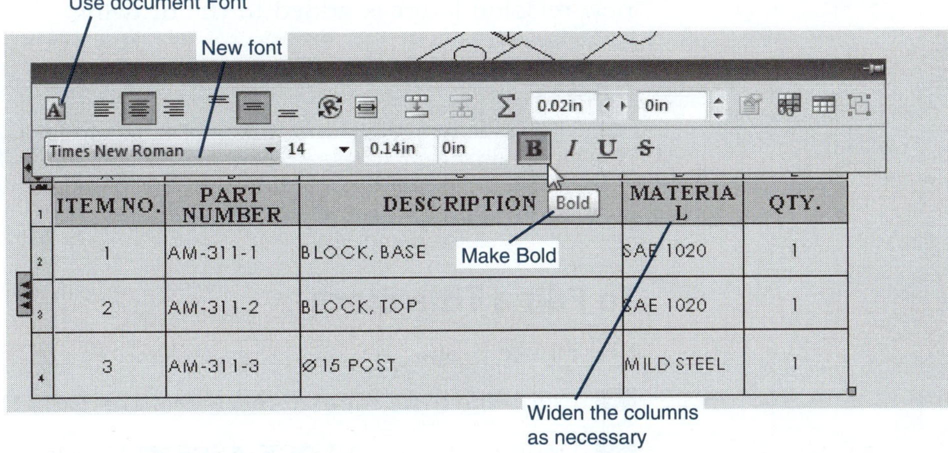

New font

ITEM NO.	PART NUMBER	DESCRIPTION	MATERIAL	QTY.
1	AM-311-1	BLOCK, BASE	SAE 1020	1
2	AM-311-2	BLOCK, TOP	SAE 1020	1
3	AM-311-3	Ø15 POST	MILD STEEL	1

1 Swipe the text to be changed.

2 Click the **Use Document Font** tool.

3 Select a new font.

4 Click the **Bold (B)** tool.

5-12 Title Blocks

A title block contains information about the drawing. See Figure 5-37. The information presented in a title block varies from company to company but usually includes the company's name, the drawing name and part number, the drawing scale, and a revision letter.

Figure 5-37

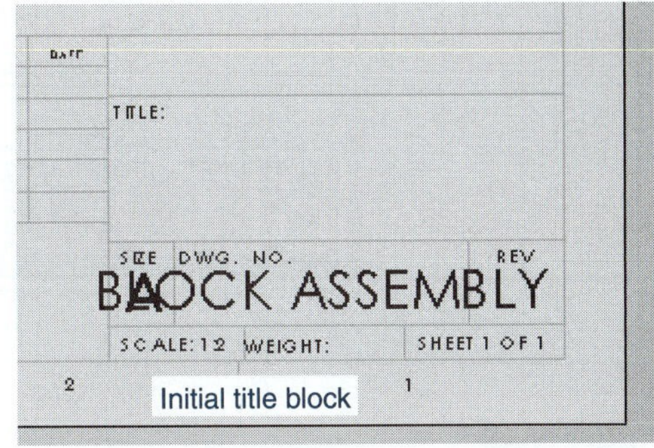

Revision Letters

As a drawing goes through its production cycle, changes are sometimes made. The changes may be because of errors but they may also be because of the availability of new materials, manufacturing techniques, or new customer requirements. As the changes are incorporated onto the drawing, a new revision letter is added to the drawing.

NOTE

SolidWorks automatically enters the file name of the document as the part number. In the example shown in Figure 5-37 the drawing number BLOCK ASSEMBLY is not the document's part number. The title block will have to be edited and the correct part number entered.

To Edit a Title Block

See Figure 5-38.

1. Right-click the mouse and select the **Edit Sheet Format** option.
2. Double-click the **BLOCK ASSEMBLY** file name.

   The **Formatting** dialog box will appear.

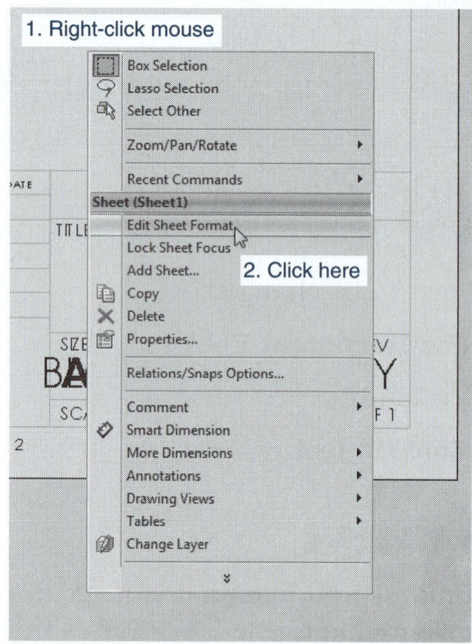

Figure 5-38

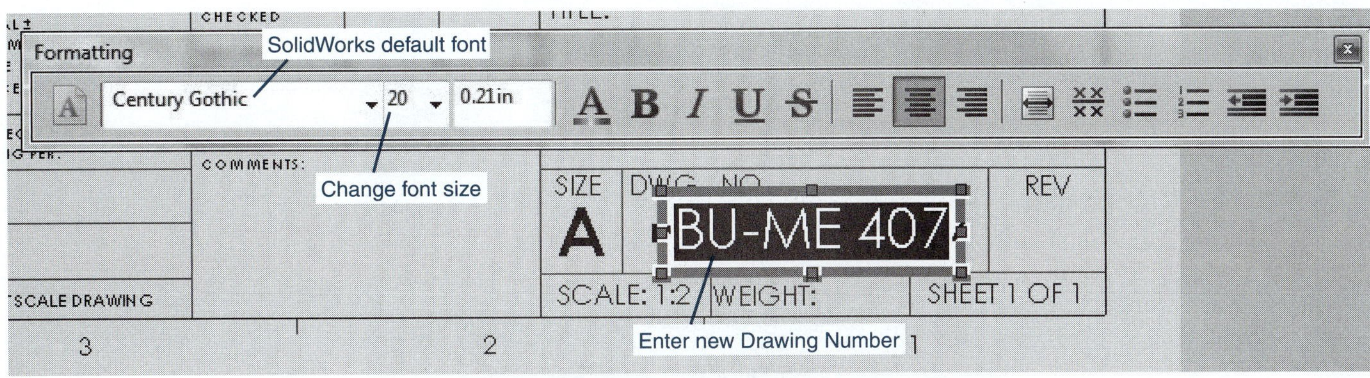

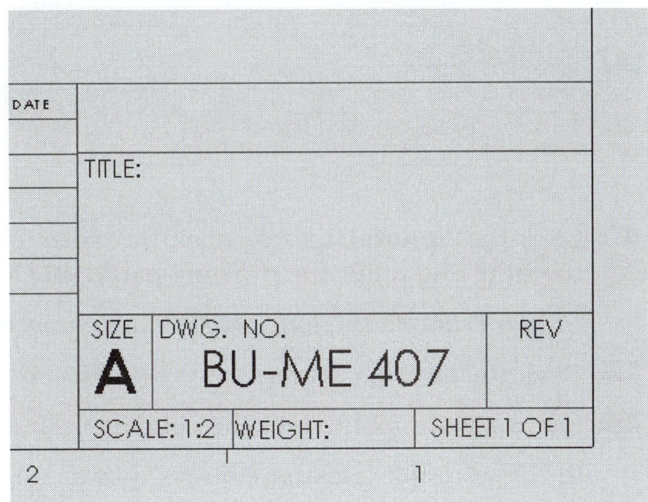

Figure 5-38
(*Continued*)

3 Change the **DWG. NO.** to **BU-ME 407**.

4 Change the font size to fit the number within the DWG. NO. box.

In this example a size of **20** was selected.

5 Click the drawing screen.

The **Notes** tool is used to add text to the other boxes in the **Title Block**. See Figure 5-39.

Figure 5-39

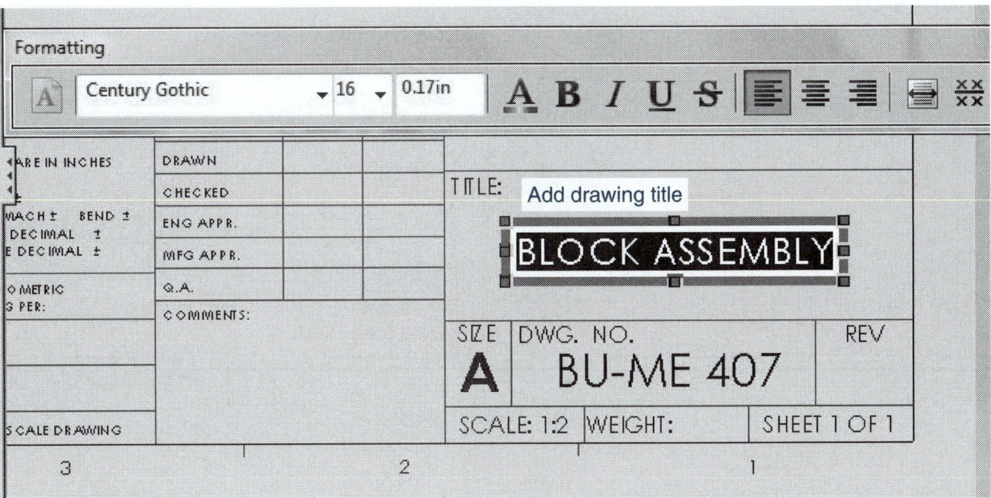

Figure 5-39
(Continued)

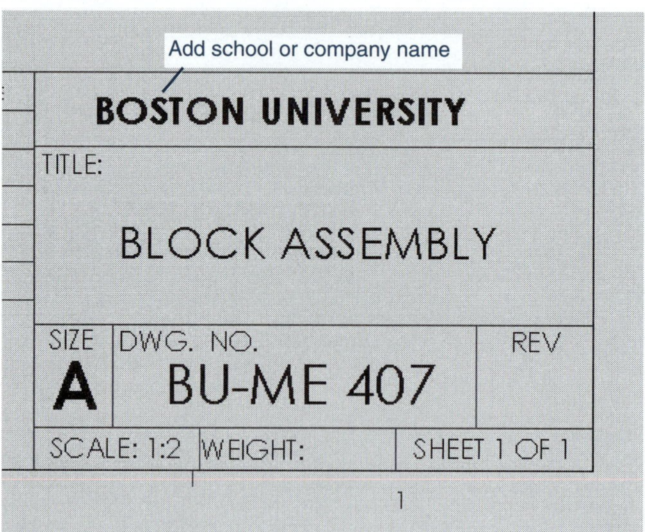

6 Click the **Annotation** tab, click the **Note** tool, locate the note within the title, and enter the drawing name (**BLOCK ASSEMBLY**).

In this example the font height was changed to **16**.

7 Click the drawing screen.

8 Use the **Note** tool to add the company name.

In this example "Boston University" was added. Consider using your own school or company name.

Release Blocks

A finished engineering drawing is a legal document that goes through a release process before it becomes final. The release block documents the release process. For example, once you have completed a drawing, you will initial and date the **DRAWN** box located just to the left of the title block. The drawing will then go to a checker, who, after reviewing and incorporating any changes, will sign and date the **CHECKED** box. See Figure 5-40.

Figure 5-40

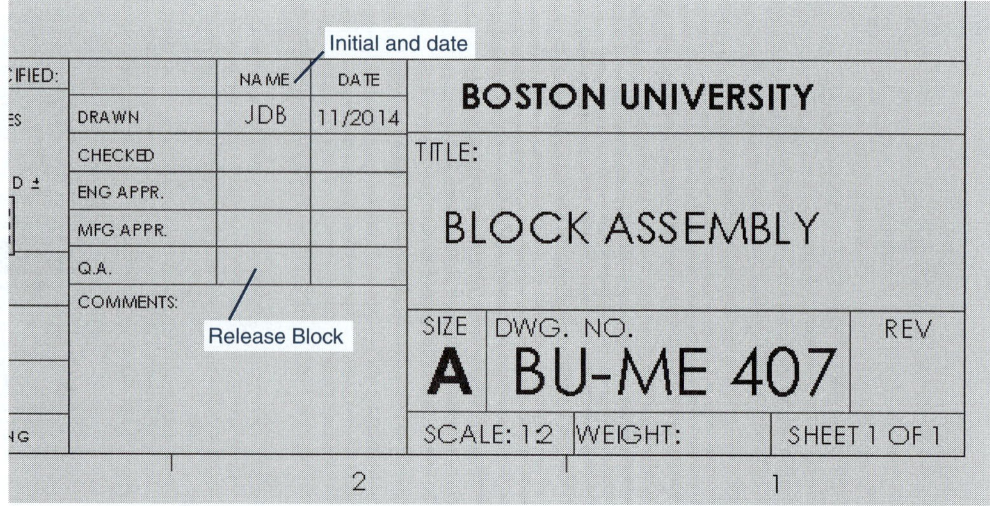

Tolerance Block

The tolerance block will be discussed in Chapter 8, Tolerancing.

Application Block

See Figure 5-41.

Figure 5-41

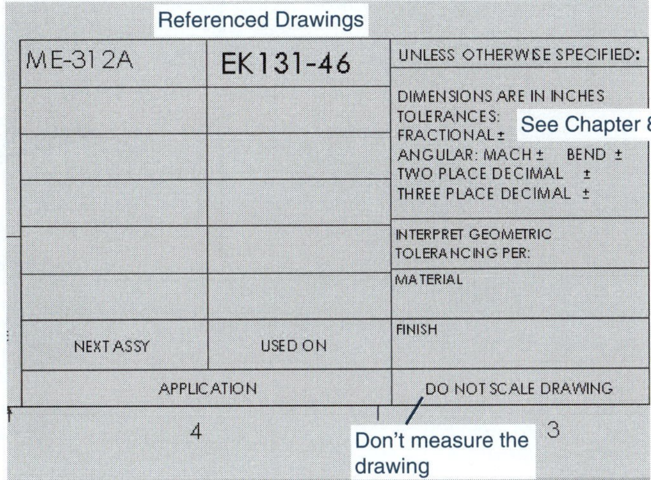

This block is used to reference closely related drawings. In this example, we know that the block assembly will be used on assembly ME-312A and that it was also used on EK131-46. This information makes it easier to access related drawings that can be checked for interfaces.

NOTE

The note "DO NOT SCALE DRAWING" located at the bottom of the tolerance block is a reminder not to measure the views on the drawing. If a dimension is missing, do not measure the distance on the drawing, because the drawing may not have been reproduced at exactly 100% of the original.

5-13 Animate Collapse

Exploded Assembly drawings can be animated. In this example the **Animate collapse** tool will be used.

1. Open the block assembly.
2. Right-click the **BLOCK ASSEMBLY** heading in the **FeatureManager**.

   See Figure 5-42.
3. Click the **Animate collapse** option.

Figure 5-42

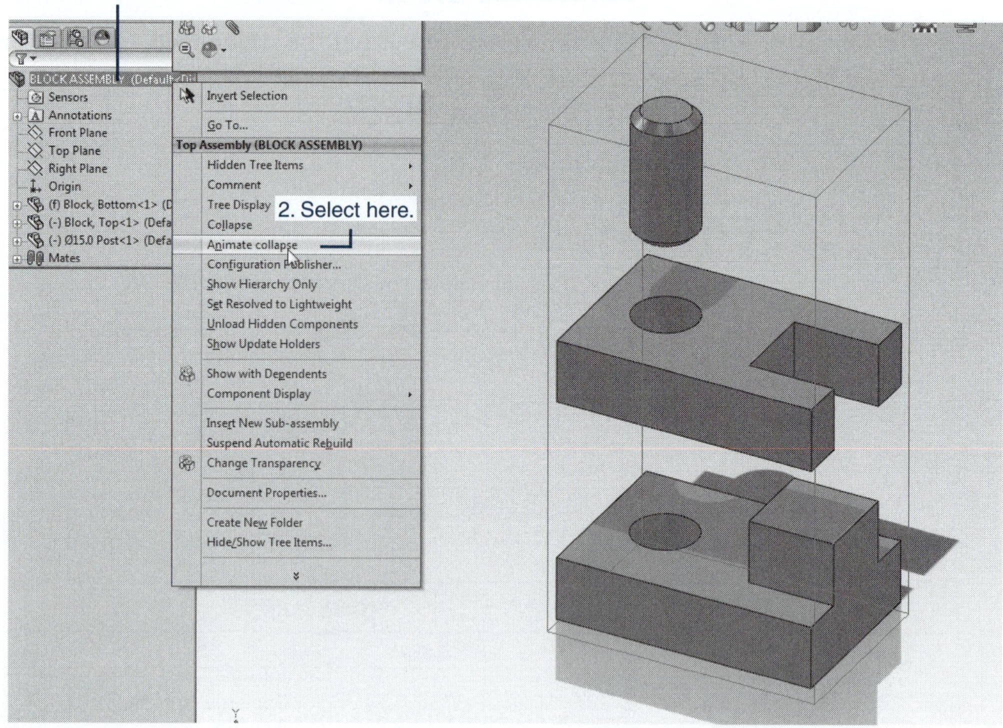

The assembly will automatically be animated and start to move. See Figure 5-43.

4 Click the **Stop** button to stop the animation and return the assembly to the exploded position.

5 Close the **Animation Controller**.

Figure 5-43

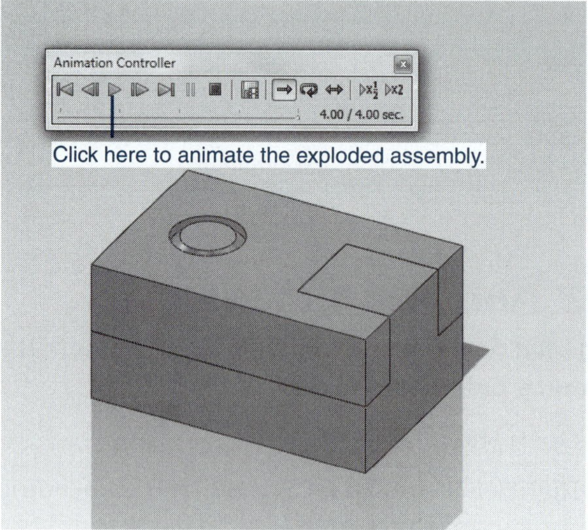

In this example the **Animate collapse** tool was used because the assembly was shown in the exploded position. Had the assembly been in a closed assembled position, the **Animate explode** option would have been used.

5-14 Sample Problem 5-1: Creating the Rotator Assembly

Figure 5-44 shows the components for the Rotator Assembly. The dimensions for the components can be found in Project P5-10 at the end of the chapter. Draw and save the four Rotator Assembly components as **Part** documents.

Figure 5-44

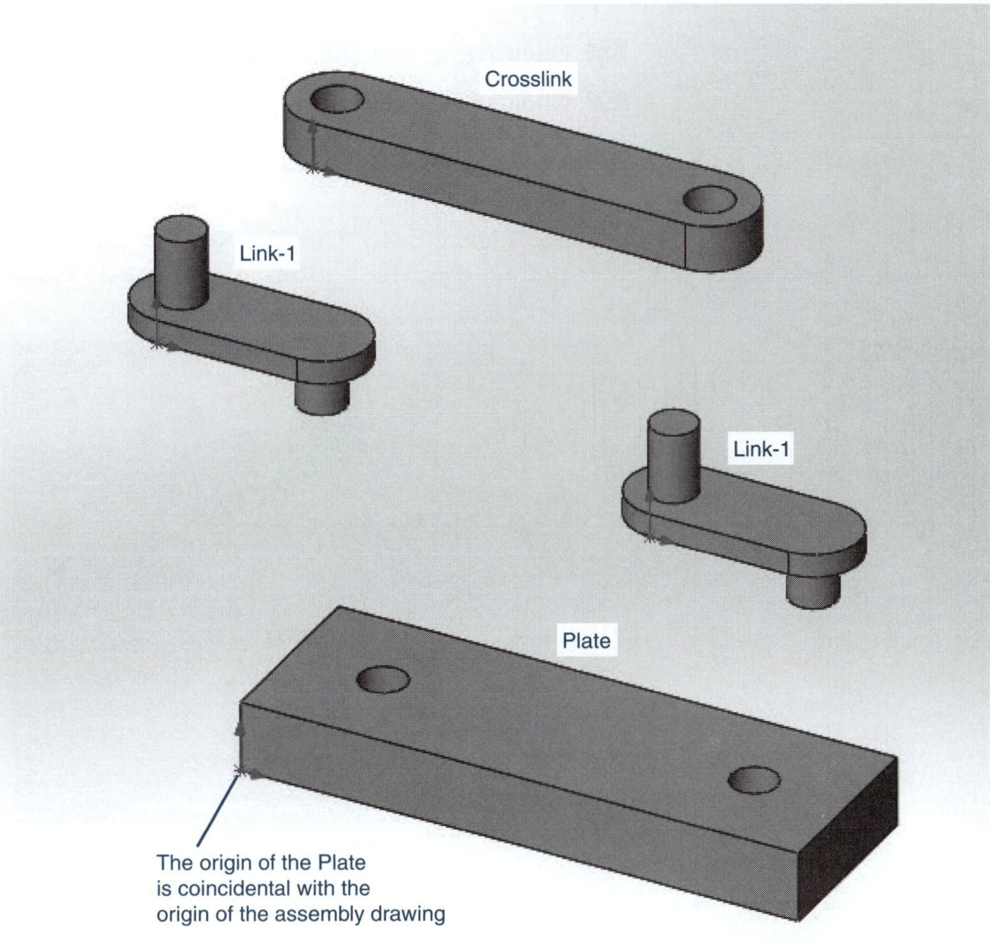

1 Start a new **Assembly** document.

2 Insert the **PLATE**, **CROSSLINK**, and two **LINK-L**s into the drawing.

> **NOTE**
>
> Insert the **PLATE** first so it will be fixed **(f)**.
>
> Insert the plate so that it is aligned with the origin of the assembly drawing.

3 Click the **Mate** tool.

4 Click the **Concentric** tool.

5 Click the side of the bottom post of one **LINK-L** and the inside of the left hole in the **PLATE**.

See Figure 5-45. The LINK-L and PLATE will align.

> **TIP**
>
> Click the surfaces of the posts and holes. Do not click the edge lines.

6 Click the green **OK** check mark.

7 Click the **Mate** tool.

8 Click the top surface of the **PLATE** and the bottom surface of the **LINK-L**.

See Figure 5-46. Use the **Rotate View** tool to manipulate the view orientation so that the bottom surface of the LINK-L is visible. The part also

Figure 5-45

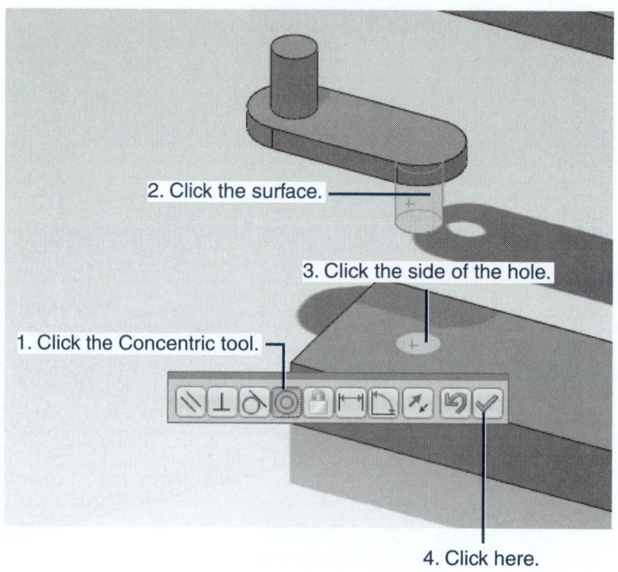

2. Click the surface.
3. Click the side of the hole.
1. Click the Concentric tool.
4. Click here.

Figure 5-46

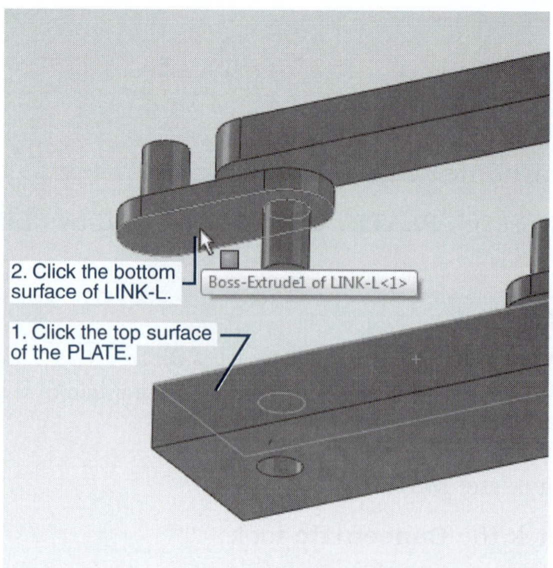

2. Click the bottom surface of LINK-L.
Boss-Extrude1 of LINK-L<1>
1. Click the top surface of the PLATE.

can be rotated by holding down the mouse wheel and moving the cursor.

9 Click the **Distance** box and enter a value. In this example a value of **2.0mm** was entered.

See Figure 5-47. The initial offset values may be in inches. Enter the new value of **2.00** followed by **mm**, and the system will automatically change to metric (millimeter) distances.

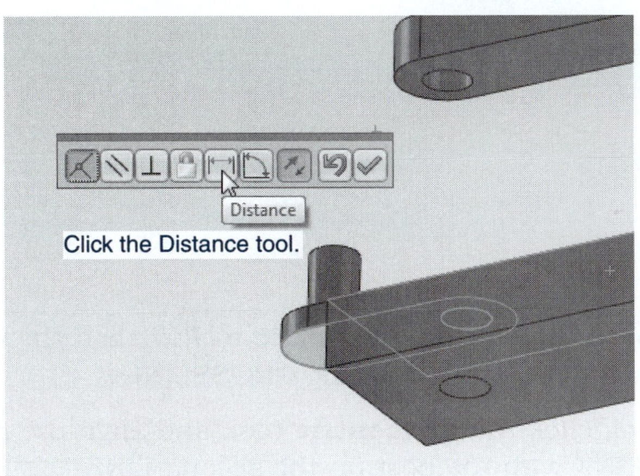

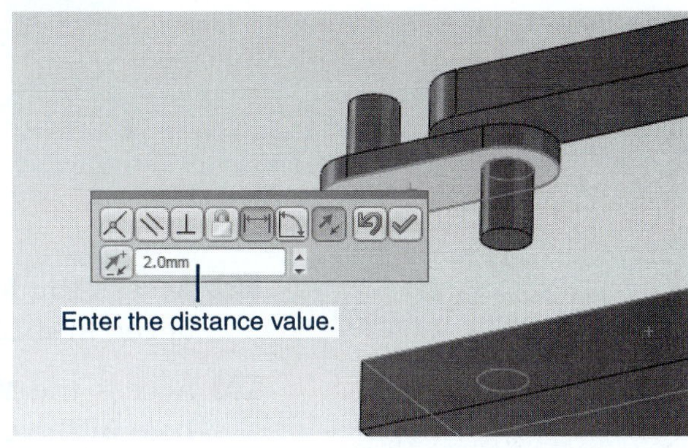

Figure 5-47

10 Click the green **OK** check mark.

The 2.00mm distance will allow for clearance between the PLATE and the LINK-L.

Figure 5-48 shows the 2-mm offset between the PLATE and the LINK-L.

11 Return the drawing to the **Isometric** orientation.

Figure 5-48

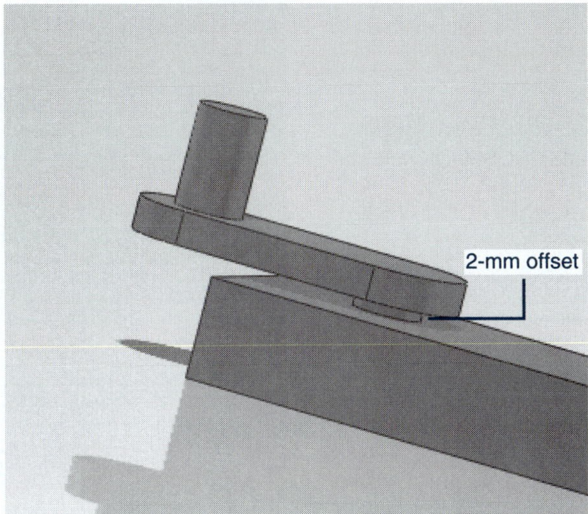

12 Repeat the procedure for the other LINK-L.

See Figure 5-49.

Figure 5-49

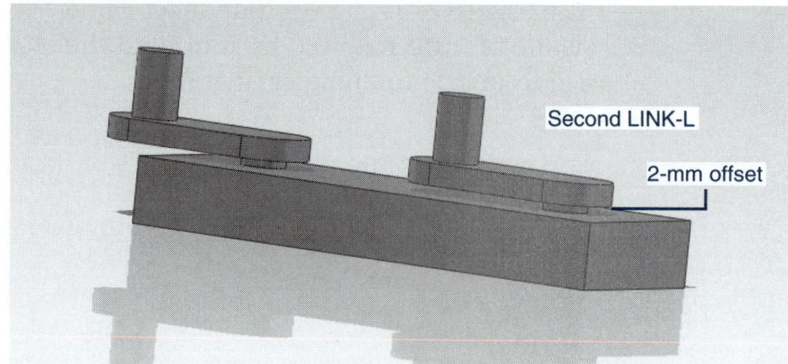

13 Access the **Mate** tool and use the **Concentric** tool to align the top post of the first LINK-L with the left hole in the CROSSLINK.

14 Access the **Mate** tool, click the **Concentric** tool, and align the right hole in the CROSSLINK with the post on the second LINK-L.

15 Use the **Mate** tool and click the top surface of the LINK-L's post and the top surface of the CROSSLINK.

16 Use the **Mate** tool and click the top surface of the second LINK-L's post and the top surface of the CROSSLINK.

See Figure 5-50.

17 Click the green **OK** check mark.

See Figure 5-51.

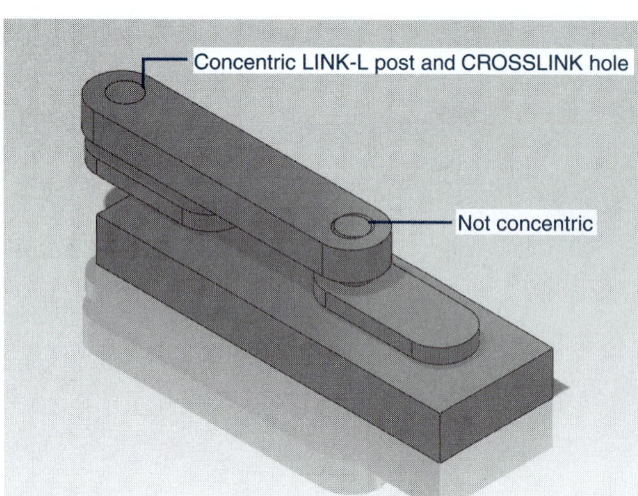

Figure 5-50

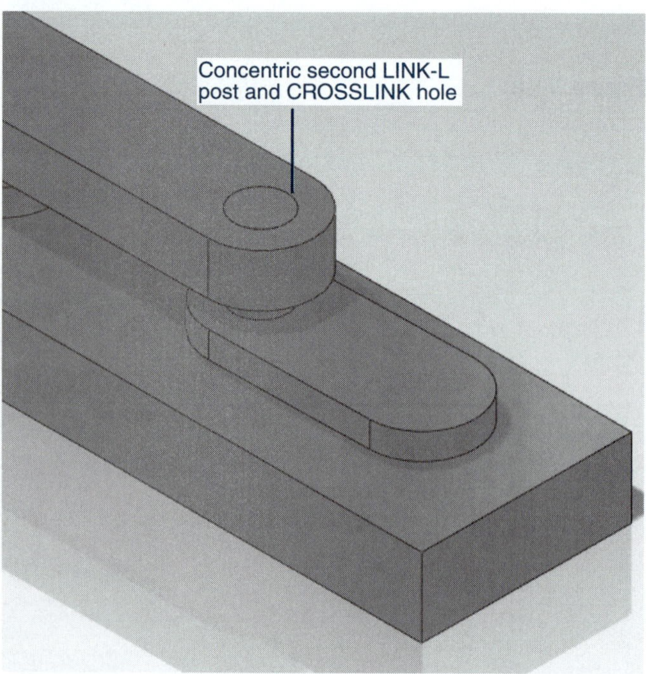

Figure 5-51

18 Locate the cursor on the **CROSSLINK** and move it around.

The CROSSLINK and two LINK-Ls should rotate about the PLATE.

19 Save the assembly as the **Rotator Assembly**.

> **TIP**
> All mates will be listed in the **FeatureManager**.

5-15 Using the SolidWorks Motion Study Tool

1 Access the **Rotator Assembly**.

2 Click the **Mate** tool, then click the **Parallel** tool.

3 Make the front surface line of the **CROSSLINK** parallel to the front edge of the **PLATE**; click the green **OK** check mark.

This step will ensure that the CROSSLINK rotates in an orientation parallel to the front edge of the PLATE. See Figure 5-52.

Figure 5-52

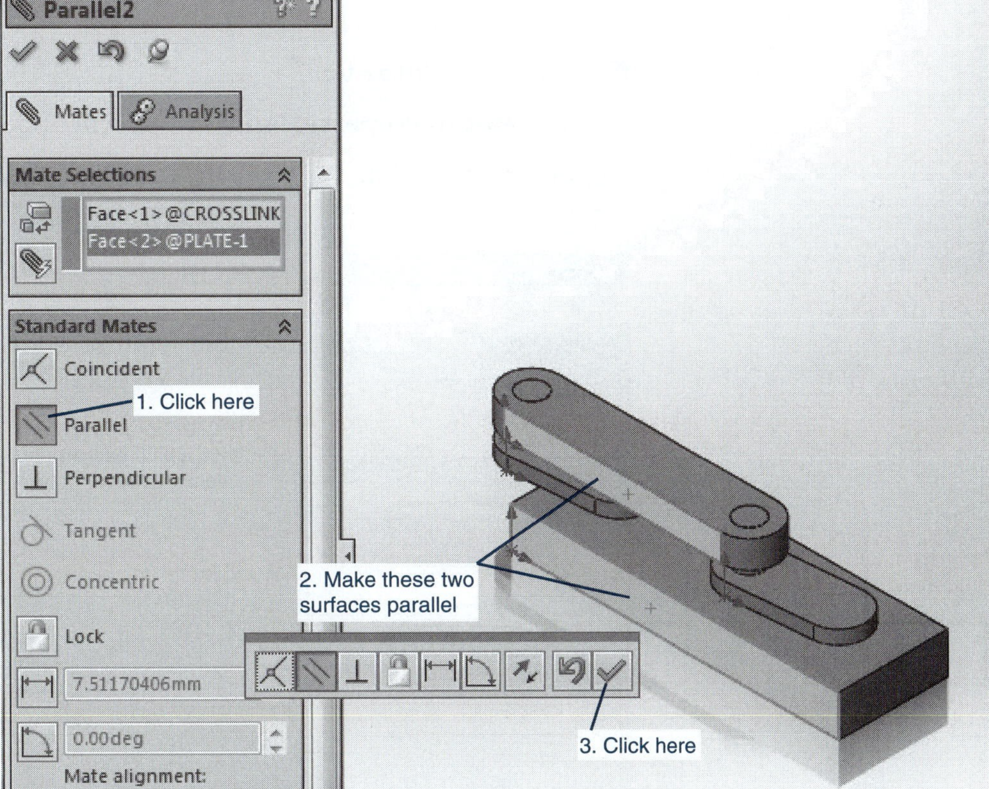

4 Click the **Motion Study** tab at the bottom of the screen.

See Figure 5-53.

Figure 5-53

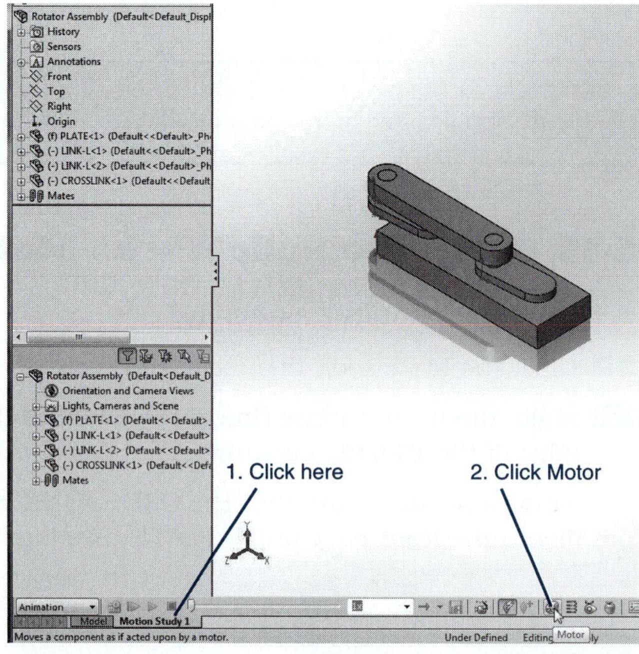

5 Click the **Motor** tool.

The **Motor PropertyManager** box will appear.

See Figure 5-54.

Figure 5-54

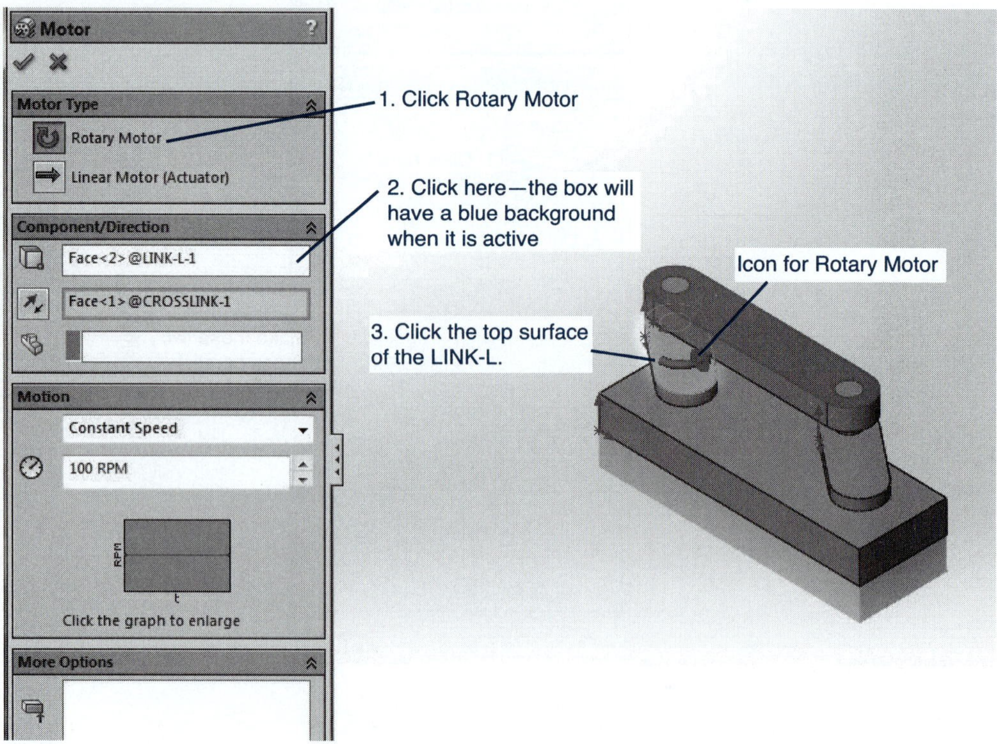

6 Click the **Rotary Motor** tool.

7 Click the box under the **Component/Direction** heading, then click the top surface of the left LINK-L.

The box will have a blue background when it is active.

The left LINK-L is now the driver link. An arrow will appear indicating the direction of motion. It will drive the other components.

Motion

1 Go to the **Motion** box on the **Motor** property manager and define the assembly's motion.

In this example the default values of **Constant Speed** and **100 RPM** were accepted.

2 Click the green **OK** check mark and return to the **MotionManager**.

See Figure 5-55.

3 Click **Play**.

Figure 5-55

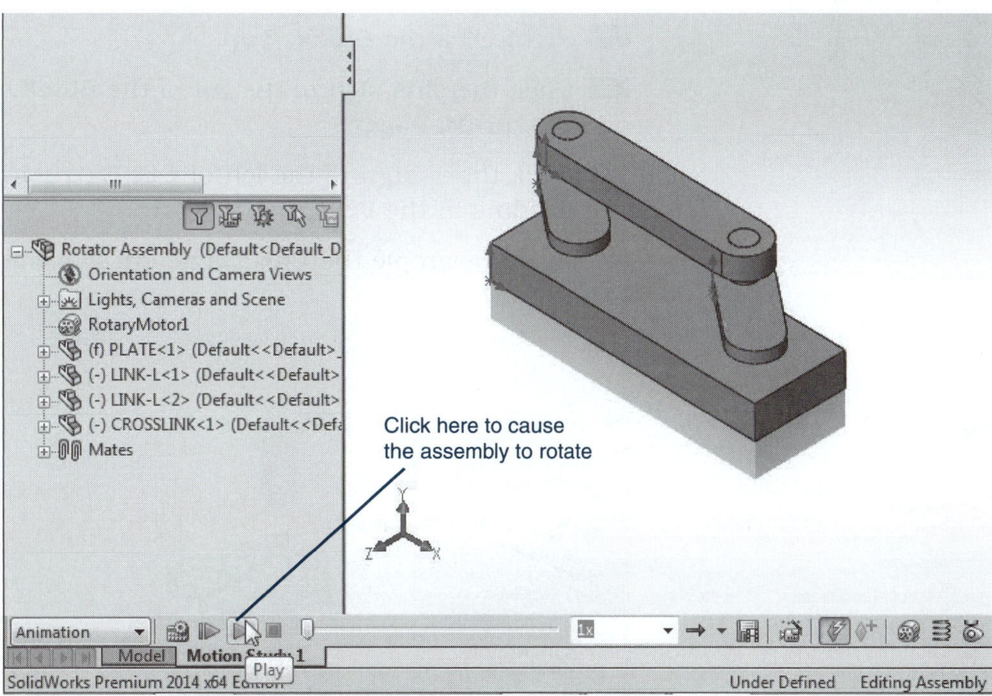

5-16 Editing a Part within an Assembly

Parts already inserted into an assembly drawing can be edited. Figure 5-56 shows the block assembly created earlier in the chapter. The general operating concept is to isolate a part, make the changes, and insert the part back into the assembly.

Say we wanted to change the hole in the top block.

Figure 5-56

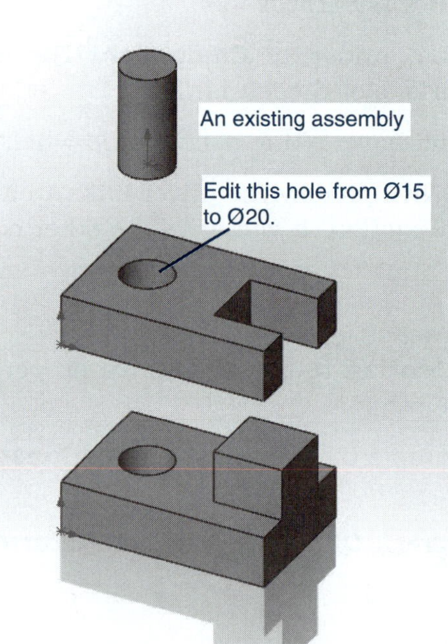

1 Right-click the **Block, Top**.

2 Click the plus sign to the left of the **Block, Top** heading in the **FeatureManager**.

3 Click the + **sign** to the left of **Cut-Extrusion 1** under the **Block, Top** heading in the **FeatureManager**.

In this example the **Cut-Extrusion 2** is the hole in the top block. See Figure 5-57.

Figure 5-57

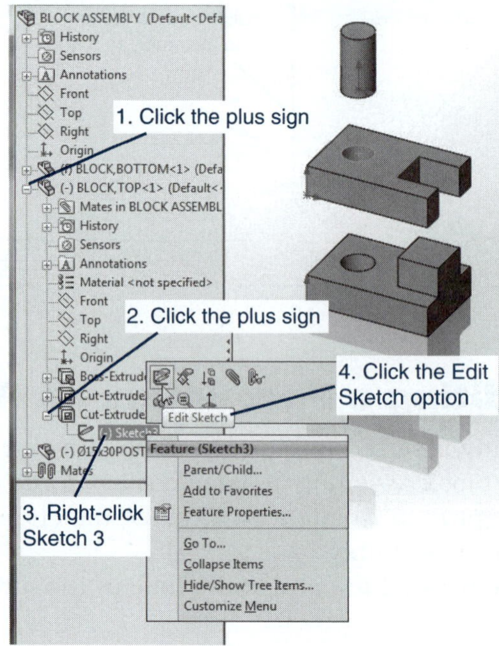

4 Right-click the **Sketch3** heading and click the **Edit Sketch** option.

5 Double-click the **Ø15** hole value and enter a new value.

In this example, the value was changed to **Ø20.0**. See Figure 5-58.

Figure 5-58

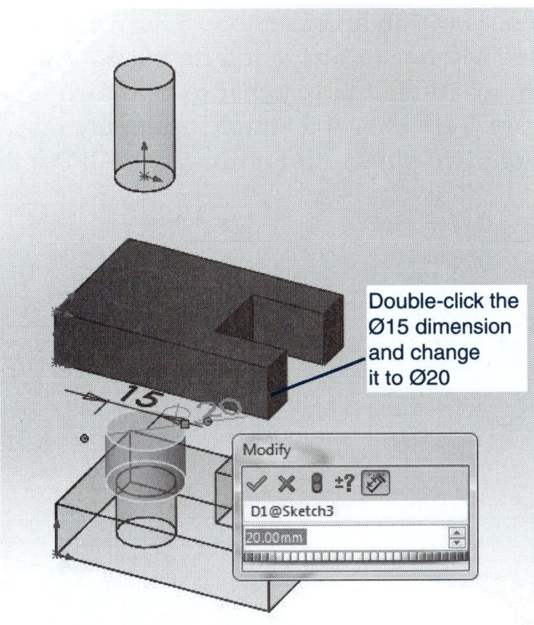

Double-click the
Ø15 dimension
and change
it to Ø20
Modify
D1@Sketch3
20.00mm

6 Click the green **OK** check mark.

7 Click the **Exit Sketch** option.

8 Click the **Edit Assembly** tool located in the upper right corner of the drawing screen.

Figure 5-59 shows the Top Block with the enlarged hole.

Figure 5-59

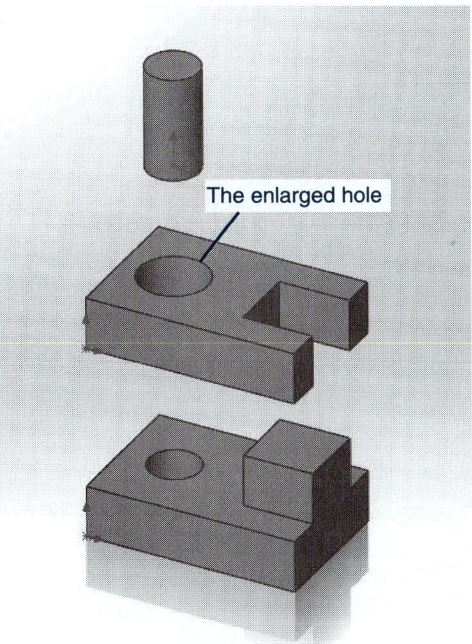

The enlarged hole

5-17 Interference Detection/Clearance Verification

Interference Detection

The **Interference Detection** tool is used to determine whether there are interferences within an assembly. SolidWorks will assemble parts as instructed without regard to interferences. It is possible for two parts that would, in reality, not fit together to be drawn together in an assembly.

Figure 5-61 shows a simple assembly made from two parts whose dimensions are shown in Figure 5-60. All the dimensions use whole num-

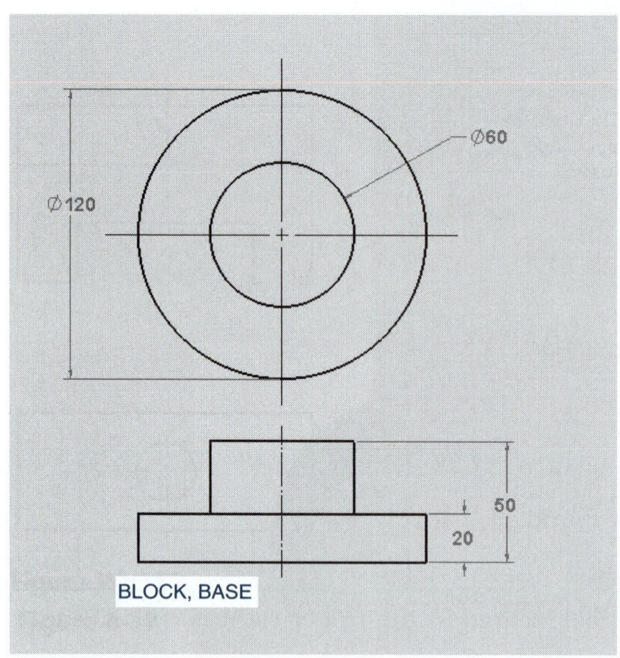

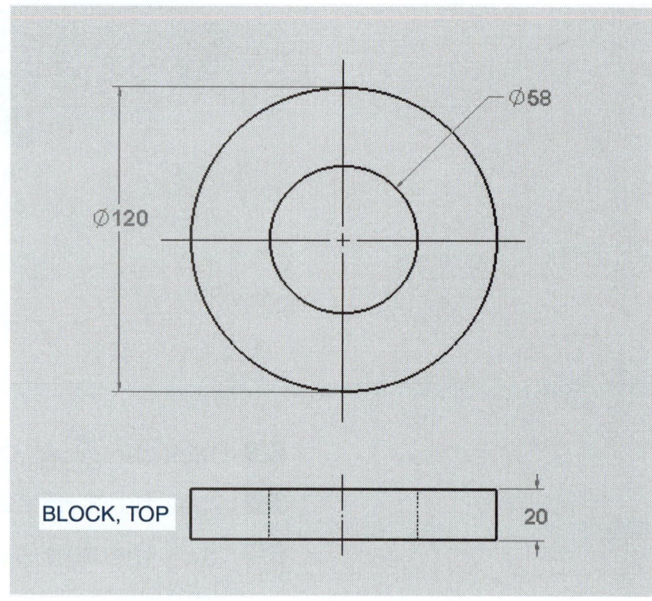

Figure 5-60

Figure 5-61

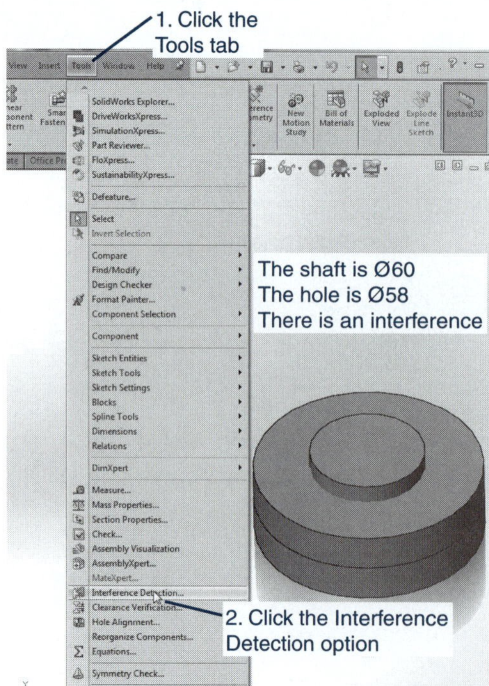

bers. There are no tolerances specified. From the given dimensions we know that there is an interference between the Base and the Top. The shaft on the Base is **Ø60** and the hole in the Top is **Ø58**. The **Interference Detection** tool can be used to highlight this interference.

To Detect an Interference

1 Click the **Tool** heading and the top of the screen and select the **Interference Detection** tool.

See Figure 5-61. The Assembly will appear enclosed in a wireframe box. A note will appear indicating that an interference has been detected. See Figure 5-62.

Figure 5-62

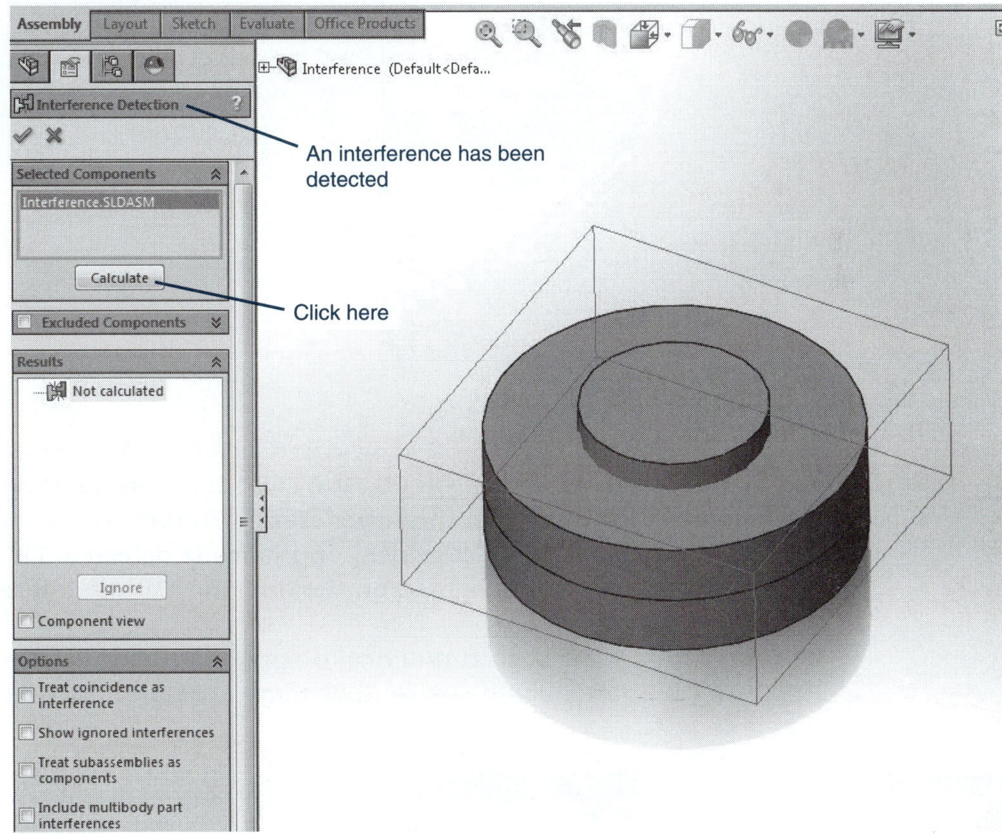

2 Click the **Calculate** box.

The interference will be highlighted using a red hollow cylinder. See Figure 5-63. There will also be a sound indicating an interference has been detected.

Figure 5-63

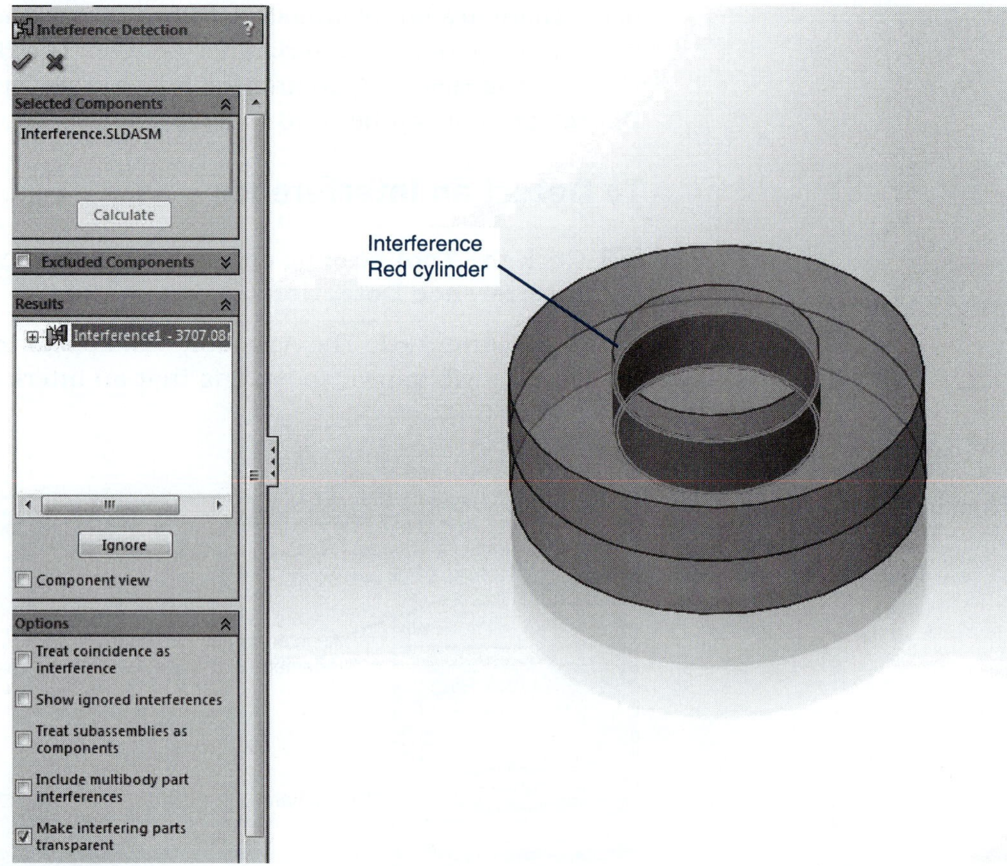

Figure 5-64 shows an assembly. The eight fasteners in the **Base** have a diameter of Ø8.00. The eight holes in the **Cover Plate** have a diameter of Ø6.00. When the **Interference Detection** tool is applied, the interference between the fasteners and holes is detected. One fastener hole interference will be highlighted on the drawing. All the interferences will be listed in the results box.

It is good practice to apply the **Interference Detection** tool to an assembly to ensure that there are no interferences.

Figure 5-64

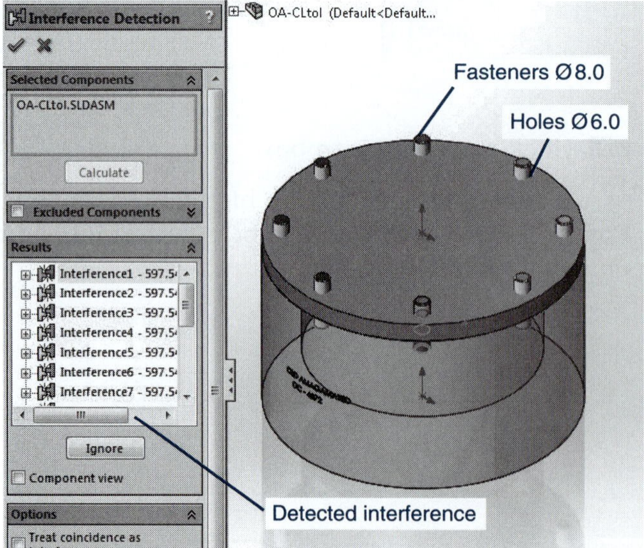

Figure 5-65 shows a similar assembly, but the eight holes in the **Cover Plate** have been enlarged to **Ø10**.

Figure 5-65

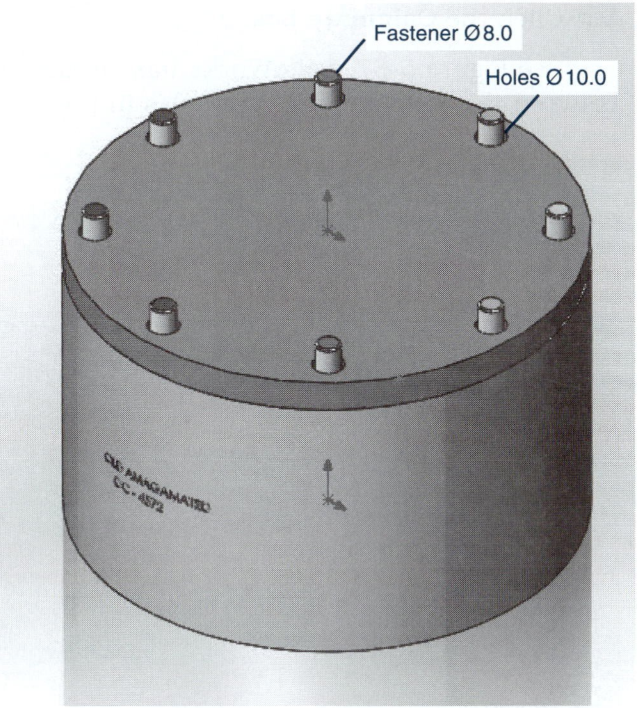

The Ø holes should create a clearance between the **Fasteners** and the **Cover Plate**'s holes.

To Verify the Clearance

1 Click the **Tools** tab and select the **Clearance Verification** tool.

See Figure 5-66.

Figure 5-66

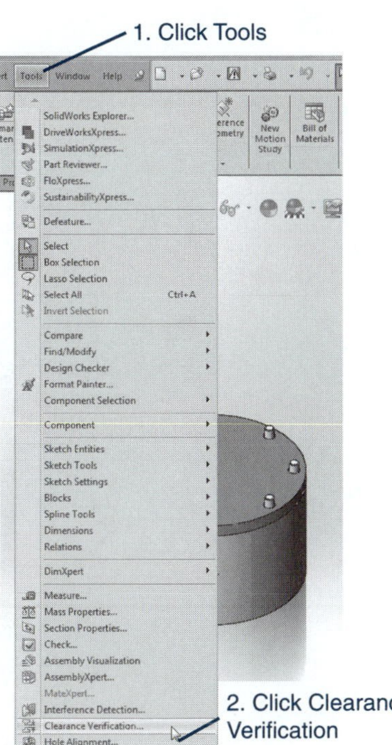

Chapter 5

2 Click a fastener and the **Cover Plate**.

The file names of the part should appear in the **Selected Components** box.

3 Click the **Calculate** box.

See Figure 5-67. SolidWorks has verified that there is a 1.0 clearance between the fasteners and the holes in the cover plate.

Figure 5-67

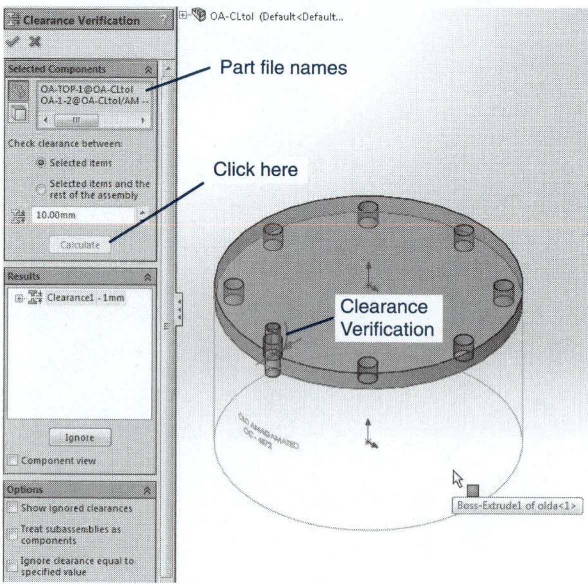

To Remove the Interference

Figure 5-68 shows the assembly originally shown in Figure 5-61. There is an interference between the post on the Base and the Hole in the Top. The

Figure 5-68

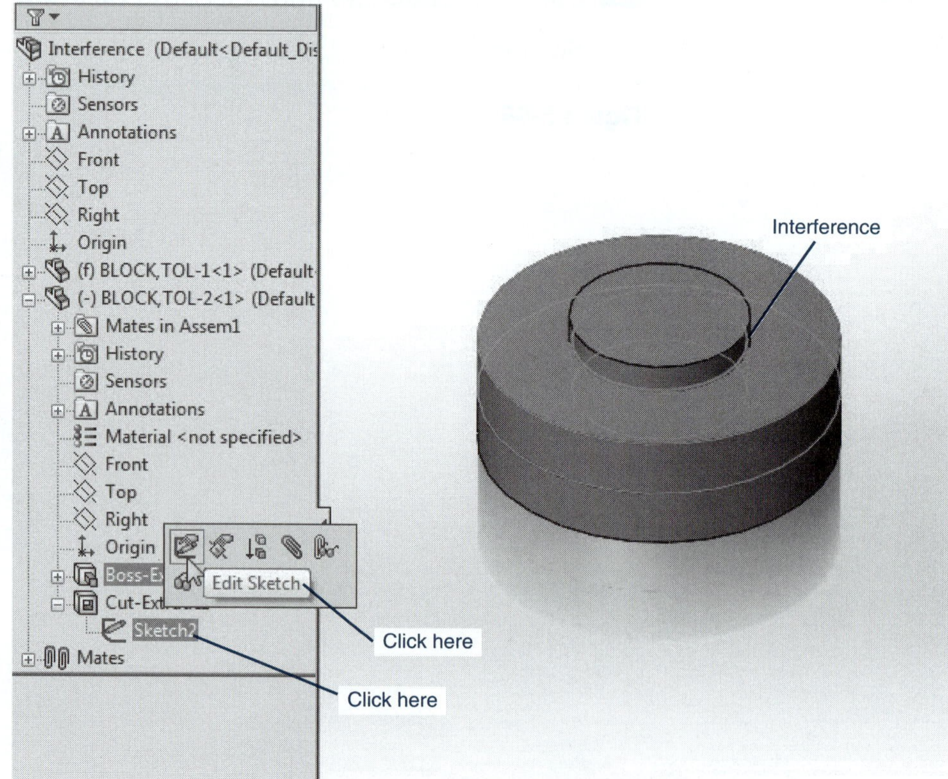

interference can be removed by going back to the original Top part drawing or it can be removed in the assembly mode.

1. Locate the **Sketch** that created the hole in the **Feature Manager Design tree**.
2. Click the **Sketch** listing.
3. Click the **Edit Sketch** option.

The hole's diameter value will appear. See Figure 5-69. In this example the diameter is **Ø58**.

Figure 5-69

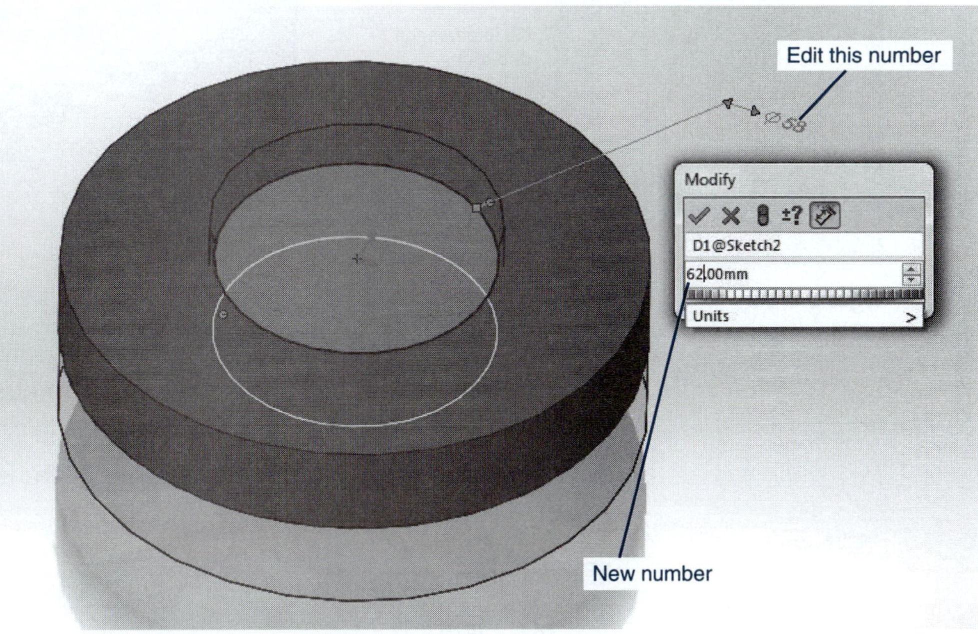

4. Double-click the **Ø58** value and change it to **Ø62**.
5. Click the **Exit Sketch** icon in the upper right corner of the drawing screen and the **Return to the Feature Manager Design tree** icon.

   There is now a clearance between the Base and Top parts.
6. Click the **Interference Detection** tool again.
7. Click the **Calculate** box.

The results box will indicate that no interferences were found. See Figure 5-70.

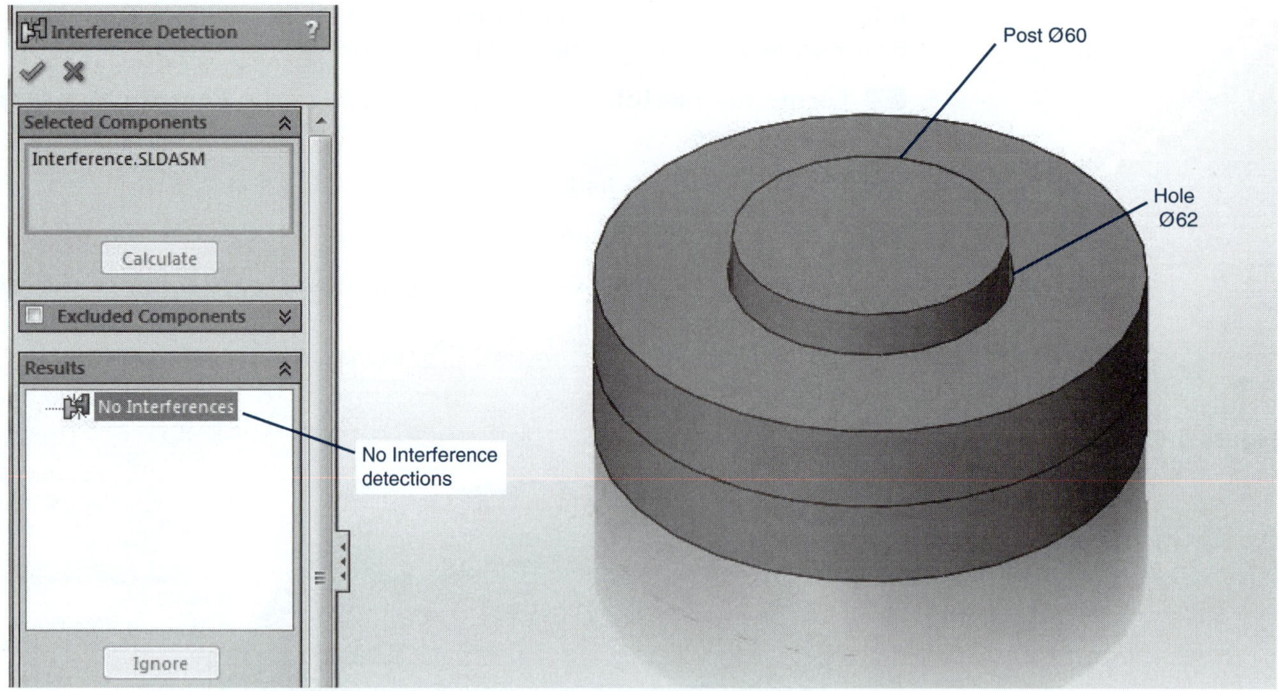

Figure 5-70

To Verify That a Clearance Exists

The **Clearance Verification** tool is used to ensure that clearances have been created.

1. Click the **Tools** heading at the top of the screen.
2. Click the **Clearance Verification** tool.
3. Click the two parts.

   See Figure 5-71.

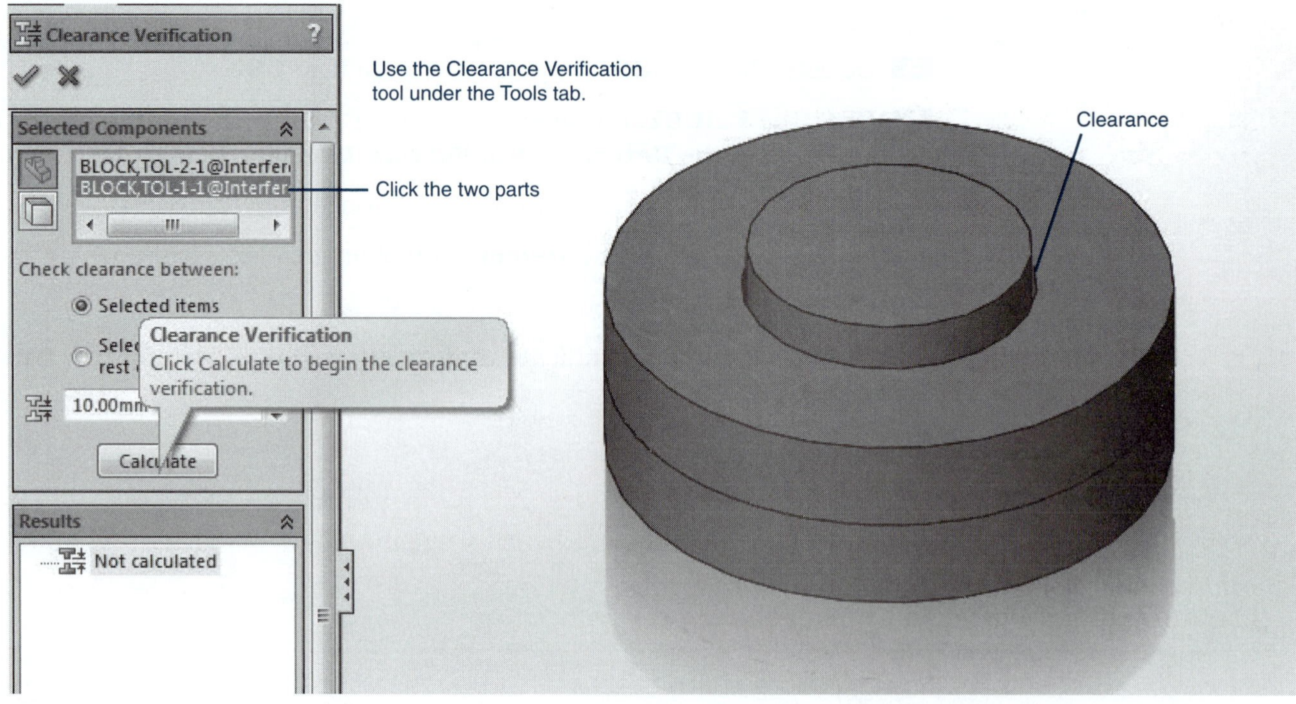

Figure 5-71

4 Click the **Calculate** box.

The **Clearance Verification** will be listed in the **Results** box and the onscreen drawing will show the clearance. See Figure 5-72.

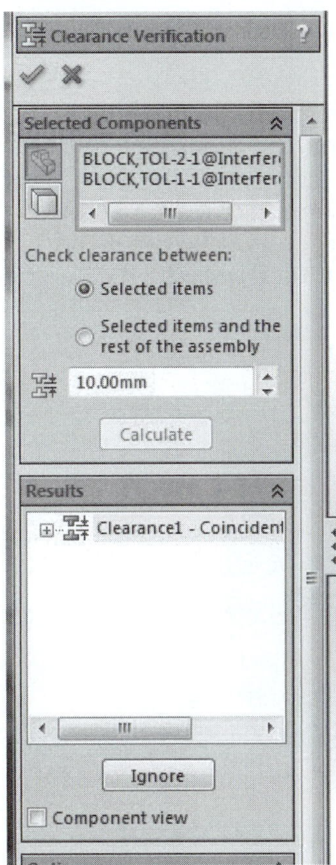

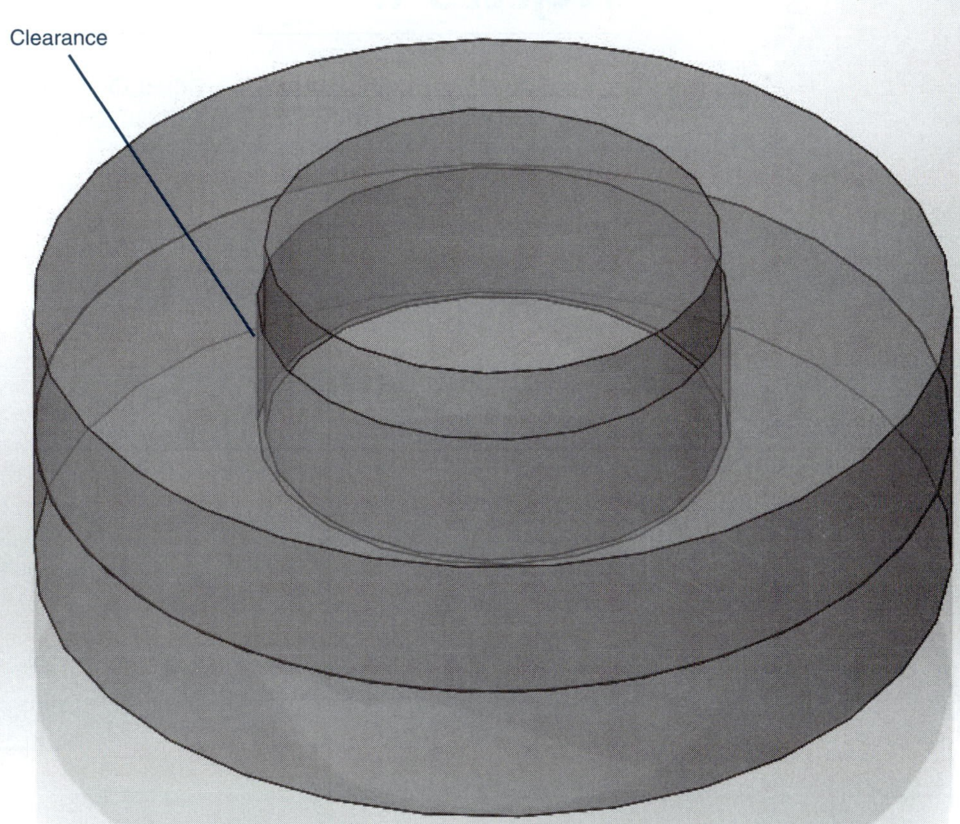

Figure 5-72

Chapter Projects

Project 5-1:

Create a **Part** document of the SQBLOCK using the given dimensions. Create assemblies using two SQBLOCKS, positioning the blocks as shown in Figures P5-1A through P5-1G.

Pages 342 through 345 show a group of parts. These parts are used to create the assemblies presented as problems in this section. Use the given descriptions, part numbers, and materials when creating BOMs for the assemblies.

Figure P5-1
MILLIMETERS

P5-1A

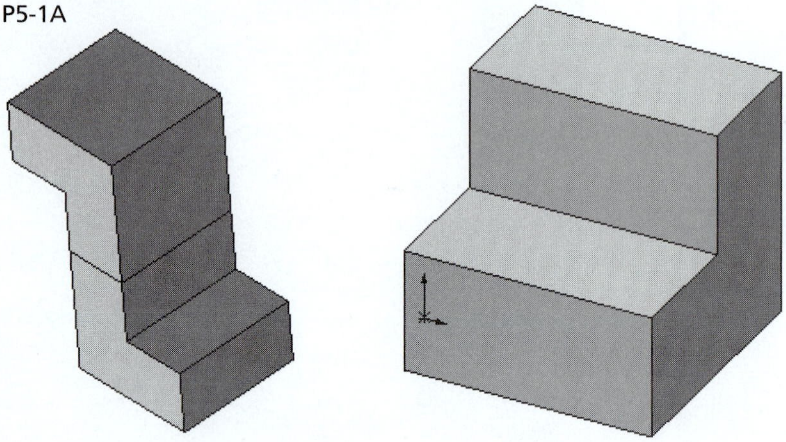

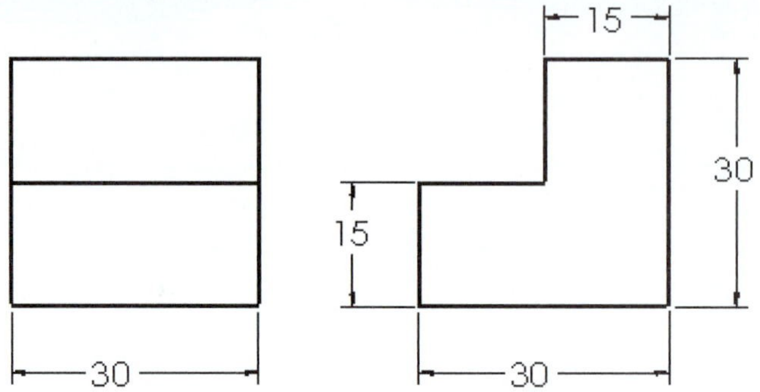

15

30

15

30

30

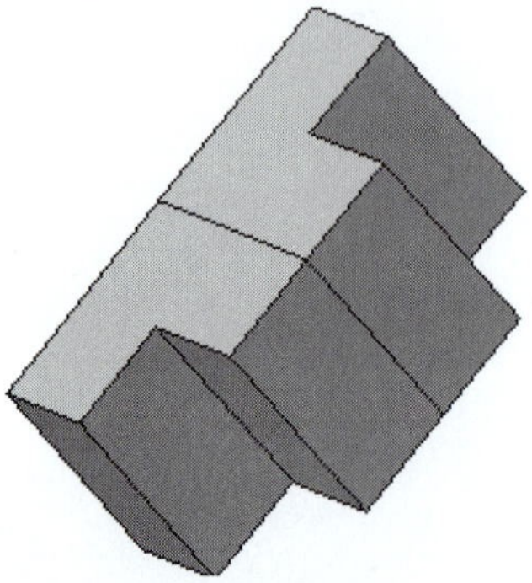

P5-1C

P5-1B

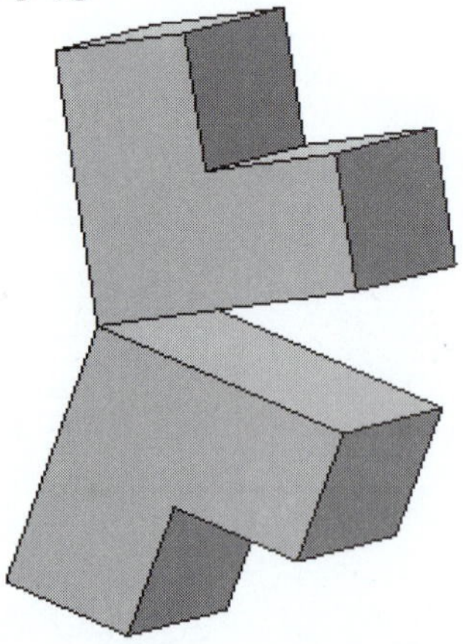

P5-1D

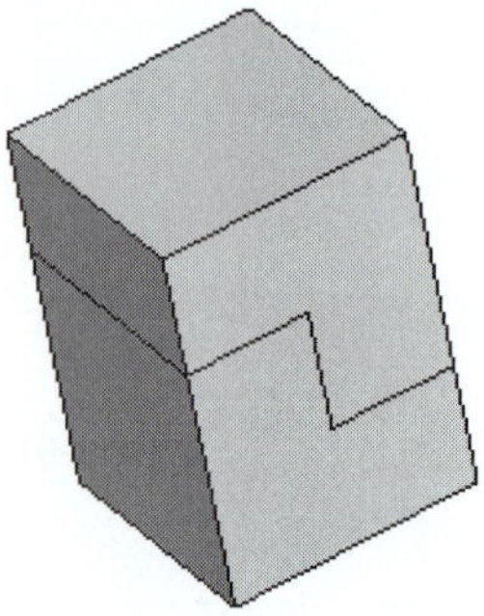

Figure P5-1
(Continued)

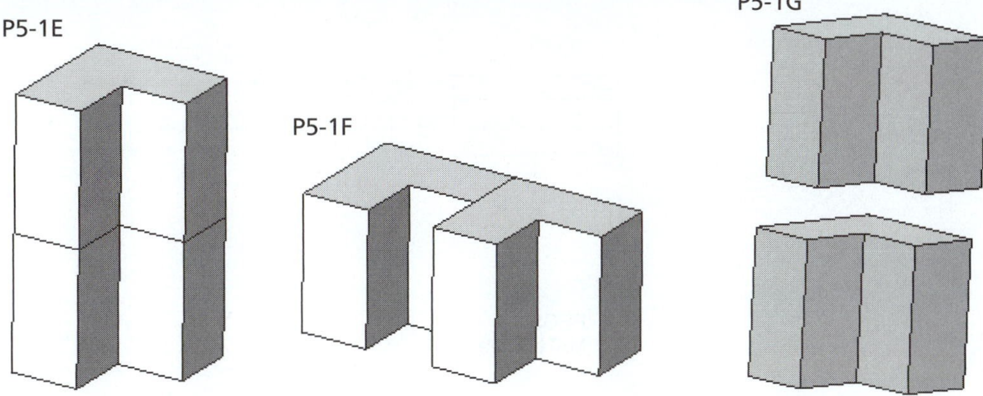

Project 5-2:

Redraw the following models and save them as **Standard (mm).ipn** files. All dimensions are in millimeters.

Figure P5-2
MILLIMETERS

SPACER
P/N ME311-1
MATL: SAE 1020 Steel

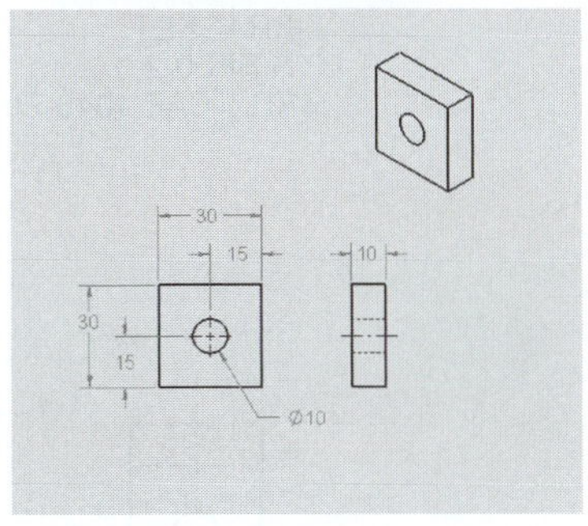

SPACER DOUBLE
P/N ME311-2
MATL: SAE 1020 Steel

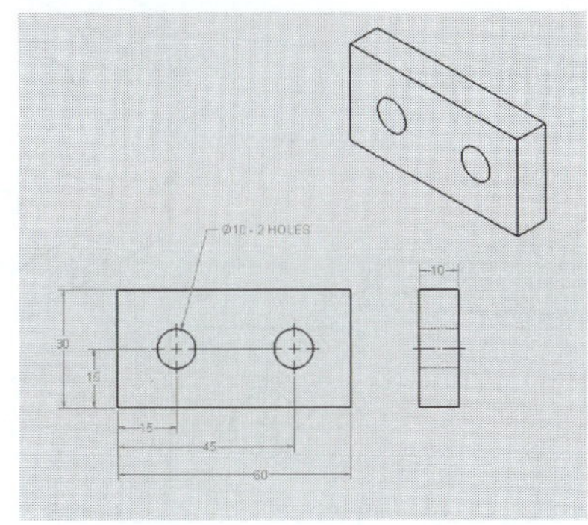

SPACER TRIPLE
P/N ME311-3
MATL: SAE 1020 Steel

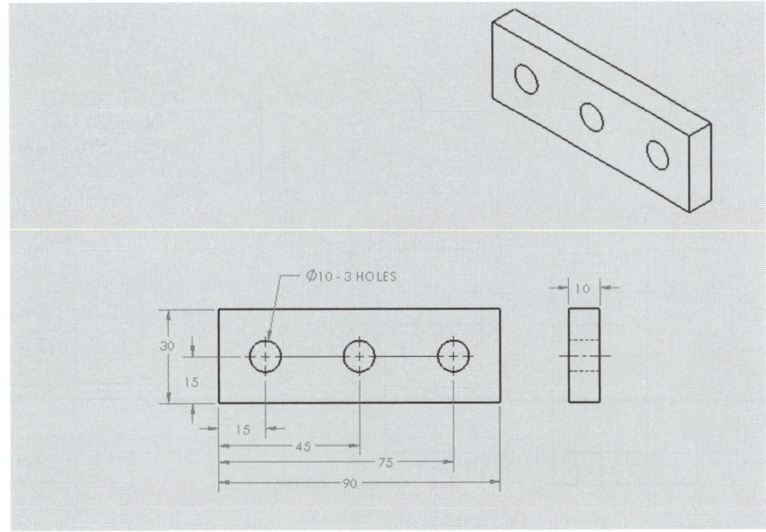

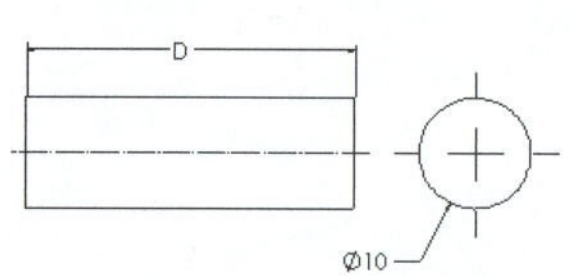

DESCRIPTION	PART NO.	D-VALUE
PEG, SHORT	PG20-1	20
PEG	PG30-1	30
PEG, LONG	PG40-1	40

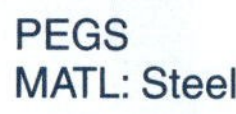

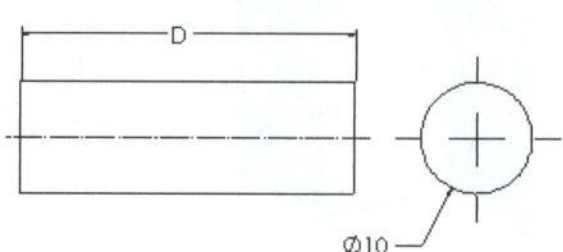

DESCRIPTION	PART NO.	D-VALUE
PEG, SHORT	PG20-1	20
PEG	PG30-1	30
PEG, LONG	PG40-1	40

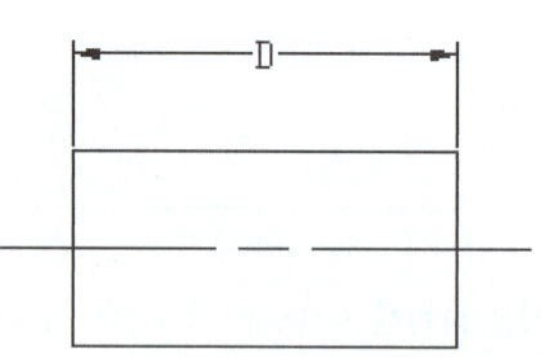

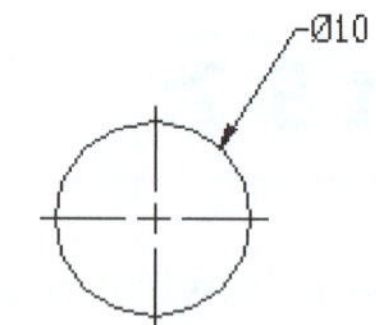

DESCRIPTION	PART NO.	D
PEG, SHORT	PG20-1	20
PEG	PG30-1	30
PEG, LONG	PG40-1	40

ALL DISTANCES IN MILLIMETERS

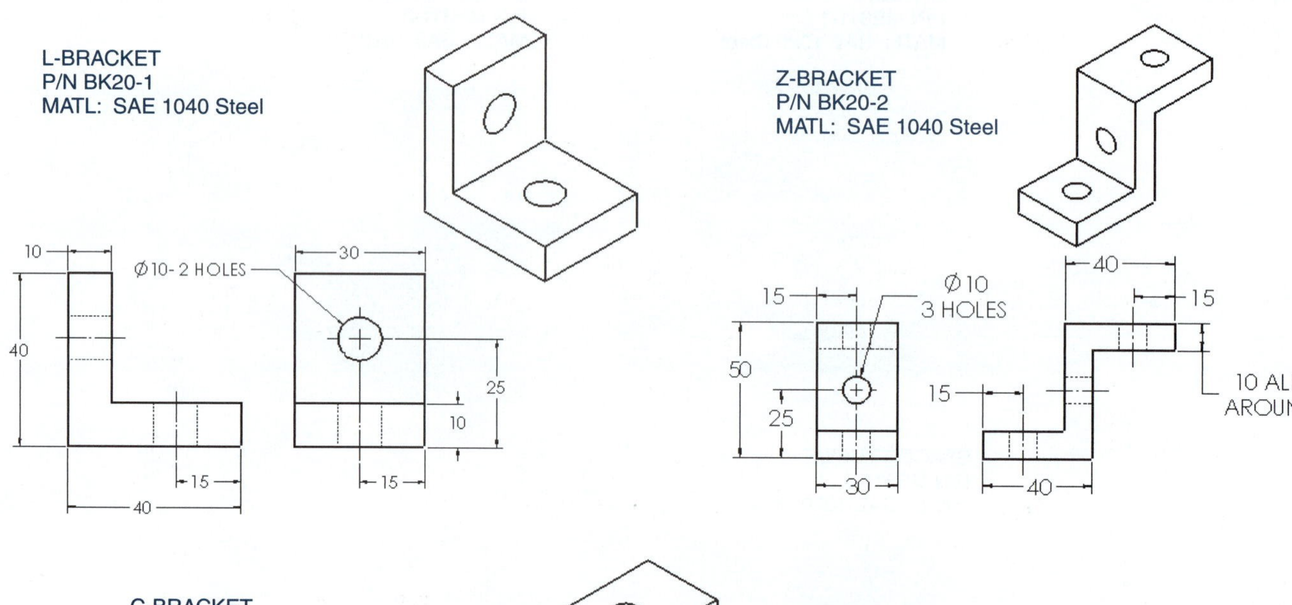

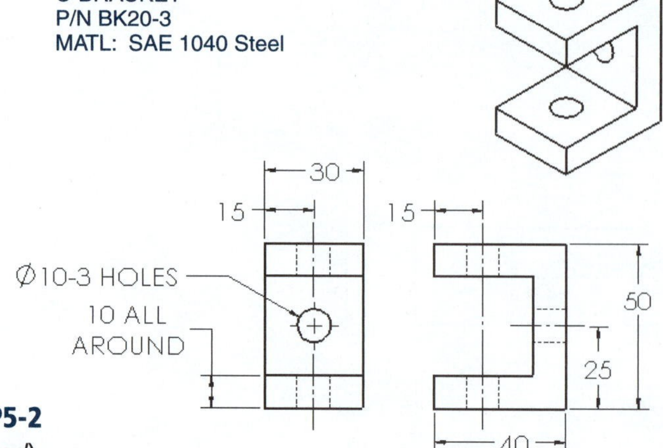

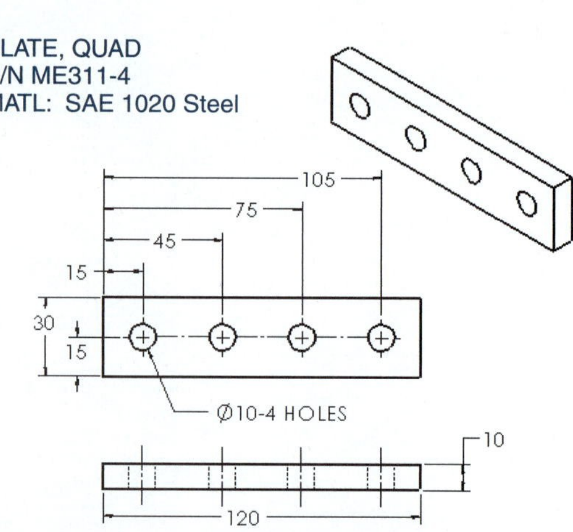

Figure P5-2
(Continued)

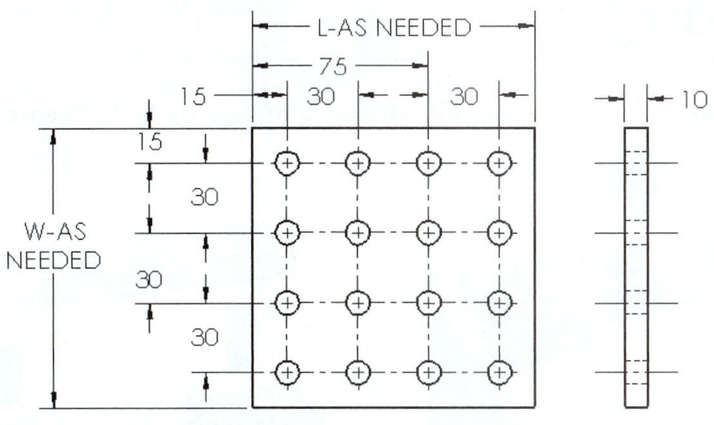

PART NO.	TOTAL NO. OF HOLES	L	W	HOLE PATTERN
PL110-9	9	90	90	3×3
PL110-16	16	120	120	4×4
PL110-6	6	60	90	2×3
PL110-8	8	60	120	2×4
PL110-4	4	60	60	2×2

PLATE, BASE
P/N ME311-5
MATL: SAE 1020 Steel

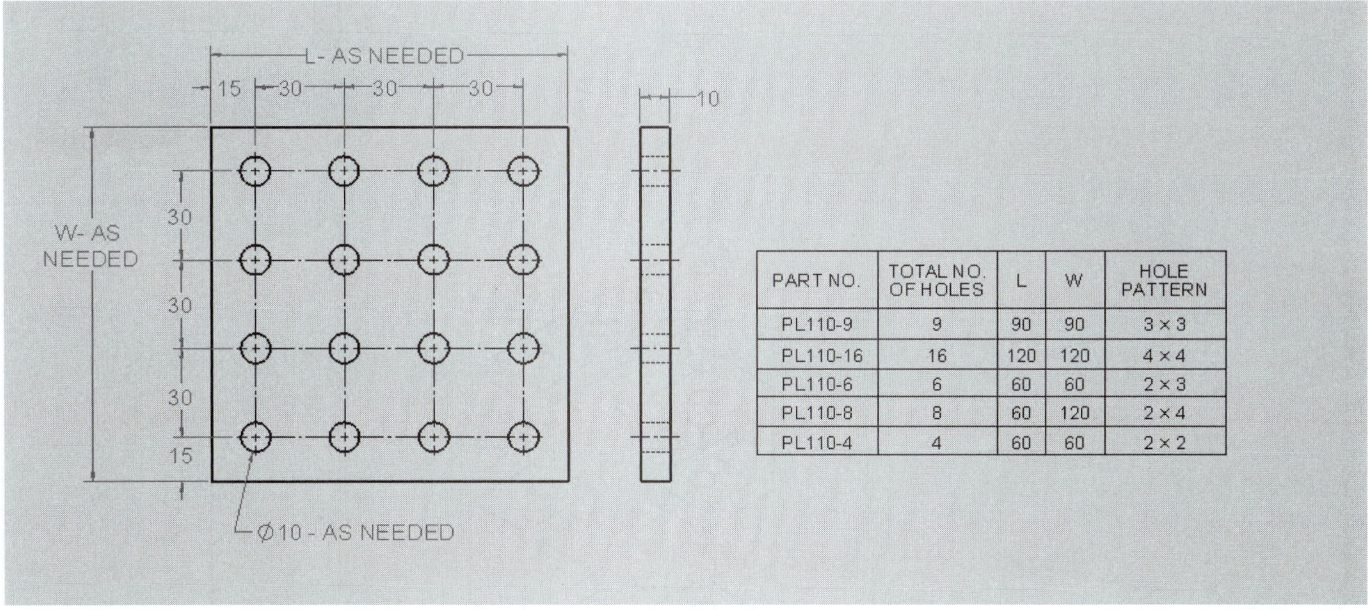

PART NO.	TOTAL NO. OF HOLES	L	W	HOLE PATTERN
PL110-9	9	90	90	3 × 3
PL110-16	16	120	120	4 × 4
PL110-6	6	60	60	2 × 3
PL110-8	8	60	120	2 × 4
PL110-4	4	60	60	2 × 2

Figure P5-2
(*Continued*)

Project 5-3:

Draw an exploded isometric assembly drawing of Assembly 1. Create a BOM.

Figure P5-3
MILLIMETERS

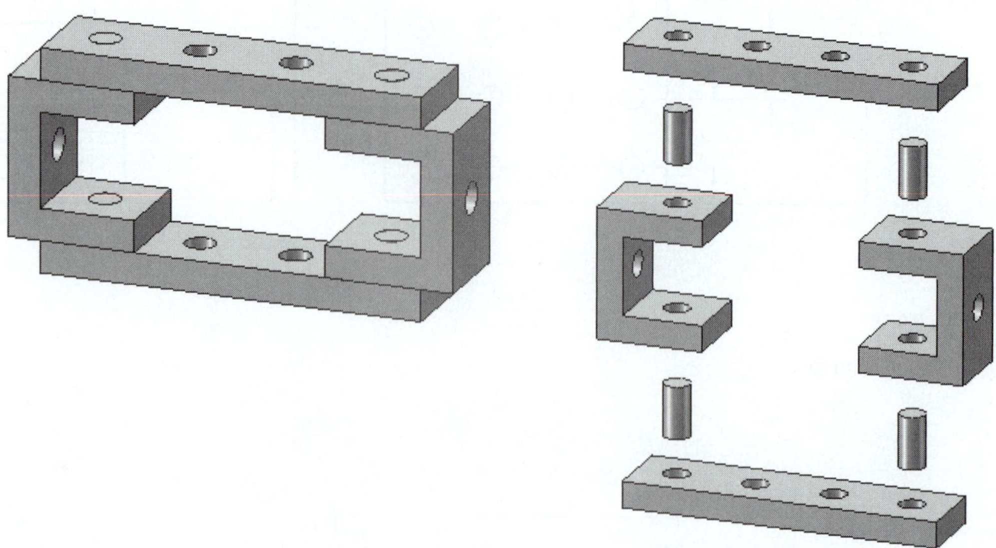

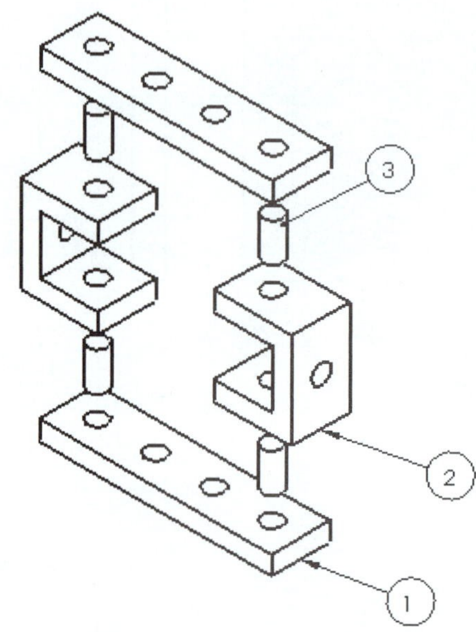

ITEM NO.	PART NUMBER	DESCRIPTION	QTY.
1	ME311-4	PLATE, QUAD	2
2	BK20-3	C-BRACKET	2
3	PG20-1	Ø12×20 PEG	4

Project 5-4:

Draw an exploded isometric assembly drawing of Assembly 2. Create a BOM.

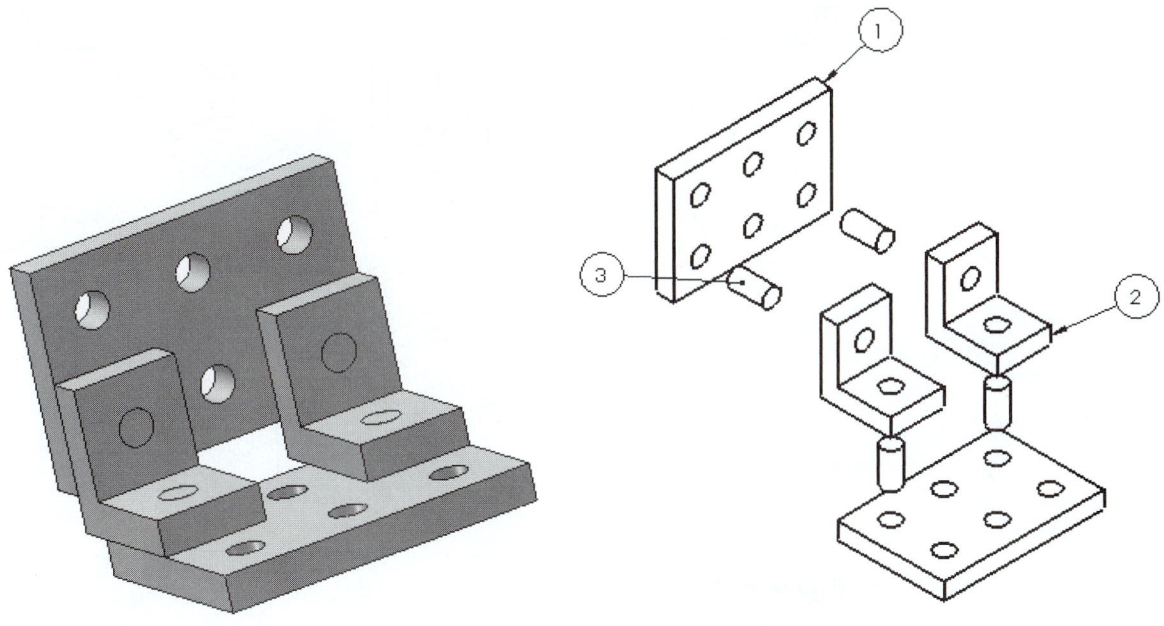

Figure P5-4
MILLIMETERS

Figure P5-5
MILLIMETERS

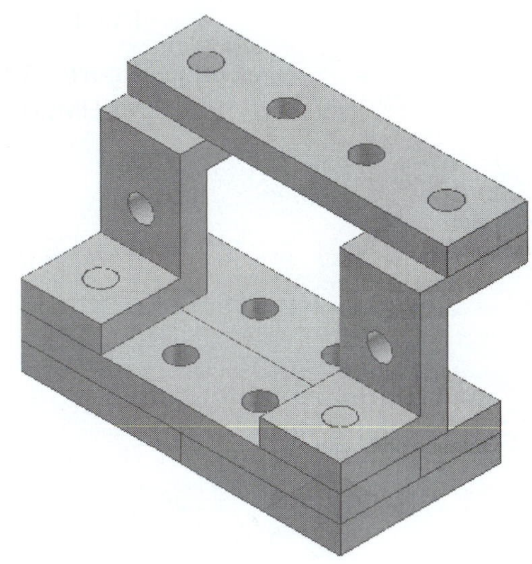

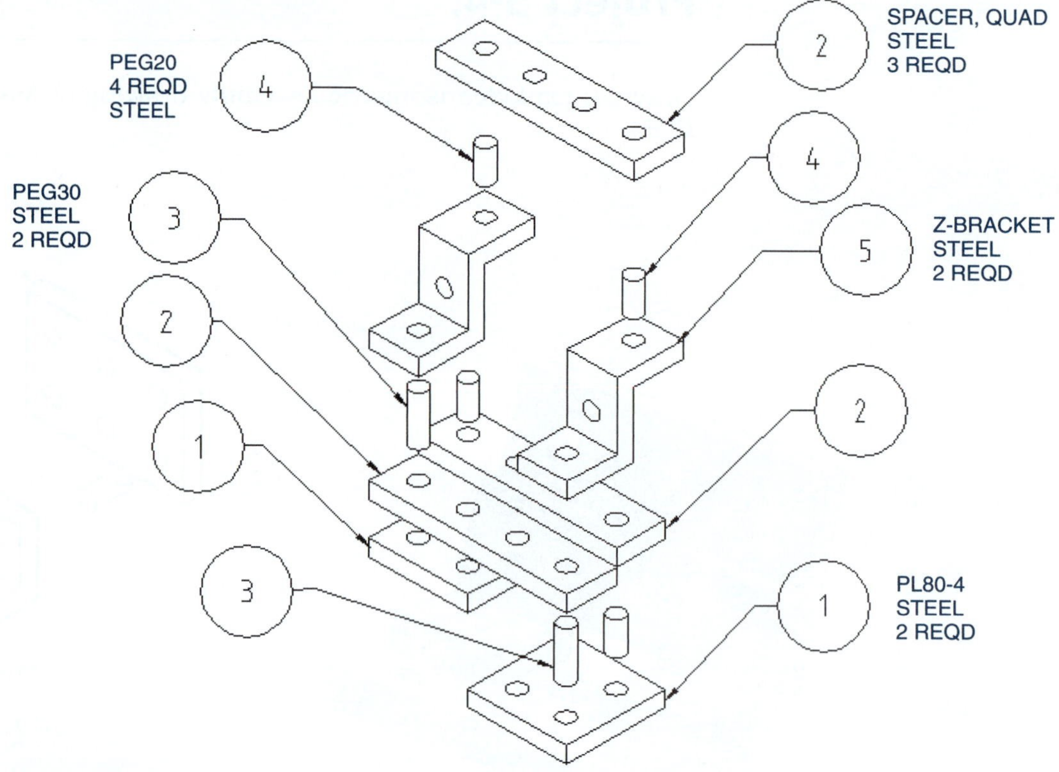

Figure P5-5
(*Continued*)

Project 5-6:

Draw an exploded isometric assembly drawing of Assembly 3. Create a BOM.

Figure P5-6
MILLIMETERS

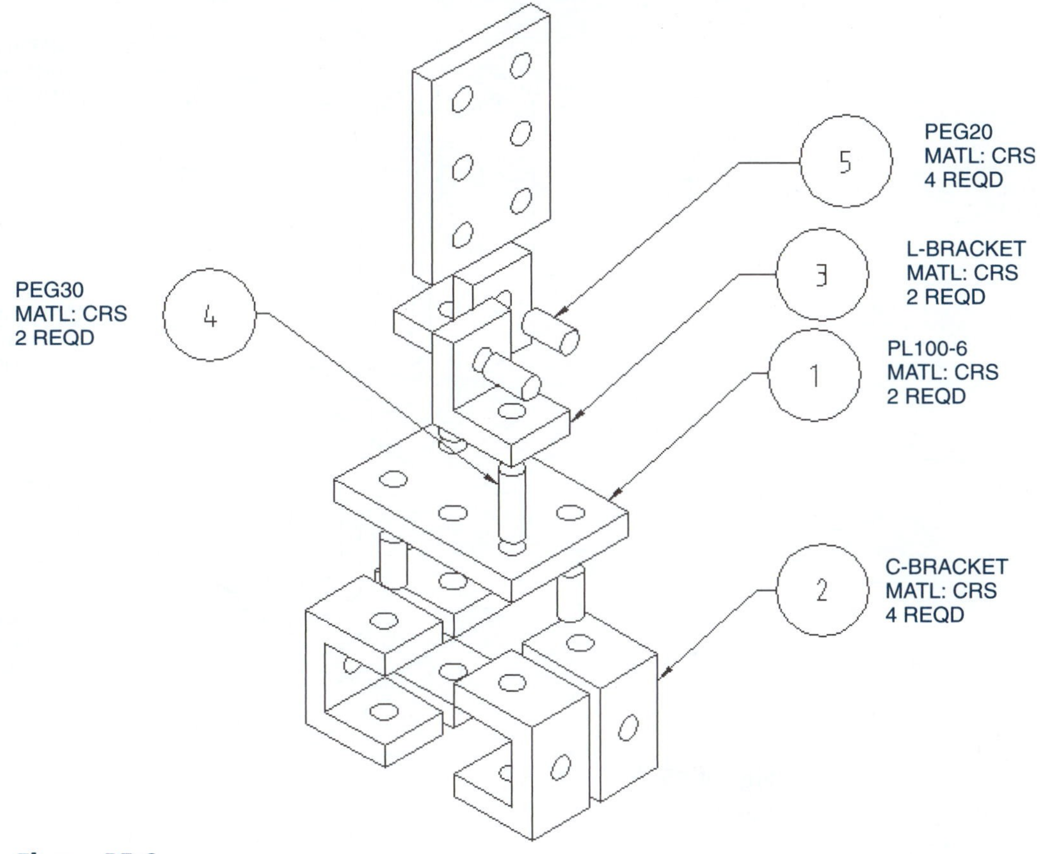

Figure P5-6
(*Continued*)

Project 5-7:

Draw an exploded isometric assembly drawing of Assembly 4. Create a BOM.

Figure P5-7
MILLIMETERS

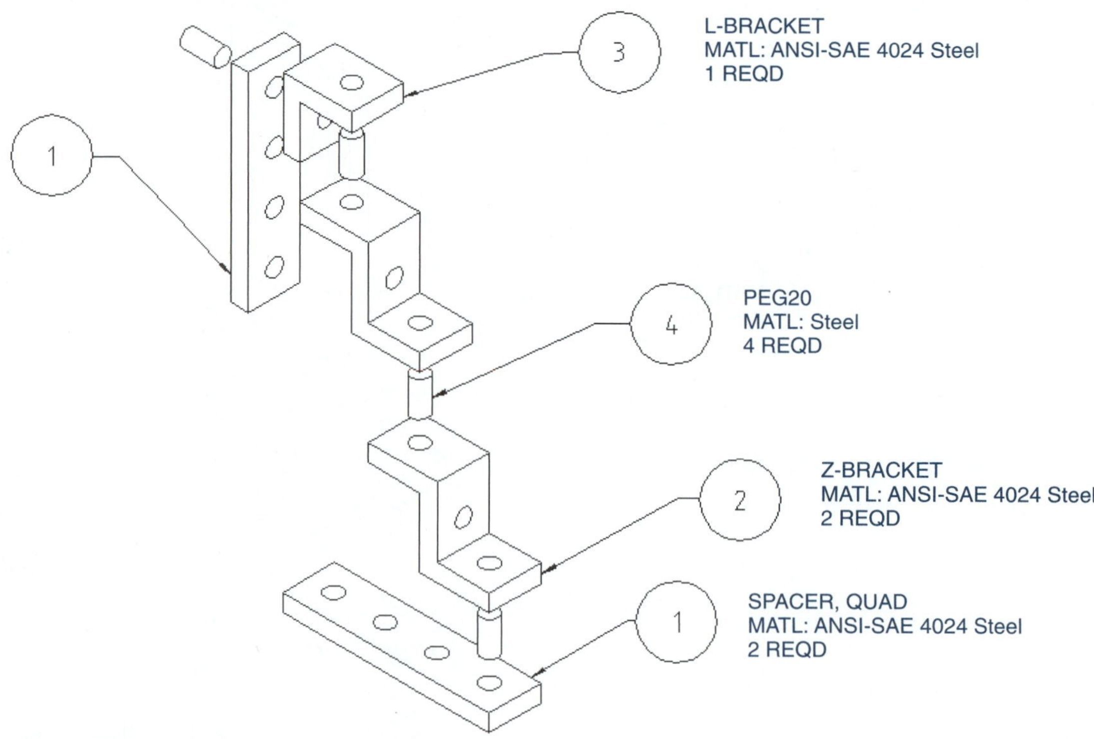

Figure P5-7
(*Continued*)

Project 5-8:

Draw an exploded isometric assembly drawing of Assembly 5. Create a BOM.

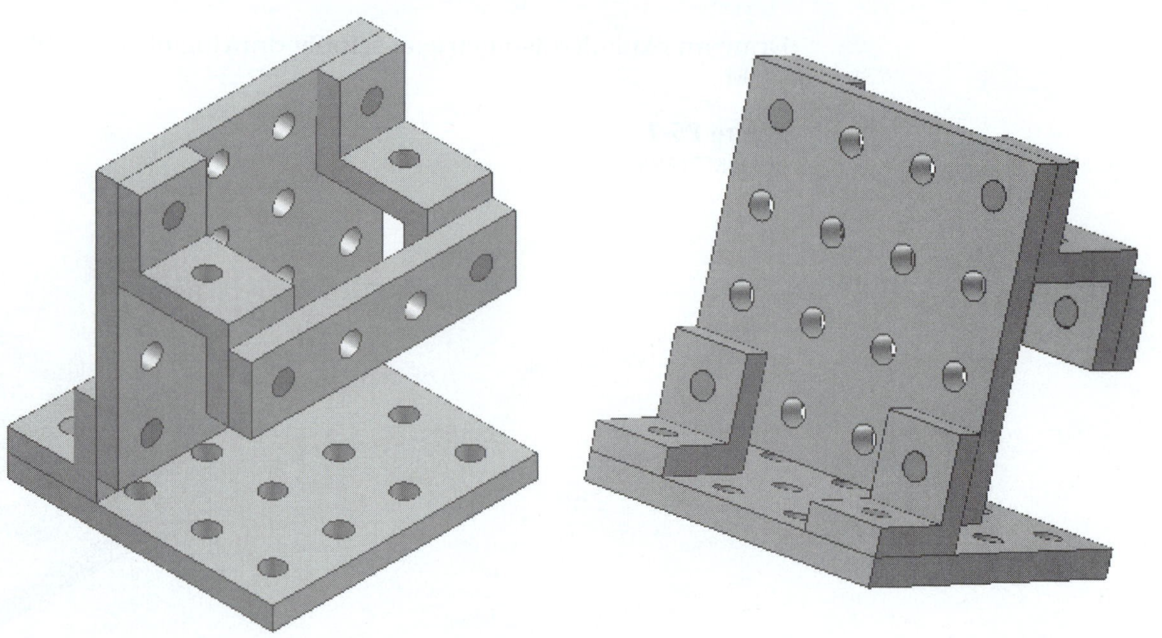

Figure P5-8
MILLIMETERS

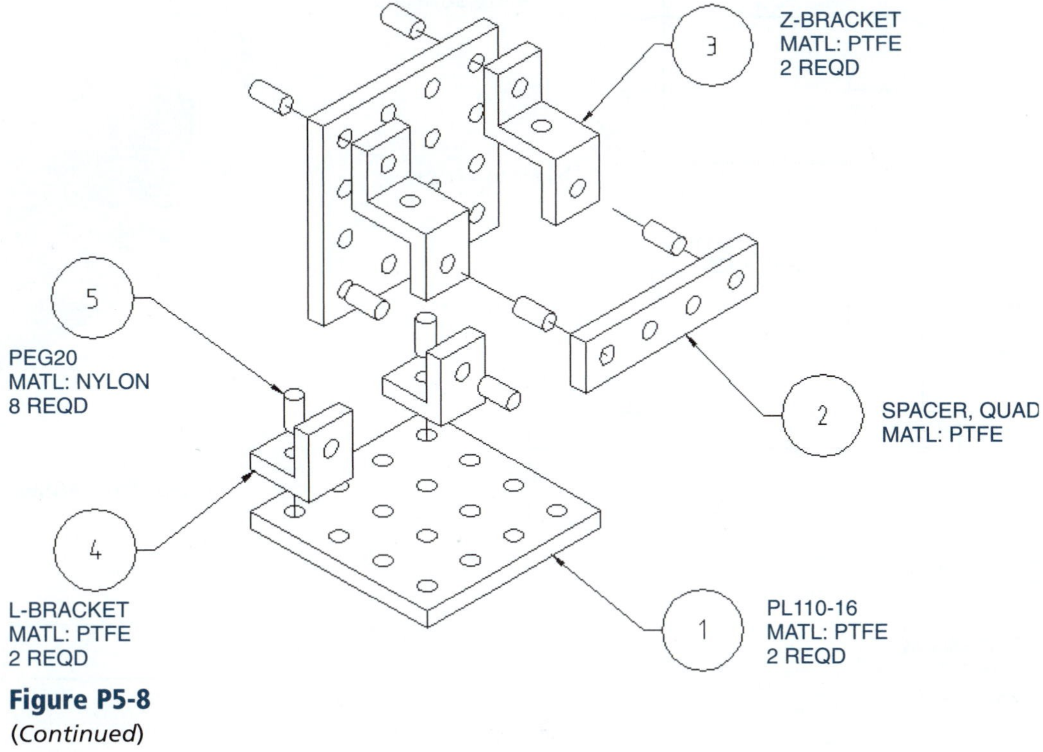

Figure P5-8
(*Continued*)

Project 5-9:

Create an original assembly based on the parts shown on pages 323–324. Include a scene, an exploded isometric drawing with assembly numbers, and a BOM. Use at least 12 parts.

Project 5-10:

Draw the ROTATOR ASSEMBLY shown. Include the following:

A. An assembly drawing

B. An exploded isometric drawing with assembly numbers

C. A parts list

D. An animated assembly drawing; the LINKs should rotate relative to the PLATE. The LINKs should carry the CROSSLINK. The CROSSLINK should remain parallel during the rotation.

NOTE

This assembly was used in the section on animating assemblies. See page 325.

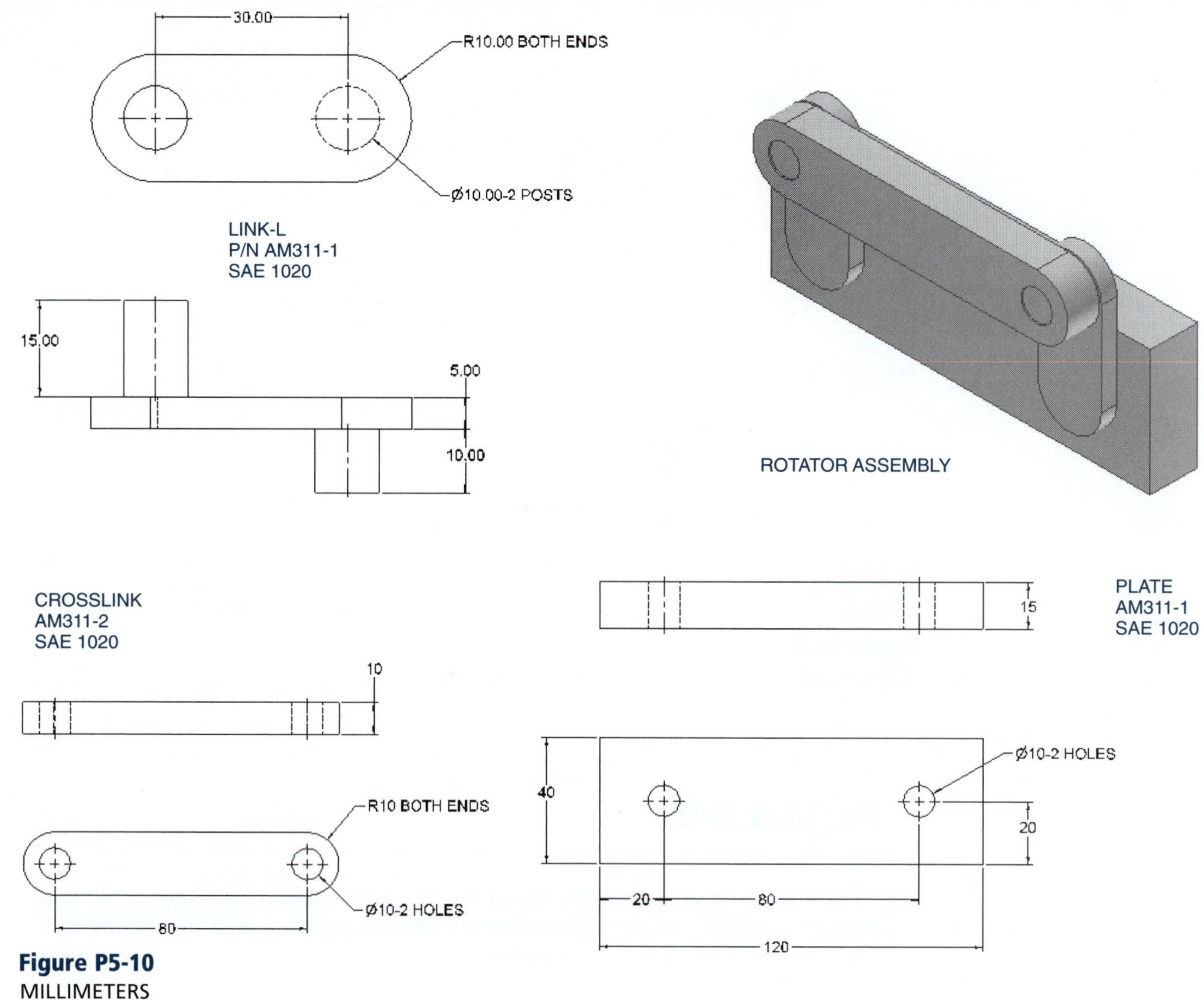

Figure P5-10
MILLIMETERS

Project 5-11:

Draw the FLY ASSEMBLY shown. Include the following:

A. An assembly drawing

B. An exploded isometric drawing with assembly numbers

C. A parts list

D. An animated assembly drawing; the FLYLINK should rotate around the SUPPORT base.

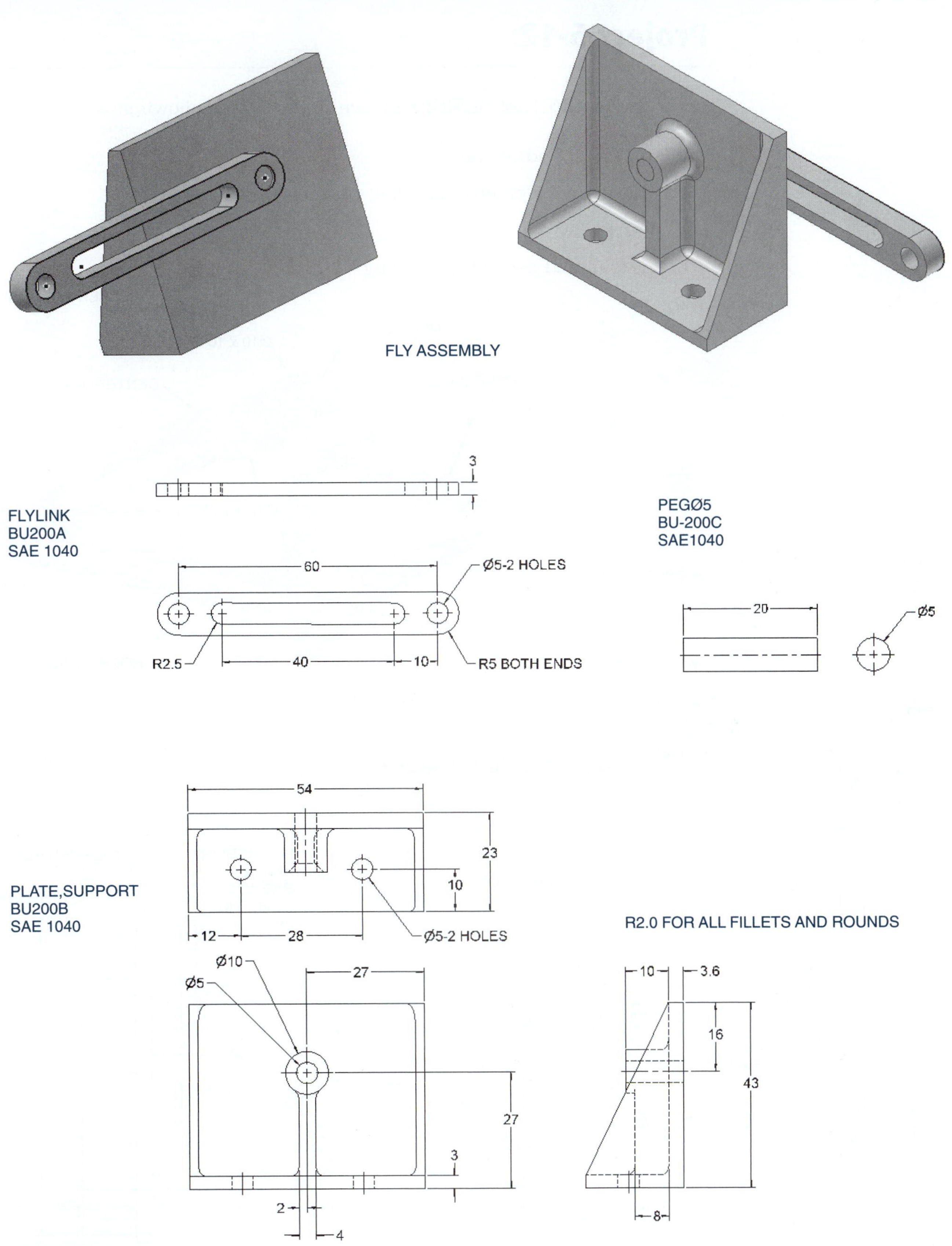

Figure P5-11
MILLIMETERS

Project 5-12:

Draw the ROCKER ASSEMBLY shown. Include the following:

A. An assembly drawing
B. An exploded isometric drawing with assembly numbers
C. A parts list
D. An animated assembly drawing

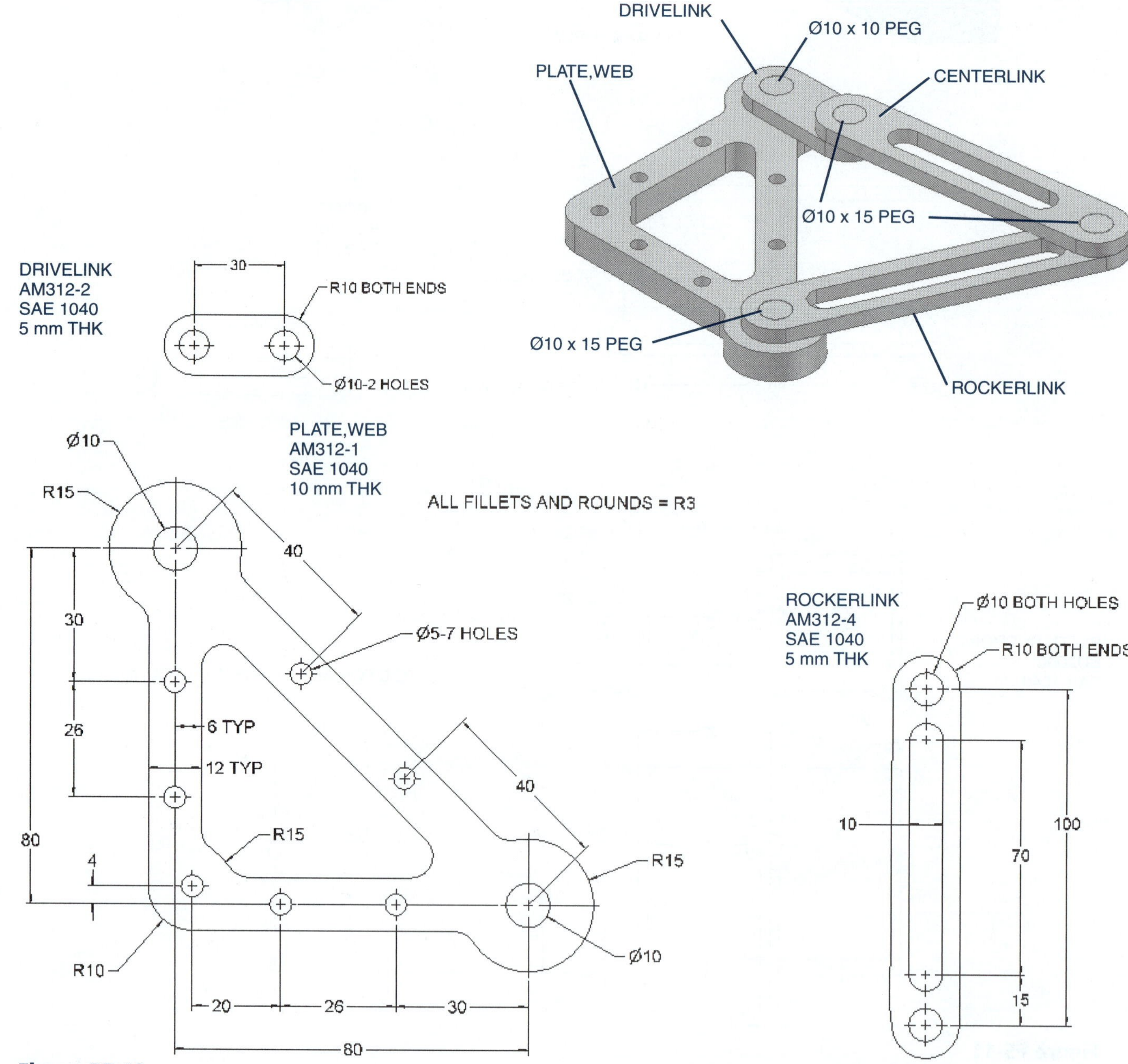

Figure P5-12
MILLIMETERS

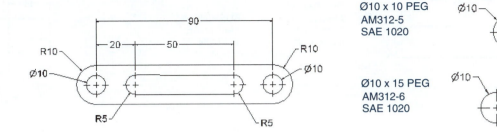

CENTERLINK
AM312-3
SAE 1040
5 mm THK

90
20
50
R10
R10
Ø10
Ø10
R5
R5

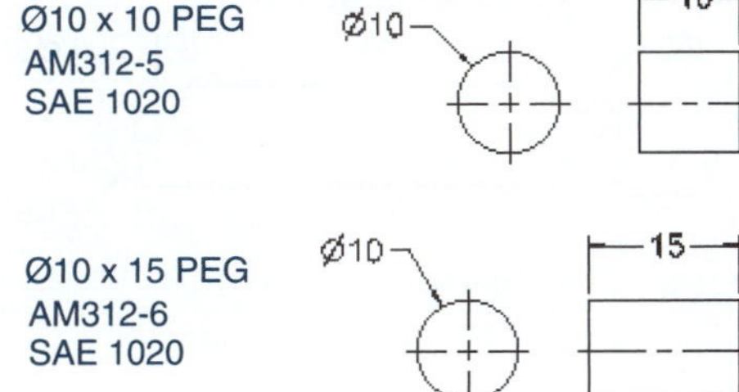

Figure P5-12
(*Continued*)

Project 5-13:

Draw the LINK ASSEMBLY shown. Include the following:

A. An assembly drawing

B. An exploded isometric drawing with assembly numbers

C. A parts list

D. An animated assembly drawing; the HOLDER ARM should rotate between –30° and +30°.

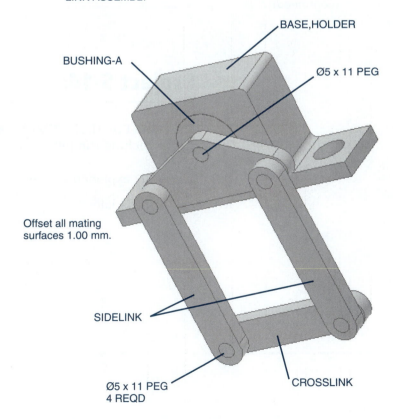

LINK ASSEMBLY
BASE,HOLDER
BUSHING-A
Ø5 x 11 PEG
Offset all mating surfaces 1.00 mm.
SIDELINK
Ø5 x 11 PEG
4 REQD
CROSSLINK

HOLDER ARM
AM-311-A3
7075-T6 AL
5 mm THK

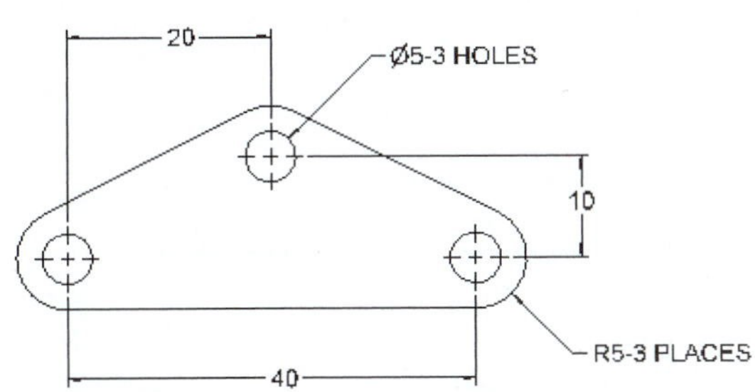

Figure P5-13

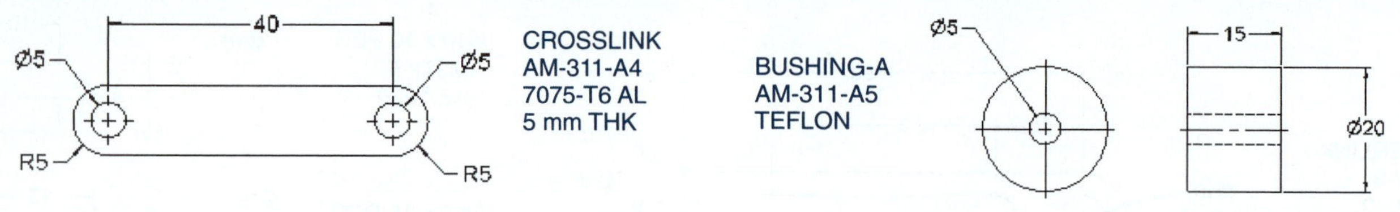

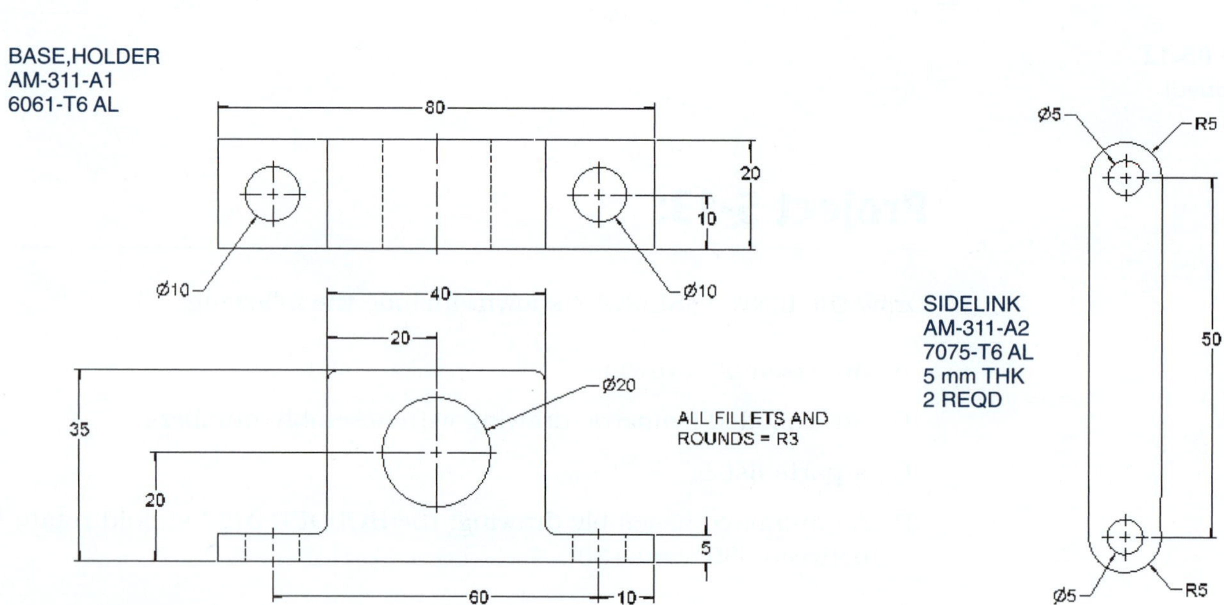

Figure P5-13
(Continued)

Project 5-14:

Draw the PIVOT ASSEMBLY shown using the dimensioned components given. Include the following:

A. A 3D exploded isometric drawing

B. A parts list

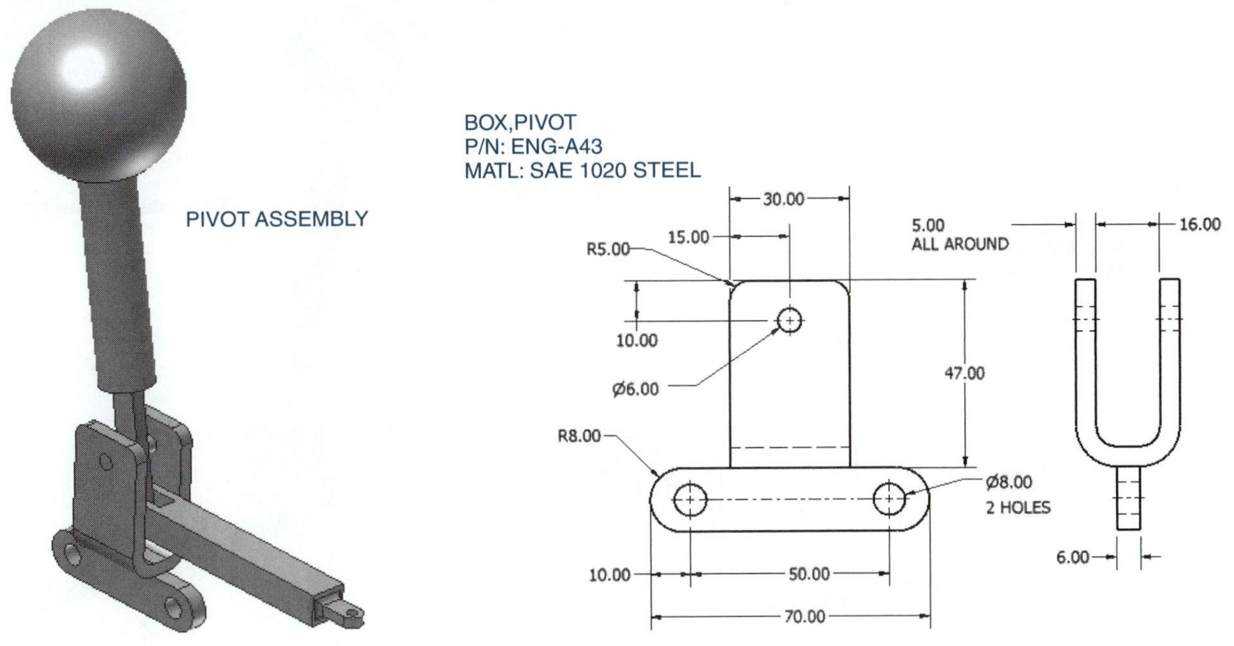
PIVOT ASSEMBLY
BOX,PIVOT
P/N: ENG-A43
MATL: SAE 1020 STEEL
30.00
15.00
R5.00
5.00
ALL AROUND
16.00
10.00
Ø6.00
47.00
R8.00
Ø8.00
2 HOLES
6.00
10.00
50.00
70.00

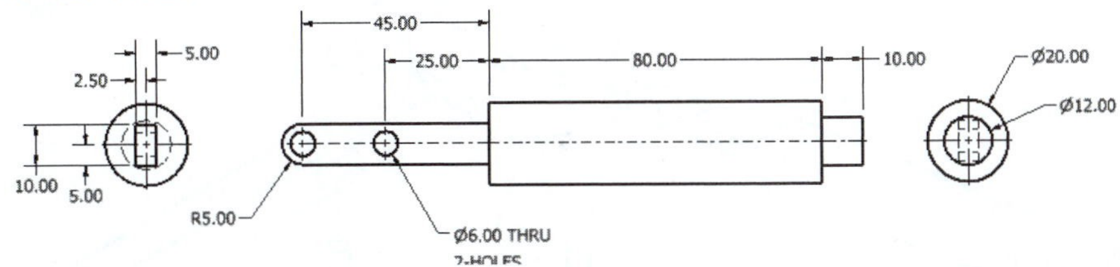
POST,HANDLE
P/N: ENG-A44
MATL: SAE 1020 STEEL
45.00
5.00
2.50
25.00
80.00
10.00
Ø20.00
Ø12.00
10.00
5.00
R5.00
Ø6.00 THRU

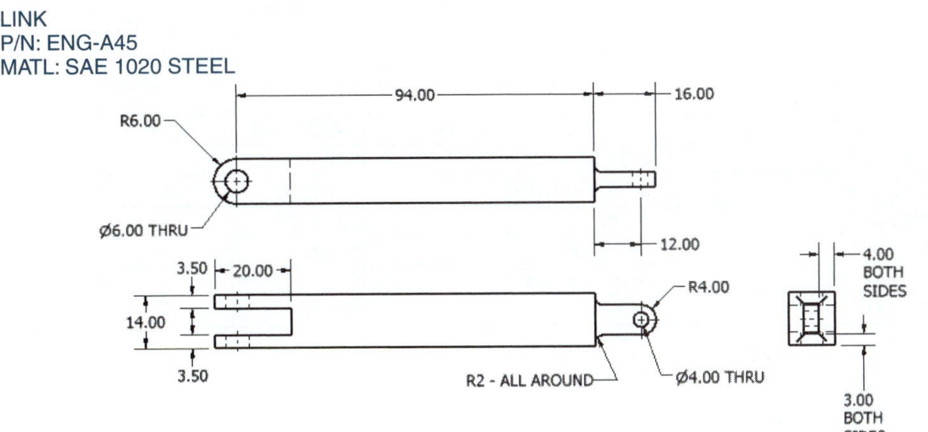
LINK
P/N: ENG-A45
MATL: SAE 1020 STEEL
94.00
16.00
R6.00
Ø6.00 THRU
12.00
3.50
20.00
14.00
3.50
R4.00
R2 - ALL AROUND
Ø4.00 THRU
4.00
BOTH
SIDES
3.00
BOTH
SIDES

Figure P5-14
MILLIMETERS

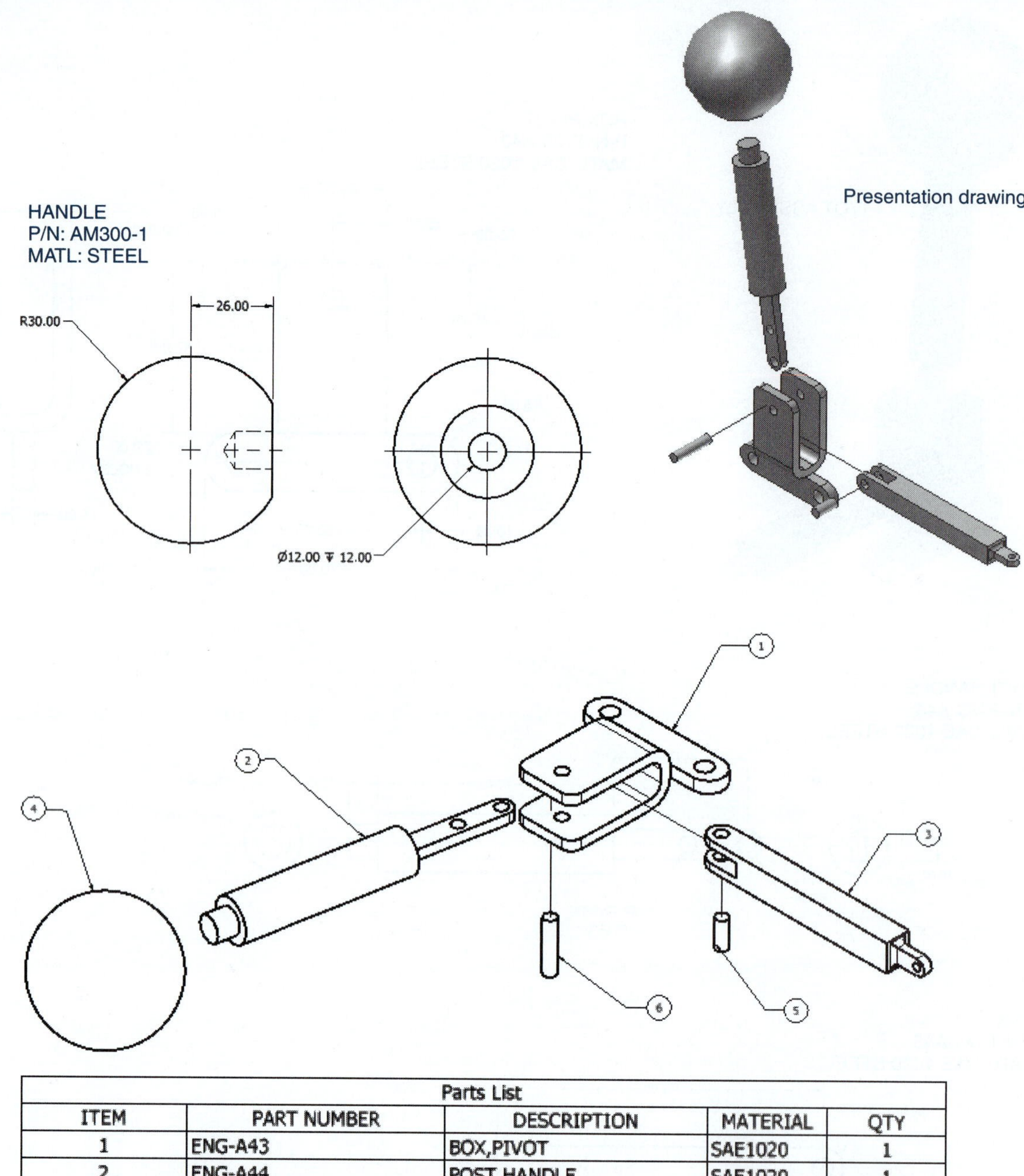

Parts List				
ITEM	PART NUMBER	DESCRIPTION	MATERIAL	QTY
1	ENG-A43	BOX,PIVOT	SAE1020	1
2	ENG-A44	POST,HANDLE	SAE1020	1
3	ENG-A45	LINK	SAE1020	1
4	AM300-1	HANDLE	STEEL	1
5	EK-132	POST-Ø6x14	STEEL	1
6	EK-131	POST-Ø6x26	STEEL	1

Figure P5-14
(Continued)

Project 5-15

Soma Cube Puzzle

The Soma Cube puzzle was invented by Piet Hein in 1933. In makes an interesting assembly drawing problem. Each of the 27 cubes used is the same size. They are combined to form seven solid shapes, each of which has an internal edge. When combined in the correct format they will form a 3 × 3 × 3 cube.

1. Specify a cube size.
2. Create the seven different shapes shown below.
3. Assemble the seven shapes to create a 3 × 3 × 3 cube.
4. As assigned by your instructor, build and assemble the seven shapes.

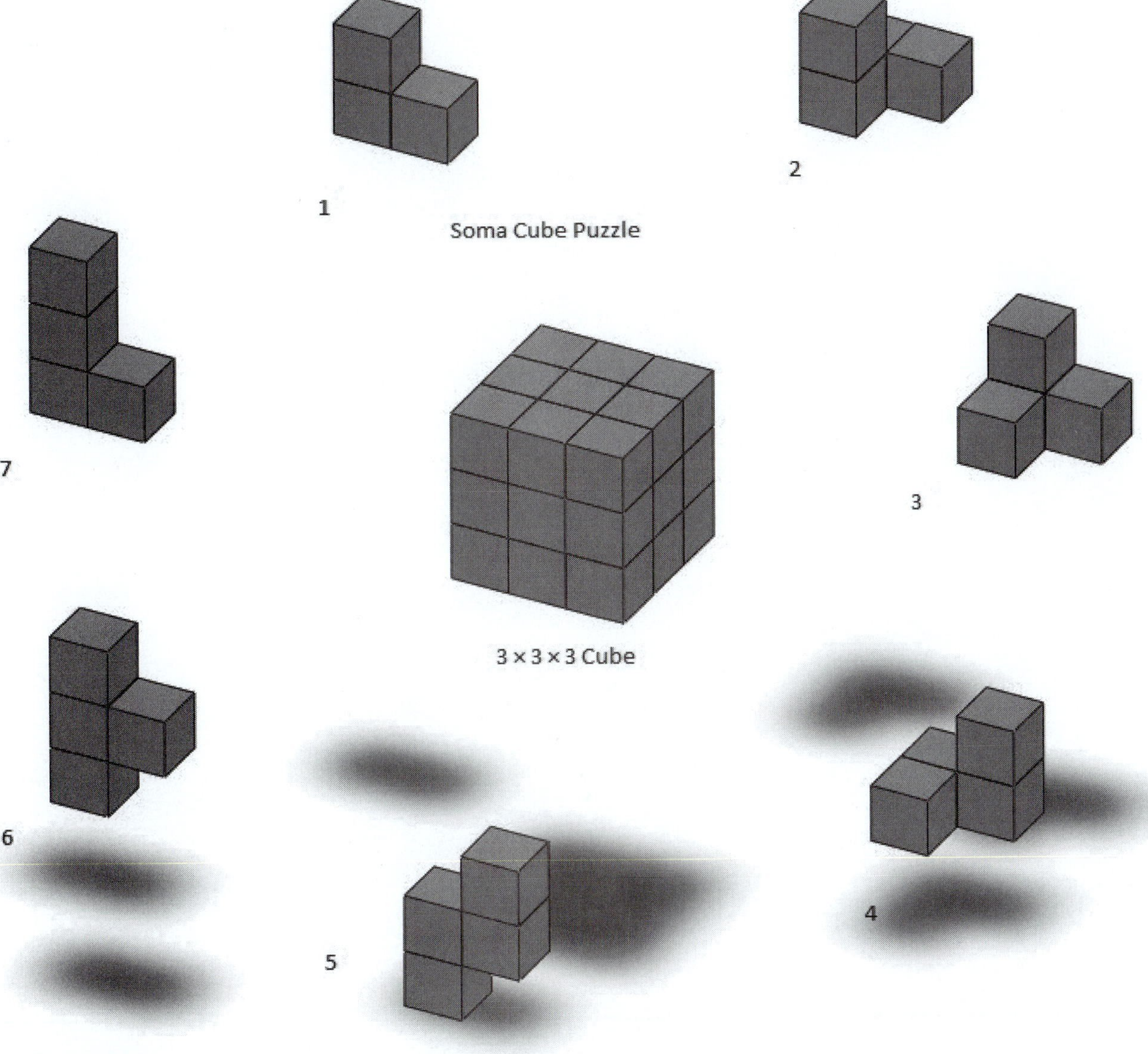

Figure P5-15
MILLIMETERS

chapter six
Threads and Fasteners

CHAPTER OBJECTIVES

- Learn thread terminology and conventions
- Learn how to draw threads
- Learn how to size both internal and external threads
- Learn how to use standard-sized threads
- Learn how to use and size washers, nuts, and setscrews

6-1 Introduction

This chapter explains how to draw threads, washers, and nuts. It also explains how to select fasteners, washers, nuts, and setscrews.

Internal threads are created using the **Hole Wizard** tool, which is located on the **Features** toolbar. Predrawn fasteners and other standard components may be accessed using the **Design Library**.

All threads in this book are in compliance with ANSI (American National Standards Institute) standards—ANSI Inch and ANSI Metric threads.

6-2 Thread Terminology

Figure 6-1 shows a thread. The peak of a thread is called the ***crest,*** and the valley portion is called the ***root***. The ***major diameter*** of a thread is the distance across the thread from crest to crest. The ***minor diameter*** is the distance across the thread from root to root.

Figure 6-1

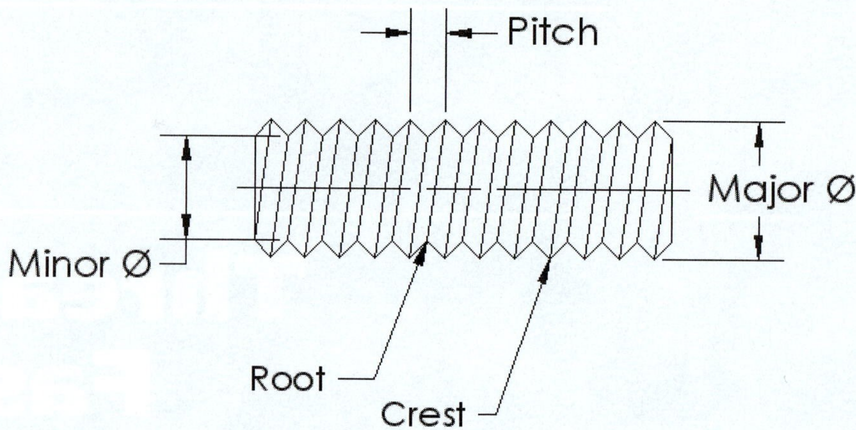

Pitch

The pitch of a thread is the distance from the center of one thread to the center of the next thread. By definition the pitch of a thread is equal to one over the number of threads per distance unit. The formula for pitch is:

Pitch = 1/Number of threads per inch (millimeters)

Number of threads per inch (millimeter) = 1/Pitch

6-3 Thread Callouts—ANSI Metric Units

Threads are specified on a drawing using drawing callouts. See Figure 6-2. The M at the beginning of a drawing callout specifies that the callout is for a metric thread. Holes that are not threaded use the Ø symbol.

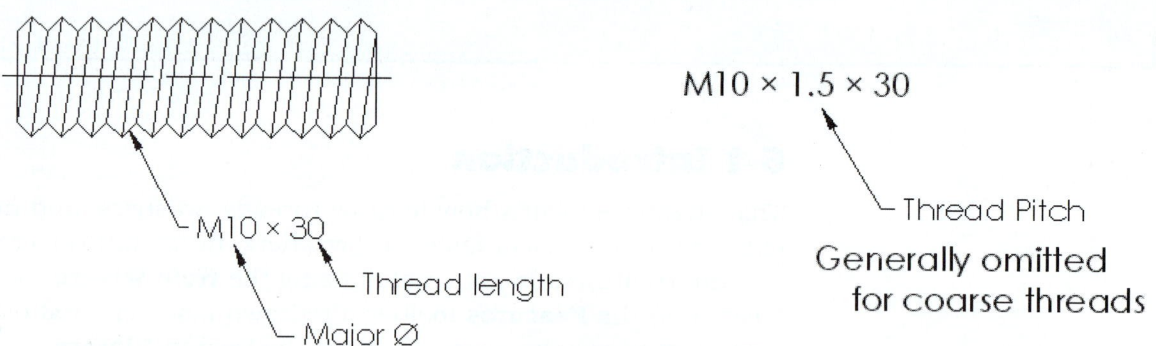

Figure 6-2

The number following the M is the major diameter of the thread. An M10 thread has a major diameter of 10 mm. The pitch of a metric thread is assumed to be a coarse thread unless otherwise stated. The callout M10 × 30 assumes a coarse thread, or a thread length of 1.5 mm per thread. The number 30 is the thread length in millimeters. The "×" is read as "by," so the thread is called a "ten by thirty."

NOTE

For metric threads the pitch is specified, not the number of threads per millimeter.

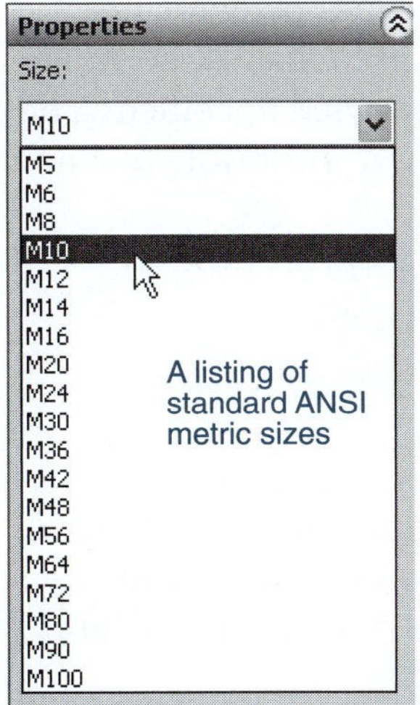

Figure 6-3

The callout M10 × 1.25 × 30 specifies a pitch of 1.25 mm per thread. This is not a standard coarse thread size, so the pitch must be specified.

Figure 6-3 shows a listing of standard metric thread sizes available in the SolidWorks **Design Library** for one type of hex head bolt. The sizes are in compliance with ANSI Metric specifications.

Whenever possible, use preferred thread sizes for designing. Preferred thread sizes are readily available and are usually cheaper than nonstandard sizes. In addition, tooling such as wrenches is also readily available for preferred sizes.

6-4 Thread Callouts—ANSI Unified Screw Threads

ANSI Unified Screw Threads (English units) always include a thread form specification. Thread form specifications are designated by capital letters, as shown in Figure 6-4, and are defined as follows:

UNC—Unified National Coarse

UNF—Unified National Fine

UNEF—Unified National Extra Fine

UN—Unified National, or constant-pitch threads

An ANSI (English units) thread callout starts by defining the major diameter of the thread followed by the pitch specification. The callout .500-13 UNC means a thread whose major diameter is .500 in. with 13 threads per inch. The thread is manufactured to the UNC standards. The pitch for a .500-13 UNC thread is 1/13 = .08 or 1/Number of threads per inch.

There are three possible classes of fit for a thread: 1, 2, and 3. The different class specifications specify a set of manufacturing tolerances. A class 1 thread is the loosest and a class 3 the most exact. A class 2 fit is the most common.

The letter A designates an external thread, B an internal thread. The symbol × means "by" as in 2 × 4, "two by four." The thread length (3.00) may be followed by the word LONG to prevent confusion about which value represents the length.

Drawing callouts for ANSI (English unit) threads are sometimes shortened, such as in Figure 6-4. The callout .500-13 UNC-2A × 3.00 LONG is shortened to .500-13 × 3.00. Only a coarse thread has 13 threads per

Figure 6-4

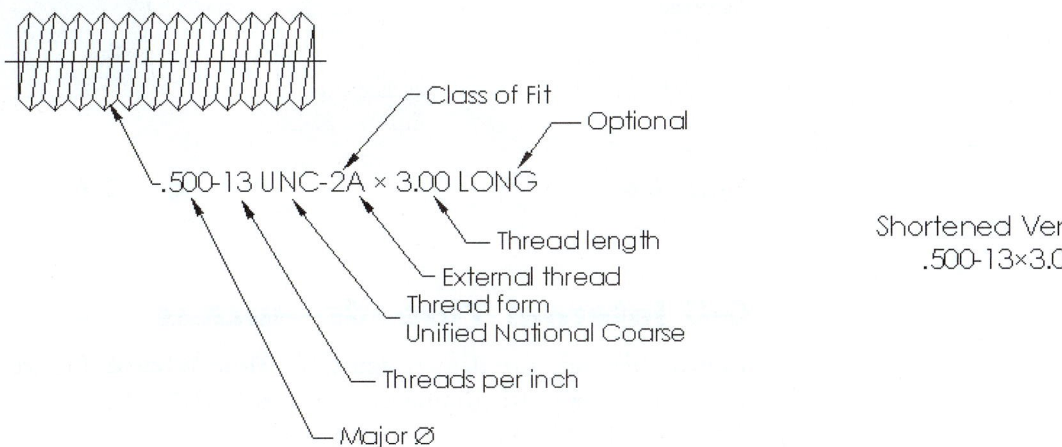

inch, and it should be obvious whether a thread is internal or external, so these specifications may be dropped. Most threads are class 2, so it is tacitly accepted that all threads are class 2 unless otherwise specified. The shortened callout form is not universally accepted. When in doubt, use a complete thread callout.

A listing of standard ANSI (English unit) threads, as presented in SolidWorks, is shown in Figure 6-5.

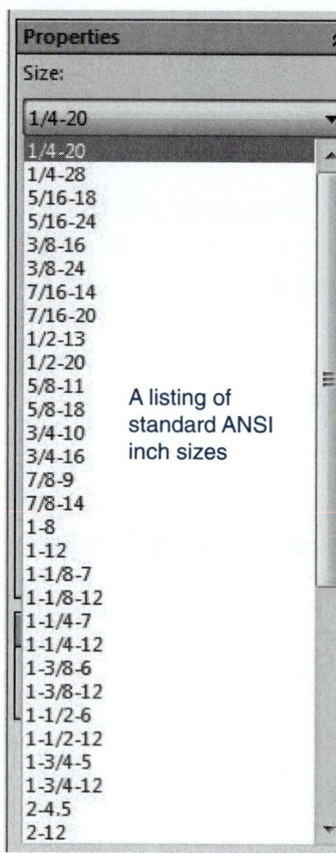

Figure 6-5

6-5 Thread Representations

There are three ways to graphically represent threads on a technical drawing: detailed, schematic, and simplified. Figure 6-6 shows an external detailed representation, and both the external and internal simplified and schematic representations.

Figure 6-7 shows an internal and an external thread created using SolidWorks. Note that no thread representations appear. This is called the **Simplified** representation. SolidWorks uses the **Simplified** thread representation as a way to minimize file size. Cosmetic thread representations can be created and will be discussed later in the chapter. Actual threads may be drawn using the **Helix** tool. SolidWorks can also draw a **Schematic** thread representation.

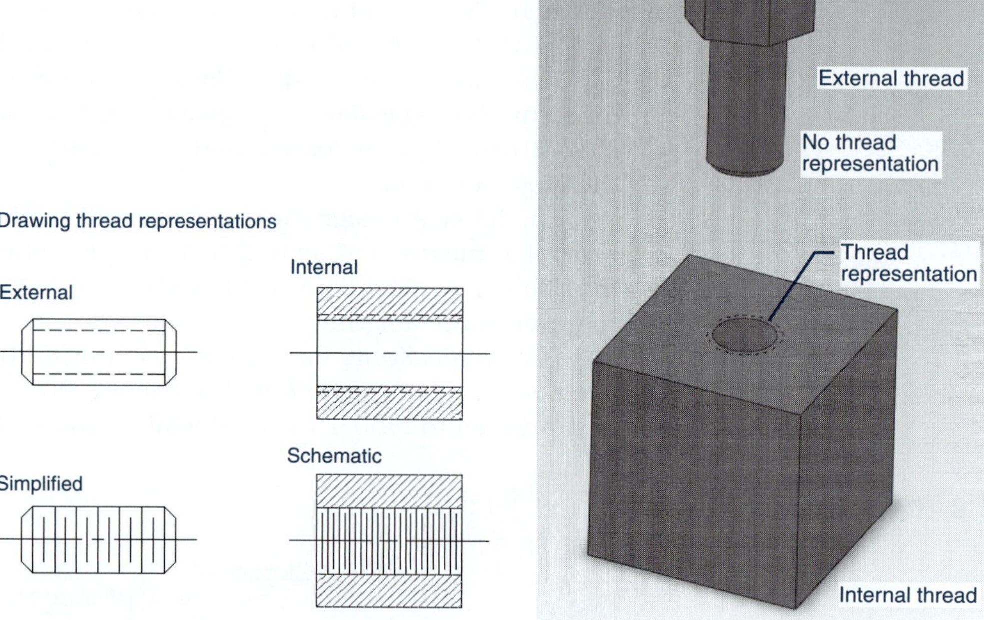

Figure 6-6

Figure 6-7

6-6 Internal Threads—Inches

Internal threads are drawn using the **Hole Wizard**. Figure 6-8 shows a 1.5 × 2.0 × 1.0 block. In this section a 3/8-16 UNC hole will be located in the center of the block.

Figure 6-8

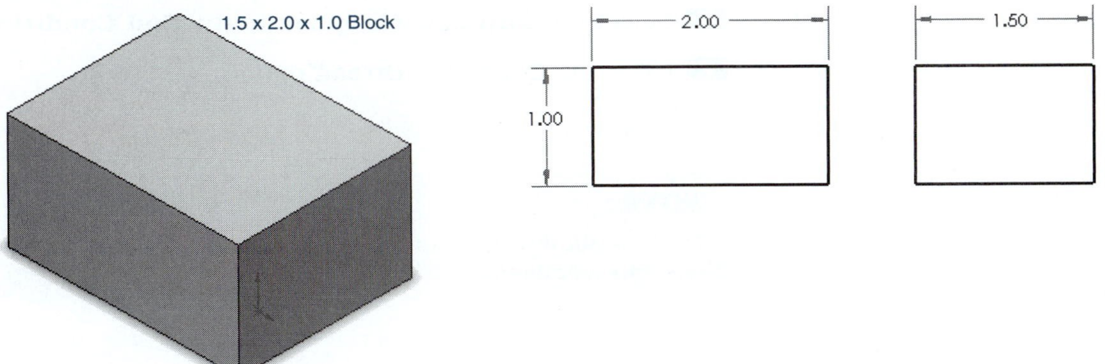

1 Draw a **1.5** × **2.0** × **1.0** block.

2 Orient the block in the **Isometric** view.

3 Click the **Hole Wizard** tool on the **Features** tab.

4 Select the **Straight Tap** option, set the **Standard** for **ANSI**, the **Type** for **Tapped hole**, and define the thread's size and length.

In this example an internal 3/8-16 UNC thread will be created. The thread will go completely through the block. See Figure 6-9.

Figure 6-9

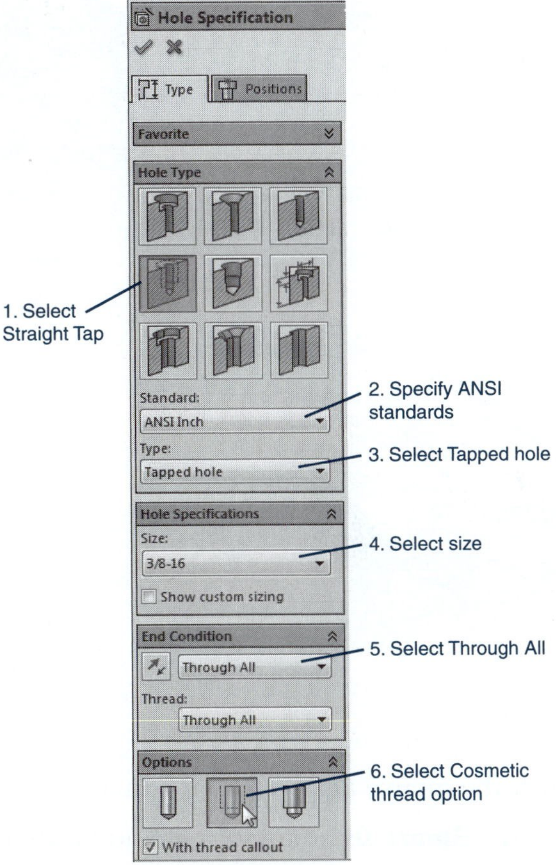

5 Select a 3/8-16 thread in the **Hole Specifications** box.

The thread will have a major diameter of 3/8 with 16 threads per inch and a pitch of 1/16 = .06.

6 Select the **Through All** option in the **End Condition** box.

7 Click the **Cosmetic thread** option.

> **NOTE**
>
> The **Cosmetic thread** option will create a hidden line around the finished hole that serves to indicate that the hole is threaded.

8 Click the **Positions** tab in the **Hole Wizard PropertyManager**.

9 Click a location near the center of the top surface of the block to identify the surface for the thread.

See Figure 6-10.

Figure 6-10

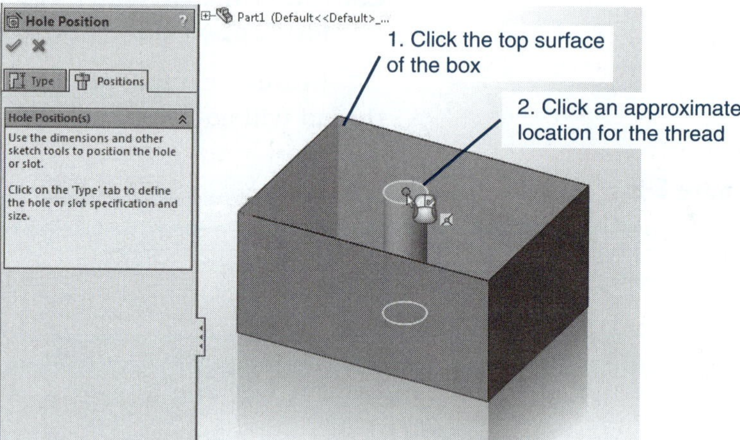

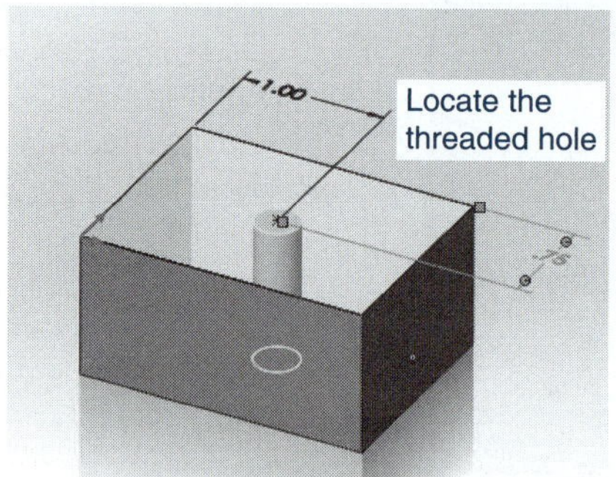

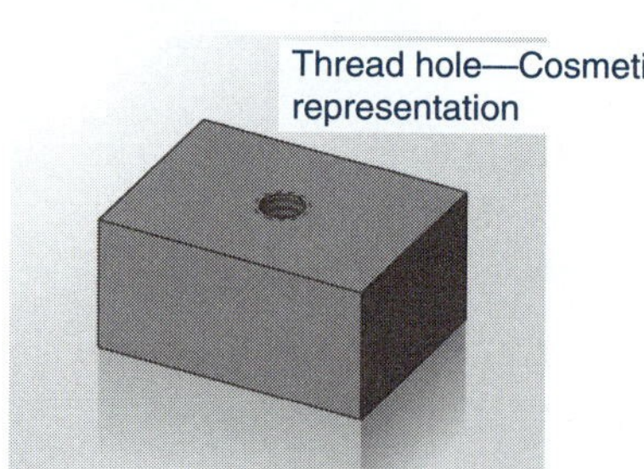

10 Click the mouse again to approximately locate the threaded hole.

11 Use the **Smart Dimension** tool and locate the center point of the hole.

12 Click the green **OK** check mark.

The hidden line surrounding the hole is a cosmetic thread and indicates that the hole is threaded. Note that there are no threads on the inside of the hole.

6-7 Threaded Blind Holes—Inches

A ***blind hole*** is one that does not go completely though an object. See Figure 6-11.

Figure 6-11

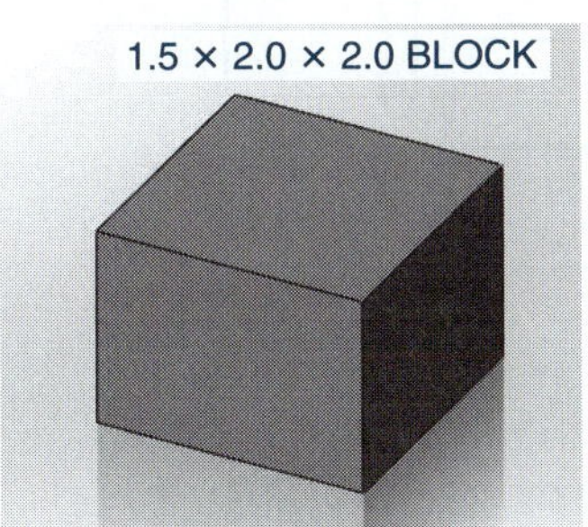

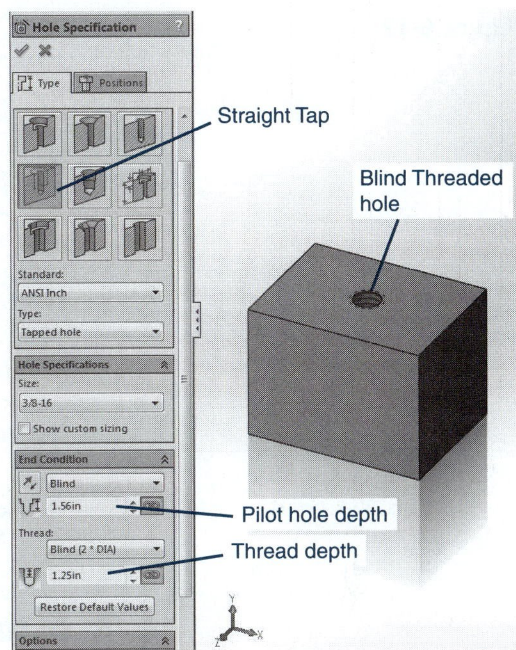

1. Use the **Undo** tool and remove the hole added to the 1.5 × 2.0 × 1.0 block.
2. Edit the block so that it is **2.00** thick.
3. Click the **Hole Wizard** tool.

   SolidWorks will remember the last hole setting used, so the hole specifications will be the same as used to create the 3/8-16 through hole: ANSI standard, tapped hole, with a cosmetic thread.

4. Size the hole to **3/8-16** and set the **Blind Thread** depth in the **End Condition** box to **1.25in**.

   In this example the tap blind hole depth was automatically calculated as 1.56 in. When a threaded hole is created, a hole (pilot hole) is first drilled and then the threads are cut (tapped hole) into the sides of the hole. The pilot hole must always be longer than the threaded portion of the hole. A tapping bit only cuts threads; it has no cutting surfaces on its bottom surface. If the bit bottoms out, that is, hits the bottom of the hole, it may break. As a general rule a distance greater than two pitches is added to the pilot hole depth beyond the threaded hole depth. In this example SolidWorks adds a distance of approximately four pitches.

5. Click the **Positions** tab and locate a center point near the center of the top surface of the block.
6. Use the **Smart Dimension** tool to locate the center point.
7. Click the green **OK** check mark.

Figure 6-12 shows an orthographic view of the block and a section view. Note that the tapped hole extends beyond the end of the threads and ends with a conical point. The depth of the tapping hole does not include the conical end point.

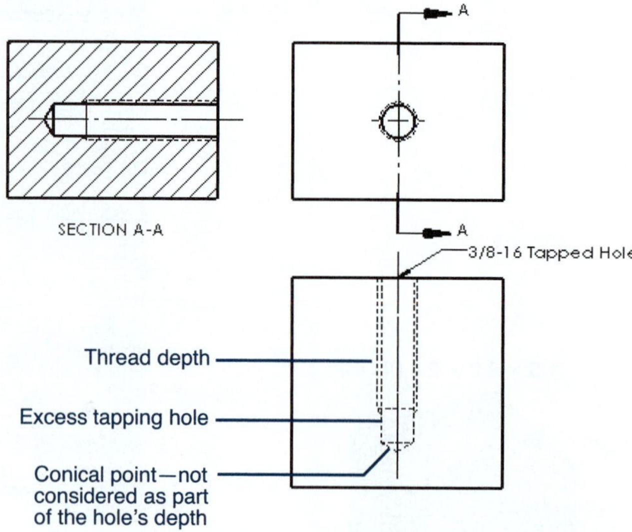

Figure 6-12

6-8 Internal Threads—Metric

Metric threads are designated by the letter M. For example, M10 × 30 is the callout for a metric thread of diameter 10 and a length of 30. The thread is assumed to be a coarse thread.

> **TIP**
>
> For metric thread drawings the symbol **Ø** indicates a hole or cylinder without threads; the symbol **M** indicates metric threads.

1. Draw a **20** × **30** × **15** block.

   See Figure 6-13.

2. Click the **Hole Wizard.**

3. Click the **Straight Tap** option, set the **Standard** for **ANSI Metric**, the hole size for **M10** × **1.0**, and the depth for **Through All.**

   The number 1.0 is the pitch of the thread. An M10 thread can be cut with several different pitch sizes.

4. Click the **Cosmetic thread** option.

5. Click the **Positions** tab and click the top surface of the block, and click the mouse again to select the approximate location for the hole.

5. Use the **Smart Dimension** tool to locate the hole's center point.

7. Click the green **OK** check mark.

Figure 6-14 shows the specifications for an M10 × 1.0 × 20 DEEP blind hole. The pilot hole depth is approximately 5 pitch deeper than the threaded portion of the hole.

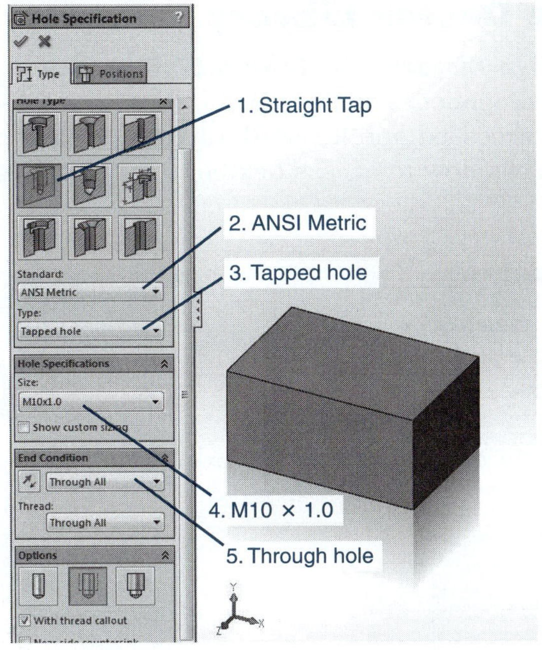

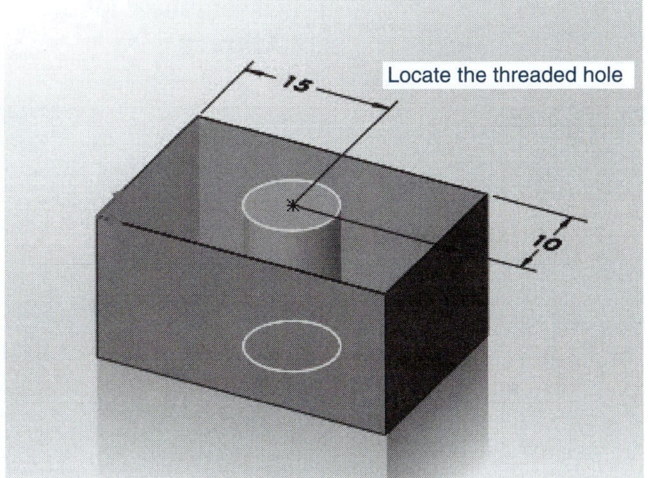

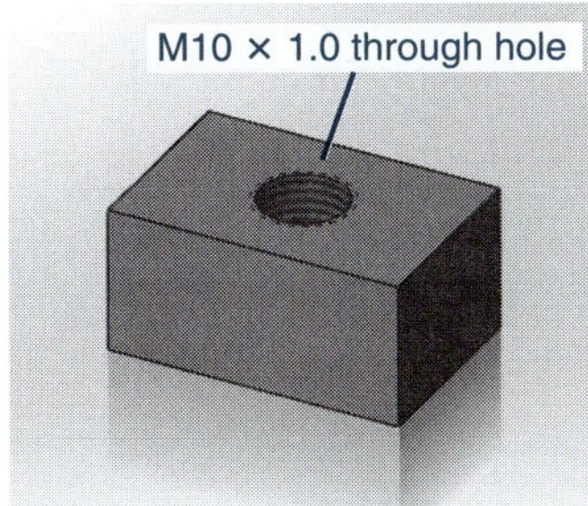

Figure 6-13

Figure 6-14

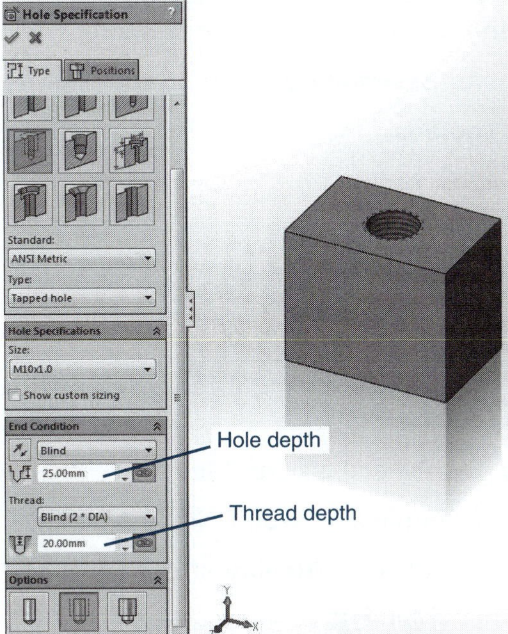

6-9 Accessing the Design Library

SolidWorks includes a **Design Library**. The **Design Library** includes a listing of predrawn standard components such as bolts, nuts, and washers. These components may be accessed and inserted into drawings to create assemblies. Figure 6-15 shows how to access hex bolts in the **Design Library**.

Figure 6-15

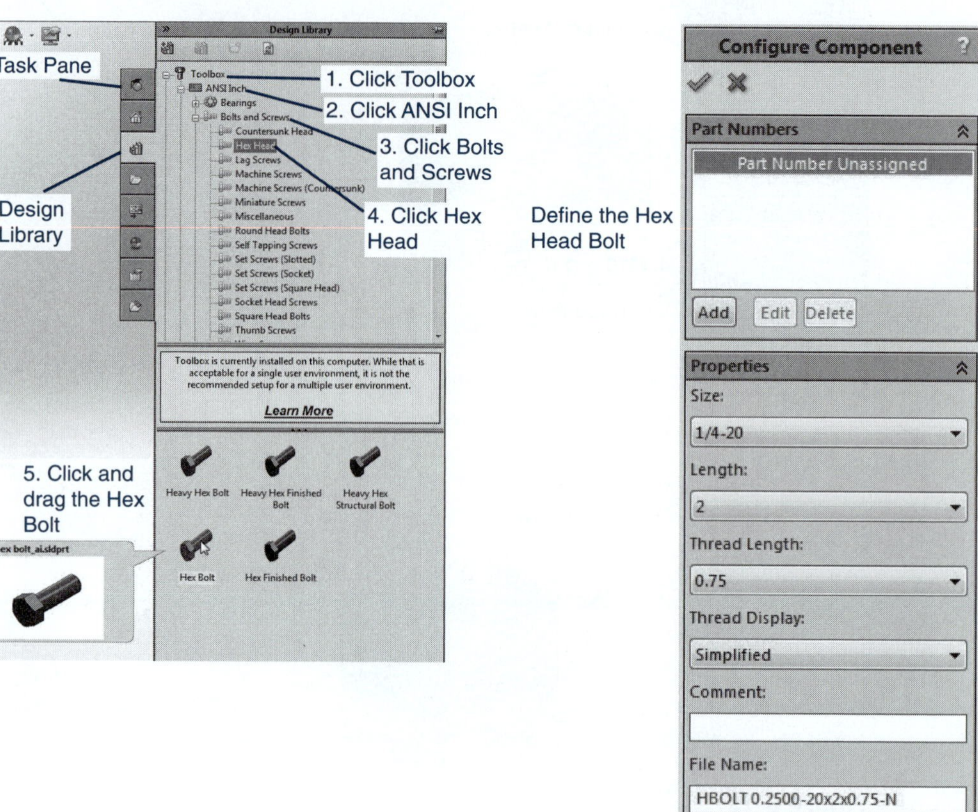

The Design Library is accessed through the Task Pane. The Design Library can only be used with an Assembly document.

1. Create an **Assembly** document and click the **Design Library** tab.
2. Click **Toolbox**.
3. Click **ANSI Inch**.
4. Click **Bolts and Screws**.
5. Click **Hex Head**.

   A listing of various types of hex head bolts will appear.
6. Click the **Hex Bolt**, drag it into the drawing screen area, and click **Create Part**.
7. Define the needed size and length.

   In this example a 1/4-20 UNC × 2.00 HEX HEAD BOLT was created.
8. Define the **Thread Display** style.
9. Click the green **OK** check mark.

> **TIP**
> There are three display styles available for threads: **Schematic, Cosmetic,** and **Simplified.** See Figure 6-16. The **Simplified** style was created to use a smaller file size.

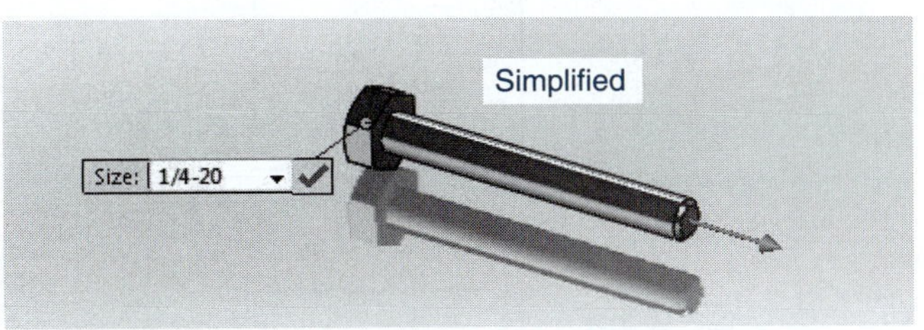

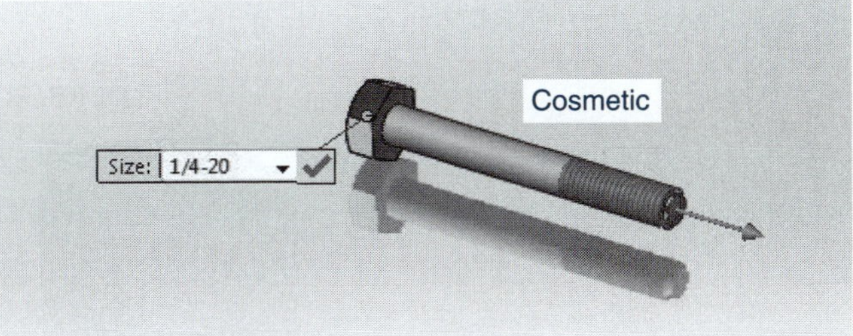

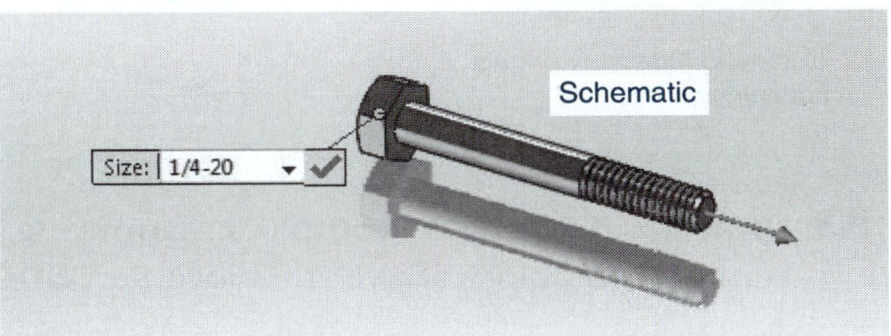

Figure 6-16

6-10 Thread Pitch

Thread pitch for an ANSI Inch fastener is defined as

$$P = \frac{1}{N}$$

where

P = pitch
N = number of threads per inch

In ANSI Inch standards a sample bolt callout is 1/4-20 UNC × length. The 20 value is the number of threads per inch, so the pitch is 1/20 or 0.05. The pitch for a 1/4-28 UNF thread would be 1/28, or 0.036.

A sample thread callout for ANSI Metric is written M10 × 1.0 × 30, where 1.0 is the pitch. No calculation is required to determine the pitch for metric threads. It is included directly in the thread callout.

> **TIP**
> Almost all threads are coarse, so ANSI metric thread callouts omit the pitch designation. A pitch size is included only when a metric thread is not coarse.

6-11 Determining an External Thread Length—Inches

Figure 6-17 shows three blocks stacked together. Their dimensions are given. They are to be held together using a hex bolt, two washers, and a nut. The bolt will be a 3/8-16 UNC. What length should be used?

Figure 6-17

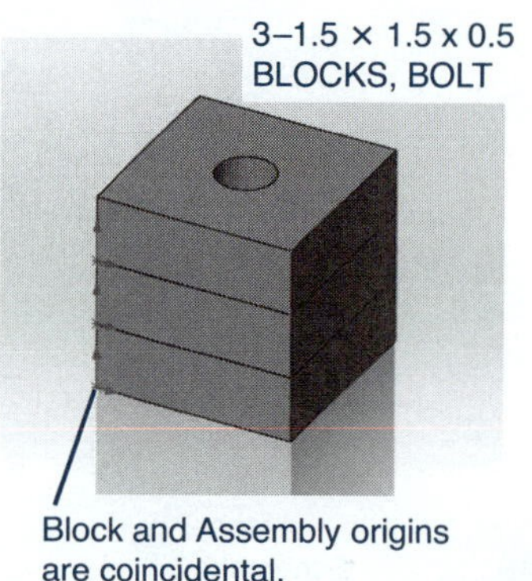
3–1.5 × 1.5 x 0.5
BLOCKS, BOLT
Block and Assembly origins
are coincidental.

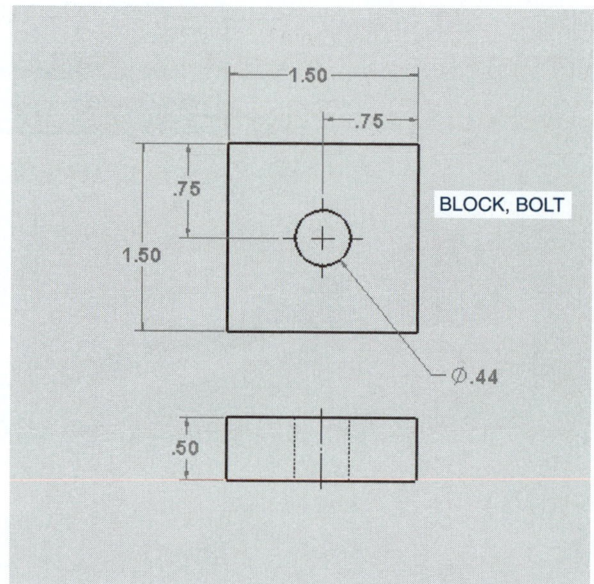
1.50
.75
.75
1.50
BLOCK, BOLT
Ø.44
.50

1 Draw a **1.5 × 1.5 × 0.50** block. Draw a **Ø.44** (7/16) hole through the center of the block. Save the block as **BLOCK, BOLT**.

The Ø.44 was selected because it allows for clearance between the bolt and the block.

2 Create an **Assembly** drawing, enter three **BLOCK, BOLT**s, and assemble them as shown.

3 Save the assembly as **3-BLOCK ASSEMBLY**.

4 Access the **Design Library, Toolbox, ANSI Inch, Washers,** and **Plain Washers (Type A)**.

See Figure 6-18.

Figure 6-18

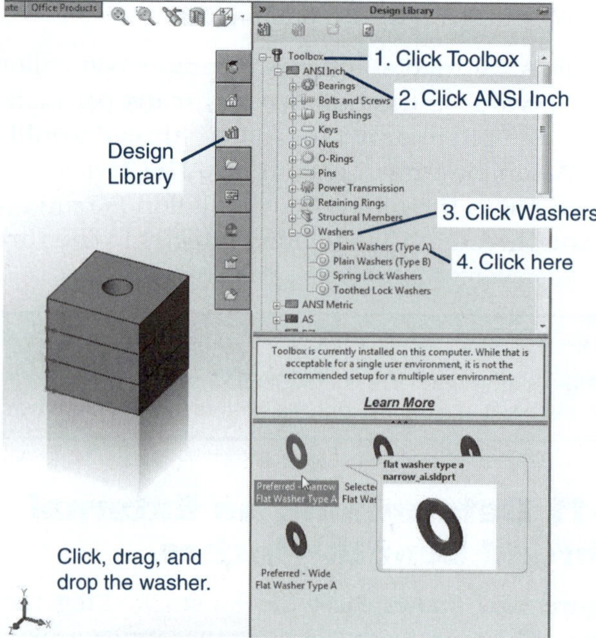
Design Library
1. Click Toolbox
2. Click ANSI Inch
Design
Library
3. Click Washers
4. Click here
Toolbox is currently installed on this computer. While that is acceptable for a single user environment, it is not the recommended setup for a multiple user environment.
Learn More
Click, drag, and
drop the washer.

5 Click and hold the **Preferred-Narrow Flat Washer Type A**.

6 Click the washer and drag-and-drop it into the field of the drawing.

The washer preview will appear in the drawing area.

7 Size the washer by clicking the arrow to the right of the initial size value and selecting a nominal value.

See Figure 6-19.

The term *nominal* refers to a starting value. In this case we are going to use a 3/8-16 UNC thread. The 3/8 value is a nominal value. We select a washer with a specified 3/8 nominal inside diameter. The actual inside diameter of the washer is .406. Clearance between the washer and the bolt is created when the washer is manufactured. The fastener will also not measure .375 but will be slightly smaller. The .375 size is the size of the bolt's shaft before the threads were cut.

Note that the washer thickness is .065 (about 1/16).

NOTE

Washers are identified using Insider diameter × Outside diameter × Thickness; for example: .406 × .812 × .065 PLAIN WASHER

Figure 6-19

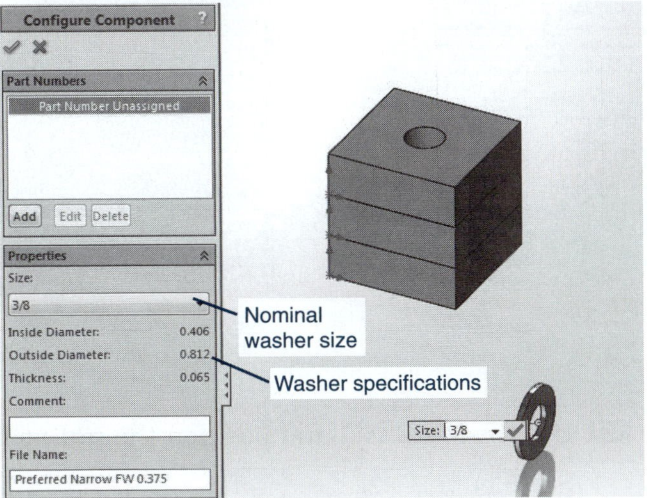

8 Click the green **OK** check mark.

9 Add a second washer into the assembly.

10 Use the **Mate Concentric** and **Mate Coincident** tools to position the washers around the block's holes as shown.

See Figure 6-20.

11 Access the **Design Library**, then click **Toolbox, ANSI Inch, Nuts, Hex Nuts, Hex Jam Nut**, and drag-and-drop the nut into the drawing area.

See Figure 6-21.

12 Size the nut to 3/8-16 UNC. It must match the 3/8-16 UNC thread on the bolt.

Figure 6-20

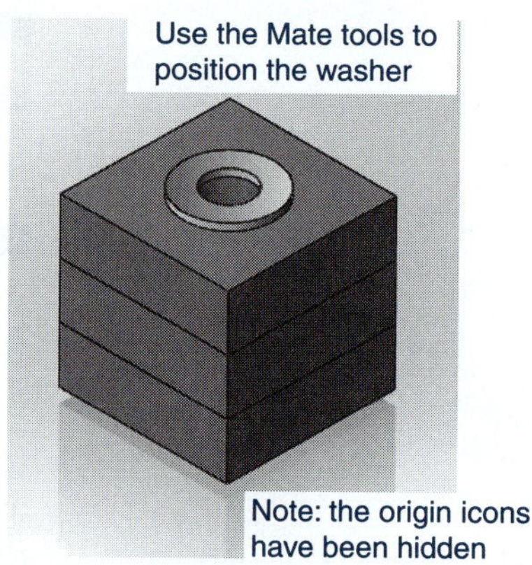

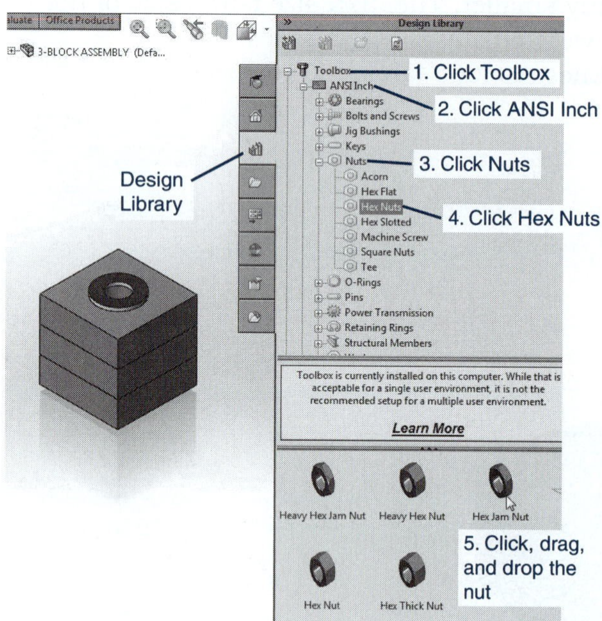

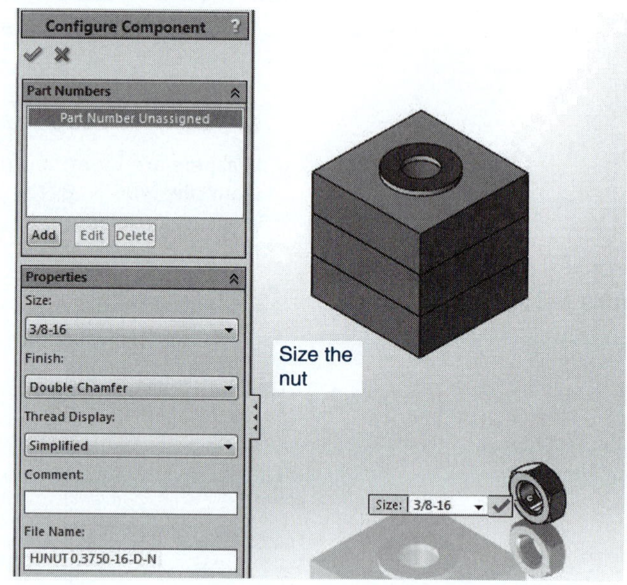

Figure 6-21

13 Use the **Mate** tool and position the nut as shown.

See Figure 6-22.

Figure 6-22

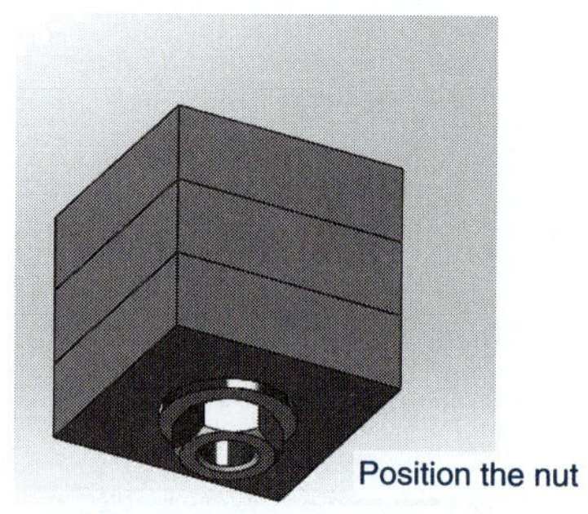

The nut thickness is .227. This value was obtained by using the **Edit Feature** tool and determining the extrusion value used to create the nut. The nut is defined as 3/8-16 UNC Hex Jam Nut.

NOTE

Bolt threads must extend beyond the nut to ensure 100% contact with the nut. The extension must be a minimum of two pitches (2*P*).

So far, the bolt must pass through three blocks (.50 × 3 = 1.50), two washers (.065 × 2 = .13), and one nut (.227). Therefore, the initial bolt length is 1.50 + .13 + .227 = 1.857.

Calculations used to determine the strength of a bolt/nut combination assume that there is 100% contact between the bolt and the nut; that is, all threads of the nut are in contact with the threads of the bolt. However, there is no assurance that the last thread on a bolt is 360°, so at least two threads must extend beyond the nut to ensure 100% contact. The 2*P* requirement is a minimum value. More than 2*P* is acceptable. SolidWorks will automatically add at least 4*P* unless defined otherwise.

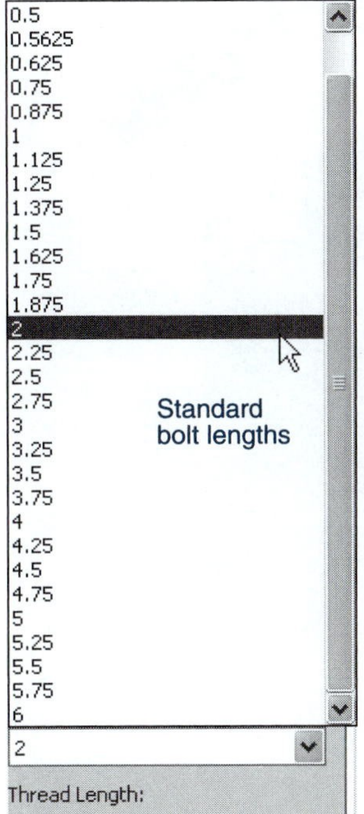

Figure 6-23

In this example the thread pitch is .0625 (1/16). Two pitches (2*P*) is .125. This value must be added to the initial thread length:

$$1.857 + .125 = 1.982$$

Therefore, the minimum bolt length is 1.982. This value must in turn be rounded up to the nearest standard size. Figure 6-23 shows a listing of standard sizes for a 1 3/8-16 UNC Hex Head bolt.

The final thread length for the given blocks, washers, nut, and 2*P* is 2.00. The bolt callout is

3/8-16 UNC × 2.00 Hex Head Bolt

14 Click the **Design Library, Toolbox, ANSI Inch, Bolts and Screws, Hex Head, Hex Bolt**, and click and drag-and-drop the bolt into the drawing.

15 Define the **Size** as **3/8-16** and the **Length** as **2**.

See Figures 6-24 and 6-25. Note how the bolt extends beyond the nut.

Figure 6-24

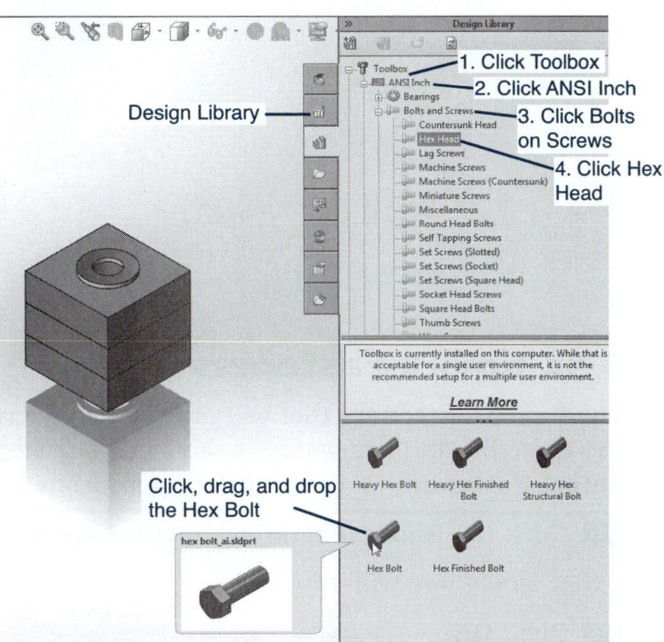

Figure 6-24
(*Continued*)

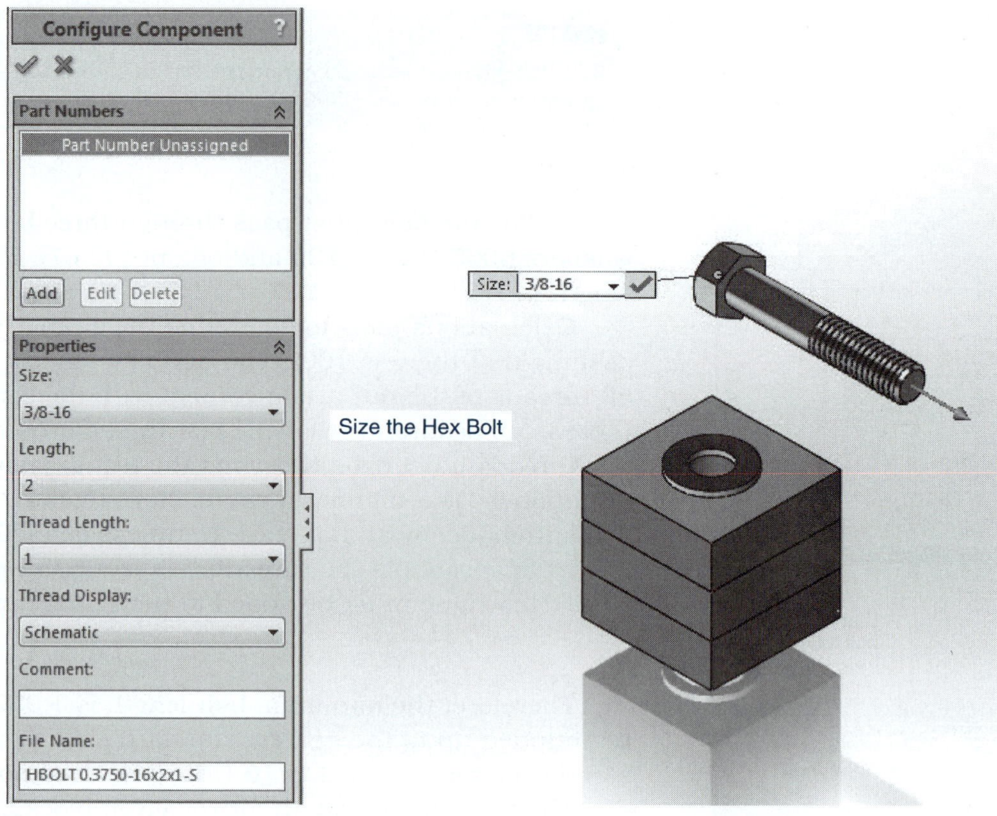

Figure 6-25

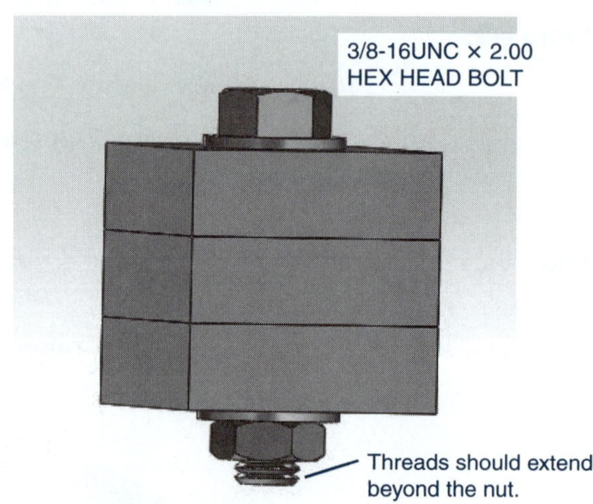

6-12 Smart Fasteners

The **Smart Fasteners** tool will automatically create the correct bolt. Given the three blocks, two washers, and nut shown in Figure 6-26, use the **Smart Fasteners** tool to add the appropriate bolt.

1 Click the **Smart Fasteners** tool located on the **Assembly** tab.

A dialog box will appear.

2 Click **OK**.

Figure 6-26

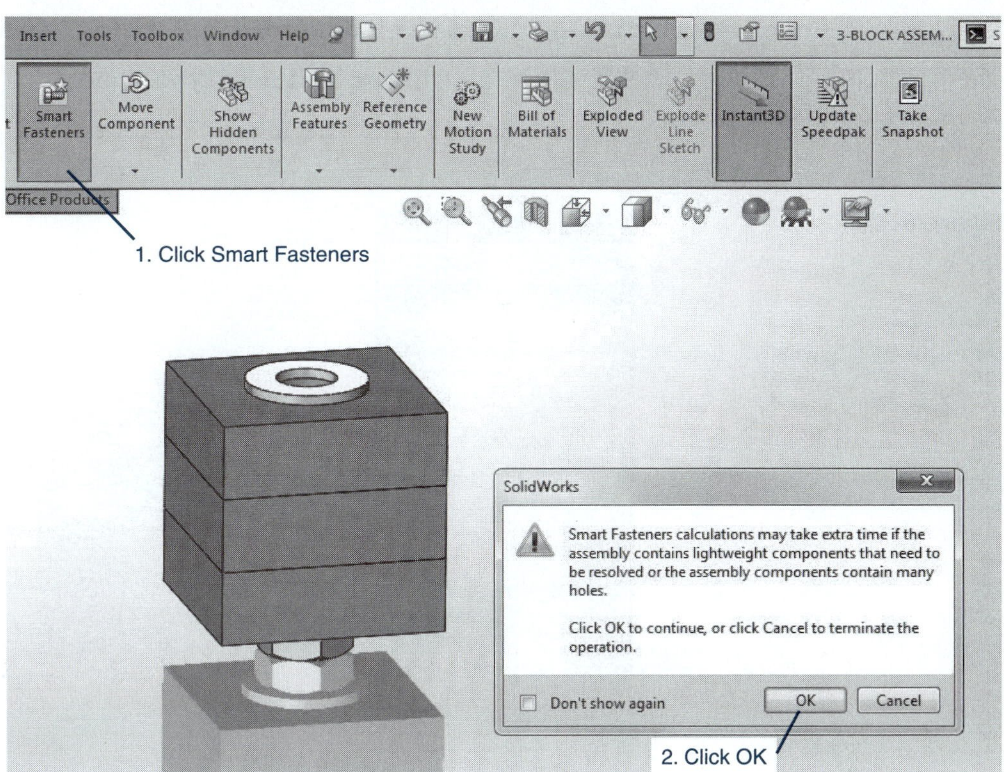

The **Smart Fasteners PropertyManager** will appear. See Figure 6-27.

3 Click the hole in the top block of the three blocks.

Figure 6-27

TIP

Click the hole, that is, a cylindrical-shaped section as shown in Figure 6-28, not the edge of the hole.

The words **Cut-Extrude 1** will appear in the **Selection** box.

4 Click **Add**.

A fastener will appear in the hole. This may take a few seconds. In this example a **Socket Head Cap Screw** appeared in the hole. Once the screw is added to the drawing it can be edited as needed.

5 Right-click the **Socket Head Cap Screw** heading in the **Fastener** box.

6 Click the **Change fastener type** option.

See Figure 6-28. The **Smart Fastener** dialog box will appear.

Figure 6-28

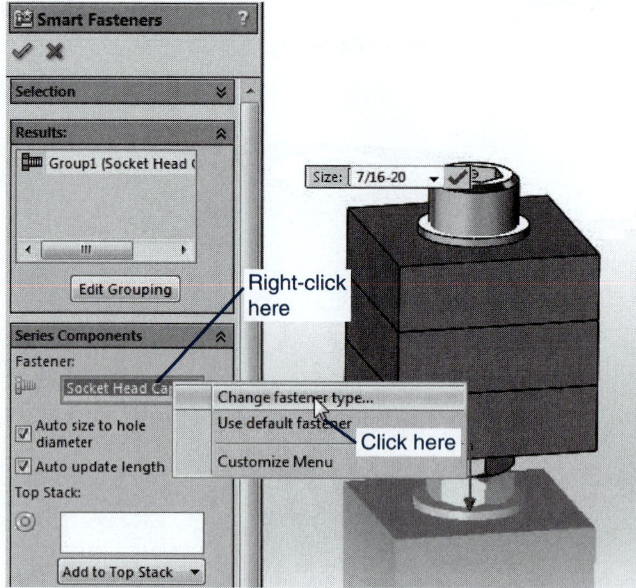

7 Scroll down the list of fasteners and select a **Hex Head** type.

See Figure 6-29.

Figure 6-29

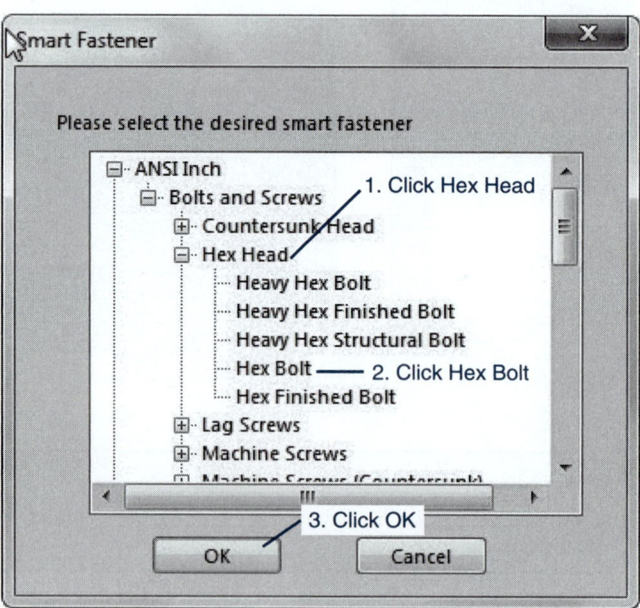

8 Click **OK**.

The **Series Components** manager will appear.
See Figure 6-30.

9 Scroll down and access the **Properties** box.

Figure 6-30

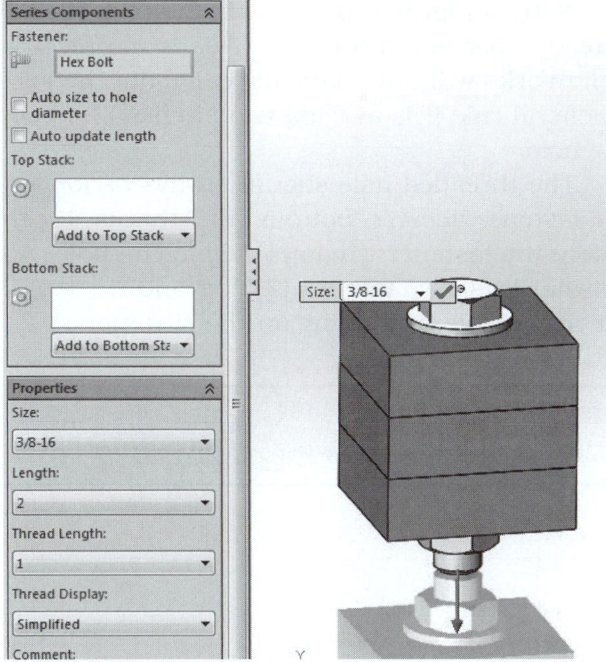

10 Define the thread size, number of threads per inch, and the thread length.

11 Click the green **OK** check mark.

Note that the fastener extends beyond the nut. See Figure 6-31.

Figure 6-31

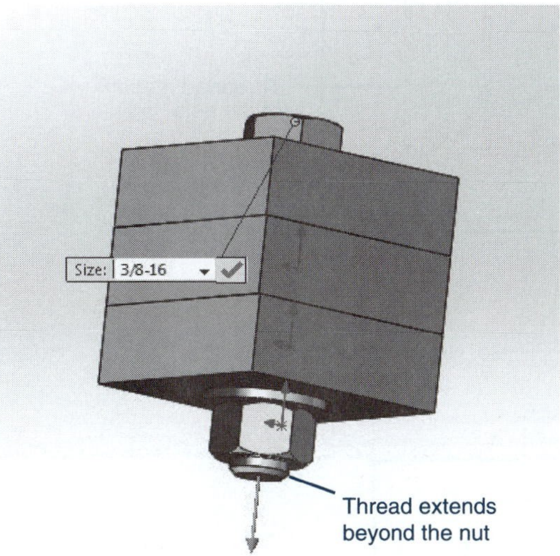

6-13 Determining an Internal Thread Length

Figure 6-32 shows two blocks: **Block, Cover** and **Block, Base**. Their dimensions are also shown. The two blocks are to be assembled and held together using an M10 × 1.25 × 25 hex head screw. What should be the threaded hole in the **Block, Base**?

Note in Figure 6-32 that the tapping hole extends beyond the threaded portion of the hole. This is to prevent damage to the tapping bit. SolidWorks will automatically calculate the excess length needed, but as a general rule it is at least two pitches (2*P*) beyond the threaded portion of the hole.

The threaded hole should always be longer than the fastener so that the fastener doesn't "bottom out," that is, hit the bottom of the threads before the fastener is completely in the hole. Again, the general rule is to allow at least two pitches (2*P*) beyond the length of the fastener, but more are acceptable depending on the situation.

> **TIP**
> Metric thread callouts give the pitch directly.

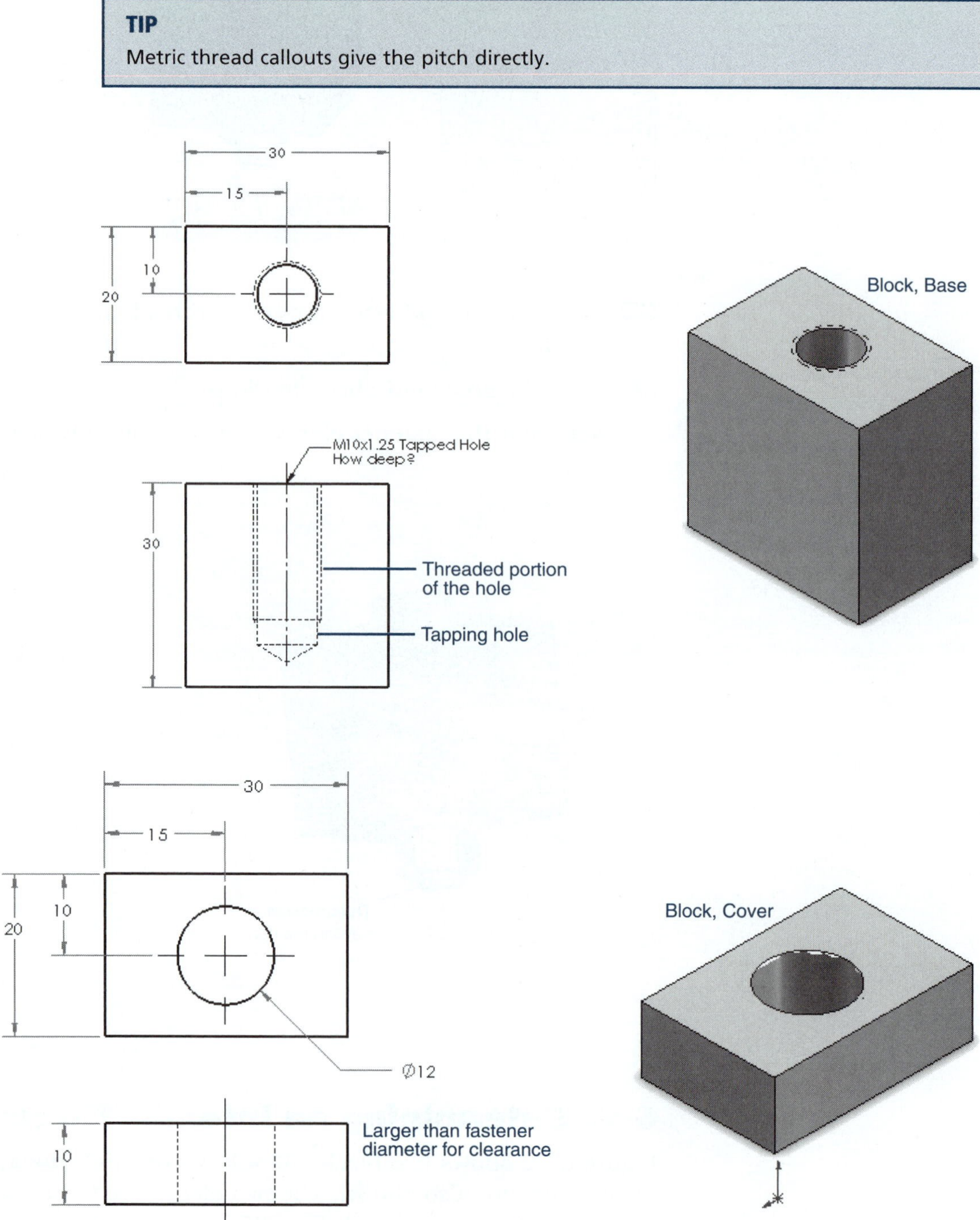

Figure 6-32

In this example the thread pitch is **1.25**. Two pitches = 2.50. The bolt length is 25, but it must initially pass through the 10-thick **Block, Cover**, so the length of the bolt in the **Block, Base** is 15. Adding 2.50 to this value yields a minimum thread length of 17.50. Rounding the value up determines that the threaded hole in the **Block, Base** should be M10 × 1.25 × 18 deep. See Figure 6-33. The 18 depth is a minimum value and can be greater.

Figure 6-33

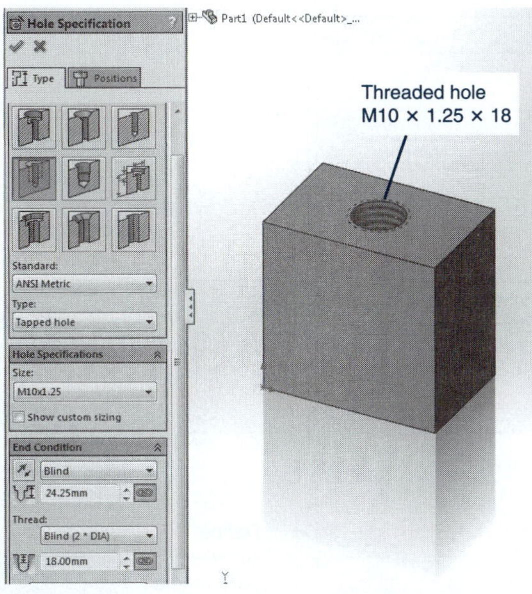

1 Draw the parts as dimensioned in Figure 6-32 and use the **Hole Wizard** to add an M10 × 18 deep hole to the **Block, Base**.

The **Ø12** hole in the **Cover** block is a clearance hole; that is, it is larger than the fastener. It has no threads.

See Figure 6-34.

Figure 6-34

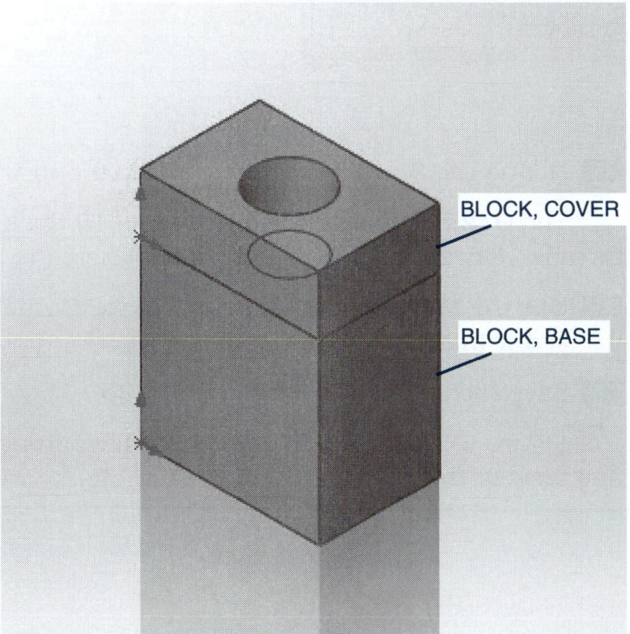

2 Assemble the parts as shown.

3 Access the **Design Library**. Click **Toolbox, ANSI Metric, Bolts and Screws, Hex Head**, and access **Formed Hex Screw ANSI B18.2.3.2M**.

4 Click and drag the bolt onto the screen.

See Figure 6-35.

Figure 6-35

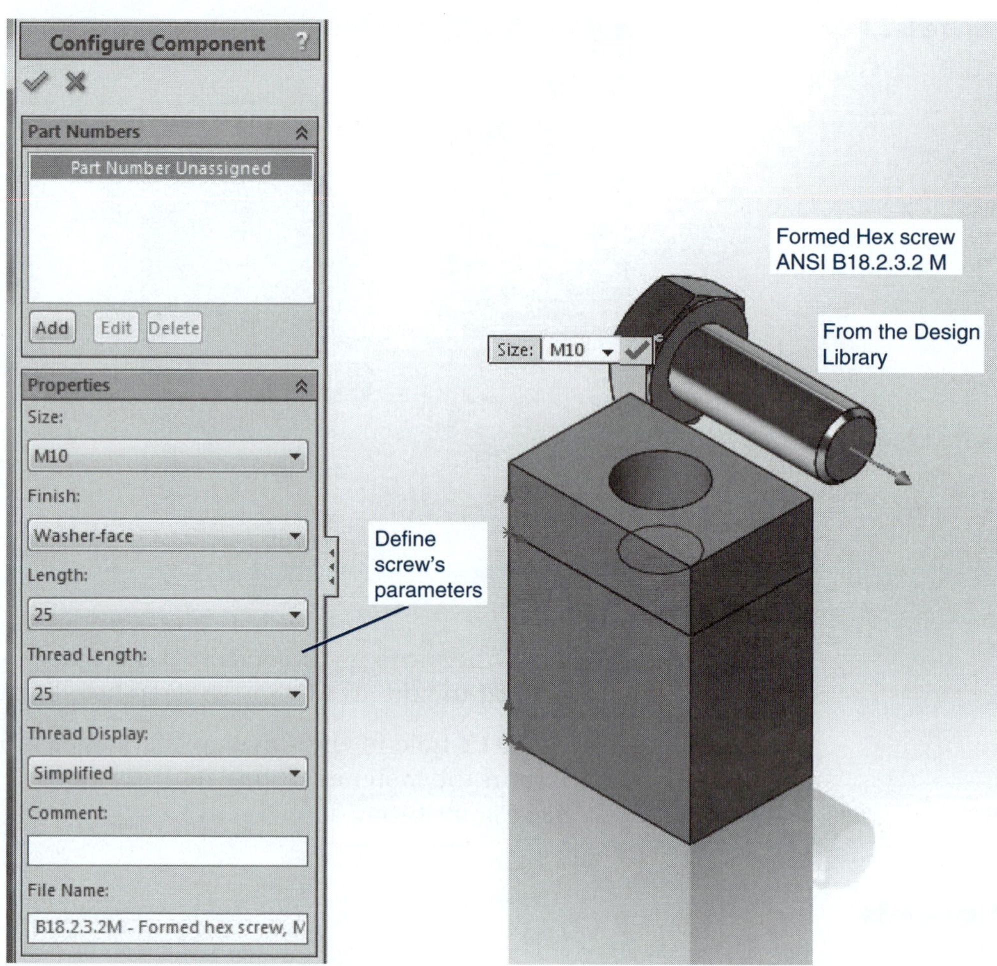

5 Define the **Size** of the screw as **M10** and the **Length** as **25**.

The **Ø12** hole in the **Cover Block** is **10** thick. The threaded hole has threads 18 deep. The screw is 25 long. There are 3 unused threads in the hole.

6 Use the **Mate Concentric** and **Mate Coincident** tools to place the screw into the assembly.

7 Save the assembly. See Figure 6-36.

Figure 6-37 shows an isometric view, an orthographic view, and a section view of the internal thread assembly.

Figure 6-36

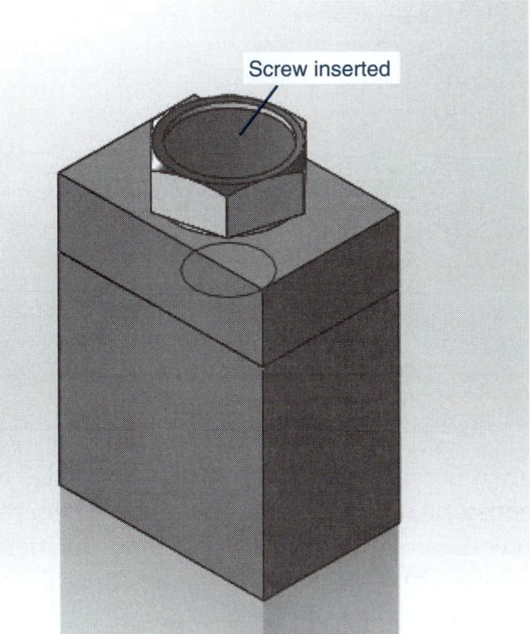

Figure 6-37

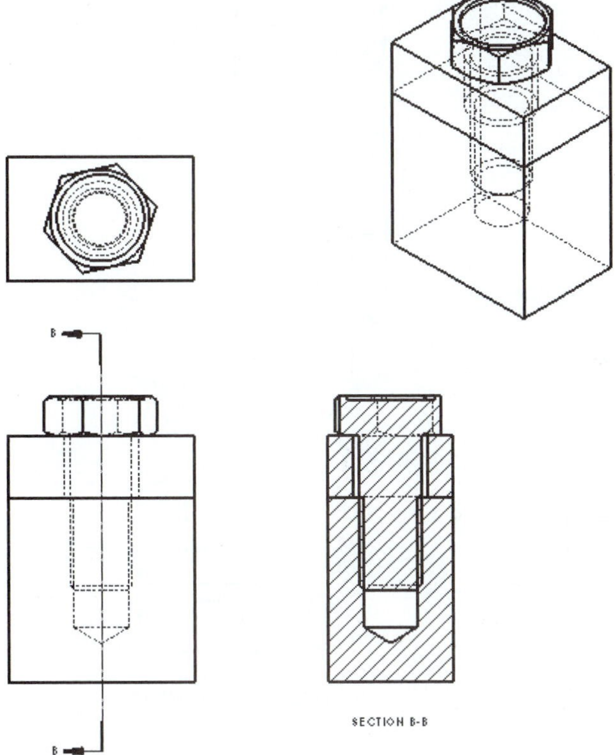

6-14 Set Screws

Set screws are fasteners used to hold parts like gears and pulleys to rotating shafts or other moving objects to prevent slippage between the two objects. See Figure 6-38.

Most set screws have recessed heads to help prevent interference with other parts.

Figure 6-38

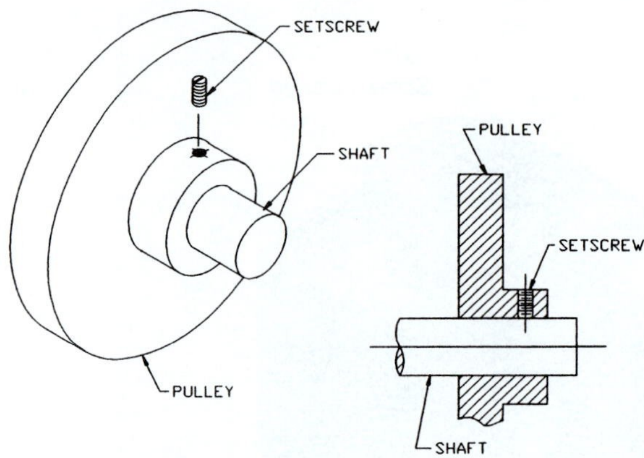

Many different head styles and point styles are available. See Figure 6-39. The dimensions shown in Figure 6-39 are general sizes for use in this book. For actual sizes, see manufacturers' specifications.

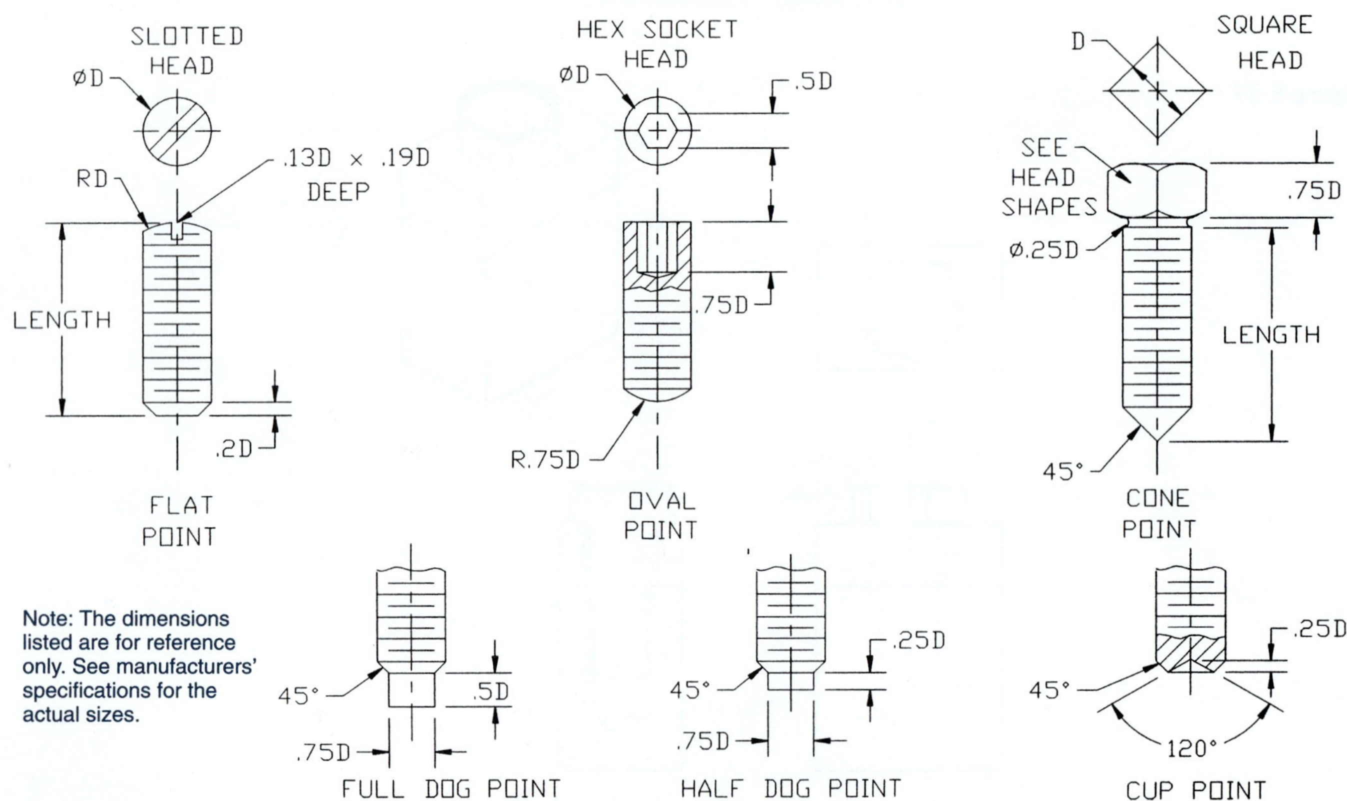

Figure 6-39

6-15 Drawing a Threaded Hole in the Side of a Cylinder

Figure 6-40 shows a Ø.75 × Ø1.00 × 1.00 collar with a #10-24 threaded hole. This section will explain how to add a threaded hole through the sides of a cylinder and insert set screws.

1 Draw the collar by drawing a **Ø1.00** × **1.00** cylinder and then drawing a **Ø.75** hole through the length of the cylinder.

Figure 6-40

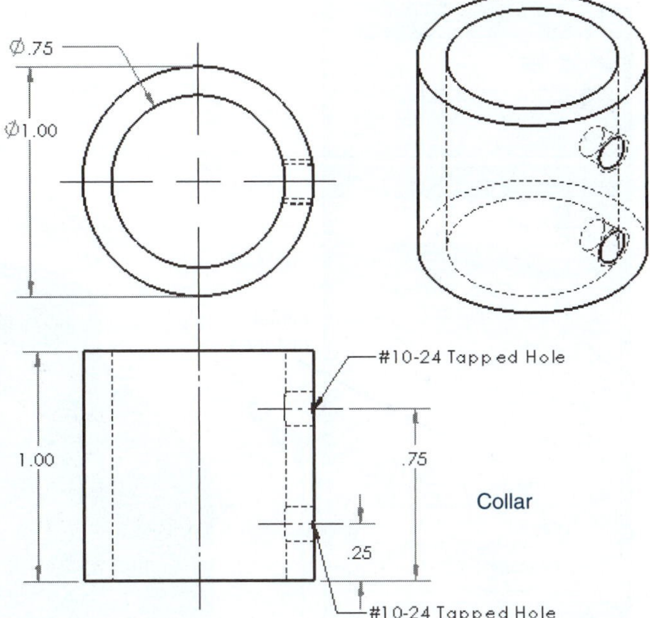

See Figure 6-41. In this example the Ø1.00 × 1.00 cylinder was centered about the origin.

Figure 6-41

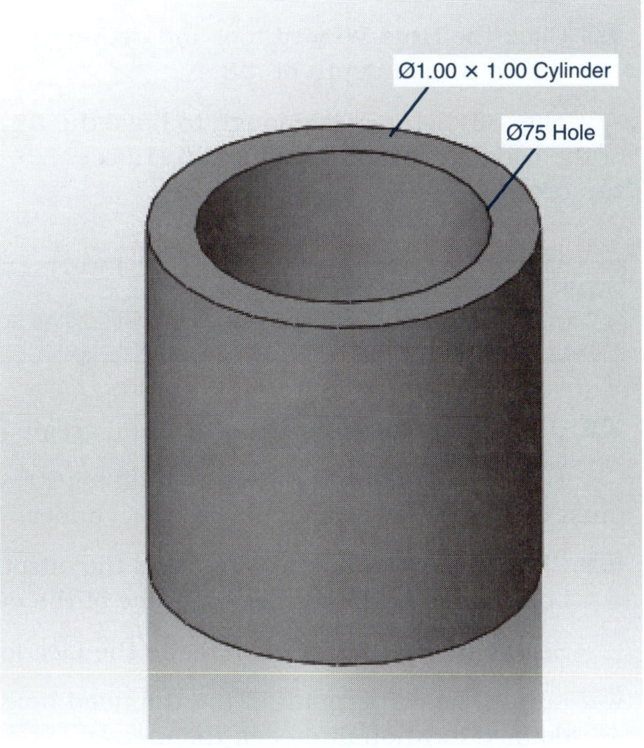

2 Use the **Plane** tool flyout from **Reference Geometry** tool and create an offset right plane tangent to the outside edge of the cylinder.

See Figure 6-42. The plane will be offset .50 from the origin.

Figure 6-42

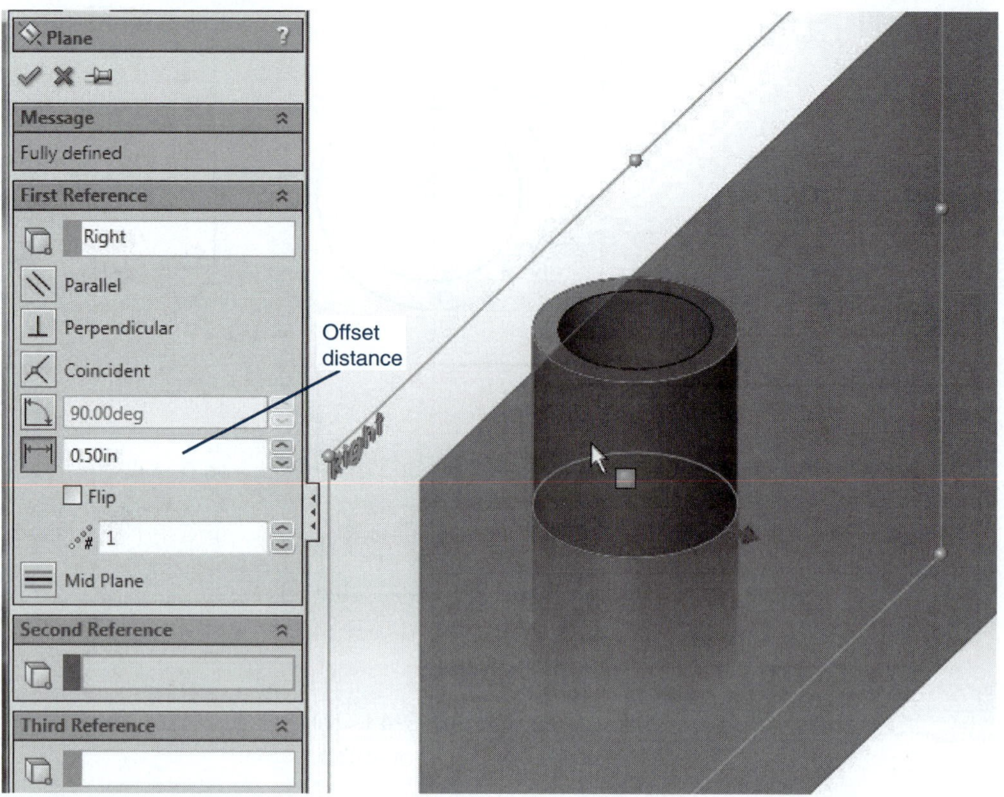

3 Click the **Hole Wizard** tool and define the **Size** of the hole as **#10-24** with a thread depth of **.38**.

The .38 distance is enough to have the threads go through the wall of the collar. The collar wall is .25 (1.00 − .75 = .25). Use the **Straight tap** option.

> **TIP**
> Do not use the **Through All** option, as this will create holes in both sides of the collar.

4 Use the **View Orientation** tool and create a **Right** view of the collar.

This view orientation will give a direct 90° view of Plane 1. The holes must be located at exactly 90° to the cylinder.

5 Click the **Positions** tab and click the offset right plane within the boundaries of the rectangular face of the cylinder.

See Figure 6-43. This will define the face for the hole location.

6 Locate the centerpoint of the threaded hole as near as possible to the defined location and click the mouse.

See Figure 6-44. This will locate the hole. The location will not be perfect, but it can easily be corrected.

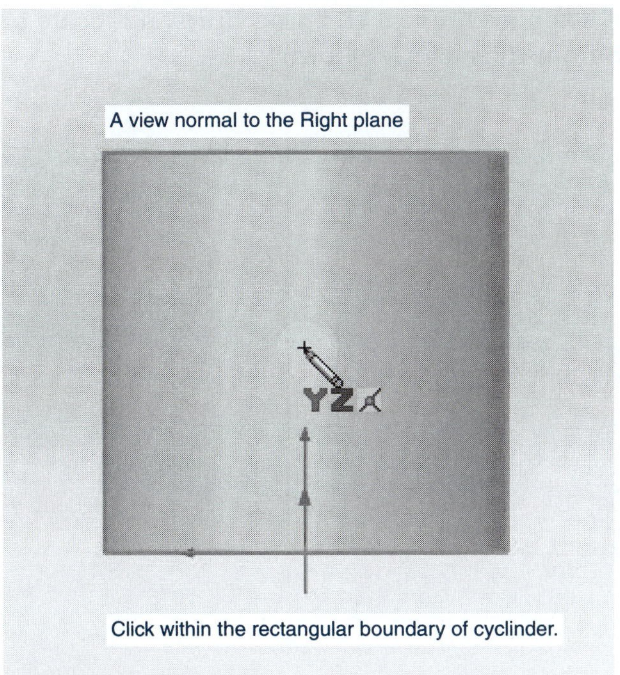

Figure 6-43

Figure 6-44

Chapter 6

7 Click the hole's center point again.

A listing of parameters will appear. See Figure 6-45.

Figure 6-45

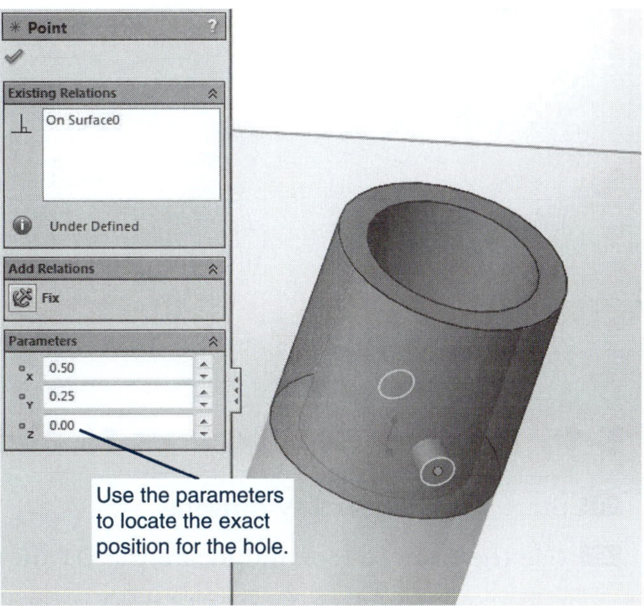

8 Change the Z value to **0.00** and the Y value to **.25**.

The Z value of 0.00 will locate the hole's center point directly on the Z-axis. Remember that the original cylinder's center point is on the origin, so this Z value will locate the hole's center point on the Z-axis. The X location will be 0.50, or a point tangent to the outside edge of the cylinder.

The .25 value comes from the given .25 dimension. See Figure 6-40.

9 Use the **Hole Wizard** tool again and repeat the procedure and locate a second threaded hole **.75** from the base as shown.

See Figure 6-46.

Figure 6-46

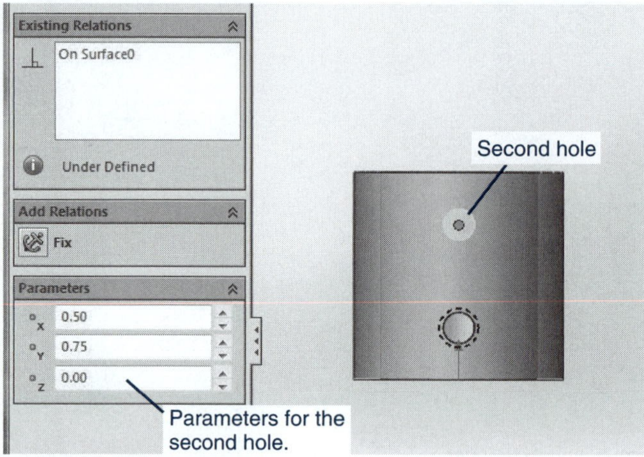

10 Hide **Plane 1**.

11 Save the collar as **Ø1.00COLLAR**.

The **Ø** symbol is created using **<Alt> 0216**. Figure 6-47 shows the resulting holes.

Figure 6-47

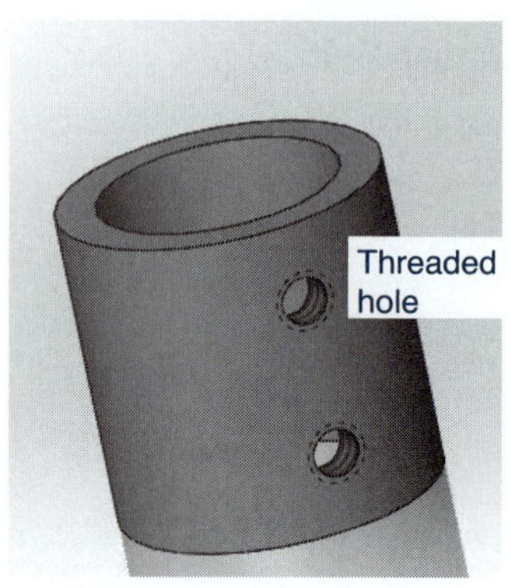

6-16 Adding Set Screws to the Collar

1 Start a new **Assembly** drawing.

2 Use the **Browse** tool and locate the **Ø1.00 COLLAR** on the screen.

Locate the origin of the **COLLAR** on the origin of the assembly drawing.

3 Access the **Design Library**, then click **Toolbox, ANSI Inch, Bolts and Screws**, and **Set Screws (Slotted)**.

See Figure 6-48.

4 Select the **Slotted Set Screw Cup Point** option and click and drag the set screw into the drawing.

See Figure 6-49.

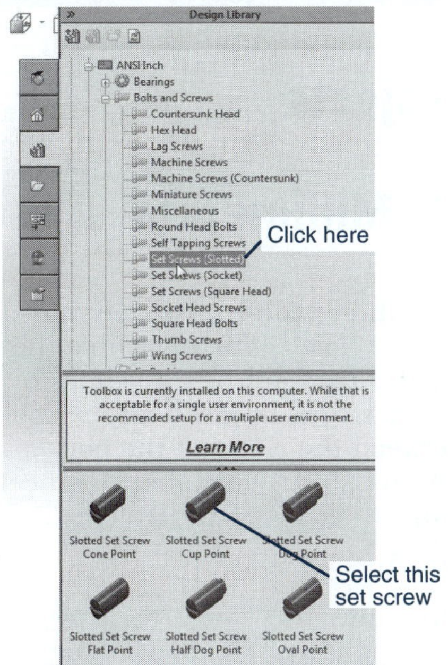

Figure 6-48

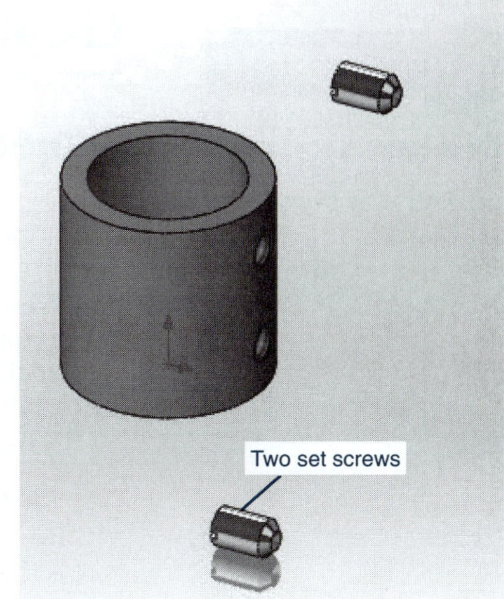

Figure 6-49

5 Define the **Size** of the set screw as **#10-24** and the **Length** as **0.315**.

The **0.315** is a standard length. It is good design practice to use standard lengths whenever possible.

6 Click the green **OK** check mark.

A second screw will automatically be attached to the cursor.

7 Add a second set screw and press the **<Esc>** key.

See Figure 6-50.

Figure 6-50

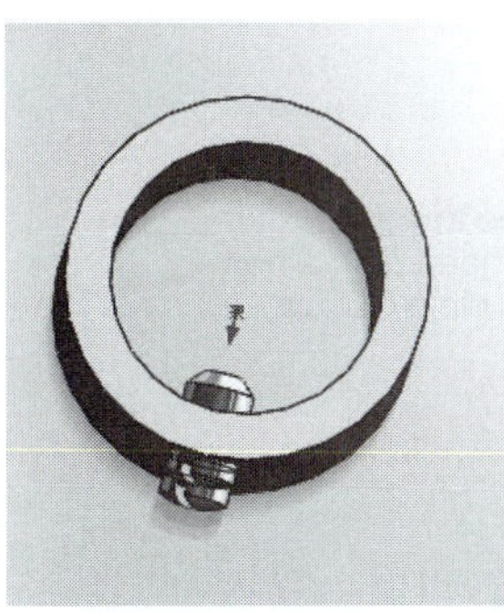

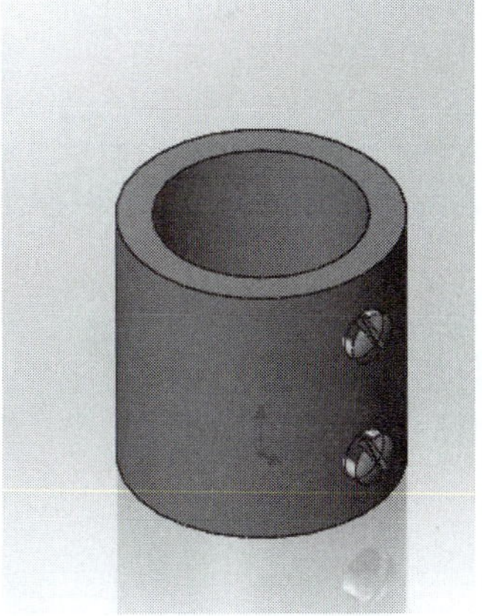

8 Use the **Mate** tool and insert the set screws into the collar.

9 Save the assembly.

Chapter Projects

Project 6-1: Millimeters

Figure P6-1 shows three blocks. Assume that the blocks are each 30 × 30 × 10 and that the hole is Ø9. Assemble the three blocks so that their holes are aligned and they are held together by a hex head bold secured by an appropriate hex nut. Locate a washer between the bolt head and the top block and between the nut and the bottom block. Create all drawings using either an A4 or A3 drawing sheet, as needed. Include a title block on all drawing sheets.

A. Define the bolt.

B. Define the nut.

C. Define the washers.

D. Draw an assembly drawing including all components.

E. Create a BOM for the assembly.

F. Create an isometric exploded drawing of the assembly.

G. Create an animation drawing of the assembly.

Figure P6-1

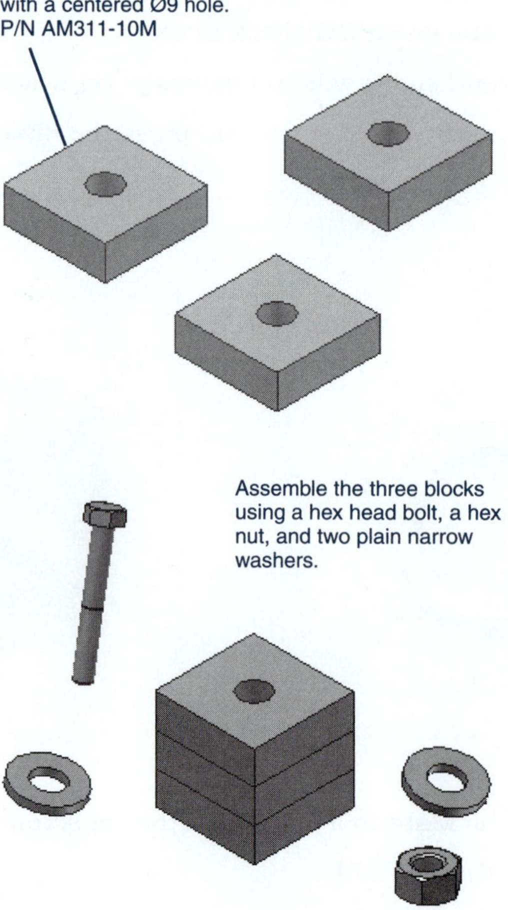
Three blocks, each 30 x 30 x 10
with a centered Ø9 hole.
P/N AM311-10M
Assemble the three blocks
using a hex head bolt, a hex
nut, and two plain narrow
washers.

Project 6-2: Millimeters

Figure P6-2 shows three blocks, one 30 × 30 × 50 block with a centered M8 threaded hole, and two 30 × 30 × 10 blocks with centered Ø9 holes. Join the two 30 × 30 × 10 blocks to the 30 × 30 × 50 block using an M8 hex head bolt. Locate a regular plain washer under the bolt head.

A. Define the bolt.

B. Define the thread depth.

C. Define the hole depth.

D. Define the washers.

E. Draw an assembly drawing including all components.

F. Create a BOM for the assembly.

G. Create an isometric exploded drawing of the assembly.

H. Create an animation drawing of the assembly.

Figure P6-2

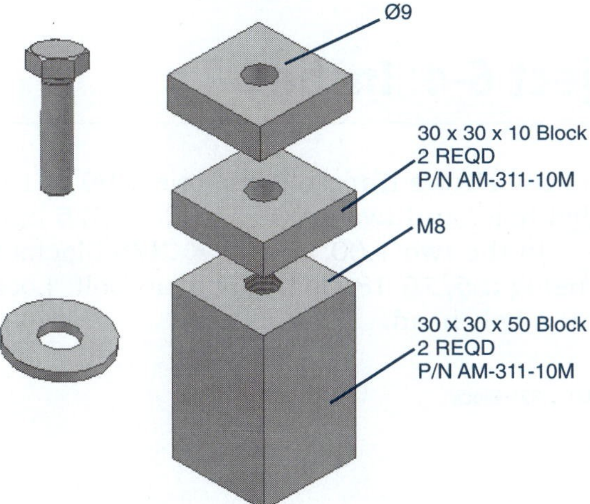

Project 6-3: Inches

Figure P6-3 shows three blocks. Assume that each block is 1.00 × 1.00 × .375 and that the hole is Ø.375. Assemble the three blocks so that their holes are aligned and that they are held together by a 5/16-18 UNC indented regular hex head bolt secured by an appropriate hex nut. Locate a washer between the bolt head and the top block and between the nut and the bottom block. Create all drawings using either an A4 or A3 drawing sheet, as needed. Include a title block on all drawing sheets.

A. Define the bolt.

B. Define the nut.

C. Define the washers.

D. Draw an assembly drawing including all components.

E. Create a BOM for the assembly.

F. Create an isometric exploded drawing of the assembly.

G. Create an animation drawing of the assembly.

Figure P6-3

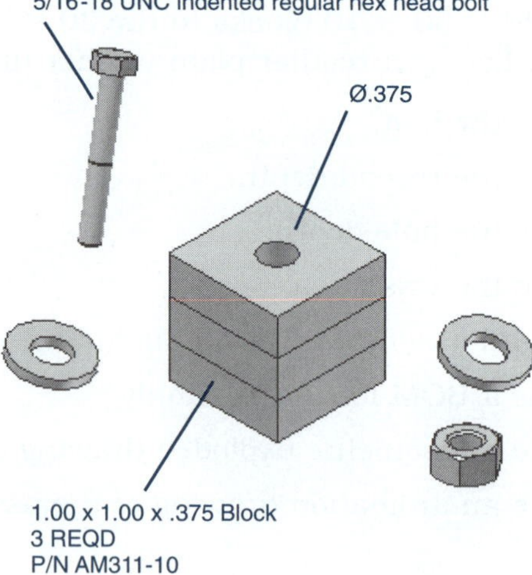

Project 6-4: Inches

Figure P6-4 shows three blocks, one 1.00 × 1.00 × 2.00 with a centered threaded hole, and two 1.00 × 1.00 × .375 blocks with centered Ø.375 holes. Join the two 1.00 × 1.00 × .375 blocks to the 1.00 × 1.00 × 2.00 block using a 5/16-18 UNC hex head bolt. Locate a regular plain washer under the bolt head.

Figure P6-4

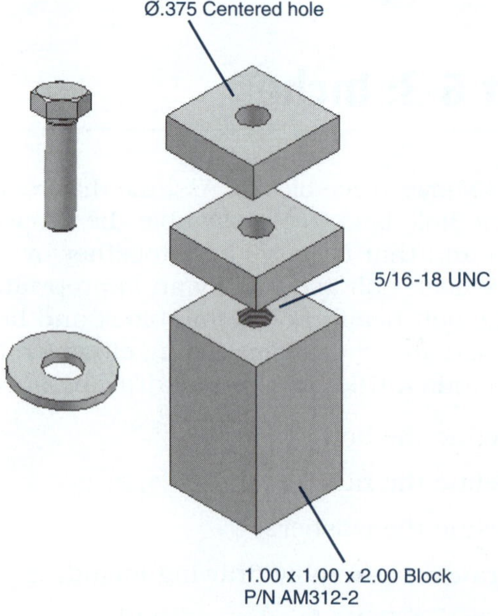

A. Define the bolt.

B. Define the thread depth.

C. Define the hole depth.

D. Define the washer.

E. Draw an assembly drawing including all components.

F. Create a BOM for the assembly.

G. Create an isometric exploded drawing of the assembly.

H. Create an animation drawing of the assembly.

Project 6-5: Inches or Millimeters

Figure P6-5 shows a centering block. Create an assembly drawing of the block and insert three set screws into the three threaded holes so that they extend at least .25 in. or 6 mm into the center hole.

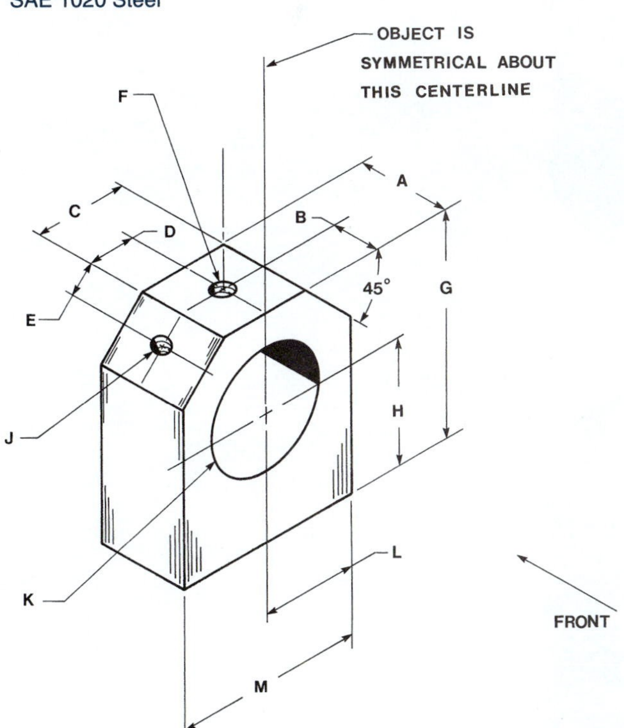

DIMENSION	INCHES	mm
A	1.00	26
B	.50	13
C	1.00	26
D	.50	13
E	.38	10
F	.190-32 UNF	M8X1
G	2.38	60
H	1.38	34
J	.164-36 UNF	M6
K	Ø1.25	Ø30
L	1.00	26
M	2.00	52

Figure P6-5

A. Use the inch dimensions.

B. Use the millimeter dimensions.

C. Define the set screws.

D. Draw an assembly drawing including all components.

E. Create a BOM for the assembly.

F. Create an isometric exploded drawing of the assembly.

G. Create an animation drawing of the assembly.

Project 6-6: Millimeters

Figure P6-6 shows two parts: a head cylinder and a base cylinder. The head cylinder has outside dimensions of Ø40 × 20, and the base cylinder has outside dimensions of Ø40 × 50. The holes in both parts are located on a Ø24 bolt circle. Assemble the two parts using hex head bolts.

Figure P6-6

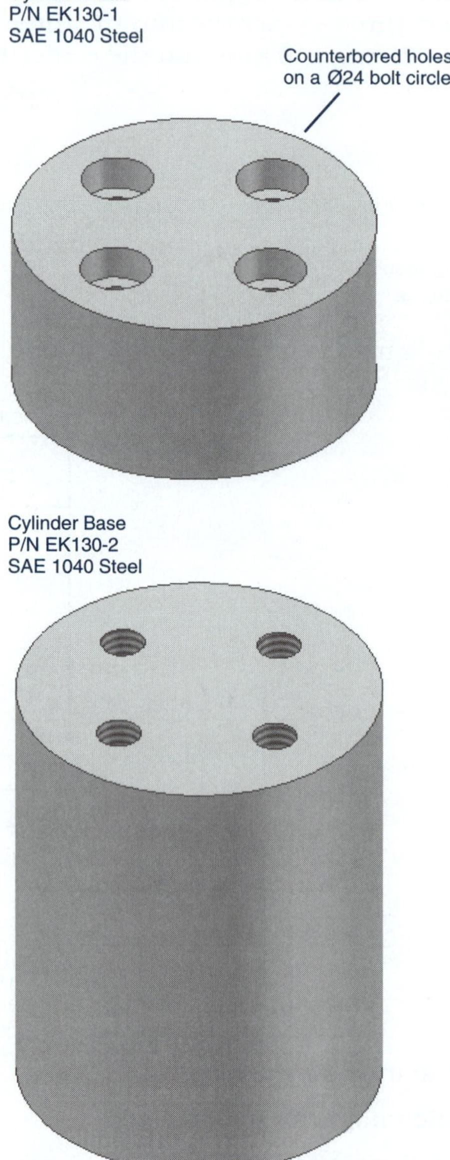

A. Define the bolt.

B. Define the holes in the head cylinder, the counterbore diameter and depth, and the clearance hole diameter.

C. Define the thread depth in the base cylinder.

D. Define the hole depth in the base cylinder.

E. Draw an assembly drawing including all components.

F. Create a BOM for the assembly.

G. Create an isometric exploded drawing of the assembly.

H. Create an animation drawing of the assembly.

Project 6-7: Millimeters

Figure P6-7 shows a pressure cylinder assembly.

Figure P6-7

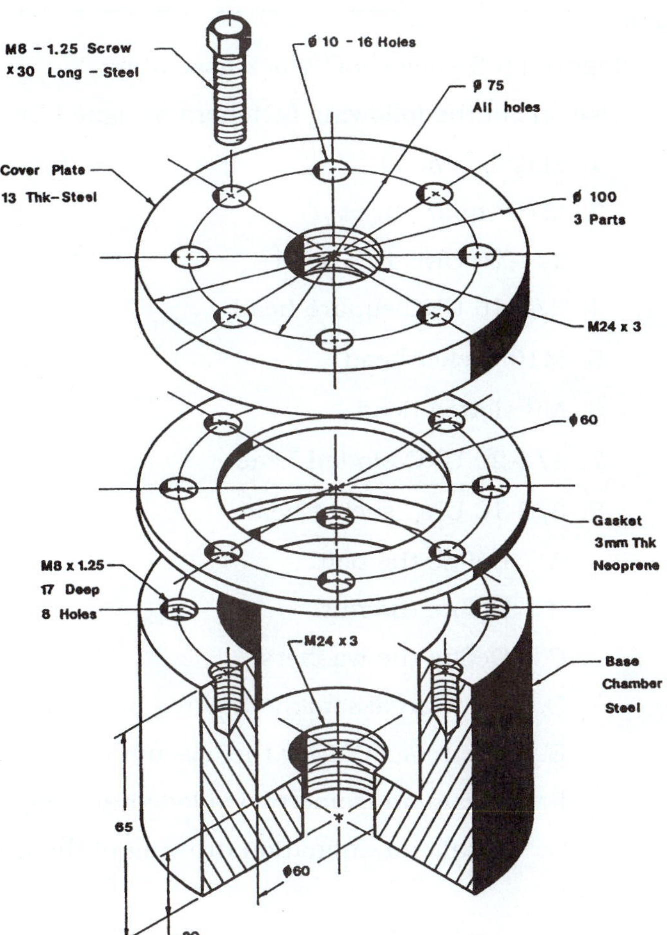

A. Draw an assembly drawing including all components.

B. Create a BOM for the assembly.

C. Create an isometric exploded drawing of the assembly.

D. Create an animation drawing of the assembly.

Project 6-8: Millimeters

Figure P6-7 shows a pressure cylinder assembly.

A. Revise the assembly so that it uses M10 × 35 hex head bolts.

B. Draw an assembly drawing including all components.

C. Create a BOM for the assembly.

D. Create an isometric exploded drawing of the assembly.

E. Create an animation drawing of the assembly.

Project 6-9: Inches and Millimeters

Figure P6-9 shows a C-block assembly.

Use one of the following fasteners assigned by your instructor.

1. M12 hex head
2. M10 square head
3. 1/4-20 UNC hex head
4. 3/8-16 UNC square head
5. M10 socket head
6. M8 slotted head
7. 1/4-20 UNC slotted head
8. 3/8-16 UNC socket head

A. Define the bolt.

B. Define the nut.

C. Define the washers.

D. Draw an assembly drawing including all components.

E. Create a BOM for the assembly.

F. Create an isometric exploded drawing of the assembly.

G. Create an animation drawing of the assembly.

Figure P6-9

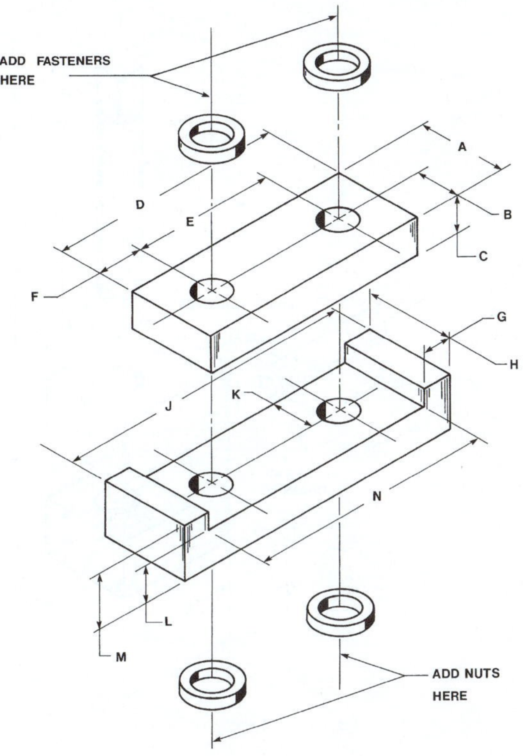

DIMENSION	INCHES	mm
A	1.25	32
B	.63	16
C	.50	13
D	3.25	82
E	2.00	50
F	.63	16
G	.38	10
H	1.25	32
J	4.13	106
K	.63	16
L	.50	13
M	.75	10
N	3.38	86

Project 6-10: Millimeters

Figure P6-10 shows an exploded assembly drawing. There are no standard parts, so each part must be drawn individually.

A. Draw an assembly drawing including all components.

B. Create a BOM for the assembly.

C. Create an isometric exploded drawing of the assembly.

D. Create an animation drawing of the assembly.

Project 6-11: Millimeters

Figure P6-11 shows an exploded assembly drawing.

A. Draw an assembly drawing including all components.

B. Create a BOM for the assembly.

C. Create an isometric exploded drawing of the assembly.

D. Create an animation drawing of the assembly.

Figure P6-10

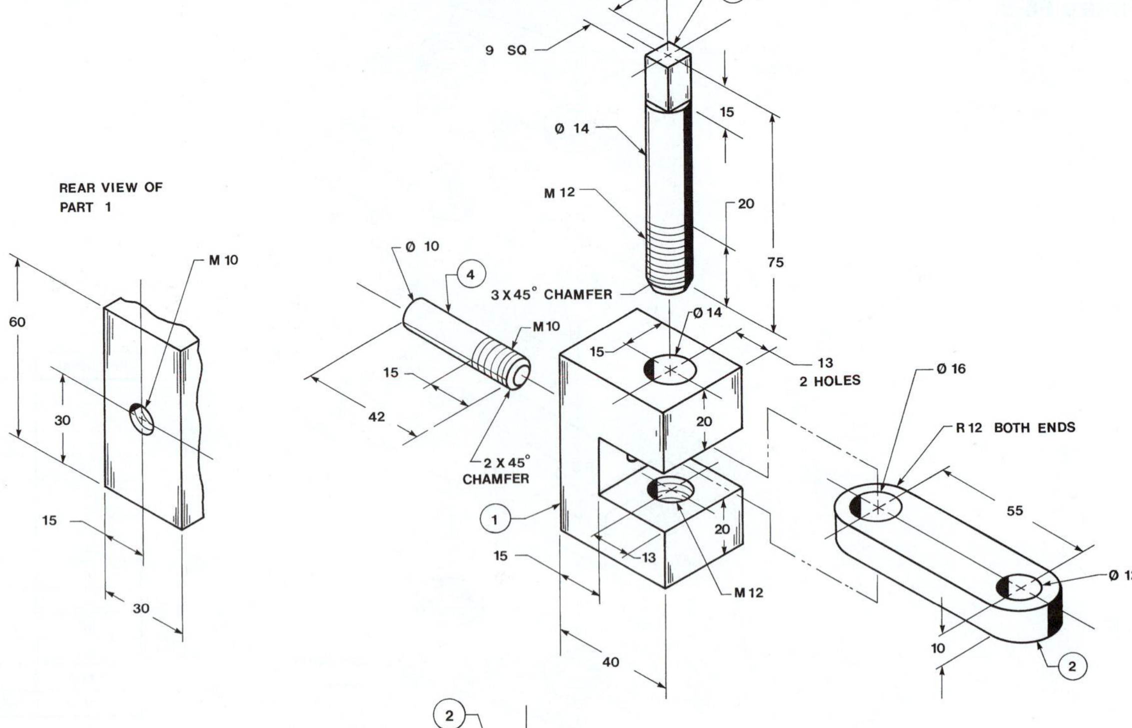

Figure P6-11

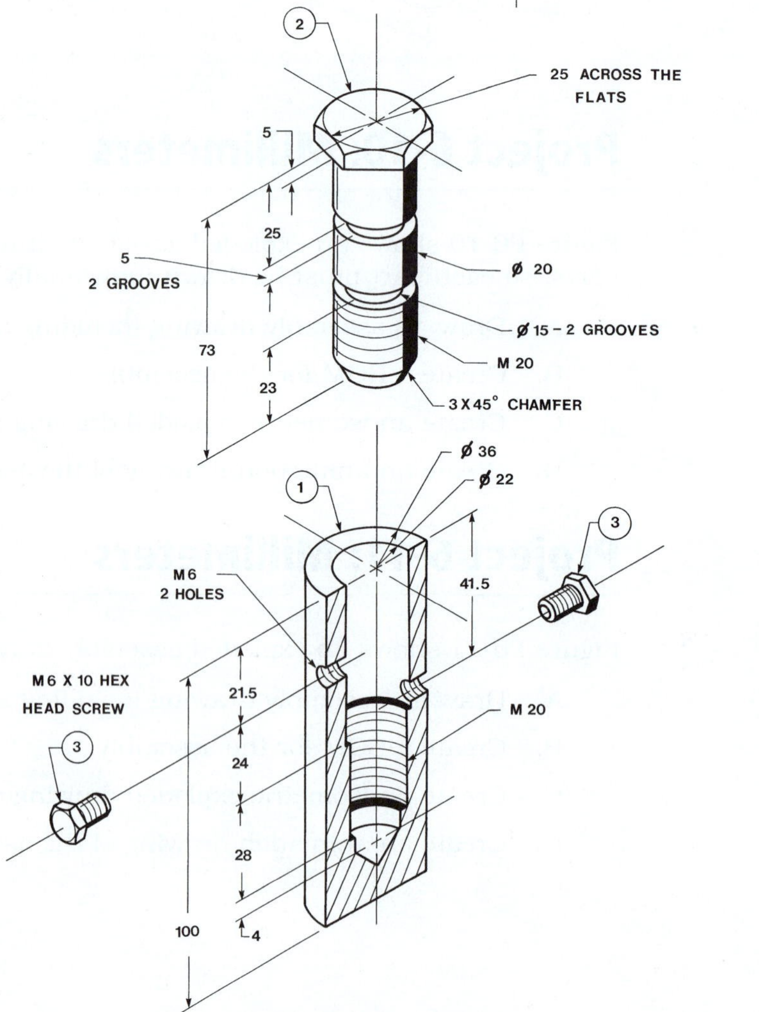

Project 6-12: Inches or Millimeters

Figure P6-12 shows an exploded assembly drawing. No dimensions are given. If parts 3 and 5 have either M10 or 3/8-16 UNC threads, size parts 1 and 2. Based on these values, estimate and create the remaining sizes and dimensions.

A. Draw an assembly drawing including all components.

B. Create a BOM for the assembly.

C. Create an isometric exploded drawing of the assembly.

D. Create an animation drawing of the assembly.

Figure P6-12

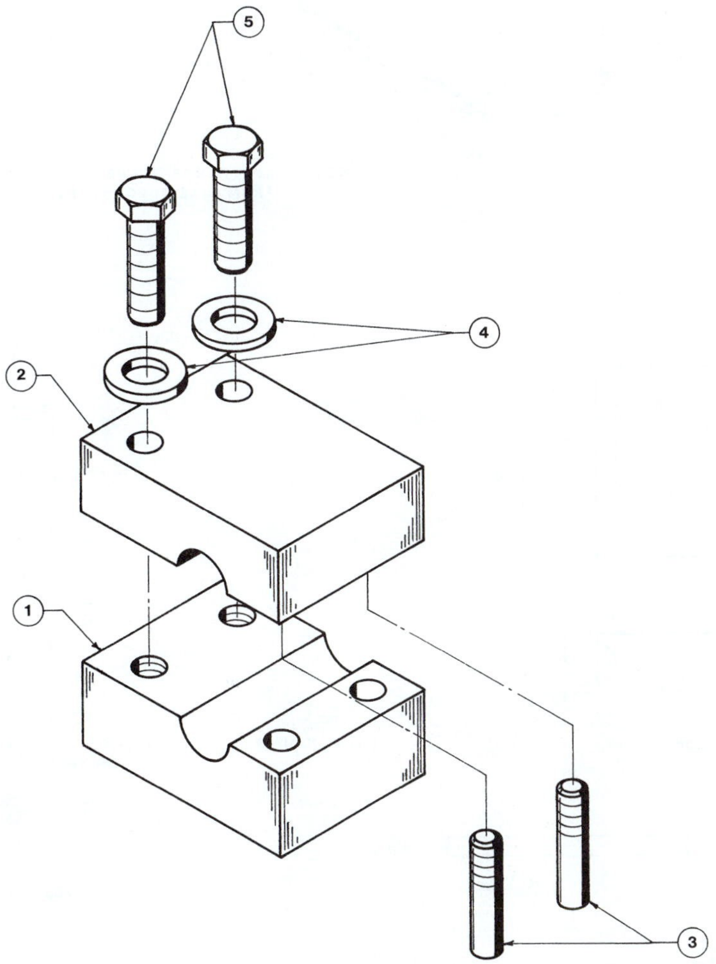

Project 6-13: Inches

Figure P6-13 shows an assembly drawing and detail drawings of a surface gauge.

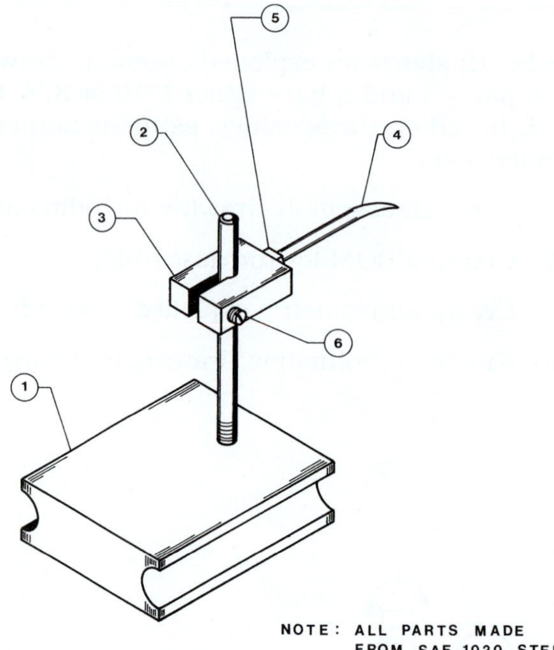

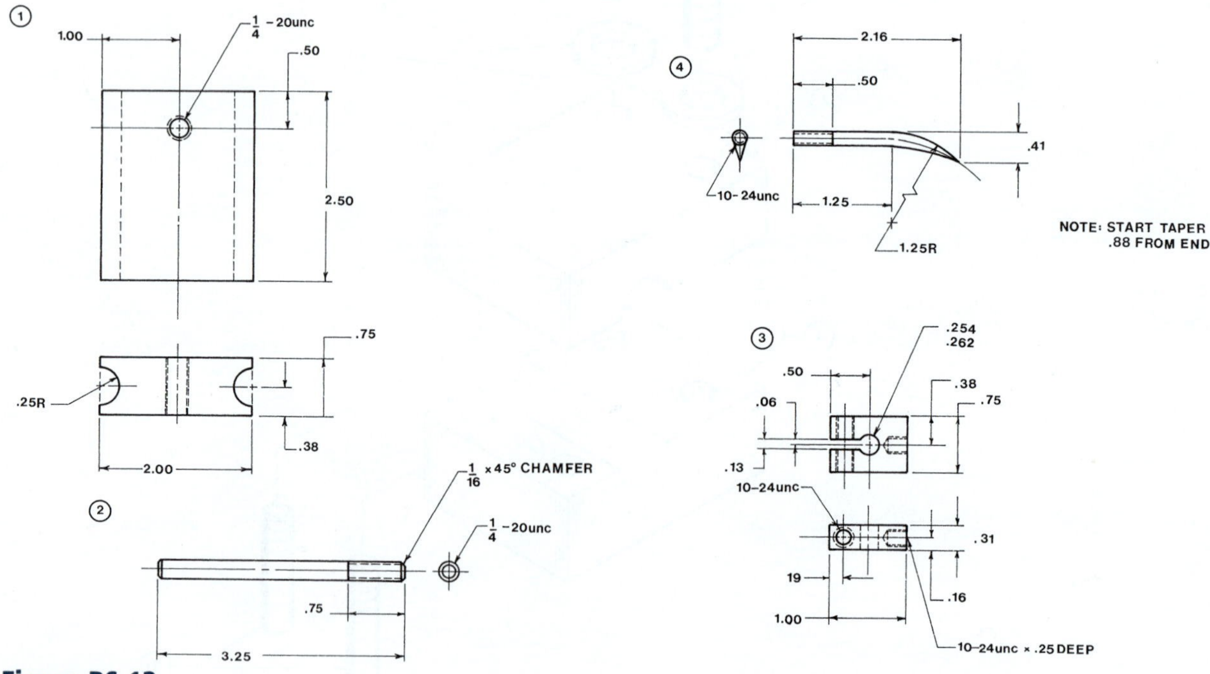

Figure P6-13

A. Draw an assembly drawing including all components.

B. Create a BOM for the assembly.

C. Create an isometric exploded drawing of the assembly.

D. Create an animation drawing of the assembly.

Project 6-14: Millimeters

Figure P6-14 shows an assembly made from parts defined on pages 343–345.

Assemble the parts using M10 threaded fasteners.

A. Define the bolt.

B. Define the nut.

C. Draw an assembly drawing including all components.

D. Create a BOM for the assembly.

E. Create an isometric exploded drawing of the assembly.

F. Create an animation drawing of the assembly.

G. Consider possible interference between the nuts and ends of the fasteners both during and after assembly. Recommend an assembly sequence.

Figure P6-14

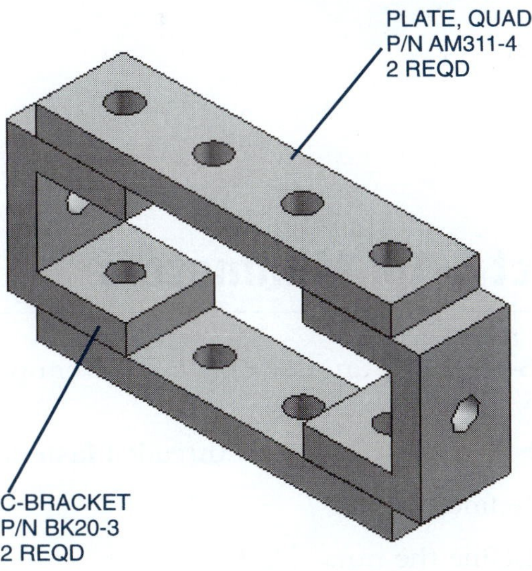

Project 6-15: Millimeters

Figure P6-15 shows an assembly made from parts defined on page XXX. Assemble the parts using M10 threaded fasteners.

A. Define the bolt.

B. Define the nut.

C. Draw an assembly drawing including all components.

D. Create a BOM for the assembly.

E. Create an isometric exploded drawing of the assembly.

F. Create an animation drawing of the assembly.

G. Consider possible interference between the nuts and ends of the fasteners both during and after assembly. Recommend an assembly sequence.

Figure P6-15

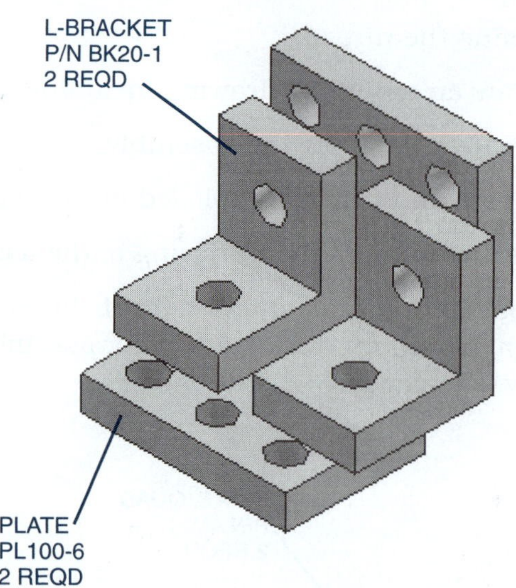

Project 6-16: Millimeters

Figure P6-16 shows an assembly made from parts defined on pages 343–345.

Assemble the parts using M10 threaded fasteners.

A. Define the bolt.

B. Define the nut.

C. Draw an assembly drawing including all components.

D. Create a BOM for the assembly.

E. Create an isometric exploded drawing of the assembly.

F. Create an animation drawing of the assembly.

G. Consider possible interference between the nuts and ends of the fasteners both during and after assembly. Recommend an assembly sequence.

Figure P6-16

Project 6-17: Access Controller

Design an access controller based on the information given in Figure P6-17. The controller works by moving an internal cylinder up and down within the base so the cylinder aligns with output holes A and B. Liquids will enter the internal cylinder from the top, then exit the base through holes A and B. Include as many holes in the internal cylinder as necessary to create the following liquid-exit combinations.

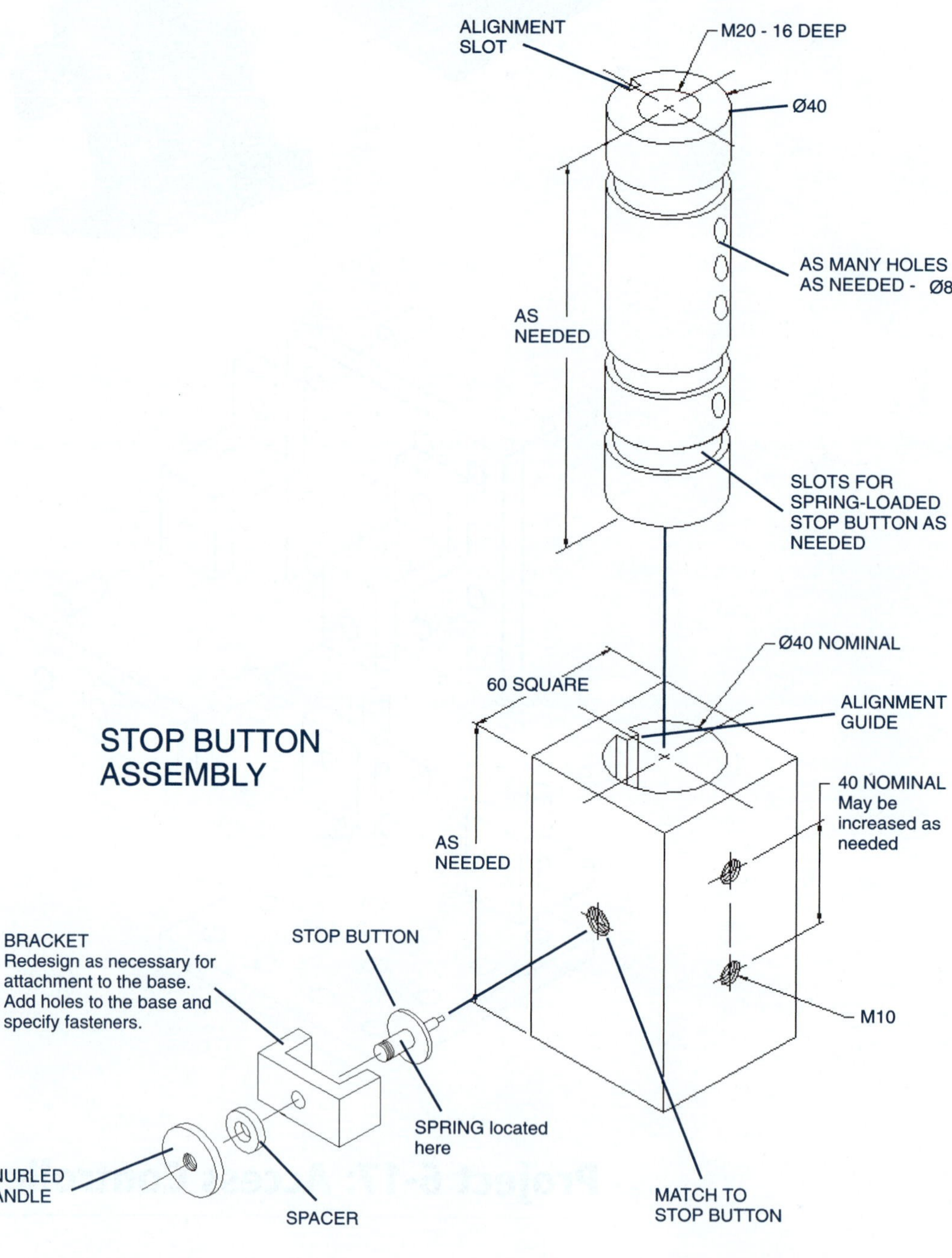

Figure P6-17

1. A open, B closed
2. A open, B open
3. A closed, B open

The internal cylinder is to be held in place by an alignment key and a stop button. The stop button is to be spring-loaded so that it will always be held in place. The internal cylinder will be moved by pulling out the stop button, repositioning the cylinder, then reinserting the stop button.
Prepare the following drawings.

A. Draw an assembly drawing.

B. Draw detail drawings of each nonstandard part. Include positional tolerances for all holes.

C. Prepare a BOM.

Project 6-18: Grinding Wheel

Design a hand-operated grinding wheel as shown in Figure P6-18 specifically for sharpening a chisel. The chisel is to be located on an adjustable rest while it is being sharpened. The mechanism should be able to be clamped to a table during operation using two thumbscrews. A standard grinding wheel is 6.00 in. and 1/2 in. thick, and has an internal mounting hole with a 50.00±.03 bore.

Prepare the following drawing.

A. Draw an assembly drawing.

B. Draw detail drawings of each nonstandard part. Include positional tolerances for all holes.

C. Prepare a BOM.

Project 6-19: Millimeters

Given the assembly shown in Figure P6-19 on page 407, add the following fasteners.

1. Create an assembly drawing.
2. Create a parts list including assembly numbers.
3. Create a dimensioned drawing of the support block and specify a dimension for each hole including the thread size and the depth required.

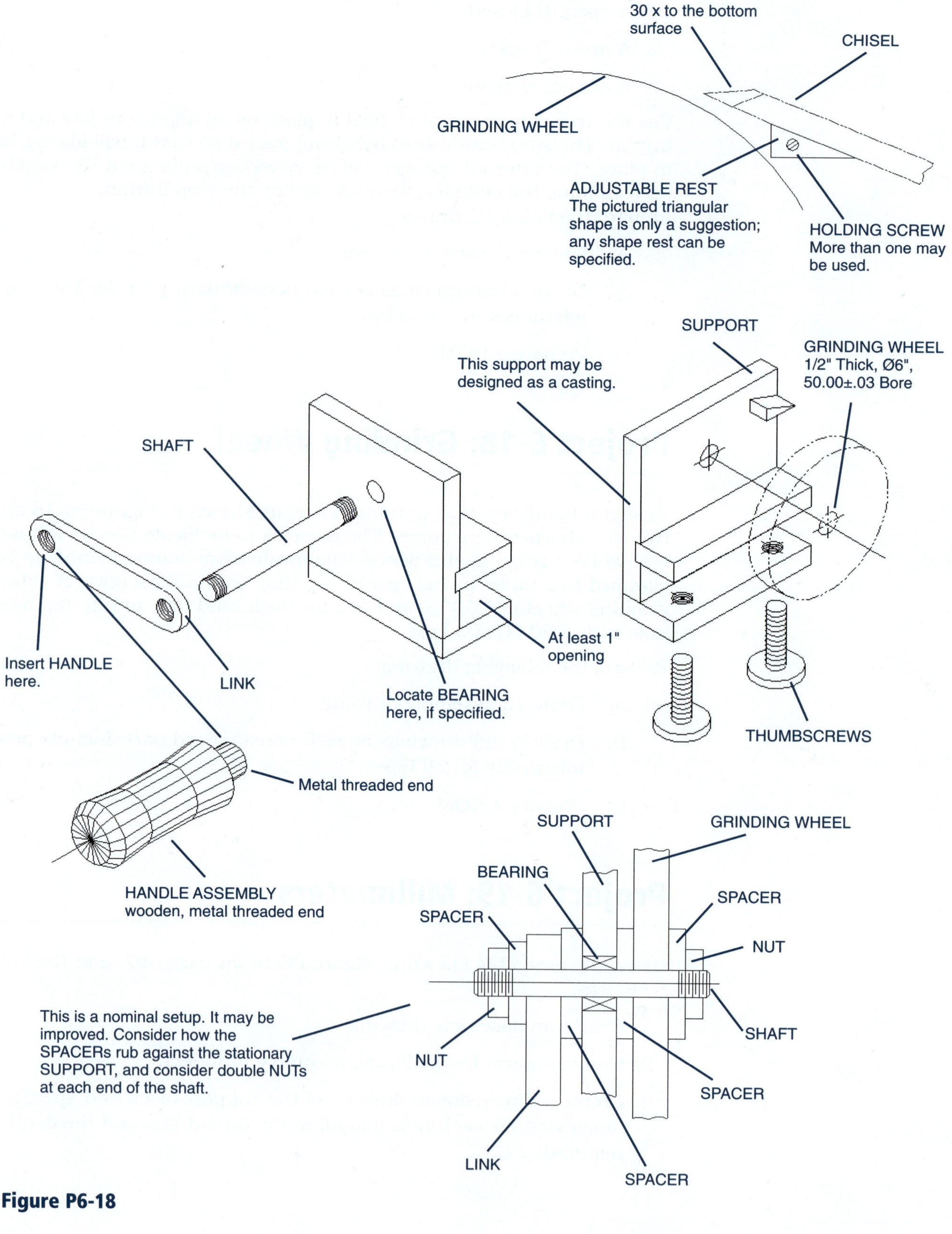

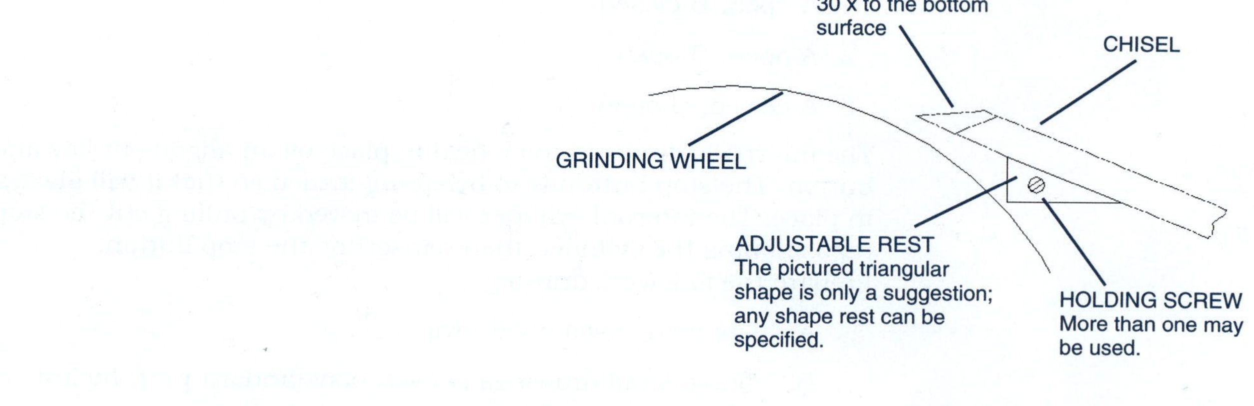

30 x to the bottom surface
CHISEL
GRINDING WHEEL
ADJUSTABLE REST The pictured triangular shape is only a suggestion; any shape rest can be specified.
HOLDING SCREW More than one may be used.
SUPPORT
This support may be designed as a casting.
GRINDING WHEEL 1/2" Thick, Ø6", 50.00±.03 Bore
SHAFT
Insert HANDLE here.
LINK
At least 1" opening
Locate BEARING here, if specified.
THUMBSCREWS
Metal threaded end
HANDLE ASSEMBLY wooden, metal threaded end
SUPPORT
GRINDING WHEEL
BEARING
SPACER
SPACER
NUT
This is a nominal setup. It may be improved. Consider how the SPACERs rub against the stationary SUPPORT, and consider double NUTs at each end of the shaft.
NUT
SHAFT
SPACER
LINK
SPACER

Figure P6-18

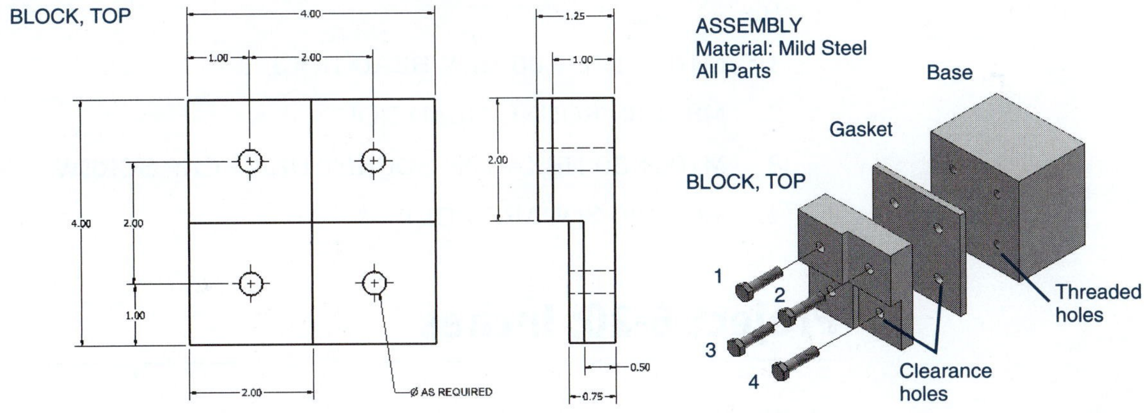

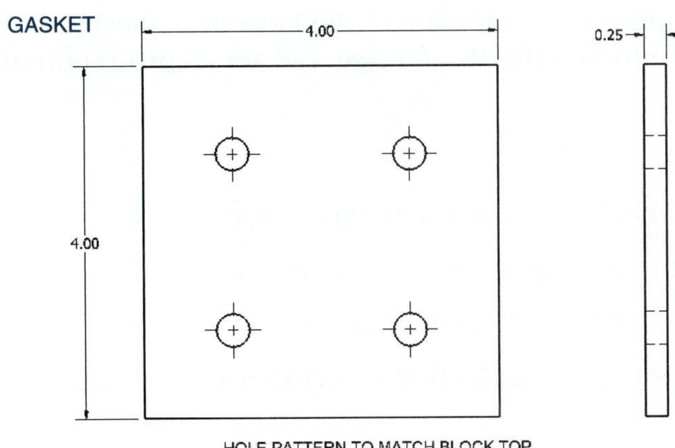

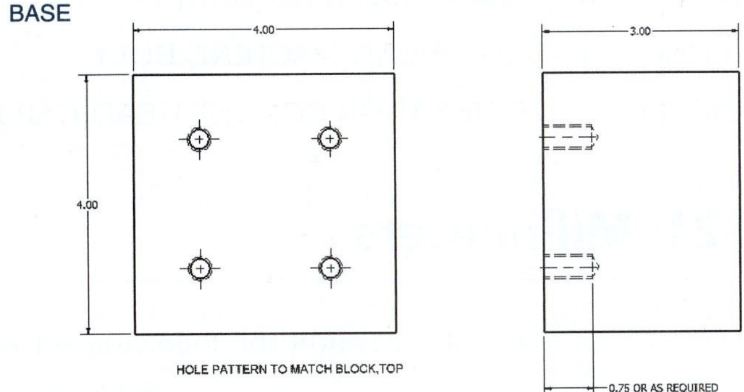

Figure P6-19

Fasteners:

A.

1. M10 × 35 HEX HEAD BOLT
2. M10 × 35 HEX HEAD BOLT
3. M10 × 30 HEX HEAD BOLT
4. M10 × 25 HEX HEAD BOLT

B.

1. M10 × 1.5 × 35 HEX HEAD BOLT
2. M8 × 35 ROUND HEAD BOLT
3. M10 × 30 HEXAGON SOCKET HEAD CAP SCREW
4. M6 × 30 SQUARE BOLT

Project 6-20: Inches

1. Create an assembly drawing.
2. Create a BOM including assembly numbers.
3. Create a dimensioned drawing of the base and specify a dimension for each hole including the thread size and the depth required.

Fasteners:

A.

1. 3/8-16 UNC × 2.50 HEX HEAD BOLT
2. 1/4-20 UNC × 2.00 HEX HEAD BOLT
3. 7/16-14 UNC × 1.75 HEX HEAD BOLT
4. 5/16-18 UNC × 2.25 HEX HEAD BOLT

B.

1. 1/4-28 UNF × 2.00 HEX HEAD BOLT
2. #8.(.164)-32 UNC × 2.00 HEX HEAD BOLT
3. 3/8-16 UNC × 1.75 PAN HEAD MACHINE BOLT
4. 5/16-18 UNC × 1.75 HEXAGON SOCKET HEAD CAP BOLT

Project 6-21: Millimeters

Given the collar shown in Figure P6-21, add the following set screws.

1. Create an assembly drawing.
2. Create a BOM.
3. Create a dimensioned drawing of the collar. Specify a thread specification for each hole as required by the designated set screw.

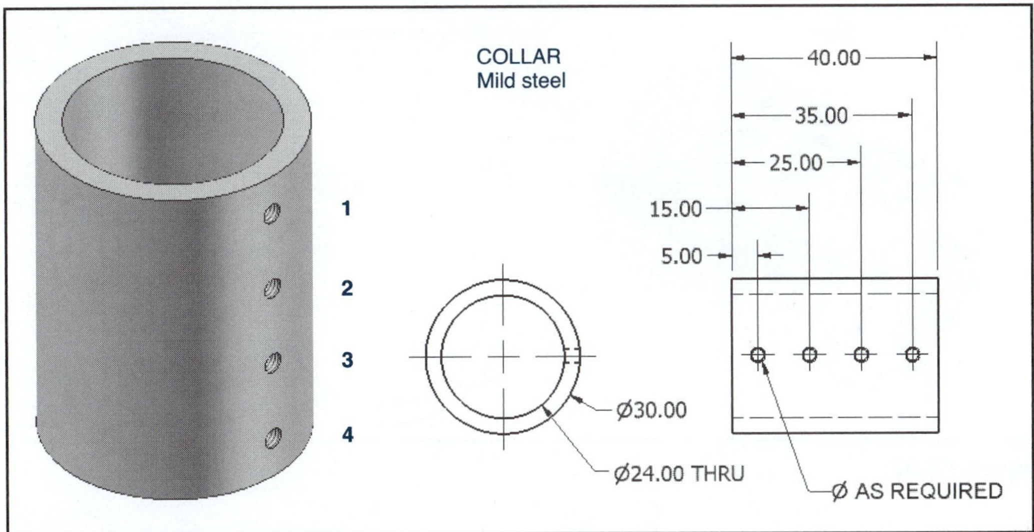

Figure P6-21

A.

1. M4 × 6 ANSI B18.3.5M SOCKET SET SCREW - HALF DOG POINT
2. M3 × 3 SOCKET SET SCREW - OVAL POINT
3. M2.5 × 4 B18.3.6M SOCKET SET SCREW - FLAT POINT
4. M4 × 5 B18.3.1M SOCKET HEAD CAP SCREW

B.

1. M2 × 4 B18.3.4M SOCKET BUTTON HEAD CAP SCREW
2. M3 × 6 B18.3.6M SOCKET SET SCREW - CONE POINT
3. M4 × 5 B18.3.6M SOCKET SET SCREW - FLAT POINT
4. M1.6 × 4 B18.3.6M SOCKET SET SCREW - CUP POINT

Project 6-22: Inches

Given the collar shown in Figure P6-22, add the following set screws.

1. Create an assembly drawing.
2. Create a BOM.
3. Create a dimensioned drawing of the collar. Specify a thread specification for each hole as required by the designated set screw.

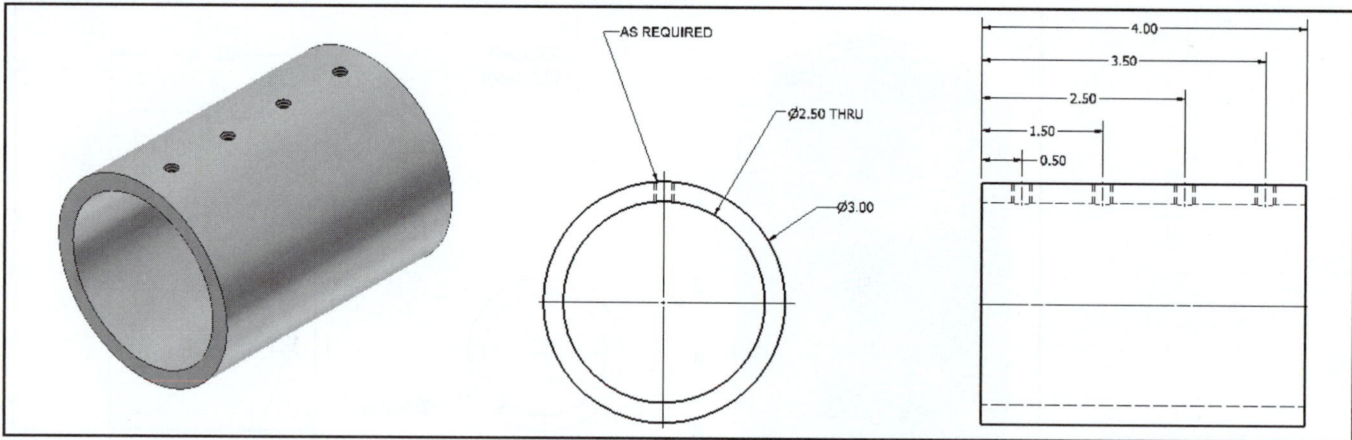

Figure P6-22

Holes:

A.

1. #10 (0.190) × .375 SQUARE HEAD SET SCREW - HALF DOG POINT-INCH
2. #6 (0.138) × .125 SLOTTED HEADLESS SET SCREW-FLAT POINT-INCH
3. #8 (0.164) × 3.75 SOCKET SET SCREW-CUP POINT-INCH
4. #5 (0.126) × .45 HEXAGON SOCKET SET SCREW-CONE POINT-INCH

B.

1. #6 (0.138) × .25 TYPE D-SOCKET SET SCREW-CUP POINT-INCH
2. #8 (0.164) × .1875 SLOTTED HEADLESS SET SCREW-DOG POINT-INCH
3. #10 (0.190) × .58 HEXAGON SOCKET SET SCREW-FLAT POINT-INCH
4. #6 (0.138) × .3125 SOCKET SET SCREW-HALF-DOG POINT-INCH

Project 6-23: Millimeters

Given the components shown in Figure P6-23:

1. Create an assembly drawing.
2. Animate the drawing.
3. Create an exploded isometric drawing.
4. Create a BOM.

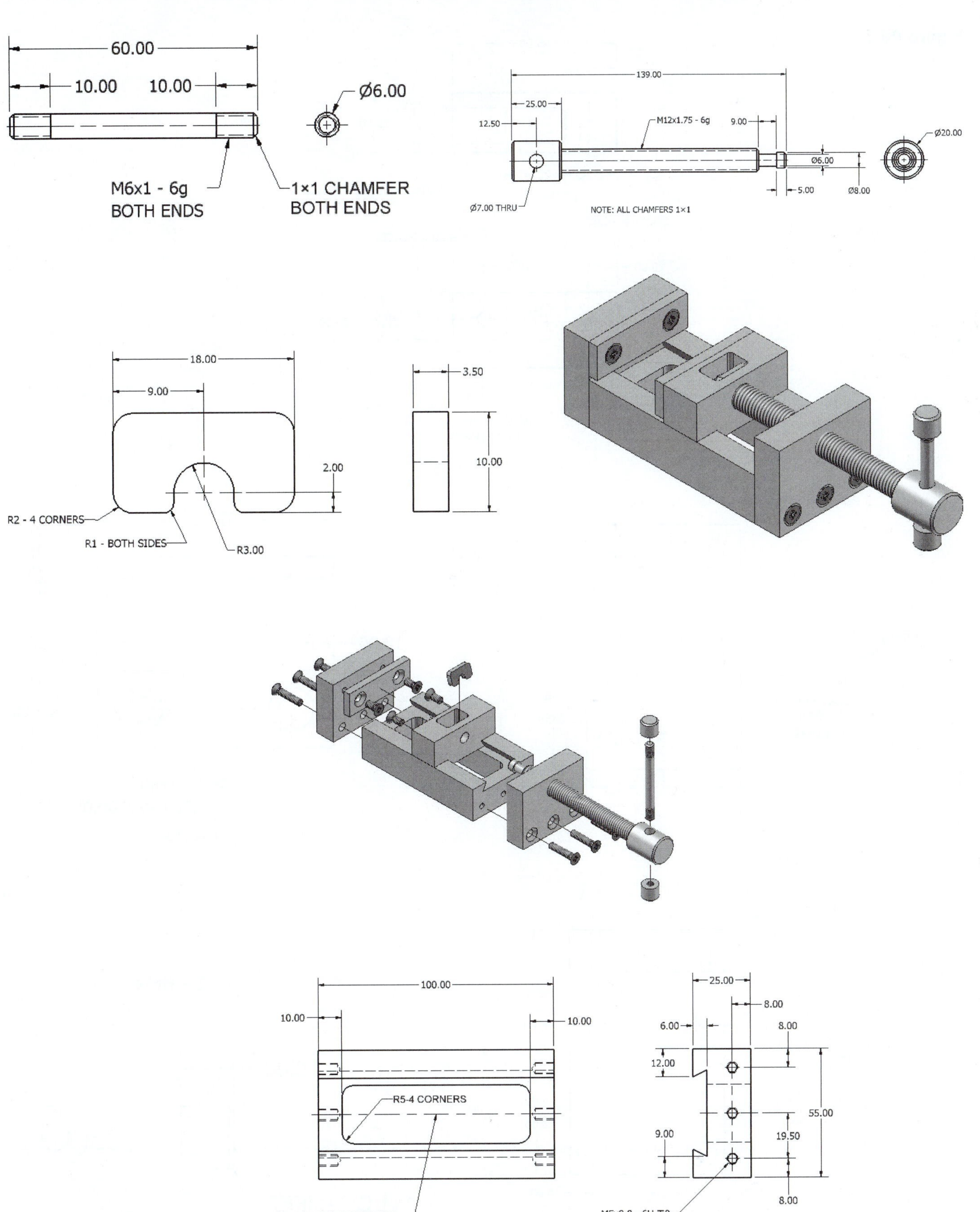
60.00
10.00
10.00
Ø6.00
M6x1 - 6g
BOTH ENDS
1×1 CHAMFER
BOTH ENDS
139.00
25.00
12.50
M12x1.75 - 6g
9.00
Ø20.00
Ø6.00
Ø7.00 THRU
5.00
Ø8.00
NOTE: ALL CHAMFERS 1×1
18.00
9.00
3.50
2.00
10.00
R2 - 4 CORNERS
R1 - BOTH SIDES
R3.00
100.00
10.00
10.00
R5-4 CORNERS
OBJECT IS SYMMETRICAL
ABOUT THIS CENTERLINE
25.00
8.00
6.00
8.00
12.00
55.00
9.00
19.50
8.00
M5x0.8 - 6H ↧8
6 HOLES

Figure P6-23

Figure P6-23
(*Continued*)

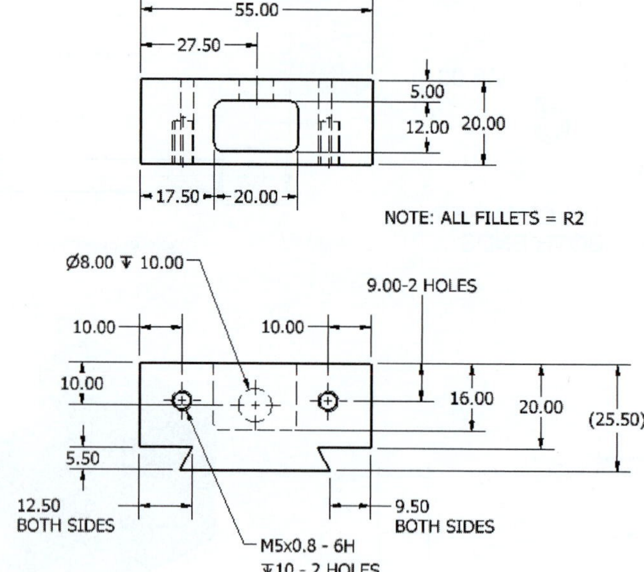

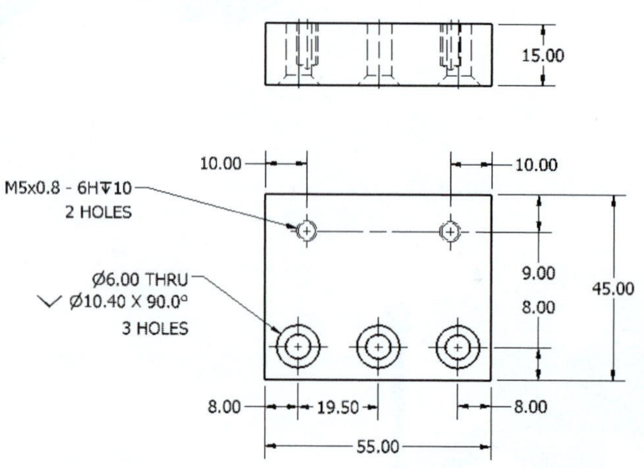

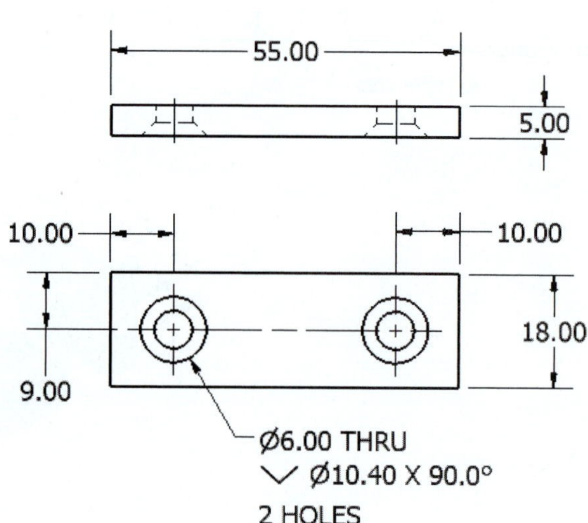

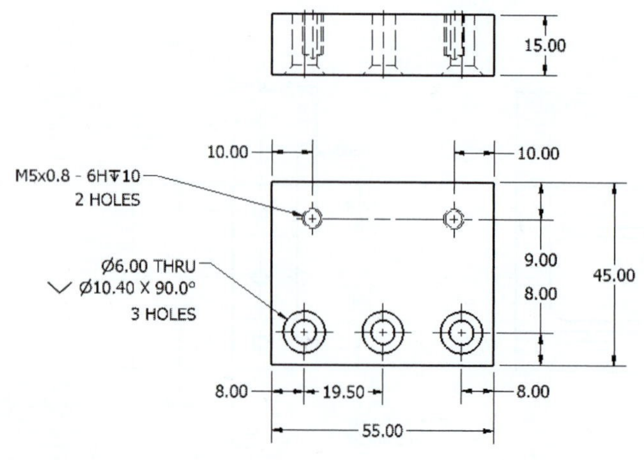

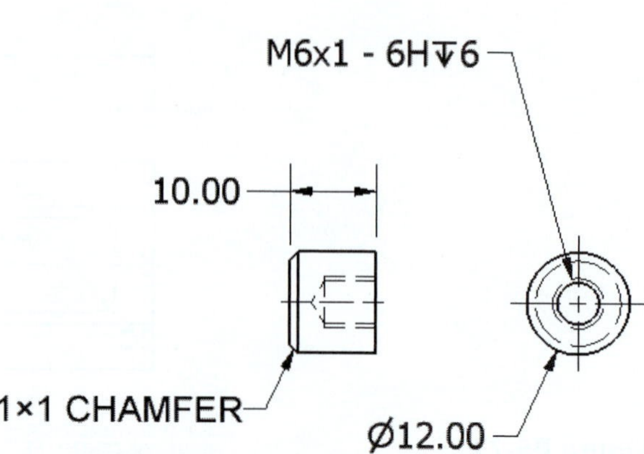

Project 6-24: Inches

Given the assembly drawing shown in Figure P6-24:

Figure P6-24

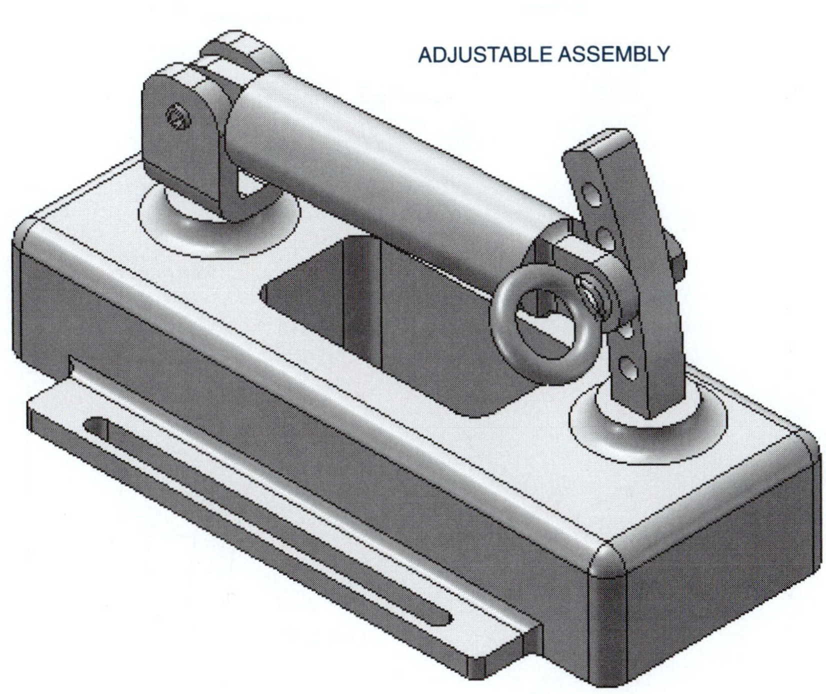

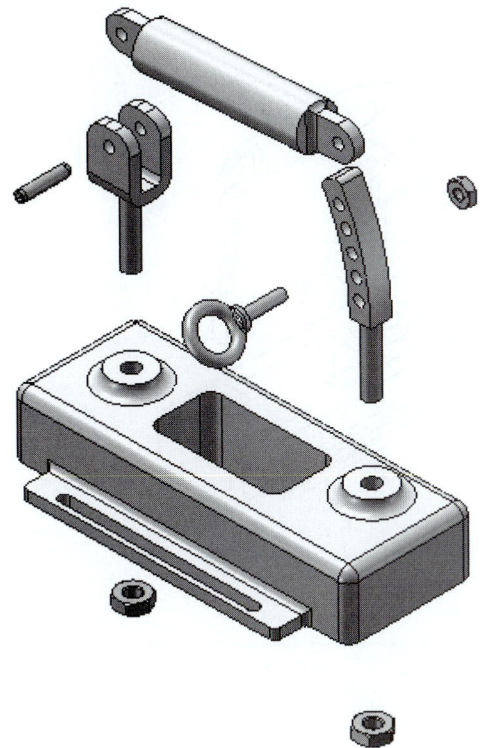

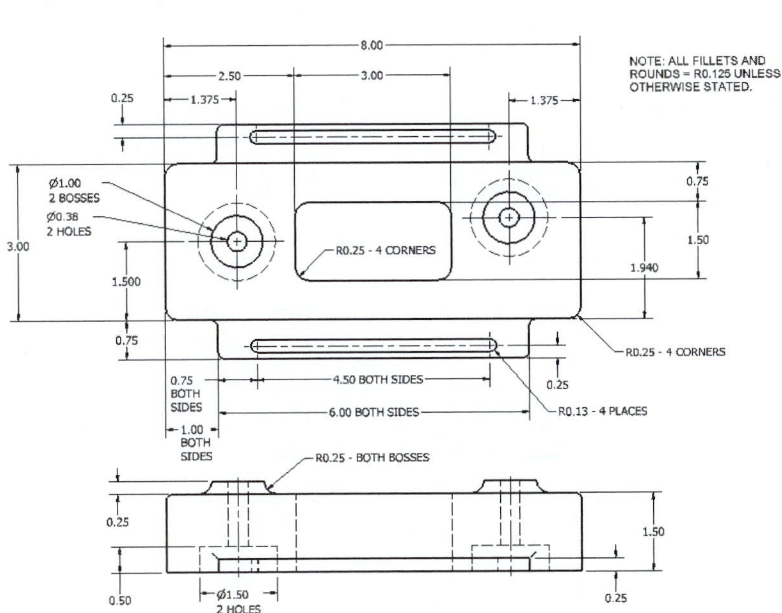

Figure P6-24
(*Continued*)

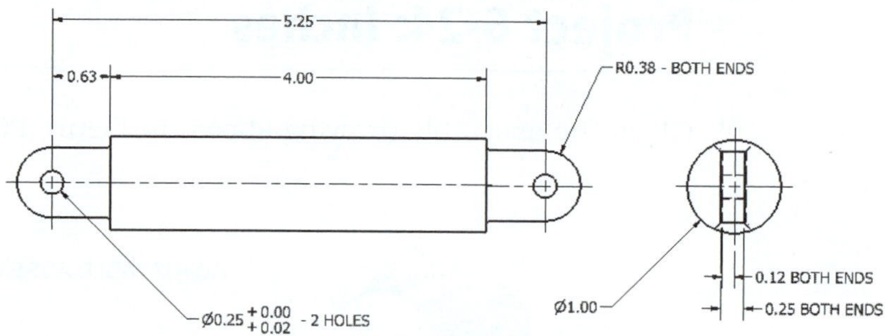
5.25
0.63
4.00
R0.38 - BOTH ENDS
Ø0.25 +0.00 +0.02 - 2 HOLES
Ø1.00
0.12 BOTH ENDS
0.25 BOTH ENDS
NOTE: ALL FILLETS = R 0.125

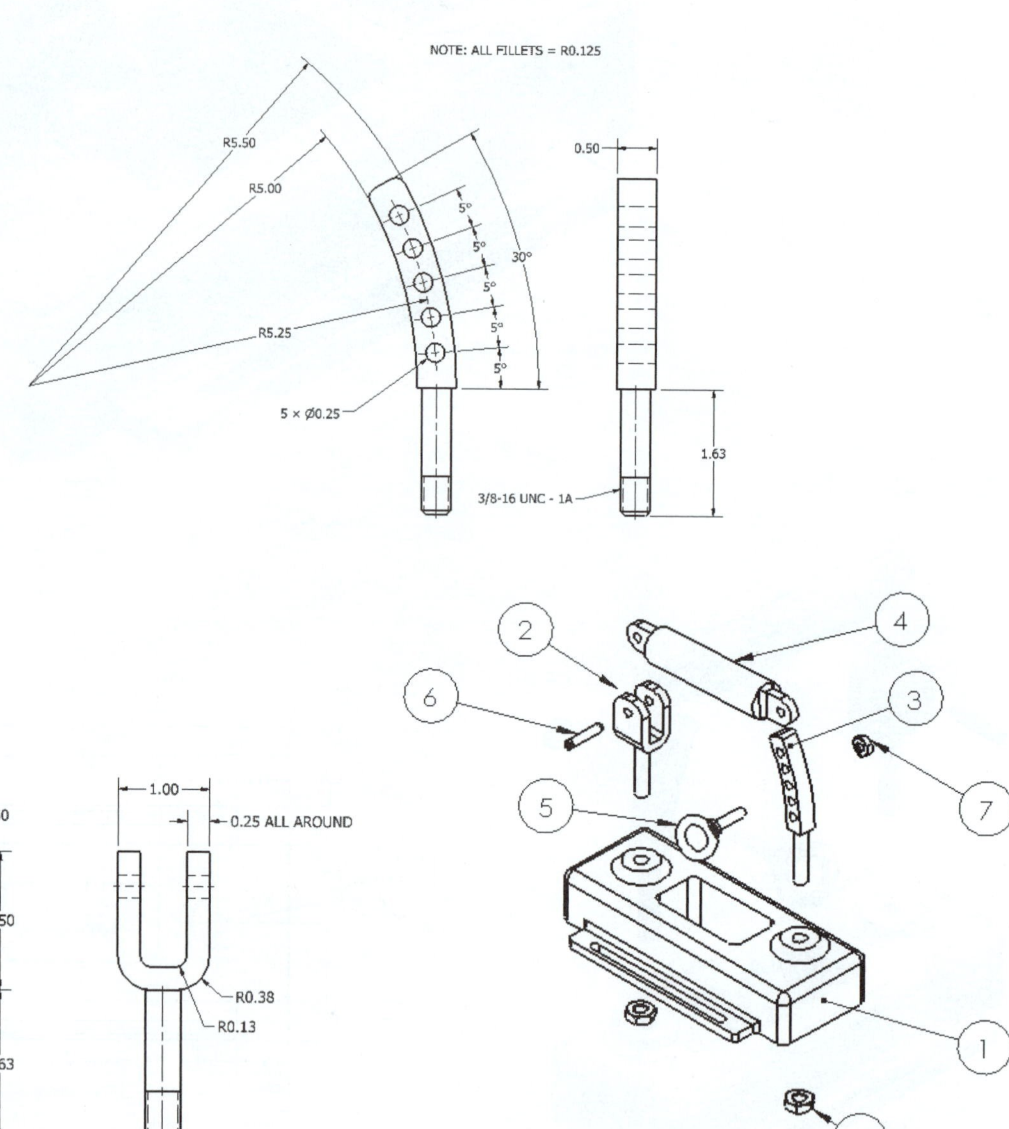
NOTE: ALL FILLETS = R0.125
R5.50
R5.00
R5.25
5°
30°
5 × Ø0.25
0.50
1.63
3/8-16 UNC - 1A
Ø0.25
2 HOLES
R0.38
BOTH
SIDES
1.00
0.50
1.50
0.53
0.25 ALL AROUND
R0.38
R0.13
1
2
3
4
5
6
7
8

Figure P6-24
(*Continued*)

ITEM NO.	PART NUMBER	DESCRIPTION	MATL	QTY.
1	SP6-24a	BASE, CAST #4	CAST IRON	1
2	SP6-24c	SUPPORT, ROUND	SAE 1020	1
3	SP6-24d	POST, ADJUSTABLE	SAE 1020	1
4	SP6-24b	YOKE	SAE 1040	1
5	AI 18.15_type2 0.25x2.22-N-0.75	EYEBOLT, TYPE 2 FORGED	STEEL	1
6	SPS 0.25x1.125	PIN, SPRING, SLOTTED	STEEL	1
7	HHJNUT 0.2500-20-D-N	HEX NUT	STEEL	1
8	HHJNUT 0.3750-16-D-N	HEX NUT	STEEL	2

1. Create an assembly drawing.
2. Animate the drawing.
3. Create an exploded isometric drawing.
4. Create a BOM.

Project 6-25: Inches

Given the assembly shown in Figure P6-25:

1. Create an assembly drawing.
2. Animate the drawing.
3. Create an exploded isometric drawing.
4. Create a BOM.

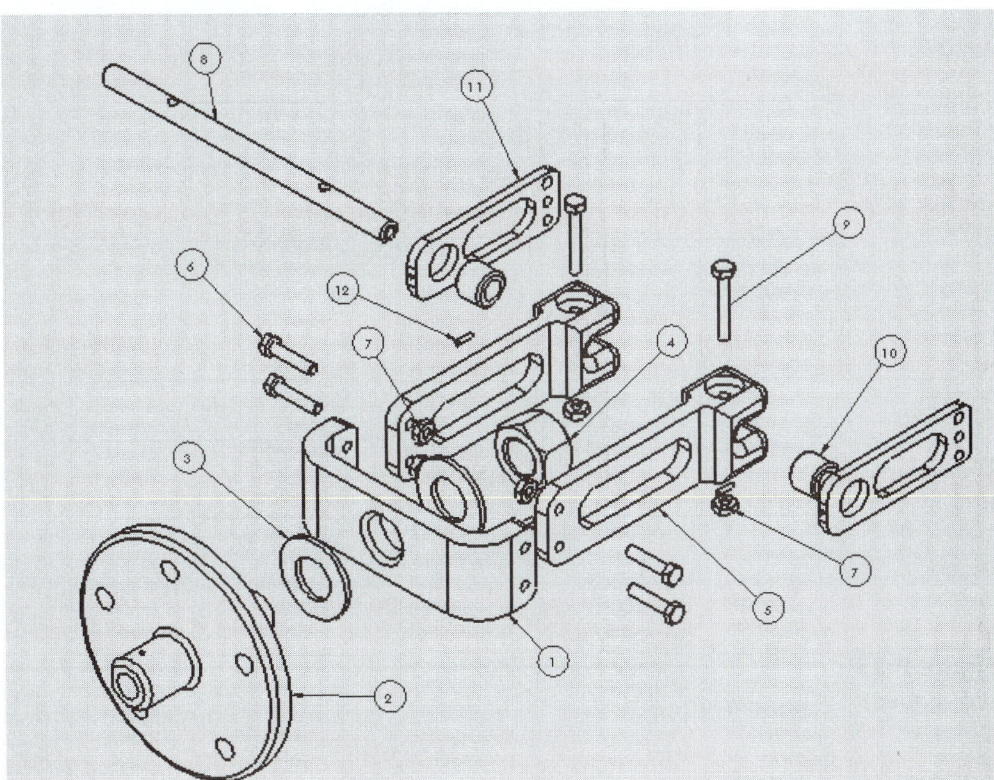

Figure P6-25

ITEM NO.	PART NUMBER	DESCRIPTION	QTY.
1	ME311-1	WHEEL BRACKET	1
2	ME311-2	WHEEL SUPPORT	1
3		1.00 × 1.75 × .06 PLAIN WASHER	2
4		1 × 8 UNC HEX NUT	1
5	ME311-3	SUPPORT ARM	2
6		1/4 - 28 UNF ×1.25 HEX HEAD	4
7		1/4 - 28 UNF × 1.75 HEX HEAD	2
8		1/4 - 28 UNF HEX NUT	6
9	ME311-4	PIVOT SHAFT	1
10		.500 × .875 .750 BEARING	2
11	ME311-5	STATIONARY ARM	2
12		#6-32 × .560 UNC SET SCREW CONE POINT	2

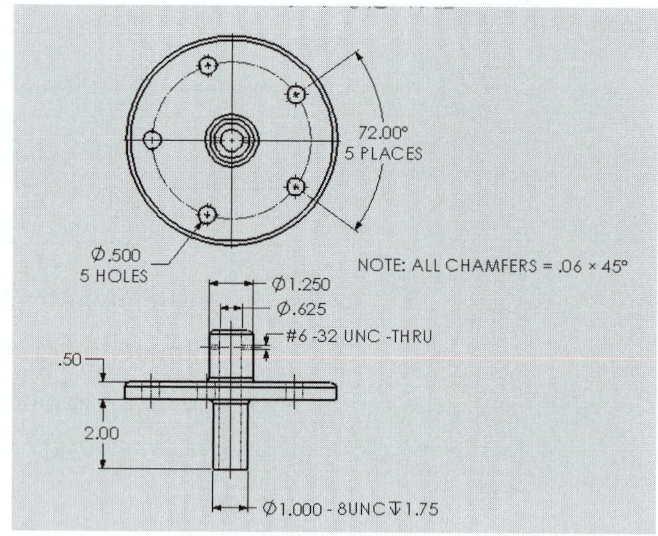

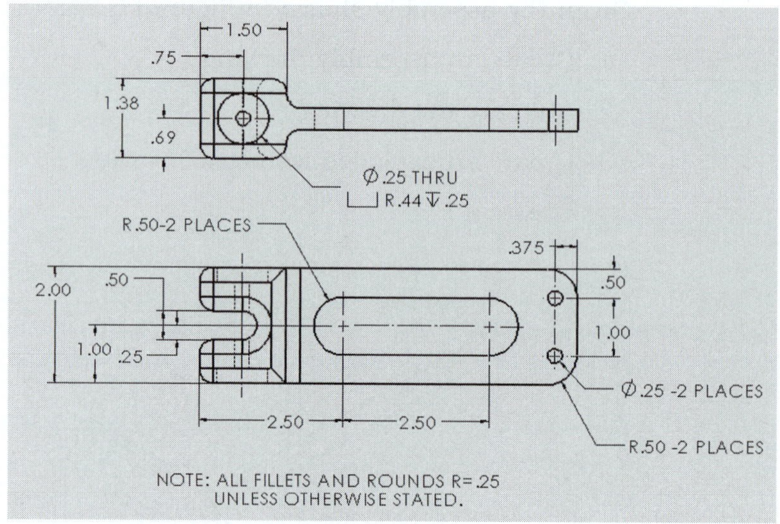

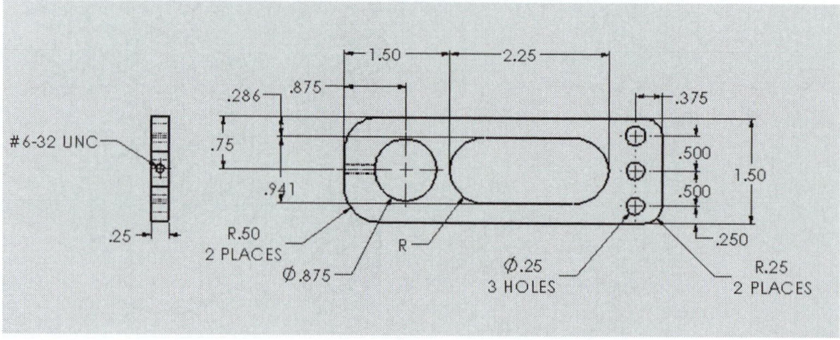

Figure P-25
(Continued)

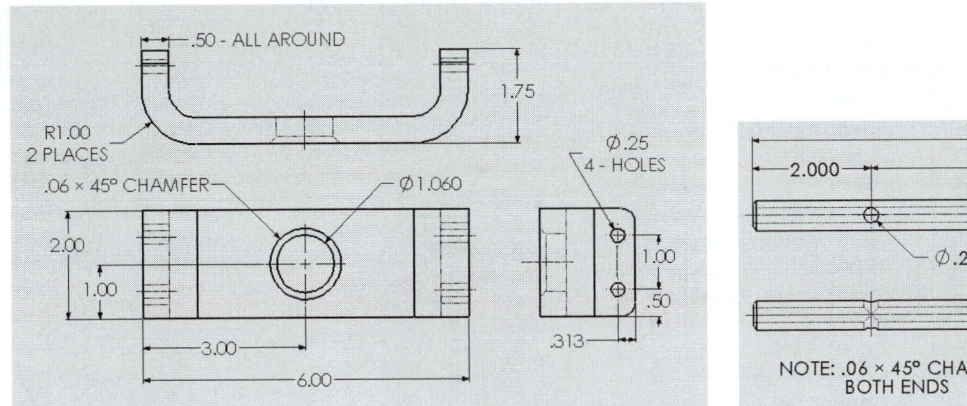

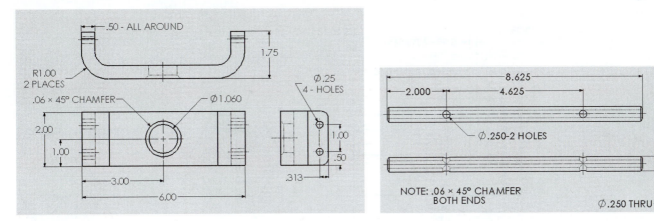

Figure P6-25
(Continued)

Project 6-26: Inches

Given the assembly shown in Figure P6-26:

1. Create an assembly drawing.
2. Animate the drawing.
3. Create an exploded isometric drawing.
4. Create a BOM.

Figure P6-26

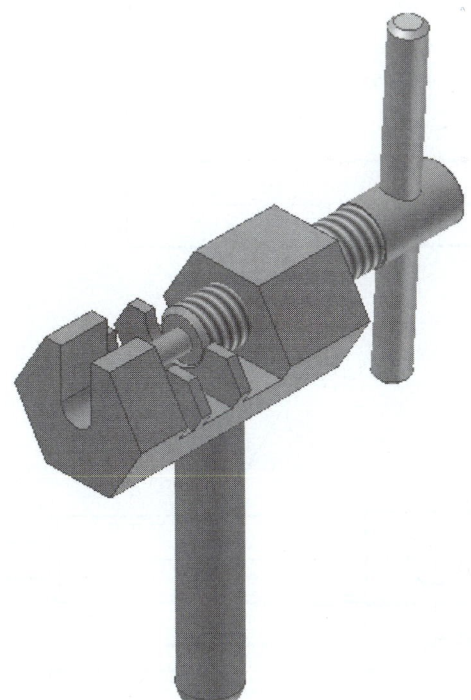

Figure P-26
(Continued)

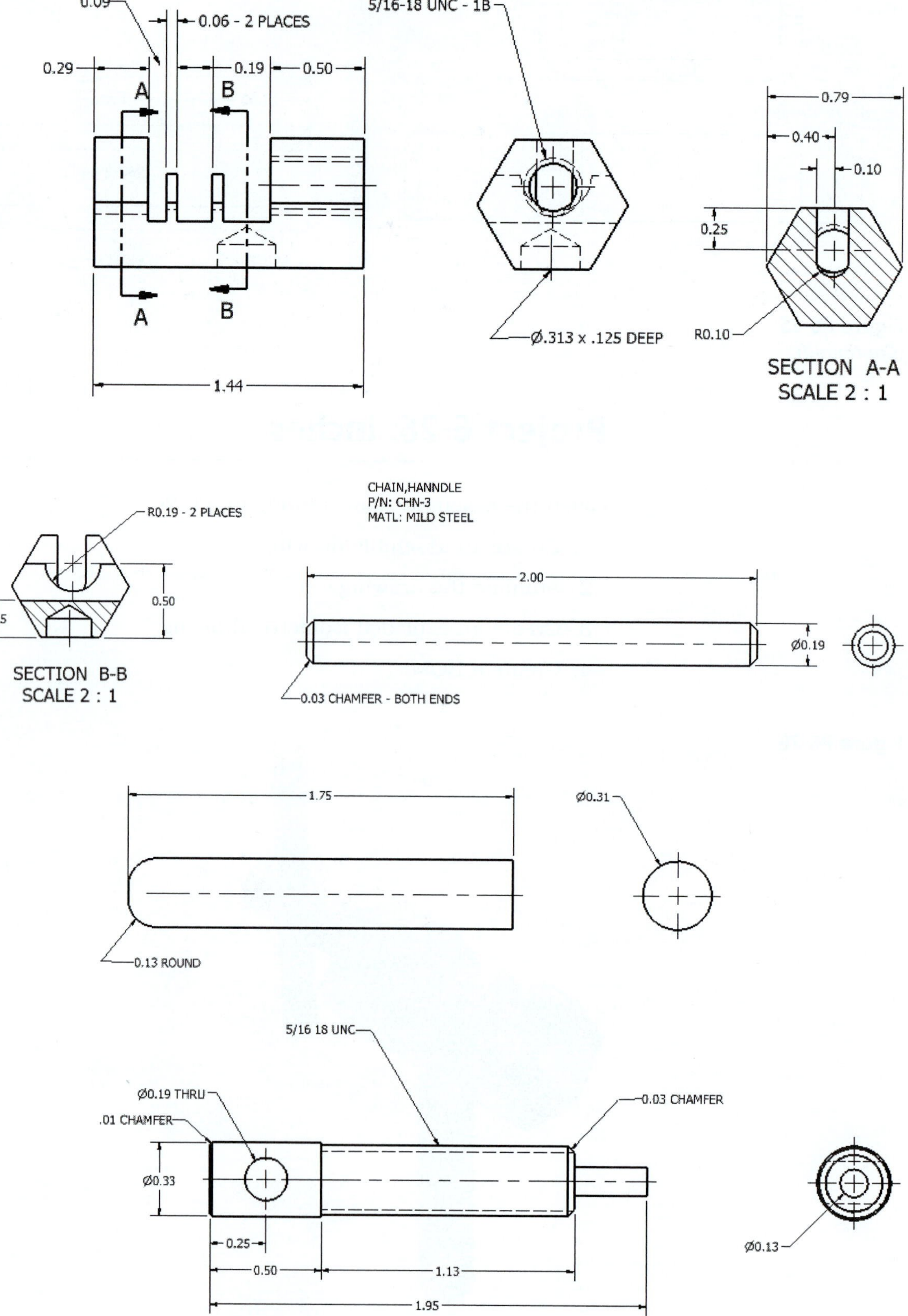
0.09
0.06 - 2 PLACES
0.29
0.19
0.50
A
B
A
B
1.44
5/16-18 UNC - 1B
Ø.313 x .125 DEEP
0.79
0.40
0.10
0.25
R0.10
SECTION A-A
SCALE 2 : 1
R0.19 - 2 PLACES
0.50
0.25
SECTION B-B
SCALE 2 : 1
CHAIN,HANNDLE
P/N: CHN-3
MATL: MILD STEEL
2.00
Ø0.19
0.03 CHAMFER - BOTH ENDS
1.75
Ø0.31
0.13 ROUND
5/16 18 UNC
Ø0.19 THRU
.01 CHAMFER
0.03 CHAMFER
Ø0.33
0.25
0.50
1.13
1.95
Ø0.13

Project 6-27: Inches

Given the assembly shown in Figure P6-27:

1. Create an assembly drawing.
2. Animate the drawing.
3. Create an exploded isometric drawing.
4. Create a BOM.

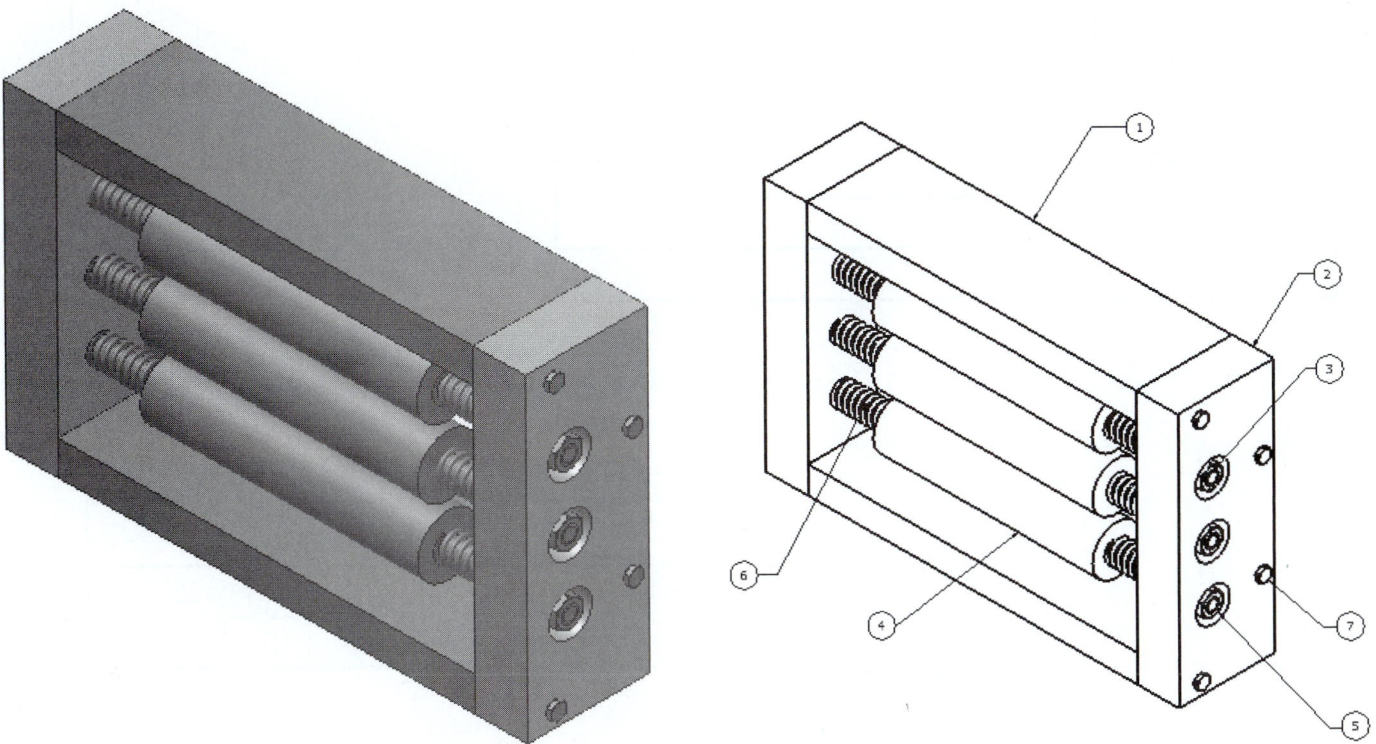

Parts List				
ITEM	PART NUMBER	DESCRIPTION	MATERIAL	QTY
1	AM311-1	BASE	Steel, Mild	1
2	AM311-2	PLATE,END	Steel, Mild	2
3	AM311-3	POST, GUIDE	Steel, Mild	3
4	EK-152	WEIGHT	Steel, Mild	3
5	AS 2465 - 1/2 UNC	HEX NUT	Steel, Mild	6
6		COMPRESSION SPRING	Steel, Mild	6
7	AS 2465 - 1/4 x 2 1/2 UNC	HEX BOLT	Steel, Mild	8

Figure P6-27

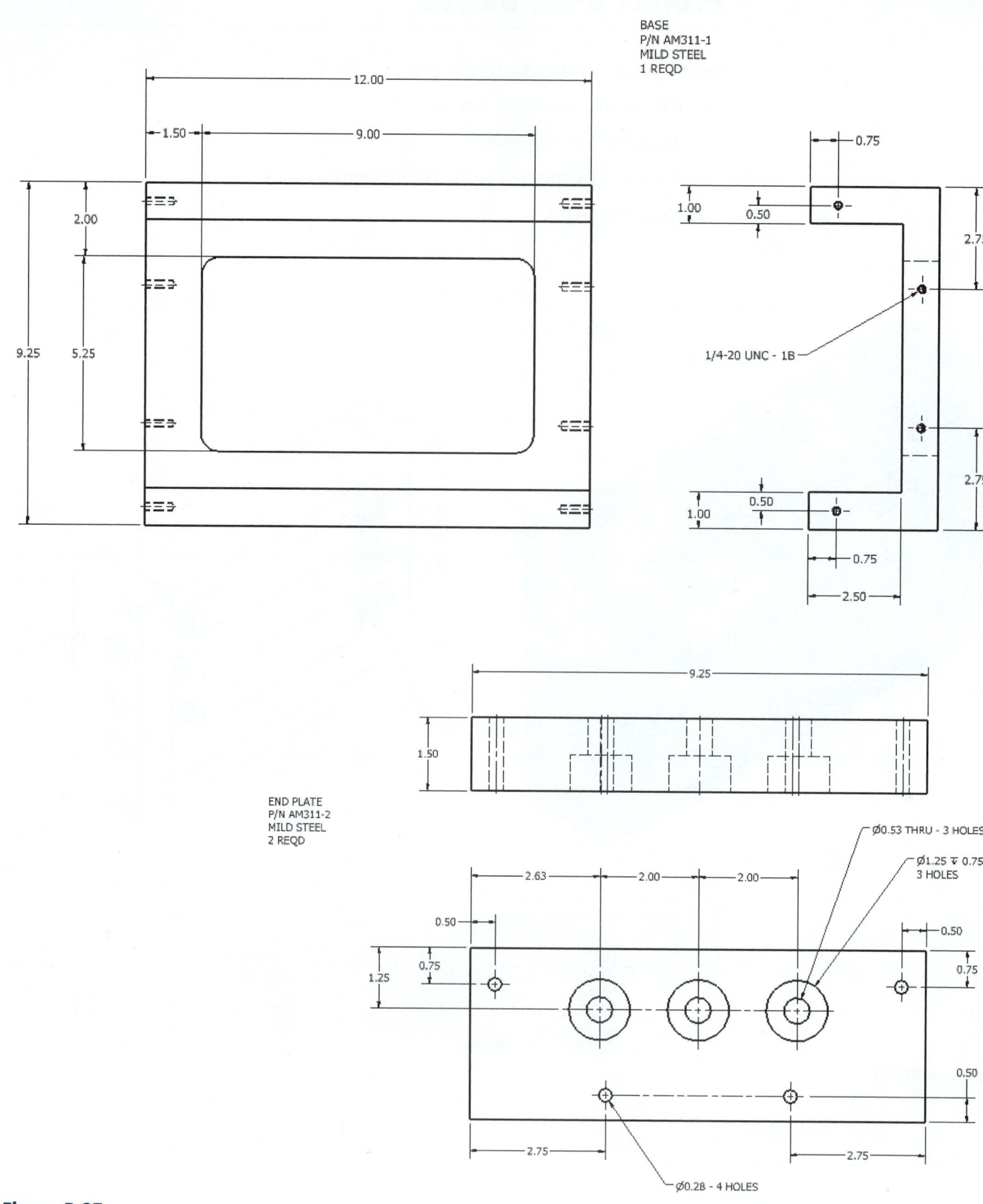

Figure P-27
(Continued)

Figure P6 -27
(*Continued*)

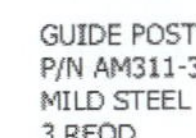

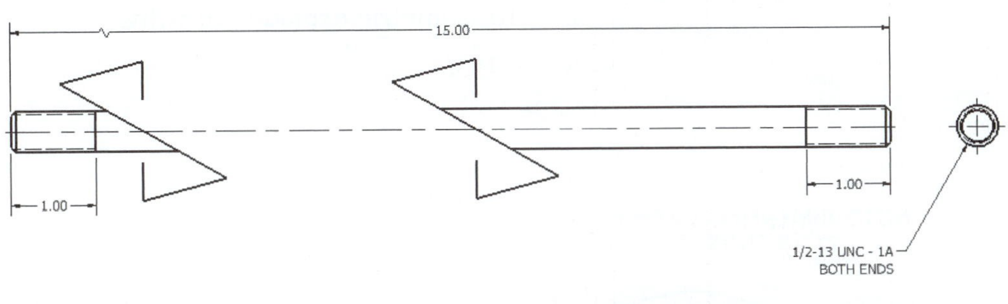

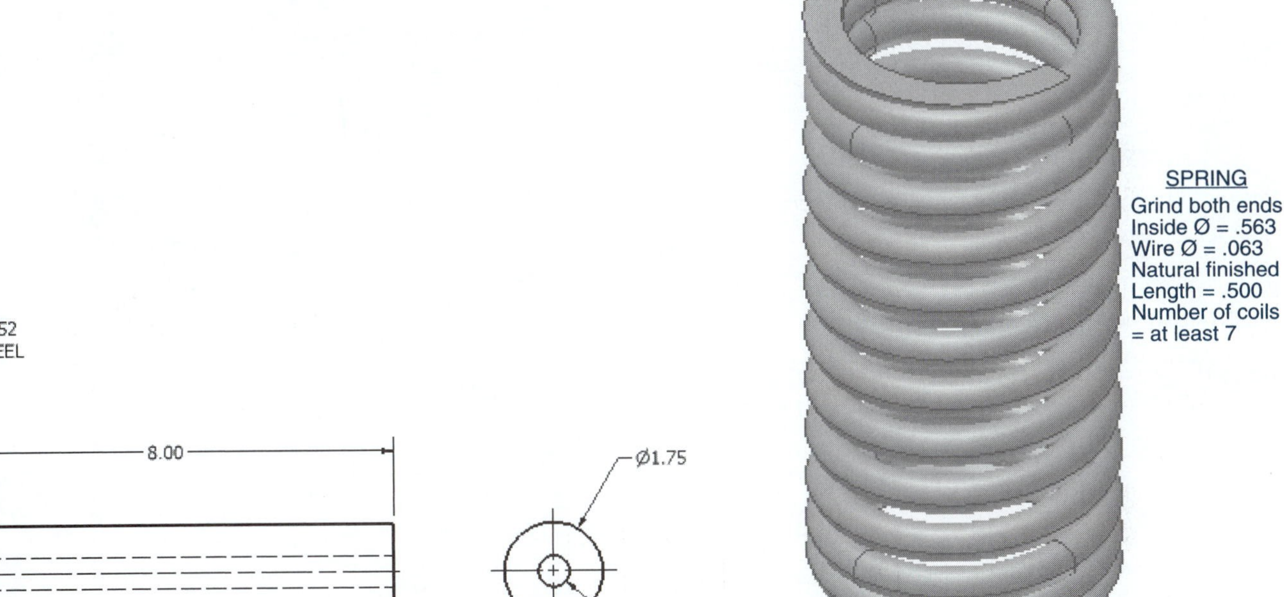

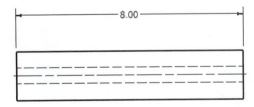

Project 6-28: Inches

Figure P6-28 shows five pipe segments. Use the segments to create the following assemblies.

A. Pipe assembly-1

B. Pipe assembly-2

C. Pipe assembly-3

D. Pipe assembly-4

E. An assembly as defined by your instructor. For each assembly join the segments using a 1/4-20 UNC × 1.375 Hex Head Screw and a 1/4-20 UNC Hex Nut.

For each assembly create the following.

1. Assembly drawing
2. An isometric exploded drawing
3. A BOM

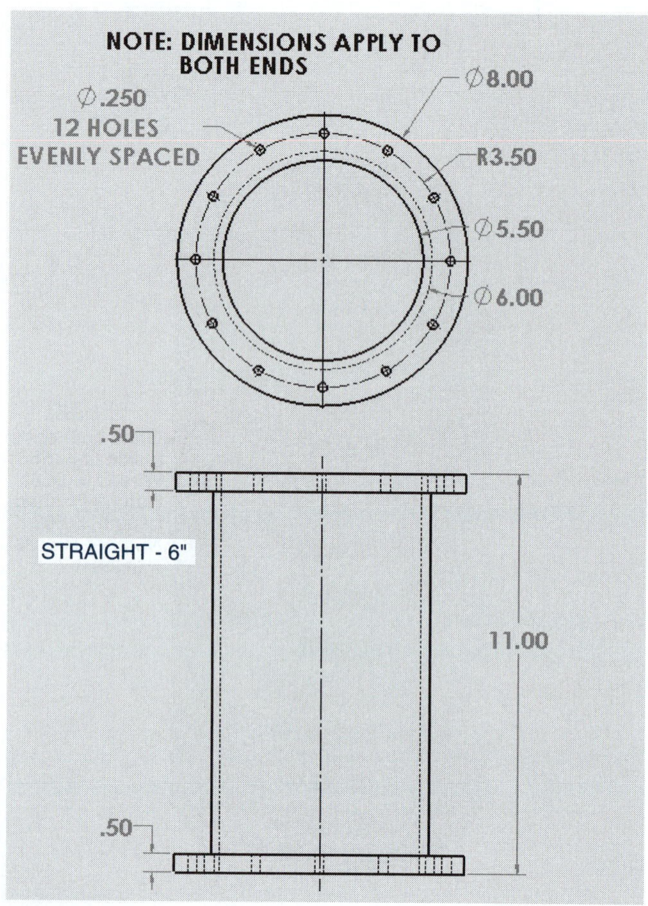

Figure P6-28

Figure P6-28
(Continued)

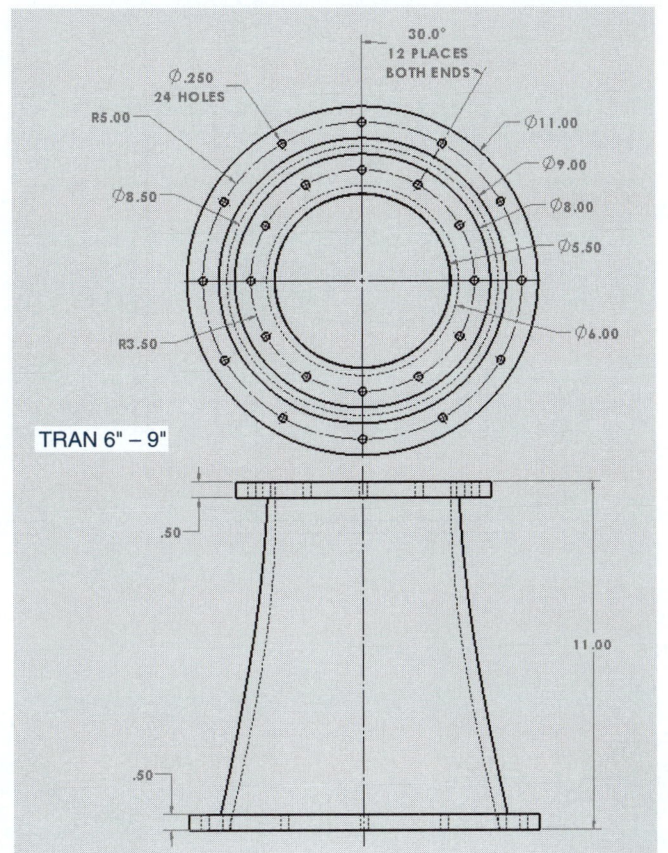

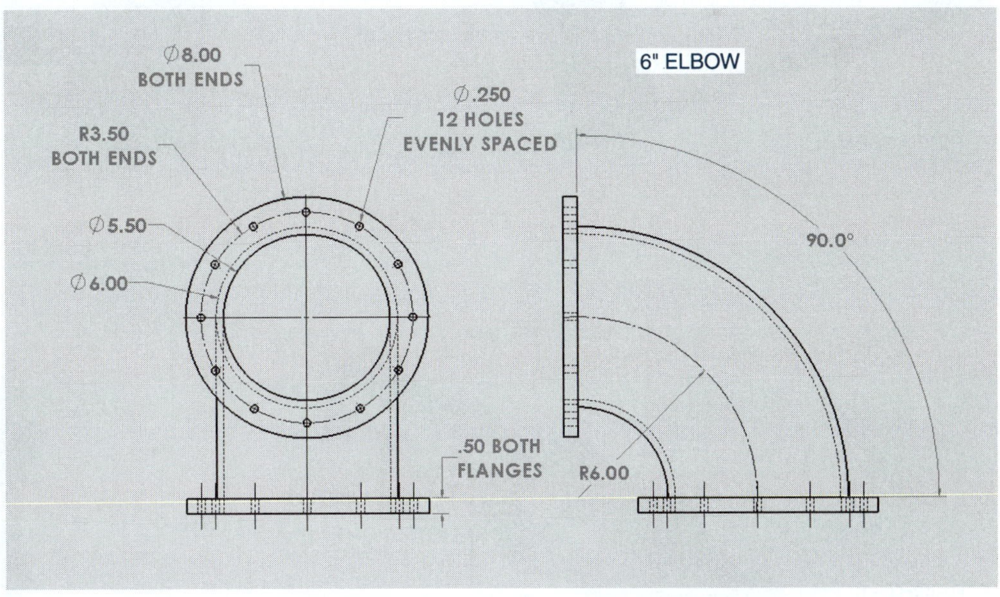

Figure P6-28
(*Continued*)

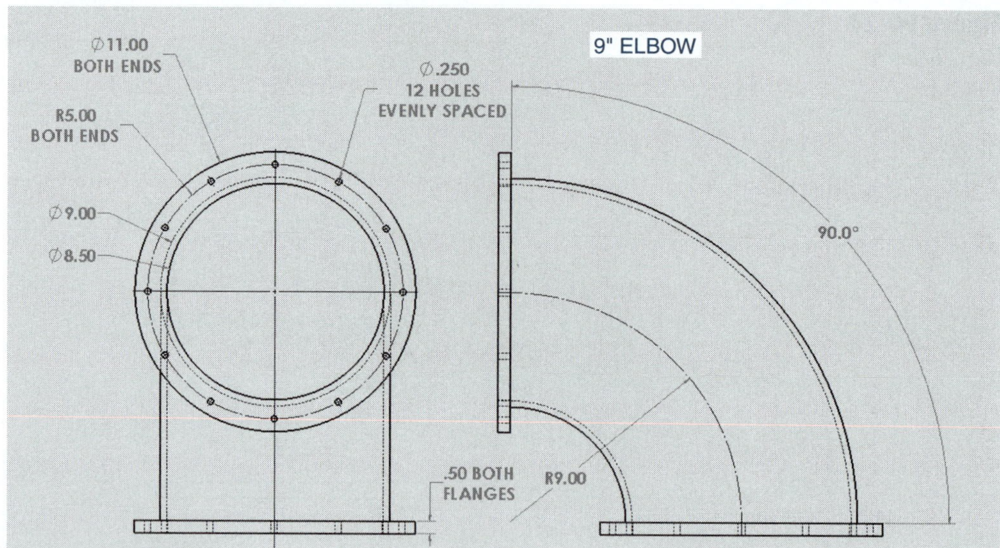

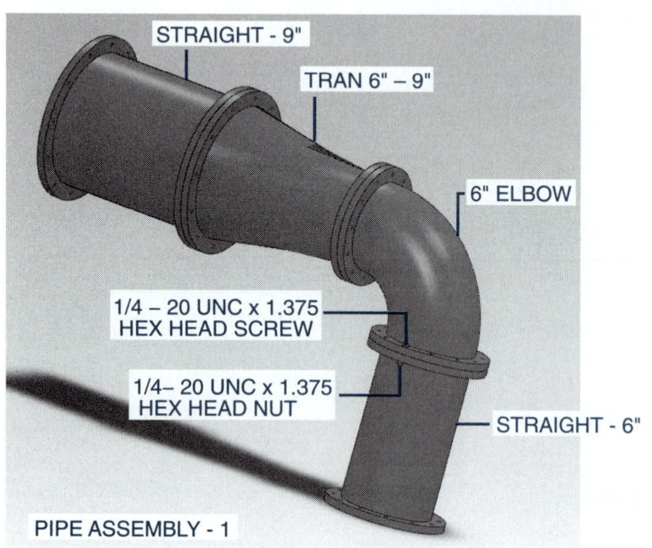

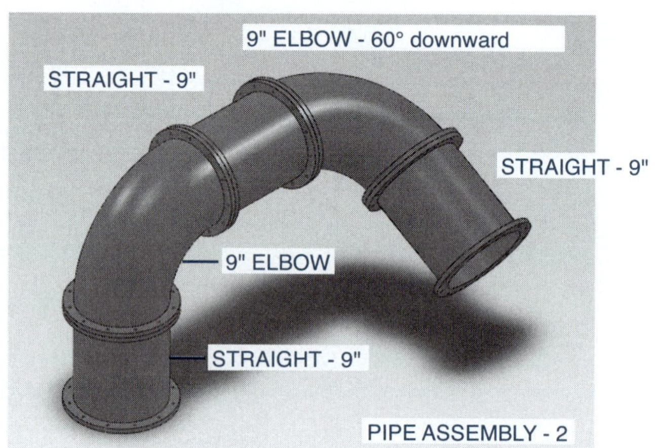

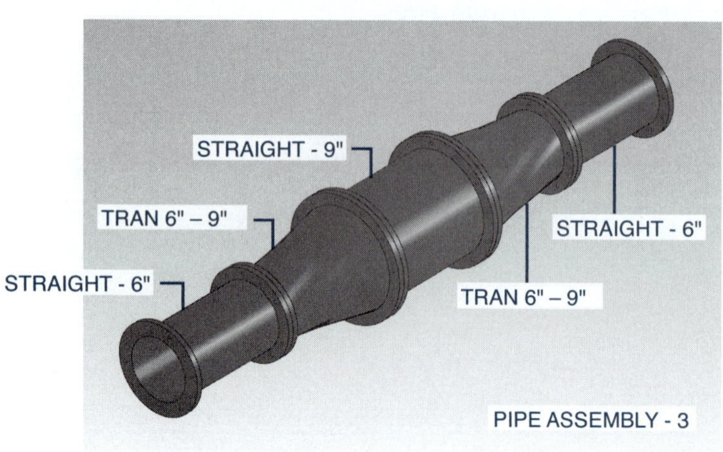

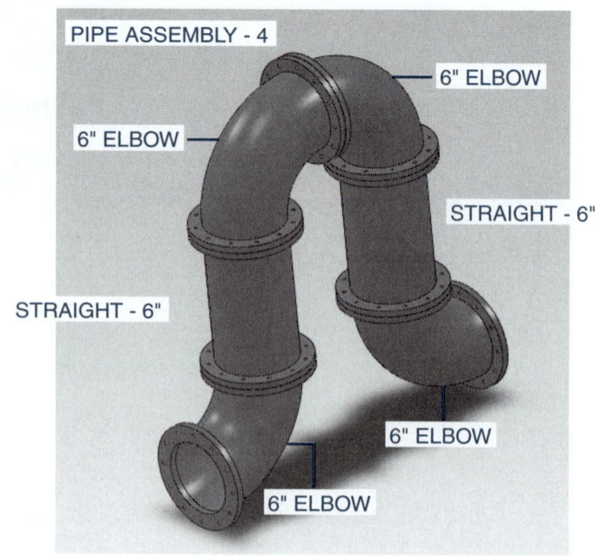

chapterseven

Dimensioning

CHAPTER OBJECTIVES

- Learn how to dimension objects
- Learn about ANSI standards and conventions
- Learn how to dimension different shapes and features
- Learn the fundamentals of 3D dimensioning

7-1 Introduction

Dimensions are added to SolidWorks on **Drawing** documents. Dimensions will appear in **Part** documents, but these are construction dimensions. These sketch dimensions are used to create a part and are used when a sketch is edited. They may be modified as the part is being created using the **Smart Dimension** tool. They will not appear on the finished model or in assembly drawings.

Figure 7-1 shows a dimensioned shape. The drawing on the left in Figure 7-1 shows the sketching dimensions that were created as the part was being created. The drawing on the right in Figure 7-1 shows dimensions that were created using the **Smart Dimension** tool in a **Drawing** document. These are defining dimensions and will appear on the working drawings. This chapter will show how to apply these types of dimensions.

SolidWorks has ANSI Inch and ANSI Metric dimensions available. Other dimensioning systems such as ISO also are available. This text is in compliance with ANSI standards.

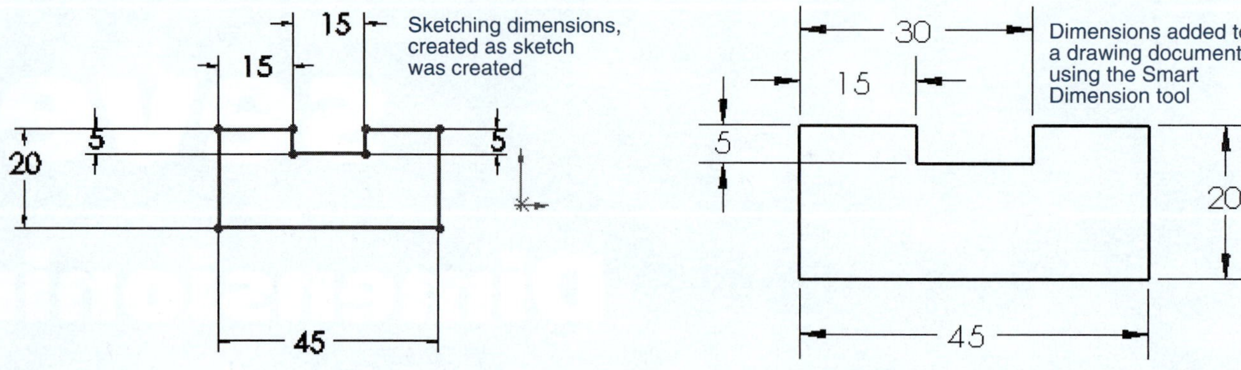

Figure 7-1

7-2 Terminology and Conventions—ANSI

Some Common Terms

Figure 7-2 shows both ANSI and ISO style dimensions. The terms apply to both styles.

Figure 7-2

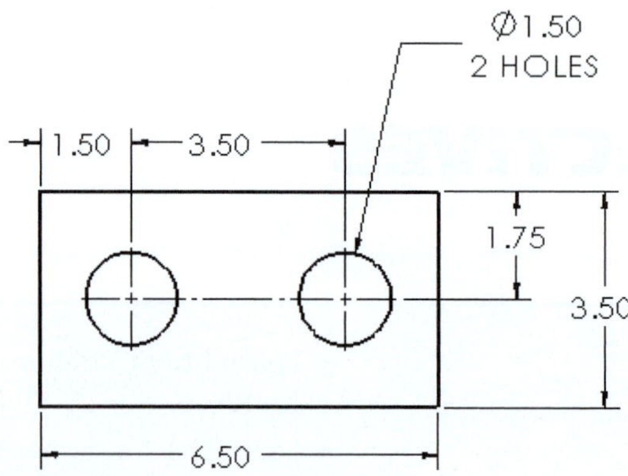

Dimension lines: In mechanical drawings, lines between extension lines that end with an arrowhead and include a numerical dimensional value located within the line.

Extension lines: Lines that extend away from an object and allow dimensions to be located off the surface of an object.

Leader lines: Lines drawn at an angle, not horizontal or vertical, that are used to dimension specific shapes such as holes. The start point of a leader line includes an arrowhead. Numerical values are drawn at the end opposite the arrowhead.

Linear dimensions: Dimensions that define the straight-line distance between two points.

Angular dimensions: Dimensions that define the angular value, measured in degrees, between two straight lines.

Some Dimensioning Conventions

See Figure 7-3.

1 Dimension lines should be drawn evenly spaced; that is, the distance between dimension lines should be uniform. A general rule of thumb is to locate dimension lines about 1/2 in. or 15 mm apart.

Figure 7-3

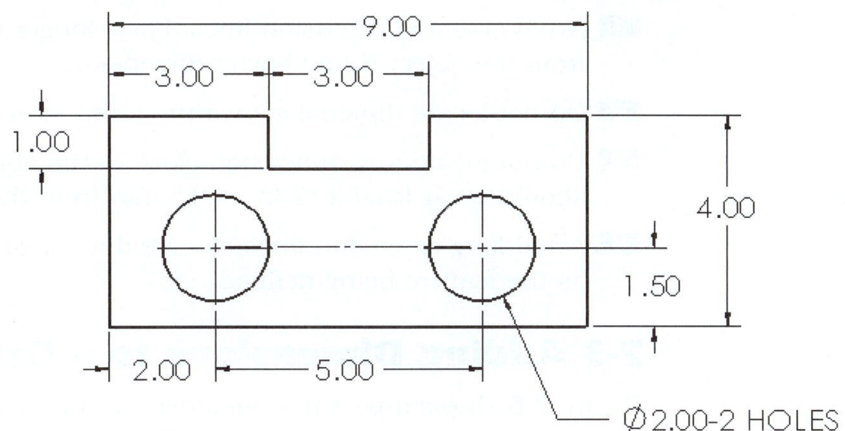

2 There should a noticeable gap between the edge of a part and the beginning of an extension line. This serves as a visual break between the object and the extension line. The visual difference between the line types can be enhanced by using different colors for the two types of lines.

3 Leader lines are used to define the size of holes and should be positioned so that the arrowhead points toward the center of the hole.

4 Centerlines may be used as extension lines. No gap is used when a centerline is extended beyond the edge lines of an object.

5 Align dimension lines whenever possible to give the drawing a neat, organized appearance.

Some Common Errors to Avoid

See Figure 7-4.

Figure 7-4

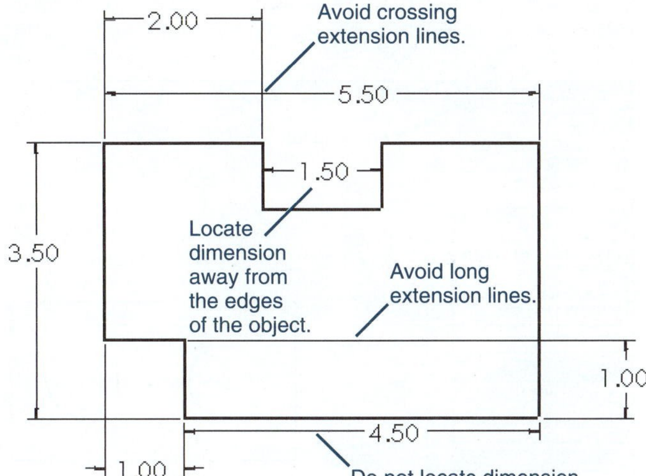

1. Avoid crossing extension lines. Place longer dimensions farther away from the object than shorter dimensions.
2. Do not locate dimensions within cutouts; always use extension lines.
3. Do not locate any dimension close to the object. Dimension lines should be at least 1/2 in. or 15 mm from the edge of the object.
4. Avoid long extension lines. Locate dimensions in the same general area as the feature being defined.

7-3 Adding Dimensions to a Drawing

Figure 7-5 shows a part that includes two holes. This section will explain how to add dimensions to the part. The part was drawn as a **Part** document and saved as **BLOCK, 2 HOLES**. See Figure 7-9 for the part's dimensions. The part is 0.50 thick. Save the part and start a new **Draw** document.

Figure 7-5

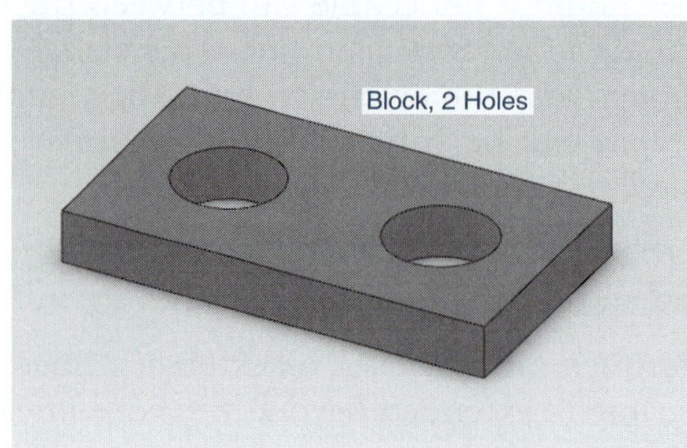

1. Click **New, Draw**, and **OK** and start a new drawing. Use a B (ANSI) Landscape sheet size.
2. Click the **View Layout** tab, **Model View**, and create a top view of the **BLOCK,2HOLES** part.

   In this example we will work with only one view. See Figure 7-6.

Figure 7-6

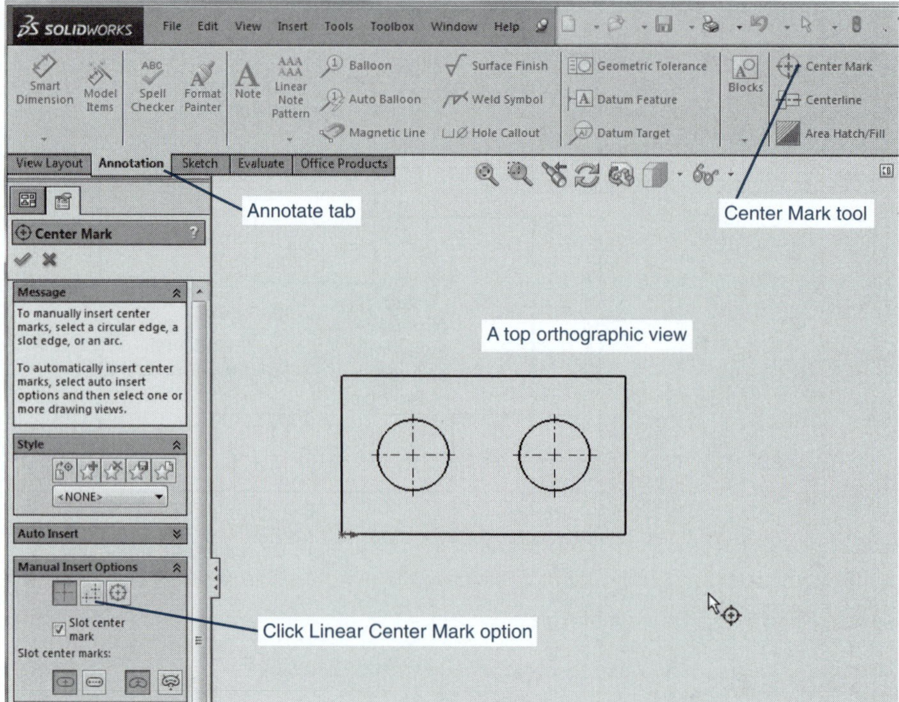

3. If the center marks do not appear, click the **Annotation** tab and select the **Center Mark** option.

4. Use the **Center Mark PropertyManager**, and the **Linear Center Mark** tool to add a centerline between the two holes by clicking the outside edge of each circle.

The holes now have the same horizontal centerline, so only one vertical dimension can be used to define the hole's location. See Figure 7-7.

Figure 7-7

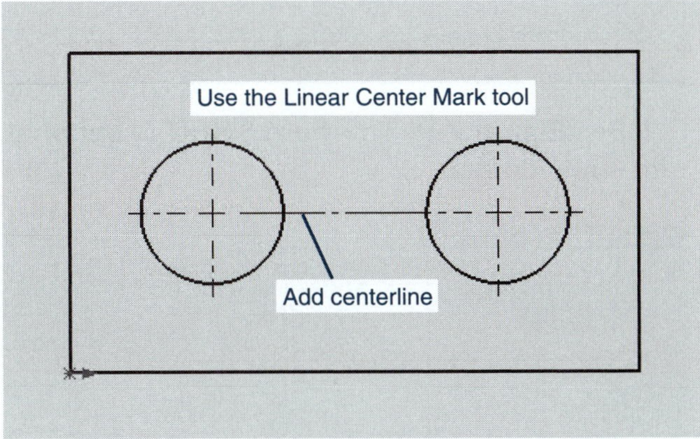

> **TIP**
>
> Centerlines can be extended by first clicking them and then dragging an endpoint to a new location.

5. Use the **Smart Dimension** tool and add the horizontal and vertical dimensions as shown.

   See Figure 7-8.

Figure 7-8

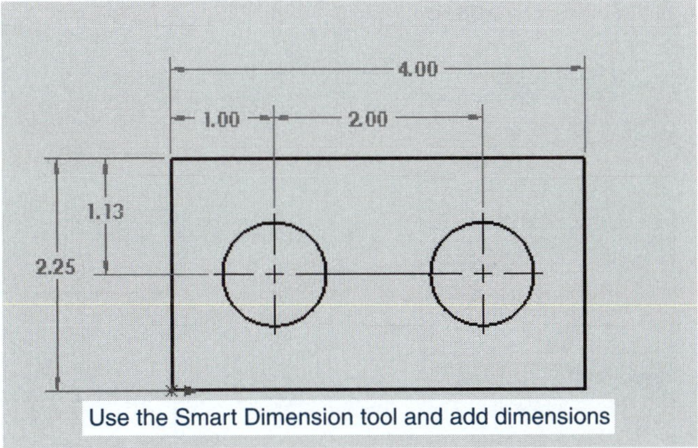

Note that the dimension values for the vertical dimensions are written horizontally. This is in compliance with ANSI standards. For this example the Century Gothic font was made bold with 14 point height.

RULE

Keep dimension lines aligned and evenly spaced.

6 Click the **Hole Callout** tool located on the **Annotate** panel, click the edge of the left hole, and move the cursor away from the hole. Note that the leader arrow always points at the center of the hole.

7 Select a location off the surface of the part and click the mouse.

RULE

Never locate dimensions on the surface of the part.

See Figure 7-9. The word THRU is optional. Some companies require it and some do not.

Figure 7-9

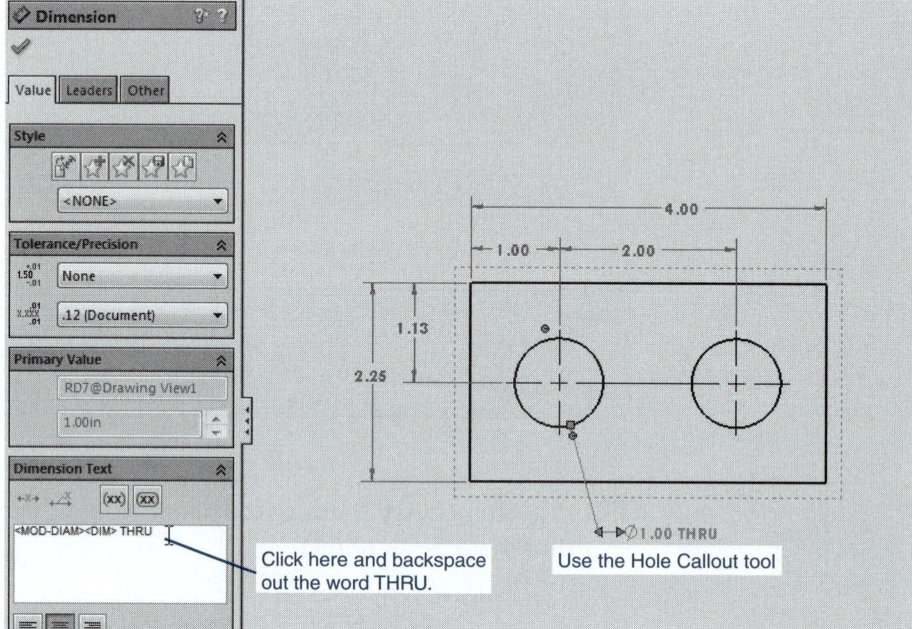

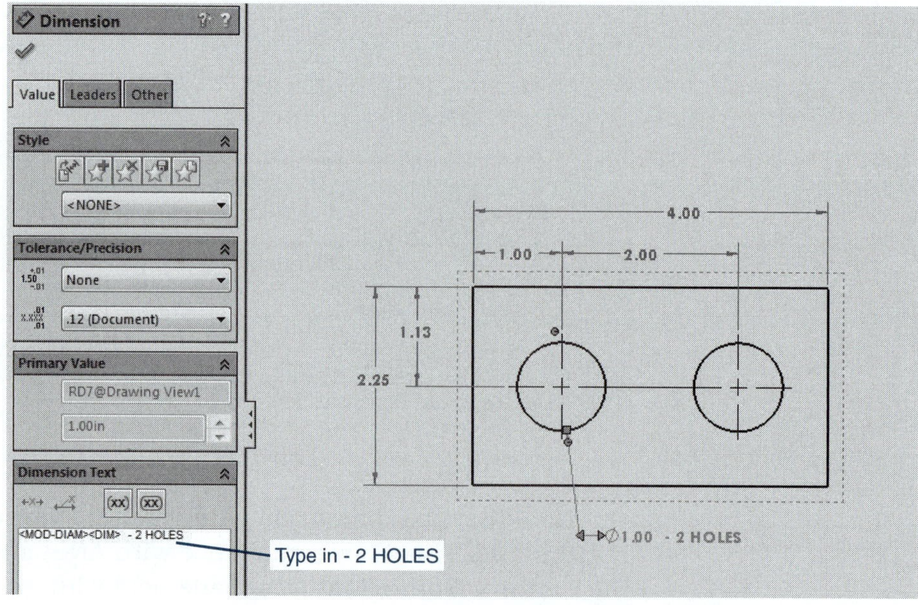

8. Go to the **Dimension PropertyManager** at the left of the screen and locate the cursor in the **Dimension Text** box, and click the mouse to the right of the word THRU and backspace the word out.

   The text already in the box defines the hole's diameter.

9. Move the cursor to the end of the existing text line, and type **2 HOLES**.

10. Click the green **OK** check mark.

    Move the dimensions if needed to create neat, uniform dimensions. See Figure 7-10.

11. Save the drawing.

Figure 7-10

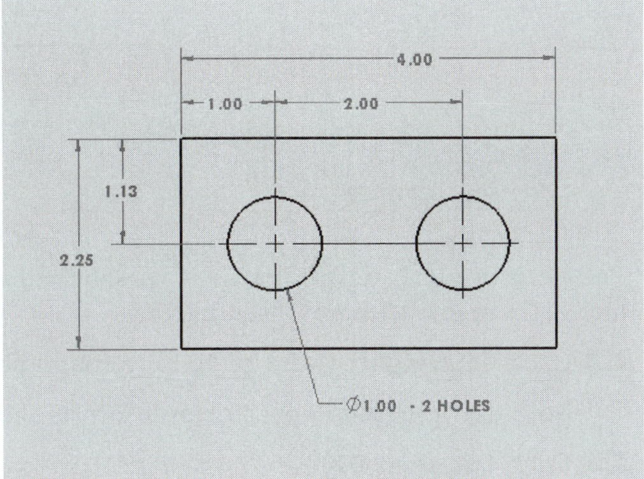

> **TIP**
> Dimensions can be relocated by clicking and dragging the dimension text.

Controlling Dimensions

Various aspects of dimensions can be edited, such as text height, arrow location, and text values.

1. Click the **Options** tool at the top of the screen.

   The **Documents Properties-Dimensions** dialog box will appear. See Figure 7-11.

2. Click the **Document Properties** tab.

3. Click the **Dimensions** option.

   The **Document Properties-Dimensions** dialog box can be used to edit the style and form of dimensions. It can also be used to change the way arrows are applied.

4. Click the **Font** option.

   The **Choose Font** dialog box will appear. See Figure 7-11. This dialog box can be used to change the font, font style, and height of dimension text. The height of text can be measured in inches, millimeters, or points.

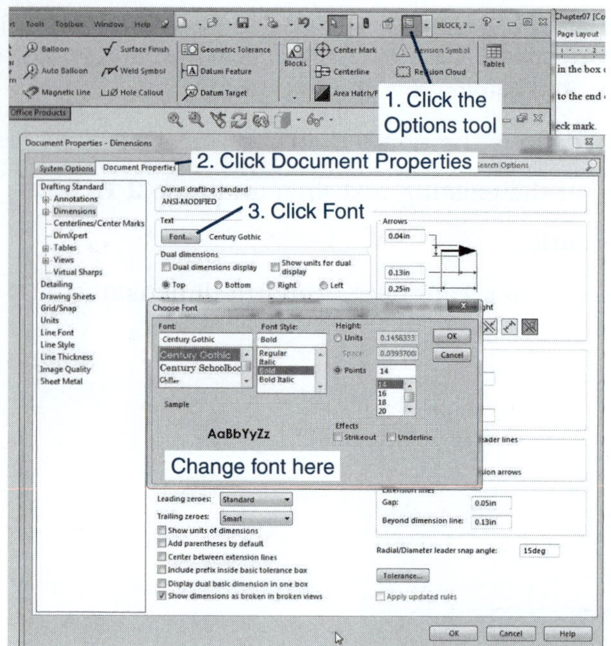

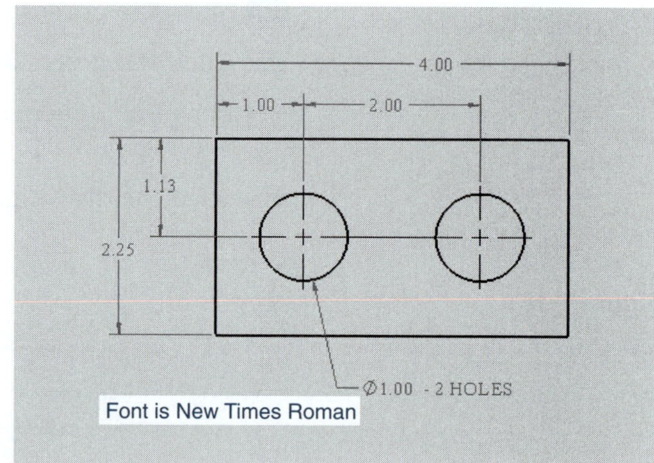

Figure 7-11

Point is a printer's term referring to a space that equals about 1/72 of an inch. (There are 12 points to a *pica*.)

5 Click the **Height: Units** radio button and change the height to **0.250in.**

Note that the SolidWorks default font is Century Gothic.

6 Click **OK**, then **OK**.

Figure 7-11 shows dimensions created using the New Times Roman font. Fonts for drawings should always be easy to read and not too stylistic.

Dimensioning Short Distances

Figure 7-12 shows an object that includes several short distances. We will start by using the standard dimensions settings and show how to edit them for a particular situation.

1 Use the **Smart Dimension** tool and add dimensions to the drawing.

Note that the arrows for the .50 dimension are aligned with the arrows for the 1.00 dimensions. Dimensions that are aligned in a single row are called ***chain dimensions***. Note that the .25 dimension is crowded between the two extension lines.

> **RULE**
>
> Never squeeze dimension values. Dimension values should always be presented clearly and should be easy to read.

There are several possible solutions to the crowded .25 value.

2 Click and drag the .25 dimension to the right outside the extension lines.

3 Add the **4.00** overall dimension.

Dimensions that define the total length and width of an object are called ***overall dimensions***. In this example the dimension 4.00 defines the total length of the part, so it is an overall dimension. Overall dimensions are located farthest away from the edge of the part.

The right edge of the part, the section below the .25, does not need a dimension. The reason for this will be discussed in the next chapter on tolerances.

> **TIP**
> To delete an existing dimension, click the dimension and press the **<Del>** key.

Figure 7-12 shows two other options for dimensioning. The first is the baseline method, in which all dimensions are taken from the same datum line. The second method is a combination of chain and baseline dimensions.

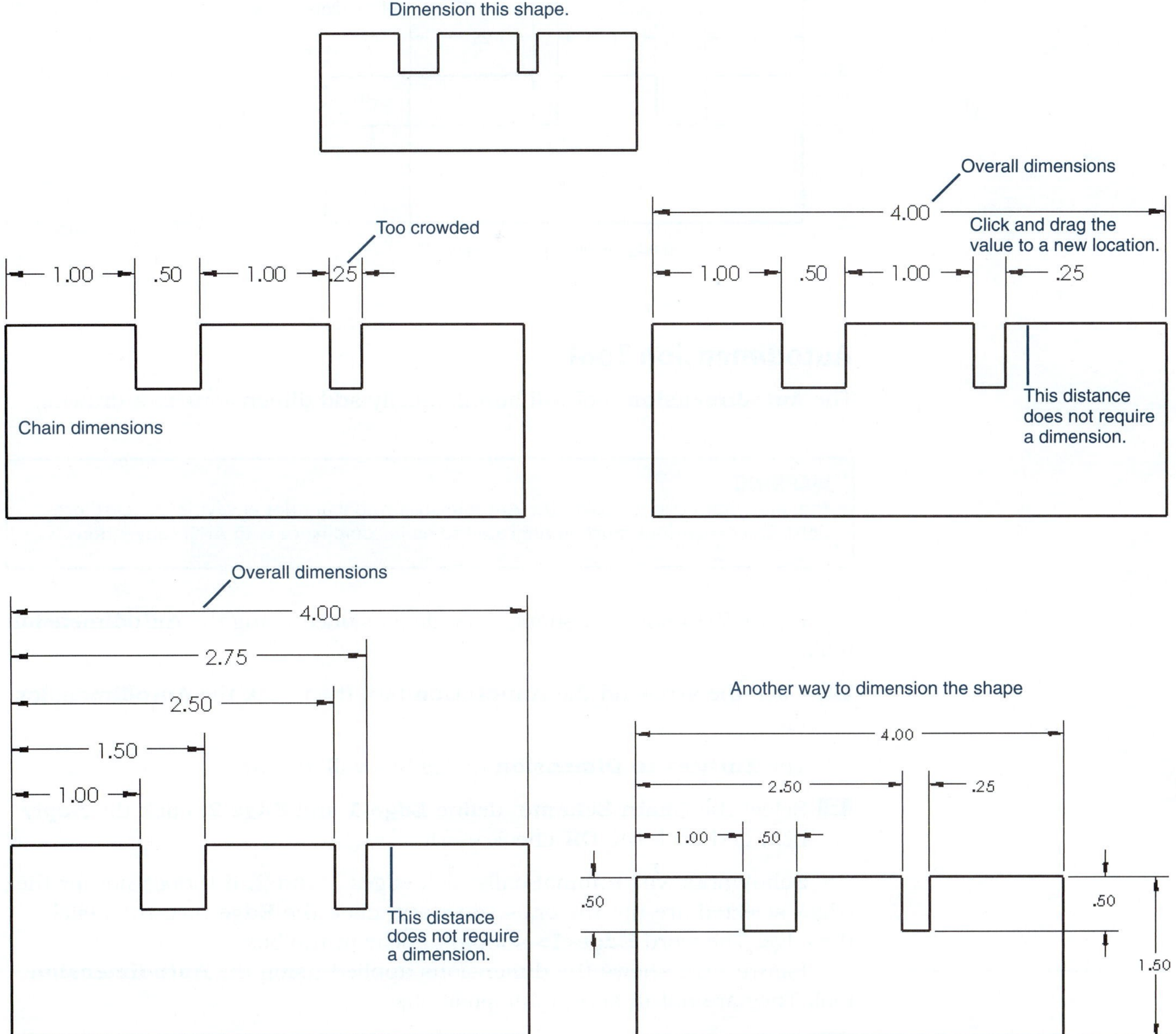

Figure 7-12

> **RULE**
> Never dimension the same distance twice. This is called *double dimensioning*.

Figure 7-13 shows an example of double dimensioning. The top edge distance is dimensioned twice: once using the 1.00 + .50 + 1.00 + .25 + 1.25 dimensions, and a second time using the 4.00 dimension. One of the dimensions must be omitted. Double dimensioning will be explained in more detail in the next chapter.

Figure 7-13

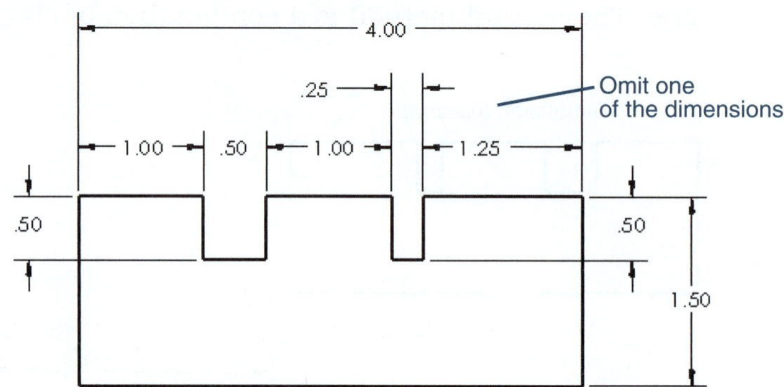

The top edge is dimensioned twice.

Autodimension Tool

The **Autodimension** tool will automatically add dimensions to a drawing.

> **WARNING**
> The dimensions created using the **Autodimension** tool are not always in the best locations. The dimensions must be relocated to be in compliance with ANSI conventions.

Figure 7-14 shows a shape to be dimensioned using the **Autodimension** tool.

1. Click the arrow on the **Annotation** tab, then click the **Autodimension** tab.

   The **Entities to Dimension** dialog box will appear.

2. Select the **Chain Scheme**, define **Edge 1** and **Edge 2**, click the **Apply** box, and click the **OK** check mark.

   SolidWorks will automatically pick edges 1 and 2. If it does not, or the edges selected are not the ones you want, click the **Edge** box, then click the edge. The word **Edge<1>** should appear in the box.
   Figure 7-14 shows the dimensions applied using the **Autodimension** tool. They are not in acceptable positions.

3. Rearrange the dimensions to comply with standard conventions.

Figure 7-14

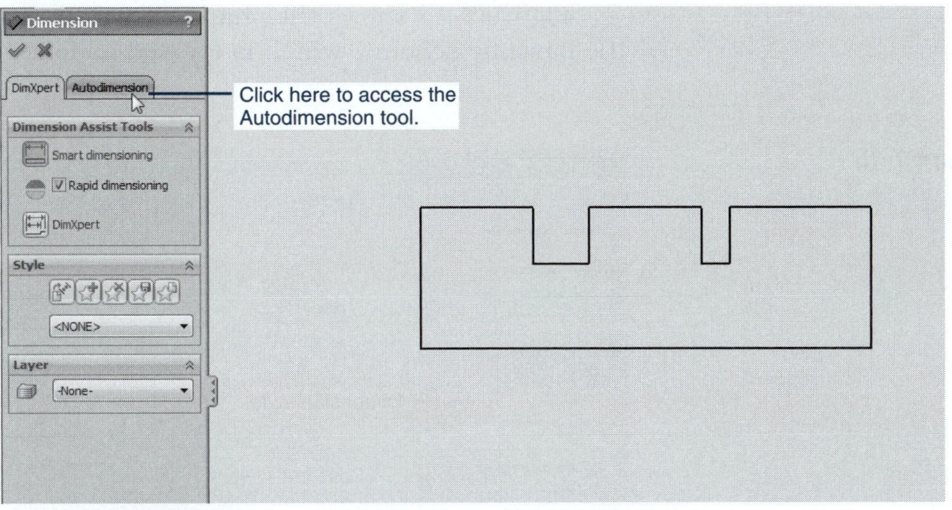
Dimension
DimXpert
Autodimension
Click here to access the Autodimension tool.
Dimension Assist Tools
Smart dimensioning
Rapid dimensioning
DimXpert
Style
<NONE>
Layer
-None-

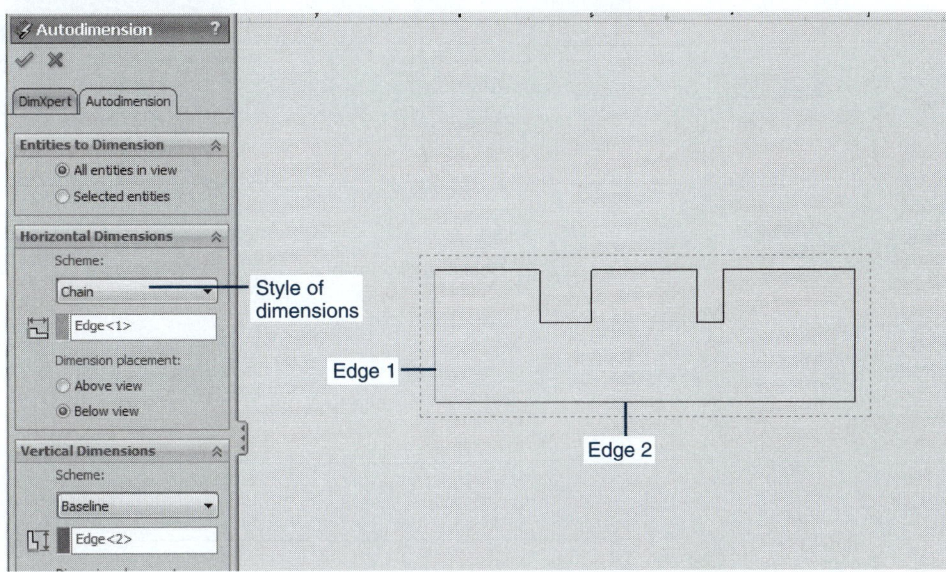
Autodimension
DimXpert
Autodimension
Entities to Dimension
All entities in view
Selected entities
Horizontal Dimensions
Scheme:
Chain
Style of dimensions
Edge<1>
Dimension placement:
Above view
Below view
Vertical Dimensions
Scheme:
Baseline
Edge<2>
Edge 1
Edge 2

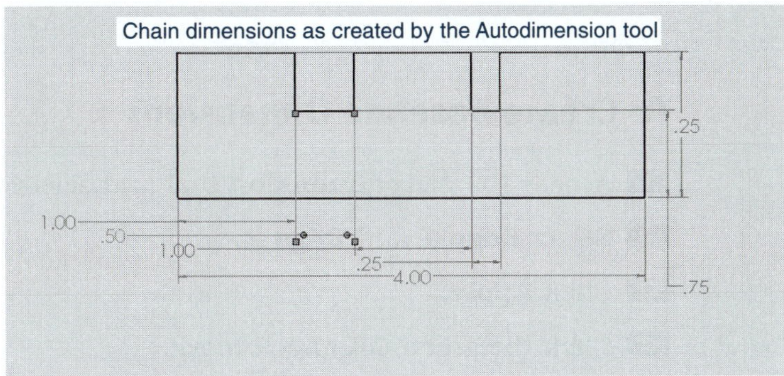
Chain dimensions as created by the Autodimension tool
.25
1.00
.50
1.00
.25
4.00
.75

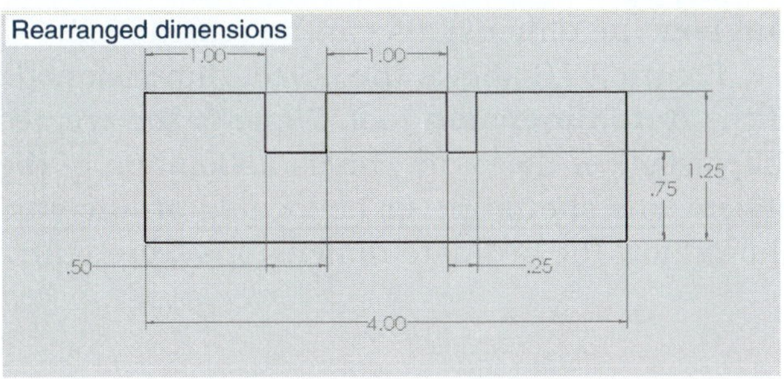
Rearranged dimensions
1.00
1.00
1.25
.75
.50
.25
4.00

Figure 7-15 shows the shape shown in Figure 7-14 dimensioned using the baseline scheme, which is created as follows.

Figure 7-15

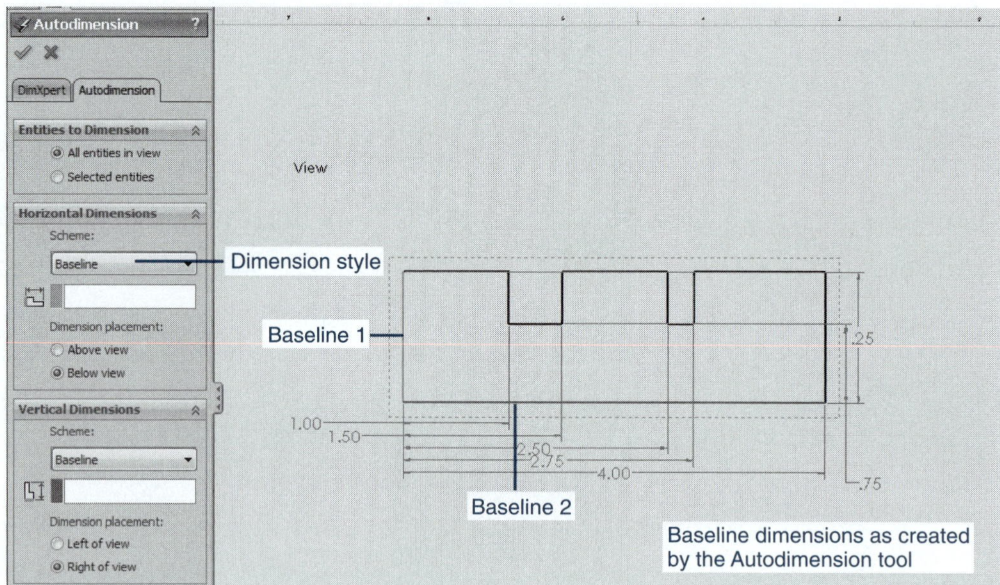

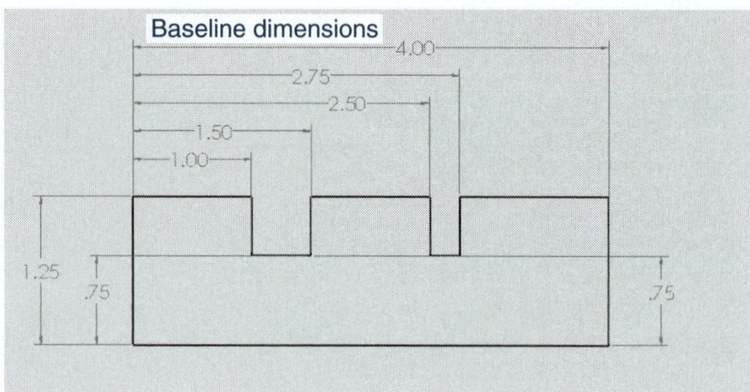

To Create Baseline Dimensions

1 Access the **Autodimension** tool and select the **Baseline Scheme**.

2 Select **Edge 1** and **Edge 2**.

3 Click **Apply**.

4 Click the green **OK** check mark.

Figure 7-15 shows the dimensions created by the **Autodimension** tool and how the dimensions can be rearranged.

Figure 7-16 shows the object dimensioned using the **Ordinate Scheme** of the **Autodimension** tool. Some of the created dimensions are located on the surface of the part. This is a violation of the convention that states that dimensions should never be located on the surface of the part. Figure 7-16 shows how the ordinate dimensions were rearranged.

Figure 7-16

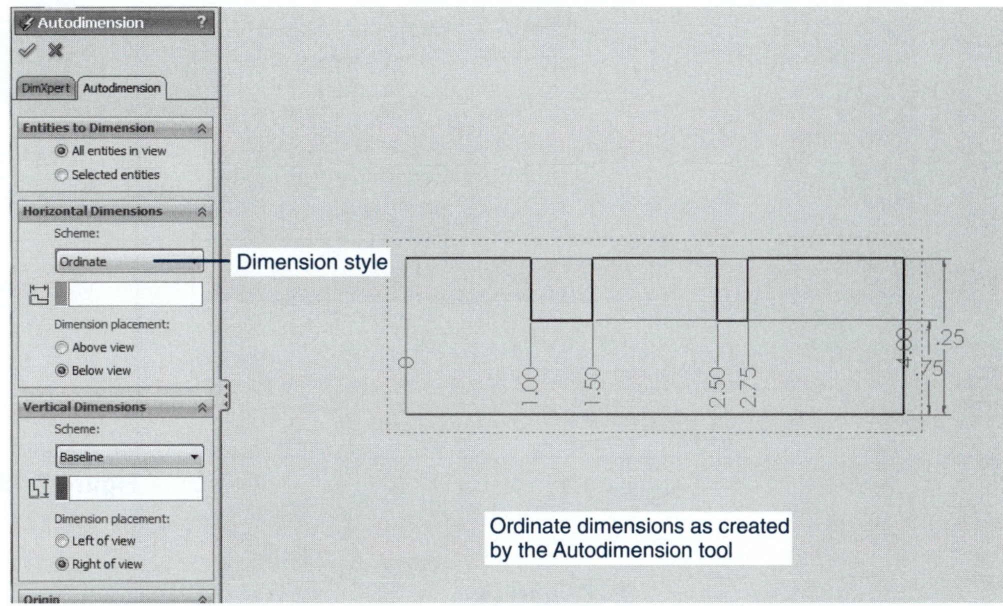

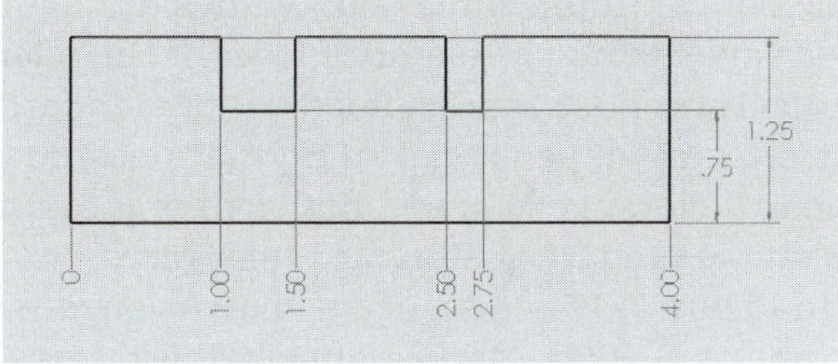

7-4 Drawing Scale

Drawings are often drawn "to scale" because the actual part is either too big to fit on a sheet of drawing paper or too small to be seen. For example, a microchip circuit must be drawn at several thousand times its actual size to be seen.

Drawing scales are written using the following formats:

SCALE: 1=1

SCALE: FULL

SCALE: 1000=1

SCALE: .25=1

In each example the value on the left indicates the scale factor. A value greater than 1 indicates that the drawing is larger than actual size. A value smaller than 1 indicates that the drawing is smaller than actual size.

Regardless of the drawing scale selected, the dimension values must be true size. Figure 7-17 shows the same rectangle drawn at two different scales. The top rectangle is drawn at a scale of 1 = 1, or its true size. The bottom rectangle is drawn at a scale of 2 = 1, or twice its true size. In both examples the 3.00 dimension remains the same.

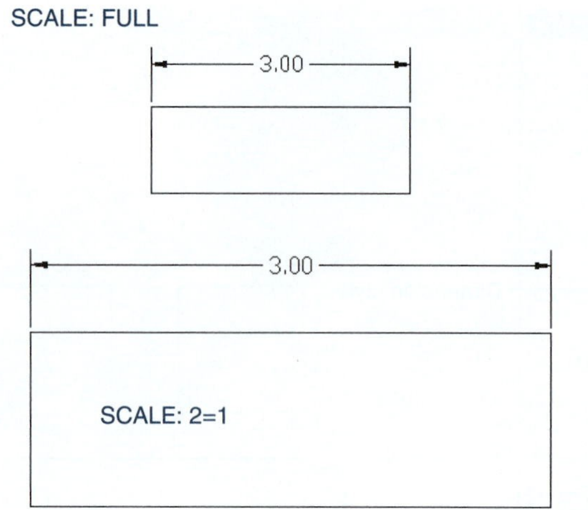

Figure 7-17

TOLERANCES UNLESS OTHERWISE STATED	
X	± 1
.X	± .1
.xx	± .01
.XXX	± .005
X°	± 1°
.X°	± .1°

Figure 7-18

7-5 Units

It is important to understand that dimension values are not the same as mathematical units. Dimension values are manufacturing instructions and always include a tolerance, even if the tolerance value is not stated. Manufacturers use a predefined set of standard dimensions that are applied to any dimensional value that does not include a written tolerance. Standard tolerance values differ from organization to organization. Figure 7-18 shows a chart of standard tolerances.

In Figure 7-19 a distance is dimensioned twice: once as 5.50 and a second time as 5.5000. Mathematically these two values are equal, but they are not the same manufacturing instruction. The 5.50 value could, for example, have a standard tolerance of ±.01, whereas the 5.5000 value could have a standard tolerance of ±.0005. A tolerance of ±.0005 is more difficult and therefore more expensive to manufacture than a tolerance of ±.01.

Figure 7-20 shows examples of units expressed in millimeters and in decimal inches. A zero is not required to the left of the decimal point for decimal inch values less than one. Millimeter values do not require zeros to the right of the decimal point. Millimeter and decimal inch values never include symbols; the units will be defined in the title block of the drawing.

These dimensions are not the same. They have different tolerance requirements.

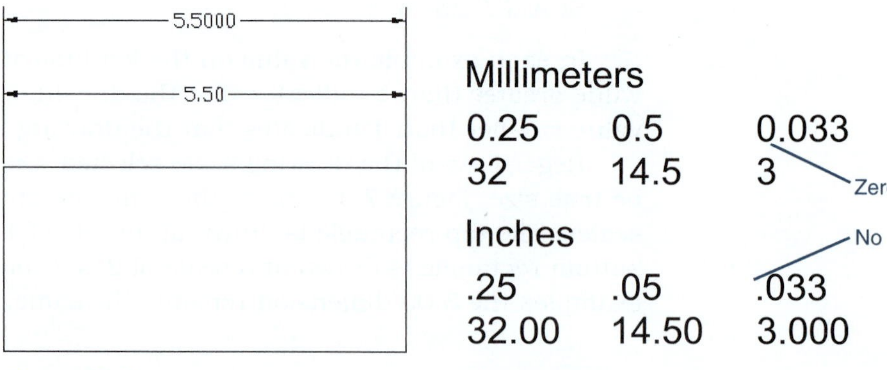

Figure 7-19

Figure 7-20

Aligned Dimensions

Aligned dimensions are dimensions that are parallel to a slanted edge or surface. They are not horizontal or vertical. The units for aligned dimensions should be written horizontally. This is called ***unidirectional dimensioning.***

Figure 7-21 shows the front, right side, and isometric views of a part with a slanted surface. The dimensions were applied using the **Smart Dimension** tool. Note that the slanted dimension, aligned with the slanted surface, has unidirectional (horizontal) text. The hole dimension was created using the **Note** tool from the **Annotation** tab.

Hole Dimensions

Figure 7-22 shows an object that has two holes, one blind, and one completely through. The object has filleted corners. In this section we will add dimensions to the views.

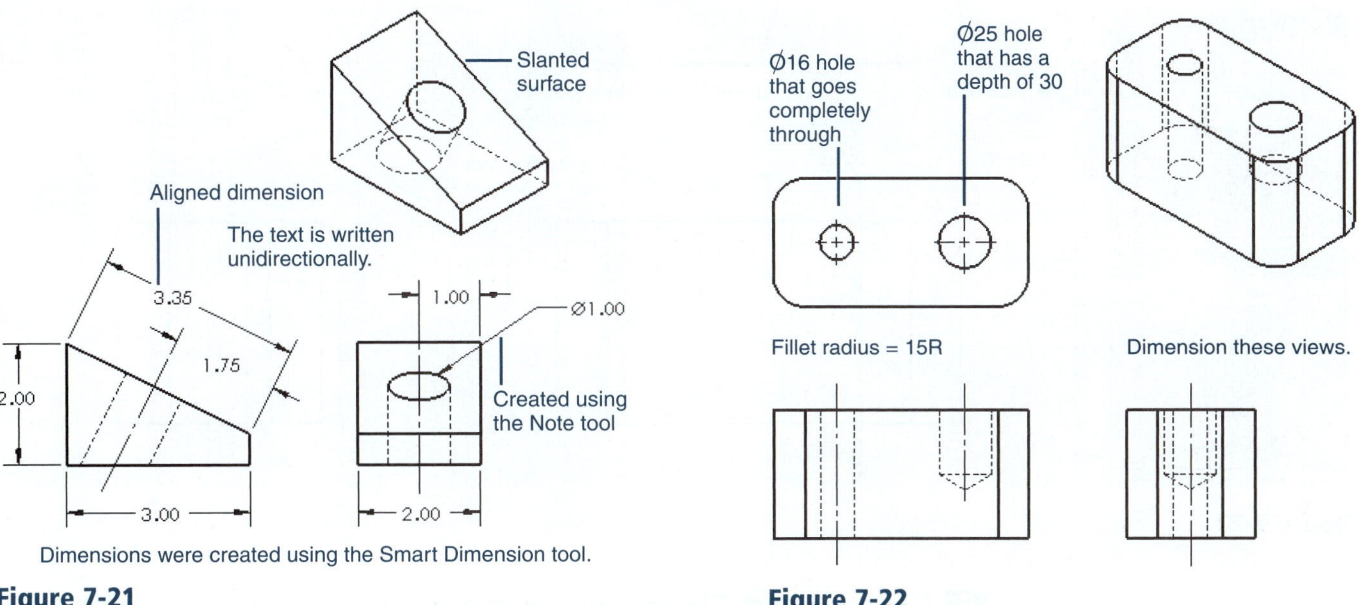

Figure 7-21

Figure 7-22

The holes were drawn using the **Hole Wizard** tool. The **Hole Wizard** tool will automatically create a conical point to a blind hole.

1 Use the **Smart Dimension** tool and locate the two holes.

See Figure 7-23. In general, dimensions are applied from the inside out, that is, starting with the features in the middle of the part and working out to the overall dimensions. Leader lines are generally applied last, as they have more freedom of location.

2 Use the **Linear Center Mark** tool to draw a centerline between the two holes and use the **Centerline** tool to add the vertical centerline in the front and side views.

The centerline between the two holes indicates that the vertical 30 dimension applies to both holes.

NOTE

Centerline should extend beyond the edges of the part. Centerlines can be extended by first clicking the centerline. Blue end boxes will appear. Click and drag the blue end boxes to a point beyond the edges of the part.

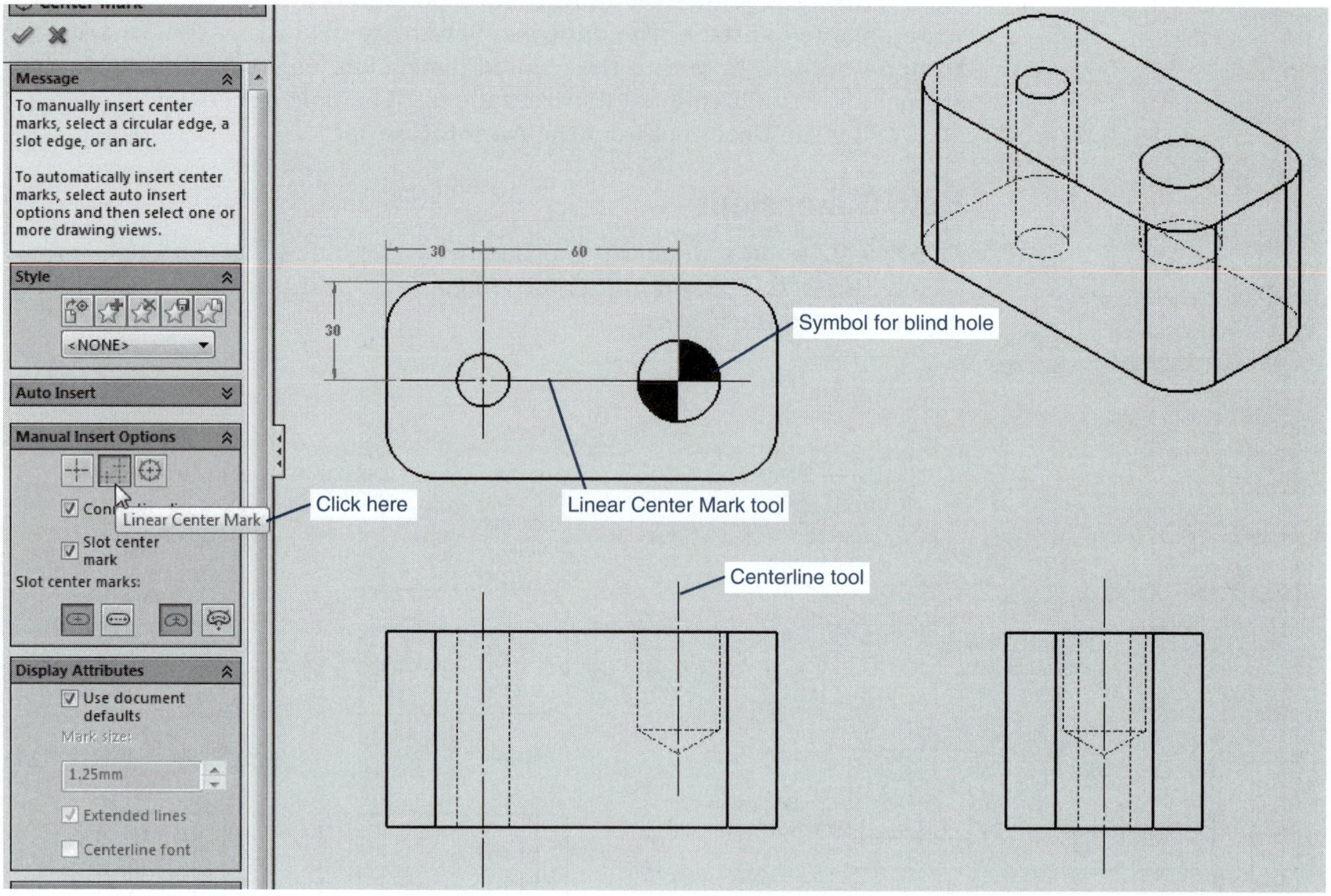

Figure 7-23

3 Use the **Smart Dimension** tool and add a dimension to one of the filleted corners.

See Figure 7-24.

Figure 7-24

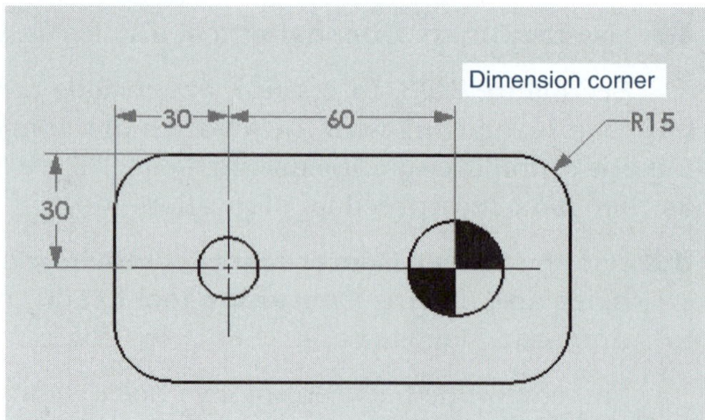

TIP

The dimension options found on the **Document PropertyManager** will change all dimensions. Clicking a dimension and using the **Dimension PropertyManager** allows you to change just that dimension.

4 Click the fillet dimension again, go to the **Dimension Text** block on the **Dimension PropertyManager**, and type **4 CORNERS** as shown.

See Figure 7-25.

Figure 7-25

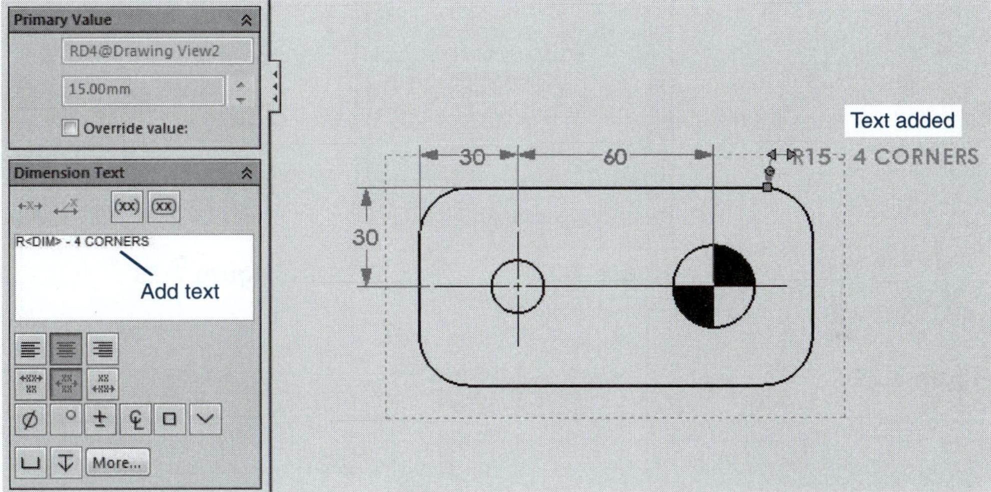

5 Click **OK, Apply**, and **OK**.

6 Use the **Hole Callout** tool on the **Annotate** panel and dimension the Ø16 hole.

The Ø16 hole goes completely through the part, so no depth specification is required. See Figure 7-26. The word THRU is optional and may be removed.

7 Use the **Hole Callout** tool and **Dimension** the Ø25 hole.

Figure 7-26

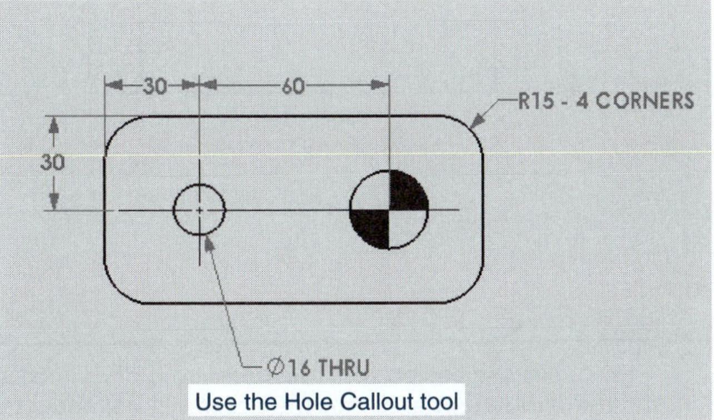

The hole callout will include the depth symbol, see Figure 7-27, and a depth value of **30**. See Figure 7-28.

8 Complete the dimensions.

See Figure 7-29.

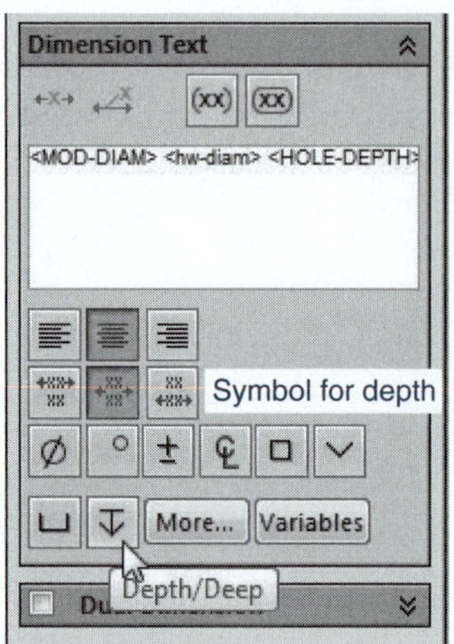

Figure 7-27

Figure 7-28

Figure 7-29

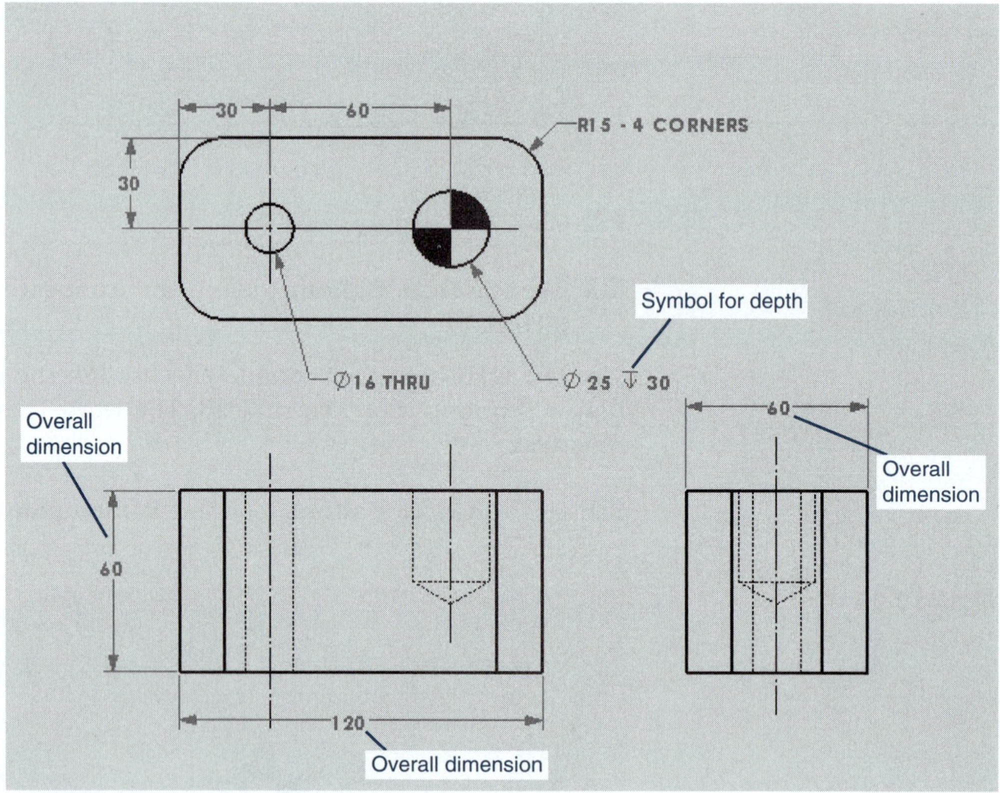

NOTE

If the **Smart Dimension** tool had been used, the dimension would have to be edited in the **Dimension text** area and the depth symbol and a numerical value added.

7-6 Dimensioning Holes and Fillets

A ***blind hole*** is a hole that does not go completely through an object. It has a depth requirement. Figure 7-30 shows a 2.00 × 2.00 × 2.00 cube with a blind Ø.50 × 1.18 DEEP hole. It was created as follows.

Dimensioning a Blind Hole

1. Draw the block.
2. Click the **Hole Wizard** tool.

   See Figure 7-30.

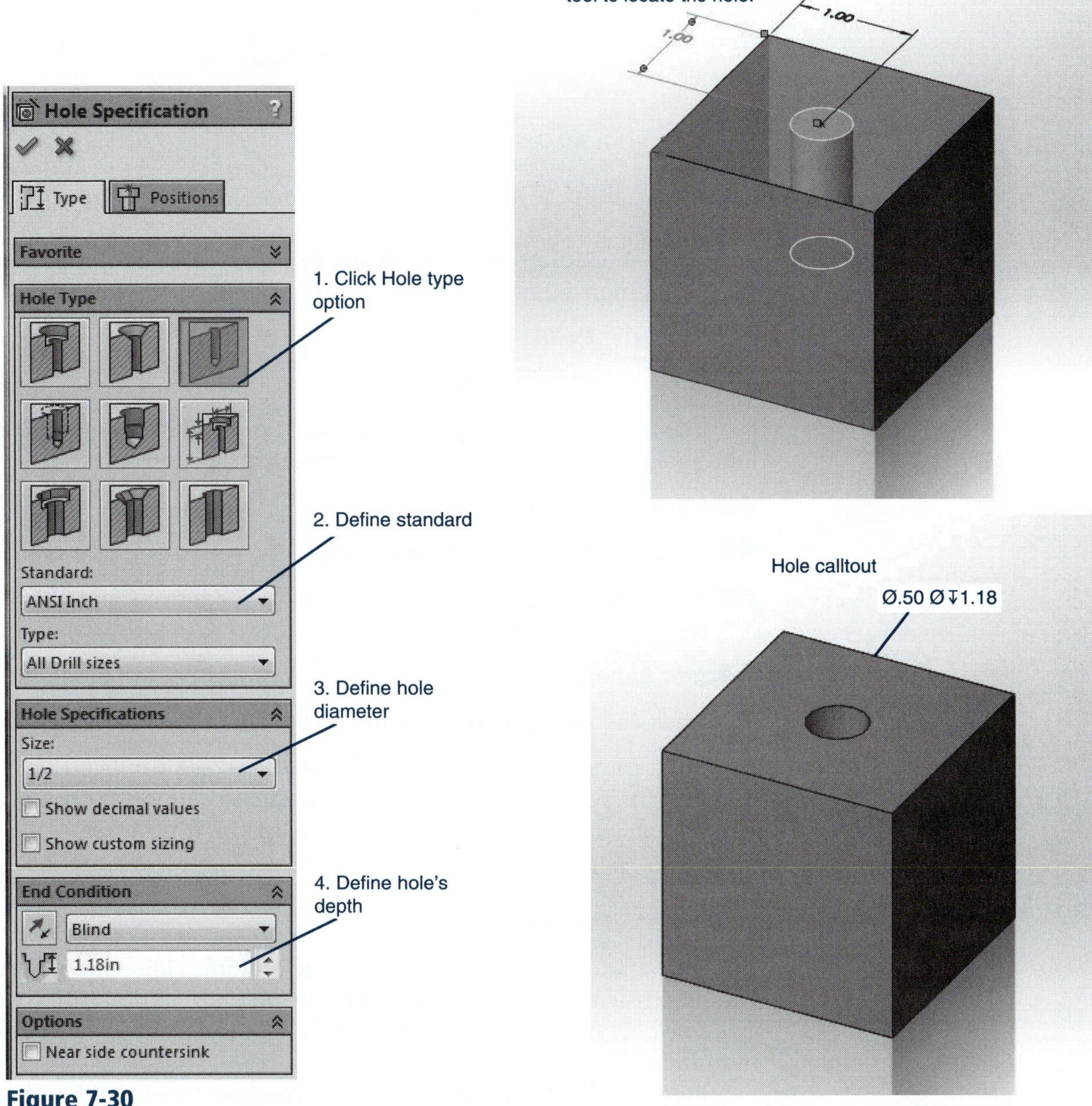

Figure 7-30

Figure 7-30
(*Continued*)

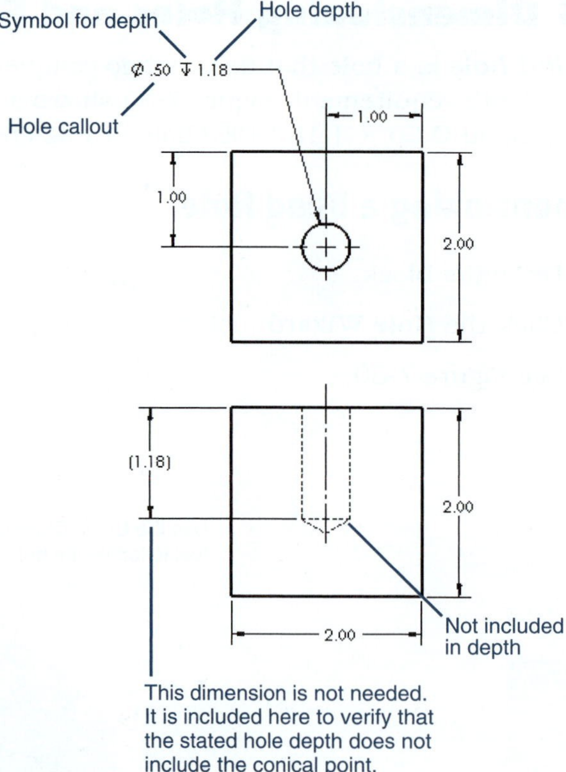

3 Click the **Hole** tool in the **Hole Type** box. Define the hole using the **ANSI Inch** standard with a diameter of **1/2** and a depth of **1.18.**

4 Click the **Positions** tab.

5 Locate the hole as shown.

The initial location is an approximation. Use the **Smart Dimension** tool to specify the exact location of the hole's centerpoint.

6 Click the green **OK** check mark.

7 Save the drawing as **Block, Blind.**

8 Start a new **Drawing** document and create a front and a top orthographic view of the **Block, Blind**.

9 Add dimensions to the views.

10 Click the **Annotation** tab and click the **Hole Callout** option.

11 Click the edge of the hole, move the cursor away from the hole, define a location for the hole callout, and click the mouse. The hole callout dimension will initially appear as a rectangular box.

Change the height of the text font if necessary.

12 Save the drawing.

Note that the hole includes a conical point. Holes manufactured using twist drills will have conical points. The conical point is not included in the hole's depth dimension. A special drill bit can be used to create a flat-bottomed hole.

Figure 7-31 shows three different methods that can be used to dimension a blind hole.

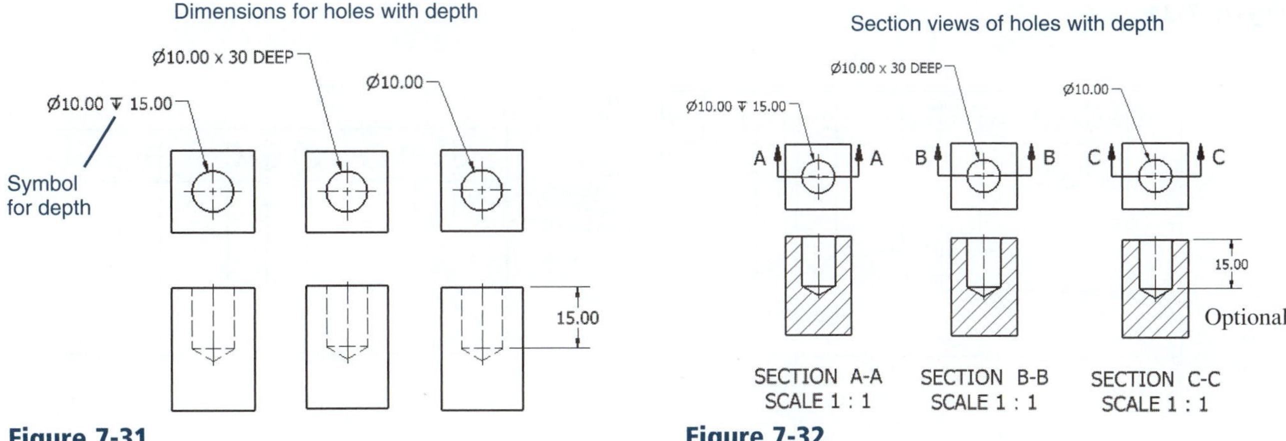

Figure 7-31

Figure 7-32

Figure 7-32 shows three methods of dimensioning holes in section views. The single line note version is the preferred method.

Dimensioning Hole Patterns

Figure 7-33 shows two different hole patterns dimensioned. The circular pattern includes the note **Ø10-4 HOLES**. This note serves to define all four holes within the object.

Figure 7-33 also shows a rectangular object that contains five holes of equal diameter, equally spaced from one another. The notation **5 × Ø10** specifies five holes of 10 diameter. The notation **4 × 20 (=80)** means

Figure 7-33

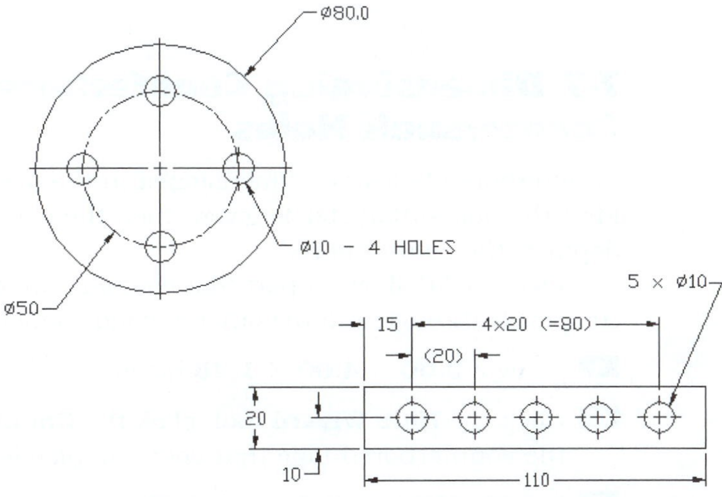

four equal spaces of 20. The notation **(=80)** is a reference dimension and is included for convenience. Reference dimensions are explained in Chapter 9.

Figure 7-34 shows two additional methods for dimensioning repeating hole patterns. Figure 7-35 shows a circular hole pattern that includes two different hole diameters. The hole diameters are not noticeably different and could be confused. One group is defined by indicating letter (A); the other is dimensioned in a normal manner.

Figure 7-34

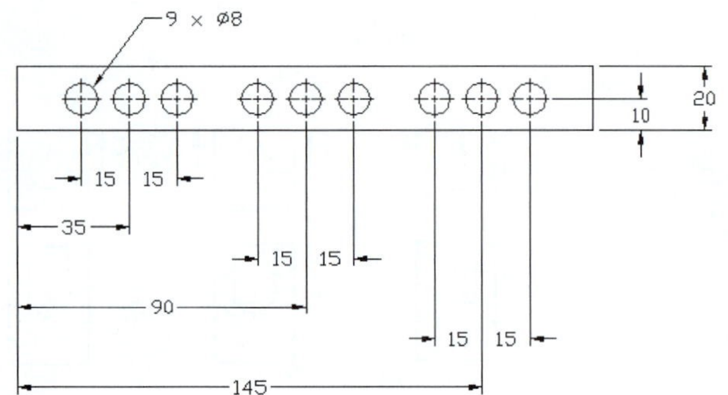

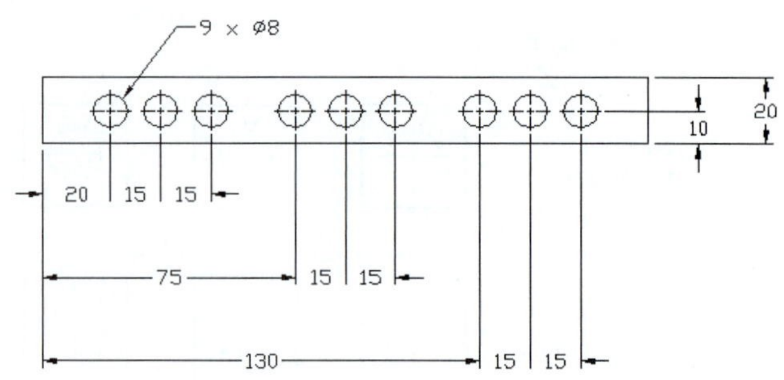

Figure 7-35

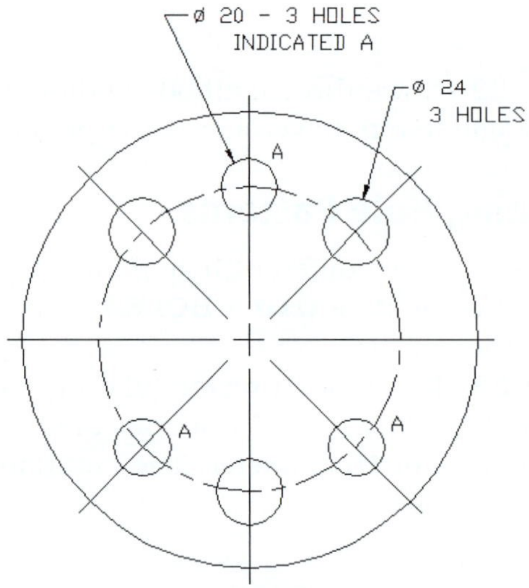

7-7 Dimensioning Counterbored and Countersunk Holes

Counterbored holes are dimensioned in the sequence of their manufacture. First the hole's diameter is given, then the counterbore diameter, then the depth of the counterbore.

Figure 7-36 shows a part that contains two counterbored holes; one goes completely through and the other is blind. Dimensions will be applied to both.

1. Draw a **3.00** × **4.00** × **1.75** block.
2. Click the **Hole Wizard** tool, click the **Counterbore** option, and draw the counterbored hole that goes completely through.
3. Specify **ANSI Inch Standard**, **Hex Screw** with a **3/8** diameter and **Through all** End condition.

See Figure 7-37. SolidWorks will automatically select the diameter for the counterbored hole that will accommodate a Ø3/8 Hex Head Screw.

Depending on your default settings, the counterbored hole may have a small chamfer added. The countersink can be removed by removing the check mark in the **Options** box on the **Hole Specification** manager.

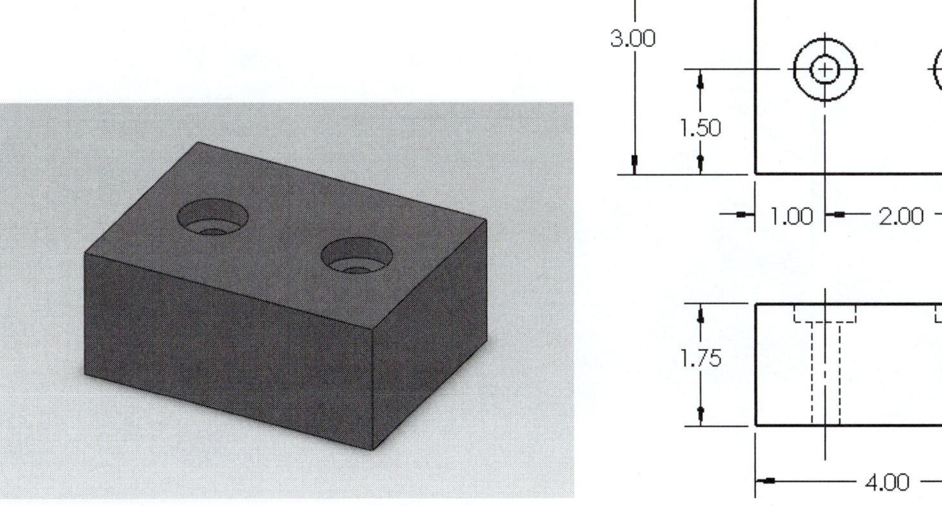

Figure 7-36

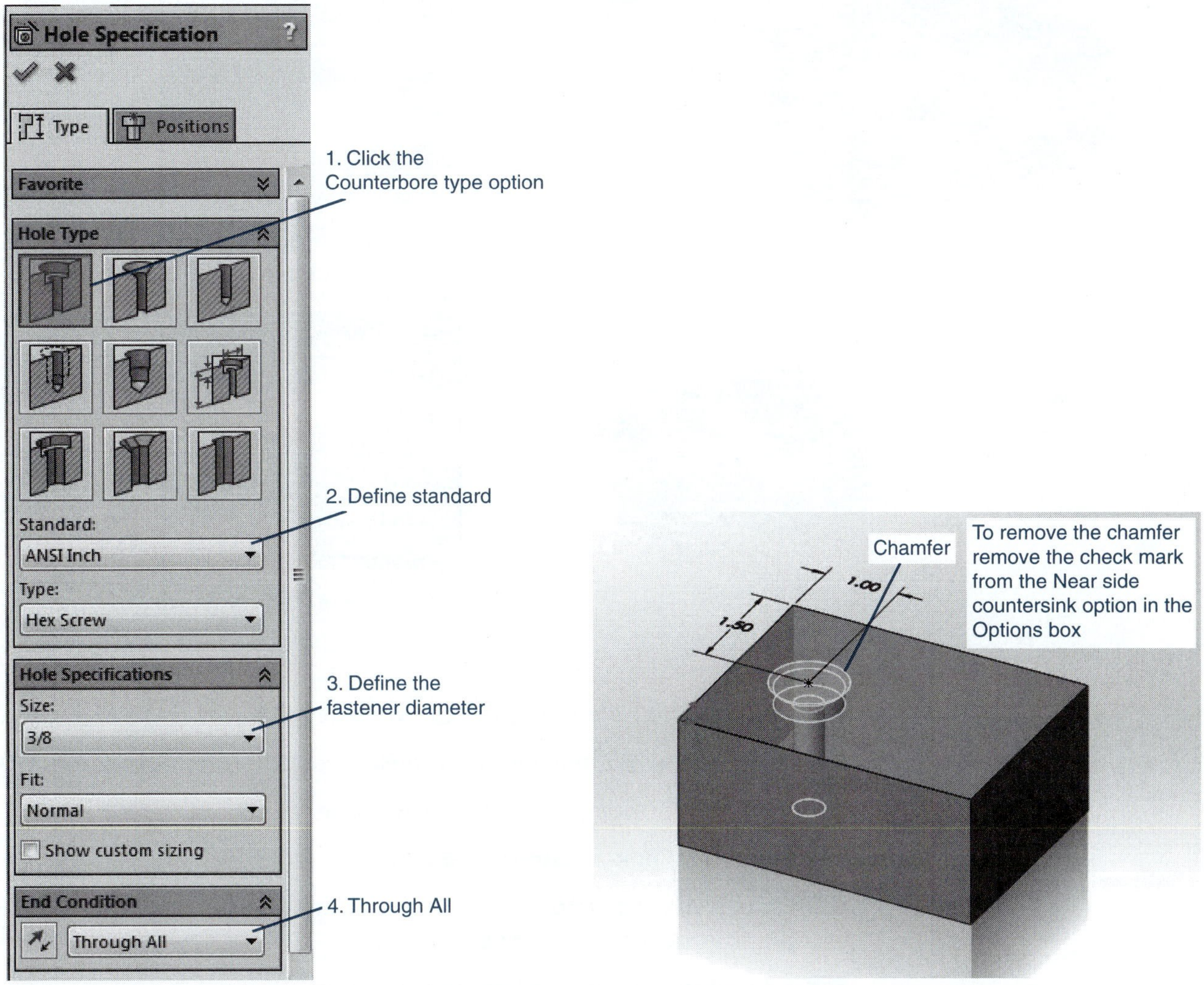

Hole Specification

Type | Positions

Favorite

Hole Type

Standard:
ANSI Inch

Type:
Hex Screw

Hole Specifications

Size:
3/8

Fit:
Normal

Show custom sizing

End Condition

Through All

1. Click the Counterbore type option

2. Define standard

3. Define the fastener diameter

4. Through All

Figure 7-37

Chamfer

To remove the chamfer remove the check mark from the Near side countersink option in the Options box

1.00

1.50

Figure 7-37
(*Continued*)

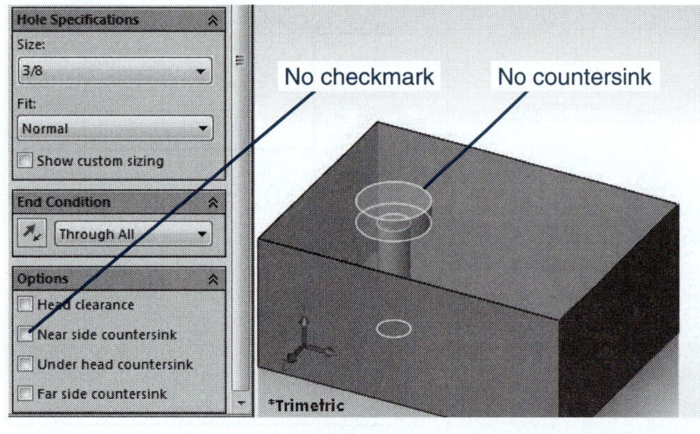

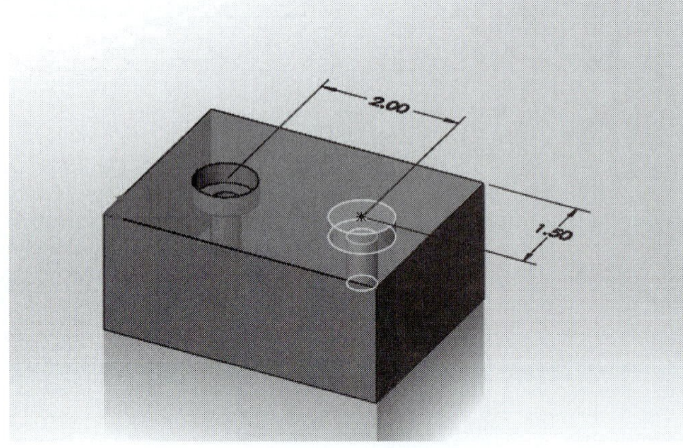

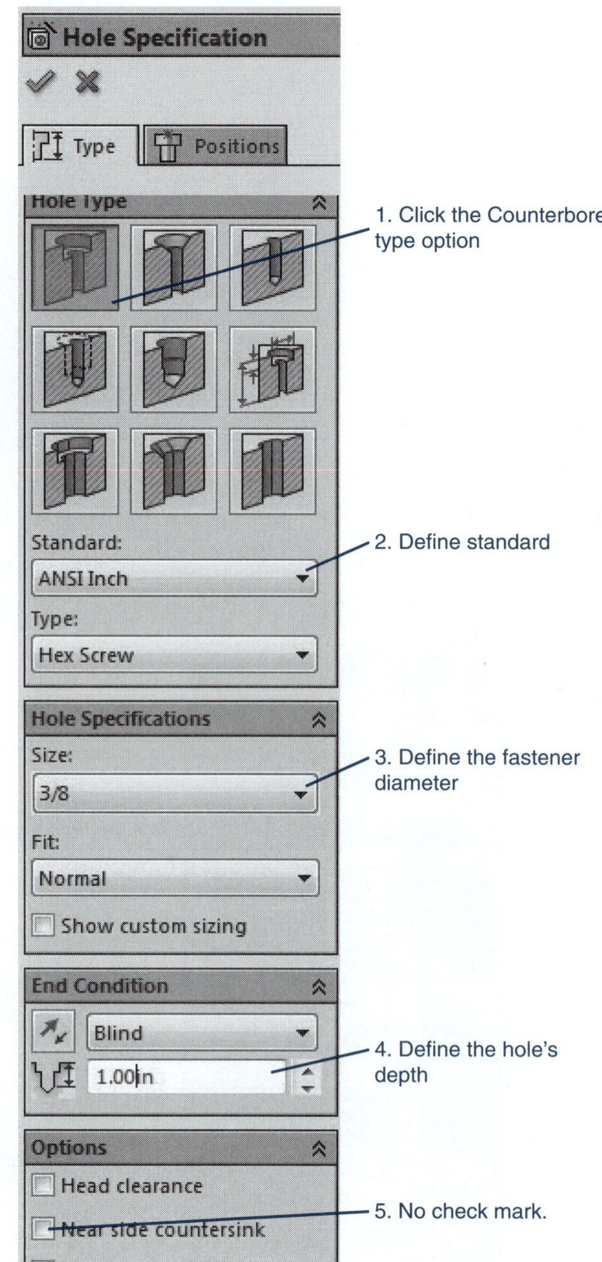

4 Position the hole using the given dimensions.

5 Add the second hole setting the **End** condition for **Blind** and **1.00 deep**.

6 Position the hole using the given dimensions.

7 Save the block as **Block, Cbore.**

8 Start a new **Drawing** document and create a front and a top orthographic view of the **Block, Cbore.**

9 Add all dimensions other than the hole dimensions.

10 Click the **Annotation** tab and select the **Hole Callout** tool.

See Figure 7-38.

11 Click the edge of each hole, move the cursor away from the hole, and click the mouse when a suitable location is found.

The counterbored hole's dimension note is interpreted as shown in Figure 7-38.

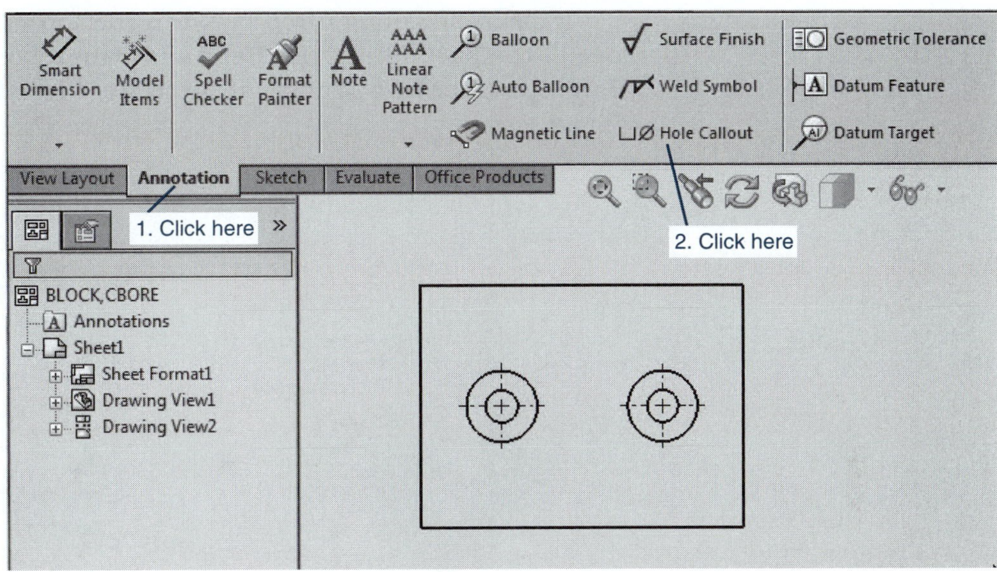

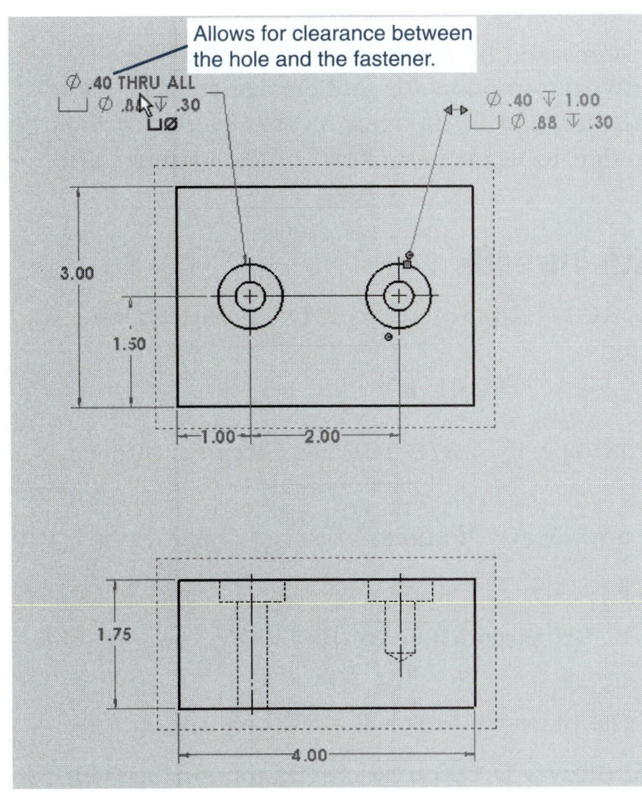

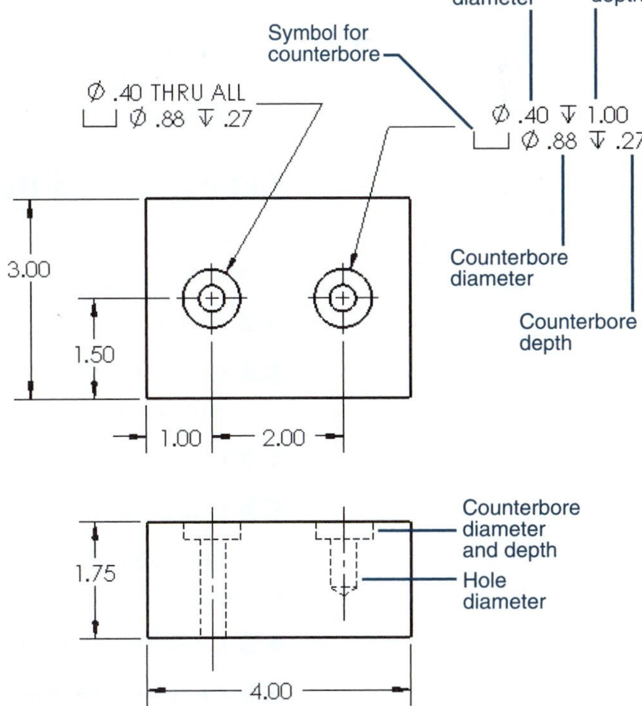

Figure 7-38

Figure 7-39 shows the **Block, Cbore** assembled with hex head screws inserted into the counterbored holes. SolidWorks will automatically generate the correct size counterbored hole for a specified screw. The counterbore depth will align the top of the screw head with the top surface of the part and will define a hole diameter that includes clearance between the fastener and the hole. In this example a clearance hole with a diameter of **Ø.40** was generated. The hole is .02 larger than the specified .38 fastener diameter

If clearance is required between the top of the screw and the top surface of the part, check the **Head clearance** box under **Options** on the **Hole Specification** section of the **Hole Wizard PropertyManager.** See Figure 7-40.

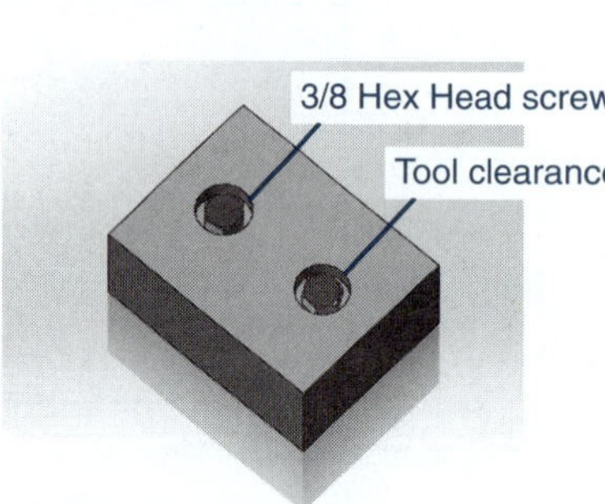

Figure 7-39

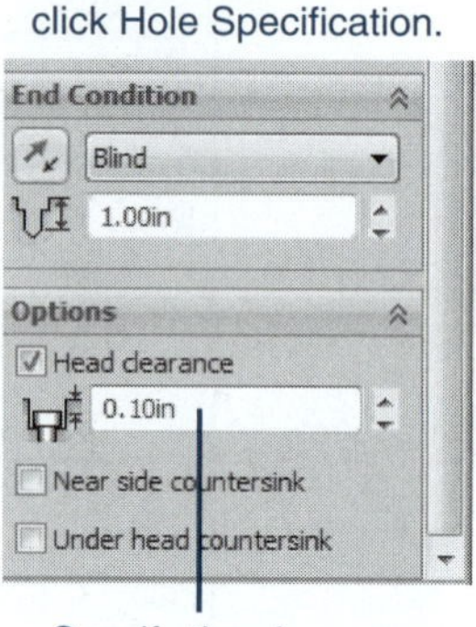

Figure 7-40

The diameter of the counterbored hole can be made larger than the clearance generated by SolidWorks to allow for tool clearance. Tool clearance allowance increases the diameter of the counterbore so that it is large enough to allow a socket wrench to fit over the head of the fastener and still fit within the hole.

Counterbored Hole with Threads

Figure 7-41 shows a 3.00 × 4.00 × 2.00 block with two counterbored holes. Both holes are threaded.

1. Draw the block.
2. Click the **Hole Wizard** and specify a **3/8-16 UNC** thread that goes completely through.
3. Click the **Positions** tab and locate the hole.
4. Click the **OK** check mark.

This will locate a 3/8-16 UNC thread hole in the block. Now, we add the counterbore.

5. Click the top surface of the block and click the **Sketch** option.
6. Click the **Circle** tool and draw a **Ø.88** circle on the top surface centered on the same center point as the Ø3/8-16 hole.

The dimensions for this example came from Figure 7-38.

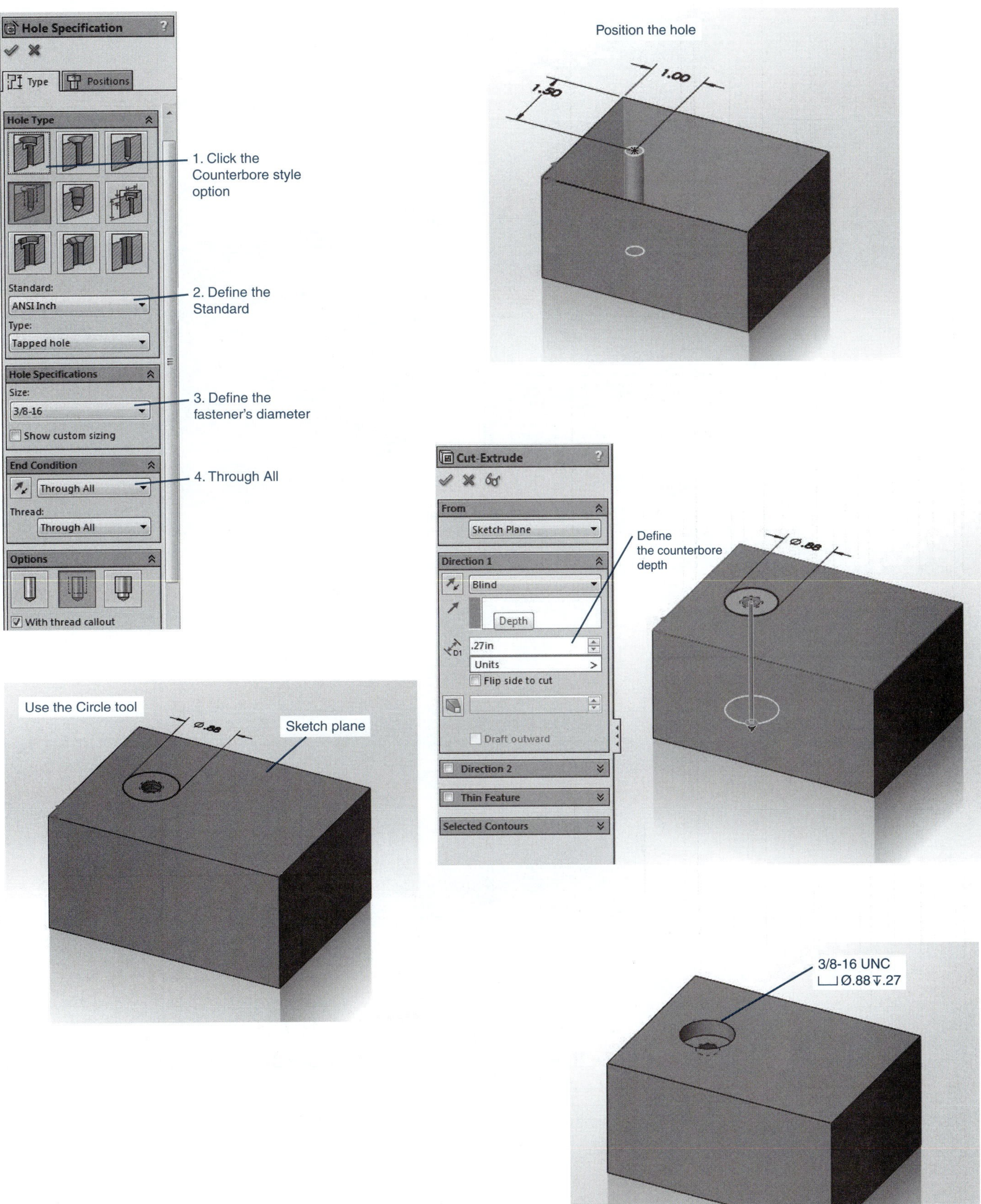
Hole Specification
Type
Positions
Hole Type
1. Click the Counterbore style option
Standard:
ANSI Inch
2. Define the Standard
Type:
Tapped hole
Hole Specifications
Size:
3/8-16
3. Define the fastener's diameter
Show custom sizing
End Condition
Through All
4. Through All
Thread:
Through All
Options
With thread callout
Position the hole
1.00
1.50
Cut-Extrude
From
Sketch Plane
Direction 1
Blind
Depth
.27in
Units
Flip side to cut
Draft outward
Direction 2
Thin Feature
Selected Contours
Define the counterbore depth
Ø.88
Use the Circle tool
Ø.88
Sketch plane
3/8-16 UNC
⌴Ø.88↧.27

Figure 7-41

Chapter 7

Figure 7-41
(*Continued*)

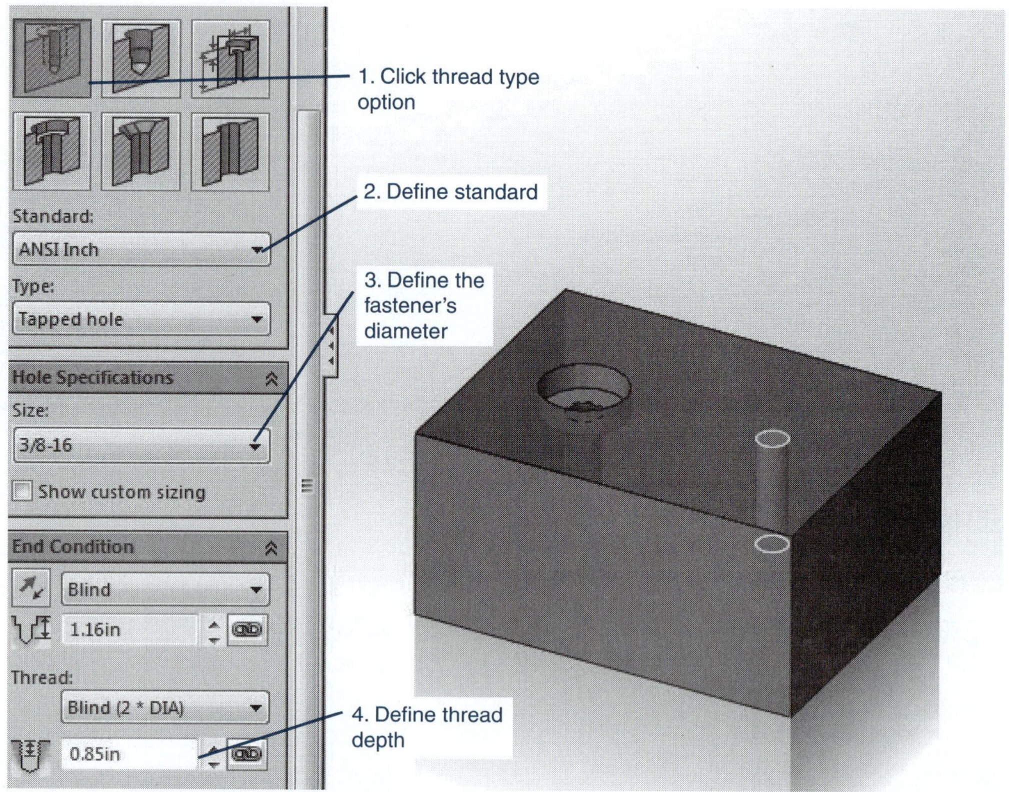

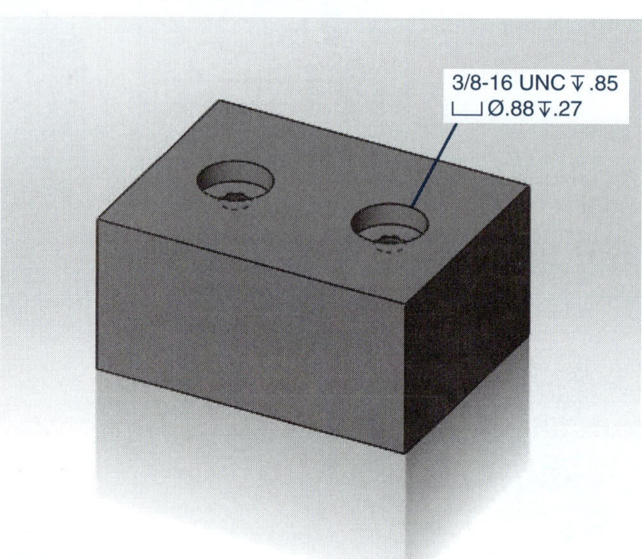

7 Click the **Features** tab, click the **Extruded Cut** tool, and specify a cut depth of **0.27**.

8 Click the green **OK** check mark.

9 Repeat the procedure adding a second hole with a thread to a depth of **0.85**.

> **TIP**
>
> For an internal thread, the thread depth is measured from the top surface of the part.

Figure 7-42

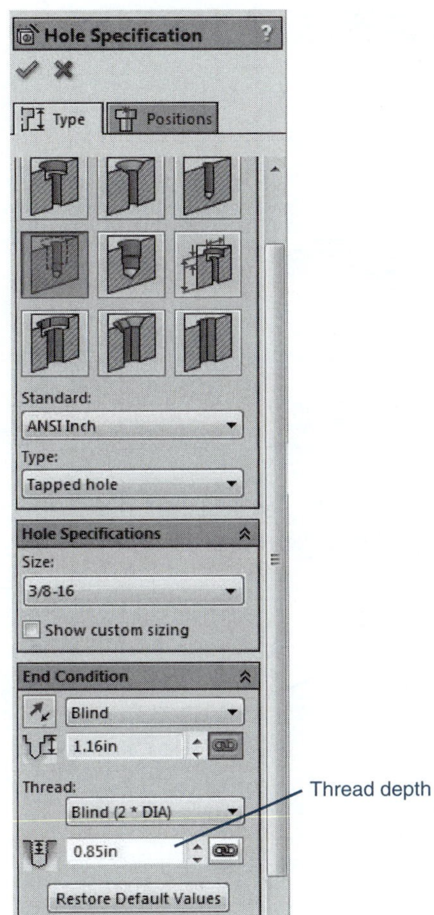

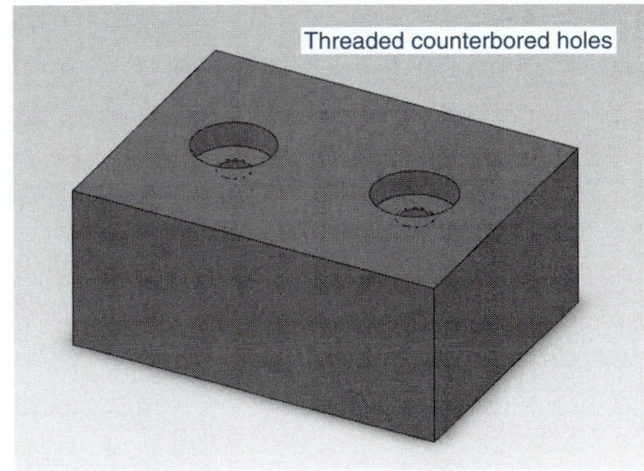

Chapter 7

See Figure 7-42.

10 Save the block as **Block, Threads**.

11 Create a new **Drawing** document and create front and top orthographic views of the **Block, Threads**.

12 Add centerlines to the front view and add dimensions as shown.

See Figure 7-43.

13 Use the **Hole Callout** tool and click the left threaded hole.

Do not click the outside of the counterbored hole. This will generate a note that includes only the counterbore.

14 Locate the text and click the mouse.

The initial note may show the counterbore callout above the thread callout. Convention calls for the note to read in the sequence of manufacture. The threaded hole is cut first and then the counterbore is added; therefore the thread callout should come before the counterbore callout.

15 Modify the callout to list the thread callout above the counterbore callout.

Access the **Dimension Text** box on the **Dimension manager** and delete the first line of text. A warning box will appear. Click the **Yes** option. Add a second line of text below the remaining line defining the counterbore. Use the symbols within the **Dimension Text** box to create the note.

Figure 7-43

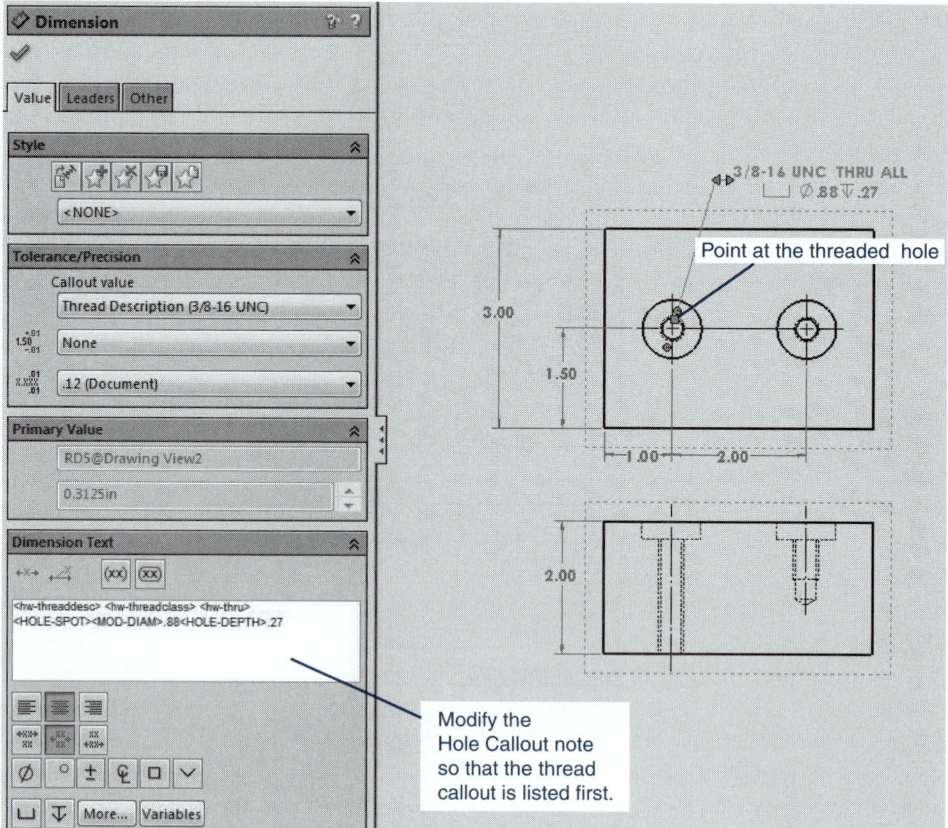

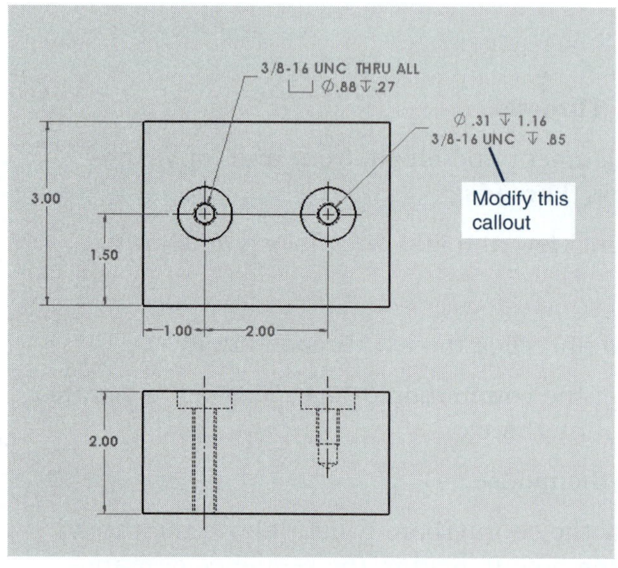

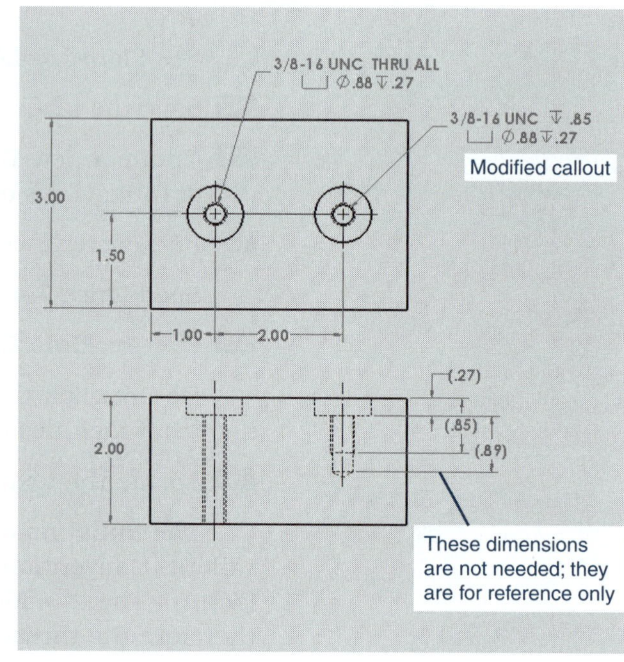

16 Click the green **OK** check mark.

17 Click the threaded portion of the right hole.

18 Locate the text and click the mouse.

19 Modify the callout as shown.

Figure 7-44 shows dimensioned counterbored holes using metric units. The **Hole Callout** tool was used to dimension the counterbored holes. Note that the hole's diameter is listed as Ø11. The fastener size was specified as M10, and the Ø11 hole is a clearance hole.

Figure 7-44

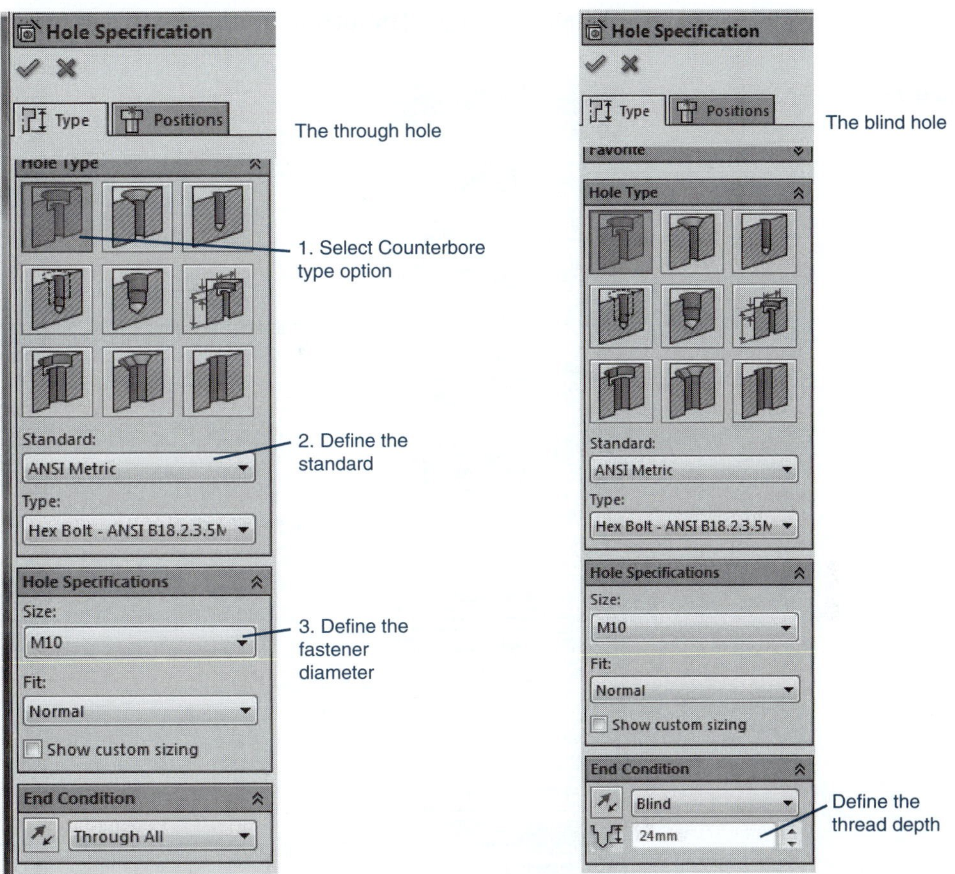

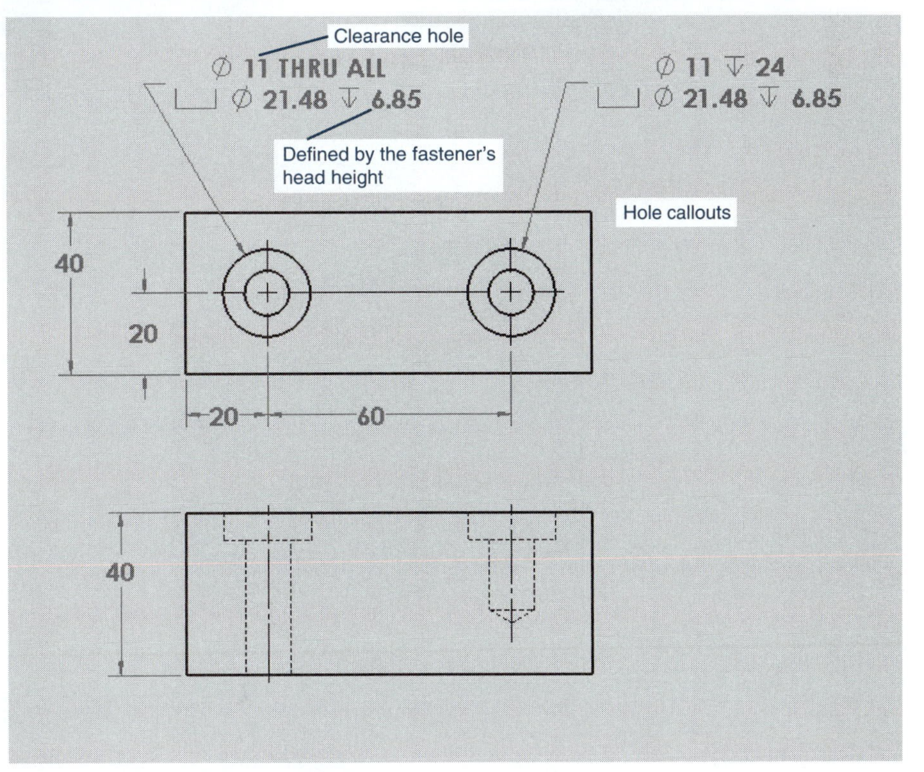

To Dimension Countersink Holes

Countersink holes are used with flat head screws to create assemblies in which the fasteners do not protrude above the surfaces.

Figure 7-45 shows a part with two countersunk holes; one goes completely through, the other has a depth specification.

Figure 7-45

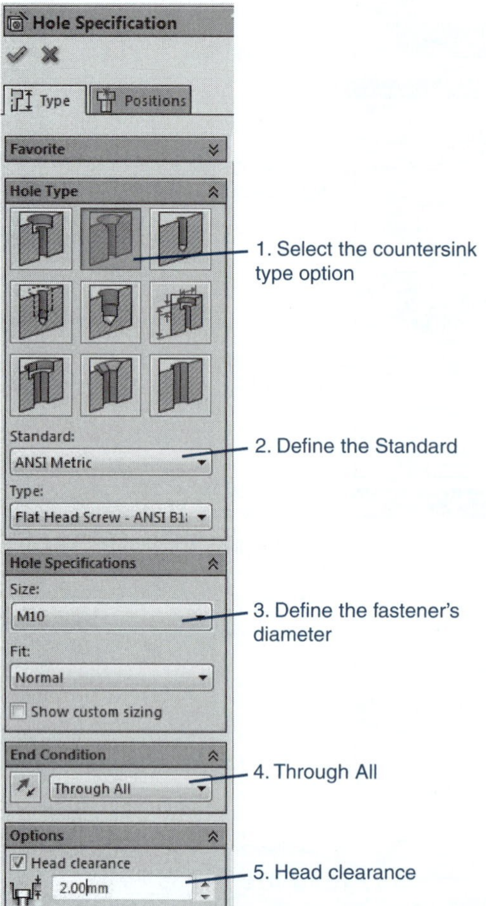

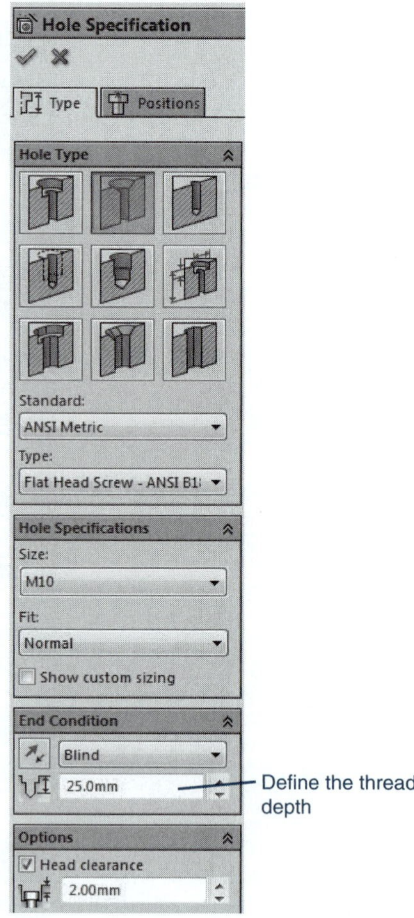

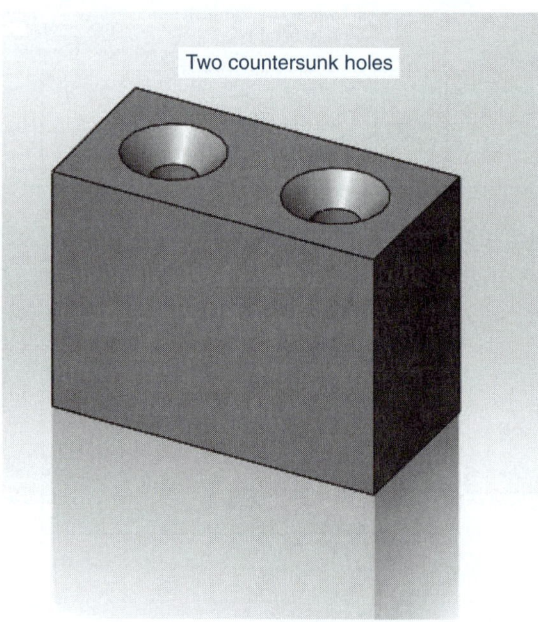

1 Draw a 40 × 80 × 60 block.

2 Use the **Hole Wizard** tool, click the **Countersink** type, specify the **ANSI Metric** standard, select an **M10** size for a flat head screw, and a hole that goes all the way through. Define a head clearance of **2.00**.

3 Click the **Positions** tab and position the countersunk hole's center point as shown using the **Smart Dimension** tool.

4 Click the green **OK** check mark.

5 Click the **Hole Wizard** tool, click the **Countersink** type, specify the **ANSI Metric** standard, select an **M10** size for a flat head screw, and specify a depth requirement of **25.0** for a blind hole. Define a head clearance of **2.00.**

6 Click the **Positions** tab and locate the hole as shown.

7 Click the green **OK** check mark.

8 Save the drawing as **Block, CSink**.

To Dimension the Block

1 Create a new **Drawing** document with a front and top orthographic view of the **Block, CSink**.

2 Use the **Smart Dimension** tool and add the appropriate dimensions.

3 Use the **Center Mark** tool to add a centerline between the two holes indicating they are aligned.

4 Click the **Annotation** tab, click the **Hole Callout** tool, and dimension the two countersunk holes.

See Figure 7-46

Figure 7-46

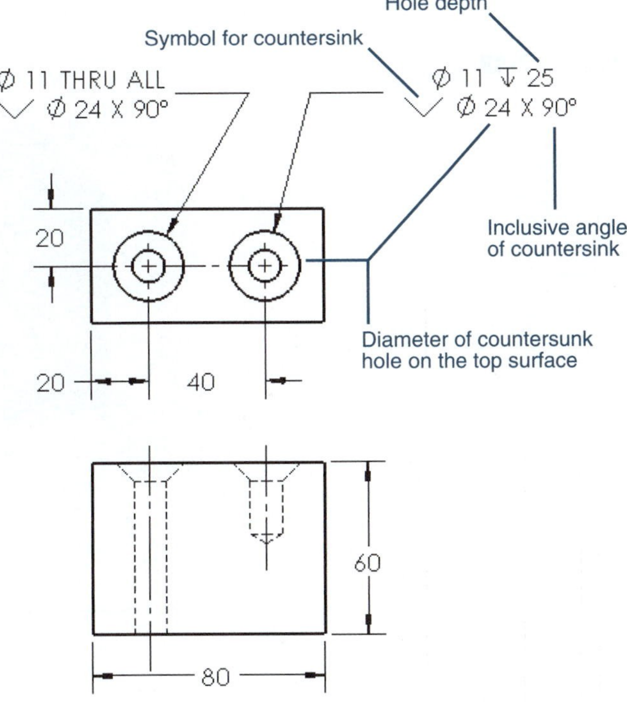

Completed dimensions

7-8 Angular Dimensions

Figure 7-47 shows a model that includes a slanted surface and dimensioned orthographic views of the model. The dimension values are located beyond the model between two extension lines. Locating dimensions between extension lines is preferred to locating the value between an extension line and the edge of the model.

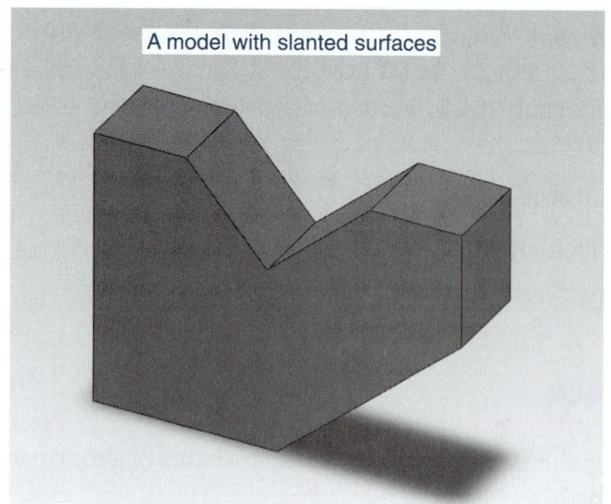

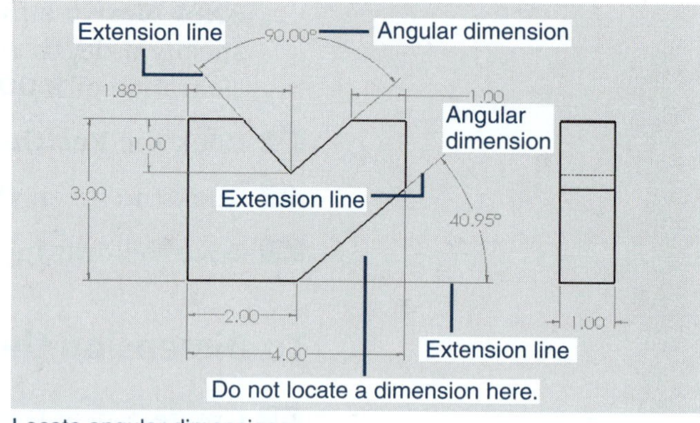

Locate angular dimensions away from the model.

Figure 7-47

Figure 7-48 shows a shape that includes a slanted surface dimensioned in two different ways. The shape on the left uses an angular dimension; the one on the right does not. Both are acceptable.

Figure 7-48

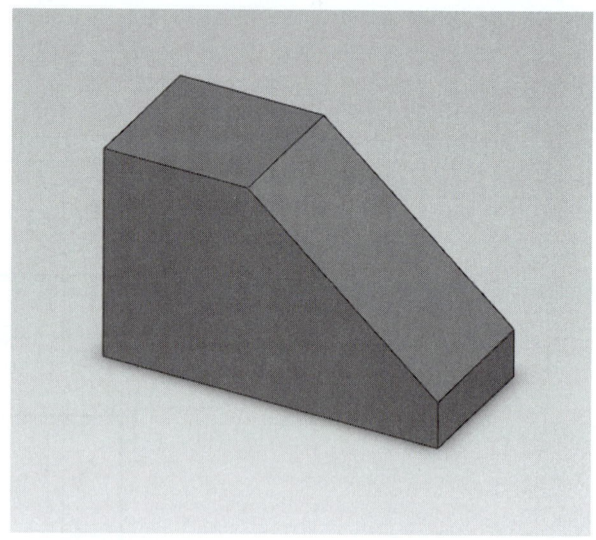

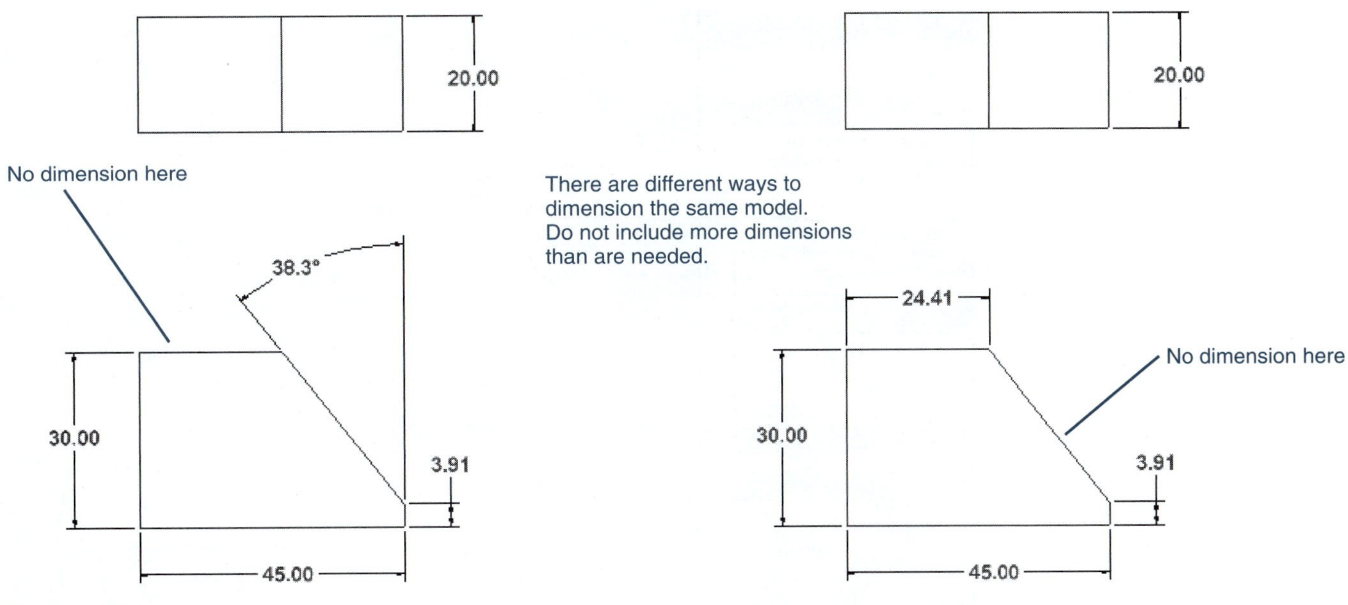
20.00
No dimension here
38.3°
30.00
3.91
45.00
20.00
There are different ways to dimension the same model. Do not include more dimensions than are needed.
24.41
No dimension here
30.00
3.91
45.00

Figure 7-48
(*Continued*)

Figure 7-49 shows two objects dimensioned using angular dimensions. One has an evenly spaced hole pattern; the other has an uneven hole pattern.

Figure 7-49

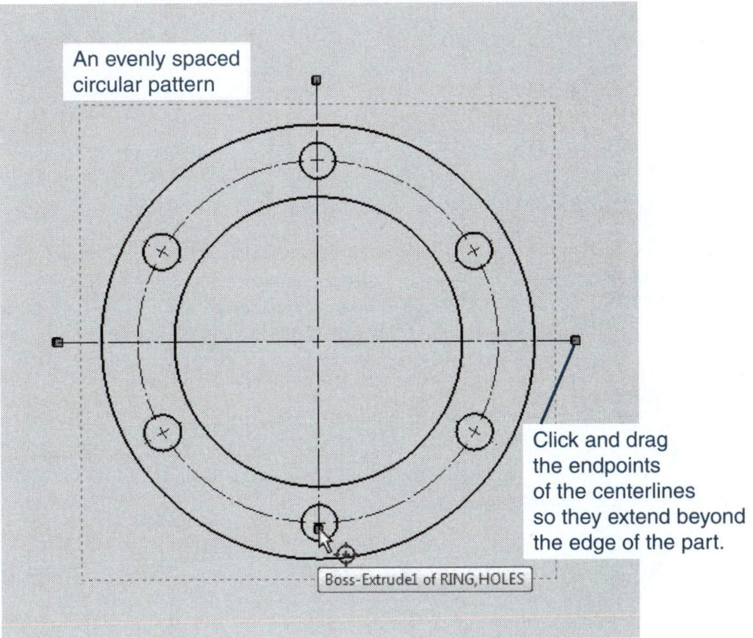
An evenly spaced circular pattern
Click and drag the endpoints of the centerlines so they extend beyond the edge of the part.
Boss-Extrude1 of RING,HOLES

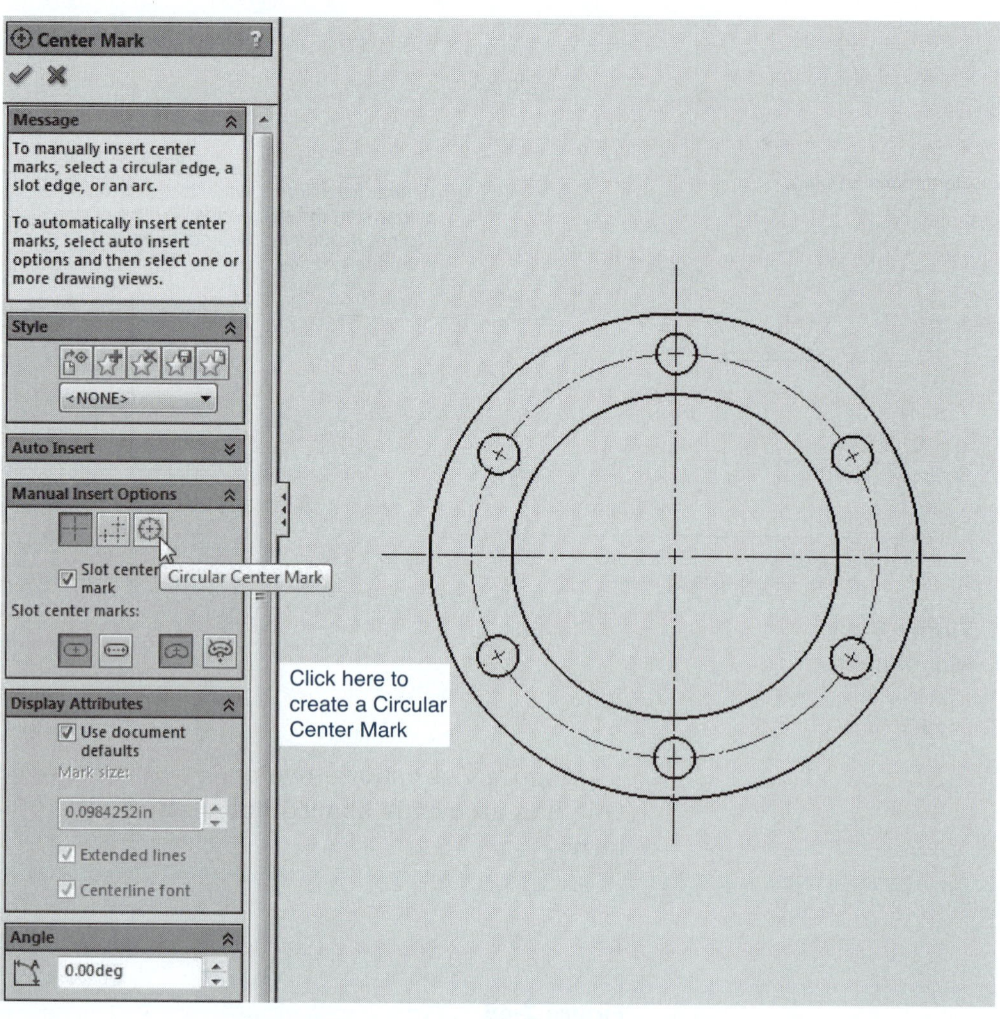

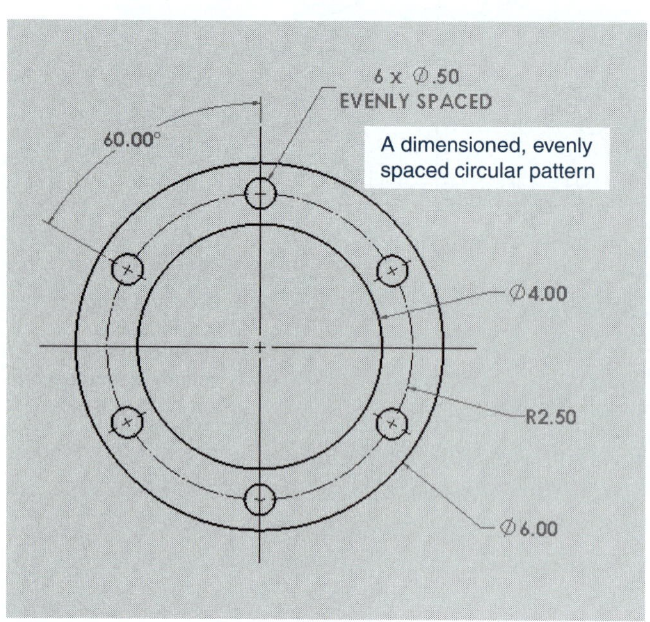

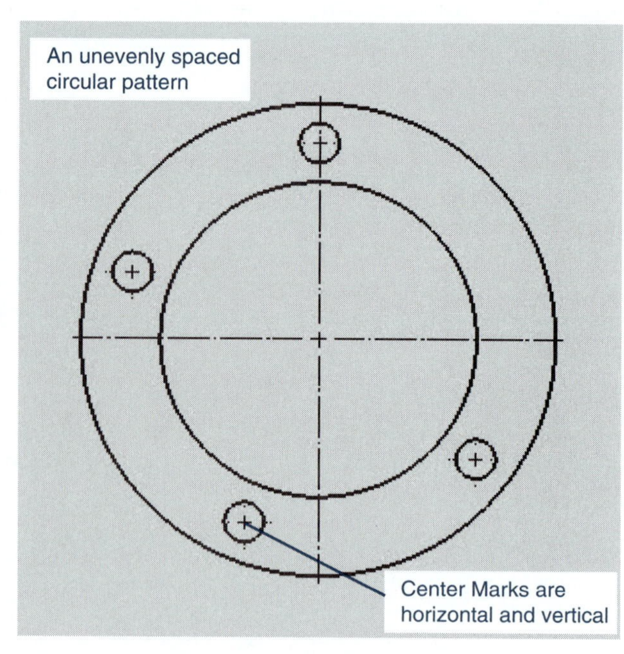

Figure 7-49
(*Continued*)

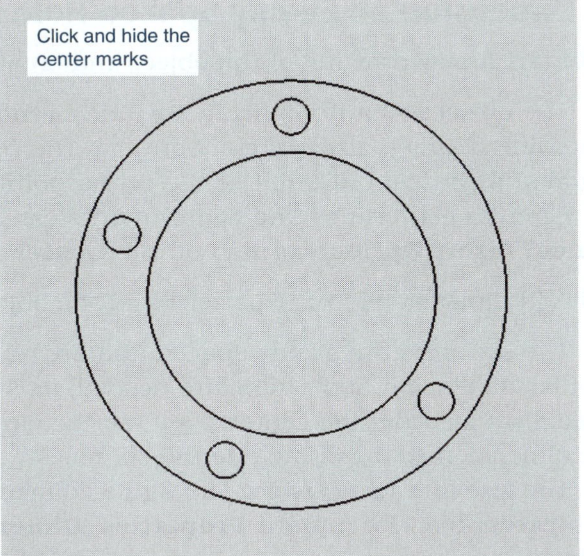

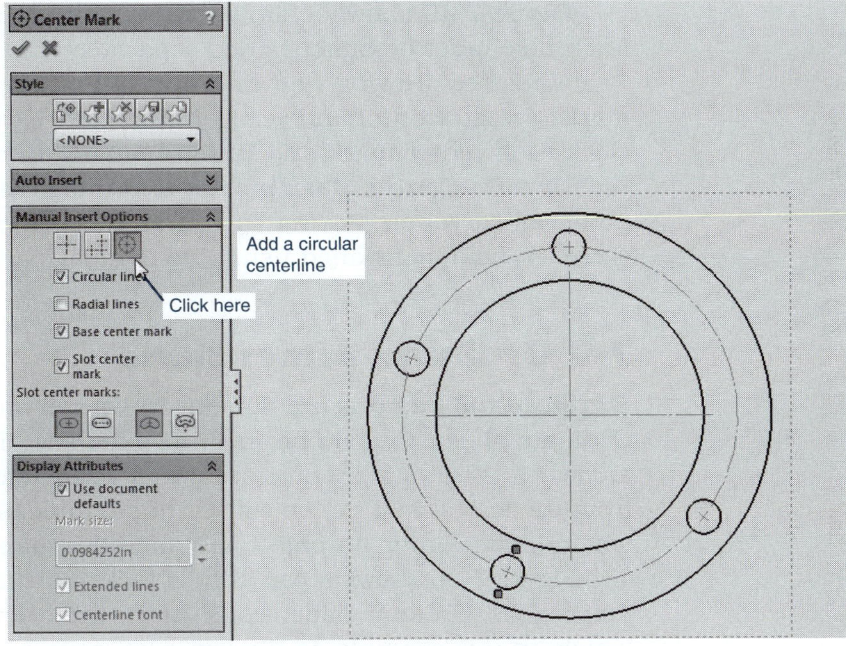

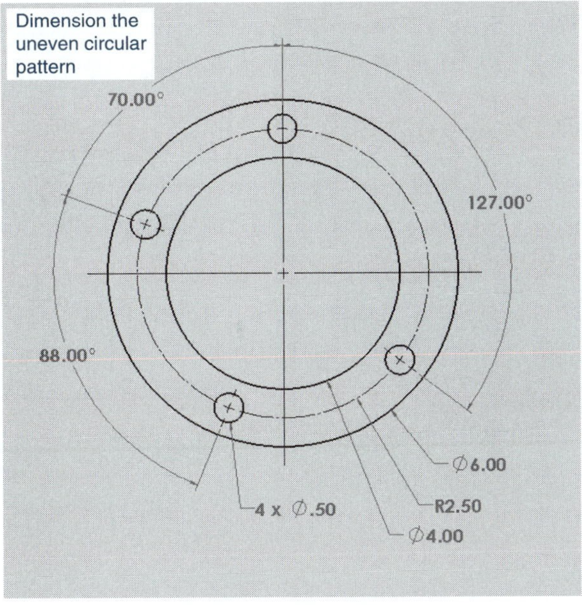

Figure 7-49
(Continued)

To Dimension an Evenly Spaced Hole Pattern

1 Start a new drawing of the object and create a view as shown.

The object will automatically include circular center lines. The circular centerline is called a *bolt circle*. Note that the center marks are not horizontal and vertical but point at the center point of the pattern.

Circular centerlines and center marks can be created using the **Manual Insert Options** located on the **Center Marks** manager.

2 Add dimensions to the pattern and the object.

The six holes are evenly spaced and are all the same size, so only one angular dimension and a note are needed, as shown. All the holes are the same distance from the center point, so the circular centerline needs only one dimension that will include the six holes.

The size and text position of angular dimension can be edited using the **System** tool, **Document Properties**, **Dimensions**, **Angle**, and entering edits.

Figure 7-49 shows a similar object but with an uneven hole pattern. Each hole must be dimensioned separately.

When the drawing view first appears on the screen, all the center marks are horizontal and vertical. A circular centerline pattern is preferred. Click each center mark and **Hide** the mark. Click the **Centermark** tool and the **Circular Center Mark** tool located under the **Manual Insert Options**, and click each hole. A circular centerline pattern will appear. The shape can then be dimensioned using the circular pattern.

7-9 Ordinate Dimensions

Ordinate dimensions are dimensions based on an X,Y coordinate system. Ordinate dimensions do not include extension lines, dimension lines, or arrowheads but simply horizontal and vertical leader lines drawn directly from the features of the object. Ordinate dimensions are particularly useful when dimensioning an object that includes many small holes.

Figure 7-50 shows a part that is to be dimensioned using ordinate dimensions. Ordinate dimensions' values are calculated from the X,Y origin, which, in this example, is the lower left corner of the front view of the model.

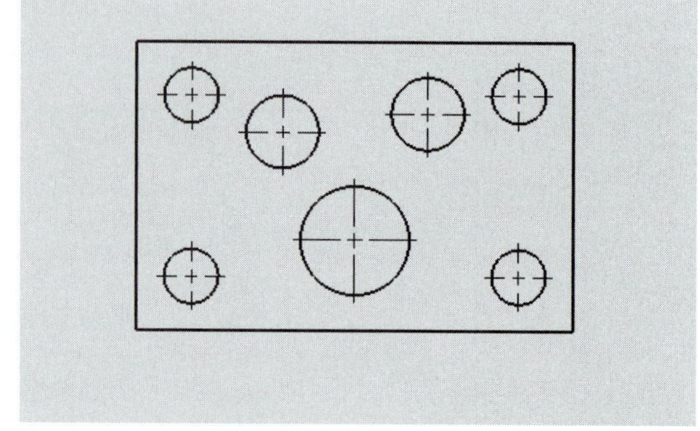

Figure 7-50

To Create Ordinate Dimensions

See Figure 7-51.

1 Start a new **Drawing** document and create a top orthographic view of the part.

Use the dimensions shown in either Figure 7-51 or 7-53 to create the drawing.

2 Click and extend the center marks and draw centerlines between the four corner holes.

Figure 7-51

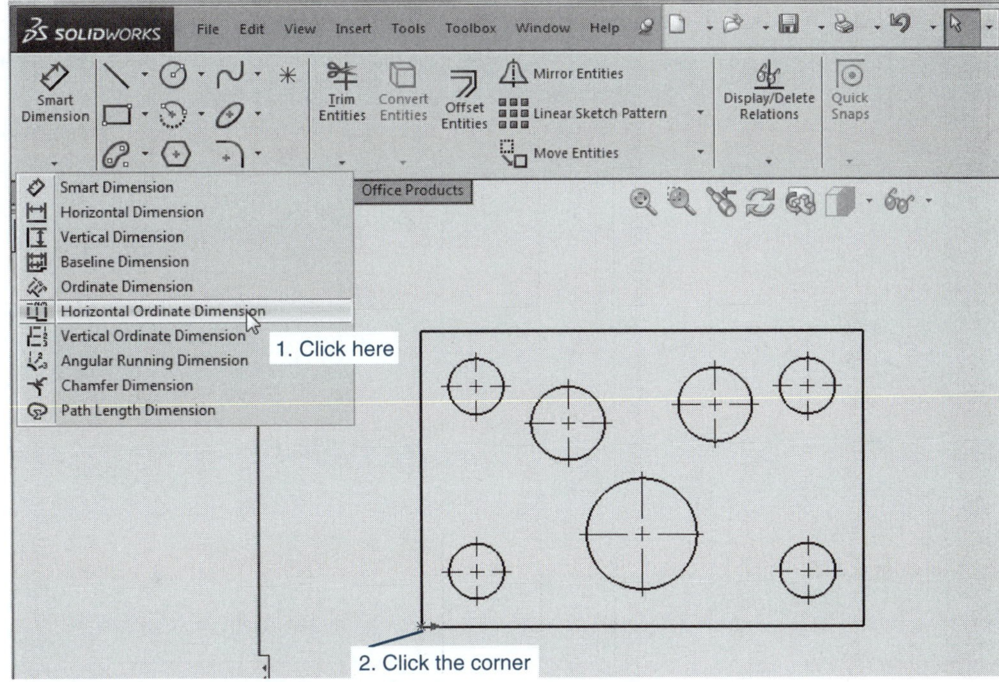

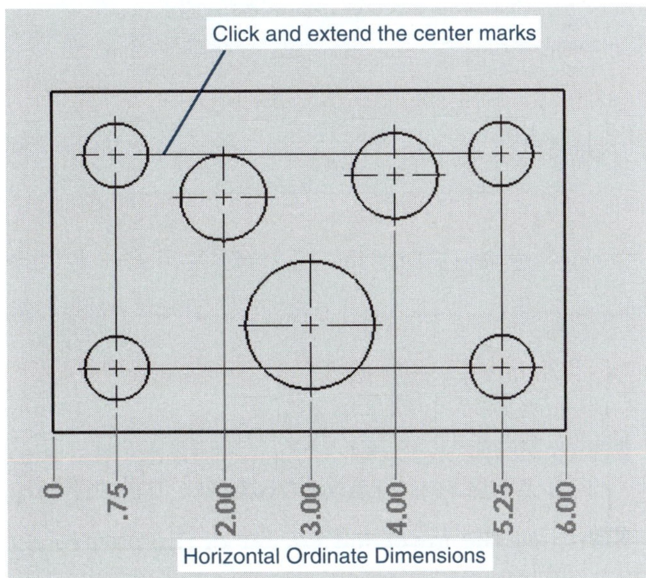

Figure 7-51
(Continued)

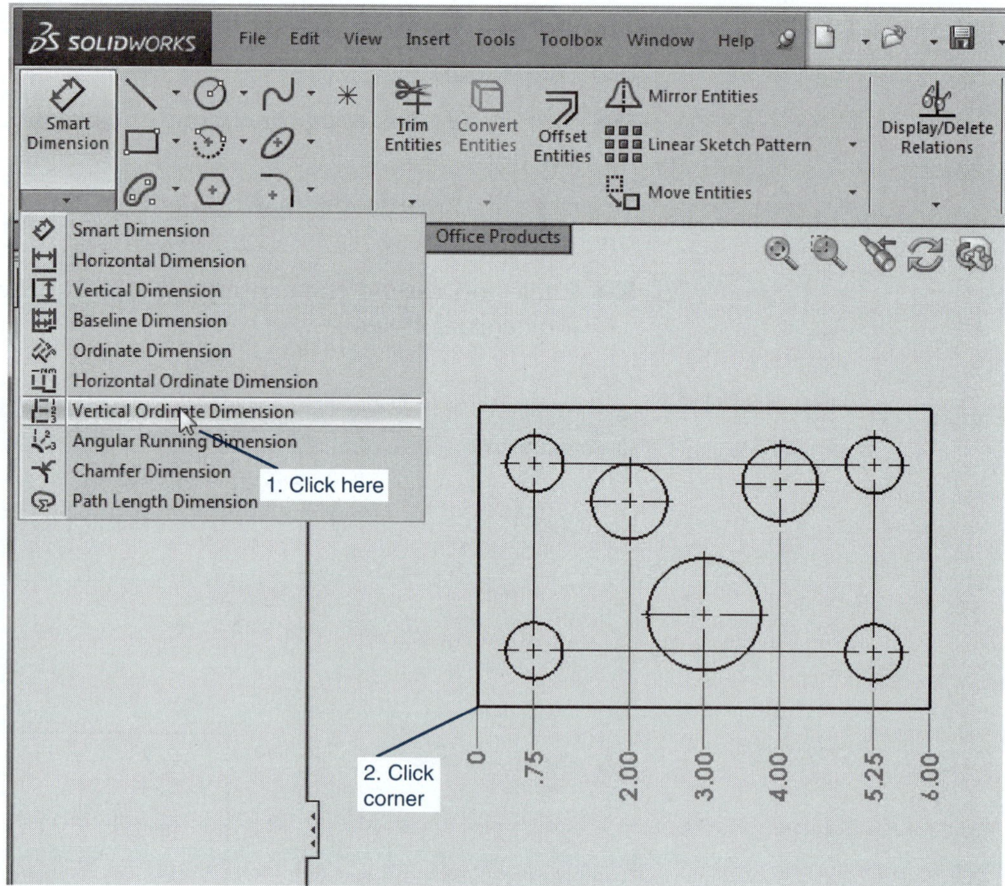

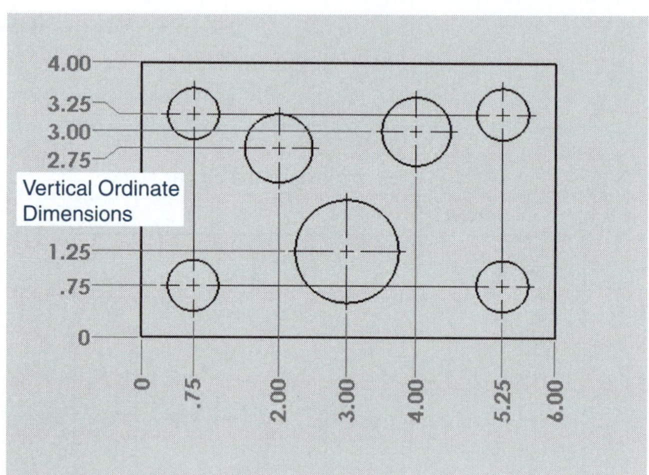

3 Click the arrowhead located under the **Smart Dimension** tool and click the **Horizontal Ordinate Dimension** option.

4 Click the lower left corner of the part to establish the origin for the dimensions.

5 Move the cursor away from the origin and define a location for the "0" dimension.

All other horizontal dimensions will align with this location.

6 Click the lower portion of each hole's vertical centerline and the lower right corner of the part.

7 Click the arrowhead located under the **Smart Dimension** tool, and click the **Vertical Ordinate Dimension** option.

8 Click the lower left corner of the part to establish the origin for the dimensions.

9 Click the left portion of each hole's horizontal centerline and the upper left corner of the part.

10 Add dimensions for the holes.

Figure 7-52 shows the dimensioned part.

Figure 7-52

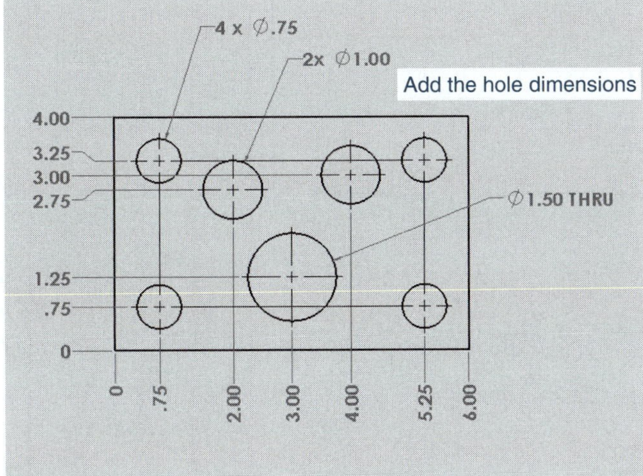

7-10 Baseline Dimensions

Baseline dimensions are a series of dimensions that originate from a common baseline or datum line. Baseline dimensions are very useful because they help eliminate the tolerance buildup that is associated with chain-type dimensions.

To Create Baseline Dimensions

See Figure 7-53.

1 Start a new **Drawing** document and create a top orthographic view of the part.

2 Use the **Linear Center Mark** tool and add connection centerlines between the four corner holes.

3 Click the arrowhead under the **Smart Dimension** tool and click the **Baseline Dimension** option.

4 Click the left vertical edge of the part and the lower portion of the first vertical centerline.

This will establish the baseline.

5 Click the lower portion of each vertical centerline and the right vertical edge line and locate the dimensions.

Figure 7-53

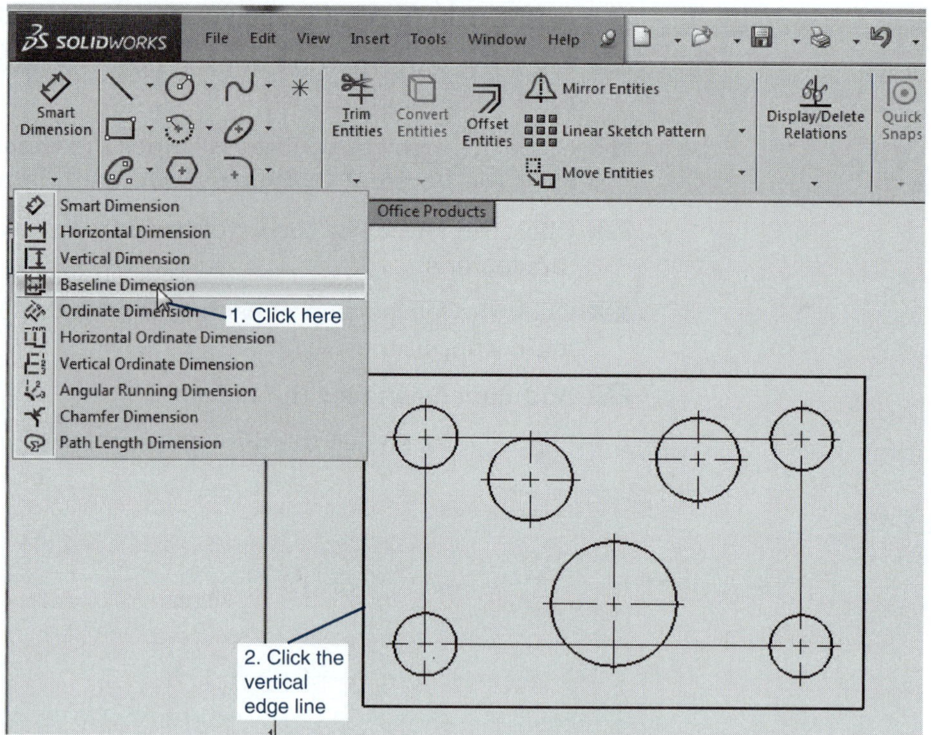
SOLIDWORKS
File Edit View Insert Tools Window Help
Smart Dimension
Trim Entities
Convert Entities
Offset Entities
Mirror Entities
Linear Sketch Pattern
Move Entities
Display/Delete Relations
Quick Snaps
Office Products
Smart Dimension
Horizontal Dimension
Vertical Dimension
Baseline Dimension
Ordinate Dimension
Horizontal Ordinate Dimension
Vertical Ordinate Dimension
Angular Running Dimension
Chamfer Dimension
Path Length Dimension
1. Click here
2. Click the vertical edge line

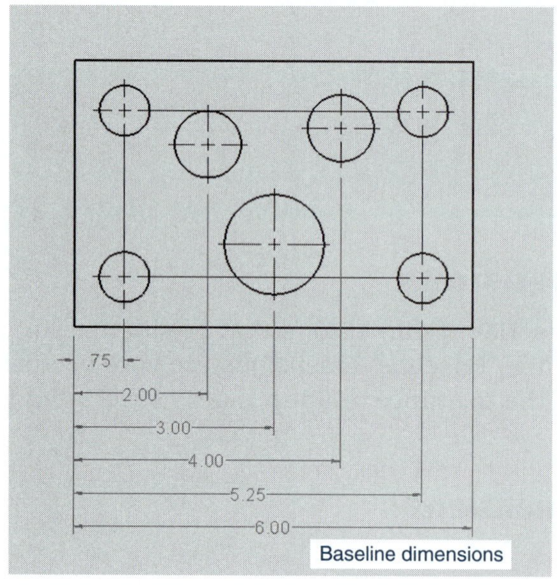
.75
2.00
3.00
4.00
5.25
6.00
Baseline dimensions

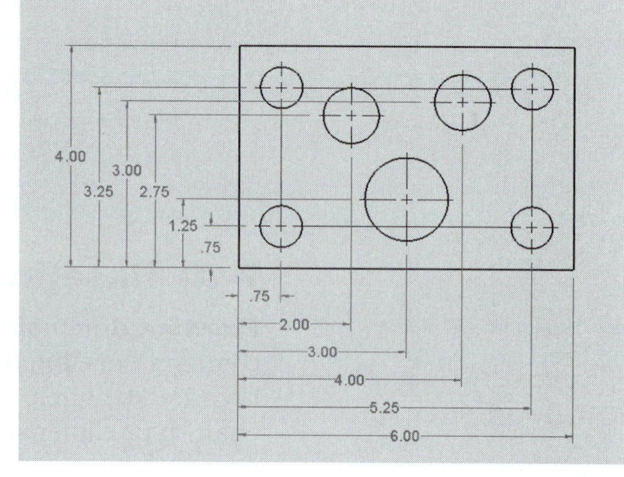
4.00
3.00
3.25
2.75
1.25
.75
.75
2.00
3.00
4.00
5.25
6.00

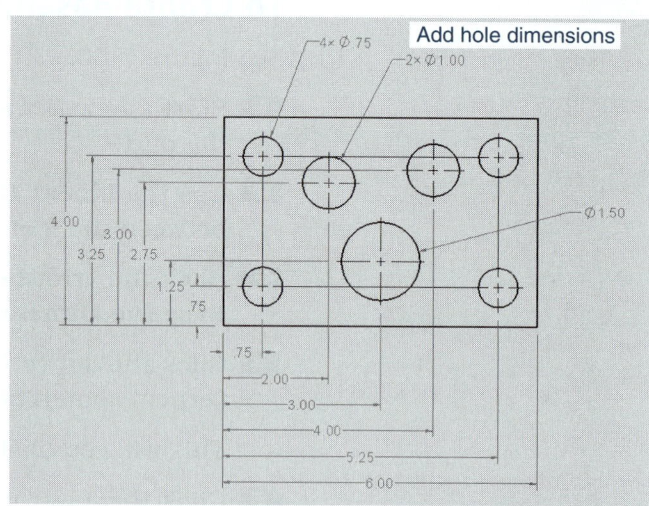
4× Ø.75
2× Ø1.00
Add hole dimensions
Ø1.50
4.00
3.00
3.25
2.75
1.25
.75
.75
2.00
3.00
4.00
5.25
6.00

NOTE

The distance between the dimension lines can be changed in the **Offset distances** box under **Dimensions** on the **Document Properties** tab of the **Options** tool.

6 Click the arrowhead under the **Smart Dimension** tool and click the **Baseline Dimension** option.

7 Click the lower horizontal edge of the part and the left end of the first horizontal centerline.

8 Click the left end of each horizontal centerline and the right top horizontal edge line.

The alignment of the vertical dimension lines can be changed by right-clicking the individual dimension and selecting the **Break Alignment** option.

9 Add the hole dimensions.

Hole Tables

Hole tables are a method for dimensioning parts that have large numbers of holes where standard dimensioning may be cluttered and difficult to read. See Figure 7-54.

Figure 7-54

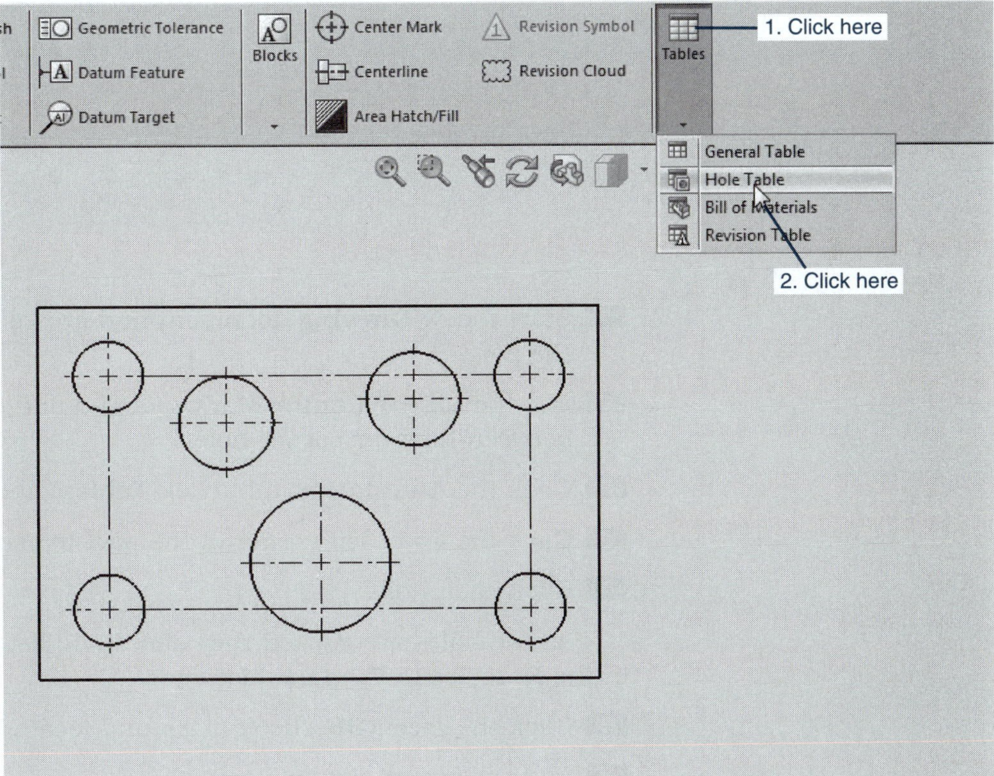

Figure 7-54
(Continued)

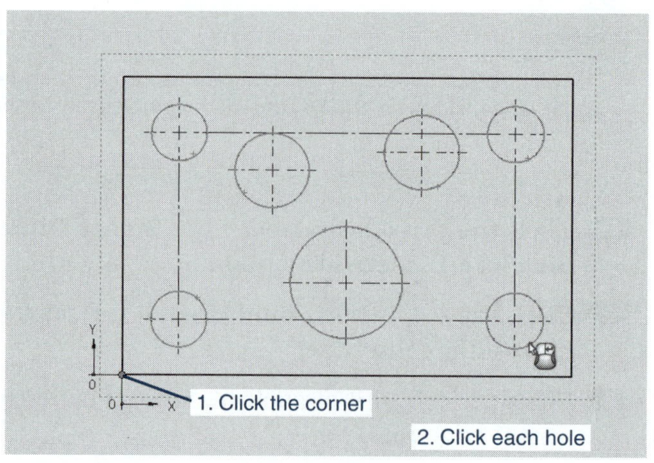

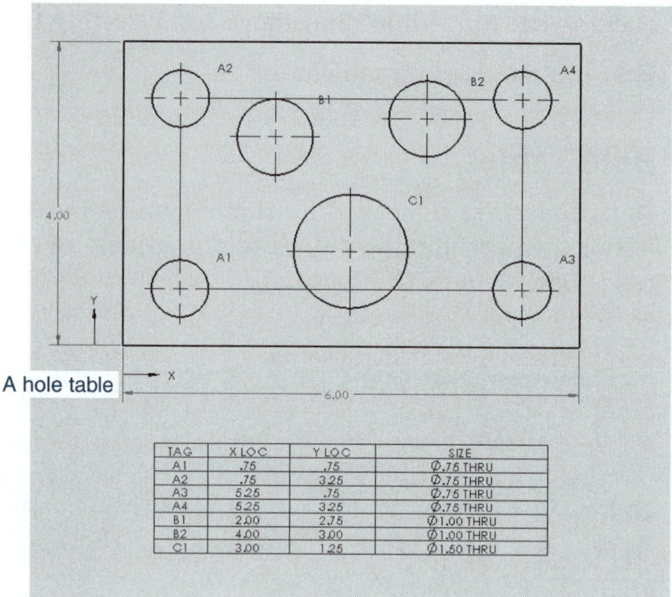

TAG	X LOC	Y LOC	SIZE
A1	.75	.75	Ø.75 THRU
A2	.75	3.25	Ø.75 THRU
A3	5.25	.75	Ø.75 THRU
A4	5.25	3.25	Ø.75 THRU
B1	2.00	2.75	Ø1.00 THRU
B2	4.00	3.00	Ø1.00 THRU
C1	3.00	1.25	Ø1.50 THRU

1. Start a new **Drawing** document and create a top orthographic view of the part.
2. Use the **Linear Center Mark** tool and add connection centerlines between the four corner holes.
3. Click the **Annotation** tab, click **Tables,** and click **Hole Table.**
4. Click the lower left corner of the part to establish an orign.
5. Click each hole.

As the holes are clicked they should be listed in the **Holes** box located in the **Hole Table PropertyManager.**

6. Click the green **OK** check mark and locate the hole table.
7. Add the overall dimensions.
8. Move the hole tags as needed to present a clear, easy-to-read drawing.

In this example all tags were located to the upper right of the holes they define. Tables can be edited using the instructions presented in Section 5-11 for BOMs.

7-11 Locating Dimensions

There are eight general rules concerning the location of dimensions. See Figure 7-55.

Figure 7-55

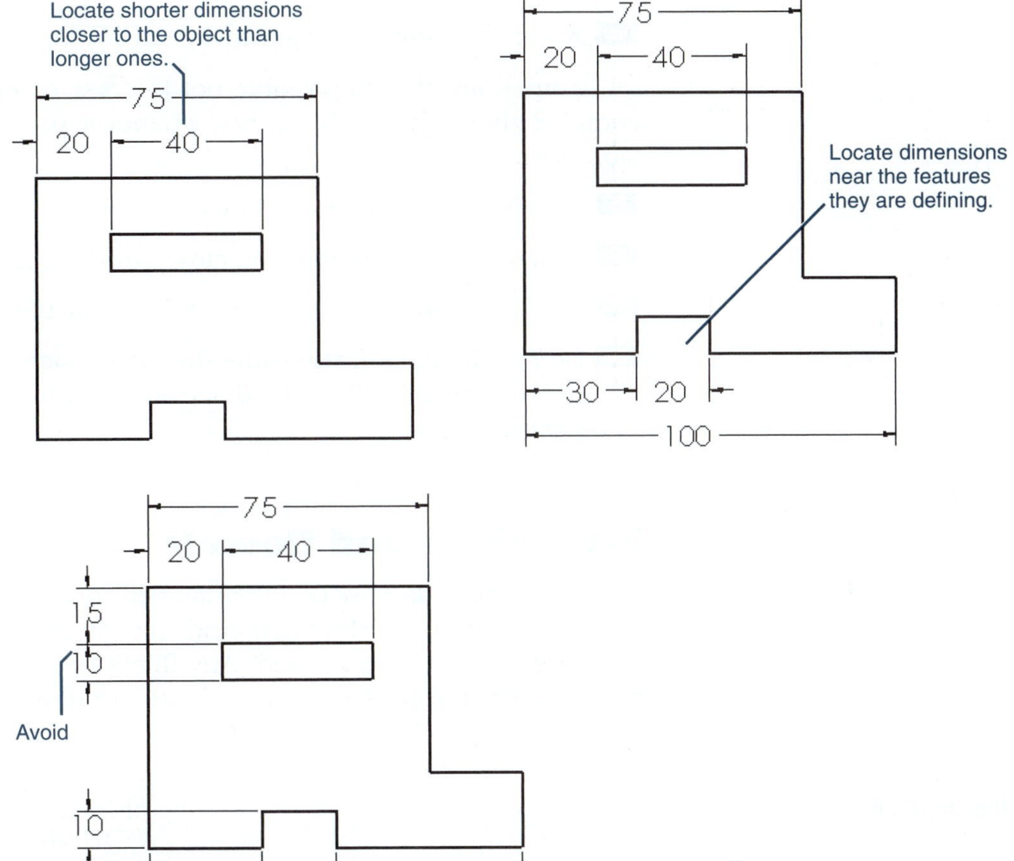

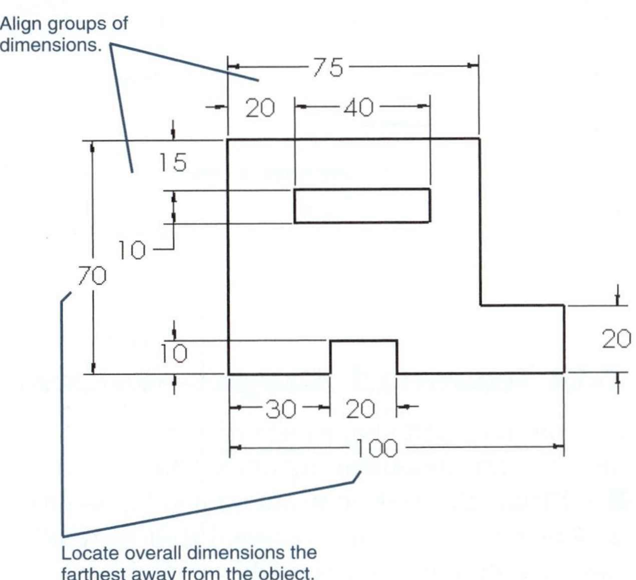

1 Locate dimensions near the features they are defining.

2 Do not locate dimensions on the surface of the object.

3 Align and group dimensions so that they are neat and easy to understand.

4 Avoid crossing extension lines.

Sometimes it is impossible not to cross extension lines because of the complex shape of the object, but whenever possible, avoid crossing extension lines.

5 Do not cross dimension lines.

6 Locate shorter dimensions closer to the object than longer ones.

7 Always locate overall dimensions the farthest away from the object.

8 Do not dimension the same distance twice. This is called *double dimensioning* and will be discussed in Chapter 8 in association with tolerancing.

7-12 Fillets and Rounds

Fillets and rounds may be dimensioned individually or by a note. In many design situations all the fillets and rounds are the same size, so a note as shown in Figure 7-56 is used. Any fillets or rounds that have a different radius from that specified by the note are dimensioned individually.

Figure 7-56

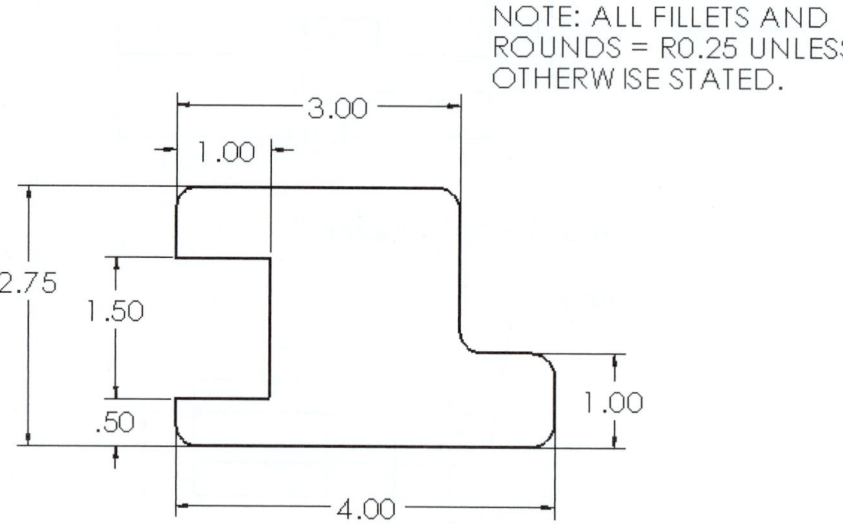

7-13 Rounded Shapes—Internal

Internal rounded shapes are called ***slots.*** Figure 7-57 shows three different methods for dimensioning slots. The end radii are indicated by the note **R - 2 PLACES**, but no numerical value is given. The width of the slot is dimensioned, and it is assumed that the radius of the rounded ends is exactly half of the stated width.

Figure 7-57

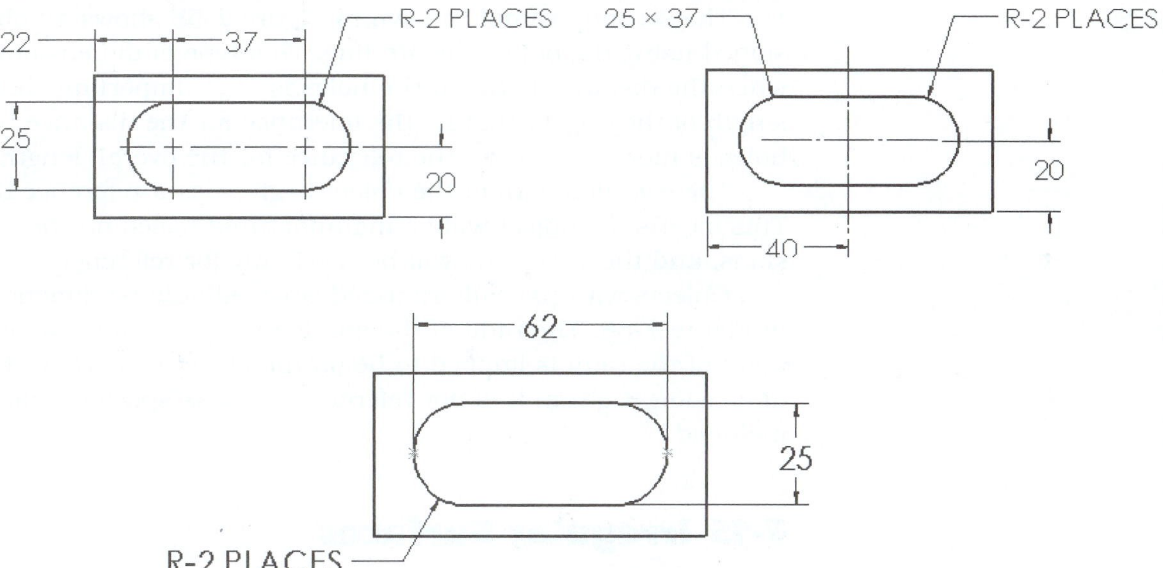

7-14 Rounded Shapes—External

Figure 7-58 shows two shapes with external rounded ends. As with internal rounded shapes, the end radii are indicated, but no value is given. The width of the object is given, and the radius of the rounded end is assumed to be exactly half of the stated width.

Figure 7-58

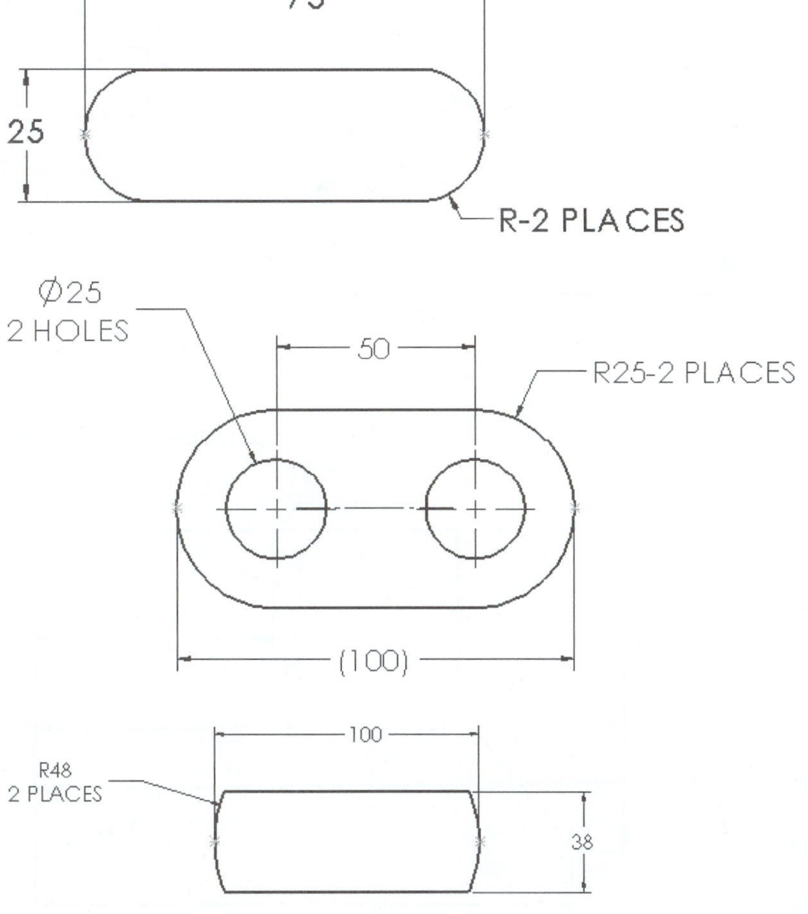

The second example shown in Figure 7-58 shows an object dimensioned using the object's centerline. This type of dimensioning is done when the distance between the holes is more important than the overall length of the object; that is, the tolerance for the distance between the holes is more exact than the tolerance for the overall length of the object.

The overall length of the object is given as a reference dimension (100). This means the object will be manufactured based on the other dimensions, and the 100 value will be used only for reference.

Objects with partially rounded edges should be dimensioned as shown in Figure 7-58. The radii of the end features are dimensioned. The center point of the radii is implied to be on the object centerline. The overall dimension is given; it is not referenced unless specific radii values are included.

7-15 Irregular Surfaces

There are three different methods for dimensioning irregular surfaces: tabular, baseline, and baseline with oblique extension lines. Figure 7-59 shows an irregular surface dimensioned using the tabular method. An XY axis is defined using the edges of the object. Points are then defined relative to the XY axis. The points are assigned reference numbers, and the reference numbers and XY coordinate values are listed in chart form as shown.

Figure 7-59

STATION	1	2	3	4	5	6
X	0	20	40	55	62	70
Y	40	38	30	16	10	0

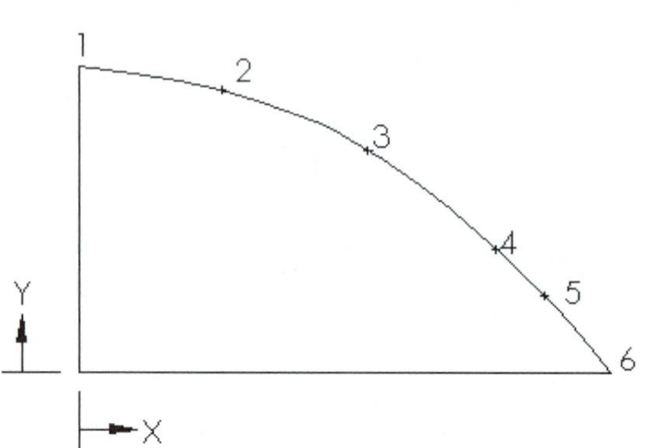

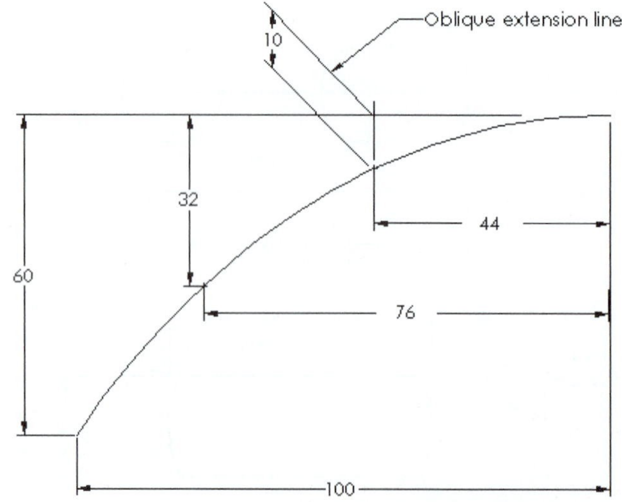

Figure 7-60 shows an irregular curve dimensioned using baseline dimensions. The baseline method references all dimensions to specified baselines. Usually there are two baselines, one horizontal and one vertical.

It is considered poor practice to use a centerline as a baseline. Centerlines are imaginary lines that do not exist on the object and would make it more difficult to manufacture and inspect the finished objects.

Baseline dimensioning is very common because it helps eliminate tolerance buildup and is easily adaptable to many manufacturing processes.

Figure 7-60

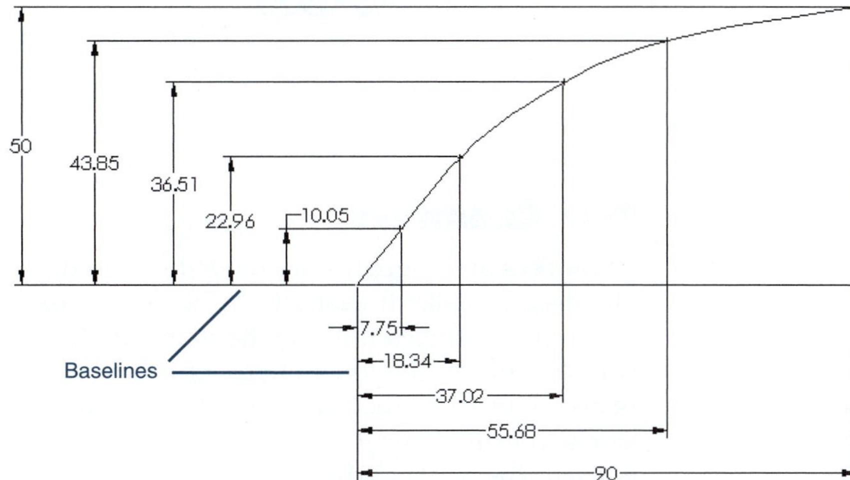

7-16 Polar Dimensions

Polar dimensions are similar to polar coordinates. A location is defined by a radius (distance) and an angle. Figure 7-61 shows an object that includes polar dimensions. The holes are located on a circular centerline, and their positions from the vertical centerline are specified using angles.

Figure 7-61

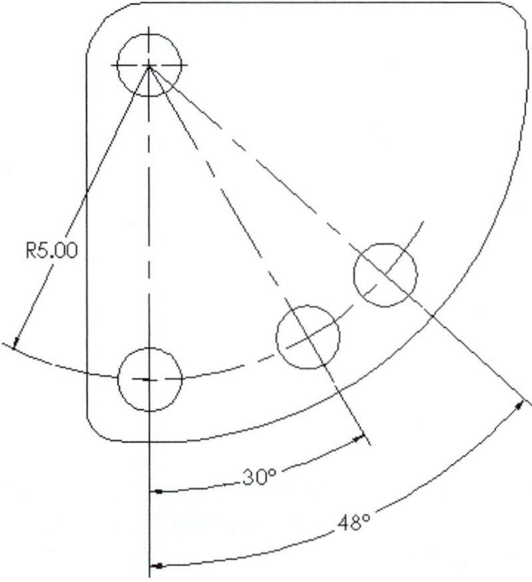

Figure 7-62 shows an example of a hole pattern dimensioned using polar dimensions.

Figure 7-62

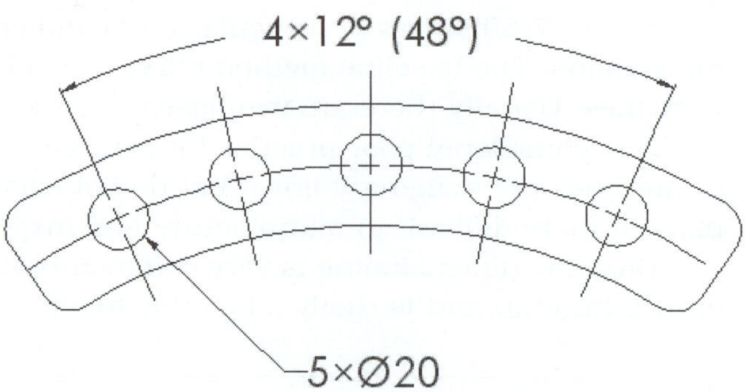

7-17 Chamfers

Chamfers are angular cuts made on the edges of objects. They are usually used to make it easier to fit two parts together. They are most often made at 45° angles but may be made at any angle. Figure 7-63 shows two objects with chamfers between surfaces 90° apart and two examples between surfaces that are not 90° apart. Either of the two types of dimensions shown for the 45° dimension may be used. If an angle other than 45° is used, the angle and setback distance must be specified.

Figure 7-63

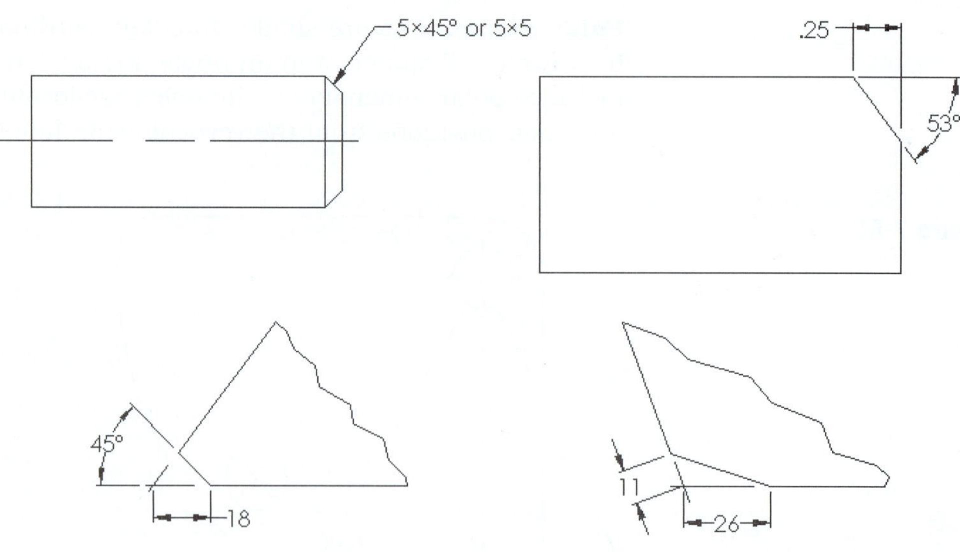

Figure 7-64 shows two examples of internal chamfers. Both define the chamfer using an angle and diameter. Internal chamfers are very similar to countersunk holes.

Figure 7-64

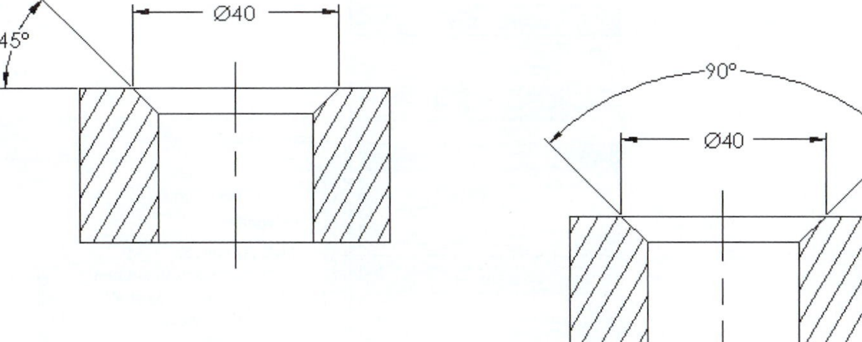

7-18 Symbols and Abbreviations

Symbols are used in dimensioning to help accurately display the meaning of the dimension. Symbols also help eliminate language barriers when reading drawings.

Abbreviations should be used very carefully on drawings. Whenever possible, write out the full word including correct punctuation. The **Dimension manager** includes a group of symbols and words commonly used on technical drawings. Figure 7-65 lists several standard abbreviations used on technical drawings.

Figure 7-65

AL = Aluminum
C'BORE = Counterbore
CRS = Cold Rolled Steel
CSK = Countersink
DIA = Diameter
EQ = Equal
HEX = Hexagon
MAT'L = Material
R = Radius
SAE = Society of Automotive Engineers
SFACE = Spotface
ST = Steel
SQ = Square
REQD = Required

Figure 7-66 shows a list of symbols available in the **Dimension Value PropertyManager**.

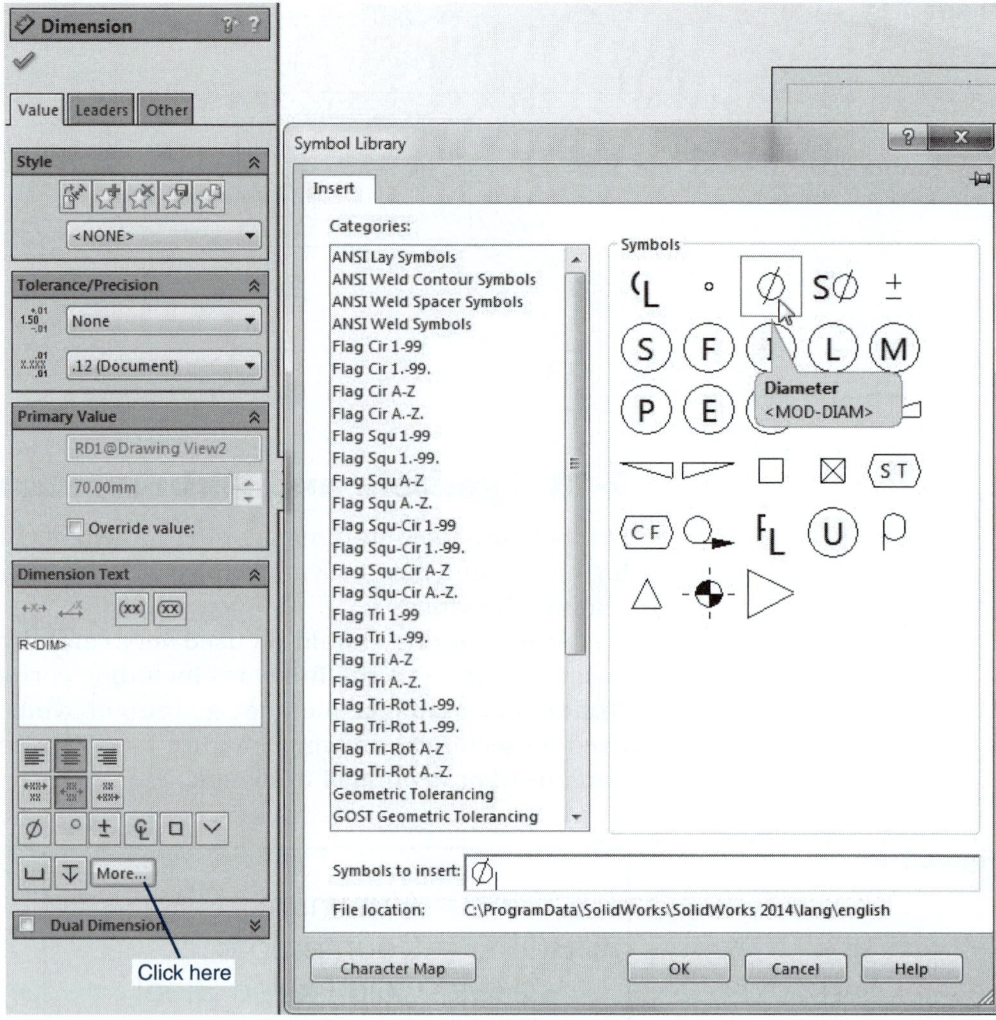

Figure 7-66

> **TIP**
>
> To access the **Dimension Value PropertyManager,** click an existing dimension.

More symbols are available by clicking the **More** box. A list of available symbols will appear. Click a new symbol. A preview of the selected symbol will appear. Click **OK** and the symbol will appear on the drawing next to the existing symbol.

7-19 Symmetrical and Centerline Symbols

An object is symmetrical about an axis when one side is the exact mirror image of the other. Figure 7-67 shows a symmetrical object. The two short parallel lines symbol or the note OBJECT IS SYMMETRICAL ABOUT THIS AXIS (centerline) may be used to designate symmetry.

If an object is symmetrical, only half the object need be dimensioned. The other dimensions are implied by the symmetry note or symbol.

Figure 7-67

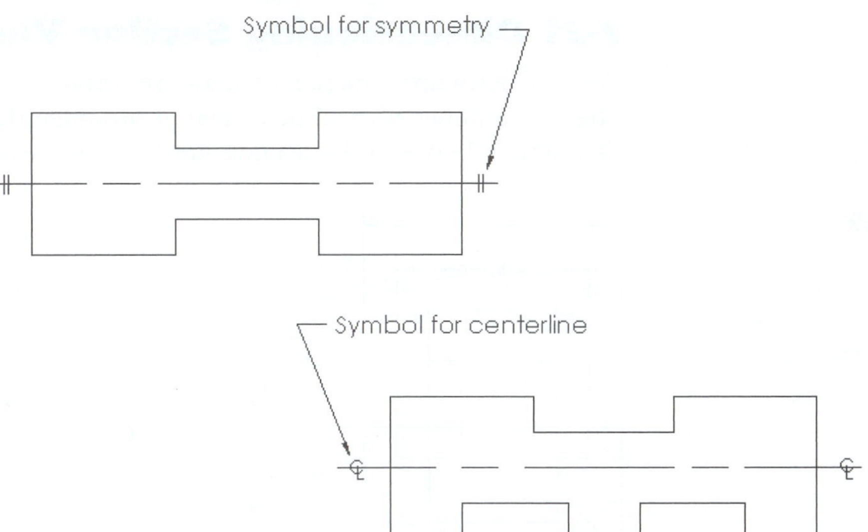

The centerline is slightly different from the axis of symmetry. An object may or may not be symmetrical about its centerline. See Figure 7-67. Centerlines are used to define the center of both individual features and entire objects. Use the centerline symbol when a line is a centerline, but do not use it in place of the symmetry symbol.

7-20 Dimensioning to a Point

Curved surfaces can be dimensioned using theoretical points. See Figure 7-68. There should be a small gap between the surface of the object and the lines used to define the theoretical point. The point should be defined by the intersection of at least two lines.

There should also be a small gap between the extension lines and the theoretical point used to locate the point.

Figure 7-68

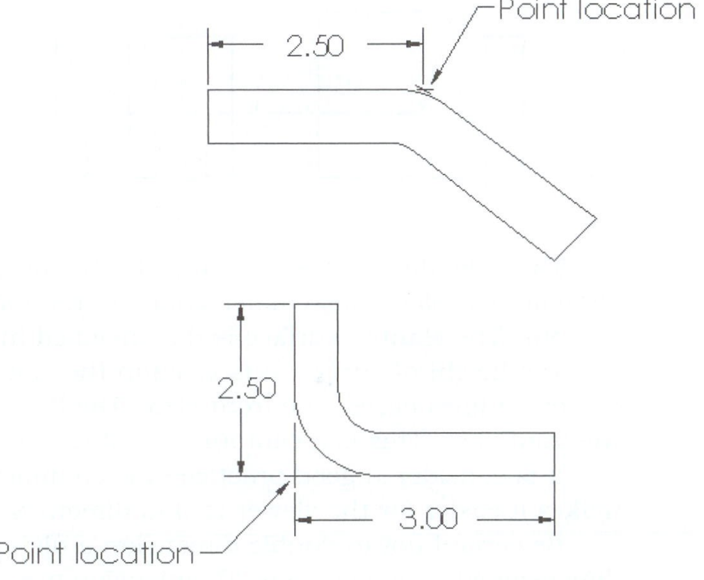

7-21 Dimensioning Section Views

Section views are dimensioned, as are orthographic views. See Figure 7-69. The section lines should be drawn at an angle that allows the viewer to clearly distinguish between the section lines and the extension lines.

Figure 7-69

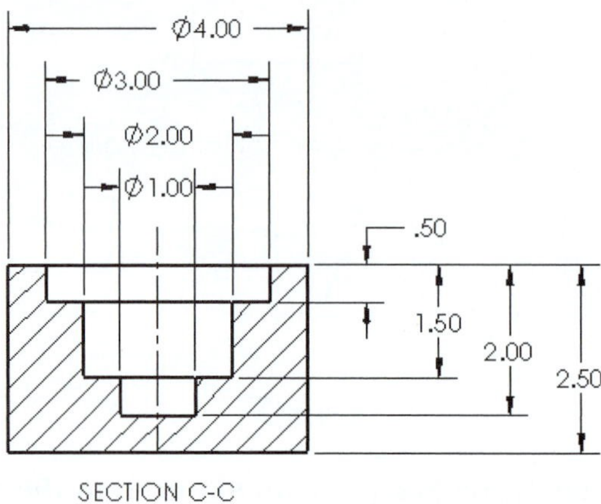

7-22 Dimensioning Orthographic Views

Dimensions should be added to orthographic views where the features appear in contour. Holes should be dimensioned in their circular views. Figure 7-70 shows three views of an object that has been dimensioned.

Figure 7-70

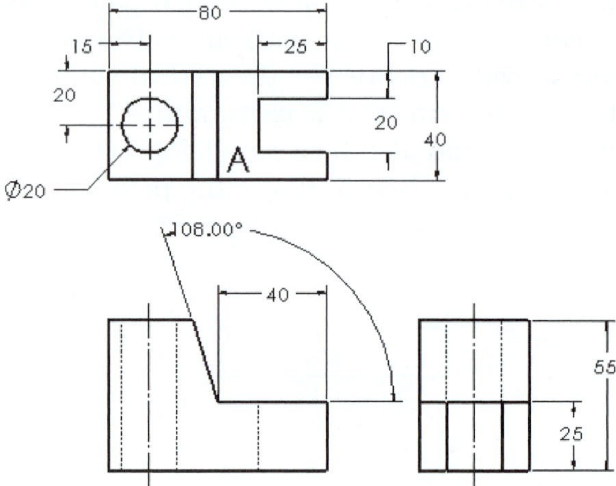

The hole dimensions are added to the top view, where the hole appears circular. The slot is also dimensioned in the top view because it appears in contour. The slanted surface is dimensioned in the front view.

The height of surface A is given in the side view rather than run along extension lines across the front view. The length of surface A is given in the front view. This is a contour view of the surface.

It is considered good practice to keep dimensions in groups. This makes it easier for the viewer to find dimensions.

Be careful not to double dimension a distance. A distance should be dimensioned only once. If a 30 dimension were added above the 25 dimension on the right-side view, it would be an error. The distance would be double dimensioned: once with the 25 + 30 dimension, and again with the

55 overall dimension. The 25 + 30 dimensions are mathematically equal to the 55 overall dimension, but there is a distinct difference in how they affect the manufacturing tolerances. Double dimensions are explained more fully in Chapter 8.

Dimensions Using Centerlines

Figure 7-71 shows an object dimensioned from its centerline. This type of dimensioning is used when the distance between the holes relative to each other is critical.

Figure 7-71

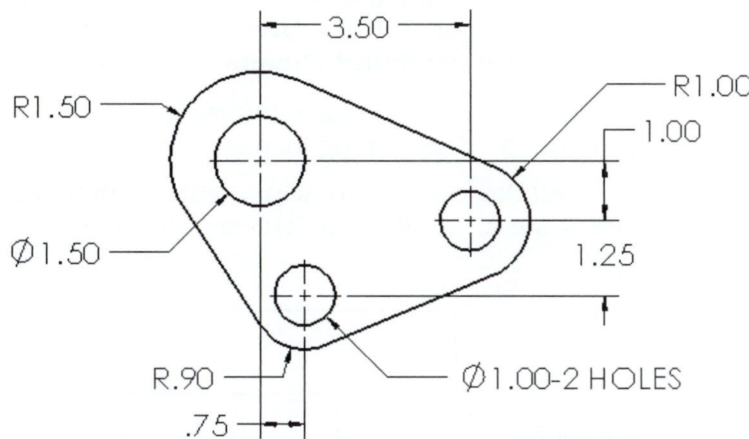

Chapter Projects

Project 7-1:

Measure and redraw the shapes in Figures P7-1 through P7-24. The dotted grid background has either .50-in. or 10-mm spacing. All holes are through holes. Specify the units and scale of the drawing. Use the **Part** template to create a model. Use the grid background pattern to determine the dimensions. Use the **Drawing** template to create the orthographic view shown. Use the **Smart Dimension** tool to dimension the view.

A. Measure using millimeters.
B. Measure using inches.

All dimensions are within either .25 in. or 5 mm. All fillets and rounds are R.50 in., R.25 in. or R10 mm, R5 mm.

THICKNESS:
40 mm
1.50 in.

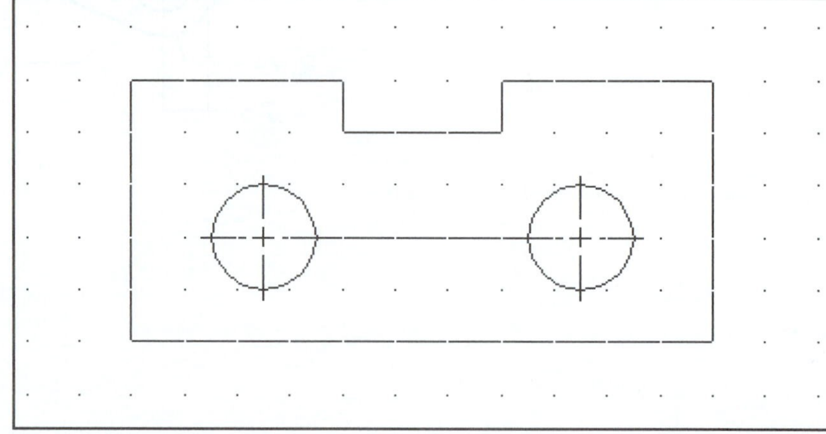

Figure P7-1

THICKNESS:
20 mm
.75 in.

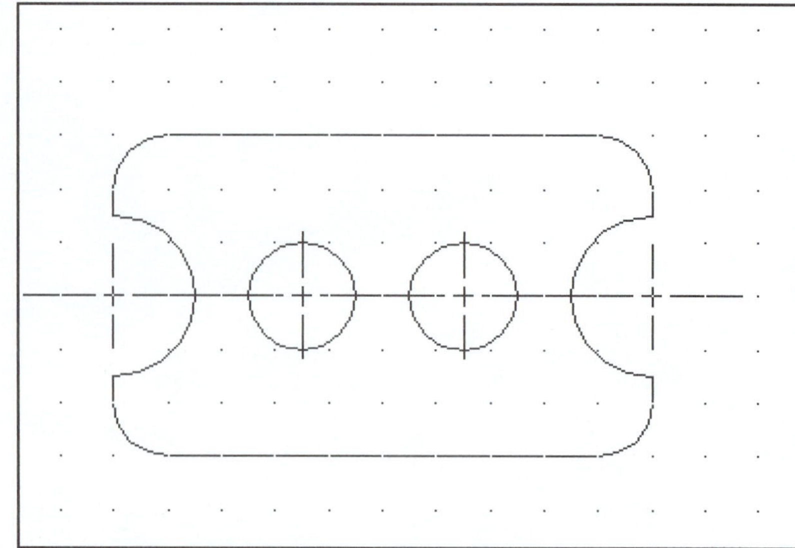

Figure P7-2

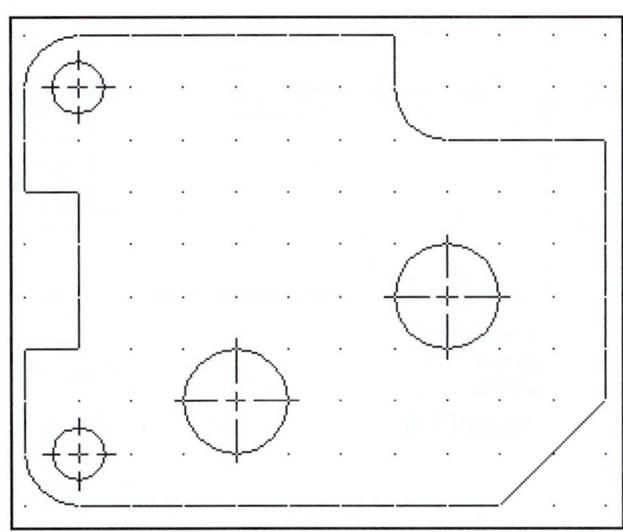

THICKNESS:
35 mm
1.25 in.

Figure P7-3

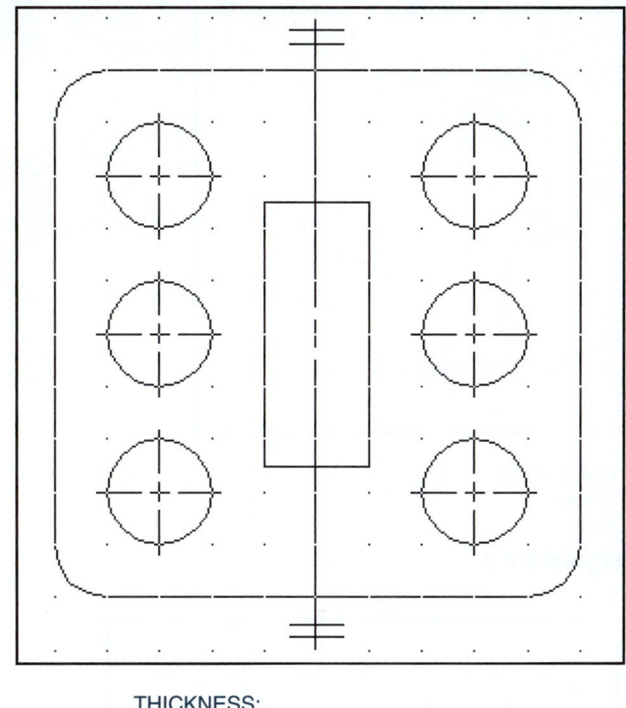

THICKNESS:
15 mm
.50 in.

Figure P7-4

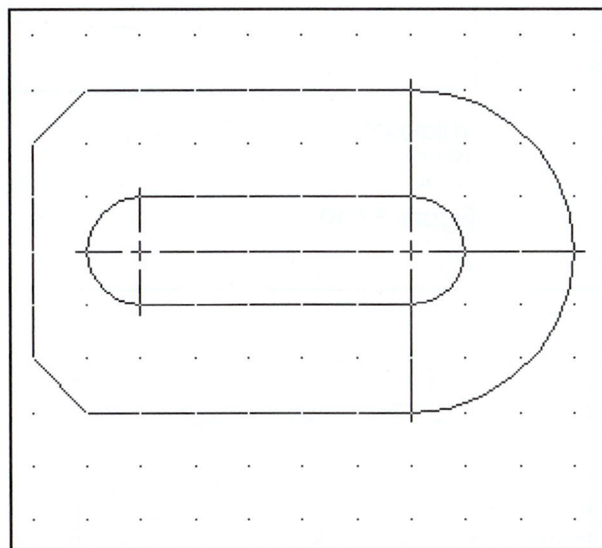

THICKNESS:
10 mm
.50 in.

Figure P7-5

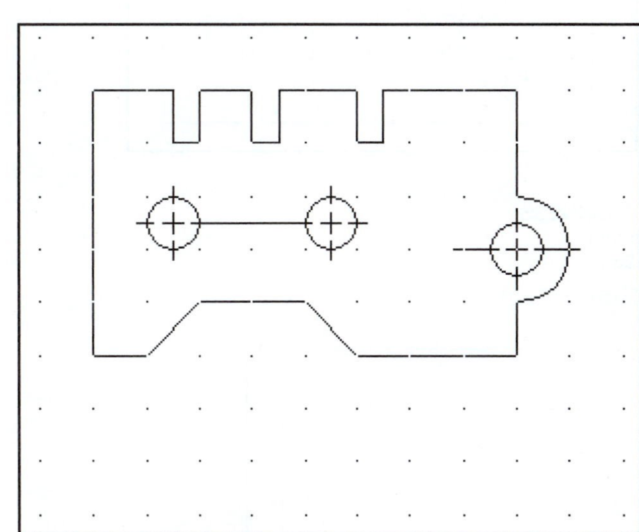

THICKNESS:
5 mm
.25 in.

Figure P7-6

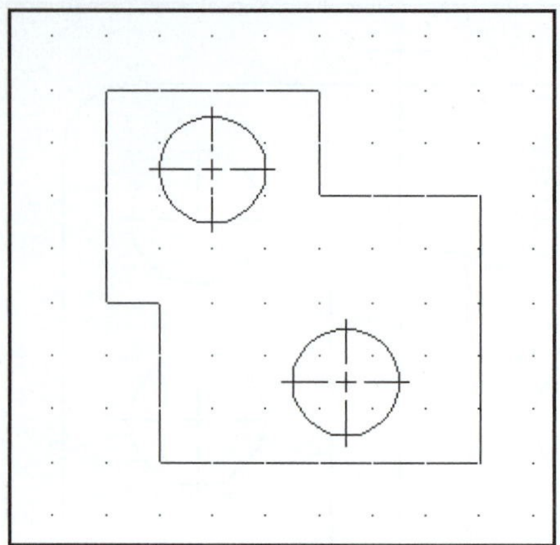

THICKNESS:
12 mm
.50 in.

Figure P7-7

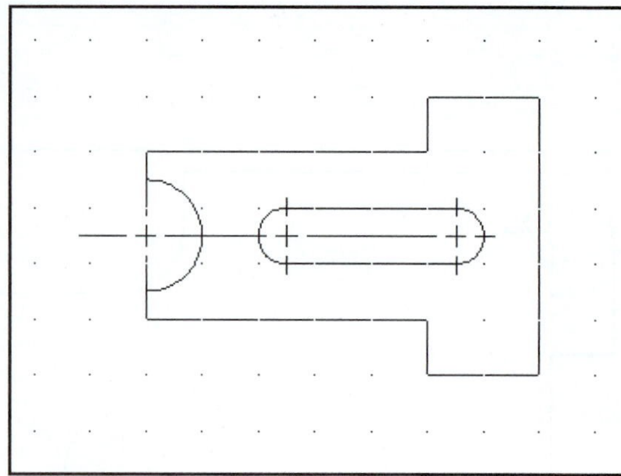

THICKNESS:
25 mm
1.00 in.

Figure P7-8

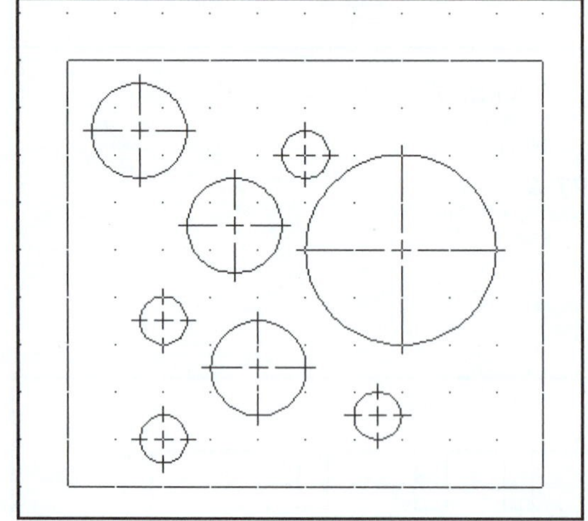

THICKNESS:
5 mm
.25 in.

Figure P7-9

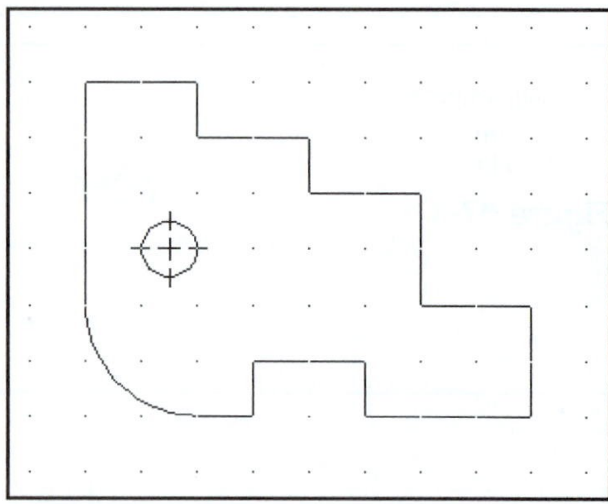

THICKNESS:
20 mm
.75 in.

Figure P7-10

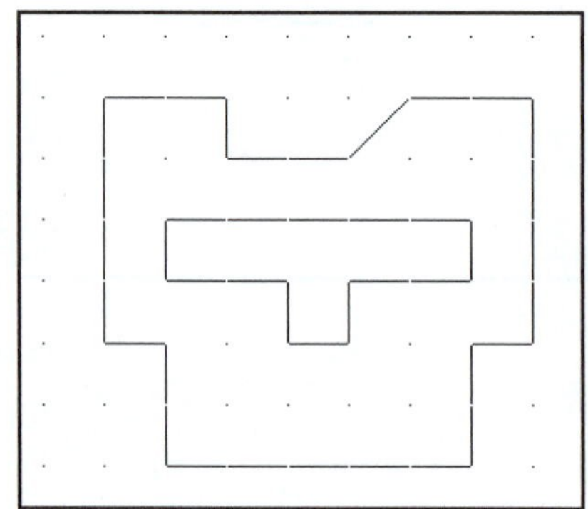

THICKNESS:
18 mm
.625 in.

Figure P7-11

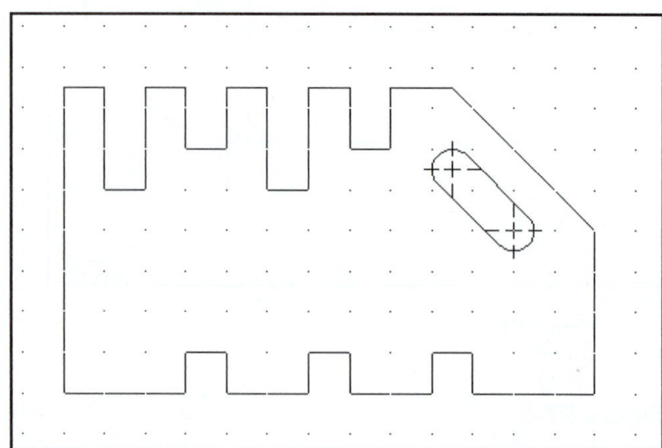

THICKNESS:
24 mm
1.00 in.

Figure P7-12

Figure P7-13

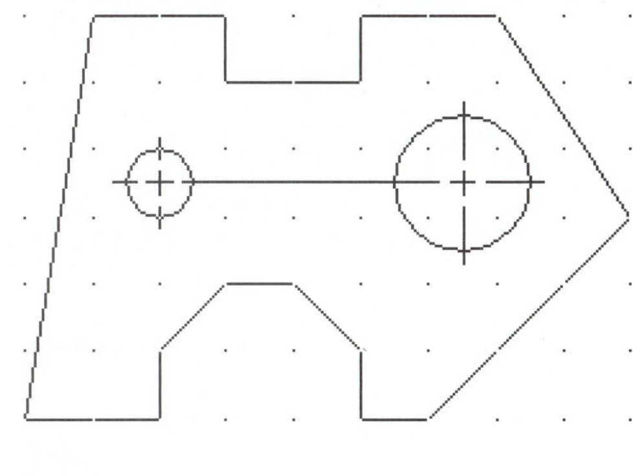

THICKNESS:
10 mm
.25 in.

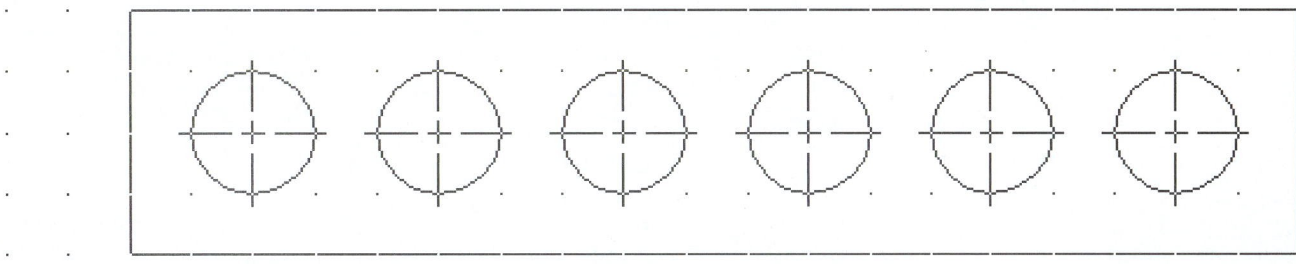

THICKNESS:
8 mm
.25 in.

Figure P7-14

Figure P7-15

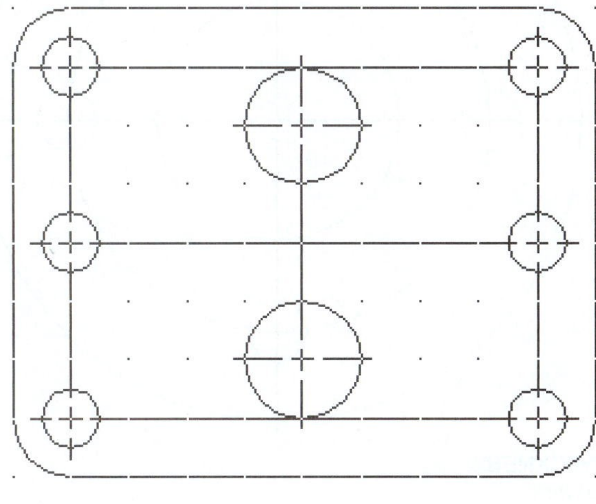

THICKNESS:
20 mm
.75 in.

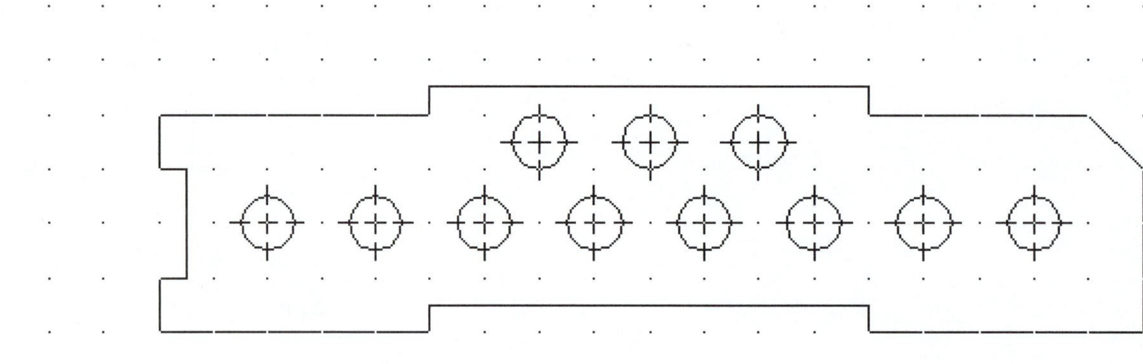

THICKNESS:

Figure P7-16

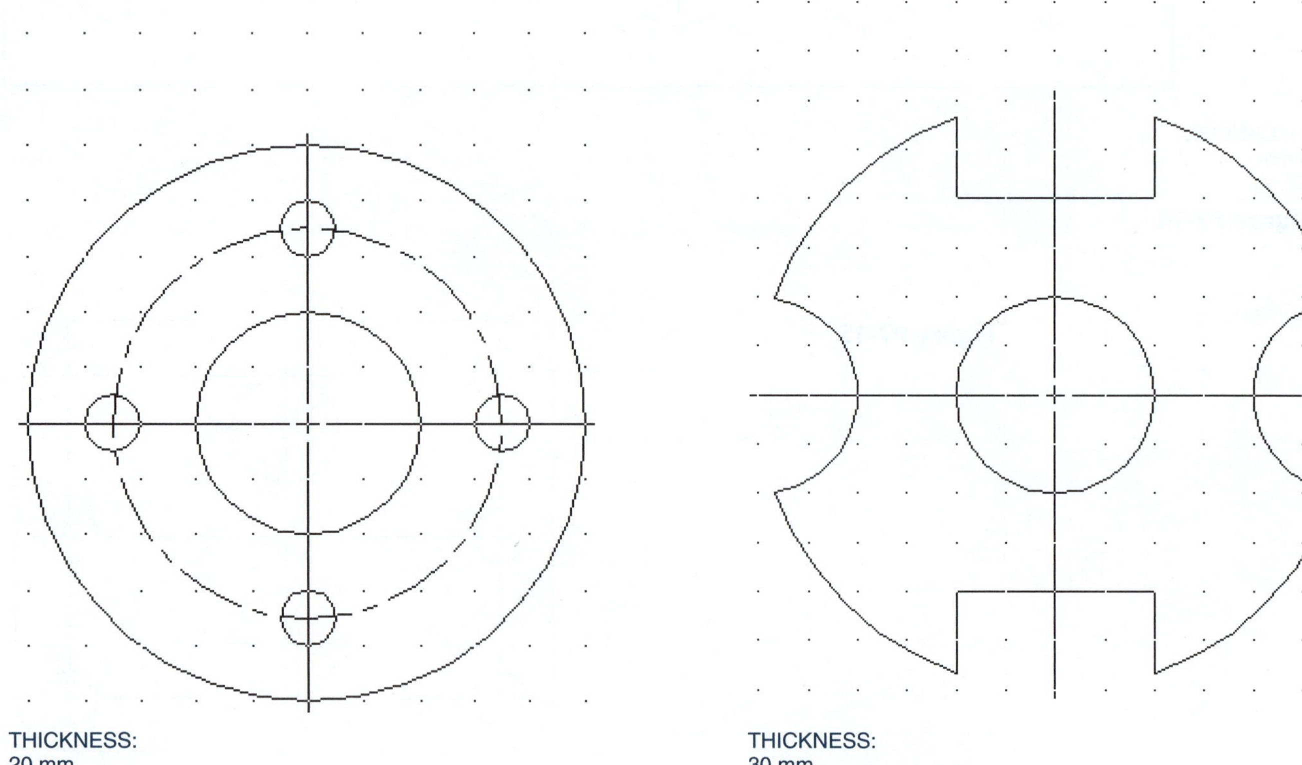

THICKNESS:
20 mm
.75 in.

Figure P7-17

THICKNESS:
30 mm
1.375 in.

Figure P7-18

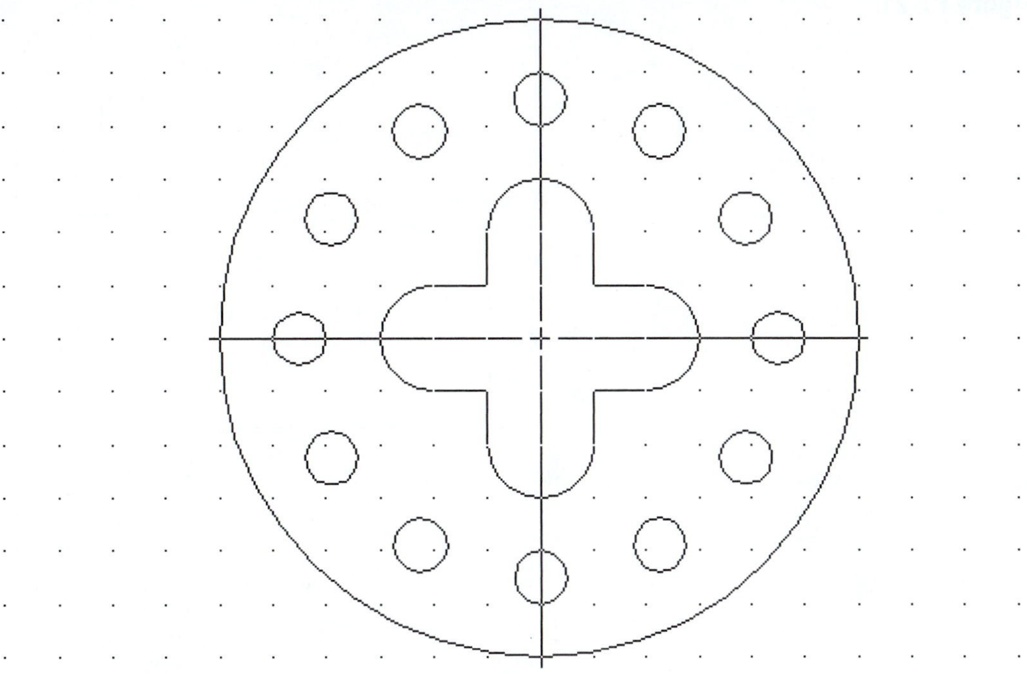

THICKNESS:
12 mm
.30 in.

Figure P7-19

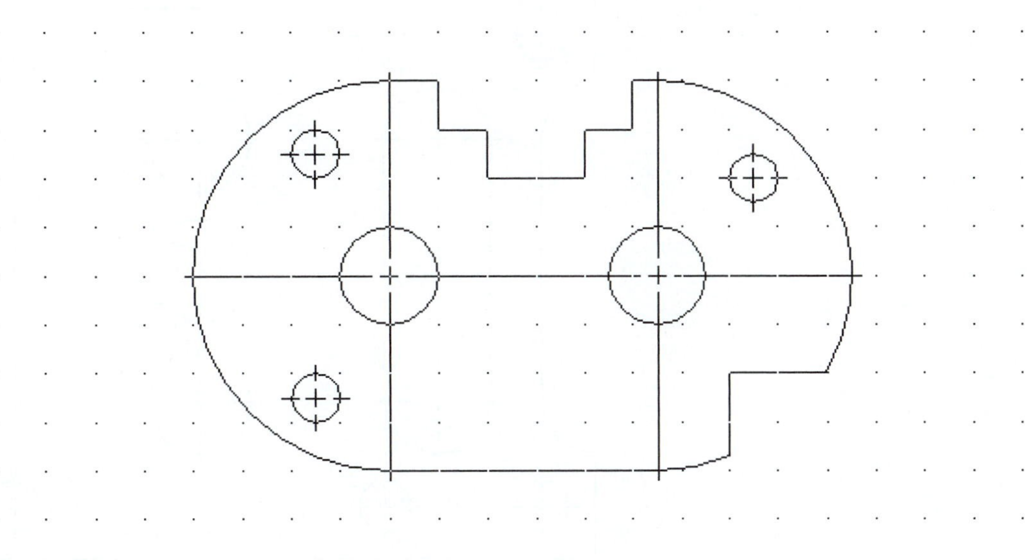

THICKNESS:
5 mm
.125 in.

Figure P7-20

Figure P7-21

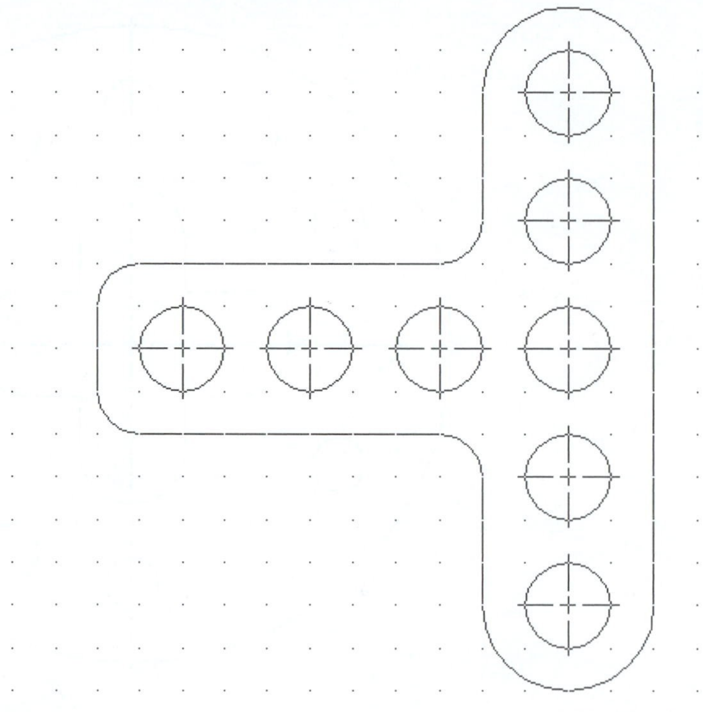

THICKNESS:
10 mm
.25 in.

Dimension using baseline dimensions.

Figure P7-22

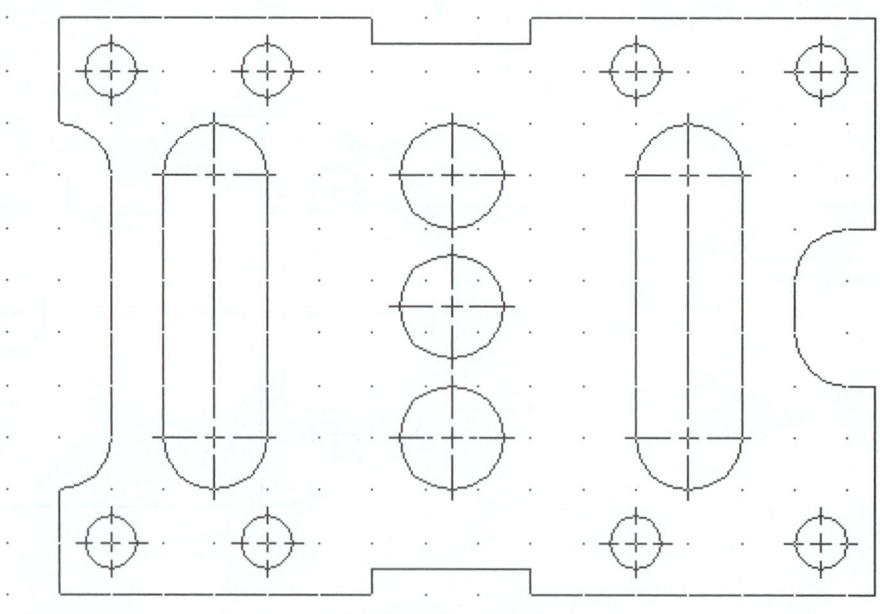

THICKNESS:
15 mm
.50 in.

Dimension using
A. Baseline dimensions.
B. Ordinate dimensions.
C. Chain dimensions.
D. Hole table.

Figure P7-23

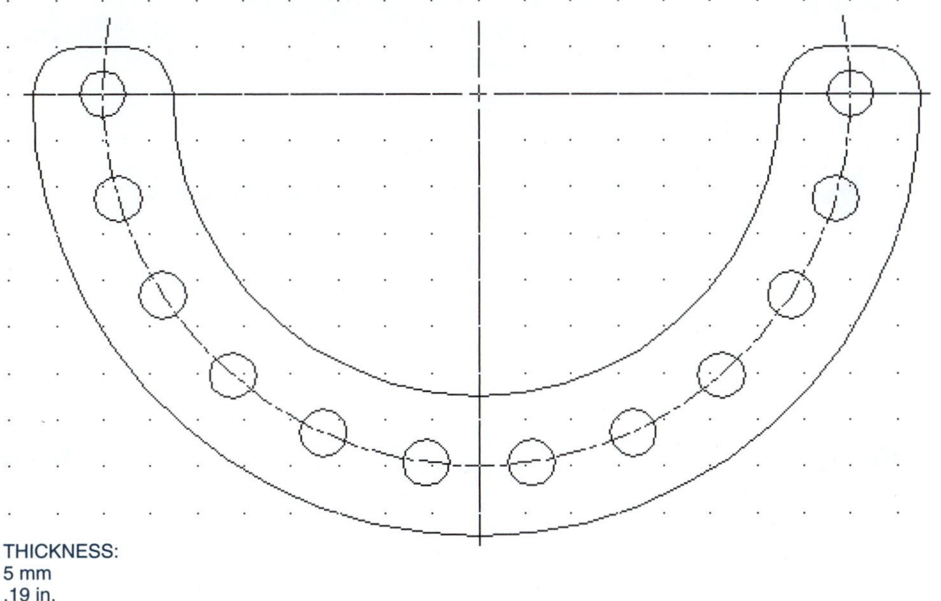

Figure P7-24

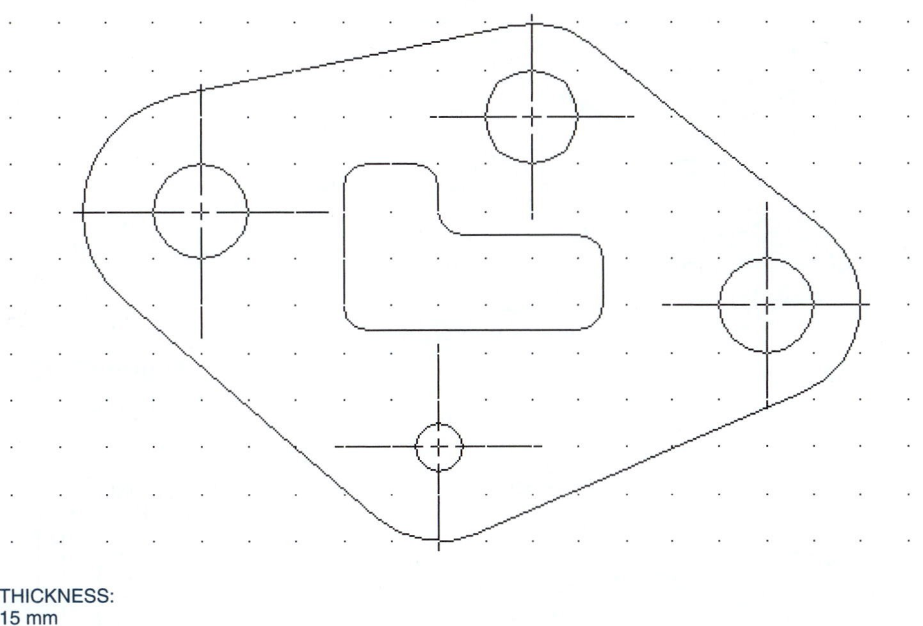

Project 7-2:

Use the **Part** template to draw models of the objects shown in Figures P7-25 through P7-42.

1. Create orthographic views of the objects. Dimension the orthographic views.
2. Create 3D models of the objects. Dimension the 3D models.

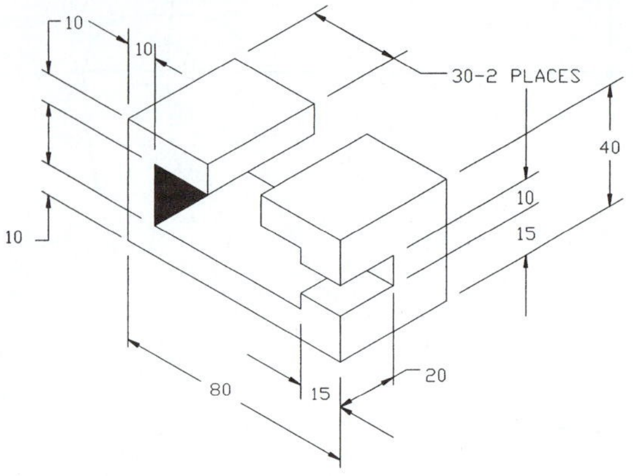

Figure P7-25
MILLIMETERS

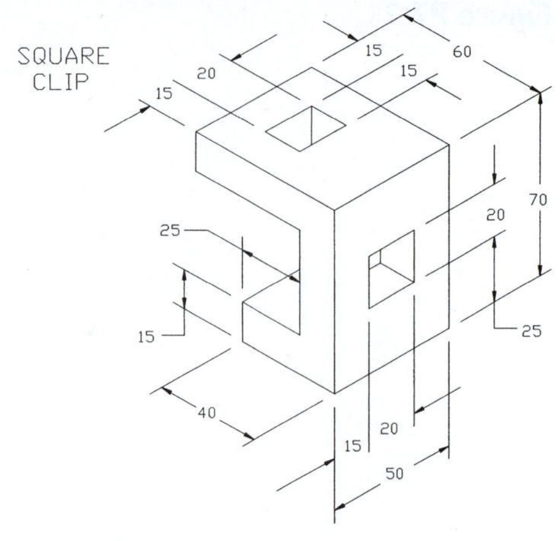

Figure P7-26
MILLIMETERS

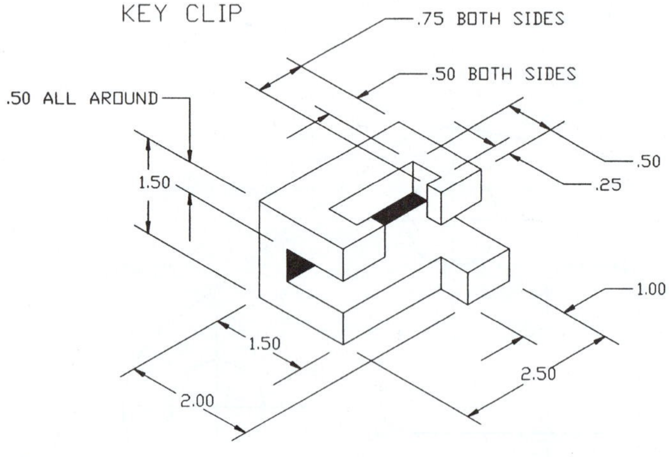

Figure P7-27
INCHES

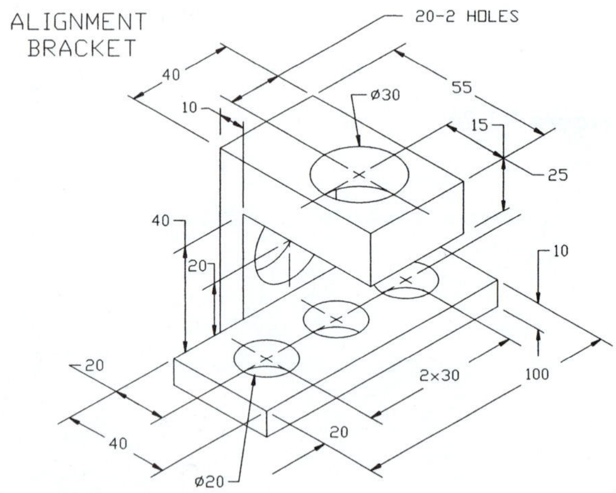

Figure P7-28
MILLIMETERS

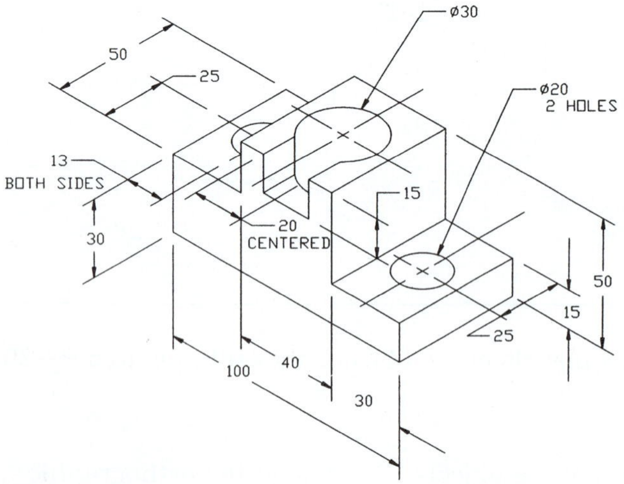

Figure P7-29
MILLIMETERS

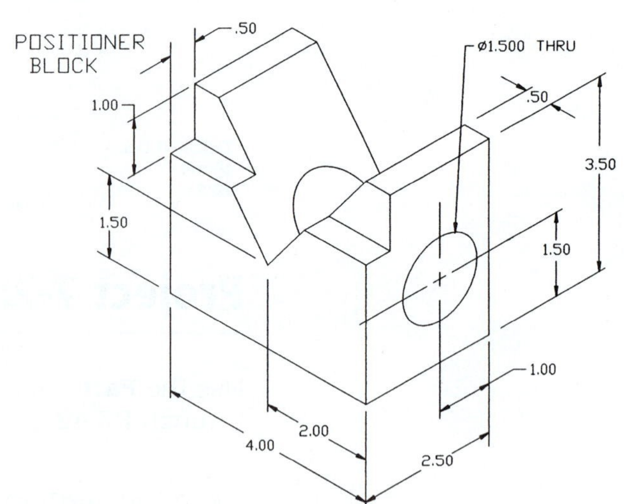

Figure P7-30
INCHES

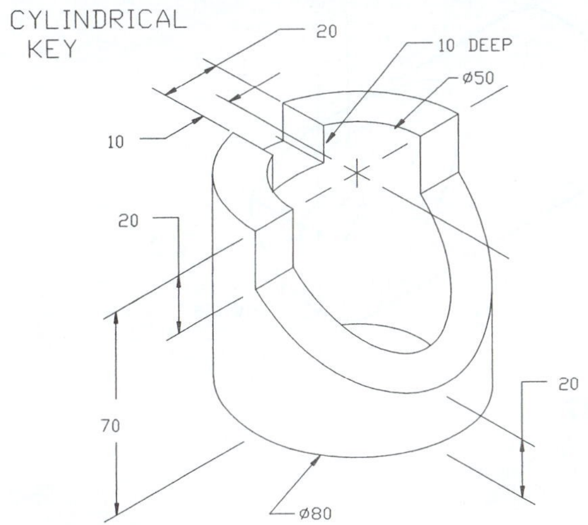

Figure P7-31
MILLIMETERS

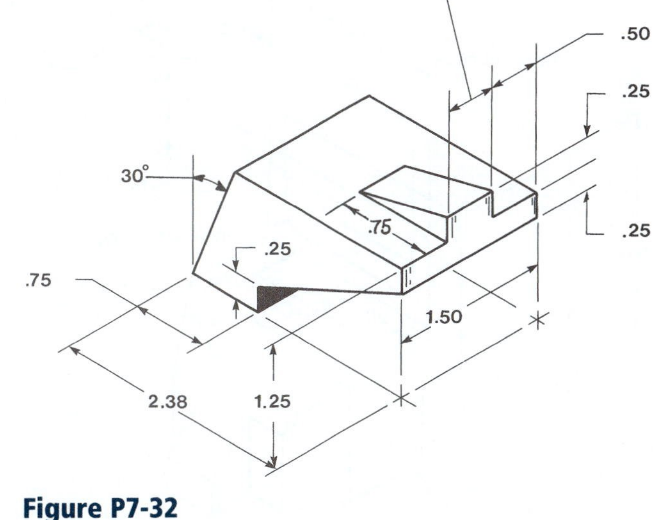

Figure P7-32
INCHES

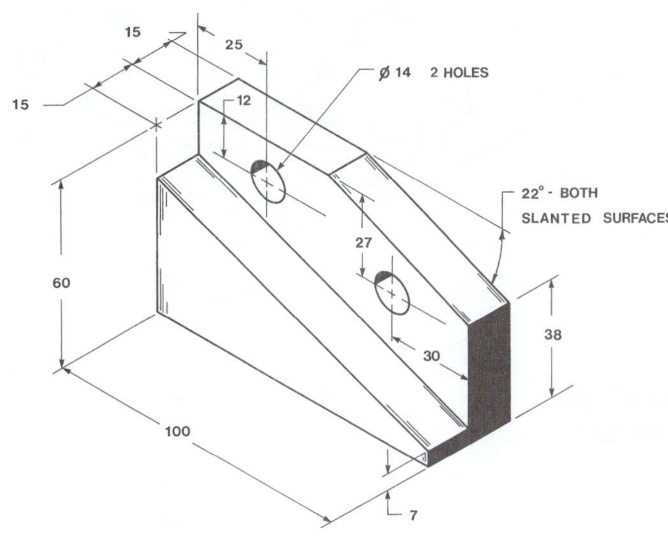

Figure P7-33
MILLIMETERS

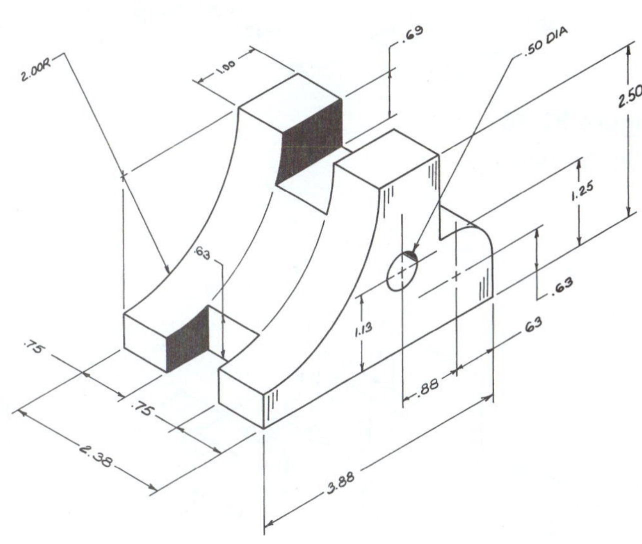

Figure P7-34
INCHES

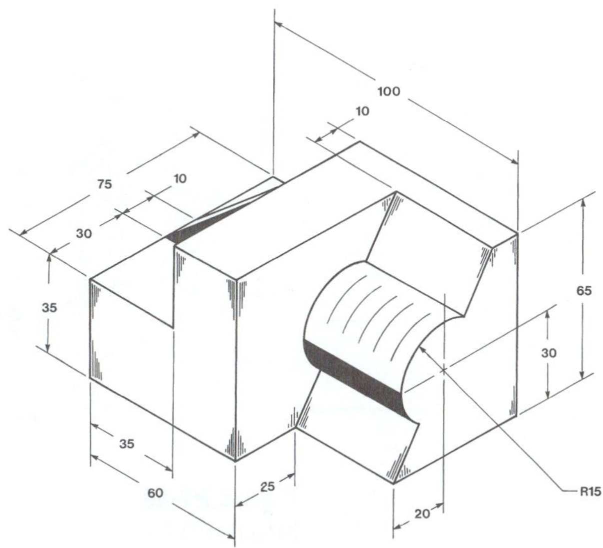

Figure P7-35
MILLIMETERS

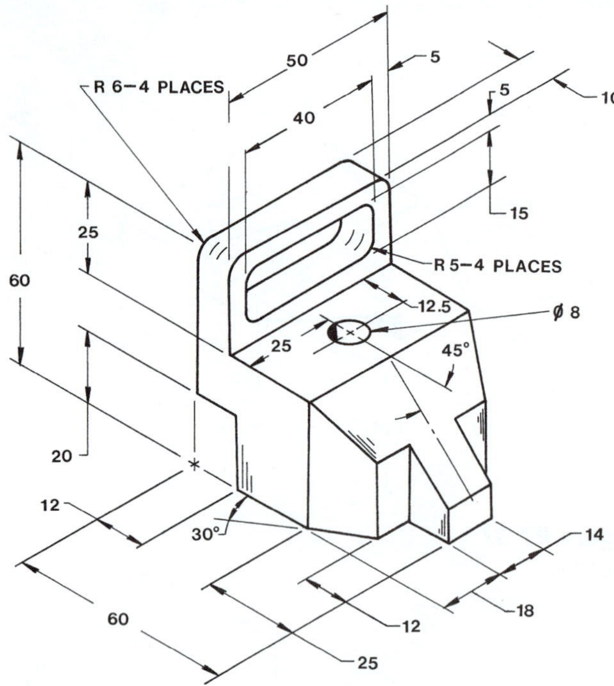

Figure P7-36
MILLIMETERS

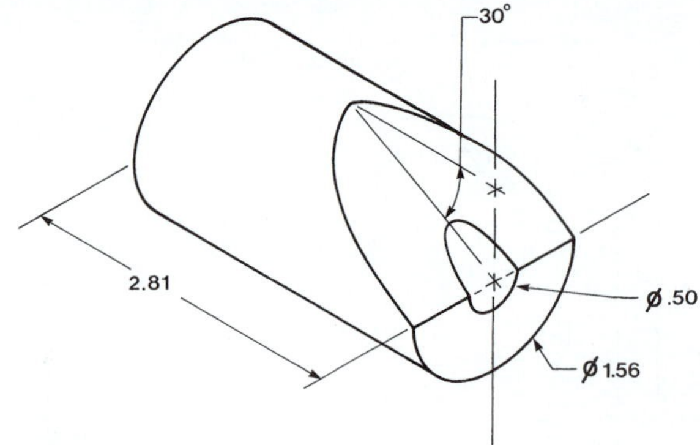

Figure P7-37
INCHES

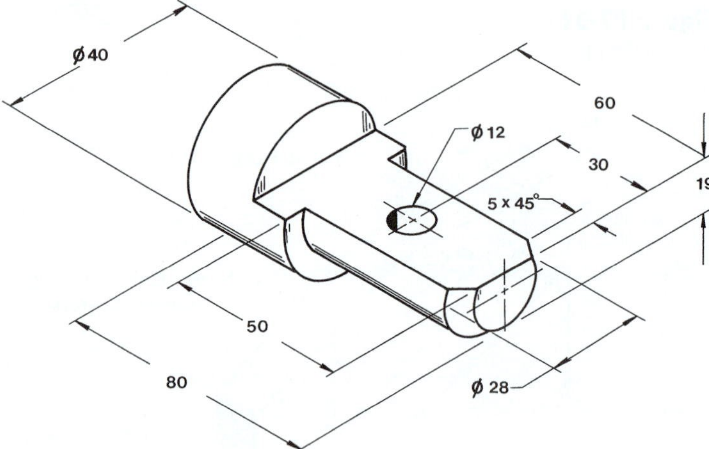

Figure P7-39
MILLIMETERS

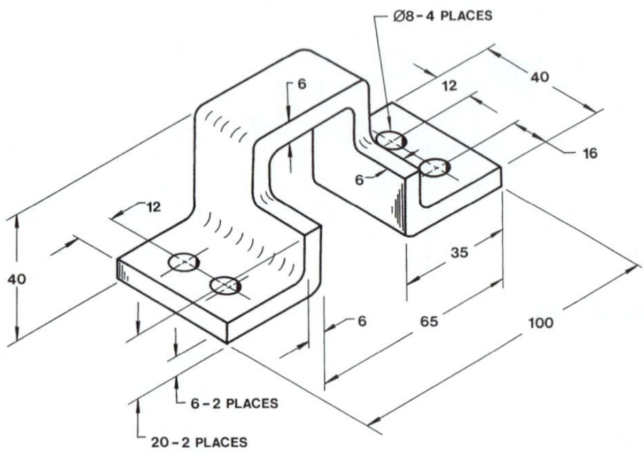

NOTE: ALL FILLET AND ROUNDS=R3

Figure P7-38
MILLIMETERS

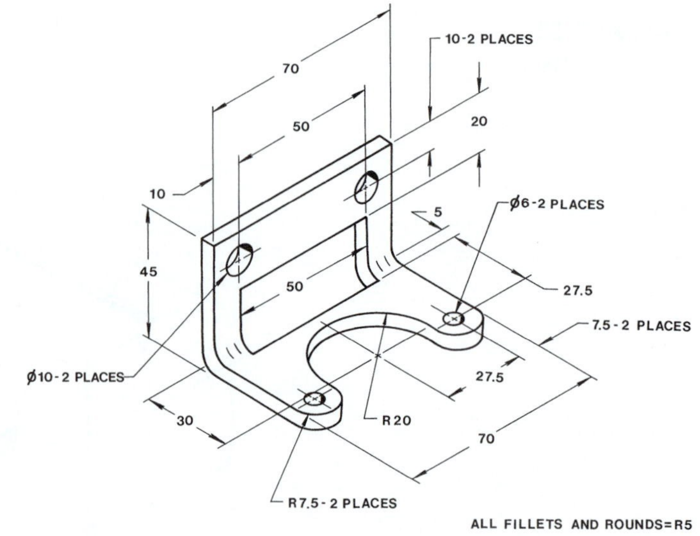

ALL FILLETS AND ROUNDS=R5
MATL 5 THK

Figure P7-40
MILLIMETERS

Figure P7-41
MILLIMETERS

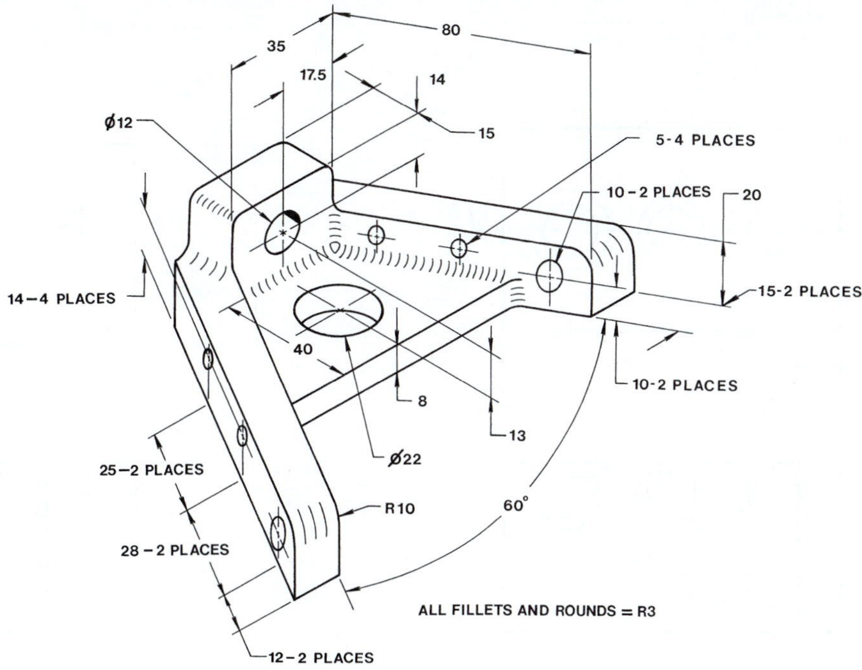

Figure P7-42
MILLIMETERS

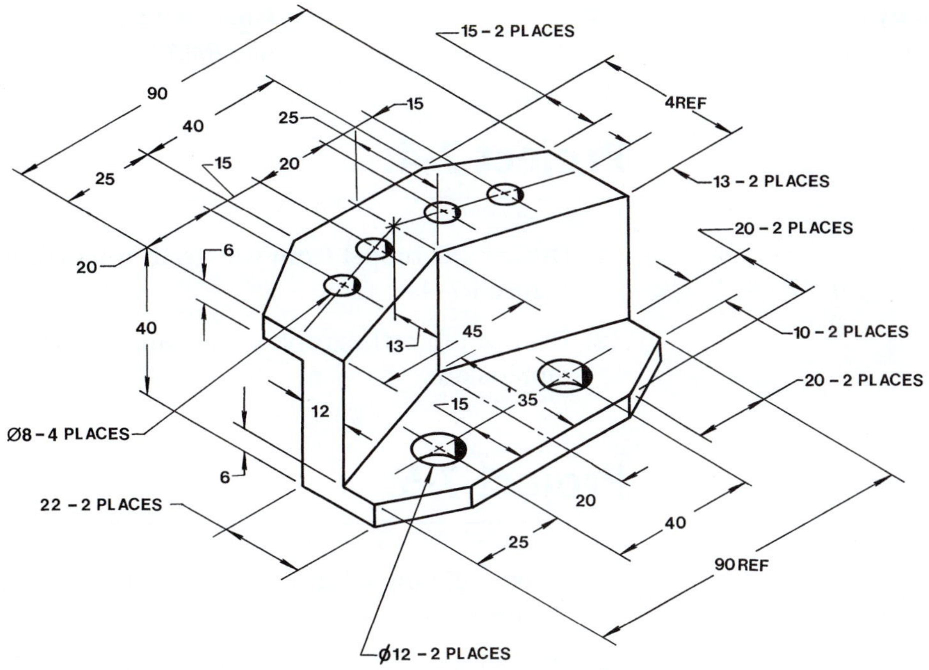

Project 7-3:

1. Draw a 3D model from the given top orthographic and section views in Figure P7-43.
2. Draw a top orthographic view and a section view of the object and add dimensions.

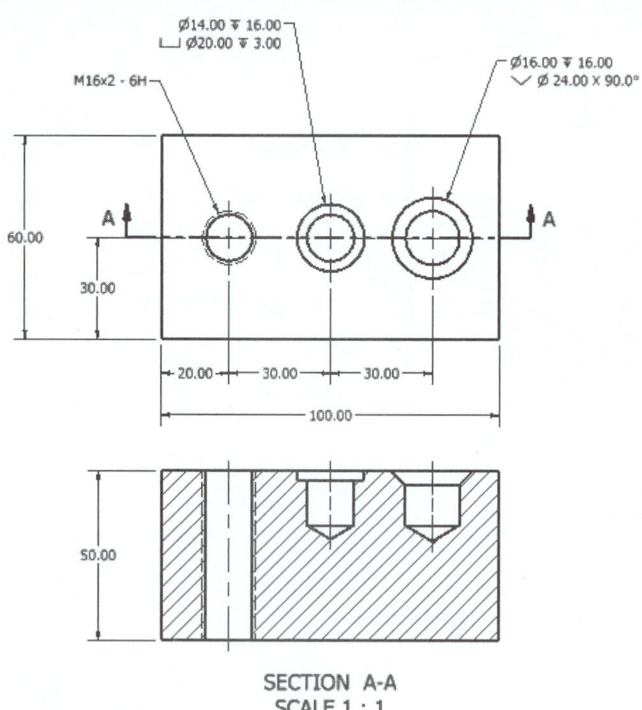

Figure P7-43
MILLIMETERS

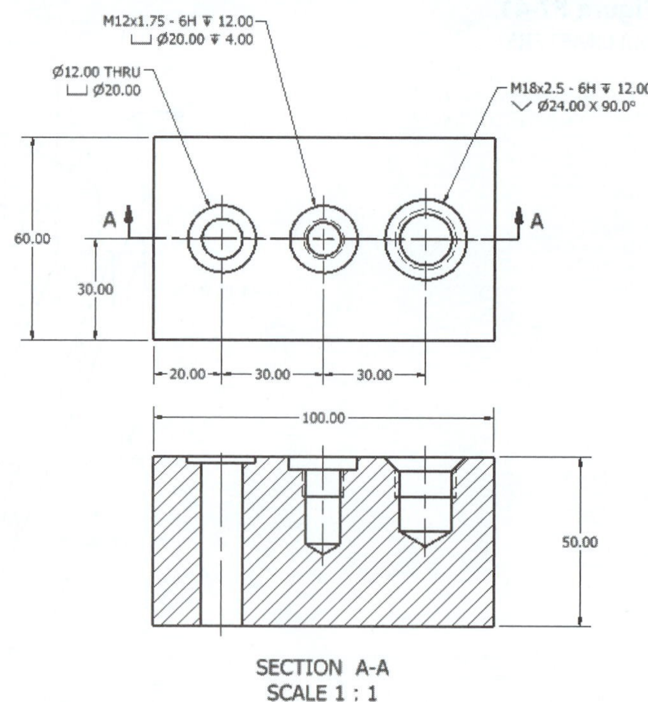

Figure P7-44
MILLIMETERS

Project 7-4:

1. Draw a 3D model from the given top orthographic and section views in Figure P7-44.
2. Draw a top orthographic view and a section view of the object and add dimensions.

Project 7-5:

1. Draw a 3D model from the given top orthographic and section views in Figure P7-45.
2. Draw a top orthographic view and a section view of the object and add dimensions.

Project 7-6:

1. Draw a 3D model from the given top orthographic and section views in Figure P7-46.
2. Draw a top orthographic view and a section view of the object and add dimensions.

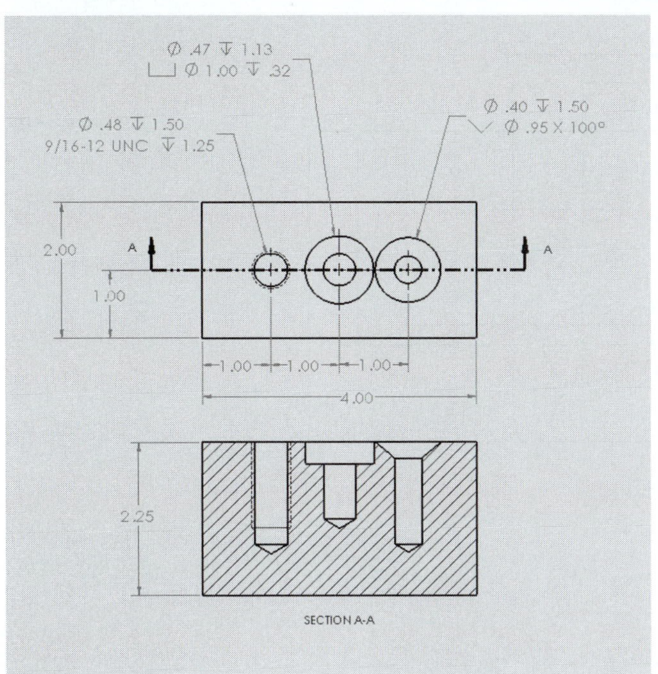

Figure P7-45
INCHES

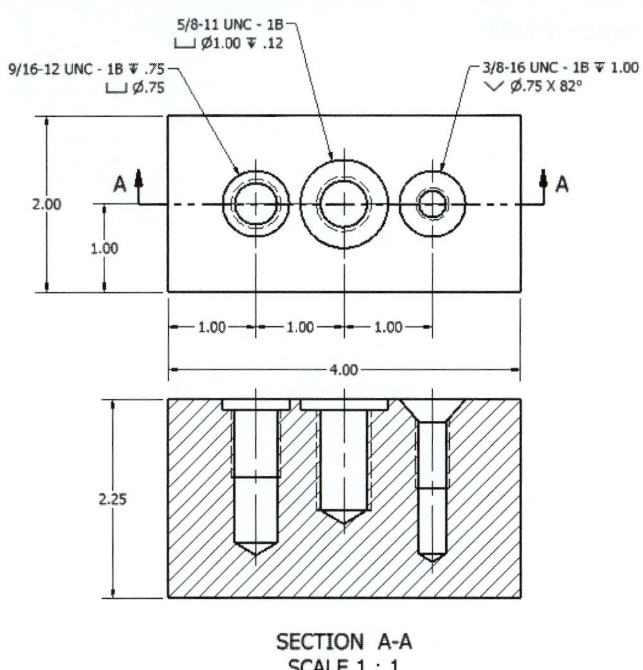

Figure P7-46
INCHES

Redraw the given shape and dimension it using the following dimension styles.

1. Baseline
2. Ordinate
3. Hole Table

Figure P7-47
INCHES
MATL = 0.50

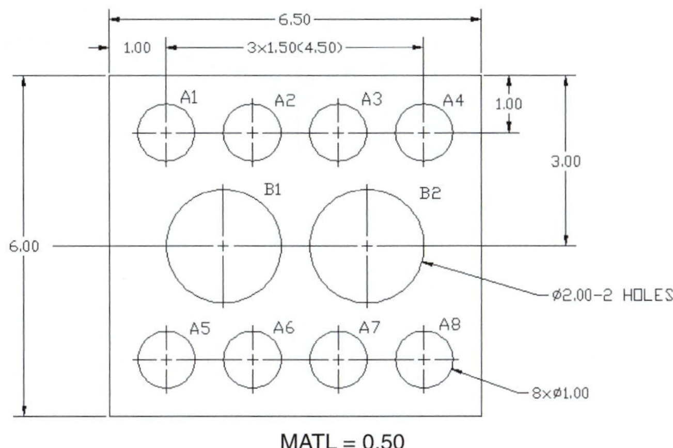

Figure P7-48
MILLIMETERS

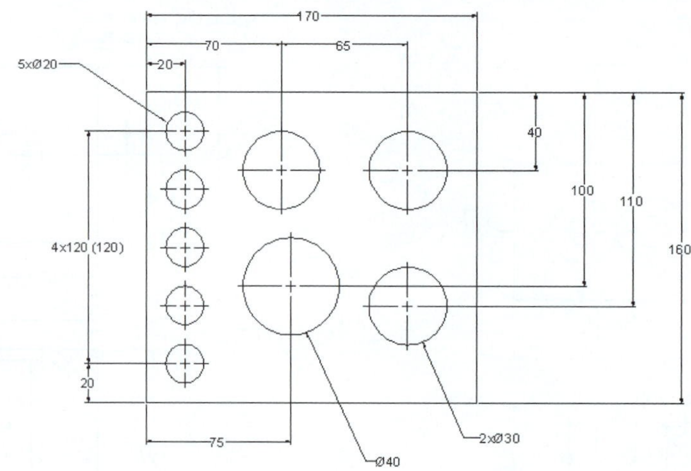

Figure P7-49
INCHES

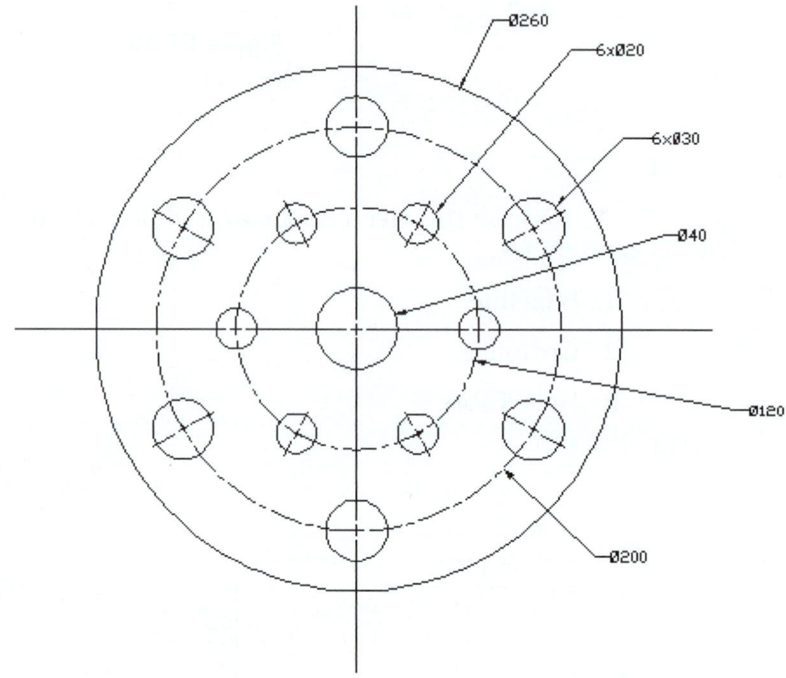

8 chapter eight
Tolerancing

CHAPTER OBJECTIVES

- Understand tolerance conventions
- Understand the meaning of tolerances
- Learn how to apply tolerances
- Understand geometric tolerances
- Understand positional tolerances

8-1 Introduction

Tolerances define the manufacturing limits for dimensions. All dimensions have tolerances either written directly on the drawing as part of the dimension or implied by a predefined set of standard tolerances that apply to any dimension that does not have a stated tolerance.

This chapter explains general tolerance conventions and how they are applied using SolidWorks. It includes a sample tolerance study and an explanation of standard fits and surface finishes.

8-2 Direct Tolerance Methods

There are two methods used to include tolerances as part of a dimension: *plus and minus*, and *limits*. Plus and minus tolerances can be expressed in either bilateral (deviation) or unilateral (symmetric) form.

A ***bilateral tolerance*** has both a plus and a minus value, whereas a ***unilateral tolerance*** has either the plus or the minus value equal to 0. Figure 8-1 shows a horizontal dimension of 60 mm that includes a bilateral tolerance of plus or minus 1 and another dimension of 60.00 mm

Figure 8-1

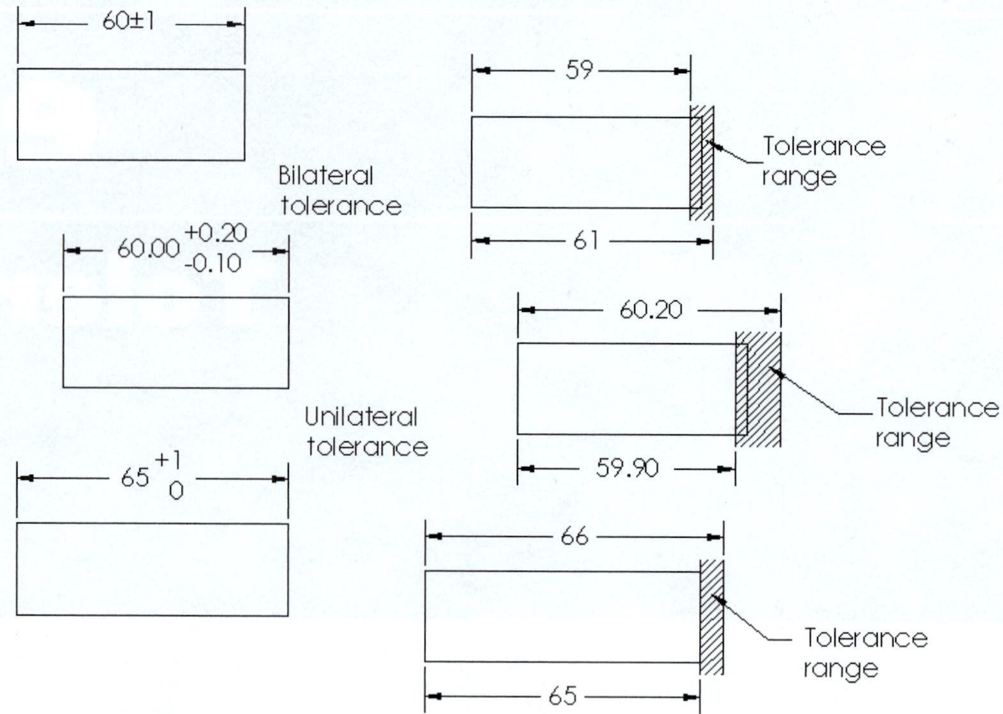

that includes a bilateral tolerance of plus 0.20 or minus 0.10. Figure 8-1 also shows a dimension of 65 mm that includes a unilateral tolerance of plus 1 or minus 0.

> **NOTE**
>
> Bilateral tolerances are called **symmetric** in SolidWorks. Unilateral tolerances are called **deviation.**

Plus or minus tolerances define a range for manufacturing. If inspection shows that all dimensioned distances on an object fall within their specified tolerance range, the object is considered acceptable; that is, it has been manufactured correctly.

Figure 8-2

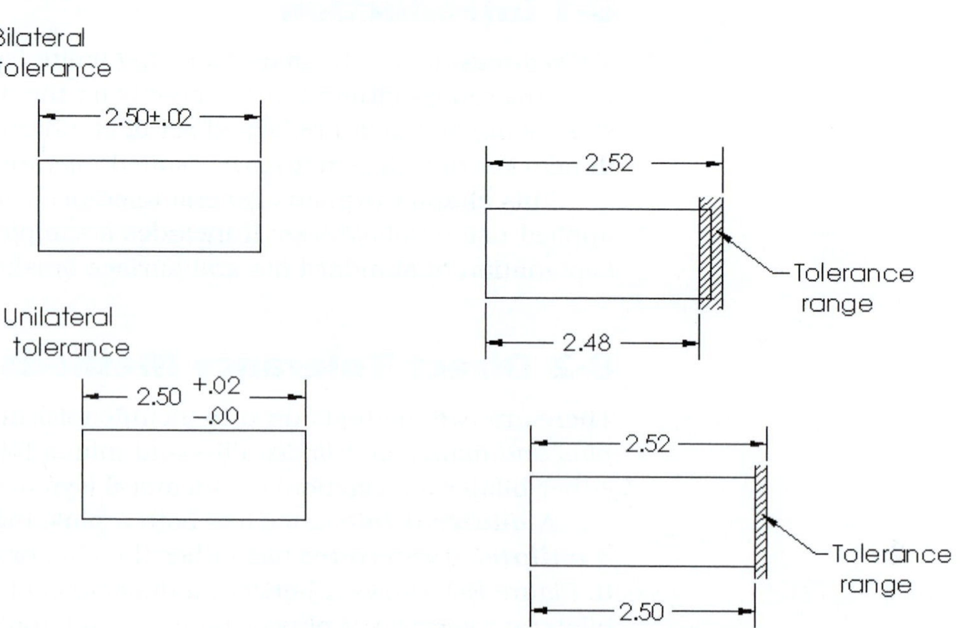

The dimension and tolerance of 60±0.1 means that the part must be manufactured within a range no greater than 60.1 nor less than 59.9. The dimension and tolerance 65 +1/−0 defines the tolerance range as 65 to 66.

Figure 8-2 shows some bilateral and unilateral tolerances applied using decimal inch values. Inch dimensions and tolerances are written using a slightly different format than millimeter dimensions and tolerances, but they also define manufacturing ranges for dimension values. The horizontal bilateral dimension and tolerance 2.50±.02 defines the longest acceptable distance as 2.52 in. and the shortest as 2.48. The unilateral dimension 2.50 +.02/−.00 defines the longest acceptable distance as 2.52 and the shortest as 2.50.

8-3 Tolerance Expressions

Dimension and tolerance values are written differently for inch and millimeter values. See Figure 8-3. Unilateral dimensions for millimeter values specify a zero limit with a single 0. A zero limit for inch values must

Figure 8-3

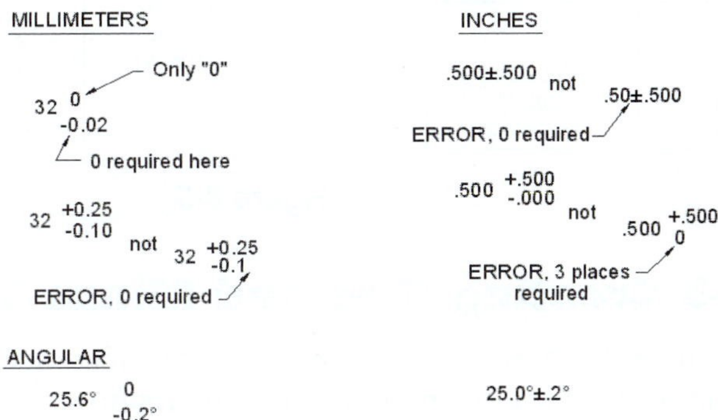

include the same number of decimal places given for the dimension value. In the example shown in Figure 8-3, the dimension value .500 has a unilateral tolerance with minus zero tolerance. The zero limit is written as .000, three decimal places for both the dimension and the tolerance.

Both values in a bilateral tolerance for inch values must contain the same number of decimal places; for millimeter values the tolerance values need not include the same number of decimal places as the dimension value. In Figure 8-3 the dimension value 32 is accompanied by tolerances of +0.25 and −0.10. This form is not acceptable for inch dimensions and tolerances. An equivalent inch dimension and tolerance would be written 32.00 +.25 −.10.

Degree values must include the same number of decimal places in both the dimension and the tolerance values for bilateral tolerances. A single 0 may be used for unilateral tolerances.

8-4 Understanding Plus and Minus Tolerances

A millimeter dimension and tolerance of 12.0 +0.2/−0.1 means that the longest acceptable distance is 12.2000 . . . 0, and the shortest is 11.9000 . . . 0. The total range is 0.3000 . . . 0.

After an object is manufactured, it is inspected to ensure that the object has been manufactured correctly. Each dimensioned distance is measured and, if it is within the specified tolerance, is accepted. If the measured distance is not within the specified tolerance, the part is rejected. Some rejected objects may be reworked to bring them into the specified tolerance range, whereas others are simply scrapped.

Figure 8-4 shows a dimension with a tolerance. Assume that five objects were manufactured using the same 12.0 +0.2/−0.1 dimension and tolerance. The objects were then inspected and the results were as listed. Inspected measurements are usually expressed to at least one more decimal place than that specified in the tolerance. Which objects are acceptable and which are not? Object 3 is too long, and object 5 is too short because their measured distances are not within the specified tolerances.

Figure 8-5 shows a dimension and tolerance of 3.50 +.02 in. Object 3 is not acceptable because it is too short, and object 4 is too long.

GIVEN (mm):

$12^{+0.2}_{-0.1}$

MEANS:
TOL MAX = 12.2
TOL MIN = 11.9
TOTAL TOL = 0.3

OBJECT	AS MEASURED	ACCEPTABLE ?
1	12.160	OK
2	12.020	OK
3	12.203	Too Long
4	11.920	OK
5	11.895	Too Short

Figure 8-4

GIVEN (inches):

3.50±.02

MEANS:
TOL MAX = 3.52
TOL MIN = 3.48
TOTAL TOL = .04

OBJECT	AS MEASURED	ACCEPTABLE ?
1	3.520	OK
2	3.486	OK
3	3.470	Too Short
4	3.521	Too Long
5	3.515	OK

Figure 8-5

8-5 Creating Plus and Minus Tolerances

Figure 8-6 shows a dimensioned view. This section will show how to add plus and minus to the existing dimensions.

Figure 8-6

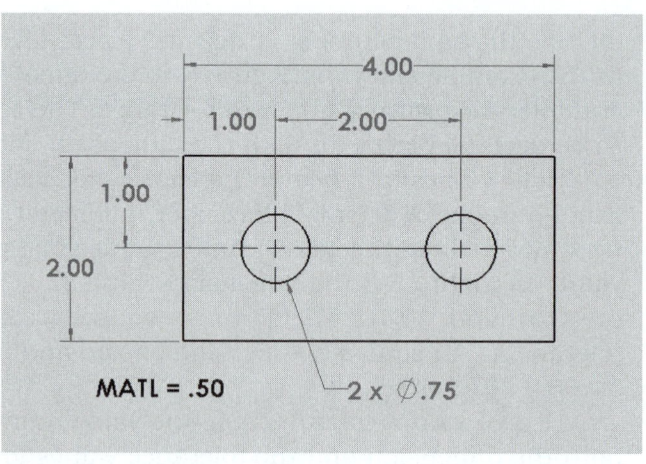

1 Create a part using the dimensions shown. The part thickness is 0.50.

2 Save the part as **BLOCK, TOL**.

3 Create a **Drawing** document of the **BLOCK, TOL** and create an orthographic view as shown.

4 Add dimensions as shown.

See Figure 8-7.

Figure 8-7

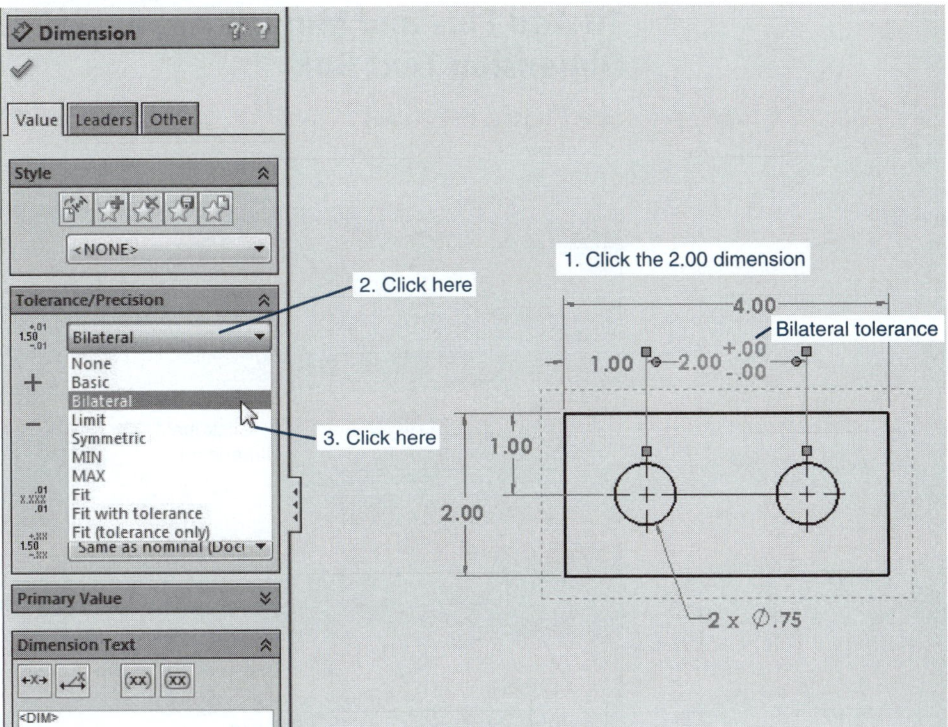

5 Click the horizontal **2.00** dimension.

Click the arrow in the **Tolerance/Precision** box as shown.

6 Select the **Bilateral** option.

7 Enter a plus tolerance of **0.02** and a minus tolerance of **0.01**.

See Figure 8-8.

8 Click the green **OK** check mark.

Figure 8-8

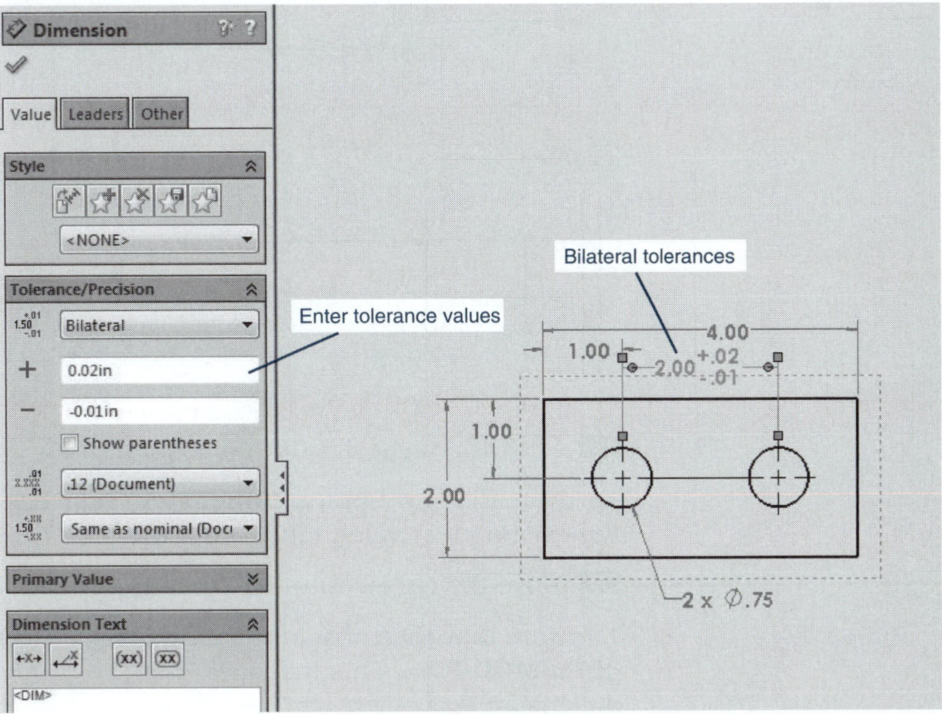

To Add Plus and Minus Symmetric Tolerances Using the Dimension Text Box

See Figure 8-9.

Figure 8-9

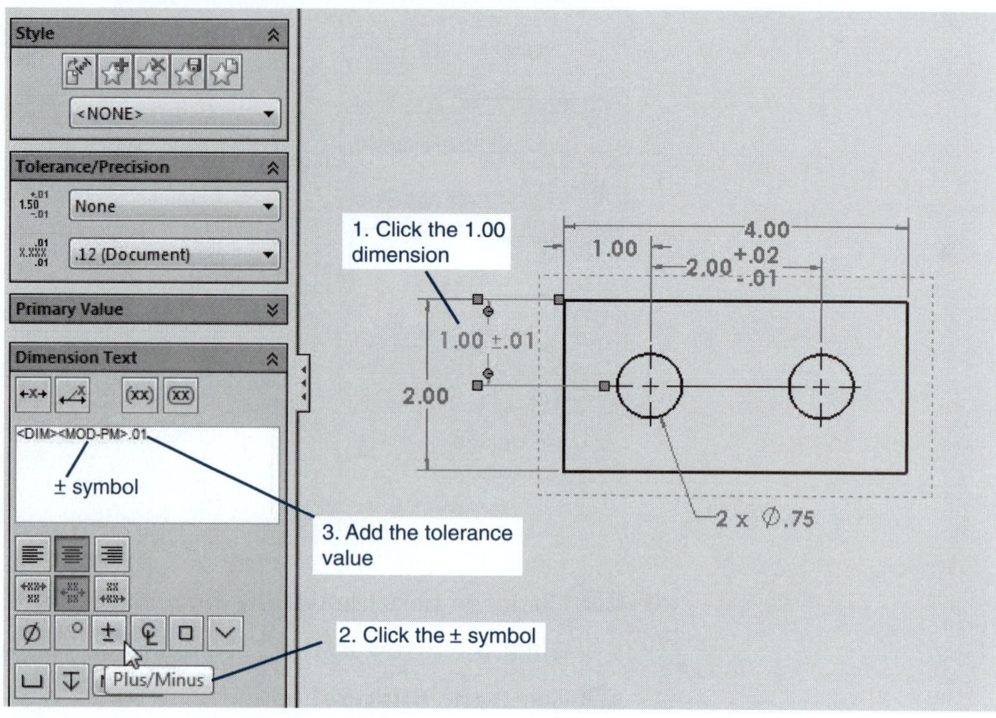

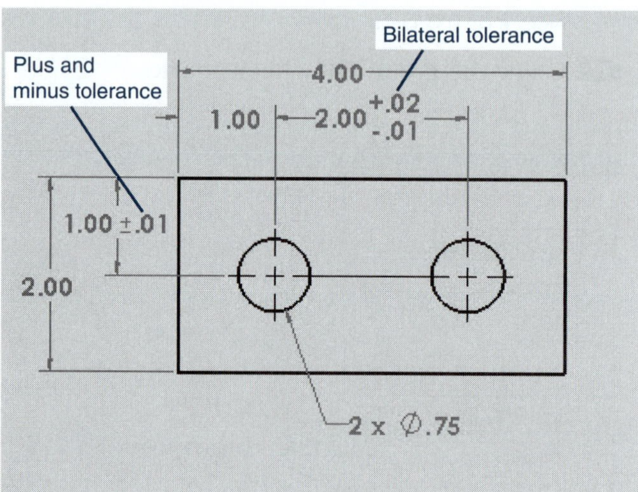

1 Click the vertical **1.00** dimension.

Note the entry in the **Dimension Text** box: **<DIM>**. This represents the existing text value taken from the part's construction dimensions.

2 Move the cursor into the **Dimension Text** box and click the ± symbol.

Note that the entry in the **Dimension Text** box now reads **<DIM MOD-PM>**. This indicates that the ± symbol has been added to the dimension text.

3 Type **.01** after <MOD-PM>.

4 Click the **OK** check mark.

5 Save the **BLOCK, TOL** drawing.

Note how the dimensions and tolerances have been aligned and moved to create a neat, uncluttered appearance.

> **TIP**
> A symmetric tolerance can also be created using the **Symmetric** option in the **Tolerance/Precision** box.

8-6 Creating Limit Tolerances

Figure 8-10 shows examples of limit tolerances. Limit tolerances replace dimension values. Two values are given: the upper and lower limits for the dimension value. The limit tolerance 62.1 and 61.9 is mathematically equal to 62±0.1, but the stated limit tolerance is considered easier to read and understand.

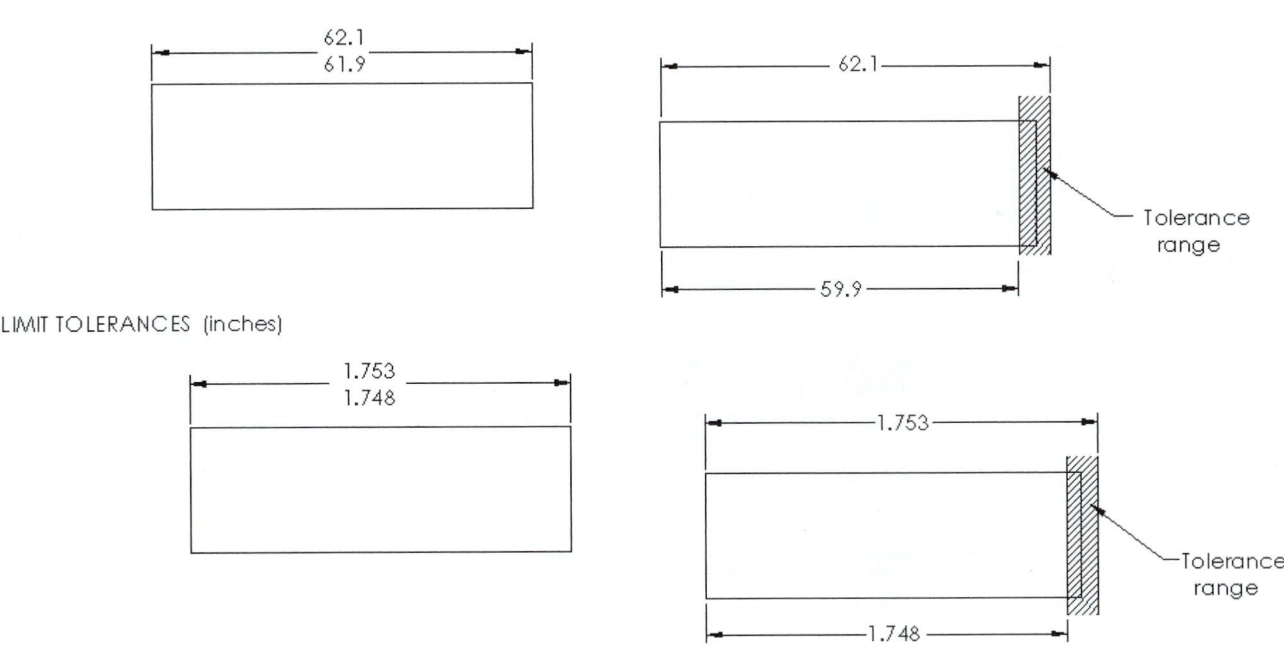

Figure 8-10

Limit tolerances define a range for manufacture. Final distances on an object must fall within the specified range to be acceptable.

This section uses the **BLOCK, TOL** drawing created in the previous section. See Figure 8-11.

1 Click the vertical **2.00** dimension.

2 Select the **Limit** option as shown.

3 Set the upper limit for **.02** and the lower limit for **.01**.

4 Click the green **OK** check mark.

Figure 8-11

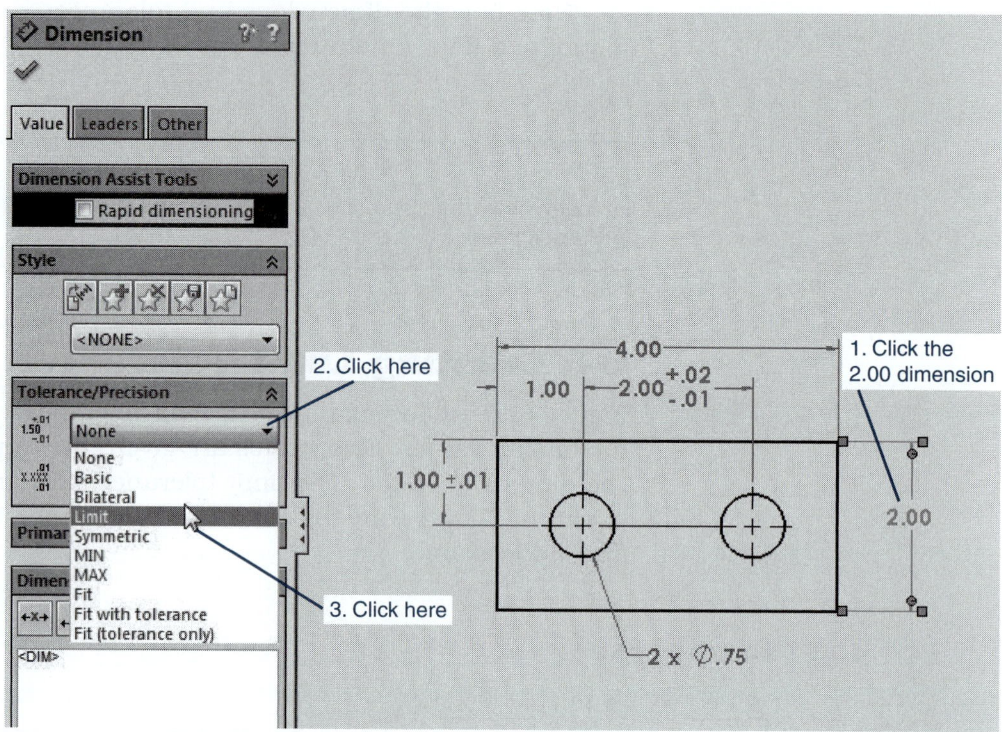

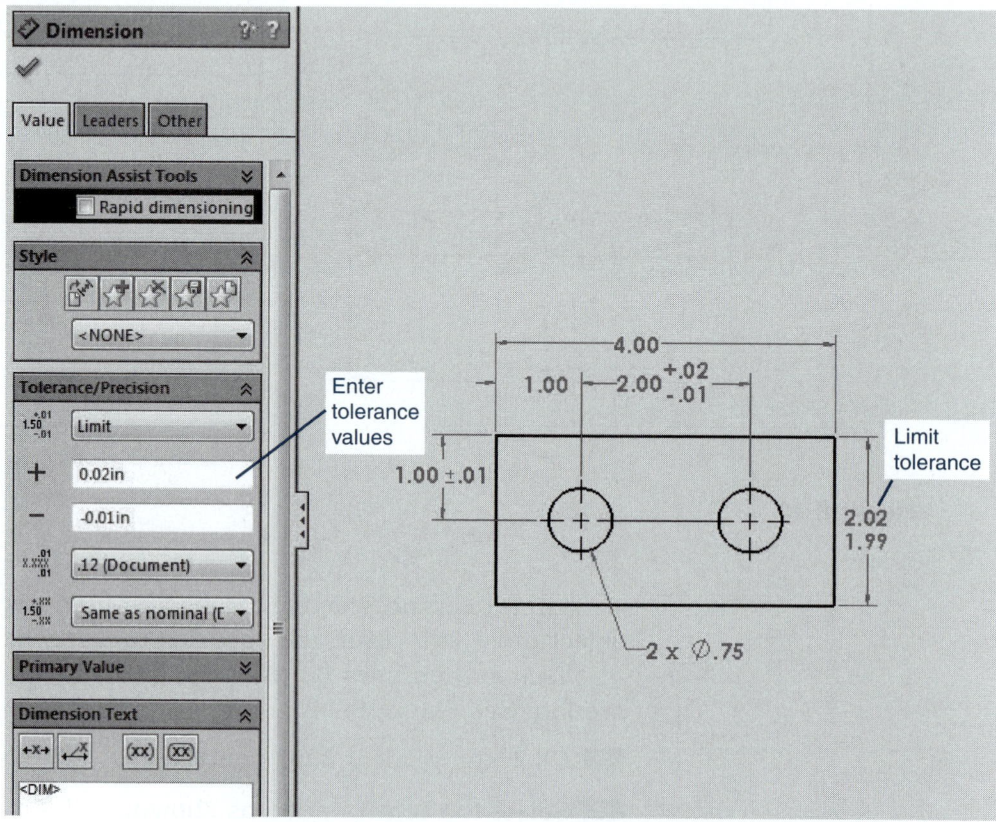

8-7 Creating Angular Tolerances

Figure 8-12 shows an example of an angular dimension with a symmetric tolerance. The procedures explained for applying different types of tolerances to linear dimensions also apply to angular dimensions.

See Figure 8-12.

1 Draw the part shown in Figure 8-12. Extrude the part to a thickness of **.50**.

2 Save the part as **BLOCK, ANGLE**

3 Create a **Drawing** document of the **BLOCK, ANGLE** and create an orthographic view as shown.

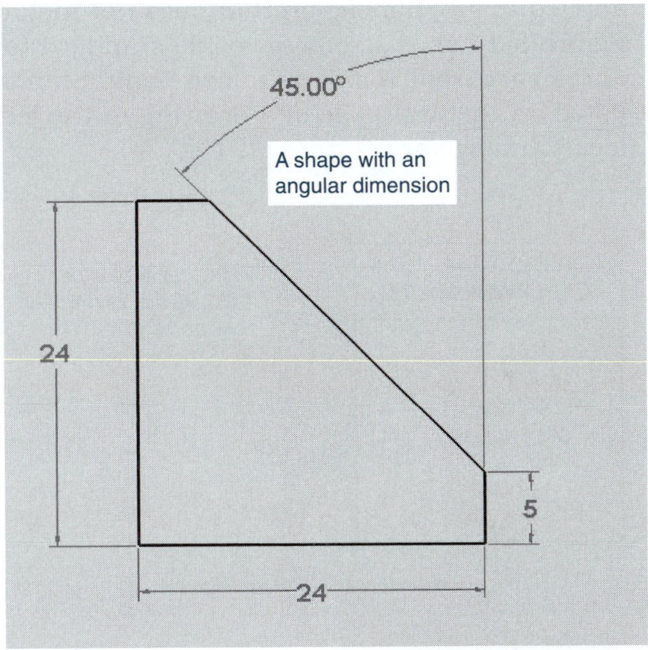

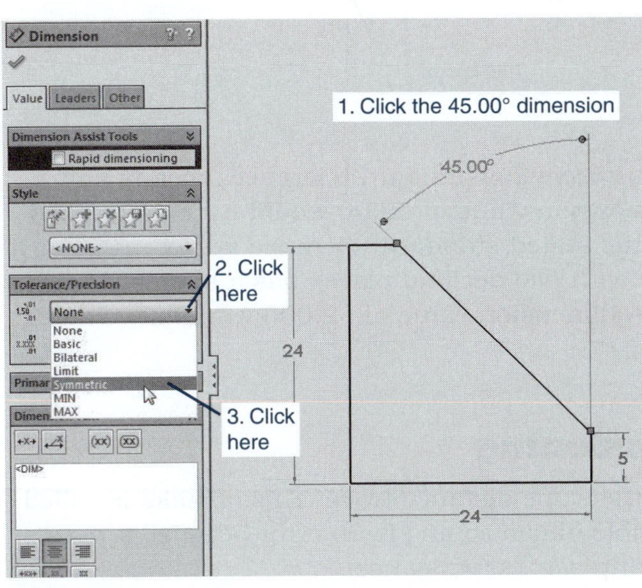

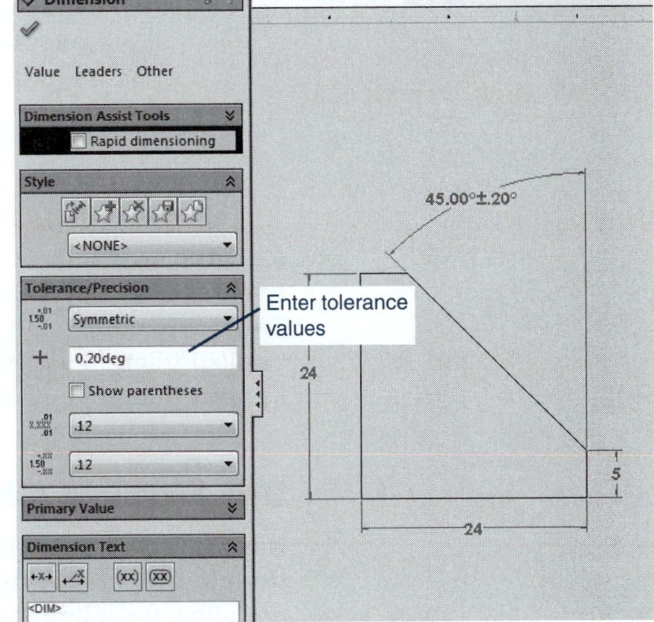

Figure 8-12

4. Dimension the view.
5. Click the **45.00°** dimension.
6. Select the **Symmetric** option.
7. Enter a value of **.20**.
8. Click the green **OK** check mark.
9. Save the drawing.

8-8 Standard Tolerances

Most manufacturers establish a set of standard tolerances that are applied to any dimension that does not include a specific tolerance. Figure 8-13 shows some possible standard tolerances. Standard tolerances vary from company to company. Standard tolerances are usually listed on the first page of a drawing to the left of the title block, but this location may vary.

Figure 8-13

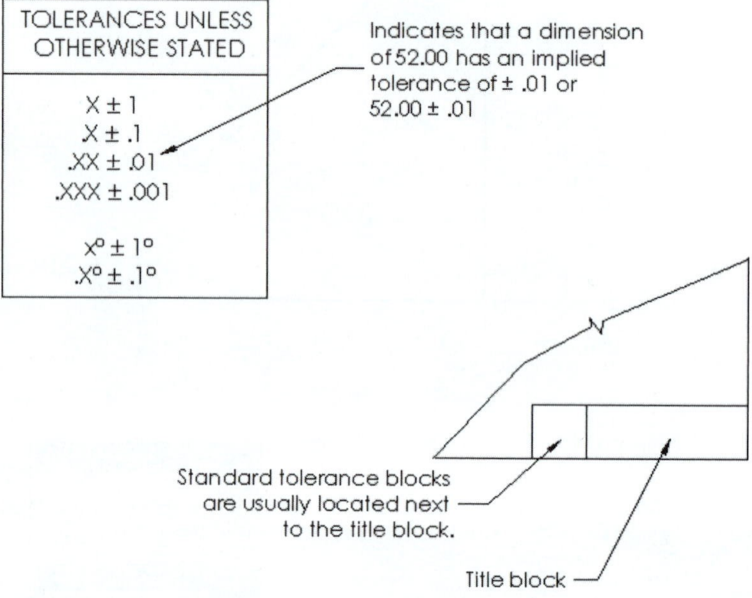

The X value used when specifying standard tolerances means any X stated in that format. A dimension value of 52.00 would have an implied tolerance of ±.01 because the stated standard tolerance is .XX ±.01. Thus, any dimension value with two decimal places has a standard implied tolerance of ±.01. A dimension value of 52.000 would have an implied tolerance of ±.001.

8-9 Double Dimensioning

It is an error to dimension the same distance twice. This mistake is called ***double dimensioning.*** Double dimensioning is an error because it does not allow for tolerance buildup across a distance.

Figure 8-14 shows an object that has been dimensioned twice across its horizontal length, once using three 30-mm dimensions and a second time using the 90-mm overall dimension. The two dimensions are mathematically equal but are not equal when tolerances are considered. Assume that each dimension has a standard tolerance of ±1 mm. The three 30-mm dimensions could create an acceptable distance of 90±3 mm, or a maximum distance of 93 and a minimum distance of 87. The overall dimension of 90 mm allows a maximum distance of 91 and a minimum distance of 89. The two dimensions yield different results when tolerances are considered.

NOTE

Never dimension the same distance twice.

The size and location of a tolerance depend on the design objectives of the object, how it will be manufactured, and how it will be inspected. Even objects that have similar shapes may be dimensioned and toleranced very differently.

One possible solution to the double dimensioning shown in Figure 8-14 is to remove one of the 30-mm dimensions and allow that distance to "float," that is, absorb the cumulated tolerances. The choice of which 30-mm dimension to eliminate depends on the design objectives of the part. For this example the far-right dimension was eliminated to remove the double-dimensioning error.

Figure 8-14

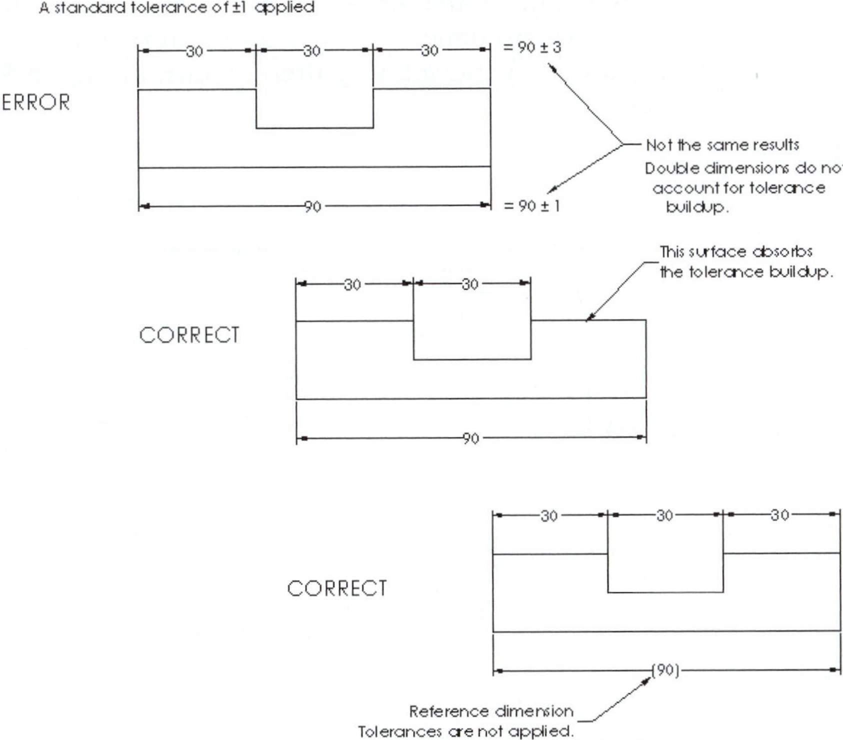

Another possible solution to the double-dimensioning error is to retain the three 30-mm dimensions and to change the 90-mm overall dimension to a reference dimension. A reference dimension is used only for mathematical convenience. It is not used during the manufacturing or inspection process. A reference dimension is designated on a drawing using parentheses: (90).

If the 90-mm dimension was referenced, then only the three 30-mm dimensions would be used to manufacture and inspect the object. This would eliminate the double-dimensioning error.

8-10 Chain Dimensions and Baseline Dimensions

There are two systems for applying dimensions and tolerances to a drawing: chain and baseline. Figure 8-15 shows examples of both systems. ***Chain dimensions*** dimension each feature to the feature next to it. ***Baseline dimensions*** dimension all features from a single baseline or datum.

Chain and baseline dimensions may be used together. Figure 8-15 also shows two objects with repetitive features; one object includes two slots, and the other, three sets of three holes. In each example, the center of the repetitive feature is dimensioned to the left side of the object, which serves as a baseline. The sizes of the individual features are dimensioned using chain dimensions referenced to centerlines.

Baseline dimensions eliminate tolerance buildup and can be related directly to the reference axis of many machines. They tend to take up much more area on a drawing than do chain dimensions.

Chain dimensions are useful in relating one feature to another, such as the repetitive hole pattern shown in Figure 8-15. In this example the

Figure 8-15

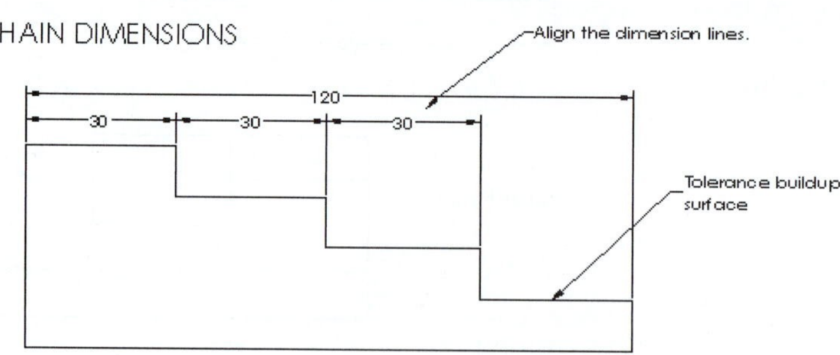

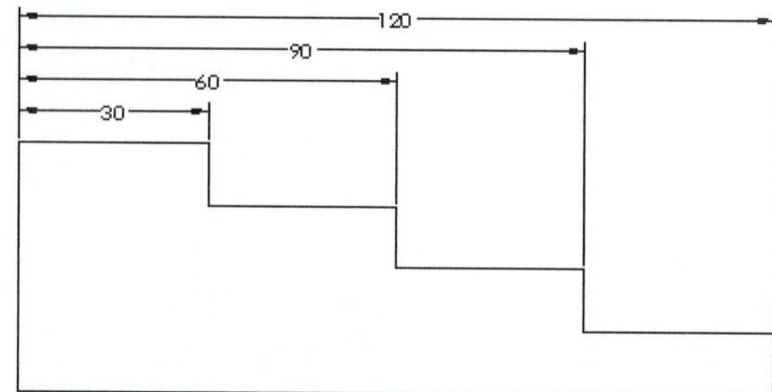

Figure 8-15
(*Continued*)

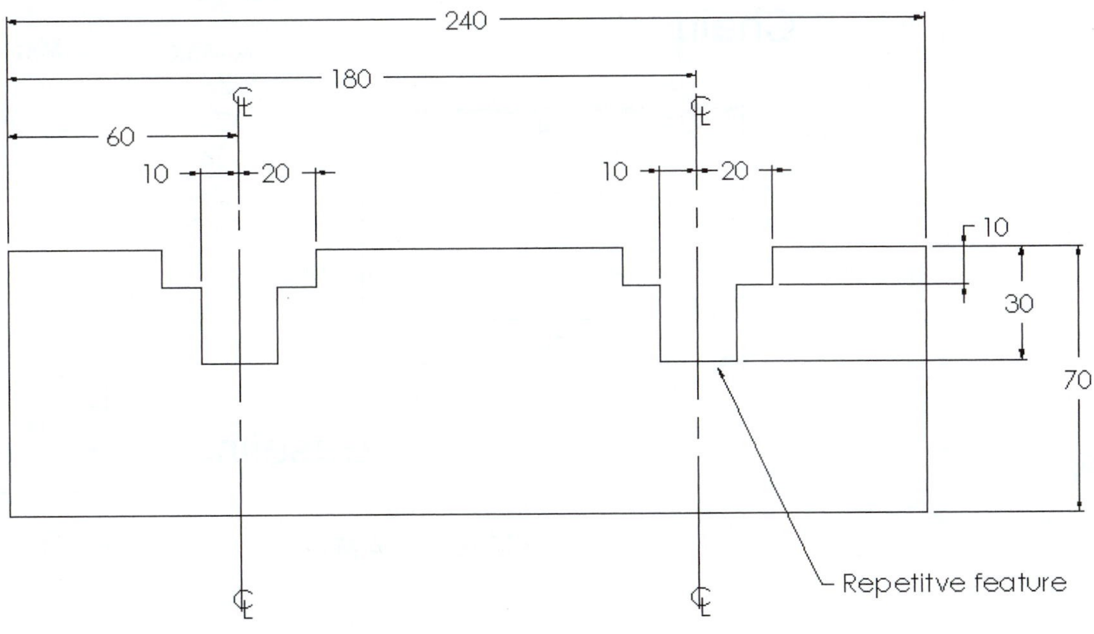

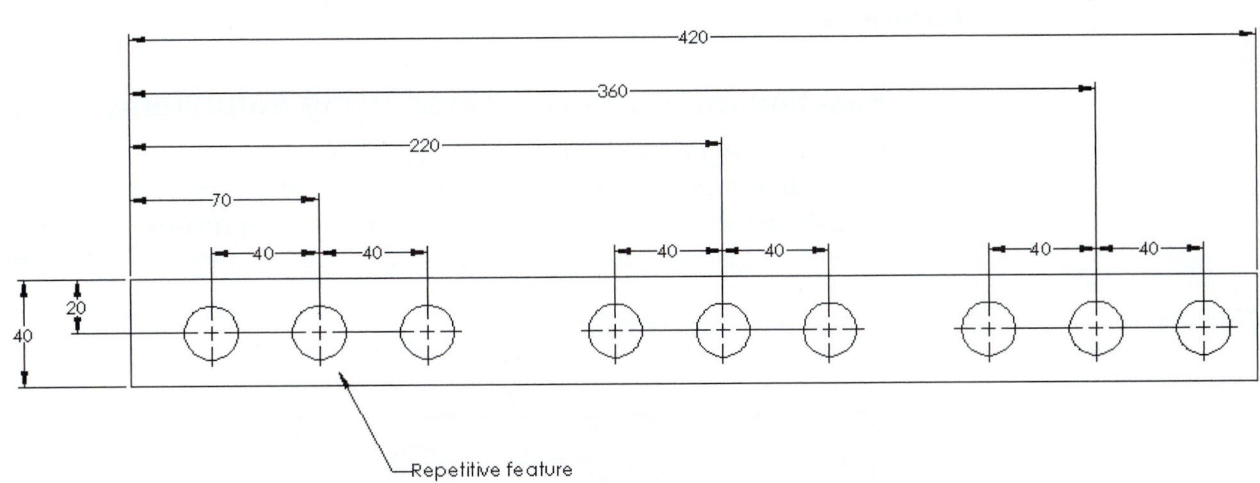

distance between the holes is more important than the individual hole's distance from the baseline.

Figure 8-16 shows the same object dimensioned twice, once using chain dimensions and once using baseline dimensions. All distances are assigned a tolerance range of 2 mm, stated using limit tolerances. The maximum distance for surface A is 28 mm using the chain system and 27 mm using the baseline system. The 1-mm difference comes from the elimination of the first 26–24 limit dimension found on the chain example but not on the baseline.

The total tolerance difference is 6 mm for the chain dimensions and 4 mm for the baseline dimensions. The baseline method reduces the tolerance variations for the object simply because it applies the tolerances and dimensions differently. So why not always use baseline dimensions? For most applications, the baseline system is probably better, but if the distance between the individual features is more critical than the distance from the feature to the baseline, use the chain system.

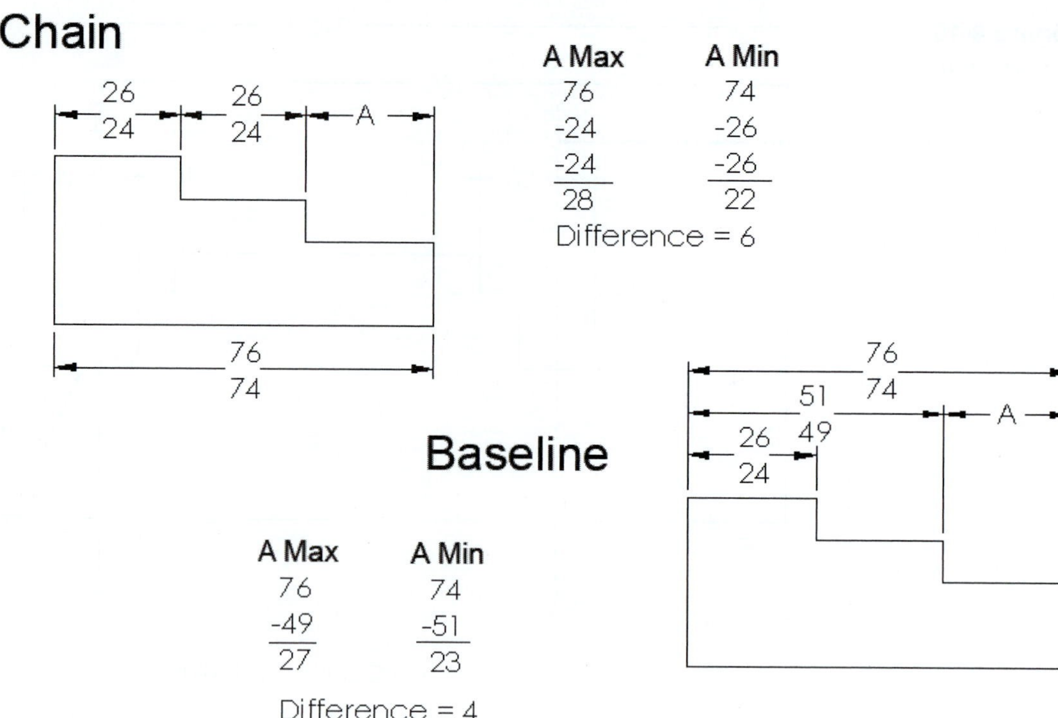

Figure 8-16

Baseline Dimensions Created Using SolidWorks

See Figure 8-17. See also Chapter 7.

Note in the example of baseline dimensioning shown in Figure 8-17 that each dimension is independent of the other. This means that if one of the dimensions is manufactured incorrectly, it will not affect the other dimensions.

Figure 8-17

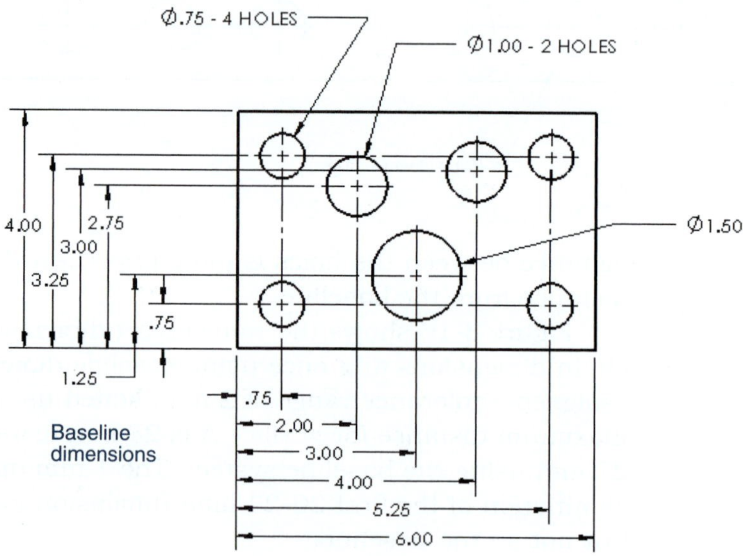

8-11 Tolerance Studies

The term ***tolerance study*** is used when analyzing the effects of a group of tolerances on one another and on an object. Figure 8-18 shows an object with two horizontal dimensions. The horizontal distance A is not dimensioned. Its length depends on the tolerances of the two horizontal dimensions.

Figure 8-18

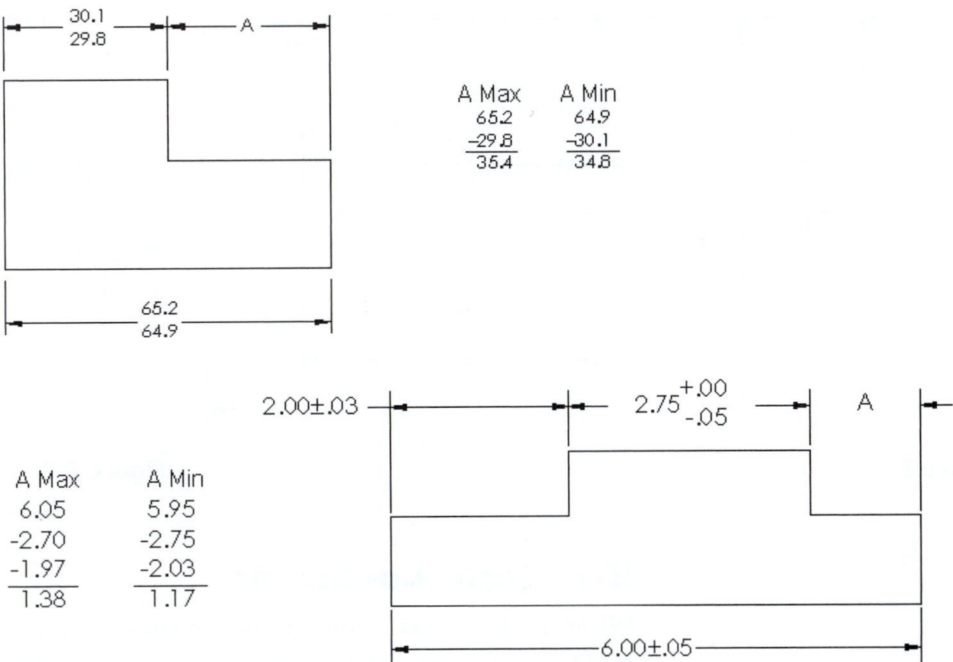

Calculating the Maximum Length of A

Distance A will be longest when the overall distance is at its longest and the other distance is at its shortest.

$$\begin{array}{r} 65.2 \\ -\underline{29.8} \\ 35.4 \end{array}$$

Calculating the Minimum Length of A

Distance A will be shortest when the overall length is at its shortest and the other length is at its longest.

$$\begin{array}{r} 64.9 \\ -\underline{30.1} \\ 34.8 \end{array}$$

NOTE

The hole locations can also be defined using polar dimensions.

8-12 Rectangular Dimensions

Figure 8-19 shows an example of rectangular dimensions referenced to baselines. Figure 8-20 shows a circular object on which dimensions are referenced to a circle's centerlines. Dimensioning to a circle's centerline is critical to accurate hole location.

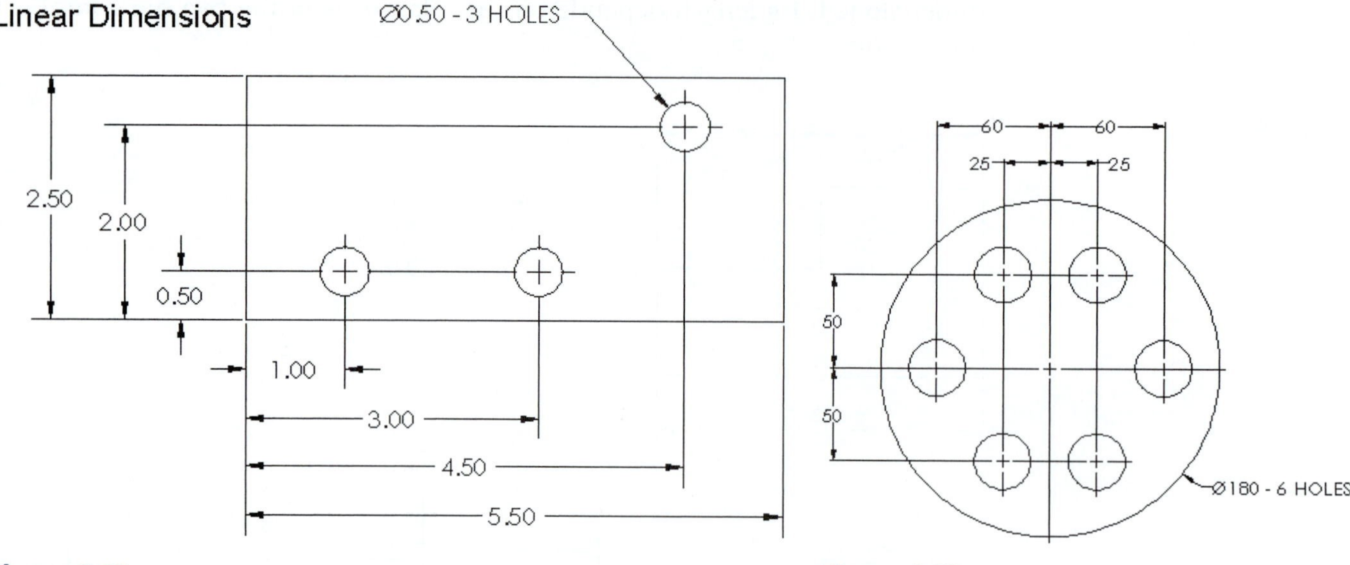

Figure 8-19

Figure 8-20

8-13 Hole Locations

When rectangular dimensions are used, the location of a hole's centerpoint is defined by two linear dimensions. The result is a rectangular tolerance zone whose size is based on the linear dimension's tolerances. The shape of the centerpoint's tolerance zone may be changed to circular using positioning tolerancing, as described later in the chapter.

Figure 8-21 shows the location and size dimensions for a hole. Also shown are the resulting tolerance zone and the overall possible hole shape. The centerpoint's tolerance is .2 by .3 based on the given linear locating tolerances.

The hole diameter has a tolerance of ±.05. This value must be added to the centerpoint location tolerances to define the maximum overall possible shape of the hole. The maximum possible hole shape is determined by drawing the maximum radius from the four corner points of the tolerance zone.

Figure 8-21

Figure 8-21
(*Continued*)

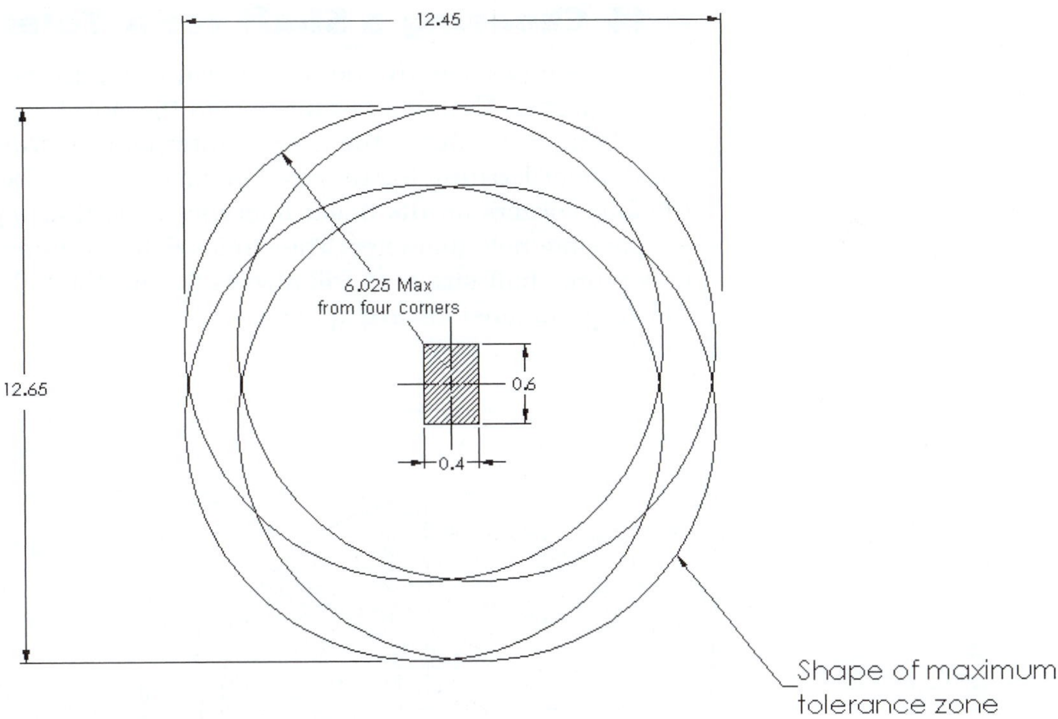

This means that the left edge of the hole could be as close to the vertical baseline as 12.75 or as far as 13.25. The 12.75 value was derived by subtracting the maximum hole diameter value 12.05 from the minimum linear distance 24.80 (24.80 − 12.05 = 12.75). The 13.25 value was derived by subtracting the minimum hole diameter 11.95 from the maximum linear distance 25.20 (25.20 − 11.95 = 13.25).

Figure 8-22 shows a hole's tolerance zone based on polar dimensions. The zone has a sector shape, and the possible hole shape is determined by locating the maximum radius at the four corner points of the tolerance zone.

Figure 8-22

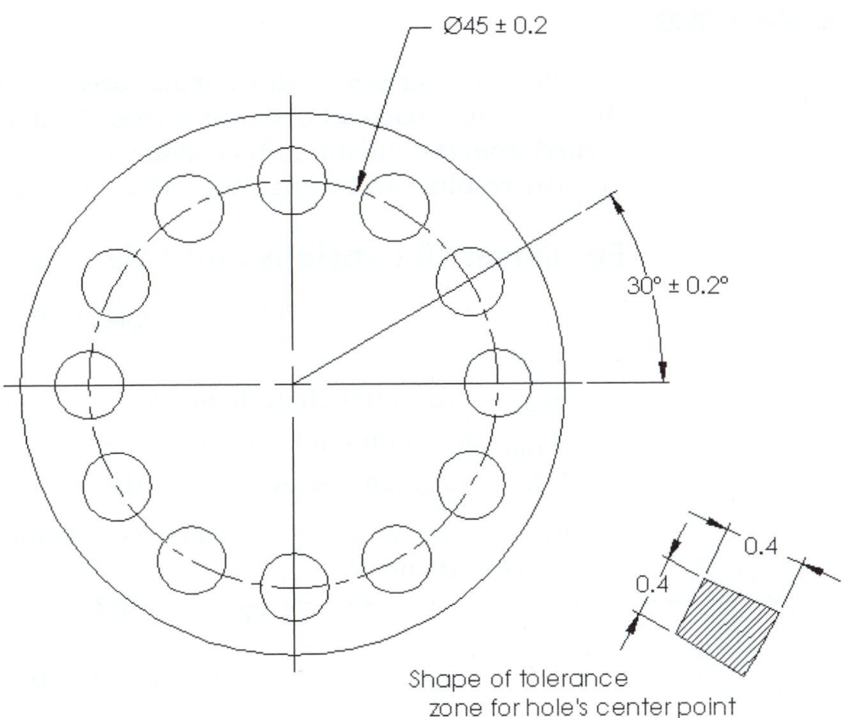

8-14 Choosing a Shaft for a Toleranced Hole

Given the hole location and size shown in Figure 8-21, what is the largest diameter shaft that will always fit into the hole?

Figure 8-23 shows the hole's centerpoint tolerance zone based on the given linear locating tolerances. Four circles have been drawn centered at the four corners on the linear tolerance zone that represents the smallest possible hole diameter. The circles define an area that represents the maximum shaft size that will always fit into the hole, regardless of how the given dimensions are applied.

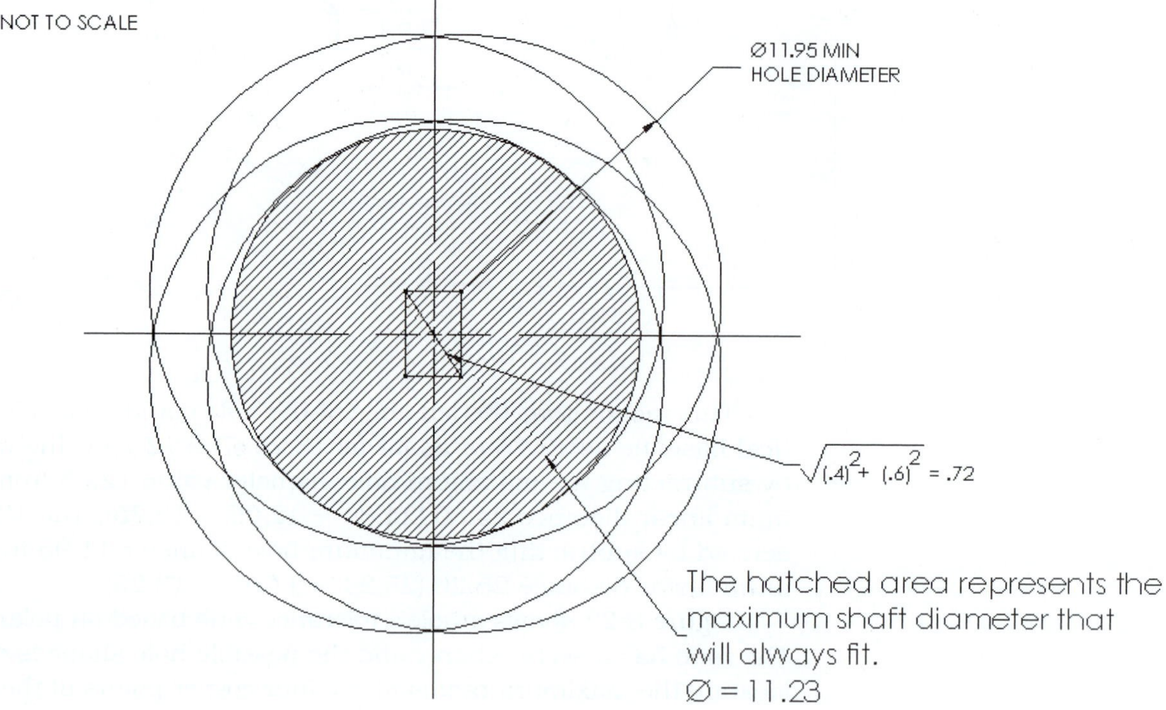

Figure 8-23

The diameter size of this circular area can be calculated by subtracting the maximum diagonal distance across the linear tolerance zone (corner to corner) from the minimum hole diameter.

The results can be expressed as a formula.

For Linear Dimensions and Tolerances

$$S_{max} = H_{min} - DTZ$$

where

S_{max} = maximum shaft diameter

H_{min} = minimum hole diamter

DTZ = diagonal distance across the tolerance zone

In the example shown the diagonal distance is determined using the Pythagorean theorem:

$$DTZ = \sqrt{(.4)^2 + (.6)^2}$$

$$= \sqrt{.16 + .36}$$

$$DTZ = .72$$

This means that the maximum shaft diameter that will always fit into the given hole is 11.43:

$$S_{max} = H_{min} - DTZ$$
$$= 11.95 - .52$$
$$S_{max} = 11.23$$

This procedure represents a restricted application of the general formula presented later in the chapter for positioning tolerances.

NOTE

Linear tolerances generate a square or rectangular tolerance zone.

Once the maximum shaft size has been established, a tolerance can be applied to the shaft. If the shaft had a total tolerance of .25, the minimum shaft diameter would be 11.43 − .25, or 11.18. Figure 8-23 shows a shaft dimensioned and toleranced using these values.

The formula presented is based on the assumption that the shaft is perfectly placed on the hole's centerpoint. This assumption is reasonable if two objects are joined by a fastener and both objects are free to move. When both objects are free to move about a common fastener, they are called ***floating objects***.

8-15 Sample Problem SP8-1

Parts A and B in Figure 8-24 are to be joined by a common shaft. The total tolerance for the shaft is to be .05. What are the maximum and minimum shaft diameters?

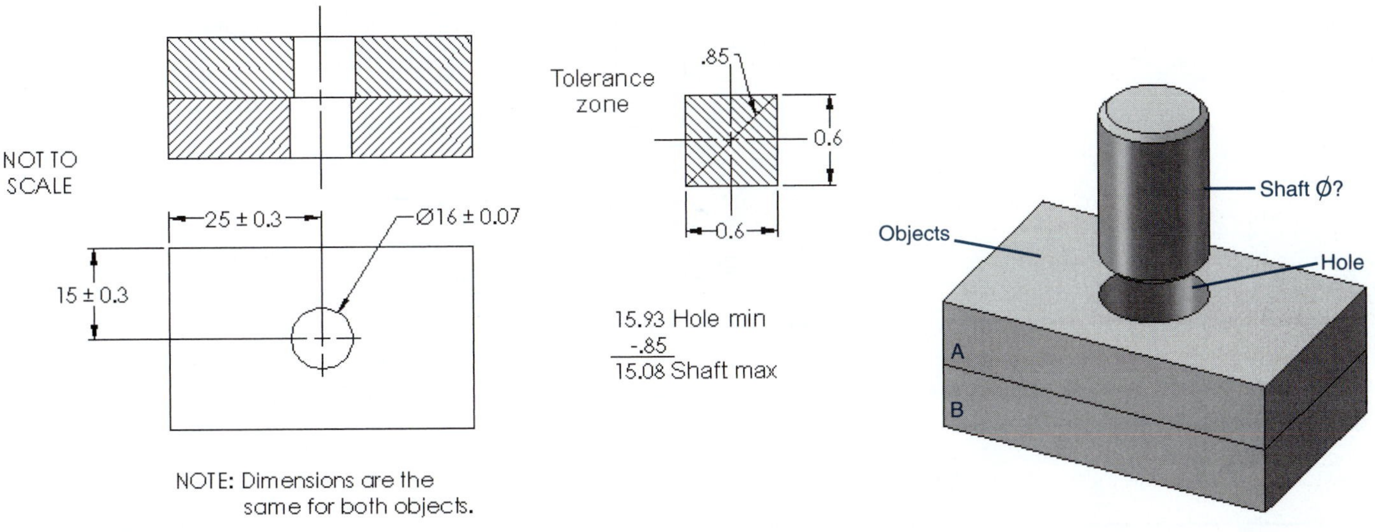

Figure 8-24

Both objects have the same dimensions and tolerances and are floating relative to each other.

$$S_{max} = H_{min} - DTZ$$
$$= 15.93 - .85$$
$$S_{max} = 15.08$$

The shaft's minimum diameter is found by subtracting the total tolerance requirement from the calculated maximum diameter:

$$15.08 - .05 = 15.03$$

Therefore,

Shaft max = 5.08

Shaft min = 5.03

8-16 Sample Problem SP8-2

The procedure presented in Sample Problem SP8-1 can be worked in reverse to determine the maximum and minimum hole size based on a given shaft size.

Objects AA and BB as shown in Figure 8-25 are to be joined using a bolt whose maximum diameter is .248. What is the minimum hole size for objects that will always accept the bolt? What is the maximum hole size if the total hole tolerance is .005?

$$S_{max} = H_{min} - DTZ$$

In this example H_{min} is the unknown factor, so the equation is rewritten as

$$H_{min} = S_{max} + DTZ$$
$$= .248 + .001$$
$$H_{min} = .249$$

Figure 8-25

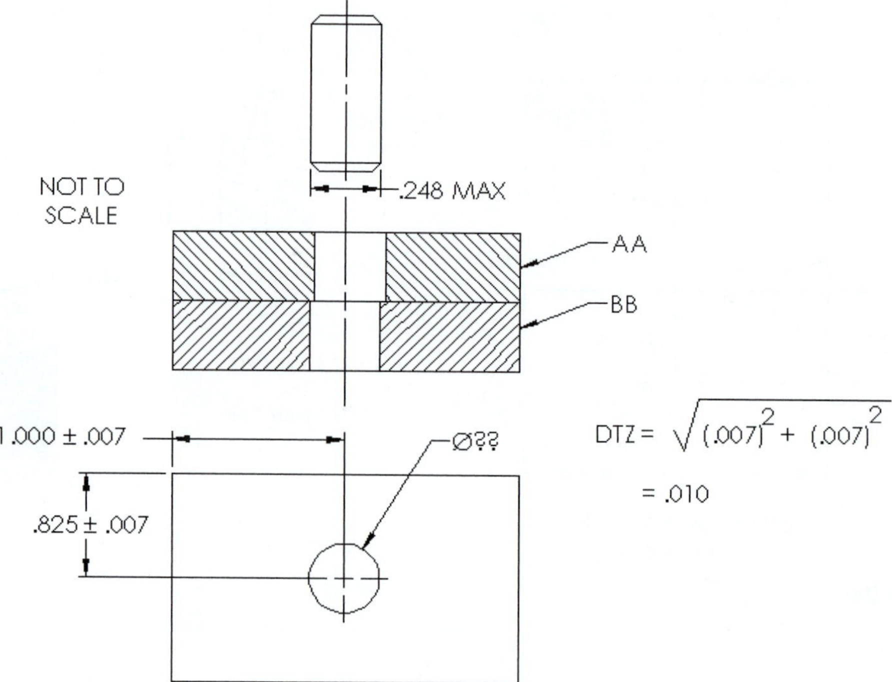

This is the minimum hole diameter, so the total tolerance requirement is added to this value:

$$.249 + .005 = .254$$

Therefore,

Hole max = .254

Hole min = .243

8-17 Nominal Sizes

The term ***nominal*** refers to the approximate size of an object that matches a common fraction or whole number. A shaft with a dimension of 1.500 + .003 is said to have a nominal size of "one and a half inches." A dimension of 1.500 +.000/−.005 is still said to have a nominal size of one and a half inches. In both examples 1.5 is the closest common fraction.

8-18 Standard Fits (Metric Values)

Calculating tolerances between holes and shafts that fit together is so common in engineering design that a group of standard values and notations has been established. These values may be calculated using the **Limits and Fits** option of the **Design Library**.

There are three possible types of fits between a shaft and a hole: clearance, transitional, and interference. There are several subclassifications within each of these categories.

A ***clearance fit*** always defines the maximum shaft diameter as smaller than the minimum hole diameter. The difference between the two diameters is the amount of clearance. It is possible for a clearance fit to be defined with zero clearance; that is, the maximum shaft diameter is equal to the minimum hole diameter.

An ***interference fit*** always defines the minimum shaft diameter as larger than the maximum hole diameter; that is, the shaft is always bigger than the hole. This definition means that an interference fit is the converse of a clearance fit. The difference between the diameter of the shaft and the hole is the amount of interference.

An interference fit is primarily used to assemble objects together. Interference fits eliminate the need for threads, welds, or other joining methods. Using an interference for joining two objects is generally limited to light load applications.

It is sometimes difficult to visualize how a shaft can be assembled into a hole with a diameter smaller than that of the shaft. It is sometimes done using a hydraulic press that slowly forces the two parts together. The joining process can be augmented by the use of lubricants or heat. The hole is heated, causing it to expand, the shaft is inserted, and the hole is allowed to cool and shrink around the shaft.

A ***transition fit*** may be either a clearance or an interference fit. It may have a clearance between the shaft and the hole or an interference.

The notations are based on Standard International Tolerance values. A specific description for each category of fit follows.

Clearance Fits

H11/c11 or C11/h11 = loose running fit

H8/d8 or D8/h8 = free running fit

H8/f7 or F8/h7 = close running fit

H7/g6 or G7/h6 = sliding fit

H7/h6 = locational clearance fit

Transitional Fits

H7/k6 or K7/h6 = locational transition fit

H7/n6 or N7/h6 = locational transition fit

Interference Fits

H7/p6 or P7/h6 = locational transition fit

H7/s6 or S7/h6 = medium drive fit

H7/u6 or U7/h6 = force fit

8-19 Standard Fits (Inch Values)

Inch values are accessed in the **Design Library** by selecting the **ANSI-Inch** standards.

Fits defined using inch values are classified as follows:

RC = running and sliding fits

LC = clearance locational fits

LT = transitional locational fits

LN = interference fits

FN = force fits

Each of these general categories has several subclassifications within it defined by a number, for example, Class RC1, Class RC2, through Class RC8. The letter designations are based on International Tolerance Standards, as are metric designations.

TIP
Charts of tolerance values can be found in the appendix.

To Add a Fit Callout to a Drawing

Figure 8-26 shows a block and post. Figure 8-27 shows a drawing containing a front and a top orthographic view of the block and post assembly.

1 Draw the block and post and create an assembly drawing as shown in Figure 8-26.

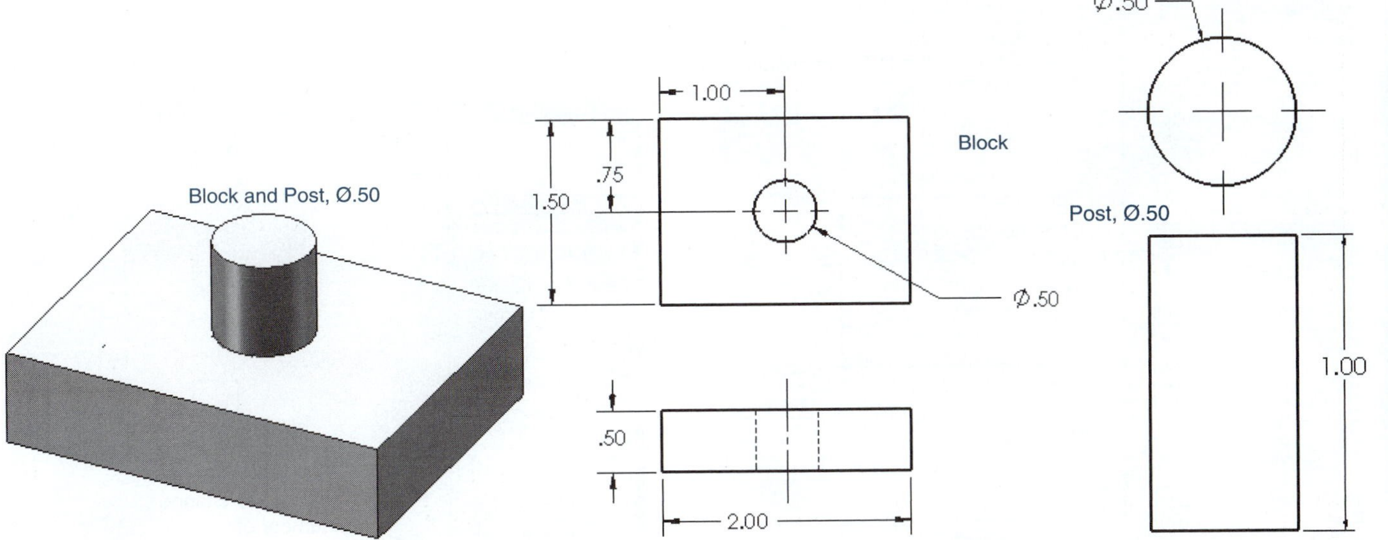

Figure 8-26

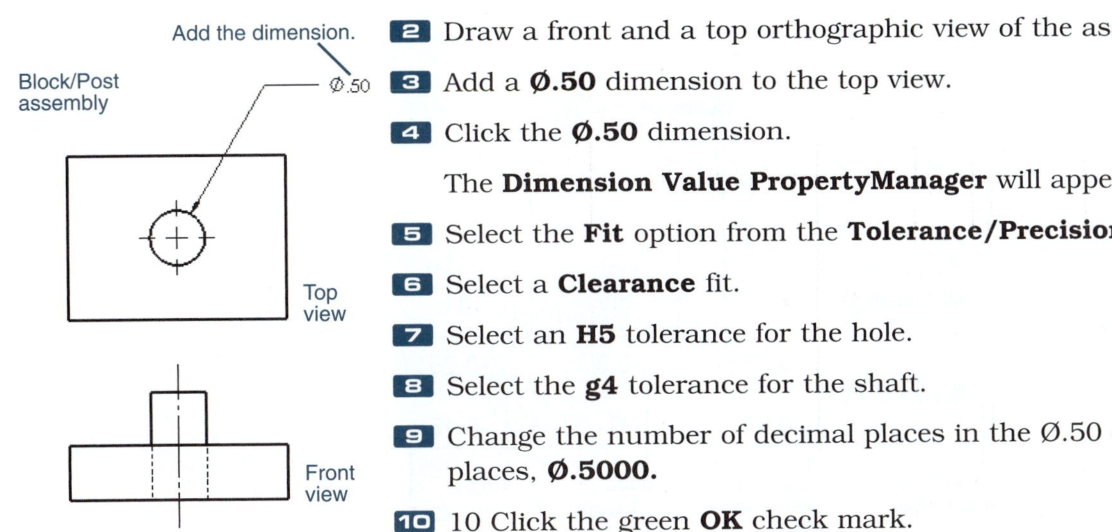

Figure 8-27

2 Draw a front and a top orthographic view of the assembly as shown.

3 Add a **Ø.50** dimension to the top view.

4 Click the **Ø.50** dimension.

The **Dimension Value PropertyManager** will appear. See Figure 8-28.

5 Select the **Fit** option from the **Tolerance/Precision** box as shown.

6 Select a **Clearance** fit.

7 Select an **H5** tolerance for the hole.

8 Select the **g4** tolerance for the shaft.

9 Change the number of decimal places in the Ø.50 dimension to four places, **Ø.5000.**

10 10 Click the green **OK** check mark.

Reading Fit Tables

There are several fit tables in the appendix for both English units and metric units. The metric tables can be read directly, as they state hole and shaft dimension. For example, the tolerance for a Preferred Clearance Fit for a 10mm nominal hole using a loose running fit is 10.090/10.000 for the hole and 9.920/9.830 for the shaft. The tolerance callout would be H11/c11. English unit tables require interpretation.

Figure 8-29 shows the table values for a .5000 nominal hole using an H5/g4 tolerance. The tolerance values are in thousandths of an inch; that is, a listed value of 0.25 equals 0.0025 in. The Ø.5000 nominal value is in the 0.40–0.71 size range, so the tolerance values are as

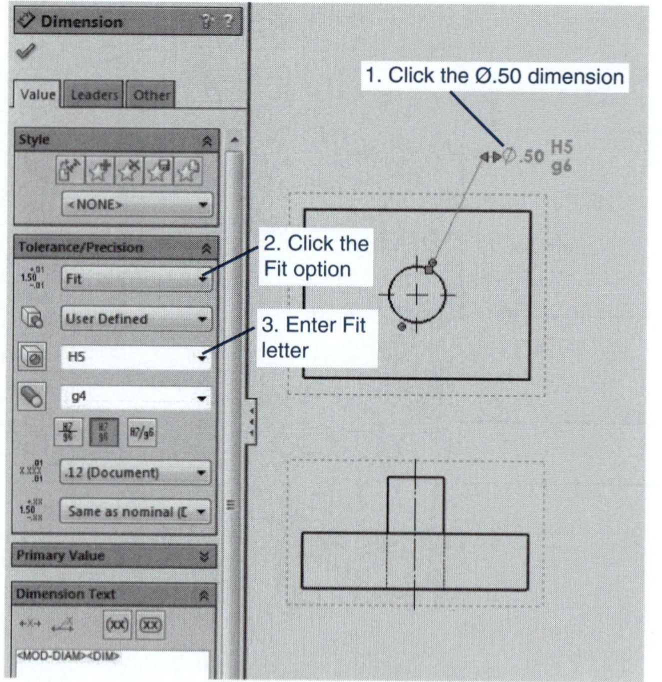

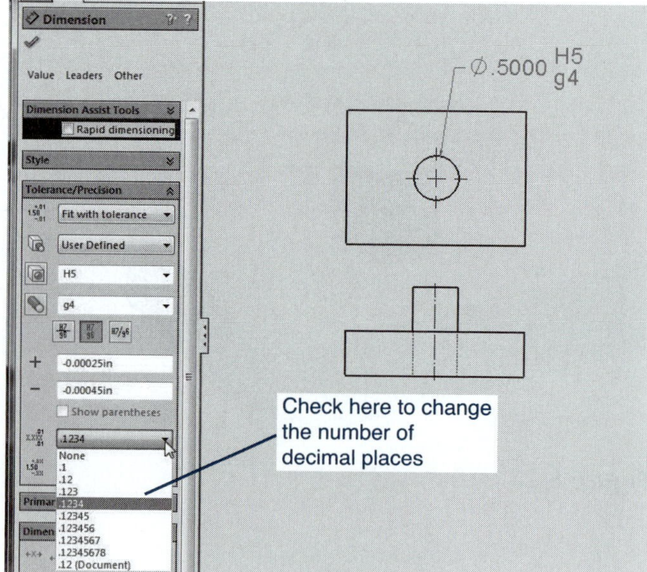

Figure 8-28

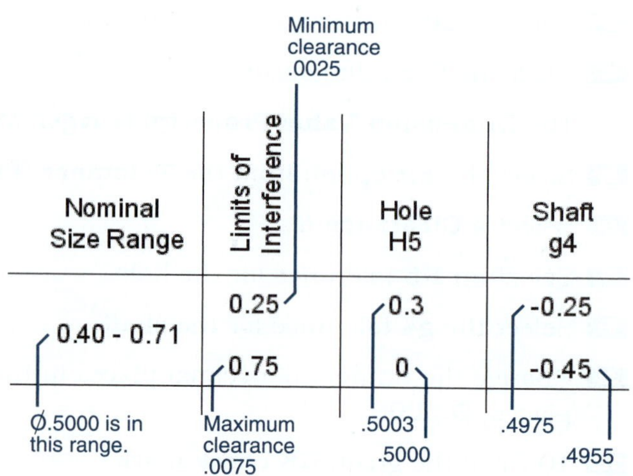

Figure 8-29

shown. These values can be applied to the detail drawings of the block and shaft. See Figure 8-30.

8-20 Preferred and Standard Sizes

It is important that designers always consider preferred and standard sizes when selecting sizes for designs. Most tooling is set up to match these sizes, so manufacturing is greatly simplified when preferred and standard sizes are specified. Figure 8-31 shows a listing of preferred sizes for metric values.

Consider the case of design calculations that call for a 42-mm-diameter hole. A 42-mm-diameter hole is not a preferred size. A diameter of 40 mm is the closest preferred size, and a 45-mm diameter is a second choice. A 42-mm hole could be manufactured but would require an unusual drill size that might not be available. It would be wise to reconsider the design to

Figure 8-30

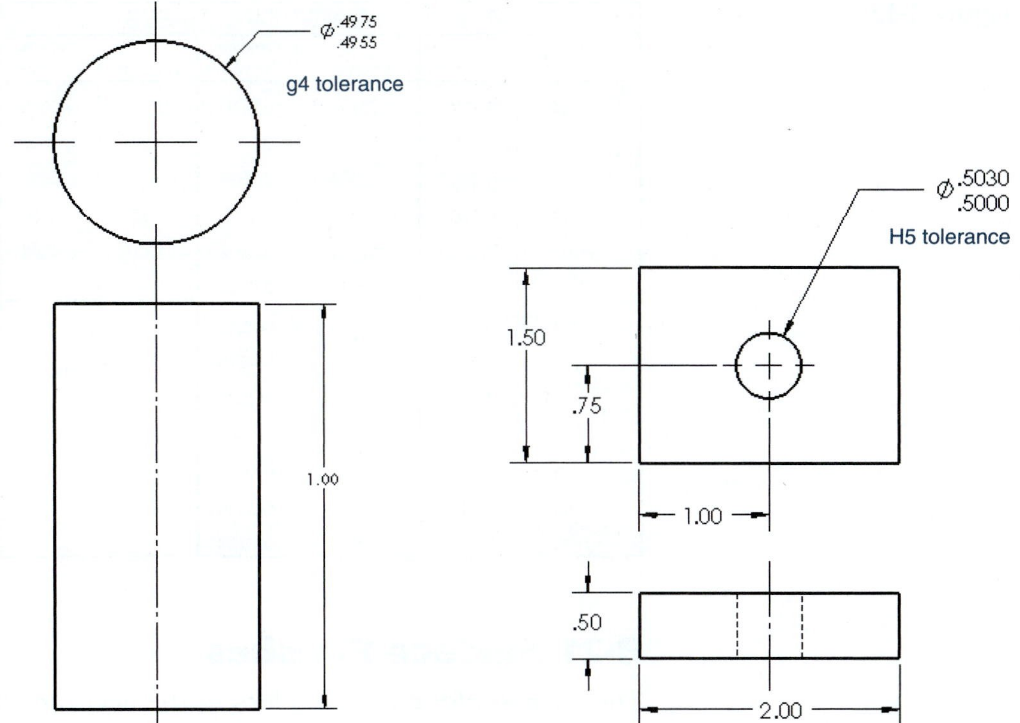

Figure 8-31

PREFERRED SIZES			
1	1.1	12	14
1.2	1.4	16	18
1.6	1.8	20	22
2	2.2	25	28
2.5	2.8	30	35
3	3.5	40	45
4	4.5	50	55
5	5.5	60	70
6	7	80	90
8	9	100	110
10	11	120	140

see if a 40-mm-diameter hole could be used, and if not, possibly a 45-mm-diameter hole.

A production run of a very large quantity could possibly justify the cost of special tooling, but for smaller runs it is probably better to use preferred sizes. Machinists will have the required drills, and maintenance people will have the appropriate tools for these sizes.

Figure 8-32 shows a listing of standard fractional drill sizes. Most companies now specify metric units or decimal inches; however, many standard items are still available in fractional sizes, and many older objects may still require fractional-sized tools and replacement parts. A more complete listing is available in the appendix.

Figure 8-32

STANDARD TWIST DRILL SIZES					
Fraction	Decimal Equivalent	Fraction	Decimal Equivalent	Fraction	Decimal Equivalent
7/16	0.1094	21/64	0.3281	11/16	0.6875
1/8	0.1250	11/64	0.3438	3/4	0.7500
9/64	0.1406	23/64	0.3594	13/16	0.8125
5/32	0.1562	3/8	0.3750	7/8	0.8750
11/64	0.1719	25/64	0.3906	15/16	0.9375
3/16	0.1875	13/32	0.4062	1	1.0000
13/64	0.2031	27/64	0.4219		
7/32	0.2188	7/16	0.4375		
1/4	0.2500	29/64	0.4531		
17/64	0.2656	15/32	0.4688		
9/32	0.2812	1/2	0.5000		
19/64	0.2969	9/16	0.5625		
5/16	0.3125	5/8	0.6250		

8-21 Surface Finishes

The term ***surface finish*** refers to the accuracy (flatness) of a surface. Metric values are measured using micrometers (μm), and inch values are measured in microinches (μin.).

The accuracy of a surface depends on the manufacturing process used to produce the surface. Figure 8-33 shows a listing of manufacturing processes and the quality of the surface finish they can be expected to produce.

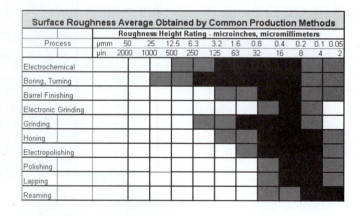

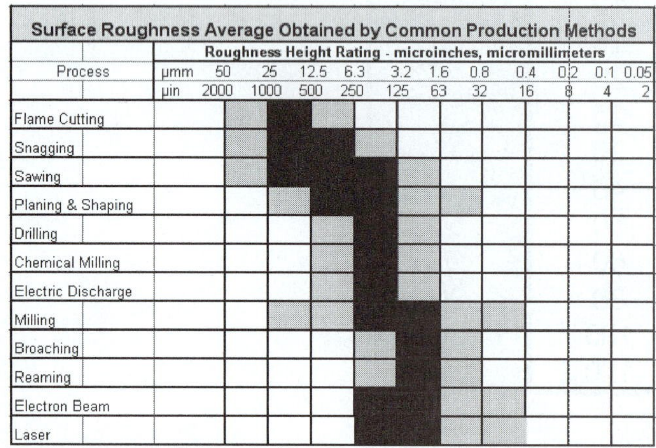

Surface Roughness Average Obtained by Common Production Methods											
	Roughness Height Rating - microinches, micromillimeters										
Process	μmm 50	25	12.5	6.3	3.2	1.6	0.8	0.4	0.2	0.1	0.05
	μin 2000	1000	500	250	125	63	32	16	8	4	2
Flame Cutting											
Snagging											
Sawing											
Planing & Shaping											
Drilling											
Chemical Milling											
Electric Discharge											
Milling											
Broaching											
Reaming											
Electron Beam											
Laser											

Surface Roughness Average Obtained by Common Production Methods											
	Roughness Height Rating - microinches, micromillimeters										
Process	μmm 50	25	12.5	6.3	3.2	1.6	0.8	0.4	0.2	0.1	0.05
	μin 2000	1000	500	250	125	63	32	16	8	4	2
Electrochemical											
Boring, Turning											
Barrel Finishing											
Electronic Grinding											
Grinding											
Honing											
Electropolishing											
Polishing											
Lapping											
Reaming											

Surface Roughness Average Obtained by Common Production Methods											
	Roughness Height Rating - microinches, micromillimeters										
Process	μmm 50	25	12.5	6.3	3.2	1.6	0.8	0.4	0.2	0.1	0.05
	μin 2000	1000	500	250	125	63	32	16	8	4	2
Sand Casting											
Hot Rolling											
Forging											
Perm Mold Casting											
Investment Casting											
Extrusion											
Cold Rolling, Drawing											
Polishing Die Casting											

Figure 8-33

Surface finishes have several design applications. ***Datum surfaces,*** or surfaces used for baseline dimensioning, should have fairly accurate surface finishes to help assure accurate measurements. Bearing surfaces should have good-quality surface finishes for better load distribution, and parts that operate at high speeds should have smooth finishes to help reduce friction. Figure 8-34 shows a screw head sitting on a very wavy surface. Note that the head of the screw is actually in contact with only two wave peaks, meaning all the bearing load is concentrated on the two peaks. This situation could cause stress cracks and greatly weaken the surface. A better-quality surface finish would increase the bearing contact area.

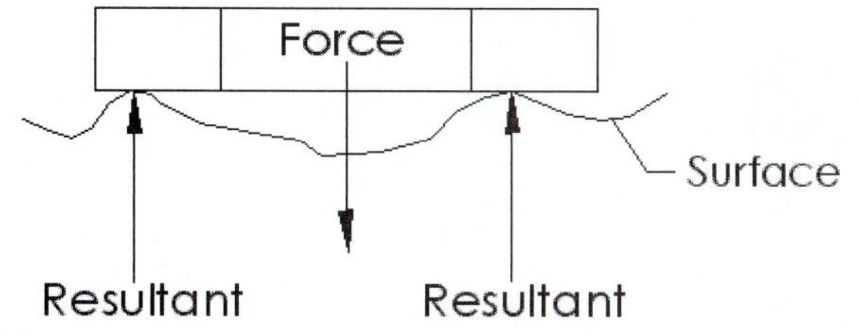

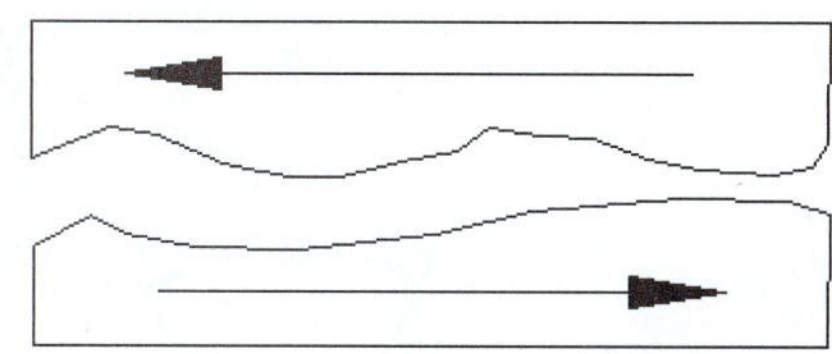

Figure 8-34

Figure 8-34 also shows two very rough surfaces moving in contact with each other. The result will be excess wear to both surfaces because the surfaces touch only on the peaks, and these peaks will tend to wear faster than flatter areas. Excess vibration can also result when interfacing surfaces are too rough.

Surface finishes are classified into three categories: surface texture, roughness, and lay. ***Surface texture*** is a general term that refers to the overall quality and accuracy of a surface.

Roughness is a measure of the average deviation of a surface's peaks and valleys. See Figure 8-35.

Figure 8-35

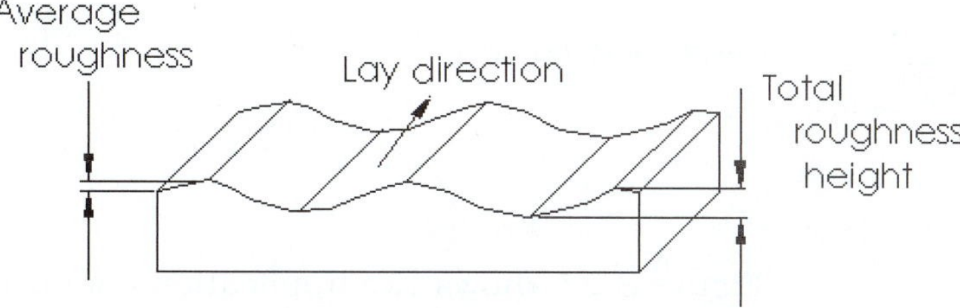

Lay refers to the direction of machine marks on a surface. See Figure 8-36. The lay of a surface is particularly important when two moving objects are in contact with each other, especially at high speeds.

8-22 Surface Control Symbols

Surface finishes are indicated on a drawing using surface control symbols. See Figure 8-37. The general surface control symbol looks like a check mark. Roughness values may be included with the symbol to specify the required accuracy. Surface control symbols can also be used to specify the manufacturing process that may or may not be used to produce a surface.

Figure 8-36

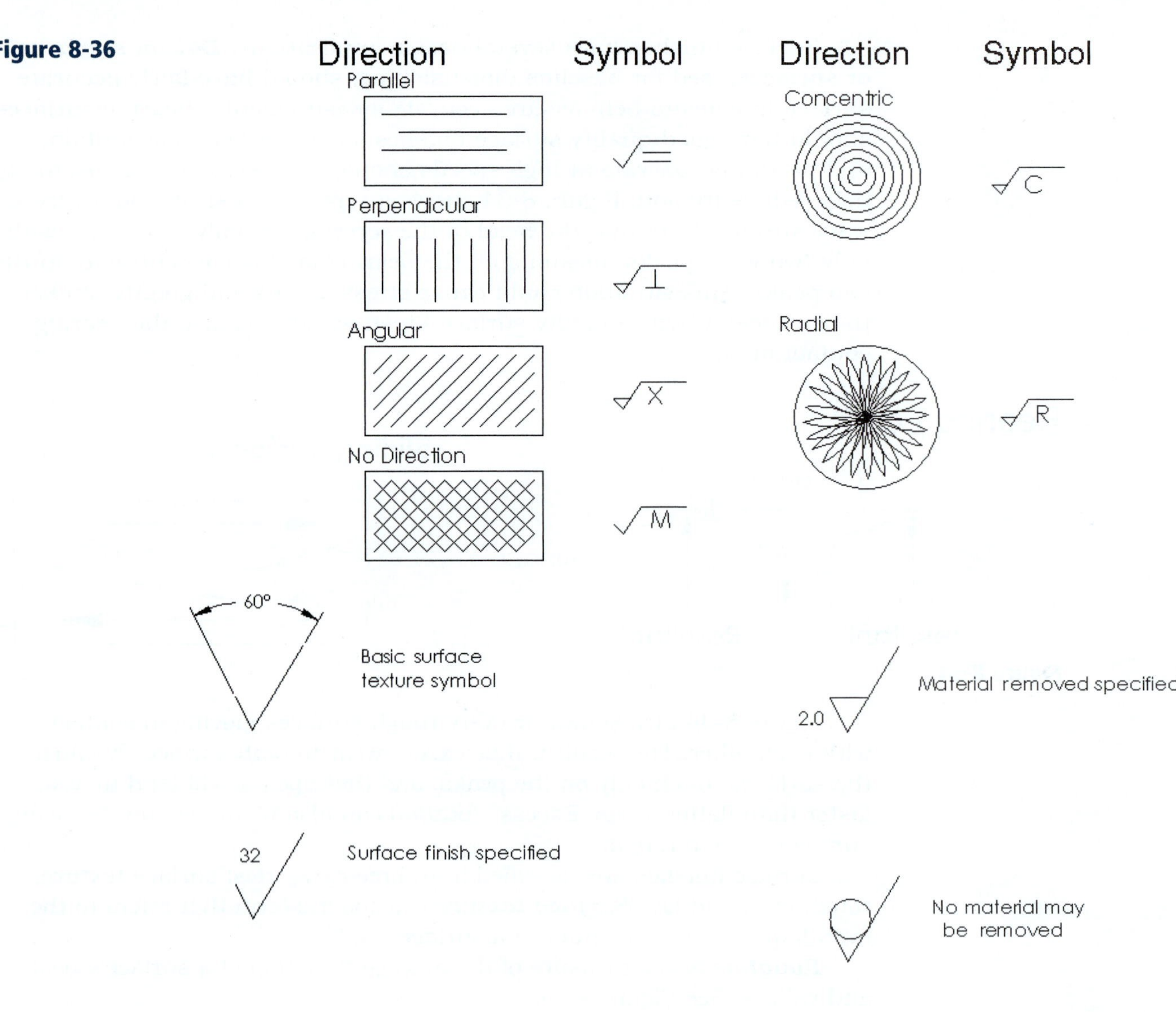

Figure 8-37 shows two applications of surface control symbols. In the first example, a 0.8-μm (32 μin.) surface finish is specified on the surface that serves as a datum for several horizontal dimensions. A 0.8-μm surface finish is generally considered the minimum acceptable finish for datums.

A second finish mark with a value of 0.4 μm is located on an extension line that refers to a surface that will be in contact with a moving object. The extra flatness will help prevent wear between the two surfaces.

8-23 Applying Surface Control Symbols

Figure 8-38 shows a dimensioned orthographic view. Surface symbols will be added to this view.

1 Click the **Annotation** tab and select the **Surface Finish** tool.

Figure 8-37

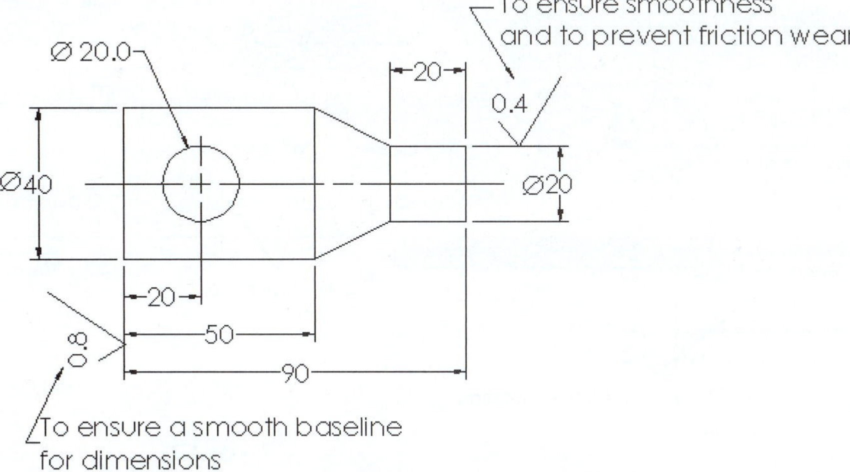

Figure 8-38

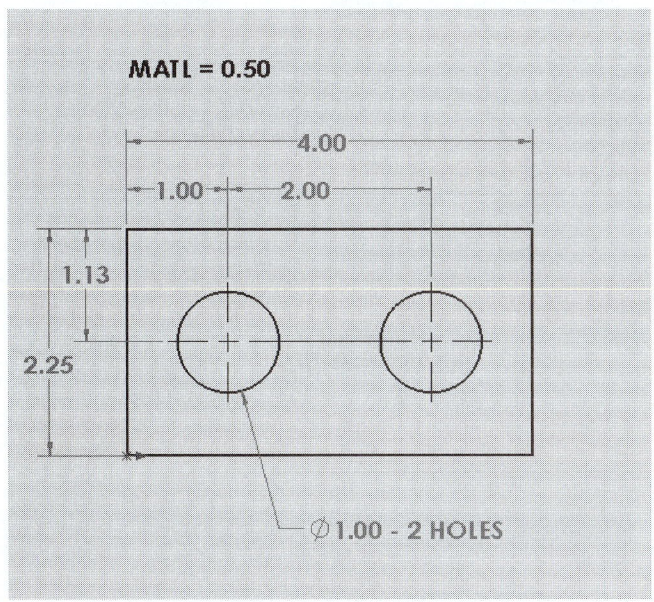

See Figure 8-39.

2 Select the **Basic** symbol and enter a value of **32**.

3 Move the cursor into the drawing area and locate the surface control symbol on an extension line as shown.

4 Click the **<Esc>** key.

5 Click the green **OK** check mark

6 Save the drawing.

To Add a Lay Symbol to a Drawing

Use the same drawing as in Figure 8-39.

1 Click the **Annotation** tab and select the **Surface Finish** tool.

See Figure 8-40.

2 Click the **Machine Required** symbol box.

3 Click the **Lay Direction** box and select the **Multi-Directional** option.

SOLIDWORKS File Edit View Insert Tools Window Help

Smart Dimension
Model Items
Spell Checker
Format Painter
Note
Linear Note Pattern
Balloon
Auto Balloon
Magnetic Line
Surface Finish
Weld Symbol
Hole Callout
Geometric Tolerance
Datum Feature
Datum Target
Blocks
Center Mark
Centerline
Area Hatch/Fill

View Layout Annotation Sketch Evaluate Office Products

BLOCK,finish
Annotations
Sheet1
Sheet Format1
Drawing View1

2. Click Surface Finish

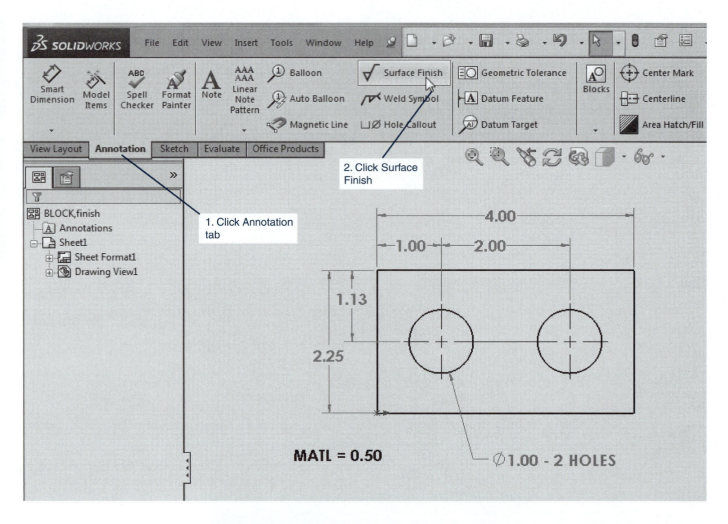

1. Click Annotation tab

4.00
1.00
2.00
1.13
2.25
MATL = 0.50
∅1.00 - 2 HOLES

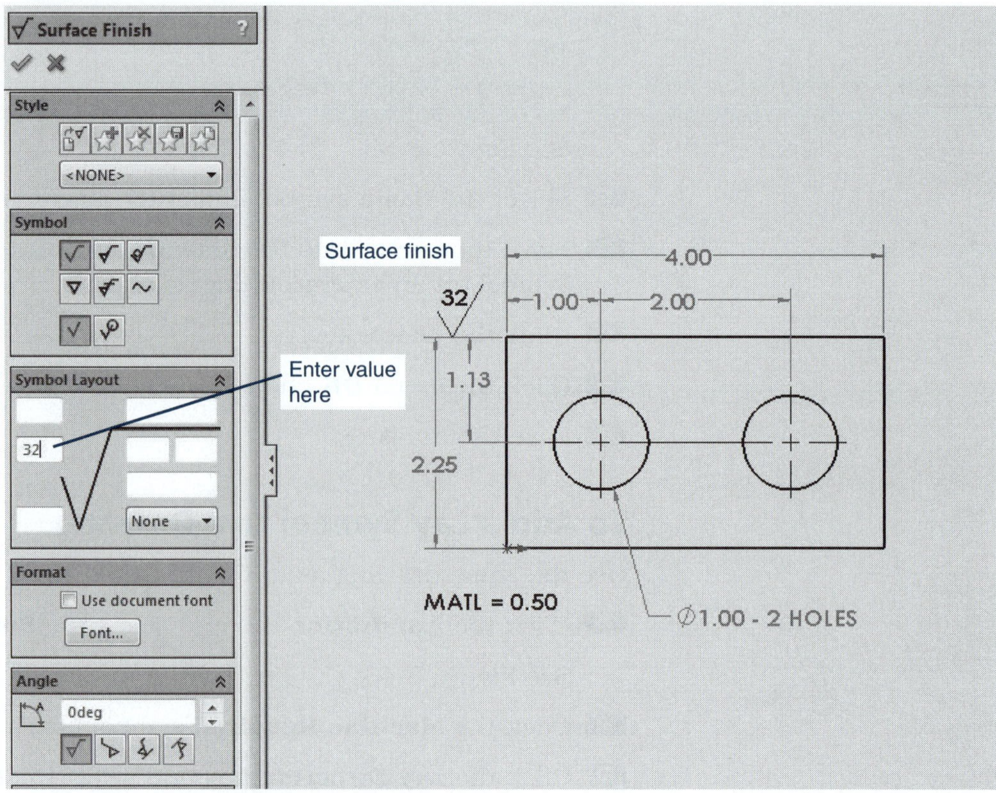

Figure 8-39

Figure 8-40

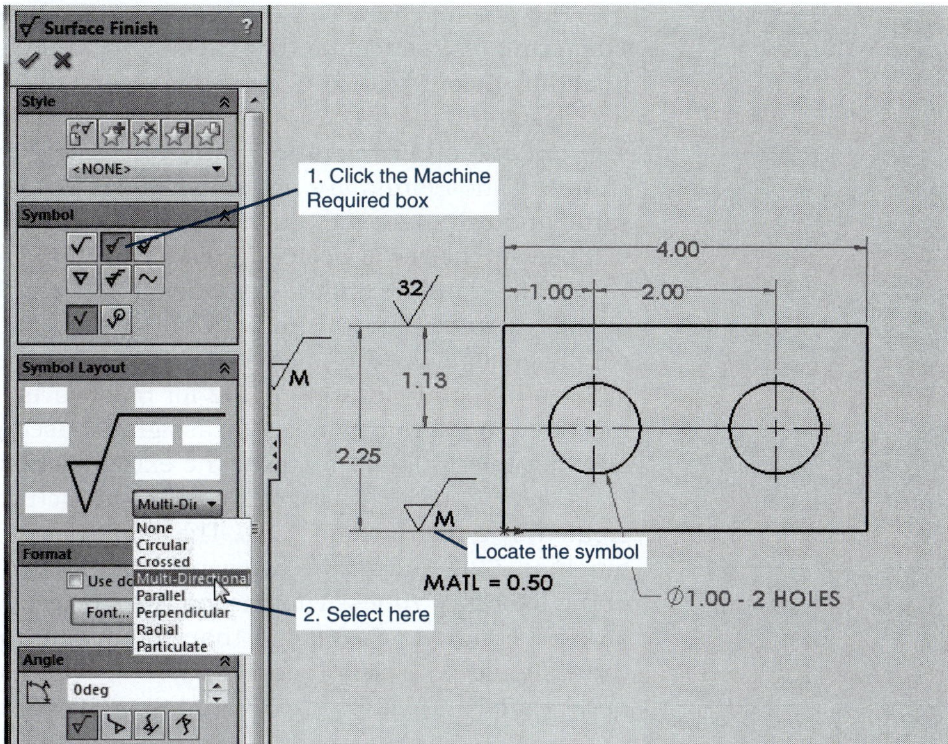

4 Move the cursor into the drawing area and locate the surface control symbol on an extension line as shown.

5 Click the **<Esc>** key.

6 Click the green **OK** check mark.

8-24 Design Problems

Figure 8-41 shows two objects that are to be fitted together using a fastener such as a screw-and-nut combination. For this example a cylinder will be used to represent a fastener. Only two nominal dimensions are given. The dimensions and tolerances were derived as follows.

Figure 8-41

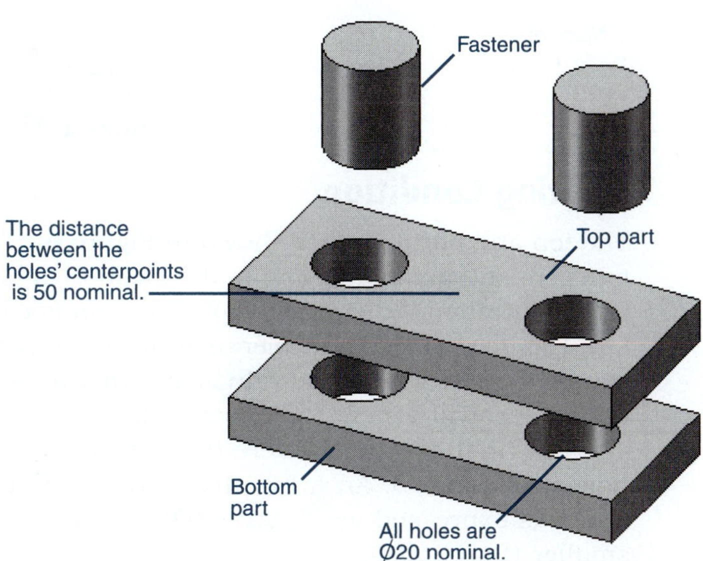

Chapter 8

The distance between the centers of the holes is given as 50 nominal. The term *nominal* means that the stated value is only a starting point. The final dimensions will be close to the given value but do not have to equal it.

Assigning tolerances is an iteration process; that is, a tolerance is selected and other tolerance values are calculated from the selected initial values. If the results are not satisfactory, go back and modify the initial value and calculate the other values again. As your experience grows you will become better at selecting realistic initial values.

In the example shown in Figure 8-41, start by assigning a tolerance of ±.01 to both the top and bottom parts for both the horizontal and vertical dimensions used to locate the holes. This means that there is a possible centerpoint variation of .02 for both parts. The parts must always fit together, so tolerances must be assigned based on the worst-case condition, or when the parts are made at the extreme ends of the assigned tolerances.

Figure 8-42 shows a greatly enlarged picture of the worst-case condition created by a tolerance of ±.01. The centerpoints of the holes could be as much as .028 apart if the two centerpoints were located at opposite corners of the tolerance zones. This means that the minimum hole diameter must always be at least .028 larger than the maximum stud diameter. In addition, there should be a clearance tolerance assigned so that the hole and stud are never exactly the same size. Figure 8-43 shows the resulting tolerances.

TIP

The tolerance zones in this section are created by line dimensions that generated square tolerance zones.

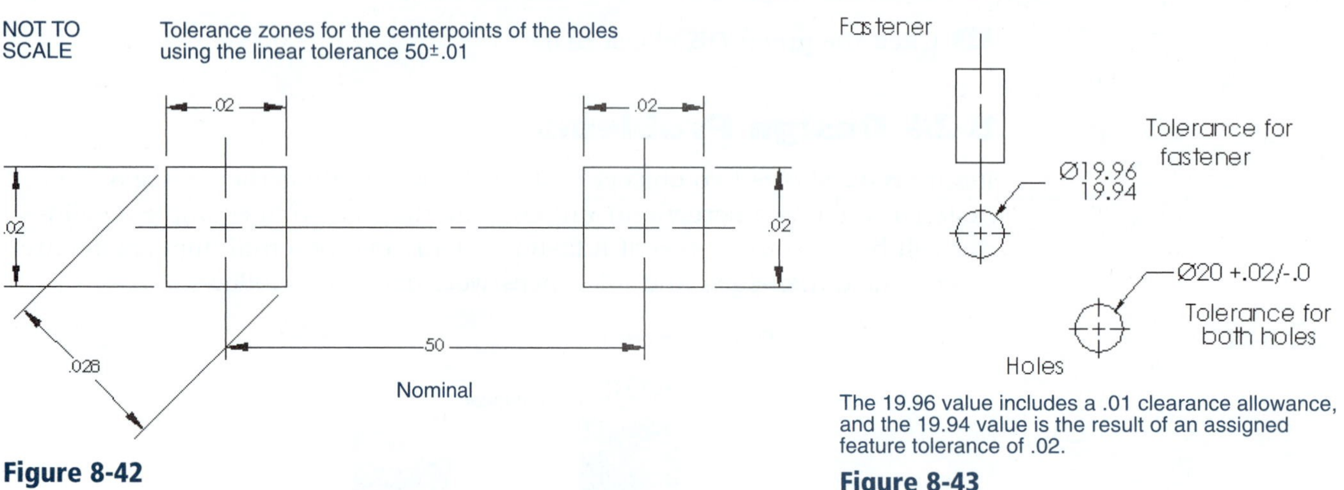

Figure 8-42

The 19.96 value includes a .01 clearance allowance, and the 19.94 value is the result of an assigned feature tolerance of .02.

Figure 8-43

Floating Condition

The top and bottom parts shown in Figure 8-41 are to be joined by two independent fasteners; that is, the location of one fastener does not depend on the location of the other. This situation is called a ***floating condition***.

This means that the tolerance zones for both the top and bottom parts can be assigned the same values and that a fastener diameter selected to fit one part will also fit the other part.

The final tolerances were developed by first defining a minimum hole size of 20.00. An arbitrary tolerance of .02 was assigned to the hole and was expressed as 20.00 +.02/−0, so the hole can never be any smaller than 20.00.

The 20.00 minimum hole diameter dictates that the maximum fastener diameter can be no greater than 19.97, or .03 (the rounded-off diagonal distance across the tolerance zone—.028) less than the minimum hole diameter. A .01 clearance was assigned. The clearance ensures that the hole and fastener are never exactly the same diameter. The resulting maximum allowable diameter for the fastener is 19.96. Again, an arbitrary tolerance of .02 was assigned to the fastener. The final fastener dimensions are therefore 19.96 to 19.94.

The assigned tolerances ensure that there will always be at least .01 clearance between the fastener and the hole. The other extreme condition occurs when the hole is at its largest possible size (20.02) and the fastener is at its smallest (19.94). This means that there could be as much as .08 clearance between the parts. If this much clearance is not acceptable, then the assigned tolerances will have to be reevaluated.

Figure 8-44 shows the top and bottom parts dimensioned and toleranced. Any dimensions that do not have assigned tolerances are assumed to have standard tolerances.

Figure 8-44

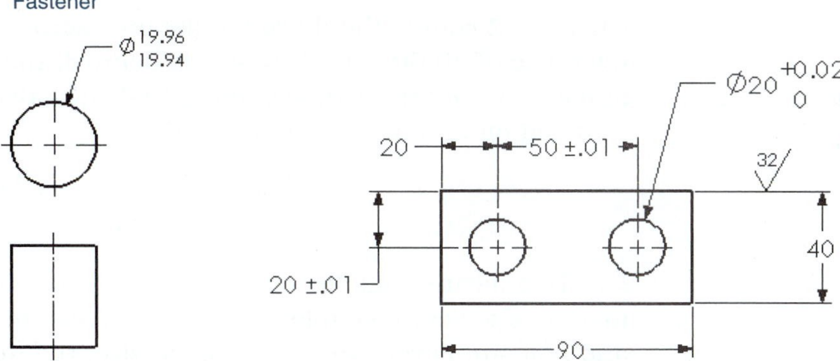

Note, in Figure 8-44, that the top edge of each part has been assigned a surface finish. This was done to help ensure the accuracy of the 20±.01 dimension. If this edge surface was rough, it could affect the tolerance measurements.

This example will be done later in the chapter using geometric tolerances. Geometric tolerance zones are circular rather than rectangular.

Fixed Condition

Figure 8-45 shows the same nominal conditions presented in Figure 8-41, but the fasteners are now fixed to the top part. This situation is called the

Figure 8-45

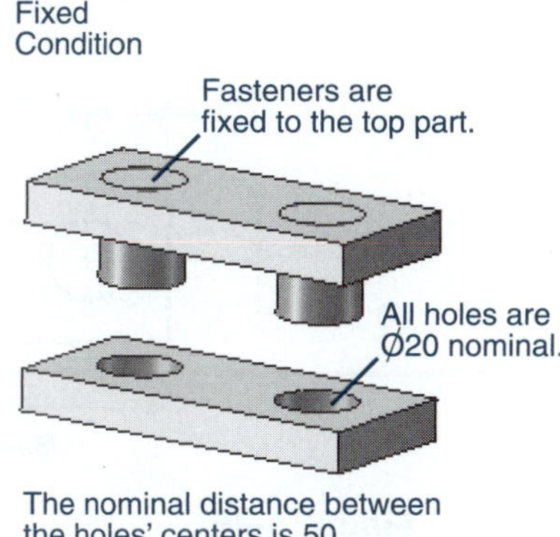

fixed condition. In analyzing the tolerance zones for the fixed condition, two positional tolerances must be considered: the positional tolerances for the holes in the bottom part, and the positional tolerances for the fixed fasteners in the top part. This relationship may be expressed in an equation, as follows:

$$S_{max} + DTSZ = H_{min} - DTZ$$

where

S_{max} = maximum shaft (fastener) diameter

H_{min} = minimum hole diameter

DTSZ = diagonal distance across the shaft's centerpoint tolerance zone

DTZ = diagonal distance across the hole's centerpoint tolerance zone

If a dimension and tolerance of 50±.01 is assigned to both the center distance between the holes and the center distance between the fixed fasteners, the values for DTSZ and DTZ will be equal. The formula can then be simplified as follows.

$$S_{max} = H_{min} - 2(DTZ)$$

where DTZ equals the diagonal distance across the tolerance zone. If a hole tolerance of 20.00 +.02/−0 also is defined, the resulting maximum shaft size can be determined, assuming that the calculated distance of .028 is rounded off to .03. See Figure 8-46.

$$S_{max} = 20.00 - 2(0.03)$$
$$= 19.94$$

This means that 19.94 is the largest possible shaft diameter that will just fit. If a clearance tolerance of .01 is assumed to ensure that the shaft and hole are never exactly the same size, the maximum shaft diameter becomes 19.93. See Figure 8-47.

A feature tolerance of .02 on the shaft will result in a minimum shaft diameter of 19.91. Note that the .01 clearance tolerance and the .02 feature tolerance were arbitrarily chosen. Other values could have been used.

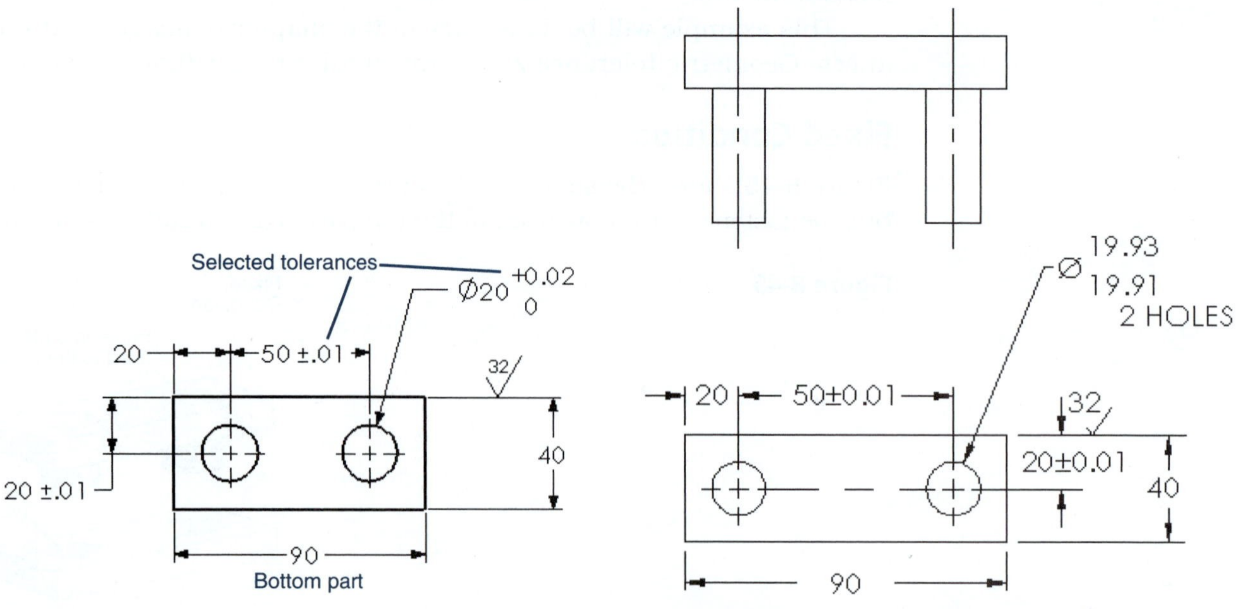

Figure 8-46

Figure 8-47

Designing a Hole Given a Fastener Size

The previous two examples started by selecting a minimum hole diameter and then calculating the resulting fastener size. Figure 8-48 shows a situation in which the fastener size is defined, and the problem is to determine the appropriate hole sizes. Figure 8-49 shows the dimensions and tolerances for both top and bottom parts.

Requirements:

Clearance, minimum = .003

Hole tolerances =.005

Positional tolerance = .002

Figure 8-48

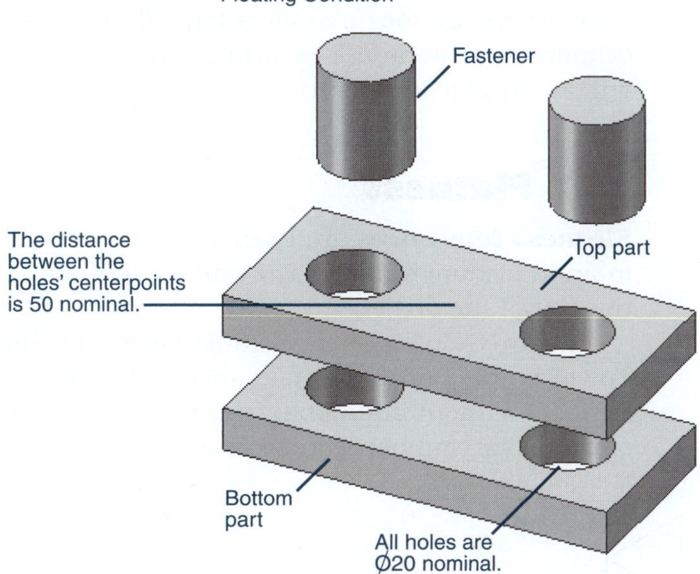
Floating Condition
Fastener
Top part
The distance between the holes' centerpoints is 50 nominal.
Bottom part
All holes are Ø20 nominal.

Figure 8-49

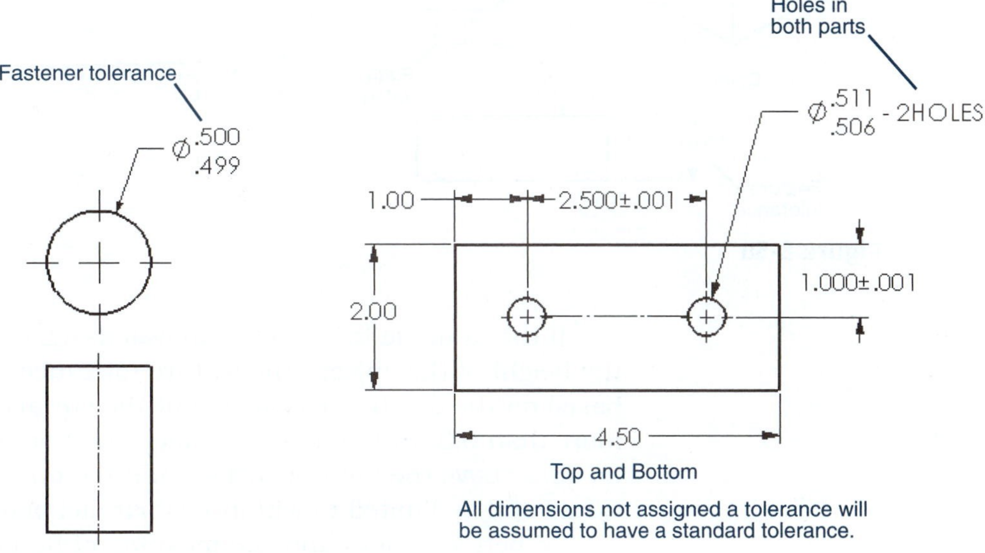
Fastener tolerance
Ø .500 .499
Holes in both parts
Ø .511 .506 - 2HOLES
1.00
2.500±.001
1.000±.001
2.00
4.50
Top and Bottom
All dimensions not assigned a tolerance will be assumed to have a standard tolerance.

8-25 Geometric Tolerances

Geometric tolerancing is a dimensioning and tolerancing system based on the geometric shape of an object. Surfaces may be defined in terms of their flatness or roundness, or in terms of how perpendicular or parallel they are to other surfaces.

Geometric tolerances allow a more exact definition of the shape of an object than do conventional coordinate-type tolerances. Objects can be toleranced in a manner more closely related to their design function or so that their features and surfaces are more directly related to each other.

8-26 Tolerances of Form

Tolerances of form are used to define the shape of a surface relative to itself. There are four classifications: flatness, straightness, roundness, and cylindricity. Tolerances of form are not related to other surfaces but apply only to an individual surface.

8-27 Flatness

Flatness tolerances are used to define the amount of variation permitted in an individual surface. The surface is thought of as a plane not related to the rest of the object.

Figure 8-50 shows a rectangular object. How flat is the top surface? The given plus or minus tolerances allow a variation of (±0.5) across the surface. Without additional tolerances the surface could look like a series of waves varying between 30.5 and 29.5.

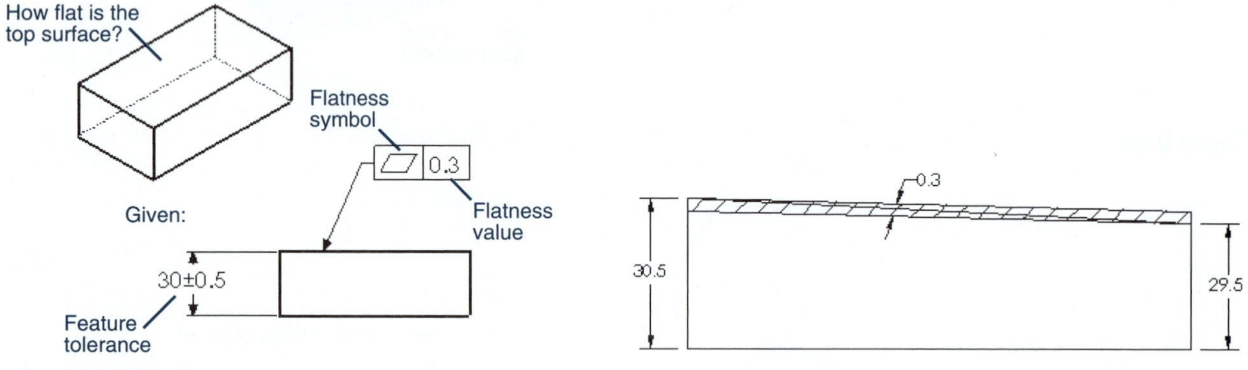

Figure 8-50

If the example in Figure 8-50 was assigned a flatness tolerance of 0.3, the height of the object—the feature tolerance—could continue to vary based on the 30±0.5 tolerance, but the surface itself could not vary by more than 0.3. In the most extreme condition, one end of the surface could be 30.5 above the bottom surface and the other end 29.5, but the surface would still be limited to within two parallel planes 0.3 apart as shown.

To better understand the meaning of flatness, consider how the surface would be inspected. The surface would be acceptable if a gauge could be moved all around the surface and never vary by more than 0.3. See Figure 8-51. Every point in the plane must be within the specified tolerance.

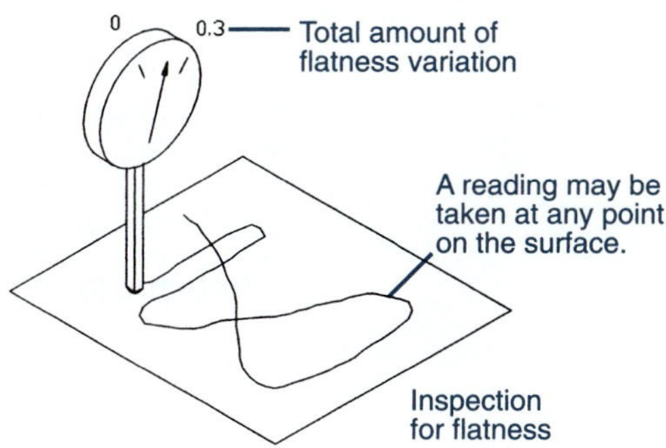

Figure 8-51

8-28 Straightness

Straightness tolerances are used to measure the variation of an individual feature along a straight line in a specified direction. Figure 8-52 shows an object with a straightness tolerance applied to its top surface. Straightness differs from flatness because straightness measurements are checked by moving a gauge directly across the surface in a single direction. The gauge is not moved randomly about the surface, as is required by flatness.

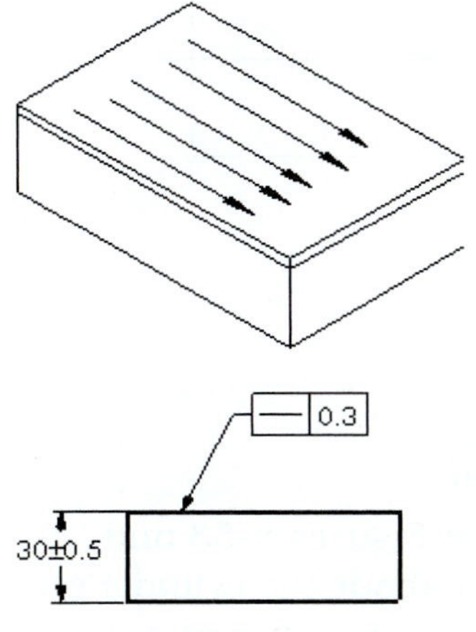

Figure 8-52

Straightness tolerances are most often applied to circular or matching objects to help ensure that the parts are not barreled or warped within the given feature tolerance range and, therefore, do not fit together well. Figure 8-53 shows a cylindrical object dimensioned and toleranced using a standard feature tolerance. The surface of the cylinder may vary within the specified tolerance range as shown.

Figure 8-53

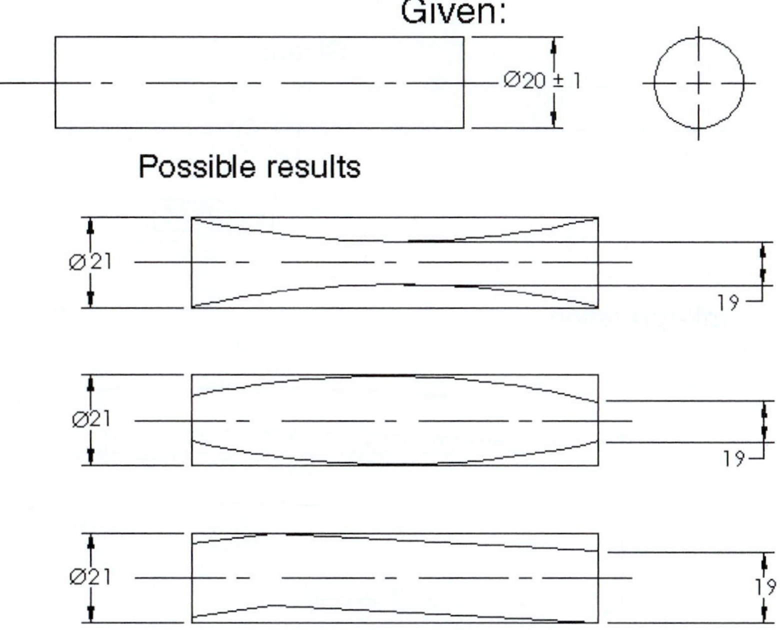

Figure 8-54 shows the same object shown in Figure 8-53 dimensioned and toleranced using the same feature tolerance but also including a 0.05 straightness tolerance. The straightness tolerance limits the surface variation to 0.05 as shown.

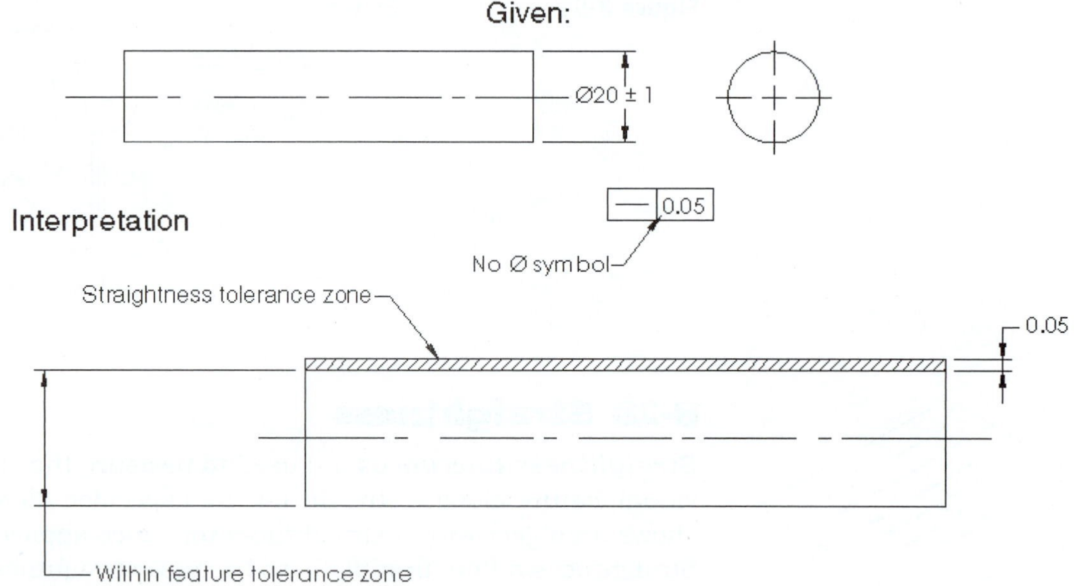

Figure 8-54

8-29 Straightness (RFS and MMC)

Figure 8-55 again shows the same cylinder shown in Figures 8-53 and 8-54. This time the straightness tolerance is applied about the cylinder's centerline. This type of tolerance permits the feature tolerance and geometric tolerance to be used together to define a *virtual condition.* A virtual condition is used to determine the maximum possible size variation of the cylinder or the smallest diameter hole that will always accept the cylinder.

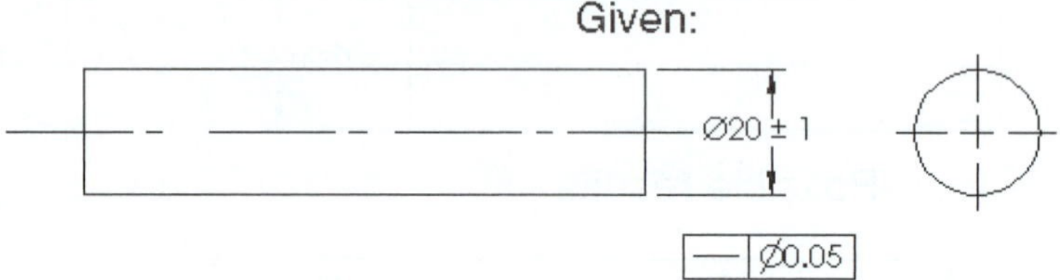

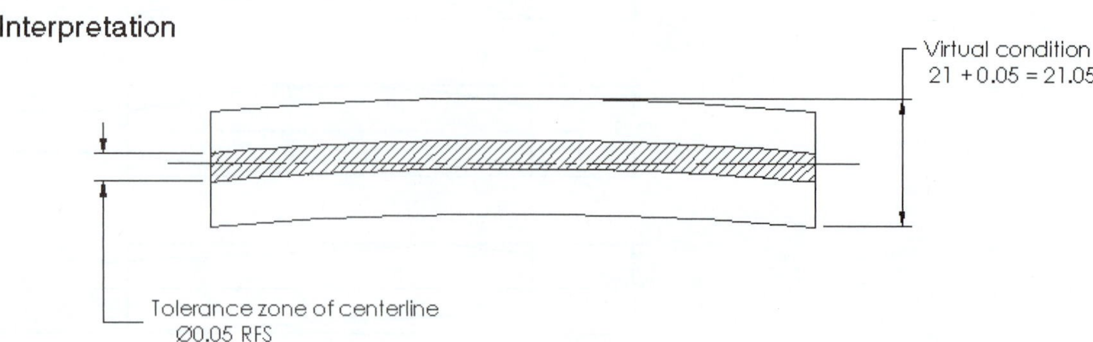

Figure 8-55

The geometric tolerance specified in Figure 8-55 is applied to any circular segment along the cylinder, regardless of the cylinder's diameter. This means that the 0.05 tolerance is applied equally when the cylinder's diameter measures 19 or when it measures 21. This application is called

RFS, ***regardless of feature size***. RFS condition applies if no material condition is specified. In Figure 8-55 no symbol is listed after the 0.05 value, so it is assumed to be applied RFS.

Figure 8-56 shows the cylinder dimensioned with an MMC condition applied to the straightness tolerance. MMC stands for ***maximum material condition*** and means that the specified straightness tolerance (0.05) is applied only at the MMC condition or when the cylinder is at its maximum diameter size (21).

Figure 8-56

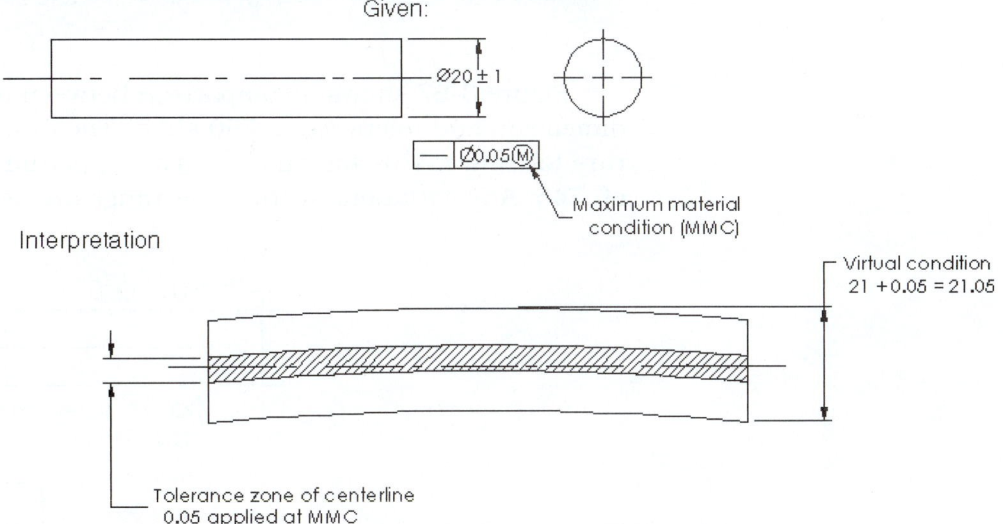

Measured Size	Allowable Tolerance Zone	Virtual Condition
21.0	0.05	21.05
20.9	0.15	21.15
20.8	0.25	21.25
.	.	.
.	.	.
.	.	.
20.0	1.05	22.05
.	.	.
.	.	.
.	.	.
19.0	2.05	23.05

A shaft is an external feature, so its largest possible size or MMC occurs when it is at its maximum diameter. A hole is an internal feature. A hole's MMC condition occurs when it is at its smallest diameter. The MMC condition for holes will be discussed later in the chapter along with positional tolerances.

Applying a straightness tolerance at MMC allows for a variation in the resulting tolerance zone. Because the 0.05 flatness tolerance is applied at MMC, the virtual condition is still 21.05, the same as with the RFS condition; however, the tolerance is applied only at MMC. As the cylinder's diameter varies within the specified feature tolerance range the acceptable tolerance zone may vary to maintain the same virtual condition.

The table in Figure 8-56 shows how the tolerance zone varies as the cylinder's diameter varies. When the cylinder is at its largest size or MMC, the tolerance zone equals 0.05, or the specified flatness variation. When the cylinder is at its smallest diameter, the tolerance zone equals 2.05, or the total feature size plus the total flatness size. In all variations the virtual size remains the same, so at any given cylinder diameter value, the size of the tolerance zone can be determined by subtracting the cylinder's diameter value from the virtual condition.

NOTE

Geometric tolerance applied at MMC allows the tolerance zone to grow.

Figure 8-57 shows a comparison between different methods used to dimension and tolerance a .750 shaft. The first example uses only a feature tolerance. This tolerance sets an upper limit of .755 and a lower limit of .745. Any variations within that range are acceptable.

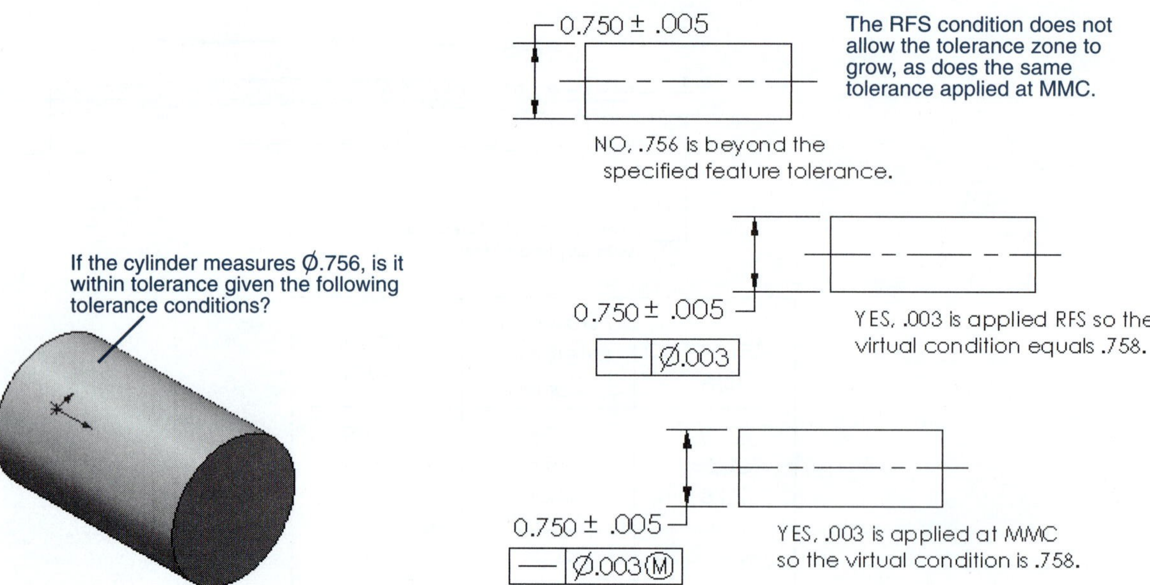

Figure 8-57

The second example in Figure 8-57 sets a straightness tolerance of .003 about the cylinder's centerline. No conditions are defined, so the tolerance is applied RFS. This limits the variations in straightness to .003 at all feature sizes. For example, when the shaft is at its smallest possible feature size of .745, the .003 still applies. This means that a shaft measuring .745 that had a straightness variation greater than .003 would be rejected. If the tolerance had been applied at MMC, the part would be accepted. This does not mean that straightness tolerances should always be applied at MMC. If straightness is critical to the design integrity or function of the part, then straightness should be applied in the RFS condition.

The third example in Figure 8-57 applies the straightness tolerance about the centerline at MMC. This tolerance creates a virtual condition of .758. The MMC condition allows the tolerance to vary as the feature tolerance varies, so when the shaft is at its smallest feature size, .745, a straightness tolerance of .003 is acceptable (.005 feature tolerance + .003 straightness tolerance).

If the tolerance specification for the cylinder shown in Figure 8-57 was 0.000 applied at MMC, it would mean that the shaft would have to be perfectly straight at MMC or when the shaft was at its maximum value (.755); however, the straightness tolerance can vary as the feature size varies, as discussed for the other tolerance conditions. A 0.000 tolerance means that the MMC and the virtual conditions are equal.

Figure 8-58 shows a very long .750 diameter shaft. Its straightness tolerance includes a length qualifier that serves to limit the straightness variations over each inch of the shaft length and to prevent excess waviness over the full length. The tolerance .002/1.000 means that the total straightness may vary over the entire length of the shaft by .003 but that the variation is limited to .002 per 1.000 of shaft length.

Figure 8-58

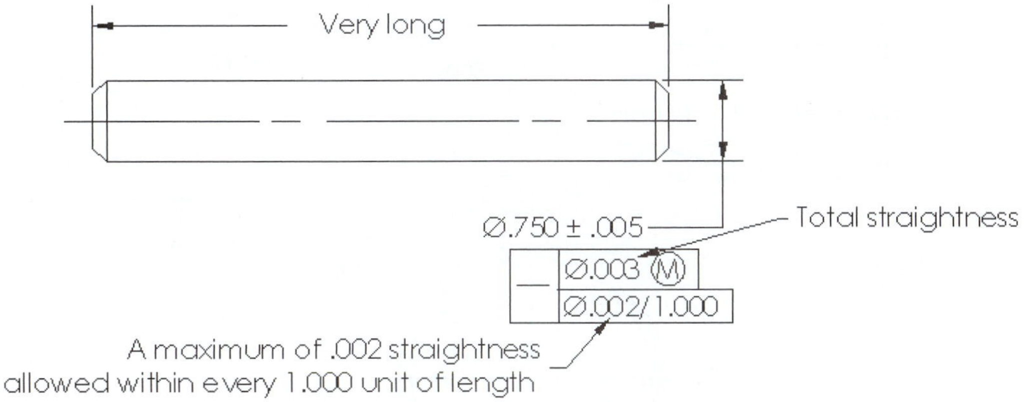

8-30 Circularity

A ***circularity tolerance*** is used to limit the amount of variation in the roundness of a surface of revolution. It is measured at individual cross sections along the length of the object. The measurements are limited to the individual cross sections and are not related to other cross sections. This means that in extreme conditions the shaft shown in Figure 8-59

Figure 8-59

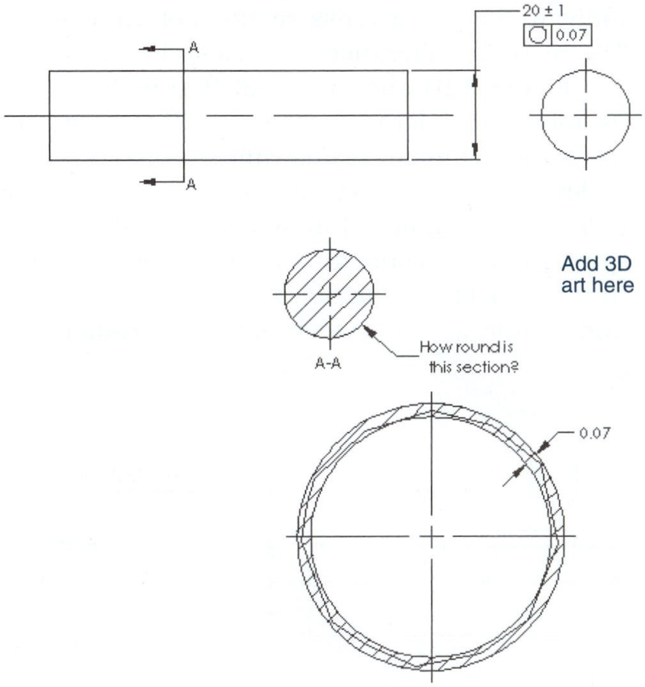

could actually taper from a diameter of 21 to a diameter of 19 and never violate the circularity requirement. It also means that qualifications such as MMC cannot be applied.

Figure 8-59 shows a shaft that includes a feature tolerance and a circularity tolerance of 0.07. To understand circularity tolerances, consider an individual cross section or slice of the cylinder. The actual shape of the outside edge of the slice varies around the slice. The difference between the maximum diameter and the minimum diameter of the slice can never exceed the stated circularity tolerance.

Circularity tolerances can be applied to tapered sections and spheres, as shown in Figure 8-60. In both applications, circularity is measured around individual cross sections, as it was for the shaft shown in Figure 8-59.

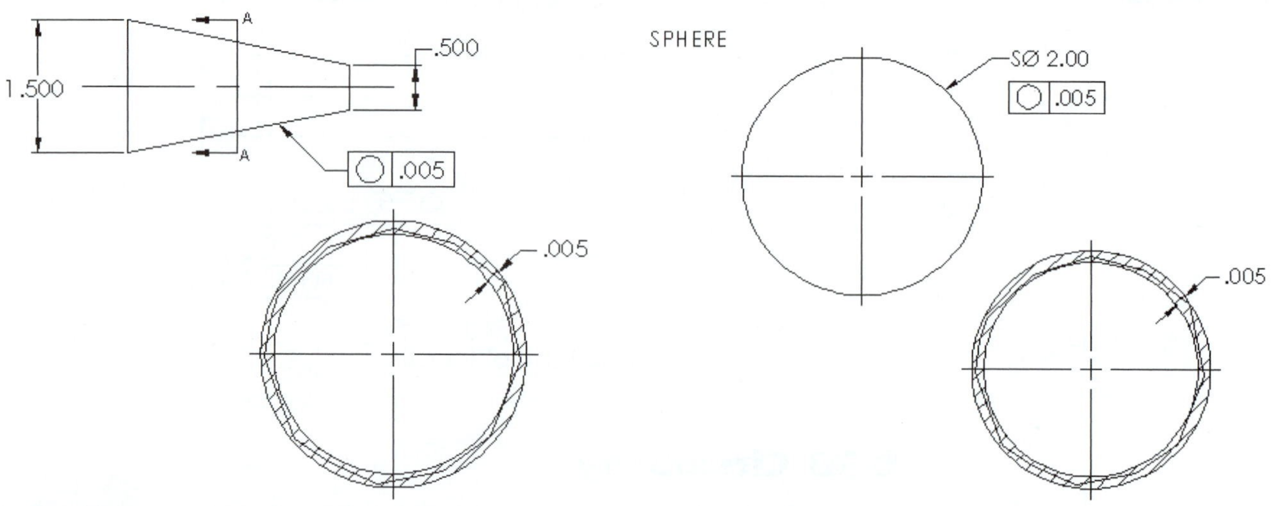

Figure 8-60

8-31 Cylindricity

Cylindricity tolerances are used to define a tolerance zone both around individual circular cross sections of an object and also along its length. The resulting tolerance zone looks like two concentric cylinders.

Figure 8-61 shows a shaft that includes a cylindricity tolerance that establishes a tolerance zone of .007. This means that if the maximum measured diameter is determined to be .755, the minimum diameter cannot be less than .748 anywhere on the cylindrical surface. Cylindricity and circularity are somewhat analogous to flatness and straightness. Flatness and cylindricity are concerned with variations across an entire surface or plane. In the case of cylindricity, the plane is shaped like a cylinder. Straightness and circularity are concerned with variations of a single element of a

Figure 8-61

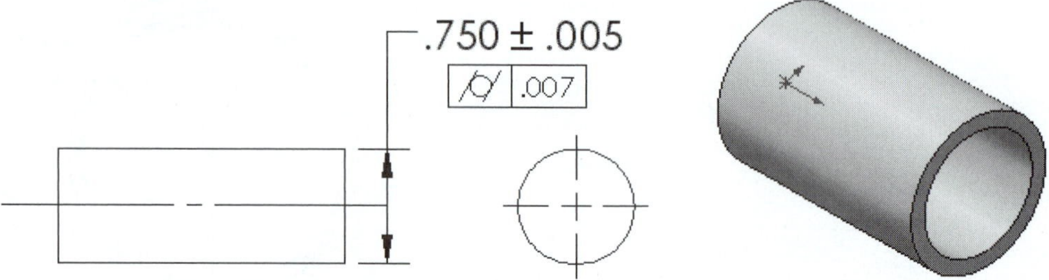

surface: a straight line across the plane in a specified direction for straightness, and a path around a single cross section for circularity.

8-32 Geometric Tolerances Using SolidWorks

Geometric tolerances are tolerances that limit dimensional variations based on the geometric properties. Figure 8-62 shows three different ways geometric tolerance boxes can be added to a drawing.

Figure 8-63 shows lists of geometric tolerance symbols.

Figure 8-62

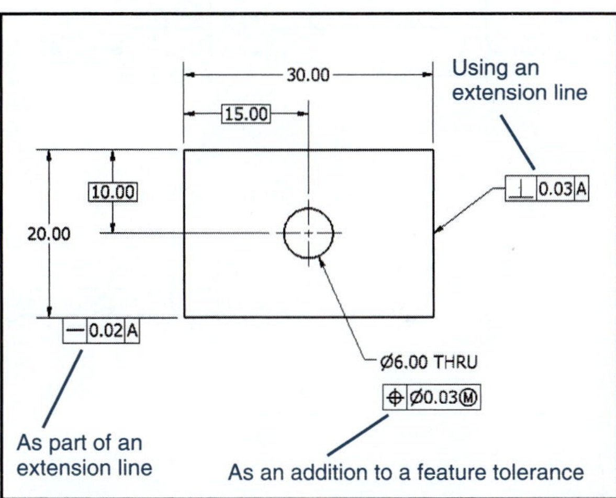

	TYPE OF TOLERANCE	CHARACTERISTIC	SYMBOL
FOR INDIVIDUAL FEATURES	FORM	STRAIGHTNESS	—
		FLATNESS	▱
		CIRCULARITY	○
		CYLINDRICITY	⌭
INDIVIDUAL OR RELATED FEATURES	PROFILE	PROFILE OF A LINE	⌒
		PROFILE OF A SURFACE	⌓
RELATED FEATURES	ORIENTATION	ANGULARITY	∠
		PERPENDICULARITY	⊥
		PARALLELISM	//
	LOCATION	POSITION	⌖
		CONCENTRICITY	◎
	RUNOUT	CIRCULAR RUNOUT	↗
		TOTAL RUNOUT	⌰

TERM	SYMBOL
AT MAXIMUM MATERIAL CONDITION	Ⓜ
REGARDLESS OF FEATURE SIZE	Ⓢ
AT LEAST MATERIAL CONDITION	Ⓛ
PROJECTED TOLERANCE ZONE	Ⓟ
DIAMETER	Ø
SPHERICAL DIAMETER	SØ
RADIUS	R
SPHERICAL RADIUS	SR
REFERENCE	( )
ARC LENGTH	⌒

Figure 8-63

8-33 Datums

A ***datum*** is a point, axis, or surface used as a starting reference point for dimensions and tolerances. Figure 8-64 and Figure 8-65 show a rectangular object with three datum planes labeled –A–, –B–, and –C–. The three datum planes are called the primary, secondary, and tertiary datums, respectively. The three datum planes are, by definition, exactly 90° to one another.

Figure 8-66 shows a cylindrical datum frame that includes three datum planes. The X and Y planes are perpendicular to each other, and the base A plane is perpendicular to the datum axis between the X and Y planes.

Figure 8-64

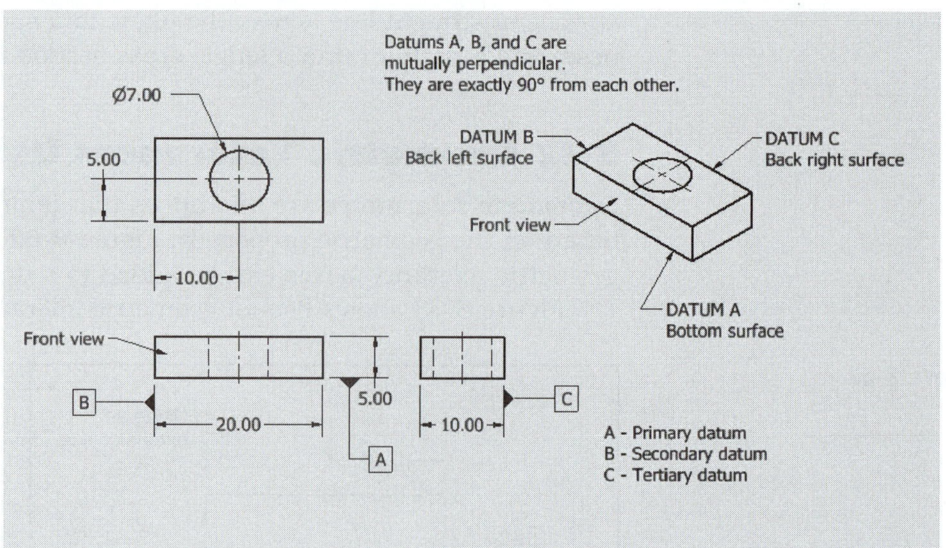

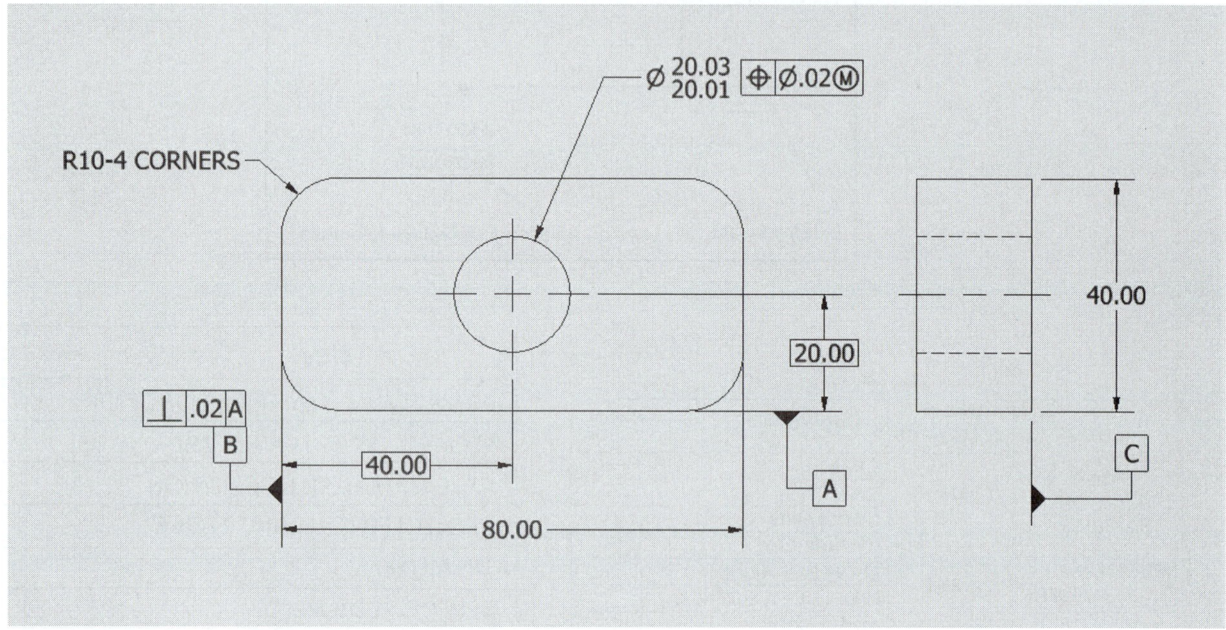

Figure 8-65

Figure 8-66

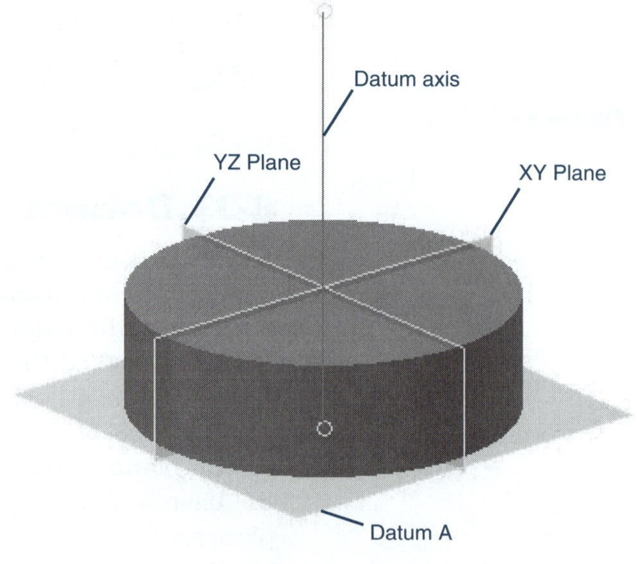

Figure 8-66
(*Continued*)

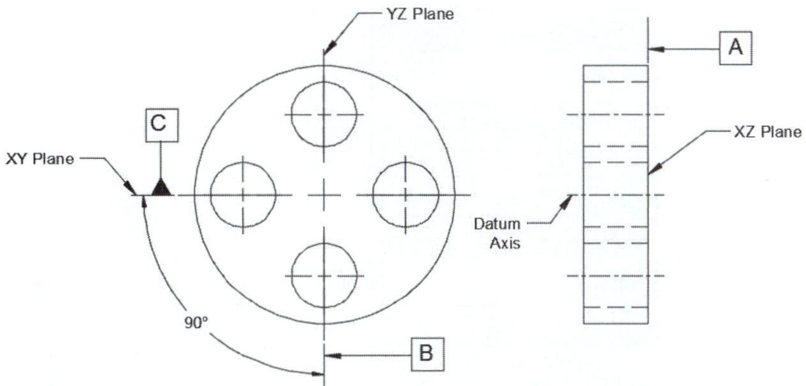

Datum planes are assumed to be perfectly flat. When assigning a datum status to a surface, be sure that the surface is reasonably flat. This means that datum surfaces should be toleranced using surface finishes, or created using machine techniques that produce flat surfaces.

To Add a Datum Indicator

Figure 8-67 shows a dimensioned orthographic view. This section shows how to define the lower surface as a datum feature, datum A. Note that a surface finish of 16 has been added to the lower surface. Datum surfaces are assumed to be flat and smooth.

1 Click the **Annotation** tab and select the **Datum Feature** tool.

The **Datum Feature PropertyManager** will appear.

2 Enter the letter **A** and select the **Filled Triangle With Shoulder** option.

Figure 8-67

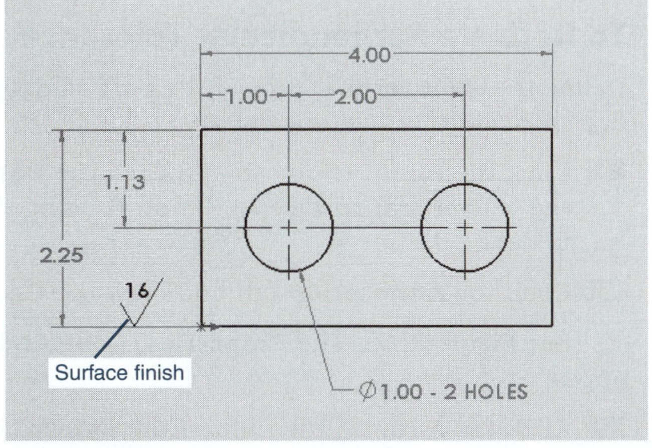

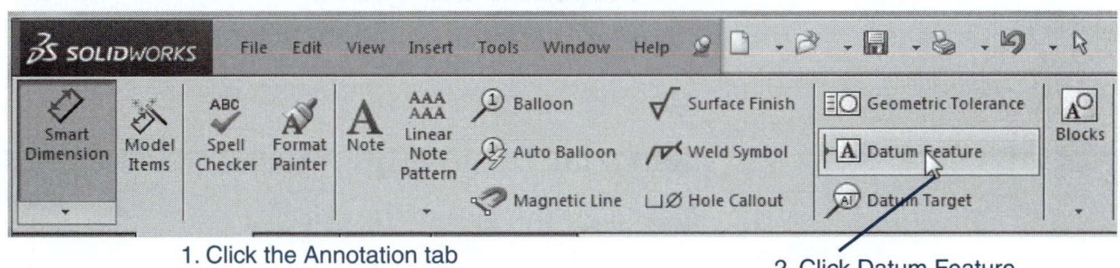

Figure 8-67
(*Continued*)

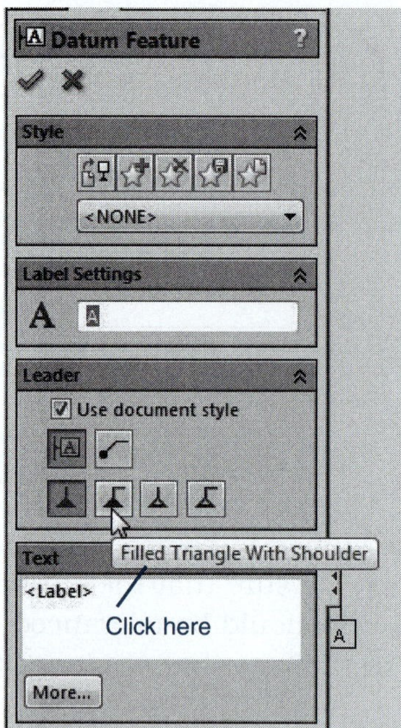

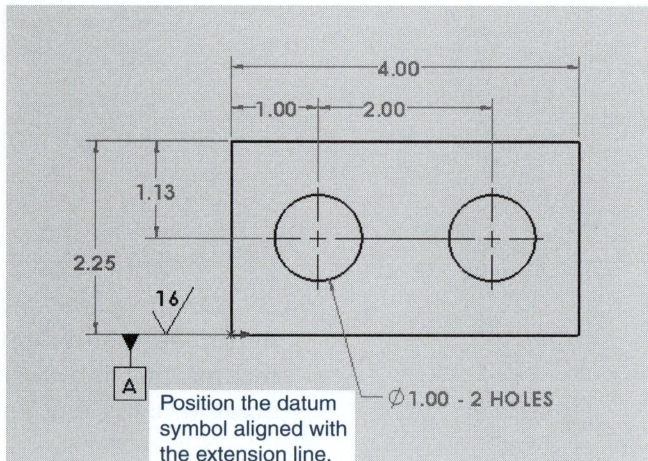

3. Locate the datum symbol on the drawing as shown.
4. Click the **<Esc>** key.
5. Click the green **OK** check mark.

> **TIP**
> Once in place, a datum symbol can be repositioned by clicking and dragging the symbol.

To Define a Perpendicular Tolerance

Define the right vertical edge of the part as datum B and perpendicular to datum A within a tolerance of .001.

1. Click the **Annotation** tab and click the **Datum Feature** tool and define the left vertical surface as datum **B** using the same procedure as used to define datum A.
2. Click the **Annotation** tab and click the **Geometric Tolerance** tool.

See Figure 8-68. The **Properties/Geometric Tolerance** dialog box will appear.

3. Click the arrow to the right of the **Symbol** box and select the **Perpendicular** symbol.
4. Enter the **Tolerance 1** and **Primary** datum information.
5. Click **OK**.
6. Locate the geometric symbol on the extension line as shown.
7. Click the green **OK** check mark.

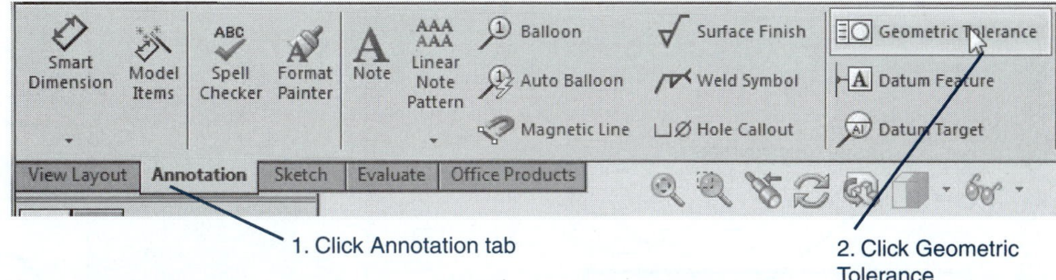

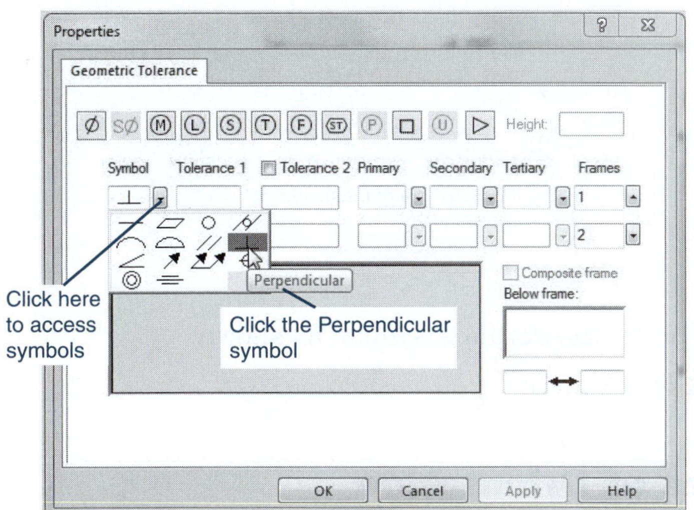

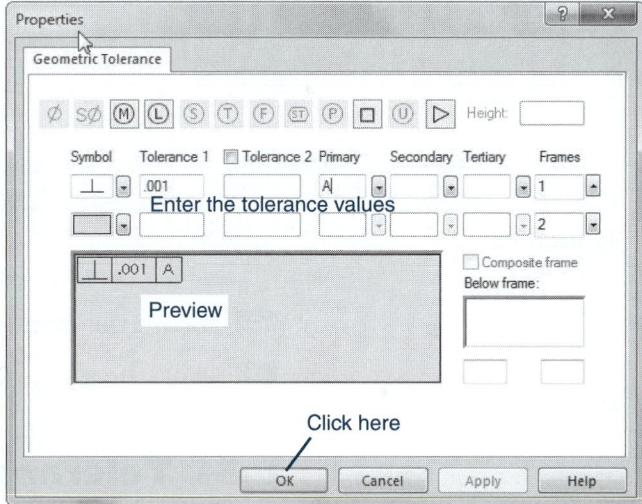

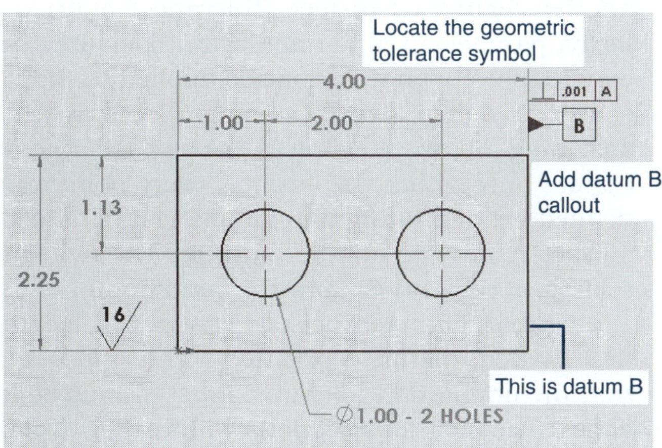

Figure 8-68

To Define a Straightness Value for Datum Surface A

The surface finish value of 16 defines the smoothness of the surface but not the straightness. Think of surface finish as waves, and straightness as an angle. A straightness value of .002 indicates that there is a band of width .002 located .001 on either side of a theoretically perfectly straight line. The surface may vary but must always be within the .002 boundary. See Figure 8-69.

1. Click the **Annotation** tab and select the **Geometric Tolerance** tool.
2. Select the **Straightness** symbol, and define the tolerance as **.002**.
3. Click **OK**.

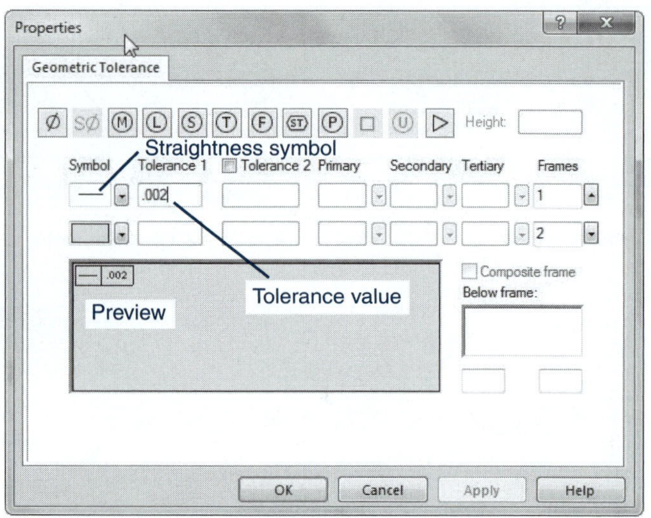

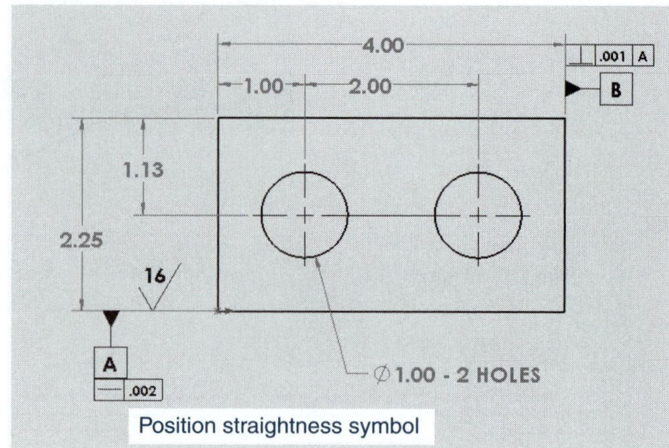

Figure 8-69

4 Locate the symbol under the datum A symbol as shown.

5 Click the green **OK** check mark.

8-34 Tolerances of Orientation

Tolerances of orientation are used to relate a feature or surface to another feature or surface. Tolerances of orientation include perpendicularity, parallelism, and angularity. They may be applied using RFS or MMC conditions, but they cannot be applied to individual features by themselves. To define a surface as parallel to another surface is very much like assigning a flatness value to the surface. The difference is that flatness applies only within the surface; every point on the surface is related to a defined set of limiting parallel planes. Parallelism defines every point in the surface relative to another surface. The two surfaces are therefore directly related to each other, and the condition of one affects the other.

Orientation tolerances are used with locational tolerances. A feature is first located, then it is oriented within the locational tolerances. This means that the orientation tolerance must always be less than the locational tolerances. The next four sections will further explain this requirement.

8-35 Perpendicularity

Perpendicularity tolerances are used to limit the amount of variation for a surface or feature within two planes perpendicular to a specified datum. Figure 8-70 shows a rectangular object. The bottom surface is assigned as datum A, and the right vertical edge is toleranced so that it must be perpendicular within a limit of 0.05 to datum A. The perpendicularity tolerance defines a tolerance zone 0.05 wide between two parallel planes that are perpendicular to datum A.

The object also includes a horizontal dimension and tolerance of 40 ± 1. This tolerance is called a *locational tolerance* because it serves to locate the right edge of the object. As with rectangular coordinate tolerances, discussed earlier in the chapter, the 40 ± 1 controls the location of the edge—how far away or how close it can be to the left edge—but does

Figure 8-70

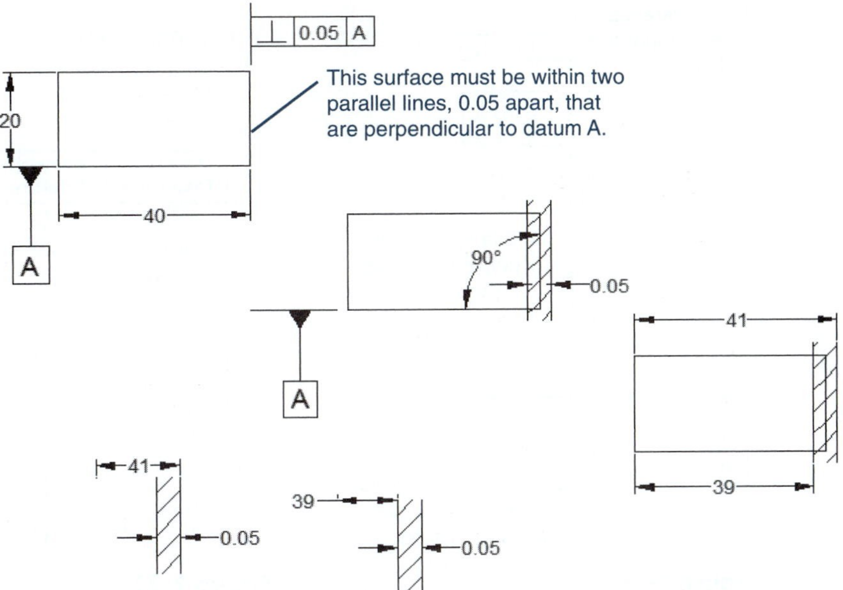

not directly control the shape of the edge. Any shape that falls within the specified tolerance range is acceptable. This may in fact be sufficient for a given design, but if a more controlled shape is required, a perpendicularity tolerance must be added. The perpendicularity tolerance works within the locational tolerance to ensure that the edge is not only within the locational tolerance but is also perpendicular to datum A.

Figure 8-70 shows the two extreme conditions for the 40±1 locational tolerance. The perpendicularity tolerance is applied by first measuring the surface and determining its maximum and minimum lengths. The difference between these two measurements must be less than 0.05. Thus, if the measured maximum distance is 41, then no other part of the surface may be less then 41 – 0.05 = 40.95.

Tolerances of perpendicularity serve to complement locational tolerances, to make the shape more exact, so tolerances of perpendicularity must always be smaller than tolerances of location. It would be of little use, for example, to assign a perpendicularity tolerance of 1.5 for the object shown in Figure 8-71. The locational tolerance would prevent the variation from ever reaching the limits specified by such a large perpendicularity tolerance.

Figure 8-72 shows a perpendicularity tolerance applied to cylindrical features: a shaft and a hole. The figure includes examples of both RFS and MMC applications. As with straightness tolerances applied at MMC, perpendicularity tolerances applied about a hole or shaft's centerline allow the tolerance zone to vary as the feature size varies.

The inclusion of the Ø symbol in a geometric tolerance is critical to its interpretation. See Figure 8-73. If the Ø symbol is not included, the tolerance applies only to the view in which it is written. This means that the tolerance zone is shaped like a rectangular slice, not a cylinder, as would be the case if the Ø symbol were included. In general it is better to always include the Ø symbol for cylindrical features because it generates a tolerance zone more like that used in positional tolerancing.

Figure 8-73 shows a perpendicularity tolerance applied to a slot, a noncylindrical feature. Again, the MMC specification is always for variations in the tolerance zone.

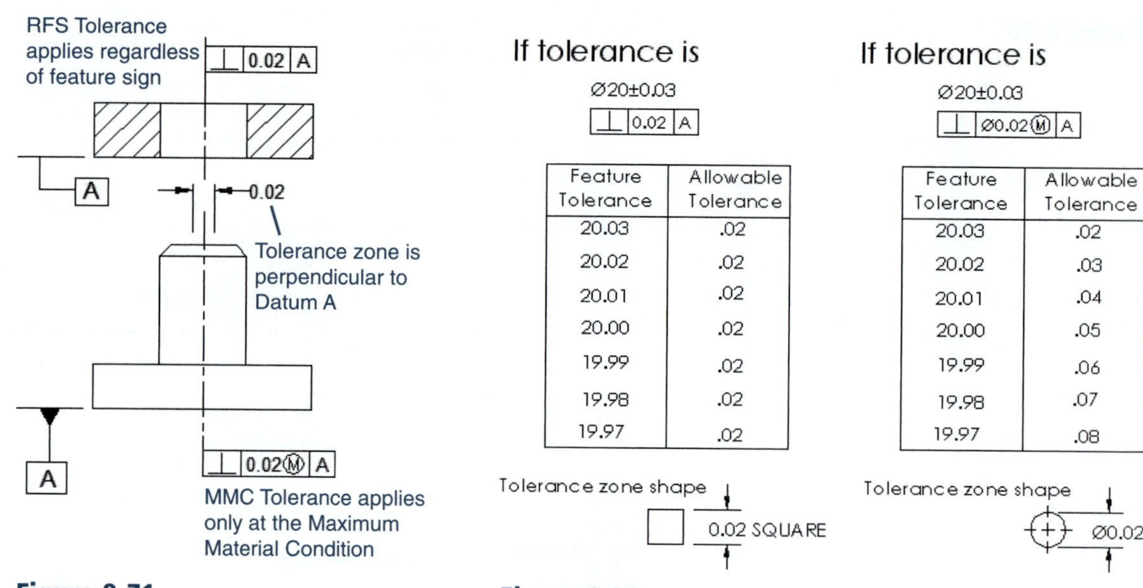

Feature Tolerance	Allowable Tolerance
20.03	.02
20.02	.02
20.01	.02
20.00	.02
19.99	.02
19.98	.02
19.97	.02

Feature Tolerance	Allowable Tolerance
20.03	.02
20.02	.03
20.01	.04
20.00	.05
19.99	.06
19.98	.07
19.97	.08

Figure 8-71

Figure 8-72

Figure 8-73

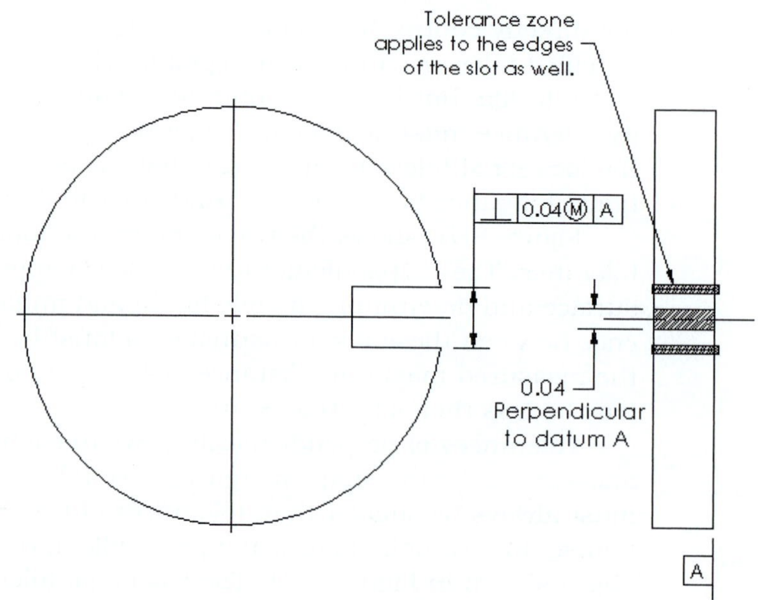

8-36 Parallelism

Parallelism tolerances are used to ensure that all points within a plane are within two parallel planes that are parallel to a referenced datum plane. Figure 8-74 shows a rectangular object that is toleranced so that its top surface is parallel to the bottom surface within 0.02. This means that every point on the top surface must be within a set of parallel planes 0.02 apart. These parallel tolerancing planes are located by determining the maximum and minimum distances from the datum surface. The difference between the maximum and minimum values may not exceed the stated 0.02 tolerance.

In the extreme condition of maximum feature size, the top surface is located 40.5 above the datum plane. The parallelism tolerance is then applied, meaning that no point on the surface may be closer than 40.3 to the datum. This is an RFS condition. The MMC condition may also be applied, thereby allowing the tolerance zone to vary as the feature size varies.

Figure 8-74

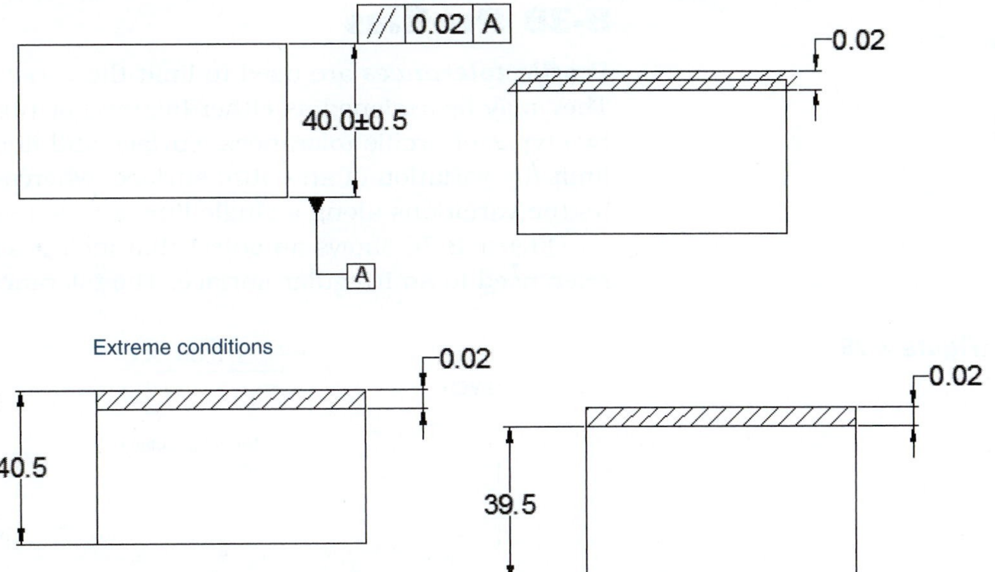

8-37 Angularity

Angularity tolerances are used to limit the variance of surfaces and axes that are at an angle relative to a datum. Angularity tolerances are applied like perpendicularity and parallelism tolerances as a way to better control the shape of locational tolerances.

Figure 8-75 shows an angularity tolerance and several ways it is interpreted at extreme conditions.

Figure 8-75

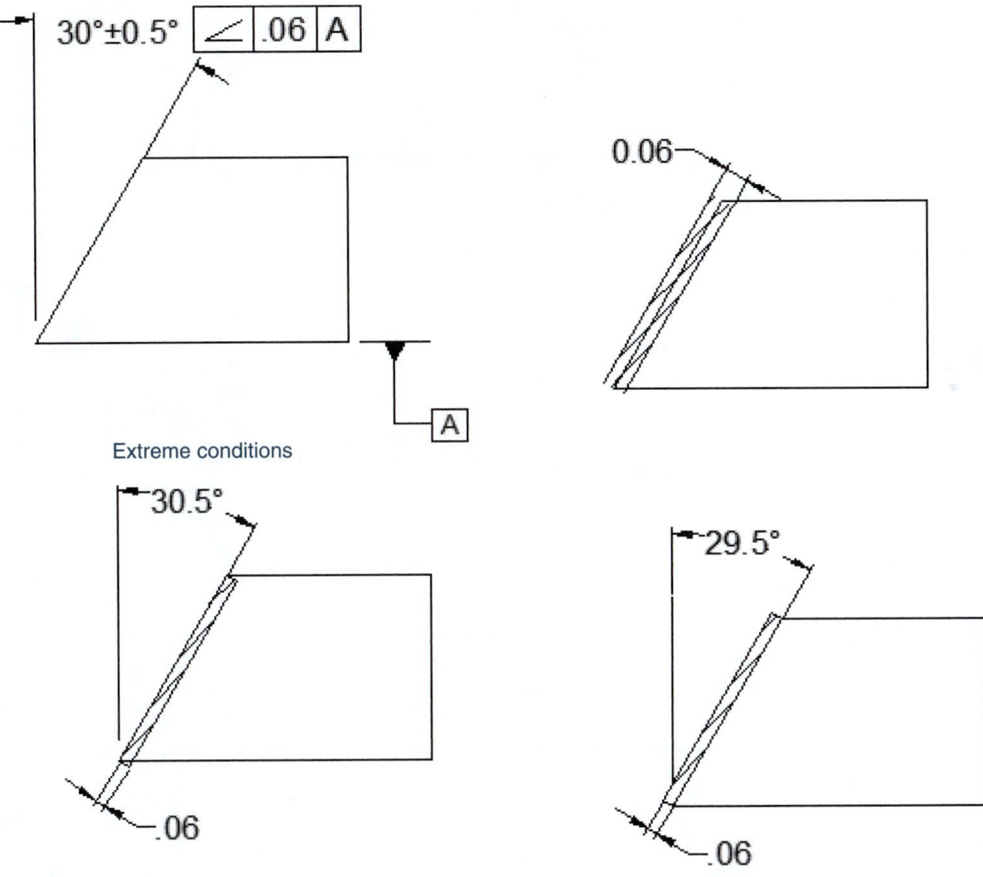

8-38 Profiles

Profile tolerances are used to limit the variations of irregular surfaces. They may be assigned as either bilateral or unilateral tolerances. There are two types of profile tolerances: surface and line. *Surface* profile tolerances limit the variation of an entire surface, whereas a *line* profile tolerance limits the variations along a single line across a surface.

Figure 8-76 shows an object that includes a surface profile tolerance referenced to an irregular surface. The tolerance is considered a bilateral

Figure 8-76

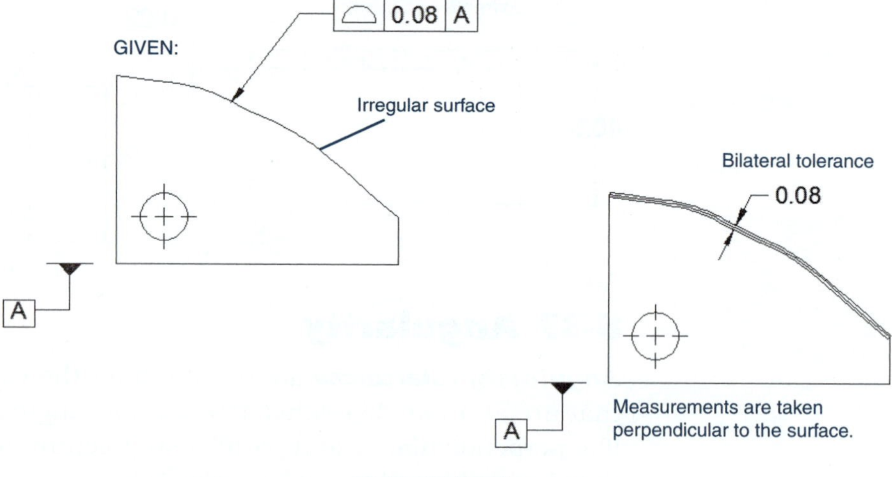

0.08 A
GIVEN:
Irregular surface
A
Bilateral tolerance
0.08
A
Measurements are taken perpendicular to the surface.

Figure 8-77

0.08 A
GIVEN:
Irregular surface
Phantom line
A
Unilateral tolerance
0.08
A
Measurements are taken perpendicular to the surface.
0.08 A
GIVEN:
Phantom line
Irregular surface
A
Unilateral tolerance
0.08
A
Measurements are taken perpendicular to the surface.

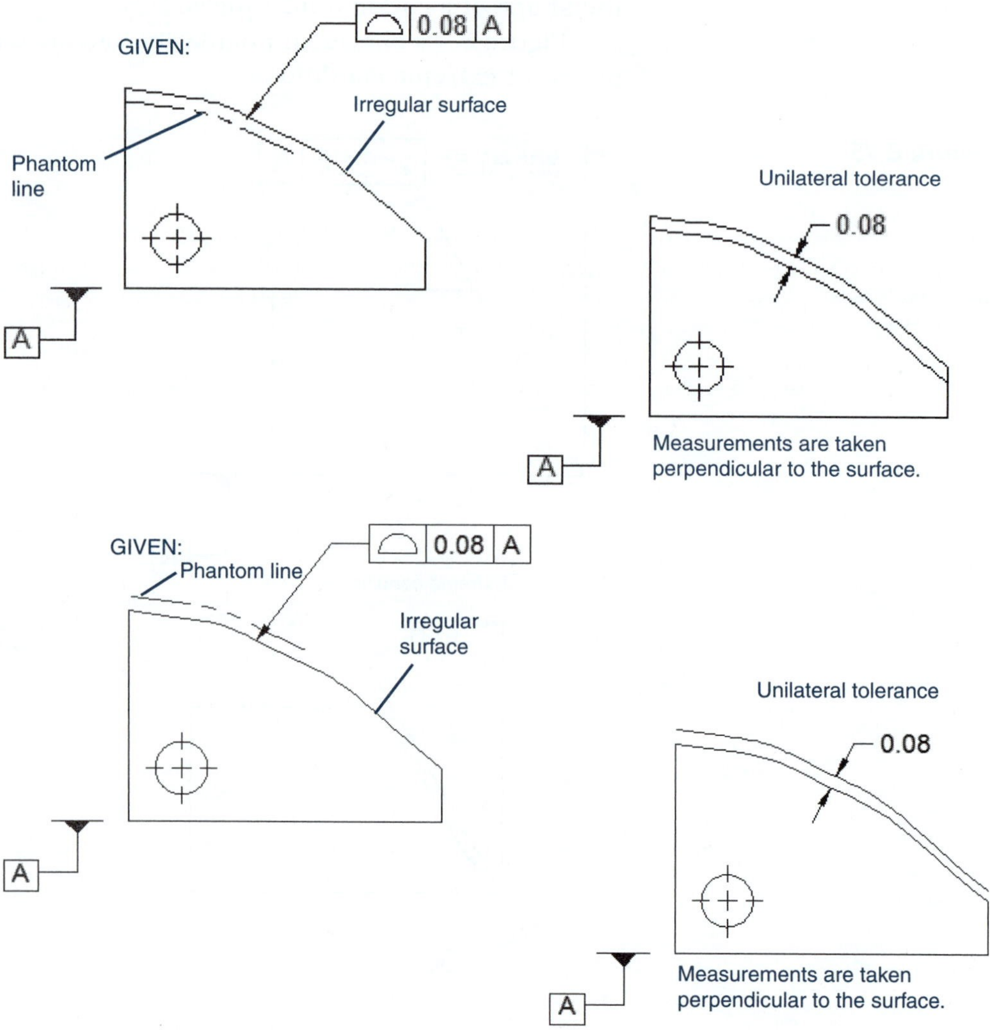

tolerance because no other specification is given. This means that all points on the surface must be located between two parallel planes 0.08 apart that are centered about the irregular surface. The measurements are taken perpendicular to the surface.

Unilateral applications of surface profile tolerances must be indicated on the drawing using phantom lines. The phantom line indicates the side of the true profile line of the irregular surface on which the tolerance is to be applied. A phantom line above the irregular surface indicates that the tolerance is to be applied using the true profile line as 0, and then the specified tolerance range is to be added above that line. See Figures 8-77 and 8-78.

Figure 8-78

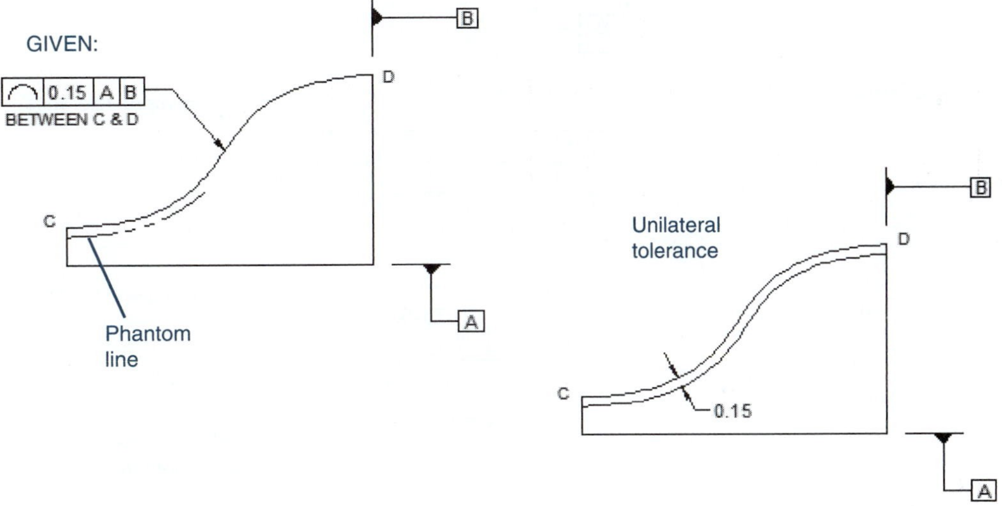

Profiles of line tolerances are applied to irregular surfaces, as shown in Figure 8-78. Profiles of line tolerances are particularly helpful when tolerancing an irregular surface that is constantly changing, such as the surface of an airplane wing.

Surface and line profile tolerances are somewhat analogous to flatness and straightness tolerances. Flatness and surface profile tolerances are applied across an entire surface, whereas straightness and line profile tolerances are applied only along a single line across the surface.

8-39 Runouts

A ***runout tolerance*** is used to limit the variations between features of an object and a datum. More specifically, they are applied to surfaces around a datum axis such as a cylinder or to a surface constructed perpendicular to a datum axis. There are two types of runout tolerances: circular and total.

Figure 8-79 shows a cylinder that includes a circular runout tolerance. The runout requirements are checked by rotating the object about its longitudinal axis or datum axis while holding an indicator gauge in a fixed position on the object's surface.

Runout tolerances may be either bilateral or unilateral. A runout tolerance is assumed to be bilateral unless otherwise indicated. If a runout tolerance is to be unilateral, a phantom line is used to indicate the side of the object's true surface to which the tolerance is to be applied. See Figure 8-80.

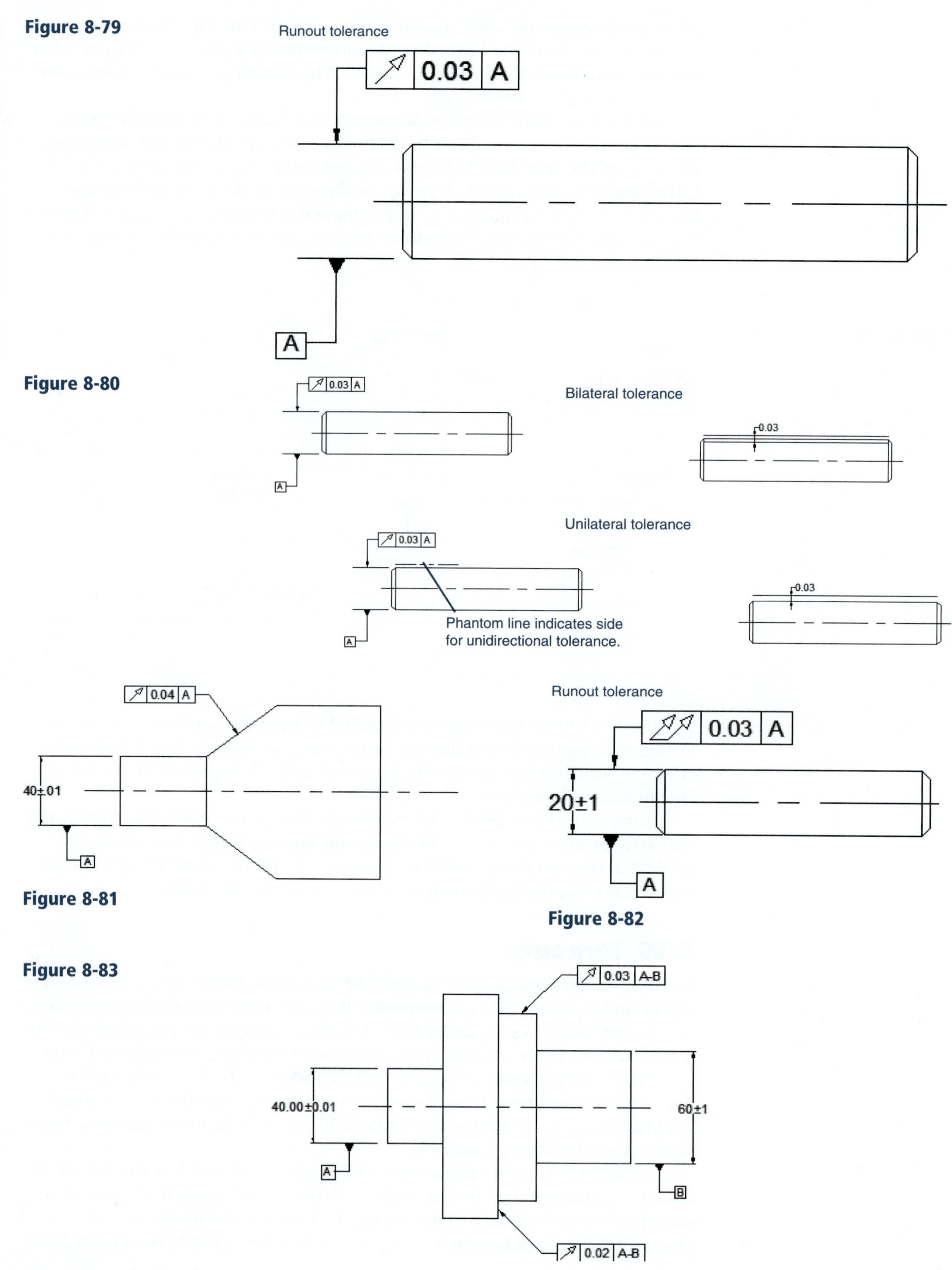

Figure 8-79

Figure 8-80

Figure 8-81

Figure 8-82

Figure 8-83

Runout tolerances may be applied to tapered areas of cylindrical objects, as shown in Figure 8-81. The tolerance is checked by rotating the object about a datum axis while holding an indicator gauge in place.

A total runout tolerance limits the variation across an entire surface. See Figure 8-82. An indicator gauge is not held in place while the object is rotated, as it is for circular runout tolerances, but is moved about the rotating surface.

Figure 8-83 shows a circular runout tolerance that references two datums. The two datums serve as one datum. The object can then be rotated about both datums simultaneously as the runout tolerances are checked.

8-40 Positional Tolerances

As defined earlier, *positional tolerances* are used to locate and tolerance holes. Positional tolerances create a circular tolerance zone for hole centerpoint locations, in contrast with the rectangular tolerance zone created by linear coordinate dimensions. See Figure 8-84. The circular tolerance zone allows for an increase in acceptable tolerance variation without compromising the design integrity of the object. Note how some of the possible hole centerpoints fall in an area outside the rectangular tolerance zone but are still within the circular tolerance zone. If the hole had been located using linear coordinate dimensions, centerpoints located beyond the rectangular tolerance zone would have been rejected as beyond tolerance, and yet holes produced using these locations would function correctly from a design standpoint. The centerpoint locations would be acceptable if positional tolerances had been specified. The finished hole is round, so a round tolerance zone is appropriate. The rectangular tolerance zone rejects some holes unnecessarily.

Holes are dimensioned and toleranced using geometric tolerances by a combination of locating dimensions, feature dimensions and tolerances, and positional tolerances. See Figure 8-85. The locating dimensions are enclosed in rectangular boxes and are called ***basic dimensions***. Basic dimensions are assumed to be exact.

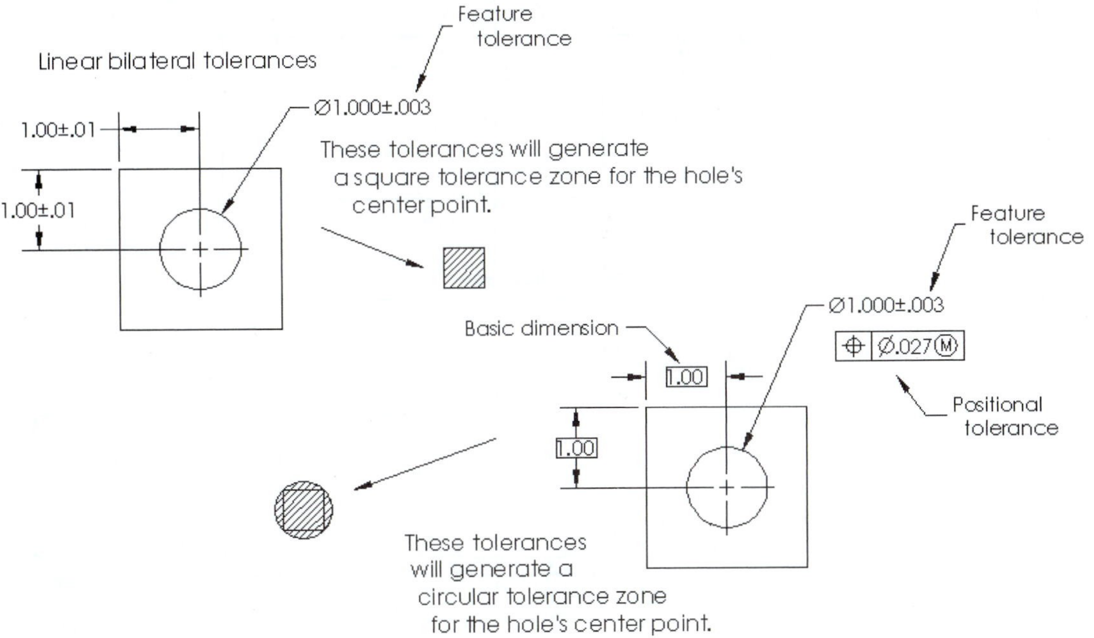

Figure 8-84

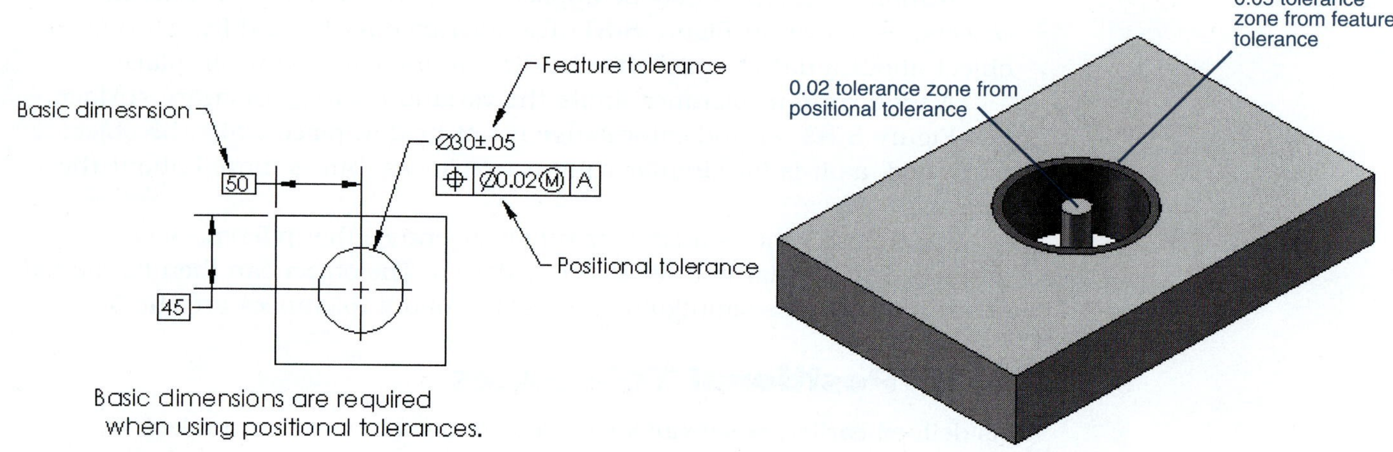

Figure 8-85

The feature tolerances for the hole are as presented earlier in the chapter. They can be presented using plus or minus or limit-type tolerances. In the example shown in Figure 8-85 the diameter of the hole is toleranced using a plus and minus 0.05 tolerance.

The basic locating dimensions of 45 and 50 are assumed to be exact. The tolerances that would normally accompany linear locational dimensions are replaced by the positional tolerance. The positional tolerance also specifies that the tolerance be applied at the centerline at maximum material condition. The resulting tolerance zones are as shown in Figure 8-85.

Figure 8-86 shows an object containing two holes that are dimensioned and toleranced using positional tolerances. There are two consecutive

Figure 8-86

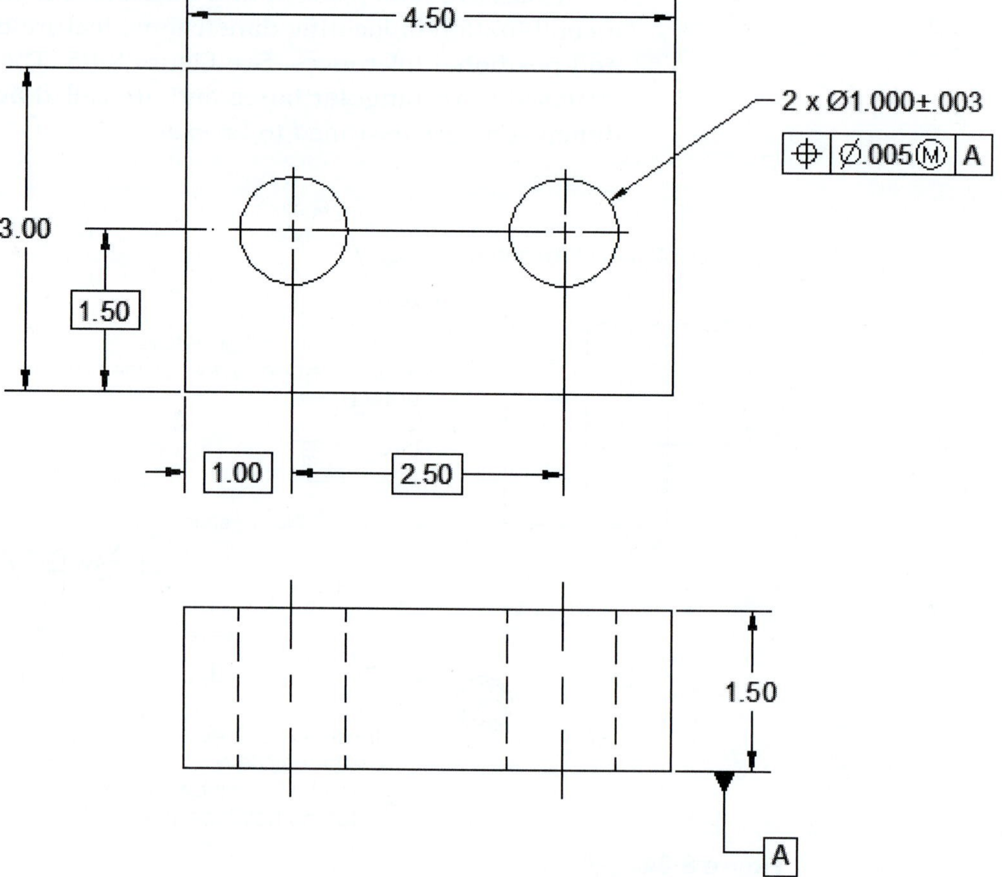

horizontal basic dimensions. Because basic dimensions are exact, they do not have tolerances that accumulate; that is, there is no tolerance buildup.

8-41 Creating Positional Tolerances Using SolidWorks

Figure 8-87 shows an orthographic view that has been dimensioned. Note that the hole has a dimension and tolerance of Ø1.000±.001. This is the hole's feature tolerance. It deals only with the hole's diameter variation. It does not tolerance the hole's location.

To Create the Positional Tolerance

1. Click the **Annotation** tab and click the **Geometric Tolerance** tool.
2. Enter the symbol for positional tolerance, the symbol for diameter, the tolerance value, and the symbol for maximum material condition.

Figure 8-87

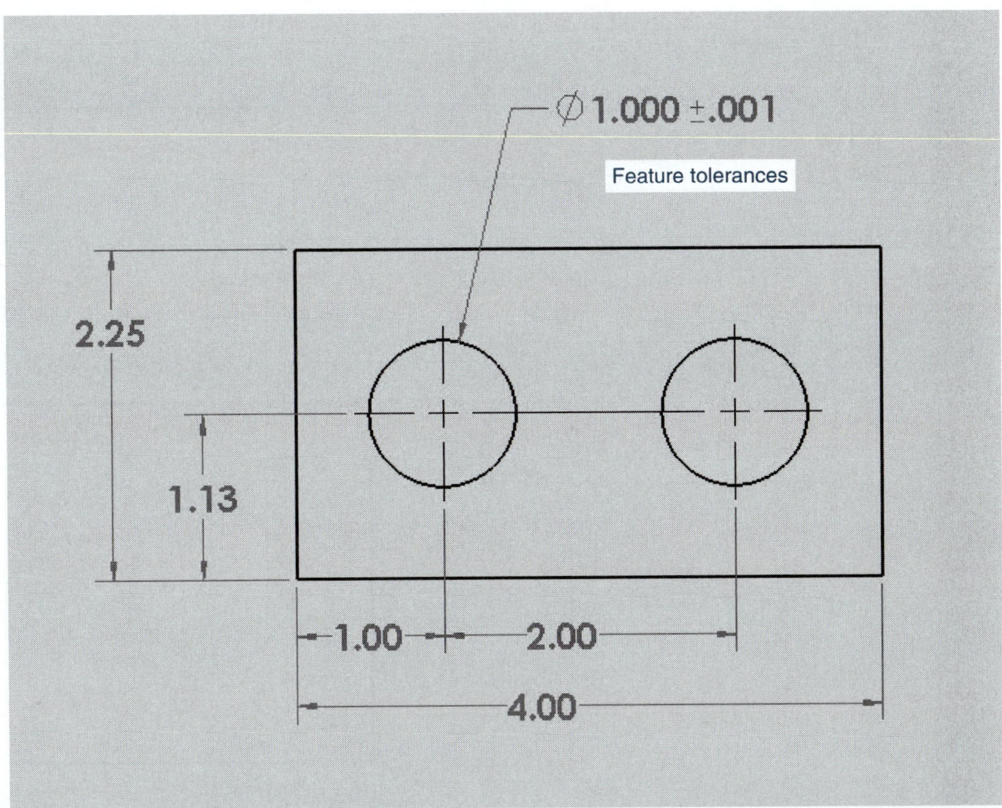

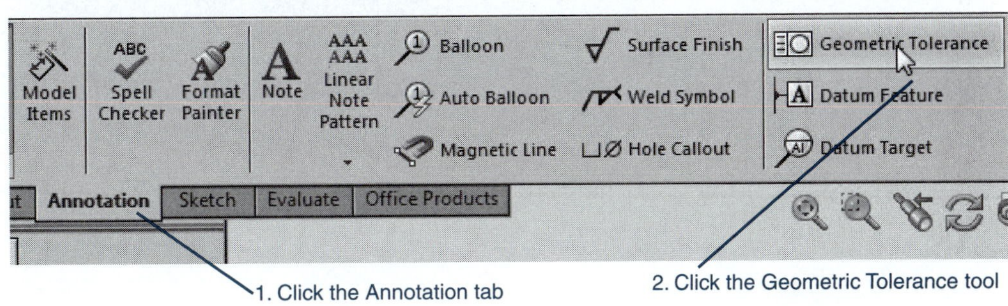

Figure 8-87
(*Continued*)

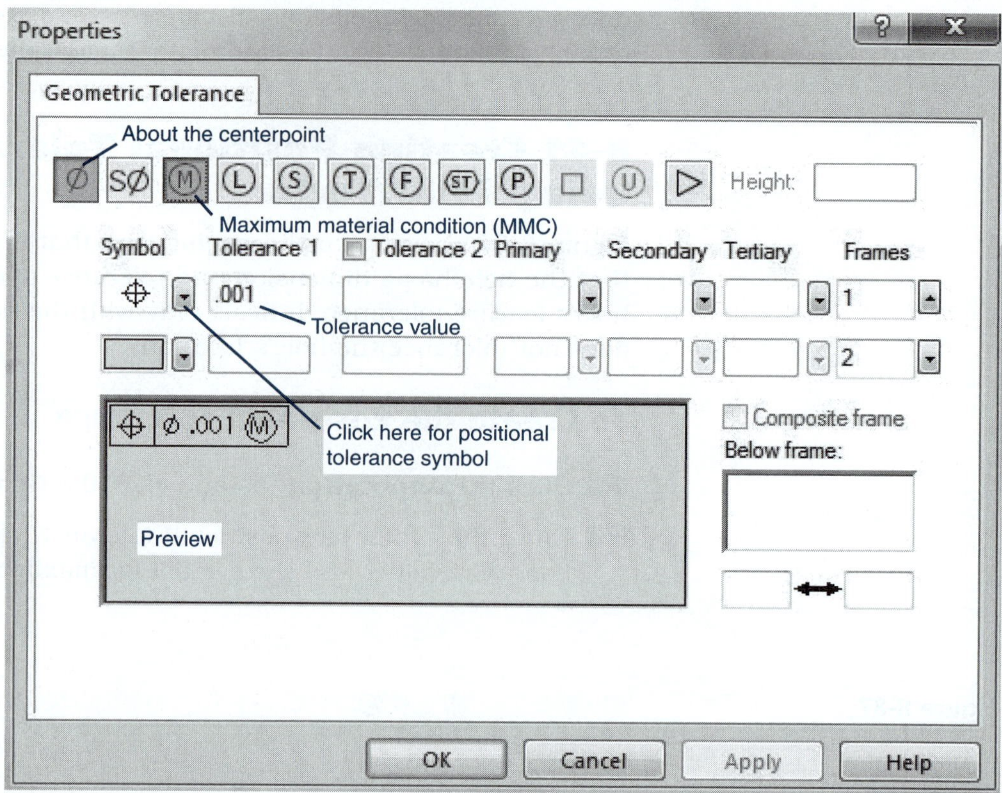

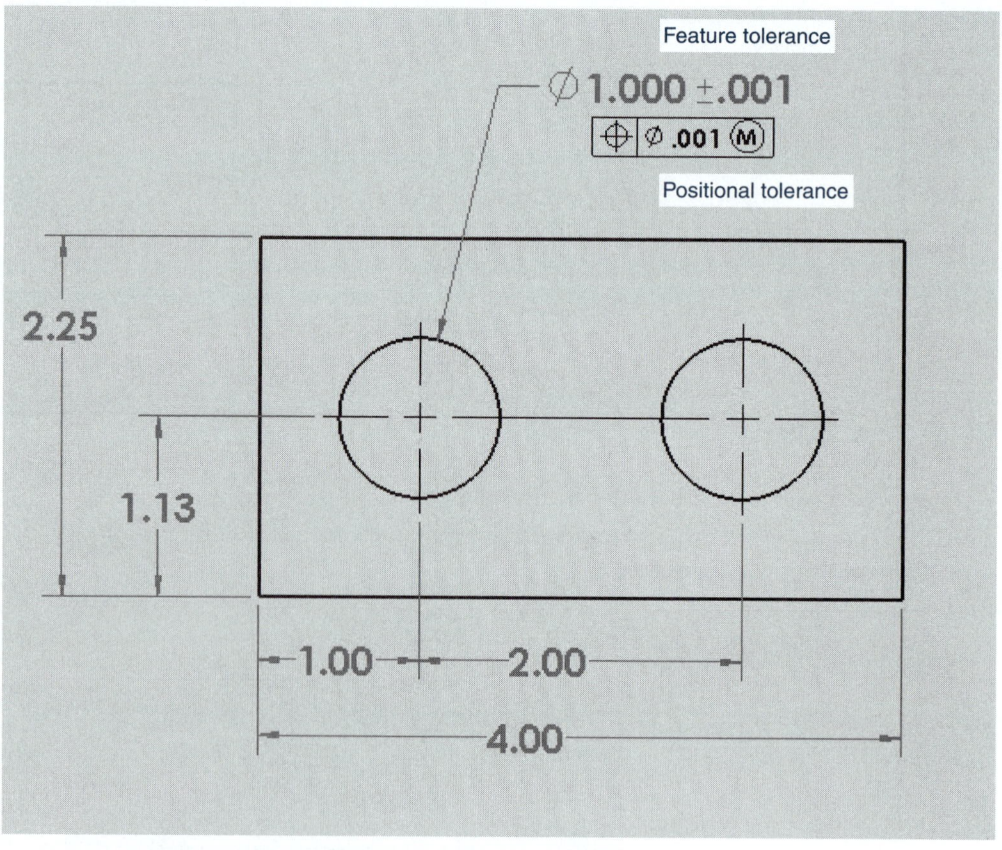

The collective symbol would read "Apply a .001 positional tolerance about the hole's centerpoint at the maximum material condition."

3 Locate the symbol on the drawing under the feature tolerance as shown.

Create basic dimensions to accompany the geometric positional tolerance. See Figure 8-88.

Figure 8-88

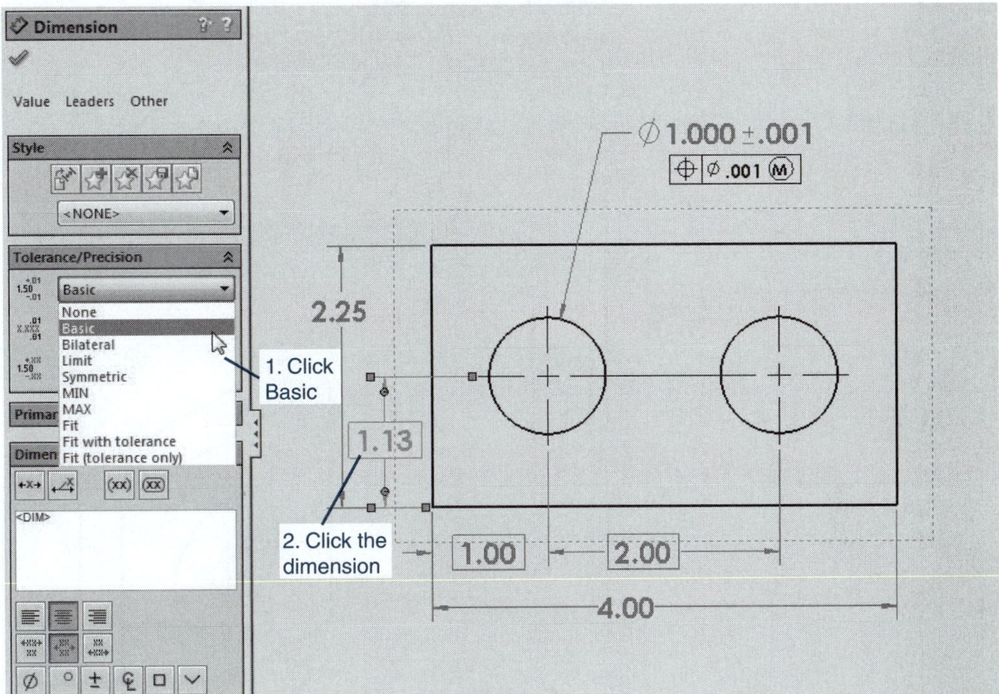

4 Click on the **2.00** dimension.

The **Dimension PropertyManager** will appear.

5 Click the **Basic** tool in the **Tolerance/Precision** box.

6 Click the **1.13** dimension and make it a basic dimension.

7 Save the drawing.

8 Click the green **OK** check mark.

> **TIP**
> Geometric positional tolerances must include basic dimensions. Basic dimensions are assumed to be perfect. The locational tolerance associated with locating dimensions has been moved to the geometric positional tolerance.

8-42 Virtual Condition

Virtual condition is a combination of a feature's MMC and its geometric tolerance. For external features (shafts) it is the MMC plus the geometric tolerance; for internal features (holes) it is the MMC minus the geometric tolerance.

The following calculations are based on the dimensions shown in Figure 8-89.

Figure 8-89

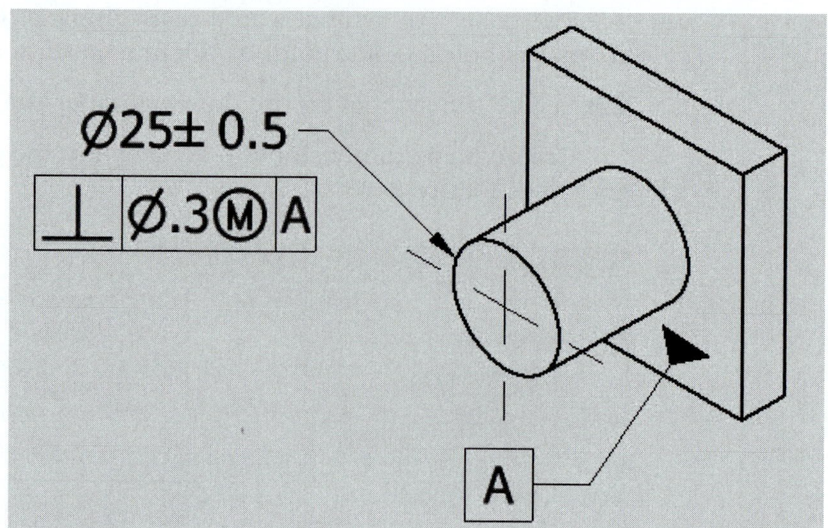

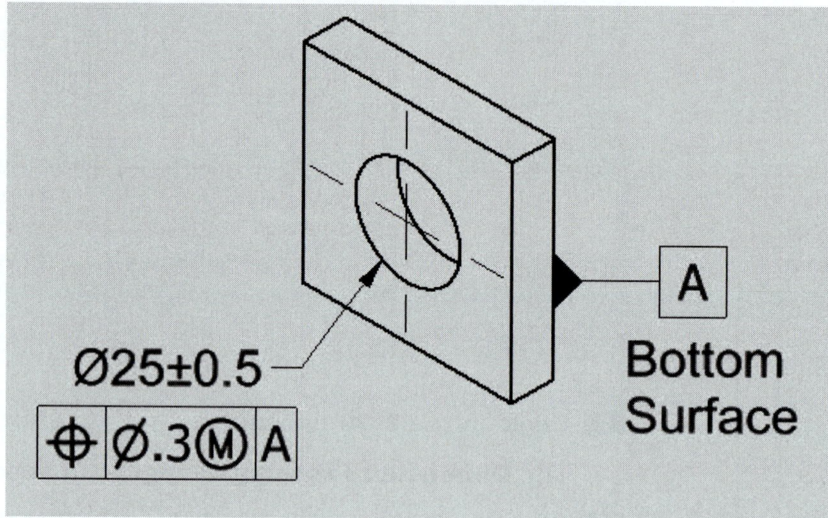

Calculating the Virtual Condition for a Shaft

25.5	MMC for shaft—maximum diameter
+0.3	Geometric tolerance
25.8	Virtual condition

Calculating the Virtual Condition for a Hole

24.5	MMC for hole—minimum diameter
−0.3	Geometric tolerance
24.2	Virtual condition

8-43 Floating Fasteners

Positional tolerances are particularly helpful when dimensioning matching parts. Because basic locating dimensions are considered exact, the sizing of mating parts is dependent only on the hole and shaft's MMC and the geometric tolerance between them.

The relationship for floating fasteners and holes in objects may be expressed as a formula:

$$H - T = F$$

where

H = hole at MMC
T = geometric tolerance
F = shaft at MMC

A ***floating fastener*** is one that passes through two or more objects, and all parts have clearance holes for the tolerance. It is not attached to either object and it does not screw into either object. Figure 8-90 shows two objects that are to be joined by a common floating shaft, such as a bolt or screw. The feature size and tolerance and the positional geometric tolerance are both given. The minimum size hole that will always just fit is determined using the preceding formula:

$$H - T = F$$

$$11.97 - .02 = 11.95$$

Therefore, the shaft's diameter at MMC, the shaft's maximum diameter, equals 11.95. Any required tolerance would have to be subtracted from this shaft size.

Figure 8-90

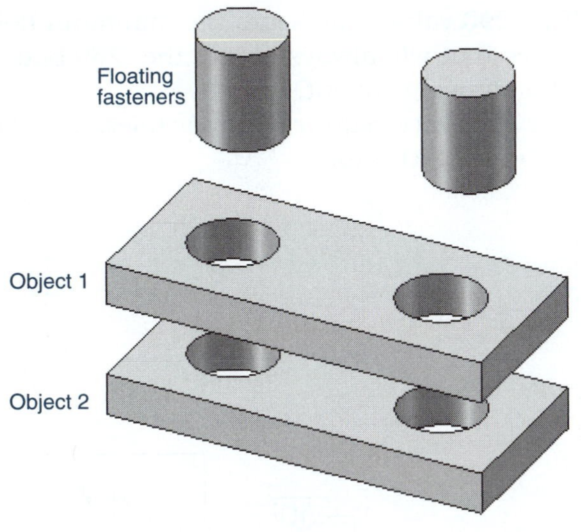

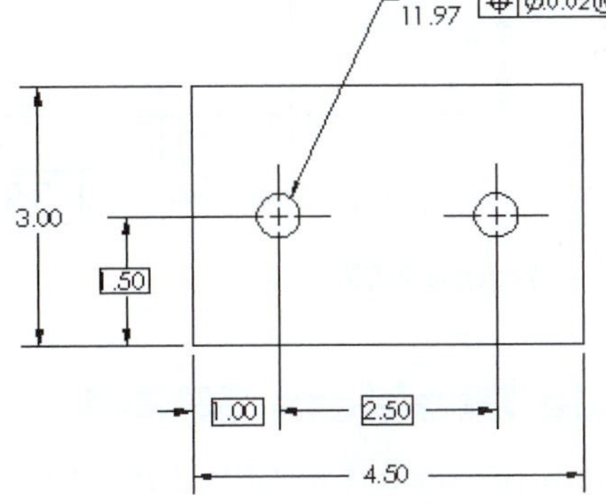

SIZE	TOLERANCE ZONE
11.97 MMC	.02
11.98	.03
11.99	.04
12.00	.05
12.01	.06
12.02	.07
12.03 LMC	.08

11.97 MMC
-0.02
11.95 virtual condition

Maximum possible
fastener diameter= 11.95

The .02 geometric tolerance is applied at the hole's MMC, so as the hole's size expands within its feature tolerance, the tolerance zone for the acceptable matching parts also expands.

8-44 Sample Problem SP8-3

The situation presented in Figure 8-90 can be worked in reverse; that is, hole sizes can be derived from given shaft sizes.

The two objects shown in Figure 8-91 are to be joined by a .250-in. bolt. The parts are floating; that is, they are both free to move, and the fastener is not joined to either object. What is the MMC of the holes if the positional tolerance is to be .030?

A manufacturer's catalog specifies that the tolerance for .250 bolts is .2500 to .2600.

Rewriting the formula

$$H - T = F$$

to isolate the H yields

$$\begin{aligned} H &= F + T \\ &= .260 + .030 \\ &= .290 \end{aligned}$$

The .290 value represents the minimum hole diameter, MMC, for all four holes that will always accept the .250 bolt. Figure 8-92 shows the resulting drawing callout.

Any clearance requirements or tolerances for the hole would have to be added to the .290 value.

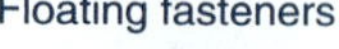

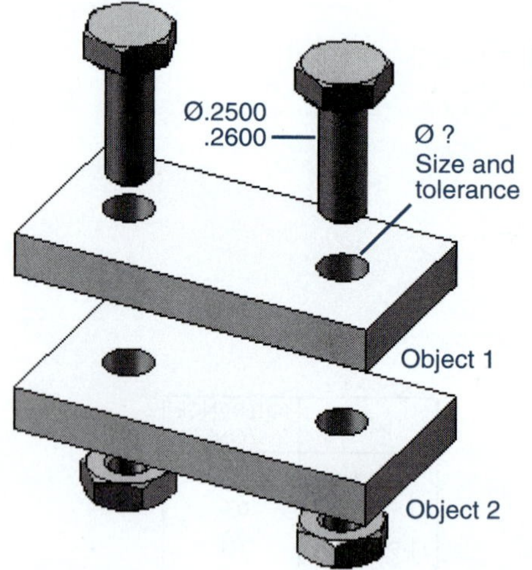

Figure 8-91

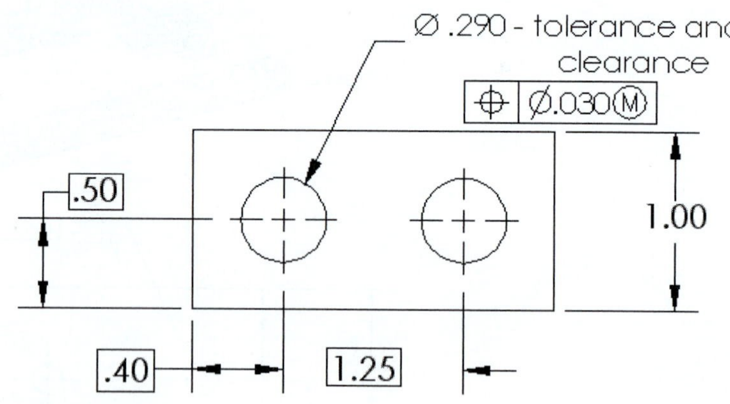

Figure 8-92

8-45 Sample Problem SP8-4

Repeat the problem presented in SP8-3 but be sure that there is always a minimum clearance of .002 between the hole and the shaft, and assign a hole tolerance of .008.

Sample problem SP8-3 determined that the maximum hole diameter that will always accept the .250 bolt was .290 based on the .030 positioning tolerance. If the minimum clearance is to be .002, the maximum hole diameter is found as follows:

.290	Minimum hole diameter that will always accept the bolt (0 clearance at MMC)
+.002	Minimum clearance
.292	Minimum hole diameter including clearance

Now, assign the tolerance to the hole:

.292	Minimum hole diameter
+.001	Tolerance
.293	Maximum hole diameter

See Figure 8-93 for the appropriate drawing callout. The choice of clearance size and hole tolerance varies with the design requirements for the objects.

Figure 8-93

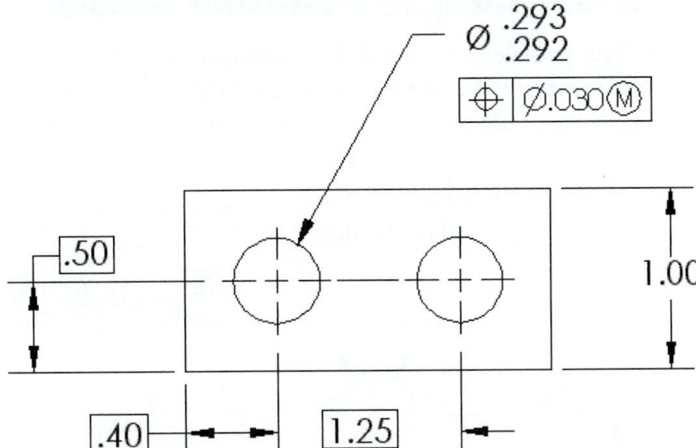

8-46 Fixed Fasteners

A ***fixed fastener*** is a fastener that is restrained in one of the parts. For example, one end of the fastener is screwed into one of the parts using threads. See Figure 8-94. Because the fastener is fixed to one of the objects, the geometric tolerance zone must be smaller than that used for floating fasteners. The fixed fastener cannot move without moving the object it is attached to. The relationship between fixed fasteners and holes in mating objects is defined by the formula

$$H - 2T = F$$

The tolerance zone is cut in half for each part. This can be demonstrated by the objects shown in Figure 8-95. The same feature sizes that were used in Figure 8-91 are assigned, but in this example the fasteners are fixed. Solving for the geometric tolerance yields a value as follows:

$$H - F = 2T$$

$$11.97 - 11.95 = 2T$$

$$.02 = 2T$$

$$.01 = T$$

The resulting positional tolerance is half that obtained for floating fasteners.

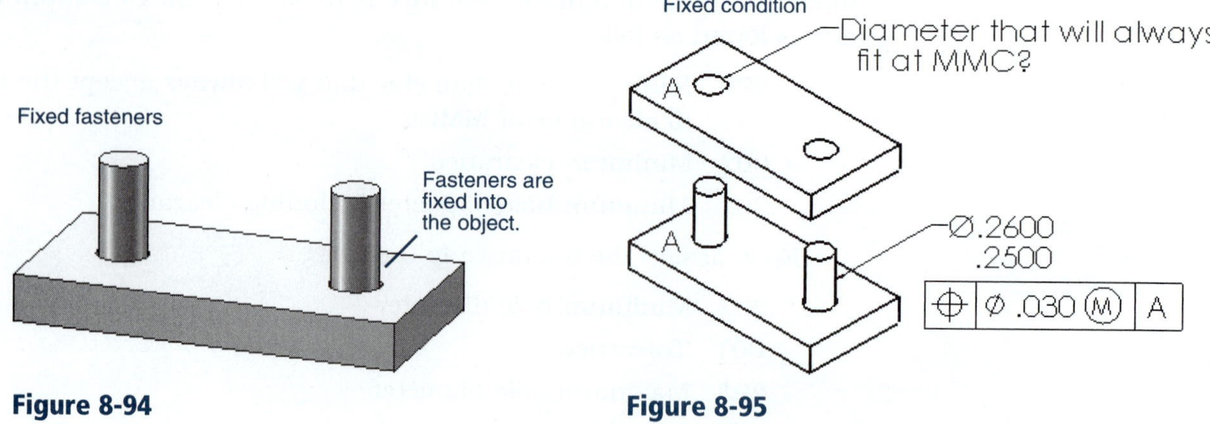

Figure 8-94

Figure 8-95

8-47 Sample Problem SP8-5

This problem is similar to sample problem SP8-3, but the given conditions are applied to fixed fasteners rather than floating fasteners. Compare the resulting shaft diameters for the two problems. See Figure 8-96.

Figure 8-96

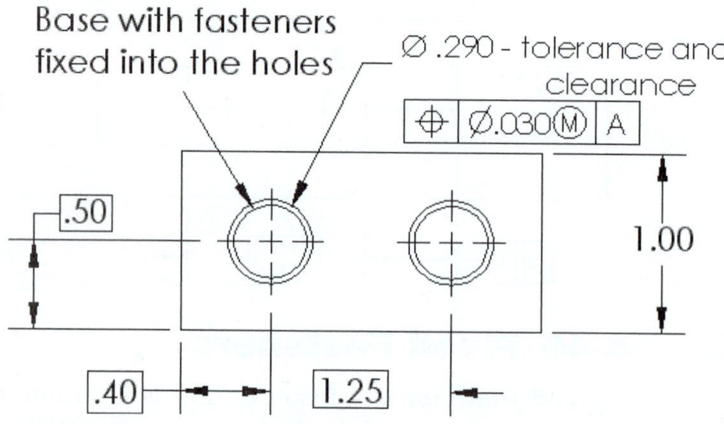

A. What is the minimum diameter hole that will always accept the fixed fasteners?

B. If the minimum clearance is .005 and the hole is to have a tolerance of .002, what are the maximum and minimum diameters of the hole?

$$H - 2T = F$$

$$H = F + 2T$$

$$= .260 + 2(.030)$$

$$= .260 + .060$$

$= .320$ Minimum diameter that will always accept the fastener

If the minimum clearance is .005 and the hole tolerance is .002,

.320	Virtual condition
+.005	Clearance
.325	Minimum hole diameter
.325	Minimum hole diameter
+.002	Tolerance
.327	Maximum hole diameter

The maximum and minimum values for the hole's diameter can then be added to the drawing of the object that fits over the fixed fasteners. See Figure 8-97.

Figure 8-97

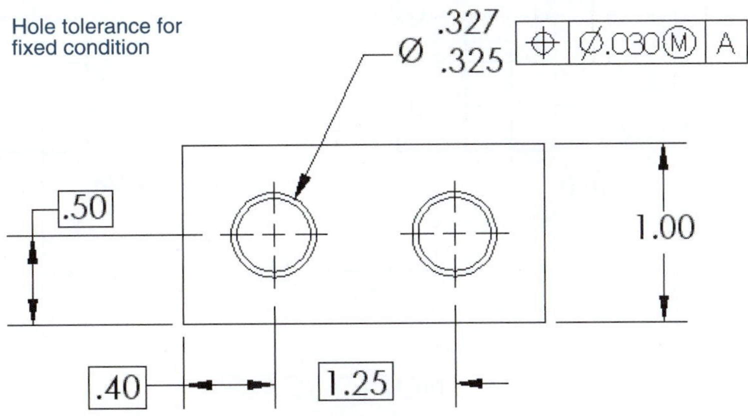

8-48 Design Problems

This problem was originally done on p. 554 using rectangular tolerances. It is done in this section using positional geometric tolerances so that the two systems can be compared. It is suggested that the previous problem be reviewed before reading this section.

Figure 8-98 shows top and bottom parts that are to be joined in the floating condition. A nominal distance of 50 between hole centers and Ø20 for the holes has been assigned. In the previous solution a rectangular tolerance of ±.01 was selected, and there was a minimum hole diameter of 20.00. Figure 8-99 shows the resulting tolerance zones.

Figure 8-98

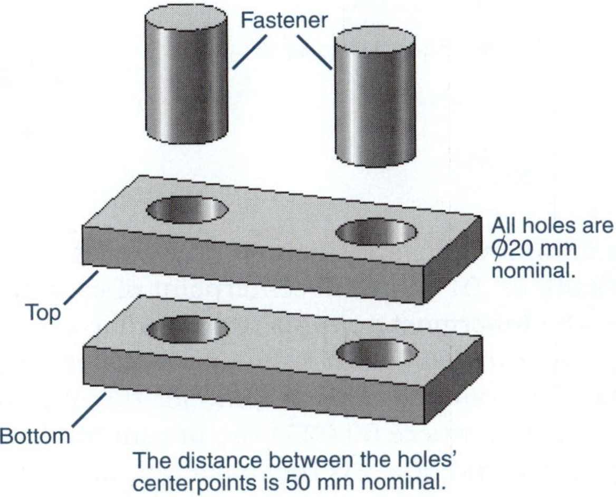

The diagonal distance across the rectangular tolerance zone is .028 and was rounded off to .03 to yield a maximum possible fastener diameter of 19.97. If the same .03 value is used to calculate the fastener diameter using positional tolerance, the results are as follows:

$$H - T = F$$

$$20.00 - .03 = 19.97$$

The results seem to be the same, but because of the circular shape of the positional tolerance zone, the manufactured results are not the same.

Figure 8-99

Linear tolerance zones based on ± tolerances.

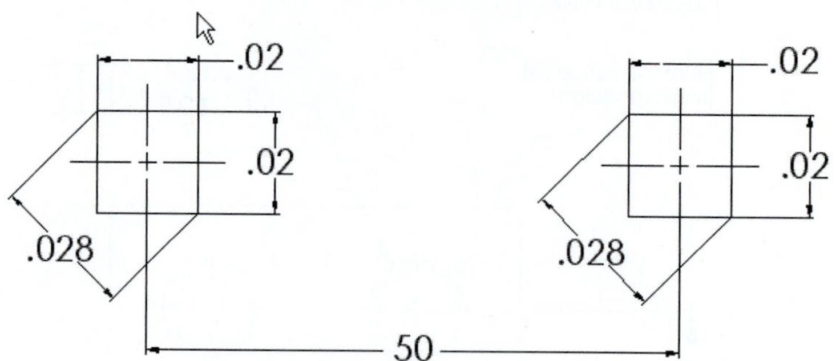
.02
.02
.028
50

NOT TO SCALE

Rectangular range: 49.98 to 50.02

Tolerance zones based on positional (⌖) tolerances.

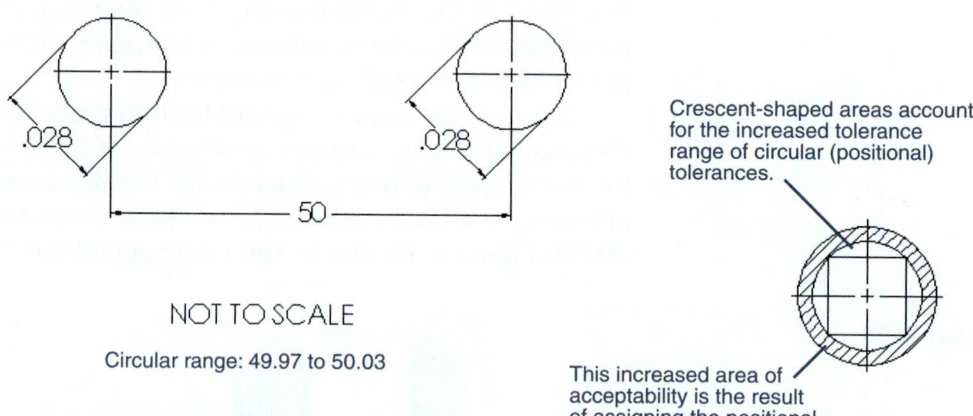
.028
.028
50
Crescent-shaped areas account for the increased tolerance range of circular (positional) tolerances.
This increased area of acceptability is the result of assigning the positional tolerance at MMC.

NOT TO SCALE

Circular range: 49.97 to 50.03

The minimum distance between the inside edges of the rectangular zones is 49.98, or .01 from the centerpoint of each hole. The minimum distance from the innermost points of the circular tolerance zones is 49.97, or .015 (half the rounded-off .03 value) from the centerpoint of each hole. The same value difference also occurs for the maximum distance between centerpoints, where 50.02 is the maximum distance for the rectangular tolerances, and 50.03 is the maximum distance for the circular tolerances. The size of the circular tolerance zone is larger because the hole tolerances are assigned at MMC. In this example the MMC for the hole is 20.00, that is, when the hole is at its smallest. As the hole's feature tolerance increases, goes from 20.00 to 20.02, the size of the tolerance zone increases.

The basic dimensions of 20 and 50 are assumed to be perfect. The original ±.01 positional tolerances assigned are still present; they are now presented as geometric positional tolerances. The hole tolerance, 20 +.02/−0, is the feature tolerance, and the geometric tolerance of .03 is the positional tolerance.

Figure 8-99 shows a comparison between the tolerance zones, and Figure 8-100 shows how the positional tolerances would be presented on a drawing of either the top or bottom part.

Figure 8-100

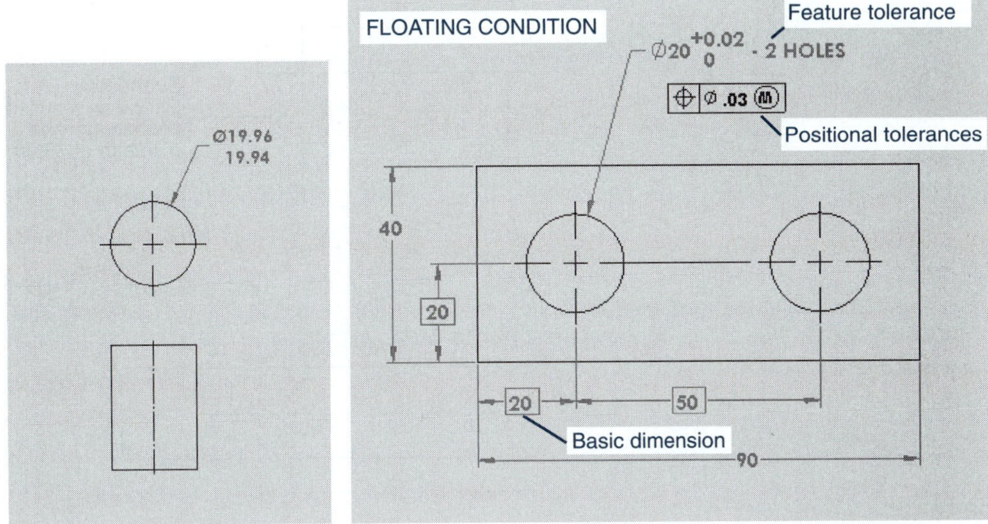

Figure 8-101 shows the same top and bottom parts joined together in the fixed condition. The initial nominal values are the same. If the same .03 diagonal value is assigned as a positional tolerance, the results are as follows:

$$H - 2T = F$$

$$20.00 - .06 = 19.94$$

These results appear to be the same as those generated by the rectangular tolerance zone, but the circular tolerance zone allows a greater variance in acceptable manufactured parts.

Figure 8-101

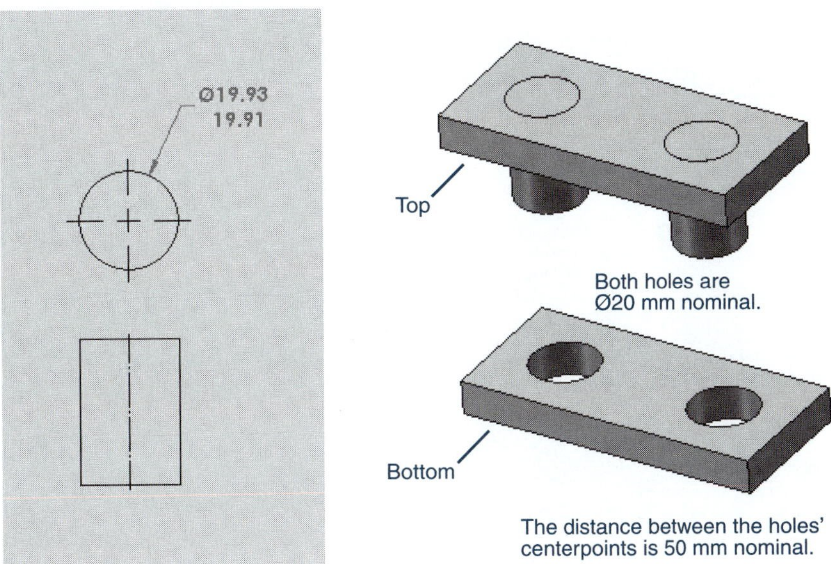

Figure 8-102 shows the parts dimensioned using baseline dimensions.

Figure 8-102

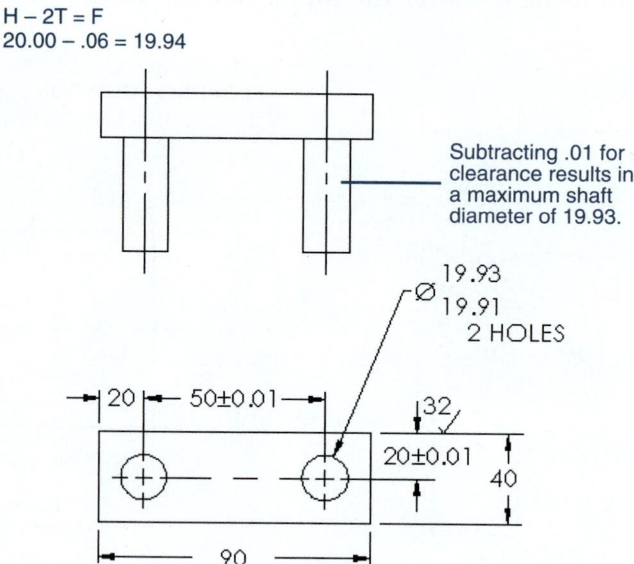
H – 2T = F
20.00 – .06 = 19.94
Subtracting .01 for clearance results in a maximum shaft diameter of 19.93.
Ø 19.93
19.91
2 HOLES
20
50±0.01
32
20±0.01
40
90
Assigning a shaft tolerance of .02 results in a maximum shaft diameter of 19.91.

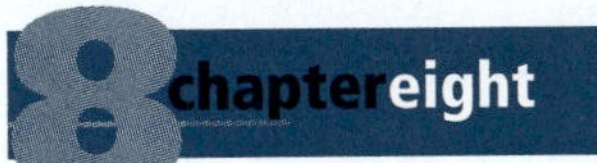

Chapter Projects

Project 8-1:

Draw a model of the objects shown in Figure P8-1A through P8-1D using the given dimensions and tolerances. Create a drawing layout with a view of the model as shown. Add the specified dimensions and tolerances.

1. 38±0.05
2. 10±0.1
3. 5±0.05
4. 45.50° / 44.50°
5. 40±0.1
6. 22±0.1
7. 12 $^{+0}_{-.1}$
8. 25 $^{+.05}_{-0}$
9. 51.50 / 50.75
10. 76±0.1

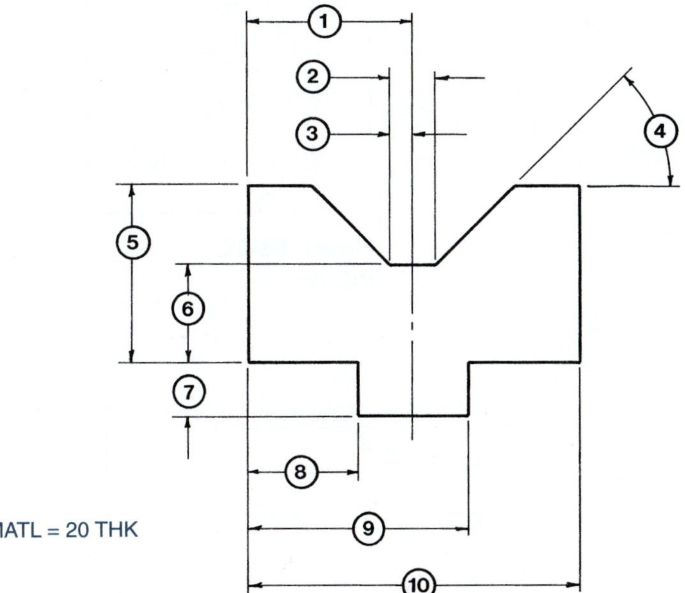

Figure P8-1A
MILLIMETERS

1. 34±0.25
2. 17±0.25
3. 25±0.05
4. 15.00 / 14.80
5. 50±0.05
6. 80±0.1
7. R5±0.1-8 PLACES
8. 45±0.25
9. 60±0.1
10. Ø14 - 3 HOLES
11. 15.00 / 14.80
12. 30.00 / 29.80

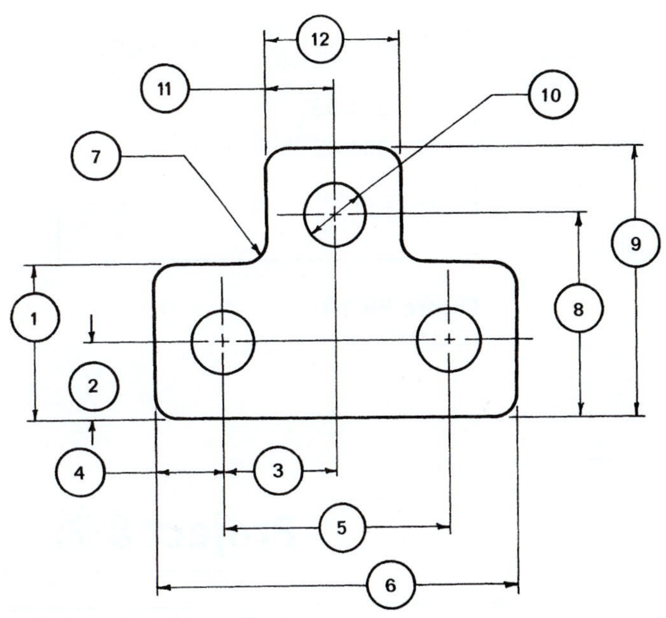

Figure P8-1B
MILLIMETERS

1. 3.00±.01
2. 1.56±.01
3. 46.50° / 45.50°
4. .750±.005
5. 2.75 / 2.70
6. 3.625±.010
7. 45°±.5°
8. 2.250±.005

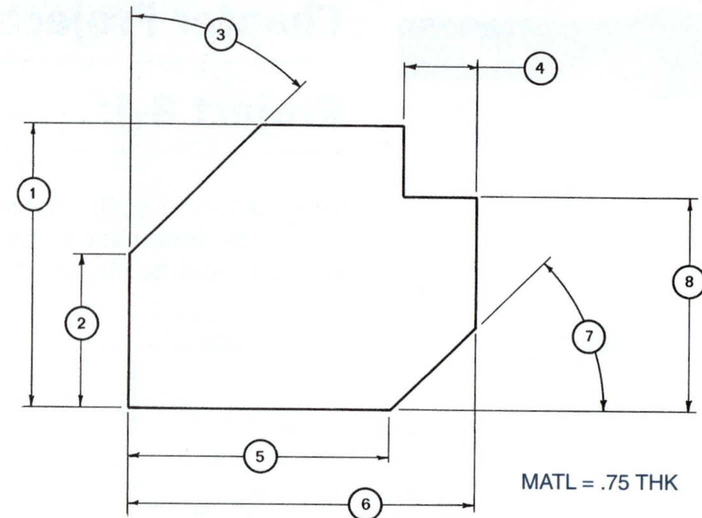

Figure P8-1C
INCHES

1. 50 $^{+.2}_{0}$
2. R45±.1 – 2 PLACES
3. 63.5 $^{0}_{-.2}$
4. 76±.1
5. 38±.1
6. Ø12.00 $^{+.05}_{0}$ – 3 HOLES
7. 30±.03
8. 30±.03
9. 100 $^{+.4}_{0}$

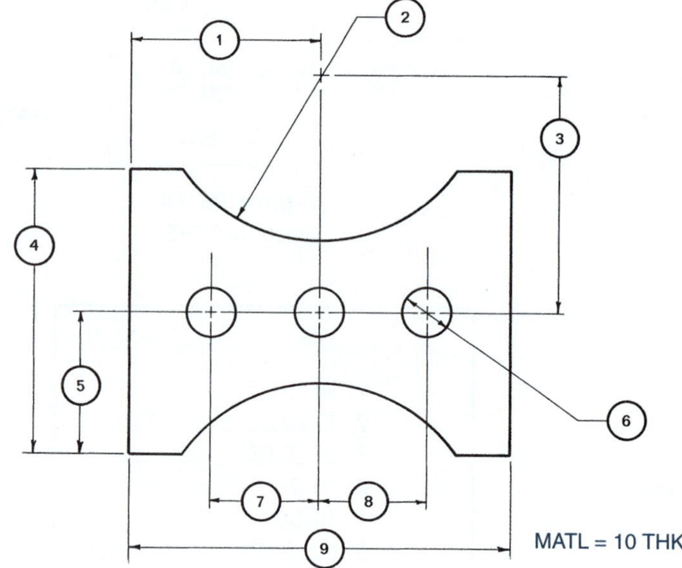

Figure P8-1D
MILLIMETERS

Project 8-2:

Redraw the object shown in Figure P8-2, including the given dimensions and tolerances. Calculate and list the maximum and minimum distances for surface A.

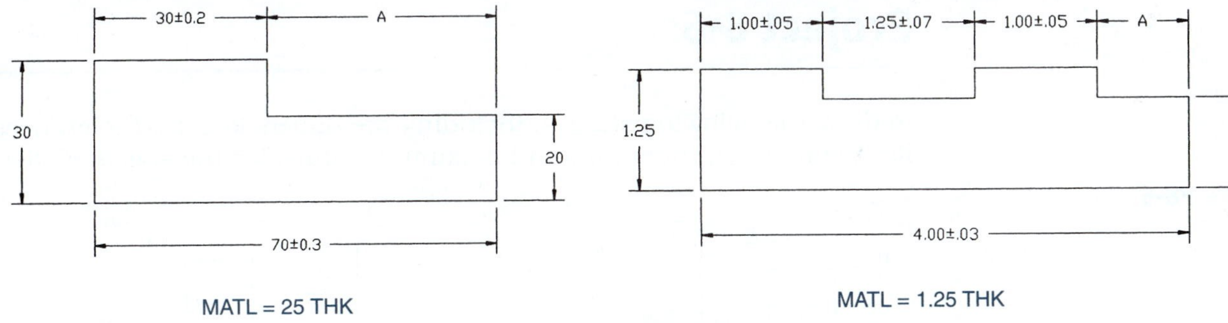

Figure P8-2
MILLIMETERS

Figure P8-3
INCHES

Project 8-3:

A. Redraw the object shown in Figure P8-3, including the dimensions and tolerances. Calculate and list the maximum and minimum distances for surface A.

B. Redraw the given object and dimension it using baseline dimensions. Calculate and list the maximum and minimum distances for surface A.

Project 8-4:

Redraw the object shown in Figure P8-4, including the dimensions and tolerances. Calculate and list the maximum and minimum distances for surfaces D and E.

Project 8-5:

Dimension the object shown in Figure P8-5 twice, once using chain dimensions and once using baseline dimensions. Calculate and list the maximum and minimum distances for surface D for both chain and baseline dimensions. Compare the results.

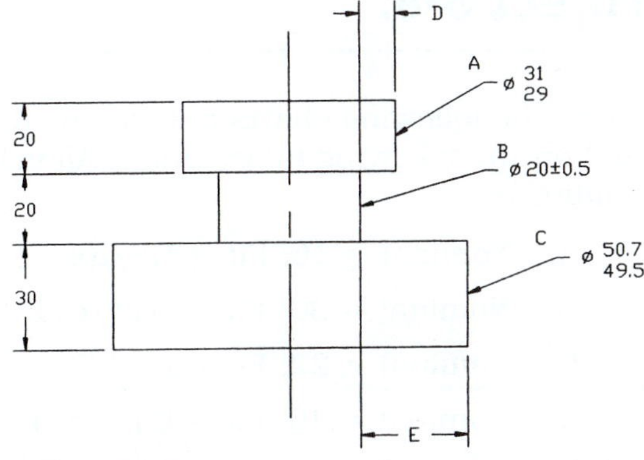

Figure P8-4
MILLIMETERS

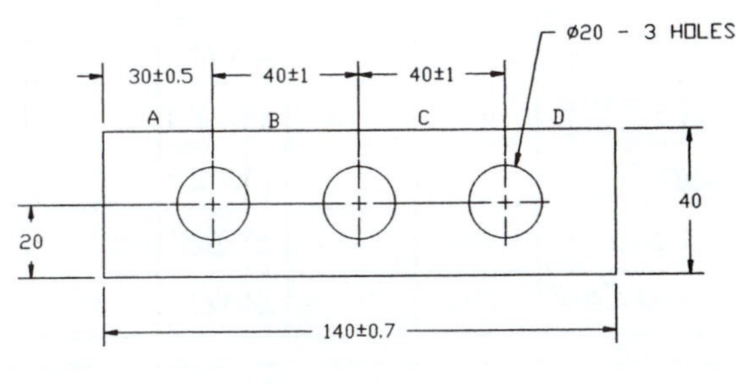

MATL = 20 THK

Figure P8-5
MILLIMETERS

Project 8-6:

Redraw the following shapes, including the dimensions and tolerances. Also list the required minimum and maximum values for the specified distances.

Figure P8-6
INCHES

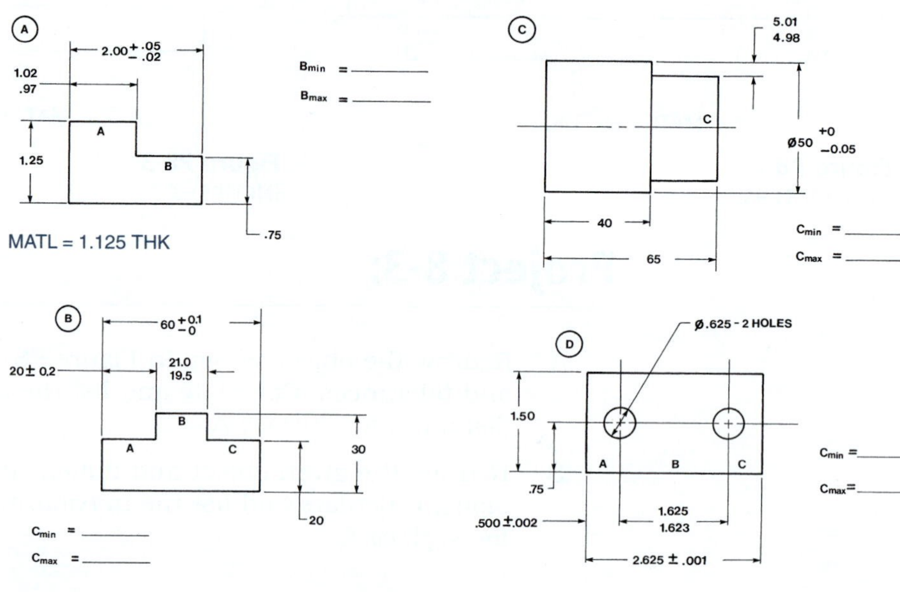

INSPECTION REPORT

PART NAME AND NO: 10755002

INSPECTOR:

DATE:

BASE DIMENSION	TOLERANCES MAX	TOLERANCES MIN	AS MEASURED	RESULTS
(1) 100 ± 0.5			99.8	
(2) $\phi\, {}^{57}_{56}$			57.01	
(3) 22 ± 0.3			21.72	
(4) ${}^{40.05}_{39.95}$			39.98	
(5) 22 ± 0.3			21.68	
(6) $R52\, {}^{+0}_{-0.2}$			51.99	
(7) $35\, {}^{+0.2}_{-0.3}$			35.20	
(8) $30\, {}^{+0.4}_{0}$			30.27	
(9) $6.0\, {}^{+.1}_{-.2}$			5.85	
(10) 12.0 ± 0.2			11.90	

1.00 3 PLACES

.50—10 PLACES

Figure P8-7
MILLIMETERS

Project 8-7:

Redraw and complete the inspection report shown in Figure P8-7. Under the RESULTS column classify each "AS MEASURED" value as OK if the value is within the stated tolerances, REWORK if the value indicates that the measured value is beyond the stated tolerance but can be reworked to bring it into the acceptable range, or SCRAP if the value is not within the tolerance range and cannot be reworked to make it acceptable.

Project 8-8:

Redraw the following charts and complete them based on the following information. All values are in millimeters.

A. Nominal = 16, Fit = H8/d8
B. Nominal = 30, Fit = H11/c11
C. Nominal = 22, Fit = H7/g6
D. Nominal = 10, Fit = C11/h11
E. Nominal = 25, Fit = F8/h7
F. Nominal = 12, Fit = H7/k6

G. Nominal = 3, Fit = H7/p6
H. Nominal = 18, Fit = H7/s6
I. Nominal = 27, Fit = H7/u6
J. Nominal = 30, Fit = N7/h6

Figure P8-8
MILLIMETERS

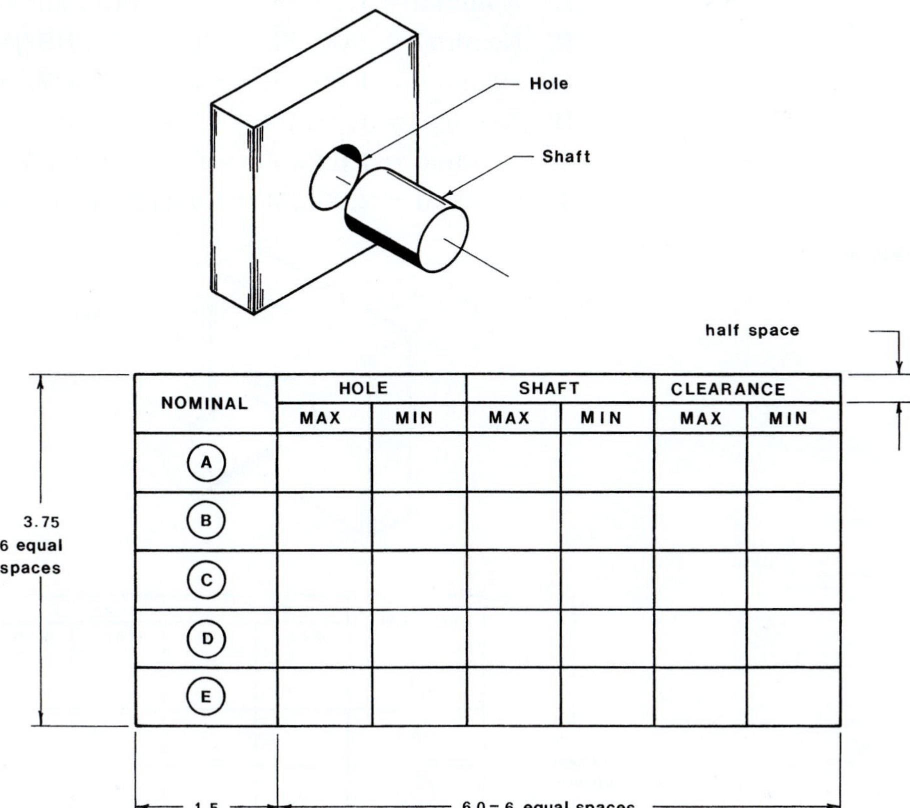

NOMINAL	HOLE		SHAFT		CLEARANCE	
	MAX	MIN	MAX	MIN	MAX	MIN
(A)						
(B)						
(C)						
(D)						
(E)						

NOMINAL	HOLE		SHAFT		INTERFERENCE	
	MAX	MIN	MAX	MIN	MAX	MIN
(F)						
(G)						
(H)						
(I)						
(J)						

Use the same dimensions given above

Project 8-9:

Redraw the following charts and complete them based on the following information. All values are in inches.

A. Nominal 0.25, Fit Class LC5, H7/g6
B. Nominal = 1.00, Fit = Class LC7, H10/e9
C. Nominal = 1.50, Fit = Class LC9, F11/h11
D. Nominal = 0.75, Fit = Class RC3, H7/f6
E. Nominal = 1.75, Fit = Class RC6, H9/e8
F. Nominal = .500, Fit = Class LT2, H8/js7
G. Nominal = 1.25, Fit = Class LT5, H7/n6
H. Nominal = 1.38, Fit = Class LN3, J7/h6
I. Nominal = 1.625, Fit = Class FN2, H7/s6
J. Nominal = 2.00, Fit = Class FN4, H7/u6

Figure P8-9
INCHES

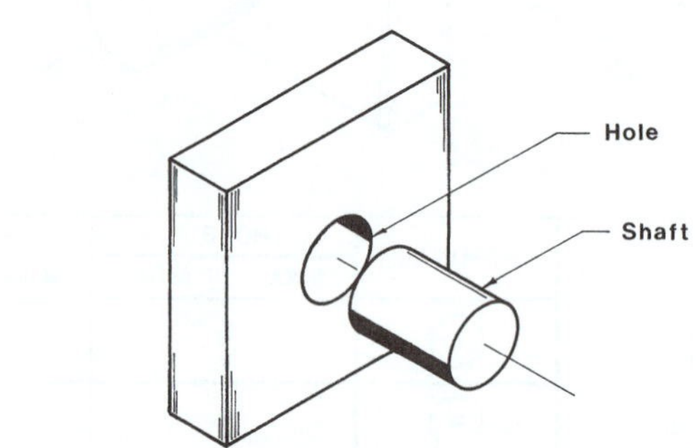

half space

3.75
6 equal
spaces

NOMINAL	HOLE		SHAFT		CLEARANCE	
	MAX	MIN	MAX	MIN	MAX	MIN
(A)						
(B)						
(C)						
(D)						
(E)						

1.5 — 6.0 – 6 equal spaces

NOMINAL	HOLE		SHAFT		INTERFERENCE	
	MAX	MIN	MAX	MIN	MAX	MIN
(F)						
(G)						
(H)						
(I)						
(J)						

Use the same dimensions given above

Project 8-10:

Draw the chart shown and add the appropriate values based on the dimensions and tolerances given in Figures P8-10A through P8-10D.

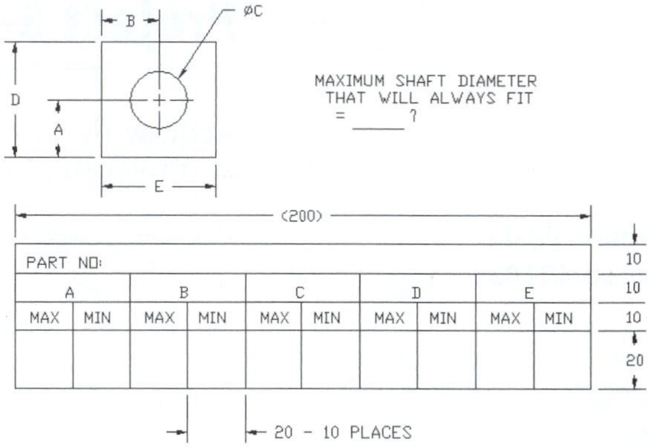

MAXIMUM SHAFT DIAMETER THAT WILL ALWAYS FIT = _____ ?

(200)

PART NO:									
A		B		C		D		E	
MAX	MIN	MAX	MIN	MAX	MIN	MAX	MIN	MAX	MIN

20 - 10 PLACES

PART NO: 9-M53A

A. 20±0.1

B. 30±0.2

C. ⌀20±0.05

D. 40

E. 60

Figure P8-10A
MILLIMETERS

PART NO: 9-M53B

A. 32.02 / 31.97

B. 47.52 / 47.50

C. ⌀18 $^{+0.05}_{0}$

D. 64±0.05

E. 100±0.05

Figure P8-10B
MILLIMETERS

PART NO: 9-E47A

A. 2.00±.02

B. 1.75±.03

C. ⌀.750±.005

D. 4.00±.05

E. 3.50±.05

Figure P8-10C
MILLIMETERS

PART NO: 9-E47B

A. 18 $^{+0}_{-0.02}$

B. 26 $^{+0}_{-0.04}$

C. ⌀ 24.03 / 23.99

D. 52±0.04

E. 36±0.02

Figure P8-10D
MILLIMETERS

Project 8-11:

Prepare front and top views of parts 4A and 4B in Figure P8-11. Based on the given dimensions. Add tolerances to produce the stated clearances.

Project 8-12:

Redraw the Box, Top and Box, Bottom in Figure P8-12 and dimensions and tolerances to meet the "UPON ASSEMBLY" requirements.

Figure P8-11
MILLIMETERS

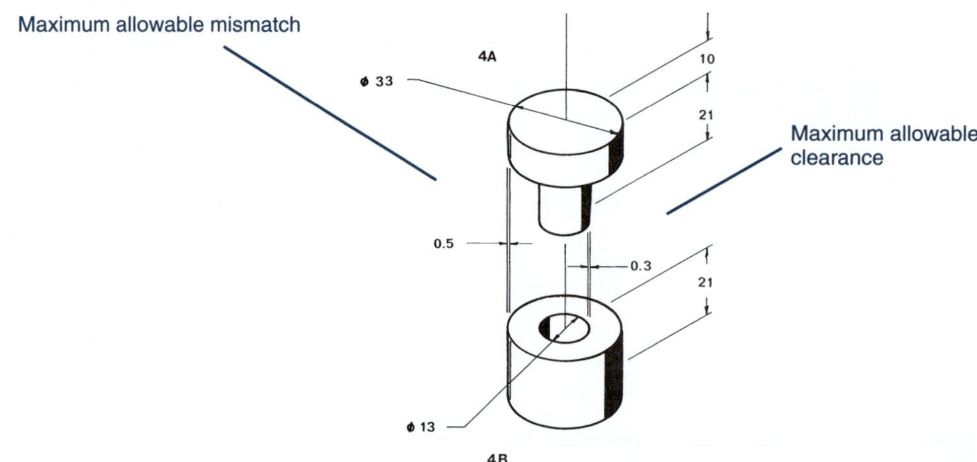

Figure P8-12
INCHES

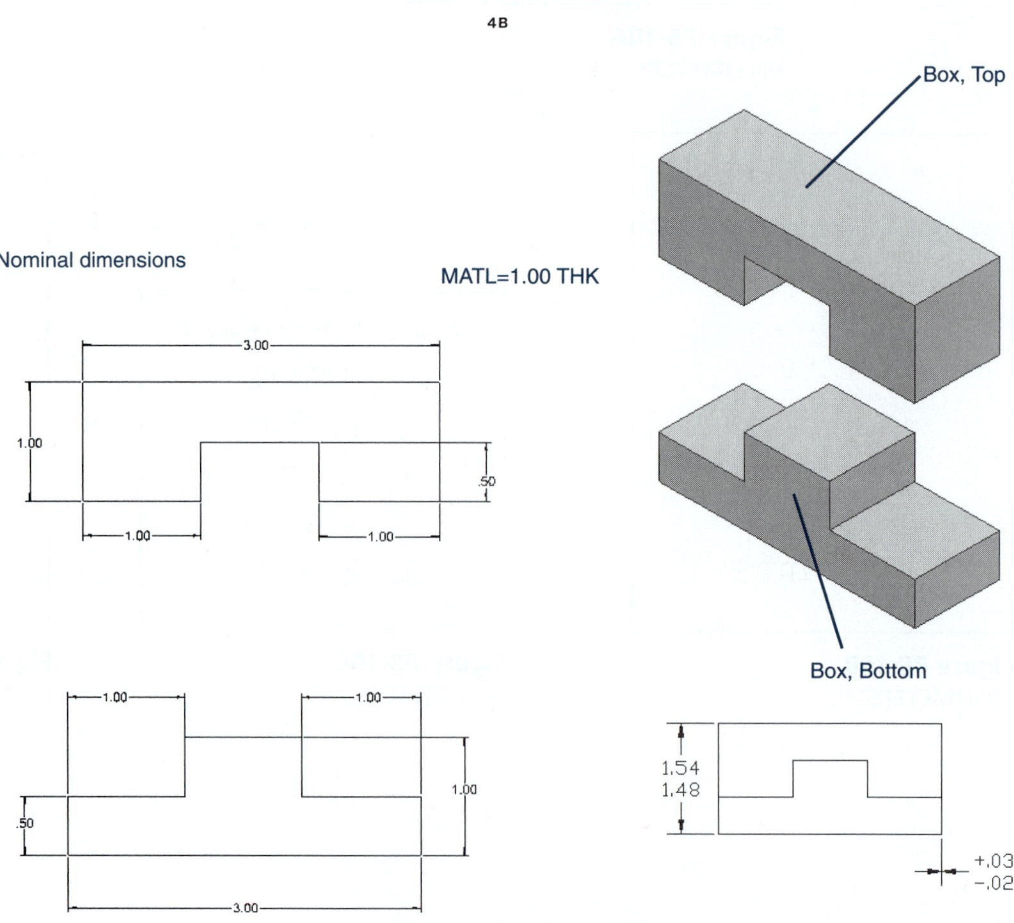

Project 8-13:

Draw a front and top view of both given objects in Figure P8-13. Add dimensions and tolerances to meet the "FINAL CONDITION" requirements.

Project 8-14:

Given the following nominal sizes, dimension and tolerance parts AM311 and AM312 in Figure P8-14 so that they always fit together regardless of orientation. Further, dimension the overall lengths of each part so that in the assembled condition they will always pass through a clearance gauge with an opening of 80.00±0.02.

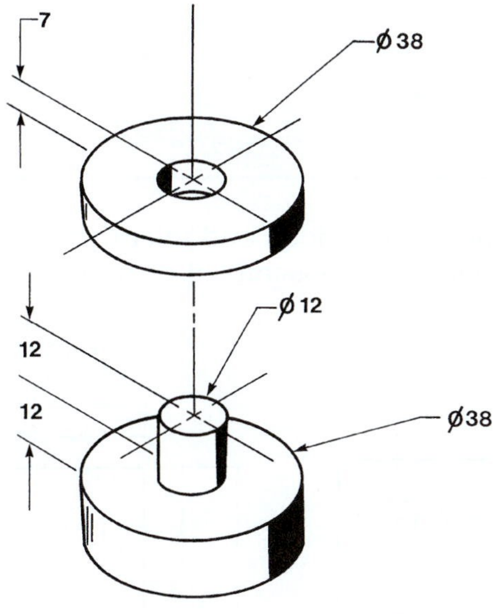

Figure P8-13
MILLIMETERS

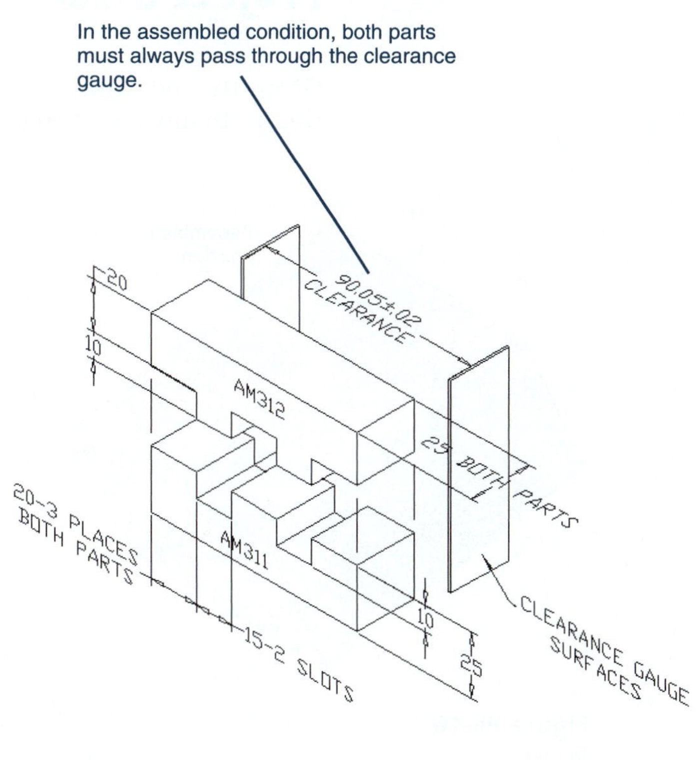

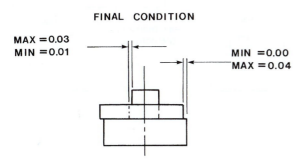

Figure P8-14
MILLIMETERS

Project 8-15:

Given the following rail assembly, add dimensions and tolerances so that the parts always fit together as shown in the assembled position.

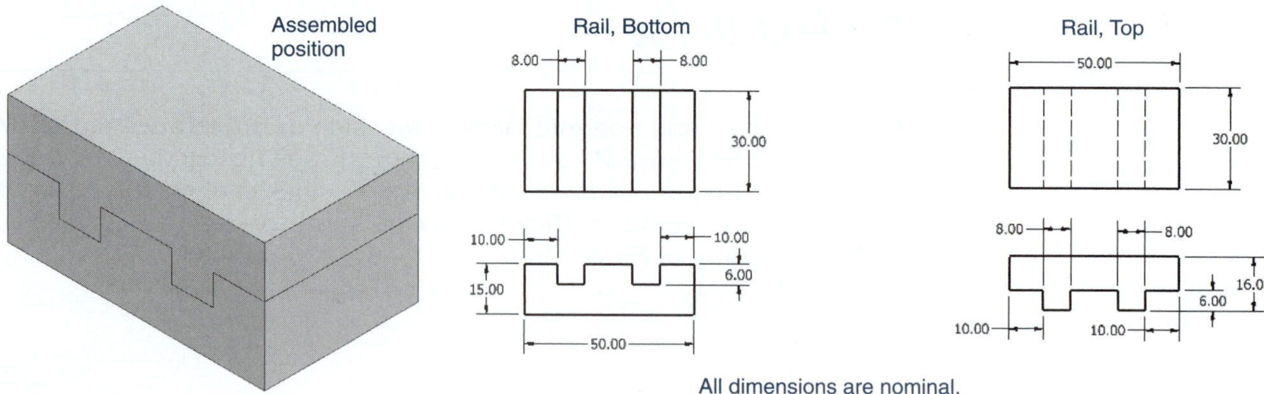

Figure P8-15
MILLIMETERS

Project 8-16:

Given the following peg assembly, add dimensions and tolerances so that the parts always fit together as shown in the assembled position.

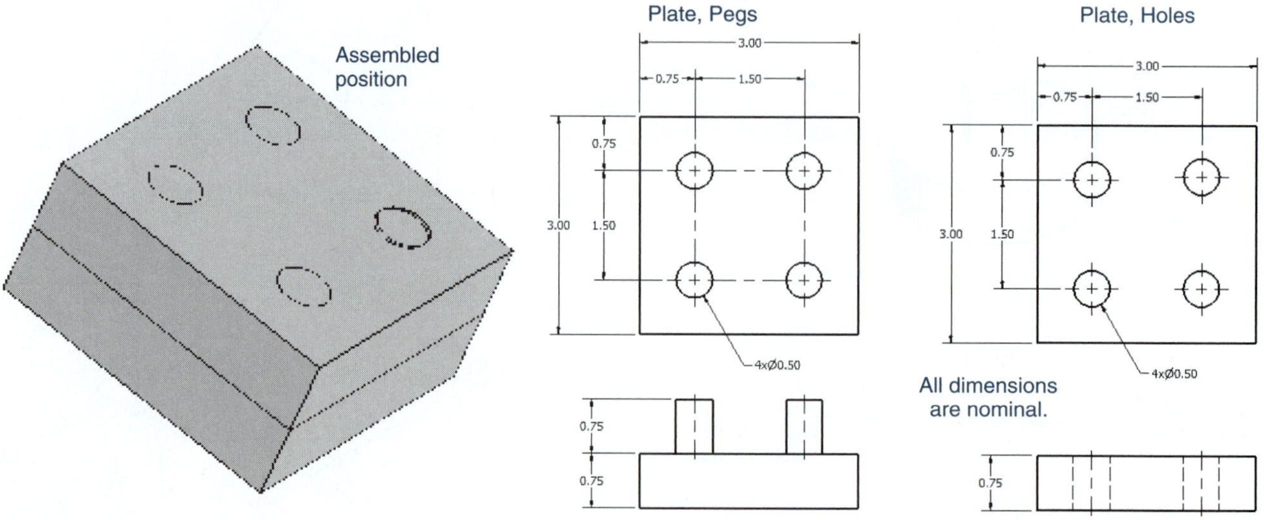

Figure P8-16
INCHES

Project 8-17:

Given the following collar assembly, add dimensions and tolerances so that the parts always fit together as shown in the assembled position.

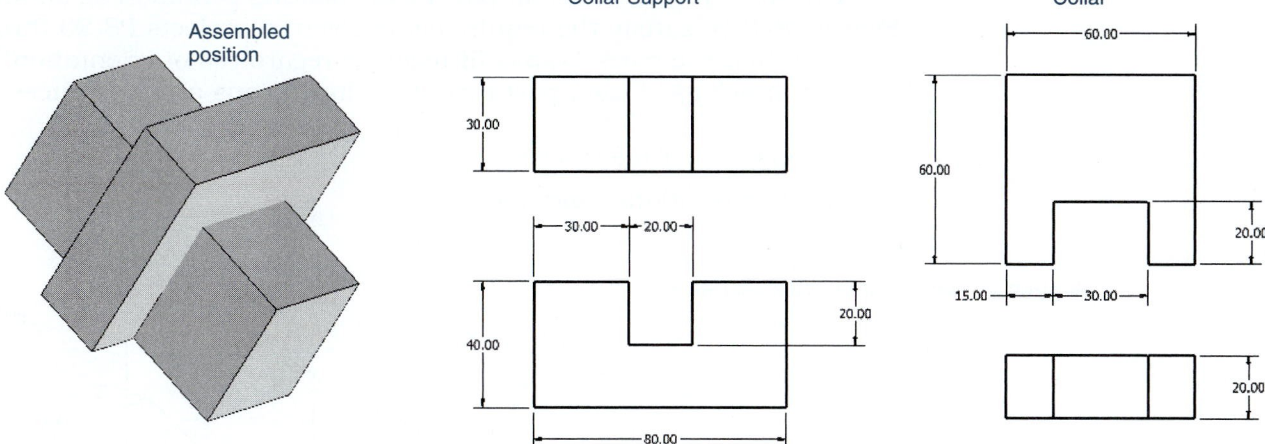

Figure P8-17
MILLIMETERS

Project 8-18:

Given the following vee-block assembly, add dimensions and tolerances so that the parts always fit together as shown in the assembled position. The total height of the assembled blocks must be between 4.45 and 4.55 in.

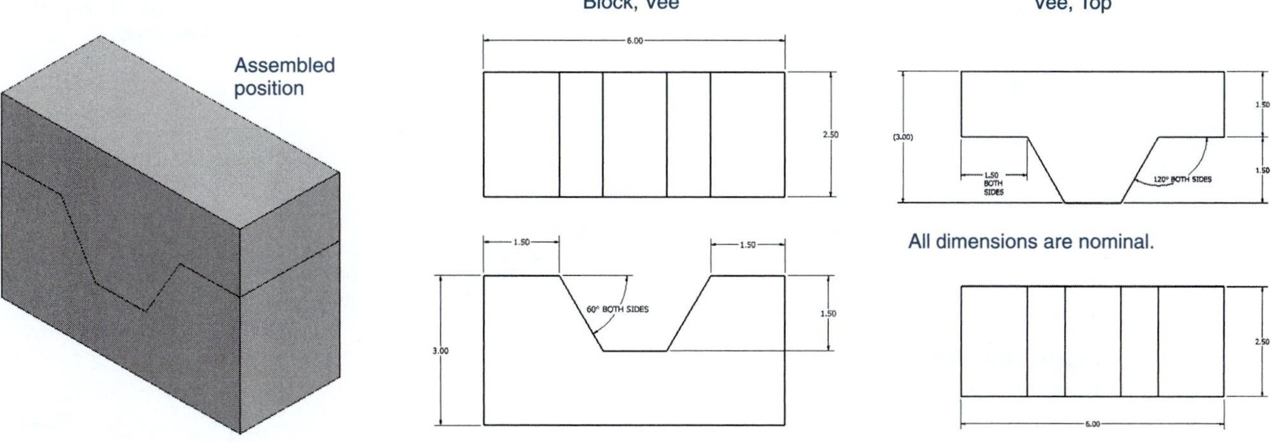

Figure P8-18
INCHES

Project 8-19:

Design a bracket that will support the three Ø100 wheels shown in Figure P8-19A. The wheels will utilize three Ø5.00±0.01 shafts attached to the bracket. The bottom of the bracket must be a minimum of 10 mm from the ground. The wall thickness of the bracket must always be at least 5 mm, and the minimum bracket opening must be at least 15 mm.

1. Prepare a front and a side view of the bracket.
2. Draw the wheels in their relative positions using phantom lines.
3. Add all appropriate dimensions and tolerances.

Given a top and a bottom part in the floating condition as shown in Figure P8-19B, satisfy the requirements given in projects P8-20 through P8-23 so that the parts always fit together regardless of orientation. Prepare drawings of each part including dimensions and tolerances.

A. Use linear tolerances.

B. Use positional tolerances.

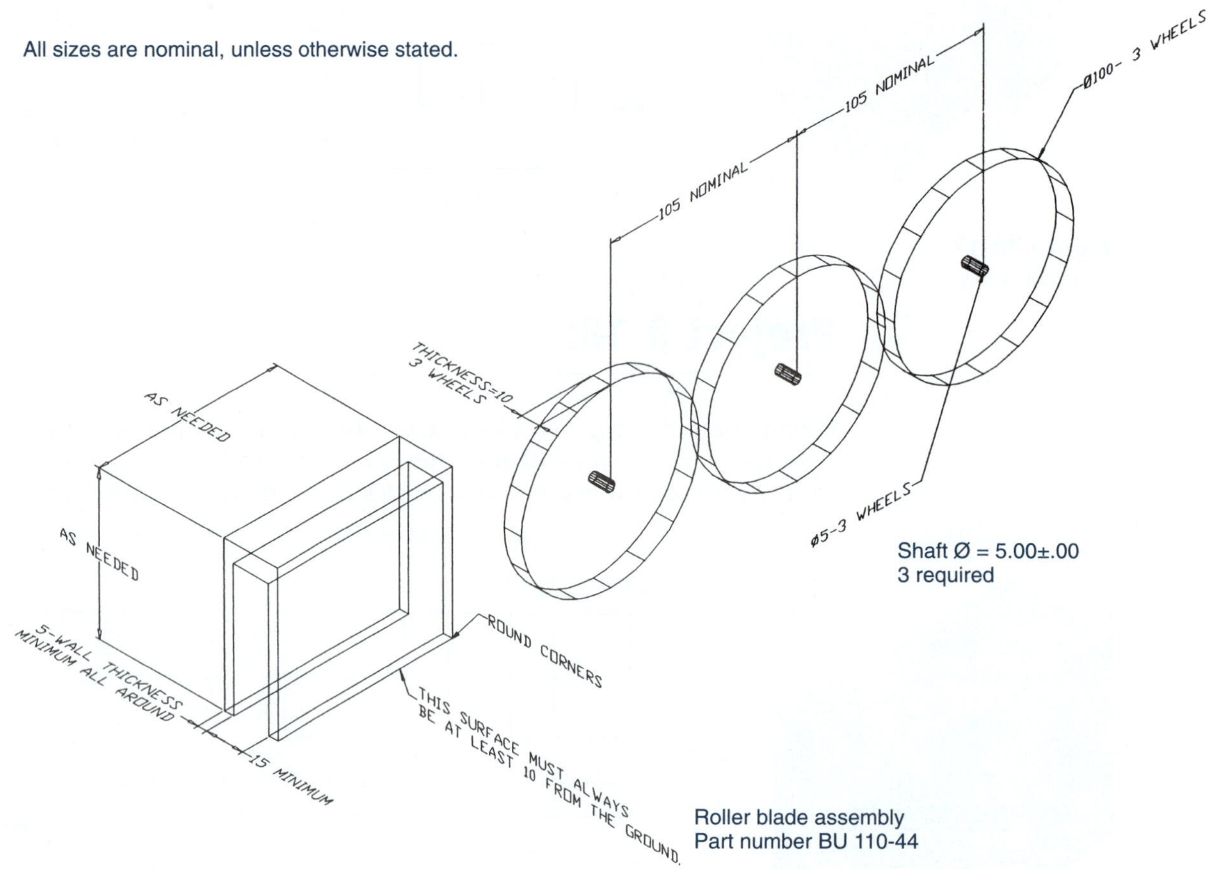

Figure P8-19A
MILLIMETERS

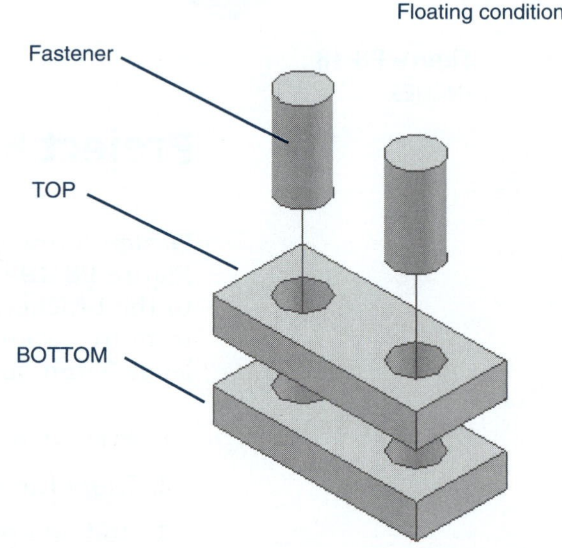

Figure P8-19B

Project 8-20: Inches

A. The distance between the holes' centerpoints is 2.00 nominal.
B. The holes are Ø.375 nominal.
C. The fasteners have a tolerance of .001.
D. The holes have a tolerance of .002.
E. The minimum allowable clearance between the fasteners and the holes is .003.
F. The positional tolerance is .001.

Project 8-21: Millimeters

A. The distance between the holes' centerpoints is 80 nominal.
B. The holes are Ø12 nominal.
C. The fasteners have a tolerance of 0.05.
D. The holes have a tolerance of 0.03.
E. The minimum allowable clearance between the fasteners and the holes is 0.02.
F. The positional tolerance is .01.

Project 8-22: Inches

A. The distance between the holes' centerpoints is 3.50 nominal.
B. The holes are Ø.625 nominal.
C. The fasteners have a tolerance of .005.
D. The holes have a tolerance of .003.
E. The minimum allowable clearance between the fasteners and the holes is .002.
F. The positional tolerance is .002.

Project 8-23: Millimeters

A. The distance between the holes' centerpoints is 65 nominal.
B. The holes are Ø16 nominal.
C. The fasteners have a tolerance of 0.03.
D. The holes have a tolerance of 0.04.
E. The minimum allowable clearance between the fasteners and the holes is 0.03.
F. The positional tolerance is .02.

Given a top and a bottom part in the fixed condition as shown in Figure P8-23, satisfy the requirements given in projects P8-24 through P8-27 so that the parts fit together regardless of orientation. Prepare drawings of each part including dimensions and tolerances.

A. Use linear tolerances.

B. Use positional tolerances.

Figure P8-23

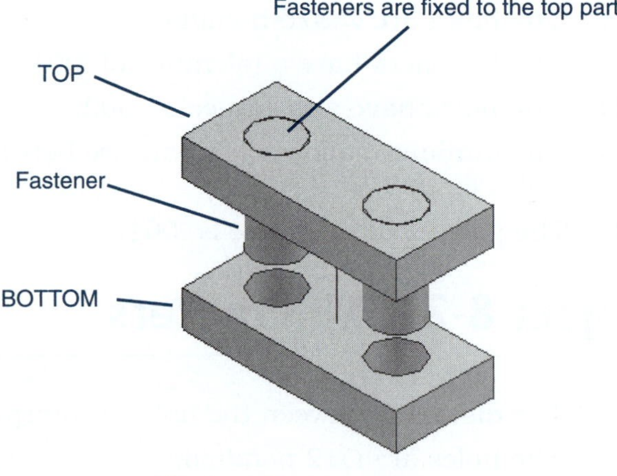

Project 8-24: Millimeters

A. The distance between the holes' centerpoints is 60 nominal.

B. The holes are Ø10 nominal.

C. The fasteners have a tolerance of 0.04.

D. The holes have a tolerance of 0.02.

E. The minimum allowable clearance between the fasteners and the holes is 0.02.

F. The positional tolerance is 0.01.

Project 8-25: Inches

A. The distance between the holes' centerpoints is 3.50 nominal.

B. The holes are Ø.563 nominal.

C. The fasteners have a tolerance of .005.

D. The holes have a tolerance of .003.

E. The minimum allowable clearance between the fasteners and the holes is .002.

F. The positional tolerance is .001.

Project 8-26: Millimeters

A. The distance between the holes' centerpoints is 100 nominal.

B. The holes are Ø18 nominal.

C. The fasteners have a tolerance of 0.02.

D. The holes have a tolerance of 0.01.

E. The minimum allowable clearance between the fasteners and the holes is 0.03.

F. The positional tolerance is 0.02.

Project 8-27: Inches

A. The distance between the holes' centerpoints is 1.75 nominal.

B. The holes are Ø.250 nominal.

C. The fasteners have a tolerance of .002.

D. The holes have a tolerance of .003.

E. The minimum allowable clearance between the fasteners and the holes is .001.

F. The positional tolerance is .002.

Project 8-28: Millimeters

Dimension and tolerance the rotator assembly shown in Figure P8-28. Use the given dimensions as nominal and add sleeve bearings between the LINKs and both the CROSS-LINK and the PLATE. Create drawings of each part. Modify the dimensions as needed and add the appropriate tolerances. Specify the selected sleeve bearing.

Project 8-29:

Dimension and tolerance the rocker assembly shown in Figure P8-29. Use the given dimensions as nominal, and add sleeve bearings between all moving parts. Create drawings of each part. Modify the dimensions as needed and add the appropriate tolerances. Specify the selected sleeve bearing.

Project 8-30:

Draw the model shown in Figure P8-30, create a drawing layout with the appropriate views, and add the specified dimensions and tolerances.

Project 8-31:

Redraw the shaft shown in Figure P8-31, create a drawing layout with the appropriate views, and add a feature dimension and tolerance of 36±0.1 and a straightness tolerance of 0.07 about the centerline at MMC.

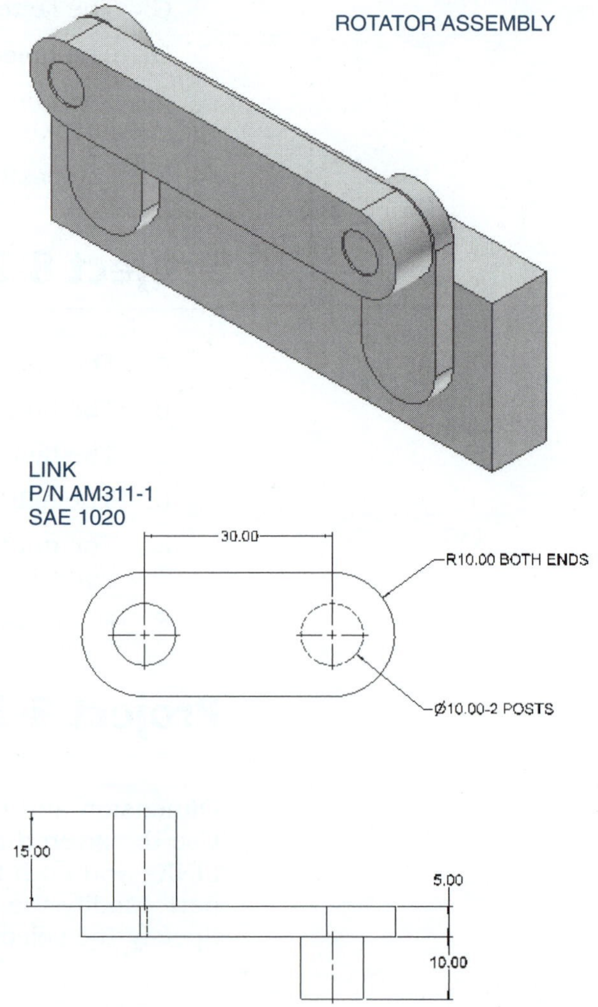
ROTATOR ASSEMBLY
LINK
P/N AM311-1
SAE 1020
30.00
R10.00 BOTH ENDS
Ø10.00-2 POSTS
15.00
5.00
10.00

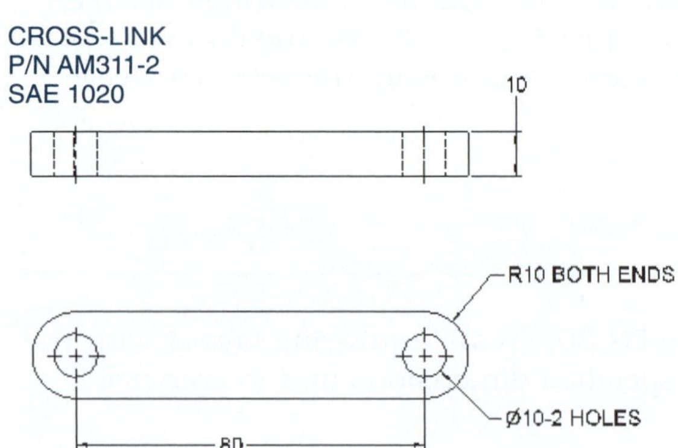
CROSS-LINK
P/N AM311-2
SAE 1020
10
R10 BOTH ENDS
Ø10-2 HOLES
80

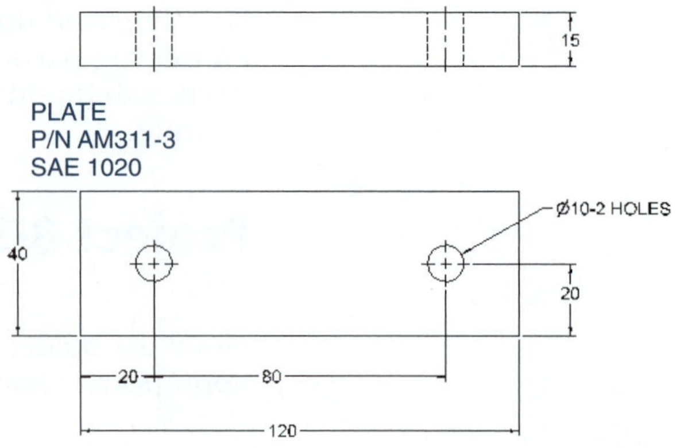
15
PLATE
P/N AM311-3
SAE 1020
Ø10-2 HOLES
40
20
20
80
120

Figure P8-28

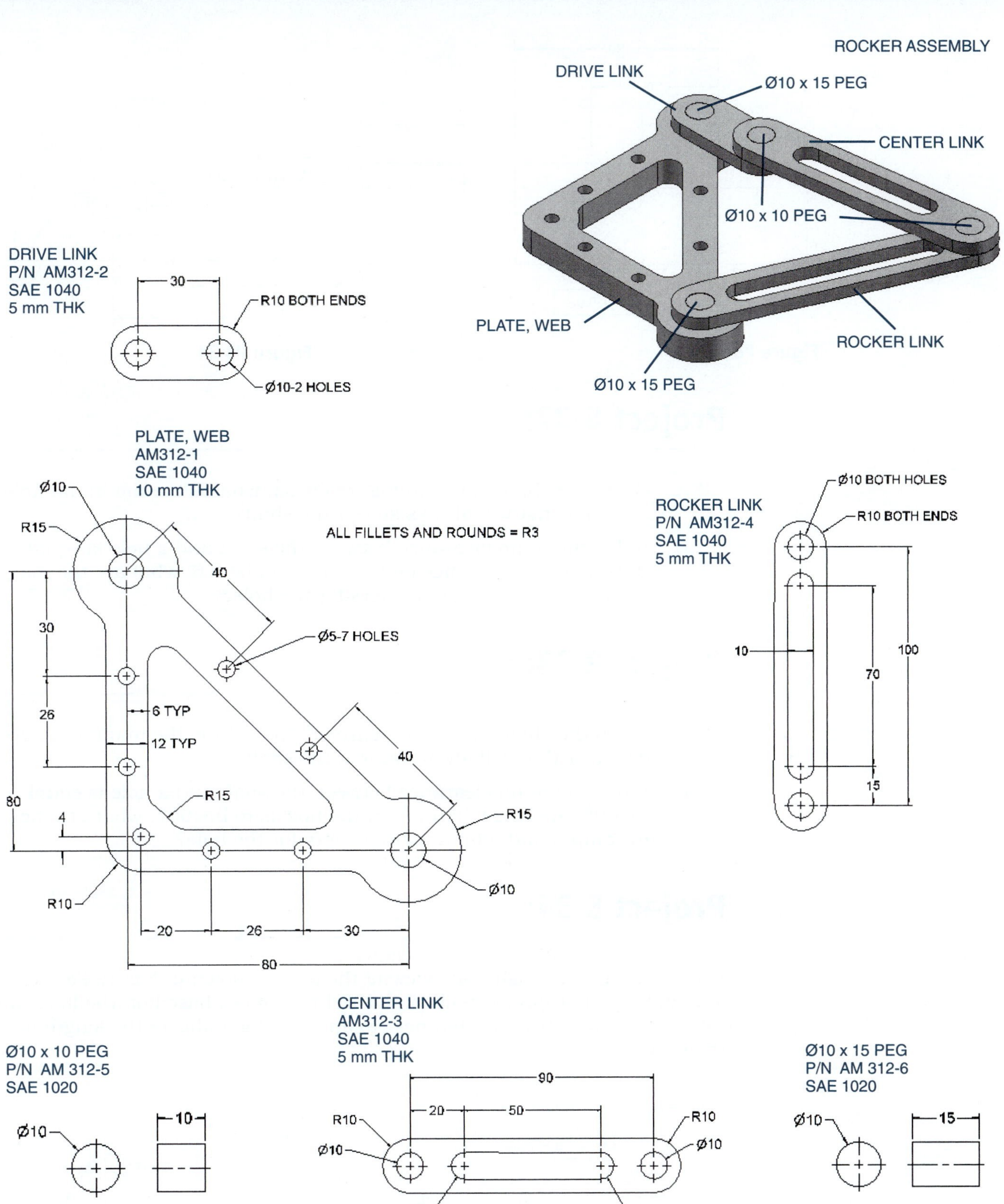
ROCKER ASSEMBLY
DRIVE LINK
Ø10 x 15 PEG
CENTER LINK
Ø10 x 10 PEG
PLATE, WEB
ROCKER LINK
Ø10 x 15 PEG
DRIVE LINK
P/N AM312-2
SAE 1040
5 mm THK
30
R10 BOTH ENDS
Ø10-2 HOLES
PLATE, WEB
AM312-1
SAE 1040
10 mm THK
Ø10
R15
40
ALL FILLETS AND ROUNDS = R3
30
Ø5-7 HOLES
26
6 TYP
12 TYP
40
80
R15
4
R15
R10
Ø10
20
26
30
80
ROCKER LINK
P/N AM312-4
SAE 1040
5 mm THK
Ø10 BOTH HOLES
R10 BOTH ENDS
10
100
70
15
Ø10 x 10 PEG
P/N AM 312-5
SAE 1020
Ø10
10
CENTER LINK
AM312-3
SAE 1040
5 mm THK
90
20
50
R10
Ø10
R10
Ø10
R5
R5
Ø10 x 15 PEG
P/N AM 312-6
SAE 1020
Ø10
15

Figure P8-29

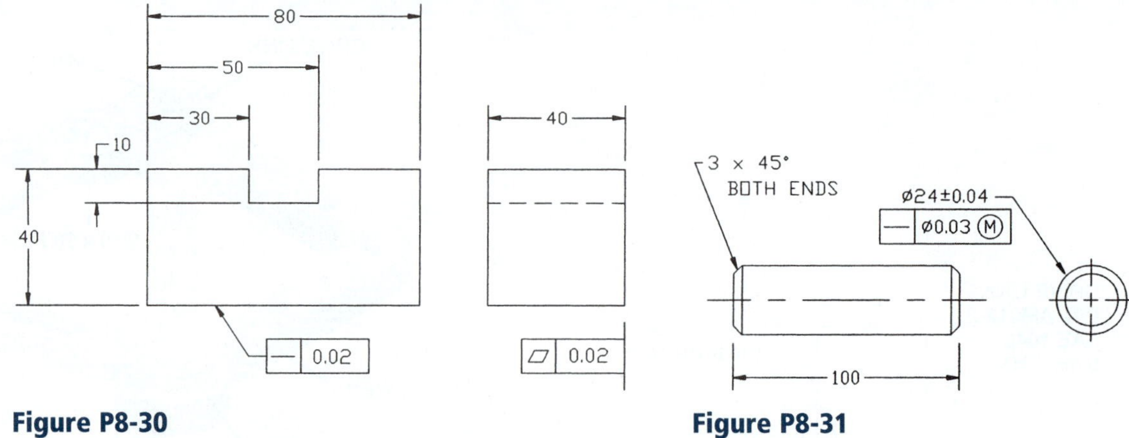

Figure P8-30

Figure P8-31

Project 8-32:

A. Given the shaft shown in Figure P8-32, what is the minimum hole diameter that will always accept the shaft?

B. If the minimum clearance between the shaft and a hole is equal to 0.02, and the tolerance on the hole is to be 0.6, what are the maximum and minimum diameters for the hole?

Project 8-33:

A. Given the shaft shown in Figure P8-33, what is the minimum hole diameter that will always accept the shaft?

B. If the minimum clearance between the shaft and a hole is equal to .005, and the tolerance on the hole is to be .007, what are the maximum and minimum diameters for the hole?

Project 8-34:

Draw a front and a right-side view of the object shown in Figure P8-34 and add the appropriate dimensions and tolerances based on the following information. Numbers located next to an edge line indicate the length of the edge.

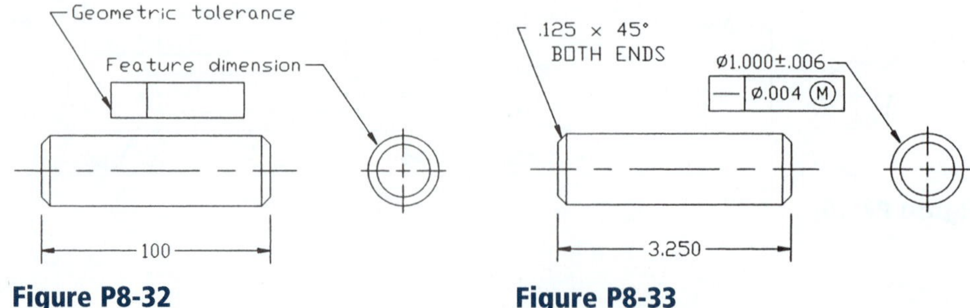

Figure P8-32

Figure P8-33

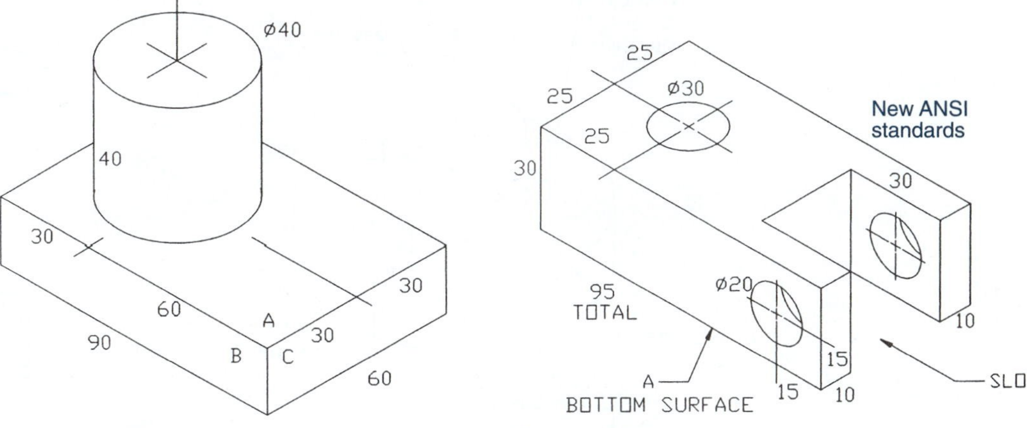

Figure P8-34

Figure P8-35

A. Define surfaces A, B, and C as primary, secondary, and tertiary datums, respectively.

B. Assign a tolerance of ±0.5 to all linear dimensions.

C. Assign a feature tolerance of 12.07−12.00 to the protruding shaft.

D. Assign a flatness tolerance of 0.01 to surface A.

E. Assign a straightness tolerance of 0.03 to the protruding shaft.

F. Assign a perpendicularity tolerance to the centerline of the protruding shaft of 0.02 at MMC relative to datum A.

Project 8-35:

Draw a front and a right-side view of the object shown in Figure P8-35 and add the following dimensions and tolerances.

A. Define the bottom surface as datum A.

B. Assign a perpendicularity tolerance of 0.4 to both sides of the slot relative to datum A.

C. Assign a perpendicularity tolerance of 0.2 to the centerline of the 30 diameter hole at MMC relative to datum A.

D. Assign a feature tolerance of ±0.8 to all three holes.

E. Assign a parallelism tolerance of 0.2 to the common centerline between the two 20 diameter holes relative to datum A.

F. Assign a tolerance of ±0.5 to all linear dimensions.

Project 8-36:

Draw a circular front and the appropriate right-side view of the object shown in Figure P8-36 and add the following dimensions and tolerances.

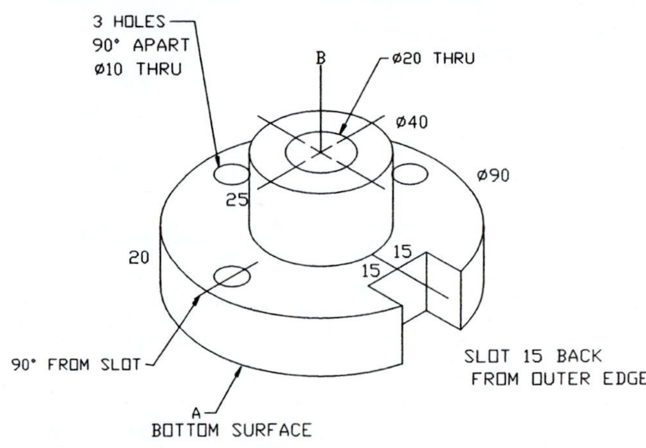

Figure P8-36

Figure P8-37

A. Assign datum A as indicated.

B. Assign the object's longitudinal axis as datum B.

C. Assign the object's centerline through the slot as datum C.

D. Assign a tolerance of ±0.5 to all linear tolerances.

E. Assign a tolerance of ±0.5 to all circular features.

F. Assign a parallelism tolerance of 0.01 to both edges of the slot.

G. Assign a perpendicularity tolerance of 0.01 to the outside edge of the protruding shaft.

Project 8-37:

Given the two objects shown in Figure P8-37, draw a front and a side view of each. Assign a tolerance of ±0.5 to all linear dimensions. Assign a feature tolerance of ±0.4 to the shaft, and also assign a straightness tolerance of 0.2 to the shaft's centerline at MMC.

Tolerance the hole so that it will always accept the shaft with a minimum clearance of 0.1 and a feature tolerance of 0.2. Assign a perpendicularity tolerance of 0.05 to the centerline of the hole at MMC.

Project 8-38:

Given the two objects shown in Figure P8-38, draw a front and a side view of each. Assign a tolerance of ±.005 to all linear dimensions. Assign a feature tolerance of ±.004 to the shaft, and also assign a straightness tolerance of .002 to the shaft's centerline at MMC.

Tolerance the hole so that it will always accept the shaft with a minimum clearance of .001 and a feature tolerance of .002.

Project 8-39:

Draw a model of the object shown in Figure P8-39, then create a drawing layout including the specified dimensions. Add the following tolerances and specifications to the drawing.

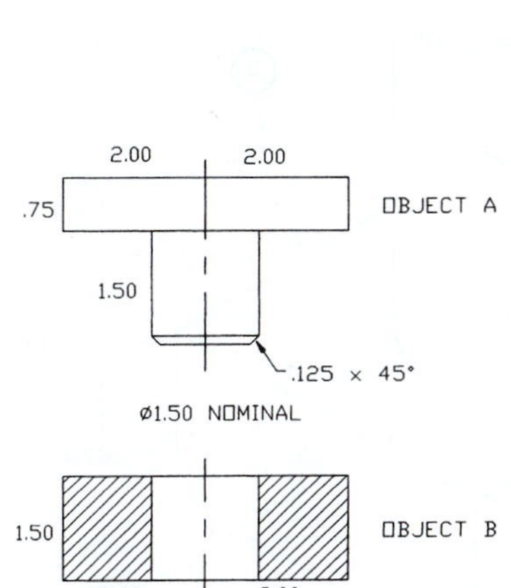

Figure P8-38

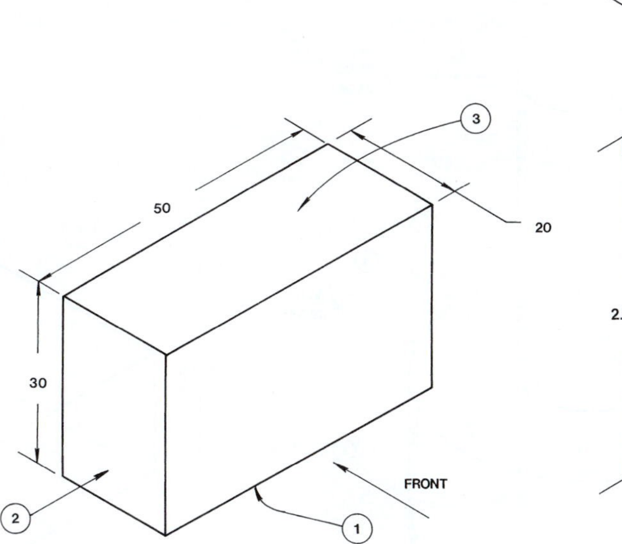

Figure P8-39

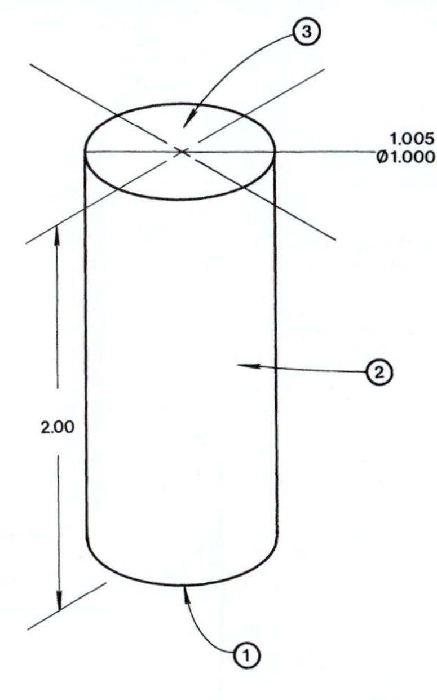

Figure P8-40

A. Surface 1 is datum A.

B. Surface 2 is datum B and is perpendicular to datum A within 0.1 mm.

C. Surface 3 is datum C and is parallel to datum A within 0.3 mm.

D. Locate a 16-mm diameter hole in the center of the front surface that goes completely through the object. Use positional tolerances to locate the hole. Assign a positional tolerance of 0.02 at MMC perpendicular to datum A.

Project 8-40:

Draw a model of the object shown in Figure P8-40, then create a drawing layout including the specified dimensions. Add the following tolerances and specifications to the drawing.

A. Surface 1 is datum A.

B. Surface 2 is datum B and is perpendicular to datum A within .003 in.

C. Surface 3 is parallel to datum A within .005 in.

D. The cylinder's longitudinal centerline is to be straight within .001 in. at MMC.

E. Surface 2 is to have circular accuracy within .002 in.

Project 8-41:

Draw a model of the object shown in Figure P8-41, then create a drawing layout including the specified dimensions. Add the following tolerances and specifications to the drawing.

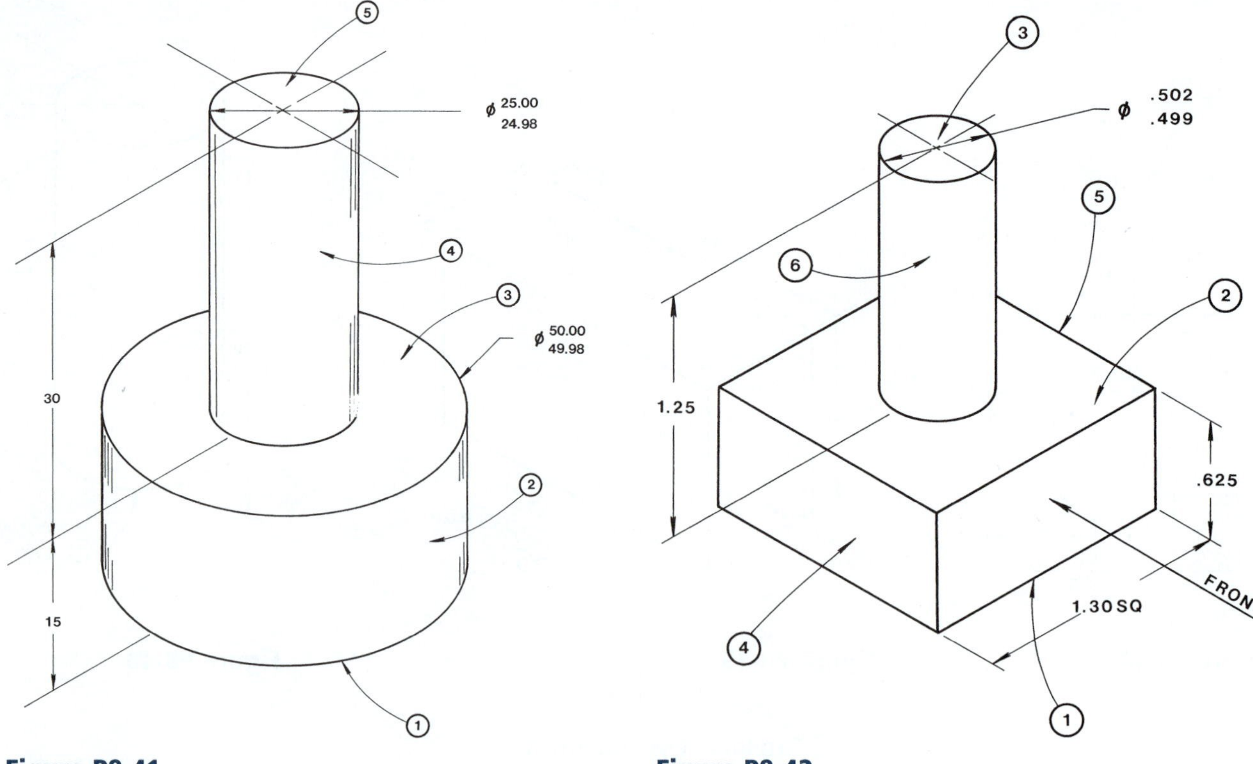

Figure P8-41

Figure P8-42

A. Surface 1 is datum A.

B. Surface 4 is datum B and is perpendicular to datum A within 0.08 mm.

C. Surface 3 is flat within 0.03 mm.

D. Surface 5 is parallel to datum A within 0.01 mm.

E. Surface 2 has a runout tolerance of 0.2 mm relative to surface 4.

F. Surface 1 is flat within 0.02 mm.

G. The longitudinal centerline is to be straight within 0.02 at MMC and perpendicular to datum A.

Project 8-42:

Draw a model of the object shown in Figure P8-42, then create a drawing layout including the specified dimensions. Add the following tolerances and specifications to the drawing.

A. Surface 2 is datum A.

B. Surface 6 is perpendicular to datum A with .000 allowable variance at MMC but with a .002 in. MAX variance limit beyond MMC.

C. Surface 1 is parallel to datum A within .005.

D. Surface 4 is perpendicular to datum A within .004 in.

Project 8-43:

Draw a model of the object shown in Figure P8-43, then create a drawing layout including the specified dimensions. Add the following tolerances and specifications to the drawing.

A. Surface 1 is datum A.

B. Surface 2 is datum B.

C. The hole is located using a true position tolerance value of 0.13 mm at MMC. The true position tolerance is referenced to datums A and B.

D. Surface 1 is to be straight within 0.02 mm.

E. The bottom surface is to be parallel to datum A within 0.03 mm.

Project 8-44:

Draw a model of the object shown in Figure P8-44, then create a drawing layout including the specified dimensions. Add the following tolerances and specifications to the drawing.

A. Surface 1 is datum A.

B. Surface 2 is datum B.

C. Surface 3 is perpendicular to surface 2 within 0.02 mm.

D. The four holes are to be located using a positional tolerance of 0.07 mm at MMC referenced to datums A and B.

E. The centerlines of the holes are to be straight within 0.01 mm at MMC.

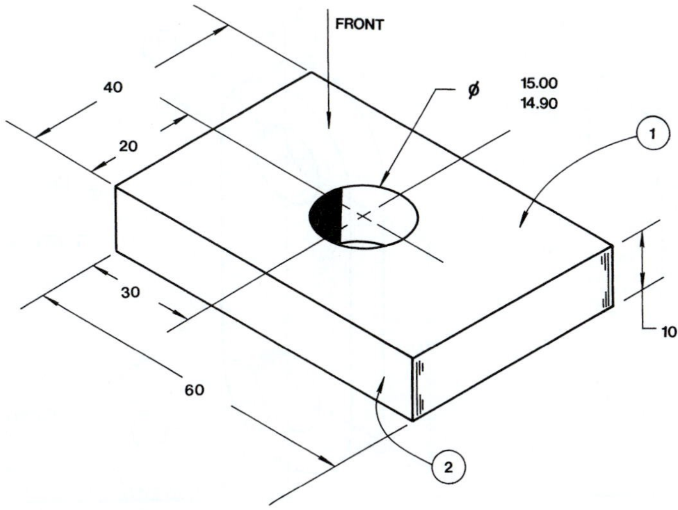

Figure P8-43

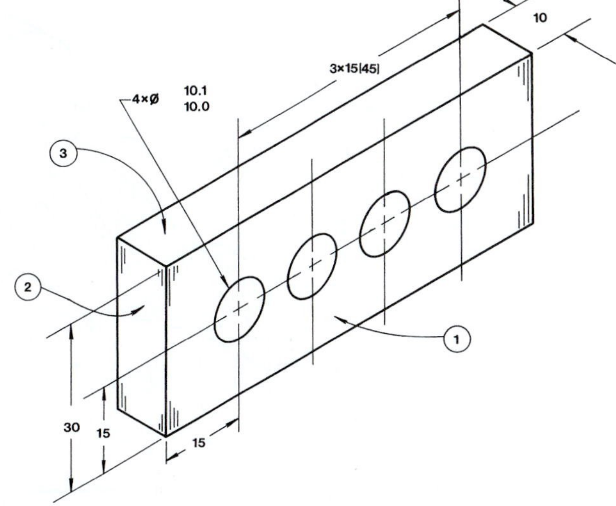

Figure P8-44

Project 8-45:

Draw a model of the object shown in Figure P8-45, then create a drawing layout including the specified dimensions. Add the following tolerances and specifications to the drawing.

A. Surface 1 has a dimension of .378−.375 in. and is datum A. The surface has a dual primary runout with datum B to within .005 in. The runout is total.

B. Surface 2 has a dimension of 1.505−1.495 in. Its runout relative to the dual primary datums A and B is .008 in. The runout is total.

C. Surface 3 has a dimension of 1.000±.005 and has no geometric tolerance.

D. Surface 4 has no circular dimension but has a total runout tolerance of .006 in. relative to the dual datums A and B.

E. Surface 5 has a dimension of .500−.495 in. and is datum B. It has a dual primary runout with datum A within .005 in. The runout is total.

Project 8-46:

Draw a model of the object shown in Figure P8-46, then create a drawing layout including the specified dimensions. Add the following tolerances and specifications to the drawing.

A. Hole 1 is datum A.

B. Hole 2 is to have its circular centerline parallel to datum A within 0.2 mm at MMC when datum A is at MMC.

C. Assign a positional tolerance of 0.01 to each hole's centerline at MMC.

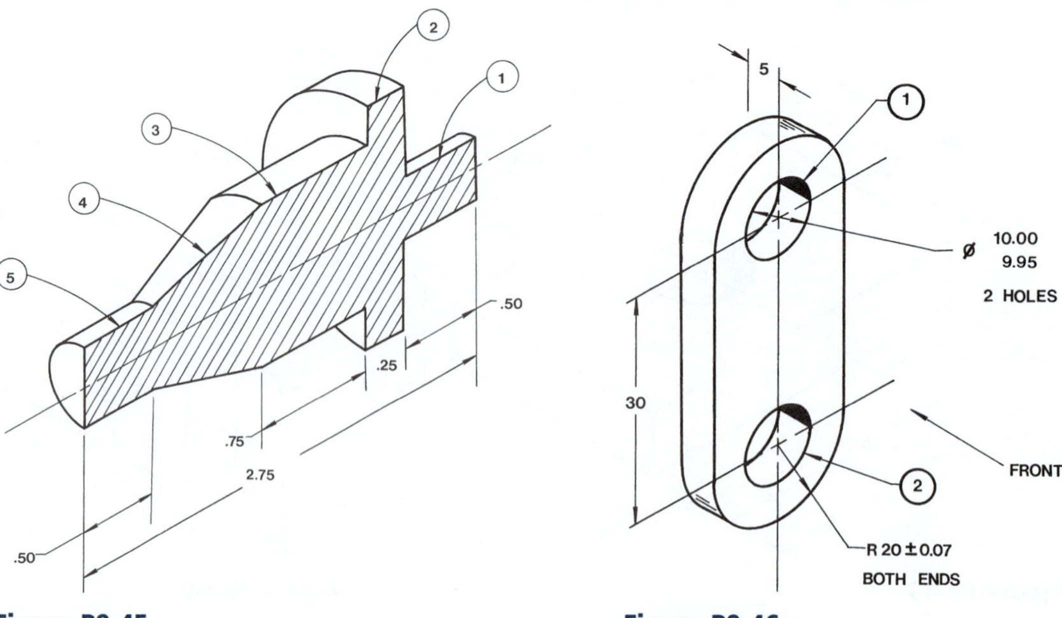

Figure P8-45

Figure P8-46

Project 8-47:

Draw a model of the object shown in Figure P8-47, then create a drawing layout including the specified dimensions. Add the following tolerances and specifications to the drawing.

A. Surface 1 is datum A.

B. Surface 2 is datum B.

C. The six holes have a diameter range of .502−.499 in. and are to be located using positional tolerances so that their centerlines are within .005 in. at MMC relative to datums A and B.

D. The back surface is to be parallel to datum A within .002 in.

Project 8-48:

Draw a model of the object shown in Figure P8-48, then create a drawing layout including the specified dimensions. Add the following tolerances and specifications to the drawing.

A. Surface 1 is datum A.

B. Hole 2 is datum B.

C. The eight holes labeled 3 have diameters of 8.4−8.3 mm with a positional tolerance of 0.15 mm at MMC relative to datums A and B. Also, the eight holes are to be counterbored to a diameter of 14.6−14.4 mm and to a depth of 5.0 mm.

D. The large center hole is to have a straightness tolerance of 0.2 at MMC about its centerline.

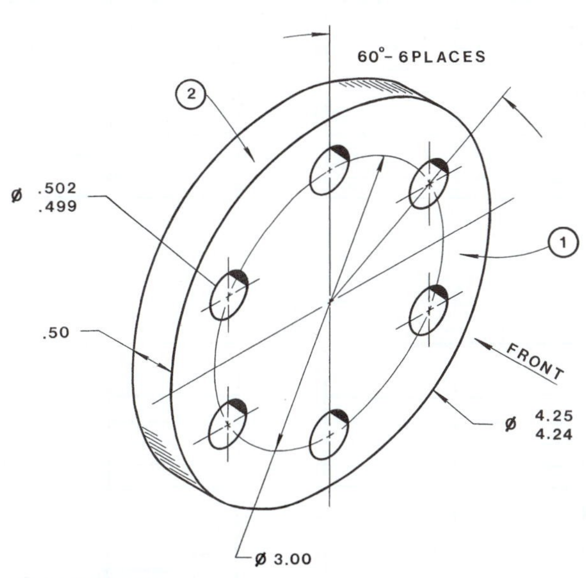

Figure P8-47

Figure P8-48

Project 8-49:

Draw a model of the object shown in Figure P8-49, then create a drawing layout including the specified dimensions. Add the following tolerances and specifications to the drawing.

A. Surface 1 is datum A.

B. Surface 2 is datum B.

C. Surface 3 is datum C.

D. The four holes labeled 4 have a dimension and tolerance of 8 +0.3/−0 mm. The holes are to be located using a positional tolerance of 0.05 mm at MMC relative to datums A, B, and C .

E. The six holes labeled 5 have a dimension and tolerance of 6 +0.2/ −0 mm. The holes are to be located using a positional tolerance of 0.01 mm at MMC relative to datums A, B, and C .

Project 8-50:

The objects in Figure P8-50B labeled A and B are to be toleranced using four different tolerances as shown. Redraw the charts shown in Figure P8-50A and list the appropriate allowable tolerance for "as measured" increments of 0.1 mm or .001 in. Also include the appropriate geometric tolerance drawing called out above each chart.

A. (Millimeters)

B. Inches

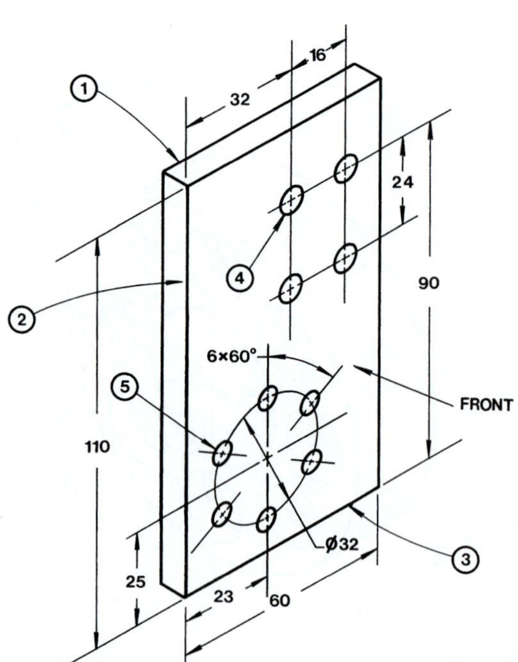

Figure P8-49

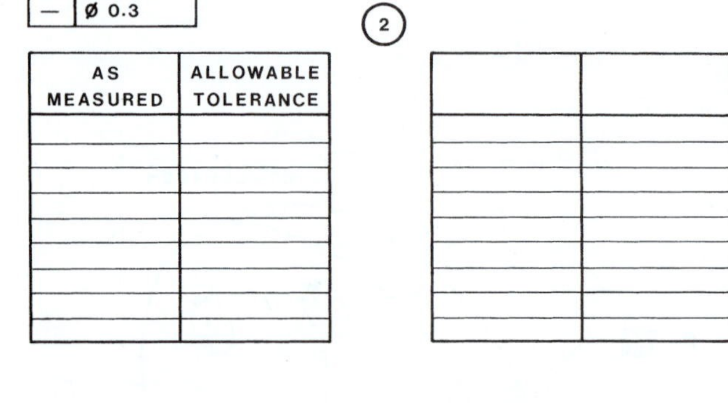

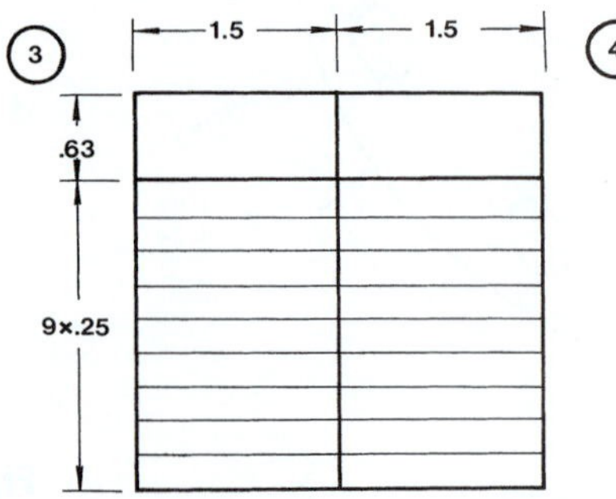

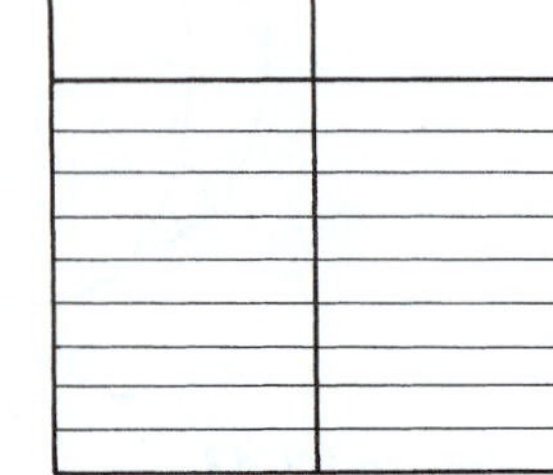

Figure P8-50A

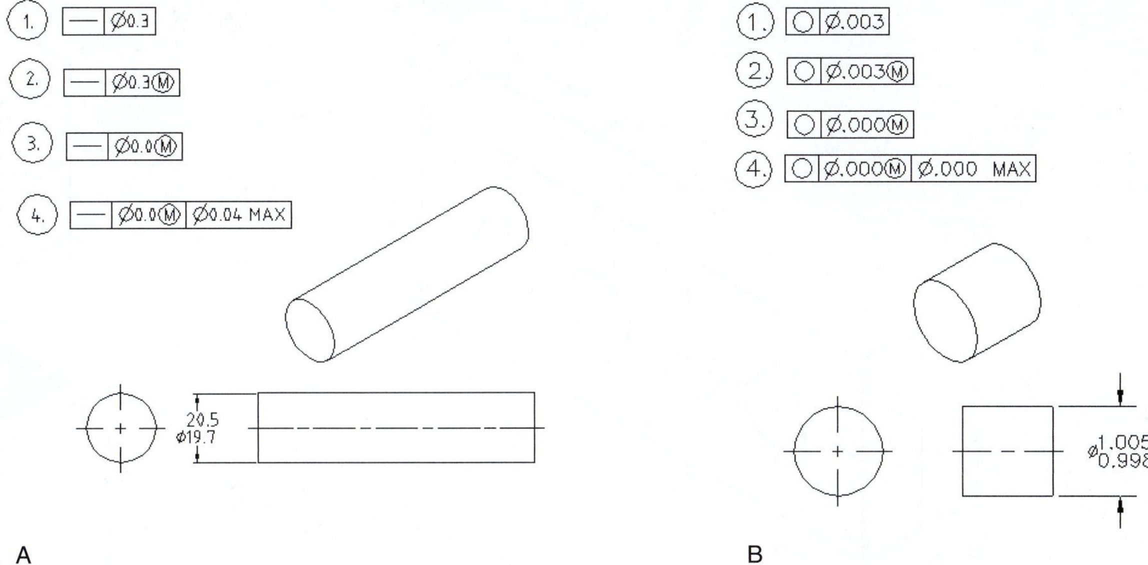

Figure P8-50B

Project 8-51:

Assume that there are two copies of the part in Figure P8-51 and that these parts are to be joined together using four fasteners in the floating condition. Draw front and top views of the object, including dimensions and tolerances. Add the following tolerances and specifications to the drawing, then draw front and top views of a shaft that can be used to join the two objects. The shaft should be able to fit into any of the four holes.

A. Surface 1 is datum A.

B. Surface 2 is datum B.

C. Surface 3 is perpendicular to surface 2 within 0.02 mm.

D. Specify the positional tolerance for the four holes applied at MMC.

E. The centerlines of the holes are to be straight within 0.01 mm at MMC.

F. The clearance between the shafts and the holes is to be 0.05 minimum and 0.10 maximum.

Project 8-52:

Dimension and tolerance parts 1 and 2 of Figure P8-52 so that part 1 always fits into part 2 with a minimum clearance of .005 in. The tolerance for part 1's outer matching surface is .006 in.

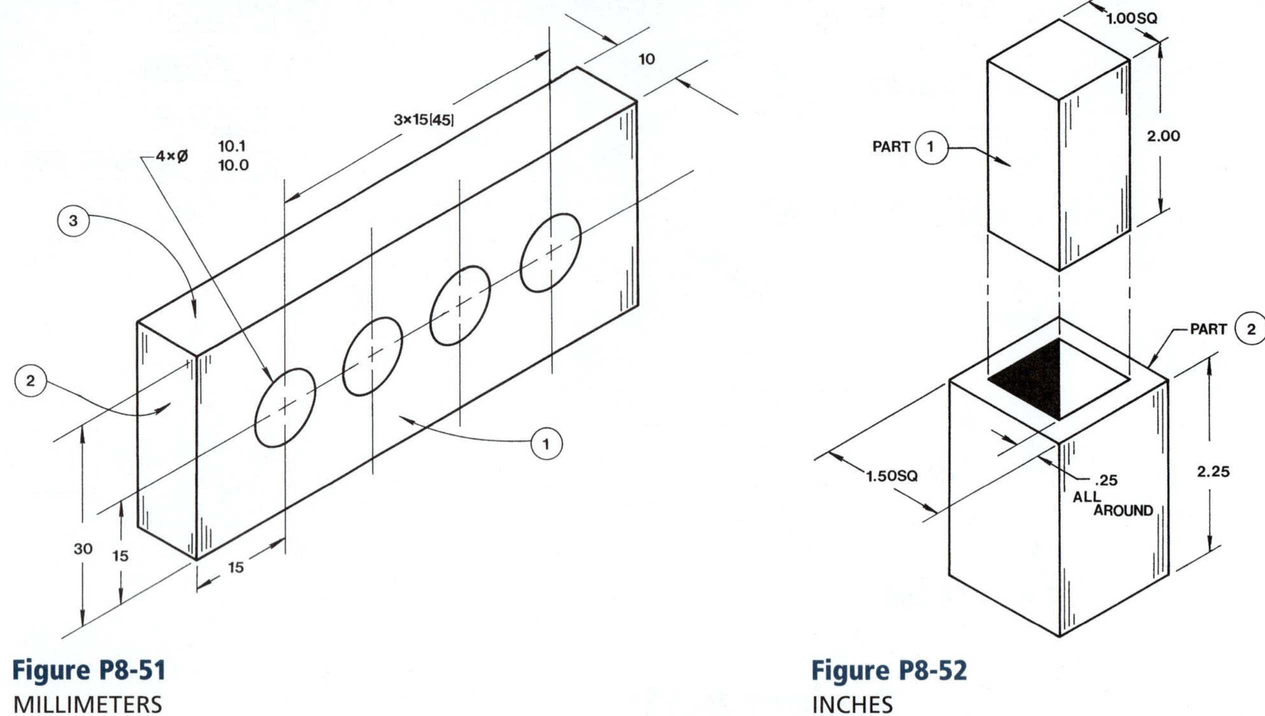

Figure P8-51
MILLIMETERS

Figure P8-52
INCHES

Project 8-53:

Dimension and tolerance parts 1 and 2 of Figure P8-53 so that part 1 always fits into part 2 with a minimum clearance of 0.03 mm. The tolerance for part 1's diameter is 0.05 mm. Take into account the fact that the interface is long relative to the diameters.

Project 8-54:

Assume that there are two copies of the part in Figure P8-54 and that these parts are to be joined together using six fasteners in the floating condition. Draw front and top views of the object, including dimensions and tolerances. Add the following tolerances and specifications to the drawing, then draw front and top views of a shaft that can be used to join the two objects. The shaft should be able to fit into any of the six holes.

A. Surface 1 is datum A.

B. Surface 2 is round within .003.

C. Specify the positional tolerance for the six holes applied at MMC.

D. The clearance between the shafts and the holes is to be .001 minimum and .003 maximum.

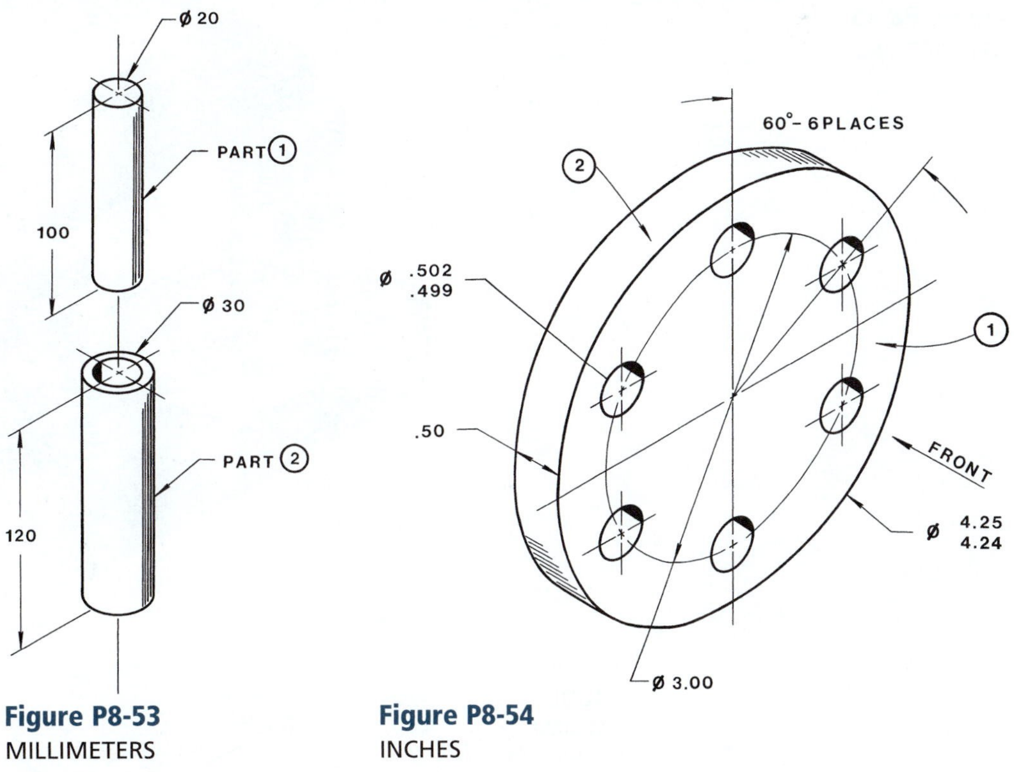

Figure P8-53
MILLIMETERS

Figure P8-54
INCHES

Project 8-55:

The assembly shown in Figure P8-55 is made from parts defined in Chapter 5.

A. Draw an exploded assembly drawing.

B. Draw a BOM.

C. Use the drawing layout mode and draw orthographic views of each part. Include dimensions and geometric tolerances. The pegs should have a minimum clearance of 0.02. Select appropriate tolerances.

Figure P8-55
MILLIMETERS

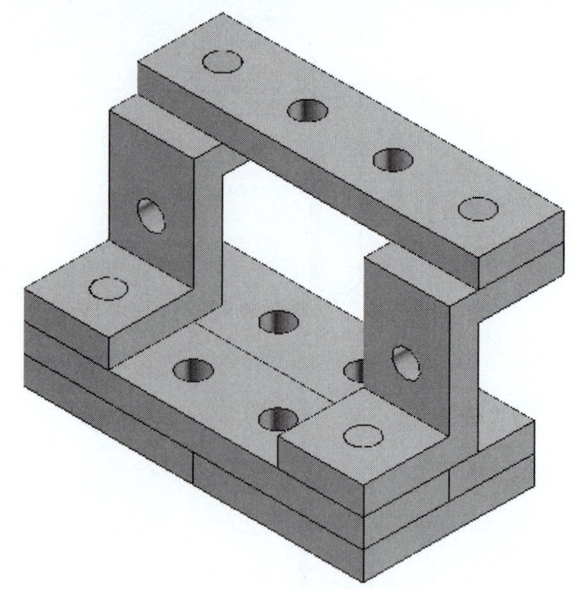

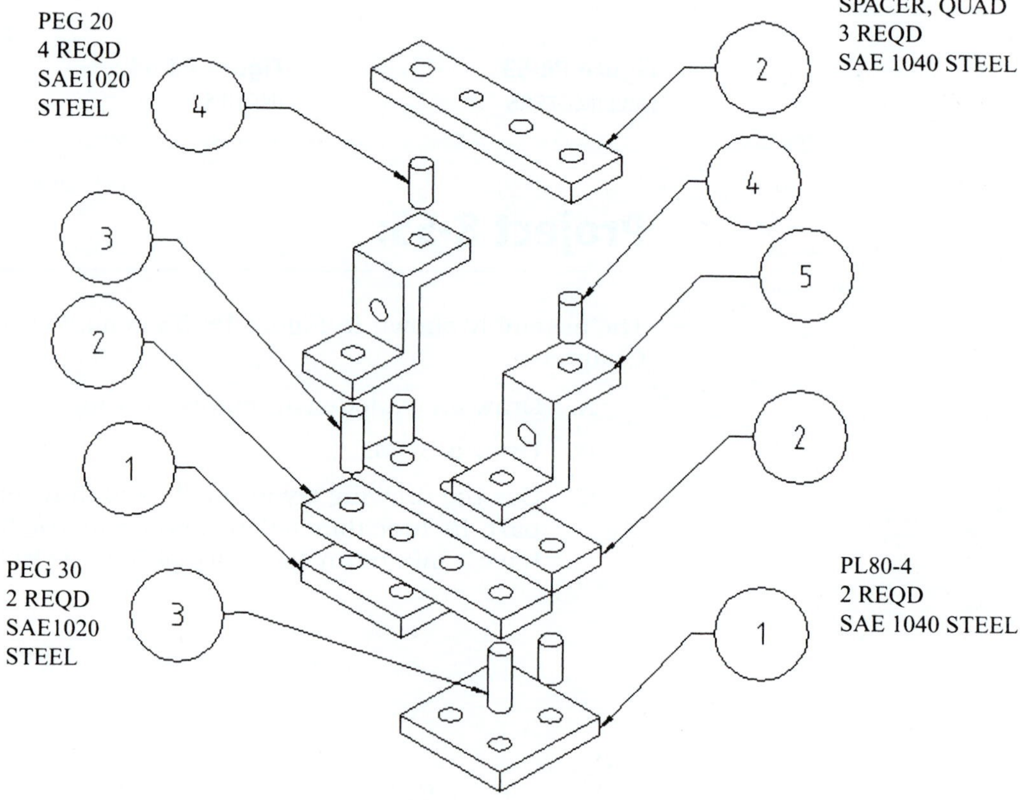

chapternine
Bearings and Fit Tolerances

CHAPTER OBJECTIVES

- Learn about sleeve and ball bearings
- Learn about fits
- Learn how fits are applied to bearings and shafts
- Learn about tolerances for bearings

9-1 Introduction

This text deals with two types of bearings: sleeve bearings or *bushings* and ball bearings. See Figure 9-1. ***Sleeve bearings***, or jig bushings, are hollow cylinders made from a low-friction material such as Teflon or impregnated bronze. Sleeve bearings may have flanges. ***Ball bearings*** include spherical bearings in an internal race that greatly reduce friction. In general, sleeve bearings are cheaper than ball bearings, but ball bearings can take heavier loads at faster speeds. A listing of ball bearings is included in the **Design Library**.

Figure 9-1

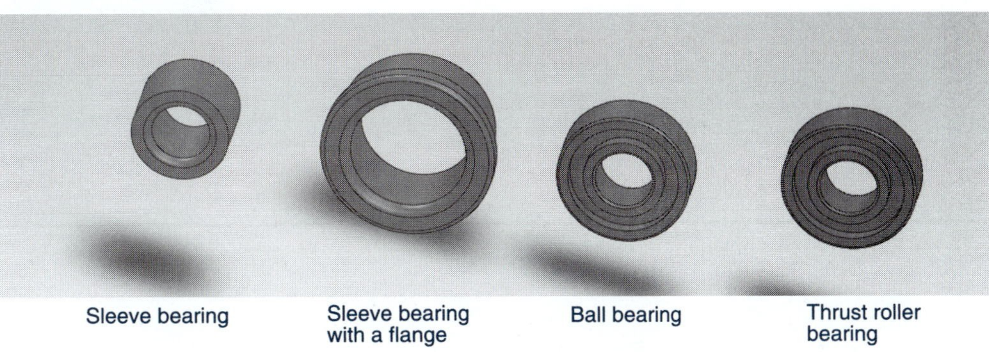

9-2 Sleeve Bearings

Sleeve bearings are identified by the following callout format:

Inside diameter × Outside Diameter × Thickness

For example,

.375 × .750 × .500 or 3/8 × 3/4 × 1/2

To Draw a Sleeve Bearing

Draw a .500 × 1.000 × 1.000 sleeve bearing. See Figure 9-2.

1. Start a new **Part** document.
2. Select the **Front Plane** orientation.

Figure 9-2

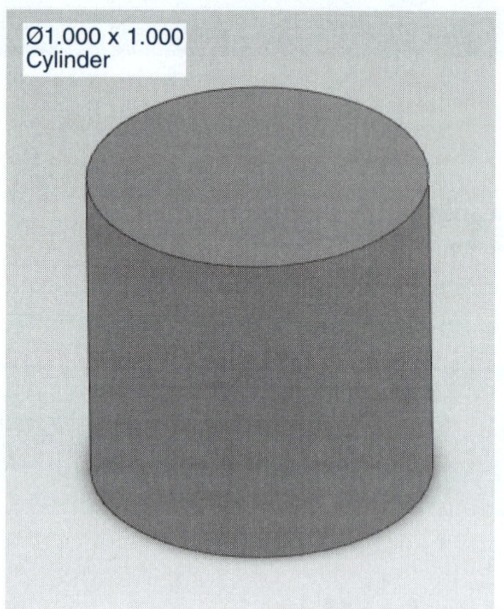
Ø1.000 x 1.000
Cylinder

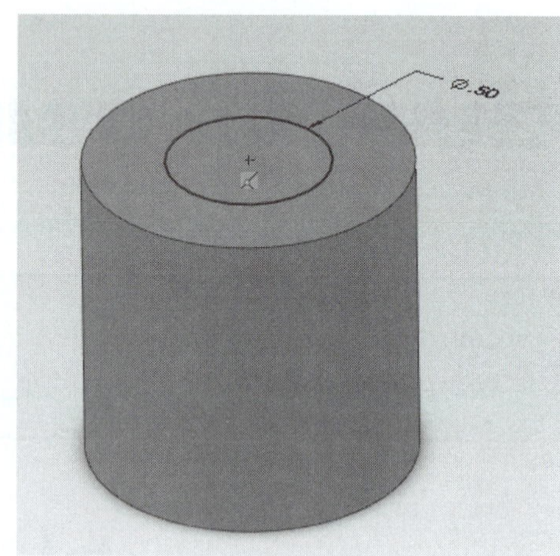
Ø.50

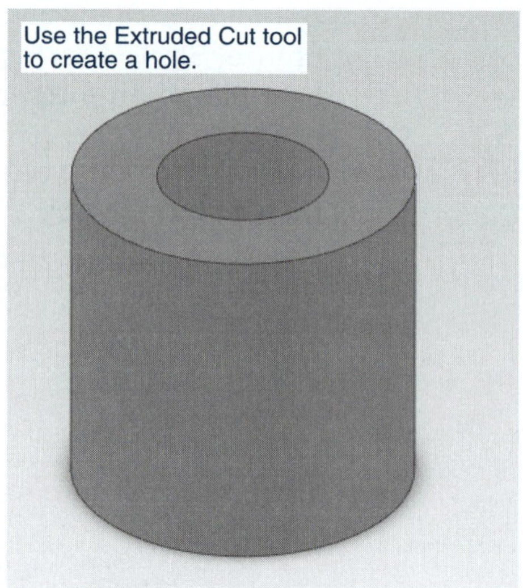
Use the Extruded Cut tool
to create a hole.

3. Click the **Sketch** group, then click the **Circle** tool.
4. Draw a **Ø1.000** circle. Use the **Smart Dimension** tool to size the circle.
5. Click the **Features** tab, then click the **Extruded Boss/Base** tool.
6. Define the thickness of the extrusion as **1.00 in**.
7. Click the green **OK** check mark.
8. Right-click the front surface of the cylinder and select the **Sketch** tool.
9. Use the **Circle** tool and the **Smart Dimension** tool to draw a **Ø.5000** circle on the front surface of the cylinder.
10. Click the **Features** tab and select the **Extruded Cut** tool.
11. Define the Ø.500 to be cut and click the green **OK** check mark.
12. Save the drawing as **.50 BEARING**.

NOTE
No tolerances are assigned to the bushing.

To Use a Sleeve Bearing in an Assembly Drawing

See Figures 9-3 and 9-4.

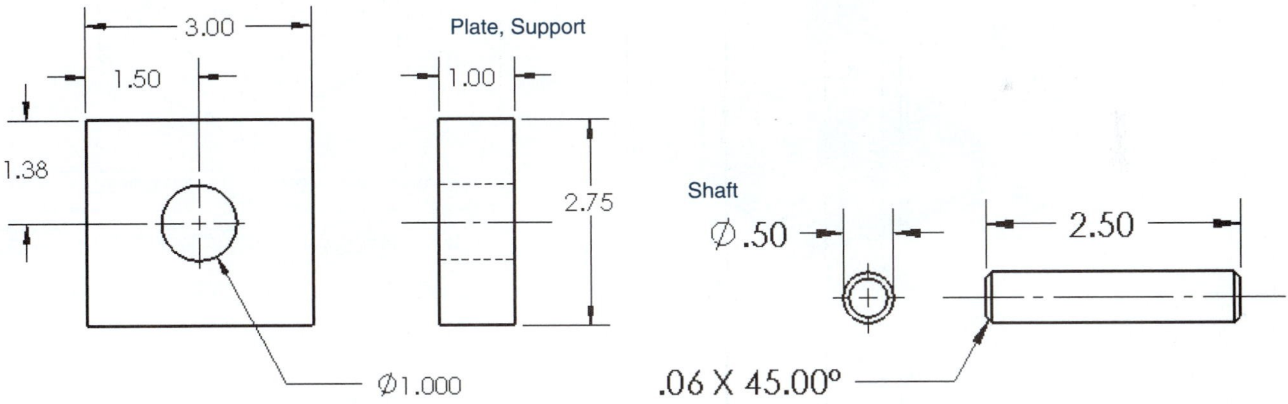

Figure 9-3

1. Draw the support plate and Ø.500 × 2.500 shaft defined in Figure 9-3. Save the drawings.
2. Create a new **Assembly** drawing.
3. Add the support plate, shaft, and sleeve bearing (created in the last section) to the assembly drawing.

   See Figure 9-4.

Figure 9-4

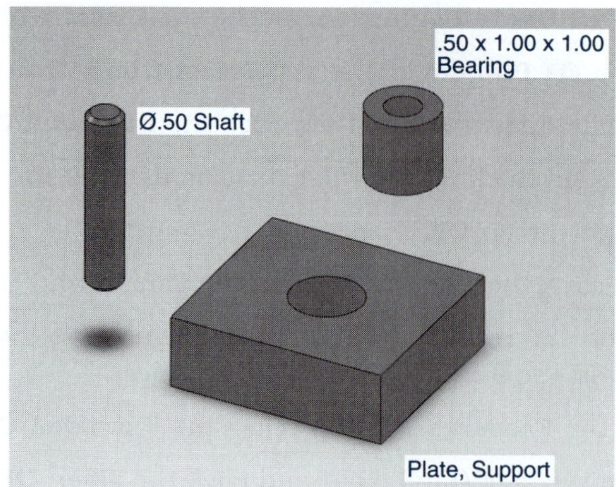

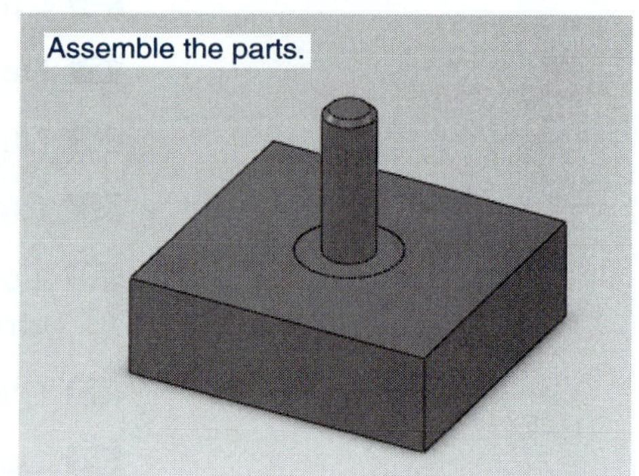

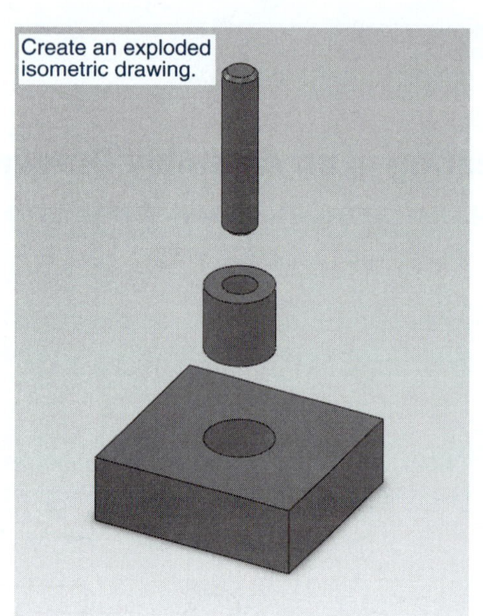

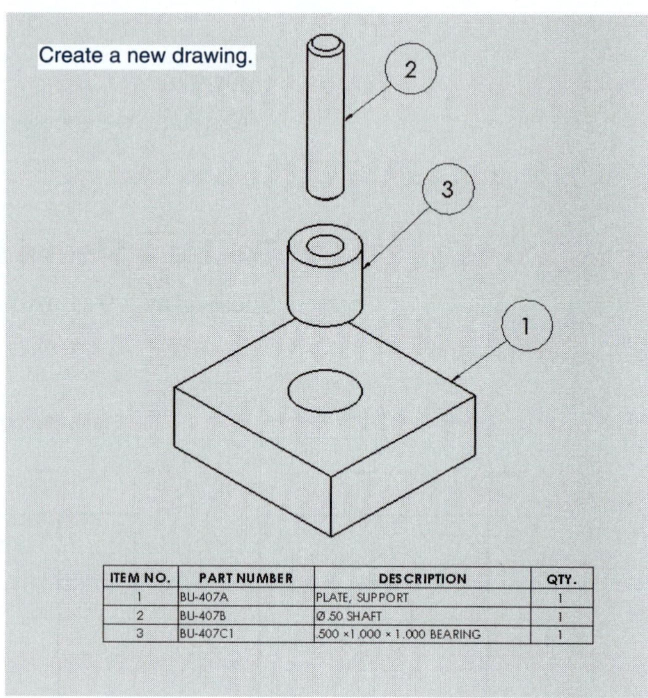

ITEM NO.	PART NUMBER	DESCRIPTION	QTY.
1	BU-407A	PLATE, SUPPORT	1
2	BU-407B	Ø.50 SHAFT	1
3	BU-407C1	.500 ×1.000 × 1.000 BEARING	1

4 Use the **Mate** tool and assemble the sleeve bearing into the support plate, and the shaft into the sleeve bearing. The front surface of the shaft should be offset **1.50** from the front surface of the support plate.

5 Save the assembly as **Sleeve Assembly**.

6 Use the **Exploded View** tool. Select a direction and pull the shaft out of the assembly.

See Figure 5-25 for instructions on how to use the **Exploded View** tool.

7 Use the **Exploded View** tool to pull the bearing away from the support plate.

8 Save the exploded drawing as **Sleeve Assembly**.

Replace the old **Sleeve Assembly** drawing.

9 Start a new drawing.

10 Click the **Annotation** tab, click the **Balloon** tool, and add the assembly numbers as shown.

11 Click the **Annotation** tab, click the **Tables** tool, click the **Bill of Materials** tool, and add a BOM to the drawing.

9-3 Bearings from the Toolbox

The SolidWorks **Toolbox** includes many different sizes and styles of bearings. In this example we will use the **PLATE, SUPPORT**, and **Ø.50 SHAFT** with a bearing from the **Toolbox** to create an assembly.

1 Create a new assembly drawing and add the **PLATE, SUPPORT**, and **Ø.50 SHAFT** to the drawing.

See Figure 9-5.

2 Save the assembly as **BEARING ASSEMBLY**.

3 Click the **Toolbox** and access the **Jig Bushings** tool.

4 Select a **Jig Bushing Type P** and drag it onto the drawing screen.

We know the shaft diameter is **.500**, so we will size the bearing's inside diameter to match the shaft diameter. Tolerances are not considered. They will be included in later examples.

5 Select a **.5000 × 1.000** bearing and click the green **OK** check mark.

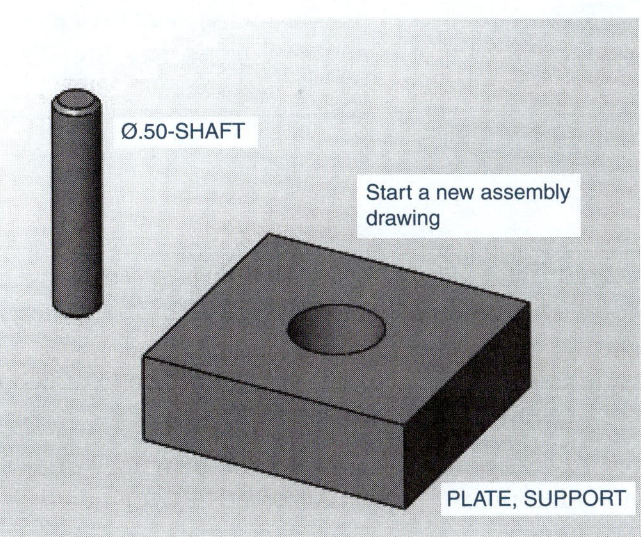

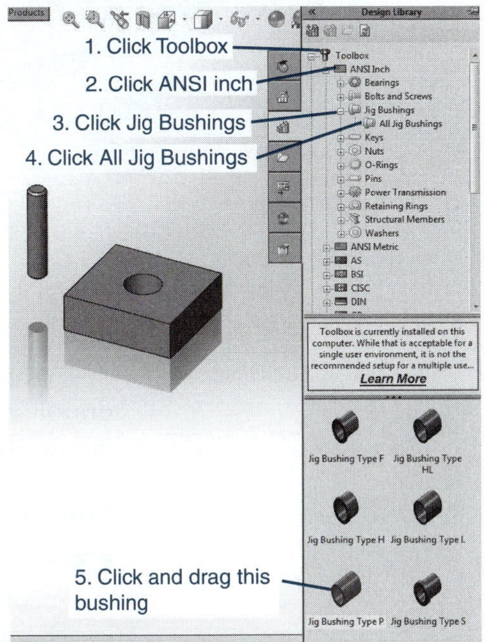

Figure 9-5

Figure 9-5
(*Continued*)

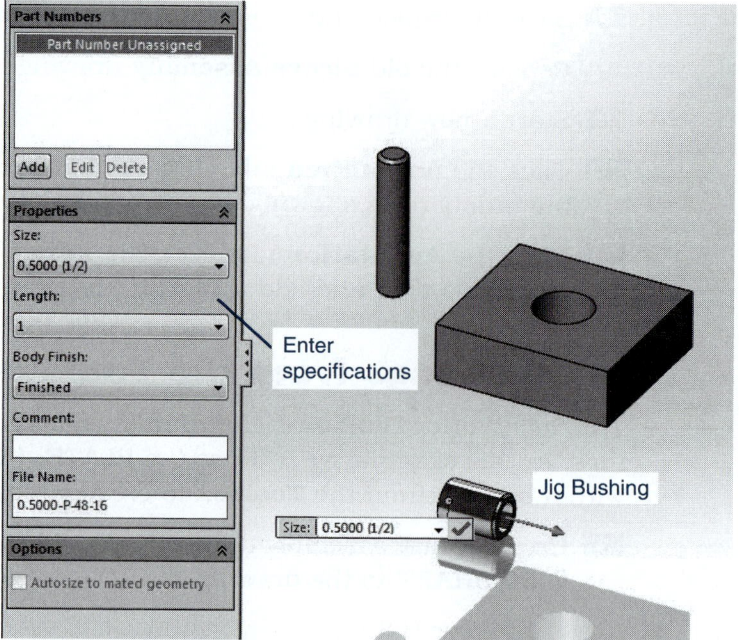

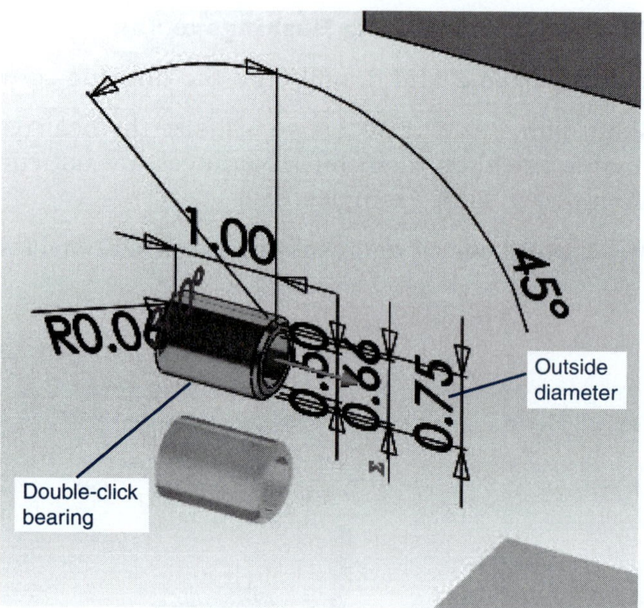

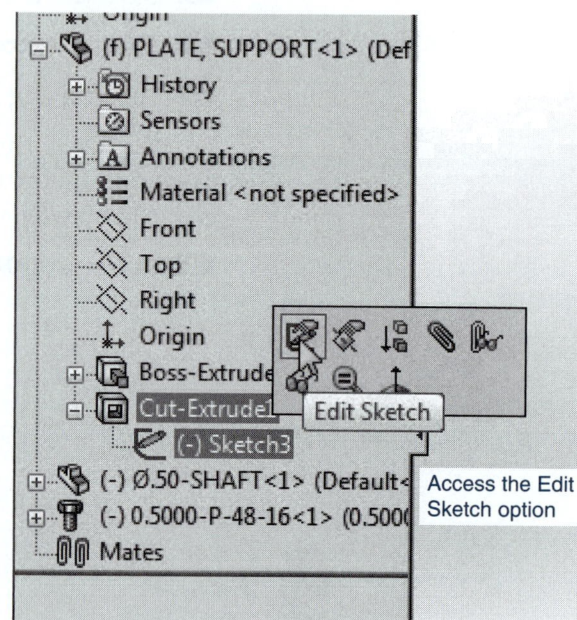

What is the outside diameter of the bearing? To answer this question, double-click the bearing on the drawing screen. The dimensions used to create the bearing will become visible.

The outside diameter is defined as **.75**, so the Ø1.00 hole in the Plate, Support is not acceptable. It must be edited.

6 Click the drawing screen to remove the bearing's dimensions and click the + sign next to the **PLATE, SUPPORT** heading in the **FeatureManager**.

7 Click the + sign next to the **Cut-Extrusion** heading and click the **Sketch 3** heading.

Figure 9-5
(Continued)

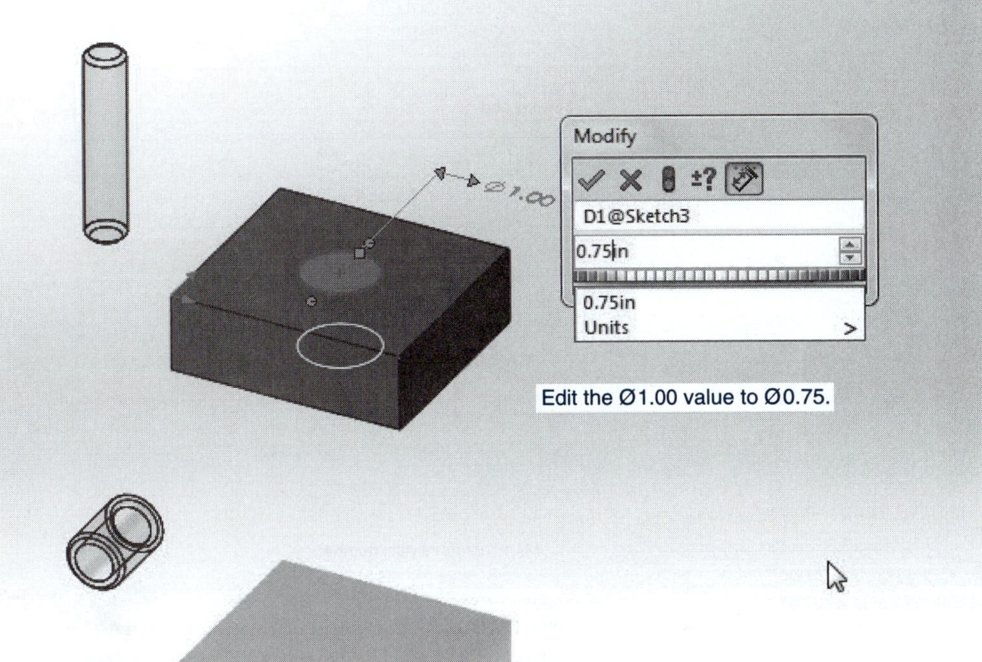
Modify
D1@Sketch3
0.75in
0.75in
Units
Edit the Ø1.00 value to Ø0.75.

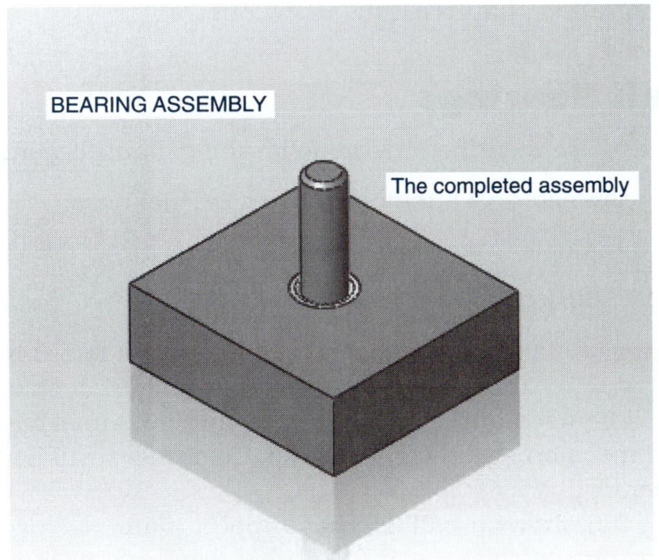
BEARING ASSEMBLY
The completed assembly

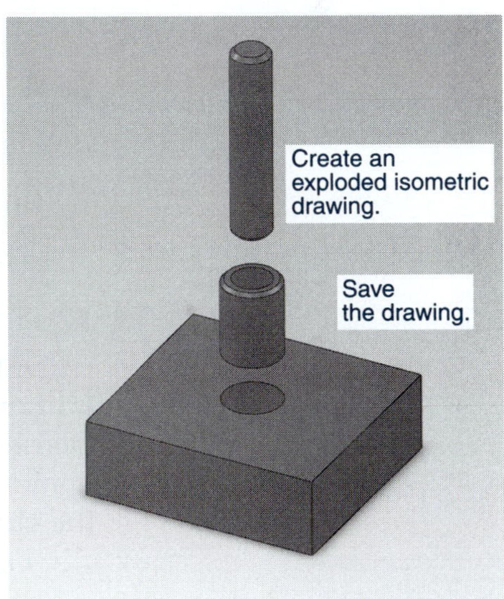
Create an exploded isometric drawing.
Save the drawing.

Your sketch number may be different.

8 Click the **Edit Sketch** option.

The dimension for the Ø1.00 hole will appear.

9 Double-click the **1.00** dimension and change the diameter from 1.00 to **.75**; click the **OK** check mark.

10 Return to the assembly drawing and assemble the components.

11 Create an exploded assembly drawing of the assembly.

12 Create a drawing of the exploded assembly and add a BOM.

Figure 9-5
(*Continued*)

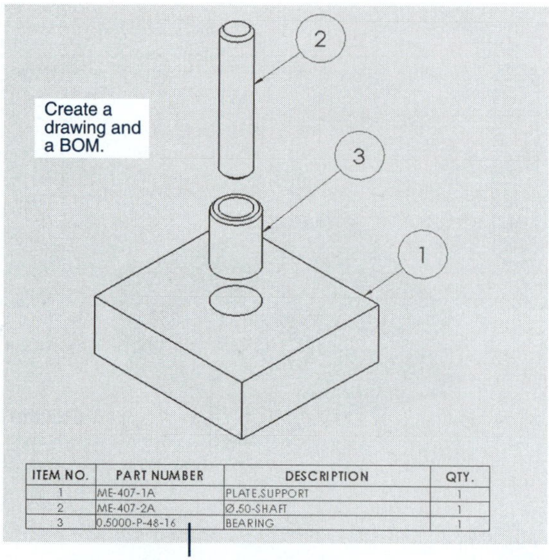

ITEM NO.	PART NUMBER	DESCRIPTION	QTY.
1	ME-407-1A	PLATE,SUPPORT	1
2	ME-407-2A	Ø.50-SHAFT	1
3	0.5000-P-48-16	BEARING	1

Note that the bearing's part number, 0.5000-P-48-16, was used as is.

This is a vendor item; that is, we did not make it but purchased it from an outside source, so we used the manufacturer's part number for the bearing rather than creating a new one.

9-4 Ball Bearings

Ball bearings are identified by the following callout format:

Inside Diameter × Outside Diameter × Thickness

For example,

.375 × .750 × .500 or 3/8 × 3/4 × 1/2

A listing of standard ball bearing sizes can be found in the **Design Library**. For this example a .5000 × .8750 × .2188 instrument ball bearing will be used and will be inserted into a counterbored hole. Only nominal dimensions will be considered. Tolerances will be defined later in the chapter.

Figure 9-6 shows a ball bearing support plate.

Figure 9-6

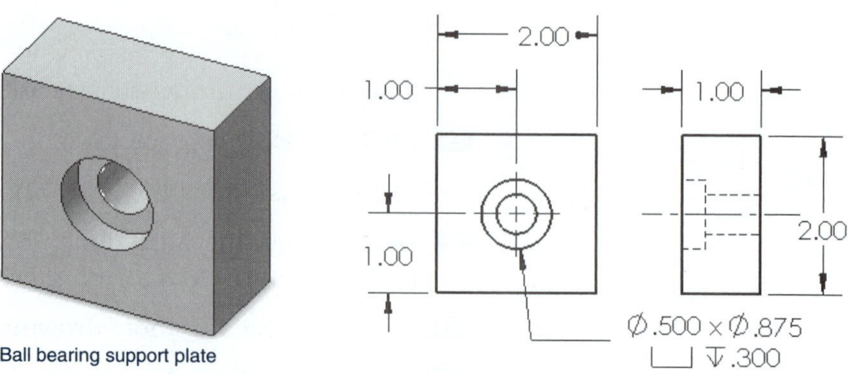

1 Draw and save the ball bearing support plate.

2 Draw and save a **Ø.500 × 2.50** shaft with **.03** chamfers at each end.

3 Start a new **Assembly** drawing and insert the ball bearing support plate and Ø.500 × 2.50 shaft.

See Figure 9-7.

4 Access the **Design Library, Toolbox, Ansi Inch, Bearings, Ball Bearings**, and select **Instrument Ball Bearing – AFBMA 12.2**.

5 Click and drag the bearing into the drawing screen and set the properties as shown.

In this example a 0.5000 – 0.8750 – 0.2188 was used. See the values in the **Instrument Ball Bearing PropertyManager**.

6 Insert the ball bearing into the counterbored hole.

7 Insert the shaft into the bearing.

8 Create a new drawing showing the assembly drawing and a BOM.

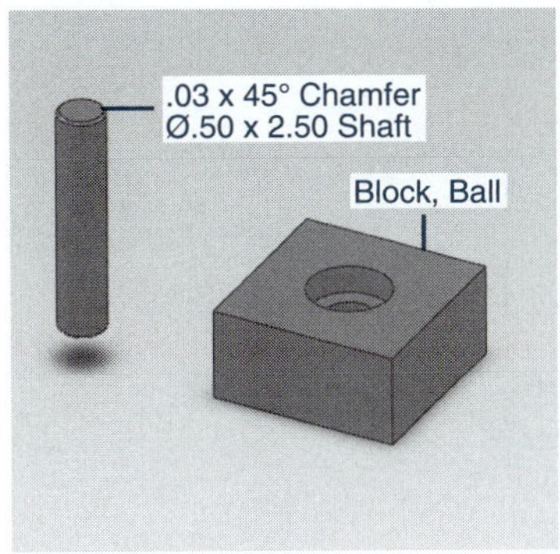

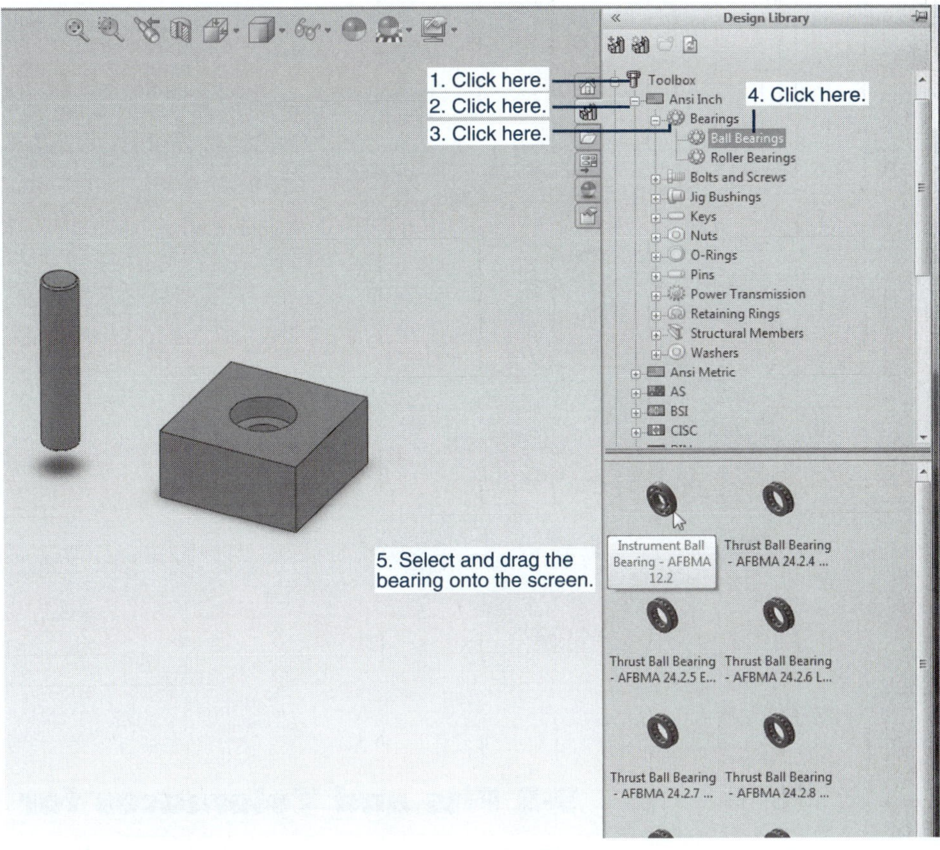

Figure 9-7

Figure 9-7
(Continued)

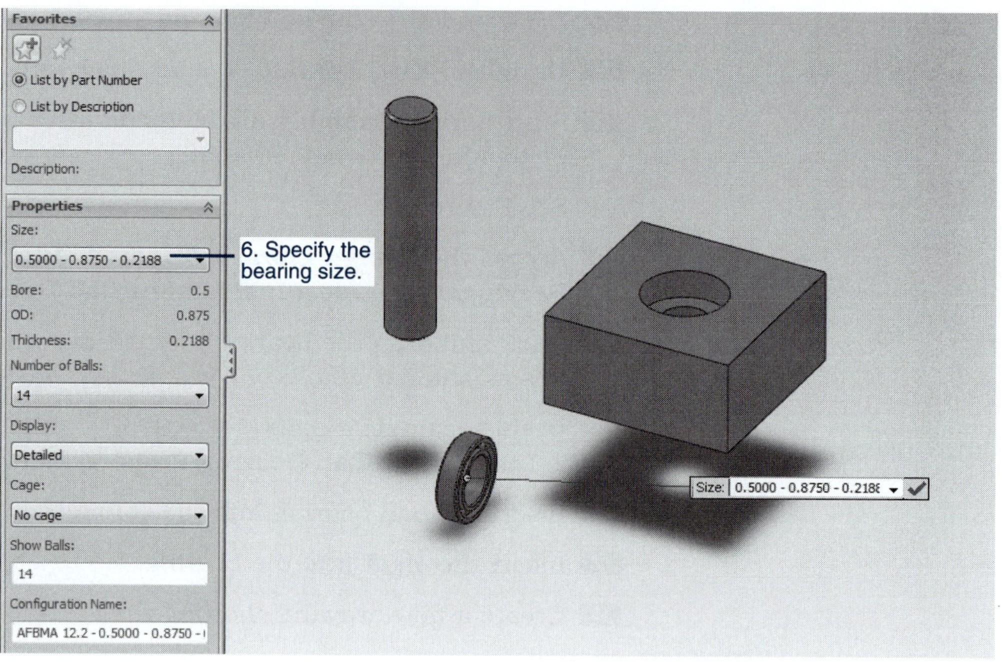

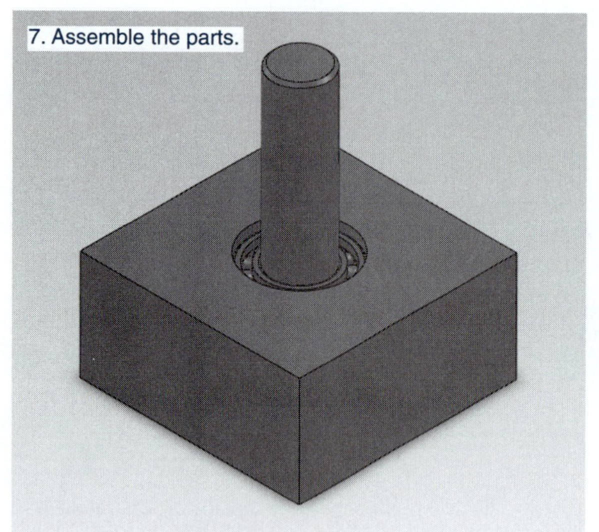

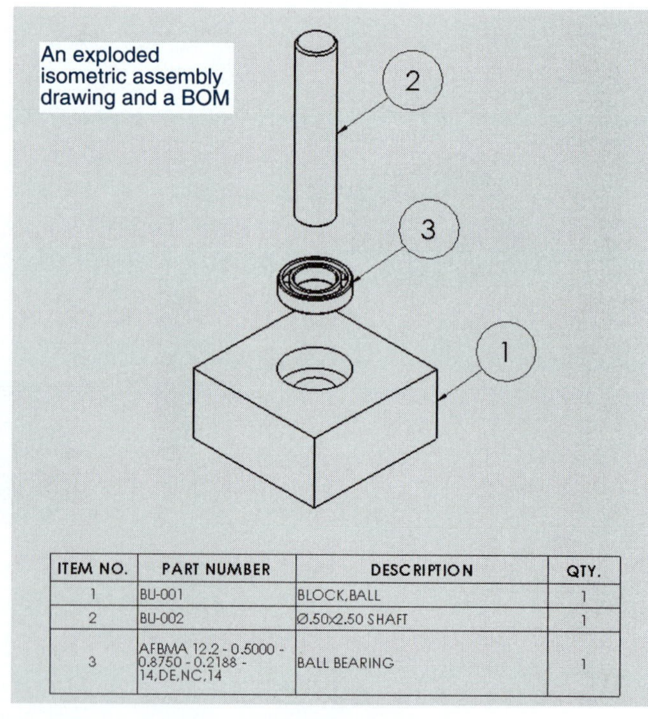

ITEM NO.	PART NUMBER	DESCRIPTION	QTY.
1	BU-001	BLOCK,BALL	1
2	BU-002	Ø.50x2.50 SHAFT	1
3	AFBMA 12.2 - 0.5000 - 0.8750 - 0.2188 - 14,DE,NC,14	BALL BEARING	1

9-5 Fits and Tolerances for Bearings

The tolerance between a shaft and a bearing and between a bearing and a support part is critical. Incorrect tolerances can cause excessive wear or vibration and affect the performance of the assembly.

In general, a clearance fit is used between the shaft and the inside diameter of the bearing, and an interference fit is used between the outside diameter of the bearing and the support structure. A listing of standard fit tolerances is included in the appendix.

9-6 Fits—Inches

Tolerances for shafts and holes have been standardized and are called ***fits***. An example of a fit callout is H7/g6. The hole tolerance is always given first using an uppercase letter, and the shaft tolerance is given second using a lowercase letter.

9-7 Clearance Fits

Say a Ø0.500 nominal shaft is to be inserted into a Ø0.500 nominal hole using an H7/g6 clearance fit, which is also referred to as a Class LC5 Clearance fit.

NOTE

The term *nominal* refers to a starting value for the shaft and hole. It is not the final dimension.

The following data are given in a table in the appendix. See Figure 9-8. The values are given in thousandths of an inch. The nominal value for the hole is 0.5000, so the +0.7 table value means .5007 in. The −0.25 table value for the 0.5000 nominal shaft means 0.49975 in. The limits of clearance values are the differences between the hole minimum value and the shaft maximum value, 0.00 and −0.25, or 0.25 absolute, and between the maximum hole value and the minimum shaft value, +0.7 and −0.65, or 1.35.

Figure 9-8

Class LC5			
Nominal Size Range	Limits of Clearance	Standard Limits	
		Hole H7	Shaft g6
0.40 – 0.71	0.25 1.35	+0.7 0	−0.25 −0.65

Hole basis

9-8 Hole Basis

The 0 value for the hole's minimum indicates that the tolerances were derived using ***hole basis*** calculations; that is, the tolerances were applied starting with the minimum hole value. Tolerances applied starting with the shaft are called ***shaft basis***.

9-9 Shaft Basis

The limits of clearance values would be applied starting with the minimum shaft diameter. If the H7/g6 tolerances were applied using the shaft basis, the resulting tolerance values for the shaft would be 0.50000 to 0.50040, and for the hole would be 0.50065 (.50040 + .00025) to 0.50135. These values maintain the limits of tolerance, 0.50135 to 0.00135, and the individual tolerances for the hole (0.50135 − 0.50065 = 0.0007) and the shaft (0.50040 − 0.50000 = .00040).

9-10 Sample Problem SP10-2

Say a shaft with nominal values of Ø.750 × 3.00 is to be fitted into a bearing with an inside diameter bore, nominal, of 0.7500 using Class LC5 fit, hole basis. What are the final dimensions for the shaft and bearing's bore? The table values for the hole are 0/+0.5, yielding a hole tolerance of .7500 to .7505, and the shaft values for the hole are 0/−0.4, yielding a shaft tolerance of .7500 to .7496.

> **NOTE**
> The fact that both the hole and the shaft could be .7500 is called ***locational fit***.

See Figure 9-9.

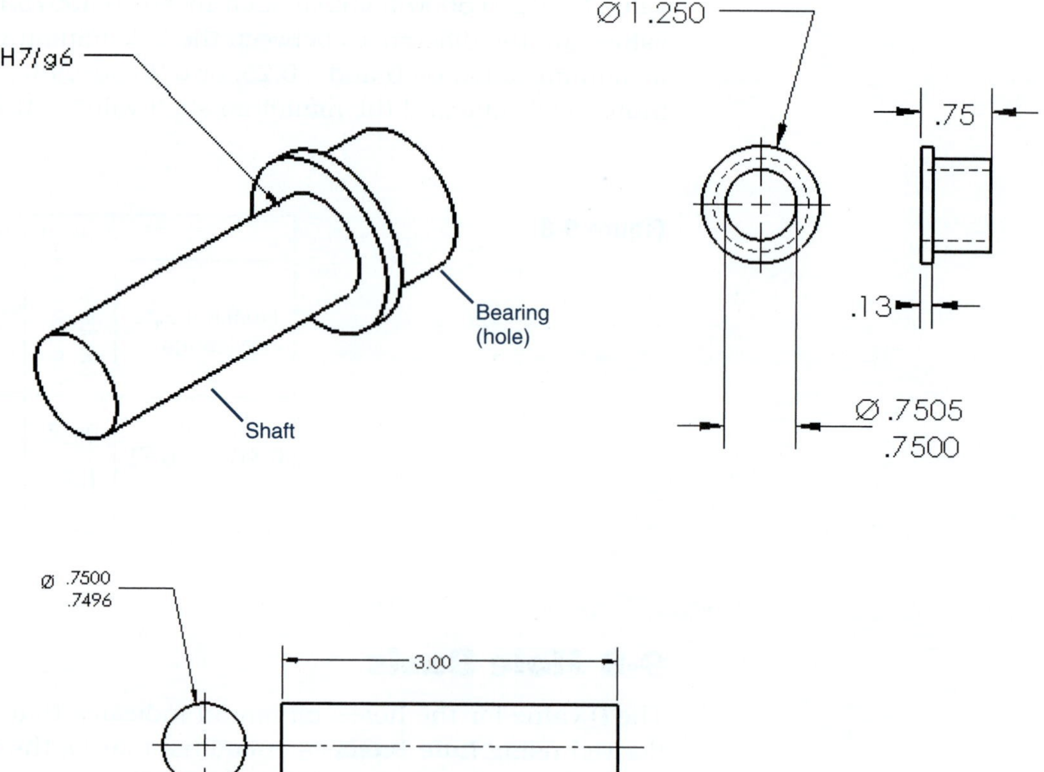

Figure 9-9

9-11 Interference Fits

When the shaft is equal to or larger in diameter than the hole, the fit is called an ***interference fit***. Interference fits are sometimes used to secure a shaft into a hole rather than using a fastener or adhesive. For example, an aluminum shaft with a nominal diameter of .250 in. inserted into a hole in a steel housing with a Ø.250 nominal hole using .0006 in. interference would require approximately 123 in.-lb of torque to turn the shaft.

In this example a sleeve bearing with a nominal outside diameter (O.D.) of .875 is to be inserted into a Ø.875 nominal hole using an LN1 Interference Locational fit. The hole and shaft (bearing O.D.) specifications are H6/n5. The following values were derived from a table in the appendix. See Figure 9-10.

Figure 9-10

Class LN1			
Nominal Size Range	Limits of Interference	Standard Limits	
		Hole	Shaft
0.71 – 1.19	0 1.0	+0.5 0	+1.0 +0.5

Hole basis

All stated values are in thousandths of an inch. The 0 in the column for the hole indicates that it is a hole basis calculation (see the explanation in the previous section). Given the .875 nominal value for both the hole and the shaft, the hole and shaft tolerances are as follows. See Figure 9-11.

Hole: Ø.8755/.875 Shaft: Ø.8760/.8755

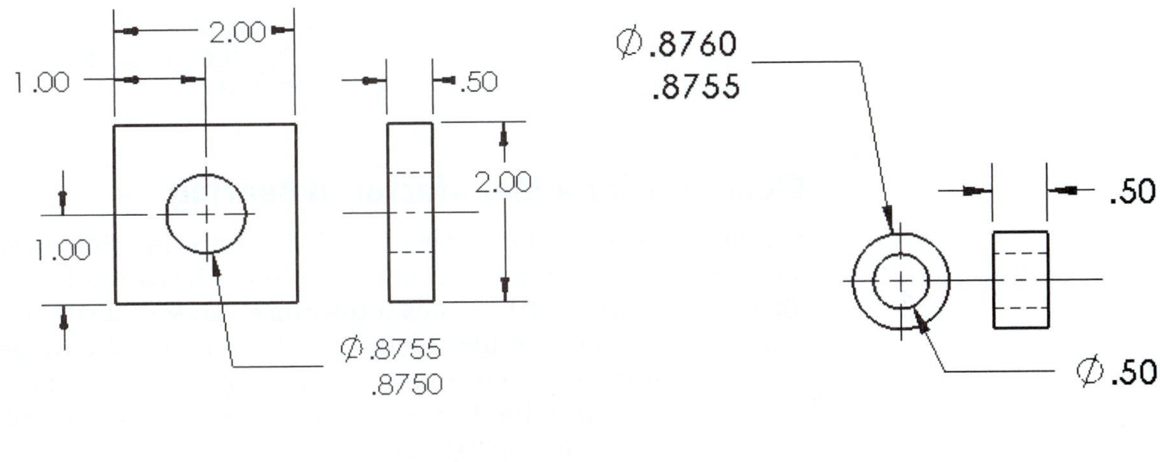

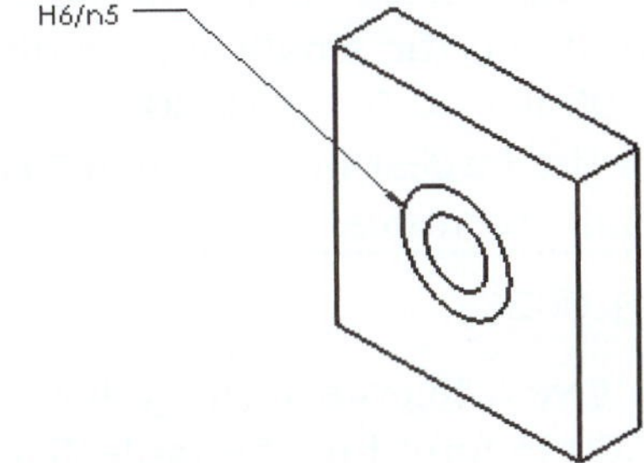

Figure 9-11

9-12 Manufactured Bearings

Most companies do not manufacture their own bearings but, rather, purchase them from a bearing manufacturer. This means that tolerances must be assigned to assemblies based on existing given tolerances for the purchased bearings.

NOTE

Companies that manufacture bearings usually also manufacture shafts that match the bearings; that is, the tolerances for the bearings and shafts are coordinated.

Figure 9-12 shows a typical manufactured sleeve bearing. The dimensions and tolerances are included. The inside diameter (bore or I.D.) of the bearing is matched to the shaft using a clearance fit, and the O.D. is to be matched to the support using an interference fit. The procedure is to find standard fits that are closest to the bearing's manufactured dimensions and apply the limits of tolerance to create the needed tolerances.

Figure 9-12

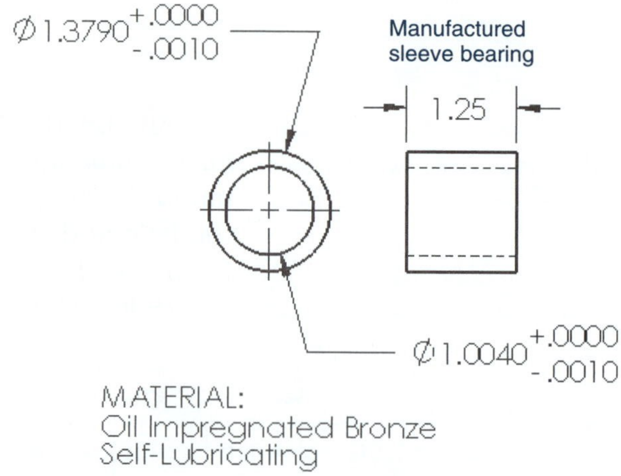

Clearance for a Manufactured Bearing

Refer to the standard fit tables in the appendix and find a tolerance range for a hole that matches or comes close to the bearing's I.D. tolerance of .001 (the +.000/−.001 creates a tolerance range of .001). The given I.D. is 1.0040, so it falls within the 0.71−1.19 nominal size range. An LC2 Clearance fit (H8/h7) has a hole tolerance specification of 0.0 to 0.0008, 0.0002 smaller than the bearing's manufactured tolerance of .0010. The given tolerance is Ø1.0040/1.0030.

The limits of clearance for the LC2 standard fit are 0.0 to 0.0013. If these limits are maintained, the smallest hole diameter is equal to the largest shaft diameter (1.0030 − 0.0 = 1.0030), and the smallest shaft diameter is .0013 less than the largest hole diameter (1.0040 − .0013 = 1.0027). Therefore, the shaft tolerances are

Shaft: Ø1.0030/1.0027

These tolerances give a tolerance range for the shaft of .0003, or .0002 less than the stated .0005 found in the table. This difference makes up for the .0002 difference between the actual hole diameter's tolerance range of 0.0010 and the standard H8 tolerance of 0.0008.

To Apply a Clearance Fit Tolerance Using SolidWorks

Figure 9-13 shows a shaft with a nominal diameter of 1.0040. Enter the required 1.0030/1.0027 tolerance.

1. Click the **Limit** option in the **Tolerance/Precision** box.
2. Set the upper limit for **+0.0010** and the lower limit for **−0.0013**.
3. Click the green **OK** check mark.

Figure 9-13

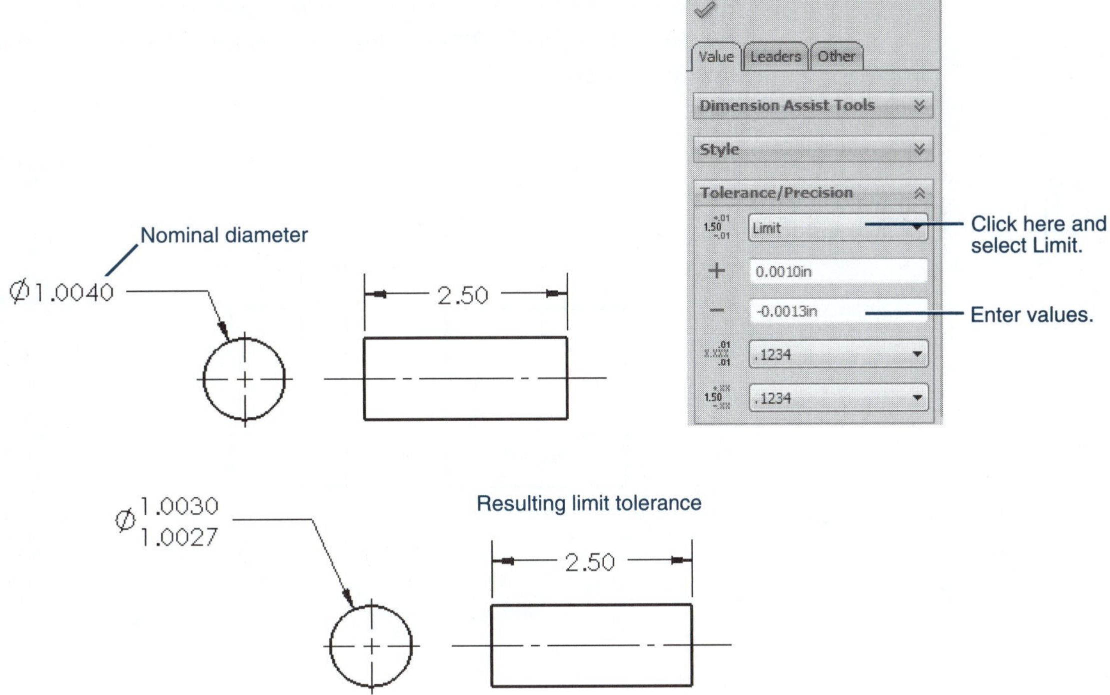

Interference for a Manufactured Bearing

The O.D. for the manufactured bearing is 1.379 +.000/−.001. Written as a limit tolerance, it is 1.3790/1.3780. The tolerance range for the O.D. is 0.001. An interference tolerance is required between the O.D. of the bearing and the hole in the support.

> **NOTE**
> An interference fit is also called a **press fit**.

A search of the Standard Fit tables in the appendix for a shaft (the O.D. of the bearing acts like a shaft in this condition) tolerance range of 0.001 finds that an LN2 (H7/p6) shaft range is .0008, or .0002 less than manufactured tolerance.

The limits of interference for the LN2 standard fit are 0.0 to 0.0013. If these limits are maintained, the smallest shaft diameter is equal to the largest hole diameter (1.3780 + 0.0 = 1.3780), and the largest shaft diameter is .0013 greater than the smallest hole diameter (1.3790 − .0013 = 1.3777).

Therefore, the shaft tolerances are

Shaft: Ø1.3780/1.3777

These tolerances give a tolerance range for the shaft of .0003, or .0002 less than the stated .0005 found in the table. This difference makes up for the .0002 difference between the actual hole diameter's tolerance range of 0.0010 and the standard H7 tolerance of 0.0008.

To Apply an Interference Fit Tolerance Using SolidWorks

Figure 9-14 shows a support with a nominal diameter of 1.3780. Enter the required 1.3780/1.3777 tolerance. See Chapter 8 for further explanation on how to apply tolerances.

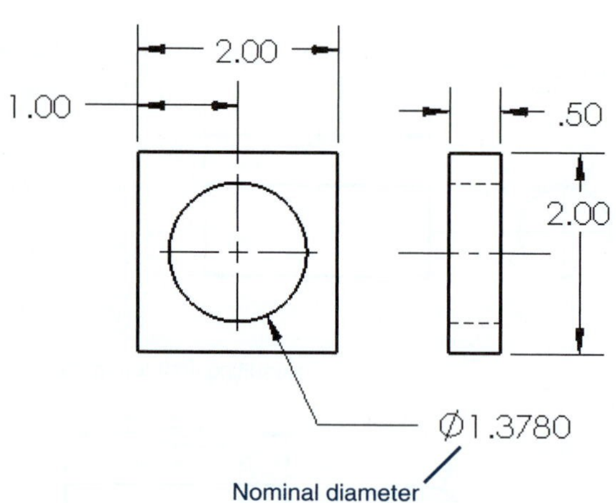

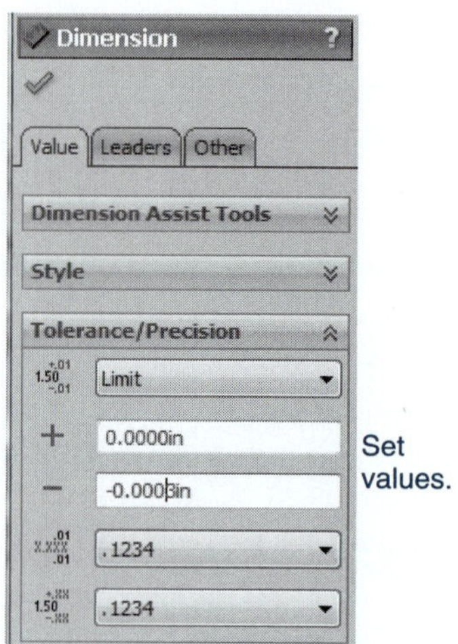

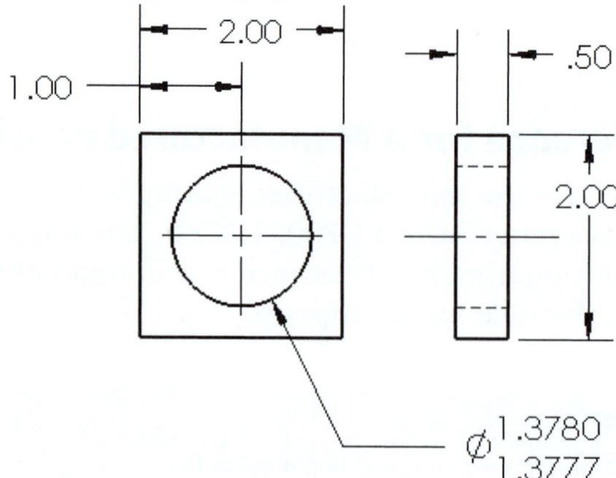

Figure 9-14

1. Click the **Limit** option in the **Tolerance/Precision** box.
2. Set the upper limit for **+0.0000** and the lower limit for **−0.0003**.
3. Click the green **OK** check mark.

Using SolidWorks to Apply Standard Fit Tolerances to an Assembly Drawing

Figure 9-15 shows the assembly of the shaft and support toleranced in the previous section with the manufactured bearing. The standard tolerance callouts are added as follows.

Figure 9-15

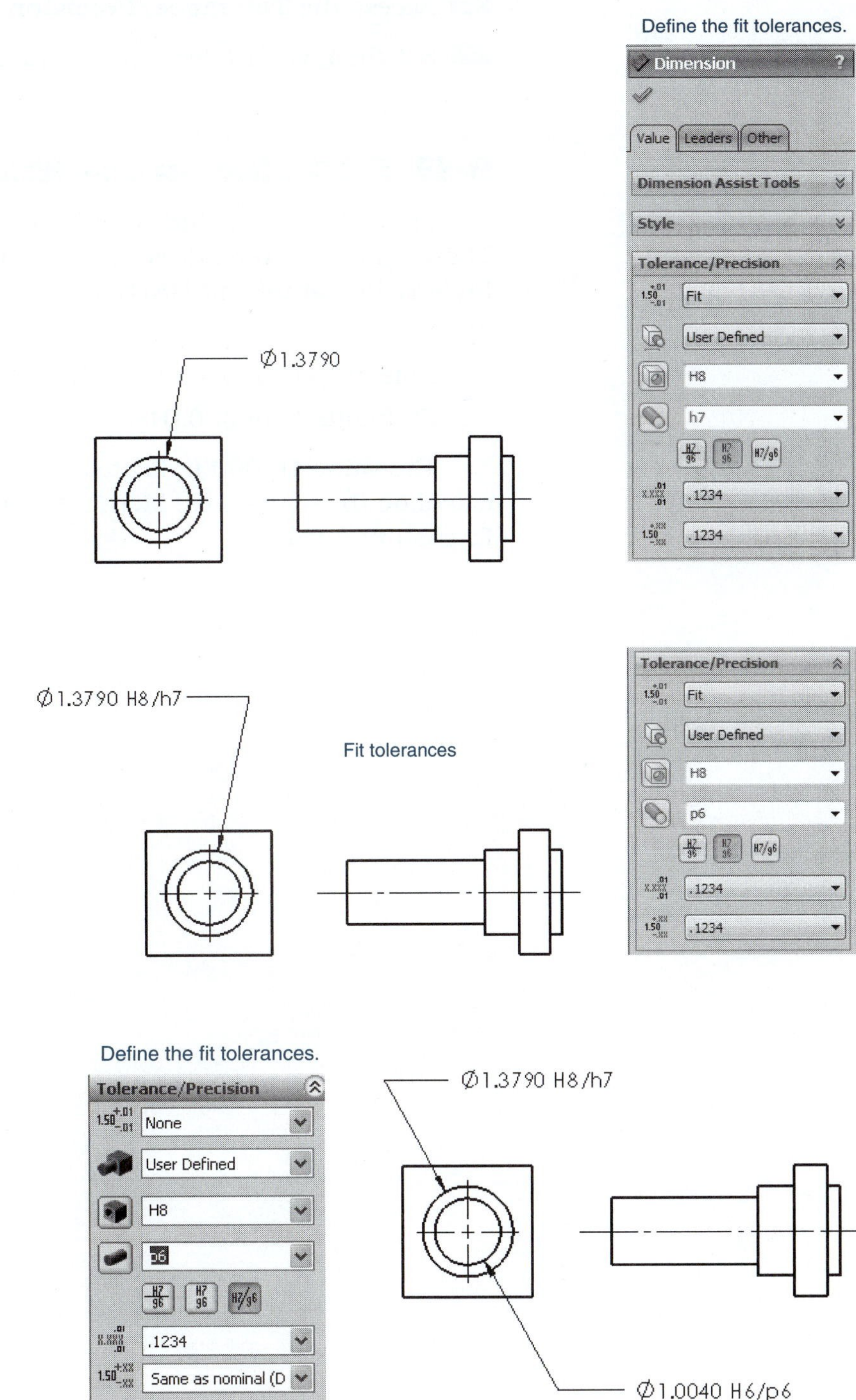

1. Use the **Smart Dimension** tool and dimension the O.D. of the bearing.
2. Access the **Tolerance/Precision** box and select the **Fit** option.
3. Set the hole fit tolerance for **H8** and the shaft fit tolerance for **h7**.
4. Click the green **OK** check mark.
5. Use the **Smart Dimension** tool and dimension the I.D. of the bearing.
6. Access the **Tolerance/Precision** box and select the **Fit** option.
7. Set the hole fit tolerance for **H6** and the shaft fit tolerance for **p6**.

9-13 Fit Tolerances—Millimeters

The appendix also includes tables for preferred fits using metric values. These tables are read directly. For example, the values for a Close Running Preferred Clearance fit H8/f7 for a nominal shaft diameter of 16 are as follows:

Hole: Ø16.027/16.00 Shaft: Ø15.984/15.966

Fit (limits of fits): 0.016

The value 16.000 indicates that the hole basis condition was used to calculate the tolerances. Metric fits are applied in the same manner as English unit values.

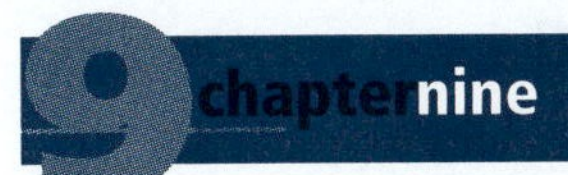

Chapter Projects

Figure P9-1 shows a support plate with three holes. A dimensioned drawing of the support plate is included. The holes are lettered. Three shafts are also shown. All shafts are 3.00 long. For Projects 9-1 to 9-8:

A. Create dimensioned and tolerances drawings for the support plate and shafts.

B. Specify tolerances for both the support plate holes and the shaft's diameters based on the given fit information.

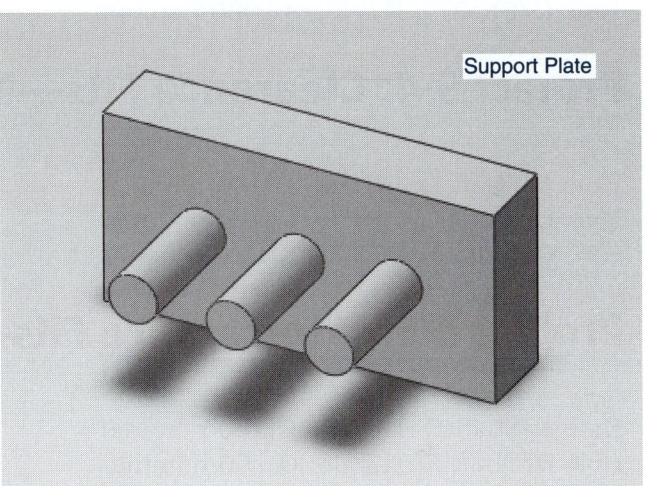

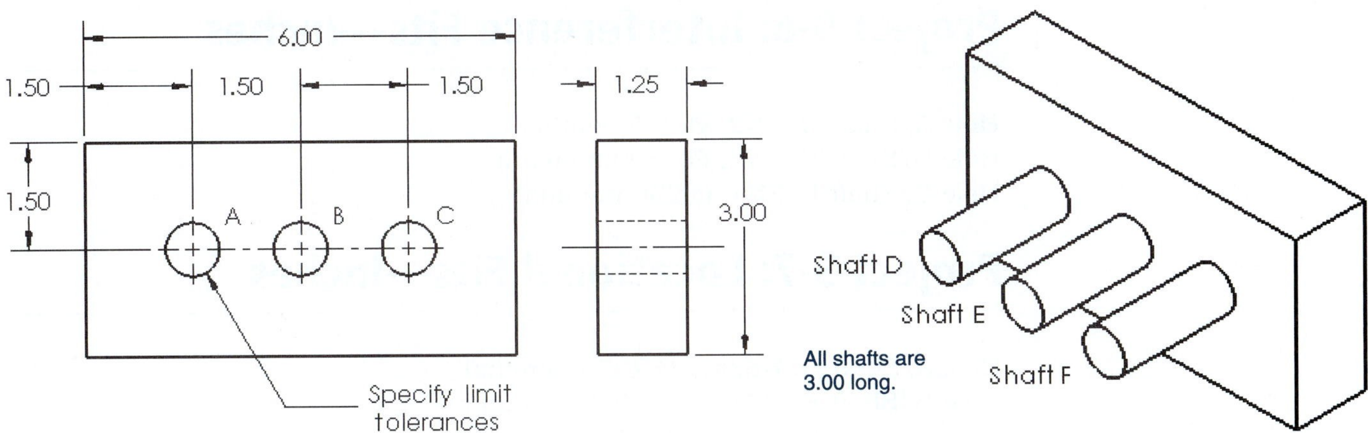

Figure P9-1

Project 9-1: Clearance Fits—Inches

Hole A/Shaft D: H7/h6, Ø.125 nominal
Hole B/Shaft E: H6/h5, Ø.750 nominal
Hole C/Shaft F: H9/f8, Ø.250 nominal

Project 9-2: Clearance Fits—Inches

Hole A/Shaft D: H10/d9, Ø1.123 nominal
Hole B/Shaft E: H7/h6, Ø.500 nominal
Hole C/Shaft F: H5/g4, Ø.625 nominal

Project 9-3: Clearance Fits—Inches

Hole A/Shaft D: H9/f8, Ø.500 nominal
Hole B/Shaft E: H8/e7, Ø.635 nominal
Hole C/Shaft F: H7/f6, Ø1.000 nominal

Project 9-4: Clearance Fits—Millimeters

Hole A/Shaft D: D9/h9, Ø10.0 nominal
Hole B/Shaft E: H7/h6, Ø16.0 nominal
Hole C/Shaft F: C11/h11, Ø20.0 nominal

Project 9-5: Interference Fits—Inches

Hole A/Shaft D: H6/n5, Ø.250 nominal
Hole B/Shaft E: H7/p6, Ø.750 nominal
Hole C/Shaft F: H7/r6, Ø.250 nominal

Project 9-6: Interference Fits—Inches

Hole A/Shaft D: FN2, Ø.375 nominal
Hole B/Shaft E: FN3, Ø1.500 nominal
Hole C/Shaft F: FN4, Ø.250 nominal

Project 9-7: Locational Fits—Inches

Hole A/Shaft D: H8/k7, Ø.4375 nominal
Hole B/Shaft E: H7/k6, Ø.7075 nominal
Hole C/Shaft F: H8/js7, Ø1.155 nominal

Project 9-8: Interference Fits—Millimeters

Hole A/Shaft D: H7/k6, Ø12.0 nominal
Hole B/Shaft E: H7/p6, Ø25.0 nominal
Hole C/Shaft F: N7/h6, Ø8.0 nominal

Figure P9-2 shows a U-bracket, four sleeve bearings, and two shafts. A dimensioned drawing of the U-bracket is also included. For Projects 9-9 to 9-12:

Figure P9-2

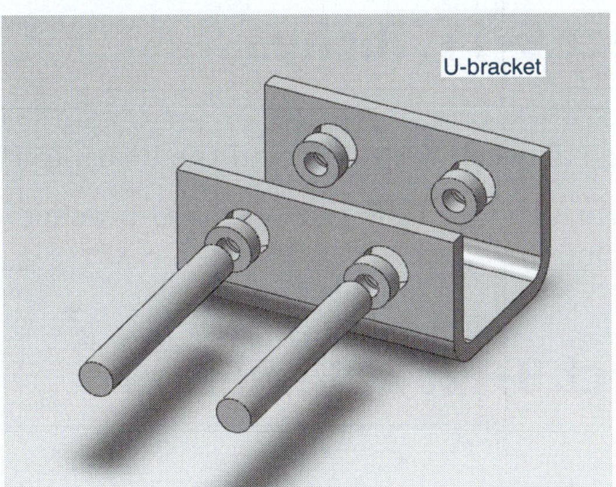

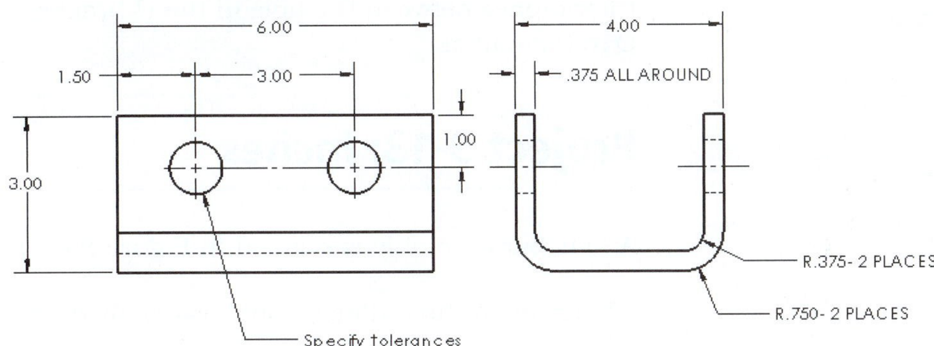

A. Create dimensioned and toleranced drawings for the U-bracket, sleeve bearings, and shafts. All shafts are 5.00 long.

B. Specify tolerances for the shaft diameter and the outside diameter of the sleeve bearings based on the given interference fits.

C. Specify tolerances for the holes in the U-bracket and the outside diameter of the sleeve bearings based on the sizes given in the appendix.

D. Create new links for parts 7, 8, and 9, and note the changes in motion created.

Project 9-9: Inches

Clearance between shaft and bearing: H9/f8, Ø.875 nominal

Interference between the hole in the U-bracket and the bearing: H7/p6, Ø.375 nominal

Project 9-10: Inches

Clearance between the shaft and the bearing: H9/f8, Ø.875 nominal

Interference between the hole in the U-bracket and the bearing: Class FN2, Ø1.125 nominal

Project 9-11: Inches

Clearance between the shaft and the bearing: H10/h9, Ø.500 nominal

Interference between the hole in the U-bracket and the bearing: H7/r6, Ø.750 nominal

Project 9-12: Millimeters

Clearance between the shaft and the bearing: H9/d9, Ø10.0 nominal

Interference between the hole in the U-bracket and the bearing: H7/p6, Ø16.0 nominal

Project 9-13: Inches

A four-bar assembly is defined in Figure P9-13.

1. Create a three-dimensional assembly drawing of the four-bar assembly.
2. Animate the links using LINK-1 as the driver.
3. Redraw the individual parts, add the appropriate dimensions, and add the following tolerances:
   A. Assign an LN1 interference fit between the links and the needle roller bearing.
   B. Assign an LC2 clearance between the holder posts, both regular and long posts, and the inside diameter of the needle roller bearing.
   C. Assign an LC3 clearance between the holder posts, both regular and long posts, and the spacers.

Project 9-14: Inches

Redraw the crank assembly shown in Figure P9-14. Create the following drawings.

A. An exploded isometric drawing with balloons and a BOM.

B. Dimensioned drawings of each part.

C. Animate the assembly drawing.

D. Change the length of the links and note the differences in motion.

Figure P9-13

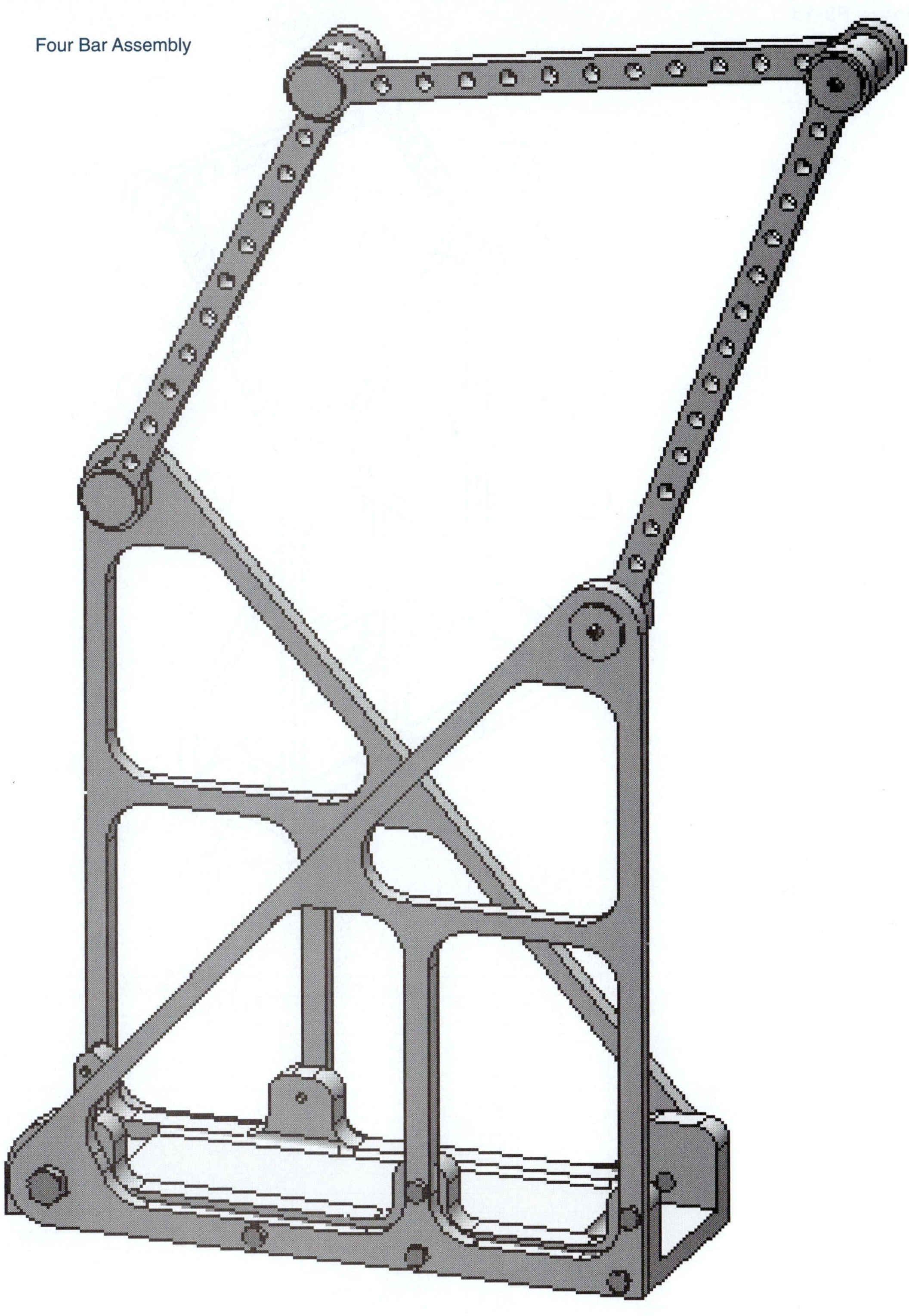

Four Bar Assembly

Figure P9-13
(Continued)

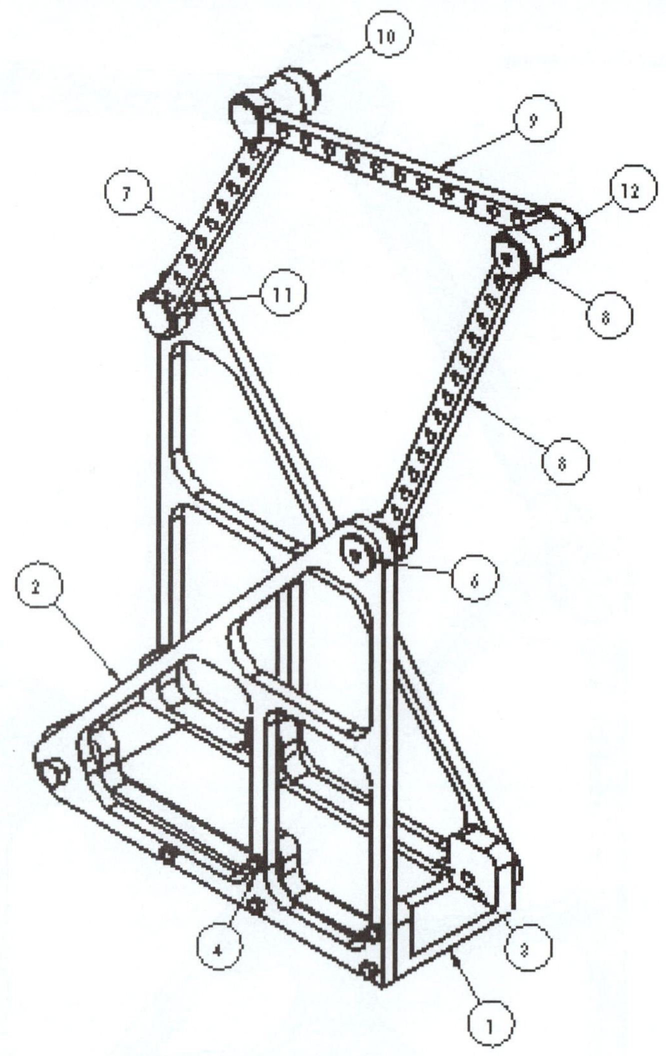

ITEM NO.	PART NUMBER	DESCRIPTION	QTY.
1	BU09-ME1	HOLDER, BASE	1
2	BU09-ME2	HOLDER, SIDE	2
3	HBOLT 0.5000-13x1.25x1.25-N		2
4	HBOLT 0.2500-20x1x1-N		11
5	HBOLT 0.2500-20x1.25x1.25-N		1
6	BU09-P1	POST, PIVOT, SHORT	2
7	LINK-01	LINK1, ASSEMBLY	1
8	LINK-03	LINK3, ASSEMBLY	1
9	LINK-02	LINK 2, ASSEMBLY	1
10	BU09-P2	POST, PIVOT, LONG	2
11	BU09-S1	SPACER, SHORT	2
12	BU09-S2	SPACER, LONG	2

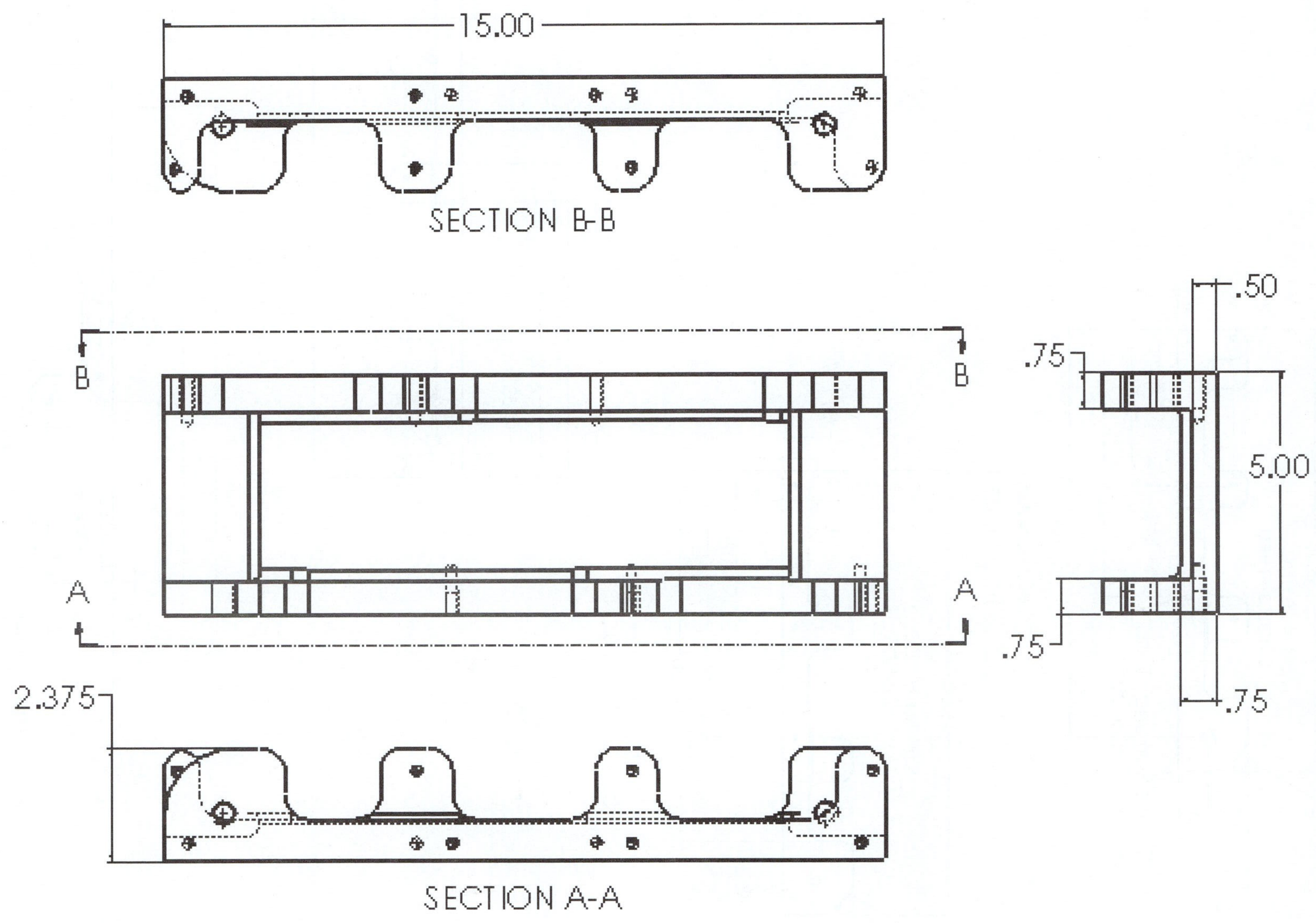

Figure P9-13
(*Continued*)

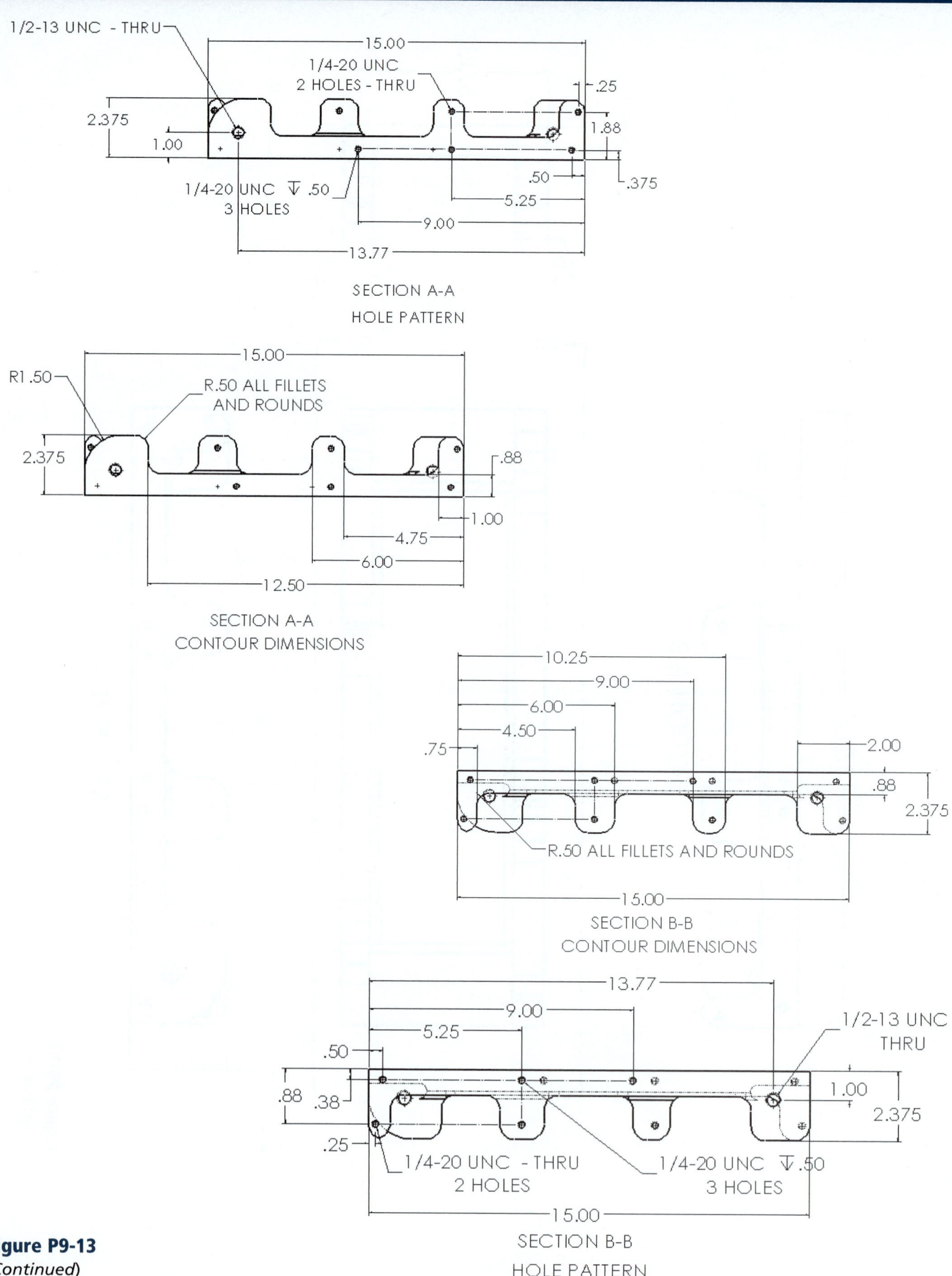

Figure P9-13
(Continued)

Holder, Side
P/N BU09-ME2

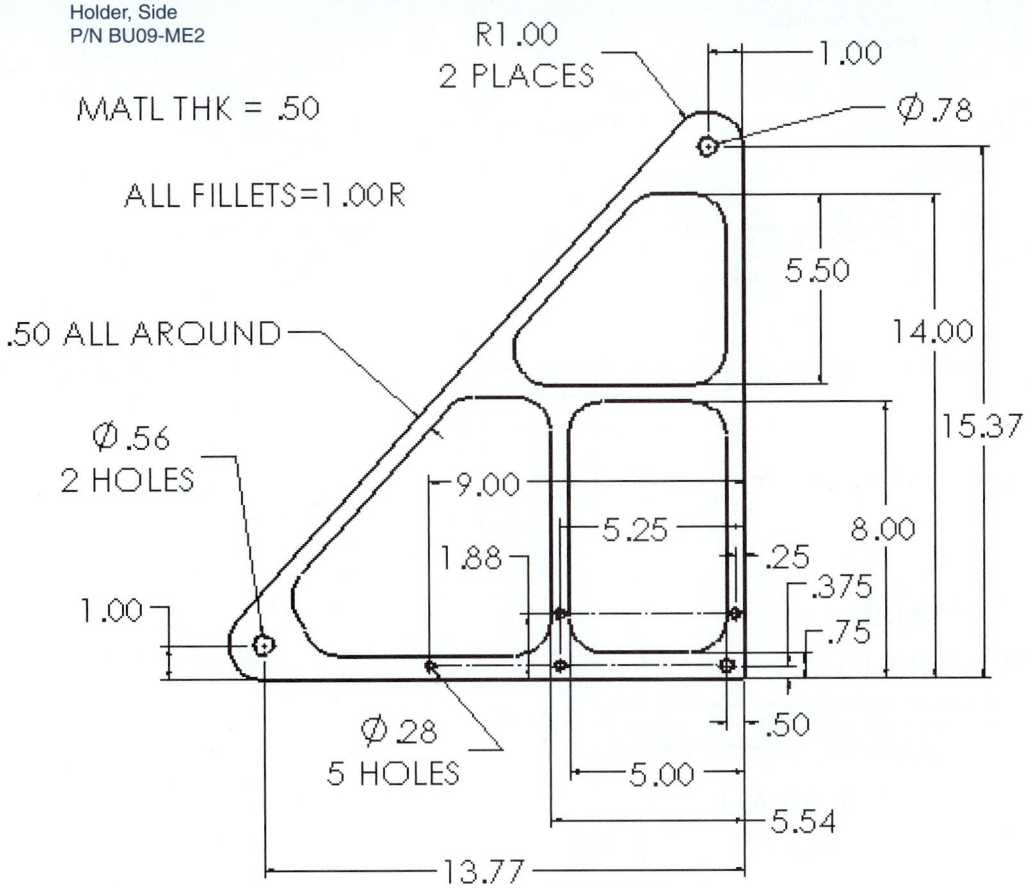

MATL THK = .50
ALL FILLETS=1.00R
R1.00
2 PLACES
1.00
∅.78
5.50
14.00
15.37
.50 ALL AROUND
∅.56
2 HOLES
9.00
5.25
1.88
.25
.375
8.00
1.00
.75
∅.28
5 HOLES
.50
5.00
5.54
13.77

LINK 1, Assembly
P/N Link-01

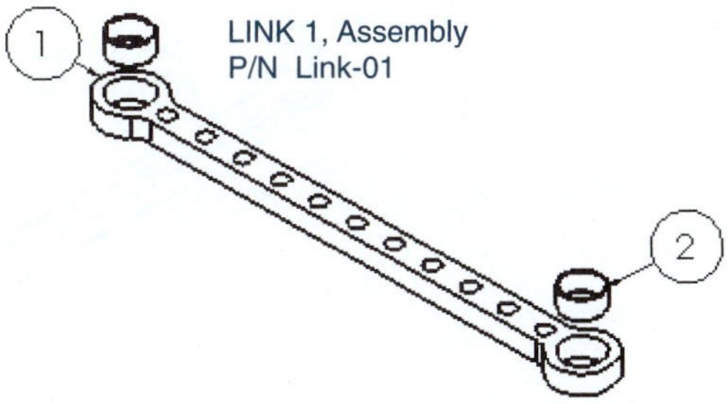

1
2

ITEM NO.	PART NUMBER	DESCRIPTION	QTY.
1	BU09-L1	LINK-1	1
2	AFBMA 18.2.3.1 - 12NIH135 - 20,SI,NC,20	NEEDLE ROLLER BEARING	2

LINK - 1
P/N BU09 - L1

12.00
1.00
10×1.00(10.00)
R.75
BOTH ENDS
∅.375 THRU
11 HOLES
R.50 ALL
FILLETS
∅1.00
2 HOLES

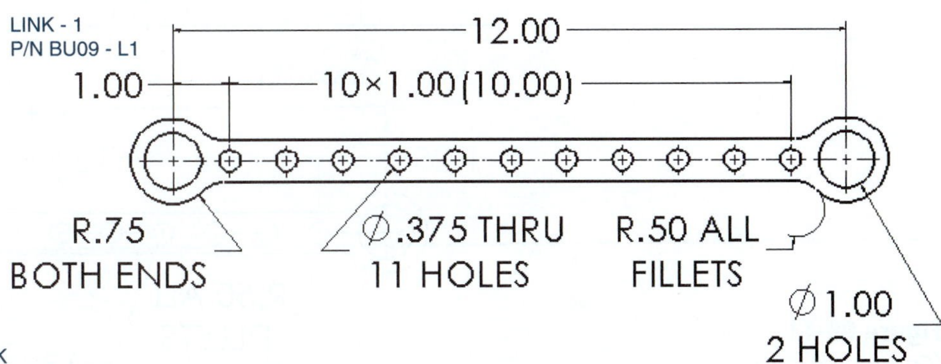

Figure P9-13
(Continued)

MATL = .50 THK

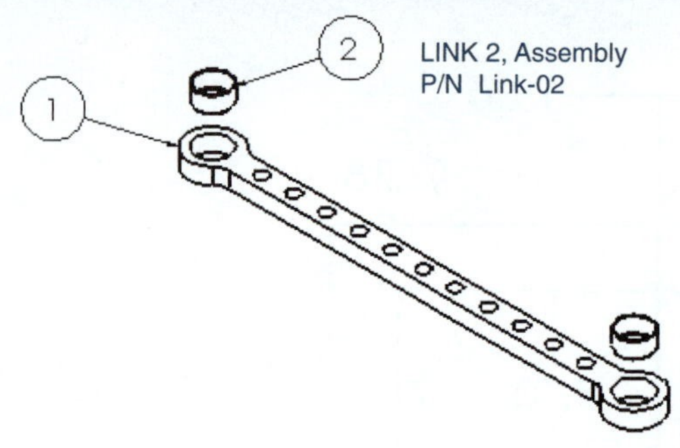

ITEM NO.	PART NUMBER	DESCRIPTION	QTY.
1	BU09-L2	LINK - 2	1
2	AFBMA 18.2.3.1 - 12NIH35 - 20,SI,NC,20	NEEDLE ROLLER BEARING	2

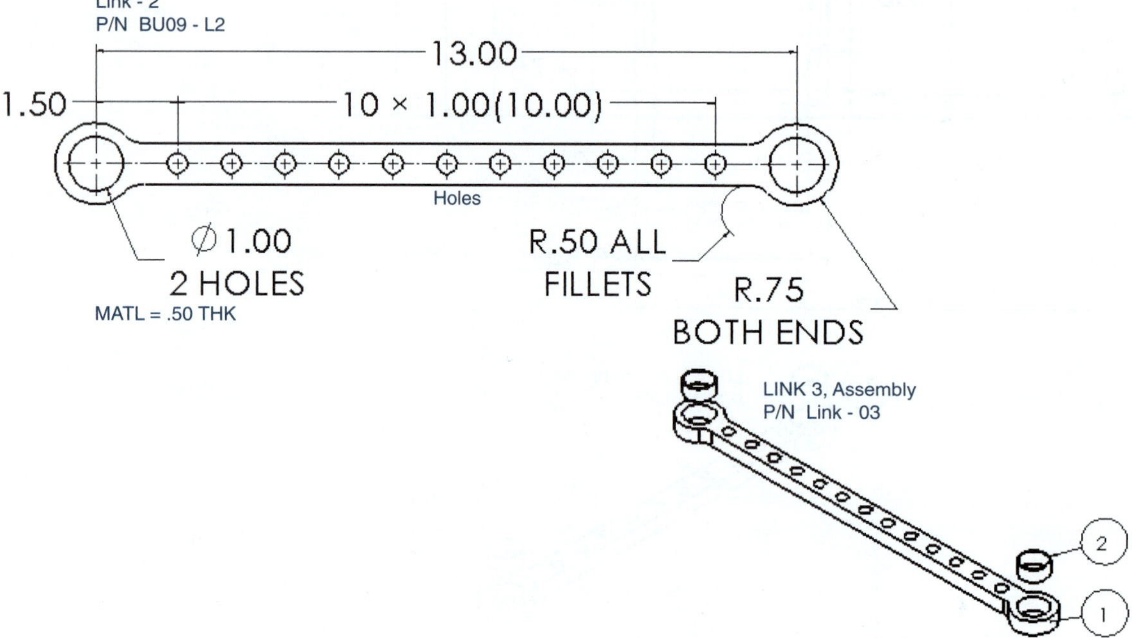

ITEM NO.	PART NUMBER	DESCRIPTION	QTY.
1	BU09-L3	LINK-3	1
2	AFBMA 18.2.3.1 - 12NIH35 - 20,SI,NC,20	NEEDLE ROLLER BEARING	2

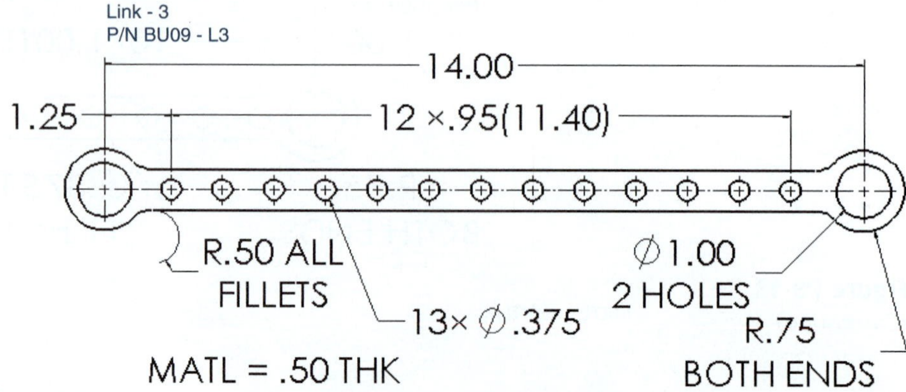

Figure P9-13
(Continued)

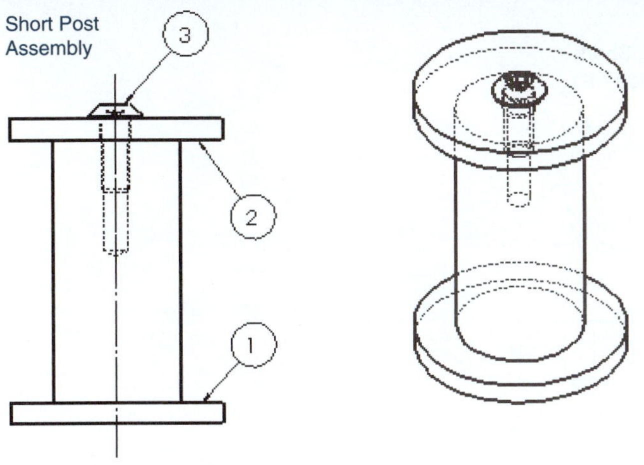

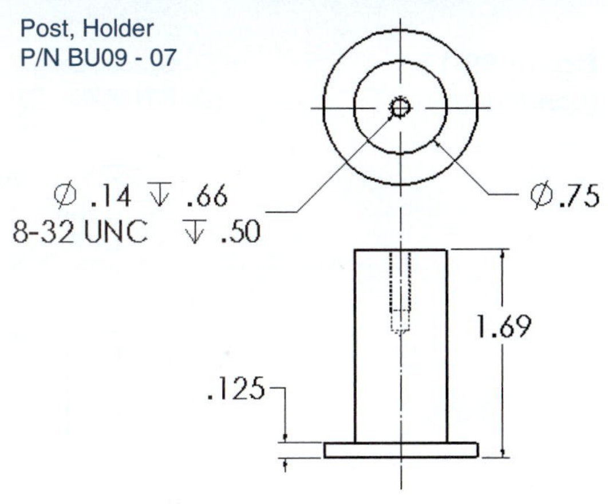

ITEM NO.	PART NUMBER	DESCRIPTION	QTY.
1	BU09-07	POST,HOLDER	1
2	BU09--10	DISK, TOP	1
3	SBHCSCREW0.164-32x0.4375-HX-N	SOCKET BUTTON HEAD CAP SCREW	1

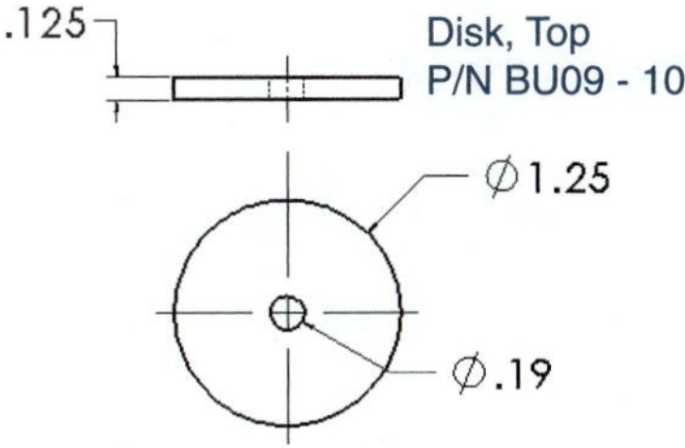

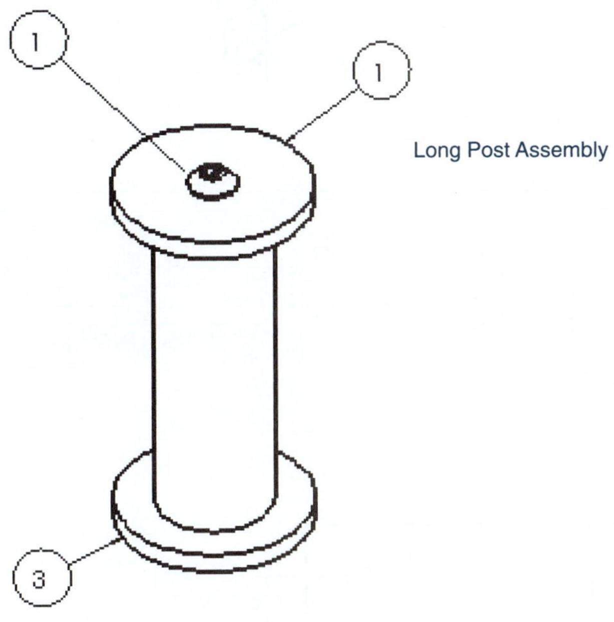

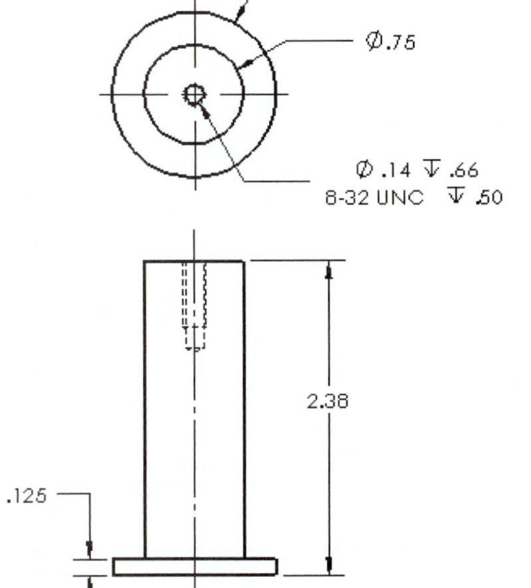

ITEM NO.	PART NUMBER	DESCRIPTION	QTY.
1	BU09-10	DISK, TOP	1
2	SBHCSCREW 0.164-32x0.4375-HX-N	SOCKET BUTTON HEAD CAP SCREW	1
3	BU09-08	POST, HOLDER, LONG	1

Figure P9-13
(Continued)

Figure P9-13
(*Continued*)

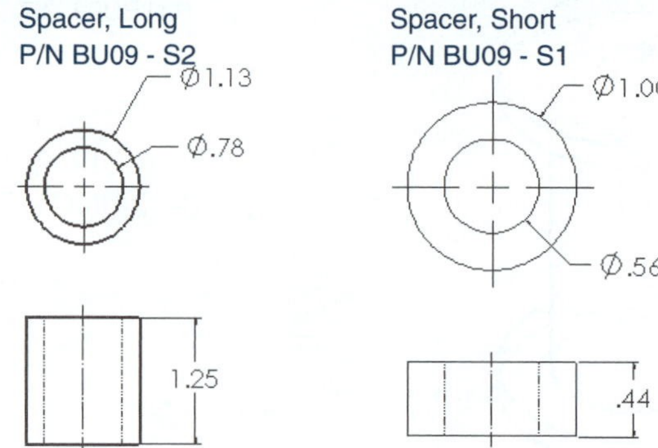

Figure P9-14

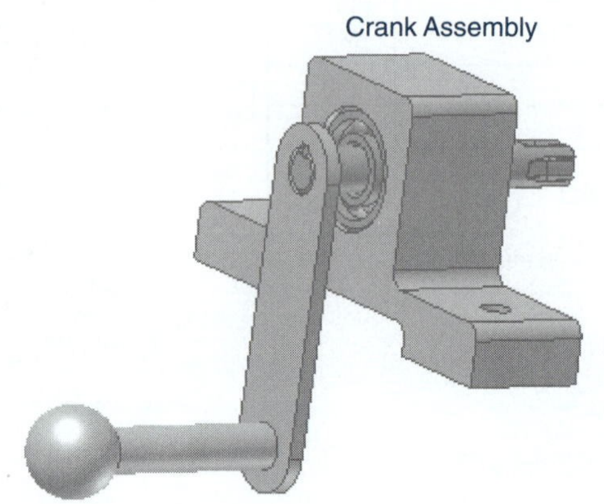

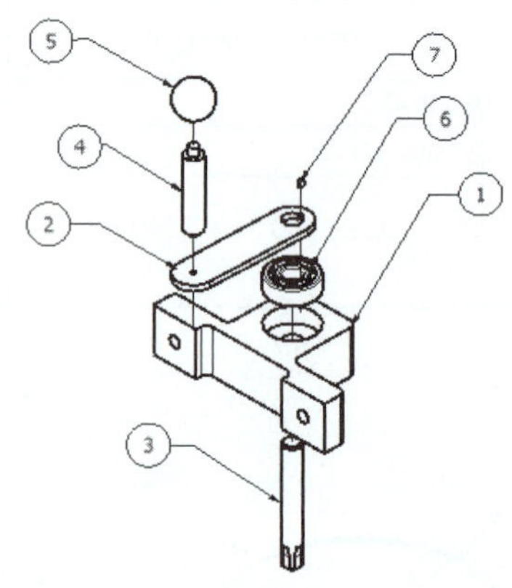

Parts List				
ITEM	PART NUMBER	DESCRIPTION	MATERIAL	QTY
1	EK131-1	SUPPORT	STEEL	1
2	EK131-2	LINK	STEEL	1
3	EK131-3	SHAFT,DRIVE	STEEL	1
4	EK131-4	POST, THREADED	STEEL	1
5	EK131-5	BALL	STEEL	1
6	BS 292 - BRM 3/4	Deep Groove Ball Bearings	STEEL,MILD	1
7	3/16x1/8x1/4	RECTANGULAR KEY	STEEL	1

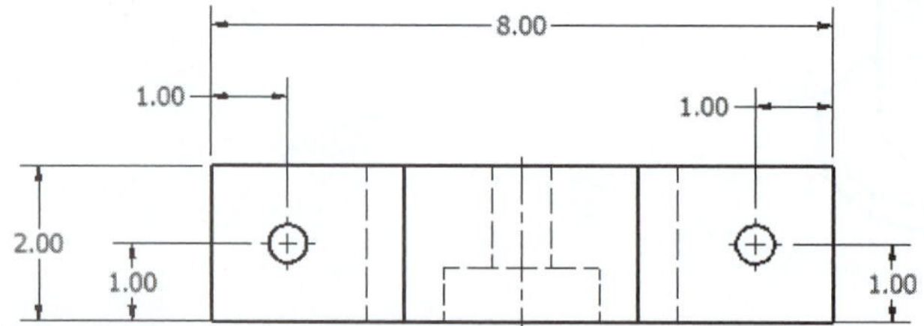

NOTE: ALL FILLETS AND ROUNDS R=0.250
UNLESS OTHERWISE STATED.

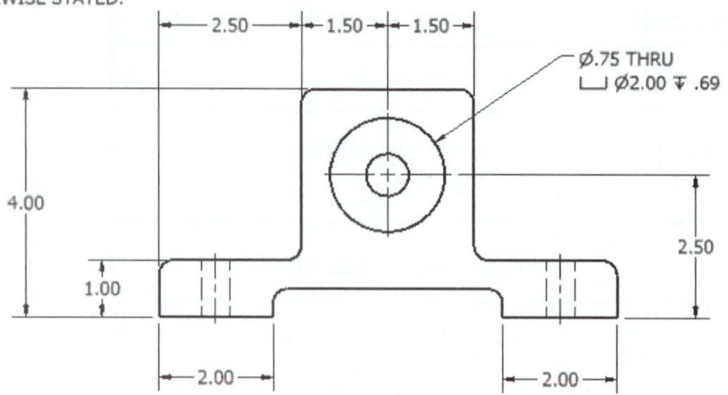

Figure P9-14
(*Continued*)

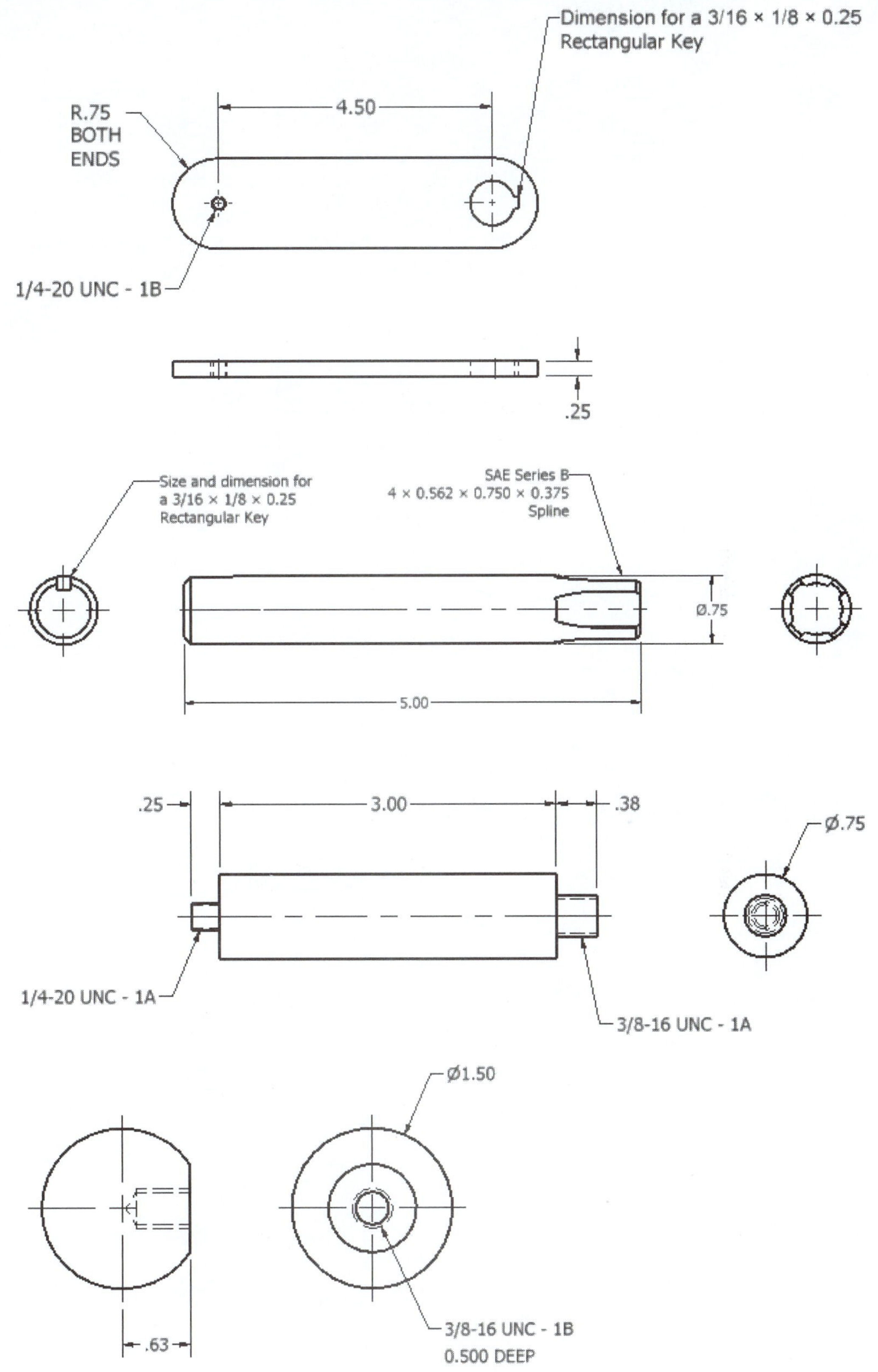

10 chapter ten Gears

CHAPTER OBJECTIVES

- Learn the concept of power transmission
- Learn the fundamentals of gears
- Learn how to draw and animate gears

10-1 Introduction

Gears, pulleys, and chains are part of a broader category called ***power transmission***. Power comes from a source such as an engine, motor, or windmill. The power is then transferred to a mechanism that performs some function. For example, an automobile engine transmits power from the engine to the wheels via a gear box. Bicyclists transmit the power of their legs to wheels via a chain and sprocket.

This section explains how gears, pulleys, and chains are drawn using SolidWorks and how the finished drawings can be animated. There is also a discussion of how speed is transferred and changed using gears, pulleys, and chains. Figure 10-1 shows a spur gear drawn using SolidWorks.

Figure 10-1

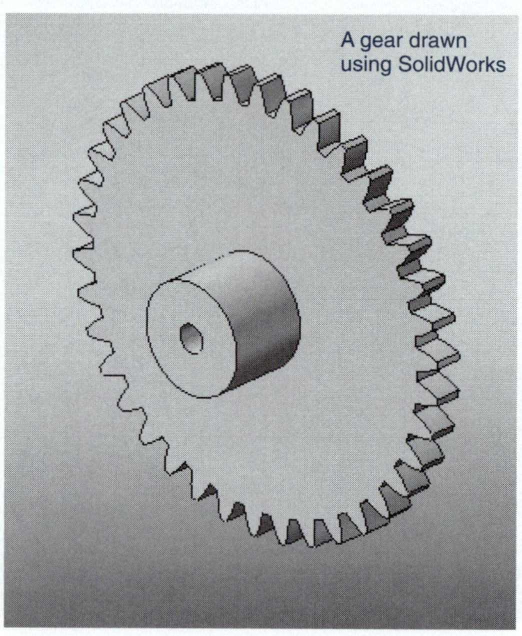

10-2 Gear Terminology

Pitch Diameter (*D*): The diameter used to define the spacing of gears. Ideally, gears are exactly tangent to each other along their pitch diameters.

Diametral Pitch (*P*): The number of teeth per inch. Meshing gears must have the same diametral pitch. Manufacturers' gear charts list gears with the same diametral pitch.

Module (*M*): The pitch diameter divided by the number of teeth. The metric equivalent of diametral pitch.

Number of Teeth (*N*): The number of teeth of a gear.

Circular Pitch (*CP*): The circular distance from a fixed point on one tooth to the same position on the next tooth as measured along the pitch circle. The circumference of the pitch circle divided by the number of teeth.

Preferred Pitches: The standard sizes available from gear manufacturers. Whenever possible, use preferred gear sizes.

Center Distance (*CD*): The distance between the center points of two meshing gears.

Backlash: The difference between a tooth width and the engaging space on a meshing gear.

Addendum (*a*): The height of a tooth above the pitch diameter.

Dedendum (*d*): The depth of a tooth below the pitch diameter.

Whole Depth: The total depth of a tooth. The addendum plus the dedendum.

Working Depth: The depth of engagement of one gear into another. Equal to the sum of the two gears' addendeums.

Circular Thickness: The distance across a tooth as measured along the pitch circle.

Face Width (*F*): The distance from front to back along a tooth as measured perpendicular to the pitch circle.

Outside Diameter: The largest diameter of the gear. Equal to the pitch diameter plus the addendum.

Root Diameter: The diameter of the base of the teeth. The pitch diameter minus the dedendum.

Clearance: The distance between the addendum of the meshing gear and the dedendum of the mating gear.

Pressure Angle: The angle between the line of action and a line tangent to the pitch circle. Most gears have pressure angles of either 14.5° or 20°.

See Figure 10-2.

Figure 10-2

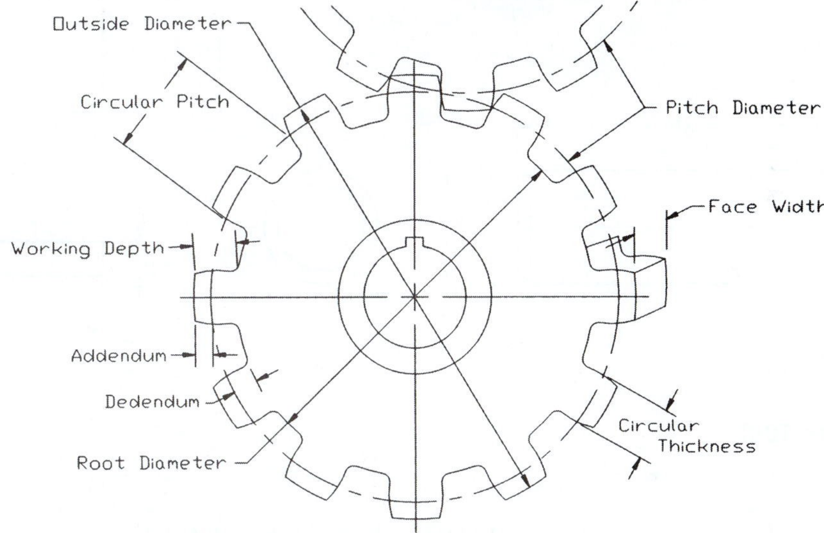

10-3 Gear Formulas

Figure 10-3 shows a chart of formulas commonly associated with gears. The formulas are for spur gears.

Figure 10-3

Diametral pitch (P)	$P = \frac{N}{D}$
Pitch diameter (D)	$D = \frac{N}{P}$
Number of teeth (N)	$N = DP$
Addendum (a)	$a = \frac{1}{P}$

Metric

Module (M)	$M = \frac{D}{N}$

10-4 Creating Gears Using SolidWorks

In this section we will create two gears and then create an assembly that includes a support plate and two posts to hold the gears in place. The specifications for the two gears are as follows. See Figure 10-4.

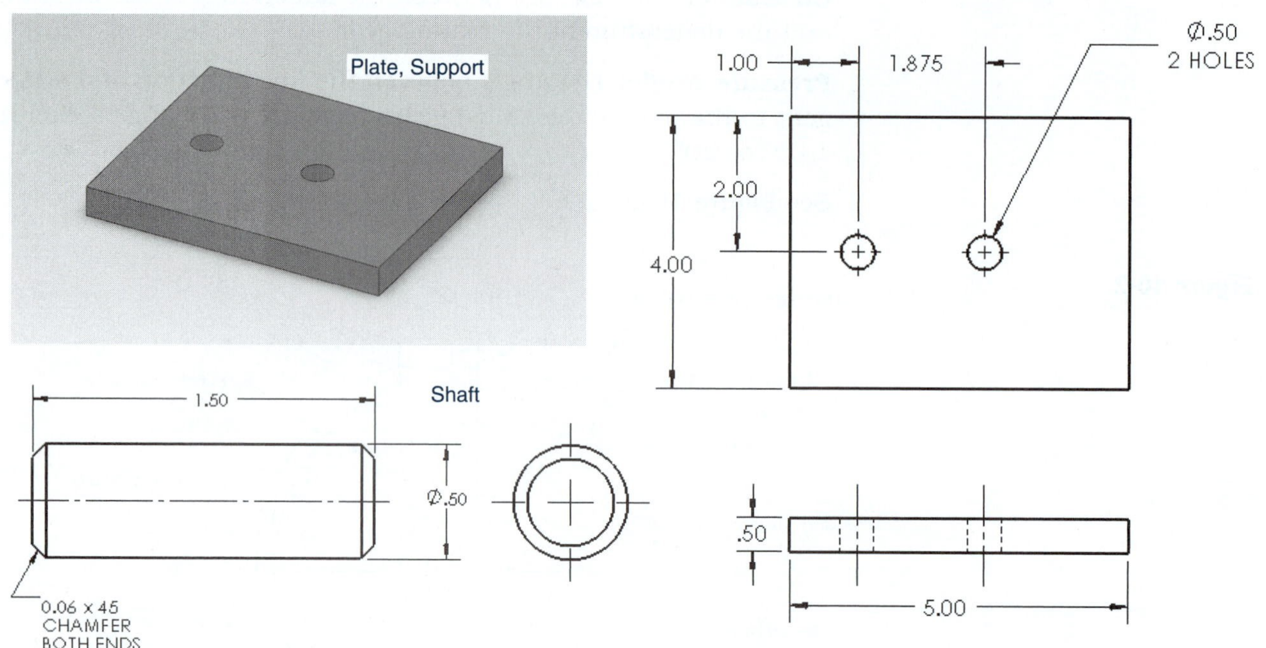

Figure 10-4

Gear 1: Diametral pitch = **24**

Number of teeth = **30**

Face thickness = **.50**

Bore = **Ø.50**

Hub = **Ø1.00**

Hub height = **.50**

Pressure angle = **20**

Gear 2: Diametral pitch = **24**

Number of teeth = **60**

Face thickness = **.50**

Bore = **Ø.50**

Hub = **Ø1.00**

Hub height = **.50**

Pressure angle = **20**

> **TIP**
> Gears must have the same diametral pitch to mesh properly.

Using the formulas presented we know that the pitch diameter is found as follows:

$$D = N/P$$

So

$$D1 = 30/24 = 1.25 \text{ in.}$$

$$D2 = 60/24 = 2.50 \text{ in.}$$

The center distance between gears is found from the relation $(D1 + D2)/2$:

$$\frac{1.25 + 2.50}{2} = 1.875 \text{ in.}$$

This center distance data was used to create the Plate, Support shown in Figure 10-4.

The bore for the gears is defined as 0.50, so shafts that hold the gears will be Ø0.50, and the holes in the Plate, Support also will be 0.50.

> **NOTE**
> Tolerances for gears, shafts, and support plates are discussed in Chapter 9, Bearings and Fit Tolerances.

The shafts will have a nominal diameter of Ø0.50 and a length of 1.50. The length was derived by allowing 0.50 for the gear thickness, 0.50 for the Plate, Support thickness, and 0.50 clearance between the gear and the plate. See Figure 10-4. In this example 0.06 × 45° chamfers were added to both ends of the shafts.

To Create a Gear Assembly

1. Draw the **Plate, Support** and **Shaft** shown in Figure 10-4.
2. Start a new **Assembly** drawing.
3. Assemble the plate and shafts as shown.

   See Figure 10-5. The top surface of the shafts is offset 1.00 from the surface of the plate.

Figure 10-5

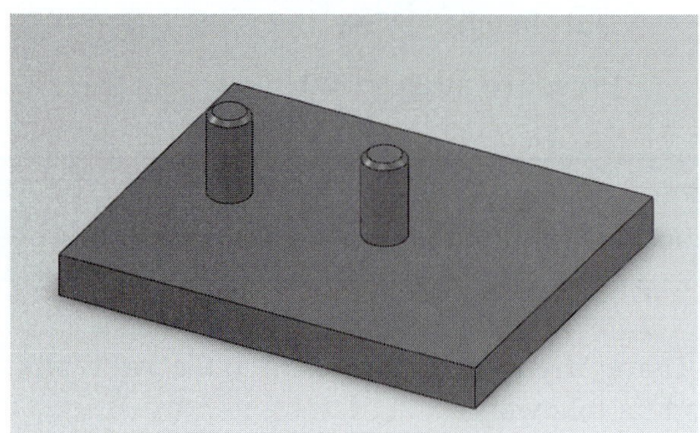

Create the gears using the **Design Library**. See Figure 10-6.

Figure 10-6

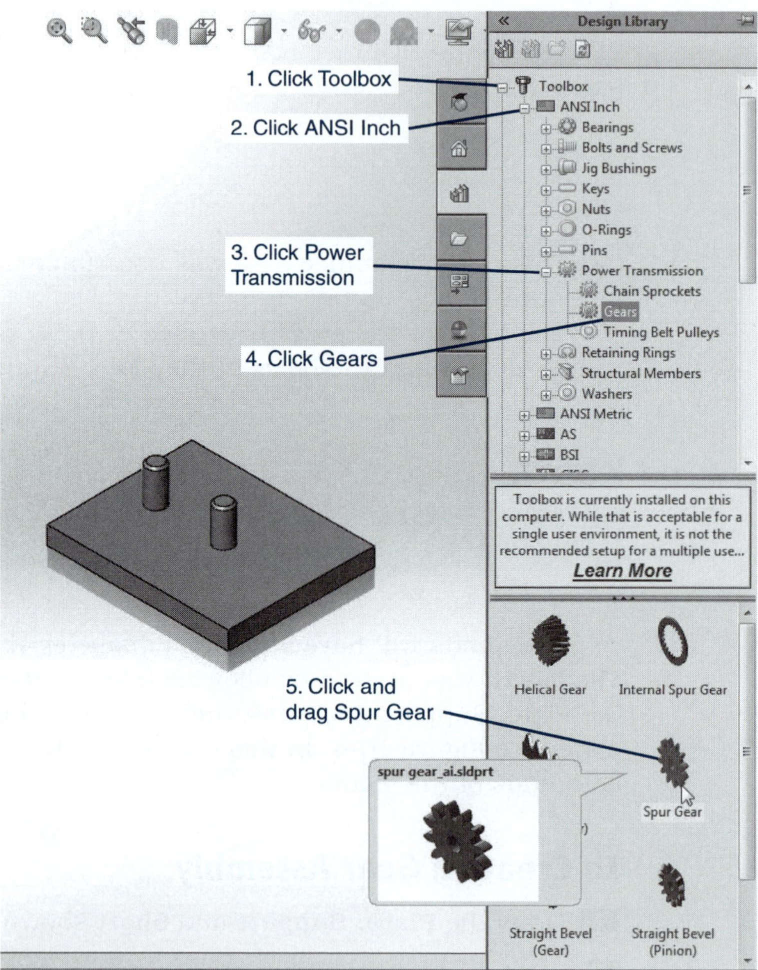

4 Click the **Design Library** tool, click **Toolbox, ANSI Inch, Power Transmission**, and **Gears.**

5 Click the **Spur Gear** tool and drag the icon into the drawing area.

The **Spur Gear PropertyManager** will appear. See Figure 10-7.

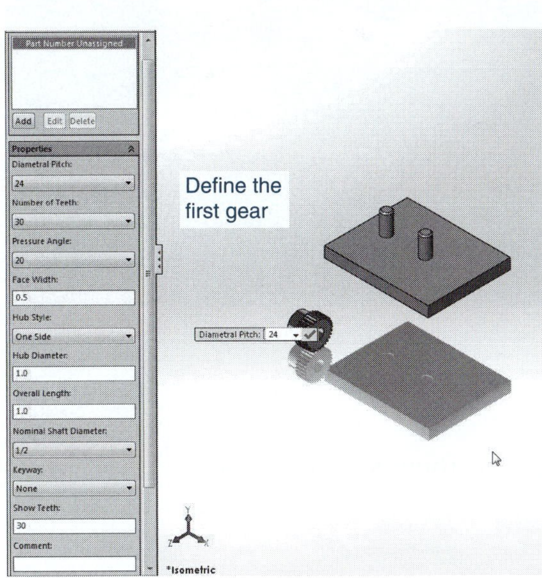

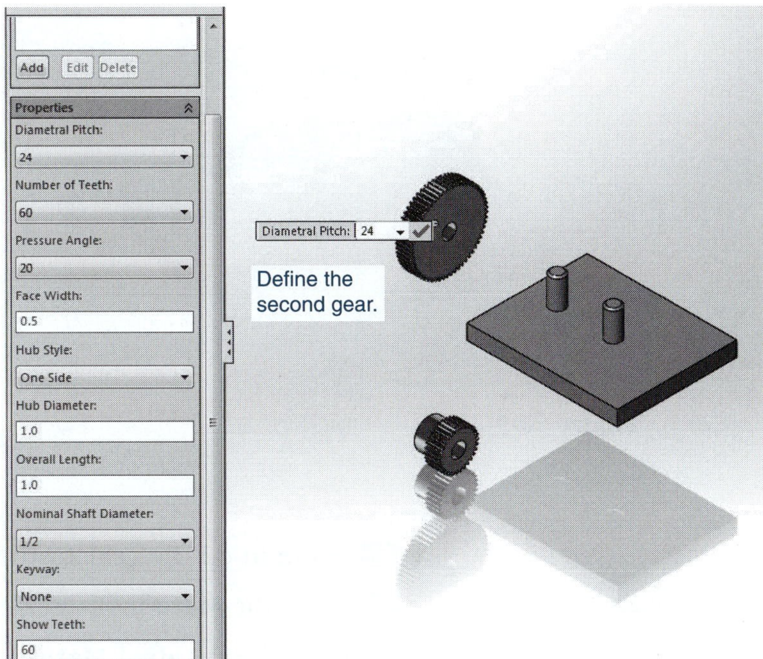

Figure 10-7

6 Enter the gear values as presented earlier for Gear 1 and create the gear.

7 Create Gear 2 using the presented values.

See Figure 10-8.

Figure 10-8

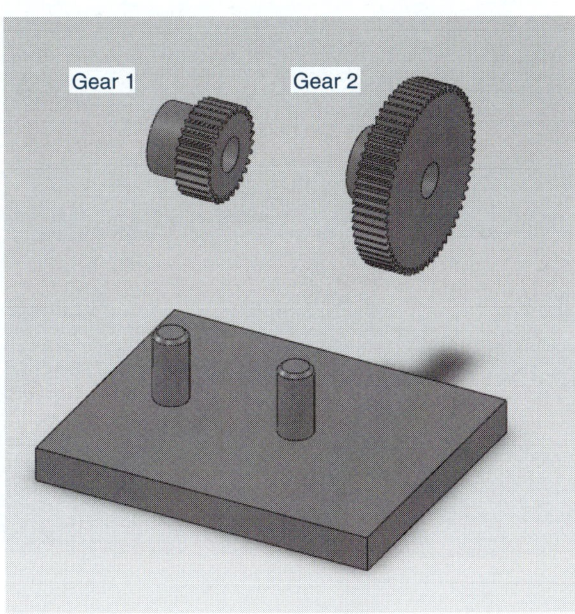

8 Use the **Mate** tool and assemble the gears onto the shafts so that they mesh. First, use the **Mate/Concentric** tool to align the gears' bores with the shafts; then use the **Mate/Parallel** tool to align the top surface of the shafts with the top surfaces of the gears.

See Figure 10-9.

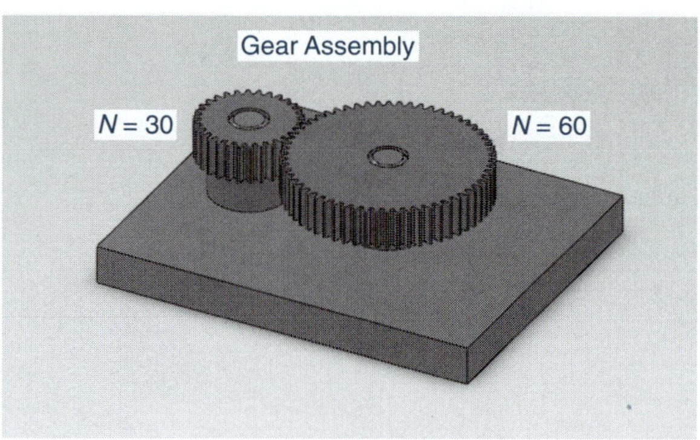

Figure 10-9

9. Zoom in on the gear teeth and align them so they mesh.
10. Click the **Mate** tool.
11. Click **Mechanical Mates.**
12. Select the **Gear** option.

    See Figure 10-10.

Figure 10-10

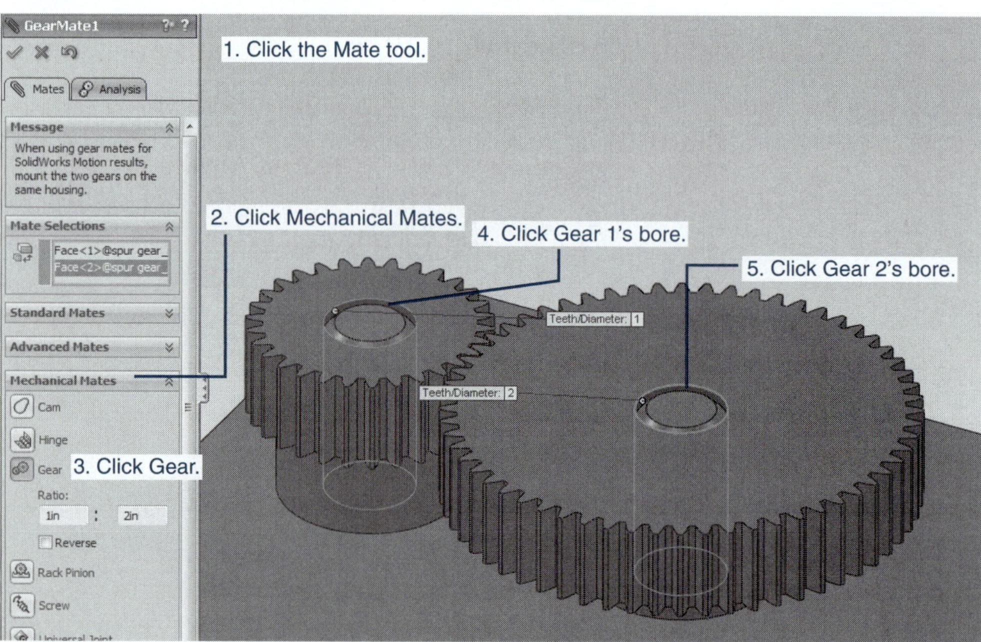

13. Define the gear mate by clicking the inside bore of the two gears.

    See Figure 10-11.
14. Define the ratio between the gears.

Figure 10-11

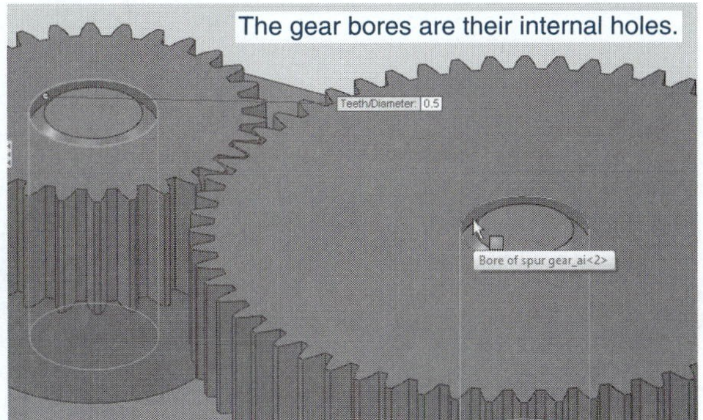

See Figure 10-12. In this example the ratio between the two gears is 2:1; that is, the smaller 30-tooth gear goes around twice for every revolution of the 60-tooth larger gear.

Figure 10-12

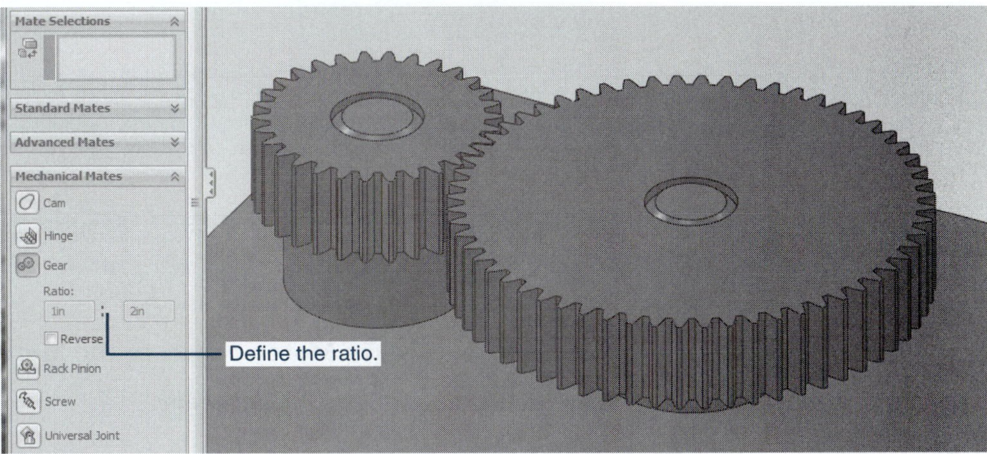

15 Click the green **OK** check mark.

16 Locate the cursor on the smaller gear and rotate the gears.

> **TIP**
>
> If the gears were not aligned (step 8) an error message would appear stating that the gears interfere with each other. The gears will turn relative to each other even if they interfere, but it is better to go back and align the gears.

To Animate the Gears

1 Click the **Motion Study** tab at the bottom of the screen.

2 Click the **Motor** tool.

See Figure 10-13.

Figure 10-13

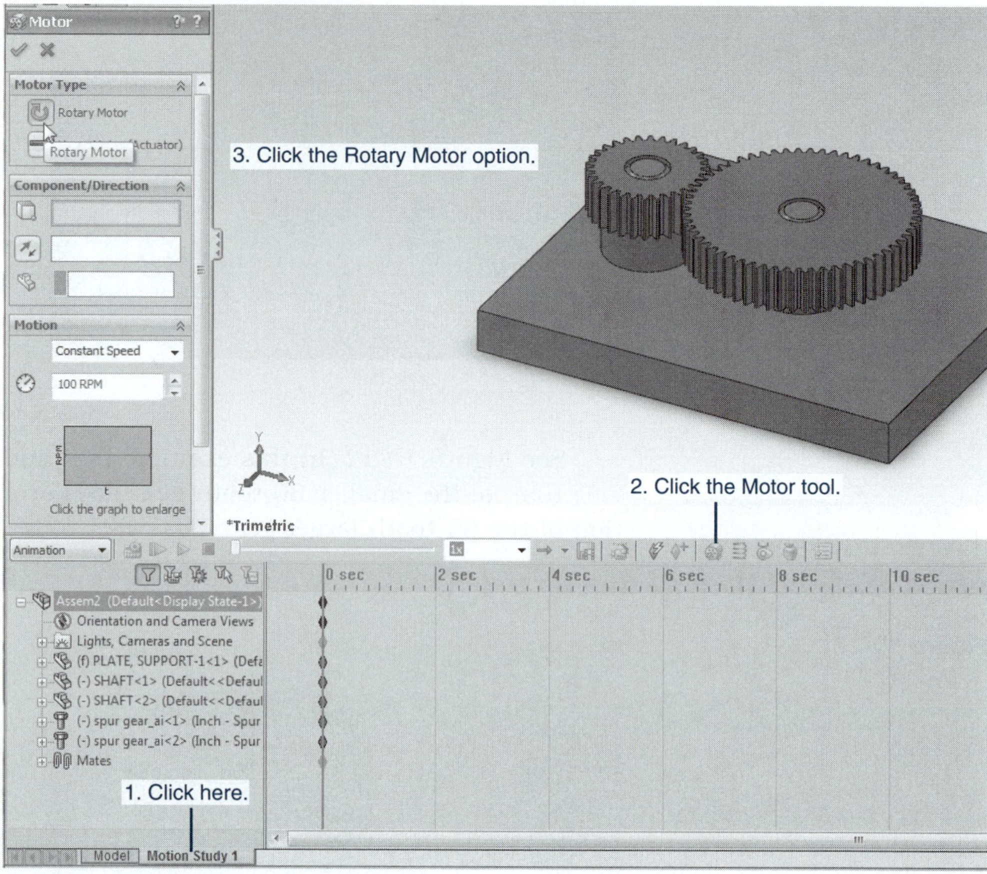

3 Click the **Rotary Motor** option in the **Motor PropertyManager**.

4 Click the smaller 30-tooth gear.

A red arrow will appear on the gear, and the gear will be identified in the **Component/Direction** box. See Figure 10-14.

Figure 10-14

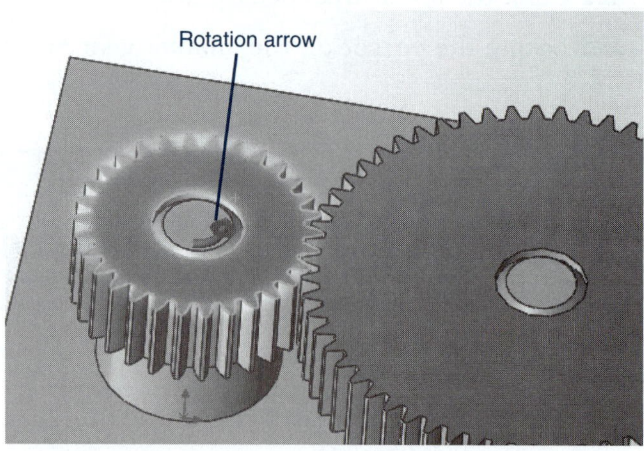

5 Click the green **OK** check mark.

Click the **Play** button and the gears will animate. See Figure 10-15.

6 Save the gear assembly.

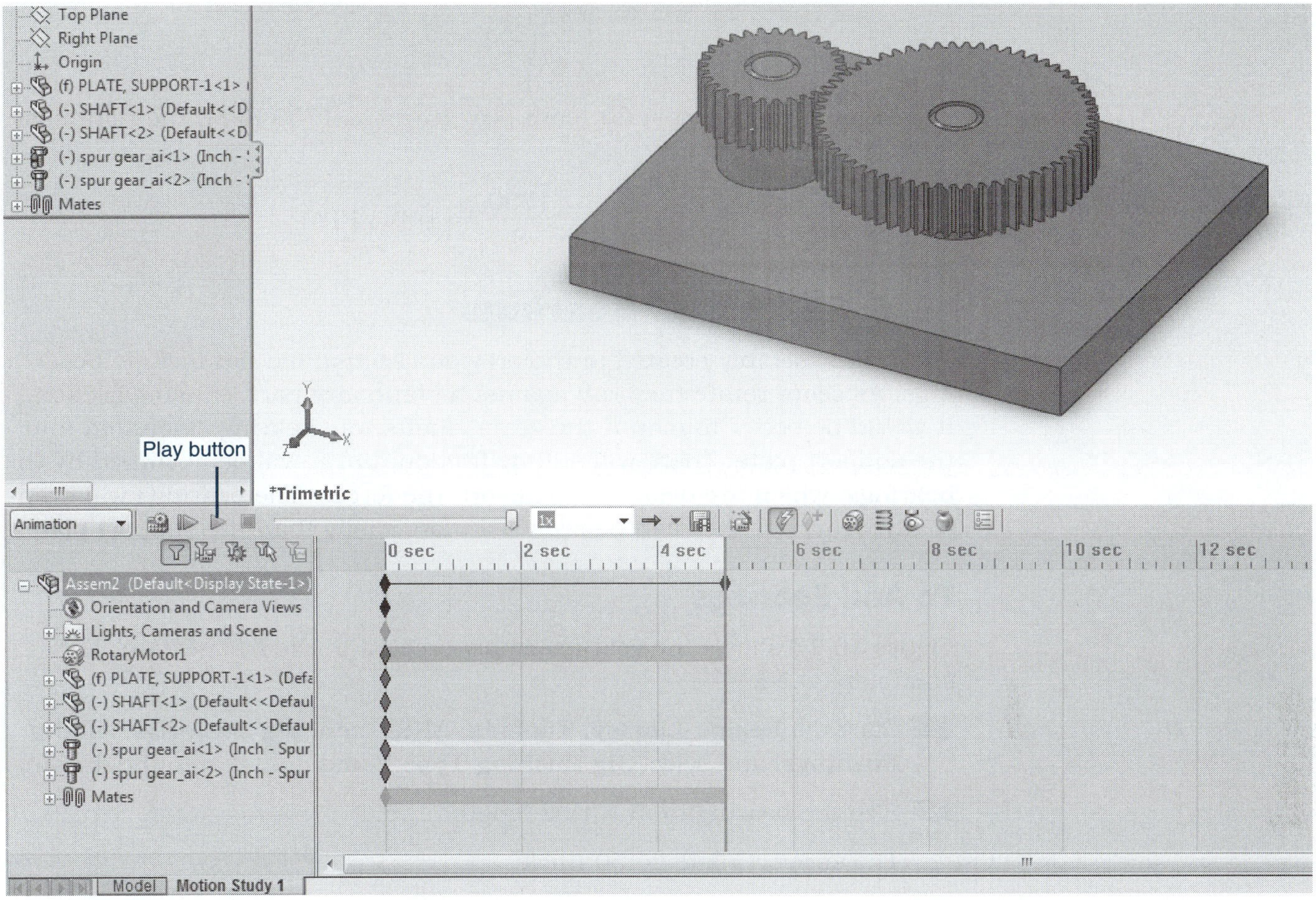

Figure 10-15

10-5 Gear Ratios

Gear ratios are determined by the number of teeth of each gear. In the previous example a gear with 30 teeth was meshed with a gear that has 60 teeth. Their gear ratio is 2:1; that is, the smaller gear turns twice for every one revolution of the larger gear. Figure 10-16 shows a group of four gears. A grouping of gears is called a ***gear train***. The gear train shown contains two gears with 30 teeth and two gears with 90 teeth. One of the 30-tooth gears is mounted on the same shaft as one of the 90-tooth gears. The gear ratio for the gear train is found as follows:

Figure 10-16

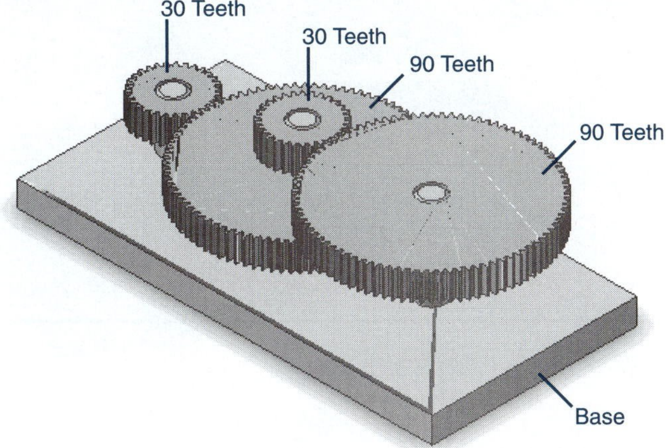

$$\left(\frac{3}{1}\right)\left(\frac{3}{1}\right) = \frac{9}{1}$$

Thus, if the leftmost 30-tooth gear turns at 1750 RPM, the rightmost 90-tooth gear turns at

$$\frac{1750}{9} = 199.4 \text{ RPM}$$

10-6 Gears and Bearings

The gear assembly created in the previous section did not include bearings. As gears rotate they rub against a stationary part, causing friction. It would be better to mount the gears' shafts into bearings mounted into the support plate. There will still be friction, but it will be absorbed by the bearings, which are designed to absorb the forces. The bearings will eventually wear, but they can be replaced more easily than the support plate.

To Add Bearings

Figure 10-17 shows the gear assembly created in the last section. The shafts are Ø.50.

1. Click the **Design Library, Toolbox, ANSI Inch, Jig Bushings, All Jig Bushings**, and select **Jig Bushing Type P** and drag it onto the drawing.
2. Size the bearing to **.5** with a length of **.5**.

   The support plate is .50 thick.
3. Add a second bearing.

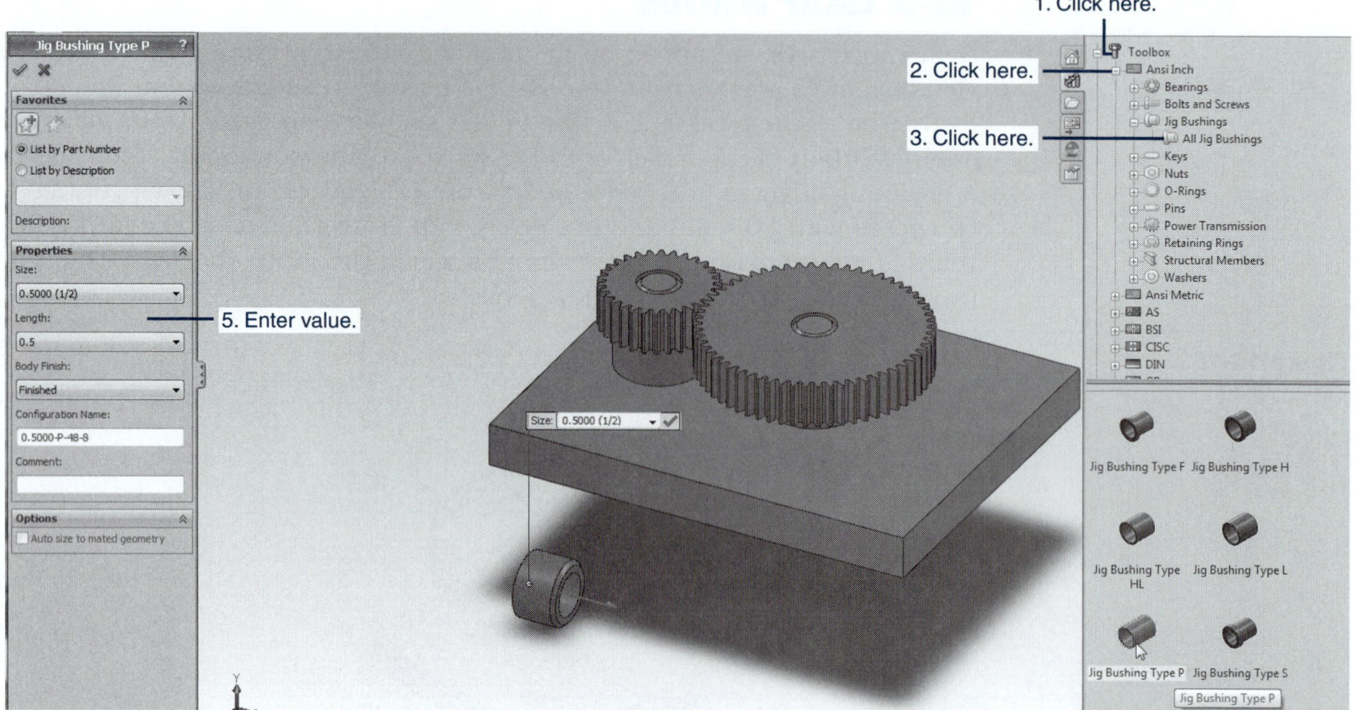

Figure 10-17

4 Double-click one of the bearings.

The dimensions used to create the bearing will appear. See Figure 10-18. The outside diameter of the bearing is .75. The holes in the support plate must be edited to Ø.75.

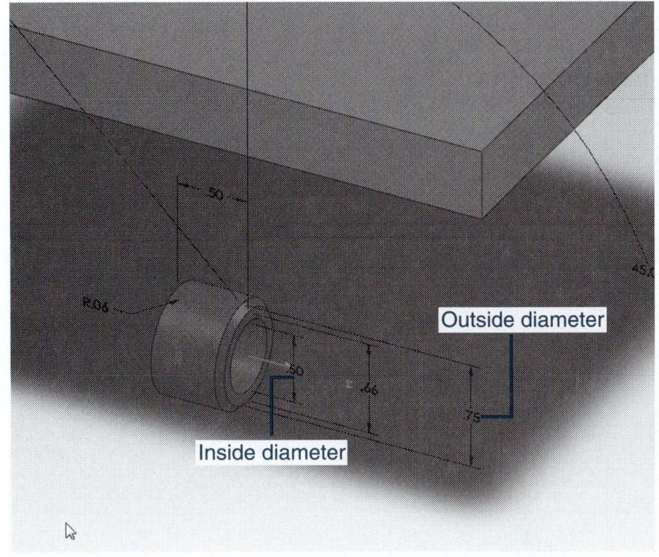

Figure 10-18

Figure 10-19

5 Click the + sign to the left of the **PLATE, SUPPORT** heading in the **FeatureManager**.

6 Right-click the **Sketch 2** heading under the **Cut-Extrude 1** heading and select the **Edit Sketch** option.

See Figure 10-19.

7 Double-click the holes' **.50** diameter dimensions and change them to **.75**.

See Figure 10-20.

8 Rotate the entire assembly so that the bottom surface of the support plate is visible.

9 Use the **Mate** tool to position the bearings onto the bottoms of the shafts.

See Figure 10-21.

Figure 10-20

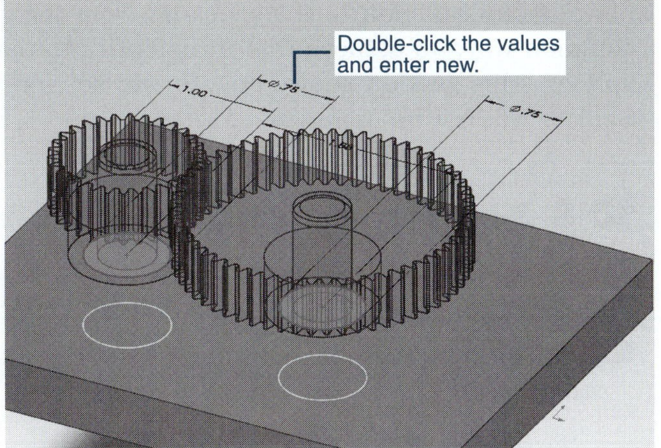

Figure 10-22 shows an exploded isometric drawing of the assembly with a BOM.

Figure 10-21

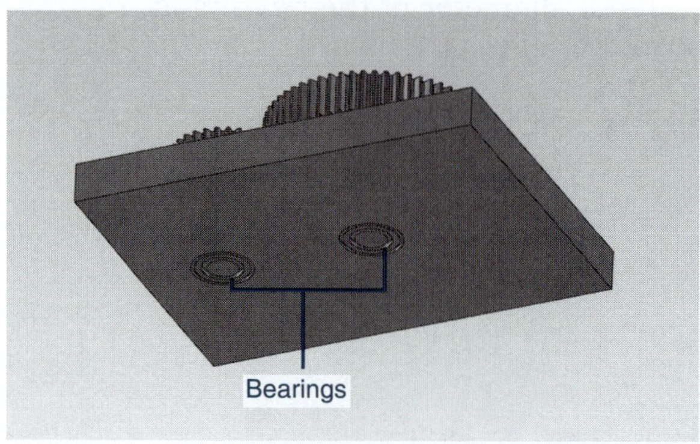

Figure 10-22

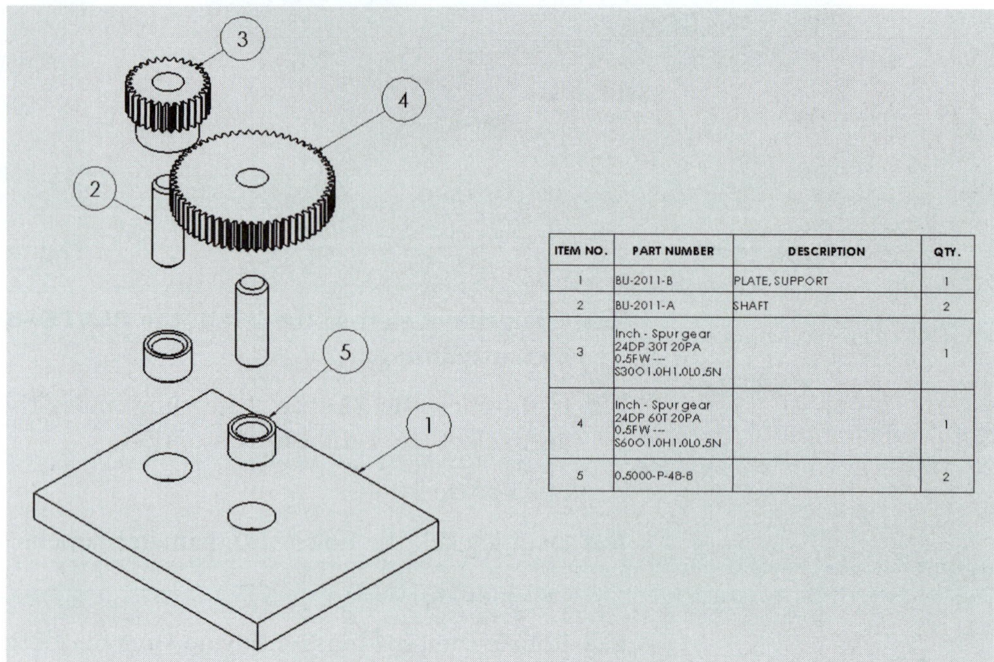

ITEM NO.	PART NUMBER	DESCRIPTION	QTY.
1	BU-2011-B	PLATE, SUPPORT	1
2	BU-2011-A	SHAFT	2
3	Inch - Spur gear 24DP 30T 20PA 0.5FW --- S30O1.0H1.0L0.5N		1
4	Inch - Spur gear 24DP 60T 20PA 0.5FW --- S60O1.0H1.0L0.5N		1
5	0.5000-P-48-8		2

10-7 Power Transmission—Shaft to Gear

When a gear is mounted on a shaft there must be a way to transfer the power from the shaft to the gear and from the gear to the shaft. Three common ways to achieve this transfer are to use set screws, keys, and splines. This section shows how to add set screws and keyways to gears. Splines will not be included.

10-8 Set Screws and Gear Hubs

This section shows how to add a hub to a gear and then how to create a threaded hole in the hub that will accept a set screw.

1 Start a new **Part** drawing and create the **Ø0.50** × **2.25** shaft shown in Figure 10-23.

Save the part as **POST, GEAR**.

2. Start a new **Assembly** drawing.
3. Use the **Insert Components** tool and add the Ø0.50 × 2.25 shaft to the drawing.
4. Access the **Design Library** and click **Toolbox, ANSI Inch, Power Transmission,** and **Gears**.
5. Select the **Spur Gear** option and click and drag a gear onto the drawing screen.

   See Figure 10-24.

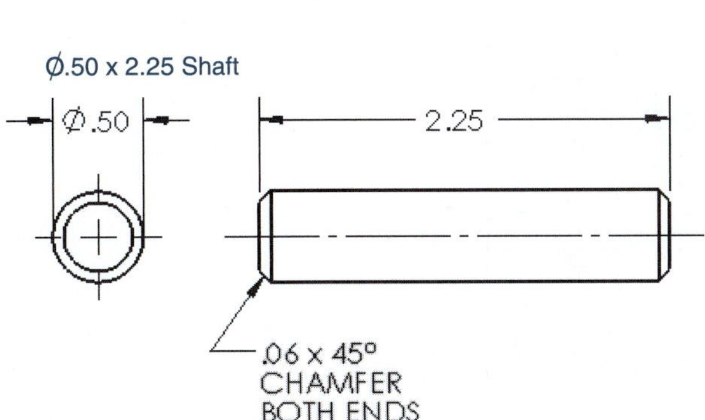

Figure 10-23

Figure 10-24

6. Set the gear's properties as follows. See Figure 10-25.

   Diametral pitch: **24**

   Number of teeth: **36**

   Pressure angle: **14.5**

   Face width: **0.5**

   Hub style: **One Side**

   Hub Diameter: **1.00**

   Overall length: **1.00**

> **TIP**
>
> The overall length is the face width plus the hub height. In this example the face width is .5, and the overall length is 1.00, so the hub height is .5.

7. Click the green **OK** check mark.

To Add a Threaded Hole to the Gear's Hub

The screen should show the shaft and gear. This is an assembly drawing.

1. Right-click the spur gear callout and click the **Open Part** option.

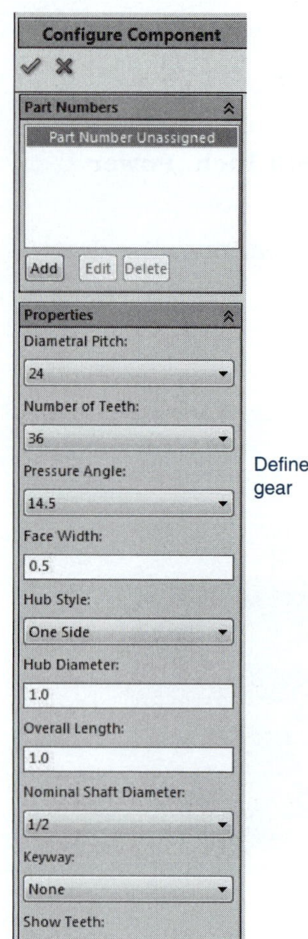

Figure 10-25

A warning dialog box will appear; click **OK**.

2 Rotate the gear so that the hub is clearly visible.

See Figure 10-26.

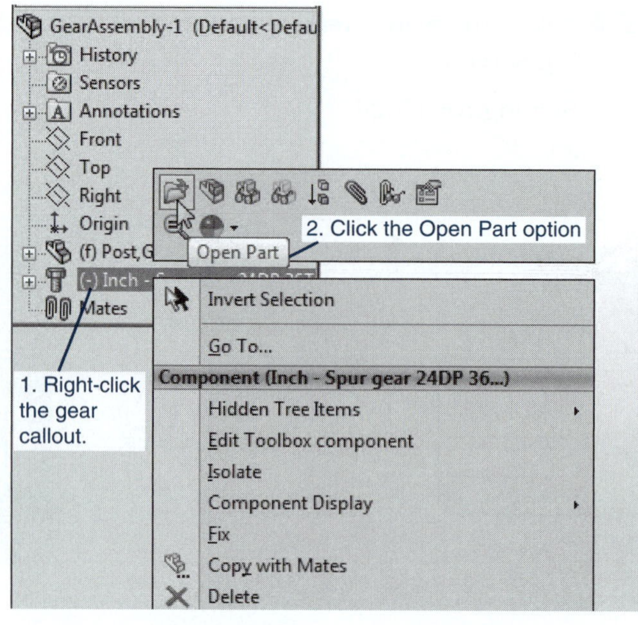

Figure 10-26

3 Click the **Hole Wizard** tool.

4 Define the **Type** of hole.

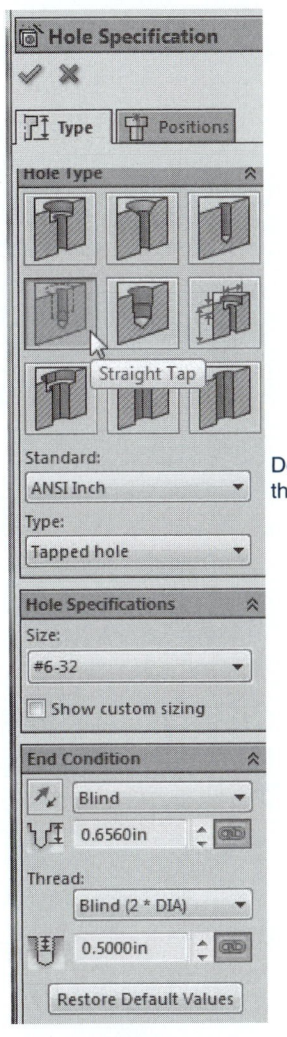

Figure 10-27

See Figure 10-27. In this example an ANSI Inch #6-32 thread was selected for the hole. The depth of the hole must exceed the wall thickness of the hub, which is 0.25. In this example a depth of 0.50 was selected.

5 Click the **Positions** tab and locate the threaded hole on the outside surface of the hub.

See Figure 10-28.

Figure 10-28

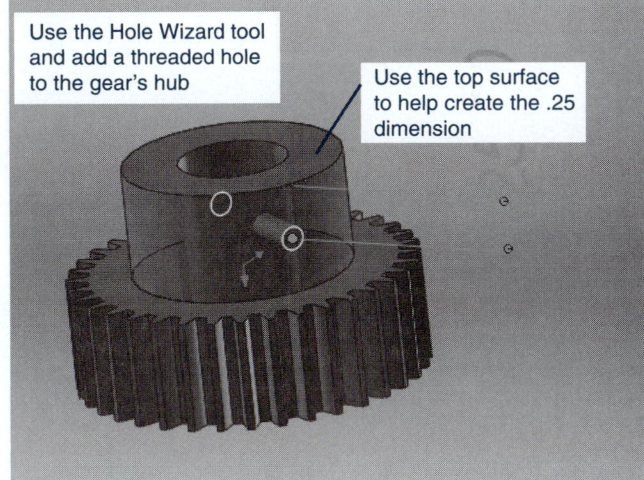

6 Click the **Smart Dimension** tool and create a **0.25** dimension between the center point of the hole and the top surface of the gear teeth.

Define the dimension between the hole's center point and the top surface of the gear's hub.

7 Click the **OK** check mark.

8 Save the reversal drawing as **GEAR, SET SCREW**.

9 Click the **Close** tool.

10 Access the **Design Library**, click **Toolbox, ANSI Inch, Bolts and Screws,** and **Set Screws (Slotted)**.

11 Select a **Slotted Set Screw Oval Point** and drag it onto the drawing screen.

12 Define the **Properties** of the set screw as **#6-32, 0.263** long; click the green **OK** check mark.

See Figure 10-29.

13 Use the **Mate** tool and assemble the shaft and set screw into the gear as shown.

See Figure 10-30.

Figure 10-29

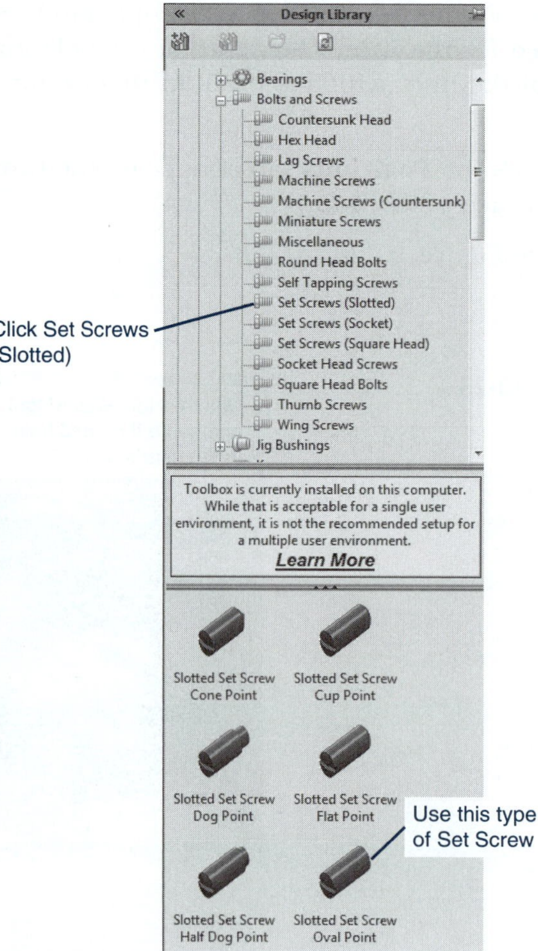
Design Library
Bearings
Bolts and Screws
Countersunk Head
Hex Head
Lag Screws
Machine Screws
Machine Screws (Countersunk)
Miniature Screws
Miscellaneous
Round Head Bolts
Self Tapping Screws
Set Screws (Slotted)
Set Screws (Socket)
Set Screws (Square Head)
Socket Head Screws
Square Head Bolts
Thumb Screws
Wing Screws
Jig Bushings
Click Set Screws (Slotted)
Toolbox is currently installed on this computer. While that is acceptable for a single user environment, it is not the recommended setup for a multiple user environment.
Learn More
Slotted Set Screw Cone Point
Slotted Set Screw Cup Point
Slotted Set Screw Dog Point
Slotted Set Screw Flat Point
Use this type of Set Screw
Slotted Set Screw Half Dog Point
Slotted Set Screw Oval Point

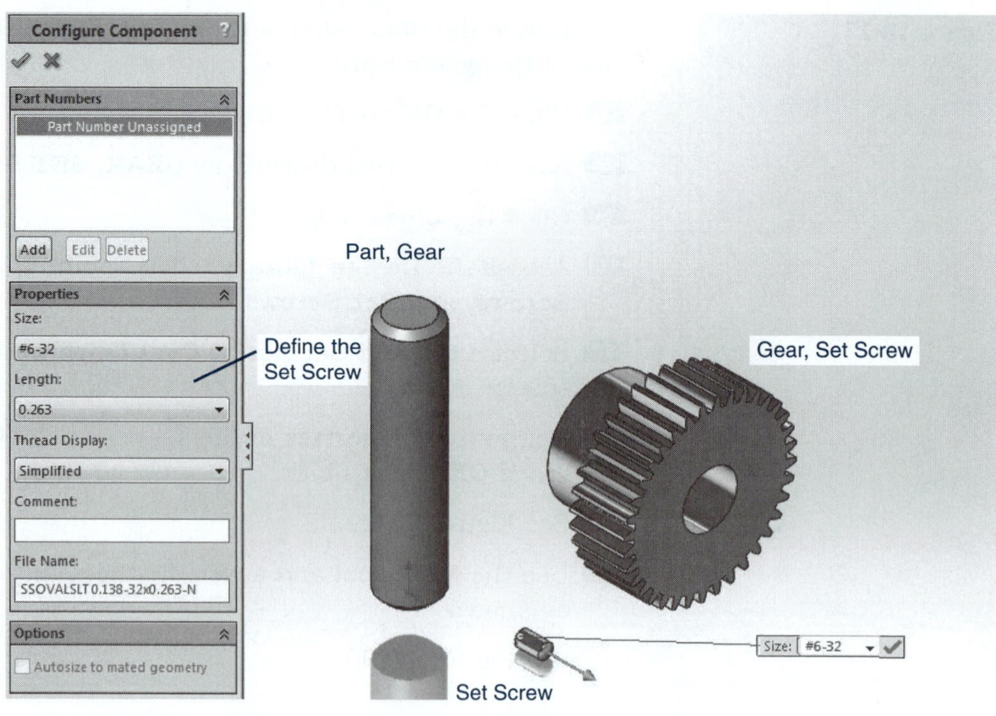
Configure Component
Part Numbers
Part Number Unassigned
Add
Edit
Delete
Properties
Size:
#6-32
Length:
0.263
Thread Display:
Simplified
Comment:
File Name:
SSOVALSLT 0.138-32x0.263-N
Options
Autosize to mated geometry
Define the Set Screw
Part, Gear
Gear, Set Screw
Size: #6-32
Set Screw

Figure 10-30

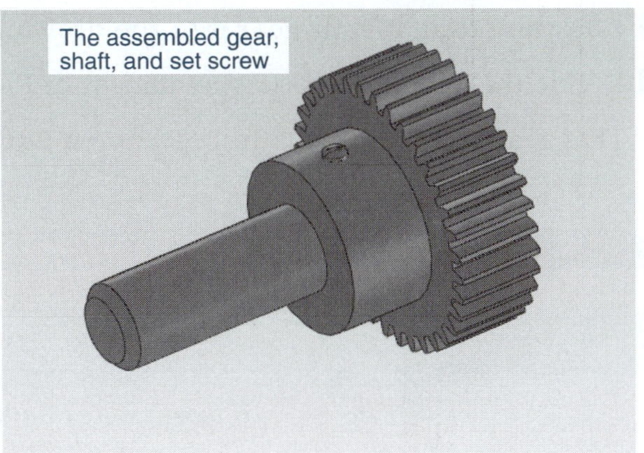

10-9 Keys, Keyseats, and Gears

Keys are used to transfer power from a drive shaft to an entity such as a gear or pulley. A ***keyseat*** is cut into both the shaft and the gear, and the key is inserted between them. See Figure 10-31. The SolidWorks **Design Library** contains two types of keys: parallel and Woodruff.

In this section we will insert a parallel key between a Ø0.50 × 3.00 shaft and a gear. Both the shaft and gear will have keyseats.

To Define and Create Keyseats in Gears

1. Draw a **Ø0.50 × 3.00** shaft and save the shaft as **Ø0.50 × 3.00 SHAFT**.
2. Start a new **Assembly** drawing and insert the shaft into the drawing.
3. Access the **Design Library, Toolbox, ANSI Inch, Power Transmission,** and **Gears** options.

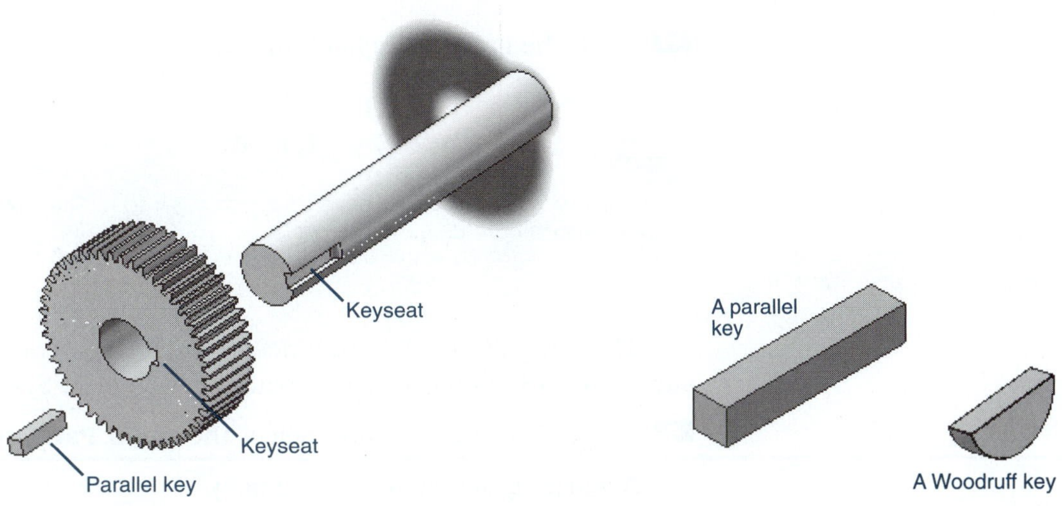

Figure 10-31

4. Click and drag a spur gear into the drawing area.
5. Define the gear's properties as shown in Figure 10-32.

   This gear will not have a hub. Define a **Square(1)** keyway.

Figure 10-32

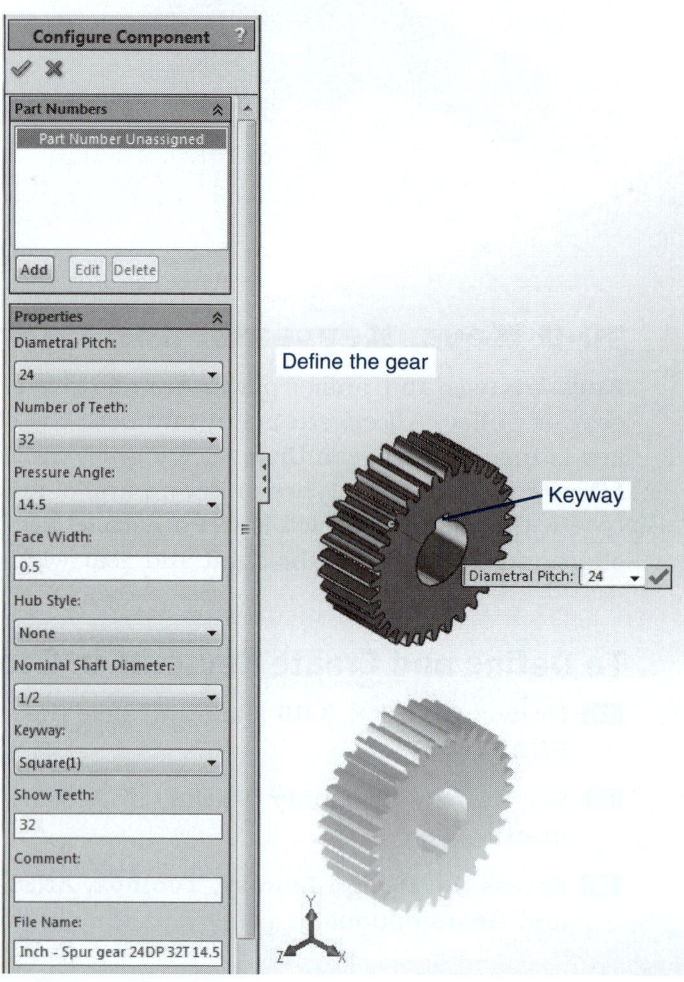

6. Click the green **OK** check mark.

> **NOTE**
> The gear will automatically have a keyseat cut into it. The size of the keyseat is based on the gear's bore diameter.

The key will also be sized according to the gear's bore diameter, but say we wish to determine the exact keyway size. See Figure 10-33.

1. Right-click the gear and select the **Open Part** option.

   A warning dialog box will appear.
2. Click **OK.**
3. Click the **Make Drawing from Part/Assembly** tool.

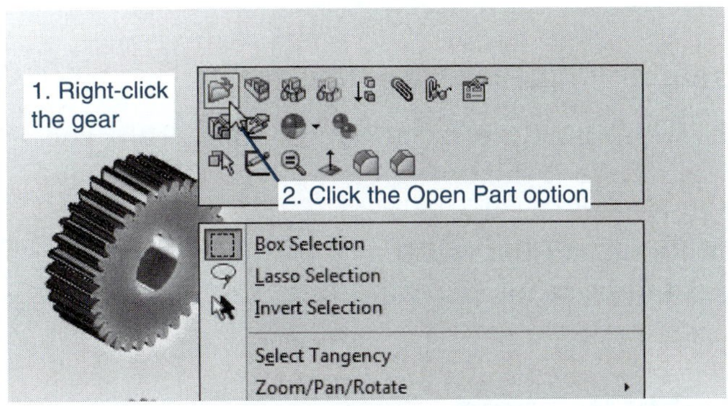

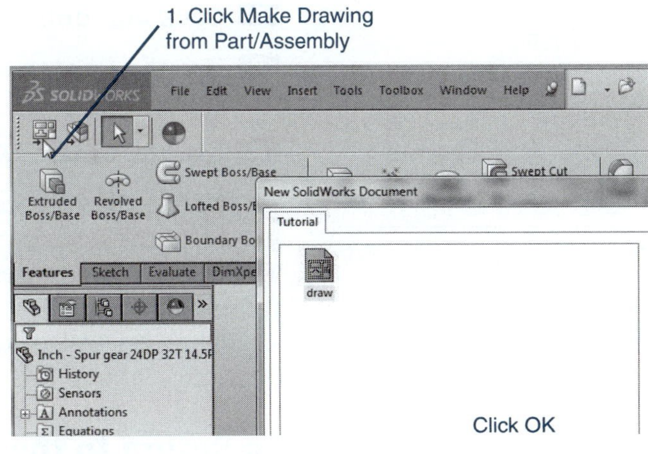

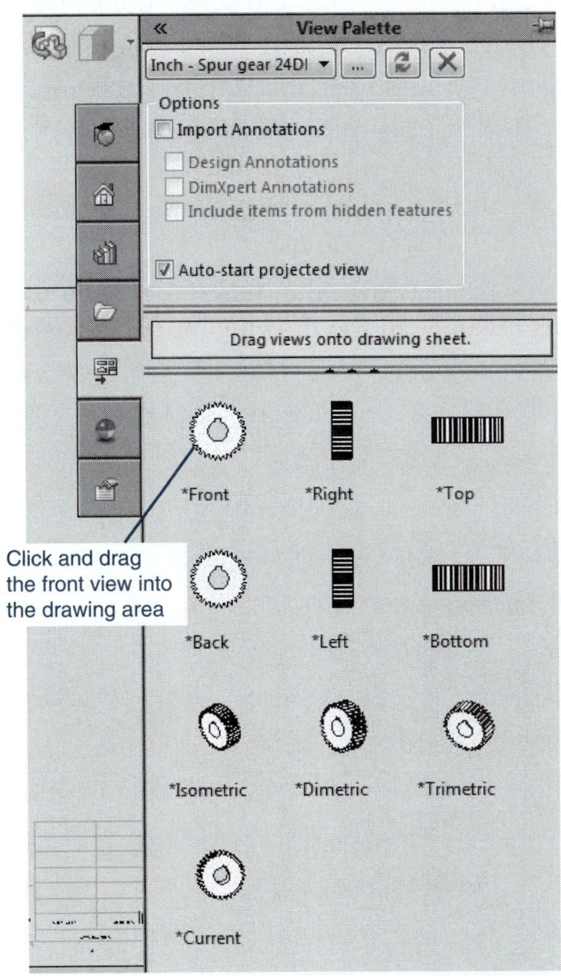

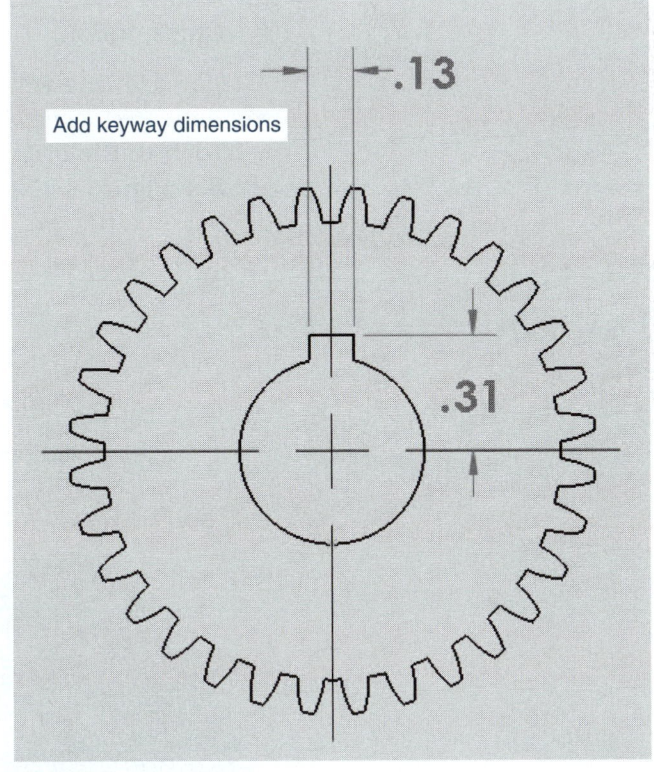

Figure 10-33

The **New SolidWorks Document** dialog box will appear.

4 Click **OK.**

The **Sheet Format/Size** dialog box will appear.

5 Select the **B(ANSI) Landscape** sheet size; click **OK.**

The system will switch to the **Drawing** format.

6. Click and drag a front view of the gear into the drawing area.

7. Use the **Center Mark** tool to add a centerline to the gear's bore. Click **<Esc>**.

8. Use the **Smart Dimension** tool and dimension the keyseat.

Notice that the keyseat's width is .13, or a little more than .125. The height of the keyseat is measured from the bore's centerpoint. The height is defined as .31. Therefore, the height of the keyseat is .31 − .25 (the radius of the bore) = .06, or about half the width.

Changes to the keyway could be made here and then saved as a **New** drawing.

To Return to the Assembly Drawing

1. Click the **File** heading at the top of the screen, and select the **Close** option.

Do not save the gear drawing. If changes were made to the drawing, they would have to be saved as a new drawing and the drawing closed to return to the original assembly.

2. Again, click the **File** heading at the top of the screen, and select the **Close** option.

The drawing will return to the assembly drawing. When the **Open** tool was activated, a new drawing was created, so the drawing must be closed to return to the original drawing.

See Figure 10-34.

Figure 10-34

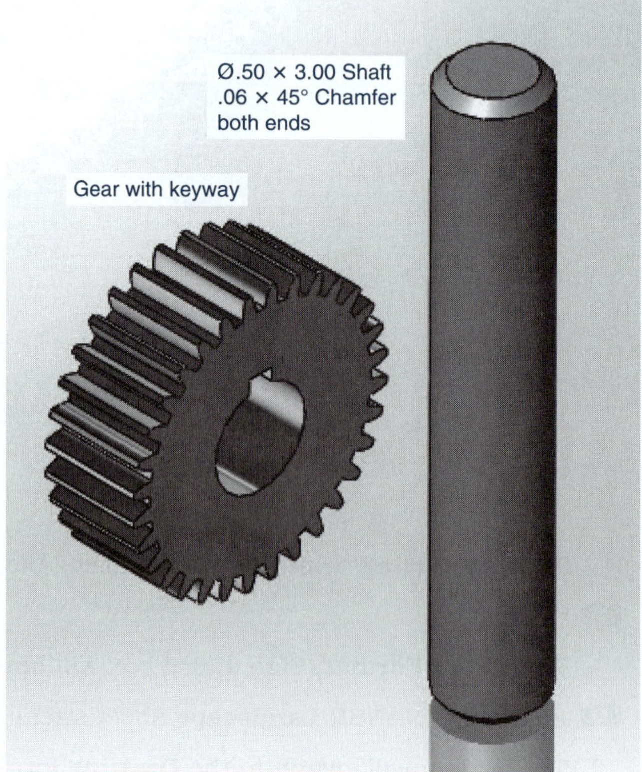

To Define and Create a Parallel Key

1 Access the **Design Library, Toolbox, ANSI Inch, Keys,** and **Parallel Keys.**

See Figure 10-35.

2 Click and drag the **Key (B17.1)** icon onto the drawing screen.

3 Enter the shaft diameter value and the key's length.

See Figure 10-36.

In this example the shaft diameter is .50, or 1/2 inch (8/16). This value is between 7/16 and 9/16. The key's length is .50.

Note that key dimensions are also given as .125 × .125 and a keyseat depth of .0625. Define the key length as .50. The key size automatically matches the keyway in the gear. The width of the gear's keyseat was .13, or .005 larger than the key. A rule of thumb is to make the height of the keyseat in both the shaft and the gear equal to a little more than the key's height. The height of the keyseat in the gear was .06, so the depth of the keyway in the shaft will be .07, for a total keyway height of .06 + .07 = .13. The calculation does not take into account the tolerances between the shaft and the gear or the tolerances between the key and the keyseat. For exact tolerance values refer to *Machinery's Handbook* (available from Industrial Press, Inc., at http://new.industrialpress.com/machinery-s-handbook-29th-edition-downloadable-ebook-in-pdf-please-see-important-ordering-information.html) or some equivalent source.

Figure 10-35

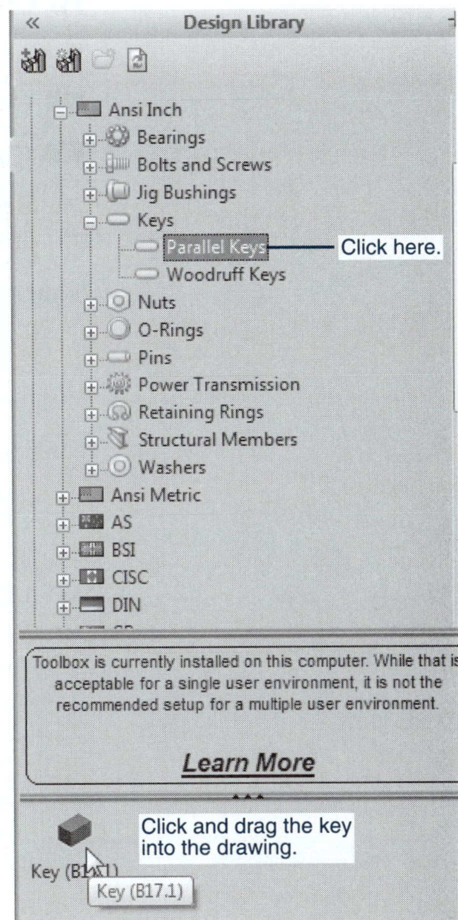

Figure 10-36

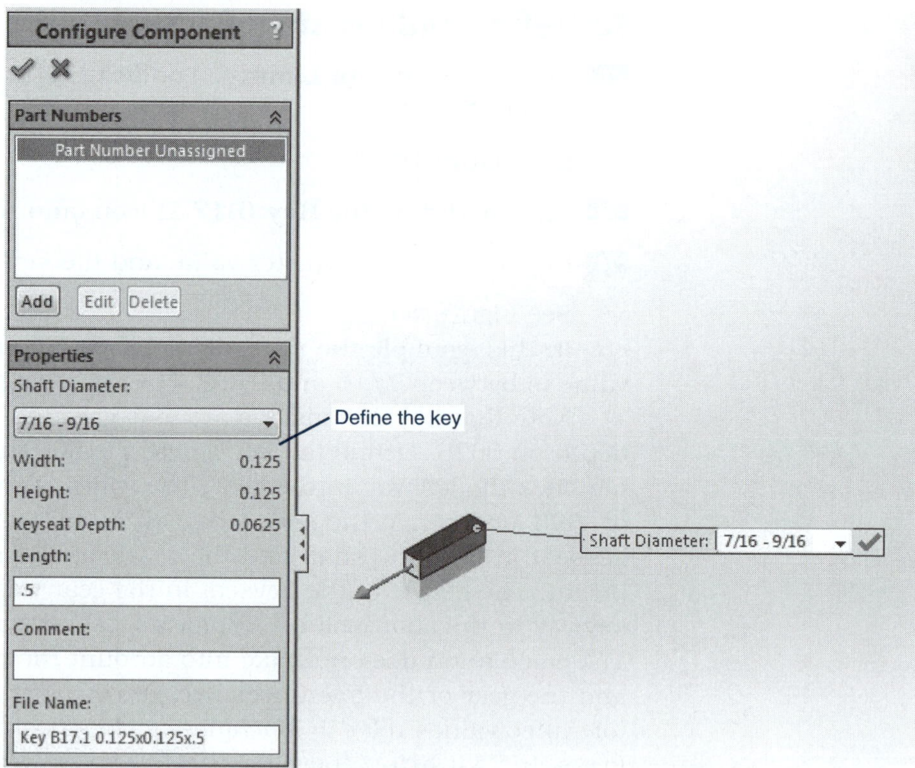

4 Click the green **OK** check mark.

To Create a Keyseat in the Shaft

The keyway for the shaft will be .13 × .07.

1 Click the top surface of the Ø.50 × 3.00 shaft.

2 Click the **Normal To** tool.

See Figure 10-37. The top surface of the shaft will become normal to the drawing screen.

Figure 10-37

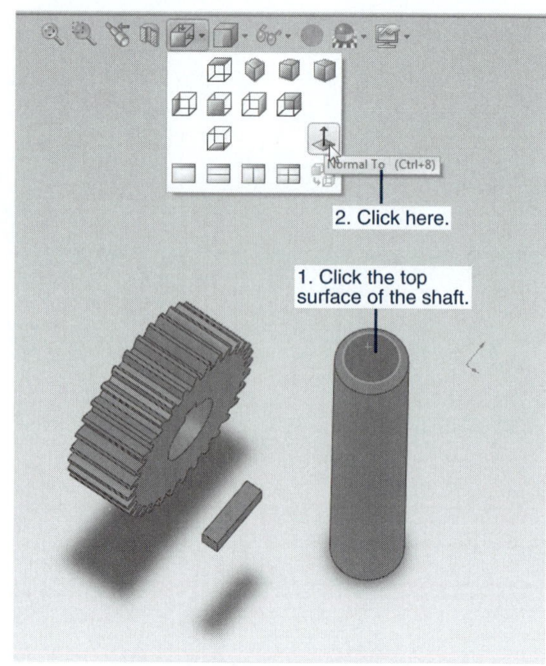

3 Right-click the normal surface again and select the **Sketch** option.

4 Click the **Sketch** tab, then click the **Line** tool and draw a vertical line from the centerpoint of the shaft as shown.

This vertical line is for construction purposes. It will be used to center the keyway.

See Figure 10-38.

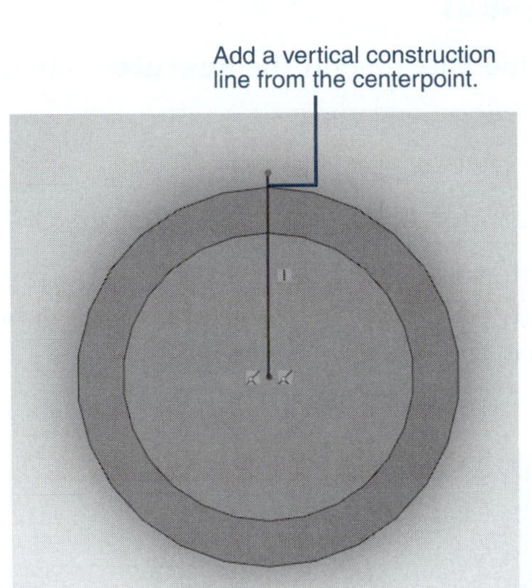
Add a vertical construction
line from the centerpoint.

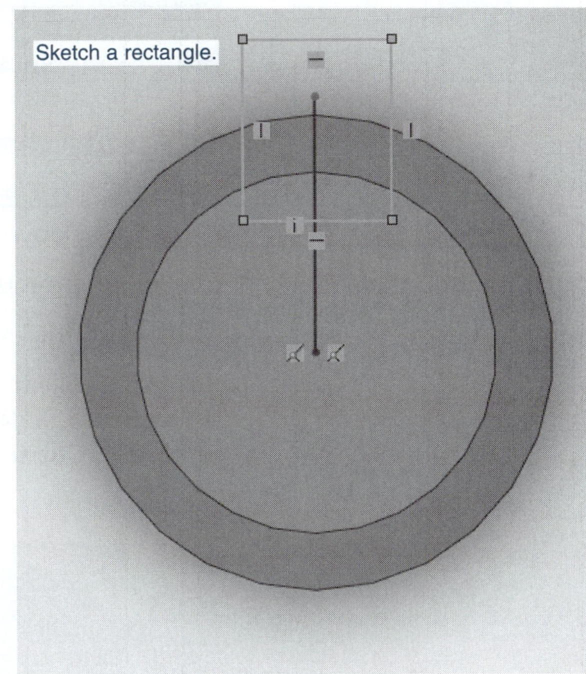
Sketch a rectangle.

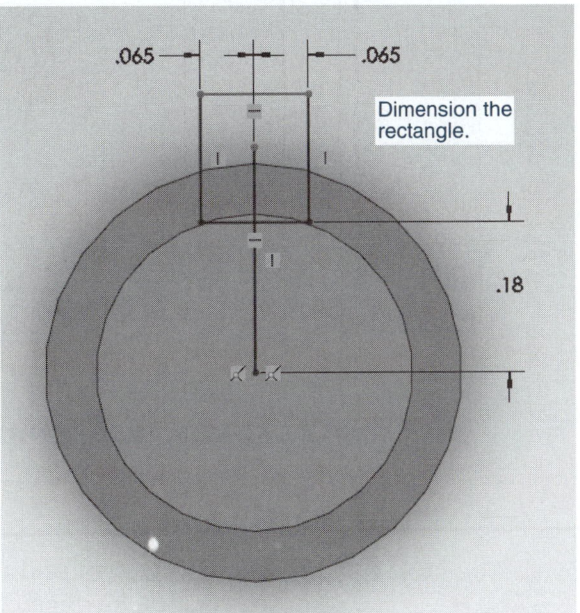
.065
.065
Dimension the
rectangle.
.18

Figure 10-38

5 Click the **Sketch** tab, then the **Rectangle** tool, and draw a rectangle as shown.

Draw the rectangle so that the top horizontal line is above the edge of the shaft.

6 Use the **Smart Dimension** tool to locate the left vertical line of the rectangle .065 from the vertical construction line.

> **TIP**
>
> If the vertical centerline moves rather than the left vertical line, left-click the vertical construction line and select the **Fix** option.

7 Use the **Smart Dimension** tool and dimension the rectangle as shown.

To Create the Keyseat

1 Access the **Extruded Cut** tool on the **Features** tab and click the dimensioned rectangle.

See Figure 10-39.

2 Define the cut length for **0.50 in**; click the green **OK** check mark.

The keyseat will end with an arc-shaped cut. The radius of the arc is equal to the depth of the keyseat. The arc shape is generated by the **Extruded Cut** tool used to create the keyseat.

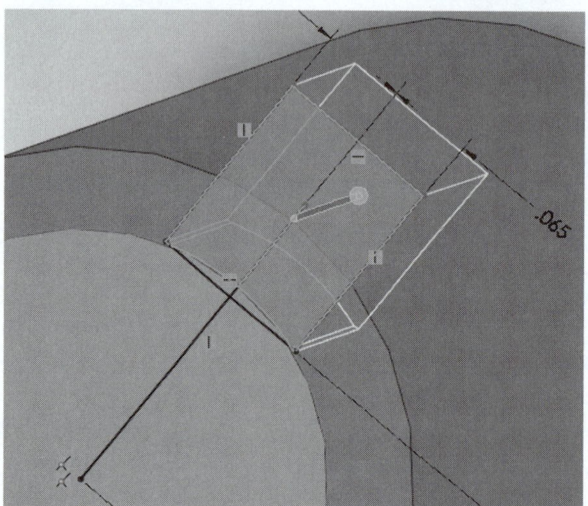

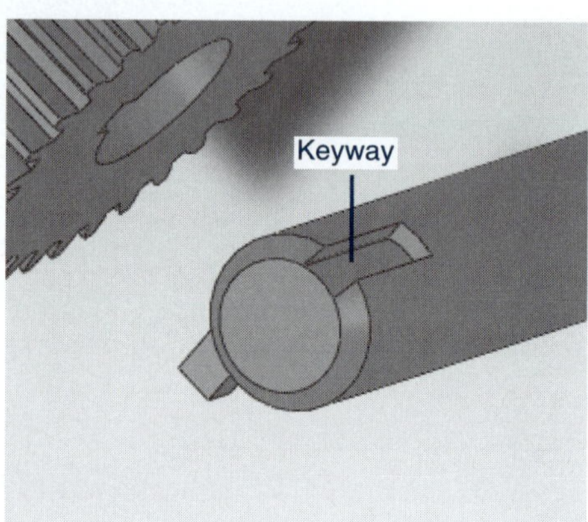

Figure 10-39

To Create the Arc-Shaped End of a Keyseat

1 Right-click the inside vertical surface of the keyseat and click the **Sketch** tab.

2 Use the **Circle** tool and draw a circle centered about the end of the keyseat on the surface of the shaft, using the lower corner of the key-way to define the radius value.

See Figure 10-40.

3 Use the **Extruded Cut** tool to cut the arc-shaped end surface.

The length of the cut equals the width of the keyway, or .13.

4 Assemble the parts as shown. See Figures 10-41 and 10-42.

Figure 10-40

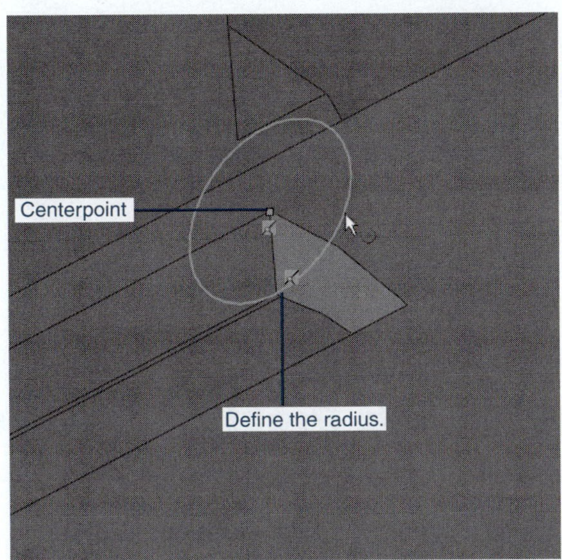

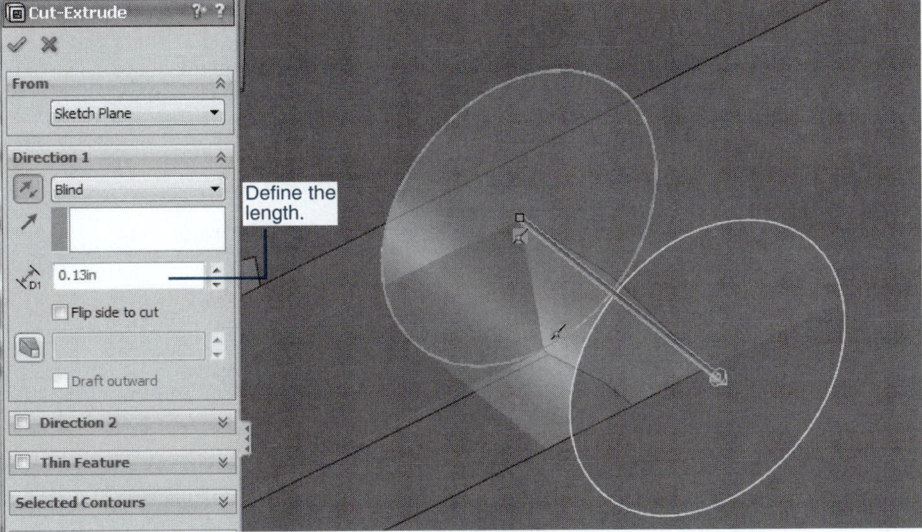

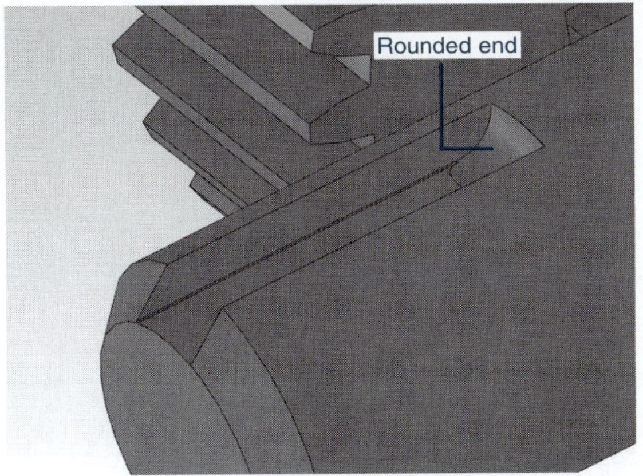

Figure 10-41

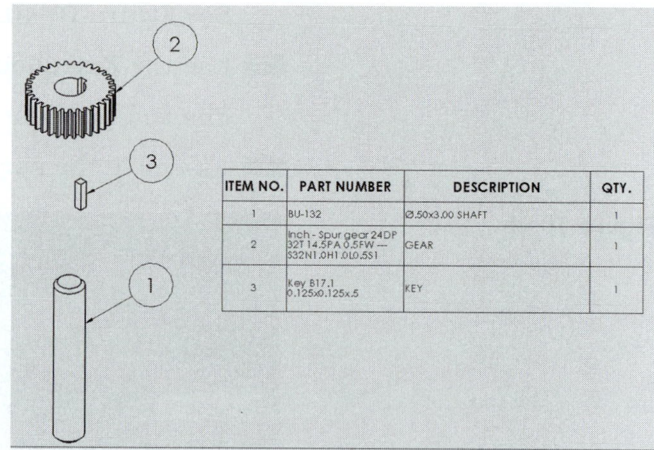

ITEM NO.	PART NUMBER	DESCRIPTION	QTY.
1	BU-132	Ø.50x3.00 SHAFT	1
2	Inch - Spur gear 24DP 32T 14.5PA 0.5FW --- S32N1.0H1.0L0.5S1	GEAR	1
3	Key 817.1 0.125x0.125x.5	KEY	1

Figure 10-42

Figure 10-43

10-10 Sample Problem 10-1—Support Plates

This exercise explains how to determine the size of plates used to support spur gears and their shafts.

Say we wish to design a support plate that will support four spur gears. The gear specifications are as follows. See Figure 10-43.

Gear 1

- Diametral Pitch = **24**
- Number of Teeth = **20**
- Pressure Angle = **14.5°**
- Face Width = **0.375**
- Hub Style = **One Side**
- Hub Diameter = **0.75**
- Overall Length = **.875**
- Nominal Shaft Diameter = **1/2**
- Keyway = **None**

Gear 2

- Diametral pitch = **24**
- Number of teeth = **60**
- Pressure angle = **14.5°**
- Face width = **0.375**
- Hub style = **One Side**
- Hub diameter = **0.75**
- Overall length = **.875**
- Nominal shaft diameter = **1/2**
- Keyway = **None**

To Determine the Pitch Diameter

The pitch diameter of the gears is determined by

$$D = \frac{N}{P}$$

where

D = pitch diameter

N = number of teeth

P = diametral pitch

Therefore, for Gear 1

$$D = \frac{20}{24} = .83$$

For Gear 2

$$D = \frac{60}{24} = 2.50$$

The center distance (CD) between the gears is calculated as follows.

$$\text{CD} = \frac{.83 + 2.50}{2} = 1.67$$

The radii of the gears are .46 and 1.25, respectively.

Figure 10-44 shows a support plate for the gears. The dimensions for the support plate were derived from the gear pitch diameters and an allowance of about .50 between the gear pitch diameters and the edge of the support plate. For example, the pitch diameter of the larger gear is 2.50. Allowing .50 between the top edge and the bottom edge gives .50 + 2.50 + .50 = 3.50.

The pitch diameter for the smaller gear is .83. Adding the .50 edge distance gives a distance of 1.33 from the smaller gear's center point. In this example the distance was rounded up to 1.50. The center distance between the gears is 1.67. Therefore, the total length of the support plate is 1.50 + 1.67 + 1.67 + 1.75 = 6.59, or about 6.75. The height will be 3.50.

Each of the gears has a 1/2 nominal shaft diameter. For this example a value of .50 will be assigned to both the gear bores and the holes in the support plate. Bearings are not included in this example.

Figure 10-45 shows a Ø.50 × 2.00 shaft that will be used to support the gears.

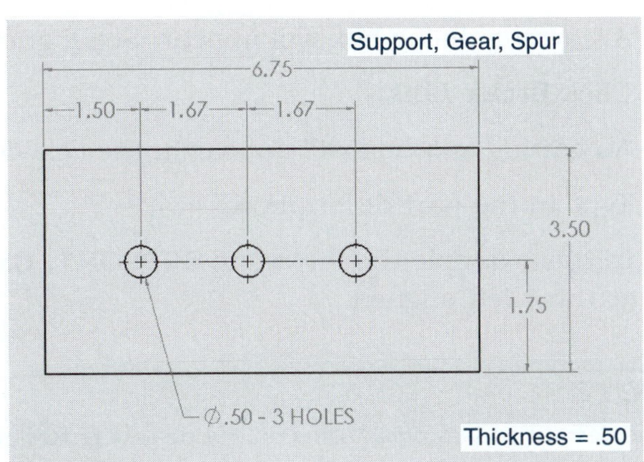

Figure 10-44

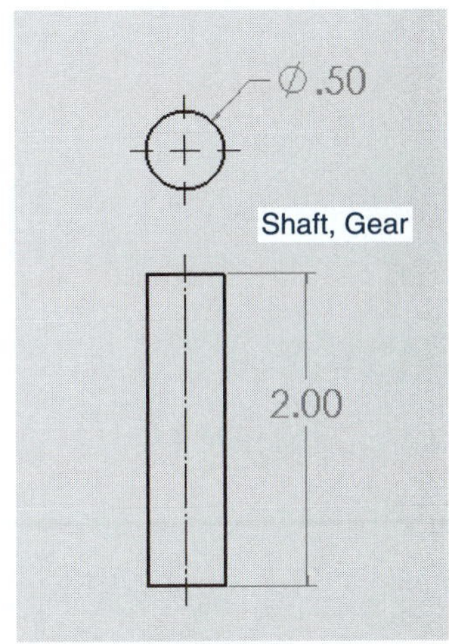

Figure 10-45

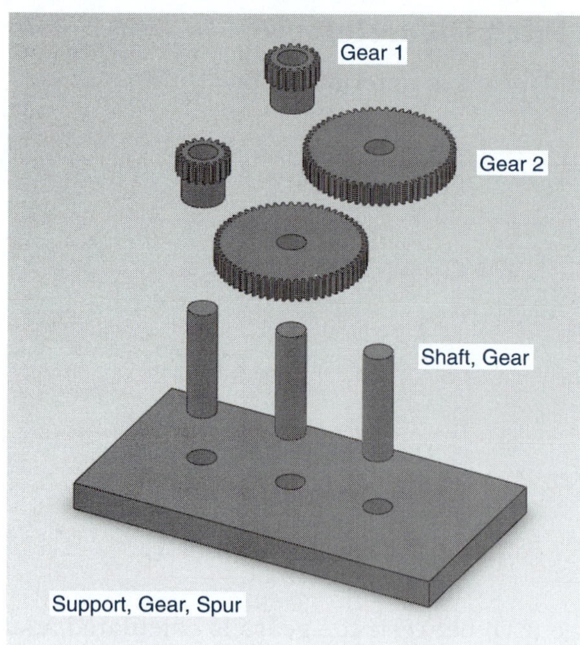

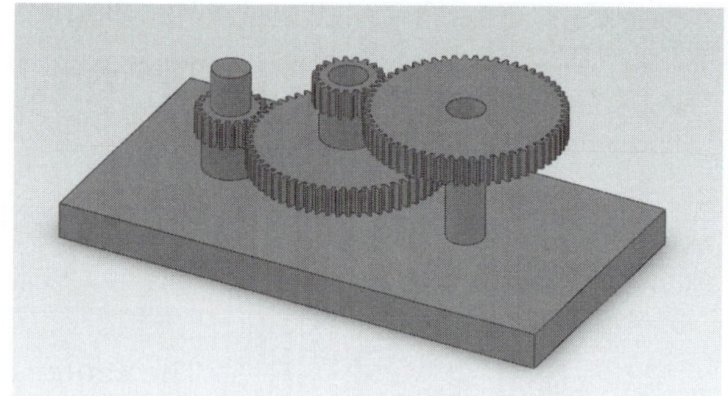

Figure 10-46

Create an **Assembly** drawing using the support plate, three Ø.50 × 2.00 shafts, and the four gears. The gears were created using the given information. Figure 10-46 shows the components in their assembled position. Save the assembly as **ASSEMBLY, THREE GEAR.** The gears are offset .60 from the support and from each other on the gear shafts.

Create an exploded drawing using the three-gear assembly in the **Isometric** orientation with no hidden lines. See Figure 10-47. Access the **Annotation** menu, add balloons, click the **Tables** tool, and select the **Bill of Materials** tool. Locate the bill of materials (BOM) as shown.

The Part Number column does not show part numbers but lists the file names assigned to each part and, in the case of the gears, a listing of gear parameters.

To Edit the Bill of Materials

See Figure 10-47. SolidWorks will automatically insert the part's file name as its part number.

1 Double-click the first cell under the heading **DESCRIPTION**.

A warning dialog box will appear. See Figure 10-48.

2 Click **Break Link**.

An editing text box will appear in the cell. See Figure 10-49.

3 Type in the part description.

In this example the file name **SUPPORT, GEAR, SPUR** was used, and the text was left aligned.

NOTE

The description was typed using only uppercase letters. Uppercase letters are the preferred convention.

4 Complete the editing of the **DESCRIPTION** column.

5 Click the **Part Number** column and add part numbers as shown.

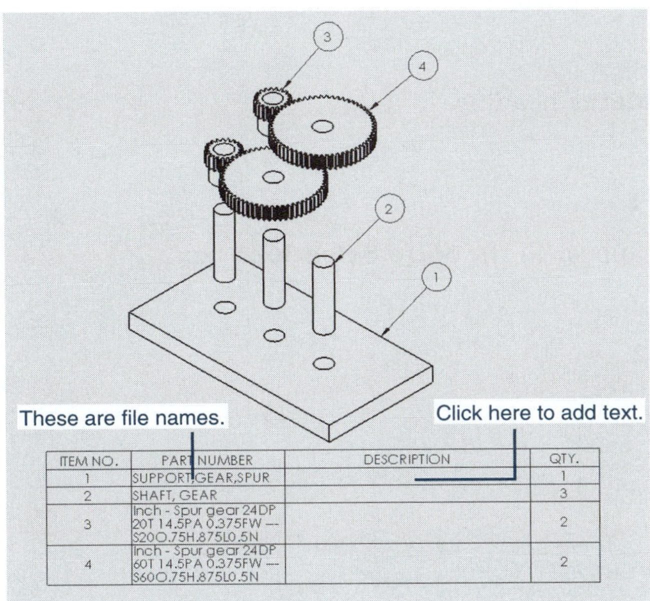

These are file names.
Click here to add text.
ITEM NO. PART NUMBER DESCRIPTION QTY.
1 SUPPORT, GEAR, SPUR 1
2 SHAFT, GEAR 3
3 Inch - Spur gear 24DP 20T 14.5PA 0.375FW --- S20O.75H.875L0.5N 2
4 Inch - Spur gear 24DP 60T 14.5PA 0.375FW --- S60O.75H.875L0.5N 2

Figure 10-47

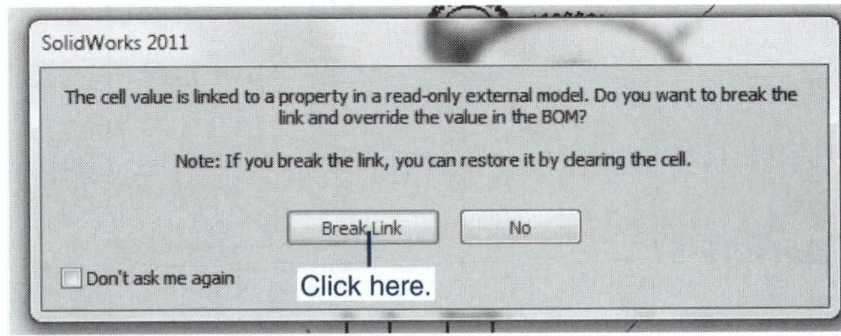

Figure 10-48

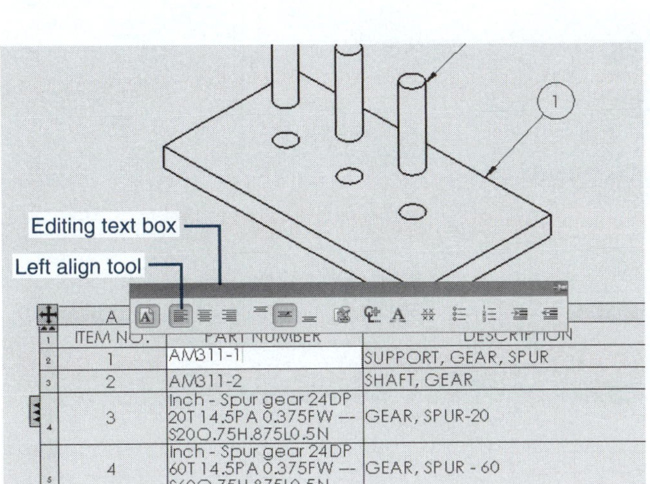

Editing text box
Left align tool
ITEM NO. PART NUMBER DESCRIPTION
1 AM311-1 SUPPORT, GEAR, SPUR
2 AM311-2 SHAFT, GEAR
3 Inch - Spur gear 24DP 20T 14.5PA 0.375FW --- S20O.75H.875L0.5N GEAR, SPUR-20
4 Inch - Spur gear 24DP 60T 14.5PA 0.375FW --- S60O.75H.875L0.5N GEAR, SPUR - 60

Figure 10-49

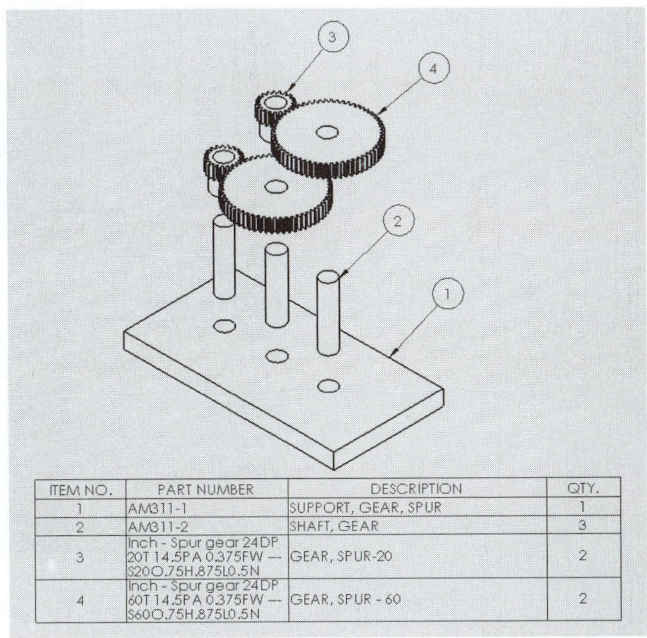

ITEM NO. PART NUMBER DESCRIPTION QTY.
1 AM311-1 SUPPORT, GEAR, SPUR 1
2 AM311-2 SHAFT, GEAR 3
3 Inch - Spur gear 24DP 20T 14.5PA 0.375FW --- S20O.75H.875L0.5N GEAR, SPUR-20 2
4 Inch - Spur gear 24DP 60T 14.5PA 0.375FW --- S60O.75H.875L0.5N GEAR, SPUR - 60 2

Figure 10-50

NOTE

Part numbers differ from item numbers (assembly numbers). The support plate is item number 1 and has a part number of AM311-1. If the plate was to be used in another assembly, it might have a different item number, but it will always have the same AM311-1 part number.

The gears' part numbers were accepted "as is" because they are standard numbers assigned to each gear. In general, manufacturers' part numbers are used directly. See Figure 10-50.

The Three Gear Assembly can be animated using the **Mechanical Mates** tool.

1. Click the **Mate** tool.
2. Click the **Mechanical Mates** heading.
3. Click the **Gear** tool.
4. Click the edge of **Gear 1**.

   The gear name should appear in the **Mate Selections** box.
5. Click **Gear 2**.
6. Set the gear ratio to **1:2**.

   See Figure 10-51.

Figure 10-51

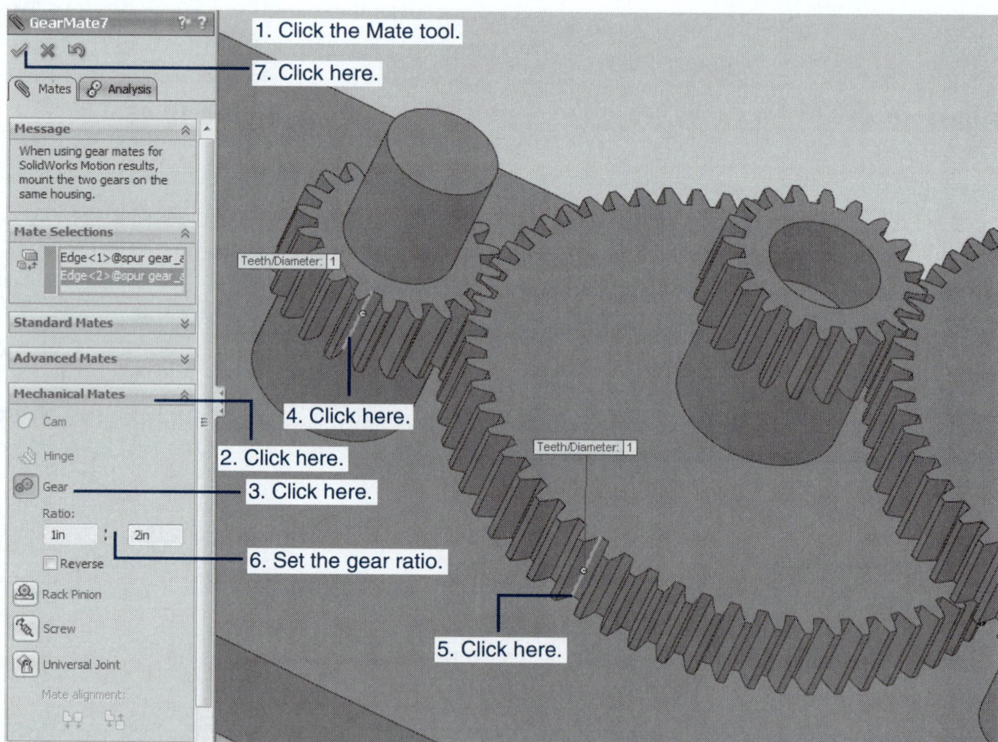

7. Create a **Gear Mate** between Gear 2 and the second Gear 1, located on the same shaft, and set the gear ratio for **1:1**.

   The gears are on the same shaft, so they will rotate at the same speed.
8. Create a **Gear Mate** between the second Gear 1 and the second Gear 2 and set the gear ratio for **1:2.**

   Figure 10-52 shows mates used to create the animated assembly.

9 Use the cursor to rotate Gear 1.

The gears should rotate.

The total gear ratio is

$$\left(\frac{3}{1}\right)\left(\frac{3}{1}\right) = \frac{9}{1}$$

Therefore, if the first Gear 1 rotates at 1750 RPM, the second Gear 2 will rotate at 194.4 RPM:

1750/9 = 194.4 RPM

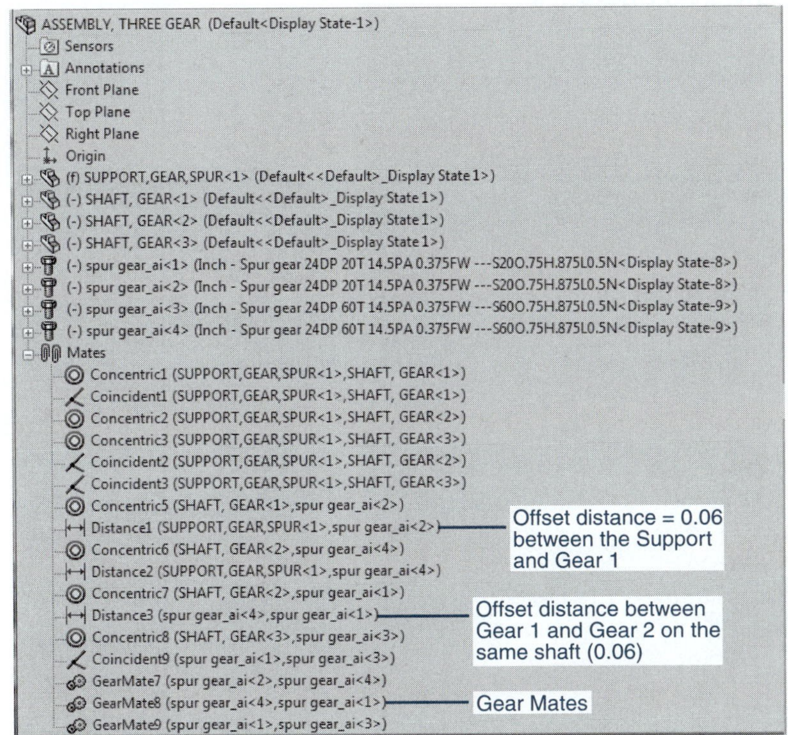

Figure 10-52

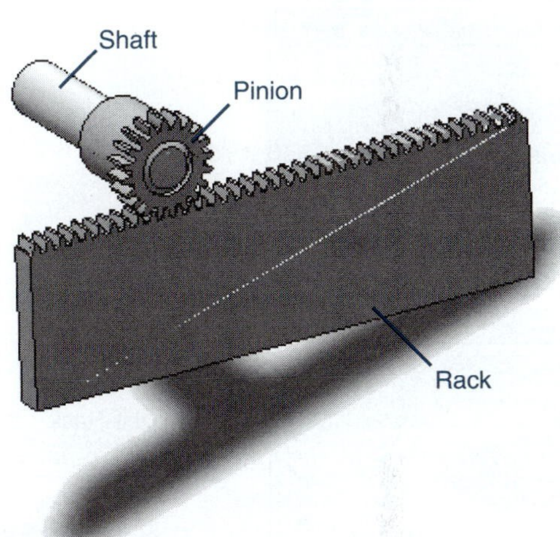

Figure 10-53

10-11 Rack and Pinion Gears

Figure 10-53 shows a rack and pinion gear setup. It was created using the **Assembly** format starting with a Ø.50 × 2.25 shaft and with a rack and pinion gear from the **Design Library.**

1 Draw a **Ø.50** × **2.25** shaft with a **0.30** chamfer at each end.

2 Start a new **Assembly** drawing.

3 Insert the Ø.50 × 2.25 shaft.

The shaft will serve as a base for the gears and a reference for animation.

4 Access the **Design Library, Toolbox, ANSI Inch, Power Transmission, Gears,** and click and drag a **Rack (Spur Rectangular)** into the drawing area.

5 Set the rack's properties as shown in Figure 10-54.

6. Click and drag a spur gear into the drawing area.

   The spur gear will become the pinion.

7. Set the pinion's properties as shown in Figure 10-55.

8. Reorient the components and use the **Mate** tool to insert the pinion onto the shaft.

   See Figure 10-56.

   Remember that the shaft was entered into the assembly first, so its position is fixed. Use the **Float** option to move the shaft.

9. Use the **Mate** tool to make the top surface of the pinion parallel with the front flat surface of the rack.

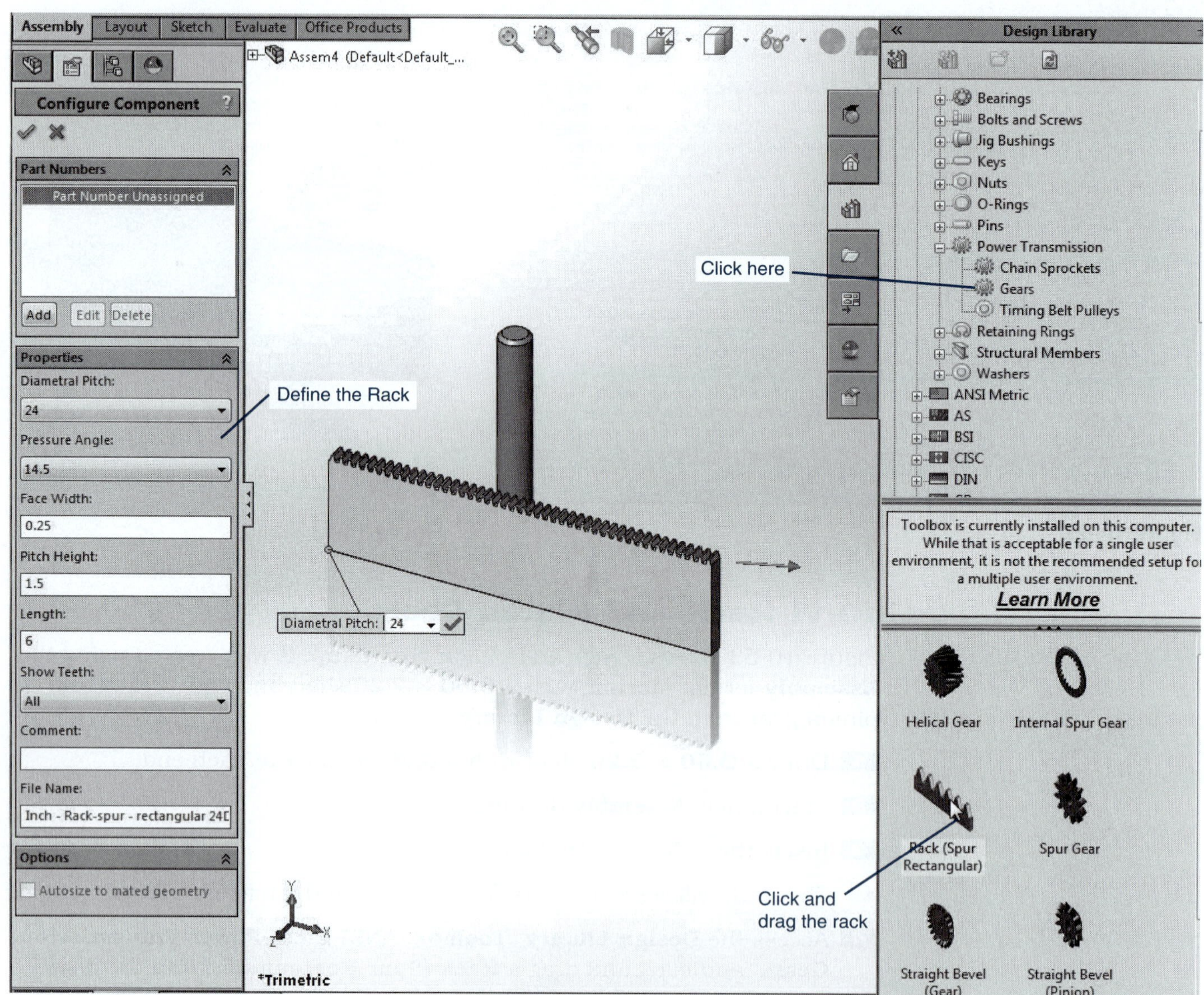

Figure 10-54

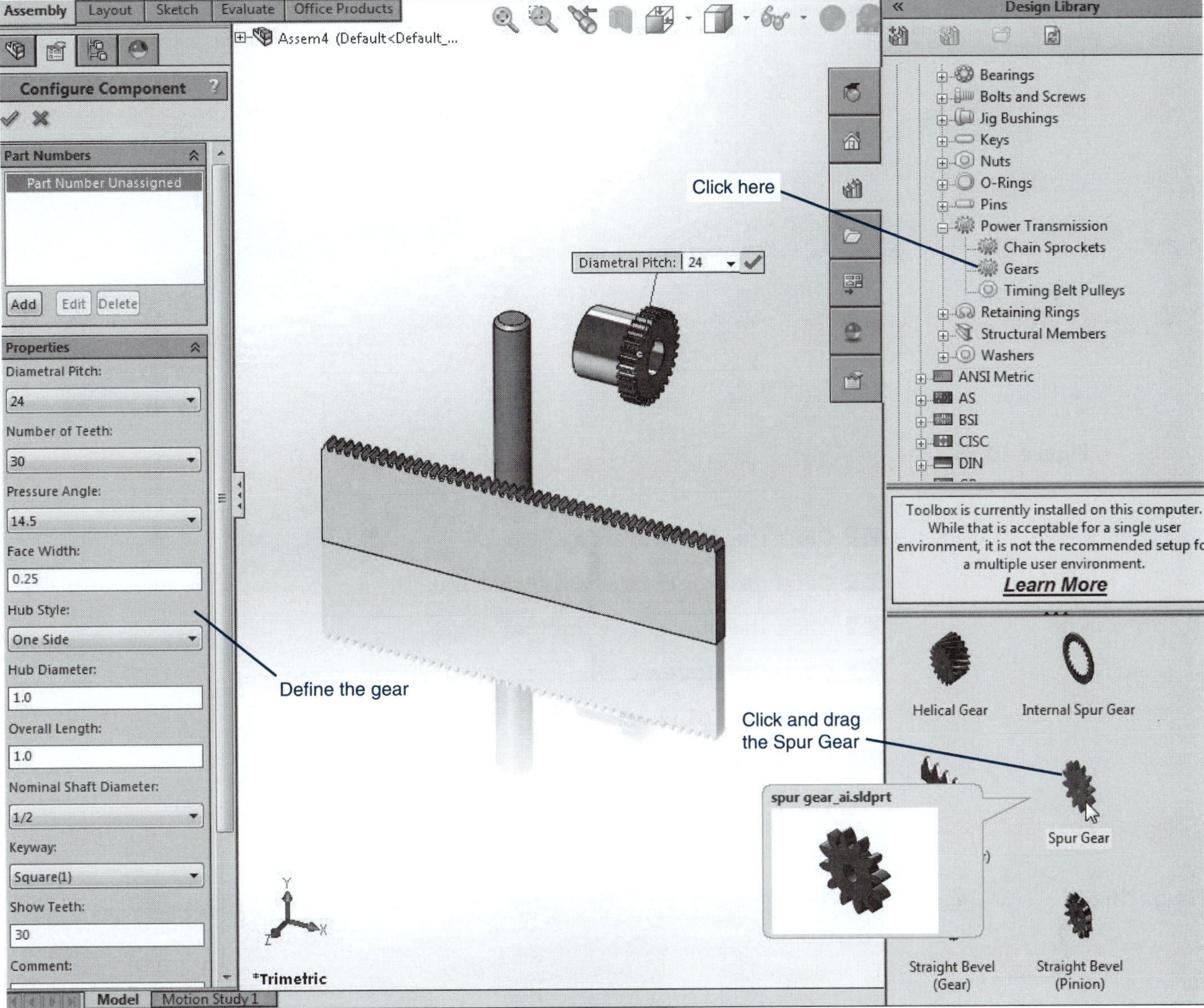

Figure 10-55

10 Use the **Mate** tool and align the top surface (not an edge) of one of the pinion's teeth with the bottom surface of one of the rack's teeth.

See Figure 10-57.

11 Adjust the pinion as needed to create a proper fit between the pinion and the rack by locating the cursor on the pinion and rotating the pinion.

To Animate the Rack and Pinion

See Figure 10-58. This animation is based on the rack and pinion setup created in the previous section.

1 Click the **Mate** tool.

2 Click **Mechanical Mates.**

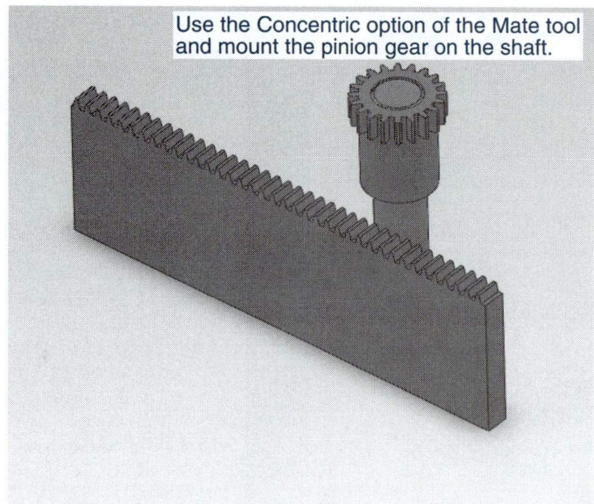

Use the Concentric option of the Mate tool
and mount the pinion gear on the shaft.

Figure 10-56

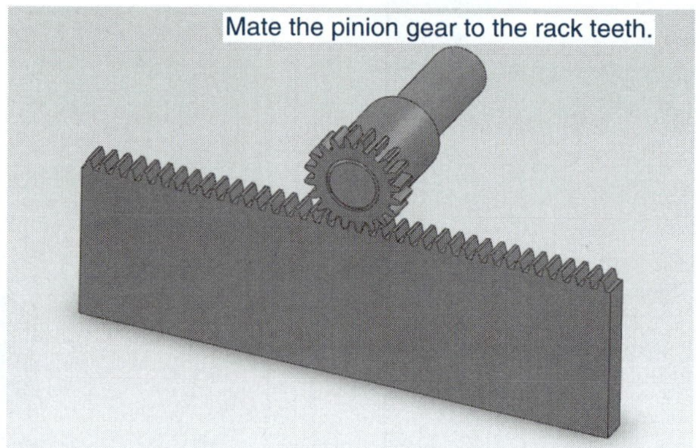

Mate the pinion gear to the rack teeth.

Figure 10-57

3 Click the **Rack Pinion** option.

4 Click the front edge of the rack's teeth.

5 Click the pinion gear (not an edge).

6 Click the **Reverse** box so that a check mark appears.

7 Click the **OK** check mark.

8 Use the cursor to rotate the pinion.

The rack will slide back and forth as the pinion is moved.

Figure 10-58

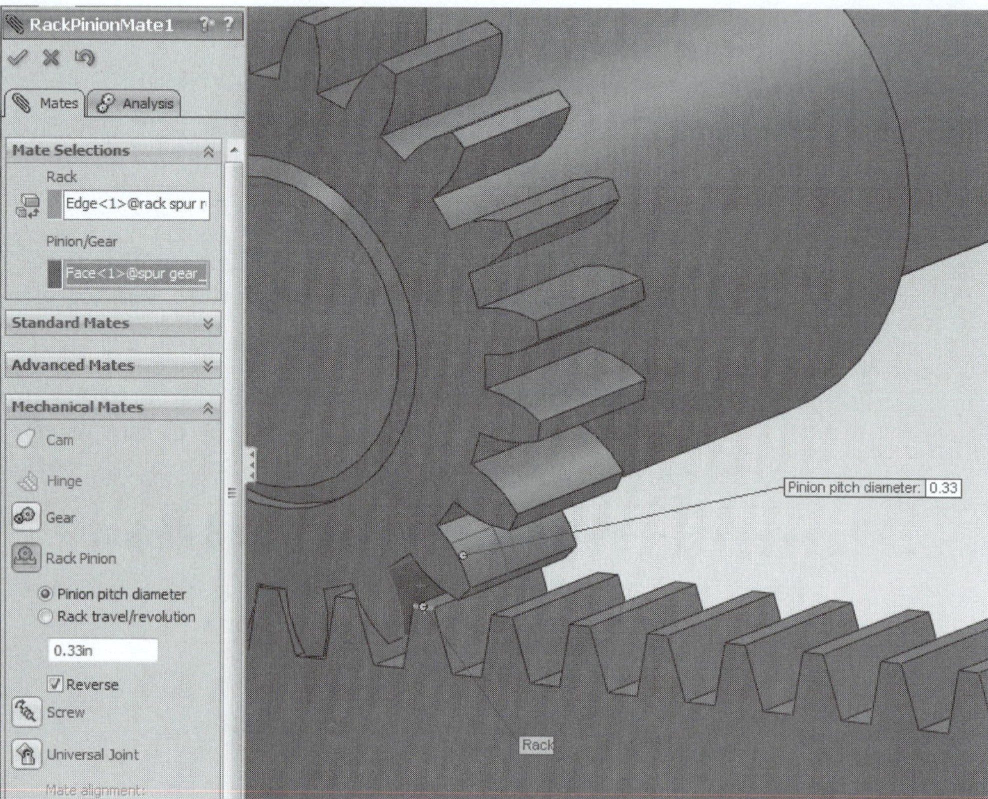

10-12 Metric Gears

Gears created using the metric system are very similar to gears created using English units with one major exception. Gears in the English system use the term *pitch* to refer to the number of teeth per inch. Gears in the metric system use the term ***module*** to refer to the pitch diameter divided by the number of teeth. As meshing gears in the English system must have the same pitch, so meshing gears in the metric system must have the same module.

To Create a Metric Gear

See Figure 10-59.

1. Start a new **Part** document and set the units for **MMGS.**
2. Draw a **Ø16 × 60** millimeter shaft. Save the shaft as **Ø16 × 60.**
3. Start a new **Assembly** drawing and insert the **Ø16 × 60** shaft.
4. Access the **Design Library.**

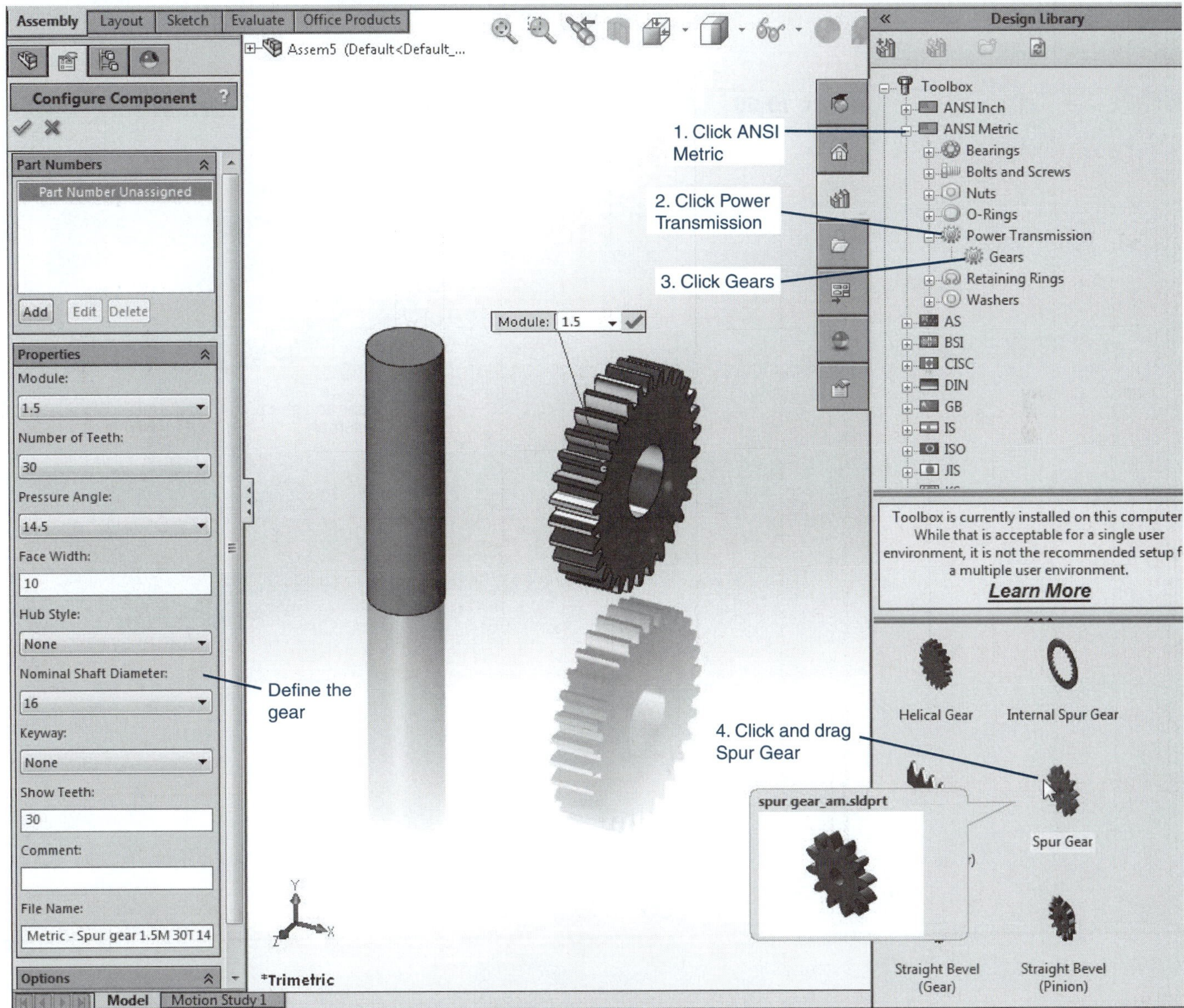

Figure 10-59

5. Click **Toolbox, ANSI Metric, Power Transmission,** and **Gears.**
6. Click and drag a spur gear into the drawing area.
7. Define the gear's properties as follows:
   a. Module = **1.5**
   b. Number of Teeth = **30**
   c. Pressure Angle = **14.5**
   d. Face Width = **10**
   e. Hub Style = **None**
   f. Nominal Shaft Diameter = **16**
   g. Keyway = **None**
8. Use the **Mate** tool and assemble the gear onto the shaft.

   Figure 10-60 shows the gear assembled onto the shaft.

Figure 10-60

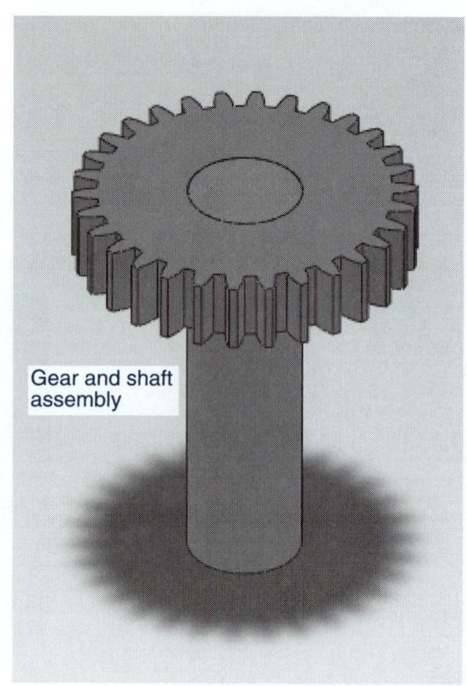

Chapter Projects

Project 10-1: Inches

Use Figure P10-1 for Projects 10-1 through 10-4.

A. Create a Ø.375 × 1.75 shaft.

B. Create a spur gear based on the following specifications:

Diametral Pitch = **32**

Number of Teeth = **36**

Pressure Angle = **14.5**

Face Width = **.250**

Hub Style = **None**

Nominal Shaft Diameter = **3/8**

Keyway = **None**

C. Assemble the gear onto the shaft with a .25 offset from the end of the shaft.

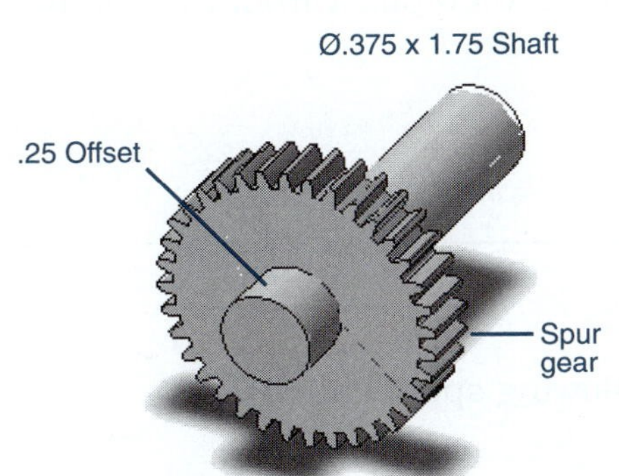

Figure P10-1

Project 10-2: Inches

A. Create a Ø.250 × 1.50 shaft.

B. Create a spur gear based on the following specifications:

Diametral Pitch = **40**

Number of Teeth = **56**

Pressure Angle = **14.5**

Face Width = **.125**

Hub Style = **None**

Nominal Shaft Diameter = **1/4**

Keyway = **None**

C. Assemble the gear onto the shaft with a .125 offset from the end of the shaft.

Project 10-3: Inches

A. Create a Ø1.00 × 4.00 shaft.

B. Create a spur gear based on the following specifications:

Diametral Pitch = **8**

Number of Teeth = **66**

Pressure Angle = **14.5**

Face Width = **.625**

Hub Style = **None**

Nominal Shaft Diameter = **1**

Keyway = **None**

C. Assemble the gear onto the shaft with a 0.00 offset from the end of the shaft.

Project 10-4: Millimeters

A. Create a Ø8.0 × 30 shaft.

B. Create a spur gear based on the following specifications:

Module = **2**

Number of Teeth = **40**

Pressure Angle = **14.5**

Face Width = **12**

Hub Style = **None**

Nominal Shaft Diameter = 8

Keyway = **None**

C. Assemble the gear onto the shaft with a 5.0 offset from the end of the shaft.

Project 10-5: Inches

Use Figure P10-5 for Projects 10-5 through 10-7.

A. Create a Ø.625 × 4.00 shaft.

B. Create a spur gear based on the following specifications:

Gear 1:

Diametral Pitch = **10**

Number of Teeth = **42**

Pressure Angle = **14.5**

Face Width = **.375**

Figure P10-5

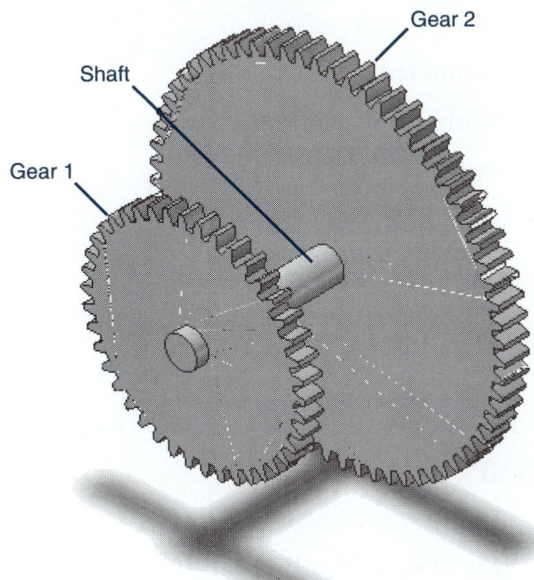

Hub Style = **None**

Nominal Shaft Diameter = **5/8**

Keyway = **None**

C. Create a second gear based on the following specifications:

Gear 2:

Diametral Pitch = **10**

Number of Teeth = **68**

Pressure Angle = **14.5**

Face Width = **.500**

Hub Style = **None**

Nominal Shaft Diameter = **5/8**

Keyway = **None**

D. Assemble the gears onto the shaft with Gear 1 offset .25 from the front end of the shaft and Gear 2 offset 3.00 from the front end of the shaft.

Project 10-6: Inches

A. Create a Ø.25 × 2.75 shaft.

B. Create a spur gear based on the following specifications:

Gear 1:

Diametral Pitch = **40**

Number of Teeth = **18**

Pressure Angle = **14.5**

Face Width = **.125**

Hub Style = **None**

Nominal Shaft Diameter = **1/4**

Keyway = **None**

C. Create a second gear based on the following specifications:

Gear 2:

Diametral Pitch = **40**

Number of Teeth = **54**

Pressure Angle = **14.5**

Face Width = **.125**

Hub Style = **None**

Nominal Shaft Diameter = **1/4**

Keyway = **None**

D. Assemble the gears onto the shaft with Gear 1 offset .125 from the front end of the shaft and Gear 2 offset 2.00 from the front end of the shaft.

Project 10-7: Millimeters

A. Create a Ø12 × 80 shaft.

B. Create a spur gear based on the following specifications:

Gear 1:

Module = **1.0**

Number of Teeth = **16**

Pressure Angle = **14.5**

Face Width = **8**

Hub Style = **None**

Nominal Shaft Diameter = **12**

Keyway = **None**

C. Create a second gear based on the following specifications:

Gear 2:

Module = **1.0** Number of Teeth = **48**

Pressure Angle = **14.5**

Face Width = **10**

Hub Style = **None**

Nominal Shaft Diameter = **12**

Keyway = **None**

D. Assemble the gears onto the shaft with Gear 1 offset 5 from the front end of the shaft and Gear 2 offset 50 from the front end of the shaft.

Project 10-8: Inches

Use Figure P10-8 for Projects 10-8 through 10-10.

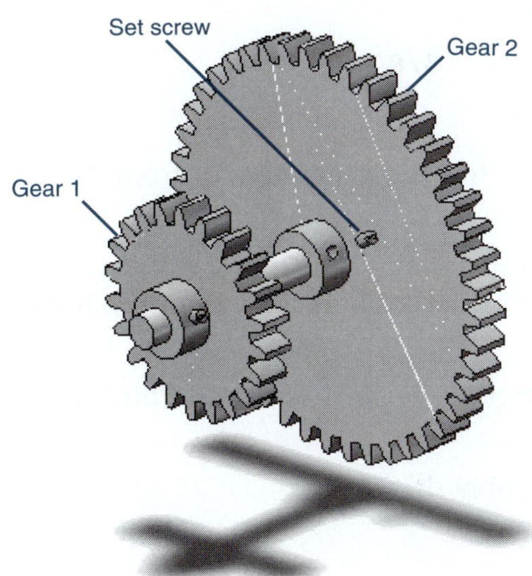

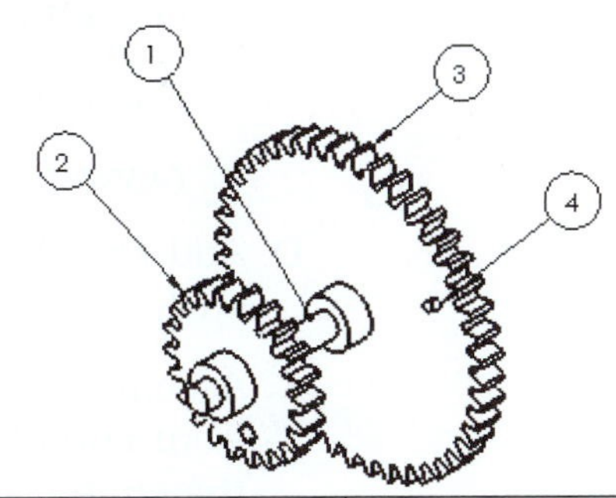

ITEM NO.	PART NUMBER	DESCRIPTION	QTY.
1	AM-311-A1	Ø.375 × 2.75 SHAFT	1
2	AM311-A2	GEAR 1 - N22	1
3	AM311-A3	GEAR 2 - N44	1
4	#6-32 UNC	SET SCREW	2

Figure P10-8

A. Create aØ.375 × 2.75 shaft. Save the shaft.

B. Create a spur gear based on the following specifications:

Gear 1:

Diametral Pitch = **10**

Number of Teeth = **22**

Pressure Angle = **14.5**

Face Width = **.375**

Hub Style = **One Side**

Hub Diameter = **.75**

Overall Length = **.75**

Nominal Shaft Diameter = **3/8**

Keyway = **None**

C. Create another spur gear based on the following specifications:

Gear 2:

Diametral Pitch = **10**

Number of Teeth = **44**

Pressure Angle = **14.5**

Face Width = **.375**

Hub Style = **One Side**

Hub Diameter = **.75**

Overall Length = **.75**

Nominal Shaft Diameter = **3/8**

Keyway = **None**

D. Add #6-32 threaded holes to each gear hub .19 from the top hub surface.

E. Assemble the gears onto the shaft with Gear 1 offset .25 from the front end of the shaft and Gear 2 offset 2.25 from the front end of the shaft.

F. Insert a #6-32 Slotted Set Screw with an Oval Point into each hole.

G. Create an exploded assembly drawing.

H. Create a bill of materials.

I. Animate the assembly.

Project 10-9: Inches

A. Create a Ø.375 × 3.25 shaft. Save the shaft.

B. Create a spur gear based on the following specifications:

Gear 1:

Diametral Pitch = **20**

Number of Teeth = **18**

Pressure Angle = **14.5**

Face Width = **.25**

Hub Style = **One Side**

Hub Diameter = **.50**

Overall Length = **.50**

Nominal Shaft Diameter = **3/8**

Keyway = **None**

C. Create another spur gear based on the following specifications:

Gear 2:

Diametral Pitch = **20**

Number of Teeth = **63**

Pressure Angle = **14.5**

Face Width = **.25**

Hub Style = **One Side**

Hub Diameter = **.50**

Overall Length = **.50**

Nominal Shaft Diameter = **3/8**

Keyway = **None**

D. Add #4-40 threaded holes to each gear hub .19 from the top hub surface.

E. Assemble the gears onto the shaft with Gear 1 offset .25 from the front end of the shaft and Gear 2 offset 2.63 from the front end of the shaft.

Project 10-10: Millimeters

A. Create a 24 × 120 shaft. Save the shaft.

B. Create a spur gear based on the following specifications:

Gear 1:

Module = **1.5**

Number of Teeth = **18**

Pressure Angle = **14.5**

Face Width = **12**

Hub Style = **One Side**

Hub Diameter = **32**

Overall Length = **30**

Nominal Shaft Diameter = **16**

Keyway = **None**

C. Create another spur gear based on the following specifications:

Gear 2:

Module = **1.5**

Number of Teeth = **70**

Pressure Angle = **14.5**

Face Width = **12**

Hub Style = **One Side**

Hub Diameter = **40**

Overall Length = **30**

Nominal Shaft Diameter = **16**

Keyway = **None**

D. Add M3.0 threaded holes to each gear hub 10 from the top hub surface.

E. Assemble the gears onto the shaft with Gear 1 offset 4.0 from the front end of the shaft and Gear 2 offset 80.0 from the front end of the shaft.

F. Insert an M3 Socket Set Screw with a Cup Point into each hole.

G. Create an exploded assembly drawing.

H. Create a bill of materials.

I. Animate the assembly.

Project 10-11: Inches

See Figure P10-11.

A. Draw three Ø.375 × 3.00 shafts.

B. Draw the support plate shown.

C. Access the **Design Library** and create two Gear 1s and two Gear 2s.

The gears are defined as follows:

Gear 1:

Diametral Pitch = **16**

Number of Teeth = **24**

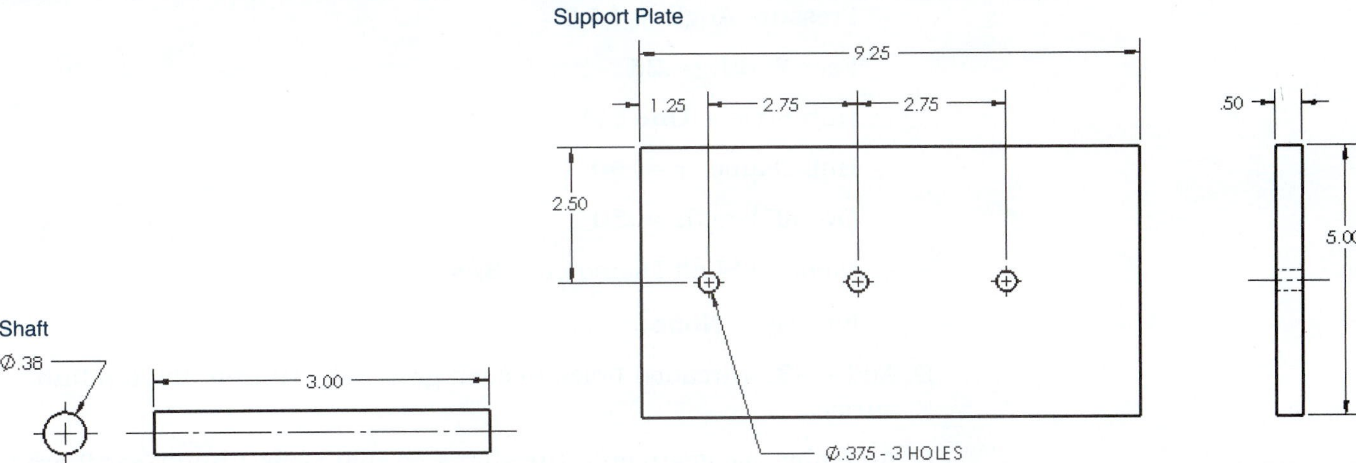

Figure P10-11

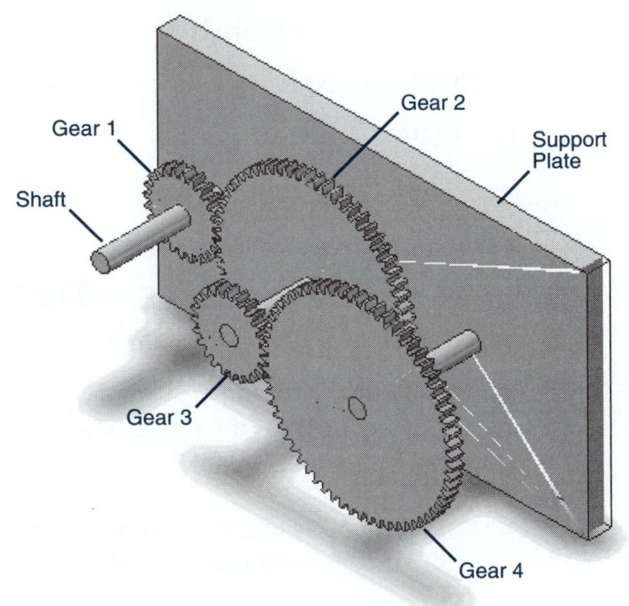

Pressure Angle = **14.5**

Face Width = **.25**

Hub Style = **One Side**

Hub Diameter = **.50**

Overall Length = **.50**

Nominal Shaft Diameter = **3/8**

Keyway = **None**

Gear 2:

Diametral Pitch = **16**

Number of Teeth = **64**

Pressure Angle = **14.5**

Face Width = **.25**

Hub Style = **One Side**

Hub Diameter = **.50**

Overall length = **.50**

Nominal Shaft Diameter = **3/8**

Keyway = **None**

D. Add #6-32 threaded holes to each gear hub .19 from the top hub surface.

E. Assemble the gears onto the shafts so that Gear 1 and Gear 2 are offset .50 from the support plate, and Gear 3 and Gear 4 are parallel to the ends of the shafts.

F. Insert a #6-32 Slotted Set Screw with an Oval Point into each hole.

G. Create an exploded assembly drawing.

H. Create a bill of materials.

I. Animate the assembly.

Project 10-12: Inches—Design Problem

Using Figure P10-11 define a support plate and shafts that support the following gears. Use two of each gear.

Parameters:

Plate: .50 thick, a distance of at least .50 beyond the other edge of the gears to the edge of the plate.

Shafts: Diameters that match the gear's bore diameter; minimum offset between the plate and the gear is .50 or greater.

Gear 1:

Diametral Pitch = **10**

Number of Teeth = **24**

Pressure Angle = **14.5**

Face Width = **.375**

Hub Style = **One Side**

Hub Diameter = **1.00**

Overall Length = **.875**

Nominal Shaft Diameter = **7/16**

Keyway = **None**

A. Create another spur gear based on the following specifications:

Gear 2:

Diametral Pitch = **10**

Number of Teeth = **96**

Pressure Angle = **14.5**

Face Width = **.375**

Hub Style = **One Side**

Hub Diameter = **1.00**

Overall Length = **.875**

Nominal Shaft Diameter = **7/16**

Keyway = **None**

B. Add #6-32 threaded holes to each gear hub .19 from the top hub surface.

C. Assemble the gears onto the shafts so that Gear 1 and Gear 2 are offset .50 from the support plate, and Gear 3 and Gear 4 are parallel to the ends of the shafts.

D. Insert a #6-32 Slotted Set Screw with an Oval Point into each hole.

E. Create an exploded assembly drawing.

F. Create a bill of materials.

G. Animate the assembly.

Project 10-13: Inches—Design Problem

Based on Figure P10-11 define a support plate and shafts that support the following gears. Use two of each gear.

Parameters:

Plate: .50 thick, a distance of at least .50 beyond the other edge of the gears to the edge of the plate.

Shafts: Diameters that match the gear's bore diameter: minimum offset between the plate and the gear is .50 or greater.

Gear 1:

Diametral Pitch = **6**

Number of Teeth = **22**

Pressure Angle = **14.5**

Face Width = **.500**

Hub Style = **One Side**

Hub Diameter = **1.50**

Overall Length = **1.25**

Nominal Shaft Diameter = **.75**

Keyway = **None**

Gear 2:

Diametral Pitch = **6**

Number of Teeth = **77**

Pressure Angle = **14.5**

Face Width = **.50**

Hub Style = **One Side**

Hub Diameter = **1.50**

Overall Length = **1.25**

Nominal Shaft Diameter = **.75**

Keyway = **None**

A. Add 1/4-20 UNC threaded holes to each gear hub 0.25 from the top hub surface.

B. Assemble the gears onto the shafts so that Gear 1 and Gear 2 are offset .50 from the support plate, and Gear 3 and Gear 4 are parallel to the ends of the shafts.

C. Insert a 1/4-20 UNC Slotted Set Screw with an Oval Point into each hole.

D. Create an exploded assembly drawing.

E. Create a bill of materials.

F. Animate the assembly.

Project 10-14: Millimeters—Design Problem

Based on Figure P10-11 define a support plate and shafts that support the following gears. Use two of each gear.

Parameters:

Plate: 20 thick, a distance of at least 25 beyond the other edge of the gears to the edge of the plate.

Shafts: Diameters that match the gear's bore diameter; minimum offset between the plate and the gear is 20 or greater.

Gear 1:

Module = **2.5**

Number of Teeth = **20**

Pressure Angle = **14.5**

Face Width = **16**

Hub Style = **One Side**

Hub Diameter = **26**

Overall Length = **30**

Nominal Shaft Diameter = **20**

Keyway = **None**

Gear 2:

Module = **2.5**

Number of Teeth = **50**

Pressure Angle = **14.5**

Face Width = **16**

Hub Style = **One Side**

Hub Diameter = **30**

Overall Length = **30**

Nominal Shaft Diameter = **20**

Keyway = **None**

A. Add M4 threaded holes to each gear hub 12 from the top hub surface.

B. Assemble the gears onto the shafts so that Gear 1 and Gear 2 are offset 10 from the support plate, and Gear 3 and Gear 4 are parallel to the ends of the shafts.

C. Insert an M4 Slotted Set Screw with an Oval Point into each hole.

D. Create an exploded assembly drawing.

E. Create a bill of materials.

F. Animate the assembly.

Project 10-15: Inches

See Figure P10-15.

A. Draw four Ø.375 × 3.00 shafts.

B. Draw the support plate shown in Figure P10-15.

C. Access the **Design Library** and create three Gear 1s and three Gear 2s.

The gears are defined as follows:

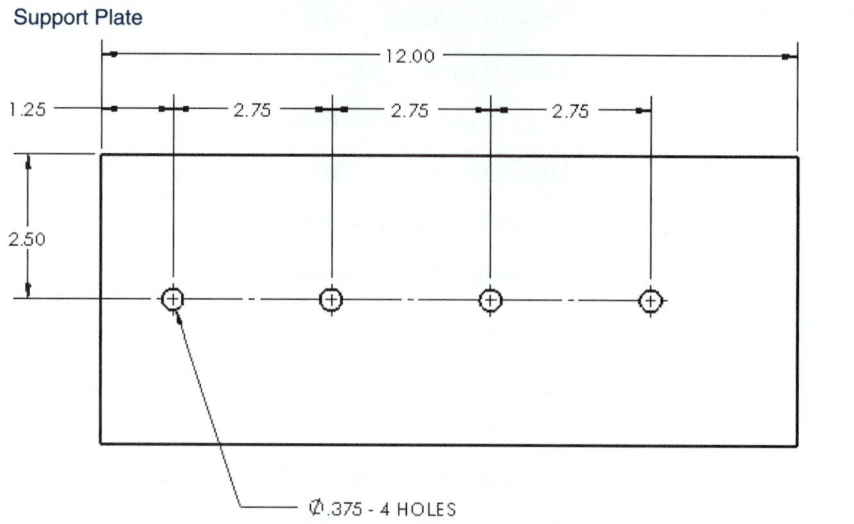

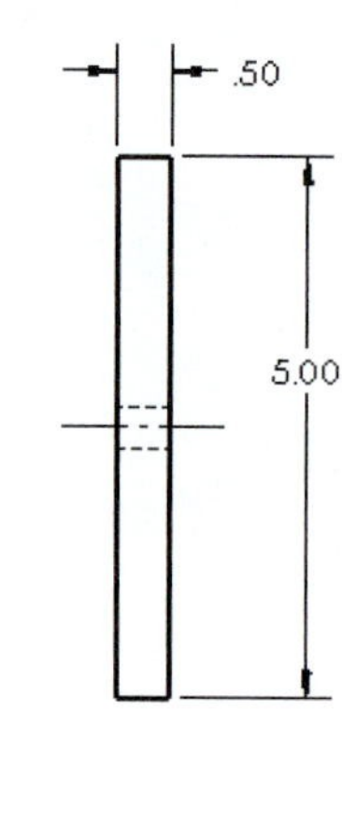

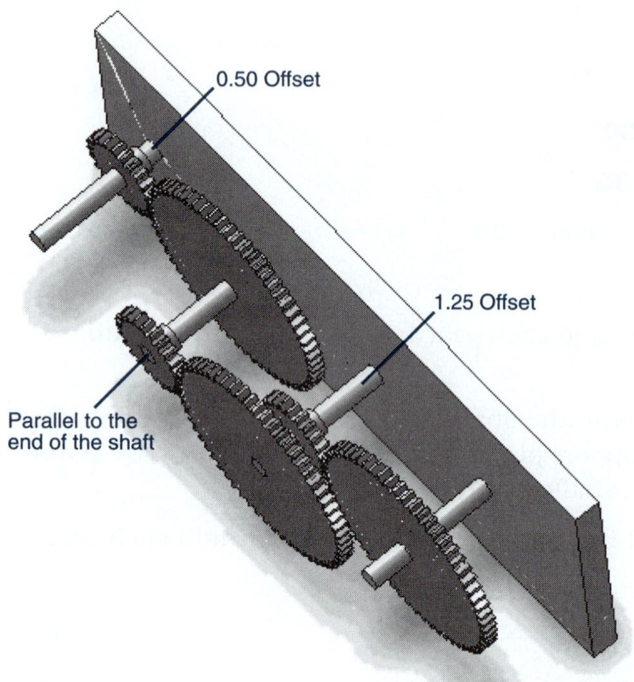

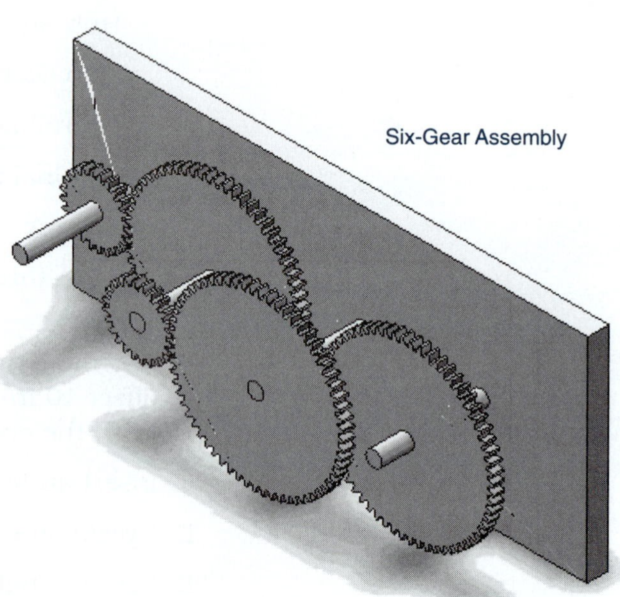

Figure P10-15

Gear 1:

Diametral Pitch = **16**

Number of Teeth = **24**

Pressure Angle = **14.5**

Face Width = **.25**

Hub Style = **One Side**

Hub Diameter = **.50**

Overall Length = **.50**

Nominal Shaft Diameter = **3/8**

Keyway = **None**

Gear 2:

Diametral Pitch = **16**

Number of Teeth = **64**

Pressure Angle = **14.5**

Face Width = **.25**

Hub Style = **One Side**

Hub Diameter = **.50**

Overall Length = **.50**

Nominal Shaft Diameter = **3/8**

Keyway = **None**

D. Add #6-32 threaded holes to each gear hub 0.19 from the top hub surface.

E. Assemble the gears onto the shafts so that the offset between the support plate and the gears is as defined in Figure P10-15.

F. Insert a #6-32 Slotted Set Screw with an Oval Point into each hole.

G. Create an exploded assembly drawing.

H. Create a bill of materials.

I. Animate the assembly.

Project 10-16: Inches—Design Problem

Based on Figure P10-15 define a support plate and shafts that support the following gears. Use two of each gear.

Parameters:

Plate: .50 thick, a distance of at least .50 beyond the other edge of the gears to the edge of the plate.

Shafts: Diameters that match the gear's bore diameter; minimum offset between the plate and the gear is .50 or greater.

Gear 1:

Diametral Pitch = **6**

Number of Teeth = **22**

Pressure Angle = **14.5**

Face Width = **.500**

Hub Style = **One Side**

Hub Diameter = **1.50**

Overall Length = **1.25**

Nominal Shaft Diameter = **.75**

Keyway = **None**

Gear 2:

Diametral Pitch = **6**

Number of Teeth = **77**

Pressure Angle = **14.5**

Face Width = **.50**

Hub Style = **One Side**

Hub Diameter = **1.50**

Overall Length = **1.25**

Nominal Shaft Diameter = **.75**

Keyway = **None**

A. Add 1/4-20 UNC threaded holes to each gear hub .25 from the top hub surface.

B. Assemble the gears onto the shafts so that Gear 1 and Gear 2 are offset .50 from the support plate, and Gear 3 and Gear 4 are parallel to the ends of the shafts.

C. Insert a 1/4-20 UNC Slotted Set Screw with an Oval Point into each hole.

D. Create an exploded assembly drawing.

E. Create a bill of materials.

F. Animate the assembly.

Project 10-17: Millimeters—Design Problem

Based on Figure P10-15 define a support plate and shafts that support the following gears. Use two of each gear.

Parameters:

Plate: 20 thick, a distance of at least 25 beyond the other edge of the gears to the edge of the plate.

Shafts: Diameters that match the gear's bore diameter; minimum offset between the plate and the gear is 20 or greater.

Gear 1:

Module = **2.5**

Number of Teeth = **20**

Pressure Angle = **14.5**

Face Width = **16**

Hub Style = **One Side**

Hub Diameter = **26**

Overall Length = **30**

Nominal Shaft Diameter = **20**

Keyway = **None**

Gear 2:

Module = **2.5**

Number of Teeth = **50**

Pressure Angle = **14.5**

Face Width = **16**

Hub Style = **One Side**

Hub Diameter = **30**

Overall Length = **30**

Nominal Shaft Diameter = **20**

Keyway = **None**

A. Add M4 threaded holes to each gear hub 12 from the top hub surface.

B. Assemble the gears onto the shafts so that Gear 1 and Gear 2 are offset 10 from the support plate, and Gear 3 and Gear 4 are parallel to the ends of the shafts.

C. Insert an M4 Slotted Set Screw with an Oval Point into each hole.

D. Create an exploded assembly drawing.

E. Create a bill of materials.

F. Animate the assembly.

Project 10-18: Inches—Design Problem

Redraw the gear assembly shown in Figure P10-18. Create the following:

A. An assembly drawing

B. An exploded isometric drawing

C. A BOM

D. Dimensioned drawings of each part

Parts from the **Toolbox** do not need drawings but should be listed on the BOM.

Figure P10-18

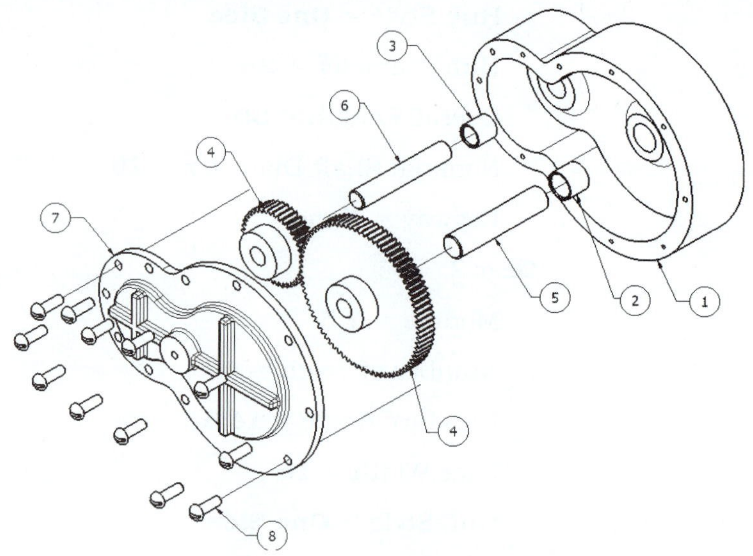

Parts List				
ITEM	PART NUMBER	DESCRIPTION	MATERIAL	QTY
1	ENG-453-A	GEAR, HOUSING	CAST IRON	1
2	BU-1123	BUSHING Ø0.75	Delrin, Black	1
3	BU-1126	BUSHING Ø0.625	Delrin, Black	1
4	ASSEMBLY-6	GEAR ASSEEMBLY	STEEL	1
5	AM-314	SHAFT, GEAR Ø.625	STEEL	1
6	AM-315	SHAFT, GEAR Ø.0.500	STEEL	1
7	ENG -566-B	COVER, GEAR	CAST IRON	1
8	ANSI B18.6.2 - 1/4-20 UNC - 0.75	Slotted Round Head Cap Screw	Steel, Mild	12

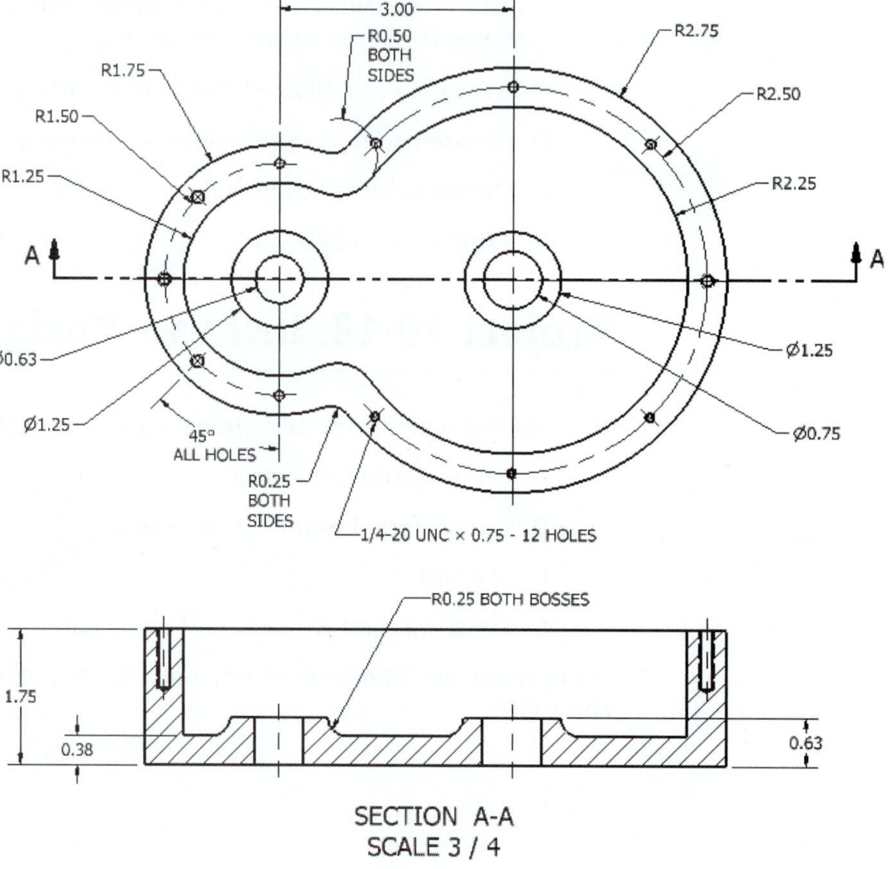

Figure P10-18
(Continued)

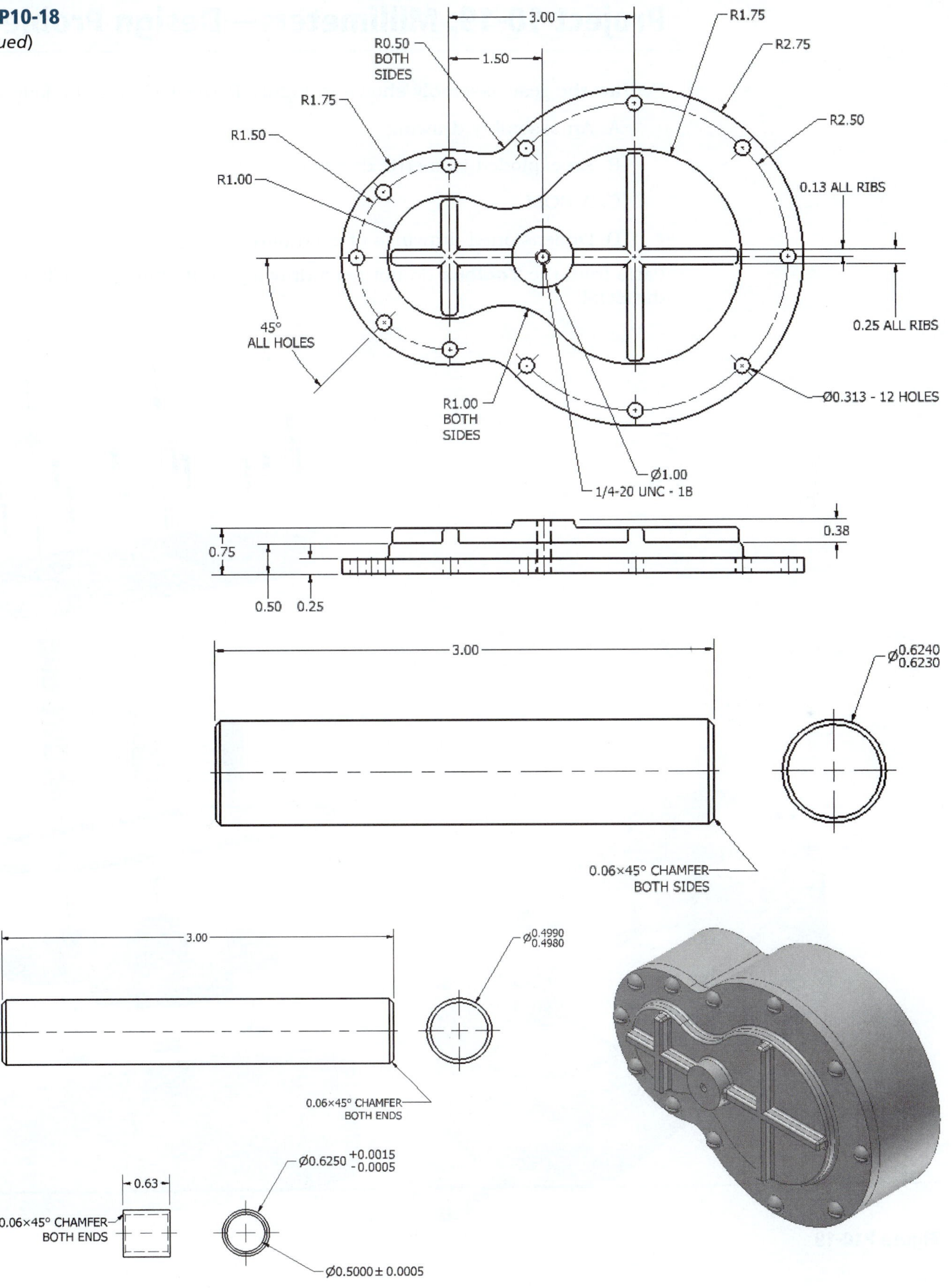
3.00
1.50
R0.50
BOTH
SIDES
R1.75
R1.50
R1.00
R1.75
R2.75
R2.50
0.13 ALL RIBS
0.25 ALL RIBS
45°
ALL HOLES
Ø0.313 - 12 HOLES
R1.00
BOTH
SIDES
Ø1.00
1/4-20 UNC - 1B
0.75
0.38
0.50
0.25
3.00
Ø0.6240
0.6230
0.06×45° CHAMFER
BOTH SIDES
3.00
Ø0.4990
0.4980
0.06×45° CHAMFER
BOTH ENDS
0.63
0.06×45° CHAMFER
BOTH ENDS
Ø0.6250 +0.0015
-0.0005
Ø0.5000 ± 0.0005

Project 10-19: Millimeters—Design Problem

Redraw the gear assembly shown in Figure P10-19. Create the following:

A. An assembly drawing

B. An exploded isometric drawing

C. A BOM

D. Dimensioned drawings of each part

Parts from the **Toolbox** do not need drawings but should be listed on the BOM.

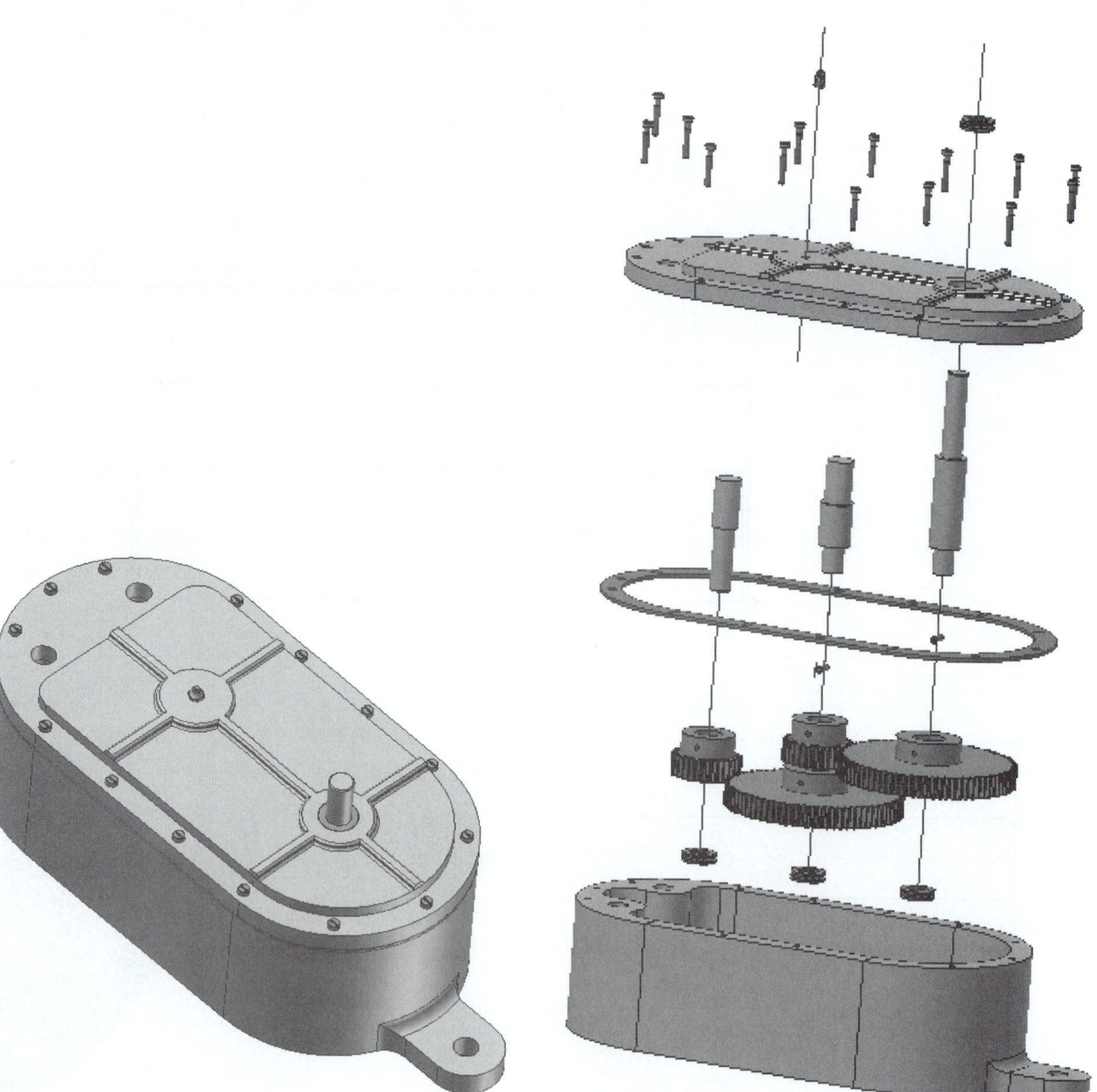

Figure P10-19

Figure P10-19
(Continued)

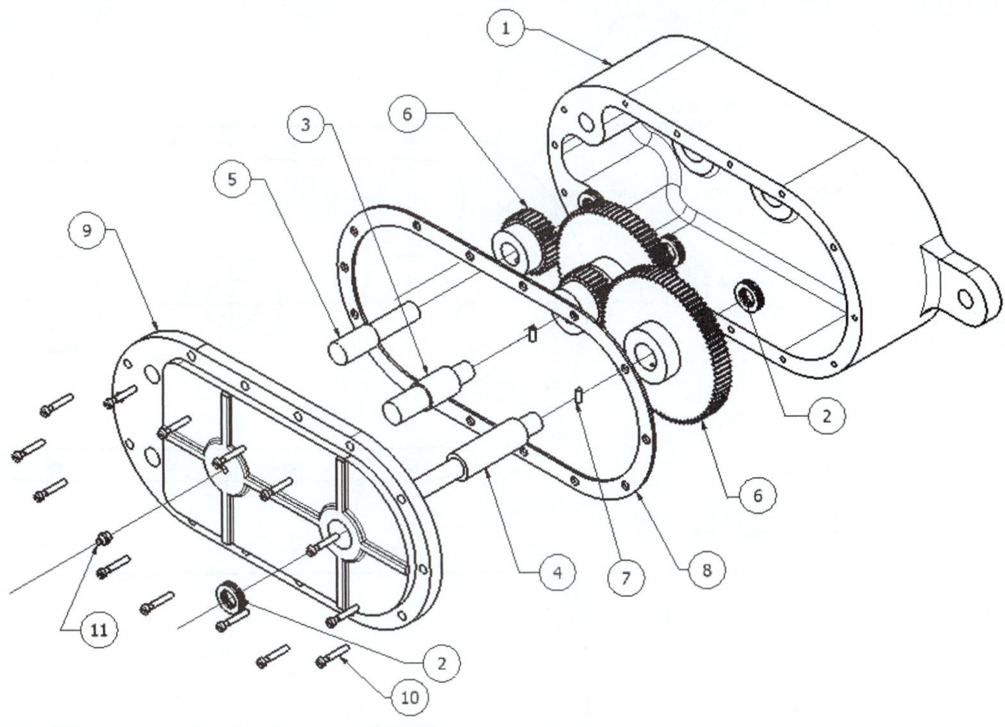

Parts List				
ITEM	PART NUMBER	DESCRIPTION	MATERIAL	QTY
1	ENG-311-1	4-GEAR HOUSING	CAST IRON	1
2	BS 5989: Part 1 - 0 10 - 20x32x8	Thrust Thrust Ball Bearing	Steel, Mild	4
3	SH-4002	SHAFT, NEUTRAL	STEEL	1
4	SH-4003	SHAFT, OUTPUT	STEEL	1
5	SH-4004A	SHAFT,INPUT	STEEL	1
6	4-GEAR-ASSEMBLY		STEEL	2
7	CSN 02 1181 - M6 x 16	Slotted Headless Set Screw - Flat Point	Steel, Mild	2
8	ENG-312-1	GASKET	Brass, Soft Yellow	1
9	COVER			1
10	CNS 4355 - M 6 x 35	Slotted Cheese Head Screw	Steel, Mild	14
11	CSN 02 7421 - M10 x 1coned short	Lubricating Nipple, coned Type A	Steel, Mild	1

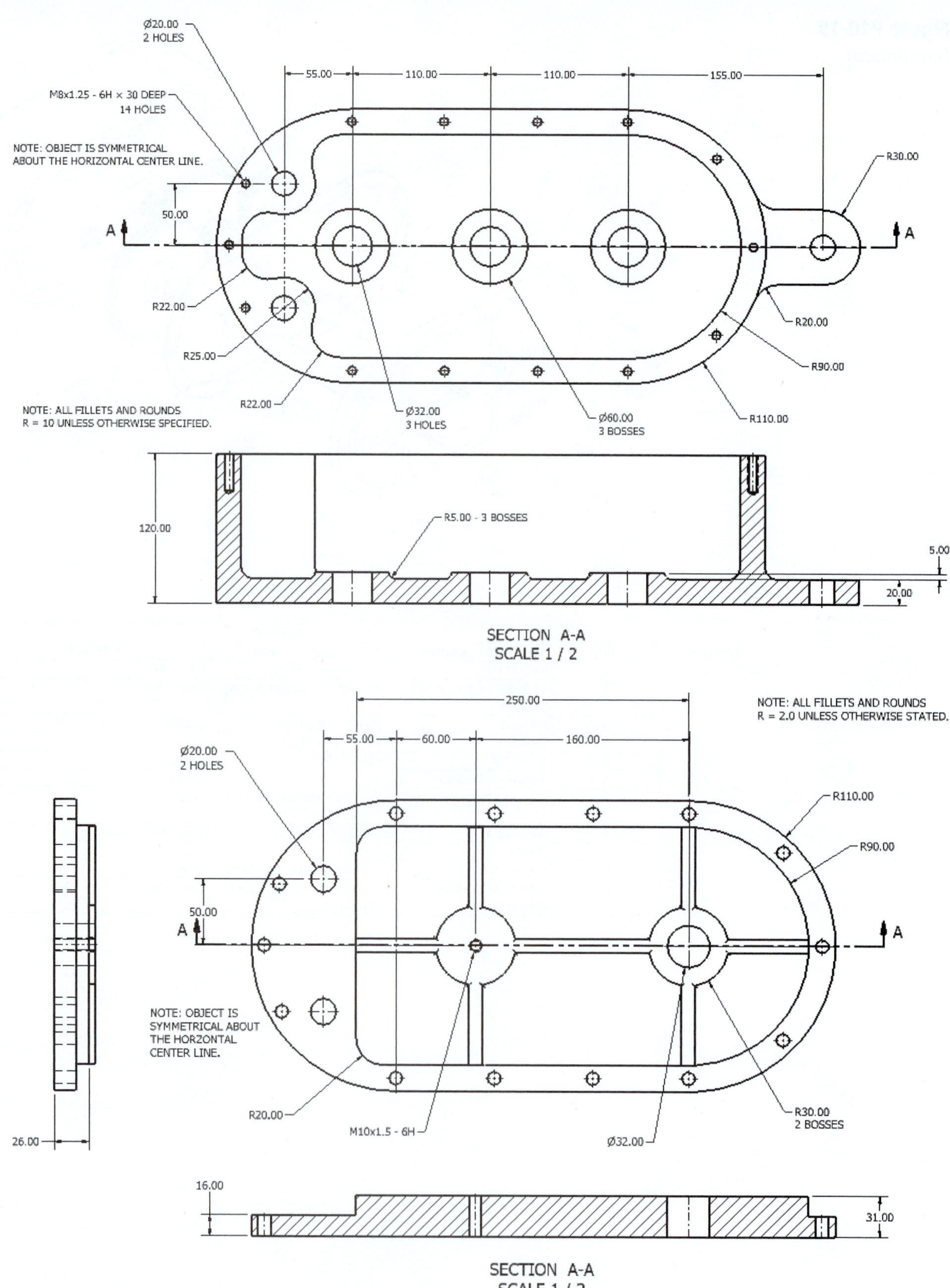
Ø20.00
2 HOLES
M8x1.25 - 6H × 30 DEEP
14 HOLES
NOTE: OBJECT IS SYMMETRICAL
ABOUT THE HORIZONTAL CENTER LINE.
55.00
110.00
110.00
155.00
R30.00
50.00
A
A
R22.00
R25.00
R22.00
R20.00
R90.00
R110.00
NOTE: ALL FILLETS AND ROUNDS
R = 10 UNLESS OTHERWISE SPECIFIED.
Ø32.00
3 HOLES
Ø60.00
3 BOSSES
120.00
R5.00 - 3 BOSSES
5.00
20.00
SECTION A-A
SCALE 1 / 2
250.00
NOTE: ALL FILLETS AND ROUNDS
R = 2.0 UNLESS OTHERWISE STATED.
55.00
60.00
160.00
Ø20.00
2 HOLES
R110.00
R90.00
50.00
A
A
NOTE: OBJECT IS
SYMMETRICAL ABOUT
THE HORZONTAL
CENTER LINE.
R20.00
M10x1.5 - 6H
Ø32.00
R30.00
2 BOSSES
26.00
16.00
31.00
SECTION A-A
SCALE 1 / 2

Figure P10-19
(Continued)

Figure P10-19
(Continued)

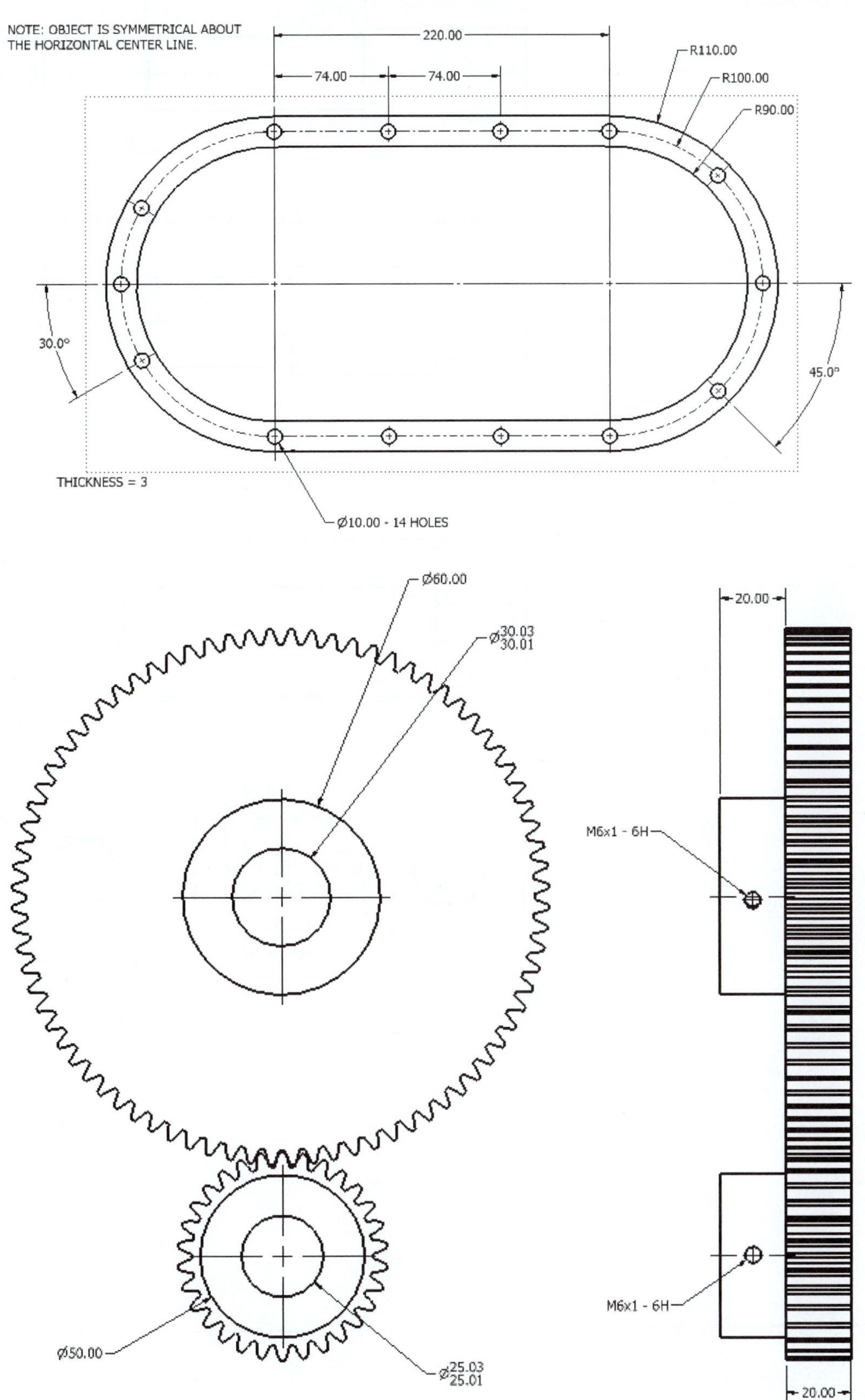
NOTE: HOLE PATTERN IS THE SAME FOR THE GASKET, GEAR HOUSING, AND GEAR COVER.
NOTE: OBJECT IS SYMMETRICAL ABOUT THE HORIZONTAL CENTER LINE.
220.00
74.00
74.00
R110.00
R100.00
R90.00
30.0°
45.0°
THICKNESS = 3
Ø10.00 - 14 HOLES
Ø60.00
Ø30.03 30.01
20.00
M6x1 - 6H
M6x1 - 6H
Ø50.00
Ø25.03 25.01
20.00

Figure P10-19
(Continued)

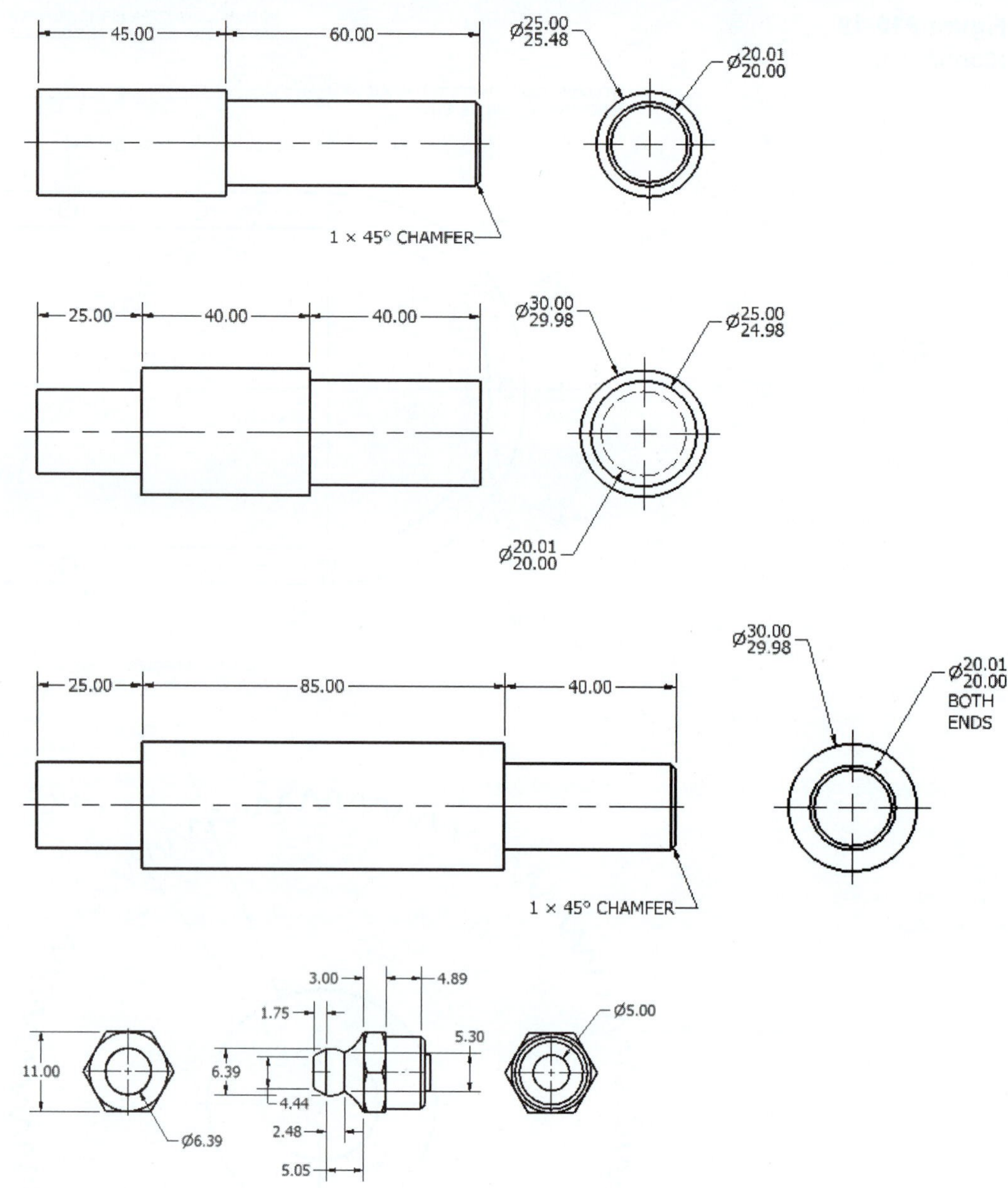

chaptereleven Belts and Pulleys

CHAPTER OBJECTIVES

- Learn to draw belts and pulleys
- Understand the use of standard sizes for designs using belts and pulleys

11-1 Introduction

Belts and pulleys are another form of power transmission. They are cheaper than gears, require less stringent tolerances, can be used to cover greater distances, and can absorb shock better. However, belts cannot take as much load as gears and can slip or creep, and operate at slower speeds.

11-2 Belt and Pulley Standard Sizes

There are many different-sized belts and pulleys. Listed here are belt designations, belt overall thicknesses, belt widths, and pulley widths that can be used to create assemblies within the context of this text. For other belt and pulley properties see manufacturers' specifications.

Standard Belt Sizes—Single-Sided Belt Thickness

Mini Extra Light: MXL (0.080)—0.045

Extra Light: XL (0.200)—0.09

Light: L (0.375)—0.14

Heavy: H (0.500)—0.16

Extra Heavy: XL (0.875)—0.44

Double Extra Heavy: XXL (1.250)—0.62

Standard Pulley Widths

MXL: 0.25

XL: 0.38

L: 0.50, 0.75, 1.00

H: 1.00, 1.50, 2.00, 3.00

XH: 2.00, 3.00, 4.00

XXH: 2.00, 3.00, 4.00, 5.00

Standard Belt Widths

XXL—0.12, 0.19, 0.25

XL—0.25, 0.38

L—0.50, 0.75, 1.00

H—0.75, 1.00, 1.50, 2.00, 3.00

XH—2.00, 3.00, 4.00

XXH—2.00, 3.00, 4.00, 5.00

To Draw a Belt and Pulley Assembly

Figure 11-1 shows dimensioned drawings of the support plate and shaft.

1 Create **Part** documents of the support plate and the shaft.

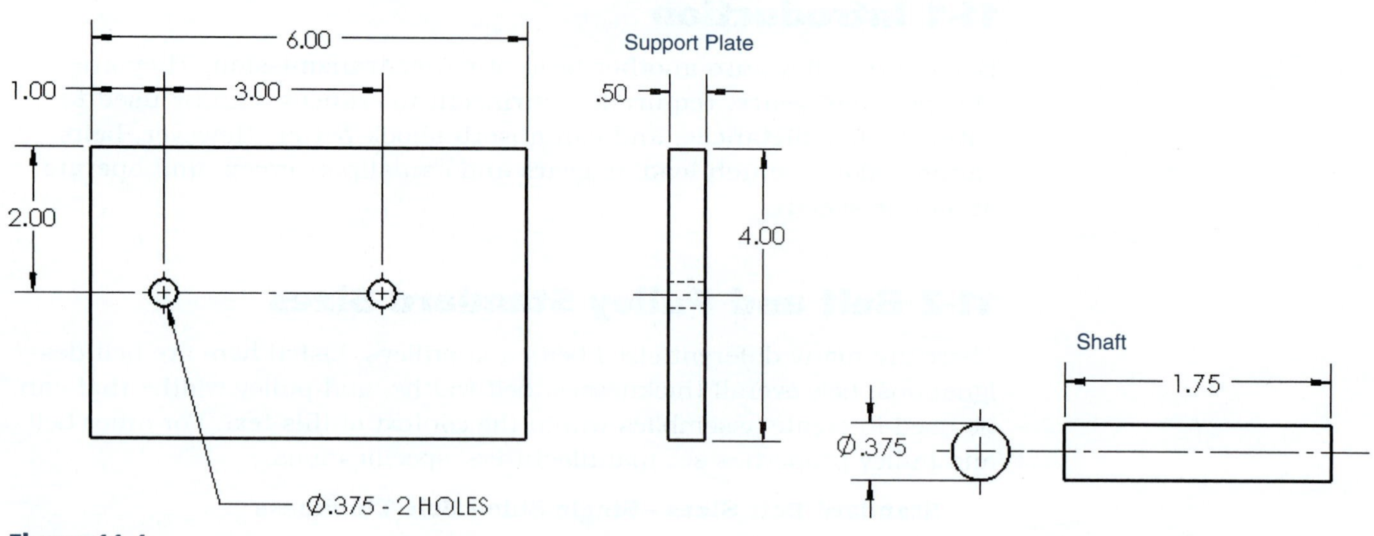

Figure 11-1

2 Assemble two shafts into the support plate. The shafts should extend **1.25** beyond the support plate.

See Figure 11-2.

Figure 11-2

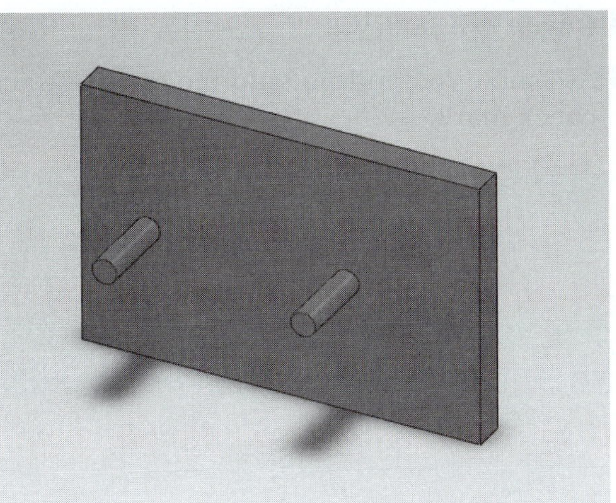

3 Access the **Design Library** and click **Toolbox, ANSI Inch, Power Transmission**, and **Timing Belts**.

4 Click and drag a **Timing Belt Pulley** into the drawing area. Nominal Shaft Diameter = 3/8.

5 Set the properties as follows: Belt Pitch = **(0.200) = XL**, Belt Width = **0.38**, Pulley Style = **Flanged**, Number of grooves = **20**, Hub Diameter = **.500**, Overall Length = **.500**, Nominal Shaft Diameter = **3/8**, and Keyway = **None**.

See Figure 11-3.

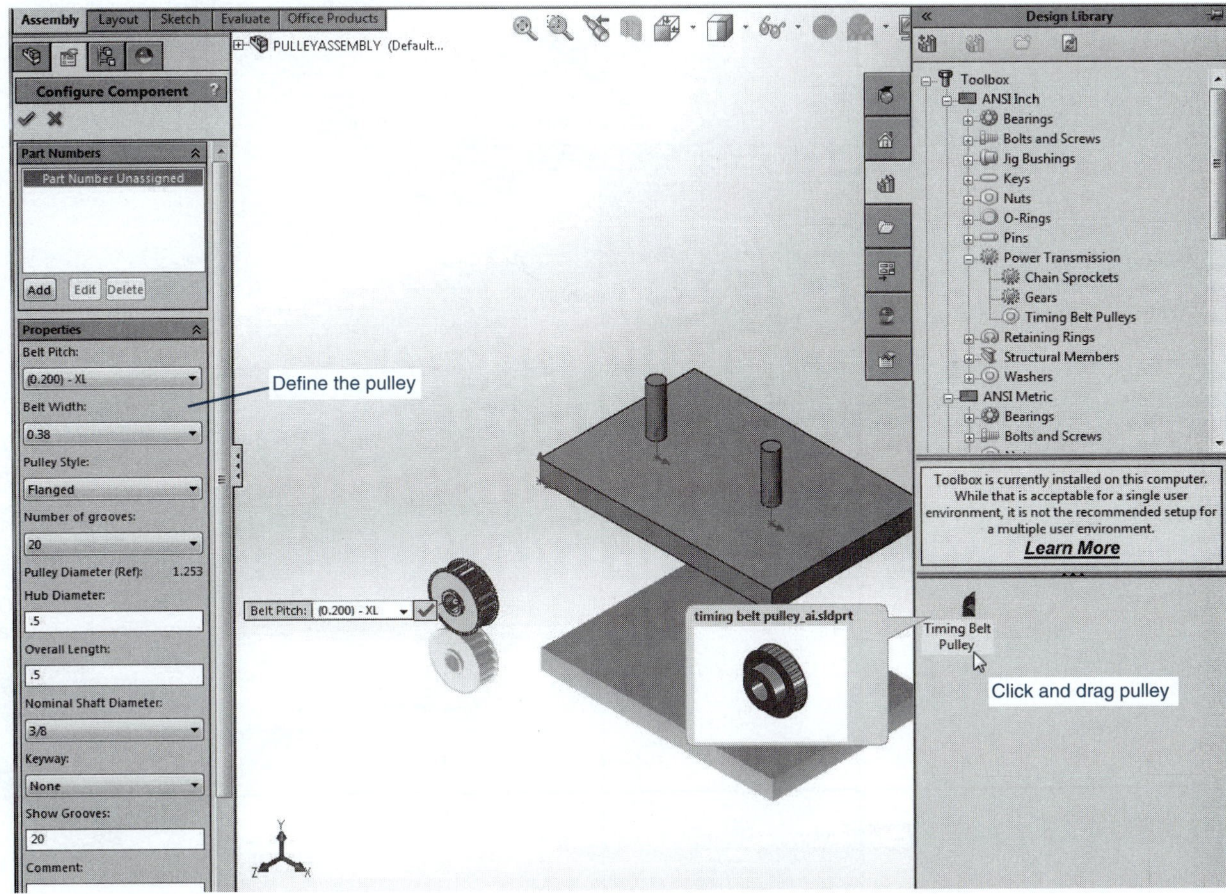

Figure 11-3

6. Create two pulleys.
7. Assemble the pulleys onto the ends of the shafts; click the green **OK** check mark.

   See Figure 11-4.

Figure 11-4

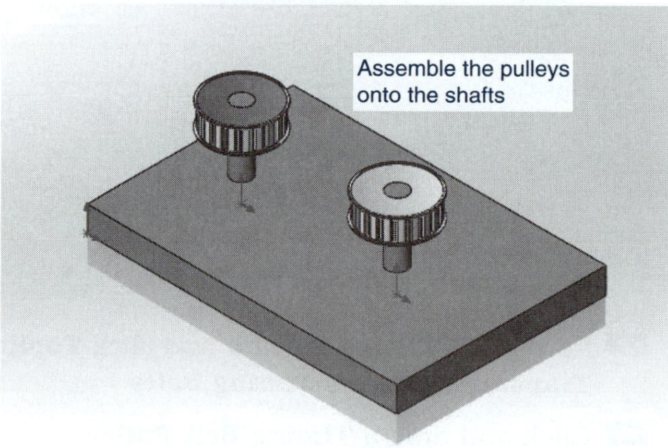

8. Click the **Insert** tool at the top of the screen, click **Assembly Feature**, then **Belt/Chain**.

   See Figure 11-5.

Figure 11-5

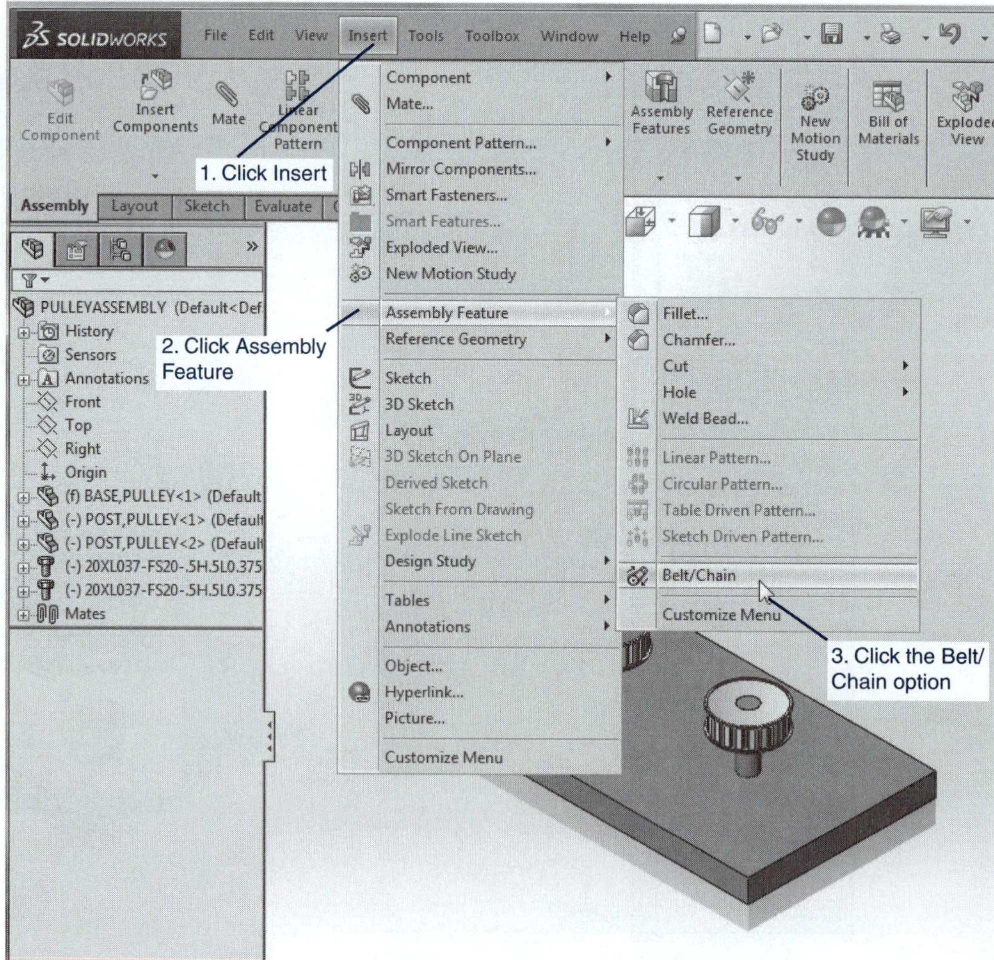

9 Select the **Belt Members** by clicking the top surfaces on the pulleys' teeth as shown.

See Figure 11-6.

Figure 11-6

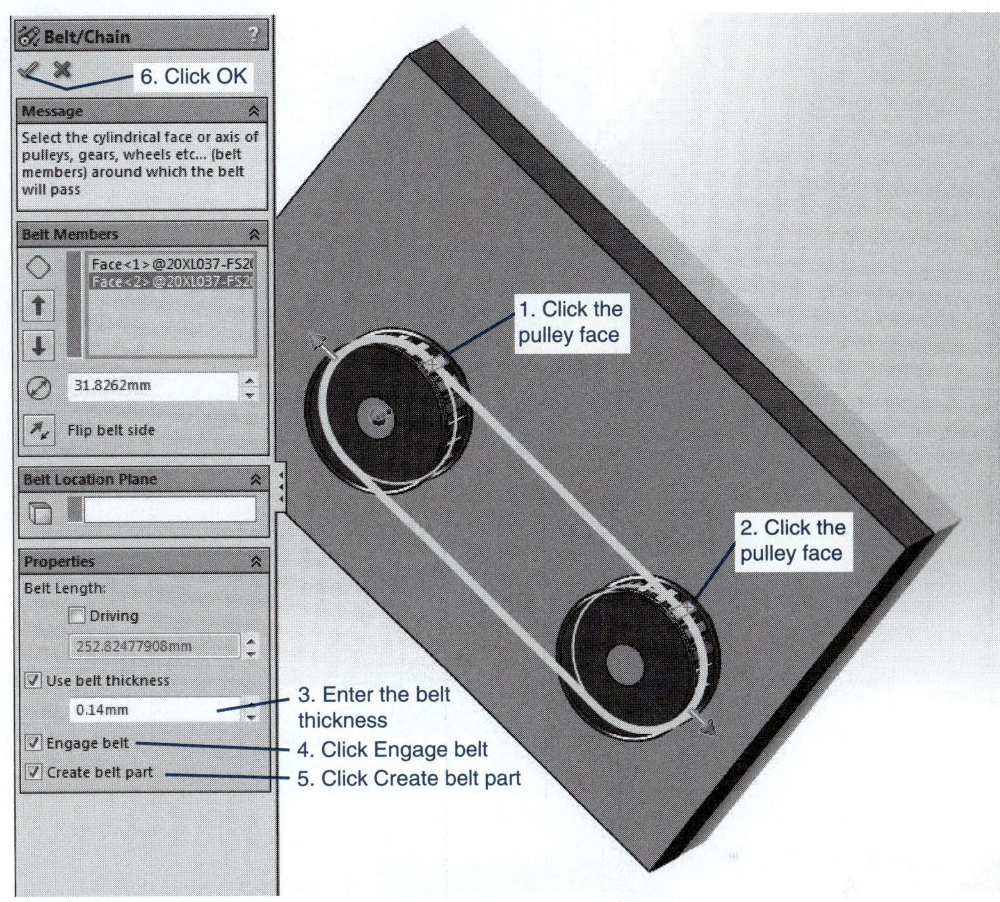

10 Scroll down the **Belt/Chain PropertyManager**, click the **Use belt thickness** box, and set the thickness for **0.14in**.

11 Click the **Engage belt** box.

12 Click the **Create belt part** box.

13 Click the green **OK** check mark.

14 Click the + sign to the left of the **Belt1** heading in the **FeatureManager**.

15 Click the + sign to the left of the **Belt1-5^ Pulley-test** heading.

Your headings may be slightly different. See Figure 11-7.

16 Right-click **Sketch 2** and select the **Edit Sketch** option.

17 Click the **Features** tab to access the **Features** tools.

18 Click the **Extruded Boss/Base** tool.

The **Extrude PropertyManager** and a preview will appear. See Figure 11-8.

19 Set **Direction 1** for **Mid Plane**.

Figure 11-7

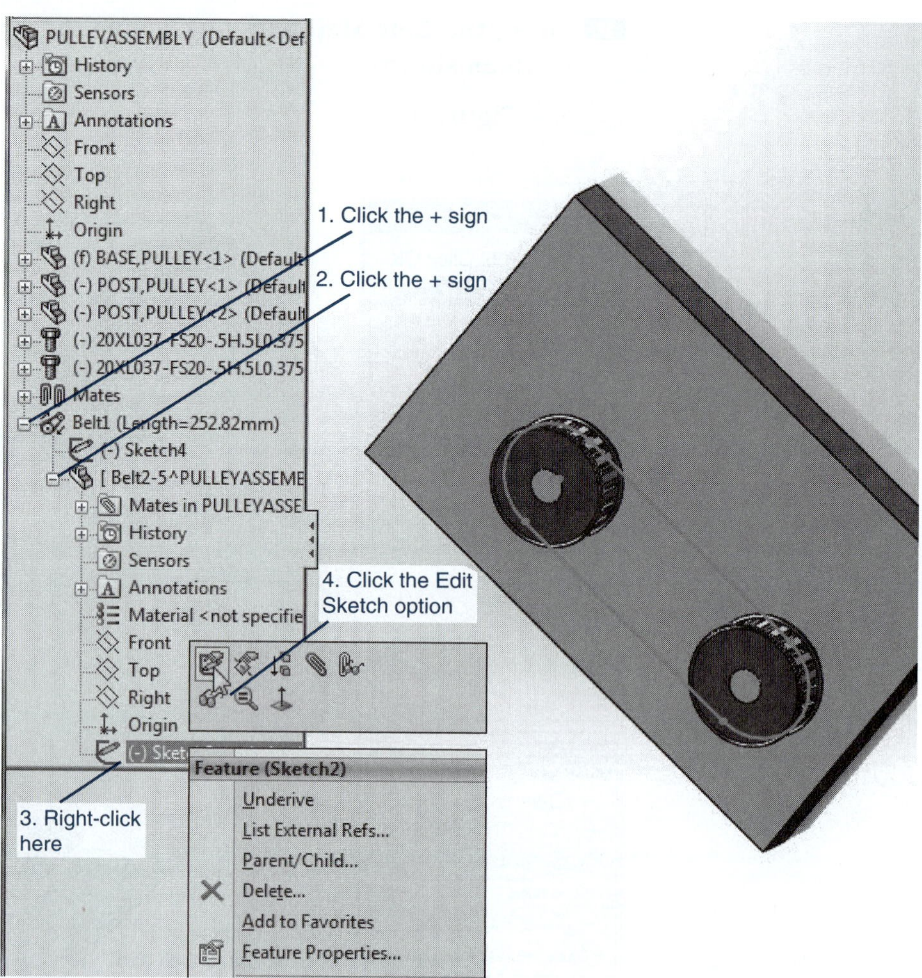

Figure 11-8

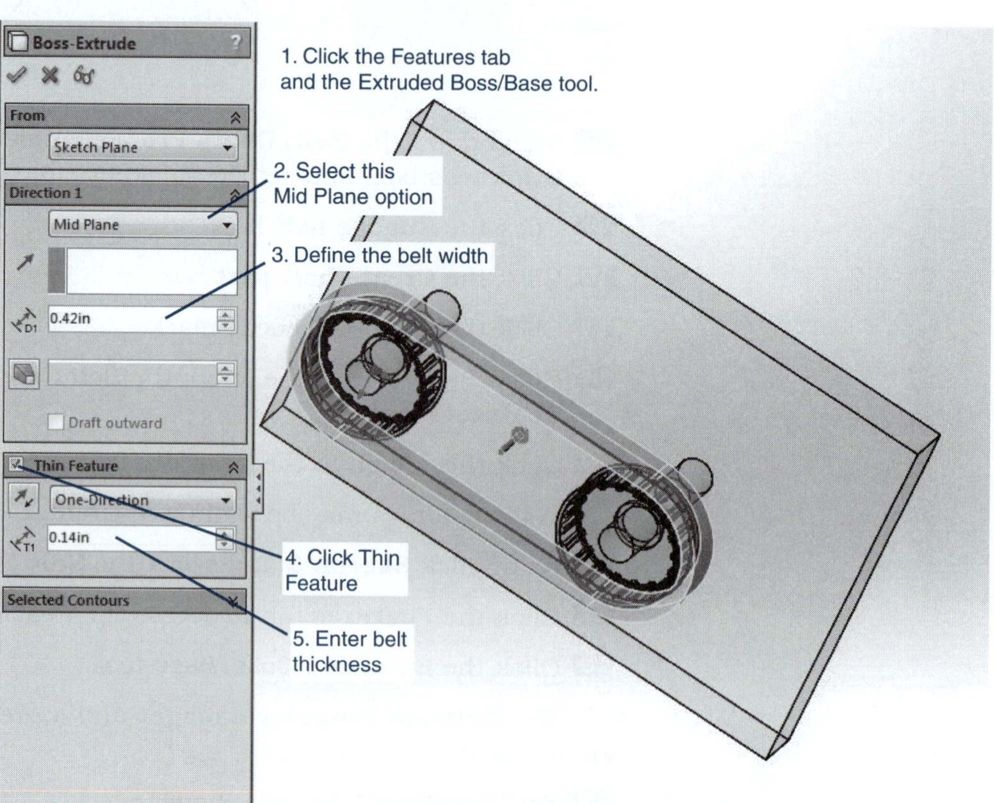

20 Set the belt width value for **0.42in**.

This width keeps the belt inside the flanges.

21 Click the **Thin Feature** option and set the thickness for **0.14in**.

22 Click the green **OK** check mark and return to the assembly drawing.

See Figure 11-9.

Figure 11-9

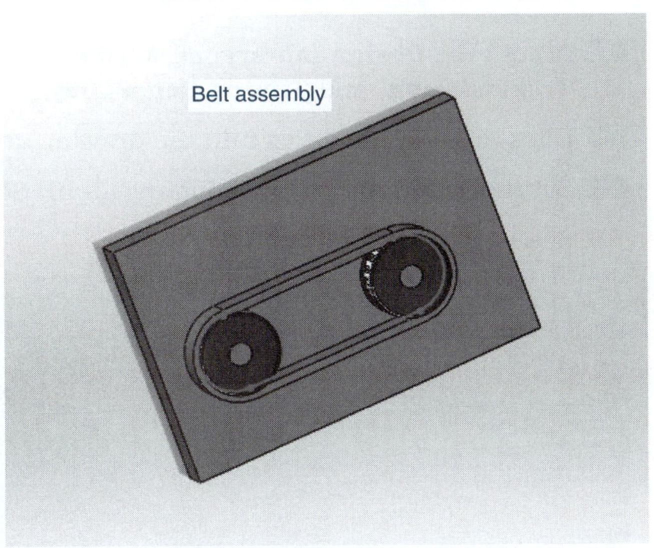

23 Save the assembly as **Belt Assembly**.

Locate the cursor on the left pulley and rotate the pulley. The right pulley also will rotate. The pulleys will rotate in the same direction. Remember, gears rotate in opposite directions.

11-3 Pulleys and Keys

Figure 11-10 shows a dimensioned drawing of the support plate defined in Figure 11-1 and a shaft. The shaft includes a keyway defined to accept a .125 × .125 × .250 square key.

Figure 11-10

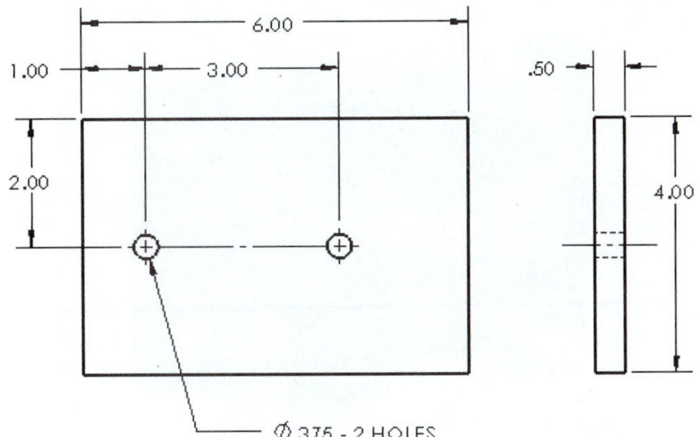

Figure 11-10
(Continued)

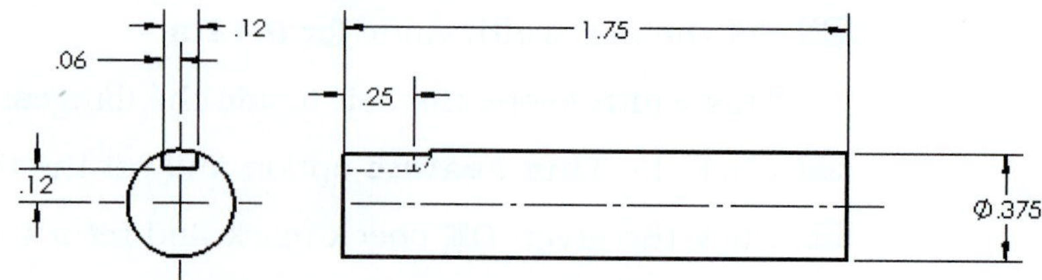

To Add a Keyway to a Pulley

1. Access the **Design Library**, click **Toolbox, ANSI Inch, Power Transmission**, and **Timing Belt Pulley**.
2. Click and drag a pulley into the drawing area.
3. Set the values as shown in Figure 11-10. Set the **Keyway** for **Square(1)**.

   See Figure 11-11.

Figure 11-11

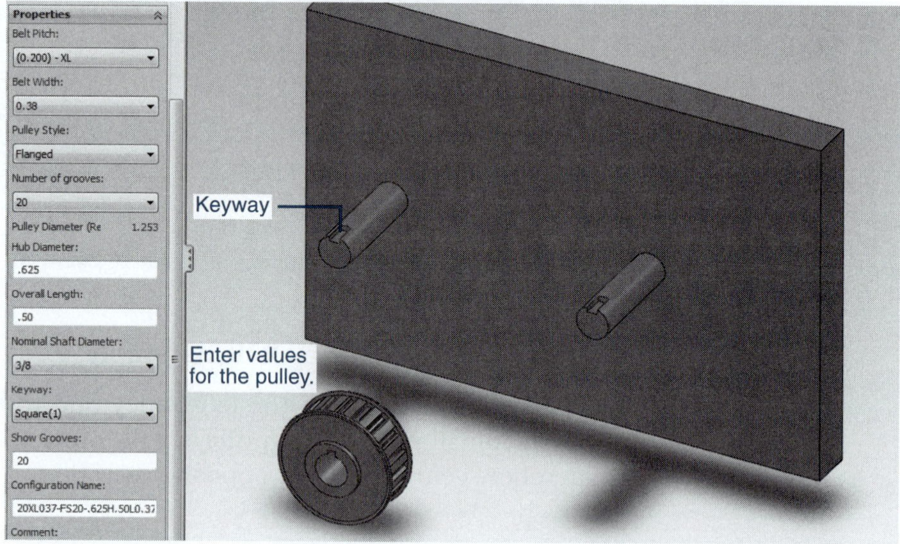

4. Create a second pulley identical to the first and click the green **OK** check mark.
5. Assemble the pulleys onto the shafts.

   See Figure 11-12.

Figure 11-12

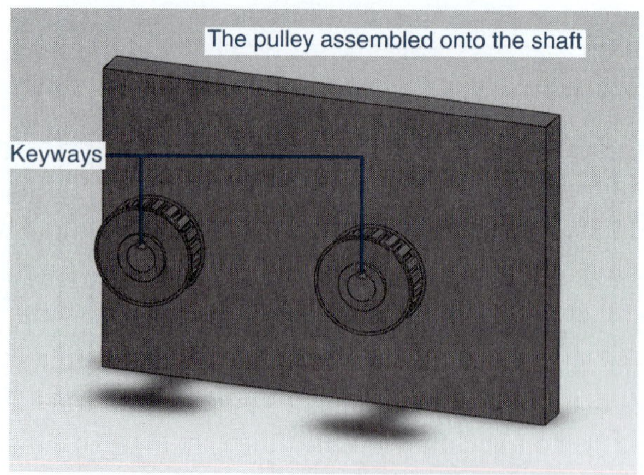

6 Access the **Design Library**, click **Toolbox, ANSI Inch, Keys**, and **Parallel Keys**.

7 Click and drag a key into the drawing area.

See Figure 11-13.

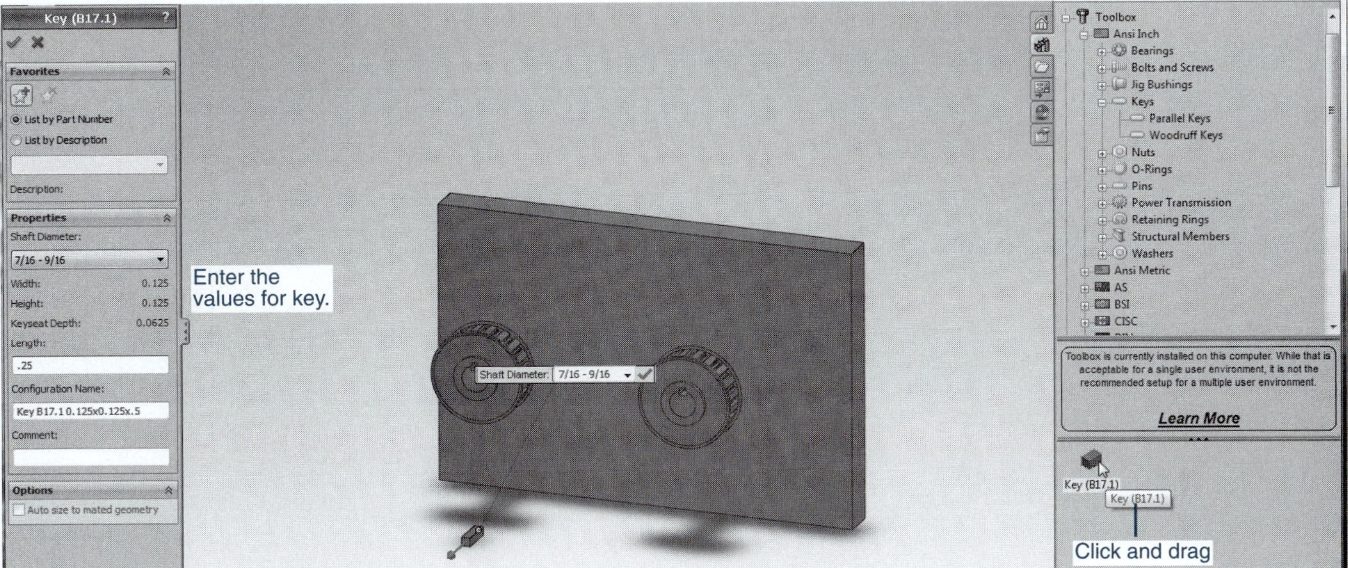

Figure 11-13

8 Set the key's **Properties** values for a shaft diameter of **7/16 – 9/16**. Set the **Length** for **.25**.

9 Create a second key and click the green **OK** check mark.

10 Assemble the keys into the keyways and add a timing belt as defined in the last section.

See Figure 11-14.

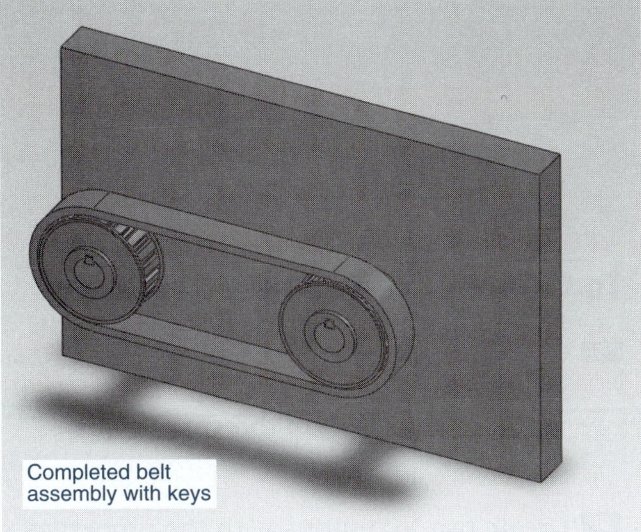

Figure 11-14

11-4 Multiple Pulleys

More than one pulley can be included in an assembly. Figure 11-15 shows drawings for a support plate and shaft.

Figure 11-15

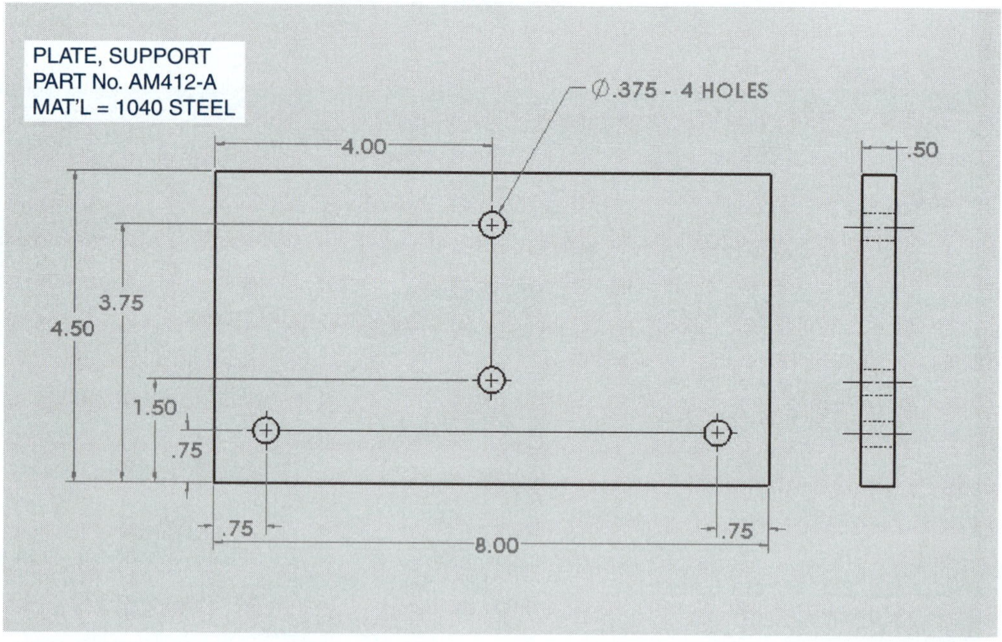

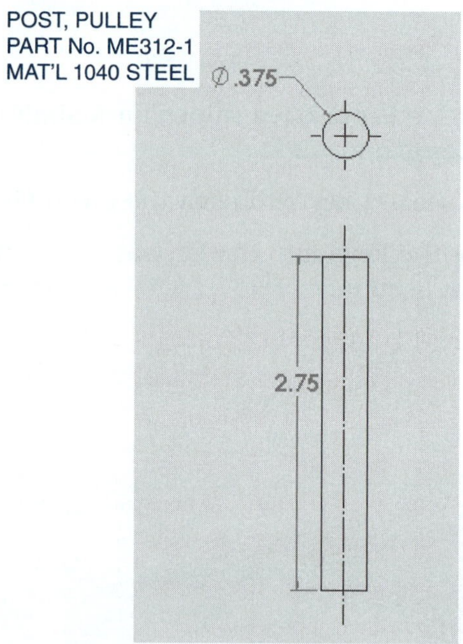

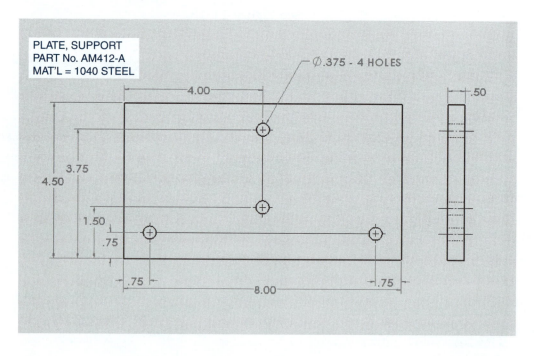

To Create a Multi-Pulley Assembly

1 Create an **Assembly** drawing of the support plate and shafts with the shafts inserted into the support plate so that the shafts extend 1.00 beyond the surface of the support plate.

See Figure 11-16.

2 Create two XL pulleys with the properties specified in Figure 11-17.

Figure 11-16

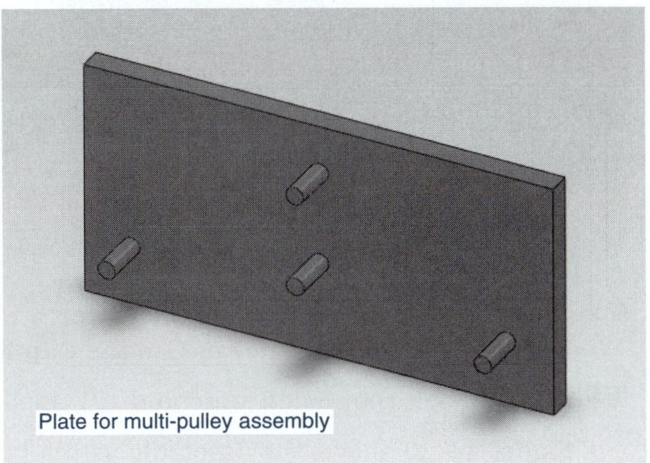

Plate for multi-pulley assembly

XL pulley properties

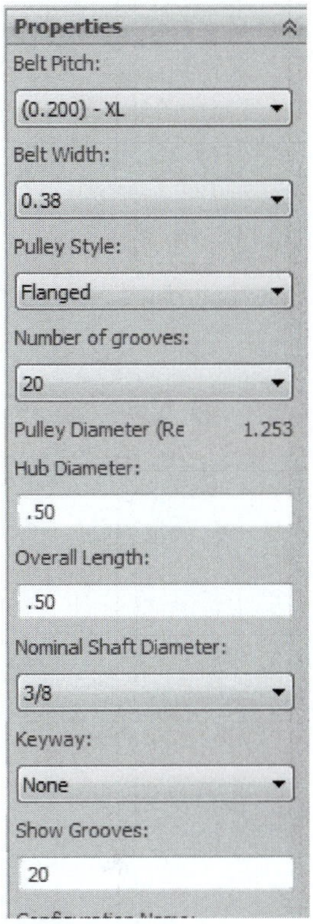

Properties
Belt Pitch: (0.200) - XL
Belt Width: 0.38
Pulley Style: Flanged
Number of grooves: 20
Pulley Diameter (Re 1.253
Hub Diameter: .50
Overall Length: .50
Nominal Shaft Diameter: 3/8
Keyway: None
Show Grooves: 20

Figure 11-17

3 Create two L pulleys with the properties specified in Figure 11-18.

Figure 11-18

L pulley properties

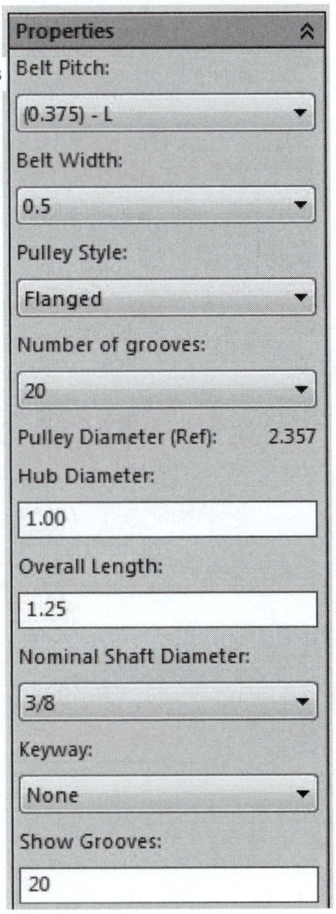

Properties
Belt Pitch: (0.375) - L
Belt Width: 0.5
Pulley Style: Flanged
Number of grooves: 20
Pulley Diameter (Ref): 2.357
Hub Diameter: 1.00
Overall Length: 1.25
Nominal Shaft Diameter: 3/8
Keyway: None
Show Grooves: 20

4 Assemble the pulleys onto the shafts as shown.

See Figure 11-19.

5 Click **Insert, Assembly Feature**, and **Belt/Chain**.

See Figure 11-19.

Figure 11-19

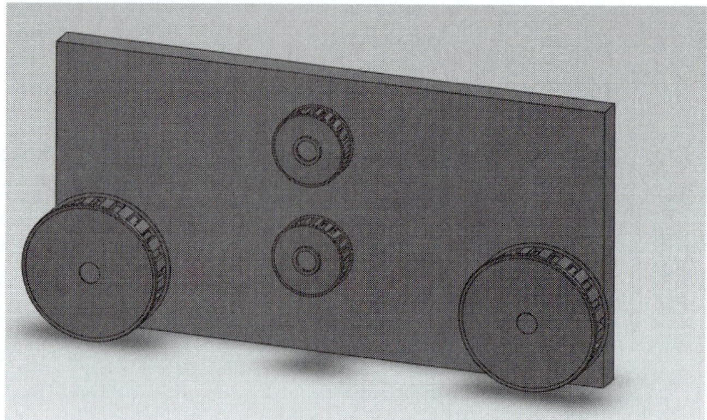

6 Click the top surface of the pulley's teeth to identify the belt location. Use the belt position shown. Use the **Flip belt side** tool if necessary.

7 Create a belt that is .42 wide and .14 thick centered on the pulleys.

Figure 11-20 shows the assembly with a belt profile.

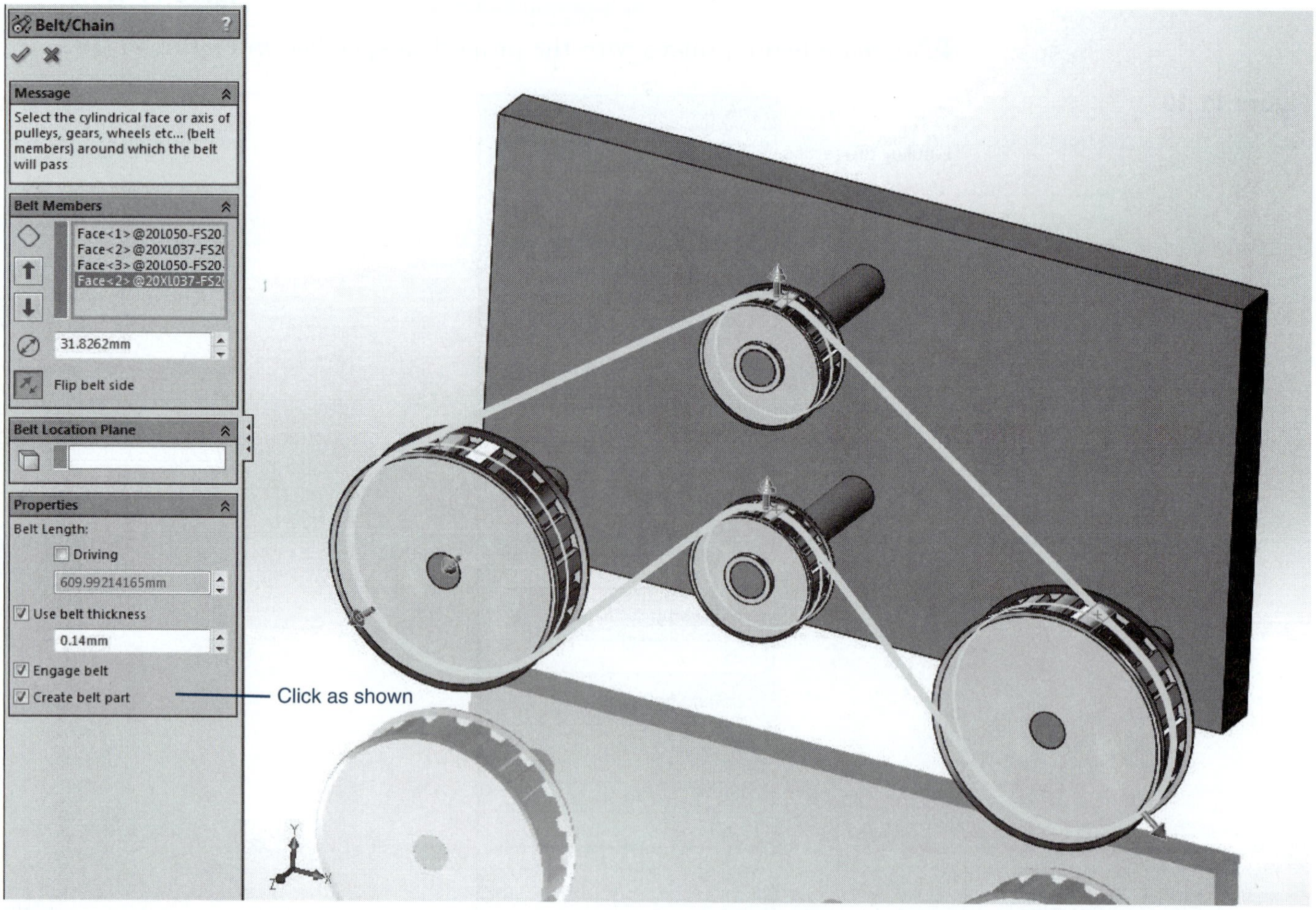

Figure 11-20

Figure 11-20
(*Continued*)

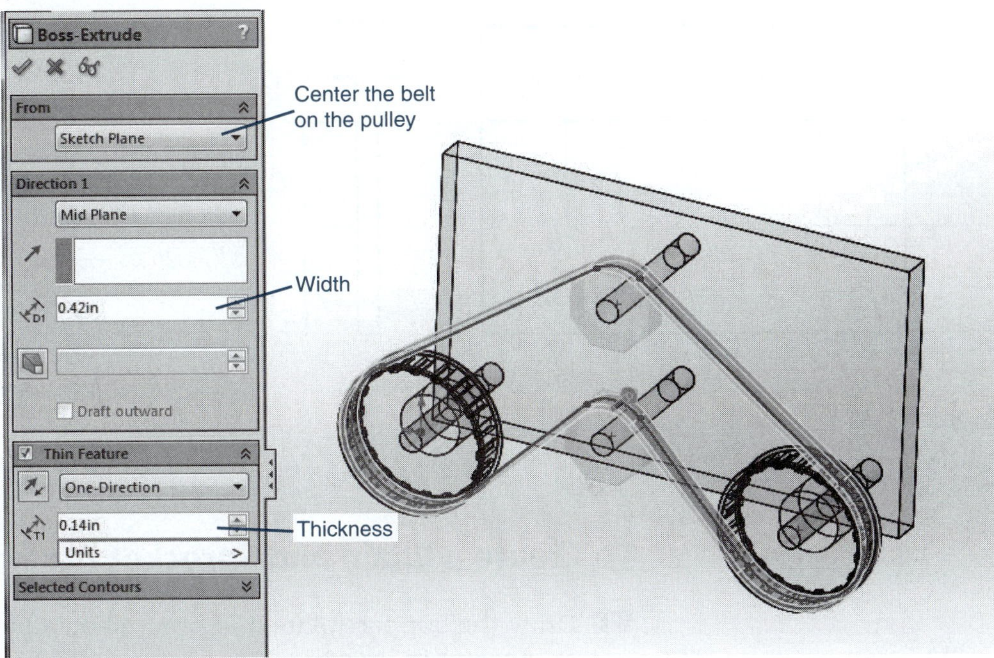

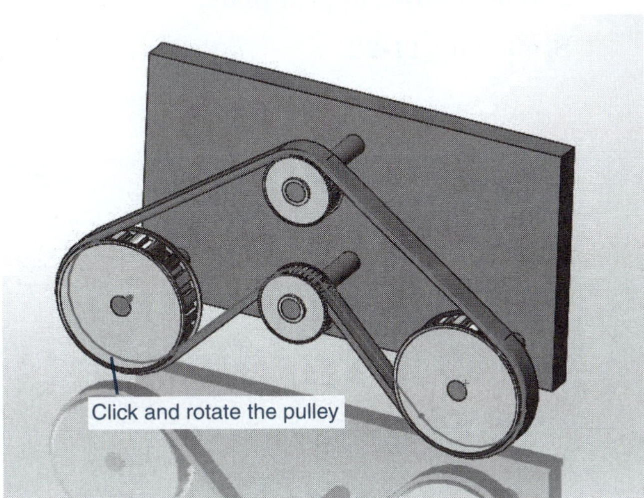

Click and rotate the lower left pulley. Note the direction of rotation for each pulley.

NOTE

If you use the cursor to rotate the edge line of one of the pulleys, they will all rotate.

11-5 Chains and Sprockets

SolidWorks creates chains and sprockets in a manner similar to that used to create belts and pulleys. The resulting chain is a representation of a chain but looks like a belt representation.

Figure 11-21 shows a support plate and a shaft that will be used to create a chain and sprocket assembly.

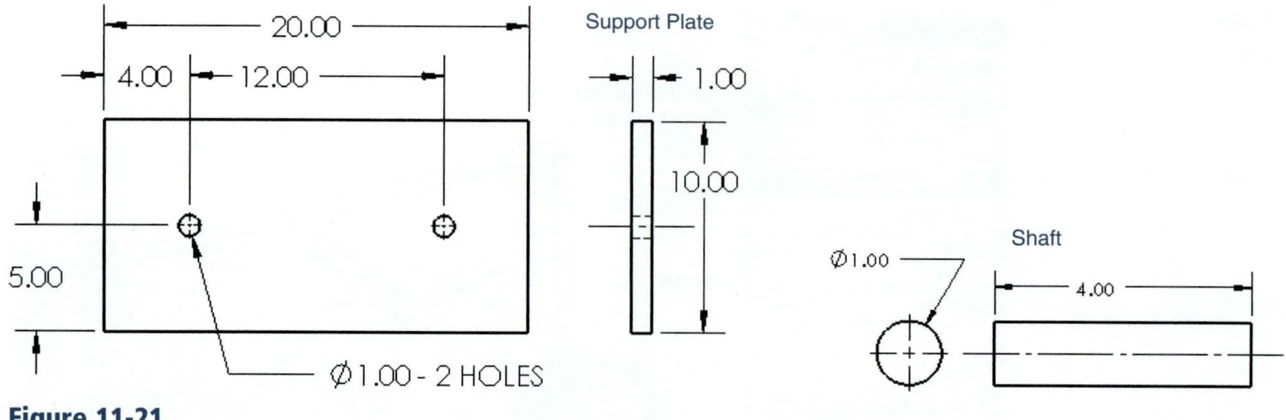

Figure 11-21

To Create a Chain and Sprocket Assembly

1. Draw the support plate and shaft shown in Figure 11-21 and create an **Assembly** drawing. The shafts should extend 3.00 beyond the top surface of the support plate.

   See Figure 11-22.

Figure 11-22

2. Access the **Design Library** and click **Toolbox, ANSI Inch, Power Transmission**, and **Chain Sprockets**.
3. Click **Silent Larger Sprocket** and drag the icon into the field of the drawing.

   See Figure 11-23.
4. Set the following chain property values:

   Chain Number = **SC610**

   Number of Teeth = **24**

Figure 11-23

Hub Style = **None**

Nominal Shaft Diameter = **1**

Keyway = **None**

5 Create a second Silent Larger Sprocket and assemble the sprockets onto the shafts.

See Figure 11-24.

Figure 11-24

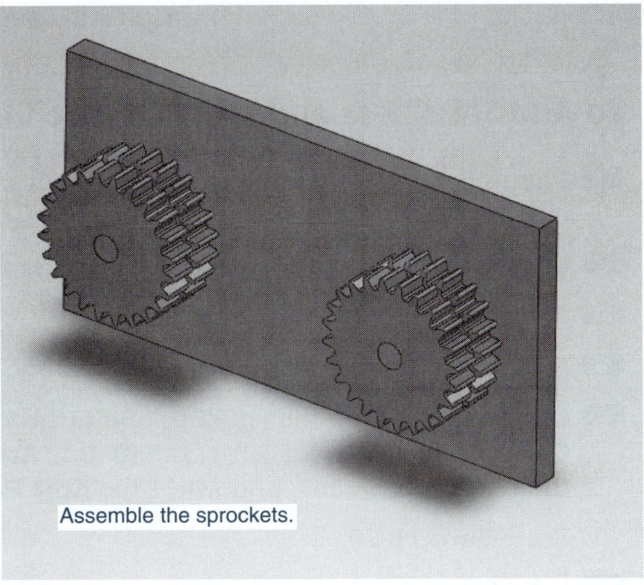

6 Click **Insert** at the top of the screen, then **Assembly Feature**, and **Belt/Chain**.

See Figure 11-25.

7 Click the bottom surface of a sprocket tooth, as shown, on both sprockets.

See Figure 11-25.

8 Click the green **OK** check mark.

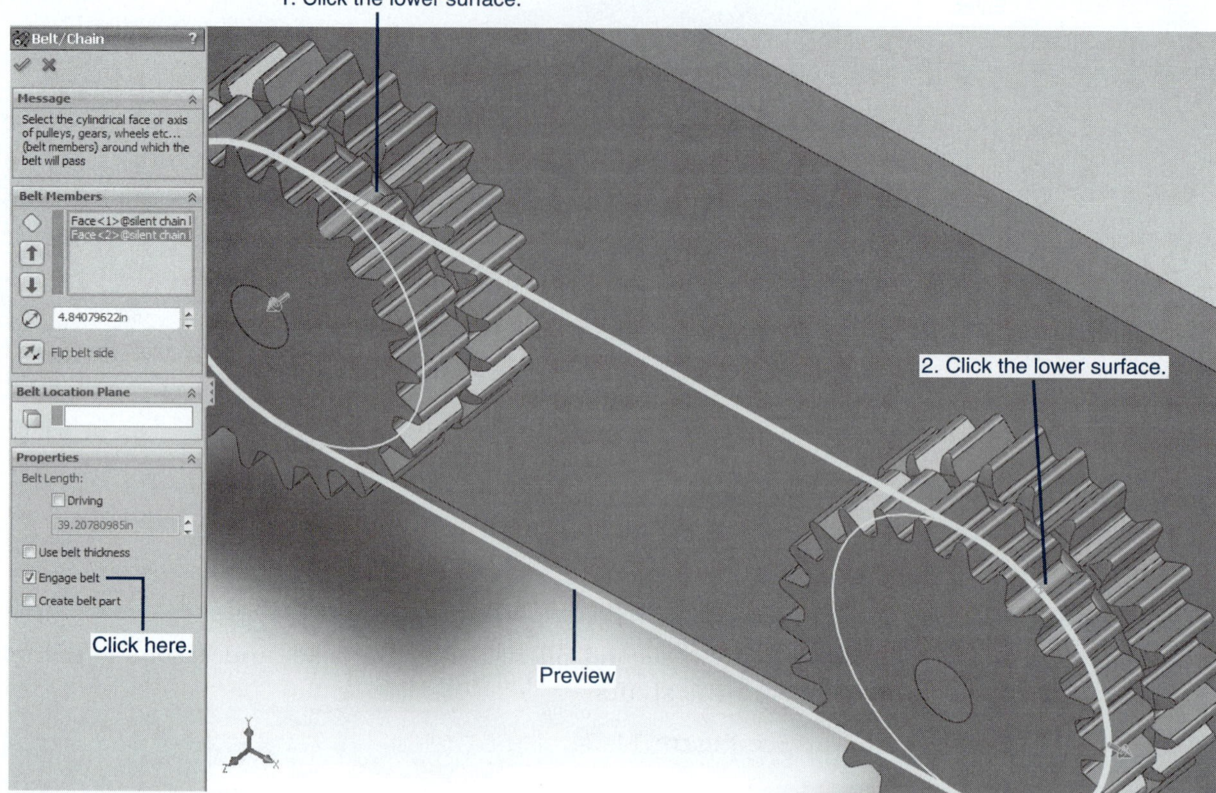

Figure 11-25

To Add Thickness and Width to the Chain

1 Right-click **Belt1** in the **FeatureManager** and select the **Edit Feature** option.

2 In the **Belt1** box scroll down and click the **Create belt part** box.

3 Click the green **OK** check mark.

4 Save the assembly as **Sprocket Assembly**.

5 Click the + sign to the left of the **Belt1** heading in the **FeatureManager**. Click the + sign to the left of the **[Belt1, Assem . .]** heading, right-click the **Sketch2** heading, and select the **Edit Part** option.

See Figure 11-26.

6 Left-click the upper horizontal segment of the belt (chain) and select the **Edit Sketch** option.

7 Click the **Extruded Boss/Base** tool on the **Features** tab.

Set **Direction 1** for **Blind, D1** for **2.00in**, and **Thin Feature** value for **0.125in**.

See Figures 11-27 and Figure 11-28.

Figure 11-26

1. Click here.
2. Click here.

(-) silent chain larger sproc
Mates
Belt1 (Length=39.21in)
(-) Sketch2
[Belt1-17^Sprocket Ass
Mates in Sprocket A
Sensors
Annotations
Material <not specif
Front Plane
Top Plane
Right Plane
Origin
(-) Sketch2

3. Right-click here and select the Edit Sketch option.

Feature (Sketch2)
Underive
List External Refs...
Parent/Child...
Delete...
Feature Properties...
Go To...

Boss-Extrude

From
Sketch Plane

Direction 1
Blind

2.00in

Draft outward

Direction 2

Thin Feature
One-Direction

0.125in

Cap ends

Selected Contours

1. Define the belt width.

3. Click here.

2. Define the thickness.

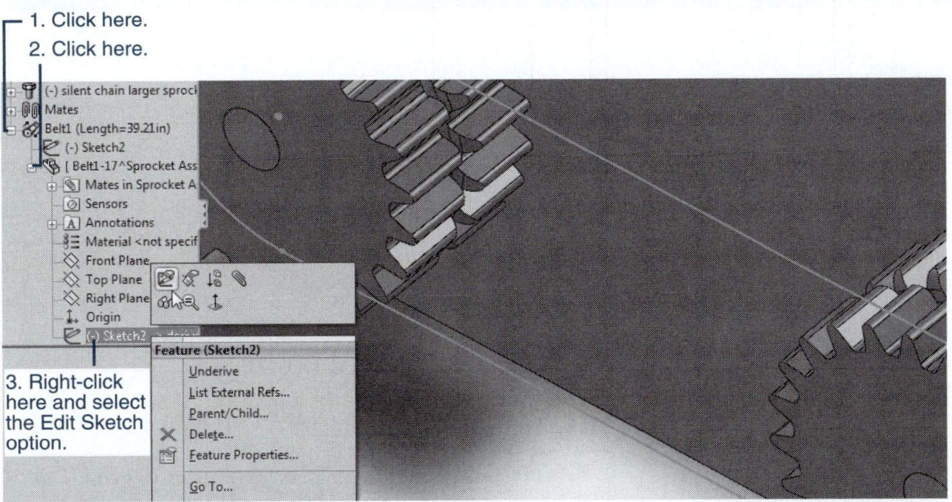

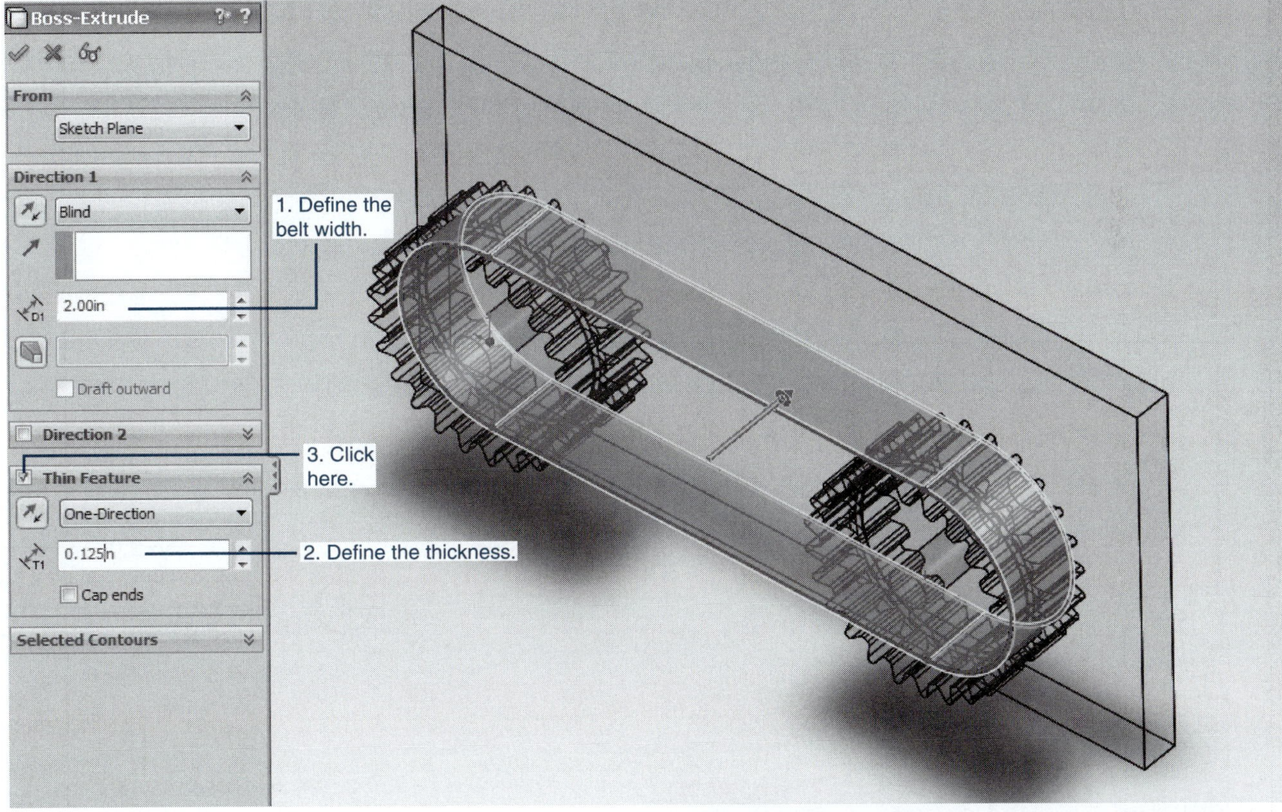

Figure 11-27

The completed assembly

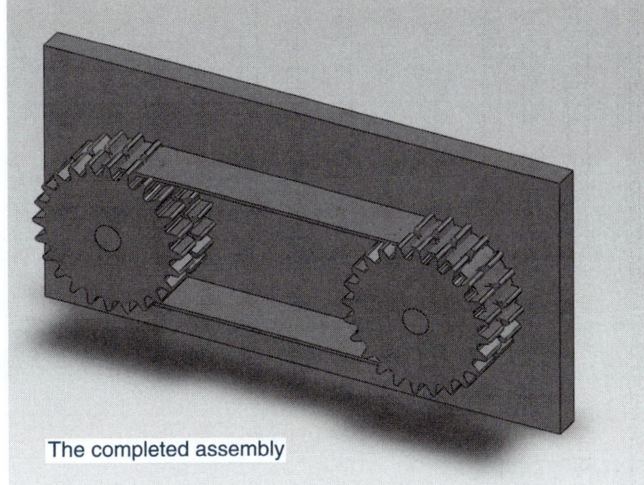

Figure 11-28

Chapter Projects

Project 11-1: Inches

See Figure P11-1.

A. Create a Ø.375 × 2.00 shaft.

B. Create the support plate shown.

C. Access the **Design Library** and create two pulleys.

The pulleys are defined as follows:

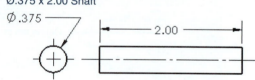

Pulley 1:
Belt Pitch = **(.375)–L**
Belt Width = **.5**

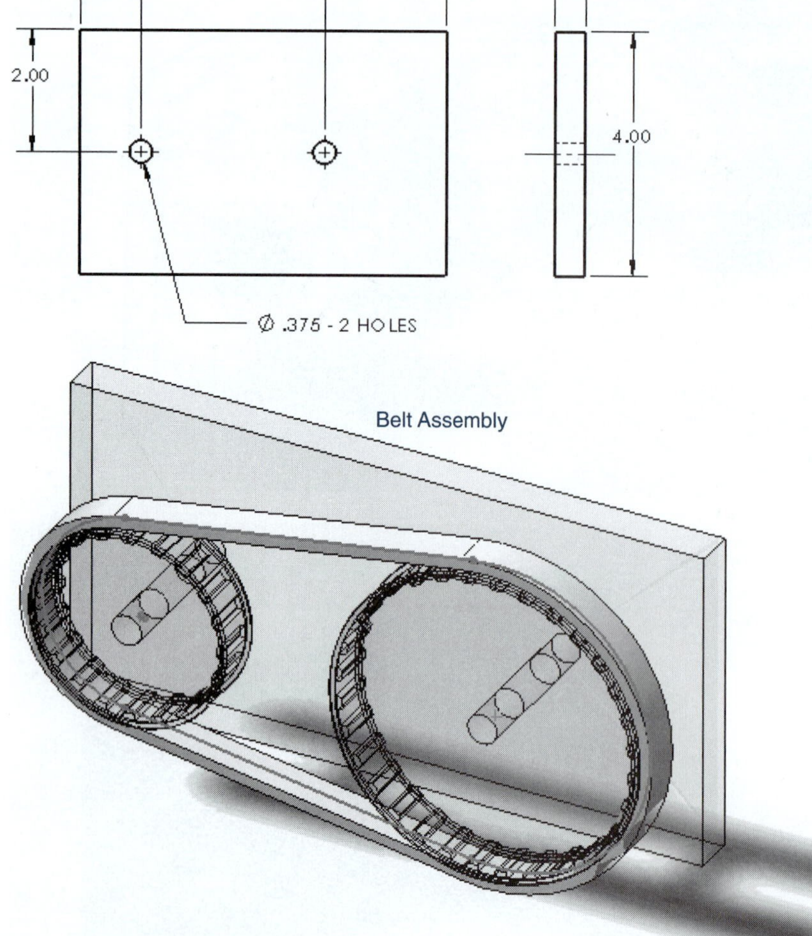

Figure P11-1

Pulley Style = **Flanged**
Number of Grooves = **20**
Hub Diameter = **.500**
Overall Length = **1.500**
Nominal Shaft Diameter = **3/8**
Keyway = **None**

Pulley 2:
Belt Pitch = **(.375)–L**
Belt Width = **.5**
Pulley Style = **Flanged**
Number of Grooves = **32**
Hub Diameter = **.500**
Overall Length = **1.500**
Nominal Shaft Diameter = **3/8**
Keyway = **None**

D. Assemble the shafts into the support plate.

E. Assemble the pulleys onto the shafts.

F. Add a timing belt between the pulleys. Make the belt a thin feature with a .14 thickness. Make the width of the belt .42.

Project 11-2: Inches—Design Problem

A. Create a Ø.500 × 3.50 shaft.

B. Create a support plate so that the center distance between the pulleys' centerpoints is 5.52 and the distance between the outside edge of the pulleys and the edge of the support plate is at least .50 but less than 1.00.

C. Access the **Design Library** and create two pulleys.

The pulleys are defined as follows:

Pulley 1:
Belt Pitch = **(.500)–H**
Belt Width = **1.5**
Pulley Style = **Unflanged**
Number of Grooves = **26**
Hub Diameter = **1.000**
Hub Length = **1.000**
Nominal Shaft Diameter = **1/2**
Keyway = **None**

Pulley 2:
Belt Pitch = **(.500)–H**
Belt Width = **1.5**
Pulley Style = **Unflanged**

Number of Grooves = **44**
Hub Diameter = **1.00**
Overall Length = **1.00**
Nominal Shaft Diameter = **1/2**
Keyway = **None**

D. Assemble the shafts into the support plate.

E. Assemble the pulleys onto the shafts.

F. Add a timing belt between the pulleys. Make the belt a thin feature with a .16 thickness. Make the width of the belt 1.50 in.

Project 11-3: Inches—Design Problem

A. Create a Ø.500 × 4.50 shaft.

B. Create a support plate so that the center distance between the pulleys' centerpoints is 21.50 and the distance between the outside edge of the pulleys and the edge of the support plate is at least .50 but less than 1.00.

C. Access the **Design Library** and create two pulleys.

The pulleys are defined as follows:

Pulley 1:
Belt Pitch = **(.875)–XH**
Belt Width = **2**
Pulley Style = **Unflanged**
Number of Grooves = **30**
Hub Diameter = **3.00**
Overall Length = **6.00**
Nominal Shaft Diameter = **1/2**
Keyway = **None**

Pulley 2:
Belt Pitch = **(.875)–XH**
Belt Width = **2.0**
Pulley Style = **Unflanged**
Number of Grooves = **48**
Hub Diameter = **3.00**
Overall Length = **6.00**
Nominal Shaft Diameter = **1/2**
Keyway = **None**

D. Assemble the shafts into the support plate.

E. Assemble the pulleys onto the shafts.

F. Add a #10-24 threaded hole to each pulley hub and insert a #10-24 UNC Slotted Set Screw Dog Point into each threaded hole.

G. Add a timing belt between the pulleys. Make the belt a thin feature with a .44 thickness. Make the width of the belt 2.00.

Project 11-4: Inches

Create a support plate and shaft as defined in Figure P11-4.

A. Create three Ø.375 × 3.00 shafts.

B. Create three identical pulleys as defined below.

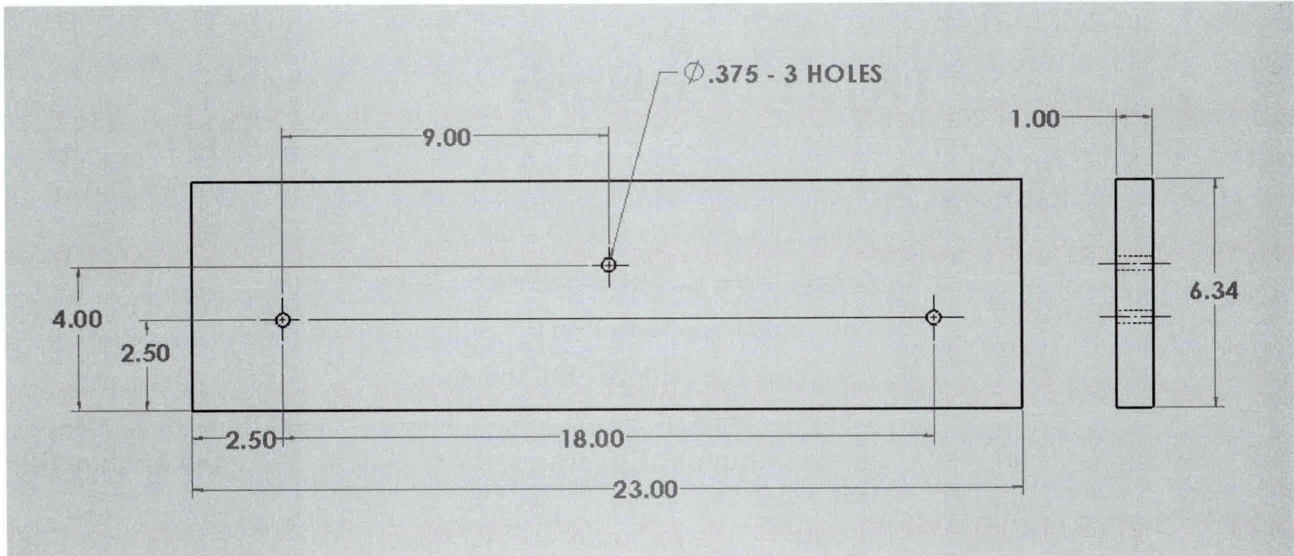

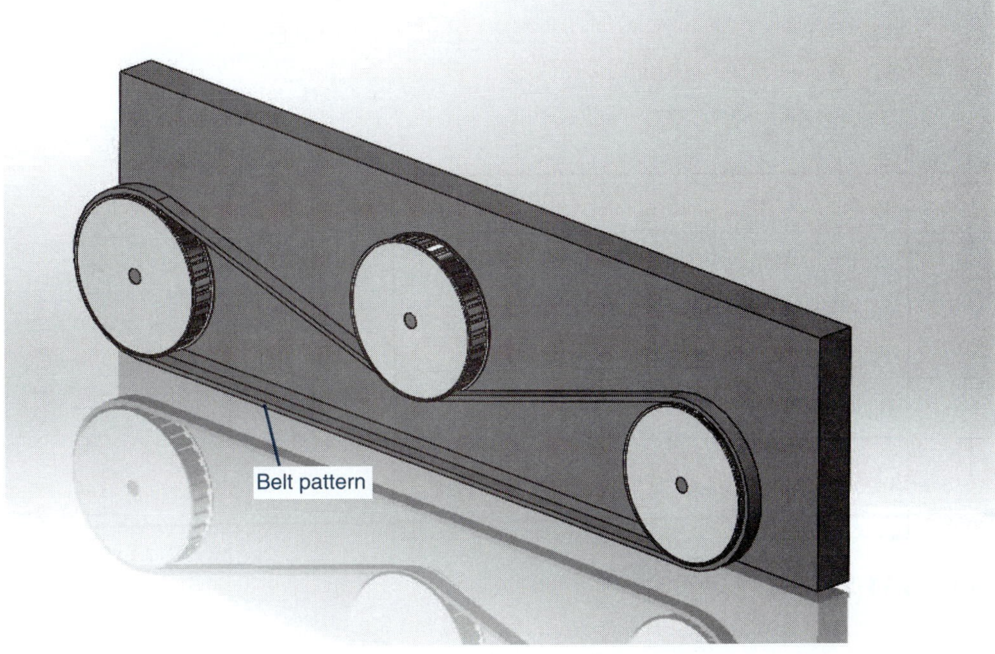

Figure P11-4

C. Assemble the shafts into the support plate.

D. Assemble the pulleys onto the shafts.

E. Add a timing belt between the pulleys in the pattern shown. Make the belt a thin feature with a .14 thickness. Make the width of the belt .42.

Pulley Properties:

Belt Pitch = **(.375)–L**
Belt Width = **.5**
Pulley Style = **Flanged**
Number of Grooves = **32**
Hub Diameter = **.875**
Overall Length = **1.000**
Nominal Shaft Diameter = **3/8**
Keyway = **None**

Project 11-5: Inches

A. Create a support plate and shaft as defined in Figure P11-5.

B. Create four Ø.375 × 3.00 shafts.

C. Create three identical pulleys as defined below.

D. Assemble the shafts into the support plate.

E. Assemble the pulleys onto the shafts.

F. Add a timing belt between the pulleys in the pattern shown. Make the belt a thin feature with a .14 thickness. Make the width of the belt .42.

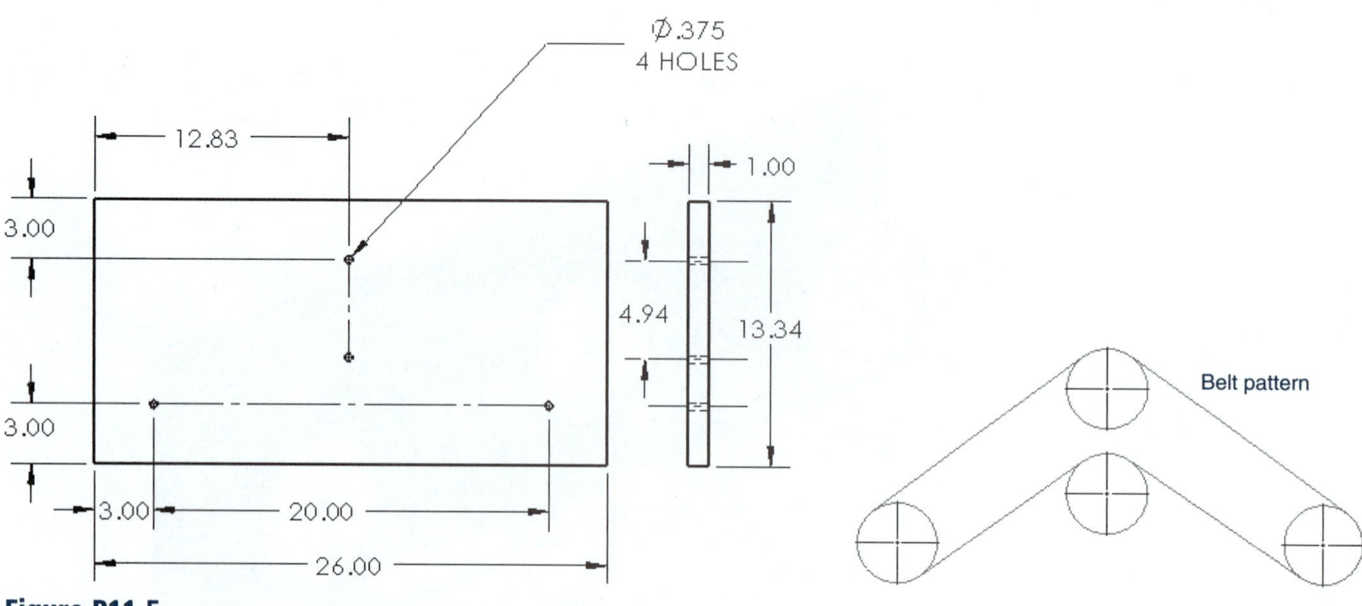

Figure P11-5

Pulley Properties:

- Belt Pitch = **(0.375)–L**
- Belt Width = **.5**
- Pulley Style = **Flanged**
- Number of Grooves = **32**
- Hub Diameter = **1.000**
- Overall Length = **2.000**
- Nominal Shaft Diameter = **3/8**
- Keyway = **None**

Project 11-6: Inches

A. Create a support plate and shaft as defined in Figure P11-6. Holes labeled A are Ø.375, and holes labeled B are Ø.500.

B. Create three Ø.375 × 3.00 shafts and two Ø.500 × 3.00 shafts.

C. Create three small pulleys and two large pulleys as defined below.

D. Assemble the shafts into the support plate.

E. Assemble the pulleys onto the shafts.

F. Add a timing belt between the pulleys in the pattern shown. Make the belt a thin feature with a .16 thickness. Make the width of the belt .92.

Pulley 1 Properties:

- Belt Pitch = **(.375)–L**
- Belt Width = **1.00**
- Pulley Style = **Flanged**
- Number of Grooves = **24**

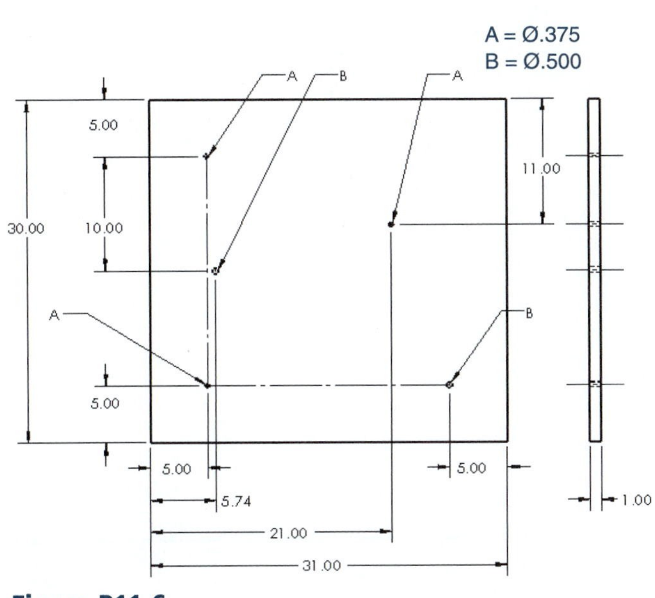

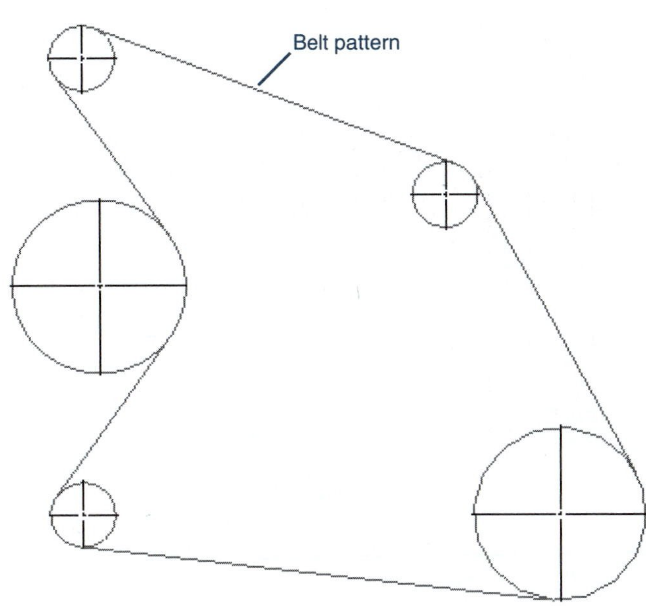

Figure P11-6

Hub Diameter = **2.000**
Overall Length = **2.000**
Nominal Shaft Diameter = **3/8**
Keyway = **None**

Pulley 2 Properties:
Belt Pitch = **(.500)–H**
Belt Width = **1.00**
Pulley Style = **Flanged**
Number of Grooves = **48**
Hub Diameter = **2.000**
Overall Length = **2.000**
Nominal Shaft Diameter = **1/2**
Keyway = **None**

chapter twelve

Cams

CHAPTER OBJECTIVES

- Learn how to draw cams using SolidWorks
- Learn how to draw displacement diagrams
- Understand the relationship between cams and followers

12-1 Introduction

Cams are mechanical devices used to translate rotary motion into linear motion. Traditionally, cam profiles are designed by first defining a displacement diagram and then transferring the displacement diagram information to a base circle. Figure 12-1 shows a cam and a displacement diagram.

12-2 Base Circle

Cam profiles are defined starting with a base circle. The diameter of the base circle will vary according to the design situation. The edge of the base circle is assumed to be the 0.0 displacement line on the displacement diagram.

12-3 Trace Point

The trace point is the centerpoint of the roller follower. SolidWorks defines the shape of the cam profile by defining the path of the trace point.

Figure 12-1

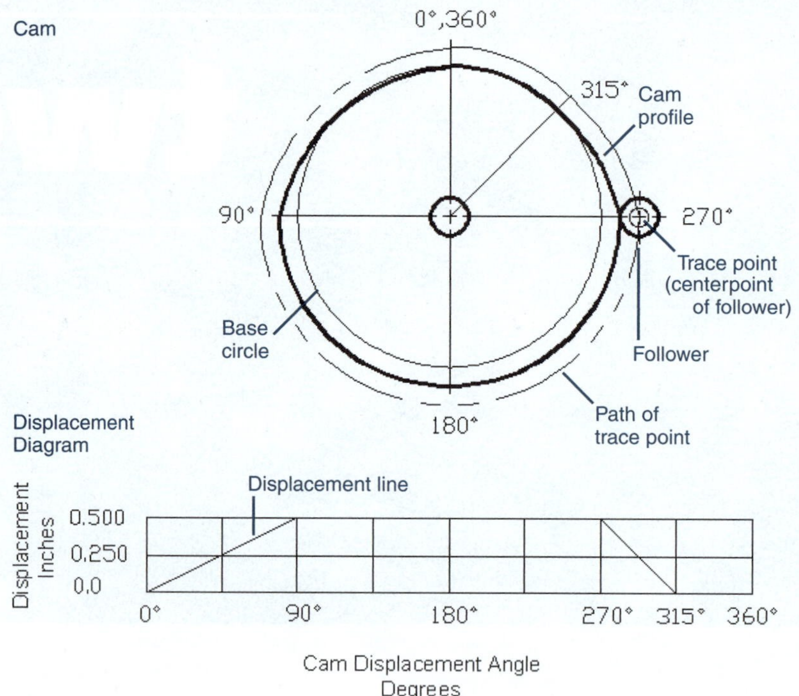

12-4 Dwell, Rise, and Fall

In the displacement diagram shown in Figure 12-1, the displacement line rises .500 in. in the first 90°. This type of motion is called *rise*.

The displacement line then remains at .500 from 90° to 270°. This type of motion is called *dwell*.

> **TIP**
>
> A circle is a shape of constant radius. If a circle was used as a cam, the follower would not go up or down but would remain in the same position, because a circle's radius is constant.

The displacement line falls 0.500 from 270° to 315°. This type of motion is called *fall*. The displacement line dwells between 315° and 360°.

Shape of the Rise and Fall Lines

The shape of the cam's surface during either a rise or a fall is an important design consideration. The shape of the profile will affect the acceleration and deceleration of the follower, and that will in turn affect the forces in both the cam and the follower. SolidWorks includes 13 different types of motions.

Cam Direction

Note that the 90° reference is located on the left side of the cam. This indicates clockwise direction.

12-5 Creating Cams in SolidWorks

Cams are created in SolidWorks by working from existing templates. There are templates for circular and linear cams and for internal and external cams. The templates allow you to work directly on the cam profile and eliminate the need for a displacement diagram. In the following example a circular cam with a 4.00-in. base circle and a profile that rises 0.5 in. in 90° using harmonic motion, dwells for 180°, falls 0.50 in. in 45° using harmonic motions, and dwells for 45°. See the approximate shape of the cam presented in Figure 12-1.

To Access the Cam Tools

1. Create a new **Part** document.
2. Select a **Front Plane** orientation and create a sketch plane.
3. Click the **Toolbox** heading at the top of the screen.

> **NOTE**
>
> You may have to configure the **Toolbox** and use the **Add-Ins . . .** feature under the **Tools** heading to add the **Cam** features to your system.

4. Click the **Cams** . . . tool.

See Figure 12-2. The **Cam - Circular** toolbox will appear. See Figure 12-3.

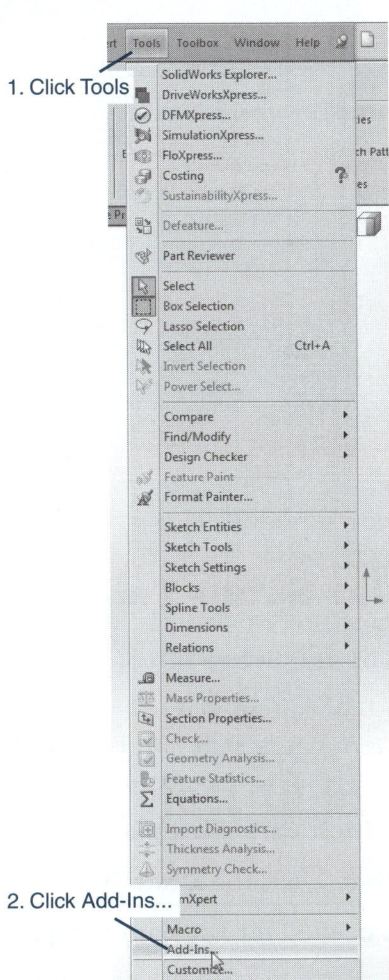

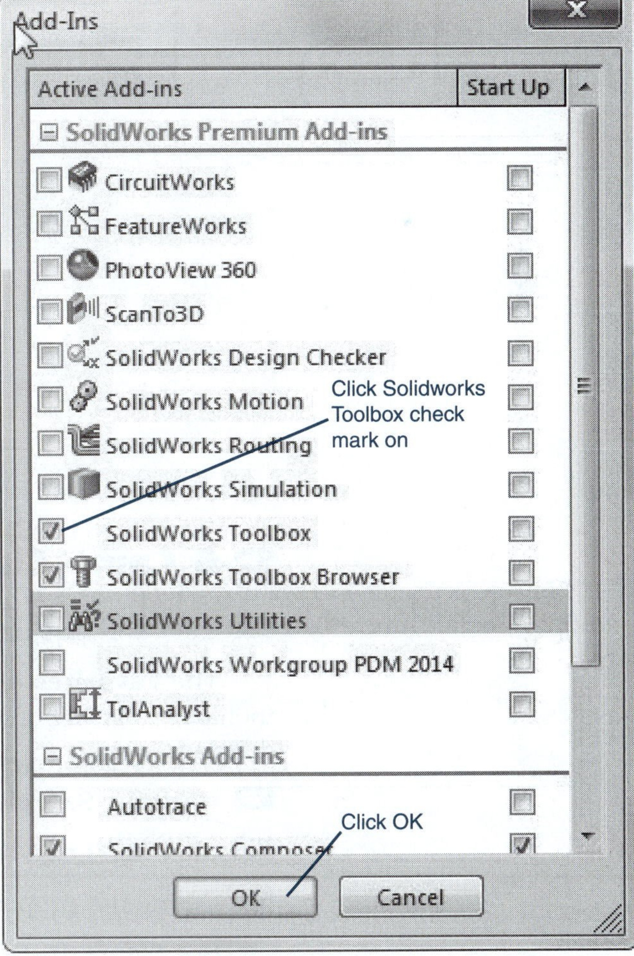

Figure 12-2

Figure 12-2
(Continued)

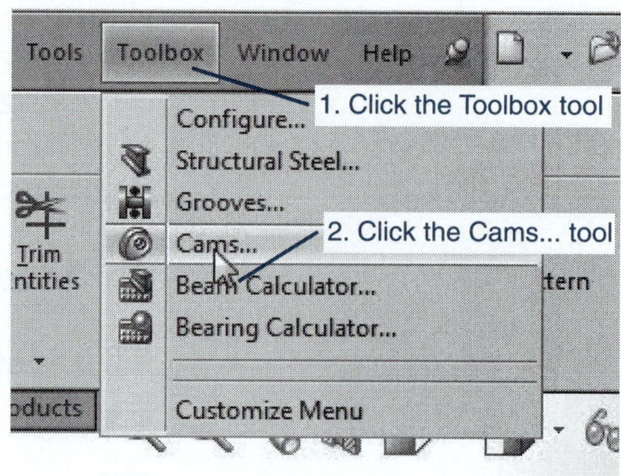

Figure 12-3

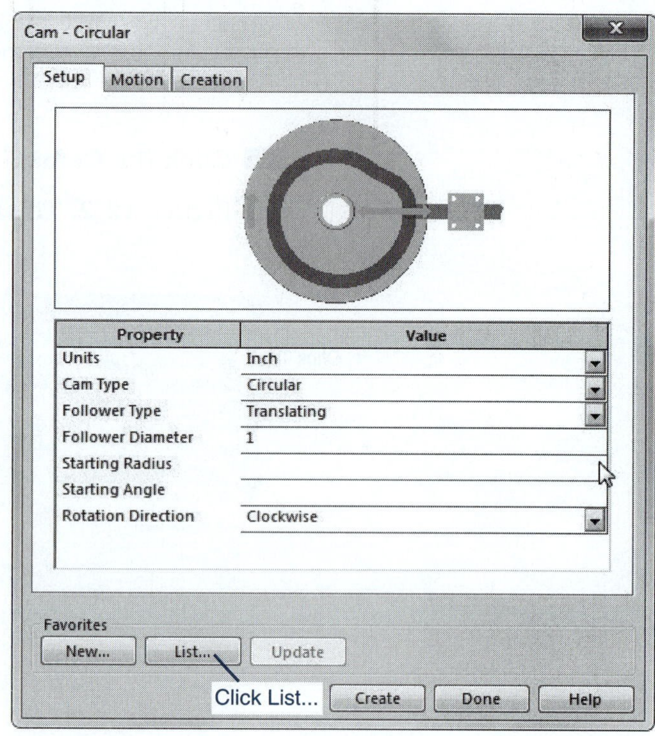

12-6 Cam - Circular Setup Tab

1. Click the **List** option on the **Setup** tab of the **Cam - Circular** toolbox.

   The **Favorites** dialog box will appear. See Figure 12-4. This box includes a listing of cam templates that can be modified to create a different cam.

2. Click the **Sample 2 - Inch Circular** option.
3. Click the **Load** box, then click the **Done** box.
4. Define the properties needed for the cam setup.

   See Figure 12-5. The properties for the example cam are as follows:

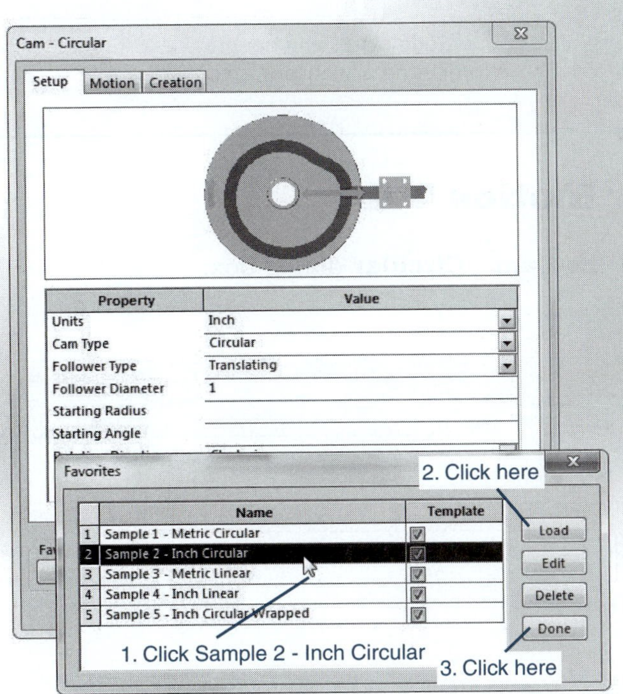

Figure 12-4

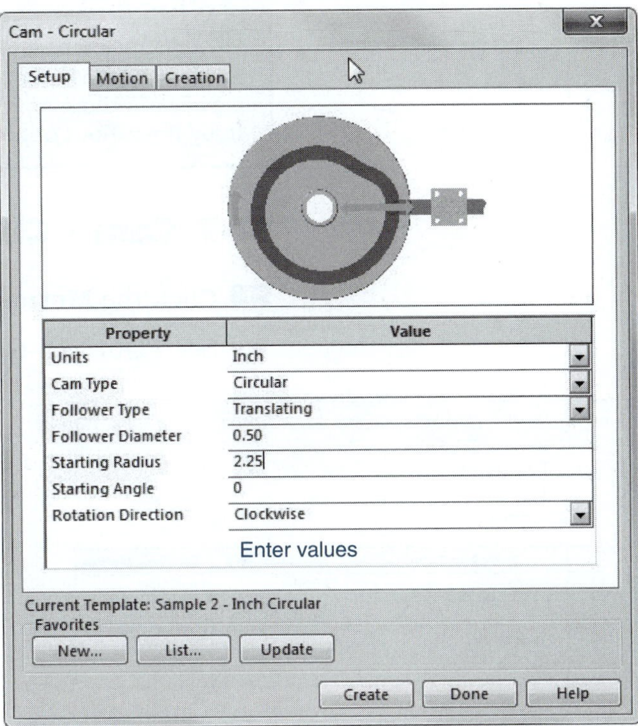

Figure 12-5

Units: **Inch**

Cam Type: **Circular**

Follower Type: **Translating**

This follower type will locate the follower directly inline with a ray from the cam's centerpoint.

Follower Diameter: **0.50**

Starting Radius: **2.25**

This property defines a Ø4.00 base circle with a radius of 2.00 plus an additional 0.25 radius value to reach the centerpoint of the follower. The follower has a diameter of 0.5.

NOTE

The cam profile is defined by the *path of the trace point*. The trace point is the centerpoint of the circular follower. In this example the trace point at the 0.0° point on the cam is located 2.00 + 0.25 from the centerpoint of the cam.

Starting Angle: **0**

This property defines the ray between the cam's centerpoint and the follower's centerpoint as 0.0°.

Rotation Direction: **Clockwise**

See Figure 12-5.

TIP

Some of the **Value** boxes include other options. SolidWorks will generate real-time previews as these options are clicked. Those interested are encouraged to click and study the various options available.

12-7 Cam - Circular Motion Tab

1 Click the **Motion** tab on the **Cam - Circular** dialog box.

See Figure 12-6.

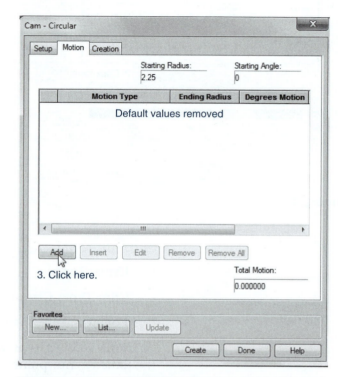

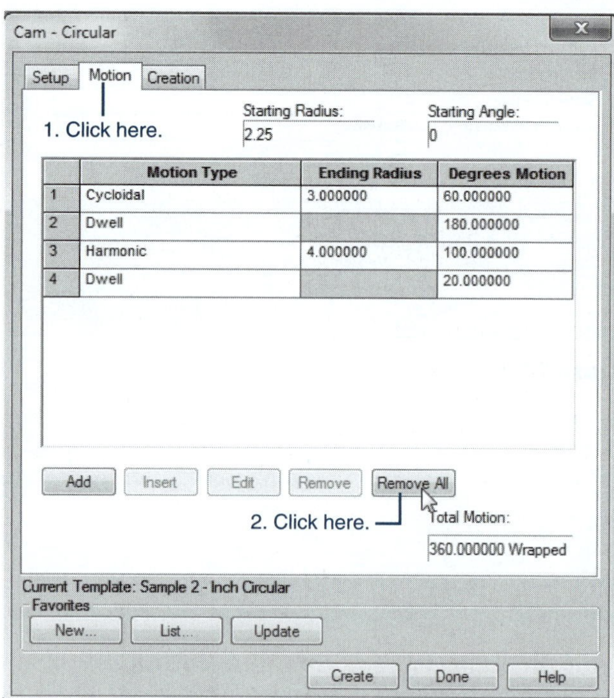

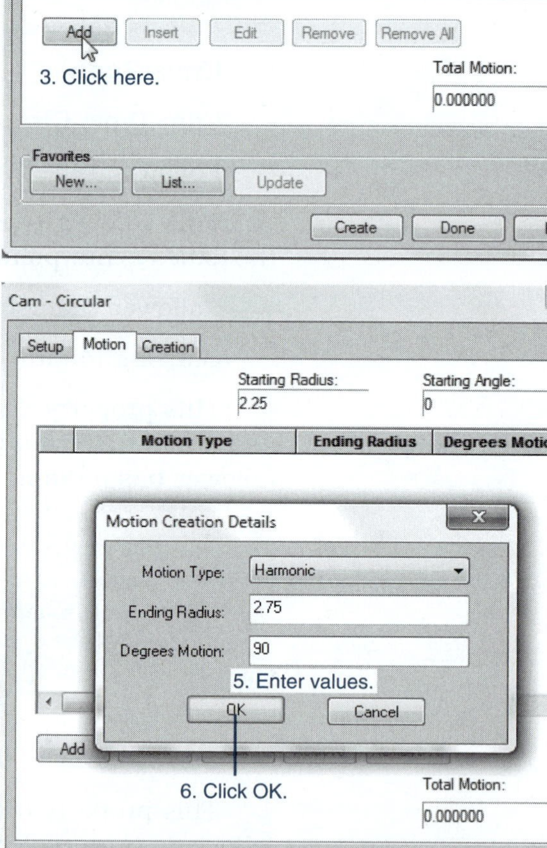

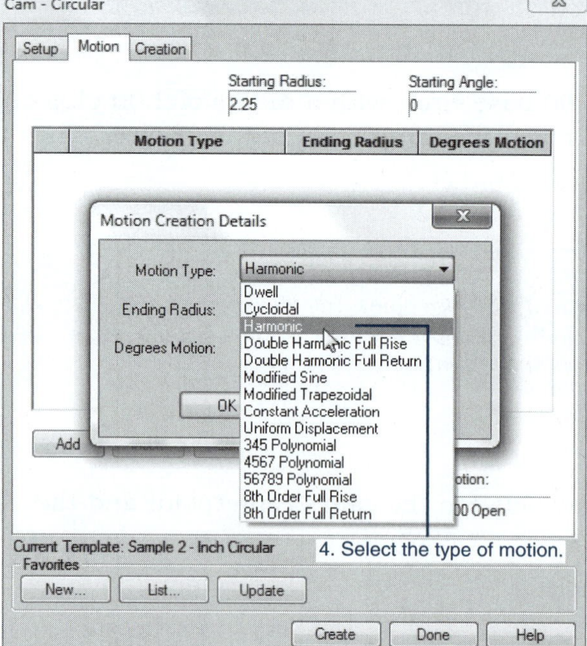

Figure 12-6

2 Click the **Remove All** box.

This step will remove all existing motion types and enable you to define a new cam.

3 Click the **Add** box.

The **Motion Creation Details** dialog box will appear.

4 Click the arrowhead to the right of the **Motion Type** box.

5 Define the **Motion Type** as **Harmonic**.

6 Define the **Ending Radius** as **2.75** and the **Degrees Motion** as **90**; click **OK**.

These values define the follower rise as 0.50 in. over a distance of 90° using harmonic motion.

7 Click the **Add** box again.

8 Define the following details:

Motion Type: **Dwell**

Degrees Motion: **180**

Because the dwell motion type was selected, the ending radius will automatically be the same as the starting radius.

9 Click **OK**.

10 Click the **Add** box again.

11 Set the motion as follows:

Motion Type: **Harmonic**

Ending Radius: **2.25**

Degrees Motion: **45.00**

This will return the follower to the base circle.

12 Click the **Add** box again.

13 Select a **Motion Type** of **dwell** and **Degrees Motion** of **45**.

Note that the **Total Motion** is 360.00. The cam profile has now returned to the original starting point of 0.0°. This is called the ***closed condition***. If the total number of degrees of motion is less than 360°, it is called an ***open condition***. If the number of degrees of motion is greater than 360°, it is called the ***wrapped condition***.

Figure 12-7 shows the finished **Motion** tab box.

12-8 Cam - Circular Creation Tab

1 Click the **Creation** tab on the **Cam - Circular** dialog box.

Enter the appropriate values as shown in Figure 12-8.

2 Define the cam's **Blank Outside Dia** as **6** and the **Thickness** as **.50 in**.

This cam will not have a hub.

3 Define both the **Near** and **Far Hub Dia & Length** as **0**.

4 Define the **Blank Fillet Rad & Chamfer** as **0**.

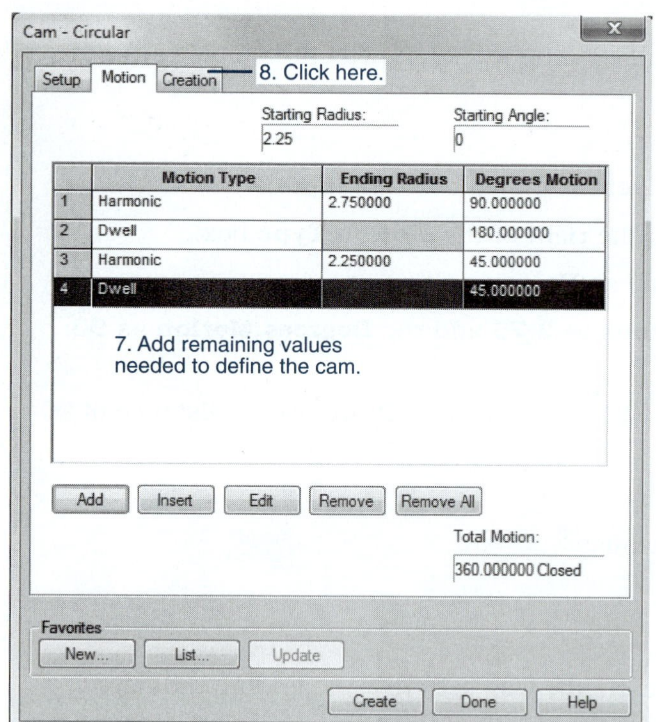

Figure 12-7

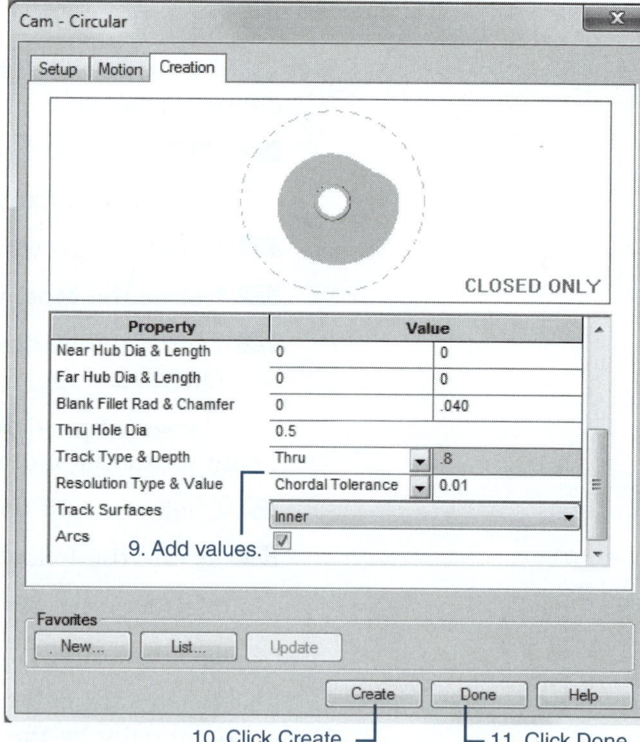

Figure 12-8

5 Define the **Thru Hole Dia** as **0.5**.

6 Define the **Track Type & Depth** as **Thru**.

7 Click the **Arcs** box so that a check mark appears.

8 Set the **Track Surfaces** for **Inner**.

Accept all the other default values.

9 Click the **Create** box.

10 Click the **Done** box.

11 Click the **Top Plane** orientation.

Accept all other default values. Figure 12-9 shows the finished cam. Figure 12-10 shows a top view of the cam. It also shows the cam with the original Ø4.00 base circle superposed onto the surface. Note how the cam profile rises, dwells, falls, and dwells. Figure 12-10 also shows the relationship between the cam profile and the path of the trace point.

Figure 12-9

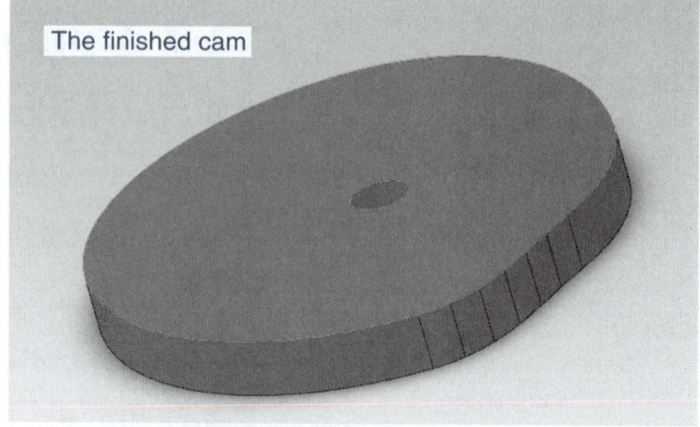

Figure 12-10

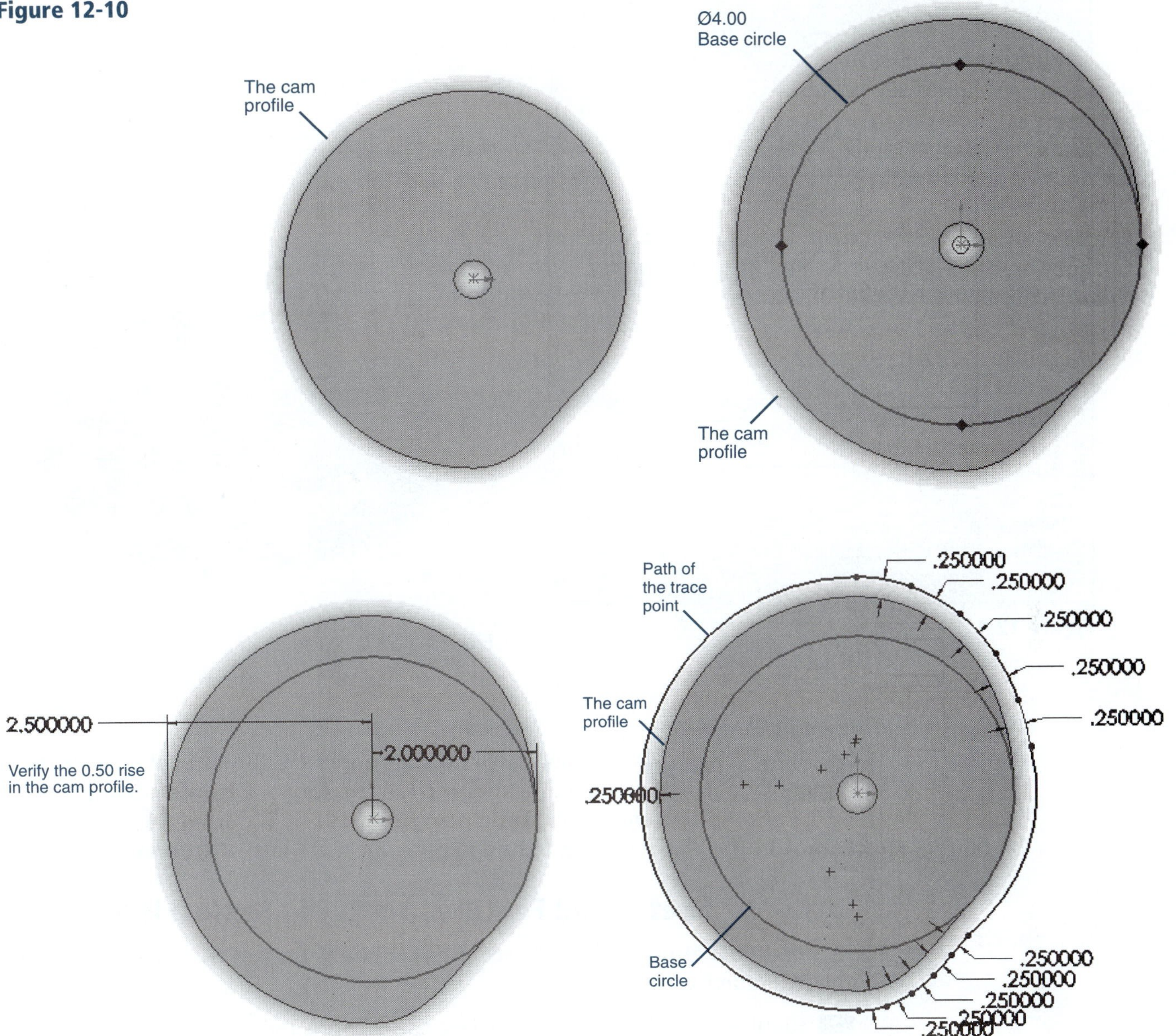

12-9 Hubs on Cams

There are two methods for creating hubs on cams: add the hub directly using the **Cam - Circular** dialog box, or create a hub on an existing cam using the **Sketch** and **Features** tools.

Using the Cam - Circular Dialog Box to Create a Hub

Figure 12-11 shows the **Creation** tab portion of the **Cam - Circular** dialog box that was originally presented in Figure 12-8. The values shown in Figure 12-11 include values for the **Near Hub**. The diameter is to be **Ø1.0** and the length **.75**. All other values are the same as those used to create the cam shown in Figure 12-9.

A value of **0.5** has also been entered in the **Thru Hole Dia** box. This will generate a Ø0.5 hole through the hub diameter. The hole will go through the hub and through the cam. All other values are the same.

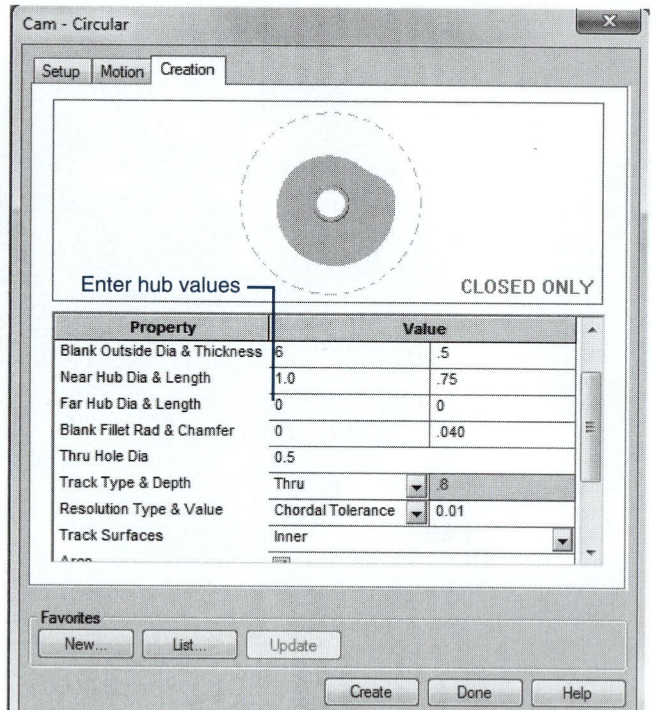

Figure 12-11

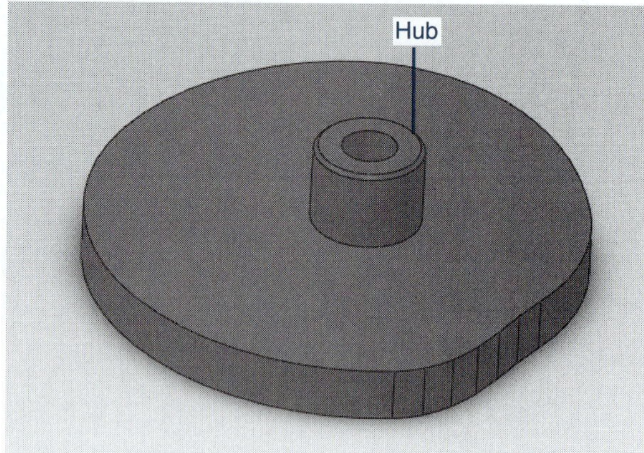

Figure 12-12

Figure 12-12 shows the modified cam that includes the hub. Note that the dimensions match those values entered in the **Cam - Circular** dialog box.

A second, far-side, hub could also be added by entering the appropriate values in the **Far Hub Dia & Length** box on the **Cam - Circular** dialog box.

Using the Sketch and Features Tools to Create a Hub

Figure 12-13 shows the cam created in the first part of this chapter. A hub can be added as follows.

1. Right-click the mouse and select the **Sketch** tool.
2. Use the **Circle** tool and add a circle on the sketch plan centered on the cam's centerpoint.
3. Use the **Smart Dimension** tool and size the circle to **Ø1.00**.
4. Use the **Extruded Boss/Base** tool to extrude the Ø1.00 circle **0.75**.
5. Right-click the mouse and add a new sketch to the top surface of the hub.
6. Use the **Sketch** and **Features** tools to create a **Ø0.50** circle on the top surface of the hub.
7. Use the **Extruded Cut** tool and cut the circle through both the hub and the cam a distance of **1.25 in**.

To Add a Threaded Hole to a Cam's Hub

Threaded holes are added to a cam's hub to accept set screws that hold a cam in place against a rotating shaft. Use the cam created in Section 12-9 with the **Cam - Circular** dialog box to create a **Hub** option.

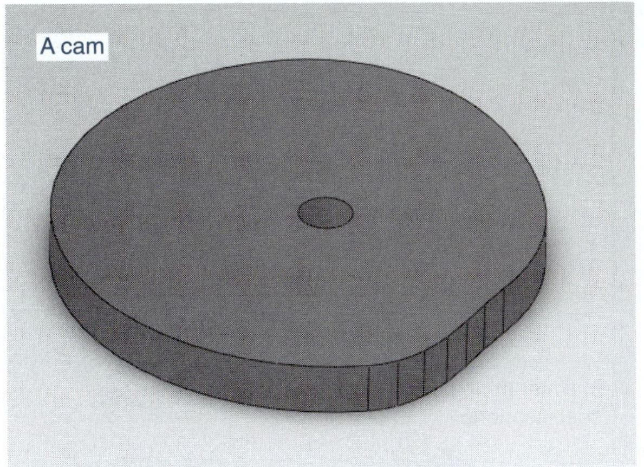
A cam

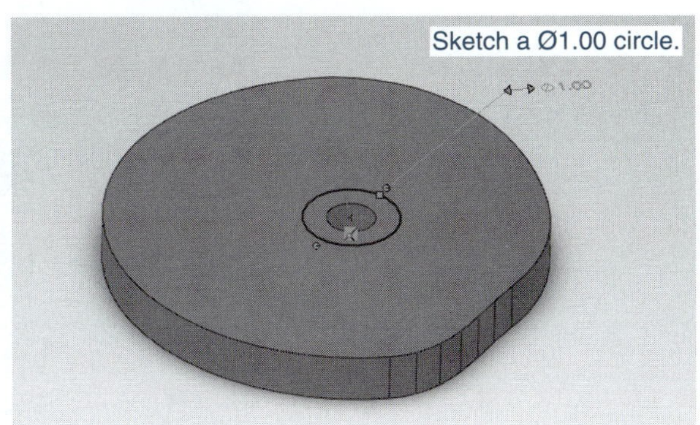
Sketch a Ø1.00 circle.

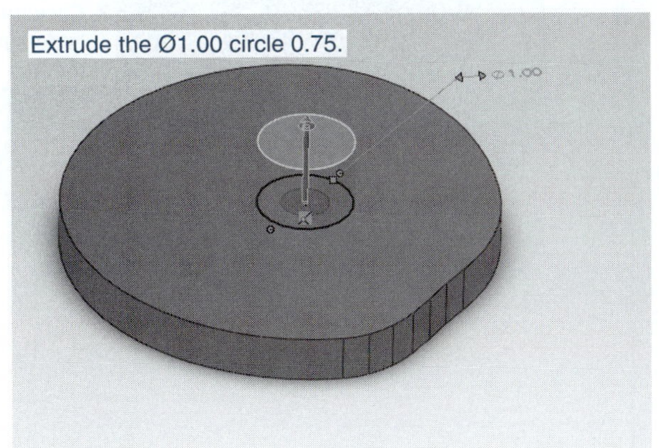
Extrude the Ø1.00 circle 0.75.

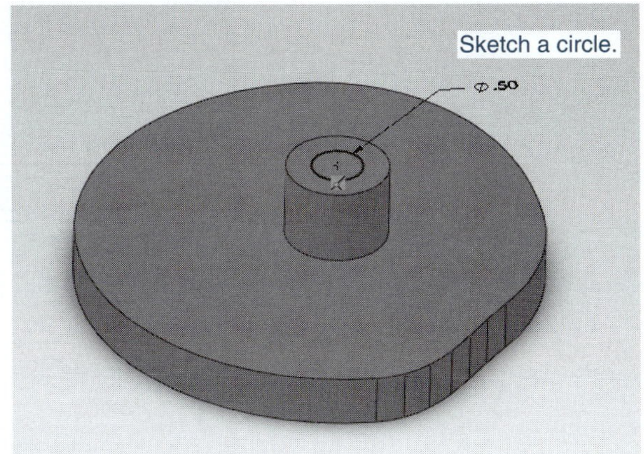
Sketch a circle.
Ø .50

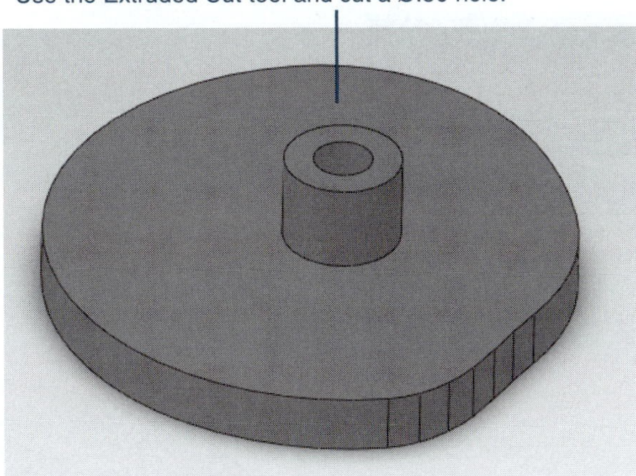
Use the Extruded Cut tool and cut a Ø.50 hole.

Figure 12-13

1. Click the **Hole Wizard** tool on the **Features** tool panel.
2. Click the **Tap** box in the **Hole Specification** box. See Figure 12-14.
3. Select the **ANSI Inch** standards.
4. Select a $\frac{1}{4}$**-20 UNC** thread.
5. Define the threaded hole's depth as **0.25**.

Figure 12-14

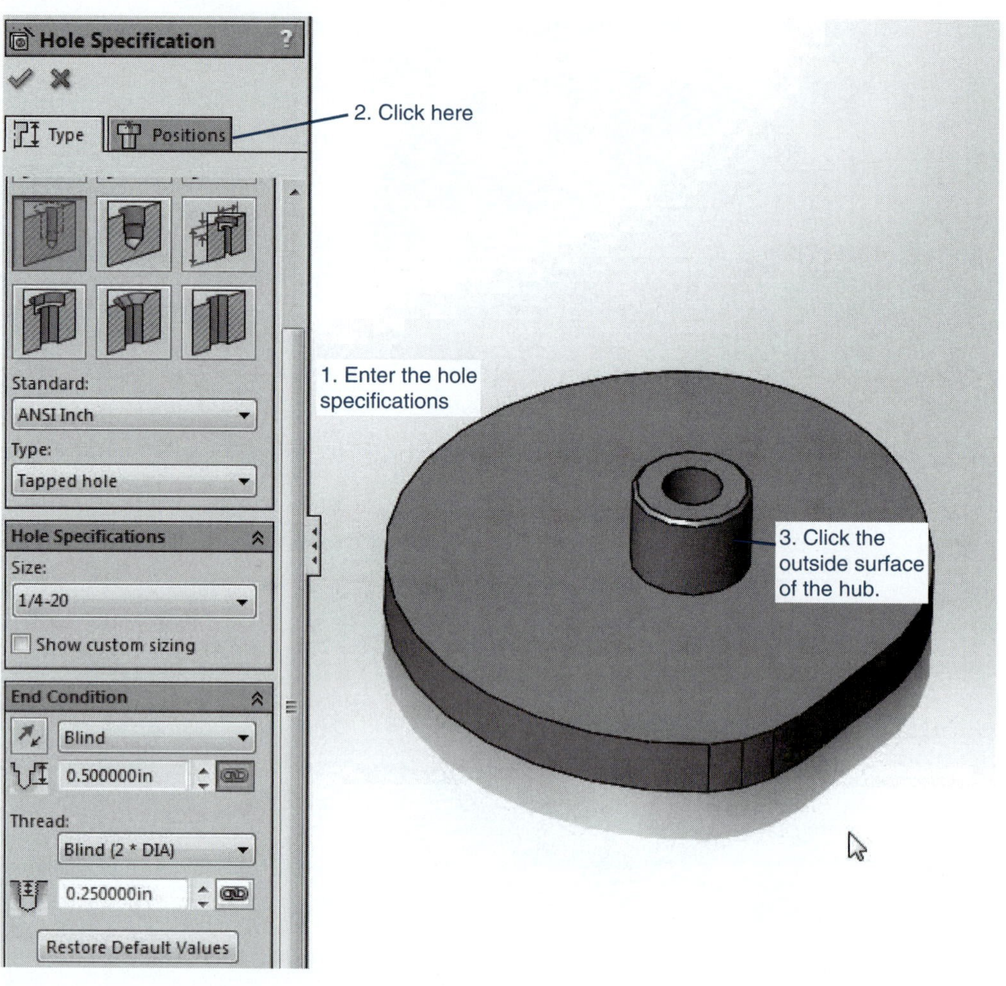
Hole Specification
Type
Positions
2. Click here
Standard:
ANSI Inch
Type:
Tapped hole
Hole Specifications
Size:
1/4-20
Show custom sizing
End Condition
Blind
0.500000in
Thread:
Blind (2 * DIA)
0.250000in
Restore Default Values
1. Enter the hole specifications
3. Click the outside surface of the hub.

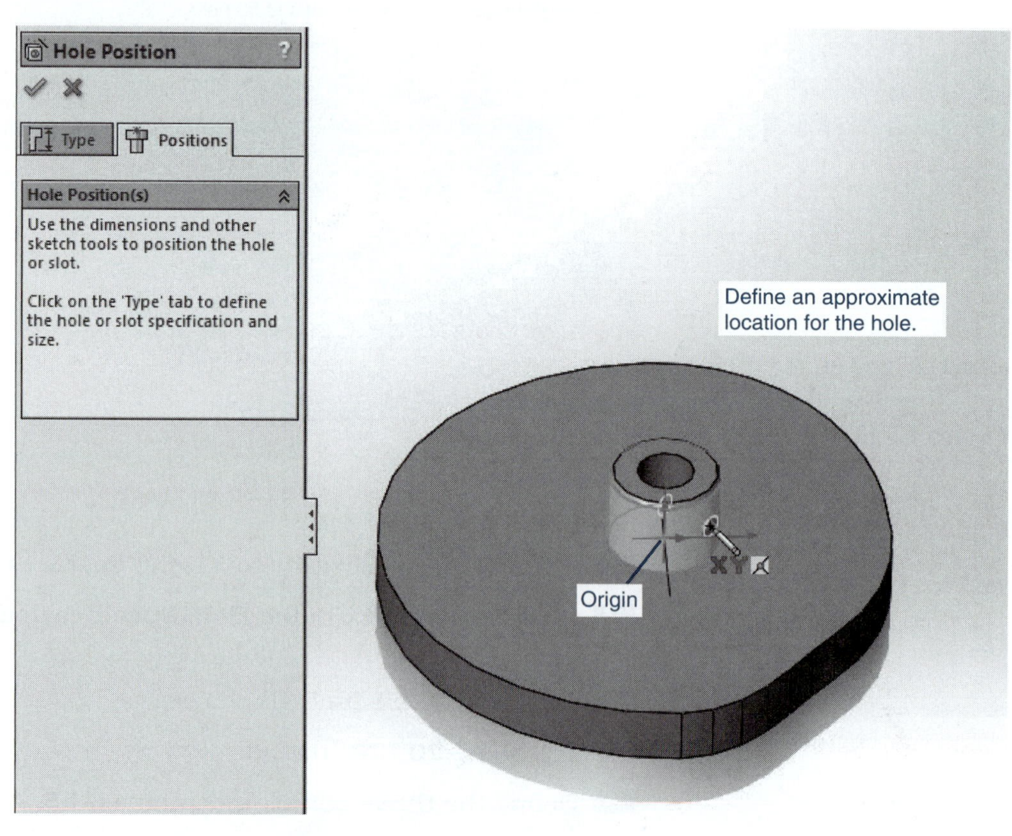
Hole Position
Type
Positions
Hole Position(s)
Use the dimensions and other sketch tools to position the hole or slot.
Click on the 'Type' tab to define the hole or slot specification and size.
Define an approximate location for the hole.
Origin

Figure 12-14
(Continued)

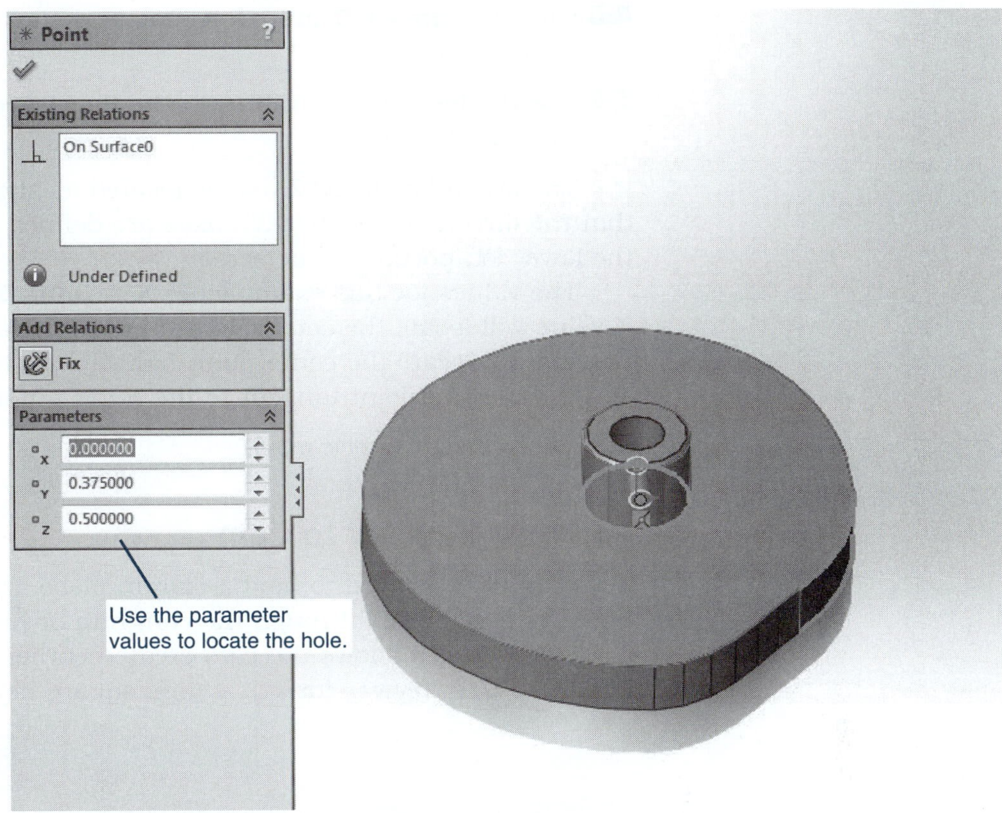

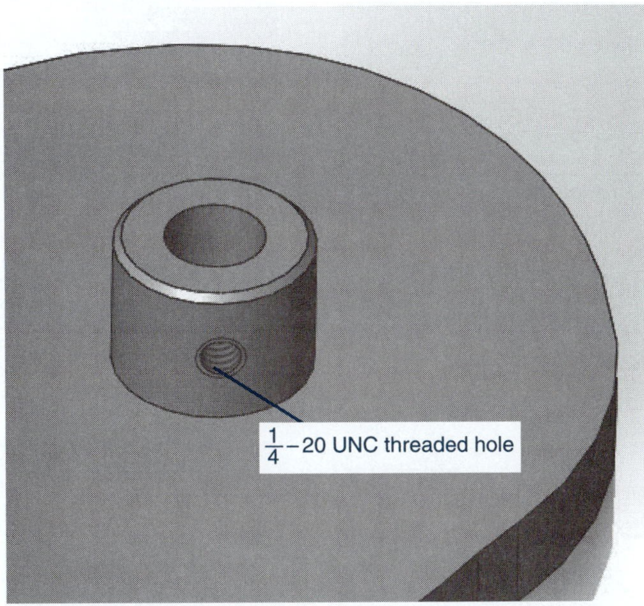

The diameter of the hub is 1.00 and that of the through hole is 0.50. Therefore, the wall thickness of the hub is 0.25. The threaded hole will pass through only one side.

The threaded hole can be added anywhere on the hub's surface by clicking the **Positions** tab on the **Hole Wizard PropertyManager** and then clicking a point on the hub's outside surface, but say we wish to locate the threaded hole at a specific location and orientation. See Figure 12-14.

6 Click the **Positions** tab and locate the threaded hole on the surface of the hub.

7 Click the **Smart Dimension** tool, and click the threaded hole's centerpoint.

8 Use the **Parameters** values in the **Point PropertyManager** to locate the threaded hole.

The origin for the XYZ axis is located as shown in Figure 12-14. Note that the directions for the XYZ axes are defined by the orientation icon in the lower left portion of the screen.

The values for this example are **X = 0.00, Y = 3.75, Z = .50**. The X-value will locate the centerpoint on the vertical construction line, the Y-value will locate the centerpoint halfway up the hub, and the Z-value will locate the centerpoint tangent to the hub's outside surface.

9 Click the **OK** check mark.

To Add a Keyway to Cam

Keys can also be used to hold a cam in place against a drive shaft. Keyways may be cut through the cam hub or just through the cam.

Figure 12-15 shows the cam created earlier in the chapter. See Figure 12-9. A keyway for a $\frac{1}{4} \times \frac{1}{4}$-in. square key is created as follows.

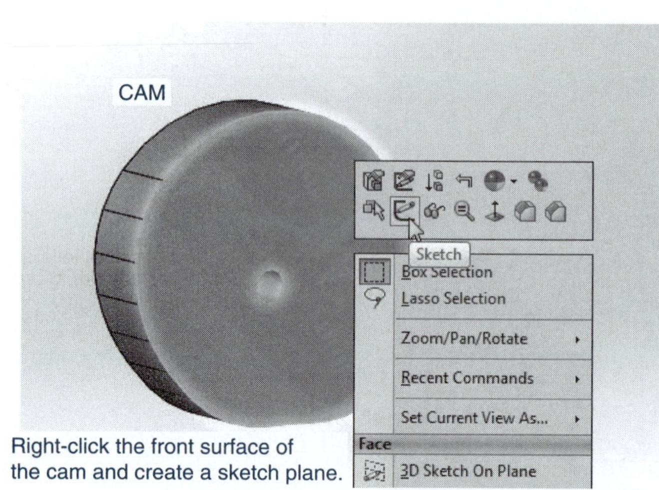

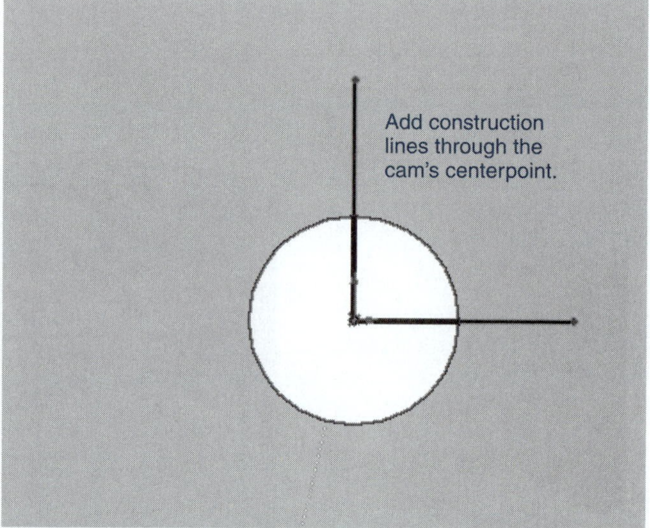

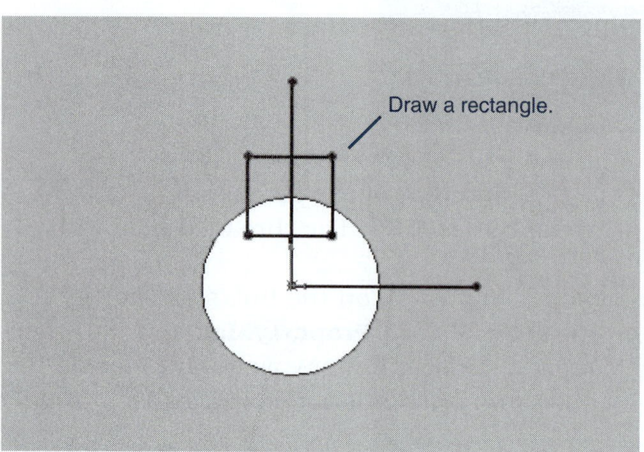

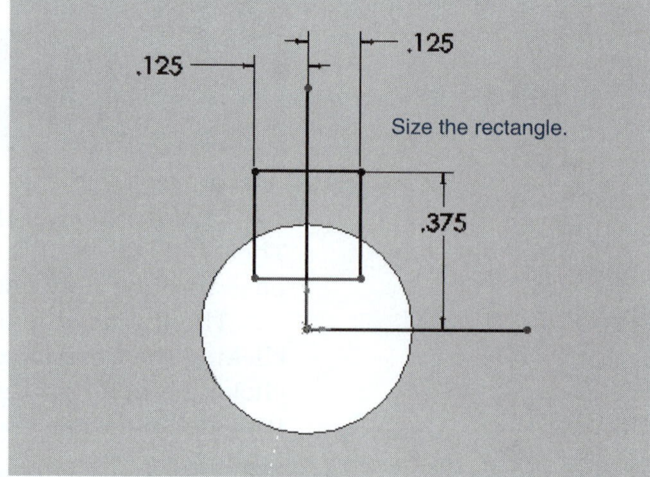

Figure 12-15

Figure 12-15
(*Continued*)

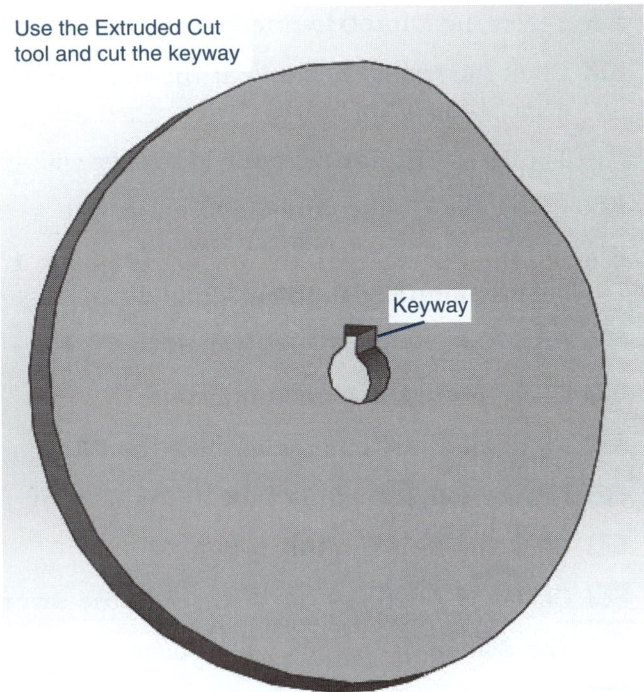

1. Right-click the front surface of the cam, and click the **Sketch** tool.
2. Change the orientation to a view looking directly at the front surface.

   In this example the **Normal** tool, located under the **View orientation** tool, was used.
3. Draw two construction lines from the cam's centerpoint, one vertical and one horizontal.
4. Use the **Rectangle** tool and draw a rectangle as shown.
5. Use the **Smart Dimension** tool and size the rectangle to accept a $\frac{1}{4} \times \frac{1}{4}$-in. square key.

   Tolerances for keys and keyways can be found in Chapter 10.
6. Use the **Extruded Cut** tool and cut the keyway into the cam.

12-10 Springs for Cams

To Draw a Spring

This section shows how to draw a spring that will be used in the next section as part of the cam assembly. See Figure 12-16.

1. Start a new **Part** document.
2. Select the top plane, create a **Sketch** plane, and draw a **Ø0.500** circle centered on the origin.

   The value Ø0.500 will define the outside diameter of the helix. A Ø0.125 circle will be swept along the helical path to form the spring. This generates an inside diameter for the spring of Ø0.437 (0.500 – 0.063 = 0.437) and an outside diameter of Ø0.563.

3. Select the **Dimetric** orientation from the axis orientation icon menu.
4. Click the **Insert** heading at the top of the screen, click **Curve**, and select the **Helix/Spiral** tool.

   The **Helix/Spiral PropertyManager** will appear.
5. Select the **Height and Revolution** option in the **Defined By** box.
6. Set the **Parameters** values for **Height = 1.75 in., 6 Revolutions**, and a **Start angle** of **0.00deg**.
7. Click the green **OK** check mark.
8. Click the **Right Plane** option.
9. Right-click the plane and click the **Sketch** option.
10. Draw a **Ø0.125** circle with its centerpoint on the end of the helix.
11. Click the **Exit Sketch** option.
12. Click the **Features** tab and select the **Swept Boss/Base** tool.

    The **Sweep** dialog box will appear.
13. Define the circle as the **Profile** and the helix as the **Path**.

Figure 12-16

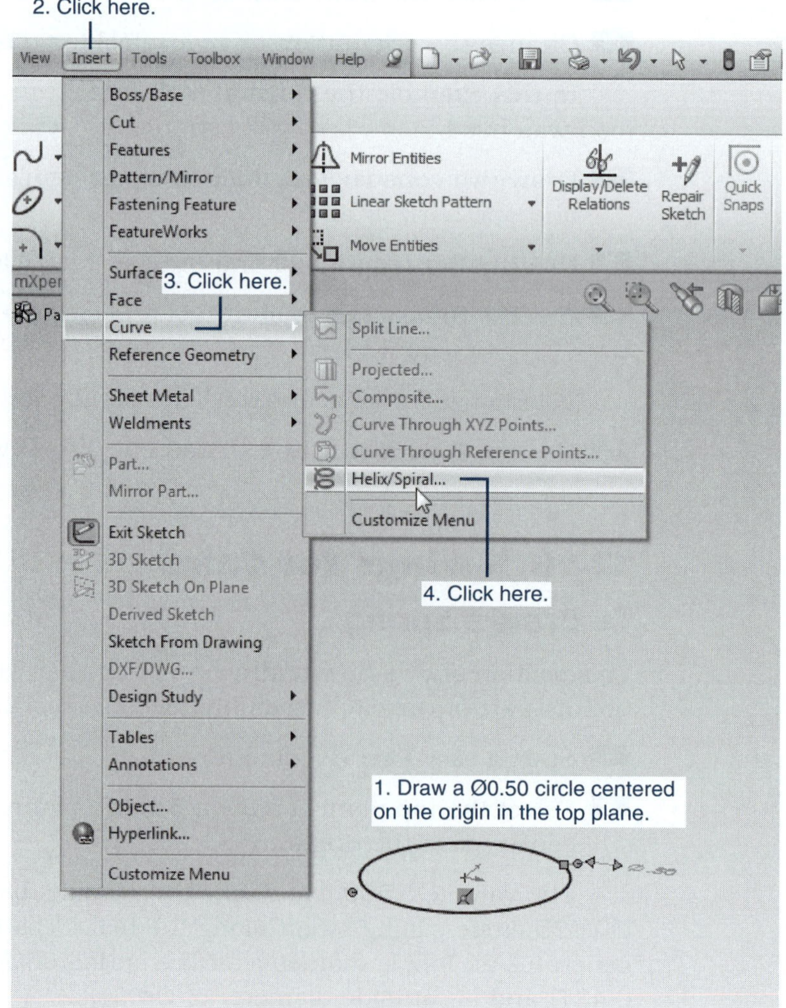

Figure 12-16
(Continued)

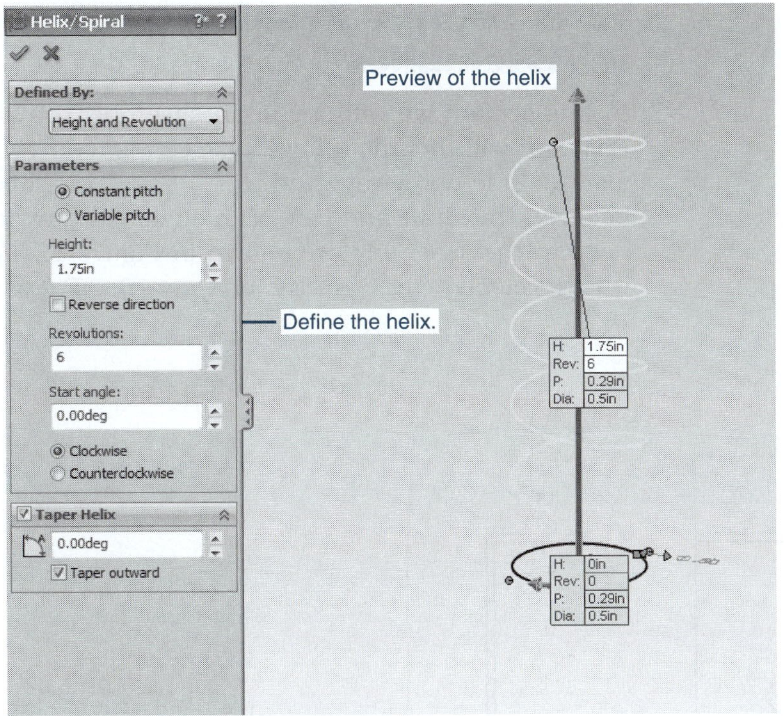

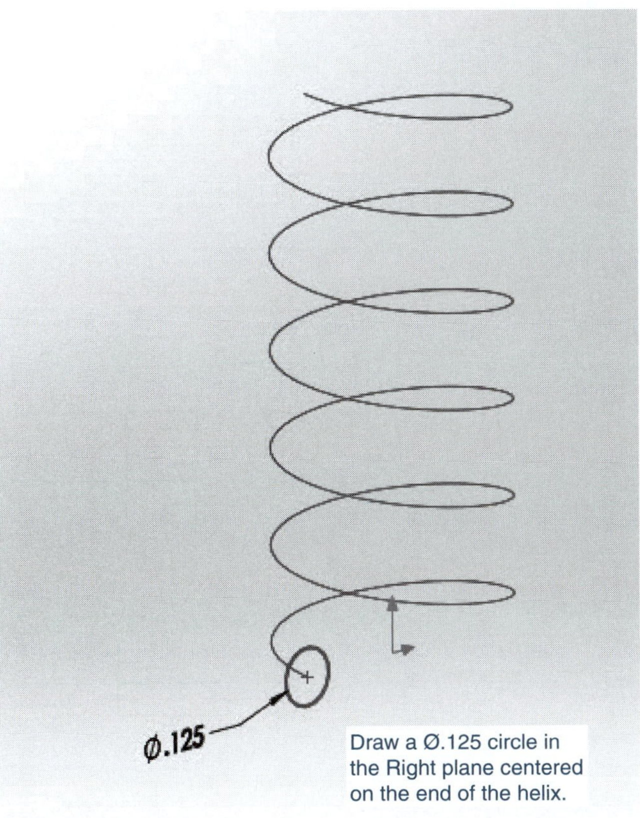

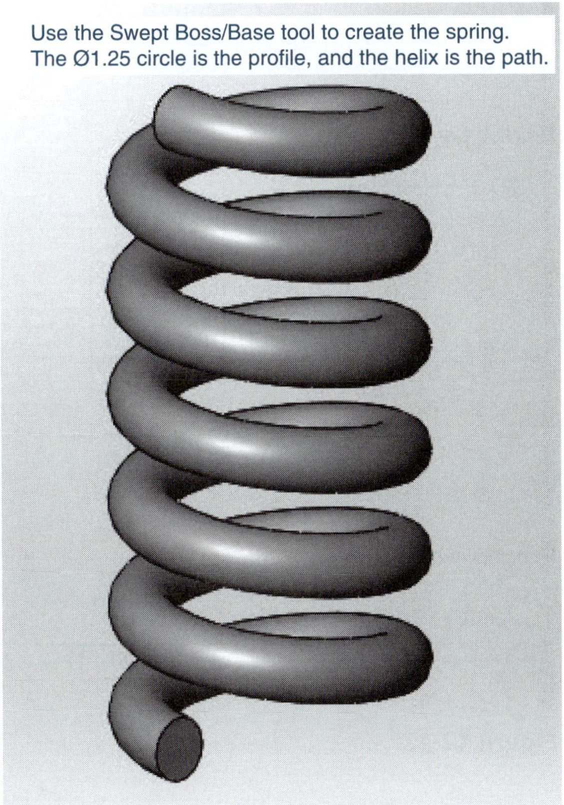

Use the Swept Boss/Base tool to create the spring. The Ø1.25 circle is the profile, and the helix is the path.

14 Click the green **OK** check mark.

15 Save the spring as **Cam Spring**.

12-11 Sample Problem SP12-1—Cams in Assemblies

In this section we will create an assembly drawing that includes a cam. The cam will include a keyway. See Figure 12-15. The support shaft will also include a keyway, and a $\frac{1}{4} \times \frac{1}{4} \times \frac{1}{2}$-in. square key will be inserted between the shaft and cam. Dimensioned drawings for the components used in the assembly are shown in Figure 12-17. The cam is the same as was developed earlier in the chapter. See Figures 12-6 to 12-9.

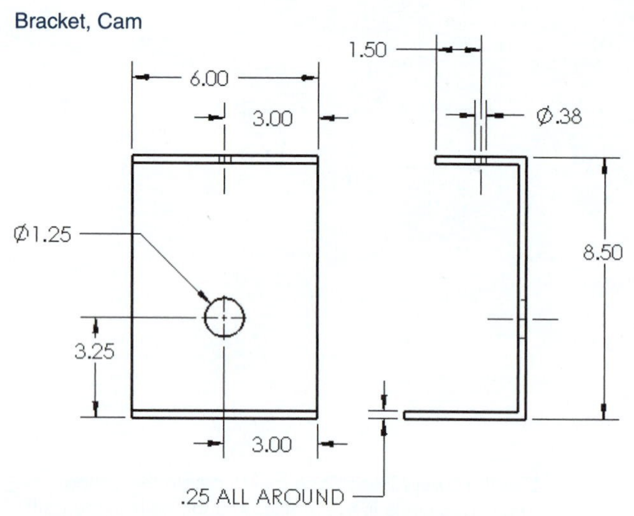

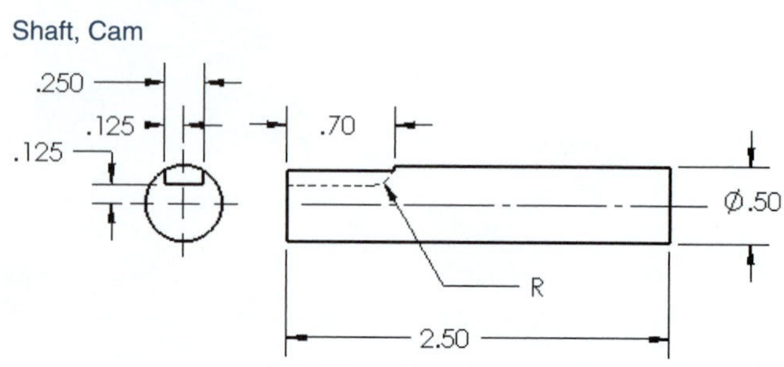

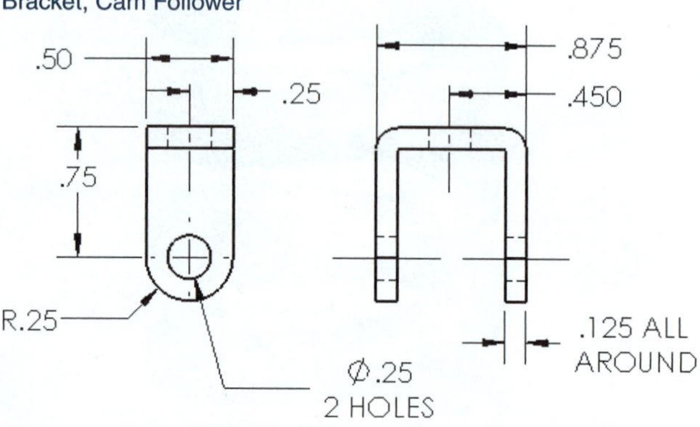

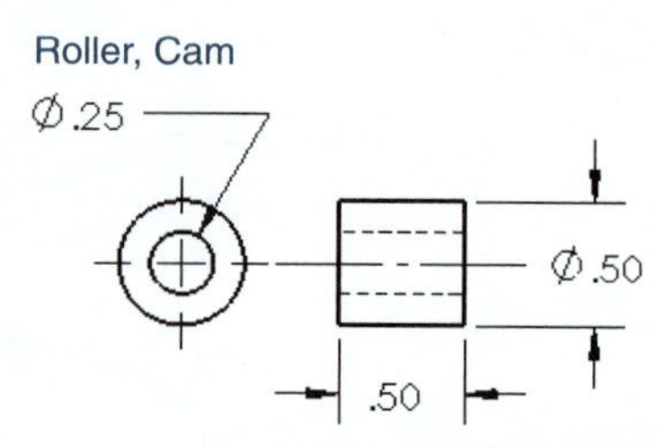

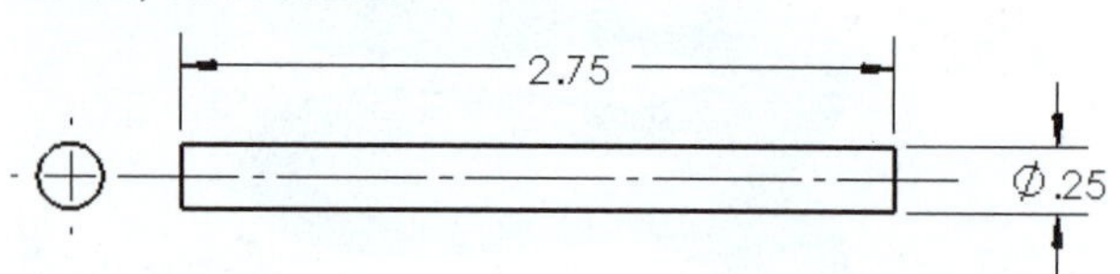

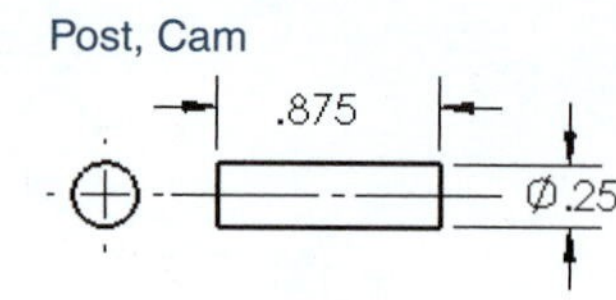

Figure 12-17

Figure 12-17
(*Continued*)

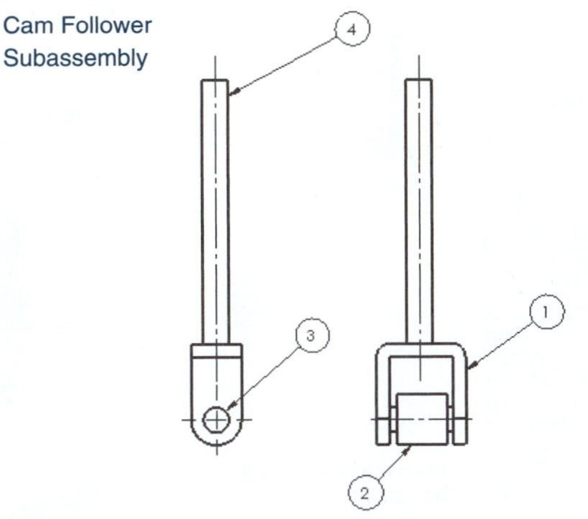

ITEM NO.	PART NUMBER	DESCRIPTION	QTY.
1	AM40 7A	BRACKET, CAM FOLLOWER	1
2	EK 40 7B	ROLLER, CAM	1
3	AM 347 A1	POST, CAM	1
4	MN78	HANDLE, CAM FOLLOWER	1

1 Start a new **Assembly** document.

2 Use the **Insert Component Browse** . . . option and insert the appropriate components.

In this example the first component entered into the assembly drawing screen is the cam bracket. The cam bracket will automatically be fixed in place so that all additional components will move to the bracket. See Figure 12-18.

Figure 12-18

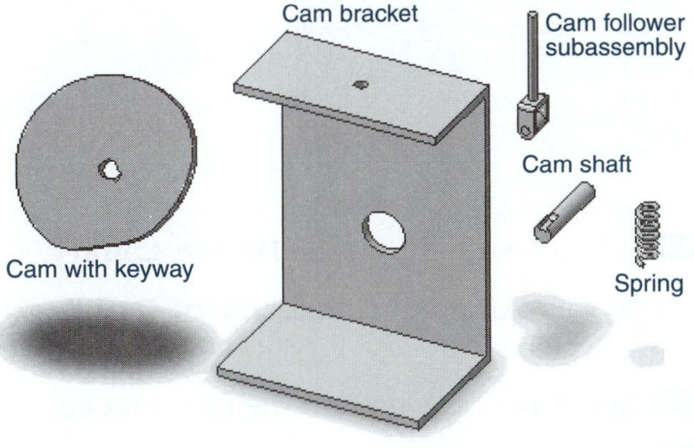

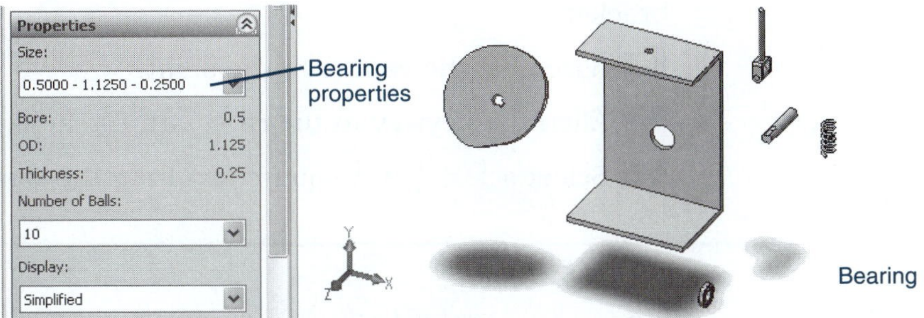

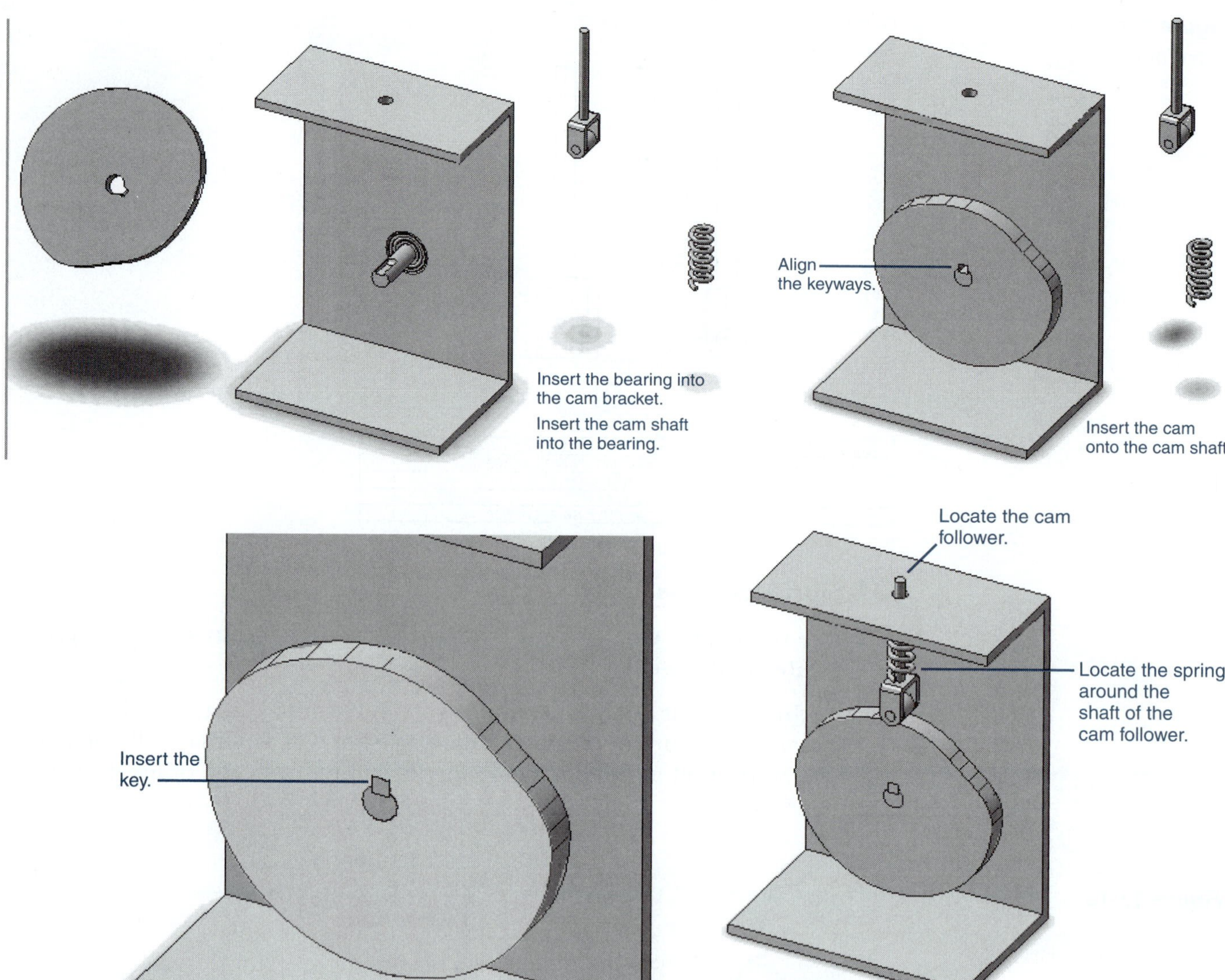

Figure 12-18
(*Continued*)

3 Add a bearing from the **Design Library**.

In this example an **Instrument Ball Bearing 0.5000-1.1250-0.2500** was selected.

4 Insert the bearing into the cam bracket.

5 Insert the cam shaft into the bearing.

Insert the shaft so that it extends 1.50 from the front surface of the bracket.

6 Assemble the cam onto the shaft.

7 Align the keyway in the cam with the keyway in the shaft.

8 Select a $\frac{1}{4} \times \frac{1}{4} \times \frac{1}{2}$ square key from the **Design Library**.

NOTE

A $\frac{1}{4} \times \frac{1}{4} \times \frac{1}{2}$ key can be drawn as an individual component.

9 Insert the key between the shaft and the cam.

10 Insert the cam follower subassembly.

11 Align the roller cam follower with the profile of the cam.

12 Locate the spring around the shaft of the cam follower subassembly.

13 Save the assembly as **Cam Assembly**.

Creating an Orthographic Drawing and a Bill of Materials

1 Start a new **Drawing** document.

2 Use third-angle projection and create a front and a right-side orthographic view of the cam assembly.

3 Click on the **Annotation** tab and add the appropriate centerlines.

See Figure 12-19.

4 Click on the **Annotation** tab and select the **Auto Balloon** tool.

Note that balloon numbers (assembly numbers) have been added to all parts including the parts of the cam follower subassembly.

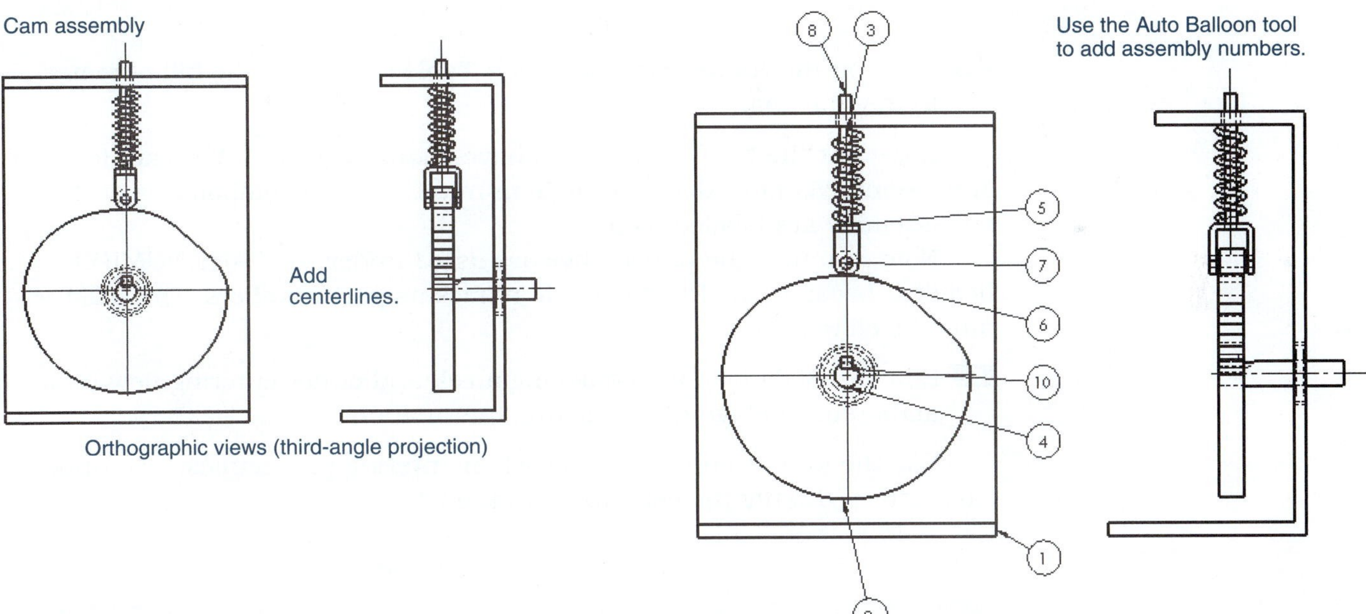

An edited BOM

ITEM NO.	PART NUMBER	DESCRIPTION	QTY.
1	EK-407A	BRACKET, CAM	1
2	EK-407B	CAM - KEYWAY	1
3	AM311-A2	SPRING, CAM	1
4	MN402-1	SHAFT, CAM	1
5	BU-2009S	FOLLOWER, CAM SUB-ASSEMBLY	1
6	AFBMA 12.2 - 0.5000 - 1.1250 - 0.2500 - 10,SI,NC,10		1
7	DR42	KEY, CAM	1

Figure 12-19

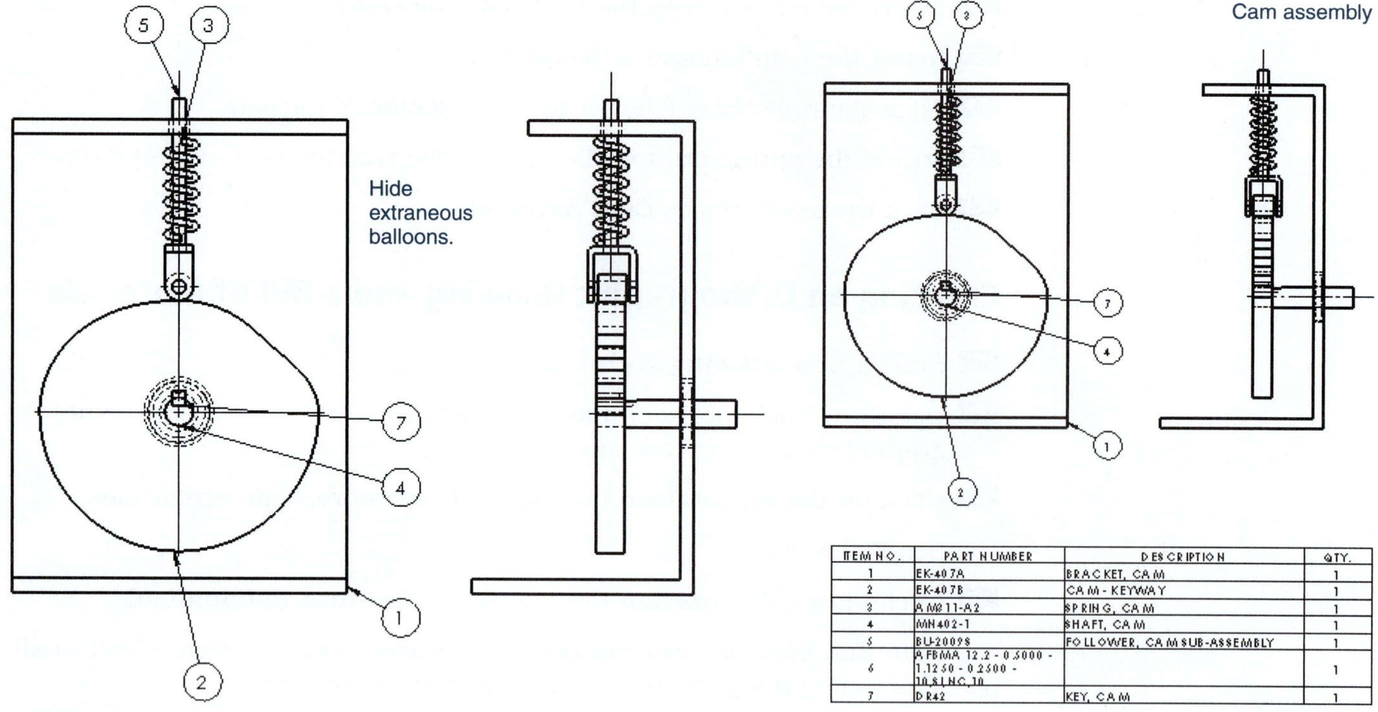

ITEM NO.	PART NUMBER	DESCRIPTION	QTY.
1	EK-407A	BRACKET, CAM	1
2	EK-407B	CAM - KEYWAY	1
3	AM811-A2	SPRING, CAM	1
4	MN402-1	SHAFT, CAM	1
5	BU-20098	FOLLOWER, CAM SUB-ASSEMBLY	1
6	AFBMA 12.2 - 0.5000 - 1.1250 - 0.2500 - 10,SI,NC,10		1
7	DR42	KEY, CAM	1

Figure 12-19
(*Continued*)

5 Click on the **Annotation** tab, then **Tables**, and add the bill of materials to the drawing.

Note that the balloon numbers have changed, so that the cam follower subassembly is now identified as item number 5. All components of the subassembly are labeled as 5.

Note also that the part names are listed under the PART NUMBER heading, because the BOM lists file names as part numbers. The BOM must be edited.

6 Edit the BOM by double-clicking a cell and either entering new information or modifying the existing information.

Use the format noun, modifier when entering part names. Use uppercase letters. Justify the cell inputs to the left.

NOTE

The part number of the bearing selected from the **Design Library** will automatically be inserted into the BOM.

7 Hide the extraneous number 5 balloons. Only one is needed.

8 Save the drawing as **Cam Assembly**.

Chapter Projects

Draw the cams as specified in Projects 12-1 through 12-6.

Project 12-1: Inches

See Figure P12-1.

Units = **Inches**
Cam Type = **Circular**
Follower Type = **Translating**
Follower Diameter = **.375**
Starting Radius = **1.4375**
Starting Angle = **0°**
Rotation Direction = **Clockwise**

Starting Radius = **1.4375**
Dwell = **45°**
Rise **0.375**, Harmonic Motion, **135°**
Dwell = 90°
Fall **0.375**, Harmonic Motion, **90°**

Blank Outside Dia = **2.8675**
Thickness = **.375**
No Hub
Thru Hole Dia = **0.625**

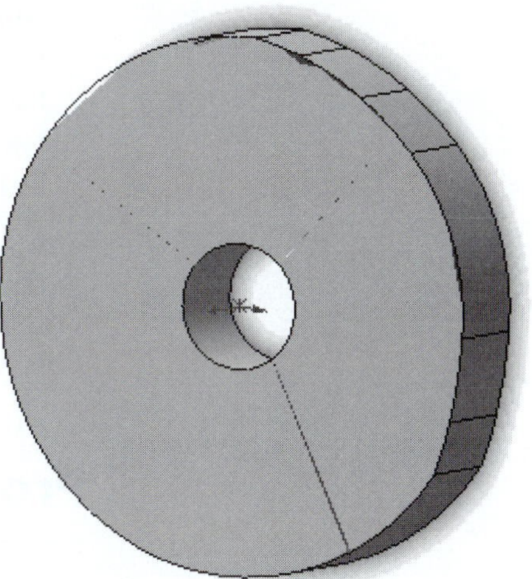

Figure P12-1

Project 12-2: Inches

See Figure P12-2.

Units = **Inches**
Cam Type = **Circular**
Follower Type = **Translating**
Follower Diameter = **.50**
Starting Radius = **2.00**
Starting Angle = **0°**
Rotation Direction = **Clockwise**

Starting Radius = **2.00**
Dwell = **45°**
Rise **0.438**, Modified Trapezoidal Motion, **90°**
Dwell = **90°**
Fall **0.375**, Modified Trapezoidal Motion, **90°**
Dwell = **45°**

Blank Outside Dia = **4.876**
Thickness = **0.500**
Near Hub Dia & Length = **1.25, 1.00**
Thru Hole Dia = **0.75**

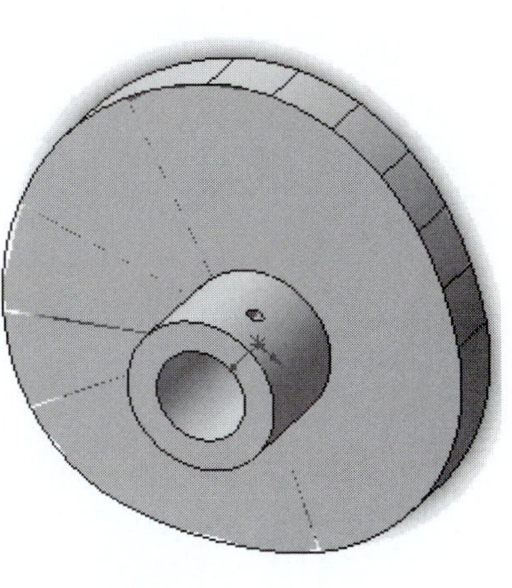

Figure P12-2

Add a #6-32 threaded hole 0.50 from the top of the hub.

Project 12-3: Design Problem

Use a base circle of Ø4.00 in.
The follower has a Ø0.50 in.
Cam motion:

Dwell = **45°**
Rise **0.25 in.** using Uniform Displacement for 45°
Dwell = **45°**
Rise **0.25 in.** using Uniform Displacement for 45°
Dwell = **45°**
Fall **0.25 in.** using Uniform Displacement for 45°
Dwell = **45°**
Fall **0.25 in.** using Uniform Displacement for 45°

The hub has a diameter of 1.50 in. and extends 1.50 in. from the surface of the cam.
The cam bore is Ø0.75 in.
The hub includes a #10-32 threaded hole.

Project 12-4: Millimeters

See Figure P12-4.

Units = **Metric**
Cam Type = **Circular**
Follower Type = **Translating**
Follower Diameter = **20**
Starting Radius = **40**
Starting Angle = **0°**
Rotation Direction = **Clockwise**

Starting Radius = **40**
Dwell = **45°**
Rise **10**, Harmonic Motion, **135°**
Dwell = **90°**
Fall **10**, Harmonic Motion, **90°**

Blank Outside Dia = **100**
Thickness = **20**
No Hub
Thru Hole Dia = **16.0**

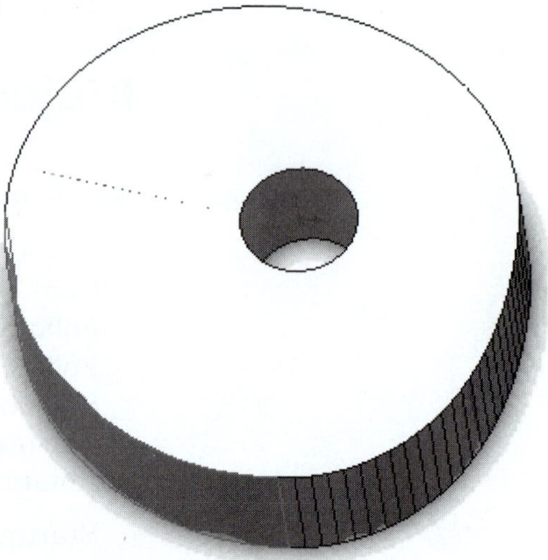

Figure P12-4

Project 12-5:

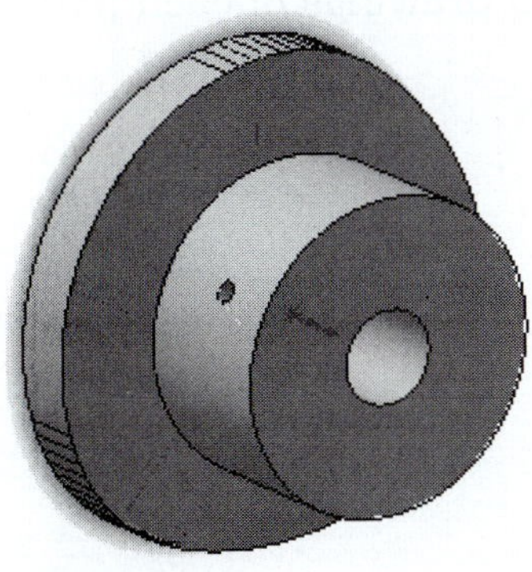

Figure P12-5

See Figure P12-5.

Units = **Metric**
Cam Type = **Circular**
Follower Type = **Translating**
Follower Diameter = **16**
Starting Radius = **50.0**
Starting Angle = **0°**
Rotation Direction = **Clockwise**

Starting Radius = **50.0**
Dwell = **45°**
Rise **8.0**, Modified Trapezoidal Motion, **90°**
Dwell = **90°**
Fall **8.0**, Modified Trapezoidal Motion, **90°**
Dwell = **45°**

Blank Outside Dia = **108**
Thickness = **12.0**
Near Hub Dia & Length = **60, 30**
Thru Hole Dia = **18.00**

Add an M4 threaded hole 15 from the top of the hub.

Project 12-6: Design Problem

See Figure P12-5.

Use a base circle of Ø80.0 mm.
The follower has a Ø12.0 mm.
Cam motion:

Dwell = **45°**
Rise **10.0 mm** using Uniform Displacement for 45°
Dwell = **45°**
Rise **5.0 mm** using Uniform Displacement for 45°
Dwell = **45°**
Fall **5.0 mm** using Uniform Displacement for 45°
Dwell = **45°**
Fall **10.0 mm** using Uniform Displacement for 45°

The hub has a diameter of 30.0 mm and extends 26.0 mm from the surface of the cam.
The cam bore is Ø16.0 mm.
The hub includes an M6 threaded hole.

Project 12-7: Design Problem

Figure P12-7 shows a cam assembly. Dimensioned drawings of each part are shown in Figure 12-17.

1. Draw the assembly.
2. Insert the following cam:
   Cam parameters:

   Base circle = **Ø5.00 in.**
   Follower Diameter = **0.50 in.**
   Select a bearing from the **Design Library**.

   Cam motion:

   Dwell = **45°**
   Rise **0.250 in.** using Uniform Displacement for 45°
   Dwell = **45°**
   Rise **0.250 in.** using Uniform Displacement for 45°
   Dwell = **45°**
   Fall **0.250 in.** using Uniform Displacement for 45°
   Dwell = **45°**
   Fall **0.250 in.** using Uniform Displacement for 45°
3. Create a keyway in both the cam and the cam shaft that will accept a 0.375 × 0.375 × 0.500-in. square key.
4. Calculate the distance between the top surface of the cam follower bracket and the underside of the top flange on the cam bracket, and create a spring to fit into the space.

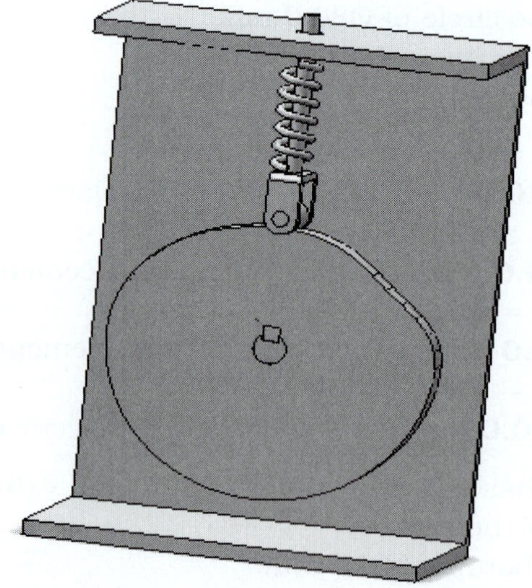

Figure P12-7

Project 12-8: Millimeters

Draw a 3D solid model of the following assembly. See Figure P12-8.

1. Create an exploded isometric assembly drawing with assembly (item) numbers.
2. Create a BOM for the assembly.
3. Create dimensioned drawings of each individual part.

Cam parameters:

Base circle = **Ø146**
Face width = **16**
Motion: rise **10** using harmonic motion over 90°, dwell for 180°, fall **10** in 90°.
Bore = **Ø16.0**
Keyway = **2.3 × 5 × 16**
Follower = **Ø16**
Follower width = **4**
Square key = **5 × 5 × 16**
Bearing overall dimensions

Select an appropriate bearing from the **Design Library** or manufacturer's website.
Bearing 1: **17 × 40 × 10 (D × OD × THK)**
Bearing 2: **4 × 13 × 4**
Bearing 3: **8 × 18 × 5**

Spring parameters:

Wire Ø = **1.5**
Inside Ø = **9.0**
Length = **20**
Coil Direction = **Right**
Coils = **10**

Consider creating a spring with ground ends. See Chapter 3.

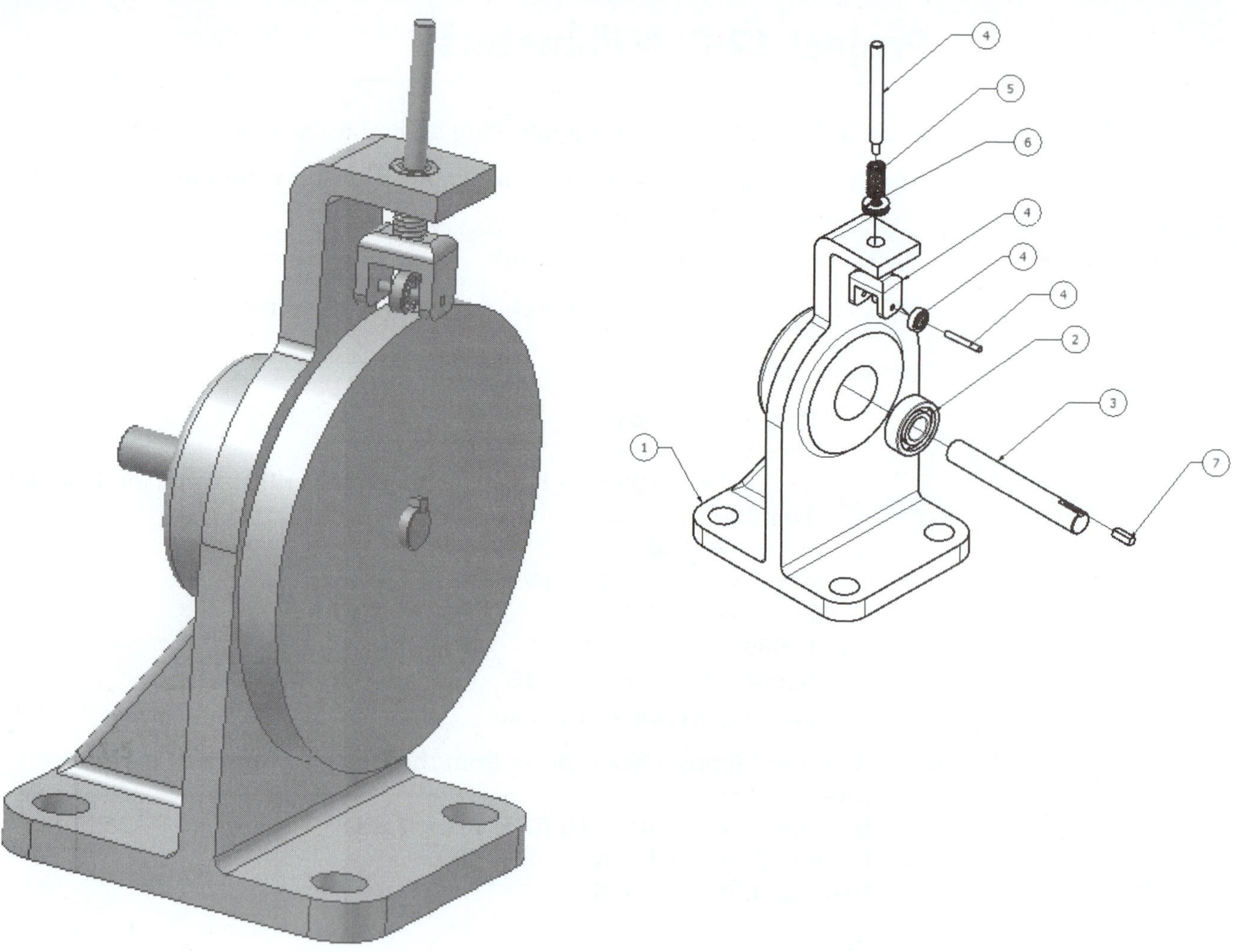

Parts List			
ITEM	QTY	PART NUMBER	DESCRIPTION
1	1	ENG-2008-A	BASE, CAST
2	1	DIN625 - SKF 6203	Single row ball bearings
3	1	SHF-4004-16	SHAFT: Ø16×120,WITH 2.3×5×16 KEYWAY
4	1		SUB-ASSEMBLY, FOLLOWER
5	1	SPR-C22	SPRING,COMPRESSION
6	1	GB 273.2-87 - 7/70 - 8 x 18 x 5	Rolling bearings - Thrust bearings - Plan of boundary dimensions
7	1	IS 2048 - 1983 - Specification for Parallel Keys and Keyways B 5 x 5 x 16	Specification for Parallel Keys and Keyways

Figure P12-8

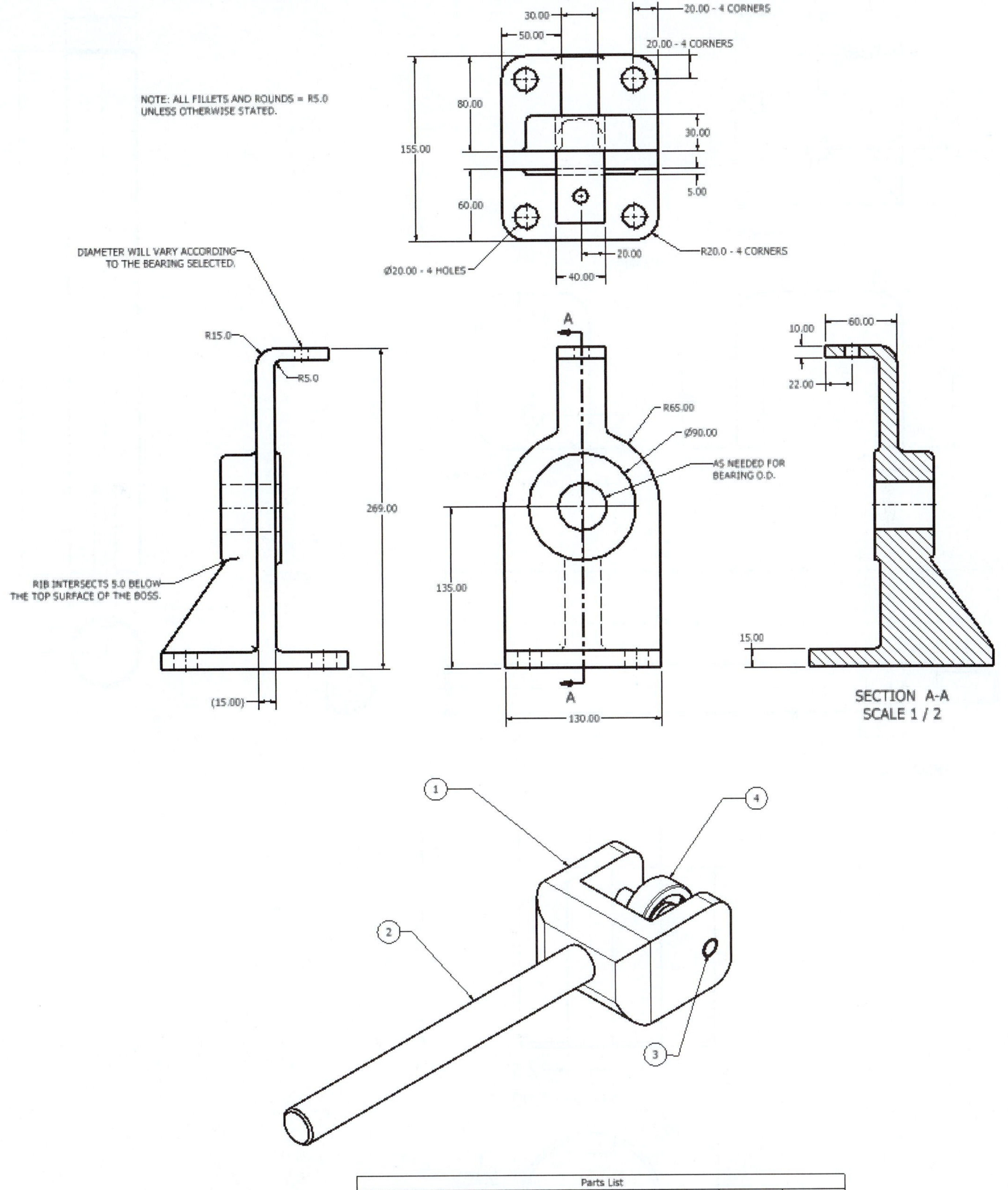

Parts List				
ITEM	PART NUMBER	DESCRIPTION	MATERIAL	QTY
1	AM-232	HOLDER	STEEL	1
2	AM-256	POST, FOLLOWER	STEEL	1
3	BS 1804-2 - 4 x 30	Parallel steel dowel pins - metric series	Steel, Mild	1
4	DIN625- SKF 634	Single row ball bearings	Steel, Mild	1

Figure P12-8
(*Continued*)

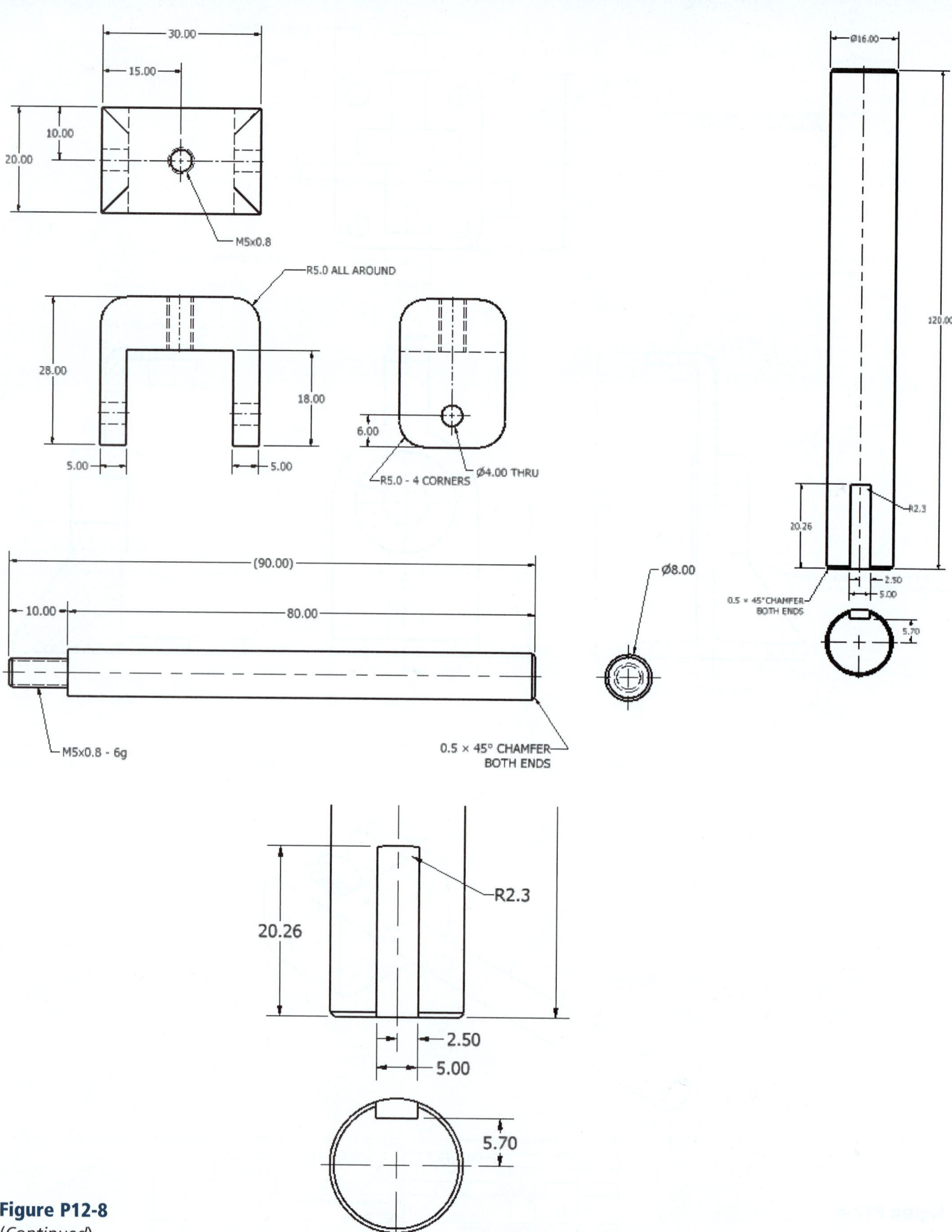

Figure P12-8
(*Continued*)

appendix

Wire and Sheet Metal Gauges

Gauge	Thickness	Gauge	Thickness
000 000	0.5800	18	0.0403
00 000	0.5165	19	0.0359
0 000	0.4600	20	0.0320
000	0.4096	21	0.0285
00	0.3648	22	0.0253
0	0.3249	23	0.0226
1	0.2893	24	0.0201
2	0.2576	25	0.0179
3	0.2294	26	0.0159
4	0.2043	27	0.0142
5	0.1819	28	0.0126
6	0.1620	29	0.0113
7	0.1443	30	0.0100
8	0.1285	31	0.0089
9	0.1144	32	0.0080
10	0.1019	33	0.0071
11	0.0907	34	0.0063
12	0.0808	35	0.0056
13	0.0720	36	0.0050
14	0.0641	37	0.0045
15	0.0571	38	0.0040
16	0.0508	39	0.0035
17	0.0453	40	0.0031

Figure A-1

American Standard Clearance Locational Fits

Nominal Size Range Inches Over – To	Class LC1			Class LC2			Class LC3			Class LC4		
	Limits of Clearance	Standard Limits Hole H6	Standard Limits Shaft h5	Limits of Clearance	Standard Limits Hole H7	Standard Limits Shaft h6	Limits of Clearance	Standard Limits Hole H8	Standard Limits Shaft h7	Limits of Clearance	Standard Limits Hole H10	Standard Limits Shaft h9
0 - 0.12	0	+0.25	0	0	+0.4	0	0	+0.6	0	0	+1.6	0
	0.45	0	-0.2	0.65	0	-0.25	1	0	-0.4	2.6	0	-1.0
0.12 - 0.24	0	+0.3	0	0	+0.5	0	0	+0.7	0	0	+1.8	0
	0.5	0	-0.2	0.8	0	-0.3	1.2	0	-0.5	3.0	0	-1.2
0.24 - 0.40	0	+0.4	0	0	+0.6	0	0	+0.9	0	0	+2.2	0
	0.65	0	-0.25	1.0	0	-0.4	1.5	0	-0.6	3.6	0	-1.4
0.40 - 0.71	0	+0.4	0	0	+0.7	0	0	+1.0	0	0	+2.8	0
	0.7	0	-0.3	1.1	0	-0.4	1.7	0	-0.7	4.4	0	-1.6
0.71 - 1.19	0	+0.5	0	0	+0.8	0	0	+1.2	0	0	+3.5	0
	0.9	0	-0.4	1.3	0	-0.5	2	0	-0.8	5.5	0	-2.0
1.19 - 1.97	0	+0.6	0	0	+1.0	0	0	+1.6	0	0	+4.0	0
	1.0	0	-0.4	1.6	0	-0.6	2.6	0	-1.0	6.5	0	-2.5

Figure A-2A

Nominal Size Range Inches Over – To	Class LC5			Class LC6			Class LC7			Class LC8		
	Limits of Clearance	Standard Limits Hole H7	Standard Limits Shaft g6	Limits of Clearance	Standard Limits Hole H9	Standard Limits Shaft f8	Limits of Clearance	Standard Limits Hole H10	Standard Limits Shaft e9	Limits of Clearance	Standard Limits Hole H10	Standard Limits Shaft d9
0 - 0.12	0.1	+0..4	-0.1	0.3	+1.0	-0.3	0.6	+1.6	-0.6	1.0	+1.6	-1.0
	0.75	0	-0.35	1.9	0	-0.9	3.2	0	-1.6	3.6	0	-2.0
0.12 - 0.24	0.15	+0.5	-0.15	0.4	+1.2	-0.4	0.8	+1.8	-0.8	1.2	+1.8	-1.2
	0.95	0	-0.45	2.3	0	-1.1	3.8	0	-2.0	4.2	0	-2.4
0.24 - 0.40	0.2	+0.6	-0.2	0.5	+1.4	-0.5	1.0	+2.2	-1.0	1.6	+2.2	-1.6
	1.2	0	-0.6	2.8	0	-1.4	4.6	0	-2.4	5.2	0	-3.0
0.40 - 0.71	0.25	+0.7	-0.25	0.6	+1.6	-0.6	1.2	+2.8	-1.2	2.0	+2.8	-2.0
	1.35	0	-0.65	3.2	0	-1.6	5.6	0	-2.8	6.4	0	-3.6
0.71 - 1.19	0.3	+0.8	-0.3	0.8	+2.0	-0.8	1.6	+3.5	-1.6	2.5	+3.5	-2.5
	1.6	0	-0.8	4.0	0	-2.0	7.1	0	-3.6	8.0	0	-4.5
1.19 - 1.97	0.4	+1.0	-0.4	1.0	+2.5	-1.0	2.0	+4.0	-2.0	3.0	+4.0	-3.0
	2.0	0	-1.0	5.1	0	-2.6	8.5	0	-4.5	9.5	0	-5.5

Figure A-2B

American Standard Running and Sliding Fits
(Hole Basis)

Nominal Size Range Inches		Class RC1			Class RC2			Class RC3			Class RC4		
		Limits of Clearance	Standard Limits		Limits of Clearance	Standard Limits		Limits of Clearance	Standard Limits		Limits of Clearance	Standard Limits	
Over	To		Hole H5	Shaft g4		Hole H6	Shaft g5		Hole H7	Shaft f6		Hole H8	Shaft f7
0	0.12	0.1 0.45	+0.2 0	−0.1 −0.25	0.1 0.55	+0.25 0	−0.1 −0.3	0.3 0.95	+0.4 0	−0.3 −0.55	0.3 1.3	+0.6 0	−0.3 −0.7
0.12	0.24	0.15 0.5	+0.2 0	−0.15 −0.3	0.15 0.65	+0.3 0	−0.15 −0.35	0.4 1.12	+0.5 0	−0.4 −0.7	0.4 1.5	+0.7 0	−0.4 −0.0
0.24	0.40	0.2 0.6	+0.25 0	−0.2 −0.35	0.2 0.85	+0.4 0	−0.2 −0.45	0.5 1.5	+0.6 0	−0.5 −0.9	0.5 2.0	+0.9 0	−0.5 −1.1
0.40	0.71	0.25 0.75	+0.3 0	−0.25 −0.45	0.25 0.95	+0.4 0	−0.25 −0.55	0.6 1.7	+0.7 0	−0.6 −1.0	0.6 2.3	+1.0 0	−0.6 −1.3
0.71	1.19	0.3 0.95	+0.4 0	−0.3 −0.55	0.3 1.2	+0.5 0	−0.3 −0.7	0.8 2.1	+0.8 0	−0.8 −1.3	0.8 2.8	+1.2 0	−0.8 −1.6
1.19	1.97	0.4 1.1	+0.4 0	−0.4 −0.7	0.4 1.4	+0.6 0	−0.4 −0.8	1.0 2.6	+1.0 0	−1.0 −1.6	1.0 3.6	+1.6 0	−1.0 −2.0

Figure A-3A

Nominal Size Range Inches		Class RC5			Class RC6			Class RC7			Class RC8		
		Limits of Clearance	Standard Limits		Limits of Clearance	Standard Limits		Limits of Clearance	Standard Limits		Limits of Clearance	Standard Limits	
Over	To		Hole H8	Shaft e7		Hole H9	Shaft e8		Hole H9	Shaft d8		Hole H10	Shaft c9
0	0.12	0.6 1.6	+0.6 0	−0.6 −1.0	0.6 2.2	+1.0 0	−0.6 −1.2	1.0 2.6	+1.0 0	−1.0 −1.6	2.5 5.1	+1.6 0	−2.5 −3.5
0.12	0.24	0.8 2.0	+0.7 0	−0.8 −1.3	0.8 2.7	+1.2 0	−0.8 −1.5	1.2 3.1	+1.2 0	−1.2 −1.9	2.8 5.8	+1.8 0	−2.8 −4.0
0.24	0.40	1.0 2.5	+0.9 0	−1.0 −1.6	1.0 3.3	+1.4 0	−1.0 −1.9	1.6 3.9	+1.4 0	−1.6 −2.5	3.0 6.6	+2.2 0	−3.0 −4.4
0.40	0.71	1.2 2.9	+1.0 0	−1.2 −1.9	1.2 3.8	+1.6 0	−1.2 −2.2	2.0 4.6	+1.6 0	−2.0 −3.0	3.5 7.9	+2.8 0	−3.5 −5.1
0.71	1.19	1.6 3.6	+1.2 0	−1.6 −2.4	1.6 4.8	+2.0 0	−1.6 −2.8	2.5 5.7	+2.0 0	−2.5 −3.7	4.5 10.0	+3.5 0	−4.5 −6.5
1.19	1.97	2.0 4.6	+1.6 0	−2.0 −3.0	2.0 6.1	+2.5 0	−2.0 −3.6	3.0 7.1	+2.5 0	−3.0 −4.6	5.0 11.5	+4.0 0	−5.0 −7.5

Figure A-3B

American Standard Transition Locational Fits

Nominal Size Range Inches Over – To	Fit	Class LT1 Standard Limits Hole H7	Shaft js6	Fit	Class LT2 Standard Limits Hole H8	Shaft js7	Fit	Class LT3 Standard Limits Hole H7	Shaft k6
0 – 0.12	−0.10 +0.50	+0.4 0	+0.10 −0.10	−0.2 +0.8	+0.6 0	+0.2 −0.2			
0.12 – 0.24	−0.15 −0.65	+0.5 0	+0.15 −0.15	−0.25 +0.95	+0.7 0	+0.25 −0.25			
0.24 – 0.40	−0.2 +0.5	+0.6 0	+0.2 −0.2	−0.3 +1.2	+0.9 0	+0.3 −0.3	−0.5 +0.5	+0.6 0	+0.5 +0.1
0.40 – 0.71	−0.2 +0.9	+0.7 0	+0.2 −0.2	−0.35 +1.35	+1.0 0	+0.35 −0.35	−0.5 +0.6	+0.7 0	+0.5 +0.1
0.71 – 1.19	−0.25 +1.05	+0.8 0	+0.25 −0.25	−0.4 +1.6	+1.2 0	+0.4 −0.4	−0.6 +0.7	+0.8 0	+0.6 +0.1
1.19 – 1.97	−0.3 +1.3	+1.0 0	+0.3 −0.3	−0.5 +2.1	+1.6 0	+0.5 −0.5	+0.7 +0.1	+1.0 0	+0.7 +0.1

Figure A-4A

Nominal Size Range Inches Over – To	Fit	Class LT4 Standard Limits Hole H8	Shaft k7	Fit	Class LT5 Standard Limits Hole H7	Shaft n6	Fit	Class LT6 Standard Limits Hole H7	Shaft n7
0 – 0.12				−0.5 +0.15	+0.4 0	+0.5 +0.25	−0.65 +0.15	+0.4 0	+0.65 +0.25
0.12 – 0.24				−0.6 +0.2	+0.5 0	+0.6 +0.3	−0.8 +0.2	+0.5 0	+0.8 +0.3
0.24 – 0.40	−0.7 +0.8	+0.9 0	+0.7 +0.1	−0.8 +0.2	+0.6 0	+0.8 +0.4	−1.0 +0.2	+0.6 0	+1.0 +0.4
0.40 – 0.71	−0.8 +0.9	+1.0 0	+0.8 +0.1	−0.9 +0.2	+0.7 0	+0.9 +0.5	−1.2 +0.2	+0.7 0	+1.2 +0.5
0.71 – 1.19	−0.9 +1.1	+1.2 0	+0.9 +0.1	−1.1 +0.2	+0.8 0	+1.1 +0.6	−1.4 +0.2	+0.8 0	+1.4 +0.6
1.19 – 1.97	−1.1 +1.5	+1.6 0	+1.1 +0.1	−1.3 +0.3	+1.0 0	+1.3 +0.7	−1.7 +0.3	+1.0 0	+1.7 +0.7

Figure A-4B

American Standard Interference Locational Fits

Nominal Size Range Inches Over To	Limits of Interference	Class LN1 Standard Limits		Limits of Interference	Class LN2 Standard Limits		Limits of Interference	Class LN3 Standard Limits	
		Hole H6	Shaft n5		Hole H7	Shaft p6		Hole H7	Shaft r6
0 - 0.12	0 0.45	+0.25 0	+0.45 +0.25	0 0.65	+0.4 0	+0.63 +0.4	0.1 0.75	+0.4 0	+0.75 +0.5
0.12 - 0.24	0 0.5	+0.3 0	+0.5 +0.3	0 0.8	+0.5 0	+0.8 +0.5	0.1 0.9	+0.5 0	+0.9 +0.6
0.24 - 0.40	0 0.65	+0.4 0	+0.65 +0.4	0 1.0	+0.6 0	+1.0 +0.6	0.2 1.2	+0.6 0	+1.2 +0.8
0.40 - 0.71	0 0.8	+0.4 0	+0.8 +0.4	0 1.1	+0.7 0	+1.1 +0.7	0.3 1.4	+0.7 0	+1.4 +1.0
0.71 - 1.19	0 1.0	+0.5 0	+1.0 +0.5	0 1.3	+0.8 0	+1.3 +0.8	0.4 1.7	+0.8 0	+1.7 +1.2
1.19 - 1.97	0 1.1	+0.6 0	+1.1 +0.6	0 1.6	+1.0 0	+1.6 +1.0	0.4 2.0	+1.0 0	+2.0 +1.4

Figure A-5

American Standard Force and Shrink Fits

Nominal Size Range Inches Over To	Limits of Interference	Class FN 1 Standard Limits		Limits of Interference	Class FN 2 Standard Limits		Limits of Interference	Class FN 3 Standard Limits		Limits of Interference	Class FN 4 Standard Limits	
		Hole	Shaft		Hole	Shaft		Hole	Shaft		Hole	Shaft
0 - 0.12	0.05 0.5	+0.25 0	+0.5 +0.3	0.2 0.85	+0.4 0	+0.85 +0.6				0.3 0.95	+0.4 0	+0.95 +0.7
0.12 - 0.24	0.1 0.6	+0.3 0	+0.6 +0.4	0.2 1.0	+0.5 0	+1.0 +0.7				0.4 1.2	+0.5 0	+1.2 +0.9
0.24 - 0.40	0.1 0.75	+0.4 0	+0.75 +0.5	0.4 1.4	+0.6 0	+1.4 +1.0				0.6 1.6	+0.6 0	+1.6 +1.2
0.40 - 0.56	0.1 0.8	+0.4 0	+0.8 +0.5	0.5 1.6	+0.7 0	+1.6 +1.2				0.7 1.8	+0.7 0	+1.8 +1.4
0.56 - 0.71	0.2 0.9	+0.4 0	+0.9 +0.6	0.5 1.6	+0.7 0	+1.6 +1.2				0.7 1.8	+0.7 0	+1.8 +1.4
0.71 - 0.95	0.2 1.1	+0.5 0	+1.1 +0.7	0.6 1.9	+0.8 0	+1.9 +1.4				0.8 2.1	+0.8 0	+2.1 +1.6
0.95 - 1.19	0.3 1.2	+0.5 0	+1.2 +0.8	0.6 1.9	+0.8 0	+1.9 +1.4	0.8 2.1	+0.8 0	+2.1 +1.6	1.0 2.3	+0.8 0	+2.1 +1.8
1.19 - 1.58	0.3 1.3	+0.6 0	+1.3 +0.9	0.8 2.4	+1.0 0	+2.4 +1.8	1.0 2.6	+1.0 0	+2.6 +2.0	1.5 3.1	+1.0 0	+3.1 +2.5
1.58 - 1.97	0.4 1.4	+0.6 0	+1.4 +1.0	0.8 2.4	+1.0 0	+2.4 +1.8	1.2 2.8	+1.0 0	+2.8 +2.2	1.8 3.4	+1.0 0	+3.4 +2.8

Figure A-6

Preferred Clearance Fits — Cylindrical Fits

(Hole Basis; ANSI B4.2)

Basic Size		Loose Running			Free Running			Close Running			Sliding			Locational Clear.		
		Hole H11	Shaft c11	Fit	Hole H9	Shaft d9	Fit	Hole H8	Shaft f7	Fit	Hole H7	Shaft g6	Fit	Hole H7	Shaft h6	Fit
4	Max	4.075	3.930	0.220	4.030	3.970	0.090	4.018	3.990	0.040	4.012	3.996	0.024	4.012	4.000	0.020
	Min	4.000	3.855	0.070	4.000	3.940	0.030	4.000	3.978	0.010	4.000	3.988	0.004	4.000	3.992	0.000
5	Max	5.075	4.930	0.220	5.030	4.970	0.090	5.018	4.990	0.040	5.012	4.996	0.024	5.012	5.000	0.020
	Min	5.000	4.855	0.070	5.000	4.940	0.030	5.000	4.978	0.010	5.000	4.988	0.004	5.000	4.992	0.000
6	Max	6.075	5.930	0.220	6.030	5.970	0.090	6.018	5.990	0.040	6.012	5.996	0.024	6.012	6.000	0.020
	Min	6.000	5.885	0.070	6.000	5.940	0.030	6.000	5.978	0.010	6.000	5.988	0.004	6.000	5.992	0.000
8	Max	8.090	7.920	0.260	8.036	7.960	0.112	8.022	7.987	0.050	8.015	7.995	0.029	8.015	8.000	0.024
	Min	8.000	7.830	0.080	8.000	7.924	0.040	8.000	7.972	0.013	8.000	7.986	0.005	8.000	7.991	0.000
10	Max	10.090	9.920	0.026	10.036	9.960	0.112	10.022	9.987	0.050	10.015	9.995	0.029	10.015	10.000	0.024
	Min	10.000	9.830	0.080	10.000	9.924	0.040	10.000	9.972	0.013	10.000	9.986	0.005	10.000	9.991	0.000
12	Max	12.112	11.905	0.315	12.043	11.950	0.136	12.027	11.984	0.061	12.018	11.994	0.035	12.018	12.000	0.029
	Min	12.000	11.795	0.095	12.000	11.907	0.050	12.000	11.966	0.016	12.000	11.983	0.006	12.000	11.989	0.000
16	Max	16.110	15.905	0.315	16.043	15.950	0.136	16.027	15.984	0.061	16.018	15.994	0.035	16.018	16.000	0.029
	Min	16.000	15.795	0.095	16.000	15.907	0.050	16.000	15.966	0.016	16.000	19.983	0.006	16.000	15.989	0.000
20	Max	20.130	19.890	0.370	20.052	19.935	0.169	20.033	19.980	0.074	20.021	19.993	0.041	20.021	20.000	0.034
	Min	20.000	19.760	0.110	20.000	19.883	0.065	20.000	19.959	0.020	20.000	19.980	0.007	20.000	19.987	0.000
25	Max	25.130	24.890	0.370	25.052	24.935	0.169	25.033	24.980	0.074	25.021	24.993	0.041	25.021	25.000	0.034
	Min	25.000	24.760	0.110	25.000	24.883	0.065	25.000	24.959	0.020	25.000	24.980	0.007	25.000	24.987	0.000
30	Max	30.130	29.890	0.370	30.052	29.935	0.169	30.033	29.980	0.074	30.021	29.993	0.041	30.021	30.000	0.034
	Min	30.000	29.760	0.110	30.000	29.883	0.065	30.000	29.959	0.020	30.000	29.980	0.007	30.000	29.987	0.000

Figure A-7

Preferred Transition and Interference Fits — Cylindrical Fits
(Hole Basis; ANSI B4.2)

Basic Size		Locational Trans.			Locational Trans.			Locational Inter.			Medium Drive			Force		
		Hole H7	Shaft k6	Fit	Hole H7	Shaft n6	Fit	Hole H7	Shaft p6	Fit	Hole H7	Shaft s6	Fit	Hole H7	Shaft u6	Fit
4	Max	4.012	4.009	0.011	4.012	4.016	0.004	4.012	4.020	0.000	4.012	4.027	-0.007	4.012	4.031	-0.011
	Min	4.000	4.001	-0.009	4.000	4.008	-0.016	4.000	4.012	-0.020	4.000	4.019	-0.027	4.000	4.023	-0.031
5	Max	5.012	5.009	0.011	5.012	5.016	0.004	5.012	5.020	0.000	5.012	5.027	-0.007	5.012	5.031	-0.011
	Min	5.000	5.001	-0.009	5.000	5.008	-0.016	5.000	5.012	-0.020	5.000	5.019	-0.027	5.000	5.023	-0.031
6	Max	6.012	6.009	0.011	6.012	6.016	0.004	6.012	6.020	0.000	6.012	6.027	-0.007	6.012	6.031	-0.011
	Min	6.000	6.001	-0.009	6.000	6.008	-0.016	6.000	6.012	-0.020	6.000	6.019	-0.027	6.000	6.023	-0.031
8	Max	8.015	8.010	0.014	8.015	8.019	0.005	8.015	8.024	0.000	8.015	8.032	-0.008	8.015	8.037	-0.013
	Min	8.000	8.001	-0.010	8.000	8.010	-0.019	8.000	8.015	-0.024	8.000	8.023	-0.032	8.000	8.028	-0.037
10	Max	10.015	10.010	0.014	10.015	10.019	0.005	10.015	10.024	0.000	10.015	10.032	-0.008	10.015	10.037	-0.013
	Min	10.000	10.001	-0.010	10.000	10.010	-0.019	10.000	10.015	-0.024	10.000	10.023	-0.032	10.000	10.028	-0.037
12	Max	12.018	12.012	0.017	12.018	12.023	0.006	12.018	12.029	0.000	12.018	12.039	-0.010	12.018	12.044	-0.015
	Min	12.000	12.001	-0.012	12.000	12.012	-0.023	12.000	12.018	-0.029	12.000	12.028	-0.039	12.000	12.033	-0.044
16	Max	16.018	16.012	0.017	16.018	16.023	0.006	16.018	16.029	0.000	16.018	16.039	-0.010	16.018	16.044	-0.015
	Min	16.000	16.001	-0.012	16.000	16.012	-0.023	16.000	16.018	-0.029	16.000	16.028	-0.039	16.000	16.033	-0.044
20	Max	20.021	20.015	0.019	20.021	20.028	0.006	20.021	20.035	-0.001	20.021	20.048	-0.014	20.021	20.054	-0.020
	Min	20.000	20.002	-0.015	20.000	20.015	-0.028	20.000	20.022	-0.035	20.000	20.035	-0.048	20.000	20.041	-0.054
25	Max	25.021	25.015	0.019	25.021	25.028	0.006	25.021	25.035	-0.001	25.021	25.048	-0.014	25.021	25.061	-0.027
	Min	25.000	25.002	-0.015	25.000	25.015	-0.028	25.000	25.022	-0.035	25.000	25.035	-0.048	25.000	25.048	-0.061
30	Max	30.021	30.015	0.019	30.021	30.028	0.006	30.021	30.035	-0.001	30.021	30.048	-0.014	30.021	30.061	-0.027
	Min	30.000	30.002	-0.015	30.000	30.015	-0.028	30.000	30.022	-0.035	30.000	30.035	-0.048	30.000	30.048	-0.061

Figure A-8

Preferred Clearance Fits — Cylindrical Fits
(Shaft Basis; ANSI B4.2)

Basic Size		Loose Running			Free Running			Close Running			Sliding			Locational Clear.		
		Hole C11	Shaft h11	Fit	Hole D9	Shaft h9	Fit	Hole F8	Shaft h7	Fit	Hole G7	Shaft h6	Fit	Hole H7	Shaft h6	Fit
4	Max	4.145	4.000	0.220	4.060	4.000	0.090	4.028	4.000	0.040	4.016	4.000	0.024	4.012	4.000	0.020
	Min	4.070	3.925	0.070	4.030	3.970	0.030	4.010	3.988	0.010	4.004	3.992	0.004	4.000	3.992	0.000
5	Max	5.145	5.000	0.220	5.060	5.000	0.090	5.028	5.000	0.040	5.016	5.000	0.024	5.012	5.000	0.020
	Min	5.070	4.925	0.070	5.030	4.970	0.030	5.010	4.988	0.010	5.004	4.992	0.004	5.000	4.992	0.000
6	Max	6.145	6.000	0.220	6.060	6.000	0.090	6.028	6.000	0.040	6.016	6.000	0.024	6.012	6.000	0.020
	Min	6.070	5.925	0.070	6.030	5.970	0.030	6.010	5.988	0.010	6.004	5.992	0.004	6.000	5.992	0.000
8	Max	8.170	8.000	0.260	8.076	8.000	0.112	8.035	8.000	0.050	8.020	8.000	0.029	8.015	8.000	0.024
	Min	8.080	7.910	0.080	8.040	7.964	0.040	8.013	7.985	0.013	8.005	7.991	0.005	8.000	7.991	0.000
10	Max	10.170	10.000	0.260	10.076	10.000	0.112	10.035	10.000	0.050	10.020	10.000	0.029	10.015	10.000	0.024
	Min	10.080	9.910	0.080	10.040	9.964	0.040	10.013	9.985	0.013	10.005	9.991	0.005	10.000	9.991	0.000
12	Max	12.205	12.000	0.315	12.093	12.000	0.136	12.043	12.000	0.061	12.024	12.000	0.035	12.018	12.000	0.029
	Min	12.095	11.890	0.095	12.050	11.957	0.050	12.016	11.982	0.016	12.006	11.989	0.006	12.000	11.989	0.000
16	Max	16.205	16.000	0.315	16.093	16.000	0.136	16.043	16.000	0.061	16.024	16.000	0.035	16.018	16.000	0.029
	Min	16.095	15.890	0.095	16.050	15.957	0.050	16.016	15.982	0.016	06.006	15.989	0.006	16.000	15.989	0.000
20	Max	20.240	20.000	0.370	20.117	20.000	0.169	20.053	20.000	0.074	20.028	20.000	0.041	20.021	20.000	0.034
	Min	20.110	19.870	0.110	20.065	19.948	0.065	20.020	19.979	0.020	20.007	19.987	0.007	20.000	19.987	0.000
25	Max	25.240	25.000	0.370	25.117	25.000	0.169	25.053	25.000	0.074	25.028	25.000	0.041	25.021	25.000	0.034
	Min	25.110	24.870	0.110	25.065	24.948	0.065	25.020	24.979	0.020	25.007	24.987	0.007	25.000	24.987	0.000
30	Max	30.240	30.000	0.370	30.117	30.000	0.169	30.053	30.000	0.074	30.028	30.000	0.041	30.021	30.000	0.034
	Min	30.110	29.870	0.110	30.065	29.948	0.065	29.979	29.979	0.020	30.007	29.987	0.007	30.000	29.987	0.000

Figure A-9

Preferred Transition and Interference Fits — Cylindrical Fits

(Shaft Basis; ANSI B4.2)

Basic Size		Locational Trans.			Locational Trans.			Locational Inter.			Medium Drive			Force		
		Hole K7	Shaft h6	Fit	Hole N7	Shaft h6	Fit	Hole F7	Shaft h6	Fit	Hole S7	Shaft h6	Fit	Hole U7	Shaft h6	Fit
4	Max	4.003	4.000	0.011	3.996	4.000	0.004	3.992	4.000	0.000	3.985	4.000	-0.007	3.981	4.000	-0.011
	Min	3.991	3.992	-0.009	3.984	3.992	-0.016	3.980	3.992	-0.020	3.973	3.992	-0.027	3.969	3.992	-0.031
5	Max	5.003	5.000	0.011	4.996	5.000	0.004	4.992	5.000	0.000	4.985	5.000	-0.007	4.981	5.000	-0.011
	Min	4.991	4.992	-0.009	4.984	4.992	-0.016	4.980	4.992	-0.020	4.973	4.992	-0.027	4.969	4.992	-0.031
6	Max	6.003	6.000	0.011	5.996	6.000	0.004	5.992	6.000	0.000	5.985	6.000	-0.007	5.981	6.000	-0.011
	Min	5.991	5.991	-0.009	5.984	5.992	-0.016	5.980	5.992	-0.020	5.973	5.992	-0.027	5.969	5.992	-0.031
8	Max	8.005	8.000	0.014	7.996	8.000	0.005	7.991	8.000	0.000	7.983	8.000	-0.008	7.978	8.000	-0.013
	Min	7.990	7.991	-0.010	7.981	7.991	-0.019	7.976	7.991	-0.024	7.968	7.991	-0.032	7.963	7.991	-0.037
10	Max	10.005	10.000	0.014	9.996	10.000	0.005	9.991	10.000	0.000	9.983	10.000	-0.008	9.978	10.000	-0.013
	Min	9.990	9.991	-0.010	9.981	9.991	-0.019	9.976	9.991	-0.024	9.968	9.991	-0.032	9.963	9.991	-0.037
12	Max	12.006	12.000	0.017	11.995	12.000	0.006	11.989	12.000	0.000	11.979	12.000	-0.010	11.974	12.000	-0.015
	Min	11.988	11.989	-0.012	11.977	11.989	-0.023	11.971	11.989	-0.029	11.961	11.989	-0.039	11.956	11.989	-0.044
16	Max	16.006	16.000	0.017	15.995	16.000	0.006	15.989	16.000	0.000	15.979	16.000	-0.010	15.974	16.000	-0.015
	Min	15.988	15.989	-0.012	15.977	15.989	-0.023	15.971	15.989	-0.029	15.961	15.989	-0.039	15.956	15.989	-0.044
20	Max	20.006	20.000	0.019	19.993	20.000	0.006	19.986	20.000	-0.001	19.973	20.000	-0.014	19.967	20.000	-0.020
	Min	19.985	19.987	-0.015	19.972	19.987	-0.028	19.965	19.987	-0.035	19.952	19.987	-0.048	19.946	19.987	-0.054
25	Max	25.006	25.000	0.019	24.993	25.000	0.006	24.986	25.000	-0.001	24.973	25.000	-0.014	24.960	25.000	-0.027
	Min	24.985	24.987	-0.015	24.972	24.987	-0.028	24.965	24.987	-0.035	24.952	24.987	-0.048	24.939	24.987	-0.061
30	Max	30.006	30.000	0.019	29.993	30.000	0.006	29.986	30.000	-0.001	29.973	30.000	-0.014	29.960	30.000	-0.027
	Min	29.985	29.987	-0.015	29.972	29.987	-0.028	29.987	29.987	-0.035	29.952	29.987	-0.048	29.939	29.987	-0.061

Figure A-10

Metric Threads—Preferred Sizes			
First Choice	Second Choice	First Choice	Second Choice
1	1.1	12	14
1.2	1.4	16	18
1.6	1.8	20	22
2	2.2	25	28
2.5	2.8	30	35
3	3.5	40	45
4	4.5	50	55
5	5.5	60	70
6	7	80	90
8	9	100	110
10	11	120	140

Figure A-11

Standard Thread Lengths—Inches														
	3/16	**1/4**	**3/8**	**1/2**	**5/8**	**3/4**	**7/8**	**1**	**1 1/4**	**1 1/2**	**1 3/4**	**2**	**2 1/2**	**3**
#2 - 5	√	√	√	√	√	√	√	√						
#4 - 40		√	√	√	√	√	√	√	√	√				
#6 - 32		√	√	√	√	√		√	√	√	√	√	√	√
#8 - 32		√	√	√	√	√		√	√	√	√	√	√	√
#10 - 24		√	√	√	√	√		√	√	√	√	√	√	√
#10 - 32		√	√	√		√		√	√	√	√	√	√	√
#12 - 24			√	√	√	√		√	√	√	√	√	√	√
1/4 20				√	√	√	√	√	√	√	√	√	√	√
5/16 18				√	√	√	√	√	√	√	√	√	√	√
3/8 16				√	√	√	√	√	√	√	√	√	√	√
1/2 13						√	√	√	√	√	√	√	√	√
5/8 11							√	√	√	√	√	√	√	√
3/4 10										√		√		√

Figure A-12

American National Standard Plain Washers					
Nominal Washer Size		Series	Inside Diameter	Outside Diameter	Thickness
No. 0	0.060	N R W	0.068 0.068 0.068	0.125 0.188 0.250	0.025 0.025 0.025
No. 1	0.073	N R W	0.084 0.084 0.084	0.156 0.219 0.281	0.025 0.025 0.032
No. 2	0.086	N R W	0.094 0.094 0.094	0.188 0.250 0.344	0.025 0.032 0.032
No. 3	0.099	N R W	0.109 0.109 0.109	0.219 0.312 0.406	0.025 0.032 0.040
No. 4	0.112	N R W	0.125 0.125 0.125	0.250 0.375 0.438	0.032 0.040 0.040
No. 5	0.125	N R W	0.141 0.141 0.141	0.281 0.406 0.500	0.032 0.040 0.040
No. 6	1.380	N R W	0.156 0.156 0.156	0.312 0.438 0.562	0.032 0.040 0.040
No. 8	0.164	N R W	0.188 0.188 0.188	0.375 0.500 0.633	0.040 0.040 0.063
No. 10	0.190	N R W	0.203 0.203 0.203	0.406 0.562 0.734	0.040 0.040 0.063
No. 12	0.216	N R W	0.234 0.234 0.234	0.438 0.625 0.875	0.040 0.063 0.063
1/4	0.250	N R W	0.281 0.281 0.281	0.500 0.734 1.000	0.063 0.063 0.063
5/16	0.312	N R W	0.344 0.344 0.344	0.625 0.875 1.125	0.063 0.063 0.063
3/8	0.375	N R W	0.406 0.406 0.406	0.734 1.000 1.250	0.063 0.063 0.100
7/16	0.438	N R W	0.469 0.469 0.469	0.875 1.125 1.469	0.063 0.063 0.100
1/2	0.500	N R W	0.531 0.531 0.531	1.000 1.2.5 1.125	0.063 0.100 0.100

Figure A-13

American National Standard Plain Washers					
Nominal Washer Size		Series	Inside Diameter	Outside Diameter	Thickness
9/16	0.562	N R W	0.594 0.594 0.594	1.125 1.469 2.000	0.063 0.100 0.100
5/8	0.625	N R W	0.656 0.656 0.656	1.250 1.750 2.250	0.100 0.100 0.160
3/4	0.750	N R W	0.812 0.812 0.812	1.375 2.000 2.500	0.100 0.100 0.160
7/8	0.875	N R W	0.938 0.938 0.938	1.469 2.250 2.750	0.100 0.160 0.160
1	1.000	N R W	1.062 1.062 1.062	1.750 2.500 3.000	0.100 0.160 0.160

Figure A-13
(*Continued*)

Index